→	1B	2B	3A	4A	5A	6A	7A	NOBLE ELEMENTS
								HELIUM 4.0026 0.179 **He** 2 $1s^2$ 3.57 HEX 1.633 ~1.0 (26 Atm) 26^{LT}
			BORON 10.81 2.34 **B** 5 $1s^2 2s^2 2p^1$ 8.73 TET 0.576 2600 1250	CARBON 12.01 2.26 **C** 6 $1s^2 2s^2 2p^2$ 3.57 DIA (4300) 1860	NITROGEN 14.007 1.03 **N** 7 $1s^2 2s^2 2p^3$ 4.039 HEX 1.651 63.3 $(\beta) 79^{LT}$	OXYGEN 15.999 1.43 **O** 8 $1s^2 2s^2 2p^4$ 6.83 CUB 54.7 $(\gamma) 46^{LT}$	FLUORINE 18.998 1.97 (α) **F** 9 $1s^2 2s^2 2p^5$ MCL 53.5	NEON 20.18 1.56 **Ne** 10 $1s^2 2s^2 2p^6$ 4.43 FCC 24.5 63
			ALUMINUM 26.982 2.70 **Al** 13 [Ne] $3s^2 3p^1$ 4.05 FCC 933 394	SILICON 28.086 2.33 **Si** 14 [Ne] $3s^2 3p^2$ 5.43 DIA 1683 625	PHOSPHORUS 30.974 1.82 (white) **P** 15 [Ne] $3s^2 3p^3$ 7.17 CUB 317.3	SULFUR 32.064 2.07 **S** 16 [Ne] $3s^2 3p^4$ 10.47 ORC 2.339 1.229 386	CHLORINE 35.453 2.09 **Cl** 17 [Ne] $3s^2 3p^5$ 6.24 ORC 1.324 0.718 172.2	ARGON 39.948 1.78 **Ar** 18 [Ne] $3s^2 3p^6$ 5.26 FCC 83.9 85
NICKEL 58.71 8.9 **Ni** 28 [Ar] $3d^8 4s^2$ 3.52 FCC 1726 375	COPPER 63.55 8.96 **Cu** 29 [Ar] $3d^{10} 4s^1$ 3.61 FCC 1356 315	ZINC 65.38 7.14 **Zn** 30 [Ar] $3d^{10} 4s^2$ 2.66 HEX 1.856 693 234	GALLIUM 69.72 5.91 **Ga** 31 [Ar] $3d^{10} 4s^2 3p^1$ 4.51 ORC 1.695 1.001 303 240	GERMANIUM 72.59 5.32 **Ge** 32 [Ar] $3d^{10} 4s^2 4p^2$ 5.66 DIA 1211 360	ARSENIC 74.922 5.72 **As** 33 [Ar] $3d^{10} 4s^2 4p^3$ 4.13 RHL 54°10′ 1090 285	SELENIUM 78.96 4.79 **Se** 34 [Ar] $3d^{10} 4s^2 4p^4$ 4.36 HEX 1.136 490 150^{LT}	BROMINE 79.91 4.10 **Br** 35 [Ar] $3d^{10} 4s^2 4p^5$ 6.67 ORC 1.307 0.672 266	KRYPTON 83.80 3.07 **Kr** 36 [Ar] $3d^{10} 4s^2 4p^6$ 5.72 FCC 116.5 73^{LT}
PALLADIUM 106.40 12.0 **Pd** 46 [Kr] $4d^{10} 5s^0$ 3.89 FCC 1825 275	SILVER 107.87 10.5 **Ag** 47 [Kr] $4d^{10} 5s^1$ 4.09 FCC 1234 215	CADMIUM 112.40 8.65 **Cd** 48 [Kr] $4d^{10} 5s^2$ 2.98 HEX 1.886 594 120	INDIUM 114.82 7.31 **In** 49 [Kr] $4d^{10} 5s^2 5p^1$ 4.59 TET 1.076 429.8 129	TIN 118.69 7.30 **Sn** 50 [Kr] $4d^{10} 5s^2 5p^2$ 5.82 TET 0.546 505 170	ANTIMONY 121.75 6.62 **Sb** 51 [Kr] $4d^{10} 5s^2 5p^3$ 4.51 RHL 57°6′ 904 200	TELLURIUM 127.60 6.24 **Te** 52 [Kr] $4d^{10} 5s^2 5p^4$ 4.45 HEX 1.330 723 139^{LT}	IODINE 126.90 4.94 **I** 53 [Kr] $4d^{10} 5s^2 5p^5$ 7.27 ORC 1.347 0.659 387	XENON 131.30 3.77 **Xe** 54 [Kr] $4d^{10} 5s^2 5p^6$ 6.20 FCC 161.3 55^{LT}
PLATINUM 195.09 21.4 **Pt** 78 [Xe] $4f^{14} 5d^{10} 6s^0$ 3.92 FCC 2045 230	GOLD 196.97 19.3 **Au** 79 [Xe] $4f^{14} 5d^{10} 6s^1$ 4.08 FCC 1337 170	MERCURY 200.59 13.6 **Hg** 80 [Xe] $4f^{14} 5d^{10} 6s^2$ 2.99 RHL 70°45′ 234.3 100	THALLIUM 204.37 11.85 **Tl** 81 [Xe] $4f^{14} 5d^{10} 6s^2 6p^1$ 3.46 HEX 1.599 577 96	LEAD 207.19 11.4 **Pb** 82 [Xe] $4f^{14} 5d^{10} 6s^2 6p^2$ 4.95 FCC 601 88	BISMUTH 208.98 9.8 **Bi** 83 [Xe] $4f^{14} 5d^{10} 6s^2 6p^3$ 4.75 RHL 57°14′ 544.5 120	POLONIUM 210 9.4 **Po** 84 [Xe] $4f^{14} 5d^{10} 6s^2 6p^4$ 3.35 SC 527	ASTATINE 210 **At** 85 [Xe] $4f^{14} 5d^{10} 6s^2 6p^5$ (575)	RADON 222 (4.4) **Rn** 86 [Xe] $4f^{14} 5d^{10} 6s^2 6p^6$ (FCC) (202)

EUROPIUM 151.96 7.90 **Eu** 63 [Xe] $4f^7 5d^0 6s^2$ 4.61 BCC 1095 107^{LT}	GADOLINIUM 157.25 8.23 **Gd** 64 [Xe] $4f^7 5d^1 6s^2$ 3.64 HEX 1.588 1585 176^{LT}	TERBIUM 158.92 8.54 **Tb** 65 [Xe] $4f^9 5d^0 6s^2$ 3.60 HEX 1.581 1633 188^{LT}	DYSPROSIUM 162.50 8.78 **Dy** 66 [Xe] $4f^{10} 5d^0 6s^2$ 3.59 HEX 1.573 1680 186^{LT}	HOLMIUM 164.93 9.05 **Ho** 67 [Xe] $4f^{11} 5d^0 6s^2$ 3.58 HEX 1.570 1743 191^{LT}	ERBIUM 167.26 9.37 **Er** 68 [Xe] $4f^{12} 5d^0 6s^2$ 3.56 HEX 1.570 1795 195^{LT}	THULIUM 168.93 9.31 **Tm** 69 [Xe] $4f^{13} 5d^0 6s^2$ 3.54 HEX 1.570 1818 200^{LT}	YTTERBIUM 173.04 6.97 **Yb** 70 [Xe] $4f^{14} 5d^0 6s^2$ 5.49 FCC 1097 118^{LT}	LUTETIUM 174.97 9.84 **Lu** 71 [Xe] $4f^{14} 5d^1 6s^2$ 3.51 HEX 1.585 1929 207^{LT}
AMERICIUM 243 11.8 **Am** 95 [Rn] $5f^7 6d^0 7s^2$ 1267	CURIUM 247 **Cm** 96 [Rn] $5f^7 6d^1 7s^2$ 1600	BERKELIUM 247 **Bk** 97 [Rn] $5f^7 6d^2 7s^2$	CALIFORNIUM 251 **Cf** 98 [Rn] $5f^9 6d^1 7s^2$	EINSTEINIUM 254 **Es** 99	FERMIUM 257 **Fm** 100	MENDELEVIUM 256 **Md** 101	NOBELIUM 254 **No** 102	LAWRENCIUM 257 **Lw** 103

Solid State Physics

Solid State Physics

Neil W. Ashcroft
N. David Mermin
Cornell University

CENGAGE

Australia • Brazil • Canada • Mexico • Singapore • United Kingdom • United States

CENGAGE

Solid State Physics
Neil W. Ashcroft,
N. David Mermin

Library of Congress Control Number: 74-9772

ISBN-13: 978-0-357-67081-1

ISBN-10: 0-357-67081-7

Cengage
200 Pier 4 Boulevard
Boston, MA 02210
USA

Printed at CLDPC, USA, 12-23

for Elizabeth, Jonathan, Robert, and Ian

Preface

We began this project in 1968 to fill a gap we each felt acutely after several years of teaching introductory solid state physics to Cornell students of physics, chemistry, engineering, and materials science. In both undergraduate and graduate courses we had to resort to a patchwork array of reading assignments, assembled from some half dozen texts and treatises. This was only partly because of the great diversity of the subject; the main problem lay in its dual nature. On the one hand an introduction to solid state physics must describe in some detail the vast range of real solids, with an emphasis on representative data and illustrative examples. On the other hand there is now a well-established basic theory of solids, with which any seriously interested student must become familiar.

Rather to our surprise, it has taken us seven years to produce what we needed: a single introductory text presenting both aspects of the subject, descriptive and analytical. Our aim has been to explore the variety of phenomena associated with the major forms of crystalline matter, while laying the foundation for a working understanding of solids through clear, detailed, and elementary treatments of fundamental theoretical concepts.

Our book is designed for introductory courses at either the undergraduate or graduate level.[1] Statistical mechanics and the quantum theory lie at the heart of solid state physics. Although these subjects are used as needed, we have tried, especially in the more elementary chapters, to recognize that many readers, particularly undergraduates, will not yet have acquired expertise. When it is natural to do so, we have clearly separated topics based entirely on classical methods from those demanding a quantum treatment. In the latter case, and in applications of statistical mechanics, we have proceeded carefully from explicitly stated first principles. The book is therefore suitable for an introductory course taken concurrently with first courses in quantum theory and statistical mechanics. Only in the more advanced chapters and appendices do we assume a more experienced readership.

The problems that follow each chapter are tied rather closely to the text, and are of three general kinds: (a) routine steps in analytical development are sometimes relegated to problems, partly to avoid burdening the text with formulas of no intrinsic interest, but, more importantly, because such steps are better understood if completed by the reader with the aid of hints and suggestions; (b) extensions of the chapter (which the spectre of a two volume work prevented us from including) are presented as problems when they lend themselves to this type of exposition; (c) further numerical and analytical applications are given as problems, either to communicate additional

---

[1] Suggestions for how to use the text in courses of varying length and level are given on pp. xviii–xxi.

information or to exercise newly acquired skills. Readers should therefore examine the problems, even if they do not intend to attempt their solution.

Although we have respected the adage that one picture is worth a thousand words, we are also aware that an uninformative illustration, though decorative, takes up the space that could usefully be filled by several hundred. The reader will thus encounter stretches of expository prose unrelieved by figures, when none are necessary, as well as sections that can profitably be perused entirely by looking at the figures and their captions.

We anticipate use of the book at different levels with different areas of major emphasis. A particular course is unlikely to follow the chapters (or even selected chapters) in the order in which they are presented here, and we have written them in a way that permits easy selection and rearrangement.[2] Our particular choice of sequence follows certain major strands of the subject from their first elementary exposition to their more advanced aspects, with a minimum of digression.

We begin the book[3] with the elementary classical [1] and quantum [2] aspects of the free electron theory of metals because this requires a minimum of background and immediately introduces, through a particular class of examples, almost all of the phenomena with which theories of insulators, semiconductors, and metals must come to grips. The reader is thereby spared the impression that nothing can be understood until a host of arcane definitions (relating to periodic structures) and elaborate quantum mechanical explorations (of periodic systems) have been mastered.

Periodic structures are introduced only after a survey [3] of those metallic properties that can and cannot be understood without investigating the consequences of periodicity. We have tried to alleviate the tedium induced by a first exposure to the language of periodic systems by (a) separating the very important consequences of purely translational symmetry [4, 5] from the remaining but rather less essential rotational aspects [7], (b) separating the description in ordinary space [4] from that in the less familiar reciprocal space [5], and (c) separating the abstract and descriptive treatment of periodicity from its elementary application to X-ray diffraction [6].

Armed with the terminology of periodic systems, readers can pursue to whatever point seems appropriate the resolution of the difficulties in the free electron model of metals or, alternatively, can embark directly upon the investigation of lattice vibrations. The book follows the first line. Bloch's theorem is described and its implications examined [8] in general terms, to emphasize that its consequences transcend the illustrative and very important practical cases of nearly free electrons [9] and tight binding [10]. Much of the content of these two chapters is suitable for a more advanced course, as is the following survey of methods used to compute real band structures [11]. The remarkable subject of semiclassical mechanics is introduced and given elementary applications [12] before being incorporated into the more elaborate semiclassical theory of transport [13]. The description of methods by which Fermi surfaces are measured [14] may be more suitable for advanced readers, but much of the survey

[2] The Table on pp. xix–xxi lists the prerequisites for each chapter, to aid those interested primarily in one aspect of the subject, or those preferring a different order of presentation.

[3] References to chapter numbers are given in brackets.

of the band structures of actual metals [15] is readily incorporated into an elementary course.

Except for the discussion of screening, an elementary course might also bypass the essays on what is overlooked by the relaxation-time approximation [16] and by the neglect of electron-electron interactions [17].

Work functions and other surface properties [18] can be taken up at any time after the discussion of translational symmetry in real space. Our description of the conventional classification of solids [19] has been separated from the analysis of cohesive energies [20]. Both have been placed after the introduction to band structure, because it is in terms of electronic structure that the categories are most clearly distinguished.

To motivate the study of lattice vibrations (at whatever point after Chapter 5 readers choose to begin the subject) a summary [21] lists those solid properties that cannot be understood without their consideration. Lattice dynamics is given an elementary introduction, with the classical [22] and quantum [23] aspects of the harmonic crystal treated separately. The ways in which phonon spectra are measured [24], the consequences of anharmonicity [25], and the special problems associated with phonons in metals [26] and ionic crystals [27] are surveyed at an elementary level, though some parts of these last four chapters might well be reserved for a more advanced course. None of the chapters on lattice vibrations rely on the use of normal mode raising and lowering operators; these are described in several appendices for readers wanting a more advanced treatment.

Homogeneous [28] and inhomogeneous [29] semiconductors can be examined at any point after the introduction of Bloch's theorem and the elementary discussion of semiclassical mechanics. Crystalline defects [30] can be studied as soon as crystals themselves have been introduced, though parts of earlier chapters are occasionally referred to.

Following a review of atomic magnetism, we examine how it is modified in a solid environment [31], explore exchange and other magnetic interactions [32], and apply the resulting models to magnetic ordering [33]. This brief introduction to magnetism and the concluding essay on superconductivity [34] are largely self-contained. They are placed at the end of the book so the phenomena can be viewed, not in terms of abstract models but as striking properties of real solids.

To our dismay, we discovered that it is impossible at the end of a seven year project, labored upon not only at Cornell, but also during extended stays in Cambridge, London, Rome, Wellington, and Jülich, to recall all the occasions when students, postdoctoral fellows, visitors, and colleagues gave us invaluable criticism, advice, and instruction. Among others we are indebted to V. Ambegaokar, B. W. Batterman, D. Beaglehole, R. Bowers, A. B. Bringer, C. di Castro, R. G. Chambers, G. V. Chester, R. M. Cotts, R. A. Cowley, G. Eilenberger, D. B. Fitchen, C. Friedli, V. Heine, R. L. Henderson, D. F. Holcomb, R. O. Jones, B. D. Josephson, J. A. Krumhansl, C. A. Kukkonen, D. C. Langreth, W. L. McLean, H. Mahr, B. W. Maxfield, R. Monnier, L. G. Parratt, O. Penrose, R. O. Pohl, J. J. Quinn, J. J. Rehr, M. V. Romerio, A. L. Ruoff, G. Russakoff, H. S. Sack, W. L. Schaich, J. R. Schrieffer, J. W. Serene, A. J. Sievers, J. Silcox, R. H. Silsbee, J. P. Straley, D. M. Straus, D. Stroud, K. Sturm, and J. W. Wilkins.

One person, however, has influenced almost every chapter. Michael E. Fisher, Horace White Professor of Chemistry, Physics, *and* Mathematics, friend and neighbor, gadfly and troubadour, began to read the manuscript six years ago and has followed ever since, hard upon our tracks, through chapter, and, on occasion, through revision and re-revision, pouncing on obscurities, condemning dishonesties, decrying omissions, labeling axes, correcting misspellings, redrawing figures, and often making our lives very much more difficult by his unrelenting insistence that we could be more literate, accurate, intelligible, and thorough. We hope he will be pleased at how many of his illegible red marginalia have found their way into our text, and expect to be hearing from him about those that have not.

One of us (NDM) is most grateful to the Alfred P. Sloan Foundation and the John Simon Guggenheim Foundation for their generous support at critical stages of this project, and to friends at Imperial College London and the Istituto di Fisica "G. Marconi," where parts of the book were written. He is also deeply indebted to R. E. Peierls, whose lectures converted him to the view that solid state physics is a discipline of beauty, clarity, and coherence. The other (NWA), having learnt the subject from J. M. Ziman and A. B. Pippard, has never been in need of conversion. He also wishes to acknowledge with gratitude the support and hospitality of the Kernforschungsanlage Jülich, the Victoria University of Wellington, and the Cavendish Laboratory and Clare Hall, Cambridge.

Ithaca
June 1975

N. W. Ashcroft
N. D. Mermin

Contents

APPENDICES

Important Tables

The more important tables of data[1] or theoretical results are listed below. To aid the reader in hunting down a particular table we have grouped them into several broad categories. Theoretical results are listed only under that heading, and data on magnetic and superconducting metals are listed under magnetism and superconductivity, rather than under metals. Accurate values of the fundamental constants are given in the end paper and on page 757.

Theoretical Results

[1] The data in the tables are presented with the aim of giving the reader an appreciation for orders of magnitude and relative sizes. We have therefore been content to quote numbers to one or two significant places and have not made special efforts to give the most precise values. Readers requiring data for fundamental research should consult the appropriate sources.

Crystal Structure

Metals

Insulators and Semiconductors

Magnetism

Superconductivity

Suggestions for Using the Book

It is in the nature of books that chapters must be linearly ordered. We have tried to select a sequence that least obscures several interwoven lines of development. The accompanying Table (pp. xix–xxi) is designed to help readers with special interests (in lattice vibrations, semiconductors, or metals, for example) or teachers of courses with particular constraints on time or level.

The prerequisites for each chapter are given in the Table according to the following conventions: (a) If "M" is listed after Chapter N, then the contents of Chapter M (as well, of course, as *its* essential prerequisites) are essential for understanding much or all of Chapter N; (b) If "(M)" appears after Chapter N, then Chapter M is not an *essential* prerequisite: either a small part of Chapter N is based on Chapter M, or a few parts of Chapter M may be of some help in reading Chapter N; (c) The absence of "M" or "(M)" after Chapter N does not mean that no reference back to Chapter M is made; however, such references as may occur are primarily because N illuminates the subject of M, rather than because M aids in the development of N.[1]

The rest of the Table indicates how the book might be used in a one semester (40 to 50 lectures) or two-semester (80 to 100 lectures) introductory undergraduate course. Chapters (or selections from a chapter) are listed for reading[2] if they are almost entirely descriptive, or, alternatively, if we felt that an introductory course should at least make students aware of a topic, even if time was not available for its careful exploration. The order of presentation is, of course, flexible. For example, although a two-semester course could follow the order of the book, one might well prefer to follow the pattern of the one-semester course for the first term, filling in the more advanced topics in the second.

An introductory course at the graduate level, or a graduate course following a one-semester undergraduate survey, would probably make use of sections omitted in the two-semester undergraduate course, as well as many of the sixteen appendices.

---

[1] Thus to proceed to Chapter 12 with a minimum of digression it is necessary to read Chapters 8, 5, 4, 2, and 1.

[2] Students would presumably be asked to read the chapters bearing on the lectures as well.

Chapter	***Prerequisites***	***One-Semester Introduction***		***Two-Semester Introduction***	
		LECTURES	READING	LECTURES	READING
1. Drude	None	All		All	
2. Sommerfeld	1	All		All	
3. Failures of free-electron model	2		All		All
4. Crystal lattices	None	Summarize	All	All	
5. Reciprocal lattice	4	All		All	
6. X-ray diffraction	5	96–104		All	
7. Crystal symmetries	4				All
8. Bloch's theorem	5	132–143		All	
9. Nearly free electrons	8 (6)	152–166		All	
10. Tight binding	8		176	176–184	184–189
11. Computing band structure	8 (9)		192–193		All
12. Semiclassical dynamics	2, 8	214–233		214–233	
13. Semiclassical transport	12				244–246

Chapter	*Prerequisites*	*One-Semester Introduction*		*Two-Semester Introduction*	
		LECTURES	READING	LECTURES	READING
14. Measuring the Fermi surface	12	264–275		264–275	
15. Band structure of metals	8 (2, 9, 10, 11, 12)		All	All	
16. Beyond relaxation-time approximation	2 (13)				All
17. Beyond independent electron approximation	2		337–342	330–344	345–351
18. Surface effects	2, 4 (6, 8)		354–364	354–364	
19. Classification of solids	2, 4 (9, 10)		All		All
20. Cohesive energy	19 (17)	396–410		All	
21. Failures of static lattice model	(2, 4)		All		All
22. Classical harmonic crystal	5	422–437		All	
23. Quantum harmonic crystal	22	452–464		All	

24. Measuring phonons	2, 23		470–481	All	
25. Anharmonic effects	23		499–505	All	
26. Phonons in metals	17, 23 (16)	523–526		512–519 523–526	
27. Dielectric properties	19, 22		534–542	All	
28. Homogeneous semiconductors	2, 8, (12)	562–580		All	
29. Inhomogeneous semiconductors	28	590–600		All	
30. Defects	4 (8, 12, 19, 22, 28, 29)		628–636		All
31. Diamagnetism, Paramagnetism	(2, 4, 14)	661–665		All	
32. Magnetic interactions	31 (2, 8, 10, 16, 17)	672–682		672–684	
33. Magnetic ordering	4, 5, 32		694–700	All	
34. Superconductivity	1, 2 (26)		726–736	All	

1

The Drude Theory of Metals

Metals occupy a rather special position in the study of solids, sharing a variety of striking properties that other solids (such as quartz, sulfur, or common salt) lack. They are excellent conductors of heat and electricity, are ductile and malleable, and display a striking luster on freshly exposed surfaces. The challenge of accounting for these metallic features gave the starting impetus to the modern theory of solids.

Although the majority of commonly encountered solids are nonmetallic, metals have continued to play a prominent role in the theory of solids from the late nineteenth century to the present day. Indeed, the metallic state has proved to be one of the great fundamental states of matter. The elements, for example, definitely favor the metallic state: over two thirds are metals. Even to understand nonmetals one must also understand metals, for in explaining why copper conducts so well, one begins to learn why common salt does not.

During the last hundred years physicists have tried to construct simple models of the metallic state that account in a qualitative, and even quantitative, way for the characteristic metallic properties. In the course of this search brilliant successes have appeared hand in hand with apparently hopeless failures, time and again. Even the earliest models, though strikingly wrong in some respects, remain, when properly used, of immense value to solid state physicists today.

In this chapter we shall examine the theory of metallic conduction put forth by P. Drude[1] at the turn of the century. The successes of the Drude model were considerable, and it is still used today as a quick practical way to form simple pictures and rough estimates of properties whose more precise comprehension may require analysis of considerable complexity. The failures of the Drude model to account for some experiments, and the conceptual puzzles it raised, defined the problems with which the theory of metals was to grapple over the next quarter century. These found their resolution only in the rich and subtle structure of the quantum theory of solids.

BASIC ASSUMPTIONS OF THE DRUDE MODEL

J. J. Thomson's discovery of the electron in 1897 had a vast and immediate impact on theories of the structure of matter, and suggested an obvious mechanism for conduction in metals. Three years after Thomson's discovery Drude constructed his theory of electrical and thermal conduction by applying the highly successful kinetic theory of gases to a metal, considered as a gas of electrons.

In its simplest form kinetic theory treats the molecules of a gas as identical solid spheres, which move in straight lines until they collide with one another.[2] The time taken up by a single collision is assumed to be negligible, and, except for the forces coming momentarily into play during each collision, no other forces are assumed to act between the particles.

Although there is only one kind of particle present in the simplest gases, in a metal there must be at least two, for the electrons are negatively charged, yet the metal is electrically neutral. Drude assumed that the compensating positive charge was at-

---

[1] *Annalen der Physik* **1**, 566 and **3**, 369 (1900).

[2] Or with the walls of the vessel containing them, a possibility generally ignored in discussing metals unless one is interested in very fine wires, thin sheets, or effects at the surface.

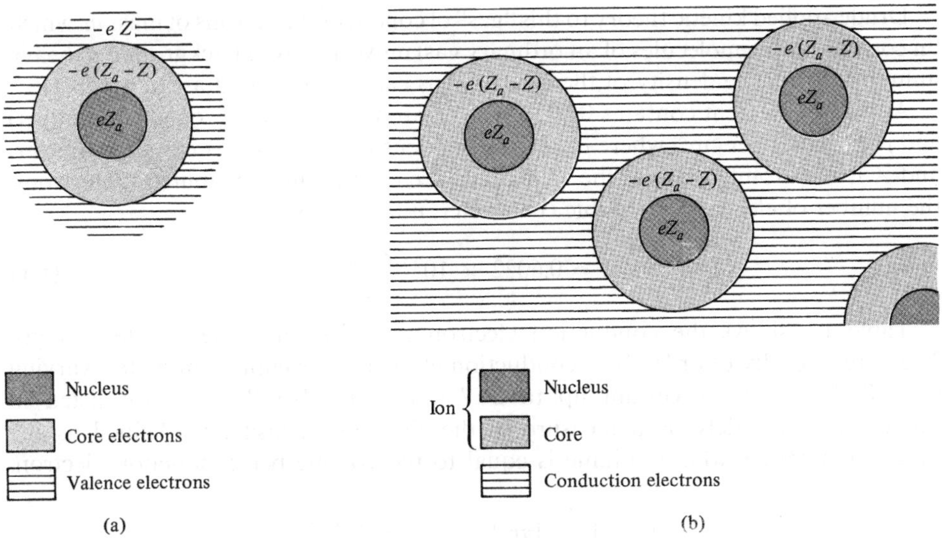

Figure 1.1
(a) Schematic picture of an isolated atom (not to scale). (b) In a metal the nucleus and ion core retain their configuration in the free atom, but the valence electrons leave the atom to form the electron gas.

tached to much heavier particles, which he considered to be immobile. At his time, however, there was no precise notion of the origin of the light, mobile electrons and the heavier, immobile, positively charged particles. The solution to this problem is one of the fundamental achievements of the modern quantum theory of solids. In this discussion of the Drude model, however, we shall simply assume (and in many metals this assumption can be justified) that when atoms of a metallic element are brought together to form a metal, the valence electrons become detached and wander freely through the metal, while the metallic ions remain intact and play the role of the immobile positive particles in Drude's theory. This model is indicated schematically in Figure 1.1. A single isolated atom of the metallic element has a nucleus of charge eZ_a, where Z_a is the atomic number and e is the magnitude of the electronic charge[3]: $e = 4.80 \times 10^{-10}$ electrostatic units (esu) $= 1.60 \times 10^{-19}$ coulombs. Surrounding the nucleus are Z_a electrons of total charge $-eZ_a$. A few of these, Z, are the relatively weakly bound valence electrons. The remaining $Z_a - Z$ electrons are relatively tightly bound to the nucleus, play much less of a role in chemical reactions, and are known as the core electrons. When these isolated atoms condense to form a metal, the core electrons remain bound to the nucleus to form the metallic ion, but the valence electrons are allowed to wander far away from their parent atoms. In the metallic context they are called conduction electrons.[4]

---

[3] We shall always take e to be a positive number.

[4] When, as in the Drude model, the core electrons play a passive role and the ion acts as an indivisible inert entity, one often refers to the conduction electrons simply as "the electrons," saving the full term for times when the distinction between conduction and core electrons is to be emphasized.

Drude applied kinetic theory to this "gas" of conduction electrons of mass m, which (in contrast to the molecules of an ordinary gas) move against a background of heavy immobile ions. The density of the electron gas can be calculated as follows:

A metallic element contains 0.6022×10^{24} atoms per mole (Avogadro's number) and ρ_m/A moles per cm^3, where ρ_m is the mass density (in grams per cubic centimeter) and A is the atomic mass of the element. Since each atom contributes Z electrons, the number of electrons per cubic centimeter, $n = N/V$, is

$$n = 0.6022 \times 10^{24} \frac{Z\rho_m}{A}. \tag{1.1}$$

Table 1.1 shows the conduction electron densities for some selected metals. They are typically of order 10^{22} conduction electrons per cubic centimeter, varying from 0.91×10^{22} for cesium up to 24.7×10^{22} for beryllium.[5] Also listed in Table 1.1 is a widely used measure of the electronic density, r_s, defined as the radius of a sphere whose volume is equal to the volume per conduction electron. Thus

$$\frac{V}{N} = \frac{1}{n} = \frac{4\pi r_s^3}{3}; \quad r_s = \left(\frac{3}{4\pi n}\right)^{1/3}. \tag{1.2}$$

Table 1.1 lists r_s both in angstroms (10^{-8} cm) and in units of the Bohr radius $a_0 = \hbar^2/me^2 = 0.529 \times 10^{-8}$ cm; the latter length, being a measure of the radius of a hydrogen atom in its ground state, is often used as a scale for measuring atomic distances. Note that r_s/a_0 is between 2 and 3 in most cases, although it ranges between 3 and 6 in the alkali metals (and can be as large as 10 in some metallic compounds).

These densities are typically a thousand times greater than those of a classical gas at normal temperatures and pressures. In spite of this and in spite of the strong electron-electron and electron-ion electromagnetic interactions, the Drude model boldly treats the dense metallic electron gas by the methods of the kinetic theory of a neutral dilute gas, with only slight modifications. The basic assumptions are these:

1. Between collisions the interaction of a given electron, both with the others and with the ions, is neglected. Thus in the absence of externally applied electromagnetic fields each electron is taken to move uniformly in a straight line. In the presence of externally applied fields each electron is taken to move as determined by Newton's laws of motion in the presence of those external fields, but neglecting the additional complicated fields produced by the other electrons and ions.[6] The neglect of electron-electron interactions between collisions is known as the *independent electron approximation*. The corresponding neglect of electron-ion interactions is known as the *free electron approximation*. We shall find in subsequent chapters that

[5] This is the range for metallic elements under normal conditions. Higher densities can be attained by application of pressure (which tends to favor the metallic state). Lower densities are found in compounds.

[6] Strictly speaking, the electron-ion interaction is not entirely ignored, for the Drude model implicitly assumes that the electrons are confined to the interior of the metal. Evidently this confinement is brought about by their attraction to the positively charged ions. Gross effects of the electron-ion and electron-electron interaction like this are often taken into account by adding to the external fields a suitably defined internal field representing the average effect of the electron-electron and electron-ion interactions.

Table 1.1
FREE ELECTRON DENSITIES OF SELECTED METALLIC ELEMENTS[a]

ELEMENT	Z	$n\ (10^{22}/\text{cm}^3)$	r_s(Å)	r_s/a_0
Li (78 K)	1	4.70	1.72	3.25
Na (5 K)	1	2.65	2.08	3.93
K (5 K)	1	1.40	2.57	4.86
Rb (5 K)	1	1.15	2.75	5.20
Cs (5 K)	1	0.91	2.98	5.62
Cu	1	8.47	1.41	2.67
Ag	1	5.86	1.60	3.02
Au	1	5.90	1.59	3.01
Be	2	24.7	0.99	1.87
Mg	2	8.61	1.41	2.66
Ca	2	4.61	1.73	3.27
Sr	2	3.55	1.89	3.57
Ba	2	3.15	1.96	3.71
Nb	1	5.56	1.63	3.07
Fe	2	17.0	1.12	2.12
Mn (α)	2	16.5	1.13	2.14
Zn	2	13.2	1.22	2.30
Cd	2	9.27	1.37	2.59
Hg (78 K)	2	8.65	1.40	2.65
Al	3	18.1	1.10	2.07
Ga	3	15.4	1.16	2.19
In	3	11.5	1.27	2.41
Tl	3	10.5	1.31	2.48
Sn	4	14.8	1.17	2.22
Pb	4	13.2	1.22	2.30
Bi	5	14.1	1.19	2.25
Sb	5	16.5	1.13	2.14

[a] At room temperature (about 300 K) and atmospheric pressure, unless otherwise noted. The radius r_s of the free electron sphere is defined in Eq. (1.2). We have arbitrarily selected one value of Z for those elements that display more than one chemical valence. The Drude model gives no theoretical basis for the choice. Values of n are based on data from R. W. G. Wyckoff, *Crystal Structures*, 2nd ed., Interscience, New York, 1963.

although the independent electron approximation is in many contexts surprisingly good, the free electron approximation must be abandoned if one is to arrive at even a qualitative understanding of much of metallic behavior.

2. Collisions in the Drude model, as in kinetic theory, are instantaneous events that abruptly alter the velocity of an electron. Drude attributed them to the electrons bouncing off the impenetrable ion cores (rather than to electron-electron collisions, the analogue of the predominant collision mechanism in an ordinary gas). We shall find later that electron-electron scattering is indeed one of the least important of the several scattering mechanisms in a metal, except under unusual conditions. However,

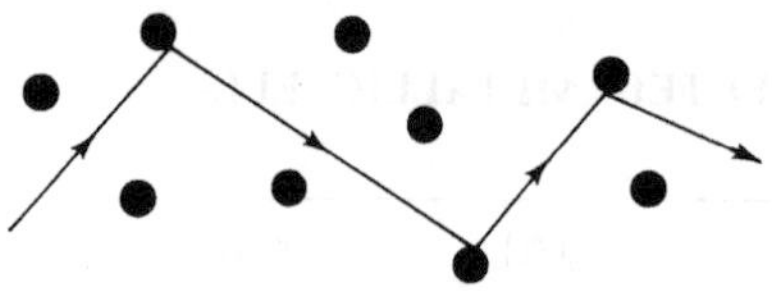

Figure 1.2
Trajectory of a conduction electron scattering off the ions, according to the naive picture of Drude.

the simple mechanical picture (Figure 1.2) of an electron bumping along from ion to ion is very far off the mark.[7] Fortunately, this does not matter for many purposes: a qualitative (and often a quantitative) understanding of metallic conduction can be achieved by simply assuming that there is *some* scattering mechanism, without inquiring too closely into just what that mechanism might be. By appealing, in our analysis, to only a few general effects of the collision process, we can avoid committing ourselves to any specific picture of how electron scattering actually takes place. These broad features are described in the following two assumptions.

3. We shall assume that an electron experiences a collision (i.e., suffers an abrupt change in its velocity) with a probability per unit time $1/\tau$. We mean by this that the probability of an electron undergoing a collision in any infinitesimal time interval of length dt is just dt/τ. The time τ is variously known as the relaxation time, the collision time, or the mean free time, and it plays a fundamental role in the theory of metallic conduction. It follows from this assumption that an electron picked at random at a given moment will, on the average, travel for a time τ before its next collision, and will, on the average, have been traveling for a time τ since its last collision.[8] In the simplest applications of the Drude model the collision time τ is taken to be independent of an electron's position and velocity. We shall see later that this turns out to be a surprisingly good assumption for many (but by no means all) applications.

4. Electrons are assumed to achieve thermal equilibrium with their surroundings only through collisions.[9] These collisions are assumed to maintain local thermodynamic equilibrium in a particularly simple way: immediately after each collision an electron is taken to emerge with a velocity that is not related to its velocity just before the collision, but randomly directed and with a speed appropriate to the temperature prevailing at the place where the collision occurred. Thus the hotter the region in which a collision occurs, the faster a typical electron will emerge from the collision.

In the rest of this chapter we shall illustrate these notions through their most important applications, noting the extent to which they succeed or fail to describe the observed phenomena.

DC ELECTRICAL CONDUCTIVITY OF A METAL

According to *Ohm's law*, the current I flowing in a wire is proportional to the potential drop V along the wire: $V = IR$, where R, the resistance of the wire, depends on its

---

[7] For some time people were led into difficult but irrelevant problems connected with the proper aiming of an electron at an ion in each collision. So literal an interpretation of Figure 1.2 is strenuously to be avoided.

[8] See Problem 1.

[9] Given the free and independent electron approximation, this is the only possible mechanism left.

dimensions, but is independent of the size of the current or potential drop. The Drude model accounts for this behavior and provides an estimate of the size of the resistance.

One generally eliminates the dependence of R on the shape of the wire by introducing a quantity characteristic only of the metal of which the wire is composed. The resistivity ρ is defined to be the proportionality constant between the electric field $\mathbf{E}$ at a point in the metal and the current density $\mathbf{j}$ that it induces[10]:

$$\mathbf{E} = \rho\mathbf{j}. \tag{1.3}$$

The current density $\mathbf{j}$ is a vector, parallel to the flow of charge, whose magnitude is the amount of charge per unit time crossing a unit area perpendicular to the flow. Thus if a uniform current I flows through a wire of length L and cross-sectional area A, the current density will be $j = I/A$. Since the potential drop along the wire will be $V = EL$, Eq. (1.3) gives $V = I\rho L/A$, and hence $R = \rho L/A$.

If n electrons per unit volume all move with velocity $\mathbf{v}$, then the current density they give rise to will be parallel to $\mathbf{v}$. Furthermore, in a time dt the electrons will advance by a distance $v\,dt$ in the direction of $\mathbf{v}$, so that $n(v\,dt)A$ electrons will cross an area A perpendicular to the direction of flow. Since each electron carries a charge $-e$, the charge crossing A in the time dt will be $-nevA\,dt$, and hence the current density is

$$\mathbf{j} = -ne\mathbf{v}. \tag{1.4}$$

At any point in a metal, electrons are always moving in a variety of directions with a variety of thermal energies. The net current density is thus given by (1.4), where $\mathbf{v}$ is the average electronic velocity. In the absence of an electric field, electrons are as likely to be moving in any one direction as in any other, $\mathbf{v}$ averages to zero, and, as expected, there is no net electric current density. In the presence of a field $\mathbf{E}$, however, there will be a mean electronic velocity directed opposite to the field (the electronic charge being negative), which we can compute as follows:

Consider a typical electron at time zero. Let t be the time elapsed since its last collision. Its velocity at time zero will be its velocity $\mathbf{v}_0$ immediately after that collision plus the additional velocity $-e\mathbf{E}t/m$ it has subsequently acquired. Since we assume that an electron emerges from a collision in a random direction, there will be no contribution from $\mathbf{v}_0$ to the average electronic velocity, which must therefore be given entirely by the average of $-e\mathbf{E}t/m$. However, the average of t is the relaxation time τ. Therefore

$$\mathbf{v}_{\text{avg}} = -\frac{e\mathbf{E}\tau}{m}; \quad \mathbf{j} = \left(\frac{ne^2\tau}{m}\right)\mathbf{E}. \tag{1.5}$$

This result is usually stated in terms of the inverse of the resistivity, the conductivity $\sigma = 1/\rho$:

$$\boxed{\mathbf{j} = \sigma\mathbf{E}; \quad \sigma = \frac{ne^2\tau}{m}.} \tag{1.6}$$

---

10 In general, $\mathbf{E}$ and $\mathbf{j}$ need not be parallel. One then defines a resistivity *tensor*. See Chapters 12 and 13.

This establishes the linear dependence of **j** on **E** and gives an estimate of the conductivity σ in terms of quantities that are all known except for the relaxation time τ. We may therefore use (1.6) and the observed resistivities to estimate the size of the relaxation time:

$$\tau = \frac{m}{\rho n e^2}. \tag{1.7}$$

Table 1.2 gives the resistivities of several representative metals at several temperatures. Note the strong temperature dependence. At room temperature the resistivity is roughly linear in T, but it falls away much more steeply as low temperatures are

Table 1.2
ELECTRICAL RESISTIVITIES OF SELECTED ELEMENTS[a]

ELEMENT	77 K	273 K	373 K	$\dfrac{(\rho/T)_{373\text{ K}}}{(\rho/T)_{273\text{ K}}}$
Li	1.04	8.55	12.4	1.06
Na	0.8	4.2	Melted	
K	1.38	6.1	Melted	
Rb	2.2	11.0	Melted	
Cs	4.5	18.8	Melted	
Cu	0.2	1.56	2.24	1.05
Ag	0.3	1.51	2.13	1.03
Au	0.5	2.04	2.84	1.02
Be		2.8	5.3	1.39
Mg	0.62	3.9	5.6	1.05
Ca		3.43	5.0	1.07
Sr	7	23		
Ba	17	60		
Nb	3.0	15.2	19.2	0.92
Fe	0.66	8.9	14.7	1.21
Zn	1.1	5.5	7.8	1.04
Cd	1.6	6.8		
Hg	5.8	Melted	Melted	
Al	0.3	2.45	3.55	1.06
Ga	2.75	13.6	Melted	
In	1.8	8.0	12.1	1.11
Tl	3.7	15	22.8	1.11
Sn	2.1	10.6	15.8	1.09
Pb	4.7	19.0	27.0	1.04
Bi	35	107	156	1.07
Sb	8	39	59	1.11

[a] Resistivities in microhm centimeters are given at 77 K (the boiling point of liquid nitrogen at atmospheric pressure), 273 K, and 373 K. The last column gives the ratio of ρ/T at 373 K and 273 K to display the approximate linear temperature dependence of the resistivity near room temperature.

Source: G. W. C. Kaye and T. H. Laby, *Table of Physical and Chemical Constants*, Longmans Green, London, 1966.

reached. Room temperature resistivities are typically of the order of microhm centimeters (μohm-cm) or, in atomic units, of order 10^{-18} statohm-cm.[11] If ρ_μ is the resistivity in microhm centimeters, then a convenient way of expressing the relaxation time implied by (1.7) is

$$\tau = \left(\frac{0.22}{\rho_\mu}\right)\left(\frac{r_s}{a_0}\right)^3 \times 10^{-14} \text{ sec.} \tag{1.8}$$

Relaxation times calculated from (1.8) and the resistivities in Table 1.2 are displayed in Table 1.3. Note that at room temperatures τ is typically 10^{-14} to 10^{-15} sec. In considering whether this is a reasonable number, it is more instructive to contemplate the mean free path, $\ell = v_0\tau$, where v_0 is the average electronic speed. The length ℓ measures the average distance an electron travels between collisions. In Drude's time it was natural to estimate v_0 from classical equipartition of energy: $\frac{1}{2}mv_0{}^2 = \frac{3}{2}k_BT$. Using the known electronic mass, we find a v_0 of order 10^7 cm/sec at room temperature, and hence a mean free path of 1 to 10 Å. Since this distance is comparable to the interatomic spacing, the result is quite consistent with Drude's original view that collisions are due to the electron bumping into the large heavy ions.

However, we shall see in Chapter 2 that this classical estimate of v_0 is an order of magnitude too small at room temperatures. Furthermore, at the lowest temperatures in Table 1.3, τ is an order of magnitude larger than at room temperature, while (as we shall see in Chapter 2) v_0 is actually temperature-independent. This can raise the low-temperature mean free path to 10^3 or more angstroms, about a thousand times the spacing between ions. Today, by working at sufficiently low temperatures with carefully prepared samples, mean free paths of the order of centimeters (i.e., 10^8 interatomic spacings) can be achieved. This is strong evidence that the electrons do not simply bump off the ions, as Drude supposed.

Fortunately, however, we may continue to calculate with the Drude model without any precise understanding of the cause of collisions. In the absence of a theory of the collision time it becomes important to find predictions of the Drude model that are independent of the value of the relaxation time τ. As it happens, there are several such τ-independent quantities, which even today remain of fundamental interest, for in many respects the precise quantitative treatment of the relaxation time remains the weakest link in modern treatments of metallic conductivity. As a result, τ-independent quantities are highly valued, for they often yield considerably more reliable information.

Two cases of particular interest are the calculation of the electrical conductivity when a spatially uniform static magnetic field is present, and when the electric field

---

[11] To convert resistivities from microhm centimeters to statohm centimeters note that a resistivity of 1 μohm-cm yields an electric field of 10^{-6} volt/cm in the presence of a current of 1 amp/cm^2. Since 1 amp is 3×10^9 esu/sec, and 1 volt is $\frac{1}{300}$ statvolt, a resistivity of 1 μohm-cm yields a field of 1 statvolt/cm when the current density is $300 \times 10^6 \times 3 \times 10^9$ esu-cm^{-2}-sec^{-1}. The statohm-centimeter is the electrostatic unit of resistivity, and therefore gives 1 statvolt/cm with a current density of only 1 esu-cm^{-2}-sec^{-1}. Thus 1 μohm-cm is equivalent to $\frac{1}{9} \times 10^{-17}$ statohm-cm. To avoid using the statohm-centimeter, one may evaluate (1.7) taking ρ in ohm meters, m in kilograms, n in electrons per cubic meter, and e in coulombs. (*Note:* The most important formulas, constants, and conversion factors from Chapters 1 and 2 are summarized in Appendix A.)

Table 1.3
DRUDE RELAXATION TIMES IN UNITS OF 10^{-14} SECOND[a]

ELEMENT	77 K	273 K	373 K
Li	7.3	0.88	0.61
Na	17	3.2	
K	18	4.1	
Rb	14	2.8	
Cs	8.6	2.1	
Cu	21	2.7	1.9
Ag	20	4.0	2.8
Au	12	3.0	2.1
Be		0.51	0.27
Mg	6.7	1.1	0.74
Ca		2.2	1.5
Sr	1.4	0.44	
Ba	0.66	0.19	
Nb	2.1	0.42	0.33
Fe	3.2	0.24	0.14
Zn	2.4	0.49	0.34
Cd	2.4	0.56	
Hg	0.71		
Al	6.5	0.80	0.55
Ga	0.84	0.17	
In	1.7	0.38	0.25
Tl	0.91	0.22	0.15
Sn	1.1	0.23	0.15
Pb	0.57	0.14	0.099
Bi	0.072	0.023	0.016
Sb	0.27	0.055	0.036

[a] Relaxation times are calculated from the data in Tables 1.1 and 1.2, and Eq. (1.8). The slight temperature dependence of n is ignored.

is spatially uniform but time-dependent. Both of these cases are most simply dealt with by the following observation:

At any time t the average electronic velocity $\mathbf{v}$ is just $\mathbf{p}(t)/m$, where $\mathbf{p}$ is the total momentum per electron. Hence the current density is

$$\mathbf{j} = -\frac{ne\mathbf{p}(t)}{m}. \tag{1.9}$$

Given that the momentum per electron is $\mathbf{p}(t)$ at time t, let us calculate the momentum per electron $\mathbf{p}(t + dt)$ an infinitesimal time dt later. An electron taken at random at time t will have a collision before time $t + dt$, with probability dt/τ, and will therefore survive to time $t + dt$ without suffering a collision with probability $1 - dt/\tau$. If it experiences no collision, however, it simply evolves under the influence of the force $\mathbf{f}(t)$ (due to the spatially uniform electric and/or magnetic fields) and will therefore acquire an additional momentum[12] $\mathbf{f}(t)\,dt + O(dt)^2$. The contribution of all those electrons that do not collide between t and $t + dt$ to the momentum per electron at time $t + dt$ is the fraction $(1 - dt/\tau)$ they constitute of all electrons, times *their* average momentum per electron, $\mathbf{p}(t) + \mathbf{f}(t)\,dt + O(dt)^2$.

Thus neglecting for the moment the contribution to $\mathbf{p}(t + dt)$ from those electrons that *do* undergo a collision in the time between t and $t + dt$, we have[13]

$$\begin{aligned}\mathbf{p}(t + dt) &= \left(1 - \frac{dt}{\tau}\right)\left[\mathbf{p}(t) + \mathbf{f}(t)dt + O(dt)^2\right]\\ &= \mathbf{p}(t) - \left(\frac{dt}{\tau}\right)\mathbf{p}(t) + \mathbf{f}(t)dt + O(dt)^2.\end{aligned} \tag{1.10}$$

The correction to (1.10) due to those electrons that have had a collision in the interval t to $t + dt$ is only of the order of $(dt)^2$. To see this, first note that such electrons constitute a fraction dt/τ of the total number of electrons. Furthermore, since the electronic velocity (and momentum) is randomly directed immediately after a collision, each such electron will contribute to the average momentum $\mathbf{p}(t + dt)$ only to the extent that it has acquired momentum from the force $\mathbf{f}$ since its last collision. Such momentum is acquired over a time no longer than dt, and is therefore of order $\mathbf{f}(t)dt$. Thus the correction to (1.10) is of order $(dt/\tau)\mathbf{f}(t)dt$, and does not affect the terms of linear order in dt. We may therefore write:

$$\mathbf{p}(t + dt) - \mathbf{p}(t) = -\left(\frac{dt}{\tau}\right)\mathbf{p}(t) + \mathbf{f}(t)\,dt + O(dt)^2, \tag{1.11}$$

where the contribution of *all* electrons to $\mathbf{p}(t + dt)$ is accounted for. Dividing this by dt and taking the limit as $dt \to 0$, we find

$$\frac{d\mathbf{p}(t)}{dt} = -\frac{\mathbf{p}(t)}{\tau} + \mathbf{f}(t). \tag{1.12}$$

This simply states that the effect of individual electron collisions is to introduce a frictional damping term into the equation of motion for the momentum per electron.

We now apply (1.12) to several cases of interest.

HALL EFFECT AND MAGNETORESISTANCE

In 1879 E. H. Hall tried to determine whether the force experienced by a current carrying wire in a magnetic field was exerted on the whole wire or only upon (what we would now call) the moving electrons in the wire. He suspected it was the latter, and his experiment was based on the argument that "if the current of electricity in a fixed conductor is itself attracted by a magnet, the current should be drawn to one side of the wire, and therefore the resistance experienced should be increased."[14] His

---

[12] By $O(dt)^2$ we mean a term of the order of $(dt)^2$.

[13] If the force on the electrons is not the same for every electron, (1.10) will remain valid provided that we interpret $\mathbf{f}$ as the *average* force per electron.

[14] *Am. J. Math.* **2**, 287 (1879).

efforts to detect this extra resistance were unsuccessful,[15] but Hall did not regard this as conclusive: "The magnet may *tend* to deflect the current without being able to do so. It is evident that in this case there would exist a state of stress in the conductor, the electricity pressing, as it were, toward one side of the wire." This state of stress should appear as a transverse voltage (known today as the Hall voltage), which Hall was able to observe.

Hall's experiment is depicted in Figure 1.3. An electric field E_x is applied to a wire extending in the x-direction and a current density j_x flows in the wire. In addition, a magnetic field $\mathbf{H}$ points in the positive z-direction. As a result the Lorentz force[16]

$$-\frac{e}{c}\mathbf{v} \times \mathbf{H} \tag{1.13}$$

acts to deflect electrons in the negative y-direction (an electron's drift velocity is *opposite* to the current flow). However the electrons cannot move very far in the y-direction before running up against the sides of the wire. As they accumulate there, an electric field builds up in the y-direction that opposes their motion and their further accumulation. In equilibrium this transverse field (or Hall field) E_y will balance the Lorentz force, and current will flow only in the x-direction.

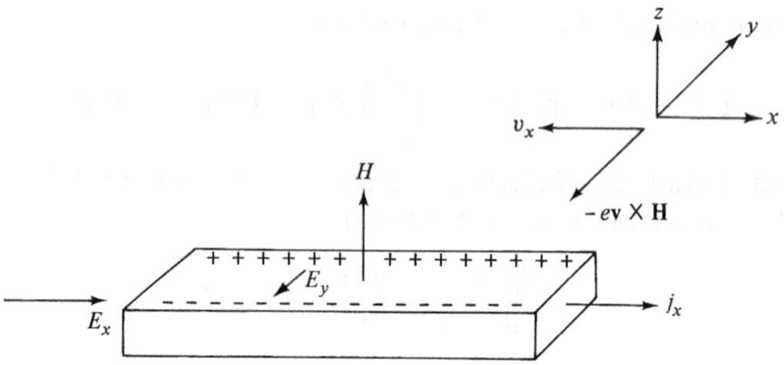

Figure 1.3
Schematic view of Hall's experiment.

There are two quantities of interest. One is the ratio of the field along the wire E_x to the current density j_x,

$$\rho(H) = \frac{E_x}{j_x}. \tag{1.14}$$

This is the magnetoresistance,[17] which Hall found to be field-independent. The other is the size of the transverse field E_y. Since it balances the Lorentz force, one might expect it to be proportional both to the applied field H and to the current along the

[15] The increase in resistance (known as the magnetoresistance) does occur, as we shall see in Chapters 12 and 13. The Drude model, however, predicts Hall's null result.

[16] When dealing with nonmagnetic (or weakly magnetic) materials, we shall always call the field **H**, the difference between **B** and **H** being extremely small.

[17] More precisely, it is the transverse magnetoresistance. There is also a longitudinal magnetoresistance, measured with the magnetic field parallel to the current.

wire j_x. One therefore defines a quantity known as the Hall coefficient by

$$R_H = \frac{E_y}{j_x H}. \tag{1.15}$$

Note that since the Hall field is in the negative y-direction (Figure 1.3), R_H should be negative. If, on the other hand, the charge carriers were positive, then the sign of their x-velocity would be reversed, and the Lorentz force would therefore be unchanged. As a consequence the Hall field would be opposite to the direction it has for negatively charged carriers. This is of great importance, for it means that a measurement of the Hall field determines the sign of the charge carriers. Hall's original data agreed with the sign of the electronic charge later determined by Thomson. One of the remarkable aspects of the Hall effect, however, is that in some metals the Hall coefficient is positive, suggesting that the carriers have a charge opposite to that of the electron. This is another mystery whose solution had to await the full quantum theory of solids. In this chapter we shall consider only the simple Drude model analysis, which though incapable of accounting for positive Hall coefficients, is often in fairly good agreement with experiment.

To calculate the Hall coefficient and magnetoresistance we first find the current densities j_x and j_y in the presence of an electric field with arbitrary components E_x and E_y, and in the presence of a magnetic field $\mathbf{H}$ along the z-axis. The (position independent) force acting on each electron is $\mathbf{f} = -e(\mathbf{E} + \mathbf{v} \times \mathbf{H}/c)$, and therefore Eq. (1.12) for the momentum per electron becomes[18]

$$\frac{d\mathbf{p}}{dt} = -e\left(\mathbf{E} + \frac{\mathbf{p}}{mc} \times \mathbf{H}\right) - \frac{\mathbf{p}}{\tau}. \tag{1.16}$$

In the steady state the current is independent of time, and therefore p_x and p_y will satisfy

$$0 = -eE_x - \omega_c p_y - \frac{p_x}{\tau}, \tag{1.17}$$

$$0 = -eE_y + \omega_c p_x - \frac{p_y}{\tau},$$

where

$$\omega_c = \frac{eH}{mc}. \tag{1.18}$$

We multiply these equations by $-ne\tau/m$ and introduce the current density components through (1.4) to find

$$\begin{aligned} \sigma_0 E_x &= \omega_c \tau j_y + j_x, \\ \sigma_0 E_y &= -\omega_c \tau j_x + j_y, \end{aligned} \tag{1.19}$$

where σ_0 is just the Drude model DC conductivity in the absence of a magnetic field, given by (1.6).

---

18 Note that the Lorentz force is not the same for each electron since it depends on the electronic velocity $\mathbf{v}$. Therefore the force $\mathbf{f}$ in (1.12) is to be taken as the average force per electron (see Footnote 13). Because, however, the force depends on the electron on which it acts only through a term *linear* in the electron's velocity, the average force is obtained simply by replacing that velocity by the average velocity, $\mathbf{p}/m$.

The Hall field E_y is determined by the requirement that there be no transverse current j_y. Setting j_y to zero in the second equation of (1.19) we find that

$$E_y = -\left(\frac{\omega_c\tau}{\sigma_0}\right)j_x = -\left(\frac{H}{nec}\right)j_x. \tag{1.20}$$

Therefore the Hall coefficient (1.15) is

$$R_H = -\frac{1}{nec}. \tag{1.21}$$

This is a very striking result, for it asserts that the Hall coefficient depends on no parameters of the metal except the density of carriers. Since we have already calculated n assuming that the atomic valence electrons become the metallic conduction electrons, a measurement of the Hall constant provides a direct test of the validity of this assumption.

In trying to extract the electron density n from measured Hall coefficients one is faced with the problem that, contrary to the prediction of (1.21), they generally do depend on magnetic field. Furthermore, they depend on temperature and on the care with which the sample has been prepared. This result is somewhat unexpected, since the relaxation time τ, which can depend strongly on temperature and the condition of the sample, does not appear in (1.21). However, at very low temperatures in very pure, carefully prepared samples at very high fields, the measured Hall constants do appear to approach a limiting value. The more elaborate theory of Chapters 12 and 13 predicts that for many (but not all) metals this limiting value is precisely the simple Drude result (1.21).

Some Hall coefficients at high and moderate fields are listed in Table 1.4. Note the occurrence of cases in which R_H is actually positive, apparently corresponding to carriers with a positive charge. A striking example of observed field dependence totally unexplained by Drude theory is shown in Figure 1.4.

The Drude result confirms Hall's observation that the resistance does not depend on field, for when $j_y = 0$ (as is the case in the steady state when the Hall field has been established), the first equation of (1.19) reduces to $j_x = \sigma_0 E_x$, the expected result for the conductivity in zero magnetic field. However, more careful experiments on a variety of metals have revealed that there is a magnetic field dependence to the resistance, which can be quite dramatic in some cases. Here again the quantum theory of solids is needed to explain why the Drude result applies in some metals and to account for some truly extraordinary deviations from it in others.

Before leaving the subject of DC phenomena in a uniform magnetic field, we note for future applications that the quantity $\omega_c\tau$ is an important, dimensionless measure of the strength of a magnetic field. When $\omega_c\tau$ is small, Eq. (1.19) gives $\mathbf{j}$ very nearly parallel to $\mathbf{E}$, as in the absence of a magnetic field. In general, however, $\mathbf{j}$ is at an angle ϕ (known as the Hall angle) to $\mathbf{E}$, where (1.19) gives $\tan\phi = \omega_c\tau$. The quantity ω_c, known as the cyclotron frequency, is simply the angular frequency of revolution[19]

[19] In a uniform magnetic field the orbit of an electron is a spiral along the field whose projection in a plane perpendicular to the field is a circle. The angular frequency ω_c is determined by the condition that the centripetal acceleration $\omega_c^2 r$ be provided by the Lorentz force, $(e/c)(\omega_c r)H$.

Table 1.4
HALL COEFFICIENTS OF SELECTED ELEMENTS IN MODERATE TO HIGH FIELDS[a]

METAL	VALENCE	$-1/R_H nec$
Li	1	0.8
Na	1	1.2
K	1	1.1
Rb	1	1.0
Cs	1	0.9
Cu	1	1.5
Ag	1	1.3
Au	1	1.5
Be	2	−0.2
Mg	2	−0.4
In	3	−0.3
Al	3	−0.3

[a] These are roughly the limiting values assumed by R_H as the field becomes very large (of order 10^4 G), and the temperature very low, in carefully prepared specimens. The data are quoted in the form n_0/n, where n_0 is the density for which the Drude form (1.21) agrees with the measured R_H: $n_0 = -1/R_H ec$. Evidently the alkali metals obey the Drude result reasonably well, the noble metals (Cu, Ag, Au) less well, and the remaining entries, not at all.

of a free electron in the magnetic field H. Thus $\omega_c\tau$ will be small if electrons can complete only a small part of a revolution between collisions, and large if they can complete many revolutions. Alternatively, when $\omega_c\tau$ is small the magnetic field deforms the electronic orbits only slightly, but when $\omega_c\tau$ is comparable to unity or larger, the effect of the magnetic field on the electronic orbits is quite drastic. A useful numerical evaluation of the cyclotron frequency is

$$\nu_c\,(10^9 \text{ hertz}) = 2.80 \times H\,(\text{kilogauss}), \qquad \omega_c = 2\pi\nu_c. \tag{1.22}$$

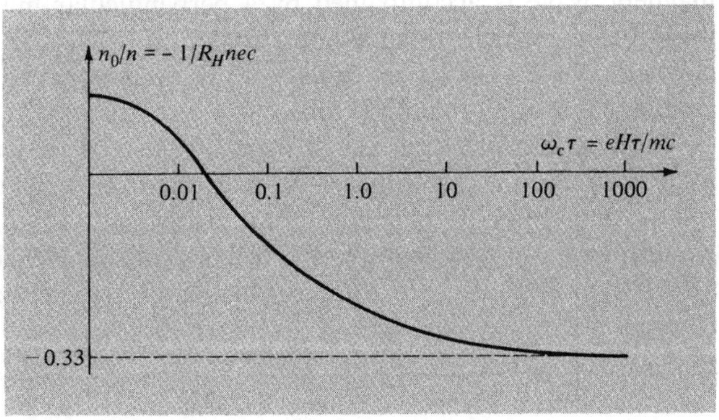

Figure 1.4
The quantity $n_0/n = -1/R_H nec$, for aluminum, as a function of $\omega_c\tau$. The free electron density n is based on a nominal chemical valence of 3. The high field value suggests only one carrier per primitive cell, with a positive charge. (From R. Lück, *Phys. Stat. Sol.* **18**, 49 (1966).)

AC ELECTRICAL CONDUCTIVITY OF A METAL

To calculate the current induced in a metal by a time-dependent electric field, we write the field in the form

$$\mathbf{E}(t) = \mathrm{Re}\,(\mathbf{E}(\omega)e^{-i\omega t}). \tag{1.23}$$

The equation of motion (1.12) for the momentum per electron becomes

$$\frac{d\mathbf{p}}{dt} = -\frac{\mathbf{p}}{\tau} - e\mathbf{E}. \tag{1.24}$$

We seek a steady-state solution of the form

$$\mathbf{p}(t) = \mathrm{Re}\,(\mathbf{p}(\omega)e^{-i\omega t}). \tag{1.25}$$

Substituting the complex **p** and **E** into (1.24), which must be satisfied by both the real and imaginary parts of any complex solution, we find that $\mathbf{p}(\omega)$ must satisfy

$$-i\omega\mathbf{p}(\omega) = -\frac{\mathbf{p}(\omega)}{\tau} - e\mathbf{E}(\omega). \tag{1.26}$$

Since $\mathbf{j} = -ne\mathbf{p}/m$, the current density is just

$$\mathbf{j}(t) = \mathrm{Re}\,(\mathbf{j}(\omega)e^{-i\omega t}),$$
$$\mathbf{j}(\omega) = -\frac{ne\mathbf{p}(\omega)}{m} = \frac{(ne^2/m)\mathbf{E}(\omega)}{(1/\tau) - i\omega}. \tag{1.27}$$

One customarily writes this result as

$$\mathbf{j}(\omega) = \sigma(\omega)\mathbf{E}(\omega), \tag{1.28}$$

where $\sigma(\omega)$, known as the frequency-dependent (or AC) conductivity, is given by

$$\sigma(\omega) = \frac{\sigma_0}{1 - i\omega\tau}, \quad \sigma_0 = \frac{ne^2\tau}{m}. \tag{1.29}$$

Note that this correctly reduces to the DC Drude result (1.6) at zero frequency.

The most important application of this result is to the propagation of electromagnetic radiation in a metal. It might appear that the assumptions we made to derive (1.29) would render it inapplicable to this case, since (a) the **E** field in an electromagnetic wave is accompanied by a perpendicular magnetic field **H** of the same magnitude,[20] which we have not included in (1.24), and (b) the fields in an electromagnetic wave vary in space as well as time, whereas Eq. (1.12) was derived by assuming a spatially uniform force.

The first complication can always be ignored. It leads to an additional term $-e\mathbf{p}/mc \times \mathbf{H}$ in (1.24), which is smaller than the term in **E** by a factor v/c, where v is the magnitude of the mean electronic velocity. But even in a current as large as 1 amp/mm^2, $v = j/ne$ is only of order 0.1 cm/sec. Hence the term in the magnetic field is typically 10^{-10} of the term in the electric field and can quite correctly be ignored.

---

[20] One of the more appealing features of CGS units.

The second point raises more serious questions. Equation (1.12) was derived by assuming that at any time the same force acts on each electron, which is not the case if the electric field varies in space. Note, however, that the current density at point **r** is entirely determined by what the electric field has done to each electron at **r** since its last collision. This last collision, in the overwhelming majority of cases, takes place no more than a few mean free paths away from **r**. Therefore if the field does not vary appreciably over distances comparable to the electronic mean free path, we may correctly calculate $\mathbf{j}(\mathbf{r}, t)$, the current density at point **r**, by taking the field everywhere in space to be given by its value $\mathbf{E}(\mathbf{r}, t)$ at the point **r**. The result,

$$\mathbf{j}(\mathbf{r}, \omega) = \sigma(\omega)\mathbf{E}(\mathbf{r}, \omega), \tag{1.30}$$

is therefore valid whenever the wavelength λ of the field is large compared to the electronic mean free path ℓ. This is ordinarily satisfied in a metal by visible light (whose wavelength is of the order of 10^3 to 10^4 Å). When it is not satisfied, one must resort to so-called nonlocal theories, of greater complexity.

Assuming, then, that the wavelength is large compared to the mean free path, we may proceed as follows: in the presence of a specified current density **j** we may write Maxwell's equations as[21]

$$\nabla \cdot \mathbf{E} = 0; \quad \nabla \cdot \mathbf{H} = 0; \quad \nabla \times \mathbf{E} = -\frac{1}{c}\frac{\partial \mathbf{H}}{\partial t};$$

$$\nabla \times \mathbf{H} = \frac{4\pi}{c}\mathbf{j} + \frac{1}{c}\frac{\partial \mathbf{E}}{\partial t}. \tag{1.31}$$

We look for a solution with time dependence $e^{-i\omega t}$, noting that in a metal we can write **j** in terms of **E** via (1.28). We then find

$$\nabla \times (\nabla \times \mathbf{E}) = -\nabla^2\mathbf{E} = \frac{i\omega}{c}\nabla \times \mathbf{H} = \frac{i\omega}{c}\left(\frac{4\pi\sigma}{c}\mathbf{E} - \frac{i\omega}{c}\mathbf{E}\right), \tag{1.32}$$

or

$$-\nabla^2\mathbf{E} = \frac{\omega^2}{c^2}\left(1 + \frac{4\pi i\sigma}{\omega}\right)\mathbf{E}. \tag{1.33}$$

This has the form of the usual wave equation,

$$-\nabla^2\mathbf{E} = \frac{\omega^2}{c^2}\epsilon(\omega)\mathbf{E}, \tag{1.34}$$

with a complex dielectric constant given by

$$\epsilon(\omega) = 1 + \frac{4\pi i\sigma}{\omega}. \tag{1.35}$$

If we are at frequencies high enough to satisfy

$$\omega\tau \gg 1, \tag{1.36}$$

---

21 We are considering here an electromagnetic wave, in which the induced charge density ρ vanishes. Below we examine the possibility of oscillations in the charge density.

then, to a first approximation, Eqs. (1.35) and (1.29) give

$$\epsilon(\omega) = 1 - \frac{\omega_p^{\ 2}}{\omega^2}, \tag{1.37}$$

where ω_p, known as the plasma frequency, is given by

$$\omega_p^{\ 2} = \frac{4\pi n e^2}{m}. \tag{1.38}$$

When ϵ is real and negative ($\omega < \omega_p$) the solutions to (1.34) decay exponentially in space; i.e., no radiation can propagate. However, when ϵ is positive ($\omega > \omega_p$) the solutions to (1.34) become oscillatory, radiation can propagate, and the metal should become transparent. This conclusion is only valid, of course, if our high-frequency assumption (1.36) is satisfied in the neighborhood of $\omega = \omega_p$. If we express τ in terms of the resistivity through Eq. (1.8), then we can use the definition (1.38) of the plasma frequency to compute that

$$\omega_p \tau = 1.6 \times 10^2 \left(\frac{r_s}{a_0}\right)^{3/2} \left(\frac{1}{\rho_\mu}\right). \tag{1.39}$$

Since the resistivity in microhm centimeters, ρ_μ, is of the order of unity or less, and since r_s/a_0 is in the range from 2 to 6, the high frequency condition (1.36) will be well satisfied at the plasma frequency.

The alkali metals have, in fact, been observed to become transparent in the ultraviolet. A numerical evaluation of (1.38) gives the frequency at which transparency should set in as

$$\nu_p = \frac{\omega_p}{2\pi} = 11.4 \times \left(\frac{r_s}{a_0}\right)^{-3/2} \times 10^{15}\ \text{Hz} \tag{1.40}$$

or

$$\lambda_p = \frac{c}{\nu_p} = 0.26 \left(\frac{r_s}{a_0}\right)^{3/2} \times 10^3\ \text{Å}. \tag{1.41}$$

In Table 1.5 we list the threshold wavelengths calculated from (1.41), along with the

Table 1.5

OBSERVED AND THEORETICAL WAVELENGTHS BELOW WHICH THE ALKALI METALS BECOME TRANSPARENT

ELEMENT	THEORETICAL[a] λ (10^3 Å)	OBSERVED λ (10^3 Å)
Li	1.5	2.0
Na	2.0	2.1
K	2.8	3.1
Rb	3.1	3.6
Cs	3.5	4.4

[a] From Eq. (1.41).

Source: M. Born and E. Wolf, *Principles of Optics*, Pergamon, New York, 1964.

observed thresholds. The agreement between theory and experiment is rather good. As we shall see, the actual dielectric constant of a metal is far more complicated than (1.37) and it is to some extent a piece of good fortune that the alkali metals so strikingly display this Drude behavior. In other metals different contributions to the dielectric constant compete quite substantially with the "Drude term" (1.37).

A second important consequence of (1.37) is that the electron gas can sustain charge density oscillations. By this we mean a disturbance in which the electric charge density[22] has an oscillatory time dependence $e^{-i\omega t}$. From the equation of continuity,

$$\nabla \cdot \mathbf{j} = -\frac{\partial \rho}{\partial t}, \quad \nabla \cdot \mathbf{j}(\omega) = i\omega\rho(\omega), \tag{1.42}$$

and Gauss's law,

$$\nabla \cdot \mathbf{E}(\omega) = 4\pi\rho(\omega), \tag{1.43}$$

we find, in view of Eq. (1.30), that

$$i\omega\rho(\omega) = 4\pi\sigma(\omega)\rho(\omega). \tag{1.44}$$

This has a solution provided that

$$1 + \frac{4\pi i\sigma(\omega)}{\omega} = 0, \tag{1.45}$$

which is precisely the condition we found above for the onset of propagation of radiation. In the present context it emerges as the condition the frequency must meet if a charge density wave is to propagate.

The nature of this charge density wave, known as a plasma oscillation or plasmon, can be understood in terms of a very simple model.[23] Imagine displacing the entire electron gas, as a whole, through a distance d with respect to the fixed positive background of the ions (Figure 1.5).[24] The resulting surface charge gives rise to an electric field of magnitude $4\pi\sigma$, where σ is the charge per unit area[25] at either end of the slab.

Figure 1.5
Simple model of a plasma oscillation.

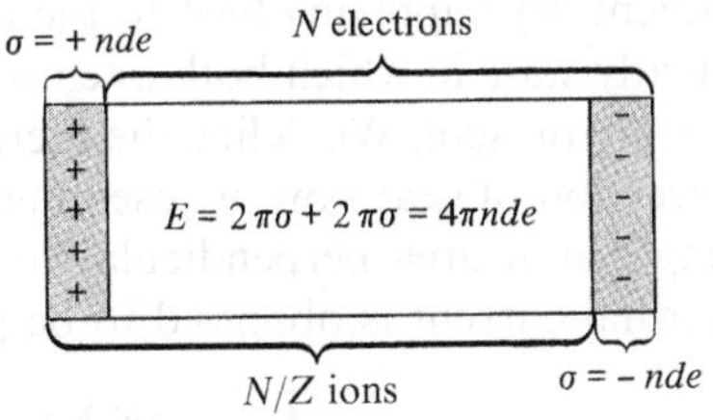

[22] The charge density ρ should not be confused with the resistivity, also generally denoted by ρ. The context will always make it clear which is being referred to.

[23] Since the field of a uniform plane of charge is independent of the distance from the plane, this crude argument, which places all of the charge density on two opposite surfaces, is not as crude as it appears at first glance.

[24] We observed earlier that the Drude model does take the electron-ion interaction into account by acknowledging that the attraction to the positively charged ions confines the electrons to the interior of the metal. In this simple model of a plasma oscillation it is precisely this attraction that provides the restoring force.

[25] The surface charge density σ should not be confused with the conductivity, also generally denoted by σ.

Consequently the electron gas as a whole will obey the equation of motion:

$$N m \ddot{d} = -Ne|4\pi\sigma| = -Ne\,(4\pi n d e) = -4\pi n e^2 N d, \tag{1.46}$$

which leads to oscillation at the plasma frequency.

Few direct observations have been made of plasmons. Perhaps the most notable is the observation of energy losses in multiples of $\hbar\omega_p$ when electrons are fired through thin, metallic films.[26] Nevertheless, the possibility of their excitation in the course of other electronic processes must always be borne in mind.

THERMAL CONDUCTIVITY OF A METAL

The most impressive success of the Drude model at the time it was proposed was its explanation of the empirical law of Wiedemann and Franz (1853). The Wiedemann-Franz law states that the ratio, κ/σ, of the thermal to the electrical conductivity of a great number of metals is directly proportional to the temperature, with a proportionality constant which is to a fair accuracy the same for all metals. This remarkable regularity can be seen in Table 1.6, where measured thermal conductivities are given for several metals at 273 K and 373 K, along with the ratios $\kappa/\sigma T$ (known as the Lorenz number) at the two temperatures.

In accounting for this the Drude model assumes that the bulk of the thermal current in a metal is carried by the conduction electrons. This assumption is based on the empirical observation that metals conduct heat much better than insulators do. Thus thermal conduction by the ions[27] (present in both metals and insulators) is much less important than thermal conduction by the conduction electrons (present only in metals).

To define and estimate the thermal conductivity, consider a metal bar along which the temperature varies slowly. If there were no sources and sinks of heat at the ends of the bar to maintain the temperature gradient, then the hot end would cool and the cool end would warm, i.e., thermal energy would flow in a sense opposite to the temperature gradient. By supplying heat to the hot end as fast as it flows away, one can produce a steady state in which both a temperature gradient and a uniform flow of thermal energy are present. We define the thermal current density $\mathbf{j}^q$ to be a vector parallel to the direction of heat flow, whose magnitude gives the thermal energy per unit time crossing a unit area perpendicular to the flow.[28] For small temperature gradients the thermal current is observed to be proportional to ∇T (Fourier's law):

$$\mathbf{j}^q = -\kappa\,\nabla T. \tag{1.47}$$

The proportionality constant κ is known as the thermal conductivity, and is positive, since the thermal current flows opposite to the direction of the temperature gradient.

---

[26] C. J. Powell and J. B. Swan, *Phys. Rev.* **115**, 869 (1959).

[27] Although the metallic ions cannot wander through the metal, there *is* a way in which they can transport thermal energy (though not electric charge): the ions can vibrate a little about their mean positions, leading to the transmission of thermal energy in the form of elastic waves propagating through the network of ions. See Chapter 25.

[28] Note the analogy to the definition of the electrical current density **j**, as well as the analogy between the laws of Ohm and Fourier.

Table 1.6
EXPERIMENTAL THERMAL CONDUCTIVITIES AND LORENZ NUMBERS OF SELECTED METALS

	273 K		373 K	
ELEMENT	κ (watt/cm-K)	$\kappa/\sigma T$ (watt-ohm/K^2)	κ (watt/cm-K)	$\kappa/\sigma T$ (watt-ohm/K^2)
Li	0.71	2.22×10^{-8}	0.73	2.43×10^{-8}
Na	1.38	2.12		
K	1.0	2.23		
Rb	0.6	2.42		
Cu	3.85	2.20	3.82	2.29
Ag	4.18	2.31	4.17	2.38
Au	3.1	2.32	3.1	2.36
Be	2.3	2.36	1.7	2.42
Mg	1.5	2.14	1.5	2.25
Nb	0.52	2.90	0.54	2.78
Fe	0.80	2.61	0.73	2.88
Zn	1.13	2.28	1.1	2.30
Cd	1.0	2.49	1.0	
Al	2.38	2.14	2.30	2.19
In	0.88	2.58	0.80	2.60
Tl	0.5	2.75	0.45	2.75
Sn	0.64	2.48	0.60	2.54
Pb	0.38	2.64	0.35	2.53
Bi	0.09	3.53	0.08	3.35
Sb	0.18	2.57	0.17	2.69

Source: G. W. C. Kaye and T. H. Laby, *Table of Physical and Chemical Constants*, Longmans Green, London, 1966.

As a concrete example let us examine a case where the temperature drop is uniform in the positive x-direction. In the steady state the thermal current will also flow in the x-direction and have a magnitude $j^q = -\kappa\, dT/dx$. To calculate the thermal current we note (assumption 4, page 6) that after each collision an electron emerges with a speed appropriate to the local temperature; the hotter the place of the collision, the more energetic the emerging electron. Consequently, even though the mean electronic velocity at a point may vanish (in contrast to the case when an electric current flows) electrons arriving at the point from the high-temperature side will have higher energies than those arriving from the low-temperature side leading to a net flow of thermal energy toward the low-temperature side (Figure 1.6).

To extract a quantitative estimate of the thermal conductivity from this picture, consider first an oversimplified "one-dimensional" model, in which the electrons can only move along the x-axis, so that at a point x half the electrons come from the high-temperature side of x, and half from the low. If $\mathcal{E}(T)$ is the thermal energy per electron in a metal in equilibrium at temperature T, then an electron whose last collision was at x' will, on the average, have a thermal energy $\mathcal{E}(T[x'])$. The electrons arriving at x from the high-temperature side will, on the average, have had their last collision at

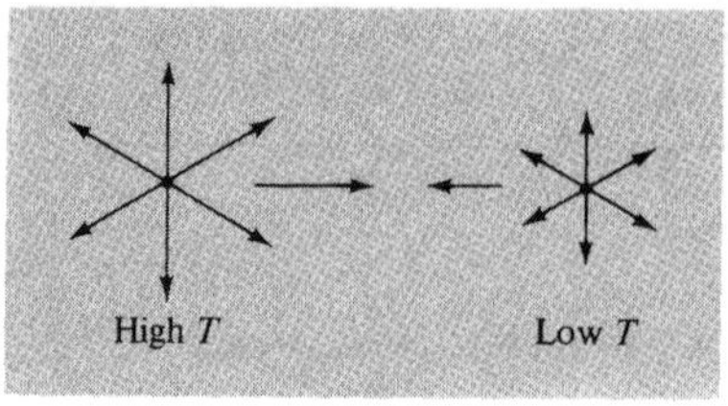

Figure 1.6
Schematic view of the relation between temperature gradient and thermal current. Electrons arriving at the center of the bar from the left had their last collision in the high-temperature region. Those arriving at the center from the right had their last collision in the low-temperature region. Hence electrons moving to the right at the center of the bar tend to be more energetic than those moving to the left, yielding a net thermal current to the right.

$x - v\tau$, and will therefore carry a thermal energy per electron of size $\mathcal{E}(T[x - v\tau])$. Their contribution to the thermal current density at x will therefore be the number of such electrons per unit volume, $n/2$, times their velocity, v, times this energy, or $(n/2)v\mathcal{E}(T[x - v\tau])$. The electrons arriving at x from the low-temperature side, on the other hand, will contribute $(n/2)(-v)[\mathcal{E}(T[x + v\tau])]$, since they have come from the positive x-direction and are moving toward negative x. Adding these together gives

$$j^q = \tfrac{1}{2}nv[\mathcal{E}(T[x - v\tau]) - \mathcal{E}(T[x + v\tau])]. \tag{1.48}$$

Provided that the variation in temperature over a mean free path ($\ell = v\tau$) is very small,[29] we may expand this about the point x to find:

$$j^q = nv^2\tau \frac{d\mathcal{E}}{dT}\left(-\frac{dT}{dx}\right). \tag{1.49}$$

To go from this to the three-dimensional case we need only replace v by the x-component v_x of the electronic velocity $\mathbf{v}$, and average over all directions. Since[30] $\langle v_x{}^2\rangle = \langle v_y{}^2\rangle = \langle v_z{}^2\rangle = \frac{1}{3}v^2$, and since $n\, d\mathcal{E}/dT = (N/V)d\mathcal{E}/dT = (dE/dT)/V = c_v$, the electronic specific heat, we have:

$$\mathbf{j}^q = \tfrac{1}{3}v^2\tau c_v(-\nabla T) \tag{1.50}$$

or

$$\kappa = \tfrac{1}{3}v^2\tau c_v = \tfrac{1}{3}\ell v c_v, \tag{1.51}$$

where v^2 is the mean square electronic speed.

We emphasize the roughness of this argument. We have spoken rather glibly about the thermal energy per electron carried by a particular group of electrons, a quantity one might be hard pressed to define with precision. We have also been quite careless about replacing quantities, at various stages of the calculation, by their thermal averages. One might object, for example, that if the thermal energy per electron depends on the direction the electrons come from, so will their average speed, since this too depends on the temperature at the place of their last collision. We shall note below that this last oversight is canceled by yet another oversight, and in Chapter 13

---

[29] Its change in ℓ is (ℓ/L) times its change in the sample length L.

[30] In equilibrium the velocity distribution is isotropic. Corrections to this due to the temperature gradient are exceedingly small.

we shall find by a more rigorous argument that the result (1.51) is quite close to (and, in special circumstances, precisely) the correct one.

Given the estimate (1.51), we can derive another result independent of the mysteries buried in the relaxation time τ, by dividing the thermal conductivity by the electrical conductivity (1.6):

$$\frac{\kappa}{\sigma} = \frac{\frac{1}{3}c_v m v^2}{ne^2}. \tag{1.52}$$

It was natural for Drude to apply the classical ideal gas laws in evaluating the electronic specific heat and mean square velocity. He therefore in effect took c_v to be $\frac{3}{2}nk_B$ and $\frac{1}{2}mv^2$ to be $\frac{3}{2}k_B T$, where k_B is Boltzmann's constant, 1.38×10^{-16} erg/K. This leads to the result

$$\frac{\kappa}{\sigma} = \frac{3}{2}\left(\frac{k_B}{e}\right)^2 T. \tag{1.53}$$

The right side of (1.53) is proportional to T and depends only on the universal constants k_B and e, in complete agreement with the law of Wiedemann and Franz. Equation (1.53) gives a Lorenz number[31]

$$\begin{aligned}\frac{\kappa}{\sigma T} = \frac{3}{2}\left(\frac{k_B}{e}\right)^2 &= 1.24 \times 10^{-13}\ (\text{erg/esu-K})^2 \\ &= 1.11 \times 10^{-8}\ \text{watt-ohm/K}^2,\end{aligned} \tag{1.54}$$

which is about half the typical value given in Table 1.6. In his original calculation of the electrical conductivity, Drude erroneously found half the correct result (1.6), as a result of which he found a value $\kappa/\sigma T = 2.22 \times 10^{-8}$ watt-ohm/K^2, in extraordinary agreement with experiment.

This success, though wholly fortuitous, was so impressive as to spur further investigations with the model. It was, however, quite puzzling, since no electronic contribution to the specific heat remotely comparable to $\frac{3}{2}nk_B$ was ever observed. Indeed, at room temperature there appeared to be no electronic contribution to the specific heat measured at all. In Chapter 2 we shall find that the classical ideal gas laws cannot be applied to the electron gas in a metal. Drude's impressive success, aside from his factor-of-two mistake, is a consequence of two errors of about 100 that cancel: at room temperature the actual electronic contribution to the specific heat is about 100 times smaller than the classical prediction, but the mean square electronic speed is about 100 times larger.

We shall examine the correct theory of the equilibrium thermal properties of the free electron gas in Chapter 2, and shall return to a more correct analysis of the thermal conductivity of a metal in Chapter 13. Before leaving the subject of thermal transport, however, we should correct one oversimplification in our analysis that obscures an important physical phenomenon:

We calculated the thermal conductivity by ignoring all manifestations of the temperature gradient except for the fact that the thermal energy carried by a group

---

[31] Since (joule/coulomb)^2 = (watt/amp)^2 = watt-ohm, the practical units in which Lorenz numbers are quoted are often called watt-ohm/K^2, instead of (joule/coulomb-K)^2.

of electrons depends on the temperature at the place of their last collision. But if electrons emerge from a collision with higher energies when the temperature is higher, they will also have higher speeds. It would therefore appear that we should let the electron's speed v as well as its contribution to the thermal energy depend on the place of the last collision. As it turns out such an additional term only alters the result by a factor of the order of unity, but we were in fact quite right to ignore such a correction. It is true that immediately after the temperature gradient is applied there will be a nonvanishing mean electronic velocity directed toward the low-temperature region. Since the electrons are charged, however, this velocity will result in an electric current. But thermal conductivity measurements are performed under open-circuit conditions, in which no electric current can flow. Therefore the electric current can continue only until enough charge has accumulated at the surface of the sample to build up a retarding electric field that opposes the further accumulation of charge, and hence precisely cancels the effect of the temperature gradient on the electronic mean velocity.[32] When the steady state is reached there will be no electric current flow, and we were therefore correct in assuming that the mean electronic velocity at a point vanished.

In this way we are led to consider another physical effect: A temperature gradient in a long, thin bar should be accompanied by an electric field directed opposite to the temperature gradient. The existence of such a field, known as the thermoelectric field, has been known for some time (the Seebeck effect). The field is conventionally written as

$$\mathbf{E} = Q\nabla T, \tag{1.55}$$

and the proportionality constant Q is known as the thermopower. To estimate the thermopower note that in our "one-dimensional" model the mean electronic velocity at a point x due to the temperature gradient is

$$\begin{aligned} v_Q &= \tfrac{1}{2}[v(x - v\tau) - v(x + v\tau)] = -\tau v \frac{dv}{dx} \\ &= -\tau \frac{d}{dx}\left(\frac{v^2}{2}\right). \end{aligned} \tag{1.56}$$

We can again generalize to three dimensions[33] by letting $v^2 \rightarrow v_x{}^2$, and noting that $\langle v_x{}^2 \rangle = \langle v_y{}^2 \rangle = \langle v_z{}^2 \rangle = \frac{1}{3}v^2$, so that

$$\mathbf{v}_Q = -\frac{\tau}{6}\frac{dv^2}{dT}(\nabla T). \tag{1.57}$$

The mean velocity due to the electric field is[34]

$$\mathbf{v}_E = -\frac{e\mathbf{E}\tau}{m}. \tag{1.58}$$

---

32 See the analogous discussion of the genesis of the Hall field on page 12.

33 Cf. the discussion leading from Eq. (1.49) to Eq. (1.50).

34 See the discussion on page 7.

To have $\mathbf{v}_Q + \mathbf{v}_E = 0$, we require that

$$Q = -\left(\frac{1}{3e}\right)\frac{d}{dT}\frac{mv^2}{2} = -\frac{c_v}{3ne}. \tag{1.59}$$

This result is also independent of the relaxation time. Drude evaluated it by another inappropriate application of classical statistical mechanics, setting c_v equal to $3nk_B/2$ to find that

$$Q = -\frac{k_B}{2e} = -0.43 \times 10^{-4} \text{ volt/K}. \tag{1.60}$$

Observed metallic thermopowers at room temperature are of the order of microvolts per degree, a factor of 100 smaller. This is the same error of 100 that appeared twice in Drude's derivation of the Wiedemann-Franz law, but being now uncompensated, it offers unambiguous evidence of the inadequacy of classical statistical mechanics in describing the metallic electron gas.

With the use of quantum statistical mechanics one removes this discrepancy. However, in some metals the sign of the thermopower—the direction of the thermoelectric field—is opposite to what the Drude model predicts. This is as mysterious as the discrepancies in the sign of the Hall coefficient. The quantum theory of solids can account for a sign reversal in the thermopower as well, but one's sense of triumph is somewhat tempered in this case, for a really quantitative theory of the thermoelectric field is still lacking. We shall note in later discussions some of the peculiarities of this phenomenon that make it particularly difficult to calculate with precision.

These last examples have made it clear that we cannot proceed very far with a free electron theory without a proper use of quantum statistics. This is the subject of Chapter 2.

PROBLEMS

1. *Poisson Distribution*

In the Drude model the probability of an electron suffering a collision in any infinitesimal interval dt is just dt/τ.

(a) Show that an electron picked at random at a given moment had no collision during the preceding t seconds with probability $e^{-t/\tau}$. Show that it will have no collision during the next t seconds with the same probability.

(b) Show that the probability that the time interval between two successive collisions of an electron falls in the range between t and $t + dt$ is $(dt/\tau)e^{-t/\tau}$.

(c) Show as a consequence of (a) that at any moment the mean time back to the last collision (or up to the next collision) averaged over all electrons is τ.

(d) Show as a consequence of (b) that the mean time between successive collisions of an electron is τ.

(e) Part (c) implies that at any moment the time T between the last and next collision averaged over all electrons is 2τ. Explain why this is not inconsistent with the result in (d). (A thorough explanation should include a derivation of the probability distribution for T.) A failure to appreciate this subtlety led Drude to a conductivity only half of (1.6). He did not make the same mistake in the thermal conductivity, whence the factor of two in his calculation of the Lorenz number (see page 23).

2. *Joule Heating*

Consider a metal at uniform temperature in a static uniform electric field $\mathbf{E}$. An electron experiences a collision, and then, after a time t, a second collision. In the Drude model, energy is not conserved in collisions, for the mean speed of an electron emerging from a collision does not depend on the energy that the electron acquired from the field since the time of the preceding collision (assumption 4, page 6).

(a) Show that the average energy lost to the ions in the second of two collisions separated by a time t is $(eEt)^2/2m$. (The average is over all directions in which the electron emerged from the first collision.)

(b) Show, using the result of Problem 1(b), that the average energy loss to the ions per electron per collision is $(eE\tau)^2/m$, and hence that the average loss per cubic centimeter per second is $(ne^2\tau/m)E^2 = \sigma E^2$. Deduce that the power loss in a wire of length L and cross section A is I^2R, where I is the current flowing and R is the resistance of the wire.

3. *Thomson Effect*

Suppose that in addition to the applied electric field in Problem 2 there is also a uniform temperature gradient ∇T in the metal. Since an electron emerges from a collision at an energy determined by the local temperature, the energy lost in collisions will depend on how far down the temperature gradient the electron travels between collisions, as well as on how much energy it has gained from the electric field. Consequently the power lost will contain a term proportional to $\mathbf{E} \cdot \nabla T$ (which is easily isolated from other terms since it is the only term in the second-order energy loss that changes sign when the sign of $\mathbf{E}$ is reversed). Show that this contribution is given in the Drude model by a term of order $(ne\tau/m)\,(d\mathcal{E}/dT)\,(\mathbf{E} \cdot \nabla T)$, where $\mathcal{E}$ is the mean thermal energy per electron. (Calculate the energy lost by a typical electron colliding at $\mathbf{r}$, which made its last collision at $\mathbf{r} - \mathbf{d}$. Assuming a fixed (that is, energy-independent) relaxation time τ, $\mathbf{d}$ can be found to linear order in the field and temperature gradient by simple kinematic arguments, which is enough to give the energy loss to second order.)

4. *Helicon Waves*

Suppose that a metal is placed in a uniform magnetic field $\mathbf{H}$ along the z-axis. Let an AC electric field $\mathbf{E}e^{-i\omega t}$ be applied perpendicular to $\mathbf{H}$.

(a) If the electric field is circularly polarized ($E_y = \pm iE_x$) show that Eq. (1.28) must be generalized to

$$j_x = \left(\frac{\sigma_0}{1 - i(\omega \mp \omega_c)\tau}\right) E_x, \quad j_y = \pm i j_x, \quad j_z = 0. \tag{1.61}$$

(b) Show that, in conjunction with (1.61), Maxwell's equations (1.31) have a solution

$$E_x = E_0 e^{i(kz - \omega t)}, \quad E_y = \pm iE_x, \quad E_z = 0, \tag{1.62}$$

provided that $k^2c^2 = \epsilon\omega^2$, where

$$\epsilon(\omega) = 1 - \frac{{\omega_p}^2}{\omega}\left(\frac{1}{\omega \mp \omega_c + i/\tau}\right). \tag{1.63}$$

(c) Sketch $\epsilon(\omega)$ for $\omega > 0$ (choosing the polarization $E_y = iE_x$) and demonstrate that solutions to $k^2c^2 = \epsilon\omega^2$ exist for arbitrary k at frequencies $\omega > \omega_p$ and $\omega < \omega_c$. (Assume the high field condition $\omega_c\tau \gg 1$, and note that even for hundreds of kilogauss, $\omega_p/\omega_c \gg 1$.)

(d) Show that when $\omega \ll \omega_c$ the relation between k and ω for the low-frequency solution is

$$\omega = \omega_c\left(\frac{k^2c^2}{\omega_p{}^2}\right). \tag{1.64}$$

This low-frequency wave, known as a helicon, has been observed in many metals.[35] Estimate the helicon frequency if the wavelength is 1 cm and the field is 10 kilogauss, at typical metallic densities.

5. *Surface Plasmons*

An electromagnetic wave that can propagate along the surface of a metal complicates the observation of ordinary (bulk) plasmons. Let the metal be contained in the half space $z > 0$, $z < 0$ being vacuum. Assume that the electric charge density ρ appearing in Maxwell's equations vanishes both inside and outside the metal. (This does not preclude a surface charge density concentrated in the plane $z = 0$.) The surface plasmon is a solution to Maxwell's equations of the form:

$$\begin{aligned} &E_x = Ae^{iqx}e^{-Kz}, \quad E_y = 0, \quad E_z = Be^{iqx}e^{-Kz}, \qquad z > 0; \\ &E_x = Ce^{iqx}e^{K'z}, \quad E_y = 0, \quad E_z = De^{iqx}e^{K'z}, \qquad z < 0; \\ &q,\ K,\ K' \text{ real}, \quad K,\ K' \text{ positive}. \end{aligned} \tag{1.65}$$

(a) Assuming the usual boundary conditions ($\mathbf{E}_\parallel$ continuous, $(\epsilon\mathbf{E})_\perp$ continuous) and using the Drude results (1.35) and (1.29) find three equations relating q, K, and K' as functions of ω.

(b) Assuming that $\omega\tau \gg 1$, plot q^2c^2 as a function of ω^2.

(c) In the limit as $qc \gg \omega$, show that there is a solution at frequency $\omega = \omega_p/\sqrt{2}$. Show from an examination of K and K' that the wave is confined to the surface. Describe its polarization. This wave is known as a surface plasmon.

---

[35] R. Bowers et al., *Phys. Rev. Letters* **7**, 339 (1961).

2
The Sommerfeld Theory of Metals

Fermi-Dirac Distribution

Free Electrons

Density of Allowed Wave Vectors

Fermi Momentum, Energy, and Temperature

Ground-State Energy and Bulk Modulus

Thermal Properties of a Free Electron Gas

Sommerfeld Theory of Conduction

Wiedemann-Franz Law

In Drude's time, and for many years thereafter, it seemed reasonable to assume that the electronic velocity distribution, like that of an ordinary classical gas of density $n = N/V$, was given in equilibrium at temperature T by the Maxwell-Boltzmann distribution. This gives the number of electrons per unit volume with velocities in the range[1] $d\mathbf{v}$ about $\mathbf{v}$ as $f_B(\mathbf{v})\,d\mathbf{v}$, where

$$f_B(\mathbf{v}) = n\left(\frac{m}{2\pi k_B T}\right)^{3/2} e^{-mv^2/2k_BT}. \tag{2.1}$$

We saw in Chapter 1 that in conjunction with the Drude model this leads to good order of magnitude agreement with the Wiedemann-Franz law, but also predicts a contribution to the specific heat of a metal of $\frac{3}{2}k_B$ per electron that was not observed.[2]

This paradox cast a shadow over the Drude model for a quarter of a century, which was only removed by the advent of the quantum theory and the recognition that for electrons[3] the Pauli exclusion principle requires the replacement of the Maxwell-Boltzmann distribution (2.1) with the Fermi-Dirac distribution:

$$f(\mathbf{v}) = \frac{(m/\hbar)^3}{4\pi^3}\frac{1}{\exp\left[(\frac{1}{2}mv^2 - k_BT_0)/k_BT\right] + 1}. \tag{2.2}$$

Here $\hbar$ is Planck's constant divided by 2π, and T_0 is a temperature that is determined by the normalization condition[4]

$$n = \int d\mathbf{v}\, f(\mathbf{v}), \tag{2.3}$$

and is typically tens of thousands of degrees. At temperatures of interest (that is, less than 10^3 K) the Maxwell-Boltzmann and Fermi-Dirac distributions are spectacularly different at metallic electronic densities (Figure 2.1).

In this chapter we shall describe the theory underlying the Fermi-Dirac distribution (2.2) and survey the consequences of Fermi-Dirac statistics for the metallic electron gas.

Shortly after the discovery that the Pauli exclusion principle was needed to account for the bound electronic states of atoms, Sommerfeld applied the same principle to the free electron gas of metals, and thereby resolved the most flagrant thermal anomalies of the early Drude model. In most applications Sommerfeld's model is nothing more than Drude's classical electron gas with the *single* modification that the electronic velocity distribution is taken to be the quantum Fermi-Dirac distribution

---

[1] We use standard vector notation. Thus by v we mean the magnitude of the vector $\mathbf{v}$; a velocity is in the range $d\mathbf{v}$ about $\mathbf{v}$ if its ith component lies between v_i and $v_i + dv_i$, for $i = x, y, z$; we also use $d\mathbf{v}$ to denote the volume of the region of velocity space in the range $d\mathbf{v}$ about $\mathbf{v}$: $d\mathbf{v} = dv_x dv_y dv_z$ (thereby following the practice common among physicists of failing to distinguish notationally between a region and its volume, the significance of the symbol being clear from context).

[2] Because, as we shall see, the actual electronic contribution is about 100 times smaller at room temperature, becoming smaller still as the temperature drops.

[3] And any other particles obeying Fermi-Dirac statistics.

[4] Note that the constants in the Maxwell-Boltzmann distribution (2.1) have already been chosen so that (2.3) is satisfied. Equation (2.2) is derived below; see Eq. (2.89). In Problem 3d the prefactor appearing in Eq. (2.2) is cast in a form that facilitates direct comparison with Eq. (2.1).

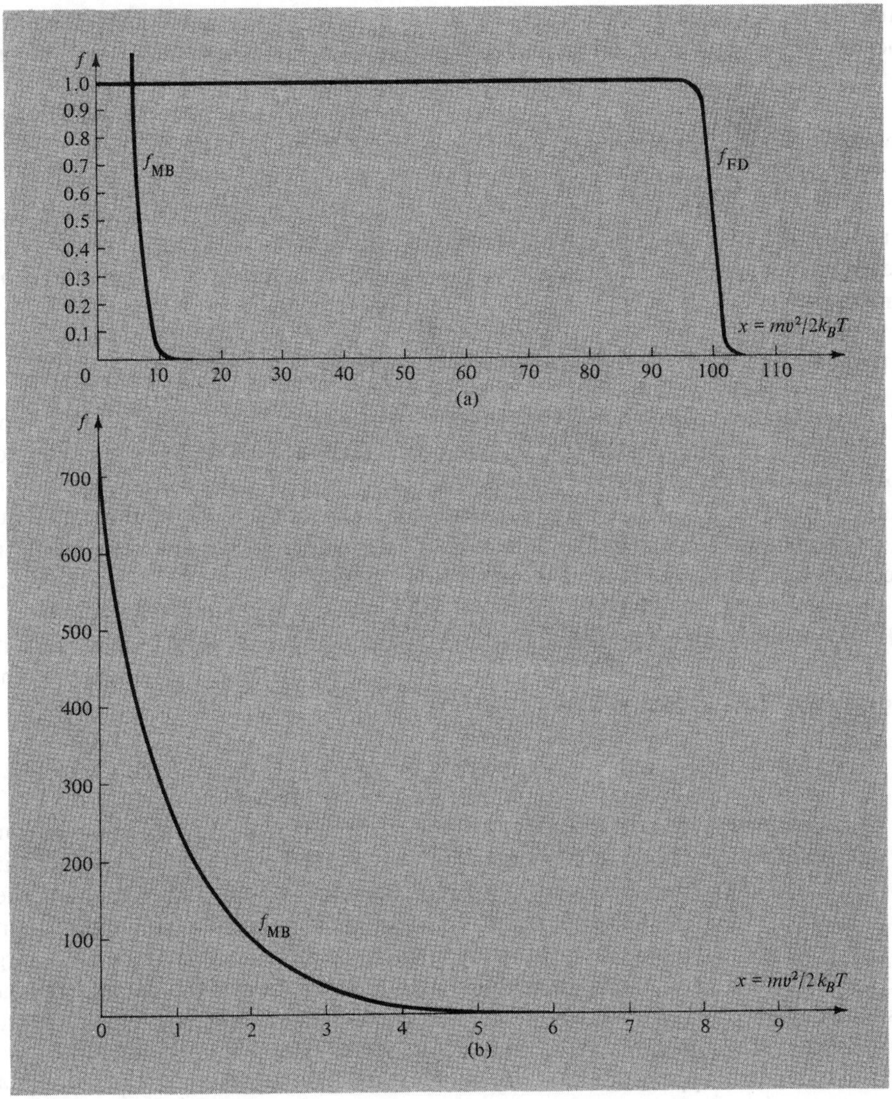

Figure 2.1
(a) The Maxwell-Boltzmann and Fermi-Dirac distributions for typical metallic densities at room temperature. (Both curves are for the density given by $T = 0.01T_0$.) The scale is the same for both distributions, and has been normalized so that the Fermi-Dirac distribution approaches 1 at low energies. Below room temperature the differences between the two distributions are even more marked. (b) A view of that part of (a) between $x = 0$ and $x = 10$. The x-axis has been stretched by about a factor of 10, and the f-axis has been compressed by about 500 to get all of the Maxwell-Boltzmann distribution in the figure. On this scale the graph of the Fermi-Dirac distribution is indistinguishable from the x-axis.

rather than the classical Maxwell-Boltzmann distribution. To justify both the use of the Fermi-Dirac distribution and its bold grafting onto an otherwise classical theory, we must examine the quantum theory of the electron gas.[5]

For simplicity we shall examine the ground state (i.e., $T = 0$) of the electron gas before studying it at nonzero temperatures. As it turns out, the properties of the ground state are of considerable interest in themselves: we shall find that room temperature, for the electron gas at metallic densities, is a very low temperature indeed, for many purposes indistinguishable from $T = 0$. Thus many (though not all) of the electronic properties of a metal hardly differ from their values at $T = 0$, even at room temperature.

GROUND-STATE PROPERTIES OF THE ELECTRON GAS

We must calculate the ground-state properties of N electrons confined to a volume V. Because the electrons do not interact with one another (independent electron approximation) we can find the ground state of the N electron system by first finding the energy levels of a single electron in the volume V, and then filling these levels up in a manner consistent with the Pauli exclusion principle, which permits at most one electron to occupy any single electron level.[6]

A single electron can be described by a wave function $\psi(\mathbf{r})$ and the specification of which of two possible orientations its spin possesses. If the electron has no interactions, the one electron wave function associated with a level of energy $\mathcal{E}$ satisfies the time-independent Schrödinger equation[7]:

$$-\frac{\hbar^2}{2m}\left(\frac{\partial^2}{\partial x^2}+\frac{\partial^2}{\partial y^2}+\frac{\partial^2}{\partial z^2}\right)\psi(\mathbf{r}) = -\frac{\hbar^2}{2m}\nabla^2\psi(\mathbf{r}) = \mathcal{E}\psi(\mathbf{r}). \tag{2.4}$$

We shall represent the confinement of the electron (by the attraction of the ions) to the volume V by a boundary condition on Eq. (2.4). The choice of boundary condition, whenever one is dealing with problems that are not explicitly concerned with effects of the metallic surface, is to a considerable degree at one's disposal and can be determined by mathematical convenience, for if the metal is sufficiently large we should expect its *bulk* properties to be unaffected by the detailed configuration of its surface.[8] In this spirit we first select the shape of the metal to suit our analytic convenience. The time-honored choice is a cube[9] of side $L = V^{1/3}$.

Next we must append a boundary condition to the Schrödinger equation (2.4),

---

[5] Throughout this chapter we shall take "electron gas" to mean a gas of free and independent electrons (see page 4) unless we are explicitly considering corrections due to electron-electron or electron-ion interactions.

[6] Note that here and later we shall reserve the term "state" for the state of the N-electron system, and the term "level" for a one-electron state.

[7] We also make the free electron approximation, so that no potential energy term appears in the Schrödinger equation.

[8] This is the approach that is almost universally followed in theories of macroscopic matter. Rigorous proofs that bulk properties are independent of the boundary conditions can now be constructed in a variety of contexts. The work most pertinent to solid state physics is by J. L. Lebowitz and E. H. Lieb, *Phys. Rev. Lett.* **22**, 631 (1969).

[9] We shall subsequently find it far more convenient to take not a cube but a parallelepiped with

reflecting the fact that the electron is confined to this cube. We also make this choice in the belief that it will not affect calculated bulk properties. One possibility is to require the wave function $\psi(\mathbf{r})$ to vanish whenever $\mathbf{r}$ is on the surface of the cube. This, however, is often unsatisfactory, for it leads to standing-wave solutions of (2.4), while the transport of charge and energy by the electrons is far more conveniently discussed in terms of running waves. A more satisfactory choice is to emphasize the inconsequence of the surface by disposing of it altogether. We can do this by imagining each face of the cube to be joined to the face opposite it, so that an electron coming to the surface is not reflected back in, but leaves the metal, simultaneously reentering at a corresponding point on the opposite surface. Thus, if our metal were one-dimensional, we would simply replace the line from 0 to L to which the electrons were confined, by a circle of circumference L. In three dimensions the geometrical embodiment of the boundary condition, in which the three pairs of opposite faces on the cube are joined, becomes topologically impossible to construct in three-dimensional space. Nevertheless, the analytic form of the boundary condition is easily generalized. In one dimension the circular model of a metal results in the boundary condition $\psi(x + L) = \psi(x)$, and the generalization to a three-dimensional cube is evidently

$$\begin{aligned} \psi(x, y, z + L) &= \psi(x, y, z), \\ \psi(x, y + L, z) &= \psi(x, y, z), \\ \psi(x + L, y, z) &= \psi(x, y, z). \end{aligned} \tag{2.5}$$

Equation (2.5) is known as the Born-von Karman (or periodic) boundary condition. We shall encounter it often (sometimes in a slightly generalized[9] form).

We now solve (2.4) subject to the boundary condition (2.5). One can verify by differentiation that a solution, neglecting the boundary condition, is

$$\psi_{\mathbf{k}}(\mathbf{r}) = \frac{1}{\sqrt{V}} e^{i\mathbf{k}\cdot\mathbf{r}}, \tag{2.6}$$

with energy

$$\mathcal{E}(\mathbf{k}) = \frac{\hbar^2 k^2}{2m}, \tag{2.7}$$

where $\mathbf{k}$ is any position independent vector. We have picked the normalization constant in (2.6) so that the probability of finding the electron *somewhere* in the whole volume V is unity:

$$1 = \int d\mathbf{r}\, |\psi(\mathbf{r})|^2. \tag{2.8}$$

To see the significance of the vector $\mathbf{k}$, note that the level $\psi_{\mathbf{k}}(\mathbf{r})$ is an eigenstate of the momentum operator,

$$\mathbf{p} = \frac{\hbar}{i}\frac{\partial}{\partial \mathbf{r}} = \frac{\hbar}{i}\nabla, \quad \left(\mathbf{p}_x = \frac{\hbar}{i}\frac{\partial}{\partial x}, \quad \text{etc.}\right), \tag{2.9}$$

edges not necessarily equal or perpendicular. For the moment we use a cube to avoid minor geometrical complexities, but it is a useful exercise to verify that all the results of this section remain valid for the parallelepiped.

with eigenvalue $\mathbf{p} = \hbar\mathbf{k}$, for

$$\frac{\hbar}{i}\frac{\partial}{\partial \mathbf{r}} e^{i\mathbf{k}\cdot\mathbf{r}} = \hbar\mathbf{k}\, e^{i\mathbf{k}\cdot\mathbf{r}}. \tag{2.10}$$

Since a particle in an eigenstate of an operator has a definite value of the corresponding observable given by the eigenvalue, an electron in the level $\psi_{\mathbf{k}}(\mathbf{r})$ has a definite momentum proportional to $\mathbf{k}$:

$$\mathbf{p} = \hbar\mathbf{k}, \tag{2.11}$$

and a velocity $\mathbf{v} = \mathbf{p}/m$ of

$$\mathbf{v} = \frac{\hbar\mathbf{k}}{m}. \tag{2.12}$$

In view of this the energy (2.7) can be written in the familiar classical form,

$$\mathcal{E} = \frac{p^2}{2m} = \tfrac{1}{2}mv^2. \tag{2.13}$$

We can also interpret $\mathbf{k}$ as a wave vector. The plane wave $e^{i\mathbf{k}\cdot\mathbf{r}}$ is constant in any plane perpendicular to $\mathbf{k}$ (since such planes are defined by the equation $\mathbf{k}\cdot\mathbf{r}$ = constant) and it is periodic along lines parallel to $\mathbf{k}$, with wavelength

$$\lambda = \frac{2\pi}{k}, \tag{2.14}$$

known as the de Broglie wavelength.

We now invoke the boundary condition (2.5). This permits only certain discrete values of $\mathbf{k}$, since (2.5) will be satisfied by the general wave function (2.6) only if

$$e^{ik_xL} = e^{ik_yL} = e^{ik_zL} = 1. \tag{2.15}$$

Since $e^z = 1$ only if $z = 2\pi in$, where n is an integer,[10] the components of the wave vector $\mathbf{k}$ must be of the form:

$$k_x = \frac{2\pi n_x}{L}, \quad k_y = \frac{2\pi n_y}{L}, \quad k_z = \frac{2\pi n_z}{L}, \qquad n_x, n_y, n_z \text{ integers.} \tag{2.16}$$

Thus in a three-dimensional space with Cartesian axes k_x, k_y, and k_z (known as k-space) the allowed wave vectors are those whose coordinates along the three axes are given by integral multiples of $2\pi/L$. This is illustrated (in two dimensions) in Figure 2.2.

Generally the only practical use one makes of the quantization condition (2.16) is this: One often needs to know how many allowed values of $\mathbf{k}$ are contained in a region of k-space that is enormous on the scale of $2\pi/L$, and that therefore contains a vast number of allowed points. If the region is very large,[11] then to an excellent approximation the number of allowed points is just the volume of k-space contained within the region, divided by the volume of k-space per point in the network of

---

[10] We shall always use the word "integer" to mean the negative integers and zero, as well as the positive integers.

[11] And not too irregularly shaped; only a negligible fraction of the points should be within $O(2\pi/L)$ of the surface.

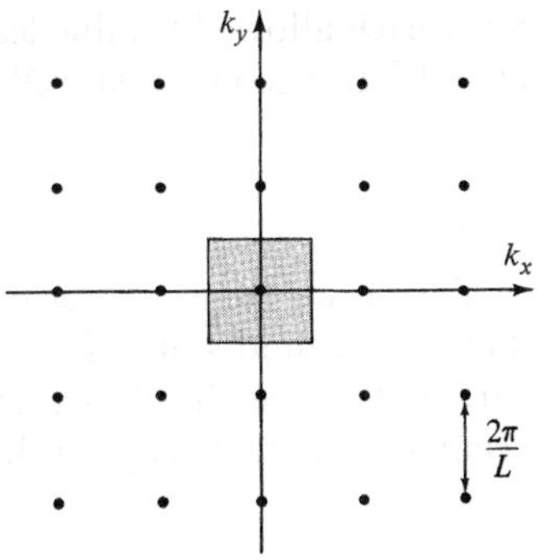

Figure 2.2
Points in a two-dimensional k-space of the form $k_x = 2\pi n_x/L$, $k_y = 2\pi n_y/L$. Note that the area per point is just $(2\pi/L)^2$. In d dimensions the volume per point is $(2\pi/L)^d$.

allowed values of **k**. That latter volume (see Figure 2.2) is just $(2\pi/L)^3$. We therefore conclude that a region of k-space of volume Ω will contain

$$\frac{\Omega}{(2\pi/L)^3} = \frac{\Omega V}{8\pi^3} \tag{2.17}$$

allowed values of **k**, or, equivalently, that the number of allowed k-values per unit volume of k-space (also known as the k-space density of levels) is just

$$\frac{V}{8\pi^3}. \tag{2.18}$$

In practice we shall deal with k-space regions so large ($\sim 10^{22}$ points) and so regular (typically spheres) that to all intents and purposes (2.17) and (2.18) can be regarded as exact. We shall begin to apply these important counting formulas shortly.

Because we assume the electrons are noninteracting we can build up the N-electron ground state by placing electrons into the allowed one-electron levels we have just found. The Pauli exclusion principle plays a vital role in this construction (as it does in building up the states of many electron atoms): we may place at most one electron in each single electron level. The one-electron levels are specified by the wave vectors **k** and by the projection of the electron's spin along an arbitrary axis, which can take either of the two values $\hbar/2$ or $-\hbar/2$. Therefore associated with each allowed wave vector **k** are *two* electronic levels, one for each direction of the electron's spin.

Thus in building up the N-electron ground state we begin by placing two electrons in the one-electron level $\mathbf{k} = 0$, which has the lowest possible one-electron energy $\mathcal{E} = 0$. We then continue to add electrons, successively filling the one-electron levels of lowest energy that are not already occupied. Since the energy of a one-electron level is directly proportional to the square of its wave vector (see (2.7)), when N is enormous the occupied region will be indistinguishable from a sphere.[12] The radius of this sphere is called k_F (F for Fermi), and its volume Ω is $4\pi k_F^3/3$. According to (2.17) the number of allowed values of **k** within the sphere is

$$\left(\frac{4\pi k_F^3}{3}\right)\left(\frac{V}{8\pi^3}\right) = \frac{k_F^3}{6\pi^2} V. \tag{2.19}$$

---

12 If it were not spherical it would not be the ground state, for we could then construct a state of lower energy by moving the electrons in levels farthest away from $\mathbf{k} = 0$ into unoccupied levels closer to the origin.

Since each allowed k-value leads to two one-electron levels (one for each spin value), in order to accommodate N electrons we must have

$$N = 2 \cdot \frac{k_F^3}{6\pi^2} V = \frac{k_F^3}{3\pi^2} V. \tag{2.20}$$

Thus if we have N electrons in a volume V (i.e., an electronic density $n = N/V$), then the ground state of the N-electron system is formed by occupying all single-particle levels with k less than k_F, and leaving all those with k greater than k_F unoccupied, where k_F is given by the condition:

$$\boxed{n = \frac{k_F^3}{3\pi^2}.} \tag{2.21}$$

This free and independent electron ground state is described by some rather unimaginative nomenclature:

The sphere of radius k_F (the *Fermi wave vector*) containing the occupied one electron levels is called the *Fermi sphere*.

The surface of the Fermi sphere, which separates the occupied from the unoccupied levels is called the *Fermi surface*. (We shall see, starting with Chapter 8, that the Fermi surface is one of the fundamental constructions in the modern theory of metals; in general it is not spherical.)

The momentum $\hbar k_F = p_F$ of the occupied one-electron levels of highest energy is known as the *Fermi momentum*; their energy, $\mathcal{E}_F = \hbar^2 k_F^2/2m$ is the *Fermi energy*; and their velocity, $v_F = p_F/m$, is the *Fermi velocity*. The Fermi velocity plays a role in the theory of metals comparable to the thermal velocity, $v = (3k_B T/m)^{1/2}$, in a classical gas.

All these quantities can be evaluated in terms of the conduction electron density, via Eq. (2.21). For estimating them numerically it is often more convenient to express them in terms of the dimensionless parameter r_s/a_0 (see page 4), which varies from about 2 to 6 in the metallic elements. Taken together, Eqs. (1.2) and (2.21) give

$$k_F = \frac{(9\pi/4)^{1/3}}{r_s} = \frac{1.92}{r_s}, \tag{2.22}$$

or

$$\boxed{k_F = \frac{3.63}{r_s/a_0} \text{Å}^{-1}.} \tag{2.23}$$

Since the Fermi wave vector is of the order of inverse angstroms, the de Broglie wavelength of the most energetic electrons is of the order of angstroms.

The Fermi velocity is

$$\boxed{v_F = \left(\frac{\hbar}{m}\right) k_F = \frac{4.20}{r_s/a_0} \times 10^8 \text{ cm/sec.}} \tag{2.24}$$

This is a substantial velocity (about 1 percent of the velocity of light). From the viewpoint of classical statistical mechanics this is quite a surprising result, for we are describing the ground state ($T = 0$), and all particles in a classical gas have zero velocity at $T = 0$. Even at room temperature the thermal (i.e., average) velocity for a classical particle with the electronic mass is only of order 10^7 cm/sec.

The Fermi energy is conveniently written in the form (since $a_0 = \hbar^2/me^2$)

$$\mathcal{E}_F = \frac{\hbar^2 k_F{}^2}{2m} = \left(\frac{e^2}{2a_0}\right)(k_F a_0)^2. \tag{2.25}$$

Here $e^2/2a_0$, known as the rydberg (Ry), is the ground-state binding energy of the hydrogen atom, 13.6 electron volts.[13] The rydberg is as convenient a unit of atomic energies as the Bohr radius is of atomic distances. Since $k_F a_0$ is of the order of unity, Eq. (2.25) demonstrates that the Fermi energy has the magnitude of a typical atomic binding energy. Using (2.23) and $a_0 = 0.529 \times 10^{-8}$ cm, we find the explicit numerical form:

$$\boxed{\mathcal{E}_F = \frac{50.1\ \text{eV}}{(r_s/a_0)^2},} \tag{2.26}$$

indicating a range of Fermi energies for the densities of metallic elements between 1.5 and 15 electron volts.

Table 2.1 lists the Fermi energy, velocity, and wave vector for the metals whose conduction electron densities are given in Table 1.1.

To calculate the ground-state energy of N electrons in a volume V we must add up the energies of all the one-electron levels inside the Fermi sphere[14]:

$$E = 2 \sum_{k<k_F} \frac{\hbar^2}{2m} k^2. \tag{2.27}$$

Quite generally, in summing any smooth function $F(\mathbf{k})$ over all allowed values of $\mathbf{k}$, one may proceed as follows:

Because the volume of k-space per allowed $\mathbf{k}$ value is $\Delta\mathbf{k} = 8\pi^3/V$ (see Eq. (2.18)) it is convenient to write

$$\sum_{\mathbf{k}} F(\mathbf{k}) = \frac{V}{8\pi^3} \sum_{\mathbf{k}} F(\mathbf{k})\,\Delta\mathbf{k}, \tag{2.28}$$

for in the limit as $\Delta\mathbf{k} \to 0$ (i.e., $V \to \infty$) the sum $\Sigma F(\mathbf{k})\,\Delta\mathbf{k}$ approaches the integral $\int d\mathbf{k}\, F(\mathbf{k})$, provided only that $F(\mathbf{k})$ does not vary appreciably[15] over distances in k-space of order $2\pi/L$. We may therefore rearrange (2.28) and write

$$\boxed{\lim_{V\to\infty} \frac{1}{V} \sum_{\mathbf{k}} F(\mathbf{k}) = \int \frac{d\mathbf{k}}{8\pi^3} F(\mathbf{k}).} \tag{2.29}$$

In applying (2.29) to finite, but macroscopically large, systems one always assumes that $(1/V)\,\Sigma F(\mathbf{k})$ differs negligibly from its infinite volume limit (for example, one

---

13 Strictly speaking, the rydberg is the binding energy in the approximation of infinite proton mass. An electron volt is the energy gained by an electron crossing a potential of 1 volt; 1 eV $= 1.602 \times 10^{-12}$ erg $= 1.602 \times 10^{-19}$ joule.

14 The factor of 2 is for the two spin levels allowed for each $\mathbf{k}$.

15 The most celebrated case in which F fails to satisfy this condition is the condensation of the ideal Bose gas. In applications to metals the problem never arises.

Table 2.1
FERMI ENERGIES, FERMI TEMPERATURES, FERMI WAVE VECTORS, AND FERMI VELOCITIES FOR REPRESENTATIVE METALS[a]

ELEMENT	r_s/a_0	$\mathcal{E}_F$	T_F	k_F	v_F
Li	3.25	4.74 eV	5.51×10^4 K	$1.12 \times 10^8\ \text{cm}^{-1}$	1.29×10^8 cm/sec
Na	3.93	3.24	3.77	0.92	1.07
K	4.86	2.12	2.46	0.75	0.86
Rb	5.20	1.85	2.15	0.70	0.81
Cs	5.62	1.59	1.84	0.65	0.75
Cu	2.67	7.00	8.16	1.36	1.57
Ag	3.02	5.49	6.38	1.20	1.39
Au	3.01	5.53	6.42	1.21	1.40
Be	1.87	14.3	16.6	1.94	2.25
Mg	2.66	7.08	8.23	1.36	1.58
Ca	3.27	4.69	5.44	1.11	1.28
Sr	3.57	3.93	4.57	1.02	1.18
Ba	3.71	3.64	4.23	0.98	1.13
Nb	3.07	5.32	6.18	1.18	1.37
Fe	2.12	11.1	13.0	1.71	1.98
Mn	2.14	10.9	12.7	1.70	1.96
Zn	2.30	9.47	11.0	1.58	1.83
Cd	2.59	7.47	8.68	1.40	1.62
Hg	2.65	7.13	8.29	1.37	1.58
Al	2.07	11.7	13.6	1.75	2.03
Ga	2.19	10.4	12.1	1.66	1.92
In	2.41	8.63	10.0	1.51	1.74
Tl	2.48	8.15	9.46	1.46	1.69
Sn	2.22	10.2	11.8	1.64	1.90
Pb	2.30	9.47	11.0	1.58	1.83
Bi	2.25	9.90	11.5	1.61	1.87
Sb	2.14	10.9	12.7	1.70	1.96

[a] The table entries are calculated from the values of r_s/a_0 given in Table 1.1 using $m = 9.11 \times 10^{-28}$ grams.

assumes that the electronic energy per unit volume in a 1-cm cube of copper is the same as in a 2-cm cube).

Using (2.29) to evaluate (2.27), we find that the energy density of the electron gas is:

$$\frac{E}{V} = \frac{1}{4\pi^3}\int_{k<k_F} d\mathbf{k}\,\frac{\hbar^2 k^2}{2m} = \frac{1}{\pi^2}\frac{\hbar^2 k_F{}^5}{10m}. \tag{2.30}$$

To find the energy per electron, E/N, in the ground state, we must divide this by $N/V = k_F{}^3/3\pi^2$, which gives

$$\frac{E}{N} = \frac{3}{10}\frac{\hbar^2 k_F{}^2}{m} = \frac{3}{5}\,\mathcal{E}_F. \tag{2.31}$$

We can also write this result as

$$\frac{E}{N} = \frac{3}{5}k_B T_F \tag{2.32}$$

where T_F, the *Fermi temperature*, is

$$\boxed{T_F = \frac{\mathcal{E}_F}{k_B} = \frac{58.2}{(r_s/a_0)^2} \times 10^4 \text{ K}.} \tag{2.33}$$

Note, in contrast to this, that the energy per electron in a classical ideal gas, $\frac{3}{2}k_BT$, vanishes at $T = 0$ and achieves a value as large as (2.32) only at $T = \frac{2}{5}T_F \approx 10^4$ K.

Given the ground-state energy E, one can calculate the pressure exerted by the electron gas from the relation $P = -(\partial E/\partial V)_N$. Since $E = \frac{3}{5}N\mathcal{E}_F$ and $\mathcal{E}_F$ is proportional to $k_F{}^2$, which depends on V only through a factor $n^{2/3} = (N/V)^{2/3}$, it follows that[16]

$$P = \frac{2}{3}\frac{E}{V}. \tag{2.34}$$

One can also calculate the compressibility, K, or bulk modulus, $B = 1/K$, defined by:

$$B = \frac{1}{K} = -V\frac{\partial P}{\partial V}. \tag{2.35}$$

Since E is proportional to $V^{-2/3}$, Eq. (2.34) shows that P varies as $V^{-5/3}$, and therefore

$$B = \frac{5}{3}P = \frac{10}{9}\frac{E}{V} = \frac{2}{3}n\mathcal{E}_F \tag{2.36}$$

or

$$B = \left(\frac{6.13}{r_s/a_0}\right)^5 \times 10^{10} \text{ dynes/cm}^2. \tag{2.37}$$

In Table 2.2 we compare the free electron bulk moduli (2.37) calculated from r_s/a_0, with the measured bulk moduli, for several metals. The agreement for the heavier alkali metals is fortuitously good, but even when (2.37) is substantially off, as it is in

Table 2.2
BULK MODULI IN 10^{10} DYNES/CM^2 FOR SOME TYPICAL METALS[a]

METAL	FREE ELECTRON B	MEASURED B
Li	23.9	11.5
Na	9.23	6.42
K	3.19	2.81
Rb	2.28	1.92
Cs	1.54	1.43
Cu	63.8	134.3
Ag	34.5	99.9
Al	228	76.0

[a] The free electron value is that for a free electron gas at the observed density of the metal, as calculated from Eq. (2.37).

16 At nonzero temperatures the pressure and energy density continue to obey this relation. See (2.101).

the noble metals, it is still of about the right order of magnitude (though it varies from three times too large to three times too small, through the table). It is absurd to expect that the free electron gas pressure alone should completely determine the resistance of a metal to compression, but Table 2.2 demonstrates that this pressure is at least as important as any other effects.

THERMAL PROPERTIES OF THE FREE ELECTRON GAS: THE FERMI-DIRAC DISTRIBUTION

When the temperature is not zero it is necessary to examine the excited states of the N-electron system as well as its ground state, for according to the basic principles of statistical mechanics, if an N-particle system is in thermal equilibrium at temperature T, then its properties should be calculated by averaging over all N-particle stationary states, assigning to each state of energy E a weight $P_N(E)$ proportional to e^{-E/k_BT}:

$$P_N(E) = \frac{e^{-E/k_BT}}{\Sigma e^{-E_\alpha{}^N/k_BT}}. \tag{2.38}$$

(Here $E_\alpha{}^N$ is the energy of the αth stationary state of the N-electron system, the sum being over all such states.)

The denominator of (2.38) is known as the partition function, and is related to the Helmholtz free energy, $F = U - TS$ (where U is the internal energy and S, the entropy) by

$$\Sigma e^{-E_\alpha{}^N/k_BT} = e^{-F_N/k_BT}. \tag{2.39}$$

We can therefore write (2.38) more compactly as:

$$P_N(E) = e^{-(E-F_N)/k_BT}. \tag{2.40}$$

Because of the exclusion principle, to construct an N-electron state one must fill N different one-electron levels. Thus each N-electron stationary state can be specified by listing which of the N one-electron levels are filled in that state. A very useful quantity to know is f_i^N, the probability of there being an electron in the particular one-electron level i, when the N-electron system is in thermal equilibrium.[17] This probability is simply the sum of the independent probabilities of finding the N-electron system in any one of those N-electron states in which the ith level is occupied:

$$f_i^N = \sum P_N(E_\alpha{}^N) \qquad \text{(summation over all } N\text{-electron states } \alpha \text{ in which there } is \text{ an electron in the one-electron level } i). \tag{2.41}$$

We can evaluate f_i^N by the following three observations:

1. Since the probability of an electron being in the level i is just one minus the probability of no electron being in the level i (those being the only two possibilities

[17] In the case we are interested in the level i is specified by the electron's wave vector $\mathbf{k}$ and the projection s of its spin along some axis.

allowed by the exclusion principle) we could equally well write (2.41) as

$$f_i^N = 1 - \sum P_N(E_\gamma{}^N) \qquad \text{(summation over all } N\text{-electron states } \gamma \text{ in which there is \textit{no} electron in the one-electron level } i). \tag{2.42}$$

2. By taking any $(N + 1)$-electron state in which there *is* an electron in the one-electron level i, we can construct an N-electron state in which there is *no* electron in the level i, by simply removing the electron in the ith level, leaving the occupation of all the other levels unaltered. Furthermore, *any* N-electron state with no electron in the one-electron level i can be so constructed from just one $(N + 1)$-electron state *with* an electron in the level i.[18] Evidently the energies of any N-electron state and the corresponding $(N + 1)$-electron state differ by just $\mathcal{E}_i$, the energy of the only one-electron level whose occupation is different in the two states. Thus the set of energies of all N-electron states with the level i unoccupied is the same as the set of energies of all $(N + 1)$-electron states with the level i occupied, provided that each energy in the latter set is reduced by $\mathcal{E}_i$. We can therefore rewrite (2.42) in the peculiar form

$$f_i^N = 1 - \sum P_N(E_\alpha^{N+1} - \mathcal{E}_i) \qquad \text{(summation over all } (N+1)\text{-electron states } \alpha \text{ in which there \textit{is} an electron in the one-electron level } i). \tag{2.43}$$

But Eq. (2.40) permits us to write the summand as

$$P_N(E_\alpha^{N+1} - \mathcal{E}_i) = e^{(\mathcal{E}_i - \mu)/k_BT} P_{N+1}(E_\alpha^{N+1}), \tag{2.44}$$

where μ, known as the chemical potential, is given at temperature T by

$$\mu = F_{N+1} - F_N. \tag{2.45}$$

Substituting this into (2.43), we find:

$$f_i^N = 1 - e^{(\mathcal{E}_i - \mu)/k_BT} \sum P_{N+1}(E_\alpha^{N+1}) \qquad \text{(summation over all } (N+1)\text{-electron states } \alpha \text{ in which there \textit{is} an electron in the one-electron level } i). \tag{2.46}$$

Comparing the summation in (2.46) with that in (2.41) one finds that (2.46) simply asserts that

$$f_i^N = 1 - e^{(\mathcal{E}_i - \mu)/k_BT} f_i^{N+1}. \tag{2.47}$$

3. Equation (2.47) gives an exact relation between the probability of the one-electron level i being occupied at temperature T in an N-electron system, and in an $(N + 1)$-electron system. When N is very large (and we are typically interested in N of the order of 10^{22}) it is absurd to imagine that by the addition of a single extra electron we could appreciably alter this probability for more than an insignificant handful of one-electron levels.[19] We may therefore replace f_i^{N+1} by f_i^N in (2.47), which

---

[18] Namely the one obtained by occupying all those levels occupied in the N-electron state *plus* the ith level.

[19] For a typical level, changing N by one alters the probability of occupation by order $1/N$. See Problem 4.

makes it possible to solve for f_i^N:

$$f_i^N = \frac{1}{e^{(\mathcal{E}_i - \mu)/k_B T} + 1}. \tag{2.48}$$

In subsequent formulas we shall drop the explicit reference to the N dependence of f_i, which is, in any event, carried through the chemical potential μ; see (2.45). The value of N can always be computed, given the f_i, by noting that f_i is the mean number of electrons in the one-electron level[20] i. Since the total number of electrons N is just the sum over all levels of the mean number in each level,

$$N = \sum_i f_i = \sum_i \frac{1}{e^{(\mathcal{E}_i - \mu)/k_B T} + 1}, \tag{2.49}$$

which determines N as a function of the temperature T and chemical potential μ. In many applications, however, it is the temperature and N (or rather the density, $n = N/V$) that are given. In such cases (2.49) is used to determine the chemical potential μ as a function of n and T, permitting it to be eliminated from subsequent formulas in favor of the temperature and density. However the chemical potential is of considerable thermodynamic interest in its own right. Some of its important properties are summarized in Appendix B.[21]

THERMAL PROPERTIES OF THE FREE ELECTRON GAS: APPLICATIONS OF THE FERMI-DIRAC DISTRIBUTION

In a gas of free and independent electrons the one-electron levels are specified by the wave vector $\mathbf{k}$ and spin quantum number s, with energies that are independent of s (in the absence of a magnetic field) and given by Eq. (2.7); i.e.,

$$\mathcal{E}(\mathbf{k}) = \frac{\hbar^2 k^2}{2m}. \tag{2.50}$$

We first verify that the distribution function (2.49) is consistent with the ground-state ($T = 0$) properties derived above. In the ground state those and only those levels are occupied with $\mathcal{E}(\mathbf{k}) \leqslant \mathcal{E}_F$, so the ground-state distribution function must be

$$\begin{aligned} f_{\mathbf{k}s} &= 1, \quad \mathcal{E}(\mathbf{k}) < \mathcal{E}_F; \\ &= 0, \quad \mathcal{E}(\mathbf{k}) > \mathcal{E}_F. \end{aligned} \tag{2.51}$$

[20] *Proof:* A level can contain either 0 or 1 electron (more than one being prohibited by the exclusion principle). The mean number of electrons is therefore 1 times the probability of 1 electron plus 0 times the probability of 0 electrons. Thus the mean number of electrons in the level is numerically equal to the probability of its being occupied. Note that this would not be so if multiple occupation of levels were permitted.

[21] The chemical potential plays a more fundamental role when the distribution (2.48) is derived in the grand canonical ensemble. See, for example, F. Reif, *Statistical and Thermal Physics*, McGraw-Hill, New York, 1965, p. 350. Our somewhat unorthodox derivation, which can also be found in Reif, uses only the canonical ensemble.

On the other hand, as $T \to 0$, the limiting form of the Fermi-Dirac distribution (2.48) is

$$\lim_{T\to 0} f_{\mathbf{k}s} = 1, \quad \mathcal{E}(\mathbf{k}) < \mu; \qquad = 0, \quad \mathcal{E}(\mathbf{k}) > \mu. \tag{2.52}$$

For these to be consistent it is necessary that

$$\lim_{T\to 0} \mu = \mathcal{E}_F. \tag{2.53}$$

We shall see shortly that for metals the chemical potential remains equal to the Fermi energy to a high degree of precision, all the way up to room temperature. As a result, people frequently fail to make any distinction between the two when dealing with metals. This, however, can be dangerously misleading. In precise calculations it is essential to keep track of the extent to which μ, the chemical potential, differs from its zero temperature value, $\mathcal{E}_F$.

The most important single application of Fermi-Dirac statistics is the calculation of the electronic contribution to the constant-volume specific heat of a metal,

$$c_v = \frac{T}{V}\left(\frac{\partial S}{\partial T}\right)_V = \left(\frac{\partial u}{\partial T}\right)_V, \qquad u = \frac{U}{V}. \tag{2.54}$$

In the independent electron approximation the internal energy U is just the sum over one-electron levels of $\mathcal{E}(\mathbf{k})$ times the mean number of electrons in the level[22]:

$$U = 2\sum_{\mathbf{k}} \mathcal{E}(\mathbf{k}) f(\mathcal{E}(\mathbf{k})). \tag{2.55}$$

We have introduced the *Fermi function* $f(\mathcal{E})$ to emphasize that $f_{\mathbf{k}}$ depends on $\mathbf{k}$ only through the electronic energy $\mathcal{E}(\mathbf{k})$:

$$\boxed{f(\mathcal{E}) = \frac{1}{e^{(\mathcal{E}-\mu)/k_B T} + 1}.} \tag{2.56}$$

If we divide both sides of (2.55) by the volume V, then (2.29) permits us to write the energy density $u = U/V$ as

$$u = \int \frac{d\mathbf{k}}{4\pi^3} \mathcal{E}(\mathbf{k}) f(\mathcal{E}(\mathbf{k})). \tag{2.57}$$

If we also divide both sides of (2.49) by V, then we can supplement (2.57) by an equation for the electronic density $n = N/V$, and use it to eliminate the chemical potential:

$$n = \int \frac{d\mathbf{k}}{4\pi^3} f(\mathcal{E}(\mathbf{k})). \tag{2.58}$$

In evaluating integrals like (2.57) and (2.58) of the form

$$\int \frac{d\mathbf{k}}{4\pi^3} F(\mathcal{E}(\mathbf{k})), \tag{2.59}$$

---

22 As usual, the factor of 2 reflects the fact that each k-level can contain two electrons of opposite spin orientations.

one often exploits the fact that the integrand depends on **k** only through the electronic energy $\mathcal{E} = \hbar^2 k^2/2m$, by evaluating the integral in spherical coordinates and changing variables from k to $\mathcal{E}$:

$$\int \frac{d\mathbf{k}}{4\pi^3} F(\mathcal{E}(\mathbf{k})) = \int_0^\infty \frac{k^2\, dk}{\pi^2} F(\mathcal{E}(\mathbf{k})) = \int_{-\infty}^{\infty} d\mathcal{E} g(\mathcal{E}) F(\mathcal{E}). \tag{2.60}$$

Here

$$\begin{aligned} g(\mathcal{E}) &= \frac{m}{\hbar^2\pi^2}\sqrt{\frac{2m\mathcal{E}}{\hbar^2}}, \quad \mathcal{E} > 0; \\ &= 0, \qquad\qquad \mathcal{E} < 0. \end{aligned} \tag{2.61}$$

Since the integral (2.59) is an evaluation of $(1/V)\sum_{\mathbf{k}s} F(\mathcal{E}(\mathbf{k}))$, the form in (2.60) shows that

$$g(\mathcal{E})\, d\mathcal{E} = \left(\frac{1}{V}\right) \times \text{[the number of one-electron levels in the energy range from } \mathcal{E} \text{ to } \mathcal{E} + d\mathcal{E}]. \tag{2.62}$$

For this reason $g(\mathcal{E})$ is known as the density of levels per unit volume (or often simply as the density of levels). A dimensionally more transparent way of writing g is

$$\begin{aligned} g(\mathcal{E}) &= \frac{3}{2}\frac{n}{\mathcal{E}_F}\left(\frac{\mathcal{E}}{\mathcal{E}_F}\right)^{1/2}, \quad \mathcal{E} > 0; \\ &= 0, \qquad\qquad \mathcal{E} < 0, \end{aligned} \tag{2.63}$$

where $\mathcal{E}_F$ and k_F are *defined* by the zero-temperature equations (2.21) and (2.25). A quantity of particular numerical importance is the density of levels at the Fermi energy, which (2.61) and (2.63) give in either of the two equivalent forms:

$$\boxed{g(\mathcal{E}_F) = \frac{mk_F}{\hbar^2\pi^2}} \tag{2.64}$$

or

$$\boxed{g(\mathcal{E}_F) = \frac{3}{2}\frac{n}{\mathcal{E}_F}.} \tag{2.65}$$

Using this notation, we rewrite (2.57) and (2.58) as:

$$u = \int_{-\infty}^{\infty} d\mathcal{E}\, g(\mathcal{E}) \mathcal{E} f(\mathcal{E}) \tag{2.66}$$

and

$$n = \int_{-\infty}^{\infty} d\mathcal{E}\, g(\mathcal{E}) f(\mathcal{E}). \tag{2.67}$$

We do this both for notational simplicity *and* because in this form the free electron approximation enters only through the particular evaluation (2.61) or (2.63) of the density of levels g. We can define a density of levels, via (2.62), in terms of which (2.66)

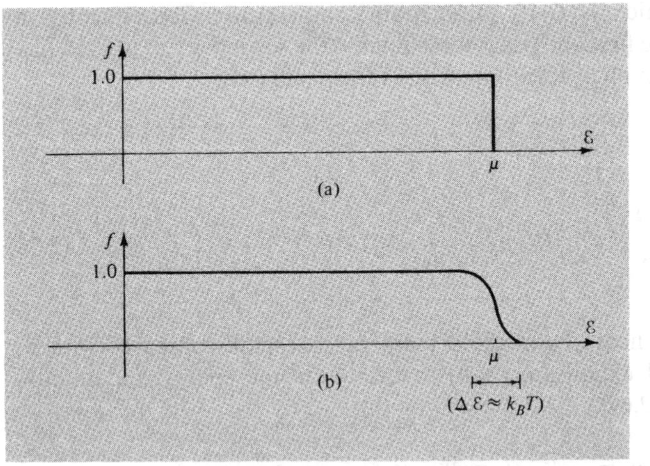

Figure 2.3
The Fermi function, $f(\mathcal{E}) = 1/[e^{\beta(\mathcal{E}-\mu)} + 1]$ versus $\mathcal{E}$ for given μ, at (a) $T = 0$ and (b) $T \approx 0.01\mu$ (of order room temperature, at typical metallic densities). The two curves differ only in a region of order k_BT about μ.

and (2.67) remain valid for any set of noninteracting (that is, independent) electrons.[23] Thus we shall later be able to apply results deduced from (2.66) and (2.67) to considerably more sophisticated models of independent electrons in metals.

In general, the integrals (2.66) and (2.67) have a rather complex structure. There is, however, a simple systematic expansion that exploits the fact that at almost all temperatures of interest in metals, T is very much smaller than the Fermi temperature (2.33). In Figure 2.3 the Fermi function $f(\mathcal{E})$ is plotted at $T = 0$ and at room temperature for typical metallic densities ($k_BT/\mu \approx 0.01$). Evidently f differs from its zero temperature form only in a small region about μ of width a few k_BT. Thus the way in which integrals of the form $\int_{-\infty}^{\infty} H(\mathcal{E})f(\mathcal{E})\,d\mathcal{E}$ differ from their zero temperature values, $\int_{-\infty}^{\mathcal{E}_F} H(\mathcal{E})\,d\mathcal{E}$, will be entirely determined by the form of $H(\mathcal{E})$ near $\mathcal{E} = \mu$. If $H(\mathcal{E})$ does not vary rapidly in the energy range of the order of k_BT about μ, the temperature dependence of the integral should be given quite accurately by replacing $H(\mathcal{E})$ by the first few terms in its Taylor expansion about $\mathcal{E} = \mu$:

$$H(\mathcal{E}) = \sum_{n=0}^{\infty} \frac{d^n}{d\mathcal{E}^n} H(\mathcal{E})\bigg|_{\mathcal{E}=\mu} \frac{(\mathcal{E}-\mu)^n}{n!}. \tag{2.68}$$

This procedure is carried out in Appendix C. The result is a series of the form:

$$\int_{-\infty}^{\infty} H(\mathcal{E})f(\mathcal{E})\,d\mathcal{E} = \int_{-\infty}^{\mu} H(\mathcal{E})\,d\mathcal{E} + \sum_{n=1}^{\infty} (k_BT)^{2n} a_n \frac{d^{2n-1}}{d\mathcal{E}^{2n-1}} H(\mathcal{E})\bigg|_{\mathcal{E}=\mu} \tag{2.69}$$

which is known as the Sommerfeld expansion.[24] The a_n are dimensionless constants of the order of unity. The functions H one typically encounters have major variations on an energy scale of the order of μ, and generally $(d/d\mathcal{E})^n H(\mathcal{E})|_{\mathcal{E}=\mu}$ is of the order of $H(\mu)/\mu^n$. When this is the case, successive terms in the Sommerfeld expansion are

---

[23] See Chapter 8.

[24] The expansion is not always exact, but is highly reliable unless $H(\mathcal{E})$ has a singularity very close to $\mathcal{E} = \mu$. If, for example, H is singular at $\mathcal{E} = 0$ (as is the free electron density of levels (2.63)) then the expansion will neglect terms of the order of $\exp(-\mu/k_BT)$, which are typically of order $e^{-100} \sim 10^{-43}$. See also Problem 1.

smaller by $O(k_BT/\mu)^2$ which is $O(10^{-4})$ at room temperature. Consequently, in actual calculations only the first and (*very* occasionally) the second terms are retained in the sum in (2.69). The explicit form for these is (Appendix C):

$$\int_{-\infty}^{\infty} H(\mathcal{E})f(\mathcal{E})\,d\mathcal{E} = \int_{-\infty}^{\mu} H(\mathcal{E})\,d\mathcal{E} + \frac{\pi^2}{6}(k_BT)^2H'(\mu) + \frac{7\pi^4}{360}(k_BT)^4H'''(\mu) + O\left(\frac{k_BT}{\mu}\right)^6. \tag{2.70}$$

To evaluate the specific heat of a metal at temperatures small compared with T_F we apply the Sommerfeld expansion (2.70) to the electronic energy and number densities (Eqs. (2.66) and (2.67)):

$$u = \int_0^{\mu} \mathcal{E}g(\mathcal{E})\,d\mathcal{E} + \frac{\pi^2}{6}(k_BT)^2[\mu g'(\mu) + g(\mu)] + O(T^4), \tag{2.71}$$

$$n = \int_0^{\mu} g(\mathcal{E})\,d\mathcal{E} + \frac{\pi^2}{6}(k_BT)^2g'(\mu) + O(T^4). \tag{2.72}$$

Equation (2.72) as we shall presently see in detail, implies that μ differs from its $T = 0$ value, $\mathcal{E}_F$, by terms of order T^2. Thus, correctly to order T^2, we may write

$$\int_0^{\mu} H(\mathcal{E})\,d\mathcal{E} = \int_0^{\mathcal{E}_F} H(\mathcal{E})\,d\mathcal{E} + (\mu - \mathcal{E}_F)H(\mathcal{E}_F). \tag{2.73}$$

If we apply this expansion to the integrals in (2.71) and (2.72), and replace μ by $\mathcal{E}_F$ in the terms already of order T^2 in these equations, we find

$$u = \int_0^{\mathcal{E}_F} \mathcal{E}g(\mathcal{E})\,d\mathcal{E} + \mathcal{E}_F\left\{(\mu - \mathcal{E}_F)g(\mathcal{E}_F) + \frac{\pi^2}{6}(k_BT)^2g'(\mathcal{E}_F)\right\} + \frac{\pi^2}{6}(k_BT)^2g(\mathcal{E}_F) + O(T^4), \tag{2.74}$$

$$n = \int_0^{\mathcal{E}_F} g(\mathcal{E})\,d\mathcal{E} + \left\{(\mu - \mathcal{E}_F)g(\mathcal{E}_F) + \frac{\pi^2}{6}(k_BT)^2g'(\mathcal{E}_F)\right\}. \tag{2.75}$$

The temperature-independent first terms on the right sides of (2.74) and (2.75) are just the values of u and n in the ground state. Since we are calculating the specific heat at constant density, n is independent of temperature, and (2.75) reduces to

$$0 = (\mu - \mathcal{E}_F)g(\mathcal{E}_F) + \frac{\pi^2}{6}(k_BT)^2g'(\mathcal{E}_F), \tag{2.76}$$

which determines the deviation of the chemical potential from $\mathcal{E}_F$:

$$\mu = \mathcal{E}_F - \frac{\pi^2}{6}(k_BT)^2\frac{g'(\mathcal{E}_F)}{g(\mathcal{E}_F)}. \tag{2.77}$$

Since for free electrons $g(\mathcal{E})$ varies as $\mathcal{E}^{1/2}$ (see Eq. (2.63)), this gives

$$\mu = \mathcal{E}_F\left[1 - \frac{1}{3}\left(\frac{\pi k_B T}{2\mathcal{E}_F}\right)^2\right], \tag{2.78}$$

which is, as we asserted above, a shift of the order of T^2 and typically only about 0.01 percent, even at room temperature.

Equation (2.76) sets the term in braces in (2.74) equal to zero, thereby simplifying the form of the thermal energy density at constant electronic density:

$$u = u_0 + \frac{\pi^2}{6}(k_B T)^2 g(\mathcal{E}_F) \tag{2.79}$$

where u_0 is the energy density in the ground state. The specific heat of the electron gas is therefore

$$\boxed{c_v = \left(\frac{\partial u}{\partial T}\right)_n = \frac{\pi^2}{3} k_B{}^2 T g(\mathcal{E}_F)} \tag{2.80}$$

or, for free electrons (see (2.65)),

$$c_v = \frac{\pi^2}{2}\left(\frac{k_B T}{\mathcal{E}_F}\right) n k_B. \tag{2.81}$$

Comparing this with the classical result for an ideal gas, $c_v = 3nk_B/2$, we see that the effect of Fermi-Dirac statistics is to depress the specific heat by a factor of $(\pi^2/3)(k_B T/\mathcal{E}_F)$, which is proportional to the temperature, and even at room temperature is only of order 10^{-2}. This explains the absence of any observable contribution of the electronic degrees of freedom to the specific heat of a metal at room temperature.

If one is willing to dispense with the precise numerical coefficient, one can understand this behavior of the specific heat quite simply from the temperature dependence of the Fermi function itself. The increase in energy of the electrons when the temperature is raised from $T = 0$ comes about entirely because some electrons with energies within $O(k_B T)$ below $\mathcal{E}_F$ (the darkly shaded region of Figure 2.4) have been excited to an energy range of $O(k_B T)$ above $\mathcal{E}_F$ (the lightly shaded region of Figure 2.4). The number of electrons per unit volume that have been so excited is the width, $k_B T$, of the energy interval times the density of levels per unit volume $g(\mathcal{E}_F)$. Furthermore, the excitation energy is of order $k_B T$, and hence the total thermal energy density is of order $g(\mathcal{E}_F)(k_B T)^2$ above the ground-state energy. This misses the precise result (2.79) by a factor of $\pi^2/6$, but it gives a simple physical picture, and is useful for rough estimates.

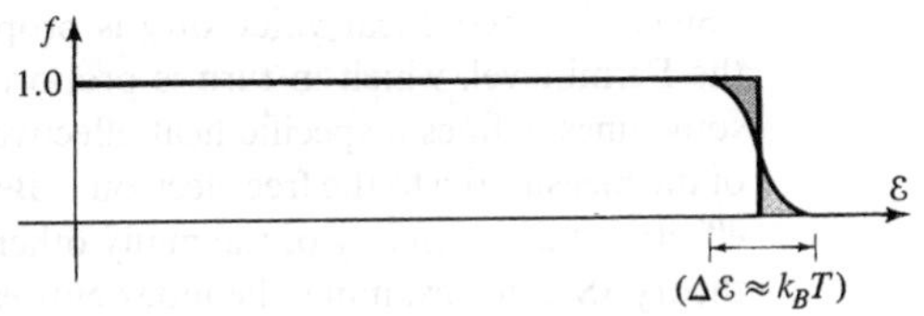

Figure 2.4
The Fermi function at nonzero T. The distribution differs from its $T = 0$ form because some electrons just below $\mathcal{E}_F$ (darkly shaded region) have been excited to levels just above $\mathcal{E}_F$ (lightly shaded region).

The prediction of a linear specific heat is one of the most important consequences of Fermi-Dirac statistics, and provides a further simple test of the electron gas theory of a metal, provided one can be sure that degrees of freedom other than the electronic ones do not make comparable or even bigger contributions. As it happens, the ionic degrees of freedom completely dominate the specific heat at high temperatures. However, well below room temperature their contribution falls off as the cube of the

Table 2.3
SOME ROUGH EXPERIMENTAL VALUES FOR THE COEFFICIENT OF THE LINEAR TERM IN T OF THE MOLAR SPECIFIC HEATS OF METALS, AND THE VALUES GIVEN BY SIMPLE FREE ELECTRON THEORY

ELEMENT	FREE ELECTRON γ (in 10^{-4} cal-mole^{-1}-K^{-2})	MEASURED γ (in 10^{-4} cal-mole^{-1}-K^{-2})	RATIO[a] (m^*/m)
Li	1.8	4.2	2.3
Na	2.6	3.5	1.3
K	4.0	4.7	1.2
Rb	4.6	5.8	1.3
Cs	5.3	7.7	1.5
Cu	1.2	1.6	1.3
Ag	1.5	1.6	1.1
Au	1.5	1.6	1.1
Be	1.2	0.5	0.42
Mg	2.4	3.2	1.3
Ca	3.6	6.5	1.8
Sr	4.3	8.7	2.0
Ba	4.7	6.5	1.4
Nb	1.6	20	12
Fe	1.5	12	8.0
Mn	1.5	40	27
Zn	1.8	1.4	0.78
Cd	2.3	1.7	0.74
Hg	2.4	5.0	2.1
Al	2.2	3.0	1.4
Ga	2.4	1.5	0.62
In	2.9	4.3	1.5
Tl	3.1	3.5	1.1
Sn	3.3	4.4	1.3
Pb	3.6	7.0	1.9
Bi	4.3	0.2	0.047
Sb	3.9	1.5	0.38

[a] Since the theoretical value of γ is proportional to the density of levels at the Fermi level, which in turn is proportional to the electronic mass m, one sometimes defines a specific heat effective mass m^* so that m^*/m is the ratio of the measured γ to the free electron γ. Beware of identifying this specific heat effective mass with any of the many other effective masses used in solid-state theory. (See, for example, the index entries under "effective mass.")

temperature (Chapter 23), and at very low temperatures it drops below the electronic contribution, which only decreases linearly with T. In order to separate out these two contributions it has become the practice to plot c_v/T against T^2, for if the electronic and ionic contributions together result in the low-temperature form,

$$c_v = \gamma T + AT^3, \tag{2.82}$$

then

$$\frac{c_v}{T} = \gamma + AT^2. \tag{2.83}$$

One can thus find γ by extrapolating the c_v/T curve linearly down to $T^2 = 0$, and noting where it intercepts the c_v/T-axis. Measured metallic specific heats typically contain a linear term that becomes comparable to the cubic one at a few degrees Kelvin.[25]

Specific heat data are usually quoted in joules (or calories) per mole per degree Kelvin. Since a mole of free electron metal contains ZN_A conduction electrons (where Z is the valence and N_A is Avogadro's number) and occupies a volume ZN_A/n, we must multiply the heat capacity per unit volume, c_v by ZN_A/n, in order to get the heat capacity per mole, C:

$$C = \frac{\pi^2}{3} ZR \frac{k_B T g(\mathcal{E}_F)}{n}, \tag{2.84}$$

where $R = k_B N_A = 8.314$ joules/mole $= 1.99$ calories/mole-K. Using the free electron density of levels (2.65) and the evaluation (2.33) of $\mathcal{E}_F/k_B$, we find a free electron contribution to the heat capacity per mole of $C = \gamma T$, where

$$\gamma = \frac{1}{2}\pi^2 R \frac{Z}{T_F} = 0.169Z\left(\frac{r_s}{a_0}\right)^2 \times 10^{-4} \text{ cal-mole}^{-1}\text{-K}^{-2}. \tag{2.85}$$

Some rough, measured values of γ are displayed in Table 2.3, together with the free electron values implied by (2.85) and the values of r_s/a_0 in Table 1.1. Note that the alkali metals continue to be reasonably well described by free electron theory, as do the noble metals (Cu, Ag, Au). Note also, however, the striking disparities in Fe and Mn (experiment of the order of 10 times theory) as well as those in Bi and Sb (experiment of the order of 0.1 times theory). These large deviations are now qualitatively understood on fairly general grounds, and we shall return to them in Chapter 15.

THE SOMMERFELD THEORY OF CONDUCTION IN METALS

To find the velocity distribution for electrons in metals, consider a small[26] volume element of k-space about a point $\mathbf{k}$, of volume $d\mathbf{k}$. Allowing for the twofold spin

---

25 Since constant density is hard to arrange experimentally, one generally measures the specific heat at constant pressure, c_p. However, one can show (Problem 2) that for the metallic free electron gas at room temperature and below, $c_p/c_v = 1 + O(k_B T/\mathcal{E}_F)^2$. Thus at temperatures where the electronic contribution to the specific heat becomes observable (a few degrees Kelvin) the two specific heats differ by a negligible amount.

26 Small enough that the Fermi function and other functions of physical interest vary negligibly throughout the volume element, but large enough that it contains very many one-electron levels.

degeneracy, the number of one-electron levels in this volume element is (see (2.18))

$$\left(\frac{V}{4\pi^3}\right) d\mathbf{k}. \tag{2.86}$$

The probability of each level being occupied is just $f(\mathcal{E}(\mathbf{k}))$, and therefore the total number of electrons in the k-space volume element is

$$\frac{V}{4\pi^3} f(\mathcal{E}(\mathbf{k}))\, d\mathbf{k}, \qquad \mathcal{E}(\mathbf{k}) = \frac{\hbar^2 \mathbf{k}^2}{2m}. \tag{2.87}$$

Since the velocity of a free electron with wave vector $\mathbf{k}$ is $\mathbf{v} = \hbar\mathbf{k}/m$ (Eq. (2.12)), the number of electrons in an element of volume $d\mathbf{v}$ about $\mathbf{v}$ is the same as the number in an element of volume $d\mathbf{k} = (m/\hbar)^3\, d\mathbf{v}$ about $\mathbf{k} = m\mathbf{v}/\hbar$. Consequently the total number of electrons per unit volume of real space in a velocity space element of volume $d\mathbf{v}$ about $\mathbf{v}$ is

$$f(\mathbf{v})\, d\mathbf{v}, \tag{2.88}$$

where

$$f(\mathbf{v}) = \frac{(m/\hbar)^3}{4\pi^3} \frac{1}{\exp\left[(\frac{1}{2}mv^2 - \mu)/k_B T\right] + 1}. \tag{2.89}$$

Sommerfeld reexamined the Drude model, replacing the classical Maxwell-Boltzmann velocity distribution (2.1) by the Fermi-Dirac distribution (2.89). Using a velocity distribution constructed from quantum-mechanical arguments in an otherwise classical theory requires some justification.[27] One can describe the motion of an electron classically if one can specify its position and momentum as accurately as necessary, without violating the uncertainty principle.[28]

A typical electron in a metal has a momentum of the order of $\hbar k_F$, so the uncertainty in its momentum, Δp, must be small compared with $\hbar k_F$ for a good classical description. Since, from (2.22), $k_F \sim 1/r_s$, the uncertainty in position must satisfy

$$\Delta x \sim \frac{\hbar}{\Delta p} \gg \frac{1}{k_F} \sim r_s, \tag{2.90}$$

where, from (1.2), r_s is of the order of the mean interelectronic distance—i.e., angstroms. Thus a classical description is impossible if one has to consider electrons localized to within atomic distances (also of the order of angstroms). However, the conduction electrons in a metal are not bound to particular ions, but can wander freely through the volume of the metal. In a macroscopic specimen, for most purposes there is no need to specify their position to an accuracy of 10^{-8} cm. The Drude model assumes a knowledge of the position of an electron primarily in only the two following contexts:

---

[27] A detailed analytical justification is fairly complicated to construct, just as it is a fairly subtle matter to specify with generality and precision when the quantum theory can be replaced by its classical limit. The underlying physics, however, is straightforward.

[28] There is also a somewhat more specialized limitation on the use of classical mechanics in describing conduction electrons. The energy of motion of an electron in the plane perpendicular to a uniform applied magnetic field is quantized in multiples of $\hbar\omega_c$ (Chapter 14). Even for fields as large as 10^4 gauss, this is a very small energy, but in suitably prepared samples at temperatures of a few degrees Kelvin, these quantum effects become observable, and are, in fact, of great practical importance.

1. When spatially varying electromagnetic fields or temperature gradients are applied, one must be able to specify the position of an electron on a scale small compared with the distance λ over which the fields or temperature gradients vary. For most applications the applied fields or temperature gradients do not vary appreciably on the scale of angstroms, and the necessary precision of definition in the electron's position need not lead to an unacceptably large uncertainty in its momentum. For example, the electric field associated with visible light varies appreciably only over a distance of order 10^3 Å. If, however, the wave length is very much shorter than this (for example, X rays), one must use quantum mechanics to describe the electronic motion induced by the field.
2. There is also an implicit assumption in the Drude model that one can localize an electron to within substantially less than a mean free path ℓ, and one should therefore be suspicious of classical arguments when mean free paths much shorter than tens of angstroms occur. Fortunately, as we shall see below, mean free paths in metals are of the order of 100 Å at room temperature, and become longer still as the temperature drops.

There is thus a wide range of phenomena in which the behavior of a metallic electron is well described by classical mechanics. It is not, however, immediately evident from this that the behavior of N such electrons can be described by classical mechanics. Since the Pauli exclusion principle so profoundly affects the statistics of N electrons, why should it not have similarly drastic effects on their dynamics? That it does not follows from an elementary theorem, which we state without proof, since the proof, though simple, is notationally rather cumbersome:

Consider a system of N electrons whose interactions with one another are ignored, and which are exposed to an arbitrary space- and time-dependent electromagnetic field. Let the N-electron state at time 0 be formed by occupying a particular group of N one-electron levels, $\psi_1(0), \ldots, \psi_N(0)$. Let $\psi_j(t)$ be the level $\psi_j(0)$ would evolve into in time t under the influence of the electromagnetic field if there were only a single electron present, which was in the level $\psi_j(0)$ at time zero. Then the correct N-electron state at time t will be the one formed by occupying the set of N one-electron levels $\psi_1(t), \ldots, \psi_N(t)$.

Thus the dynamical behavior of N noninteracting electrons is completely determined by considering N independent one-electron problems. In particular, if the classical approximation is valid for each of these one-electron problems, it will also be valid for the whole N-electron system.[29]

The use of Fermi-Dirac statistics affects only those predictions of the Drude model that require some knowledge of the electronic velocity distribution for their evaluation. If the rate $1/\tau$ at which an electron experiences collisions does not depend on its energy, then only our estimate of the electronic mean free path and our calculation of the thermal conductivity and thermopower are at all affected by a change in the equilibrium distribution function.

---

[29] Note that this implies that any classical configuration consistent with the exclusion principle at time $t = 0$ (i.e., having less than one electron of each spin per unit volume, in any momentum space region of volume $d\mathbf{p} = (2\pi\hbar)^3/V$) will remain consistent with the exclusion principle at all future times. This result can also be proved by purely classical reasoning as a direct corollary of Liouville's theorem. See Chapter 12.

Mean Free Path Using v_F (Eq. (2.24)) as a measure of the typical electronic speed, we can evaluate the mean free path $\ell = v_F\tau$ from Eq. (1.8) as follows:

$$\ell = \frac{(r_s/a_0)^2}{\rho_\mu} \times 92 \text{ Å}. \tag{2.91}$$

Since the resistivity in microhm centimeters, ρ_μ, is typically 1 to 100 at room temperature, and since r_s/a_0 is typically 2 to 6, mean free paths of order a hundred angstroms long are possible even at room temperature.[30]

Thermal Conductivity We continue to estimate the thermal conductivity by Eq. (1.51):

$$\kappa = \tfrac{1}{3}v^2\tau c_v. \tag{2.92}$$

The correct specific heat (2.81) is smaller than the classical guess of Drude by a factor of order $k_BT/\mathcal{E}_F$; the correct estimate of v^2 is not the classical thermal mean square velocity of order k_BT/m, but $v_F{}^2 = 2\mathcal{E}_F/m$, which is larger than the classical value by a factor of order $\mathcal{E}_F/k_BT$. Inserting these values in (2.92) and eliminating the relaxation time in favor of the conductivity through (1.6), we find

$$\frac{\kappa}{\sigma T} = \frac{\pi^2}{3}\left(\frac{k_B}{e}\right)^2 = 2.44 \times 10^{-8} \text{ watt-ohm/K}^2. \tag{2.93}$$

This is remarkably close to Drude's fortuitously good value, thanks to the two compensating corrections of order $k_BT/\mathcal{E}_F$, and in excellent agreement with the data in Table 1.6. We shall see (Chapter 13) that this value of the Lorenz number is a far better one than the very rough derivation of (2.93) would suggest.

Thermopower Drude's overestimate of the thermopower is also resolved by the use of Fermi-Dirac statistics. Substituting the specific heat, from Eq. (2.81), into Eq. (1.59) we find

$$Q = -\frac{\pi^2}{6}\frac{k_B}{e}\left(\frac{k_BT}{\mathcal{E}_F}\right) = -1.42\left(\frac{k_BT}{\mathcal{E}_F}\right) \times 10^{-4} \text{ volt/K}, \tag{2.94}$$

which is smaller than Drude's estimate (Eq. (1.60)) by $O(k_BT/\mathcal{E}_F) \sim 0.01$ at room temperature.

Other Properties Since the form of the electronic velocity distribution did not play a role in the calculation of the DC or AC conductivities, the Hall coefficient, or the magnetoresistance, the estimates given in Chapter 1 remain the same whether one uses Maxwell-Boltzmann or Fermi-Dirac statistics.

This is not the case, however, if one uses an energy-dependent relaxation time. If, for example, one thought the electrons collided with fixed scattering centers, then it would be natural to take an energy-independent mean free path, and hence a relaxation time $\tau = \ell/v \sim \ell/\mathcal{E}^{1/2}$. Shortly after Drude set forth the electron gas model of a

---

[30] It is perhaps just as well that Drude estimated ℓ using the very much lower classical thermal velocity, or he might have been sufficiently baffled by such long mean free paths to abandon further investigation.

metal, H. A. Lorentz showed, using the classical Maxwell-Boltzmann velocity distribution, that an energy-dependent relaxation time would lead to temperature dependence in the DC and AC conductivities, as well as to a nonvanishing magnetoresistance and a field- and temperature-dependent Hall coefficient. As one might now expect from the inappropriateness of the classical velocity distribution, none of these corrections were in any way able to bring the discrepancies of the Drude model into better alignment with the observed facts about metals.[31] Furthermore, we shall see (Chapter 13) that when the correct Fermi-Dirac velocity distribution is used, adding an energy dependence to the relaxation time has little significant effect on most of the quantities of interest in a metal.[32] If one calculates the DC or AC conductivities, the magnetoresistance, or the Hall coefficient assuming an energy-dependent $\tau(\mathcal{E})$, the results one finds are the same as those one would have calculated assuming an energy-independent τ, equal to $\tau(\mathcal{E}_F)$. In metals these quantities are determined almost entirely by the way in which electrons near the Fermi level are scattered.[33] This is another very important consequence of the Pauli exclusion principle, the justification of which will be given in Chapter 13.

PROBLEMS

1. *The Free and Independent Electron Gas in Two Dimensions*

(a) What is the relation between n and k_F in two dimensions?

(b) What is the relation between k_F and r_s in two dimensions?

(c) Prove that in two dimensions the free electron density of levels $g(\mathcal{E})$ is a constant independent of $\mathcal{E}$ for $\mathcal{E} > 0$, and 0 for $\mathcal{E} < 0$. What is the constant?

(d) Show that because $g(\mathcal{E})$ is constant, every term in the Sommerfeld expansion for n vanishes except the $T = 0$ term. Deduce that $\mu = \mathcal{E}_F$ at any temperature.

(e) Deduce from (2.67) that when $g(\mathcal{E})$ is as in (c), then

$$\mu + k_B T \ln\left(1 + e^{-\mu/k_B T}\right) = \mathcal{E}_F. \tag{2.95}$$

(f) Estimate from (2.95) the amount by which μ differs from $\mathcal{E}_F$. Comment on the numerical significance of this "failure" of the Sommerfeld expansion, and on the mathematical reason for the "failure."

2. *Thermodynamics of the Free and Independent Electron Gas*

(a) Deduce from the thermodynamic identities

$$c_v = \left(\frac{\partial u}{\partial T}\right)_n = T\left(\frac{\partial s}{\partial T}\right)_n, \tag{2.96}$$

[31] The Lorentz model is, however, of considerable importance in the description of semiconductors (Chapter 29).

[32] The thermopower is a notable exception.

[33] These assertions are correct to leading order in $k_B T/\mathcal{E}_F$, but in metals this is always a good expansion parameter.

from Eqs. (2.56) and (2.57), and from the third law of thermodynamics ($s \to 0$ as $T \to 0$) that the entropy density, $s = S/V$ is given by:

$$s = -k_B \int \frac{d\mathbf{k}}{4\pi^3} [f \ln f + (1 - f) \ln (1 - f)], \tag{2.97}$$

where $f(\mathcal{E}(\mathbf{k}))$ is the Fermi function (Eq. (2.56)).

(b) Since the pressure P satisfies Eq. (B.5) in Appendix B, $P = -(u - Ts - \mu n)$, deduce from (2.97) that

$$P = k_B T \int \frac{d\mathbf{k}}{4\pi^3} \ln \left(1 + \exp\left[-\frac{(\hbar^2 k^2/2m) - \mu}{k_B T}\right]\right). \tag{2.98}$$

Show that (2.98) implies that P is a homogeneous function of μ and T of degree 5/2; that is,

$$P(\lambda\mu, \lambda T) = \lambda^{5/2} P(\mu, T) \tag{2.99}$$

for any constant λ.

(c) Deduce from the thermodynamic relations in Appendix B that

$$\left(\frac{\partial P}{\partial \mu}\right)_T = n, \quad \left(\frac{\partial P}{\partial T}\right)_\mu = s. \tag{2.100}$$

(d) By differentiating (2.99) with respect to λ show that the ground-state relation (2.34) holds at any temperature, in the form

$$P = \tfrac{2}{3} u. \tag{2.101}$$

(e) Show that when $k_B T \ll \mathcal{E}_F$, the ratio of the constant-pressure to constant-volume specific heats satisfies

$$\left(\frac{c_p}{c_v}\right) - 1 = \frac{\pi^2}{3}\left(\frac{k_B T}{\mathcal{E}_F}\right)^2 + O\left(\frac{k_B T}{\mathcal{E}_F}\right)^4.$$

(f) Show, by retaining further terms in the Sommerfeld expansions of u and n, that correct to order T^3 the electronic heat capacity is given by

$$\begin{aligned} c_v &= \frac{\pi^2}{3} k_B^2 T g(\mathcal{E}_F) \\ &\quad - \frac{\pi^4}{90} k_B^4 T^3 g(\mathcal{E}_F) \left[15 \left(\frac{g'(\mathcal{E}_F)}{g(\mathcal{E}_F)}\right)^2 - 21 \frac{g''(\mathcal{E}_F)}{g(\mathcal{E}_F)}\right]. \end{aligned} \tag{2.102}$$

3. *The Classical Limit of Fermi-Dirac Statistics*

The Fermi-Dirac distribution reduces to the Maxwell-Boltzmann distribution, provided that the Fermi function (2.56) is much less than unity for every positive $\mathcal{E}$, for in that case we must have

$$f(\mathcal{E}) \approx e^{-(\mathcal{E}-\mu)/k_B T}. \tag{2.103}$$

The necessary and sufficient condition for (2.103) to hold for all positive $\mathcal{E}$ is

$$e^{-\mu/k_B T} \gg 1. \tag{2.104}$$

(a) Assuming that (2.104) holds, show that

$$r_s = e^{-\mu/3k_B T} 3^{1/3} \pi^{1/6} \hbar (2mk_B T)^{-1/2}. \tag{2.105}$$

In conjunction with (2.104) this requires that

$$r_s \gg \left(\frac{\hbar^2}{2mk_BT}\right)^{1/2}, \tag{2.106}$$

which can also be taken as the condition for the validity of classical statistics.

(b) What is the significance of the length r_s must exceed?

(c) Show that (2.106) leads to the numerical condition

$$\frac{r_s}{a_0} \gg \left(\frac{10^5\ \text{K}}{T}\right)^{1/2}. \tag{2.107}$$

(d) Show that the normalization constant $m^3/4\pi^3\hbar^3$ appearing in the Fermi-Dirac velocity distribution (2.2) can also be written as $(3\sqrt{\pi}/4)n(m/2\pi k_BT_F)^{3/2}$ so that $f_B(0)/f(0) = (4/3\sqrt{\pi})(T_F/T)^{3/2}$.

4. *Insensitivity of the Distribution Function to Small Changes in the Total Number of Electrons*

In deriving the Fermi distribution (page 41) we argued that the probability of a given level being occupied should not change appreciably when the total number of electrons is changed by one. Verify that the Fermi function (2.56) is compatible with this assumption as follows:

(a) Show, when $k_BT \ll \mathcal{E}_F$, that when the number of electrons changes by one at fixed temperature, the chemical potential changes by

$$\Delta\mu = \frac{1}{Vg(\mathcal{E}_F)}, \tag{2.108}$$

where $g(\mathcal{E})$ is the density of levels.

(b) Show, as a consequence of this, that the most the probability of any level being occupied can change by is

$$\Delta f = \frac{1}{6}\frac{\mathcal{E}_F}{k_BT}\frac{1}{N}. \tag{2.109}$$

[Use the free electron evaluation (2.65) of $g(\mathcal{E}_F)$.] Although temperatures of millidegrees Kelvin can be reached, at which $\mathcal{E}_F/k_BT \approx 10^8$, when N is of order 10^{22} then Δf is still negligibly small.

3

Failures of the Free Electron Model

Free electron theory successfully accounts for a wide range of metallic properties. In the form originally put forth by Drude the most striking deficiencies of the model were due to the use of classical statistical mechanics in describing the conduction electrons. As a result, predicted thermoelectric fields and heat capacities were hundreds of times too large, even at room temperature. The difficulty was obscured by the fact that classical statistics fortuitously gave a form for the Wiedemann-Franz law that was not in such gross error. Sommerfeld's application of Fermi-Dirac statistics to the conduction electrons eliminated this class of difficulties while retaining all of the other basic assumptions of the free electron model.

However, the Sommerfeld free electron model still makes many quantitative predictions that are quite unambiguously contradicted by observation, and leaves many fundamental questions of principle unresolved. We list below those inadequacies of the free electron model that have emerged from the applications made in the preceding two chapters.[1]

DIFFICULTIES WITH THE FREE ELECTRON MODEL

1. Inadequacies in the Free Electron Transport Coefficients

(a) ***The Hall Coefficient*** Free electron theory predicts a Hall coefficient which at metallic densities of electrons has the constant value $R_H = -1/nec$, independent of the temperature, the relaxation time, or the strength of the magnetic field. Although observed Hall coefficients have this order of magnitude, generally speaking they depend on both the magnetic field strength and the temperature (and presumably on the relaxation time, which is rather harder to control experimentally). Often this dependence is quite dramatic. In aluminum, for example, R_H (see Figure 1.4) never gets within a factor of three of the free electron value, depends strongly on the strength of the field, and at high fields does not even have the sign predicted by free electron theory. Such cases are not atypical. Only the Hall coefficients of the alkali metals come even close to behaving in accordance with the predictions of free electron theory.

(b) ***The Magnetoresistance*** Free electron theory predicts that the resistance of a wire perpendicular to a uniform magnetic field should not depend on the strength of the field. In almost all cases it does. In some cases (notably the noble metals, copper, silver, and gold) it can be made to increase apparently without limit as the field increases. In most metals the behavior of the resistance in a field depends quite drastically on the manner in which the metallic specimen is prepared and, for suitable specimens, on the orientation of the specimen with respect to the field.

---

[1] These examples and the remarks making up the rest of this brief chapter are not intended to give a detailed picture of the limitations of the free electron model. That will emerge in the chapters that follow, together with the solutions to the difficulties posed by the model. Our purpose in this chapter is only to emphasize how varied and extensive the shortcomings are, thereby indicating why one must resort to a considerably more elaborate analysis.

(c) ***The Thermoelectric Field*** The sign of the thermoelectric field, like the sign of the Hall constant, is not always what free electron theory predicts it should be. Only the order of magnitude is right.

(d) ***The Wiedemann-Franz Law*** That great triumph of free electron theory, the Wiedemann-Franz law, is obeyed beautifully at high (room) temperatures and also quite probably at very low (a few degrees K) temperatures. At intermediate temperatures it fails, and $\kappa/\sigma T$ depends on the temperature.

(e) ***Temperature Dependence of the DC Electrical Conductivity*** Nothing in free electron theory can account for the temperature dependence of the DC conductivity (revealed, for example, in Table 1.2). It has to be mechanically inserted into the theory as an *ad hoc* temperature dependence in the relaxation time τ.

(f) ***Directional Dependence of the DC Electrical Conductivity*** In some (but by no means all) metals the DC conductivity depends on the orientation of the specimen (if suitably prepared) with respect to the field. In such specimens the current $\mathbf{j}$ need not even be parallel to the field.

(g) ***AC Conductivity*** There is a far more subtle frequency dependence to the optical properties of metals than the simple free electron dielectric constant can hope to produce. Even sodium, in other respects a fairly good free electron metal, appears to fail this test in the detailed frequency dependence of its reflectivity. In other metals the situation is far worse. We cannot begin to explain the colors of copper and gold in terms of reflectivities calculated from the free electron dielectric constant.

2. Inadequacies in the Static Thermodynamic Predictions

(a) ***Linear Term in the Specific Heat*** The Sommerfeld theory accounts reasonably well for the size of the term linear in T in the low-temperature specific heat of the alkali metals, rather less well for the noble metals, and very poorly indeed for transition metals such as iron and manganese (much too small a prediction) as well as for bismuth and antimony (much too large a prediction).

(b) ***Cubic Term in the Specific Heat*** There is nothing in the free electron model to explain why the low-temperature specific heat should be dominated by the electronic contribution. However, it is evident from experiment that the T^3 correction to the linear term is very definitely dominated by something else, since the simple Sommerfeld theory for the electronic contribution to the T^3 term has the wrong sign and is millions of times too small.

(c) ***The Compressibility of Metals*** Although free electron theory does miraculously well in estimating the bulk moduli (or compressibilities) of many metals, it is clear that more attention must be paid to the ions and to electron-electron interactions if one is to achieve a more accurate estimate of the equation of state of a metal.

3. Fundamental Mysteries

(a) ***What Determines the Number of Conduction Electrons?*** We have assumed that all valence electrons become conduction electrons, while the others remain bound to the ions. We have given no thought to the question of why this should be, or how it is to be interpreted in the case of elements, like iron, that display more than one chemical valence.

(b) ***Why Are Some Elements Nonmetals?*** A more acute inadequacy of our rule of thumb for determining the number of conduction electrons is posed by the existence of insulators. Why, for example, is boron an insulator while its vertical neighbor in the periodic table, aluminum, an excellent metal? Why is carbon an insulator when in the form of diamond and a conductor when in the form of graphite? Why are bismuth and antimony such very poor conductors?

REVIEW OF BASIC ASSUMPTIONS

To make further progress with any of these problems we must reexamine the basic assumptions on which free electron theory rests. The most notable are these:

1. **Free Electron Approximation**[2] The metallic ions play a very minor role. In between collisions they have no effect at all on the motion of an electron, and though Drude invoked them as a source of collisions, the quantitative information we have been able to extract about the collision rate has made no sense when interpreted in terms of electrons colliding with fixed ions. The only thing the ions really seem to do properly in the models of Drude and Sommerfeld is to maintain overall charge neutrality.
2. **Independent Electron Approximation**[3] The interactions of the electrons with one another are ignored.
3. **Relaxation-Time Approximation**[4] The outcome of a collision is assumed not to depend on the configuration of the electrons at the moment of collision.

All these oversimplifications must be abandoned if we are to achieve an accurate model of a solid. However, a remarkable amount of progress can be made by first concentrating entirely on improving some aspects of the free electron approximation while continuing to use the independent electron and relaxation time approximations. We shall return to a critical examination of these last two approximations in Chapters 16 and 17, limiting ourselves here to the following general observations:

There is a surprisingly wide range of circumstances in which the independent electron approximation does not drastically diminish the validity of the analysis. In resolving the problems of free electron theory listed above, improving on the independent electron approximation plays a major role only in the calculation of metallic compressibilities (2c).[5,6] An indication of why we apparently ignore electron-electron interactions is given in Chapter 17, together with further examples in which electron-electron interactions do play a direct and crucial role.

---

[2] See page 4.

[3] See page 4.

[4] See page 6.

[5] Numbers in parentheses refer to numbered paragraphs at the beginning of this chapter.

[6] There are also some cases where a failure of the independent electron approximation (Chapter 10, p. 186 and Chapter 32) invalidates the simple distinction between metals and insulators that we shall draw in Chapters 8 and 12.

As for the relaxation time approximation, even in Drude's time there were methods in kinetic theory for correcting this oversimplification. They lead to a much more complex analysis and in many cases are primarily important in understanding metallic phenomena with greater precision. Of the difficulties described previously, only the problem of the Wiedemann-Franz law at intermediate temperatures (1d) has a resolution that requires abandoning the relaxation time approximation even at the most gross qualitative level of explanation.[7] In Chapter 16 we shall describe the form a theory must take if it is to go beyond the relaxation time approximation, together with further examples of problems requiring such a theory for their resolution.

The free electron approximation is the major source of the difficulties in the theories of Drude and Sommerfeld. It makes several simplifications:

(i) The effect of the ions on the dynamics of an electron between collisions is ignored.
(ii) What role the ions play as a source of collisions is left unspecified.
(iii) The possibility that the ions themselves, as independent dynamical entities, contribute to physical phenomena (such as the specific heat or thermal conductivity) is ignored.

The failures of assumptions (ii) and (iii) play an essential role in accounting for deviations from the Wiedemann-Franz law at intermediate temperatures (1d) and the temperature dependence of the electrical conductivity (1e). The failure of assumption (iii) accounts for the cubic term in the specific heat (2b). Relaxing these two assumptions is also essential in accounting for a variety of phenomena yet to be discussed. Such phenomena are briefly described in Chapter 21, and the consequences of abandoning assumptions (ii) and (iii) are explored in detail in Chapters 22 to 26.

It is assumption (i), that the ions have no significant effect on the motion of electrons between collisions, that is responsible for most of the deficiencies of the Drude and Sommerfeld theories described above. The reader may well be perplexed at how one can distinguish between assumptions (i) and (ii), for it is far from clear that the effect of the ions on the electrons can be unambiguously resolved into "collisional" and "noncollisional" aspects. We shall find, however (especially in Chapters 8 and 12), that a theory that takes into account the detailed field produced by an appropriate static array of ions but ignores the possibility of ionic motion (the "static ion approximation") reduces under a wide range of circumstances to a relatively simple modification of the Drude and Sommerfeld free electron theories, in which collisions are entirely absent! It is only when one allows for ionic motion that their role as a source of collisions can be properly understood.

We shall therefore relax the free electron approximation in two stages. First we shall examine the wealth of new structure and the subsequent elucidation that emerges when the electrons are considered to move not in empty space, but in the presence of a specified static potential due to a fixed array of stationary ions. Only after that (from Chapter 21 onward) will we examine the consequences of the dynamical deviations of the ionic positions from that static array.

The single most important fact about the ions is that they are not distributed at random, but are arranged in a regular periodic array, or "lattice." This was first

---

[7] It must also be abandoned to explain the detailed temperature dependence of the DC conductivity (1e).

suggested by the macroscopic crystalline forms assumed by many solids (including metals), first directly confirmed by X-ray diffraction experiments (Chapter 6) and subsequently reconfirmed by neutron diffraction, electron microscopy, and many other direct measurements.

The existence of a periodic lattice of ions lies at the heart of modern solid state physics. It provides the basis for the entire analytic framework of the subject, and without it comparatively little progress would have been made. If there is one reason why the theory of solids is so much more highly developed than the theory of liquids, even though both forms of matter have comparable densities, it is that the ions are arranged periodically in the solid state but are spatially disordered in liquids. It is the lack of a periodic array of ions that has left the subject of amorphous solids in so primitive a state compared with the highly developed theory of crystalline solids.[8]

To make further progress in the theory of solids, whether metallic or insulating, we must therefore turn to the subject of periodic arrays. The fundamental properties of such arrays are developed in Chapters 4, 5, and 7, without regard to particular physical applications. In Chapter 6 these concepts are applied to an elementary discussion of X-ray diffraction, which provides a direct demonstration of the periodicity of solids and is a paradigm for the wide variety of other wave phenomena in solids we shall subsequently encounter. Chapters 8 to 11 explore the direct consequences of the periodicity of the array of ions on the electronic structure of any solid, whether insulating or metallic. In Chapters 12 to 15 the resulting theory is used to reexplore the properties of metals described in Chapters 1 and 2. Many of the anomalies of free electron theory are thereby removed, and its mysteries are in large part resolved.

---

[8] Although there has been a great burst of interest in amorphous solids (starting in the late 1960s), the subject has yet to develop any unifying principles of a power even remotely comparable to that provided by the consequences of a periodic array of ions. Many of the concepts used in the theory of amorphous solids are borrowed, with little if any justification, from the theory of crystalline solids, even though they are only well understood as consequences of lattice periodicity. Indeed, the term "solid state physics," if defined as the subject matter of solid state physics textbooks (including this one) is currently confined almost entirely to the theory of crystalline solids. This is in large part because the normal condition of solid matter is crystalline, and also because in its present form the subject of amorphous solids still lacks the kind of broad basic principles suitable for inclusion in an elementary text.

4

Crystal Lattices

Bravais Lattice and Primitive Vectors

Simple, Body-Centered, and Face-Centered Cubic Lattices

Primitive Unit Cell, Wigner-Seitz Cell, and Conventional Cell

Crystal Structures and Lattices with Bases

Hexagonal Close-Packed and Diamond Structures

Sodium Chloride, Cesium Chloride, and Zincblende Structures

Those who have not wandered amidst the mineralogical departments of natural history museums are often surprised to learn that metals, like most other solids, are crystalline, for although one is used to the very obvious crystalline features of quartz, diamond, and rock salt, the characteristic plane faces at sharp angles with one another are absent from metals in their most commonly encountered forms. However, those metals that occur naturally in the metallic state are quite often found in crystalline forms, which are completely disguised in finished metal products by the great malleability of metals, which permits them to be fashioned into whatever macroscopic shape one wishes.

The true test of crystallinity is not the superficial appearance of a large specimen, but whether on the microscopic level the ions are arranged in a periodic array.[1] This underlying microscopic regularity of crystalline matter was long hypothesized as the obvious way to account for the simple geometric regularities of macroscopic crystals, in which plane faces make only certain definite angles with each other. It received direct experimental confirmation in 1913 through the work of W. and L. Bragg, who founded the subject of X-ray crystallography and began the investigation of how atoms are arranged in solids.

Before we describe how the microscopic structure of solids is determined by X-ray diffraction and how the periodic structures so revealed affect fundamental physical properties, it is useful to survey some of the most important geometrical properties of periodic arrays in three-dimensional space. These purely geometrical considerations are implicit in almost all the analysis one encounters throughout solid state physics, and shall be pursued in this chapter and in Chapters 5 and 7. The first of many applications of these concepts will be made to X-ray diffraction in Chapter 6.

BRAVAIS LATTICE

A fundamental concept in the description of any crystalline solid is that of the *Bravais lattice*, which specifies the periodic array in which the repeated units of the crystal are arranged. The units themselves may be single atoms, groups of atoms, molecules, ions, etc., but the Bravais lattice summarizes only the geometry of the underlying periodic structure, regardless of what the actual units may be. We give two equivalent definitions of a Bravais lattice[2]:

(a) A Bravais lattice is an infinite array of discrete points with an arrangement and orientation that appears *exactly* the same, from whichever of the points the array is viewed.

(b) A (three-dimensional) Bravais lattice consists of all points with position vectors **R** of the form

$$\mathbf{R} = n_1\mathbf{a}_1 + n_2\mathbf{a}_2 + n_3\mathbf{a}_3, \tag{4.1}$$

---

[1] Often a specimen is made up of many small pieces, each large on the microscopic scale and containing large numbers of periodically arranged ions. This "polycrystalline" state is more commonly encountered than a single macroscopic crystal, in which the periodicity is perfect, extending through the entire specimen.

[2] Why the name Bravais appears is explained in Chapter 7.

where $\mathbf{a}_1$, $\mathbf{a}_2$, and $\mathbf{a}_3$ are any three vectors not all in the same plane, and n_1, n_2, and n_3 range through all integral values.[3] Thus the point $\Sigma n_i \mathbf{a}_i$ is reached by moving n_i steps[4] of length a_i in the direction of $\mathbf{a}_i$ for $i = 1, 2$, and 3.

The vectors $\mathbf{a}_i$ appearing in definition (b) of a Bravais lattice are called *primitive vectors* and are said to *generate* or *span* the lattice.

It takes some thought to see that the two definitions of a Bravais lattice are equivalent. That any array satisfying (b) also satisfies (a) becomes evident as soon as both definitions are understood. The argument that *any* array satisfying definition (a) can be generated by an appropriate set of three vectors is not as obvious. The proof consists of an explicit recipe for constructing three primitive vectors. The construction is given in Problem 8a.

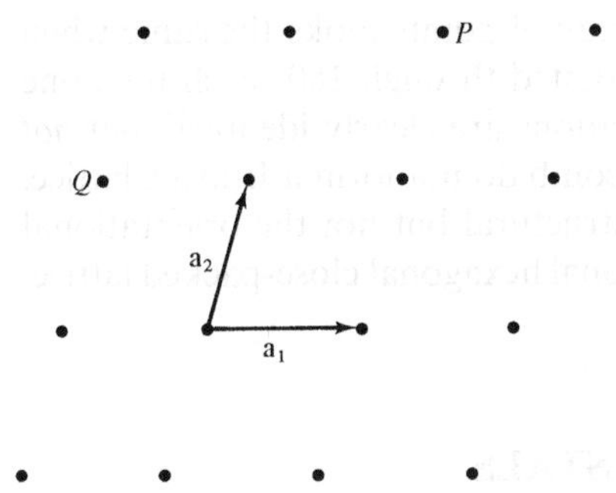

Figure 4.1
A general two-dimensional Bravais lattice of no particular symmetry: the oblique net. Primitive vectors $\mathbf{a}_1$ and $\mathbf{a}_2$ are shown. All points in the net are linear combinations of these with integral coefficients; for example, $P = \mathbf{a}_1 + 2\mathbf{a}_2$, and $Q = -\mathbf{a}_1 + \mathbf{a}_2$.

Figure 4.1 shows a portion of a two-dimensional Bravais lattice.[5] Clearly definition (a) is satisfied, and the primitive vectors $\mathbf{a}_1$ and $\mathbf{a}_2$ required by definition (b) are indicated in the figure. Figure 4.2 shows one of the most familiar of three-dimensional Bravais lattices, the simple cubic. It owes its special structure to the fact that it can be spanned by three mutually perpendicular primitive vectors of equal length.

Figure 4.2
A simple cubic three-dimensional Bravais lattice. The three primitive vectors can be taken to be mutually perpendicular, and with a common magnitude.

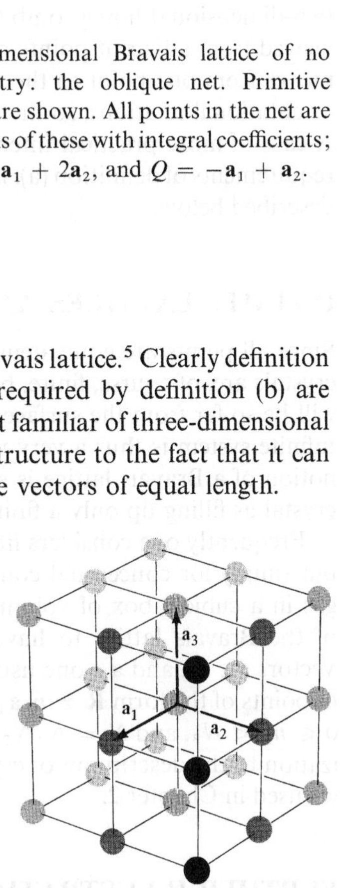

[3] We continue with the convention that "integer" means a negative integer or zero, as well as a positive integer.

[4] When n is negative, n steps in a direction means n steps in the opposite direction. The point reached does not, of course, depend on the order in which the $n_1 + n_2 + n_3$ steps are taken.

[5] A two-dimensional Bravais lattice is also known as a *net*.

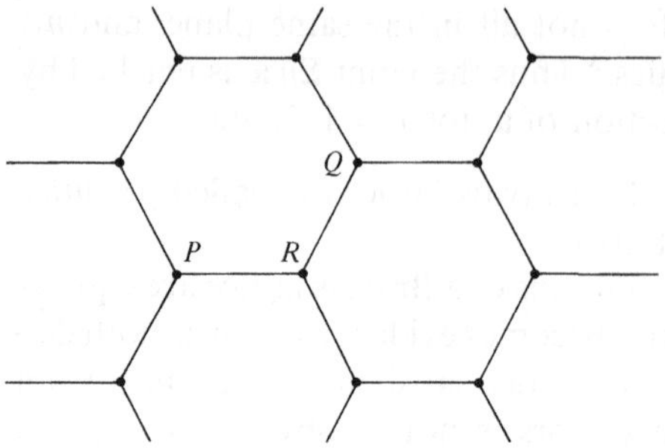

Figure 4.3
The vertices of a two-dimensional honeycomb do *not* form a Bravais lattice. The array of points has the same appearance whether viewed from point P or point Q. However, the view from point R is rotated through 180°.

It is important that not only the arrangement, but also the orientation must appear the same from every point in a Bravais lattice. Consider the vertices of a two-dimensional honeycomb (Figure 4.3). The array of points looks the same when viewed from adjacent points only if the page is rotated through 180° each time one moves from one point to the next. Structural relations are clearly identical, but *not* orientational relations, so the vertices of a honeycomb do not form a Bravais lattice. A case of more practical interest, satisfying the structural but not the orientational requirements of definition (a), is the three-dimensional hexagonal close-packed lattice, described below.

INFINITE LATTICES AND FINITE CRYSTALS

Since all points are equivalent, the Bravais lattice must be infinite in extent. Actual crystals are, of course, finite, but if they are large enough the vast majority of points will be so far from the surface as to be unaffected by its existence. The fiction of an infinite system is thus a very useful idealization. If surface effects are of interest the notion of a Bravais lattice is still relevant, but now one must think of the physical crystal as filling up only a finite portion of the ideal Bravais lattice.

Frequently one considers finite crystals, not because surface effects are important, but simply for conceptual convenience, just as in Chapter 2 we placed the electron gas in a cubical box of volume $V = L^3$. One then generally picks the finite region of the Bravais lattice to have the simplest possible form. Given three primitive vectors $\mathbf{a}_1$, $\mathbf{a}_2$, and $\mathbf{a}_3$, one usually considers the finite lattice of N sites to be the set of points of the form $\mathbf{R} = n_1\mathbf{a}_1 + n_2\mathbf{a}_2 + n_3\mathbf{a}_3$, where $0 \leqslant n_1 < N_1$, $0 \leqslant n_2 < N_2$, $0 \leqslant n_3 < N_3$, and $N = N_1N_2N_3$. This artifact is closely connected with the generalization to the description of crystalline systems[6] of the periodic boundary condition we used in Chapter 2.

FURTHER ILLUSTRATIONS AND IMPORTANT EXAMPLES

Of the two definitions of a Bravais lattice, definition (b) is mathematically more precise and is the obvious starting point for any analytic work. It has, however, two

[6] We shall make particular use of it in Chapters 8 and 22.

minor shortcomings. First, for any given Bravais lattice the set of primitive vectors is not unique—indeed, there are infinitely many nonequivalent choices (see Figure 4.4)—and it is distasteful (and sometimes misleading) to rely too heavily on a definition that emphasizes a particular choice. Second, when presented with a particular array of points one usually can tell at a glance whether the first definition is satisfied, although the existence of a set of primitive vectors or a proof that there is no such set can be rather more difficult to perceive immediately.

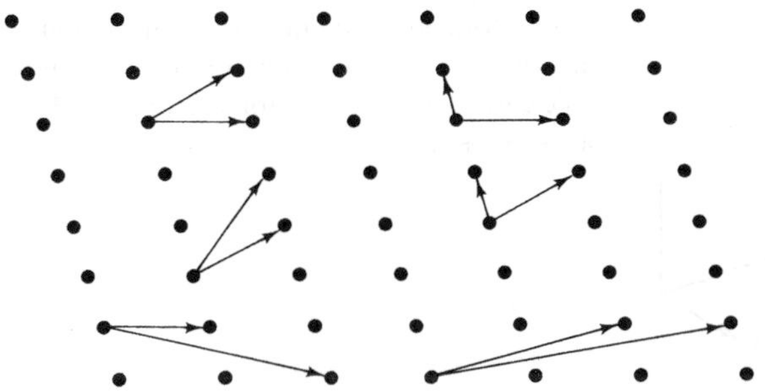

Figure 4.4
Several possible choices of pairs of primitive vectors for a two-dimensional Bravais lattice. They are drawn, for clarity, from different origins.

Consider, for example, the *body-centered cubic* (bcc) lattice, formed by adding to the simple cubic lattice of Figure 4.2 (whose sites we now label A) an additional point, B, at the center of each little cube (Figure 4.5). One might at first feel that the center points B bear a different relation to the whole than the corner points A. However, the center point B can be thought of as corner points of a second simple cubic array.

Figure 4.5
A few sites from a body-centered cubic Bravais lattice. Note that it can be regarded either as a simple cubic lattice formed from the points A with the points B at the cube centers, or as a simple cubic lattice formed from the points B with the points A at the cube centers. This observation establishes that it is indeed a Bravais lattice.

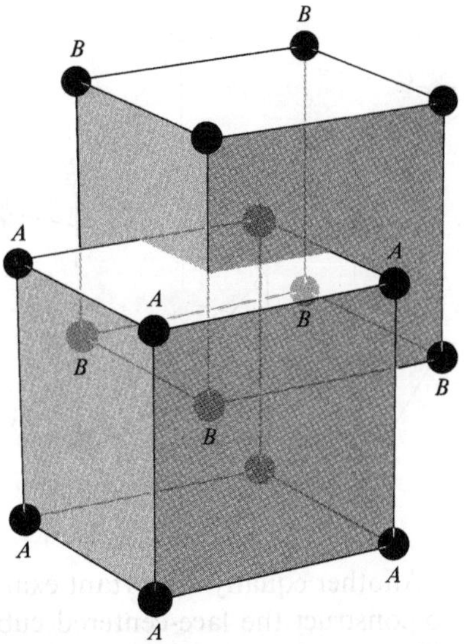

In this new array the corner points A of the original cubic array are center points. Thus all points do have identical surroundings, and the body-centered cubic lattice is a Bravais lattice. If the original simple cubic lattice is generated by primitive vectors

$$a\hat{\mathbf{x}}, \quad a\hat{\mathbf{y}}, \quad a\hat{\mathbf{z}}, \tag{4.2}$$

where $\hat{\mathbf{x}}$, $\hat{\mathbf{y}}$, and $\hat{\mathbf{z}}$ are three orthogonal unit vectors, then a set of primitive vectors for the body-centered cubic lattice could be (Figure 4.6)

$$\mathbf{a}_1 = a\hat{\mathbf{x}}, \quad \mathbf{a}_2 = a\hat{\mathbf{y}}, \quad \mathbf{a}_3 = \frac{a}{2}(\hat{\mathbf{x}} + \hat{\mathbf{y}} + \hat{\mathbf{z}}). \tag{4.3}$$

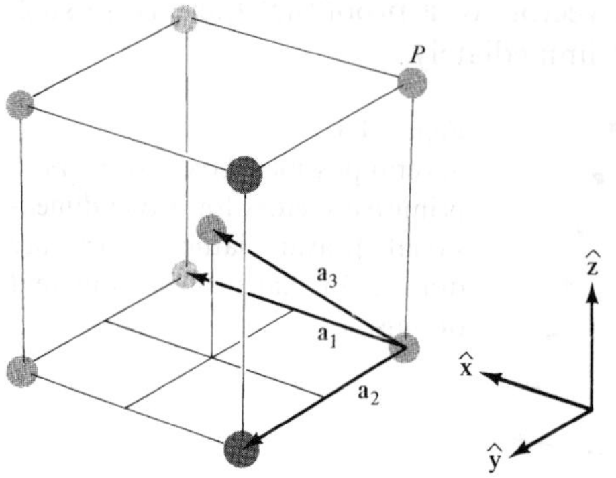

Figure 4.6
Three primitive vectors, specified in Eq. (4.3), for the body-centered cubic Bravais lattice. The lattice is formed by taking all linear combinations of the primitive vectors with integral coefficients. The point P, for example, is $P = -\mathbf{a}_1 - \mathbf{a}_2 + 2\mathbf{a}_3$.

A more symmetric set (see Figure 4.7) is

$$\mathbf{a}_1 = \frac{a}{2}(\hat{\mathbf{y}} + \hat{\mathbf{z}} - \hat{\mathbf{x}}), \quad \mathbf{a}_2 = \frac{a}{2}(\hat{\mathbf{z}} + \hat{\mathbf{x}} - \hat{\mathbf{y}}), \quad \mathbf{a}_3 = \frac{a}{2}(\hat{\mathbf{x}} + \hat{\mathbf{y}} - \hat{\mathbf{z}}). \tag{4.4}$$

It is important to convince oneself both geometrically and analytically that these sets do indeed generate the bcc Bravais lattice.

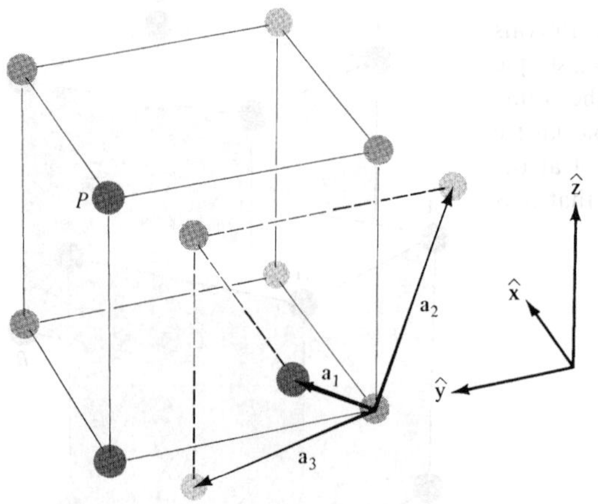

Figure 4.7
A more symmetric set of primitive vectors, specified in Eq. (4.4), for the body-centered cubic Bravais lattice. The point P, for example, has the form $P = 2\mathbf{a}_1 + \mathbf{a}_2 + \mathbf{a}_3$.

Another equally important example is the *face-centered cubic* (fcc) Bravais lattice. To construct the face-centered cubic Bravais lattice add to the simple cubic lattice of Figure 4.2 an additional point in the center of each square face (Figure 4.8). For ease in description think of each cube in the simple cubic lattice as having horizontal bottom and top faces, and four vertical side faces facing north, south, east, and west. It may sound as if all points in this new array are not equivalent, but in fact they are. One can, for example, consider the *new* simple cubic lattice formed by the points added

Figure 4.8
Some points from a face-centered cubic Bravais lattice.

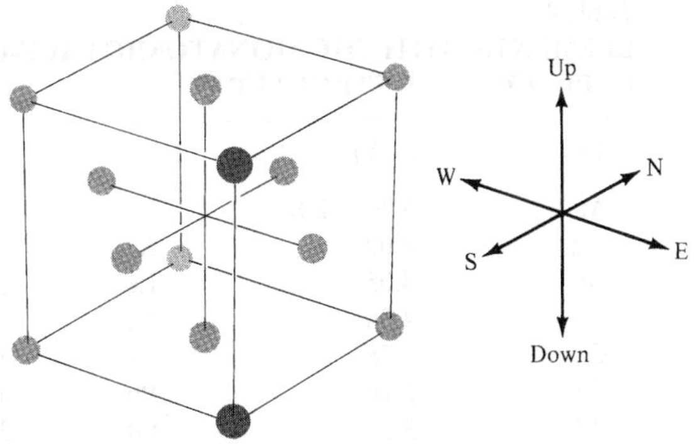

to the centers of all the horizontal faces. The original simple cubic lattice points are now centering points on the horizontal faces of the new simple cubic lattice, whereas the points that were added to the centers of the north-south faces of the original cubic lattice are in the centers of the east-west faces of the new one, and vice versa.

In the same way one can also regard the simple cubic lattice as being composed of all points centering the north-south faces of the original simple cubic lattice, or all points centering the east-west faces of the original cubic lattice. In either case the remaining points will be found centered on the faces of the new simple cubic framework. Thus any point can be thought of either as a corner point or as a face-centering point for any of the three kinds of faces, and the face-centered cubic lattice is indeed a Bravais lattice.

A symmetric set of primitive vectors for the face-centered cubic lattice (see Figure 4.9) is

$$\mathbf{a}_1 = \frac{a}{2}(\hat{\mathbf{y}} + \hat{\mathbf{z}}), \quad \mathbf{a}_2 = \frac{a}{2}(\hat{\mathbf{z}} + \hat{\mathbf{x}}), \quad \mathbf{a}_3 = \frac{a}{2}(\hat{\mathbf{x}} + \hat{\mathbf{y}}). \tag{4.5}$$

Figure 4.9
A set of primitive vectors, as given in Eq. (4.5), for the face-centered cubic Bravais lattice. The labeled points are $P = \mathbf{a}_1 + \mathbf{a}_2 + \mathbf{a}_3$, $Q = 2\mathbf{a}_2$, $R = \mathbf{a}_2 + \mathbf{a}_3$, and $S = -\mathbf{a}_1 + \mathbf{a}_2 + \mathbf{a}_3$.

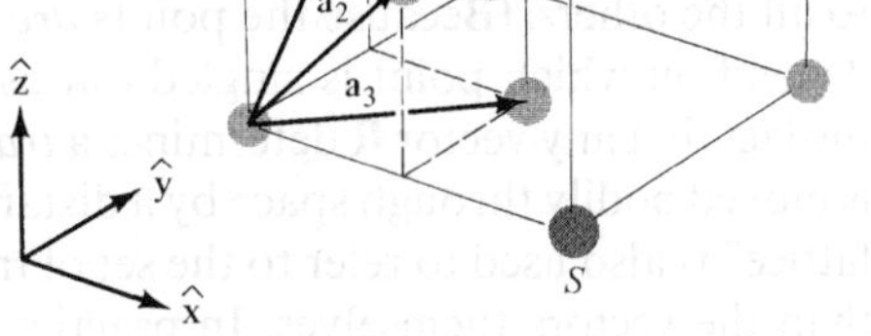

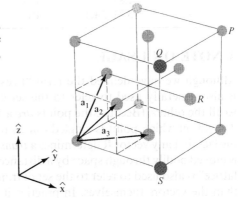

The face-centered cubic and body-centered cubic Bravais lattices are of great importance, since an enormous variety of solids crystallize in these forms with an atom (or ion) at each lattice site (see Tables 4.1 and 4.2). (The corresponding simple cubic form, however, is very rare, the alpha phase of polonium being the only known example among the elements under normal conditions.)

Table 4.1
ELEMENTS WITH THE MONATOMIC FACE-CENTERED CUBIC CRYSTAL STRUCTURE

ELEMENT	a (Å)	ELEMENT	a (Å)	ELEMENT	a (Å)
Ar	5.26 (4.2 K)	Ir	3.84	Pt	3.92
Ag	4.09	Kr	5.72 (58 K)	δ-Pu	4.64
Al	4.05	La	5.30	Rh	3.80
Au	4.08	Ne	4.43 (4.2 K)	Sc	4.54
Ca	5.58	Ni	3.52	Sr	6.08
Ce	5.16	Pb	4.95	Th	5.08
β-Co	3.55	Pd	3.89	Xe (58 K)	6.20
Cu	3.61	Pr	5.16	Yb	5.49

Data in Tables 4.1 to 4.7 are from R. W. G. Wyckoff, *Crystal Structures*, 2nd ed., Interscience, New York, 1963. In most cases, the data are taken at about room temperature and normal atmospheric pressure. For elements that exist in many forms the stable room temperature form (or forms) is given. For more detailed information, more precise lattice constants, and references, the Wyckoff work should be consulted.

Table 4.2
ELEMENTS WITH THE MONATOMIC BODY-CENTERED CUBIC CRYSTAL STRUCTURE

ELEMENT	a (Å)	ELEMENT	a (Å)	ELEMENT	a (Å)
Ba	5.02	Li	3.49 (78 K)	Ta	3.31
Cr	2.88	Mo	3.15	Tl	3.88
Cs	6.05 (78 K)	Na	4.23 (5 K)	V	3.02
Fe	2.87	Nb	3.30	W	3.16
K	5.23 (5 K)	Rb	5.59 (5 K)		

A NOTE ON USAGE

Although we have defined the term "Bravais lattice" to apply to a set of points, it is also generally used to refer to the set of vectors joining any one of these points to all the others. (Because the points *are* a Bravais lattice, this set of vectors does not depend on which point is singled out as the origin.) Yet another usage comes from the fact that any vector **R** determines a *translation* or *displacement*, in which everything is moved bodily through space by a distance R in the direction of **R**. The term "Bravais lattice" is also used to refer to the set of translations determined by the vectors, rather than the vectors themselves. In practice it is always clear from the context whether it is the points, the vectors, or the translations that are being referred to.[7]

[7] The more general use of the term provides an elegant definition of a Bravais lattice with the precision of definition (b) and the nonprejudicial nature of definition (a): A Bravais lattice is a discrete set of vectors not all in a plane, closed under vector addition and subtraction (i.e., the sum and difference of any two vectors in the set are also in the set).

COORDINATION NUMBER

The points in a Bravais lattice that are closest to a given point are called its *nearest neighbors.* Because of the periodic nature of a Bravais lattice, each point has the same number of nearest neighbors. This number is thus a property of the lattice, and is referred to as the *coordination number* of the lattice. A simple cubic lattice has coordination number 6; a body-centered cubic lattice, 8; and a face-centered cubic lattice, 12. The notion of a coordination number can be extended in the obvious way to some simple arrays of points that are not Bravais lattices, provided that each point in the array has the same number of nearest neighbors.

PRIMITIVE UNIT CELL

A volume of space that, when translated through all the vectors in a Bravais lattice, just fills all of space without either overlapping itself or leaving voids is called a *primitive cell* or *primitive unit cell* of the lattice.[8] There is no unique way of choosing a primitive cell for a given Bravais lattice. Several possible choices of primitive cells for a two-dimensional Bravais lattice are illustrated in Figure 4.10.

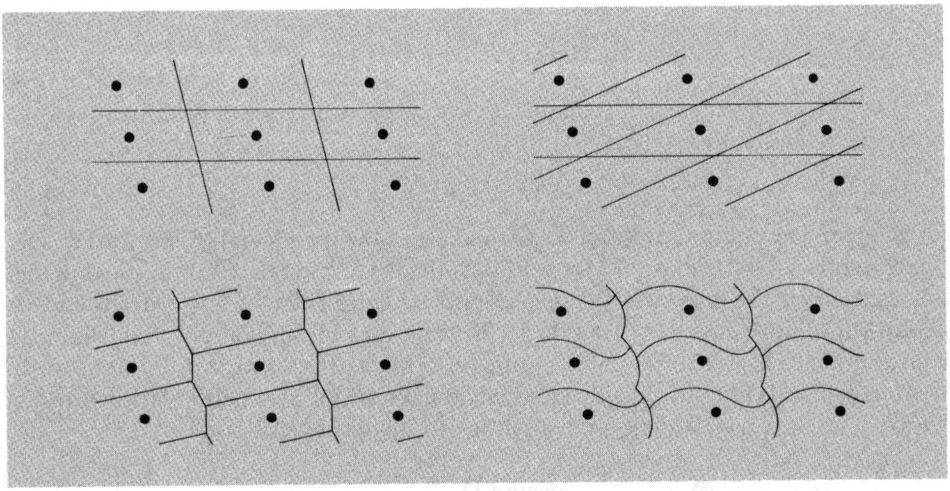

Figure 4.10
Several possible choices of primitive cell for a single two-dimensional Bravais lattice.

A primitive cell must contain precisely one lattice point (unless it is so positioned that there are points on its surface). It follows that if n is the density of points in the lattice[9] and v is the volume of the primitive cell, then $nv = 1$. Thus $v = 1/n$. Since

[8] Translations of the primitive cell may possess common surface points; the nonoverlapping proviso is only intended to prohibit overlapping regions of nonzero volume.

[9] The density n of Bravais lattice points need not, of course, be identical to the density of conduction electrons in a metal. When the possibility of confusion is present, we shall specify the two densities with different symbols.

this result holds for any primitive cell, the volume of a primitive cell is independent of the choice of cell.

It also follows from the definition of a primitive cell that, given any two primitive cells of arbitrary shape, it is possible to cut the first up into pieces, which, when translated through appropriate lattice vectors, can be reassembled to give the second. This is illustrated in Figure 4.11.

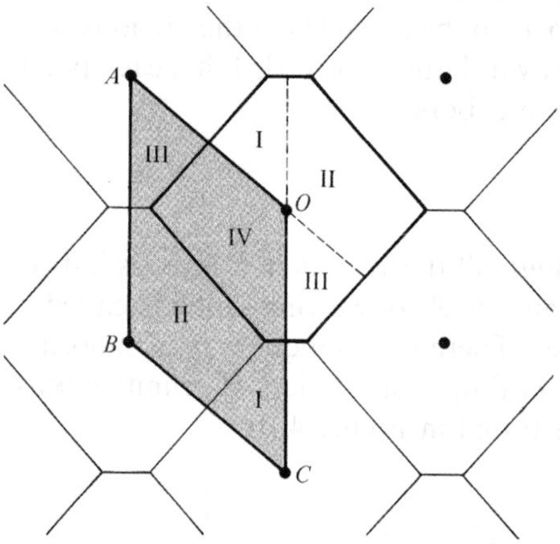

Figure 4.11
Two possible primitive cells for a two-dimensional Bravais lattice. The parallelogram cell (shaded) is obviously primitive; additional hexagonal cells are indicated to demonstrate that the hexagonal cell is also primitive. The parallelogram can be cut into pieces, which, when translated through lattice vectors, reassemble to form the hexagon. The translations for the four regions of the parallelogram are: Region I—$\overrightarrow{CO}$; Region II—$\overrightarrow{BO}$; Region III—$\overrightarrow{AO}$; Region IV—no translation.

The obvious primitive cell to associate with a particular set of primitive vectors, $\mathbf{a}_1, \mathbf{a}_2, \mathbf{a}_3$, is the set of all points $\mathbf{r}$ of the form

$$\mathbf{r} = x_1\mathbf{a}_1 + x_2\mathbf{a}_2 + x_3\mathbf{a}_3 \tag{4.6}$$

for all x_i ranging continuously between 0 and 1; i.e., the parallelipiped spanned by the three vectors $\mathbf{a}_1$, $\mathbf{a}_2$, and $\mathbf{a}_3$. This choice has the disadvantage of not displaying the full symmetry of the Bravais lattice. For example (Figure 4.12), the unit cell (4.6) for the choice of primitive vectors (4.5) of the fcc Bravais lattice is an oblique parallelipiped, which does not have the full cubic symmetry of the lattice in which it is embedded. It is often important to work with cells that do have the full symmetry of their Bravais lattice. There are two widely used solutions to this problem:

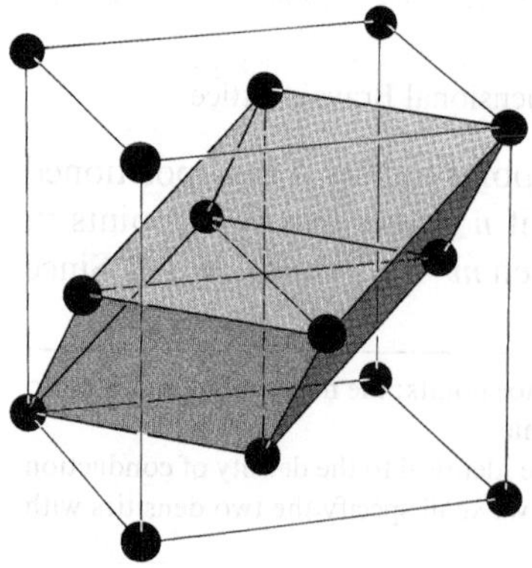

Figure 4.12
Primitive and conventional unit cells for the face-centered cubic Bravais lattice. The conventional cell is the large cube. The primitive cell is the figure with six parallelogram faces. It has one quarter the volume of the cube, and rather less symmetry.

UNIT CELL; CONVENTIONAL UNIT CELL

One can fill space up with nonprimitive unit cells (known simply as *unit cells* or *conventional unit cells*). A unit cell is a region that just fills space without any overlapping when translated through some *subset* of the vectors of a Bravais lattice. The conventional unit cell is generally chosen to be bigger than the primitive cell and to have the required symmetry. Thus one frequently describes the body-centered cubic lattice in terms of a cubic unit cell (Figure 4.13) that is twice as large as a primitive bcc unit cell, and the face-centered cubic lattice in terms of a cubic unit cell (Figure 4.12) that has four times the volume of a primitive fcc unit cell. (That the conventional cells are two and four times bigger than the primitive cells is easily seen by asking how many lattice points the conventional cubic cell must contain when it is so placed that no points are on its surface.) Numbers specifying the size of a unit cell (such as the single number a in cubic crystals) are called *lattice constants.*

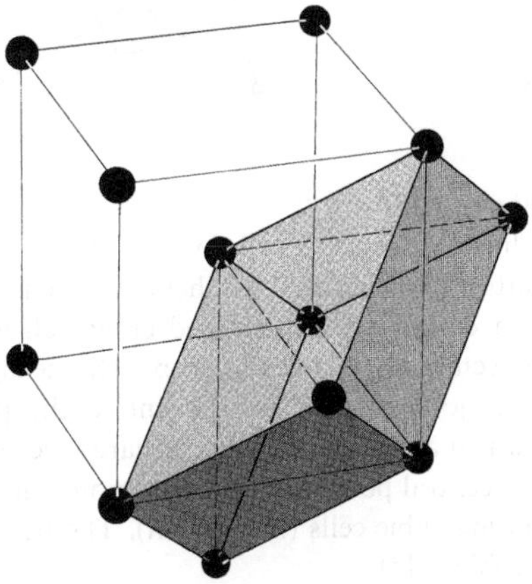

Figure 4.13
Primitive and conventional unit cells for the body-centered cubic Bravais lattice. The primitive cell (shaded) has half the volume of the conventional cubic cell.

WIGNER-SEITZ PRIMITIVE CELL

One can always choose a *primitive* cell with the full symmetry of the Bravais lattice. By far the most common such choice is the *Wigner-Seitz cell.* The Wigner-Seitz cell about a lattice point is the region of space that is closer to that point than to any other lattice point.[10] Because of the translational symmetry of the Bravais lattice, the Wigner-Seitz cell about any one lattice point must be taken into the Wigner-Seitz cell about any other, when translated through the lattice vector that joins the two points. Since any point in space has a unique lattice point, as its nearest neighbor[11] it will belong to the Wigner-Seitz cell of precisely one lattice point. It follows that a

---

[10] Such a cell can be defined for any set of discrete points that do not necessarily form a Bravais lattice. In this broader context the cell is known as a Voronoy polyhedron. In contrast to the Wigner-Seitz cell, the structure and orientation of a general Voronoy polyhedron will depend on which point of the array it encloses.

[11] Except for points on the common surface of two or more Wigner-Seitz cells.

Wigner-Seitz cell, when translated through all lattice vectors, will just fill space without overlapping; i.e., the Wigner-Seitz cell is a primitive cell.

Since there is nothing in the definition of the Wigner-Seitz cell that refers to any particular choice of primitive vectors, the Wigner-Seitz cell will be as symmetrical as the Bravais lattice.[12]

The Wigner-Seitz unit cell is illustrated for a two-dimensional Bravais lattice in Figure 4.14 and for the three-dimensional body-centered cubic and face-centered cubic Bravais lattices in Figures 4.15 and 4.16.

Note that the Wigner-Seitz unit cell about a lattice point can be constructed by drawing lines connecting the point to all others[13] in the lattice, bisecting each line

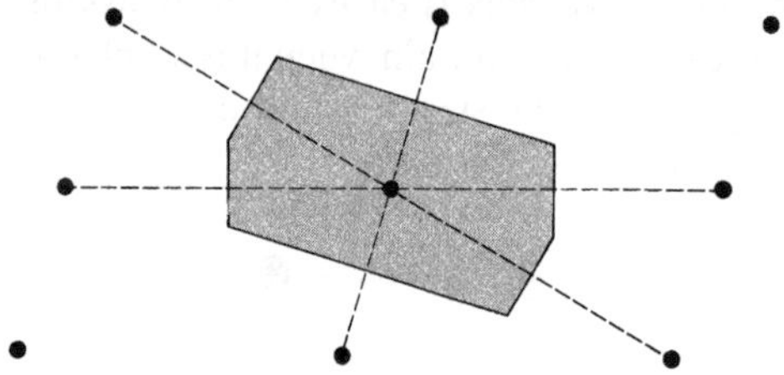

Figure 4.14
The Wigner-Seitz cell for a two-dimensional Bravais lattice. The six sides of the cell bisect the lines joining the central points to its six nearest neighboring points (shown as dashed lines). In two dimensions the Wigner-Seitz cell is always a hexagon unless the lattice is rectangular (see Problem 4a).

Figure 4.15
The Wigner-Seitz cell for the body-centered cubic Bravais lattice (a "truncated octahedron"). The surrounding cube is a conventional body-centered cubic cell with a lattice point at its center and on each vertex. The hexagonal faces bisect the lines joining the central point to the points on the vertices (drawn as solid lines). The square faces bisect the lines joining the central point to the central points in each of the six neighboring cubic cells (not drawn). The hexagons are regular (see Problem 4d).

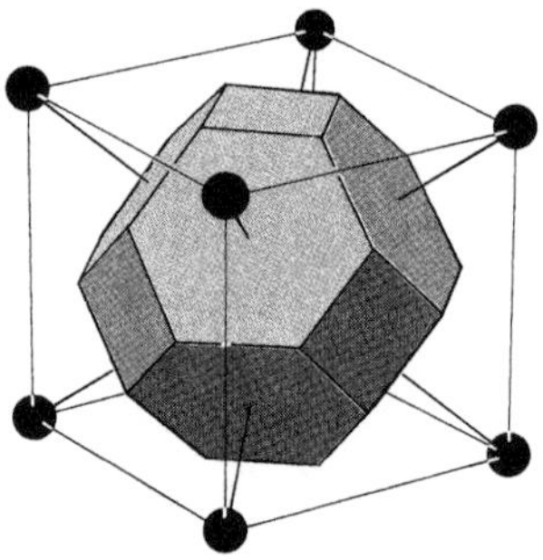

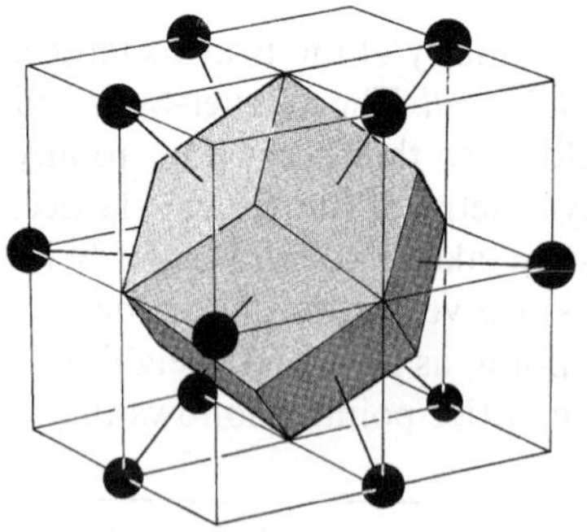

Figure 4.16
Wigner-Seitz cell for the face-centered cubic Bravais lattice (a "rhombic dodecahedron"). The surrounding cube is *not* the conventional cubic cell of Figure 4.12, but one in which lattice points are at the center of the cube and at the center of the 12 edges. Each of the 12 (congruent) faces is perpendicular to a line joining the central point to a point on the center of an edge.

---

12 A precise definition of "as symmetrical as" is given in Chapter 7.

13 In practice only a fairly small number of nearby points actually yield planes that bound the cell.

with a plane, and taking the smallest polyhedron containing the point bounded by these planes.

CRYSTAL STRUCTURE; LATTICE WITH A BASIS

A physical crystal can be described by giving its underlying Bravais lattice, together with a description of the arrangement of atoms, molecules, ions, etc., within a particular primitive cell. When emphasizing the difference between the abstract pattern of points composing the Bravais lattice and an actual physical crystal[14] embodying the lattice, the technical term "crystal structure" is used. A *crystal structure* consists of identical copies of the same physical unit, called the *basis*, located at all the points of a Bravais lattice (or, equivalently, translated through all the vectors of a Bravais lattice). Sometimes the term *lattice with a basis* is used instead. However, "lattice with a basis" is also used in a more general sense to refer to what results even when the basic unit is *not* a physical object or objects, but another set of points. For example, the vertices of a two-dimensional honeycomb, though not a Bravais lattice, can be represented as a two-dimensional triangular Bravais lattice[15] with a two-point basis (Figure 4.17). A crystal structure with a basis consisting of a single atom or ion is often called a monatomic Bravais lattice.

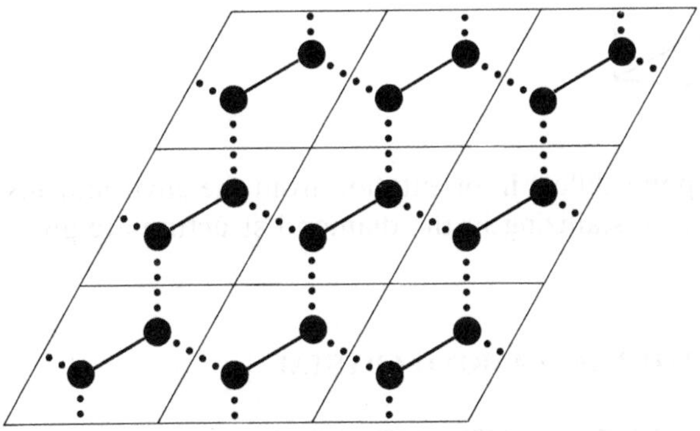

Figure 4.17
The honeycomb net, drawn so as to emphasize that it is a Bravais lattice with a two-point basis. The pairs of points joined by heavy solid lines are identically placed in the primitive cells (parallelograms) of the underlying Bravais lattice.

One also can describe a Bravais lattice as a lattice with a basis by choosing a non-primitive conventional unit cell. This is often done to emphasize the cubic symmetry of the bcc and fcc Bravais lattices, which are then described respectively, as simple cubic lattices spanned by $a\hat{\mathbf{x}}$, $a\hat{\mathbf{y}}$, and $a\hat{\mathbf{z}}$, with a two-point basis

$$\mathbf{0}, \quad \frac{a}{2}(\hat{\mathbf{x}} + \hat{\mathbf{y}} + \hat{\mathbf{z}}) \qquad \text{(bcc)} \tag{4.7}$$

or a four-point basis

$$\mathbf{0}, \quad \frac{a}{2}(\hat{\mathbf{x}} + \hat{\mathbf{y}}), \quad \frac{a}{2}(\hat{\mathbf{y}} + \hat{\mathbf{z}}), \quad \frac{a}{2}(\hat{\mathbf{z}} + \hat{\mathbf{x}}) \qquad \text{(fcc).} \tag{4.8}$$

14 But still idealized in being infinite in extent.

15 Spanned by two primitive vectors of equal length, making an angle of 60°.

SOME IMPORTANT EXAMPLES OF CRYSTAL STRUCTURES AND LATTICES WITH BASES

Diamond Structure

The diamond lattice[16] (formed by the carbon atoms in a diamond crystal) consists of two interpenetrating face-centered cubic Bravais lattices, displaced along the body diagonal of the cubic cell by one quarter the length of the diagonal. It can be regarded as a face-centered cubic lattice with the two-point basis **0** and $(a/4)(\hat{\mathbf{x}} + \hat{\mathbf{y}} + \hat{\mathbf{z}})$. The coordination number is 4 (Figure 4.18). The diamond lattice is not a Bravais lattice, because the environment of any point differs in orientation from the environments of its nearest neighbors. Elements crystallizing in the diamond structure are given in Table 4.3.

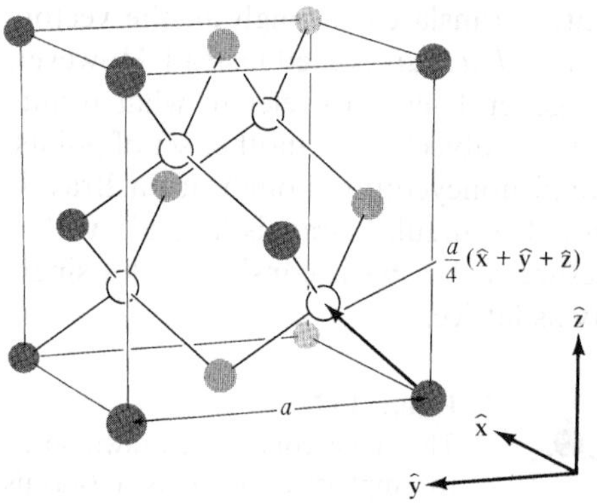

Figure 4.18
Conventional cubic cell of the diamond lattice. For clarity, sites corresponding to one of the two interpenetrating face-centered cubic lattices are unshaded. (In the zincblende structure the shaded sites are occupied by one kind of ion, and the unshaded by another.) Nearest-neighbor bonds have been drawn in. The four nearest neighbors of each point form the vertices of a regular tetrahedron.

Table 4.3
ELEMENTS WITH THE DIAMOND CRYSTAL STRUCTURE

ELEMENT	CUBE SIDE a (Å)
C (diamond)	3.57
Si	5.43
Ge	5.66
α-Sn (grey)	6.49

Hexagonal Close-Packed Structure

Though not a Bravais lattice, the *hexagonal close-packed* (hcp) structure ranks in importance with the body-centered cubic and face-centered cubic Bravais lattices; about 30 elements crystallize in the hexagonal close-packed form (Table 4.4).

---

16 We use the word "lattice," without qualifications, to refer either to a Bravais lattice or a lattice with a basis.

Table 4.4
ELEMENTS WITH THE HEXAGONAL CLOSE-PACKED CRYSTAL STRUCTURE

ELEMENT	a (Å)	c	c/a	ELEMENT	a (Å)	c	c/a
Be	2.29	3.58	1.56	Os	2.74	4.32	1.58
Cd	2.98	5.62	1.89	Pr	3.67	5.92	1.61
Ce	3.65	5.96	1.63	Re	2.76	4.46	1.62
α-Co	2.51	4.07	1.62	Ru	2.70	4.28	1.59
Dy	3.59	5.65	1.57	Sc	3.31	5.27	1.59
Er	3.56	5.59	1.57	Tb	3.60	5.69	1.58
Gd	3.64	5.78	1.59	Ti	2.95	4.69	1.59
He (2 K)	3.57	5.83	1.63	Tl	3.46	5.53	1.60
Hf	3.20	5.06	1.58	Tm	3.54	5.55	1.57
Ho	3.58	5.62	1.57	Y	3.65	5.73	1.57
La	3.75	6.07	1.62	Zn	2.66	4.95	1.86
Lu	3.50	5.55	1.59	Zr	3.23	5.15	1.59
Mg	3.21	5.21	1.62		—	—	
Nd	3.66	5.90	1.61	"Ideal"			1.63

Underlying the hcp structure is a *simple hexagonal* Bravais lattice, given by stacking two-dimensional triangular nets[15] directly above each other (Figure 4.19). The direction of stacking ($\mathbf{a}_3$, below) is known as the c-axis. Three primitive vectors are

$$\mathbf{a}_1 = a\hat{\mathbf{x}}, \quad \mathbf{a}_2 = \frac{a}{2}\hat{\mathbf{x}} + \frac{\sqrt{3}a}{2}\hat{\mathbf{y}}, \quad \mathbf{a}_3 = c\hat{\mathbf{z}}. \tag{4.9}$$

The first two generate a triangular lattice in the x-y plane, and the third stacks the planes a distance c above one another.

The hexagonal close-packed structure consists of two interpenetrating simple hexagonal Bravais lattices, displaced from one another by $\mathbf{a}_1/3 + \mathbf{a}_2/3 + \mathbf{a}_3/2$ (Figure 4.20). The name reflects the fact that close-packed hard spheres can be arranged in

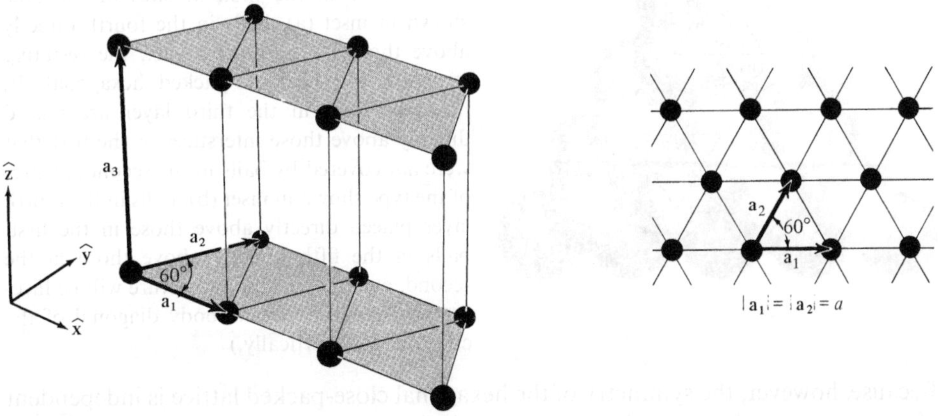

Figure 4.19
The simple hexagonal Bravais lattice. Two-dimensional triangular nets (shown in inset) are stacked directly above one another, a distance c apart.

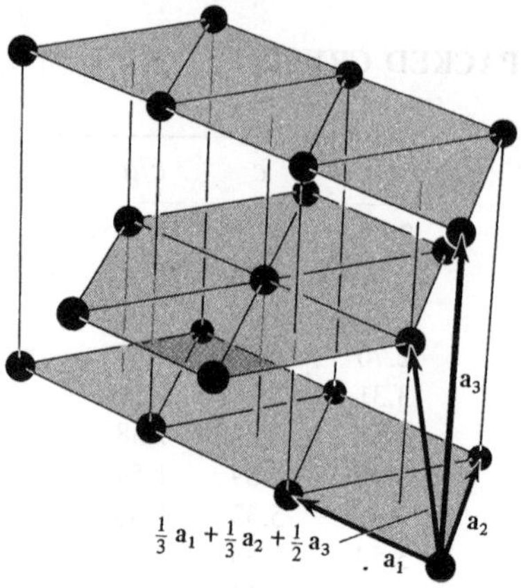

Figure 4.20
The hexagonal close-packed crystal structure. It can be viewed as two interpenetrating simple hexagonal Bravais lattices, displaced vertically by a distance $c/2$ along the common c-axis, and displaced horizontally so that the points of one lie directly above the centers of the triangles formed by the points of the other.

such a structure. Consider stacking cannonballs (Figure 4.21), starting with a close-packed triangular lattice as the first layer. The next layer is formed by placing a ball in the depressions left in the center of every other triangle in the first layer, thereby forming a second triangular layer, shifted with respect to the first. The third layer is formed by placing balls in alternate depressions in the second layer, so that they lie directly over the balls in the first layer. The fourth layer lies directly over the second, and so on. The resulting lattice is hexagonal close-packed with the particular value (see Problem 5):

$$c = \sqrt{\frac{8}{3}}\,a = 1.63299a. \quad \textbf{(4.10)}$$

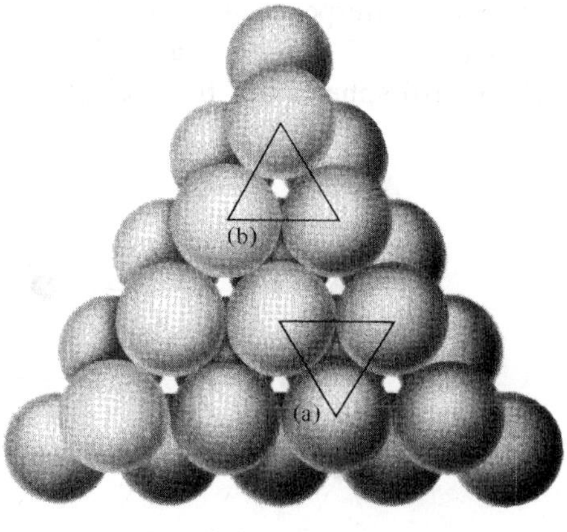

Figure 4.21
View from above of the first two layers in a stack of cannonballs. The first layer is arranged in a plane triangular lattice. Balls in the second layer are placed above alternate interstices in the first. If balls in the third layer are placed directly above those in the first, at sites of the type shown in inset (a), balls in the fourth directly above those in the second, etc., the resulting structure will be close-packed hexagonal. If, however, balls in the third layer are placed directly above those interstices in the first that were *not* covered by balls in the second, at sites of the type shown in inset (b), balls in the fourth layer placed directly above those in the first, balls in the fifth directly above those in the second, etc., the resulting structure will be face-centered cubic (with the body diagonal of the cube oriented vertically.)

Because, however, the symmetry of the hexagonal close-packed lattice is independent of the c/a ratio, the name is not restricted to this case. The value $c/a = \sqrt{8/3}$ is sometimes called "ideal," and the truly close-packed structure, with the ideal value of c/a, is known as an ideal hcp structure. Unless, however, the physical units in the hcp structure are actually close-packed spheres, there is no reason why c/a should be ideal (see Table 4.4).

Note, as in the case of the diamond structure, that the hcp lattice is not a Bravais lattice, because the orientation of the environment of a point varies from layer to layer along the c-axis. Note also that, when viewed along the c-axis, the two types of planes merge to form the two-dimensional honeycomb array of Figure 4.3, which is not a Bravais lattice.

Other Close-Packing Possibilities

Note that the hcp structure is not the only way to close-pack spheres. If the first two layers are laid down as described above, but the third is placed in the *other* set of depressions in the second—i.e., those lying above unused depressions in *both* the first and second layers (see Figure 4.21)—and then the fourth layer is placed in depressions in the third directly above the balls in the first, the fifth above the second, and so on, one generates a Bravais lattice. This Bravais lattice turns out to be nothing but the face-centered cubic lattice, with the cube diagonal perpendicular to the triangular planes (Figures 4.22 and 4.23).

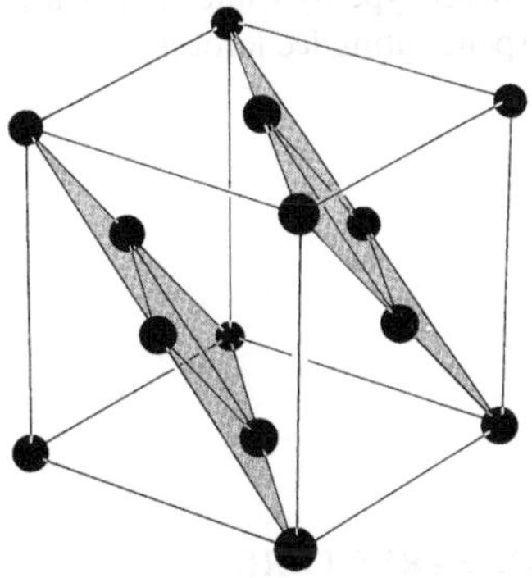

Figure 4.22
How to section the face-centered cubic Bravais lattice to get the layers pictured in Figure 4.21.

Figure 4.23
A cubic section of some face-centered cubic close-packed spheres.

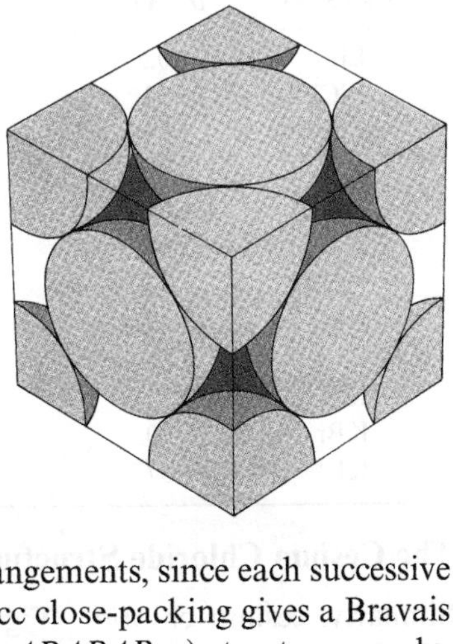

There are infinitely many other close-packing arrangements, since each successive layer can be placed in either of two positions. Only fcc close-packing gives a Bravais lattice, and the fcc (...*ABCABCABC*...) and hcp (...*ABABAB*...) structures are by far the most commonly encountered. Other close-packed structures are observed, however. Certain rare earth metals, for example, take on a structure of the form (...*ABACABACABAC*...).

The Sodium Chloride Structure

We are forced to describe the hexagonal close-packed and diamond lattices as lattices with bases by the intrinsic geometrical arrangement of the lattice points. A lattice with a basis is also necessary, however, in describing crystal structures in which the atoms or ions are located only at the points of a Bravais lattice, but in which the crystal structure nevertheless lacks the full translational symmetry of the Bravais lattice because more than one kind of atom or ion is present. For example, sodium chloride (Figure 4.24) consists of equal numbers of sodium and chlorine ions placed at alternate points of a simple cubic lattice, in such a way that each ion has six of the other kind of ions as its nearest neighbors.[17] This structure can be described as a face-centered cubic Bravais lattice with a basis consisting of a sodium ion at **0** and a chlorine ion at the center of the conventional cubic cell, $(a/2)(\hat{\mathbf{x}} + \hat{\mathbf{y}} + \hat{\mathbf{z}})$.

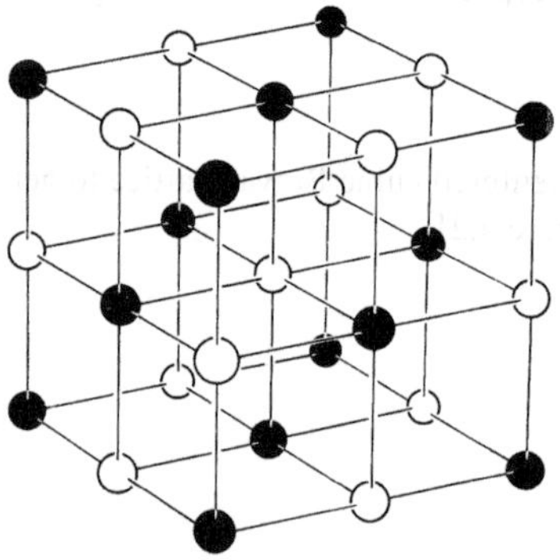

Figure 4.24
The sodium chloride structure. One type of ion is represented by black balls, the other type by white. The black and white balls form interpenetrating fcc lattices.

Table 4.5
SOME COMPOUNDS WITH THE SODIUM CHLORIDE STRUCTURE

CRYSTAL	*a* (Å)	CRYSTAL	*a* (Å)	CRYSTAL	*a* (Å)
LiF	4.02	RbF	5.64	CaS	5.69
LiCl	5.13	RbCl	6.58	CaSe	5.91
LiBr	5.50	RbBr	6.85	CaTe	6.34
LiI	6.00	RbI	7.34	SrO	5.16
NaF	4.62	CsF	6.01	SrS	6.02
NaCl	5.64	AgF	4.92	SrSe	6.23
NaBr	5.97	AgCl	5.55	SrTe	6.47
NaI	6.47	AgBr	5.77	BaO	5.52
KF	5.35	MgO	4.21	BaS	6.39
KCl	6.29	MgS	5.20	BaSe	6.60
KBr	6.60	MgSe	5.45	BaTe	6.99
KI	7.07	CaO	4.81		

The Cesium Chloride Structure

Similarly, cesium chloride (Figure 4.25) consists of equal numbers of cesium and chlorine ions, placed at the points of a body-centered cubic lattice so that each ion

[17] For examples see Table 4.5.

has eight of the other kind as its nearest neighbors.[18] The translational symmetry of this structure is that of the simple cubic Bravais lattice, and it is described as a simple cubic lattice with a basis consisting of a cesium ion at the origin **0** and a chlorine ion at the cube center $(a/2)(\hat{\mathbf{x}} + \hat{\mathbf{y}} + \hat{\mathbf{z}})$.

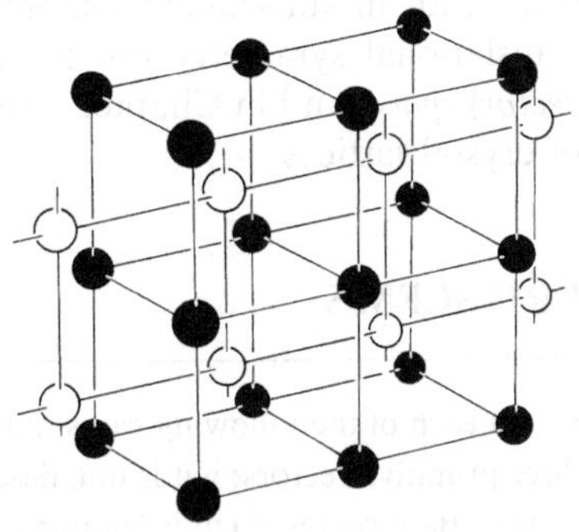

Figure 4.25
The cesium chloride structure. One type of ion is represented by black balls, the other type by white. The black and white balls form interpenetrating simple cubic lattices.

Table 4.6
SOME COMPOUNDS WITH THE CESIUM CHLORIDE STRUCTURE

CRYSTAL	a (Å)	CRYSTAL	a (Å)
CsCl	4.12	TlCl	3.83
CsBr	4.29	TlBr	3.97
CsI	4.57	TlI	4.20

The Zincblende Structure

Zincblende has equal numbers of zinc and sulfur ions distributed on a diamond lattice so that each has four of the opposite kind as nearest neighbors (Figure 4.18). This structure[19] is an example of a lattice with a basis, which must be so described both because of the geometrical position of the ions and because two types of ions occur.

Table 4.7
SOME COMPOUNDS WITH THE ZINCBLENDE STRUCTURE

CRYSTAL	a (Å)	CRYSTAL	a (Å)	CRYSTAL	a (Å)
CuF	4.26	ZnS	5.41	AlSb	6.13
CuCl	5.41	ZnSe	5.67	GaP	5.45
CuBr	5.69	ZnTe	6.09	GaAs	5.65
CuI	6.04	CdS	5.82	GaSb	6.12
AgI	6.47	CdTe	6.48	InP	5.87
BeS	4.85	HgS	5.85	InAs	6.04
BeSe	5.07	HgSe	6.08	InSb	6.48
BeTe	5.54	HgTe	6.43	SiC	4.35
MnS (red)	5.60	AlP	5.45		
MnSe	5.82	AlAs	5.62		

18 For examples see Table 4.6.
19 For examples see Table 4.7.

OTHER ASPECTS OF CRYSTAL LATTICES

This chapter has concentrated on the description of the *translational* symmetry of crystal lattices in *real physical space.* Two other aspects of periodic arrays will be dealt with in subsequent chapters: in Chapter 5 we examine the consequences of translational symmetry not in real space, but in the so-called *reciprocal* (or *wave vector*) *space*, and in Chapter 7 we describe some features of the *rotational* symmetry of crystal lattices.

PROBLEMS

1. In each of the following cases indicate whether the structure is a Bravais lattice. If it is, give three primitive vectors; if it is not, describe it as a Bravais lattice with as small as possible a basis.

(a) Base-centered cubic (simple cubic with additional points in the centers of the horizontal faces of the cubic cell).

(b) Side-centered cubic (simple cubic with additional points in the centers of the vertical faces of the cubic cell).

(c) Edge-centered cubic (simple cubic with additional points at the midpoints of the lines joining nearest neighbors).

2. What is the Bravais lattice formed by all points with Cartesian coordinates (n_1, n_2, n_3) if:

(a) The n_i are either all even or all odd?

(b) The sum of the n_i is required to be even?

3. Show that the angle between any two of the lines (bonds) joining a site of the diamond lattice to its four nearest neighbors is $\cos^{-1}(-1/3) = 109°28'$.

4. (a) Prove that the Wigner-Seitz cell for any two-dimensional Bravais lattice is either a hexagon or a rectangle.

(b) Show that the ratio of the lengths of the diagonals of each parallelogram face of the Wigner-Seitz cell for the face-centered cubic lattice (Figure 4.16) is $\sqrt{2}$:1.

(c) Show that every edge of the polyhedron bounding the Wigner-Seitz cell of the body-centered cubic lattice (Figure 4.15) is $\sqrt{2}/4$ times the length of the conventional cubic cell.

(d) Prove that the hexagonal faces of the bcc Wigner-Seitz cell are all regular hexagons. (Note that the axis perpendicular to a hexagonal face passing through its center has only threefold symmetry, so this symmetry alone is not enough.)

5. (a) Prove that the ideal c/a ratio for the hexagonal close-packed structure is $\sqrt{8/3} = 1.633$.

(b) Sodium transforms from bcc to hcp at about 23K (the "martensitic" transformation). Assuming that the density remains fixed through this transition, find the lattice constant a of the hexagonal phase, given that $a = 4.23$ Å in the cubic phase and that the c/a ratio is indistinguishable from its ideal value.

6. The face-centered cubic is the most dense and the simple cubic is the least dense of the three cubic Bravais lattices. The diamond structure is less dense than any of these. One measure of this is that the coordination numbers are: fcc, 12; bcc, 8; sc, 6; diamond, 4. Another is the following: Suppose identical solid spheres are distributed through space in such a way that their centers lie on the points of each of these four structures, and spheres on neighboring points just touch, without overlapping. (Such an arrangement of spheres is called a close-packing arrangement.) Assuming that the spheres have unit density, show that the density of a set of close-packed spheres on each of the four structures (the "packing fraction") is:

$$\begin{aligned}
&\text{fcc:} && \sqrt{2}\pi/6 = 0.74 \\
&\text{bcc:} && \sqrt{3}\pi/8 = 0.68 \\
&\text{sc:} && \pi/6 = 0.52 \\
&\text{diamond:} && \sqrt{3}\pi/16 = 0.34.
\end{aligned}$$

7. Let N_n be the number of nth nearest neighbors of a given Bravais lattice point (e.g., in a simple cubic Bravais lattice $N_1 = 6$, $N_2 = 12$, etc.). Let r_n be the distance to the nth nearest neighbor expressed as a multiple of the nearest neighbor distance (e.g., in a simple cubic Bravais lattice $r_1 = 1$, $r_2 = \sqrt{2} = 1.414$). Make a table of N_n and r_n for $n = 1, \ldots, 6$ for the fcc, bcc, and sc Bravais lattices.

8. (a) Given a Bravais lattice, let $\mathbf{a}_1$ be a vector joining a particular point P to one of its nearest neighbors. Let P' be a lattice point not on the line through P in the direction of $\mathbf{a}_1$ that is as close to the line as any other lattice point, and let $\mathbf{a}_2$ join P to P'. Let P'' be a lattice point not on the plane through P determined by $\mathbf{a}_1$ and $\mathbf{a}_2$ that is as close to the plane as any other lattice point, and let $\mathbf{a}_3$ join P to P''. Prove that $\mathbf{a}_1$, $\mathbf{a}_2$, and $\mathbf{a}_3$ are a set of primitive vectors for the Bravais lattice.

(b) Prove that a Bravais lattice can be defined as a discrete set of vectors, not all in a plane, closed under addition and subtraction (as described on page 70).

5

The Reciprocal Lattice

The reciprocal lattice plays a fundamental role in most analytic studies of periodic structures. One is led to it from such diverse avenues as the theory of crystal diffraction, the abstract study of functions with the periodicity of a Bravais lattice, or the question of what can be salvaged of the law of momentum conservation when the full translational symmetry of free space is reduced to that of a periodic potential. In this brief chapter we shall describe some important elementary features of the reciprocal lattice from a general point of view not tied to any particular application.

DEFINITION OF RECIPROCAL LATTICE

Consider a set of points **R** constituting a Bravais lattice, and a plane wave, $e^{i\mathbf{k}\cdot\mathbf{r}}$. For general **k**, such a plane wave will not, of course, have the periodicity of the Bravais lattice, but for certain special choices of wave vector it will. *The set of all wave vectors* **K** *that yield plane waves with the periodicity of a given Bravais lattice is known as its reciprocal lattice.* Analytically, **K** belongs to the reciprocal lattice of a Bravais lattice of points **R**, provided that the relation

$$e^{i\mathbf{K}\cdot(\mathbf{r}+\mathbf{R})} = e^{i\mathbf{K}\cdot\mathbf{r}} \tag{5.1}$$

holds for any **r**, and for all **R** in the Bravais lattice. Factoring out $e^{i\mathbf{K}\cdot\mathbf{r}}$, we can characterize the reciprocal lattice as the set of wave vectors **K** satisfying

$$e^{i\mathbf{K}\cdot\mathbf{R}} = 1 \tag{5.2}$$

for all **R** in the Bravais lattice.

Note that a reciprocal lattice is defined with reference to a particular Bravais lattice. The Bravais lattice that determines a given reciprocal lattice is often referred to as the *direct lattice*, when viewed in relation to its reciprocal. Note also that although one could define a set of vectors **K** satisfying (5.2) for an arbitrary set of vectors **R**, such a set of **K** is called a reciprocal lattice only if the set of vectors **R** is a Bravais lattice.[1]

THE RECIPROCAL LATTICE IS A BRAVAIS LATTICE

That the reciprocal lattice is itself a Bravais lattice follows most simply from the definition of a Bravais lattice given in footnote 7 of Chapter 4, along with the fact that if $\mathbf{K}_1$ and $\mathbf{K}_2$ satisfy (5.2), so, obviously, will their sum and difference.

It is worth considering a more clumsy proof of this fact, which provides an explicit algorithm for constructing the reciprocal lattice. Let $\mathbf{a}_1$, $\mathbf{a}_2$, and $\mathbf{a}_3$ be a set of primitive vectors for the direct lattice. Then the reciprocal lattice can be generated by the three primitive vectors

$$\begin{aligned}
\mathbf{b}_1 &= 2\pi \frac{\mathbf{a}_2 \times \mathbf{a}_3}{\mathbf{a}_1 \cdot (\mathbf{a}_2 \times \mathbf{a}_3)}, \\
\mathbf{b}_2 &= 2\pi \frac{\mathbf{a}_3 \times \mathbf{a}_1}{\mathbf{a}_1 \cdot (\mathbf{a}_2 \times \mathbf{a}_3)}, \\
\mathbf{b}_3 &= 2\pi \frac{\mathbf{a}_1 \times \mathbf{a}_2}{\mathbf{a}_1 \cdot (\mathbf{a}_2 \times \mathbf{a}_3)}.
\end{aligned} \tag{5.3}$$

---

[1] In particular, in working with a lattice with a basis one uses the reciprocal lattice determined by the underlying Bravais lattice, rather than a set of **K** satisfying (5.2) for vectors **R** describing both the Bravais lattice and the basis points.

To verify that (5.3) gives a set of primitive vectors for the reciprocal lattice, one first notes that the $\mathbf{b}_i$ satisfy[2]

$$\mathbf{b}_i \cdot \mathbf{a}_j = 2\pi\delta_{ij}, \tag{5.4}$$

where δ_{ij} is the Kronecker delta symbol:

$$\begin{aligned} \delta_{ij} &= 0, \qquad i \neq j; \\ \delta_{ij} &= 1, \qquad i = j. \end{aligned} \tag{5.5}$$

Now any vector $\mathbf{k}$ can be written as a linear combination[3] of the $\mathbf{b}_i$:

$$\mathbf{k} = k_1\mathbf{b}_1 + k_2\mathbf{b}_2 + k_3\mathbf{b}_3. \tag{5.6}$$

If $\mathbf{R}$ is any direct lattice vector, then

$$\mathbf{R} = n_1\mathbf{a}_1 + n_2\mathbf{a}_2 + n_3\mathbf{a}_3, \tag{5.7}$$

where the n_i are integers. It follows from (5.4) that

$$\mathbf{k} \cdot \mathbf{R} = 2\pi(k_1n_1 + k_2n_2 + k_3n_3). \tag{5.8}$$

For $e^{i\mathbf{k}\cdot\mathbf{R}}$ to be unity for all $\mathbf{R}$ (Eq. (5.2)) $\mathbf{k} \cdot \mathbf{R}$ must be 2π times an integer for any choices of the integers n_i. This requires the coefficients k_i to be integers. Thus the condition (5.2) that $\mathbf{K}$ be a reciprocal lattice vector is satisfied by just those vectors that are linear combinations (5.6) of the $\mathbf{b}_i$ with integral coefficients. Thus (compare Eq. (4.1)) the reciprocal lattice is a Bravais lattice and the $\mathbf{b}_i$ can be taken as primitive vectors.

THE RECIPROCAL OF THE RECIPROCAL LATTICE

Since the reciprocal lattice is itself a Bravais lattice, one can construct *its* reciprocal lattice. This turns out to be nothing but the original direct lattice.

One way to prove this is by constructing $\mathbf{c}_1$, $\mathbf{c}_2$, and $\mathbf{c}_3$ out of the $\mathbf{b}_i$ according to the same formula (5.3) by which the $\mathbf{b}_i$ were constructed from the $\mathbf{a}_i$. It then follows from simple vector identities (Problem 1) that $\mathbf{c}_i = \mathbf{a}_i$, $i = 1, 2, 3$.

A simpler proof follows from the observation that according to the basic definition (5.2), the reciprocal of the reciprocal lattice is the set of all vectors $\mathbf{G}$ satisfying

$$e^{i\mathbf{G}\cdot\mathbf{K}} = 1 \tag{5.9}$$

for all $\mathbf{K}$ in the reciprocal lattice. Since any direct lattice vector $\mathbf{R}$ has this property (again by (5.2)), all direct lattice vectors are in the lattice reciprocal to the reciprocal lattice. Furthermore, no other vectors can be, for a vector not in the direct lattice has the form $\mathbf{r} = x_1\mathbf{a}_1 + x_2\mathbf{a}_2 + x_3\mathbf{a}_3$ with at least one nonintegral x_i. For that value of i, $e^{i\mathbf{b}_i\cdot\mathbf{r}} = e^{2\pi i x_i} \neq 1$, and condition (5.9) is violated for the reciprocal lattice vector $\mathbf{K} = \mathbf{b}_i$.

---

[2] When $i \neq j$, Eq. (5.4) follows because the cross product of two vectors is normal to both. When $i = j$, it follows because of the vector identity

$$\mathbf{a}_1 \cdot (\mathbf{a}_2 \times \mathbf{a}_3) = \mathbf{a}_2 \cdot (\mathbf{a}_3 \times \mathbf{a}_1) = \mathbf{a}_3 \cdot (\mathbf{a}_1 \times \mathbf{a}_2).$$

[3] This is true for any three vectors not all in one plane. It is easy to verify that the $\mathbf{b}_i$ are not all in a plane as long as the $\mathbf{a}_i$ are not.

IMPORTANT EXAMPLES

The *simple cubic* Bravais lattice, with cubic primitive cell of side a, has as its reciprocal a simple cubic lattice with cubic primitive cell of side $2\pi/a$. This can be seen, for example, from the construction (5.3), for if

$$\mathbf{a}_1 = a\hat{\mathbf{x}}, \quad \mathbf{a}_2 = a\hat{\mathbf{y}}, \quad \mathbf{a}_3 = a\hat{\mathbf{z}}, \tag{5.10}$$

then

$$\mathbf{b}_1 = \frac{2\pi}{a}\hat{\mathbf{x}}, \quad \mathbf{b}_2 = \frac{2\pi}{a}\hat{\mathbf{y}}, \quad \mathbf{b}_3 = \frac{2\pi}{a}\hat{\mathbf{z}}. \tag{5.11}$$

The *face-centered cubic* Bravais lattice with conventional cubic cell of side a has as its reciprocal a body-centered cubic lattice with conventional cubic cell of side $4\pi/a$. This can be seen by applying the construction (5.3) to the fcc primitive vectors (4.5). The result is

$$\mathbf{b}_1 = \frac{4\pi}{a}\frac{1}{2}(\hat{\mathbf{y}} + \hat{\mathbf{z}} - \hat{\mathbf{x}}), \quad \mathbf{b}_2 = \frac{4\pi}{a}\frac{1}{2}(\hat{\mathbf{z}} + \hat{\mathbf{x}} - \hat{\mathbf{y}}), \quad \mathbf{b}_3 = \frac{4\pi}{a}\frac{1}{2}(\hat{\mathbf{x}} + \hat{\mathbf{y}} - \hat{\mathbf{z}}) \tag{5.12}$$

This has precisely the form of the bcc primitive vectors (4.4), provided that the side of the cubic cell is taken to be $4\pi/a$.

The *body-centered cubic* lattice with conventional cubic cell of side a has as its reciprocal a face-centered cubic lattice with conventional cubic cell of side $4\pi/a$. This can again be proved from the construction (5.3), but it also follows from the above result for the reciprocal of the fcc lattice, along with the theorem that the reciprocal of the reciprocal is the original lattice.

It is left as an exercise for the reader to verify (Problem 2) that the reciprocal to a *simple hexagonal* Bravais lattice with lattice constants c and a (Figure 5.1a) is another

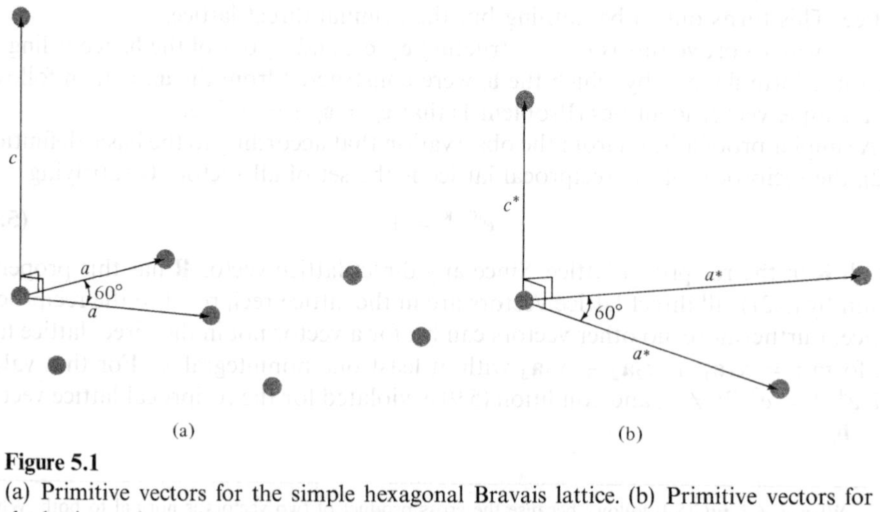

Figure 5.1
(a) Primitive vectors for the simple hexagonal Bravais lattice. (b) Primitive vectors for the lattice reciprocal to that generated by the primitive vectors in (a). The c and c^* axes are parallel. The a^* axes are rotated by 30° with respect to the a axes in the plane perpendicular to the c or c^* axes. The reciprocal lattice is also simple hexagonal.

simple hexagonal lattice with lattice constants $2\pi/c$ and $4\pi/\sqrt{3}a$ (Figure 5.1b), rotated through 30° about the c-axis with respect to the direct lattice.[4]

VOLUME OF THE RECIPROCAL LATTICE PRIMITIVE CELL

If v is the volume[5] of a primitive cell in the direct lattice, then the primitive cell of the reciprocal lattice has a volume $(2\pi)^3/v$. This is proved in Problem 1.

FIRST BRILLOUIN ZONE

The Wigner-Seitz primitive cell (page 73) of the reciprocal lattice is known as the *first Brillouin zone.* As the name suggests, one also defines higher Brillouin zones, which are primitive cells of a different type that arise in the theory of electronic levels in a periodic potential. They are described in Chapter 9.

Although the terms "Wigner-Seitz cell" and "first Brillouin zone" refer to identical geometrical constructions, in practice the latter term is applied only to the k-space cell. In particular, when reference is made to the first Brillouin zone of a particular r-space Bravais lattice (associated with a particular crystal structure), what is always meant is the Wigner-Seitz cell of the associated reciprocal lattice. Thus, because the reciprocal of the body-centered cubic lattice is face-centered cubic, the first Brillouin zone of the bcc lattice (Figure 5.2a) is just the fcc Wigner-Seitz cell (Figure 4.16). Conversely, the first Brillouin zone of the fcc lattice (Figure 5.2b) is just the bcc Wigner-Seitz cell (Figure 4.15).

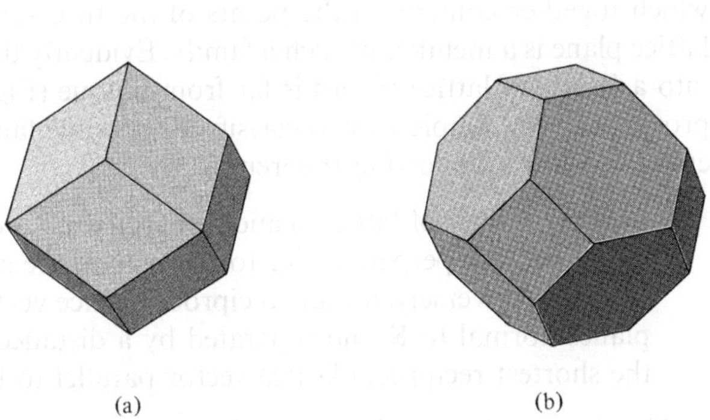

Figure 5.2
(a) The first Brillouin zone for the body-centered cubic lattice. (b) The first Brillouin zone for the face-centered cubic lattice.

LATTICE PLANES

There is an intimate relation between vectors in the reciprocal lattice and planes of points in the direct lattice. This relation is of some importance in understanding the fundamental role the reciprocal lattice plays in the theory of diffraction, and will be applied to that problem in the next chapter. Here we shall describe the relation in general geometrical terms.

---

[4] The hexagonal close-packed *structure* is not a Bravais lattice, and therefore the reciprocal lattice used in the analysis of hcp solids is that of the simple hexagonal lattice (see footnote 1).

[5] The primitive cell volume is independent of the choice of cell, as proved in Chapter 4.

Given a particular Bravais lattice, a *lattice plane* is defined to be any plane containing at least three noncollinear Bravais lattice points. Because of the translational symmetry of the Bravais lattice, any such plane will actually contain infinitely many lattice points, which form a two-dimensional Bravais lattice within the plane. Some lattice planes in a simple cubic Bravais lattice are pictured in Figure 5.3.

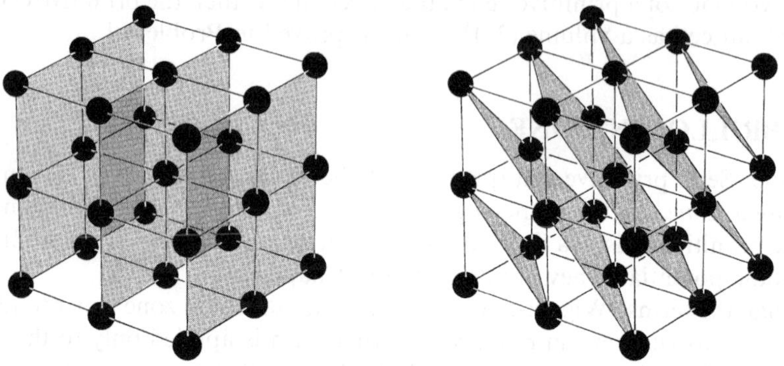

Figure 5.3
Some lattice planes (shaded) in a simple cubic Bravais lattice; (a) and (b) show two different ways of representing the lattice as a family of lattice planes.

By a *family of lattice planes* we mean a set of parallel, equally spaced lattice planes, which together contain all the points of the three-dimensional Bravais lattice. Any lattice plane is a member of such a family. Evidently the resolution of a Bravais lattice into a family of lattice planes is far from unique (Figure 5.3). The reciprocal lattice provides a very simple way to classify all possible families of lattice planes, which is embodied in the following theorem:

> For any family of lattice planes separated by a distance d, there are reciprocal lattice vectors perpendicular to the planes, the shortest of which have a length of $2\pi/d$. Conversely, for any reciprocal lattice vector $\mathbf{K}$, there is a family of lattice planes normal to $\mathbf{K}$ and separated by a distance d, where $2\pi/d$ is the length of the shortest reciprocal lattice vector parallel to $\mathbf{K}$.

The theorem is a straightforward consequence of (a) the definition (5.2) of reciprocal lattice vectors as the wave vectors of plane waves that are unity at all Bravais lattice sites and (b) the fact that a plane wave has the same value at all points lying in a family of planes that are perpendicular to its wave vector and separated by an integral number of wavelengths.

To prove the first part of the theorem, given a family of lattice planes, let $\hat{\mathbf{n}}$ be a unit vector normal to the planes. That $\mathbf{K} = 2\pi\hat{\mathbf{n}}/d$ is a reciprocal lattice vector follows from the fact that the plane wave $e^{i\mathbf{K}\cdot\mathbf{r}}$ is constant in planes perpendicular to $\mathbf{K}$ and has the same value in planes separated by $\lambda = 2\pi/K = d$. Since one of the lattice planes contains the Bravais lattice point $\mathbf{r} = \mathbf{0}$, $e^{i\mathbf{K}\cdot\mathbf{r}}$ must be unity for any point $\mathbf{r}$ in any of the planes. Since the planes contain all Bravais lattice points, $e^{i\mathbf{K}\cdot\mathbf{r}} = 1$ for all $\mathbf{R}$, so that $\mathbf{K}$ is indeed a reciprocal lattice vector. Furthermore, $\mathbf{K}$ is the shortest reciprocal lattice vector normal to the planes, for any wave vector shorter than **K** will give a plane wave with wavelength greater than $2\pi/K = d$. Such a plane wave cannot have the same value on all planes in the family, and therefore cannot give a plane wave that is unity at all Bravais lattice points.

To prove the converse of the theorem, given a reciprocal lattice vector, let **K** be the shortest parallel reciprocal lattice vector. Consider the set of real space planes on which the plane wave $e^{i\mathbf{K}\cdot\mathbf{r}}$ has the value unity. These planes (one of which contains the point $\mathbf{r} = \mathbf{0}$) are perpendicular to **K** and separated by a distance $d = 2\pi/K$. Since the Bravais lattice vectors **R** all satisfy $e^{i\mathbf{K}\cdot\mathbf{R}} = 1$ for any reciprocal lattice vector **K**, they must all lie within these planes; i.e., the family of planes must contain within it a family of lattice planes. Furthermore the spacing between the lattice planes is also d (rather than some integral multiple of d), for if only every nth plane in the family contained Bravais lattice points, then according to the first part of the theorem, the vector normal to the planes of length $2\pi/nd$, i.e., the vector $\mathbf{K}/n$, would be a reciprocal lattice vector. This would contradict our original assumption that no reciprocal lattice vector parallel to **K** is shorter than **K**.

MILLER INDICES OF LATTICE PLANES

The correspondence between reciprocal lattice vectors and families of lattice planes provides a convenient way to specify the orientation of a lattice plane. Quite generally one describes the orientation of a plane by giving a vector normal to the plane. Since we know there are reciprocal lattice vectors normal to any family of lattice planes, it is natural to pick a reciprocal lattice vector to represent the normal. To make the choice unique, one uses the shortest such reciprocal lattice vector. In this way one arrives at the *Miller indices* of the plane:

The Miller indices of a lattice plane are the coordinates of the shortest reciprocal lattice vector normal to that plane, with respect to a specified set of primitive reciprocal lattice vectors. Thus a plane with Miller indices h, k, l, is normal to the reciprocal lattice vector $h\mathbf{b}_1 + k\mathbf{b}_2 + l\mathbf{b}_3$.

As so defined, the Miller indices are integers, since any reciprocal lattice vector is a linear combination of three primitive vectors with integral coefficients. Since the normal to the plane is specified by the shortest perpendicular reciprocal lattice vector, the integers h, k, l can have no common factor. Note also that the Miller indices depend on the particular choice of primitive vectors.

In simple cubic Bravais lattices the reciprocal lattice is also simple cubic and the Miller indices are the coordinates of a vector normal to the plane in the obvious cubic coordinate system. As a general rule, face-centered and body-centered cubic Bravais lattice are described in terms of a conventional cubic cell, i.e., as simple cubic lattices with bases. Since any lattice plane in a fcc or bcc lattice is also a lattice plane in the underlying simple cubic lattice, the same elementary cubic indexing can be used to specify lattice planes. In practice, it is only in the description of noncubic crystals that one must remember that the Miller indices are the coordinates of the normal in a system given by the reciprocal lattice, rather than the direct lattice.

The Miller indices of a plane have a geometrical interpretation in the direct lattice, which is sometimes offered as an alternative way of defining them. Because a lattice plane with Miller indices h, k, l is perpendicular to the reciprocal lattice vector $\mathbf{K} = h\mathbf{b}_1 + k\mathbf{b}_2 + l\mathbf{b}_3$, it will be contained in the continuous plane $\mathbf{K} \cdot \mathbf{r} = A$, for suitable choice of the constant A. This plane intersects the axes determined by the direct lattice primitive vectors $\mathbf{a}_i$ at the points $x_1\mathbf{a}_1$, $x_2\mathbf{a}_2$, and $x_3\mathbf{a}_3$ (Figure 5.4), where the x_i are determined by the condition that $x_i\mathbf{a}_i$ indeed satisfy the equation of the plane: $\mathbf{K} \cdot (x_i\mathbf{a}_i) = A$. Since $\mathbf{K} \cdot \mathbf{a}_1 = 2\pi h$, $\mathbf{K} \cdot \mathbf{a}_2 = 2\pi k$, and $\mathbf{K} \cdot \mathbf{a}_3 = 2\pi l$, it follows that

$$x_1 = \frac{A}{2\pi h}, \quad x_2 = \frac{A}{2\pi k}, \quad x_3 = \frac{A}{2\pi l}. \tag{5.13}$$

Thus the intercepts with the crystal axes of a lattice plane are inversely proportional to the Miller indices of the plane.

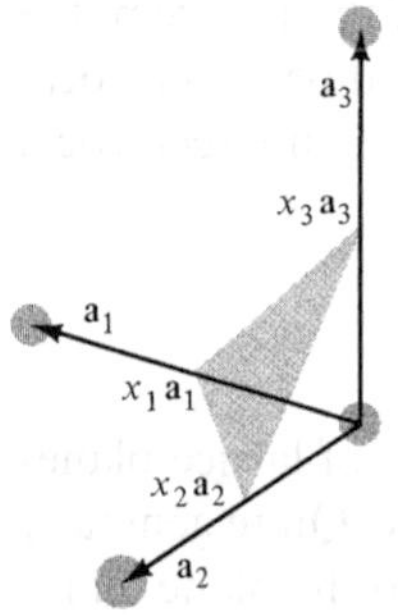

Figure 5.4
An illustration of the crystallographic definition of the Miller indices of a lattice plane. The shaded plane can be a portion of the continuous plane in which the points of the lattice plane lie, or any plane parallel to the lattice plane. The Miller indices are inversely proportional to the x_i.

Crystallographers put the cart before the horse, *defining* the Miller indices to be a set of integers with no common factors, inversely proportional to the intercepts of the crystal plane along the crystal axes:

$$h{:}k{:}l = \frac{1}{x_1}:\frac{1}{x_2}:\frac{1}{x_3}. \tag{5.14}$$

SOME CONVENTIONS FOR SPECIFYING DIRECTIONS

Lattice planes are usually specified by giving their Miller indices in parentheses: (h, k, l). Thus, in a cubic system, a plane with a normal $(4, -2, 1)$ (or, from the crystallographic viewpoint, a plane with intercepts $(1, -2, 4)$ along cubic axes) is called a $(4, -2, 1)$ plane. The commas are eliminated without confusion by writing $\bar{n}$ instead of $-n$, simplifying the description to $(4\bar{2}1)$. One must know what set of axes is being used to interpret these symbols unambiguously. Simple cubic axes are invariably used when the crystal has cubic symmetry. Some examples of planes in cubic crystals are shown in Figure 5.5.

A similar convention is used to specify directions in the direct lattice, but to avoid confusion with the Miller indices (directions in the reciprocal lattice) square brackets are used instead of parentheses. Thus the body diagonal of a simple cubic lattice lies in the [111] direction and, in general the lattice point $n_1\mathbf{a}_1 + n_2\mathbf{a}_2 + n_3\mathbf{a}_3$ lies in the direction $[n_1n_2n_3]$ from the origin.

There is also a notation specifying both a family of lattice planes and all those other families that are equivalent to it by virtue of the symmetry of the crystal. Thus

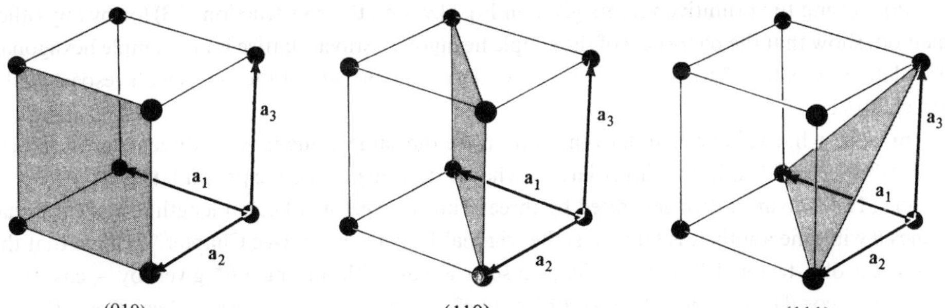

(010) (110) (111)

Figure 5.5
Three lattice planes and their Miller indices in a simple cubic Bravais lattice.

the (100), (010), and (001) planes are all equivalent in a cubic crystal. One refers to them collectively as the $\{100\}$ planes, and in general one uses $\{hkl\}$ to refer to the (hkl) planes and all those that are equivalent to them by virtue of the crystal symmetry. A similar convention is used with directions: the $[100]$, $[010]$, $[001]$, $[\bar{1}00]$, $[0\bar{1}0]$, and $[00\bar{1}]$ directions in a cubic crystal are referred to, collectively, as the $\langle 100 \rangle$ directions.

This concludes our general geometrical discussion of the reciprocal lattice. In Chapter 6 we shall see an important example of the utility and the power of the concept in the theory of the diffraction of X rays by a crystal.

PROBLEMS

1. (a) Prove that the reciprocal lattice primitive vectors defined in (5.3) satisfy

$$\mathbf{b}_1 \cdot (\mathbf{b}_2 \times \mathbf{b}_3) = \frac{(2\pi)^3}{\mathbf{a}_1 \cdot (\mathbf{a}_2 \times \mathbf{a}_3)}. \tag{5.15}$$

(*Hint:* Write $\mathbf{b}_1$ (but not $\mathbf{b}_2$ or $\mathbf{b}_3$) in terms of the $\mathbf{a}_i$, and use the orthogonality relations (5.4).)

(b) Suppose primitive vectors are constructed from the $\mathbf{b}_i$ in the same manner (Eq. (5.3)) as the $\mathbf{b}_i$ are constructed from the $\mathbf{a}_i$. Prove that these vectors are just the $\mathbf{a}_i$ themselves; i.e., show that

$$2\pi \frac{\mathbf{b}_2 \times \mathbf{b}_3}{\mathbf{b}_1 \cdot (\mathbf{b}_2 \times \mathbf{b}_3)} = \mathbf{a}_1, \quad \text{etc.} \tag{5.16}$$

(*Hint:* Write $\mathbf{b}_3$ in the numerator (but not $\mathbf{b}_2$) in terms of the $\mathbf{a}_i$, use the vector identity $\mathbf{A} \times (\mathbf{B} \times \mathbf{C}) = \mathbf{B}(\mathbf{A} \cdot \mathbf{C}) - \mathbf{C}(\mathbf{A} \cdot \mathbf{B})$, and appeal to the orthogonality relations (5.4) and the result (5.15) above.)

(c) Prove that the volume of a Bravais lattice primitive cell is

$$v = |\mathbf{a}_1 \cdot (\mathbf{a}_2 \times \mathbf{a}_3)|, \tag{5.17}$$

where the $\mathbf{a}_i$ are three primitive vectors. (In conjunction with (5.15) this establishes that the volume of the reciprocal lattice primitive cell is $(2\pi)^3/v$.)

2. (a) Using the primitive vectors given in Eq. (4.9) and the construction (5.3) (or by any other method) show that the reciprocal of the simple hexagonal Bravais lattice is also simple hexagonal, with lattice constants $2\pi/c$ and $4\pi/\sqrt{3}a$, rotated through 30° about the c-axis with respect to the direct lattice.

(b) For what value of c/a does the ratio have the same value in both direct and reciprocal lattices? If c/a is ideal in the direct lattice, what is its value in the reciprocal lattice?

(c) The Bravais lattice generated by three primitive vectors of equal length a, making equal angles θ with one another, is known as the trigonal Bravais lattice (see Chapter 7). Show that the reciprocal of a trigonal Bravais lattice is also trigonal, with an angle θ^* given by $-\cos\theta^* = \cos\theta/[1 + \cos\theta]$, and a primitive vector length a^*, given by $a^* = (2\pi/a)(1 + 2\cos\theta\cos\theta^*)^{-1/2}$.

3. (a) Show that the density of lattice points (per unit area) in a lattice plane is d/v, where v is the primitive cell volume and d the spacing between neighboring planes in the family to which the given plane belongs.

(b) Prove that the lattice planes with the greatest densities of points are the {111} planes in a face-centered cubic Bravais lattice and the {110} planes in a body-centered cubic Bravais lattice. (*Hint:* This is most easily done by exploiting the relation between families of lattice planes and reciprocal lattice vectors.)

4. Prove that any reciprocal lattice vector $\mathbf{K}$ is an integral multiple of the shortest parallel reciprocal lattice vector $\mathbf{K}_0$. (Hint: Assume the contrary, and deduce that since the reciprocal lattice is a Bravais lattice, there must be a reciprocal lattice vector parallel to $\mathbf{K}$ shorter than $\mathbf{K}_0$.)

6 Determination of Crystal Structures by X-ray Diffraction

Formulation of Bragg and von Laue

The Laue Condition and Ewald's Construction

Experimental Methods: Laue, Rotating Crystal, Powder

Geometrical Structure Factor

Atomic Form Factor

Typical interatomic distances in a solid are on the order of an angstrom (10^{-8} cm). An electromagnetic probe of the microscopic structure of a solid must therefore have a wavelength at least this short, corresponding to an energy of order

$$\hbar\omega = \frac{hc}{\lambda} = \frac{hc}{10^{-8}\ \text{cm}} \approx 12.3 \times 10^3\ \text{eV}. \tag{6.1}$$

Energies like this, on the order of several thousands of electron volts (kilovolts or keV), are characteristic X-ray energies.

In this chapter we shall describe how the distribution of X rays scattered by a rigid,[1] periodic[2] array of ions reveals the locations of the ions within that structure. There are two equivalent ways to view the scattering of X rays by a perfect periodic structure, due to Bragg and to von Laue. Both viewpoints are still widely used. The von Laue approach, which exploits the reciprocal lattice, is closer to the spirit of modern solid state physics, but the Bragg approach is still in wide use by X-ray crystallographers. Both are described below, together with a proof of their equivalence.

BRAGG FORMULATION OF X-RAY DIFFRACTION BY A CRYSTAL

In 1913 W. H. and W. L. Bragg found that substances whose macroscopic forms were crystalline gave remarkably characteristic patterns of reflected X-radiation, quite unlike those produced by liquids. In crystalline materials, for certain sharply defined wavelengths and incident directions, intense peaks of scattered radiation (now known as Bragg peaks) were observed.

W. L. Bragg accounted for this by regarding a crystal as made out of parallel planes of ions, spaced a distance d apart (i.e., the lattice planes described in Chapter 5). The conditions for a sharp peak in the intensity of the scattered radiation were: (1) that the X rays should be specularly reflected[3] by the ions in any one plane and (2) that the reflected rays from successive planes should interfere constructively. Rays specularly reflected from adjoining planes are shown in Figure 6.1. The path difference between the two rays is just $2d \sin \theta$, where θ is the angle of incidence.[4] For the rays to interfere constructively, this path difference must be an integral number of wavelengths, leading to the celebrated Bragg condition:

$$n\lambda = 2d \sin \theta. \tag{6.2}$$

The integer n is known as the order of the corresponding reflection. For a beam of X rays containing a range of different wavelengths ("white radiation") many different reflections are observed. Not only can one have higher-order reflections from a given set of lattice planes, but in addition one must recognize that there are

---

[1] Actually the ions vibrate about their ideal equilibrium sites (Chapters 21–26). This does not affect the conclusions reached in this chapter (though in the early days of X-ray diffraction it was not clear why such vibrations did not obliterate the pattern characteristic of a periodic structure). It turns out that the vibrations have two main consequences (see Appendix N): (a) the intensity in the characteristic peaks that reveal the crystal structure is diminished, but not eliminated; and (b) a much weaker continuous background of radiation (the "diffuse background") is produced.

[2] Amorphous solids and liquids have about the same density as crystalline solids, and are therefore also susceptible to probing with X rays. However, the discrete, sharp peaks of scattered radiation characteristic of crystals are not found.

[3] In specular reflection the angle of incidence equals the angle of reflection.

[4] The angle of incidence in X-ray crystallography is conventionally measured from the plane of reflection rather than from the normal to that plane (as in classical optics). Note that θ is just half the angle of deflection of the incident beam (Figure 6.2).

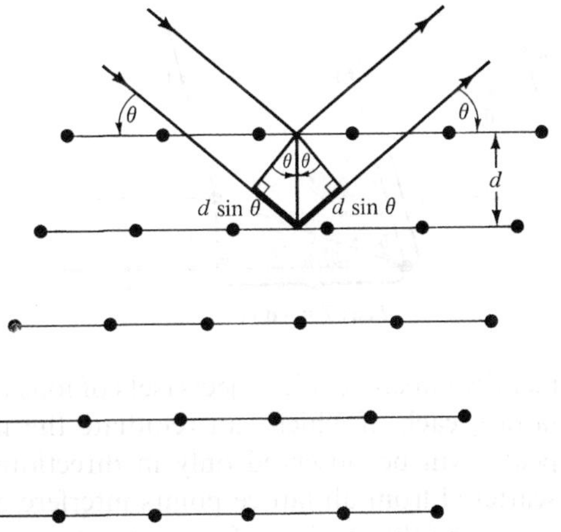

Figure 6.1
A Bragg reflection from a particular family of lattice planes, separated by a distance d. Incident and reflected rays are shown for the two neighboring planes. The path difference is $2d \sin \theta$.

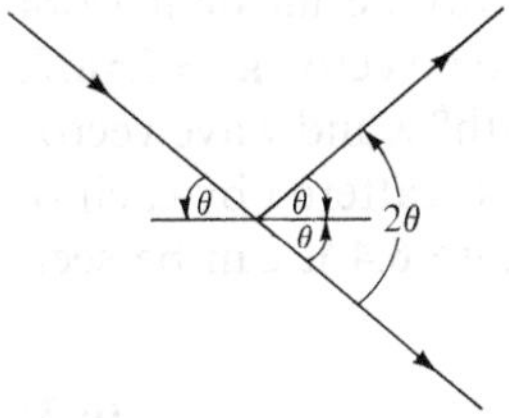

Figure 6.2
The Bragg angle θ is just half the total angle by which the incident beam is deflected.

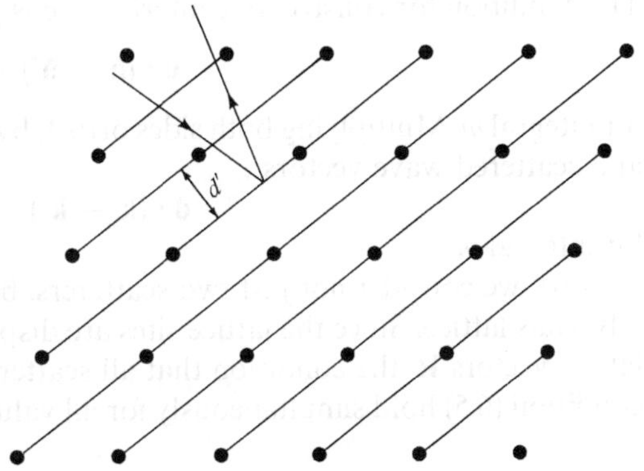

Figure 6.3
The same portion of Bravais lattice shown in Figure 6.1, with a different resolution into lattice planes indicated. The incident ray is the same as in Figure 6.1, but both the direction (shown in the figure) and wavelength (determined by the Bragg condition (6.2) with d replaced by d') of the reflected ray are different from the reflected ray in Figure 6.1. Reflections are possible, in general, for any of the infinitely many ways of resolving the lattice into planes.

many different ways of sectioning the crystal into planes, each of which will itself produce further reflections (see, for example, Figure 5.3 or Figure 6.3).

VON LAUE FORMULATION OF X-RAY DIFFRACTION BY A CRYSTAL

The von Laue approach differs from the Bragg approach in that no particular sectioning of the crystal into lattice planes is singled out, and no *ad hoc* assumption of specular reflection is imposed.[5] Instead one regards the crystal as composed of

[5] The Bragg assumption of specular reflection is, however, equivalent to the assumption that rays scattered from individual ions within each lattice plane interfere constructively. Thus both the Bragg and the von Laue approaches are based on the same physical assumptions, and their precise equivalence (see page 99) is to be expected.

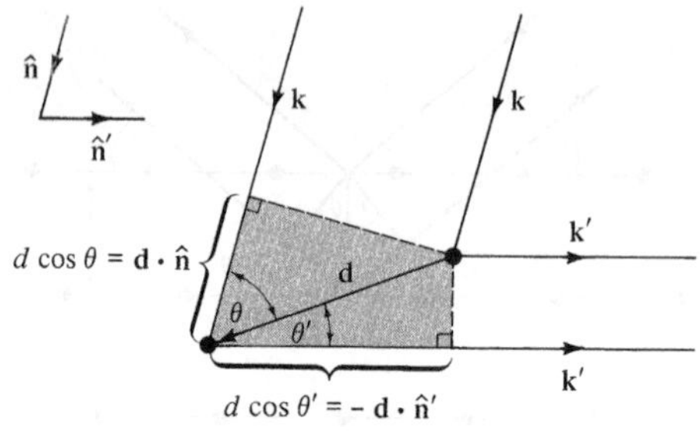

Figure 6.4
Illustrating that the path difference for rays scattered from two points separated by **d** is given by Eq. (6.3) or (6.4).

identical microscopic objects (sets of ions or atoms) placed at the sites **R** of a Bravais lattice, each of which can reradiate the incident radiation in all directions. Sharp peaks will be observed only in directions and at wavelengths for which the rays scattered from all lattice points interfere constructively.

To find the condition for constructive interference, consider first just two scatterers, separated by a displacement vector **d** (Figure 6.4). Let an X ray be incident from very far away, along a direction $\hat{\mathbf{n}}$, with wavelength λ, and wave vector $\mathbf{k} = 2\pi\hat{\mathbf{n}}/\lambda$. A scattered ray will be observed in a direction $\hat{\mathbf{n}}'$ with wavelength[6] λ and wave vector $\mathbf{k}' = 2\pi\hat{\mathbf{n}}'/\lambda$, provided that the path difference between the rays scattered by each of the two ions is an integral number of wavelengths. From Figure 6.4 it can be seen that this path difference is just

$$d \cos \theta + d \cos \theta' = \mathbf{d} \cdot (\hat{\mathbf{n}} - \hat{\mathbf{n}}'). \qquad \textbf{(6.3)}$$

The condition for constructive interference is thus

$$\mathbf{d} \cdot (\hat{\mathbf{n}} - \hat{\mathbf{n}}') = m\lambda, \qquad \textbf{(6.4)}$$

for integral m. Multiplying both sides of (6.4) by $2\pi/\lambda$ yields a condition on the incident and scattered wave vectors:

$$\mathbf{d} \cdot (\mathbf{k} - \mathbf{k}') = 2\pi m, \qquad \textbf{(6.5)}$$

for integral m.

Next, we consider not just two scatterers, but an array of scatterers, at the sites of a Bravais lattice. Since the lattice sites are displaced from one another by the Bravais lattice vectors **R**, the condition that all scattered rays interfere constructively is that condition (6.5) hold simultaneously for all values of **d** that are Bravais lattice vectors:

$$\mathbf{R} \cdot (\mathbf{k} - \mathbf{k}') = 2\pi m, \quad \text{for integral } m \text{ and all Bravais lattice vectors } \mathbf{R}. \qquad \textbf{(6.6)}$$

This can be written in the equivalent form

$$e^{i(\mathbf{k}' - \mathbf{k}) \cdot \mathbf{R}} = 1, \quad \text{for all Bravais lattice vectors } \mathbf{R}. \qquad \textbf{(6.7)}$$

---

[6] Here (and in the Bragg picture) we assume that the incident and scattered radiation has the same wavelength. In terms of photons this means that no energy has been lost in the scattering, i.e., that the scattering is elastic. To a good approximation the bulk of the scattered radiation *is* elastically scattered, though there is much to be learned from the study of that small component of the radiation that is inelastically scattered (Chapter 24 and Appendix N).

Comparing this condition with the definition (5.2) of the reciprocal lattice, we arrive at the Laue condition that *constructive interference will occur provided that the change in wave vector,* $\mathbf{K} = \mathbf{k}' - \mathbf{k}$, *is a vector of the reciprocal lattice.*

It is sometimes convenient to have an alternative formulation of the Laue condition, stated entirely in terms of the incident wave vector **k**. First note that because the reciprocal lattice is a Bravais lattice, if $\mathbf{k}' - \mathbf{k}$ is a reciprocal lattice vector, so is $\mathbf{k} - \mathbf{k}'$. Calling the latter vector **K**, the condition that **k** and **k**′ have the same magnitude is

$$k = |\mathbf{k} - \mathbf{K}|. \tag{6.8}$$

Squaring both sides of (6.8) yields the condition

$$\mathbf{k} \cdot \hat{\mathbf{K}} = \tfrac{1}{2}K; \tag{6.9}$$

i.e., the component of the incident wave vector **k** along the reciprocal lattice vector **K** must be half the length of **K**.

Thus an incident wave vector **k** will satisfy the Laue condition if and only if the tip of the vector lies in a plane that is the perpendicular bisector of a line joining the origin of k-space to a reciprocal lattice point **K** (Figure 6.5). Such k-space planes are called *Bragg planes.*

Figure 6.5
The Laue condition. If the sum of **k** and −**k**′ is a vector **K**, and if **k** and **k**′ have the same length, then the tip of the vector **k** is equidistant from the origin O and the tip of the vector **K**, and therefore it lies in the plane bisecting the line joining the origin to the tip of **K**.

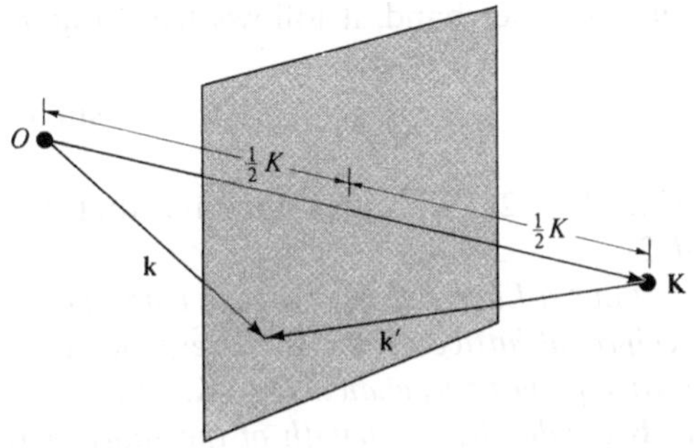

It is a consequence of the equivalence of the Bragg and von Laue points of view, demonstrated in the following section, that the k-space Bragg plane associated with a particular diffraction peak in the Laue formulation is parallel to the family of direct lattice planes responsible for the peak in the Bragg formulation.

EQUIVALENCE OF THE BRAGG AND VON LAUE FORMULATIONS

The equivalence of these two criteria for constructive interference of X rays by a crystal follows from the relation between vectors of the reciprocal lattice and families of direct lattice planes (see Chapter 5). Suppose the incident and scattered wave vectors, **k** and **k**′, satisfy the Laue condition that $\mathbf{K} = \mathbf{k}' - \mathbf{k}$ be a reciprocal lattice vector. Because the incident and scattered waves have the same wavelength,[6] **k**′ and **k** have the same magnitudes. It follows (see Figure 6.6) that **k**′ and **k** make the same angle θ with the plane perpendicular to **K**. Therefore the scattering can be viewed as a Bragg reflection, with Bragg angle θ, from the family of direct lattice planes perpendicular to the reciprocal lattice vector **K**.

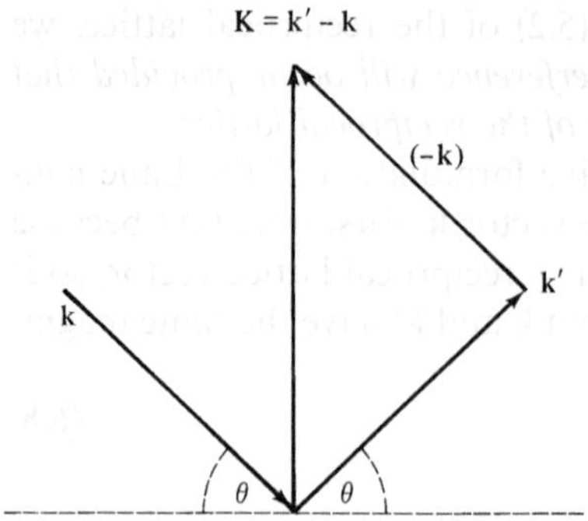

Figure 6.6
The plane of the paper contains the incident wave vector **k**, the reflected wave vector **k**′, and their difference **K** satisfying the Laue condition. Since the scattering is elastic ($k' = k$), the direction of **K** bisects the angle between **k** and **k**′. The dashed line is the intersection of the plane perpendicular to **K** with the plane of the paper.

To demonstrate that this reflection satisfies the Bragg condition (6.2), note that the vector **K** is an integral multiple[7] of the shortest reciprocal lattice vector $\mathbf{K}_0$ parallel to **K**. According to the theorem on page 90, the magnitude of $\mathbf{K}_0$ is just $2\pi/d$, where d is the distance between successive planes in the family perpendicular to $\mathbf{K}_0$ or to **K**. Thus

$$K = \frac{2\pi n}{d}. \tag{6.10}$$

On the other hand, it follows from Figure 6.6 that $K = 2k \sin \theta$, and thus

$$k \sin \theta = \frac{\pi n}{d}. \tag{6.11}$$

Since $k = 2\pi/\lambda$, Eq. (6.11) implies that the wavelength satisfies the Bragg condition (6.2).

Thus *a Laue diffraction peak corresponding to a change in wave vector given by the reciprocal lattice vector* **K** *corresponds to a Bragg reflection from the family of direct lattice planes perpendicular to* **K**. *The order, n, of the Bragg reflection is just the length of* **K** *divided by the length of the shortest reciprocal lattice vector parallel to* **K**.

Since the reciprocal lattice associated with a given Bravais lattice is far more easily visualized than the set of all possible planes into which the Bravais lattice can be resolved, the Laue condition for diffraction peaks is far more simple to work with than the Bragg condition. In the rest of this chapter we shall apply the Laue condition to a description of three of the most important ways in which X-ray crystallographic analyses of real samples are performed, and to a discussion of how one can extract information not only about the underlying Bravais lattice, but also about the arrangement of ions within the primitive cell.

EXPERIMENTAL GEOMETRIES SUGGESTED BY THE LAUE CONDITION

An incident wave vector **k** will lead to a diffraction peak (or "Bragg reflection") if and only if the tip of the wave vector lies on a k-space Bragg plane. Since the set of all

[7] This is an elementary consequence of the fact that the reciprocal lattice is a Bravais lattice. See Chapter 5, Problem 4.

Bragg planes is a discrete family of planes, it cannot begin to fill up three-dimensional k-space, and in general the tip of $\mathbf{k}$ will not lie on a Bragg plane. Thus for a fixed incident wave vector—i.e., for a fixed X-ray wavelength and fixed incident direction relative to the crystal axes—there will be in general no diffraction peaks at all.

If one wishes to search experimentally for Bragg peaks one must therefore relax the constraint of fixed $\mathbf{k}$, either varying the magnitude of $\mathbf{k}$ (i.e., varying the wavelength of the incident beam) or varying its direction (in practice, varying the orientation of the crystal with respect to the incident direction).

The Ewald Construction

A simple geometric construction due to Ewald is of great help in visualizing these various methods and in deducing the crystal structure from the peaks so observed. We draw in k-space a sphere centered on the tip of the incident wave vector $\mathbf{k}$ of radius k (so that it passes through the origin). Evidently (see Figure 6.7) there will be *some* wave vector $\mathbf{k}'$ satisfying the Laue condition if and only if some reciprocal lattice point (in addition to the origin) lies on the surface of the sphere, in which case there will be a Bragg reflection from the family of direct lattice planes perpendicular to that reciprocal lattice vector.

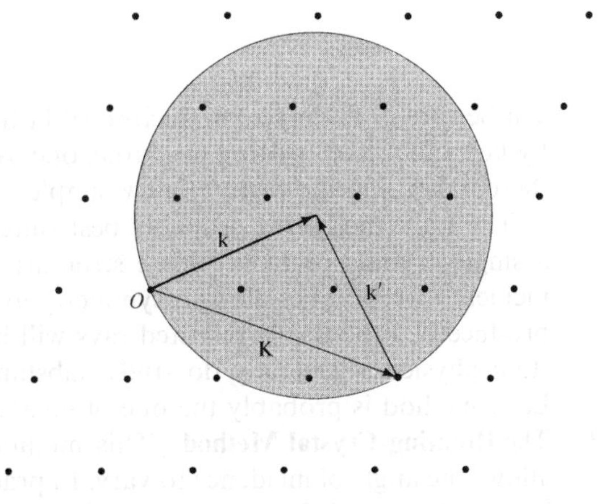

Figure 6.7
The Ewald construction. Given the incident wave vector $\mathbf{k}$, a sphere of radius k is drawn about the point $\mathbf{k}$. Diffraction peaks corresponding to reciprocal lattice vectors $\mathbf{K}$ will be observed only if $\mathbf{K}$ gives a reciprocal lattice point on the surface of the sphere. Such a reciprocal lattice vector is indicated in the figure, together with the wave vector $\mathbf{k}'$ of the Bragg reflected ray.

In general, a sphere in k-space with the origin on its surface will have no other reciprocal lattice points on its surface, and therefore the Ewald construction confirms our observation that for a general incident wave vector there will be no Bragg peaks. One can, however, ensure that some Bragg peaks will be produced by several techniques:

1. **The Laue Method** One can continue to scatter from a single crystal of fixed orientation from a fixed incident direction $\hat{\mathbf{n}}$, but can search for Bragg peaks by using not a monochromatic X-ray beam, but one containing wavelengths from λ_1 up to λ_0. The Ewald sphere will then expand into the region contained between the two spheres determined by $\mathbf{k}_0 = 2\pi\hat{\mathbf{n}}/\lambda_0$ and $\mathbf{k}_1 = 2\pi\hat{\mathbf{n}}/\lambda_1$, and Bragg peaks will be observed corresponding to any reciprocal lattice vectors lying within this region (Figure 6.8). By making the spread in wavelengths sufficiently large, one

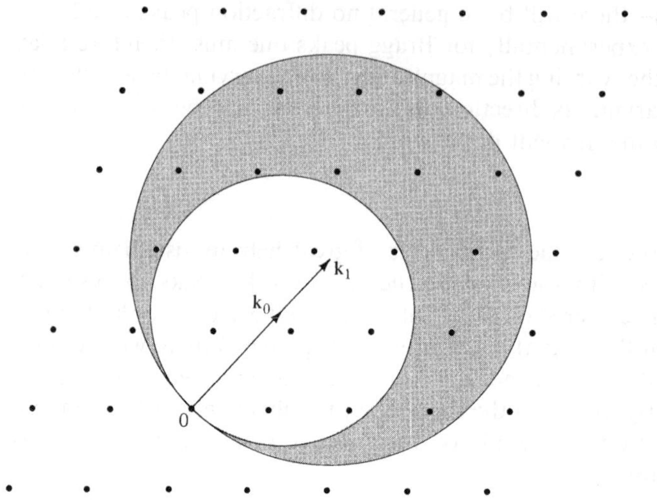

Figure 6.8
The Ewald construction for the Laue method. The crystal and incident X-ray direction are fixed, and a continuous range of wavelengths, corresponding to wave vectors between k_0 and k_1 in magnitude, is present. The Ewald spheres for all incident wave vectors fill the shaded region between the sphere centered on the tip of the vector $\mathbf{k}_0$ and that centered on the tip of $\mathbf{k}_1$. Bragg peaks will be observed corresponding to all reciprocal lattice points lying within the shaded region. (For simplicity in illustration, the incident direction has been taken to lie in a lattice plane, and only reciprocal lattice points lying in that plane are shown.)

can be sure of finding some reciprocal lattice points within the region; whereas by keeping it from getting too large, one can avoid too many Bragg reflections, thereby keeping the picture fairly simple.

The Laue method is probably best suited for determining the orientation of a single crystal specimen whose structure is known, since, for example, if the incident direction lies along a symmetry axis of the crystal, the pattern of spots produced by the Bragg reflected rays will have the same symmetry. Since solid state physicists generally do study substances of known crystal structure, the Laue method is probably the one of greatest practical interest.

2. **The Rotating-Crystal Method** This method uses monochromatic X rays, but allows the angle of incidence to vary. In practice the direction of the X-ray beam is kept fixed, and the orientation of the crystal varied instead. In the rotating crystal method the crystal is rotated about some fixed axis, and all Bragg peaks that occur during the rotation are recorded on a film. As the crystal rotates, the reciprocal lattice it determines will rotate by the same amount about the same axis. Thus the Ewald sphere (which is determined by the fixed incident wave vector $\mathbf{k}$) is fixed in k-space, while the entire reciprocal lattice rotates about the axis of rotation of the crystal. During this rotation each reciprocal lattice point traverses a circle about the rotation axis, and a Bragg reflection occurs whenever this circle intersects the Ewald sphere. This is illustrated in Figure 6.9 for a particularly simple geometry.
3. **The Powder or Debye-Scherrer Method** This is equivalent to a rotating crystal experiment in which, in addition, the axis of rotation is varied over all possible orientations. In practice this isotropic averaging of the incident direction is

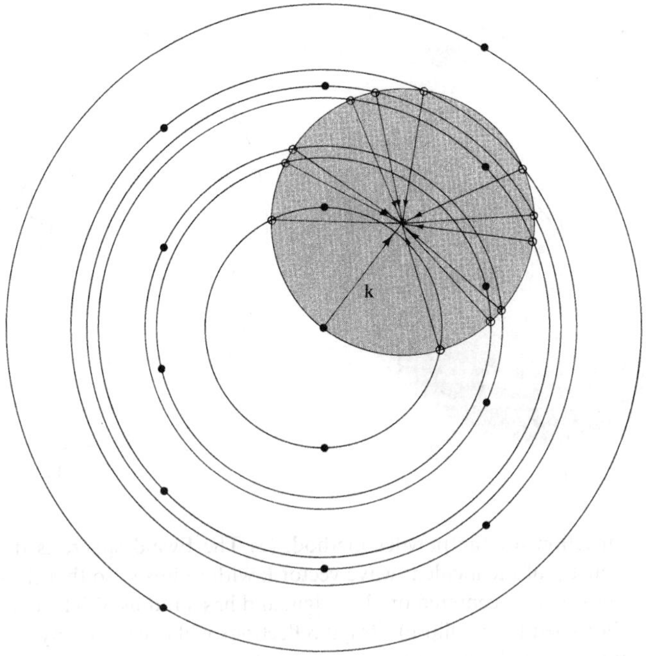

Figure 6.9
The Ewald construction for the rotating-crystal method. For simplicity a case is shown in which the incident wave vector lies in a lattice plane, and the axis of rotation is perpendicular to that plane. The concentric circles are the orbits swept out under the rotation by the reciprocal lattice vectors lying in the plane perpendicular to the axis containing **k**. Each intersection of such a circle with the Ewald sphere gives the wave vector of a Bragg reflected ray. (Additional Bragg reflected wave vectors associated with reciprocal lattice vectors in other planes are not shown.)

achieved by using a polycrystalline sample or a powder, grains of which are still enormous on the atomic scale and therefore capable of diffracting X rays. Because the crystal axes of the individual grains are randomly oriented, the diffraction pattern produced by such a powder is what one would produce by combining the diffraction patterns for all possible orientations of a single crystal.

The Bragg reflections are now determined by fixing the incident **k** vector, and with it the Ewald sphere, and allowing the reciprocal lattice to rotate through all possible angles about the origin, so that each reciprocal lattice vector **K** generates a sphere of radius K about the origin. Such a sphere will intersect the Ewald sphere in a circle (Figure 6.10a) provided that K is less than $2k$. The vector joining any point on such a circle with the tip of the incident vector **k** is a wave vector **k**′, for which scattered radiation will be observed. Thus each reciprocal lattice vector of length less than $2k$ generates a cone of scattered radiation at an angle ϕ to the forward direction, where (Figure 6.10b)

$$K = 2k \sin \tfrac{1}{2}\phi. \tag{6.12}$$

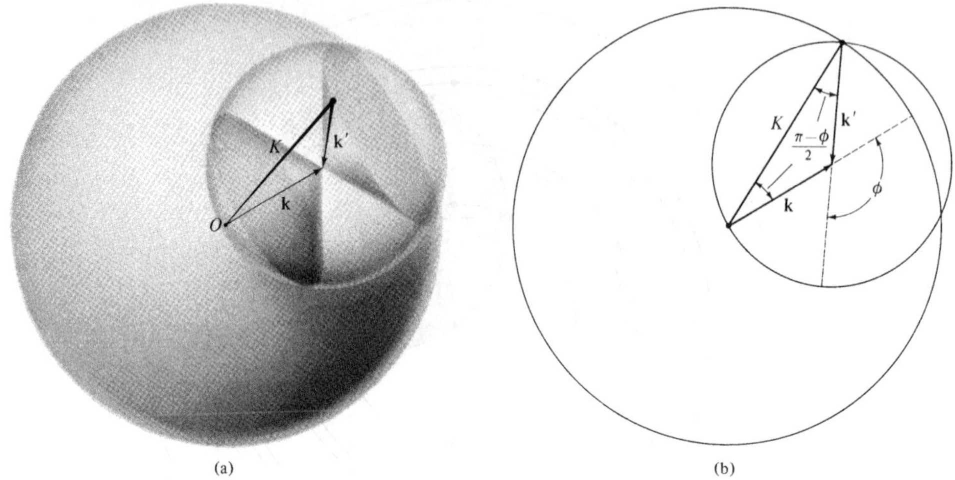

Figure 6.10
The Ewald construction for the powder method. (a) The Ewald sphere is the smaller sphere. It is centered on the tip of the incident wave vector **k** with radius k, so that the origin O is on its surface. The larger sphere is centered on the origin and has a radius K. The two spheres intersect in a circle (foreshortened to an ellipse). Bragg reflections will occur for any wave vector $\mathbf{k}'$ connecting any point on the circle of intersection to the tip of the vector **k**. The scattered rays therefore lie on the cone that opens in the direction opposite to **k**. (b) A plane section of (a), containing the incident wave vector. The triangle is isosceles, and thus $K = 2k \sin \frac{1}{2}\phi$.

By measuring the angles ϕ at which Bragg reflections are observed, one therefore learns the lengths of all reciprocal lattice vectors shorter than $2k$. Armed with this information, some facts about the macroscopic crystal symmetry, and the fact that the reciprocal lattice is a Bravais lattice, one can usually construct the reciprocal lattice itself (see, for example, Problem 1).

DIFFRACTION BY A MONATOMIC LATTICE WITH A BASIS; THE GEOMETRICAL STRUCTURE FACTOR

The preceding discussion was based on the condition (6.7) that rays scattered from each primitive cell should interfere constructively. If the crystal structure is that of a monatomic lattice with an n-atom basis (for example, carbon in the diamond structure or hexagonal close-packed beryllium, both of which have $n = 2$), then the contents of each primitive cell can be further analyzed into a set of identical scatterers at positions $\mathbf{d}_1, \ldots, \mathbf{d}_n$ within the cell. The intensity of radiation in a given Bragg peak will depend on the extent to which the rays scattered from these basis sites interfere with one another, being greatest when there is complete constructive interference and vanishing altogether should there happen to be complete destructive interference.

If the Bragg peak is associated with a change in wave vector $\mathbf{k}' - \mathbf{k} = \mathbf{K}$, then the phase difference (Figure 6.4) between the rays scattered at $\mathbf{d}_i$ and $\mathbf{d}_j$ will be $\mathbf{K} \cdot (\mathbf{d}_i - \mathbf{d}_j)$ and the amplitudes of the two rays will differ by a factor $e^{i\mathbf{K}\cdot(\mathbf{d}_i - \mathbf{d}_j)}$. Thus the amplitudes of the rays scattered at $\mathbf{d}_1, \ldots, \mathbf{d}_n$ are in the ratios $e^{i\mathbf{K}\cdot\mathbf{d}_1}, \ldots, e^{i\mathbf{K}\cdot\mathbf{d}_n}$. The net ray scattered by the entire primitive cell is the sum of the individual rays, and will therefore have an amplitude containing the factor

$$S_{\mathbf{K}} = \sum_{j=1}^{n} e^{i\mathbf{K}\cdot\mathbf{d}_j}. \tag{6.13}$$

The quantity $S_{\mathbf{K}}$, known as the *geometrical structure factor*, expresses the extent to which interference of the waves scattered from identical ions within the basis can diminish the intensity of the Bragg peak associated with the reciprocal lattice vector **K**. The intensity in the Bragg peak, being proportional to the square of the absolute value of the amplitude, will contain a factor $|S_{\mathbf{K}}|^2$. It is important to note that this is not the only source of **K** dependence to the intensity. Further dependence on the change in wave vector comes both from the ordinary angular dependence of any electromagnetic scattering, together with the influence on the scattering of the detailed internal structure of each individual ion in the basis. Therefore the structure factor alone cannot be used to predict the absolute intensity in a Bragg peak.[8] It can, however, lead to a characteristic dependence on **K** that is easily discerned even though other less distinctive **K** dependences have been superimposed upon it. The one case, in which the structure factor can be used with assurance is when it vanishes. This occurs when the elements of the basis are so arranged that there is complete destructive interference for the **K** in question; in that case no features of the rays scattered by the individual basis elements can prevent the net ray from vanishing.

We illustrate the importance of a vanishing structure factor in two cases[9]:

1. ***Body-Centered Cubic Considered as Simple Cubic with a Basis*** Since the body-centered cubic lattice is a Bravais lattice, we know that Bragg reflections will occur when the change in wave vector **K** is a vector of the reciprocal lattice, which is face-centered cubic. Sometimes, however, it is convenient to regard the bcc lattice as a simple cubic lattice generated by primitive vectors $a\hat{\mathbf{x}}$, $a\hat{\mathbf{y}}$, and $a\hat{\mathbf{z}}$, with a two-point basis consisting of $\mathbf{d}_1 = \mathbf{0}$ and $\mathbf{d}_2 = (a/2)(\hat{\mathbf{x}} + \hat{\mathbf{y}} + \hat{\mathbf{z}})$. From this point of view the reciprocal lattice is also simple cubic, with a cubic cell of side $2\pi/a$. However, there will now be a structure factor $S_{\mathbf{K}}$ associated with each Bragg reflection. In the present case, (6.13) gives

$$S_{\mathbf{K}} = 1 + \exp\left[i\mathbf{K}\cdot\tfrac{1}{2}a(\hat{\mathbf{x}} + \hat{\mathbf{y}} + \hat{\mathbf{z}})\right]. \tag{6.14}$$

A general vector in the simple cubic reciprocal lattice has the form

$$\mathbf{K} = \frac{2\pi}{a}(n_1\hat{\mathbf{x}} + n_2\hat{\mathbf{y}} + n_3\hat{\mathbf{z}}). \tag{6.15}$$

Substituting this into (6.14), we find a structure factor

$$\begin{aligned} S_{\mathbf{K}} &= 1 + e^{i\pi(n_1+n_2+n_3)} = 1 + (-1)^{n_1+n_2+n_3} \\ &= \begin{cases} 2, & n_1 + n_2 + n_3 \text{ even,} \\ 0, & n_1 + n_2 + n_3 \text{ odd.} \end{cases} \end{aligned} \tag{6.16}$$

---

8 A brief but thorough discussion of the scattering of electromagnetic radiation by crystals, including the derivation of detailed intensity formulas for the various experimental geometries described above, is given by Landau and Lifshitz, *Electrodynamics of Continuous Media*, Chapter 15, Addison-Wesley, Reading, Mass., 1966.

9 Further examples are given in Problems 2 and 3.

Thus those points in the simple cubic reciprocal lattice, the sum of whose coordinates with respect to the cubic primitive vectors are odd, will actually have no Bragg reflection associated with them. This converts the simple cubic reciprocal lattice into the face-centered cubic structure that we would have had if we had treated the body-centered cubic direct lattice as a Bravais lattice rather than as a lattice with a basis (see Figure 6.11).

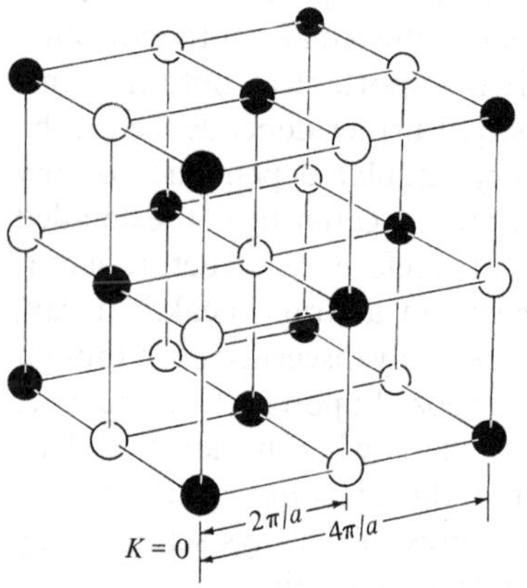

Figure 6.11
Points in the simple cubic reciprocal lattice of side $2\pi/a$, for which the structure factor (6.16) vanishes, are those (white circles) that can be reached from the origin by moving along an odd number of nearest-neighbor bonds. When such sites are eliminated, the remaining sites (black circles) constitute a face-centered cubic lattice with cubic cell of side $4\pi/a$.

Thus if, either inadvertently or for reasons of greater symmetry in description, one chooses to describe a Bravais lattice as a lattice with a basis, one still recovers the correct description of X-ray diffraction, provided that the vanishing of the structure factor is taken into account.

2. ***Monatomic Diamond Lattice*** The monatomic diamond lattice (carbon, silicon, germanium, or grey tin) is not a Bravais lattice and must be described as a lattice with a basis. The underlying Bravais lattice is face-centered cubic, and the basis can be taken to be $\mathbf{d}_1 = \mathbf{0}$, $\mathbf{d}_2 = (a/4)(\hat{\mathbf{x}} + \hat{\mathbf{y}} + \hat{\mathbf{z}})$, where $\hat{\mathbf{x}}$, $\hat{\mathbf{y}}$, and $\hat{\mathbf{z}}$, are along the cubic axes and a is the side of the conventional cubic cell. The reciprocal lattice is body-centered cubic with conventional cubic cell of side $4\pi/a$. If we take as primitive vectors

$$\mathbf{b}_1 = \frac{2\pi}{a}(\hat{\mathbf{y}} + \hat{\mathbf{z}} - \hat{\mathbf{x}}), \quad \mathbf{b}_2 = \frac{2\pi}{a}(\hat{\mathbf{z}} + \hat{\mathbf{x}} - \hat{\mathbf{y}}), \quad \mathbf{b}_3 = \frac{2\pi}{a}(\hat{\mathbf{x}} + \hat{\mathbf{y}} - \hat{\mathbf{z}}), \tag{6.17}$$

then the structure factor (6.13) for $\mathbf{K} = \Sigma n_i \mathbf{b}_i$ is

$$\begin{aligned} S_{\mathbf{K}} &= 1 + \exp\left[\tfrac{1}{2} i\pi(n_1 + n_2 + n_3)\right] \\ &= \begin{cases} 2, & n_1 + n_2 + n_3 \text{ twice an even number,} \\ 1 \pm i, & n_1 + n_2 + n_3 \text{ odd,} \\ 0, & n_1 + n_2 + n_3 \text{ twice an odd number.} \end{cases} \end{aligned} \tag{6.18}$$

To interpret these conditions on Σn_i geometrically, note that if we substitute (6.17) into $\mathbf{K} = \Sigma n_i \mathbf{b}_i$, we can write the general reciprocal lattice vector in the form

$$\mathbf{K} = \frac{4\pi}{a}(\nu_1 \hat{\mathbf{x}} + \nu_2 \hat{\mathbf{y}} + \nu_3 \hat{\mathbf{z}}), \tag{6.19}$$

where

$$\nu_j = \tfrac{1}{2}(n_1 + n_2 + n_3) - n_j, \quad \sum_{j=1}^{3} \nu_j = \tfrac{1}{2}(n_1 + n_2 + n_3). \tag{6.20}$$

We know (see Chapter 5) that the reciprocal to the fcc lattice with cubic cell of side a is a bcc lattice with cubic cell of side $4\pi/a$. Let us regard this as composed of two simple cubic lattices of side $4\pi/a$. The first, containing the origin ($\mathbf{K} = 0$), must have all ν_i integers (according to (6.19)) and must therefore be given by $\mathbf{K}$ with $n_1 + n_2 + n_3$ even (according to (6.20)). The second, containing the "body-centered point" $(4\pi/a)\frac{1}{2}(\hat{\mathbf{x}} + \hat{\mathbf{y}} + \hat{\mathbf{z}})$, must have all ν_i integers $+ \frac{1}{2}$ (according to (6.19)) and must therefore be given by $\mathbf{K}$ with $n_1 + n_2 + n_3$ odd (according to (6.20)).

Comparing this with (6.18), we find that the points with structure factor $1 \pm i$ are those in the simple cubic sublattice of "body-centered" points. Those whose structure factor S is 2 or 0 are in the simple cubic sublattice containing the origin, where $\Sigma\nu_i$ is even when $S = 2$ and odd when $S = 0$. Thus the points with zero structure factor are again removed by applying the construction illustrated in Figure 6.11 to the simple cubic sublattice containing the origin, converting it to a face-centered cubic structure (Figure 6.12).

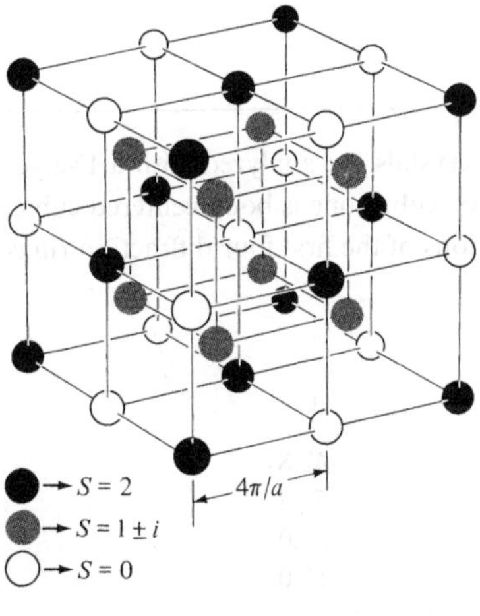

Figure 6.12
The body-centered cubic lattice with cubic cell side $4\pi/a$ that is reciprocal to a face-centered cubic lattice with cubic cell side a. When the fcc lattice is that underlying the diamond structure, then the white circles indicate sites with zero structure factor. (The black circles are sites with structure factor 2, and the gray ones are sites with structure factor $1 \pm i$.)

DIFFRACTION BY A POLYATOMIC CRYSTAL; THE ATOMIC FORM FACTOR

If the ions in the basis are not identical, the structure factor (6.13) assumes the form

$$S_{\mathbf{K}} = \sum_{j=1}^{n} f_j(\mathbf{K}) e^{i\mathbf{K}\cdot\mathbf{d}_j}, \tag{6.21}$$

where f_j, known as the *atomic form factor*, is entirely determined by the internal structure of the ion that occupies position $\mathbf{d}_j$ in the basis. Identical ions have identical form factors (regardless of where they are placed), so (6.21) reduces back to (6.13), multiplied by the common value of the form factors, in the monatomic case.

In elementary treatments the atomic form factor associated with a Bragg reflection given by the reciprocal lattice vector **K** is taken to be proportional to the Fourier transform of the electronic charge distribution of the corresponding ion[10]:

$$f_j(\mathbf{K}) = -\frac{1}{e}\int d\mathbf{r}\, e^{i\mathbf{K}\cdot\mathbf{r}}\, \rho_j(\mathbf{r}). \tag{6.22}$$

Thus the atomic form factor f_j depends on **K** and on the detailed features of the charge distribution of the ion that occupies position $\mathbf{d}_j$ in the basis. As a result, one would not expect the structure factor to vanish for any **K** unless there is some fortuitous relation between form factors of different types. By making reasonable assumptions about the **K** dependence of the different form factors, one can often distinguish quite conclusively between various possible crystal structures on the basis of the variation with **K** of the Bragg peak intensities (see, for example, Problem 5).

This concludes our discussion of the Bragg reflection of X rays. Our analysis has exploited no properties of the X rays other than their wave nature.[11] Consequently we shall find many of the concepts and results of this chapter reappearing in subsequent discussions of other wave phenomena in solids, such as electrons (Chapter 9) and neutrons (Chapter 24).[12]

PROBLEMS

1. Powder specimens of three different monatomic cubic crystals are analyzed with a Debye-Scherrer camera. It is known that one sample is face-centered cubic, one is body-centered cubic, and one has the diamond structure. The approximate positions of the first four diffraction rings in each case are (see Figure 6.13):

VALUES OF ϕ FOR SAMPLES

A	*B*	*C*
42.2°	28.8°	42.8°
49.2	41.0	73.2
72.0	50.8	89.0
87.3	59.6	115.0

(a) Identify the crystal structures of *A*, *B*, and *C*.

(b) If the wavelength of the incident X-ray beam is 1.5 Å, what is the length of the side of the conventional cubic cell in each case?

(c) If the diamond structure were replaced by a zincblende structure with a cubic unit cell of the same side, at what angles would the first four rings now occur?

[10] The electronic charge density $\rho_j(\mathbf{r})$ is that of an ion of type j placed at $\mathbf{r} = \mathbf{0}$; thus the contribution of the ion at $\mathbf{R} + \mathbf{d}_j$ to the electronic charge density of the crystal is $\rho_j(\mathbf{r} - [\mathbf{R} + \mathbf{d}_j])$. (The electronic charge is usually factored out of the atomic form factor to make it dimensionless.)

[11] As a result we have been unable to make precise statements about the absolute intensity of the Bragg peaks, or about the diffuse background of radiation in directions not allowed by the Bragg condition.

[12] Considered quantum mechanically, a particle of momentum p can be viewed as a wave of wavelength $\lambda = h/p$.

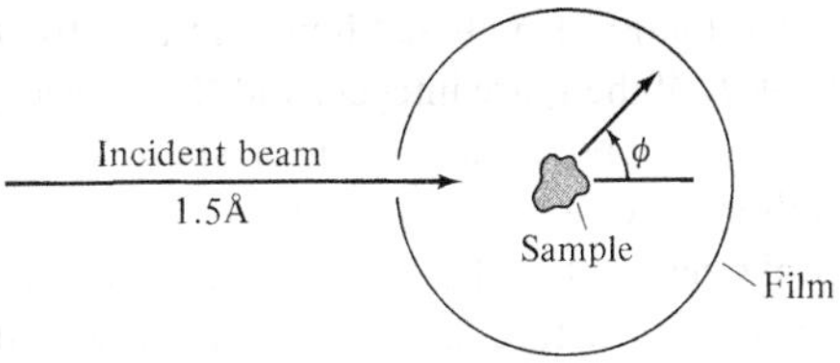

Figure 6.13
Schematic view of a Debye-Scherrer camera. Diffraction peaks are recorded on the film strip.

2. It is often convenient to represent a face-centered cubic Bravais lattice as simple cubic, with a cubic primitive cell of side a and a four-point basis.

(a) Show that the structure factor (6.13) is then either 4 or 0 at all points of the simple cubic reciprocal lattice.

(b) Show that when points with zero structure factor are removed, the remaining points of the reciprocal lattice make up a body-centered cubic lattice with conventional cell of side $4\pi/a$. Why is this to be expected?

3. (a) Show that the structure factor for a monatomic hexagonal close-packed crystal structure can take on any of the six values $1 + e^{in\pi/3}$, $n = 1, \ldots, 6$, as $\mathbf{K}$ ranges through the points of the simple hexagonal reciprocal lattice.

(b) Show that all reciprocal lattice points have nonvanishing structure factor in the plane perpendicular to the c-axis containing $\mathbf{K} = \mathbf{0}$.

(c) Show that points of zero structure factor are found in alternate planes in the family of reciprocal lattice planes perpendicular to the c-axis.

(d) Show that in such a plane the point that is displaced from $\mathbf{K} = \mathbf{0}$ by a vector parallel to the c-axis has zero structure factor.

(e) Show that the removal of all points of zero structure factor from such a plane reduces the triangular network of reciprocal lattice points to a honeycomb array (Figure 4.3).

4. Consider a lattice with an n-ion basis. Suppose that the ith ion in the basis, when translated to $\mathbf{r} = \mathbf{0}$, can be regarded as composed of m_i point particles of charge $-z_{ij}e$, located at positions $\mathbf{b}_{ij}$, $j = 1, \ldots, m_i$.

(a) Show that the atomic form factor f_i is given by

$$f_i = \sum_{j=1}^{m_i} z_{ij} e^{i\mathbf{K}\cdot\mathbf{b}_{ij}}. \tag{6.23}$$

(b) Show that the total structure factor (6.21) implied by (6.23) is identical to the structure factor one would have found if the lattice were equivalently described as having a basis of $m_1 + \cdots + m_n$ point ions.

5. (a) The sodium chloride structure (Figure 4.24) can be regarded as an fcc Bravais lattice of cube side a, with a basis consisting of a positively charged ion at the origin and a negatively charged ion at $(a/2)\hat{\mathbf{x}}$. The reciprocal lattice is body-centered cubic, and the general reciprocal lattice vector has the form (6.19), with all the coefficients v_i either integers or integers $+\frac{1}{2}$. If the atomic form factors for the two ions are f_+ and f_-, show that the structure factor is $S_{\mathbf{K}} = f_+ + f_-$, if the v_i are integers, and $f_+ - f_-$, if the v_i are integers $+\frac{1}{2}$. (Why does S vanish in the latter case when $f_+ = f_-$?)

(b) The zincblende structure (Figure 4.18) is also a face-centered cubic Bravais lattice of cube side a, with a basis consisting of a positively charged ion at the origin and a negatively charged ion at $(a/4)(\hat{\mathbf{x}} + \hat{\mathbf{y}} + \hat{\mathbf{z}})$. Show that the structure factor $S_{\mathbf{K}}$ is $f_+ \pm if_-$ if the ν_i are integers $+\frac{1}{2}$, $f_+ + f_-$ if the ν_i are integers and $\Sigma\nu_i$ is even, and $f_+ - f_-$ if the ν_i are integers and $\Sigma\nu_i$ is odd.

(c) Suppose that a cubic crystal is known to be composed of closed-shell (and hence spherically symmetric) ions, so that $f_\pm(\mathbf{K})$ depends only on the magnitude of $\mathbf{K}$. The positions of the Bragg peaks reveal that the Bravais lattice is face-centered cubic. Discuss how one might determine, from the structure factors associated with the Bragg peaks, whether the crystal structure was likely to be of the sodium chloride or zincblende type.

7

Classification of Bravais Lattices and Crystal Structures

Symmetry Operations and the Classification of Bravais Lattices

The Seven Crystal Systems and Fourteen Bravais Lattices

Crystallographic Point Groups and Space Groups

Schoenflies and International Notations

Examples from the Elements

In Chapters 4 and 5, only the *translational* symmetries of Bravais lattices were described and exploited. For example, the existence and basic properties of the reciprocal lattice depend only on the existence of three primitive direct lattice vectors $\mathbf{a}_i$, and not on any special relations that may hold among them.[1] The translational symmetries are by far the most important for the general theory of solids. It is nevertheless clear from examples already described that Bravais lattices do fall naturally into categories on the basis of symmetries other than translational. Simple hexagonal Bravais lattices, for example, regardless of the c/a ratio, bear a closer resemblance to one another than they do to any of the three types of cubic Bravais lattice we have described.

It is the subject of crystallography to make such distinctions systematic and precise.[2] Here we shall only indicate the basis for the rather elaborate crystallographic classifications, giving some of the major categories and the language by which they are described. In most applications what matters are the features of particular cases, rather than a systematic general theory, so few solid state physicists need master the full analysis of crystallography. Indeed, the reader with little taste for the subject can skip this chapter entirely with little loss in understanding what follows, referring back to it on occasion for the elucidation of arcane terms.

THE CLASSIFICATION OF BRAVAIS LATTICES

The problem of classifying all possible crystal structures is too complex to approach directly, and we first consider only the classification of Bravais lattices.[3] From the point of view of symmetry, a Bravais lattice is characterized by the specification of all rigid operations[4] that take the lattice into itself. This set of operations is known as the *symmetry group* or *space group* of the Bravais lattice.[5]

The operations in the symmetry group of a Bravais lattice include all translations through lattice vectors. In addition, however, there will in general be rotations, reflections, and inversions[6] that take the lattice into itself. A cubic Bravais lattice, for example, is taken into itself by a rotation through 90° about a line of lattice points in a $\langle 100\rangle$ direction, a rotation through 120° about a line of lattice points in a $\langle 111\rangle$ direction, reflection of all points in a $\{100\}$ lattice plane, etc.; a simple hexagonal Bravais lattice is taken into itself by a rotation through 60° about a line of lattice points parallel to the c-axis, reflection in a lattice plane perpendicular to the c-axis, etc.

---

1 An example of such a relation is the orthonormality condition $\mathbf{a}_i \cdot \mathbf{a}_j = a^2\delta_{ij}$, holding for the appropriate primitive vectors in a simple cubic Bravais lattice.

2 A detailed view of the subject can be found in M. J. Buerger, *Elementary Crystallography*, Wiley, New York, 1963.

3 In this chapter a Bravais lattice is viewed as the crystal structure formed by placing at each point of an abstract Bravais lattice a basis of maximum possible symmetry (such as a sphere, centered on the lattice point) so that no symmetries of the point Bravais lattice are lost because of the insertion of the basis.

4 Operations that preserve the distance between all lattice points.

5 We shall avoid the language of mathematical group theory, since we shall make no use of the analytical conclusions to which it leads.

6 Reflection in a plane replaces an object by its mirror image in that plane; inversion in a point P takes the point with coordinates $\mathbf{r}$ (with respect to P as origin) into $-\mathbf{r}$. All Bravais lattices have inversion symmetry in any lattice point (Problem 1).

Any symmetry operation of a Bravais lattice can be compounded out of a translation $T_{\mathbf{R}}$ through a lattice vector **R** and a rigid operation leaving at least one lattice point fixed.[7] This is not immediately obvious. A simple cubic Bravais lattice, for example, is left fixed by a rotation through 90° about a $\langle 100 \rangle$ axis that passes through the center of a cubic primitive cell with lattice points at the eight vertices of the cube. This is a rigid operation that leaves no lattice point fixed. However, it can be compounded out of a translation through a Bravais lattice vector and a rotation

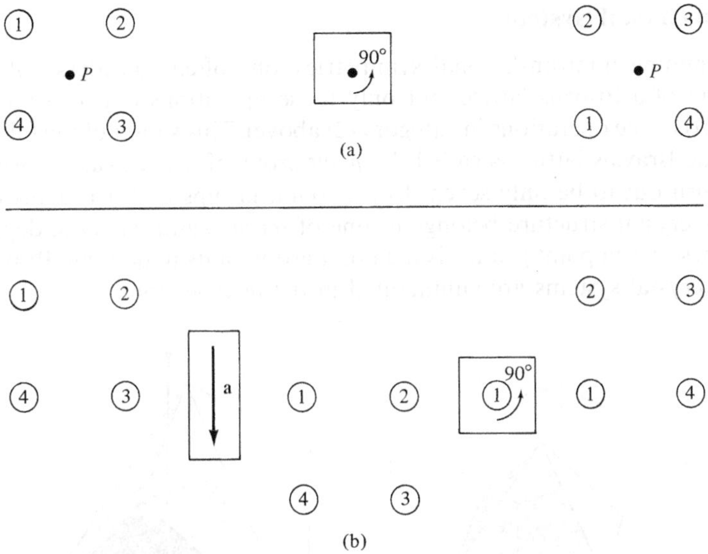

Figure 7.1
(a) A simple cubic lattice is carried into itself by a rotation through 90° about an axis that contains no lattice points. The rotation axis is perpendicular to the page, and only the four lattice points closest to the axis in a single lattice plane are shown. (b) Illustrating how the same final result can be compounded out of (at left) a translation through a lattice constant and (at right) a rotation about the lattice point numbered 1.

about a line of lattice points, as illustrated in Figure 7.1. That such a representation is always possible can be seen as follows:

Consider a symmetry operation S that leaves *no* lattice point fixed. Suppose it takes the origin of the lattice **O** into the point **R**. Consider next the operation one gets by first applying S, and then applying a translation through $-\mathbf{R}$, which we denote by $T_{-\mathbf{R}}$. The composite operation, which we call $T_{-\mathbf{R}}S$, is also a symmetry of the lattice, but it leaves the origin fixed, since S transports the origin to **R** while $T_{-\mathbf{R}}$ carries **R** back to the origin. Thus $T_{-\mathbf{R}}S$ is an operation that leaves at least one lattice point (namely the origin) fixed. If, however, after performing the operation $T_{-\mathbf{R}}S$ we then perform the operation $T_{\mathbf{R}}$, the result is equivalent to the operation S alone, since the final application of $T_{\mathbf{R}}$ just undoes the preceding application of $T_{-\mathbf{R}}$. Therefore S can be compounded out of $T_{-\mathbf{R}}S$, which leaves a point fixed, and $T_{\mathbf{R}}$, which is a pure translation.

7 Note that translation through a lattice vector (other than **O**) leaves no point fixed.

Thus the full symmetry group of a Bravais lattice[8] contains only operations of the following form:

1. Translations through Bravais lattice vectors;
2. Operations that leave a particular point of the lattice fixed;
3. Operations that can be constructed by successive applications of the operations of type (1) or (2).

The Seven Crystal Systems

When examining nontranslational symmetries, one often considers not the entire space group of a Bravais lattice, but only those operations that leave a particular point fixed (i.e., the operations in category (2) above). This subset of the full symmetry group of the Bravais lattice is called the *point group* of the Bravais lattice.

There turn out to be only seven distinct point groups that a Bravais lattice can have.[9] Any crystal structure belongs to one of *seven crystal systems*, depending on which of these seven point groups is the point group of its underlying Bravais lattice. The seven crystal systems are enumerated in the next section.

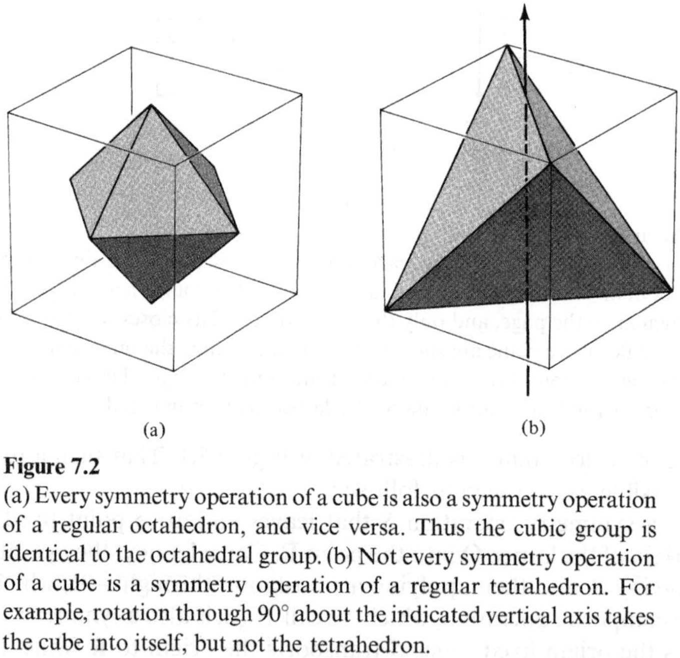

Figure 7.2
(a) Every symmetry operation of a cube is also a symmetry operation of a regular octahedron, and vice versa. Thus the cubic group is identical to the octahedral group. (b) Not every symmetry operation of a cube is a symmetry operation of a regular tetrahedron. For example, rotation through 90° about the indicated vertical axis takes the cube into itself, but not the tetrahedron.

---

[8] We shall see below that a general crystal structure can have additional symmetry operations that are not of types (1), (2), or (3). They are known as "screw axes" and "glide planes."

[9] Two point groups are identical if they contain precisely the same operations. For example, the set of all symmetry operations of a cube is identical to the set of all symmetry operations of a regular octahedron, as can readily be seen by inscribing the octahedron suitably in the cube (Figure 7.2a). On the other hand, the symmetry group of the cube is not equivalent to the symmetry group of the regular tetrahedron. The cube has more symmetry operations (Figure 7.2b).

The Fourteen Bravais Lattices

When one relaxes the restriction to point operations and considers the full symmetry group of the Bravais lattice, there turn out to be fourteen distinct space groups that a Bravais lattice can have.[10] Thus, from the point of view of symmetry, there are fourteen different kinds of Bravais lattice. This enumeration was first done by M. L. Frankenheim (1842). Frankenheim miscounted, however, reporting fifteen possibilities. A. Bravais (1845) was the first to count the categories correctly.

Enumeration of the Seven Crystal Systems and Fourteen Bravais Lattices

We list below the seven crystal systems and the Bravais lattices belonging to each. The number of Bravais lattices in a system is given in parentheses after the name of the system:

***Cubic* (3)** The cubic system contains those Bravais lattices whose point group is just the symmetry group of a cube (Figure 7.3a). Three Bravais lattices with nonequivalent space groups all have the cubic point group. They are the *simple cubic*, *body-centered cubic*, and *face-centered cubic*. All three have been described in Chapter 4.

***Tetragonal* (2)** One can reduce the symmetry of a cube by pulling on two opposite faces to stretch it into a rectangular prism with a square base, but a height not equal to the sides of the square (Figure 7.3b). The symmetry group of such an object is the tetragonal group. By so stretching the simple cubic Bravais lattice one constructs the *simple tetragonal* Bravais lattice, which can be characterized as a Bravais lattice generated by three mutually perpendicular primitive vectors, only two of which are of equal length. The third axis is called the c-axis. By similarly stretching the body-centered and face-centered cubic lattices only one more Bravais lattice of the tetragonal system is constructed, the *centered tetragonal*.

To see why there is no distinction between body-centered and face-centered tetragonal, consider Figure 7.4a, which is a representation of a centered tetragonal Bravais lattice viewed along the c-axis. The points 2 lie in a lattice plane a distance

---

10 The equivalence of two Bravais lattice space groups is a somewhat more subtle notion than the equivalence of two point groups (although both reduce to the concept of "isomorphism" in abstract group theory). It is no longer enough to say that two space groups are equivalent if they have the same operations, for the operations of identical space groups can differ in inconsequential ways. For example, two simple cubic Bravais lattices with different lattice constants, a and a', are considered to have the same space groups even though the translations in one are in steps of a, whereas the translations in the other are in steps of a'. Similarly, we would like to regard all simple hexagonal Bravais lattices as having identical space groups, regardless of the value of c/a, which is clearly irrelevant to the total symmetry of the structure.

We can get around this problem by noting that in such cases one can continuously deform a structure of a given type into another of the same type without ever losing any of the symmetry operations along the way. Thus one can uniformly expand the cube axes from a to a', always maintaining the simple cubic symmetry, or one can stretch (or shrink) the c-axis (or a-axis), always maintaining the simple hexagonal symmetry. Therefore two Bravais lattices can be said to have the same space group if it is possible continuously to transform one into the other in such a way that every symmetry operation of the first is continuously transformed into a symmetry operation of the second, and there are no additional symmetry operations of the second not so obtained from symmetry operations of the first.

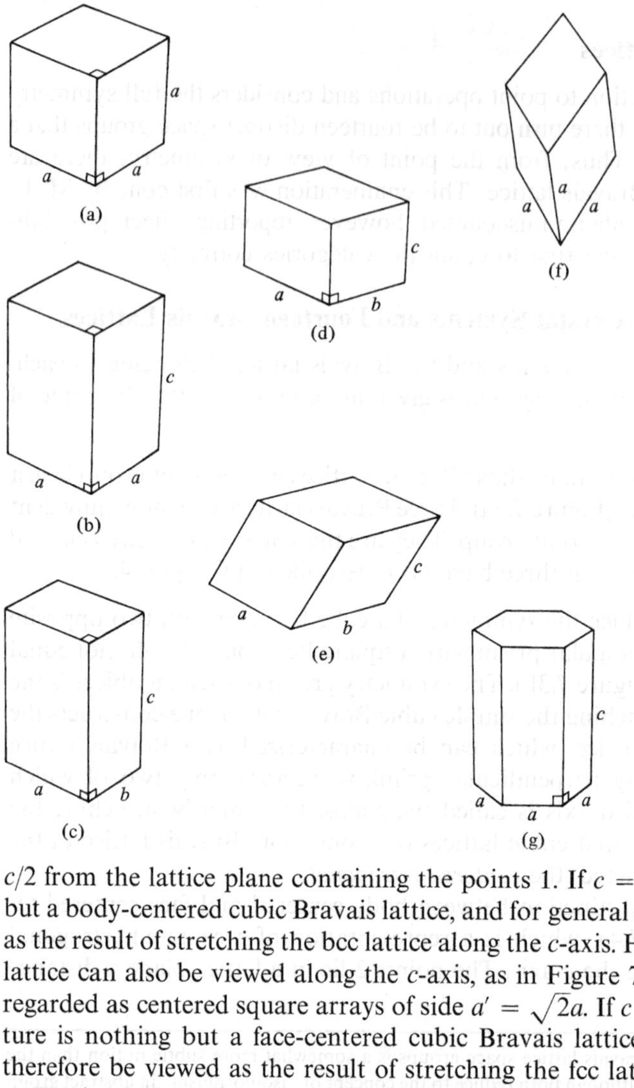

Figure 7.3
Objects whose symmetries are the point-group symmetries of Bravais lattices belonging to the seven crystal systems: (a) cubic; (b) tetragonal; (c) orthorhombic; (d) monoclinic; (e) triclinic; (f) trigonal; (g) hexagonal.

$c/2$ from the lattice plane containing the points 1. If $c = a$, the structure is nothing but a body-centered cubic Bravais lattice, and for general c it can evidently be viewed as the result of stretching the bcc lattice along the c-axis. However, precisely the same lattice can also be viewed along the c-axis, as in Figure 7.4b, with the lattice planes regarded as centered square arrays of side $a' = \sqrt{2}a$. If $c = a'/2 = a/\sqrt{2}$, the structure is nothing but a face-centered cubic Bravais lattice, and for general c it can therefore be viewed as the result of stretching the fcc lattice along the c-axis.

Putting it the other way around, face-centered cubic and body-centered cubic are both special cases of centered tetragonal, in which the particular value of the c/a ratio introduces extra symmetries that are revealed most clearly when one views the lattice as in Figure 7.4a (bcc) or Figure 7.4b (fcc).

***Orthorhombic* (4)** Continuing to still less symmetric deformations of the cube, one can reduce tetragonal symmetry by deforming the square faces of the object in Figure 7.3b into rectangles, producing an object with mutually perpendicular sides of three unequal lengths (Figure 7.3c). The orthorhombic group is the symmetry group of such an object. By stretching a simple tetragonal lattice along one of the a-axes (Figure 7.5a and b), one produces the *simple orthorhombic* Bravais lattice. However, by stretching the simple tetragonal lattice along a square diagonal (Figure 7.5c and d) one produces a second Bravais lattice of orthorhombic point group symmetry, the *base-centered* orthorhombic.

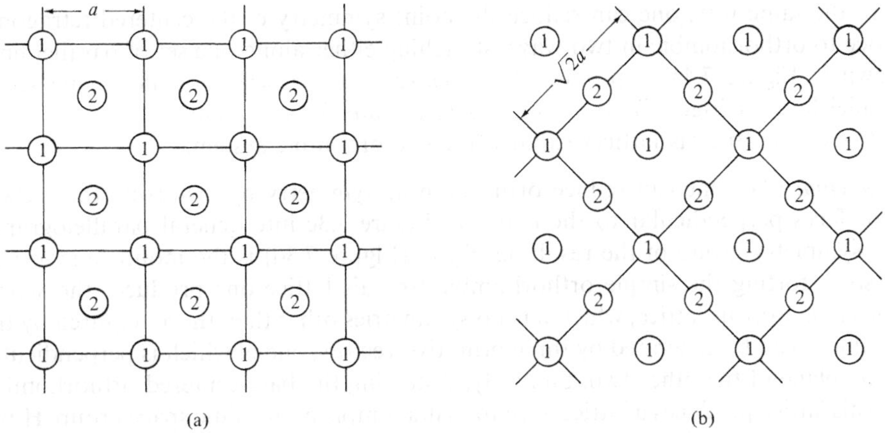

Figure 7.4
Two ways of viewing the same centered tetragonal Bravais lattice. The view is along the c-axis. The points labeled 1 lie in a lattice plane perpendicular to the c-axis, and the points labeled 2 lie in a parallel lattice plane a distance $c/2$ away. In (a) the points 1 are viewed as a simple square array, stressing that centered tetragonal is a distortion of body-centered cubic. In (b) the points 1 are viewed as a centered square array, stressing that centered tetragonal is also a distortion of face-centered cubic.

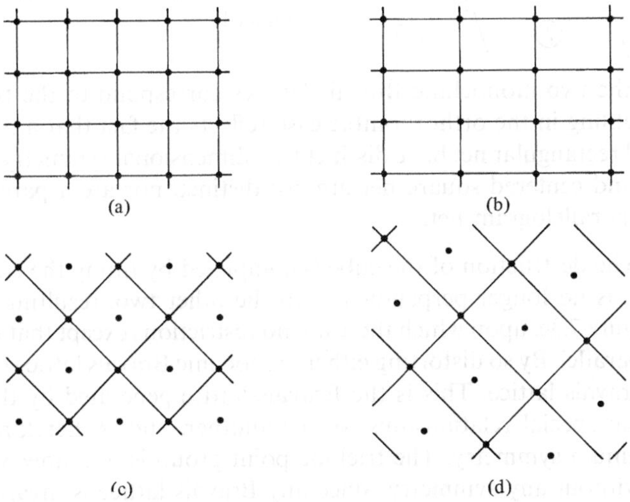

Figure 7.5
Two ways of deforming the same simple tetragonal Bravais lattice. The view is along the c-axis, and a single lattice plane is shown. In (a) bonds are drawn to emphasize that the points in the plane can be viewed as a simple square array. Stretching along a side of that array leads to the rectangular nets (b), stacked directly above one another. The resulting Bravais lattice is simple orthorhombic. In (c) lines are drawn to emphasize that the same array of points as shown in (a) can also be viewed as a centered square array. Stretching along a side of that array (i.e., along a diagonal of the square array emphasized in (a)) yields the centered rectangular nets (d), stacked directly above one another. The resulting Bravais lattice is base-centered orthorhombic.

In the same way, one can reduce the point symmetry of the centered tetragonal lattice to orthorhombic in two ways, stretching either along one set of parallel lines drawn in Figure 7.4a to produce *body-centered orthorhombic*, or along one set of parallel lines in Figure 7.4b, producing *face-centered orthorhombic*.

These four Bravais lattices exhaust the orthorhombic system.

***Monoclinic* (2)** One can reduce orthorhombic symmetry by distorting the rectangular faces perpendicular to the *c*-axis in Figure 7.3c into general parallelograms. The symmetry group of the resulting object (Figure 7.3d) is the monoclinic group. By so distorting the simple orthorhombic Bravais lattice one produces the *simple monoclinic* Bravais lattice, which has no symmetries other than those required by the fact that it can be generated by three primitive vectors, one of which is perpendicular to the plane of the other two. Similarly, distorting the base-centered orthorhombic Bravais lattice produces a lattice with the same simple monoclinic space group. However, so distorting either the face-centered or body-centered orthorhombic Bravais lattices produces the *centered monoclinic* Bravais lattice (Figure 7.6).

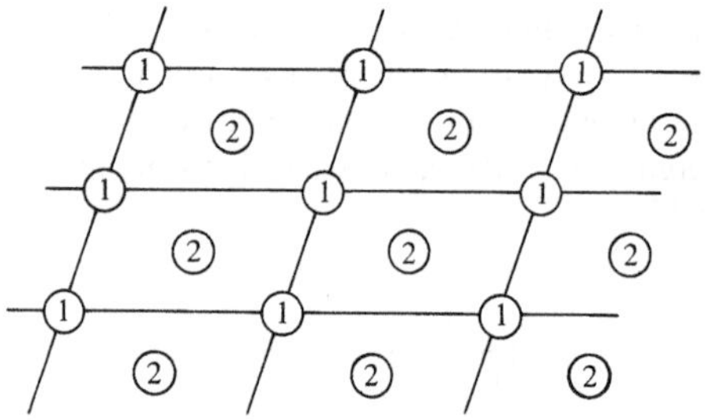

Figure 7.6
View along the *c*-axis of a centered monoclinic Bravais lattice. The points labeled 1 lie in a lattice plane perpendicular to the *c*-axis. The points labeled 2 lie in a parallel lattice plane a distance $c/2$ away, and are directly above the centers of the parallelograms formed by the points 1.

Note that the two monoclinic Bravais lattices correspond to the two tetragonal ones. The doubling in the orthorhombic case reflects the fact that a rectangular net and a centered rectangular net have distinct two-dimensional symmetry groups, while a square net and centered square net are not distinct, nor are a parallelogram net and centered parallelogram net.

***Triclinic* (*1*)** The destruction of the cube is completed by tilting the *c*-axis in Figure 7.3d so that it is no longer perpendicular to the other two, resulting in the object pictured in Figure 7.3e, upon which there are no restrictions except that pairs of opposite faces are parallel. By so distorting either monoclinic Bravais lattice one constructs the *triclinic* Bravais lattice. This is the Bravais lattice generated by three primitive vectors with no special relationships to one another, and is therefore the Bravais lattice of minimum symmetry. The triclinic point group is not, however, the group of an object without any symmetry, since any Bravais lattice is invariant under an inversion in a lattice point. That, however, is the only symmetry required by the general definition of a Bravais lattice, and therefore the only operation[11] in the triclinic point group.

By so torturing a cube we have arrived at twelve of the fourteen Bravais lattices and five of the seven crystal systems. We can find the thirteenth and sixth by returning to the original cube and distorting it differently:

[11] Other than the identity operation (which leaves the lattice where it is), which is always counted among the members of a symmetry group.

Trigonal **(*1*)** The trigonal point group describes the symmetry of the object one produces by stretching a cube along a body diagonal (Figure 7.3f). The lattice made by so distorting any of the three cubic Bravais lattices is the *rhombohedral* (or *trigonal*) Bravais lattice. It is generated by three primitive vectors of equal length that make equal angles with one another.[12]

Finally, unrelated to the cube, is:

Hexagonal **(*1*)** The hexagonal point group is the symmetry group of a right prism with a regular hexagon as base (Figure 7.3g). The *simple hexagonal* Bravais lattice (described in Chapter 4) has the hexagonal point group and is the only Bravais lattice in the hexagonal system.[13]

The seven crystal systems and fourteen Bravais lattices described above exhaust the possibilities. This is far from obvious (or the lattices would have been known as Frankenheim lattices). However, it is of no practical importance to understand why these are the only distinct cases. It is enough to know why the categories exist, and what they are.

THE CRYSTALLOGRAPHIC POINT GROUPS AND SPACE GROUPS

We next describe the results of a similar analysis, applied not to Bravais lattices but to general crystal structures. We consider the structure obtained by translating an arbitrary object through the vectors of any Bravais lattice, and try to classify the symmetry groups of the arrays so obtained. These depend both on the symmetry of the object and the symmetry of the Bravais lattice. Because the objects are no longer required to have maximum (e.g., spherical) symmetry, the number of symmetry groups is greatly increased: there turn out to be 230 different symmetry groups that a lattice with a basis can have, known as the 230 *space groups.* (This is to be compared with the fourteen space groups that result when the basis is required to be completely symmetric.)

The possible point groups of a general crystal structure have also been enumerated. These describe the symmetry operations that take the crystal structure into itself while leaving one point fixed (i.e., the nontranslational symmetries). There are thirty-two distinct point groups that a crystal structure can have, known as the *thirty-two crystallographic point groups.* (This is to be compared with the seven point groups one can have when the basis is required to have full symmetry.)

These various numbers and their relations to one another are summarized in Table 7.1.

The thirty-two crystallographic point groups can be constructed out of the seven Bravais lattice point groups by systematically considering all possible ways of reducing the symmetry of the objects (Figure 7.3) characterized by these groups.

Each of the twenty-five new groups constructed in this way is associated with one

---

12 Special values of that angle may introduce extra symmetries, in which case the lattice may actually be one of the three cubic types. See, for example, Problem 2(a).

13 If one tries to produce further Bravais lattices from distortions of the simple hexagonal, one finds that changing the angle between the two primitive vectors of equal length perpendicular to the *c*-axis yields a base-centered orthorhombic lattice, changing their magnitudes as well leads to monoclinic, and tilting the *c*-axis from the perpendicular leads, in general, to triclinic.

Table 7.1
POINT AND SPACE GROUPS OF BRAVAIS LATTICES AND CRYSTAL STRUCTURES

	BRAVAIS LATTICE (BASIS OF SPHERICAL SYMMETRY)	CRYSTAL STRUCTURE (BASIS OF ARBITRARY SYMMETRY)
Number of point groups:	7 ("the 7 crystal systems")	32 ("the 32 crystallographic point groups")
Number of space groups:	14 ("the 14 Bravais lattices")	230 ("the 230 space groups")

of the seven crystal systems according to the following rule: Any group constructed by reducing the symmetry of an object characterized by a particular crystal system continues to belong to that system until the symmetry has been reduced so far that all of the remaining symmetry operations of the object are also found in a less symmetrical crystal system; when this happens the symmetry group of the object is assigned to the less symmetrical system. Thus the crystal system of a crystallographic point group is that of the least symmetric[14] of the seven Bravais lattice point groups containing every symmetry operation of the crystallographic group.

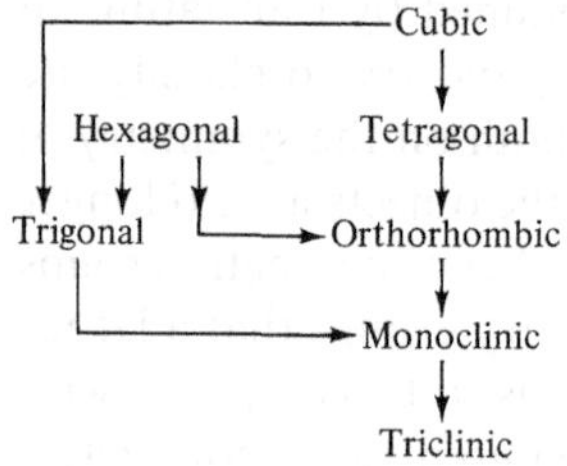

Figure 7.7
The hierarchy of symmetries among the seven crystal systems. Each Bravais lattice point group contains all those that can be reached from it by moving in the direction of the arrows.

Objects with the symmetries of the five crystallographic point groups in the cubic system are pictured in Table 7.2. Objects with the symmetries of the twenty-seven noncubic crystallographic groups are shown in Table 7.3.

Crystallographic point groups may contain the following kinds of symmetry operations:

1. **Rotations through Integral Multiples of $2\pi/n$ about Some Axis** The axis is called an n-fold rotation axis. It is easily shown (Problem 6) that a Bravais lattice can

[14] The notion of a hierarchy of crystal system symmetries needs some elaboration. In Figure 7.7 each crystal system is more symmetric than any that can be reached from it by moving along arrows; i.e., the corresponding Bravais lattice point group has no operations that are not also in the groups from which it can be so reached. There appears to be some ambiguity in this scheme since the four pairs cubic-hexagonal, tetragonal-hexagonal, tetragonal-trigonal, and orthorhombic-trigonal are not ordered by the arrows. Thus one might imagine an object all of whose symmetry operations belonged to both the tetragonal and trigonal groups but to no group lower than both of these. The symmetry group of such an object could be said to belong to either the tetragonal or trigonal systems, since there would be no unique system of lowest symmetry. It turns out, however, both in this and the other three ambiguous cases, that all symmetry elements common to both groups in a pair also belong to a group that is lower than both in the hierarchy. (For example, any element common to both the tetragonal and the trigonal groups also belongs to the monoclinic group.) There is therefore always a unique group of lowest symmetry.

Table 7.2
OBJECTS WITH THE SYMMETRY OF THE FIVE CUBIC CRYSTALLOGRAPHIC POINT GROUPS[a]

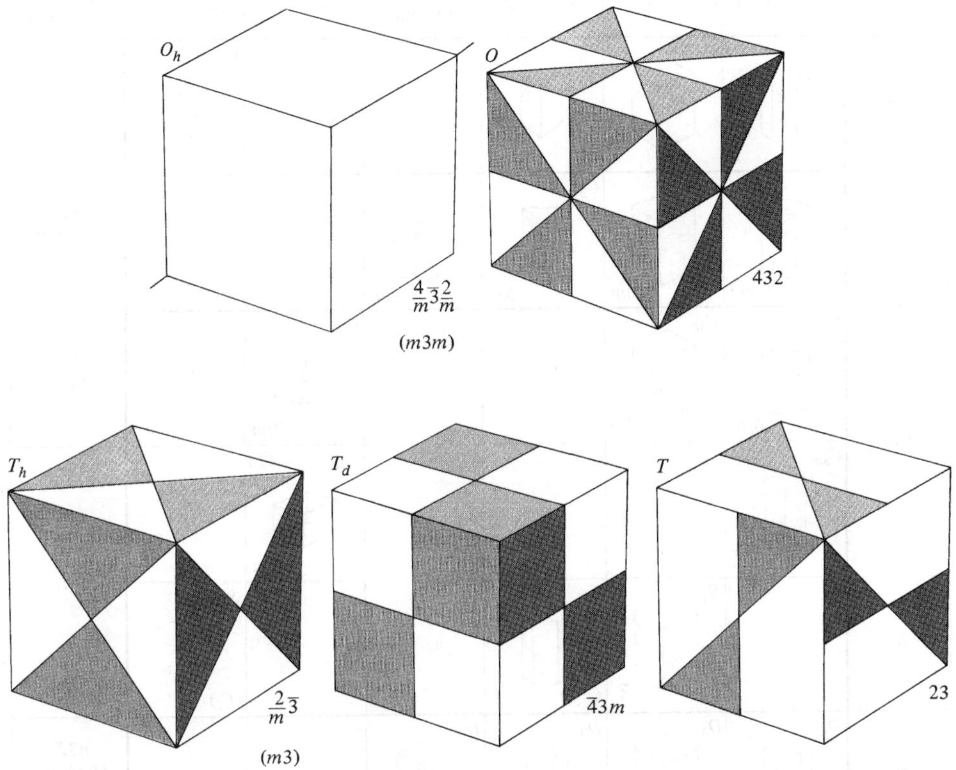

[a] To the left of each object is the Schoenflies name of its symmetry group and to the right is the international name. The unpictured faces may be deduced from the fact that rotation about a body diagonal through 120° is a symmetry operation for all five objects. (Such an axis is shown on the undecorated cube.)

contain only 2-, 3-, 4-, or 6-fold axes. Since the crystallographic point groups are contained in the Bravais lattice point groups, they too can only have these axes.

2. **Rotation-Reflections** Even when a rotation through $2\pi/n$ is not a symmetry element, sometimes such a rotation followed by a reflection in a plane perpendicular to the axis may be. The axis is then called an n-fold rotation-reflection axis. For example, the groups S_6 and S_4 (Table 7.3) have 6- and 4-fold rotation-reflection axes.
3. **Rotation-Inversions** Similarly, sometimes a rotation through $2\pi/n$ followed by an inversion in a point lying on the rotation axis is a symmetry element, even though such a rotation by itself is not. The axis is then called an n-fold rotation-inversion axis. The axis in S_4 (Table 7.3), for example, is also a 4-fold rotation-inversion axis. However, the axis in S_6 is only a 3-fold rotation-inversion axis.
4. **Reflections** A reflection takes every point into its mirror image in a plane, known as a mirror plane.

Table 7.3
THE NONCUBIC CRYSTALLOGRAPHIC POINT GROUPS[a]

SCHOEN-FLIES	HEXAGONAL	TETRAGONAL	TRIGONAL	ORTHO-RHOMBIC	MONOCLINIC	TRICLINIC	INTER-NATIONAL
C_n	C_6 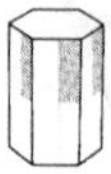 6	C_4 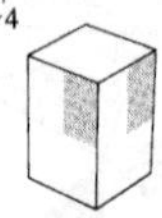 4	C_3  3		C_2 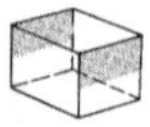 2	C_1 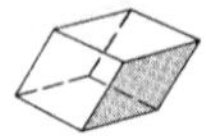 1	n
C_{nv}	C_{6v} 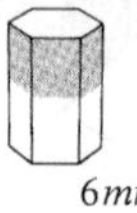 $6mm$	C_{4v} 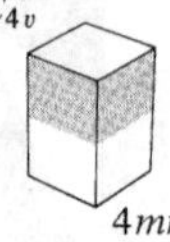 $4mm$	C_{3v}  $3m$	C_{2v} 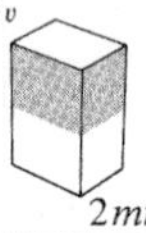 $2mm$			nmm (n even) nm (n odd)
C_{nh}	C_{6h} 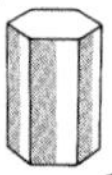 $6/m$	C_{4h}  $4/m$			C_{2h} 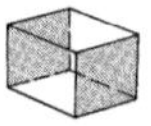 $2/m$		n/m
	C_{3h}  $\bar{6}$				C_{1h} $(\bar{2})$ 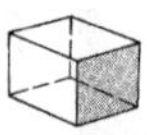 m		$\bar{n}$
S_n		S_4 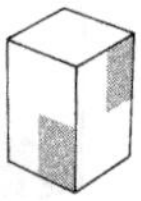 $\bar{4}$	S_6 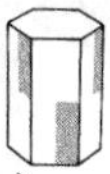 (C_{3i}) $\bar{3}$			S_2 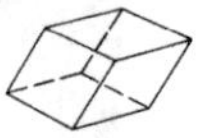 (C_i) $\bar{1}$	
D_n	D_6 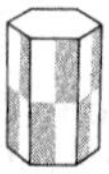 622	D_4 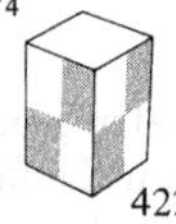 422	D_3  32	D_2  (V) 222			$n22$ (n even) $n2$ (n odd)
D_{nh}	D_{6h} 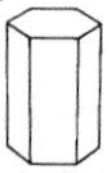 $6/mmm$	D_{4h} 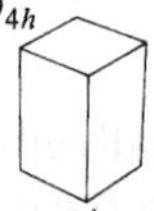 $4/mmm$		D_{2h} (mmm)  (V_h) $2/mmm$			$\frac{n}{m}\frac{2}{m}\frac{2}{m}$ (n/mmm)
	D_{3h}  $\bar{6}2m$						$\bar{n}2m$ (n even)
D_{nd}		D_{2d} 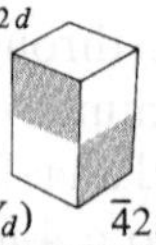 (V_d) $\bar{4}2m$	D_{3d} $(\bar{3}m)$ 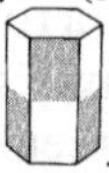 $\bar{3}\frac{2}{m}$				$\bar{n}\frac{2}{m}$ (n odd)

[a] Table caption on p. 123.

Table 7.3 (*continued*)
The unpictured faces can be deduced by imagining the representative objects to be rotated about the n-fold axis, which is always vertical. The Schoenflies name of the group is given to the left of the representative object, and the international designation the right. The groups are organized into vertical columns by crystal system, and into horizontal rows by the Schoenflies or international type. Note that the Schoenflies categories (given on the extreme left of the table) divide up the groups somewhat differently from the international categories (given on the extreme right). In most (but not all) cases the representative objects have been made by simply decorating in the appropriate symmetry reducing manner the faces of the objects used to represent the crystal systems (Bravais lattice point groups) in Figure 7.3. Exceptions are the trigonal groups and two of the hexagonal groups, where the figures have been changed to emphasize the similarity within the (horizontal) Schoenflies categories. For a representation of the trigonal groups by decorations of the object in Figure 7.3f, see Problem 4.

5. **Inversions** An inversion has a single fixed point. If that point is taken as the origin, then every other point $\mathbf{r}$ is taken into $-\mathbf{r}$.

Point-Group Nomenclature

Two nomenclatural systems, the Schönflies and the international, are in wide use. Both designations are given in Tables 7.2 and 7.3.

Schoenflies Notation for the Noncubic Crystallographic Point Groups The Schoenflies categories are illustrated by grouping the rows in Table 7.3 according to the labels given on the left side. They are:[15]

C_n: These groups contain only an n-fold rotation axis.
C_{nv}: In addition to the n-fold axis, these groups have a mirror plane that contains the axis of rotation, plus as many additional mirror planes as the existence of the n-fold axis requires.
C_{nh}: These groups contain in addition to the n-fold axis, a single mirror plane that is perpendicular to the axis.
S_n: These groups contain only an n-fold rotation-reflection axis.
D_n: In addition to an n-fold rotation axis, these groups contain a 2-fold axis perpendicular to the n-fold axis, plus as many additional 2-fold axes as are required by the existence of the n-fold axis.
D_{nh}: These (the most symmetric of the groups) contain all the elements of D_n plus a mirror plane perpendicular to the n-fold axis.
D_{nd}: These contain the elements of D_n plus mirror planes containing the n-fold axis, which bisect the angles between the 2-fold axes.

It is instructive to verify that the objects shown in Table 7.3 do indeed have the symmetries required by their Schoenflies names.

International Notation for the Noncubic Crystallographic Point Groups The international categories are illustrated by grouping the rows in Table 7.3 according to

15 C stands for "cyclic," D for "dihedral," and S for "Spiegel" (mirror). The subscripts h, v, and d stand for "horizontal," "vertical," and "diagonal," and refer to the placement of the mirror planes with respect to the n-fold axis, considered to be vertical. (The "diagonal" planes in D_{nd} are vertical and bisect the angles between the 2-fold axes.)

the labels given on the right side. Three categories are identical to the Schoenflies categories:

n is the same as C_n.

nmm is the same as C_{nv}. The two m's refer to two distinct types of mirror planes containing the n-fold axis. What they are is evident from the objects illustrating $6mm$, $4mm$, and $2mm$. These demonstrate that a $2j$-fold axis takes a vertical mirror plane into j mirror planes, but in addition j others automatically appear, bisecting the angles between adjacent planes in the first set. However, a $(2j + 1)$-fold axis takes a mirror plane into $2j + 1$ equivalent ones, and therefore[16] C_{3v} is only called $3m$.

$n22$ is the same as D_n. The discussion is the same as for nmm, but now perpendicular 2-fold axes are involved instead of vertical mirror planes.

The other international categories and their relation to those of Schoenflies are as follows:

n/m is the same as C_{nh}, *except* that the international system prefers to regard C_{3h} as containing a 6-fold rotation-inversion axis, making it $\overline{6}$ (see the next category). Note also that C_{1h} becomes simply m, rather than $1/m$.

$\bar{n}$ is a group with an n-fold rotation-inversion axis. This category contains C_{3h}, disguised as $\overline{6}$. It also contains S_4, which goes nicely into $\overline{4}$. However, S_6 becomes $\overline{3}$ and S_2 becomes $\overline{1}$ by virtue of the difference between rotation-reflection and rotation-inversion axes.

$\frac{n}{m}\frac{2}{m}\frac{2}{m}$, abbreviated n/mmm, is just D_{nh}, *except* that the international system prefers to regard D_{3h} as containing a 6-fold rotation-inversion axis, making it $\overline{6}2m$ (see the next category, and note the similarity to the ejection of C_{3h} from n/m into $\bar{n}$). Note also that $2/mmm$ is conventionally abbreviated further into mmm. The full-blown international title is supposed to remind one that D_{nh} can be viewed as an n-fold axis with a perpendicular mirror plane, festooned with two sets of perpendicular 2-fold axes, each with its own perpendicular mirror planes.

$\bar{n}2m$ is the same as D_{nd}, *except* that D_{3h} is included as $\overline{6}2m$. The name is intended to suggest an n-fold rotation-inversion axis with a perpendicular 2-fold axis and a vertical mirror plane. The $n = 3$ case is again exceptional, the full name being $\overline{3}\frac{2}{m}$ (abbreviated $\overline{3}m$) to emphasize the fact that in this case the vertical mirror plane is perpendicular to the 2-fold axis.

Nomenclature for the Cubic Crystallographic Point Groups The Schoenflies and international names for the five cubic groups are given in Table 7.2. O_h is the full symmetry group of the cube (or octahedron, whence the O) including improper operations,[17] which the horizontal reflection plane (h) admits. O is the cubic (or octahedral) group without improper operations. T_d is the full symmetry group of the regular tetrahedron including all improper operations, T is the group of the regular tetrahedron excluding all improper operations, and T_h is what results when an inversion is added to T.

---

[16] In emphasizing the differences between odd- and even-fold axes, the international system, unlike the Schoenflies, treats the 3-fold axis as a special case.

[17] Any operation that takes a right-handed object into a left-handed one is called *improper*. All others are proper. Operations containing an odd number of inversions or mirrorings are improper.

The international names for the cubic groups are conveniently distinguished from those of the other crystallographic point groups by containing 3 as a second number, referring to the 3-fold axis present in all the cubic groups.

The 230 Space Groups

We shall have mercifully little to say about the 230 space groups, except to point out that the number is larger than one might have guessed. For each crystal system one can construct a crystal structure with a different space group by placing an object with the symmetries of each of the point groups of the system into each of the Bravais lattices of the system. In this way, however, we find only 61 space groups, as shown in Table 7.4.

Table 7.4
ENUMERATION OF SOME SIMPLE SPACE GROUPS

SYSTEM	NUMBER OF POINT GROUPS	NUMBER OF BRAVAIS LATTICES	PRODUCT
Cubic	5	3	15
Tetragonal	7	2	14
Orthorhombic	3	4	12
Monoclinic	3	2	6
Triclinic	2	1	2
Hexagonal	7	1	7
Trigonal	5	1	5
Totals	32	14	61

We can eke out five more by noting that an object with trigonal symmetry yields a space group not yet enumerated, when placed in a hexagonal Bravais lattice.[18]

[18] Although the trigonal point group is contained in the hexagonal point group, the trigonal Bravais lattice cannot be obtained from the simple hexagonal by an infinitesimal distortion. (This is in contrast to all other pairs of systems connected by arrows in the symmetry hierarchy of Figure 7.7.) The trigonal point group is contained in the hexagonal point group because the trigonal Bravais lattice can be viewed as simple hexagonal with a three-point basis consisting of

$$\mathbf{0};\quad \tfrac{1}{3}\mathbf{a}_1, \tfrac{1}{3}\mathbf{a}_2, \tfrac{1}{3}\mathbf{c};\quad \text{and}\quad \tfrac{2}{3}\mathbf{a}_1, \tfrac{2}{3}\mathbf{a}_2, \tfrac{2}{3}\mathbf{c}.$$

As a result, placing a basis with a trigonal point group into a hexagonal Bravais lattice results in a different space group from that obtained by placing the same basis into a trigonal lattice. In no other case is this so. For example, a basis with tetragonal symmetry, when placed in a simple cubic lattice, yields exactly the same space group as it would if placed in a simple tetragonal lattice (unless there happens to be a special relation between the dimensions of the object and the length of the c-axis). This is reflected physically in the fact that there are crystals that have trigonal bases in hexagonal Bravais lattices, but none with tetragonal bases in cubic Bravais lattices. In the latter case there would be nothing in the structure of such an object to require the c-axis to have the same length as the a-axes; if the lattice did remain cubic it would be a mere coincidence. In contrast, a simple hexagonal Bravais lattice cannot distort continuously into a trigonal one, and can therefore be held in its simple hexagonal form even by a basis with only trigonal symmetry.

Because trigonal point groups can characterize a crystal structure with a hexagonal Bravais lattice, crystallographers sometimes maintain that there are only six crystal systems. This is because crystallography emphasizes the point symmetry rather than the translational symmetry. From the point of view of the Bravais lattice point groups, however, there are unquestionably seven crystal systems: the point groups D_{3d} and D_{6h} are both the point groups of Bravais lattices, and are not equivalent.

Another seven arise from cases in which an object with the symmetry of a given point group can be oriented in more than one way in a given Bravais lattice so that more than one space group arises. These 73 space groups are called *symmorphic.*

The majority of the space groups are *nonsymmorphic*, containing additional operations that cannot be simply compounded out of Bravais lattice translations and point-group operations. For there to be such additional operations it is essential that there be some special relation between the dimensions of the basis and the dimensions of the Bravais lattice. When the basis does have a size suitably matched to the primitive vectors of the lattice, two new types of operations may arise:

1. **Screw Axes** A crystal structure with a screw axis is brought into coincidence with itself by translation through a vector not in the Bravais lattice, followed by a rotation about the axis defined by the translation.
2. **Glide Planes** A crystal structure with a glide plane is brought into coincidence with itself by translation through a vector not in the Bravais lattice, followed by a reflection in a plane containing that vector.

The hexagonal close-packed structure offers examples of both types of operation, as shown in Figure 7.8. They occur only because the separation of the two basis points along the c-axis is precisely half the distance between lattice planes.

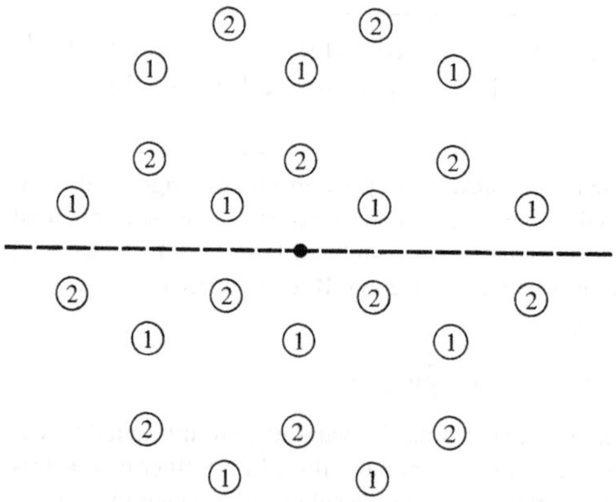

Figure 7.8
The hexagonal close-packed structure viewed along the c-axis. Lattice planes perpendicular to the c-axis are separated by $c/2$ and contain, alternately, points of type 1 and points of type 2. The line parallel to the c-axis passing through the dot in the center of the figure is a screw axis: the structure is invariant under a translation through $c/2$ along the axis followed by a rotation of 60° (but it is not invariant under either the translation or rotation alone.) The plane parallel to the c-axis that intersects the figure in the dashed line is a glide plane: the structure is invariant under a translation through $c/2$ along the c-axis followed by a reflection in the glide plane (but is not invariant under either the translation or reflection alone).

There are both Schoenflies and international systems of space-group nomenclature, which can be found, on the few occasions they may be needed, in the book by Buerger cited in footnote 2.

EXAMPLES AMONG THE ELEMENTS

In Chapter 4 we listed those elements with face-centered cubic, body-centered cubic, hexagonal close-packed, or diamond crystal structures. Over 70 percent of the elements fall into these four categories. The remaining ones are scattered among a variety of crystal structures, most with polyatomic primitive cells that are sometimes quite complex. We conclude this chapter with a few further examples listed in Table 7.5, 7.6, and 7.7. Data are from Wyckoff (cited on page 70) and are for room temperature and normal atmospheric pressure, unless stated otherwise.

Table 7.5
ELEMENTS WITH RHOMBOHEDRAL (TRIGONAL) BRAVAIS LATTICES[a]

ELEMENT	a (Å)	θ	ATOMS IN PRIMITIVE CELL	BASIS
Hg (5 K)	2.99	70°45′	1	$x = 0$
As	4.13	54°10′	2	$x = \pm 0.226$
Sb	4.51	57°6′	2	$x = \pm 0.233$
Bi	4.75	57°14′	2	$x = \pm 0.237$
Sm	9.00	23°13′	3	$x = 0, \pm 0.222$

[a] The common length of the primitive vectors is a, and the angle between any two of them is θ. In all cases the basis points expressed in terms of these primitive vectors have the form $x(\mathbf{a}_1 + \mathbf{a}_2 + \mathbf{a}_3)$. Note (Problem 2(b)) that arsenic, antimony, and bismuth are quite close to a simple cubic lattice, distorted along a body diagonal.

Table 7.6
ELEMENTS WITH TETRAGONAL BRAVAIS LATTICES[a]

ELEMENT	a (Å)	c (Å)	BASIS
In	4.59	4.94	At face-centered positions of the conventional cell
Sn (white)	5.82	3.17	At 000, $0\frac{1}{2}\frac{1}{4}$, $\frac{1}{2}0\frac{3}{4}$, $\frac{1}{2}\frac{1}{2}\frac{1}{2}$, with respect to the axes of the conventional cell

[a] The common length of two perpendicular primitive vectors is a, and the length of the third, perpendicular to these, is c. Both examples have centered tetragonal Bravais lattices, indium with a one-atom and white tin with a two-atom basis. However, both are more commonly described as simple tetragonal with bases. The conventional cell for indium is chosen to stress that it is a slightly distorted (along a cube edge) fcc structure. The white tin structure can be viewed as a diamond structure compressed along one of the cube axes.

Table 7.7
ELEMENTS WITH ORTHORHOMBIC BRAVAIS LATTICES[a]

ELEMENT	a (Å)	b (Å)	c (Å)
Ga	4.511	4.517	7.645
P (black)	3.31	4.38	10.50
Cl (113 K)	6.24	8.26	4.48
Br (123 K)	6.67	8.72	4.48
I	7.27	9.79	4.79
S (rhombic)	10.47	12.87	24.49

[a] The lengths of the three mutually perpendicular primitive vectors are a, b, and c. The structure of rhombic sulfur is complex, with 128 atoms per unit cell. The others can be described in terms of an eight-atom unit cell. For details the reader is referred to Wyckoff.

PROBLEMS

1. (a) Prove that any Bravais lattice has inversion symmetry in a lattice point. (*Hint:* Express the lattice translations as linear combinations of primitive vectors with integral coefficients.)

(b) Prove that the diamond structure is invariant under an inversion in the midpoint of any nearest neighbor bond.

2. (a) If three primitive vectors for a trigonal Bravais lattice are at angles of 90° to one another, the lattice obviously has more than trigonal symmetry, being simple cubic. Show that if the angles are 60° or arc cos $(-\frac{1}{3})$ the lattice again has more than trigonal symmetry, being face-centered cubic or body-centered cubic.

(b) Show that the simple cubic lattice can be represented as a trigonal lattice with primitive vectors $\mathbf{a}_i$ at 60° angles to one another, with a two-point basis $\pm\frac{1}{4}(\mathbf{a}_1 + \mathbf{a}_2 + \mathbf{a}_3)$. (Compare these numbers with the crystal structures in Table 7.5.)

(c) What structure results if the basis in the same trigonal lattice is taken to be $\pm\frac{1}{8}(\mathbf{a}_1 + \mathbf{a}_2 + \mathbf{a}_3)$?

3. If two systems are connected by arrows in the symmetry hierarchy of Figure 7.7, then a Bravais lattice in the more symmetric system can be reduced to one of lower symmetry by an infinitesimal distortion, except for the pair hexagonal-trigonal. The appropriate distortions have been fully described in the text in all cases except hexagonal-orthorhombic and trigonal-monoclinic.

(a) Describe an infinitesimal distortion that reduces a simple hexagonal Bravais lattice to one in the orthorhombic system.

(b) What kind of orthorhombic Bravais lattice can be reached in this way?

(c) Describe an infinitesimal distortion that reduces a trigonal Bravais lattice to one in the monoclinic system.

(d) What kind of monoclinic Bravais lattice can be reached in this way?

4. (a) Which of the trigonal point groups described in Table 7.3 is the point group of the Bravais lattice? That is, which of the representative objects has the symmetry of the object shown in Figure 7.3f?

(b) In Figure 7.9 the faces of the object of Figure 7.3f are decorated in various symmetry-reducing ways to produce objects with the symmetries of the remaining four trigonal point groups. Referring to Table 7.3, indicate the point-group symmetry of each object.

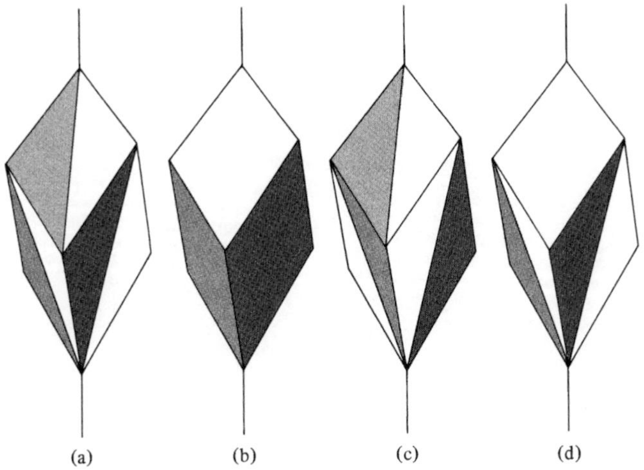

Figure 7.9
Objects with the symmetries of the trigonal groups of lower symmetry. Which is which?

5. Which of the 14 Bravais lattices other than face-centered cubic and body-centered cubic do not have reciprocal lattices of the same kind?

6. (a) Show that there is a family of lattice planes perpendicular to any n-fold rotation axis of a Bravais lattice, $n \geqslant 3$. (The result is also true when $n = 2$, but requires a somewhat more elaborate argument (Problem 7).)

(b) Deduce from (a) that an n-fold axis cannot exist in any three-dimensional Bravais lattice unless it can exist in some two-dimensional Bravais lattice.

(c) Prove that no two-dimensional Bravais lattice can have an n-fold axis with $n = 5$ or $n \geqslant 7$. (*Hint:* First show that the axis can be chosen to pass through a lattice point. Then argue by *reductio ad absurdum*, using the set of points into which the nearest neighbor of the fixed point is taken by the n rotations to find a pair of points closer together than the assumed nearest neighbor distance. (Note that the case $n = 5$ requires slightly different treatment from the others).)

7. (a) Show that if a Bravais lattice has a mirror plane, then there is a family of lattice planes parallel to the mirror plane. (*Hint:* Show from the argument on page 113 that the existence of a mirror plane implies the existence of a mirror plane containing a lattice point. It is then enough to prove that that plane contains two other lattice points not collinear with the first.)

(b) Show that if a Bravais lattice has a 2-fold rotation axis then there is a family of lattice planes perpendicular to the axis.

8

Electron Levels in a Periodic Potential: General Properties

Because the ions in a perfect crystal are arranged in a regular periodic array, we are led to consider the problem of an electron in a potential $U(\mathbf{r})$ with the periodicity of the underlying Bravais lattice; i.e.,

$$U(\mathbf{r} + \mathbf{R}) = U(\mathbf{r}) \tag{8.1}$$

for all Bravais lattice vectors **R**.

Since the scale of periodicity of the potential U ($\sim 10^{-8}$ cm) is the size of a typical de Broglie wavelength of an electron in the Sommerfeld free electron model, it is essential to use quantum mechanics in accounting for the effect of periodicity on electronic motion. In this chapter we shall discuss those properties of the electronic levels that depend only on the periodicity of the potential, without regard to its particular form. The discussion will be continued in Chapters 9 and 10 in two limiting cases of great physical interest that provide more concrete illustrations of the general results of this chapter. In Chapter 11 some of the more important methods for the detailed calculation of electronic levels are summarized. In Chapters 12 and 13 we shall discuss the bearing of these results on the problems of electronic transport theory first raised in Chapters 1 and 2, indicating how many of the anomalies of free electron theory (Chapter 3) are thereby removed. In Chapters 14 and 15 we shall examine the properties of specific metals that illustrate and confirm the general theory.

We emphasize at the outset that perfect periodicity is an idealization. Real solids are never absolutely pure, and in the neighborhood of the impurity atoms the solid is not the same as elsewhere in the crystal. Furthermore, there is always a slight temperature-dependent probability of finding missing or misplaced ions (Chapter 30) that destroy the perfect translational symmetry of even an absolutely pure crystal. Finally, the ions are not in fact stationary, but continually undergo thermal vibrations about their equilibrium positions.

These imperfections are all of great importance. They are, for example, ultimately responsible for the fact that the electrical conductivity of metals is not infinite. Progress is best made, however, by artificially dividing the problem into two parts: (a) the ideal fictitious perfect crystal, in which the potential is genuinely periodic, and (b) the effects on the properties of a hypothetical perfect crystal of all deviations from perfect periodicity, treated as small perturbations.

We also emphasize that the problem of electrons in a periodic potential does not arise only in the context of metals. Most of our general conclusions apply to all crystalline solids, and will play an important role in our subsequent discussions of insulators and semiconductors.

THE PERIODIC POTENTIAL

The problem of electrons in a solid is in principle a many-electron problem, for the full Hamiltonian of the solid contains not only the one-electron potentials describing the interactions of the electrons with the massive atomic nuclei, but also pair potentials describing the electron-electron interactions. In the independent electron approximation these interactions are represented by an effective one-electron potential $U(\mathbf{r})$. The problem of how best to choose this effective potential is a complicated one, which we shall return to in Chapters 11 and 17. Here we merely observe that whatever detailed form the one-electron effective potential may have, if the crystal is perfectly periodic it must satisfy (8.1). From this fact alone many important conclusions can already be drawn.

Qualitatively, however, a typical crystalline potential might be expected to have the form shown in Figure 8.1, resembling the individual atomic potentials as the ion is approached closely and flattening off in the region between ions.

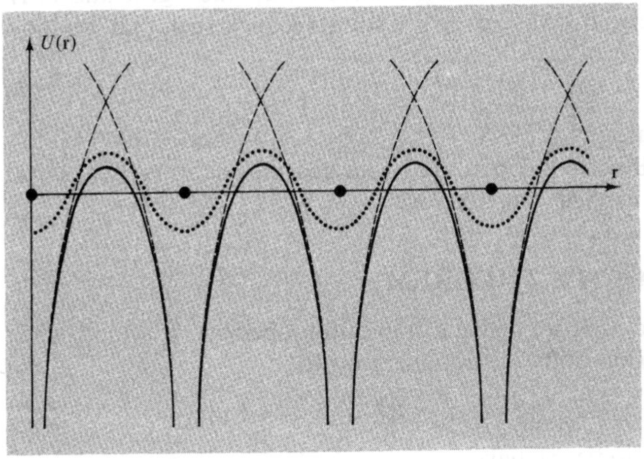

Figure 8.1
A typical crystalline periodic potential, plotted along a line of ions and along a line midway between a plane of ions. (Closed circles are the equilibrium ion sites; the solid curves give the potential along the line of ions; the dotted curves give the potential along a line between planes of ions; the dashed curves give the potential of single isolated ions.)

We are thus led to examine general properties of the Schrödinger equation for a single electron,

$$H\psi = \left(-\frac{\hbar^2}{2m}\nabla^2 + U(\mathbf{r})\right)\psi = \mathcal{E}\psi, \tag{8.2}$$

that follow from the fact that the potential U has the periodicity (8.1). The free electron Schrödinger equation (2.4) is a special case of (8.2) (although, as we shall see, in some respects a very pathological one), zero potential being the simplest example of a periodic one.

Independent electrons, each of which obeys a one electron Schrödinger equation with a periodic potential, are known as *Bloch electrons* (in contrast to "free electrons," to which Bloch electrons reduce when the periodic potential is identically zero). The stationary states of Bloch electrons have the following very important property as a general consequence of the periodicity of the potential U:

BLOCH'S THEOREM

Theorem.[1] The eigenstates ψ of the one-electron Hamiltonian $H = -\hbar^2\nabla^2/2m + U(\mathbf{r})$, where $U(\mathbf{r} + \mathbf{R}) = U(\mathbf{r})$ for all $\mathbf{R}$ in a Bravais lattice, can be chosen to have the form of a plane wave times a function with the periodicity of the Bravais lattice:

$$\boxed{\psi_{n\mathbf{k}}(\mathbf{r}) = e^{i\mathbf{k}\cdot\mathbf{r}}u_{n\mathbf{k}}(\mathbf{r}),} \tag{8.3}$$

[1] The theorem was first proved by Floquet in the one-dimensional case, where it is frequently called *Floquet's theorem.*

where

$$u_{n\mathbf{k}}(\mathbf{r} + \mathbf{R}) = u_{n\mathbf{k}}(\mathbf{r}) \tag{8.4}$$

for all **R** in the Bravais lattice.[2]

Note that Eqs. (8.3) and (8.4) imply that

$$\psi_{n\mathbf{k}}(\mathbf{r} + \mathbf{R}) = e^{i\mathbf{k}\cdot\mathbf{R}}\psi_{n\mathbf{k}}(\mathbf{r}). \tag{8.5}$$

Bloch's theorem is sometimes stated in this alternative form:[3] the eigenstates of H can be chosen so that associated with each ψ is a wave vector **k** such that

$$\boxed{\psi(\mathbf{r} + \mathbf{R}) = e^{i\mathbf{k}\cdot\mathbf{R}}\psi(\mathbf{r}),} \tag{8.6}$$

for every **R** in the Bravais lattice.

We offer two proofs of Bloch's theorem, one from general quantum-mechanical considerations and one by explicit construction.[4]

FIRST PROOF OF BLOCH'S THEOREM

For each Bravais lattice vector **R** we define a translation operator $T_{\mathbf{R}}$ which, when operating on any function $f(\mathbf{r})$, shifts the argument by **R**:

$$T_{\mathbf{R}}f(\mathbf{r}) = f(\mathbf{r} + \mathbf{R}). \tag{8.7}$$

Since the Hamiltonian is periodic, we have

$$T_{\mathbf{R}}H\psi = H(\mathbf{r} + \mathbf{R})\psi(\mathbf{r} + \mathbf{R}) = H(\mathbf{r})\psi(\mathbf{r} + \mathbf{R}) = HT_{\mathbf{R}}\psi. \tag{8.8}$$

Because (8.8) holds identically for any function ψ, we have the operator identity

$$T_{\mathbf{R}}H = HT_{\mathbf{R}}. \tag{8.9}$$

In addition, the result of applying two successive translations does not depend on the order in which they are applied, since for any $\psi(\mathbf{r})$

$$T_{\mathbf{R}}T_{\mathbf{R}'}\psi(\mathbf{r}) = T_{\mathbf{R}'}T_{\mathbf{R}}\psi(\mathbf{r}) = \psi(\mathbf{r} + \mathbf{R} + \mathbf{R}'). \tag{8.10}$$

Therefore

$$T_{\mathbf{R}}T_{\mathbf{R}'} = T_{\mathbf{R}'}T_{\mathbf{R}} = T_{\mathbf{R}+\mathbf{R}'}. \tag{8.11}$$

Equations (8.9) and (8.11) assert that the $T_{\mathbf{R}}$ for all Bravais lattice vectors **R** and the Hamiltonian H form a set of commuting operators. It follows from a fundamental theorem of quantum mechanics[5] that the eigenstates of H can therefore be chosen to be simultaneous eigenstates of all the $T_{\mathbf{R}}$:

$$\begin{aligned} H\psi &= \mathcal{E}\psi, \\ T_{\mathbf{R}}\psi &= c(\mathbf{R})\psi. \end{aligned} \tag{8.12}$$

---

[2] The index n is known as the *band index* and occurs because for a given **k**, as we shall see, there will be many independent eigenstates.

[3] Equation (8.6) implies (8.3) and (8.4), since it requires the function $u(\mathbf{r}) = \exp(-i\mathbf{k}\cdot\mathbf{r})\,\psi(\mathbf{r})$ to have the periodicity of the Bravais lattice.

[4] The first proof relies on some formal results of quantum mechanics. The second is more elementary, but also notationally more cumbersome.

[5] See, for example, D. Park, *Introduction to the Quantum Theory*, McGraw-Hill, New York, 1964, p. 123.

The eigenvalues $c(\mathbf{R})$ of the translation operators are related because of the condition (8.11), for on the one hand

$$T_{\mathbf{R}'}T_{\mathbf{R}}\psi = c(\mathbf{R})T_{\mathbf{R}'}\psi = c(\mathbf{R})c(\mathbf{R}')\psi, \tag{8.13}$$

while, according to (8.11),

$$T_{\mathbf{R}'}T_{\mathbf{R}}\psi = T_{\mathbf{R}+\mathbf{R}'}\psi = c(\mathbf{R} + \mathbf{R}')\psi. \tag{8.14}$$

It follows that the eigenvalues must satisfy

$$c(\mathbf{R} + \mathbf{R}') = c(\mathbf{R})c(\mathbf{R}'). \tag{8.15}$$

Now let $\mathbf{a}_i$ be three primitive vectors for the Bravais lattice. We can always write the $c(\mathbf{a}_i)$ in the form

$$c(\mathbf{a}_i) = e^{2\pi i x_i} \tag{8.16}$$

by a suitable choice[6] of the x_i. It then follows by successive applications of (8.15) that if $\mathbf{R}$ is a general Bravais lattice vector given by

$$\mathbf{R} = n_1\mathbf{a}_1 + n_2\mathbf{a}_2 + n_3\mathbf{a}_3, \tag{8.17}$$

then

$$c(\mathbf{R}) = c(\mathbf{a}_1)^{n_1}c(\mathbf{a}_2)^{n_2}c(\mathbf{a}_3)^{n_3}. \tag{8.18}$$

But this is precisely equivalent to

$$c(\mathbf{R}) = e^{i\mathbf{k}\cdot\mathbf{R}}, \tag{8.19}$$

where

$$\mathbf{k} = x_1\mathbf{b}_1 + x_2\mathbf{b}_2 + x_3\mathbf{b}_3 \tag{8.20}$$

and the $\mathbf{b}_i$ are the reciprocal lattice vectors satisfying Eq. (5.4): $\mathbf{b}_i \cdot \mathbf{a}_j = 2\pi\delta_{ij}$.

Summarizing, we have shown that we can choose the eigenstates ψ of H so that for every Bravais lattice vector $\mathbf{R}$,

$$T_{\mathbf{R}}\psi = \psi(\mathbf{r} + \mathbf{R}) = c(\mathbf{R})\psi = e^{i\mathbf{k}\cdot\mathbf{R}}\psi(\mathbf{r}). \tag{8.21}$$

This is precisely Bloch's theorem, in the form (8.6).

THE BORN–VON KARMAN BOUNDARY CONDITION

By imposing an appropriate boundary condition on the wave functions we can demonstrate that the wave vector $\mathbf{k}$ must be real, and arrive at a condition restricting the allowed values of $\mathbf{k}$. The condition generally chosen is the natural generalization of the condition (2.5) used in the Sommerfeld theory of free electrons in a cubical box. As in that case, we introduce the volume containing the electrons into the theory through a Born–von Karman boundary condition of macroscopic periodicity (page 33). Unless, however, the Bravais lattice is cubic and L is an integral multiple of the lattice constant a, it is not convenient to continue to work in a cubical volume of side L. Instead, it is more convenient to work in a volume commensurate with a

---

[6] We shall see that for suitable boundary conditions the x_i must be real, but for now they can be regarded as general complex numbers.

primitive cell of the underlying Bravais lattice. We therefore generalize the periodic boundary condition (2.5) to

$$\psi(\mathbf{r} + N_i\mathbf{a}_i) = \psi(\mathbf{r}), \qquad i = 1, 2, 3, \tag{8.22}$$

where the $\mathbf{a}_i$ are three primitive vectors and the N_i are all integers of order $N^{1/3}$, where $N = N_1N_2N_3$ is the total number of primitive cells in the crystal.

As in Chapter 2, we adopt this boundary condition under the assumption that the bulk properties of the solid will not depend on the choice of boundary condition, which can therefore be dictated by analytical convenience.

Applying Bloch's theorem (8.6) to the boundary condition (8.22) we find that

$$\psi_{n\mathbf{k}}(\mathbf{r} + N_i\mathbf{a}_i) = e^{iN_i\mathbf{k}\cdot\mathbf{a}_i}\psi_{n\mathbf{k}}(\mathbf{r}), \qquad i = 1, 2, 3, \tag{8.23}$$

which requires that

$$e^{iN_i\mathbf{k}\cdot\mathbf{a}_i} = 1, \qquad i = 1, 2, 3. \tag{8.24}$$

When $\mathbf{k}$ has the form (8.20), Eq. (8.24) requires that

$$e^{2\pi i N_i x_i} = 1, \tag{8.25}$$

and consequently we must have

$$x_i = \frac{m_i}{N_i}, \qquad m_i \text{ integral}. \tag{8.26}$$

Therefore the general form for allowed Bloch wave vectors is[7]

$$\mathbf{k} = \sum_{i=1}^{3} \frac{m_i}{N_i}\mathbf{b}_i, \qquad m_i \text{ integral}. \tag{8.27}$$

It follows from (8.27) that the volume $\Delta\mathbf{k}$ of k-space per allowed value of $\mathbf{k}$ is just the volume of the little parallelepiped with edges $\mathbf{b}_i/N_i$:

$$\Delta\mathbf{k} = \frac{\mathbf{b}_1}{N_1}\cdot\left(\frac{\mathbf{b}_2}{N_2}\times\frac{\mathbf{b}_3}{N_3}\right) = \frac{1}{N}\mathbf{b}_1\cdot(\mathbf{b}_2\times\mathbf{b}_3). \tag{8.28}$$

Since $\mathbf{b}_1\cdot(\mathbf{b}_2\times\mathbf{b}_3)$ is the volume of a reciprocal lattice primitive cell, Eq. (8.28) asserts that *the number of allowed wave vectors in a primitive cell of the reciprocal lattice is equal to the number of sites in the crystal.*

The volume of a reciprocal lattice primitive cell is $(2\pi)^3/v$, where $v = V/N$ is the volume of a direct lattice primitive cell, so Eq. (8.28) can be written in the alternative form:

$$\boxed{\Delta\mathbf{k} = \frac{(2\pi)^3}{V}.} \tag{8.29}$$

This is precisely the result (2.18) we found in the free electron case.

---

[7] Note that (8.27) reduces to the form (2.16) used in free electron theory when the Bravais lattice is simple cubic, the $\mathbf{a}_i$ are the cubic primitive vectors, and $N_1 = N_2 = N_3 = L/a$.

SECOND PROOF OF BLOCH'S THEOREM[8]

This second proof of Bloch's theorem illuminates its significance from a rather different point of view, which we shall exploit further in Chapter 9. We start with the observation that one can always expand any function obeying the Born–von Karman boundary condition (8.22) in the set of all plane waves that satisfy the boundary condition and therefore have wave vectors of the form (8.27):[9]

$$\psi(\mathbf{r}) = \sum_{\mathbf{q}} c_{\mathbf{q}} e^{i\mathbf{q}\cdot\mathbf{r}}. \tag{8.30}$$

Because the potential $U(\mathbf{r})$ is periodic in the lattice, its plane wave expansion will only contain plane waves with the periodicity of the lattice and therefore with wave vectors that are vectors of the reciprocal lattice:[10]

$$U(\mathbf{r}) = \sum_{\mathbf{K}} U_{\mathbf{K}} e^{i\mathbf{K}\cdot\mathbf{r}}. \tag{8.31}$$

The Fourier coefficients $U_{\mathbf{K}}$ are related to $U(\mathbf{r})$ by[11]

$$U_{\mathbf{K}} = \frac{1}{v}\int_{\text{cell}} d\mathbf{r}\, e^{-i\mathbf{K}\cdot\mathbf{r}} U(\mathbf{r}). \tag{8.32}$$

Since we are at liberty to change the potential energy by an additive constant, we fix this constant by requiring that the spatial average U_0 of the potential over a primitive cell vanish:

$$U_0 = \frac{1}{v}\int_{\text{cell}} d\mathbf{r}\, U(\mathbf{r}) = 0. \tag{8.33}$$

Note that because the potential $U(\mathbf{r})$ is real, it follows from (8.32) that the Fourier coefficients satisfy

$$U_{-\mathbf{K}} = U_{\mathbf{K}}^{*}. \tag{8.34}$$

If we assume that the crystal has inversion symmetry[12] so that, for a suitable choice of origin, $U(\mathbf{r}) = U(-\mathbf{r})$, then (8.32) implies that $U_{\mathbf{K}}$ is real, and thus

$$U_{-\mathbf{K}} = U_{\mathbf{K}} = U_{\mathbf{K}}^{*} \quad \text{(for crystals with inversion symmetry)}. \tag{8.35}$$

We now place the expansions (8.30) and (8.31) into the Schrödinger equation (8.2). The kinetic energy term gives

$$\frac{p^2}{2m}\psi = -\frac{\hbar^2}{2m}\nabla^2\psi = \sum_{\mathbf{q}} \frac{\hbar^2}{2m} q^2 c_{\mathbf{q}} e^{i\mathbf{q}\cdot\mathbf{r}}. \tag{8.36}$$

---

8 Although more elementary than the first proof, the second is also notationally more complicated, and of importance primarily as a starting point for the approximate calculations of Chapter 9. The reader may therefore wish to skip it at this point.

9 We shall subsequently understand unspecified summations over **k** to be over all wave vectors of the form (8.27) allowed by the Born–von Karman boundary condition.

10 A sum indexed by **K** shall always be understood to run over all reciprocal lattice vectors.

11 See Appendix D, where the relevance of the reciprocal lattice to Fourier expansions of periodic functions is discussed.

12 The reader is invited to pursue the argument of this section (and Chapter 9) without the assumption of inversion symmetry, which is made solely to avoid inessential complications in the notation.

The term in the potential energy can be written[13]

$$U\psi = \left(\sum_{\mathbf{K}} U_{\mathbf{K}} e^{i\mathbf{K}\cdot\mathbf{r}}\right)\left(\sum_{\mathbf{q}} c_{\mathbf{q}} e^{i\mathbf{q}\cdot\mathbf{r}}\right)$$

$$= \sum_{\mathbf{K}\mathbf{q}} U_{\mathbf{K}} c_{\mathbf{q}} e^{i(\mathbf{K}+\mathbf{q})\cdot\mathbf{r}} = \sum_{\mathbf{K}\mathbf{q}'} U_{\mathbf{K}} c_{\mathbf{q}'-\mathbf{K}} e^{i\mathbf{q}'\cdot\mathbf{r}}. \tag{8.37}$$

We change the names of the summation indices in (8.37)—from **K** and **q**′, to **K**′ and **q**—so that the Schrödinger equation becomes

$$\sum_{\mathbf{q}} e^{i\mathbf{q}\cdot\mathbf{r}} \left\{ \left(\frac{\hbar^2}{2m} q^2 - \mathcal{E} \right) c_{\mathbf{q}} + \sum_{\mathbf{K}'} U_{\mathbf{K}'} c_{\mathbf{q}-\mathbf{K}'} \right\} = 0. \tag{8.38}$$

Since the plane waves satisfying the Born–von Karman boundary condition are an orthogonal set, the coefficient of each separate term in (8.38) must vanish,[14] and therefore for all allowed wave vectors **q**,

$$\boxed{\left(\frac{\hbar^2}{2m} q^2 - \mathcal{E} \right) c_{\mathbf{q}} + \sum_{\mathbf{K}'} U_{\mathbf{K}'} c_{\mathbf{q}-\mathbf{K}'} = 0.} \tag{8.39}$$

It is convenient to write **q** in the form **q** = **k** − **K**, where **K** is a reciprocal lattice vector chosen so that **k** lies in the first Brillouin zone. Equation (8.39) becomes

$$\left(\frac{\hbar^2}{2m} (\mathbf{k}-\mathbf{K})^2 - \mathcal{E} \right) c_{\mathbf{k}-\mathbf{K}} + \sum_{\mathbf{K}'} U_{\mathbf{K}'} c_{\mathbf{k}-\mathbf{K}-\mathbf{K}'} = 0, \tag{8.40}$$

or, if we make the change of variables $\mathbf{K}' \to \mathbf{K}' - \mathbf{K}$,

$$\boxed{\left(\frac{\hbar^2}{2m} (\mathbf{k}-\mathbf{K})^2 - \mathcal{E} \right) c_{\mathbf{k}-\mathbf{K}} + \sum_{\mathbf{K}'} U_{\mathbf{K}'-\mathbf{K}} c_{\mathbf{k}-\mathbf{K}'} = 0.} \tag{8.41}$$

We emphasize that Eqs. (8.39) and (8.41) are nothing but restatements of the original Schrödinger equation (8.2) in momentum space, simplified by the fact that because of the periodicity of the potential, $U_{\mathbf{k}}$ is nonvanishing only when **k** is a vector of the reciprocal lattice.

For fixed **k** in the first Brillouin zone, the set of equations (8.41) for all reciprocal lattice vectors **K** couples only those coefficients $c_{\mathbf{k}}, c_{\mathbf{k}-\mathbf{K}}, c_{\mathbf{k}-\mathbf{K}'}, c_{\mathbf{k}-\mathbf{K}''}, \ldots$ whose wave vectors differ from **k** by a reciprocal lattice vector. Thus the original problem has separated into N independent problems: one for each allowed value of **k** in the first Brillouin zone. Each such problem has solutions that are superpositions of plane waves containing only the wave vector **k** and wave vectors differing from **k** by a reciprocal lattice vector.

---

[13] The last step follows from making the substitution **K** + **q** = **q**′, and noting that because **K** is a reciprocal lattice vector, summing over all **q** of the form (8.27) is the same as summing over all **q**′ of that form.

[14] This can also be deduced from Eq. (D.12), Appendix D, by multiplying (8.38) by the appropriate plane wave and integrating over the volume of the crystal.

Putting this information back into the expansion (8.30) of the wave function ψ, we see that if the wave vector $\mathbf{q}$ only assumes the values $\mathbf{k}$, $\mathbf{k} - \mathbf{K}'$, $\mathbf{k} - \mathbf{K}''$, . . . , then the wave function will be of the form

$$\psi_{\mathbf{k}} = \sum_{\mathbf{K}} c_{\mathbf{k}-\mathbf{K}} e^{i(\mathbf{k}-\mathbf{K})\cdot\mathbf{r}}. \tag{8.42}$$

If we write this as

$$\psi_{\mathbf{k}}(\mathbf{r}) = e^{i\mathbf{k}\cdot\mathbf{r}} \Big(\sum_{\mathbf{K}} c_{\mathbf{k}-\mathbf{K}} e^{-i\mathbf{K}\cdot\mathbf{r}}\Big), \tag{8.43}$$

then this is of the Bloch form (8.3) with the periodic function $u(\mathbf{r})$ given by[15]

$$u(\mathbf{r}) = \sum_{\mathbf{K}} c_{\mathbf{k}-\mathbf{K}} e^{-i\mathbf{K}\cdot\mathbf{r}}. \tag{8.44}$$

GENERAL REMARKS ABOUT BLOCH'S THEOREM

1. Bloch's theorem introduces a wave vector $\mathbf{k}$, which turns out to play the same fundamental role in the general problem of motion in a periodic potential that the free electron wave vector $\mathbf{k}$ plays in the Sommerfeld theory. Note, however, that although the free electron wave vector is simply $\mathbf{p}/\hbar$, where $\mathbf{p}$ is the momentum of the electron, in the Bloch case $\mathbf{k}$ is not proportional to the electronic momentum. This is clear on general grounds, since the Hamiltonian does not have complete translational invariance in the presence of a nonconstant potential, and therefore its eigenstates will not be simultaneous eigenstates of the momentum operator. This conclusion is confirmed by the fact that the momentum operator, $\mathbf{p} = (\hbar/i)\,\nabla$, when acting on $\psi_{n\mathbf{k}}$ gives

$$\begin{aligned} \frac{\hbar}{i}\nabla\,\psi_{n\mathbf{k}} &= \frac{\hbar}{i}\nabla\,(e^{i\mathbf{k}\cdot\mathbf{r}}u_{n\mathbf{k}}(\mathbf{r})) \\ &= \hbar\mathbf{k}\psi_{n\mathbf{k}} + e^{i\mathbf{k}\cdot\mathbf{r}}\frac{\hbar}{i}\nabla\,u_{n\mathbf{k}}(\mathbf{r}), \end{aligned} \tag{8.45}$$

which is not, in general, just a constant times $\psi_{n\mathbf{k}}$; i.e., $\psi_{n\mathbf{k}}$ is not a momentum eigenstate.

Nevertheless, in many ways $\hbar\mathbf{k}$ is a natural extension of $\mathbf{p}$ to the case of a periodic potential. It is known as the *crystal momentum* of the electron, to emphasize this similarity, but one should not be misled by the name into thinking that $\hbar\mathbf{k}$ is a momentum, for it is not. An intuitive understanding of the dynamical significance of the wave vector $\mathbf{k}$ can only be acquired when one considers the response of Bloch electrons to externally applied electromagnetic fields (Chapter 12). Only then does its full resemblance to $\mathbf{p}/\hbar$ emerge. For the present, the reader should view $\mathbf{k}$ as a quantum number characteristic of the translational symmetry of a periodic potential, just as the momentum $\mathbf{p}$ is a quantum number characteristic of the fuller translational symmetry of free space.

2. The wave vector $\mathbf{k}$ appearing in Bloch's theorem can always be confined to the first Brillouin zone (or to any other convenient primitive cell of the reciprocal

---

15 Note that there will be (infinitely) many solutions to the (infinite) set of equations (8.41) for a given $\mathbf{k}$. These are classified by the band index n (see footnote 2).

lattice). This is because any $\mathbf{k}'$ not in the first Brillouin zone can be written as

$$\mathbf{k}' = \mathbf{k} + \mathbf{K} \tag{8.46}$$

where $\mathbf{K}$ is a reciprocal lattice vector and $\mathbf{k}$ does lie in the first zone. Since $e^{i\mathbf{K}\cdot\mathbf{R}} = 1$ for any reciprocal lattice vector, if the Bloch form (8.6) holds for $\mathbf{k}'$, it will also hold for $\mathbf{k}$.

3. The index n appears in Bloch's theorem because for given $\mathbf{k}$ there are many solutions to the Schrödinger equation. We noted this in the second proof of Bloch's theorem, but it can also be seen from the following argument:

Let us look for all solutions to the Schrödinger equation (8.2) that have the Bloch form

$$\psi(\mathbf{r}) = e^{i\mathbf{k}\cdot\mathbf{r}}u(\mathbf{r}), \tag{8.47}$$

where $\mathbf{k}$ is fixed and u has the periodicity of the Bravais lattice. Substituting this into the Schrödinger equation, we find that u is determined by the eigenvalue problem

$$\begin{aligned} H_{\mathbf{k}}u_{\mathbf{k}}(\mathbf{r}) &= \left(\frac{\hbar^2}{2m}\left(\frac{1}{i}\nabla + \mathbf{k}\right)^2 + U(\mathbf{r})\right)u_{\mathbf{k}}(\mathbf{r}) \\ &= \mathcal{E}_{\mathbf{k}}u_{\mathbf{k}}(\mathbf{r}) \end{aligned} \tag{8.48}$$

with boundary condition

$$u_{\mathbf{k}}(\mathbf{r}) = u_{\mathbf{k}}(\mathbf{r} + \mathbf{R}). \tag{8.49}$$

Because of the periodic boundary condition we can regard (8.48) as a Hermitian eigenvalue problem restricted to a single primitive cell of the crystal. Because the eigenvalue problem is set in a fixed finite volume, we expect on general grounds to find an infinite family of solutions with *discretely* spaced eigenvalues,[16] which we label with the band index n.

Note that in terms of the eigenvalue problem specified by (8.48) and (8.49), the wave vector $\mathbf{k}$ appears only as a parameter in the Hamiltonian $H_{\mathbf{k}}$. We therefore expect each of the energy levels, for given $\mathbf{k}$, to vary continuously as $\mathbf{k}$ varies.[17] In this way we arrive at a description of the levels of an electron in a periodic potential in terms of a family of continuous[18] functions $\mathcal{E}_n(\mathbf{k})$.

4. Although the full set of levels can be described with $\mathbf{k}$ restricted to a single primitive cell, it is often useful to allow $\mathbf{k}$ to range through all of k-space, even though this gives a highly redundant description. Because the set of all wave functions and energy levels for two values of $\mathbf{k}$ differing by a reciprocal lattice vector must be

---

[16] Just as the problem of a free electron in a box of fixed finite dimensions has a set of discrete energy levels, the vibrational normal modes of a finite drumhead have a set of discrete frequencies, etc.

[17] This expectation is implicit, for example, in ordinary perturbation theory, which is possible only because small changes in parameters in the Hamiltonian lead to small changes in the energy levels. In Appendix E the changes in the energy levels for small changes in $\mathbf{k}$ are calculated explicitly.

[18] The fact that the Born–von Karman boundary condition restricts $\mathbf{k}$ to discrete values of the form (8.27) has no bearing on the continuity of $\mathcal{E}_n(\mathbf{k})$ as a function of a continuous variable $\mathbf{k}$, for the eigenvalue problem given by (8.48) and (8.49) makes no reference to the size of the whole crystal and is well defined for any $\mathbf{k}$. One should also note that the set of $\mathbf{k}$ of the form (8.27) becomes dense in k-space in the limit of an infinite crystal.

identical, we can assign the indices n to the levels in such a way that *for given n, the eigenstates and eigenvalues are periodic functions of* **k** *in the reciprocal lattice:*

$$\boxed{\begin{aligned}\psi_{n,\mathbf{k}+\mathbf{K}}(\mathbf{r}) &= \psi_{n\mathbf{k}}(\mathbf{r}),\\ \mathcal{E}_{n,\mathbf{k}+\mathbf{K}} &= \mathcal{E}_{n\mathbf{k}}.\end{aligned}} \tag{8.50}$$

This leads to a description of the energy levels of an electron in a periodic potential in terms of a family of continuous functions $\mathcal{E}_{n\mathbf{k}}$ (or $\mathcal{E}_n(\mathbf{k})$), each with the periodicity of the reciprocal lattice. The information contained in these functions is referred to as the *band structure* of the solid.

For each n, the set of electronic levels specified by $\mathcal{E}_n(\mathbf{k})$ is called an *energy band*. The origin of the term "band" will emerge in Chapter 10. Here we only note that because each $\mathcal{E}_n(\mathbf{k})$ is periodic in **k** and continuous, it has an upper and lower bound, so that all the levels $\mathcal{E}_n(\mathbf{k})$ lie in the band of energies lying between these limits.

5. It can be shown quite generally (Appendix E) that an electron in a level specified by band index n and wave vector **k** has a nonvanishing mean velocity, given by

$$\boxed{\mathbf{v}_n(\mathbf{k}) = \frac{1}{\hbar}\nabla_{\mathbf{k}}\,\mathcal{E}_n(\mathbf{k}).} \tag{8.51}$$

This is a most remarkable fact. It asserts that there are stationary (i.e., time-independent) levels for an electron in a periodic potential in which, in spite of the interaction of the electron with the fixed lattice of ions, it moves forever without any degradation of its mean velocity. This is in striking contrast to the idea of Drude that collisions were simply encounters between the electron and a static ion. Its implications are of fundamental importance, and will be explored in Chapters 12 and 13.

THE FERMI SURFACE

The ground state of N free electrons[19] is constructed by occupying all one-electron levels **k** with energies $\mathcal{E}(\mathbf{k}) = \hbar^2k^2/2m$ less than $\mathcal{E}_F$, where $\mathcal{E}_F$ is determined by requiring the total number of one-electron levels with energies less than $\mathcal{E}_F$ to be equal to the total number of electrons (Chapter 2).

The ground state of N Bloch electrons is similarly constructed, except that the one-electron levels are now labeled by the quantum numbers n and **k**, $\mathcal{E}_n(\mathbf{k})$ does not have the simple explicit free electron form, and **k** must be confined to a single primitive cell of the reciprocal lattice if each level is to be counted only once. When the lowest of these levels are filled by a specified number of electrons, two quite distinct types of configuration can result:

---

[19] We shall not distinguish notationally between the number of conduction electrons and the number of primitive cells when it is clear from the context which is meant; they are equal, however, only in a monovalent monatomic Bravais lattice(e.g., the alkali metals).

1. A certain number of bands may be completely filled, all others remaining empty. The difference in energy between the highest occupied level and the lowest unoccupied level (i.e., between the "top" of the highest occupied band and the "bottom" of the lowest empty band) is known as the *band gap*. We shall find that solids with a band gap greatly in excess of k_BT (T near room temperature) are insulators (Chapter 12). If the band gap is comparable to k_BT, the solid is known as an *intrinsic semiconductor* (Chapter 28). Because the number of levels in a band is equal to the number of primitive cells in the crystal (page 136) and because each level can accommodate two electrons (one of each spin), *a configuration with a band gap can arise* (*though it need not*) *only if the number of electrons per primitive cell is even.*
2. A number of bands may be partially filled. When this occurs, the energy of the highest occupied level, the Fermi energy $\mathcal{E}_F$, lies within the energy range of one or more bands. For each partially filled band there will be a surface in k-space separating the occupied from the unoccupied levels. The set of all such surfaces is known as the *Fermi surface*, and is the generalization to Bloch electrons of the free electron Fermi sphere. The parts of the Fermi surface arising from individual partially filled bands are known as *branches* of the Fermi surface.[20] We shall see (Chapter 12) that a solid has metallic properties provided that a Fermi surface exists.

Analytically, the branch of the Fermi surface in the nth band is that surface in k-space (if there is one) determined by[21]

$$\mathcal{E}_n(\mathbf{k}) = \mathcal{E}_F. \tag{8.52}$$

Thus the Fermi surface is a constant energy surface (or set of constant energy surfaces) in k-space, just as the more familiar equipotentials of electrostatic theory are constant energy surfaces in real space.

Since the $\mathcal{E}_n(\mathbf{k})$ are periodic in the reciprocal lattice, the complete solution to (8.52) for each n is a k-space surface with the periodicity of the reciprocal lattice. When a branch of the Fermi surface is represented by the full periodic structure, it is said to be described in a *repeated zone scheme*. Often, however, it is preferable to take just enough of each branch of the Fermi surface so that every physically distinct level is represented by just one point of the surface. This is achieved by representing each branch by that portion of the full periodic surface contained within a single primitive cell of the reciprocal lattice. Such a representation is described as a *reduced zone*

---

[20] In many important cases the Fermi surface is entirely within a single band, and generally it is found to lie within a fairly small number of bands (Chapter 15).

[21] If $\mathcal{E}_F$ is generally defined as the energy separating the highest occupied from the lowest unoccupied level, then it is not uniquely specified in a solid with an energy gap, since any energy in the gap meets this test. People nevertheless speak of "*the* Fermi energy" of an intrinsic semiconductor. What they mean is the chemical potential, which is well defined at any nonzero temperature (Appendix B). As $T \to 0$, the chemical potential of a solid with an energy gap approaches the energy at the middle of the gap (page 575), and one sometimes finds it asserted that this is the "Fermi energy" of a solid with a gap. With either the correct (undetermined) or colloquial definition of $\mathcal{E}_F$, Eq. (8.52) asserts that solids with a gap have no Fermi surface.

scheme. The primitive cell chosen is often, but not always, the first Brillouin zone.

Fermi surface geometry and its physical implications will be illustrated in many of the following chapters, particularly Chapters 9 and 15.

DENSITY OF LEVELS[22]

One must often calculate quantities that are weighted sums over the electronic levels of various one-electron properties. Such quantities are of the form[23]

$$Q = 2 \sum_{n,\mathbf{k}} Q_n(\mathbf{k}), \tag{8.53}$$

where for each n the sum is over all allowed $\mathbf{k}$ giving physically distinct levels, i.e., all $\mathbf{k}$ of the form (8.27) lying in a single primitive cell.[24]

In the limit of a large crystal the allowed values (8.27) of $\mathbf{k}$ get very close together, and the sum may be replaced with an integral. Since the volume of k-space per allowed $\mathbf{k}$ (Eq. (8.29)) has the same value as in the free electron case, the prescription derived in the free electron case (Eq. (2.29)) remains valid, and we find that[25]

$$q = \lim_{V \to \infty} \frac{Q}{V} = 2 \sum_n \int \frac{d\mathbf{k}}{(2\pi)^3} Q_n(\mathbf{k}), \tag{8.54}$$

where the integral is over a primitive cell.

If, as is often the case,[26] $Q_n(\mathbf{k})$ depends on n and $\mathbf{k}$ only through the energy $\mathcal{E}_n(\mathbf{k})$, then in further analogy to the free electron case one can define a density of levels per unit volume (or "density of levels" for short) $g(\mathcal{E})$ so that q has the form (cf. (2.60)):

$$q = \int d\mathcal{E}\, g(\mathcal{E}) Q(\mathcal{E}). \tag{8.55}$$

Comparing (8.54) and (8.55) we find that

$$g(\mathcal{E}) = \sum_n g_n(\mathcal{E}), \tag{8.56}$$

where $g_n(\mathcal{E})$, the density of levels in the nth band, is given by

$$g_n(\mathcal{E}) = \int \frac{d\mathbf{k}}{4\pi^3} \delta(\mathcal{E} - \mathcal{E}_n(\mathbf{k})), \tag{8.57}$$

where the integral is over any primitive cell.

---

22 The reader can, without loss of continuity, skip this section at a first reading, referring back to it in subsequent chapters when necessary.

23 The factor 2 is because each level specified by n and $\mathbf{k}$ can accommodate two electrons of opposite spin. We assume that $Q_n(\mathbf{k})$ does not depend on the electron spin s. If it does, the factor 2 must be replaced by a sum on s.

24 The functions $Q_n(\mathbf{k})$ usually have the periodicity of the reciprocal lattice, so the choice of primitive cell is immaterial.

25 See page 37 for the appropriate cautionary remarks.

26 For example, if q is the electronic number density n, then $Q(\mathcal{E}) = f(\mathcal{E})$, where f is the Fermi function; if q is the electronic energy density u, then $Q(\mathcal{E}) = \mathcal{E} f(\mathcal{E})$.

An alternative representation of the density of levels can be constructed by noting that, as in the free electron case (Eq. (2.62)):

$$g_n(\mathcal{E})\,d\mathcal{E} = (2/V) \times \begin{array}{l}\text{(the number of allowed wave vectors} \\ \text{in the } n\text{th band in the energy range} \\ \text{from } \mathcal{E} \text{ to } \mathcal{E} + d\mathcal{E}).\end{array} \tag{8.58}$$

The number of allowed wave vectors in the nth band in this energy range is just the volume of a k-space primitive cell, with $\mathcal{E} \leqslant \mathcal{E}_n(\mathbf{k}) \leqslant \mathcal{E} + d\mathcal{E}$, divided by the volume per allowed wave vector, $\Delta\mathbf{k} = (2\pi)^3/V$. Thus

$$g_n(\mathcal{E})\,d\mathcal{E} = \int \frac{d\mathbf{k}}{4\pi^3} \times \begin{Bmatrix} 1, & \mathcal{E} \leqslant \mathcal{E}_n(\mathbf{k}) \leqslant \mathcal{E} + d\mathcal{E}, \\ 0, & \text{otherwise} \end{Bmatrix}. \tag{8.59}$$

Since $d\mathcal{E}$ is infinitesimal, this can also be expressed as a surface integral. Let $S_n(\mathcal{E})$ be the portion of the surface $\mathcal{E}_n(\mathbf{k}) = \mathcal{E}$ lying within the primitive cell, and let $\delta k(\mathbf{k})$ be the perpendicular distance between the surfaces $S_n(\mathcal{E})$ and $S_n(\mathcal{E} + d\mathcal{E})$ at the point $\mathbf{k}$. Then (Figure 8.2):

$$g_n(\mathcal{E})\,d\mathcal{E} = \int_{S_n(\mathcal{E})} \frac{dS}{4\pi^3}\,\delta k(\mathbf{k}). \tag{8.60}$$

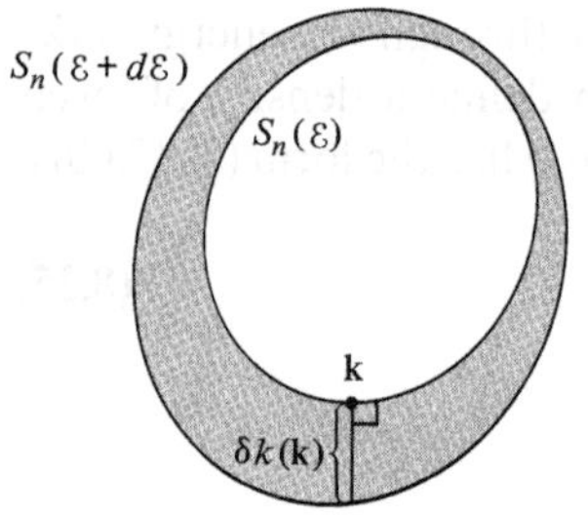

Figure 8.2
An illustration in two dimensions of the construction expressed in Eq. (8.60). The closed curves are the two constant-energy surfaces, the required area is that lying between them (shaded), and the distance $\delta k(\mathbf{k})$ is indicated for a particular $\mathbf{k}$.

To find an explicit expression for $\delta k(\mathbf{k})$ note that, since $S_n(\mathcal{E})$ is a surface of constant energy, the k-gradient of $\mathcal{E}_n(\mathbf{k})$, $\nabla\mathcal{E}_n(\mathbf{k})$, is a vector normal to that surface whose magnitude is equal to the rate of change of $\mathcal{E}_n(\mathbf{k})$ in the normal direction; i.e.,

$$\mathcal{E} + d\mathcal{E} = \mathcal{E} + |\nabla\mathcal{E}_n(\mathbf{k})|\,\delta k(\mathbf{k}), \tag{8.61}$$

and hence

$$\delta k(\mathbf{k}) = \frac{d\mathcal{E}}{|\nabla\mathcal{E}_n(\mathbf{k})|}. \tag{8.62}$$

Substituting (8.62) into (8.60), we arrive at the form

$$\boxed{g_n(\mathcal{E}) = \int_{S_n(\mathcal{E})} \frac{dS}{4\pi^3} \frac{1}{|\nabla\mathcal{E}_n(\mathbf{k})|}} \tag{8.63}$$

which gives an explicit relation between the density of levels and the band structure.

Equation (8.63) and the analysis leading to it will be applied in subsequent chapters.[27] Here we only note the following quite general property of the density of levels:

Because $\mathcal{E}_n(\mathbf{k})$ is periodic in the reciprocal lattice, bounded above and below for each n, and, in general, everywhere differentiable, there must be values of $\mathbf{k}$ in each primitive cell at which $|\nabla\mathcal{E}| = 0$. For example, the gradient of a differentiable function vanishes at local maxima and minima, but the boundedness and periodicity of each $\mathcal{E}_n(\mathbf{k})$ insure that for each n there will be at least one maximum and minimum in each primitive cell.[28]

When the gradient of $\mathcal{E}_n$ vanishes, the integrand in the density of levels (8.63) diverges. It can be shown that in three dimensions[29] such singularities are integrable, yielding finite values for g_n. However, they do result in divergences of the slope, $dg_n/d\mathcal{E}$. These are known as *van Hove singularities.*[30] They occur at values of $\mathcal{E}$ for which the constant energy surface $S_n(\mathcal{E})$ contains points at which $\nabla\mathcal{E}_n(\mathbf{k})$ vanishes. Since derivatives of the density of levels at the Fermi energy enter into all terms but the first in the Sommerfeld expansion,[31] one must be on guard for anomalies in low-temperature behavior if there are points of vanishing $\nabla\mathcal{E}_n(\mathbf{k})$ on the Fermi surface.

Typical van Hove singularities are shown in Figure 8.3 and are examined in Problem 2, Chapter 9.

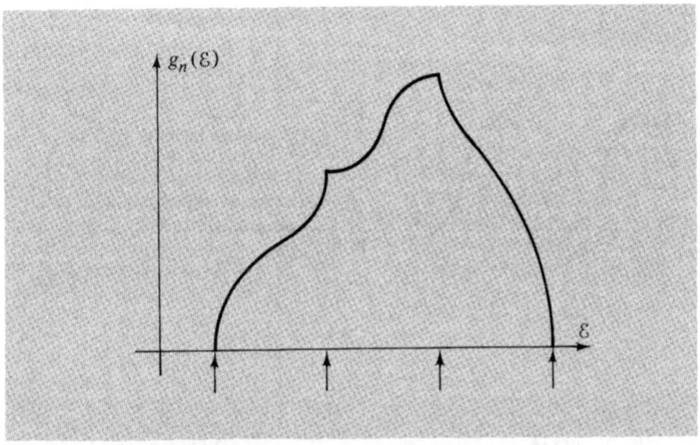

Figure 8.3
Characteristic van Hove singularities in the density of levels, indicated by arrows at right angles to the $\mathcal{E}$-axis.

This concludes our discussion of the general features of one-electron levels in a periodic potential.[32] In the following two chapters we consider two very important, but quite different, limiting cases, which provide concrete illustrations of the rather abstract discussions in this chapter.

---

27 See also Problem 2.

28 A very general analysis of how many points of vanishing gradient must occur is fairly complex. See, for example, G. Weinreich, *Solids*, Wiley, New York, 1965, pp. 73–79.

29 In one dimension $g_n(\mathcal{E})$ itself will be infinite at a van Hove singularity.

30 Essentially the same singularities occur in the theory of lattice vibrations. See Chapter 23.

31 See, for example, Problem 2f, Chapter 2.

32 Problem 1 pursues the general analysis somewhat further in the tractable but somewhat misleading case of a one-dimensional periodic potential.

PROBLEMS

1. *Periodic Potentials in One Dimension*

The general analysis of electronic levels in a periodic potential, independent of the detailed features of that potential, can be carried considerably further in one dimension. Although the one-dimensional case is in many respects atypical (there is no need for a concept of a Fermi surface) or misleading (the possibility—indeed, in two and three dimensions the likelihood—of band overlap disappears), it is nevertheless reassuring to see some of the features of three-dimensional band structure we shall describe through approximate calculations, in Chapters 9, 10, and 11, emerging from an exact treatment in one dimension.

Consider, then, a one-dimensional periodic potential $U(x)$ (Figure 8.4). It is convenient to view the ions as residing at the minima of U, which we take to define the zero of energy. We choose to view the periodic potential as a superposition of potential barriers $v(x)$ of width a, centered at the points $x = \pm na$ (Figure 8.5):

$$U(x) = \sum_{n=-\infty}^{\infty} v(x - na). \tag{8.64}$$

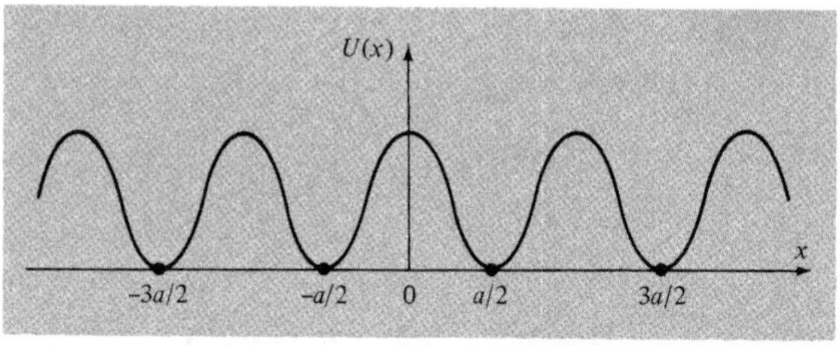

Figure 8.4
A one-dimensional periodic potential $U(x)$. Note that the ions occupy the positions of a Bravais lattice of lattice constant a. It is convenient to take these points as having coordinates $(n + \frac{1}{2})a$, and to choose the zero of potential to occur at the position of the ion.

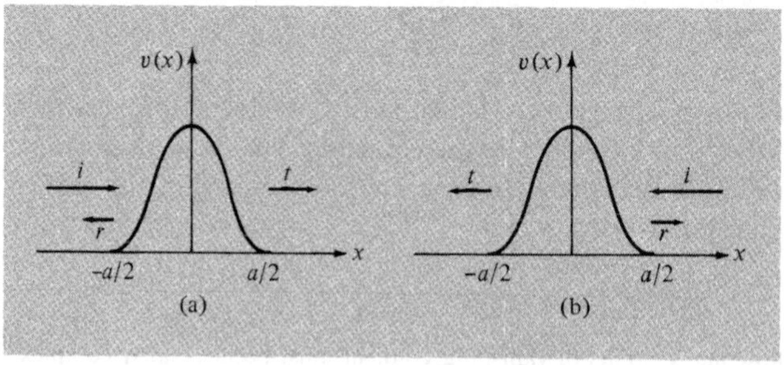

Figure 8.5
Illustrating particles incident from the left (a) and right (b) on a single one of the barriers separating neighboring ions in the periodic potential of Figure 8.4. The incident, transmitted, and reflected waves are indicated by arrows along the direction of propagation, proportional to the corresponding amplitudes.

The term $v(x - na)$ represents the potential barrier against an electron tunneling between the ions on opposite sides of the point na. For simplicity we assume that $v(x) = v(-x)$ (the one-dimensional analogue of the inversion symmetry we assumed above), but we make no other assumptions about v, so the form of the periodic potential U is quite general.

The band structure of the one-dimensional solid can be expressed quite simply in terms of the properties of an electron in the presence of a single-barrier potential $v(x)$. Consider therefore an electron incident from the left on the potential barrier $v(x)$ with energy[33] $\mathcal{E} = \hbar^2K^2/2m$. Since $v(x) = 0$ when $|x| \geqslant a/2$, in these regions the wave function $\psi_l(x)$ will have the form

$$\begin{aligned} \psi_l(x) &= e^{iKx} + re^{-iKx}, \qquad x \leqslant -\frac{a}{2}, \\ &= te^{iKx}, \qquad x \geqslant \frac{a}{2}. \end{aligned} \tag{8.65}$$

This is illustrated schematically in Figure 8.5a.

The transmission and reflection coefficients t and r give the probability amplitude that the electron will tunnel through or be reflected from the barrier; they depend on the incident wave vector K in a manner determined by the detailed features of the barrier potential v. However, one can deduce many properties of the band structure of the periodic potential U by appealing only to very general properties of t and r. Because v is even, $\psi_r(x) = \psi_l(-x)$ is also a solution to the Schrödinger equation with energy $\mathcal{E}$. From (8.65) it follows that $\psi_r(x)$ has the form

$$\begin{aligned} \psi_r(x) &= te^{-iKx}, \qquad x \leqslant -\frac{a}{2}, \\ &= e^{-iKx} + re^{iKx}, \qquad x \geqslant \frac{a}{2}. \end{aligned} \tag{8.66}$$

Evidently this describes a particle incident on the barrier from the right, as depicted in Figure 8.5b.

Since ψ_l and ψ_r are two independent solutions to the single-barrier Schrödinger equation with the same energy, any other solution with that energy will be a linear combination[34] of these two: $\psi = A\psi_l + B\psi_r$. In particular, since the crystal Hamiltonian is identical to that for a single ion in the region $-a/2 \leqslant x \leqslant a/2$, any solution to the crystal Schrödinger equation with energy $\mathcal{E}$ must be a linear combination of ψ_l and ψ_r in that region:

$$\psi(x) = A\psi_l(x) + B\psi_r(x), \qquad -\frac{a}{2} \leqslant x \leqslant \frac{a}{2}. \tag{8.67}$$

Now Bloch's theorem asserts that ψ can be chosen to satisfy

$$\psi(x + a) = e^{ika}\psi(x), \tag{8.68}$$

for suitable k. Differentiating (8.68) we also find that $\psi' = d\psi/dx$ satisfies

$$\psi'(x + a) = e^{ika}\psi'(x). \tag{8.69}$$

(a) By imposing the conditions (8.68) and (8.69) at $x = -a/2$, and using (8.65) to (8.67), show that the energy of the Bloch electron is related to its wave vector k by:

$$\cos ka = \frac{t^2 - r^2}{2t} e^{iKa} + \frac{1}{2t} e^{-iKa}, \qquad \mathcal{E} = \frac{\hbar^2K^2}{2m}. \tag{8.70}$$

Verify that this gives the right answer in the free electron case ($v \equiv 0$).

---

[33] *Note:* in this problem K is a continuous variable and has nothing to do with the reciprocal lattice.

[34] A special case of the general theorem that there are n independent solutions to an nth-order linear differential equation.

Equation (8.70) is more informative when one supplies a little more information about the transmission and reflection coefficients. We write the complex number t in terms of its magnitude and phase:

$$t = |t|\, e^{i\delta}. \tag{8.71}$$

The real number δ is known as the phase shift, since it specifies the change in phase of the transmitted wave relative to the incident one. Electron conservation requires that the probability of transmission plus the probability of reflection be unity:

$$1 = |t|^2 + |r|^2. \tag{8.72}$$

This, and some other useful information, can be proved as follows. Let ϕ_1 and ϕ_2 be any two solutions to the one-barrier Schrödinger equation with the same energy:

$$-\frac{\hbar^2}{2m}\phi_i'' + v\phi_i = \frac{\hbar^2 K^2}{2m}\phi_i, \qquad i = 1, 2. \tag{8.73}$$

Define $w(\phi_1, \phi_2)$ (the "Wronskian") by

$$w(\phi_1, \phi_2) = \phi_1'(x)\phi_2(x) - \phi_1(x)\phi_2'(x). \tag{8.74}$$

(b) Prove that w is independent of x by deducing from (8.73) that its derivative vanishes.

(c) Prove (8.72) by evaluating $w(\psi_l, \psi_l^*)$ for $x \leqslant -a/2$ and $x \geqslant a/2$, noting that because $v(x)$ is real ψ_l^* will be a solution to the same Schrödinger equation as ψ_l.

(d) By evaluating $w(\psi_l, \psi_r^*)$ prove that rt^* is pure imaginary, so r must have the form

$$r = \pm i\,|r|\, e^{i\delta}, \tag{8.75}$$

where δ is the same as in (8.71).

(e) Show as a consequence of (8.70), (8.72), and (8.75) that the energy and wave vector of the Bloch electron are related by

$$\boxed{\frac{\cos(Ka + \delta)}{|t|} = \cos ka, \qquad \mathcal{E} = \frac{\hbar^2 K^2}{2m}.} \tag{8.76}$$

Since $|t|$ is always less than one, but approaches unity for large K (the barrier becomes increasingly less effective as the incident energy grows), the left side of (8.76) plotted against K has the structure depicted in Figure 8.6. For a given k, the allowed values of K (and hence the allowed energies $\mathcal{E}(k) = \hbar^2 K^2/2m$) are given by the intersection of the curve in Figure 8.6 with the horizontal line of height $\cos(ka)$. Note that values of K in the neighborhood of those satisfying

$$Ka + \delta = n\pi \tag{8.77}$$

give $|\cos(Ka + \delta)|/|t| > 1$, and are therefore not allowed for any k. The corresponding regions of energy are the energy gaps. If δ is a bounded function of K (as is generally the case), then there will be infinitely many regions of forbidden energy, and also infinitely many regions of allowed energies for each value of k.

(f) Suppose the barrier is very weak (so that $|t| \approx 1$, $|r| \approx 0$, $\delta \approx 0$). Show that the energy gaps are then very narrow, the width of the gap containing $K = n\pi/a$ being

$$\mathcal{E}_{\text{gap}} \approx 2\pi n \frac{\hbar^2}{ma^2}|r|. \tag{8.78}$$

(g) Suppose the barrier is very strong, so that $|t| \approx 0$, $|r| \approx 1$. Show that the allowed bands

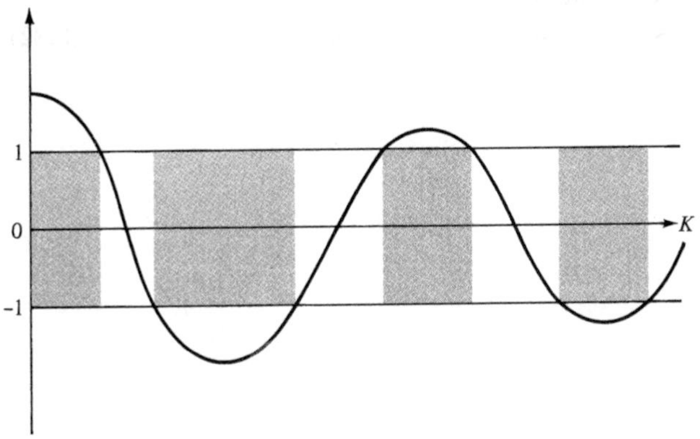

Figure 8.6
Characteristic form of the function $\cos(Ka + \delta)/|t|$. Because $|t(K)|$ is always less than unity the function will exceed unity in magnitude in the neighborhood of solutions to $Ka + \delta(K) = n\pi$. Equation (8.76) can be satisfied for real k if and only if the function is less than unity in magnitude. Consequently there will be allowed (unshaded) and forbidden (shaded) regions of K (and therefore of $\mathcal{E} = \hbar^2K^2/2m$). Note that when $|t|$ is very near unity (weak potential) the forbidden regions will be narrow, but if $|t|$ is very small (strong potential) the allowed regions will be narrow.

of energies are then very narrow, with widths

$$\mathcal{E}_{\max} - \mathcal{E}_{\min} = O(|t|). \tag{8.79}$$

(h) As a concrete example, one often considers the case in which $v(x) = g\delta(x)$, where $\delta(x)$ is the Dirac delta function (a special case of the "Kronig-Penney model"). Show that in this case

$$\cot\delta = -\frac{\hbar^2 K}{mg}, \qquad |t| = \cos\delta. \tag{8.80}$$

This model is a common textbook example of a one-dimensional periodic potential. Note, however, that most of the structure we have established is, to a considerable degree, independent of the particular functional dependence of $|t|$ and δ on K.

2. *Density of Levels*

(a) In the free electron case the density of levels at the Fermi energy can be written in the form (Eq. (2.64)) $g(\mathcal{E}_F) = mk_F/\hbar^2\pi^2$. Show that the general form (8.63) reduces to this when $\mathcal{E}_n(\mathbf{k}) = \hbar^2k^2/2m$ and the (spherical) Fermi surface lies entirely within a primitive cell.

(b) Consider a band in which, for sufficiently small k, $\mathcal{E}_n(\mathbf{k}) = \mathcal{E}_0 + (\hbar^2/2)(k_x^2/m_x + k_y^2/m_y + k_z^2/m_z)$ (as might be the case in a crystal of orthorhombic symmetry) where m_x, m_y, and m_z are positive constants. Show that if $\mathcal{E}$ is close enough to $\mathcal{E}_0$ that this form is valid, then $g_n(\mathcal{E})$ is proportional to $(\mathcal{E} - \mathcal{E}_0)^{1/2}$, so its derivative becomes infinite (van Hove singularity) as $\mathcal{E}$ approaches the band minimum. (*Hint*: Use the form (8.57) for the density of levels.) Deduce from this that if the quadratic form for $\mathcal{E}_n(\mathbf{k})$ remains valid up to $\mathcal{E}_F$, then $g_n(\mathcal{E}_F)$ can be written in the obvious generalization of the free electron form (2.65):

$$g_n(\mathcal{E}_F) = \frac{3}{2}\frac{n}{\mathcal{E}_F - \mathcal{E}_0} \tag{8.81}$$

where n is the contribution of the electrons in the band to the total electronic density.

(c) Consider the density of levels in the neighborhood of a saddle point, where $\mathcal{E}_n(\mathbf{k}) = \mathcal{E}_0 + (\hbar^2/2)(k_x^2/m_x + k_y^2/m_y - k_z^2/m_z)$ where m_x, m_y, and m_z are positive constants. Show that when $\mathcal{E} \approx \mathcal{E}_0$, the derivative of the density of levels has the form

$$\begin{aligned} g_n'(\mathcal{E}) &\approx \text{constant}, & \mathcal{E} > \mathcal{E}_0; \\ &\approx (\mathcal{E}_0 - \mathcal{E})^{-1/2}, & \mathcal{E} < \mathcal{E}_0. \end{aligned} \tag{8.82}$$

9

Electrons in a Weak Periodic Potential

Perturbation Theory and Weak Periodic Potentials

Energy Levels Near a Single Bragg Plane

Illustration of Extended-, Reduced-, and Repeated-Zone Schemes in One Dimension

Fermi Surface and Brillouin Zones

Geometrical Structure Factor

Spin-Orbit Coupling

One can gain substantial insight into the structure imposed on the electronic energy levels by a periodic potential, if that potential is very weak. This approach might once have been regarded as an instructive, but academic, exercise. We now know, however, that in many cases this apparently unrealistic assumption gives results surprisingly close to the mark. Modern theoretical and experimental studies of the metals found in groups I, II, III, and IV of the periodic table (i.e., metals whose atomic structure consists of *s* and *p* electrons outside of a closed-shell noble gas configuration) indicate that the conduction electrons can be described as moving in what amounts to an almost constant potential. These elements are often referred to as "nearly free electron" metals, because the starting point for their description is the Sommerfeld free electron gas, modified by the presence of a *weak* periodic potential. In this chapter we shall examine some of the broad general features of band structure from the almost free electron point of view. Applications to particular metals will be examined in Chapter 15.

It is by no means obvious why the conduction bands of these metals should be so free-electron-like. There are two fundamental reasons why the strong interactions of the conduction electrons with each other and with the positive ions can have the net effect of a very weak potential.

1. The electron-ion interaction is strongest at small separations, but the conduction electrons are forbidden (by the Pauli principle) from entering the immediate neighborhood of the ions because this region is already occupied by the core electrons.
2. In the region in which the conduction electrons are allowed, their mobility further diminishes the net potential any single electron experiences, for they can *screen* the fields of positively charged ions, diminishing the total effective potential.

These remarks offer only the barest indication of why the following discussion has extensive practical application. We shall return later to the problem of justifying the nearly free electron approach, taking up point 1 in Chapter 11 and point 2 in Chapter 17.

GENERAL APPROACH TO THE SCHRÖDINGER EQUATION WHEN THE POTENTIAL IS WEAK

When the periodic potential is zero, the solutions to Schrödinger's equation are plane waves. A reasonable starting place for the treatment of weak periodic potentials is therefore the expansion of the exact solution in plane waves described in Chapter 8. The wave function of a Bloch level with crystal momentum **k** can be written in the form given in Eq. (8.42):

$$\psi_{\mathbf{k}}(\mathbf{r}) = \sum_{\mathbf{K}} c_{\mathbf{k}-\mathbf{K}} e^{i(\mathbf{k}-\mathbf{K})\cdot\mathbf{r}}, \qquad \textbf{(9.1)}$$

where the coefficients $c_{\mathbf{k}-\mathbf{K}}$ and the energy of the level $\mathcal{E}$ are determined by the set of Eqs. (8.41):

$$\left[\frac{\hbar^2}{2m}(\mathbf{k}-\mathbf{K})^2 - \mathcal{E}\right] c_{\mathbf{k}-\mathbf{K}} + \sum_{\mathbf{K}'} U_{\mathbf{K}'-\mathbf{K}} c_{\mathbf{k}-\mathbf{K}'} = 0. \qquad \textbf{(9.2)}$$

The sum in (9.1) is over all reciprocal lattice vectors **K**, and for fixed **k** there is an equation of the form (9.2) for each reciprocal lattice vector **K**. The (infinitely many) different solutions to (9.2) for a given **k** are labeled with the band index n. The wave vector **k** can (but need not) be considered to lie in the first Brillouin zone of k-space.

In the free electron case, all the Fourier components $U_{\mathbf{K}}$ are precisely zero. Equation (9.2) then becomes

$$(\mathcal{E}^0_{\mathbf{k}-\mathbf{K}} - \mathcal{E})c_{\mathbf{k}-\mathbf{K}} = 0, \tag{9.3}$$

where we have introduced the notation:

$$\mathcal{E}^0_q = \frac{\hbar^2}{2m}q^2. \tag{9.4}$$

Equation (9.3) requires for each **K** that either $c_{\mathbf{k}-\mathbf{K}} = 0$ or $\mathcal{E} = \mathcal{E}^0_{\mathbf{k}-\mathbf{K}}$. The latter possibility can occur for only a single **K**, unless it happens that some of the $\mathcal{E}^0_{\mathbf{k}-\mathbf{K}}$ are equal for several different choices of **K**. If such degeneracy does *not* occur, then we have the expected class of free electron solutions:

$$\mathcal{E} = \mathcal{E}^0_{\mathbf{k}-\mathbf{K}}, \quad \psi_{\mathbf{k}} \propto e^{i(\mathbf{k}-\mathbf{K})\cdot\mathbf{r}}. \tag{9.5}$$

If, however, there is a group of reciprocal lattice vectors $\mathbf{K}_1, \ldots, \mathbf{K}_m$ satisfying

$$\mathcal{E}^0_{\mathbf{k}-\mathbf{K}_1} = \cdots = \mathcal{E}^0_{\mathbf{k}-\mathbf{K}_m}, \tag{9.6}$$

then when $\mathcal{E}$ is equal to the common value of these free electron energies there are m independent degenerate plane wave solutions. Since any linear combination of degenerate solutions is also a solution, one has complete freedom in choosing the coefficients $c_{\mathbf{k}-\mathbf{K}}$ for $\mathbf{K} = \mathbf{K}_1, \ldots, \mathbf{K}_m$.

These simple observations acquire more substance when the $U_{\mathbf{K}}$ are not zero, but very small. The analysis still divides naturally into two cases, corresponding to the nondegenerate and degenerate cases for free electrons. Now, however, the basis for the distinction is not the exact equality[1] of two or more distinct free electron levels, but only whether they are equal aside from terms of order U.

Case 1 Fix **k** and consider a particular reciprocal lattice vector $\mathbf{K}_1$ such that the free electron energy $\mathcal{E}^0_{\mathbf{k}-\mathbf{K}_1}$ is far from the values of $\mathcal{E}^0_{\mathbf{k}-\mathbf{K}}$ (for all other **K**) compared with U (see Figure 9.1)[2]:

$$\left|\mathcal{E}^0_{\mathbf{k}-\mathbf{K}_1} - \mathcal{E}^0_{\mathbf{k}-\mathbf{K}}\right| \gg U, \qquad \text{for fixed } \mathbf{k} \text{ and all } \mathbf{K} \neq \mathbf{K}_1. \tag{9.7}$$

We wish to investigate the effect of the potential on that free electron level given by:

$$\mathcal{E} = \mathcal{E}^0_{\mathbf{k}-\mathbf{K}_1}, \quad c_{\mathbf{k}-\mathbf{K}} = 0, \qquad \mathbf{K} \neq \mathbf{K}_1. \tag{9.8}$$

---

[1] The reader familiar with stationary perturbation theory may think that if there is no *exact* degeneracy, we can always make all level differences large compared with U by considering sufficiently small U. That is indeed true *for any given* **k**. However, once we are given a definite U, no matter how small, we want a procedure valid for all **k** in the first Brillouin zone. We shall see that no matter how small U is we can always find some values of **k** for which the unperturbed levels are closer together than U. Therefore what we are doing is more subtle than conventional degenerate perturbation theory.

[2] In inequalities of this form we shall use U to refer to a typical Fourier component of the potential.

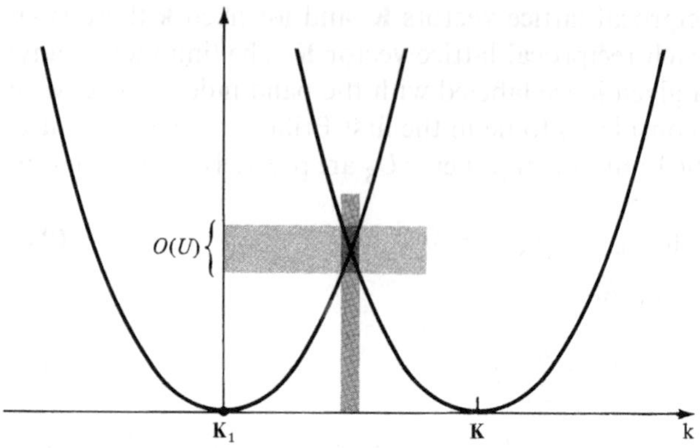

Figure 9.1
For the range of **k** within limits indicated by the dark band the free electron levels $\mathcal{E}_{\mathbf{k}-\mathbf{K}_1}$ and $\mathcal{E}_{\mathbf{k}-\mathbf{K}}$ differ by an energy $O(U)$.

Setting $\mathbf{K} = \mathbf{K}_1$ in Eq. (9.2) (and using the short notation (9.4)) we have (dropping the prime from the summation index):

$$(\mathcal{E} - \mathcal{E}^0_{\mathbf{k}-\mathbf{K}_1})c_{\mathbf{k}-\mathbf{K}_1} = \sum_{\mathbf{K}} U_{\mathbf{K}-\mathbf{K}_1} c_{\mathbf{k}-\mathbf{K}}. \tag{9.9}$$

Because we have picked the additive constant in the potential energy so that $U_{\mathbf{K}} = 0$ when $\mathbf{K} = \mathbf{0}$ (see page 137), only terms with $\mathbf{K} \neq \mathbf{K}_1$ appear on the right-hand side of (9.9). Since we are examining that solution for which $c_{\mathbf{k}-\mathbf{K}}$ vanishes when $\mathbf{K} \neq \mathbf{K}_1$ in the limit of vanishing U, we expect the right-hand side of (9.9) to be of second order in U. This can be explicitly confirmed by writing Eq. (9.2) for $\mathbf{K} \neq \mathbf{K}_1$ as

$$c_{\mathbf{k}-\mathbf{K}} = \frac{U_{\mathbf{K}_1-\mathbf{K}} c_{\mathbf{k}-\mathbf{K}_1}}{\mathcal{E} - \mathcal{E}^0_{\mathbf{k}-\mathbf{K}}} + \sum_{\mathbf{K}' \neq \mathbf{K}_1} \frac{U_{\mathbf{K}'-\mathbf{K}} c_{\mathbf{k}-\mathbf{K}'}}{\mathcal{E} - \mathcal{E}^0_{\mathbf{k}-\mathbf{K}}}. \tag{9.10}$$

We have separated out of the sum in (9.10) the term containing $c_{\mathbf{k}-\mathbf{K}_1}$, since it will be an order of magnitude larger than the remaining terms, which involve $c_{\mathbf{k}-\mathbf{K}'}$ for $\mathbf{K}' \neq \mathbf{K}_1$. This conclusion depends on the assumption (9.7) that the level $\mathcal{E}^0_{\mathbf{k}-\mathbf{K}_1}$ is not nearly degenerate to some other $\mathcal{E}^0_{\mathbf{k}-\mathbf{K}}$. Such a near degeneracy could cause some of the denominators in (9.10) to be of order U, canceling the explicit U in the numerator and resulting in additional terms in the sum in (9.10) comparable to the $\mathbf{K} = \mathbf{K}_1$ term.

Therefore, provided there is no near degeneracy,

$$c_{\mathbf{k}-\mathbf{K}} = \frac{U_{\mathbf{K}_1-\mathbf{K}} c_{\mathbf{k}-\mathbf{K}_1}}{\mathcal{E} - \mathcal{E}^0_{\mathbf{k}-\mathbf{K}}} + O(U^2). \tag{9.11}$$

Placing this in (9.9), we find:

$$(\mathcal{E} - \mathcal{E}^0_{\mathbf{k}-\mathbf{K}_1})c_{\mathbf{k}-\mathbf{K}_1} = \sum_{\mathbf{K}} \frac{U_{\mathbf{K}-\mathbf{K}_1} U_{\mathbf{K}_1-\mathbf{K}}}{\mathcal{E} - \mathcal{E}^0_{\mathbf{k}-\mathbf{K}}} c_{\mathbf{k}-\mathbf{K}_1} + O(U^3). \tag{9.12}$$

Thus the perturbed energy level $\mathcal{E}$ differs from the free electron value $\mathcal{E}^0_{\mathbf{k}-\mathbf{K}_1}$ by terms of order U^2. To solve Eq. (9.12) for $\mathcal{E}$ to this order, it therefore suffices to replace the $\mathcal{E}$ appearing in the denominator on the right-hand side by $\mathcal{E}^0_{\mathbf{k}-\mathbf{K}_1}$, leading to the following expression[3] for $\mathcal{E}$, correct to second order in U:

$$\mathcal{E} = \mathcal{E}^0_{\mathbf{k}-\mathbf{K}_1} + \sum_{\mathbf{K}} \frac{|U_{\mathbf{K}-\mathbf{K}_1}|^2}{\mathcal{E}^0_{\mathbf{k}-\mathbf{K}_1} - \mathcal{E}^0_{\mathbf{k}-\mathbf{K}}} + O(U^3). \tag{9.13}$$

Equation (9.13) asserts that weakly perturbed nondegenerate bands repel each other, for every level $\mathcal{E}^0_{\mathbf{k}-\mathbf{K}}$ that lies below $\mathcal{E}^0_{\mathbf{k}-\mathbf{K}_1}$ contributes a term in (9.13) that raises the value of $\mathcal{E}$, while every level that lies above $\mathcal{E}^0_{\mathbf{k}-\mathbf{K}_1}$ contributes a term that lowers the energy. However, the most important feature to emerge from this analysis of the case of no near degeneracy, is simply the gross observation that the shift in energy from the free electron value is second order in U. In the nearly degenerate case (as we shall now see) the shift in energy can be linear in U. Therefore, to leading order in the weak periodic potential, it is only the nearly degenerate free electron levels that are significantly shifted, and we must devote most of our attention to this important case.

Case 2 Suppose the value of $\mathbf{k}$ is such that there are reciprocal lattice vectors $\mathbf{K}_1, \ldots, \mathbf{K}_m$ with $\mathcal{E}^0_{\mathbf{k}-\mathbf{K}_1}, \ldots, \mathcal{E}^0_{\mathbf{k}-\mathbf{K}_m}$ all within order U of each other,[4] but far apart from the other $\mathcal{E}^0_{\mathbf{k}-\mathbf{K}}$ on the scale of U:

$$\left|\mathcal{E}^0_{\mathbf{k}-\mathbf{K}} - \mathcal{E}^0_{\mathbf{k}-\mathbf{K}_i}\right| \gg U, \qquad i = 1, \ldots, m, \qquad \mathbf{K} \neq \mathbf{K}_1, \ldots, \mathbf{K}_m. \tag{9.14}$$

In this case we must treat separately those equations given by (9.2) when $\mathbf{K}$ is set equal to any of the m values $\mathbf{K}_1, \ldots, \mathbf{K}_m$. This gives m equations corresponding to the single equation (9.9) in the nondegenerate case. In these m equations we separate from the sum those terms containing the coefficients $c_{\mathbf{k}-\mathbf{K}_j}$, $j = 1, \ldots, m$, which need not be small in the limit of vanishing interaction, from the remaining $c_{\mathbf{k}-\mathbf{K}}$, which will be at most of order U. Thus we have

$$(\mathcal{E} - \mathcal{E}^0_{\mathbf{k}-\mathbf{K}_i})c_{\mathbf{k}-\mathbf{K}_i} = \sum_{j=1}^{m} U_{\mathbf{K}_j-\mathbf{K}_i} c_{\mathbf{k}-\mathbf{K}_j} + \sum_{\mathbf{K}\neq\mathbf{K}_1,\ldots,\mathbf{K}_m} U_{\mathbf{K}-\mathbf{K}_i} c_{\mathbf{k}-\mathbf{K}}, \qquad i = 1, \ldots, m. \tag{9.15}$$

Making the same separation in the sum, we can write Eq. (9.2) for the remaining levels as

$$c_{\mathbf{k}-\mathbf{K}} = \frac{1}{\mathcal{E} - \mathcal{E}^0_{\mathbf{k}-\mathbf{K}}} \left(\sum_{j=1}^{m} U_{\mathbf{K}_j-\mathbf{K}} c_{\mathbf{k}-\mathbf{K}_j} + \sum_{\mathbf{K}'\neq\mathbf{K}_1,\ldots,\mathbf{K}_m} U_{\mathbf{K}'-\mathbf{K}} c_{\mathbf{k}-\mathbf{K}'} \right), \qquad \mathbf{K} \neq \mathbf{K}_1, \ldots, \mathbf{K}_m, \tag{9.16}$$

(which corresponds to equation (9.10) in the case of no near degeneracy).

Since $c_{\mathbf{k}-\mathbf{K}}$ will be at most of order U when $\mathbf{K} \neq \mathbf{K}_1, \ldots, \mathbf{K}_m$, Eq. (9.16) gives

$$c_{\mathbf{k}-\mathbf{K}} = \frac{1}{\mathcal{E} - \mathcal{E}^0_{\mathbf{k}-\mathbf{K}}} \sum_{j=1}^{m} U_{\mathbf{K}_j-\mathbf{K}} c_{\mathbf{k}-\mathbf{K}_j} + O(U^2). \tag{9.17}$$

---

[3] We use Eq. (8.34), $U_{-\mathbf{K}} = U^*_{\mathbf{K}}$.

[4] In one dimension m cannot exceed 2, but in three dimensions it can be quite large.

Placing this in (9.15), we find that

$$(\mathcal{E} - \mathcal{E}^0_{\mathbf{k}-\mathbf{K}_i})c_{\mathbf{k}-\mathbf{K}_i} = \sum_{j=1}^{m} U_{\mathbf{K}_j-\mathbf{K}_i} c_{\mathbf{k}-\mathbf{K}_j}$$

$$+ \sum_{j=1}^{m} \left(\sum_{\mathbf{K} \neq \mathbf{K}_1, \ldots, \mathbf{K}_m} \frac{U_{\mathbf{K}-\mathbf{K}_i} U_{\mathbf{K}_j-\mathbf{K}}}{\mathcal{E} - \mathcal{E}^0_{\mathbf{k}-\mathbf{K}}} \right) c_{\mathbf{k}-\mathbf{K}_j} + O(U^3). \quad \textbf{(9.18)}$$

Compare this with the result (9.12) in the case of no near degeneracy. There we found an explicit expression for the shift in energy to order U^2 (to which the set of equations (9.18) reduces when $m = 1$). Now, however, we find that to an accuracy of order U^2 the determination of the shifts in the m nearly degenerate levels reduces to the solution of m coupled equations[5] for the $c_{\mathbf{k}-\mathbf{K}_i}$. Furthermore, the coefficients in the second term on the right-hand side of these equations are of higher order in U than those in the first.[6] Consequently, to find the *leading* corrections in U we can replace (9.18) by the far simpler equations:

$$(\mathcal{E} - \mathcal{E}^0_{\mathbf{k}-\mathbf{K}_i})c_{\mathbf{k}-\mathbf{K}_i} = \sum_{i=1}^{m} U_{\mathbf{K}_j-\mathbf{K}_i} c_{\mathbf{k}-\mathbf{K}_j}, \qquad i = 1, \ldots, m, \quad \textbf{(9.19)}$$

which are just the general equations for a system of m quantum levels.[7]

ENERGY LEVELS NEAR A SINGLE BRAGG PLANE

The simplest and most important example of the preceding discussion is when two free electron levels are within order U of each other, but far compared with U from all other levels. When this happens, Eq. (9.19) reduces to the two equations:

$$\begin{aligned} (\mathcal{E} - \mathcal{E}^0_{\mathbf{k}-\mathbf{K}_1})c_{\mathbf{k}-\mathbf{K}_1} &= U_{\mathbf{K}_2-\mathbf{K}_1} c_{\mathbf{k}-\mathbf{K}_2}, \\ (\mathcal{E} - \mathcal{E}^0_{\mathbf{k}-\mathbf{K}_2})c_{\mathbf{k}-\mathbf{K}_2} &= U_{\mathbf{K}_1-\mathbf{K}_2} c_{\mathbf{k}-\mathbf{K}_1}. \end{aligned} \quad \textbf{(9.20)}$$

When only two levels are involved, there is little point in continuing with the notational convention that labels them symmetrically. We therefore introduce variables particularly convenient for the two-level problem:

$$\mathbf{q} = \mathbf{k} - \mathbf{K}_1 \quad \text{and} \quad \mathbf{K} = \mathbf{K}_2 - \mathbf{K}_1, \quad \textbf{(9.21)}$$

and write (9.20) as

$$\begin{aligned} (\mathcal{E} - \mathcal{E}^0_{\mathbf{q}})c_{\mathbf{q}} &= U_{\mathbf{K}} c_{\mathbf{q}-\mathbf{K}}, \\ (\mathcal{E} - \mathcal{E}^0_{\mathbf{q}-\mathbf{K}})c_{\mathbf{q}-\mathbf{K}} &= U_{-\mathbf{K}} c_{\mathbf{q}} = U^*_{\mathbf{K}} c_{\mathbf{q}}. \end{aligned} \quad \textbf{(9.22)}$$

---

[5] These are rather closely related to the equations of *second-order degenerate* perturbation theory, to which they reduce when all the $\mathcal{E}^0_{\mathbf{k}-\mathbf{K}_i}$ are rigorously equal, $i = 1, \ldots, m$. (See L. D. Landau and E. M. Lifshitz, *Quantum Mechanics*, Addison-Wesley, Reading Mass., 1965, p. 134.)

[6] The numerator is explicitly of order U^2, and since only **K**-values different from $\mathbf{K}_1, \ldots, \mathbf{K}_m$ appear in the sum, the denominator is not of order U when $\mathcal{E}$ is close to the $\mathcal{E}^0_{\mathbf{k}-\mathbf{K}_i}$, $i = 1, \ldots, m$.

[7] Note that the rule of thumb for going from (9.19) back to the more accurate form in (9.18) is simply that U should be replaced by U', where

$$U'_{\mathbf{K}_j-\mathbf{K}_i} = U_{\mathbf{K}_j-\mathbf{K}_i} + \sum_{\mathbf{K} \neq \mathbf{K}_1, \ldots, \mathbf{K}_m} \frac{U_{\mathbf{K}_j-\mathbf{K}} U_{\mathbf{K}-\mathbf{K}_i}}{\mathcal{E} - \mathcal{E}^0_{\mathbf{k}-\mathbf{K}}}.$$

We have:

$$\mathcal{E}^0_{\mathbf{q}} \approx \mathcal{E}^0_{\mathbf{q}-\mathbf{K}}, \quad \left|\mathcal{E}^0_{\mathbf{q}} - \mathcal{E}^0_{\mathbf{q}-\mathbf{K}'}\right| \gg U, \qquad \text{for } \mathbf{K}' \neq \mathbf{K}, \mathbf{0}. \tag{9.23}$$

Now $\mathcal{E}^0_{\mathbf{q}}$ is equal to $\mathcal{E}^0_{\mathbf{q}-\mathbf{K}}$ for some reciprocal lattice vector only when $|\mathbf{q}| = |\mathbf{q} - \mathbf{K}|$. This means (Figure 9.2a) that $\mathbf{q}$ must lie on the Bragg plane (see Chapter 6) bisecting the line joining the origin of $\mathbf{k}$ space to the reciprocal lattice point $\mathbf{K}$. The assertion that $\mathcal{E}^0_{\mathbf{q}} = \mathcal{E}_{\mathbf{q}-\mathbf{K}'}$ only for $\mathbf{K}' = \mathbf{K}$ requires that $\mathbf{q}$ lie *only* on this Bragg plane, and on no other.

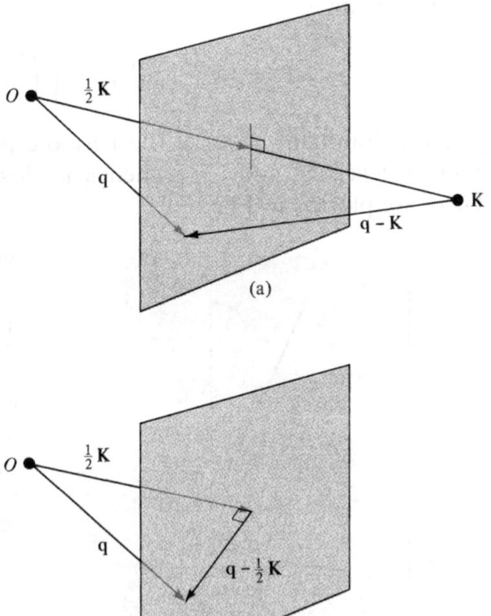

Figure 9.2
(a) If $|\mathbf{q}| = |\mathbf{q} - \mathbf{K}|$, then the point $\mathbf{q}$ must lie in the Bragg plane determined by $\mathbf{K}$. (b) If the point $\mathbf{q}$ lies in the Bragg plane, then the vector $\mathbf{q} - \frac{1}{2}\mathbf{K}$ is parallel to the plane.

Thus conditions (9.23) have the geometric significance of requiring $\mathbf{q}$ to be close to a Bragg plane (but not close to a place where *two* or more Bragg planes intersect). Therefore the case of two nearly degenerate levels applies to an electron whose wave vector very nearly satisfies the condition for a single Bragg scattering.[8] Correspondingly, the general case of many nearly degenerate levels applies to the treatment of a free electron level whose wave vector is close to one at which many simultaneous Bragg reflections can occur. Since the nearly degenerate levels are the most strongly affected by a weak periodic potential, we conclude that *a weak periodic potential has its major effects on only those free electron levels whose wave vectors are close to ones at which Bragg reflections can occur.*

We discuss systematically on pages 162 to 166 when free electron wave vectors do, or do not, lie on Bragg planes, as well as the general structure this imposes on the energy levels in a weak potential. First, however, we examine the level structure

[8] An incident X-ray beam undergoes Bragg reflection only if its wave vector lies on a Bragg plane (see Chapter 6).

when only a single Bragg plane is nearby, as determined by (9.22). These equations have a solution when:

$$\begin{vmatrix} \mathcal{E} - \mathcal{E}^0_{\mathbf{q}} & -U_{\mathbf{K}} \\ -U^*_{\mathbf{K}} & \mathcal{E} - \mathcal{E}^0_{\mathbf{q}-\mathbf{K}} \end{vmatrix} = 0. \tag{9.24}$$

This leads to a quadratic equation

$$(\mathcal{E} - \mathcal{E}^0_{\mathbf{q}})(\mathcal{E} - \mathcal{E}^0_{\mathbf{q}-\mathbf{K}}) = |U_{\mathbf{K}}|^2. \tag{9.25}$$

The two roots

$$\mathcal{E} = \tfrac{1}{2}(\mathcal{E}^0_{\mathbf{q}} + \mathcal{E}^0_{\mathbf{q}-\mathbf{K}}) \pm \left[\left(\frac{\mathcal{E}^0_{\mathbf{q}} - \mathcal{E}^0_{\mathbf{q}-\mathbf{K}}}{2}\right)^2 + |U_{\mathbf{K}}|^2\right]^{1/2} \tag{9.26}$$

give the dominant effect of the periodic potential on the energies of the two free electron levels $\mathcal{E}^0_{\mathbf{q}}$ and $\mathcal{E}^0_{\mathbf{q}-\mathbf{K}}$ when $\mathbf{q}$ is close to the Bragg plane determined by $\mathbf{K}$. These are plotted in Figure 9.3.

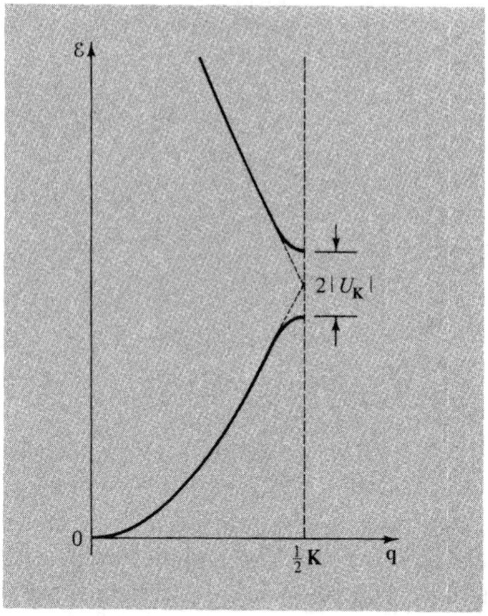

Figure 9.3
Plot of the energy bands given by Eq. (9.26) for $\mathbf{q}$ parallel to $\mathbf{K}$. The lower band corresponds to the choice of a minus sign in (9.26) and the upper band to a plus sign. When $\mathbf{q} = \frac{1}{2}\mathbf{K}$, the two bands are separated by a band gap of magnitude $2|U_{\mathbf{K}}|$. When $\mathbf{q}$ is far removed from the Bragg plane, the levels (to leading order) are indistinguishable from their free electron values (denoted by dotted lines).

The result (9.26) is particularly simple for points lying *on* the Bragg plane since, when $\mathbf{q}$ is on the Bragg plane, $\mathcal{E}^0_{\mathbf{q}} = \mathcal{E}^0_{\mathbf{q}-\mathbf{K}}$. Hence

$$\mathcal{E} = \mathcal{E}^0_{\mathbf{q}} \pm |U_{\mathbf{K}}|, \quad \mathbf{q} \text{ on a single Bragg plane.} \tag{9.27}$$

Thus, at all points on the Bragg plane, one level is uniformly raised by $|U_{\mathbf{K}}|$ and the other is uniformly lowered by the same amount.

It is also easily verified from (9.26) that when $\mathcal{E}^0_{\mathbf{q}} = \mathcal{E}^0_{\mathbf{q}-\mathbf{K}}$,

$$\frac{\partial \mathcal{E}}{\partial \mathbf{q}} = \frac{\hbar^2}{m}(\mathbf{q} - \tfrac{1}{2}\mathbf{K}); \tag{9.28}$$

i.e., when the point $\mathbf{q}$ is on the Bragg plane the gradient of $\mathcal{E}$ is parallel to the plane (see Figure 9.2b). Since the gradient is perpendicular to the surfaces on which a function is constant, the constant-energy surfaces at the Bragg plane are perpendicular to the plane.[9]

When $\mathbf{q}$ lies on a single Bragg plane we may also easily determine the form of the wave functions corresponding to the two solutions $\mathcal{E} = \mathcal{E}_{\mathbf{q}}^0 \pm |U_{\mathbf{K}}|$. From (9.22), when $\mathcal{E}$ is given by (9.27), the two coefficients $c_{\mathbf{q}}$ and $c_{\mathbf{q}-\mathbf{K}}$ satisfy[10]

$$c_{\mathbf{q}} = \pm \operatorname{sgn}(U_{\mathbf{K}}) c_{\mathbf{q}-\mathbf{K}}. \tag{9.29}$$

Since these two coefficients are the dominant ones in the plane-wave expansion (9.1), it follows that if $U_{\mathbf{K}} > 0$, then

$$\begin{aligned} |\psi(\mathbf{r})|^2 &\propto (\cos \tfrac{1}{2}\mathbf{K}\cdot\mathbf{r})^2, \quad \mathcal{E} = \mathcal{E}_{\mathbf{q}}^0 + |U_{\mathbf{K}}|, \\ |\psi(\mathbf{r})|^2 &\propto (\sin \tfrac{1}{2}\mathbf{K}\cdot\mathbf{r})^2, \quad \mathcal{E} = \mathcal{E}_{\mathbf{q}}^0 - |U_{\mathbf{K}}|, \end{aligned}$$

while if $U_{\mathbf{K}} < 0$, then

$$\begin{aligned} |\psi(\mathbf{r})|^2 &\propto (\sin \tfrac{1}{2}\mathbf{K}\cdot\mathbf{r})^2, \quad \mathcal{E} = \mathcal{E}_{\mathbf{q}}^0 + |U_{\mathbf{K}}|, \\ |\psi(\mathbf{r})|^2 &\propto (\cos \tfrac{1}{2}\mathbf{K}\cdot\mathbf{r})^2, \quad \mathcal{E} = \mathcal{E}_{\mathbf{q}}^0 - |U_{\mathbf{K}}|. \end{aligned} \tag{9.30}$$

Sometimes the two types of linear combination are called "*p*-like" ($|\psi|^2 \sim \sin^2 \frac{1}{2}\mathbf{K}\cdot\mathbf{r}$) and "*s*-like" ($|\psi|^2 \sim \cos^2 \frac{1}{2}\mathbf{K}\cdot\mathbf{r}$) because of their position dependence near lattice points. The *s*-like combination, like an atomic *s*-level, does not vanish at the ion; in the *p*-like combination the charge density vanishes as the square of the distance from the ion for small distances, which is also a characteristic of atomic *p*-levels.

ENERGY BANDS IN ONE DIMENSION

We can illustrate these general conclusions in one dimension, where twofold degeneracy is the most that can ever occur. In the absence of any interaction the electronic energy levels are just a parabola in k (Figure 9.4a). To leading order in the weak one-dimensional periodic potential this curve remains correct except near Bragg "planes" (which are points in one dimension). When q is near a Bragg "plane" corresponding to the reciprocal lattice vector K (i.e., the point $\frac{1}{2}K$) the corrected energy levels are determined by drawing another free electron parabola centered around K (Figure 9.4b), noting that the degeneracy at the point of intersection is split by $2|U_K|$ in such a way that both curves have zero slope at that point, and redrawing Figure 9.4b to get Figure 9.4c. The original free electron curve is therefore modified as in Figure 9.4d. When all Bragg planes and their associated Fourier components are included, we end up with a set of curves such as those shown in Figure 9.4e. This particular way of depicting the energy levels is known as the *extended-zone scheme.*

If we insist on specifying all the levels by a wave vector k in the first Brillouin zone, then we must translate the pieces of Figure 9.4e, through reciprocal lattice vectors, into the first Brillouin zone. The result is shown in Figure 9.4f. The representation is that of the *reduced-zone scheme* (see page 142).

---

9 This result is often, but not always, true even when the periodic potential is not weak, because the Bragg planes occupy positions of fairly high symmetry.

10 For simplicity we assume here that $U_{\mathbf{K}}$ is real (the crystal has inversion symmetry).

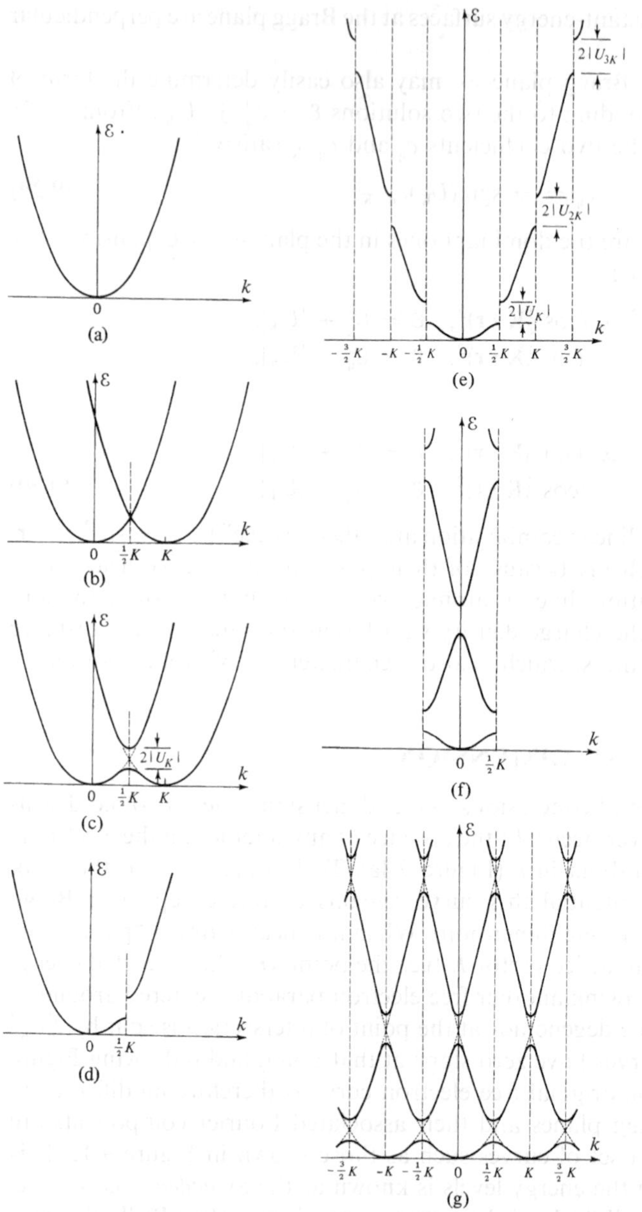

Figure 9.4
(a) The free electron ℰ vs. k parabola in one dimension. (b) Step 1 in the construction to determine the distortion in the free electron parabola in the neighborhood of a Bragg "plane," due to a weak periodic potential. If the Bragg "plane" is that determined by K, a second free electron parabola is drawn, centered on K. (c) Step 2 in the construction to determine the distortion in the free electron parabola in the neighborhood of a Bragg "plane." The degeneracy of the two parabolas at $K/2$ is split. (d) Those portions of part (c) corresponding to the original free electron parabola given in (a). (e) Effect of all additional Bragg "planes" on the free electron parabola. This particular way of displaying the electronic levels in a periodic potential is known as the *extended-zone scheme.* (f) The levels of (e), displayed in a *reduced-zone scheme.* (g) Free electron levels of (e) or (f) in a *repeated-zone scheme.*

One can also emphasize the periodicity of the labeling in k-space by periodically extending Figure 9.4f throughout all of k-space to arrive at Figure 9.4g, which emphasizes that a particular level at k can be described by any wave vector differing from k by a reciprocal lattice vector. This representation is the *repeated-zone scheme* (see page 142). The reduced-zone scheme indexes each level with a k lying in the first zone, while the extended-zone scheme uses a labeling emphasizing continuity with the free electron levels. The repeated-zone scheme is the most general representation, but is highly redundant, since the same level is shown many times, for all equivalent wave vectors $k, k \pm K, k \pm 2K, \ldots$.

ENERGY-WAVE-VECTOR CURVES IN THREE DIMENSIONS

In three dimensions the structure of the energy bands is sometimes displayed by plotting $\mathcal{E}$ vs. **k** along particular straight lines in k-space. Such curves are generally shown in a reduced-zone scheme, since for general directions in k-space they are not periodic. Even in the completely free electron approximation these curves are surprisingly complex. An example is shown in Figure 9.5, which was constructed by plotting, as **k** varied along the particular lines shown, the values of $\mathcal{E}^0_{\mathbf{k}-\mathbf{K}} = \hbar^2(\mathbf{k} - \mathbf{K})^2/2m$ for all reciprocal lattice vectors **K** close enough to the origin to lead to energies lower than the top of the vertical scale.

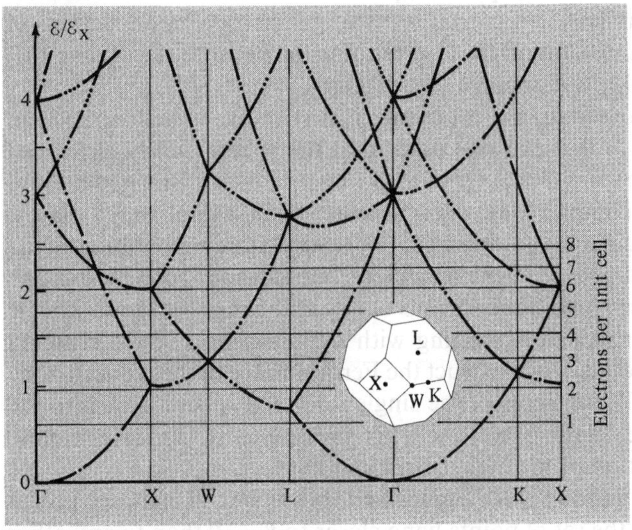

Figure 9.5
Free electron energy levels for an fcc Bravais lattice. The energies are plotted along lines in the first Brillouin zone joining the points $\Gamma(\mathbf{k} = \mathbf{0})$, K, L, W, and X. $\mathcal{E}_X$ is the energy at point X ($[\hbar^2/2m][2\pi/a]^2$). The horizontal lines give Fermi energies for the indicated numbers of electrons per primitive cell. The number of dots on a curve specifies the number of degenerate free electron levels represented by the curve. (From F. Herman, in *An Atomistic Approach to the Nature and Properties of Materials*, J. A. Pask, ed., Wiley, New York, 1967.)

Note that most of the curves are highly degenerate. This is because the directions along which the energy has been plotted are all lines of fairly high symmetry, so points along them are likely to be as far from several other reciprocal lattice vectors as they are from any given one. The addition of a weak periodic potential will in general remove some, but not necessarily all, of this degeneracy. The mathematical theory of groups is often used to determine how such degeneracies will be split.

THE ENERGY GAP

Quite generally, a weak periodic potential introduces an "energy gap" at Bragg planes. By this we mean the following:

When $U_{\mathbf{K}} = 0$, as $\mathbf{k}$ crosses a Bragg plane the energy changes continuously from the lower root of (9.26) to the upper, as illustrated in Figure 9.4b. When $U_{\mathbf{K}} \neq 0$, this is no longer so. The energy only changes continuously with $\mathbf{k}$, as the Bragg plane is crossed, if one stays with the lower (or upper) root, as illustrated in Figure 9.4c. To change branches as $\mathbf{k}$ varies continuously it is now necessary for the energy to change *discontinuously* by at least $2|U_{\mathbf{K}}|$.

We shall see in Chapter 12 that this mathematical separation of the two bands is reflected in a physical separation: When the action of an external field changes an electron's wave vector, the presence of the energy gap requires that upon crossing the Bragg plane, the electron must emerge in a level whose energy remains in the original branch of $\mathcal{E}(\mathbf{k})$. It is this property that makes the energy gap of fundamental importance in electronic transport properties.

BRILLOUIN ZONES

Using the theory of electrons in a weak periodic potential to determine the complete band structure of a three-dimensional crystal leads to geometrical constructions of great complexity. It is often most important to determine the Fermi surface (page 141) and the behavior of the $\mathcal{E}_n(\mathbf{k})$ in its immediate vicinity.

In doing this for weak potentials, the procedure is first to draw the free electron Fermi sphere centered at $\mathbf{k} = \mathbf{0}$. Next, one notes that the sphere will be deformed in a manner of which Figure 9.6 is characteristic[11] when it crosses a Bragg plane and in a correspondingly more complex way when it passes near several Bragg planes. When the effects of all Bragg planes are inserted, this leads to a representation of the Fermi surface as a fractured sphere in the extended-zone scheme. To construct the portions of the Fermi surface lying in the various bands in the repeated-zone scheme one can make a similar construction, starting with free electron spheres centered about all reciprocal lattice points. To construct the Fermi surface in the reduced-zone scheme, one can translate all the pieces of the single fractured sphere back into the first zone through reciprocal lattice vectors. This procedure is made systematic through the geometrical notion of the higher Brillouin zones.

Recall that the first Brillouin zone is the Wigner-Seitz primitive cell of the reciprocal lattice (pages 73 and 89), i.e. the set of points lying closer to $\mathbf{K} = \mathbf{0}$ than to any other

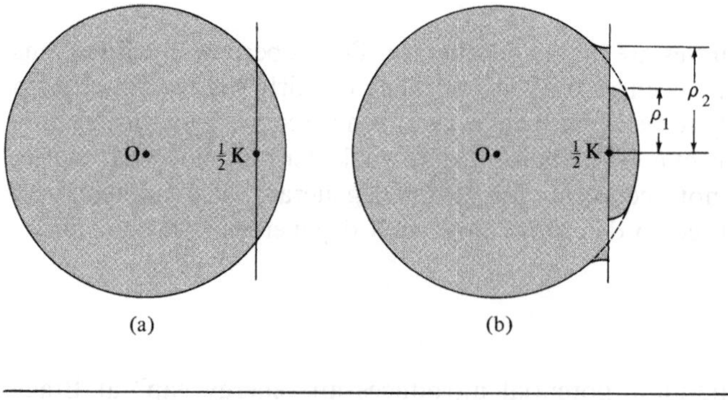

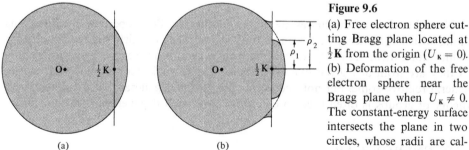

Figure 9.6
(a) Free electron sphere cutting Bragg plane located at $\frac{1}{2}\mathbf{K}$ from the origin ($U_{\mathbf{K}} = 0$). (b) Deformation of the free electron sphere near the Bragg plane when $U_{\mathbf{K}} \neq 0$. The constant-energy surface intersects the plane in two circles, whose radii are calculated in Problem 1.

[11] This follows from the demonstration on page 159 that a constant-energy surface is perpendicular to a Bragg plane when they intersect, in the nearly free electron approximation.

reciprocal lattice point. Since Bragg planes bisect the lines joining the origin to points of the reciprocal lattice, one can equally well define the first zone as the set of points that can be reached from the origin without crossing any Bragg planes.[12]

Higher Brillouin zones are simply other regions bounded by the Bragg planes, defined as follows:

The *first Brillouin zone* is the set of points in k-space that can be reached from the origin without crossing *any* Bragg plane. The *second Brillouin zone* is the set of points that can be reached from the first zone by crossing only one Bragg plane. The $(n+1)$*th Brillouin zone* is the set of points not in the $(n-1)$th zone that can be reached from the nth zone by crossing only one Bragg plane.

Alternatively, the *nth Brillouin zone* can be defined as the set of points that can be reached from the origin by crossing $n-1$ Bragg planes, but no fewer.

These definitions are illustrated in two dimensions in Figure 9.7. The surface of the first three zones for the fcc and bcc lattices are shown in Figure 9.8. Both definitions emphasize the physically important fact that the zones are bounded by Bragg planes. Thus they are regions at whose surfaces the effects of a weak periodic potential are important (i.e., first order), but in whose interior the free electron energy levels are only perturbed in second order.

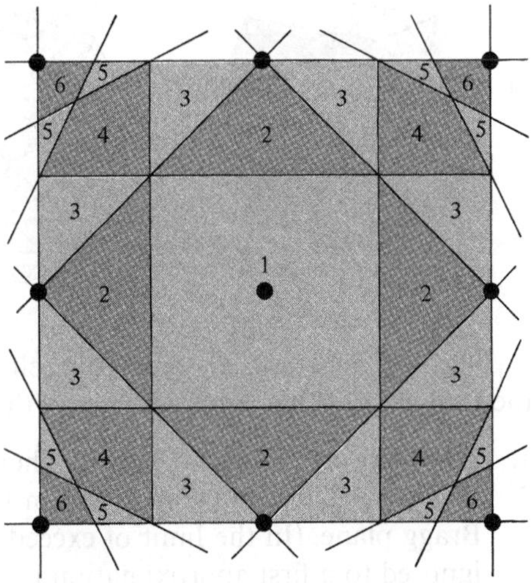

Figure 9.7
Illustration of the definition of the Brillouin zones for a two-dimensional square Bravais lattice. The reciprocal lattice is also a square lattice of side b. The figure shows all Bragg planes (lines, in two dimensions) that lie within the square of side $2b$ centered on the origin. These Bragg planes divide that square into regions belonging to zones 1 to 6. (Only zones 1, 2, and 3 are entirely contained within the square, however.)

It is very important to note that each Brillouin zone is a primitive cell of the reciprocal lattice. This is because the nth Brillouin zone is simply the set of points that have the origin as the nth nearest reciprocal lattice point (a reciprocal lattice point $\mathbf{K}$ is nearer to a point $\mathbf{k}$ than $\mathbf{k}$ is to the origin if and only if $\mathbf{k}$ is separated from the origin by the Bragg plane determined by $\mathbf{K}$). Given this, the proof that the nth Brillouin zone is a primitive cell is identical to the proof on page 73 that the Wigner-Seitz cell (i.e., the first Brillouin zone) is primitive, provided that the phrase "nth nearest neighbor" is substituted for "nearest neighbor" throughout the argument.

---

12 We exclude from consideration points lying *on* Bragg planes, which turn out to be points common to the surface of two or more zones. We define the zones in terms of their interior points.

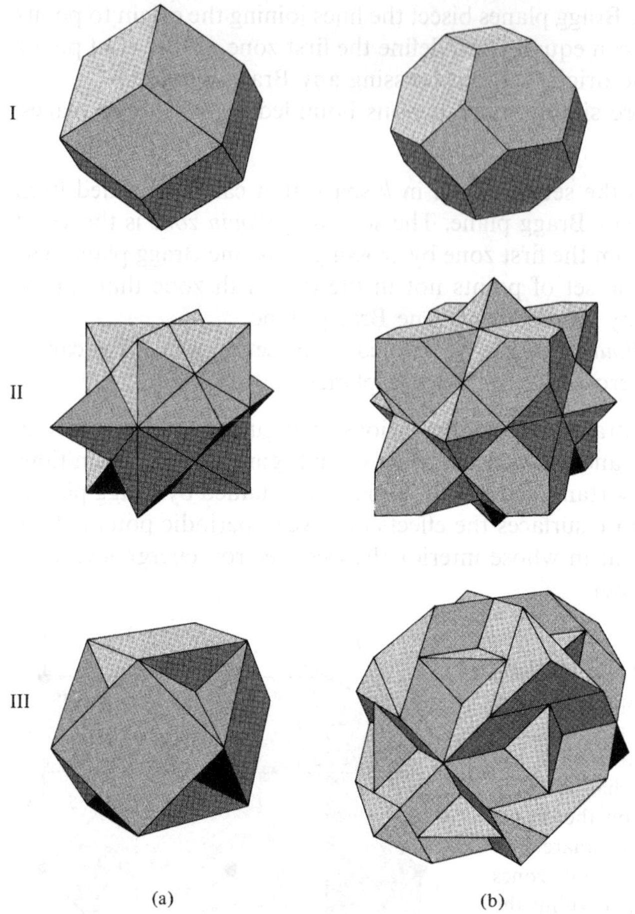

Figure 9.8
Surfaces of the first, second, and third Brillouin zones for (a) body-centered cubic and (b) face-centered cubic crystals. (Only the *exterior* surfaces are shown. It follows from the definition on page 163 that the *interior* surface of the nth zone is identical to the exterior surface of the $(n-1)$th zone.) Evidently the surfaces bounding the zones become increasingly complex as the zone number increases. In practice it is often simplest to construct free electron Fermi surfaces by procedures (such as those described in Problem 4) that avoid making use of the explicit form of the Brillouin zones. (After R. Lück, doctoral dissertation, Technische Hochschule, Stuttgart, 1965.)

Because each zone is a primitive cell, there is a simple algorithm for constructing the branches of the Fermi surface in the repeated-zone scheme[13]:

1. Draw the free electron Fermi sphere.
2. Deform it slightly (as illustrated in Figure 9.6) in the immediate vicinity of every Bragg plane. (In the limit of exceedingly weak potentials this step is sometimes ignored to a first approximation.)
3. Take that portion of the surface of the free electron sphere lying within the nth Brillouin zone, and translate it through all reciprocal lattice vectors. The resulting surface is the branch of the Fermi surface (conventionally assigned to the nth band) in the repeated-zone scheme.[14]

---

[13] The representation of the Fermi surface in the repeated-zone scheme is the most general. After surveying each branch in its full periodic splendor, one can pick that primitive cell which most lucidly represents the topological structure of the whole (which is often, but by no means always, the first Brillouin zone).

[14] An alternative procedure is to translate the pieces of the Fermi surface in the nth zone through those reciprocal lattice vectors that take the pieces of the nth zone in which they are contained, into the first zone. (Such translations exist because the nth zone is a primitive cell.) This is illustrated in Figure 9.9. The Fermi surface in the repeated-zone scheme is then constructed by translating the resulting first zone structures through all reciprocal lattice vectors.

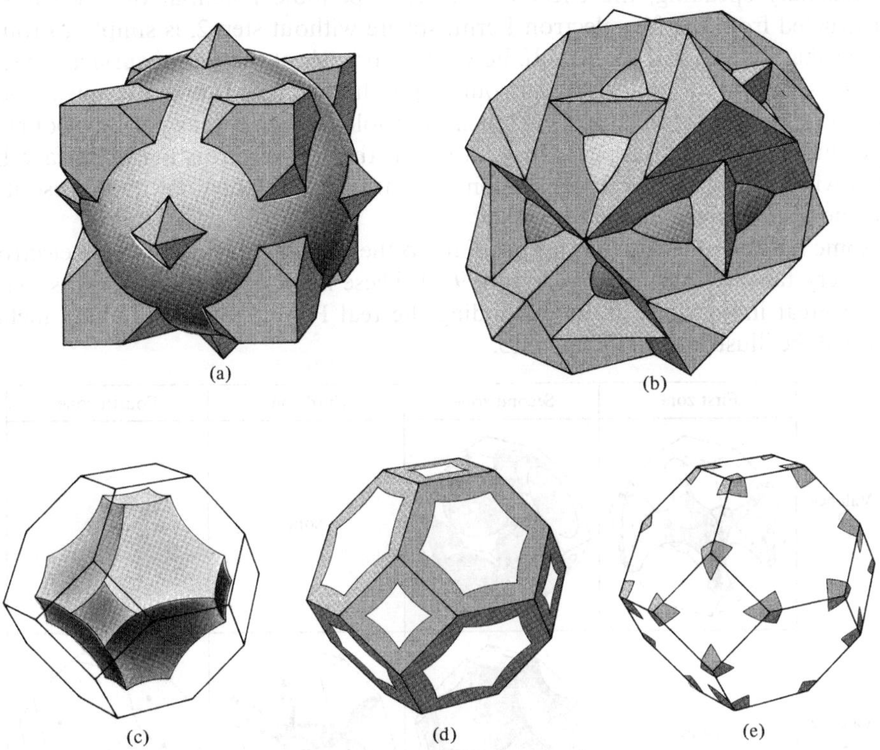

Figure 9.9
The free electron Fermi sphere for a face-centered cubic metal of valence 4. The first zone lies entirely within the interior of the sphere, and the sphere does not extend beyond the fourth zone. Thus the only zone surfaces intersected by the surface of the sphere are the (exterior) surfaces of the second and third zones (cf. Figure 9.8b). The second-zone Fermi surface consists of those parts of the surface of the sphere lying entirely within the polyhedron bounding the second zone (i.e., all of the sphere except the parts extending beyond the polyhedron in (a)). When translated through reciprocal lattice vectors into the first zone, the pieces of the second-zone surface give the simply connected figure shown in (c). (It is known as a "hole surface"; the levels it encloses have higher energies than those outside). The third-zone Fermi surface consists of those parts of the surface of the sphere lying outside of the second zone (i.e., the parts extending beyond the polyhedron in (a)) that do not lie outside the third zone (i.e., that are contained within the polyhedron shown in (b)). When translated through reciprocal lattice vectors into the first zone, these pieces of sphere give the multiply connected structure shown in (d). The fourth-zone Fermi surface consists of the remaining parts of the surface of the sphere that lie outside the third zone (as shown in (b)). When translated through reciprocal lattice vectors into the first zone they form the "pockets of electrons" shown in (e). For clarity (d) and (e) show only the intersection of the third and fourth zone Fermi surfaces with the surface of the first zone. (From R. Lück, *op. cit.*)

Generally speaking, the effect of the weak periodic potential on the surfaces constructed from the free electron Fermi sphere without step 2, is simply to round off the sharp edges and corners. If, however, a branch of the Fermi surface consists of very small pieces of surface (surrounding either occupied or unoccupied levels, known as "pockets of electrons" or "pockets of holes"), then a weak periodic potential may cause these to disappear. In addition, if the free electron Fermi surface has parts with a very narrow cross section, a weak periodic potential may cause it to become disconnected at such points.

Some further constructions appropriate to the discussion of almost free electrons in fcc crystals are illustrated in Figure 9.10. These free-electron-like Fermi surfaces are of great importance in understanding the real Fermi surfaces of many metals. This will be illustrated in Chapter 15.

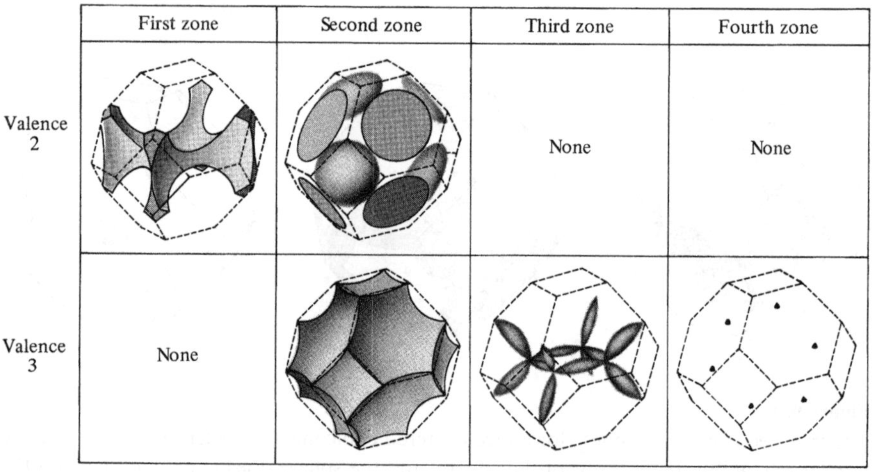

First zone
Second zone
Third zone
Fourth zone
Valence 2
None
None
Valence 3
None

Figure 9.10
The free electron Fermi surfaces for face-centered cubic metals of valence 2 and 3. (For valence 1 the surface lies entirely within the interior of the first zone and therefore remains a sphere to lowest order; the surface for valence 4 is shown in Figure 9.9.) All branches of the Fermi surface are shown. The primitive cells in which they are displayed have the shape and orientation of the first Brillouin zone. However, the cell is actually the first zone (i.e., is centered on $\mathbf{K} = \mathbf{0}$) only in the figures illustrating the second zone surfaces. In the first and third zone figures $\mathbf{K} = \mathbf{0}$ lies at the center of one of the horizontal faces, while for the fourth zone figure it lies at the center of the hexagonal face on the upper right (or the parallel face opposite it (hidden)). The six tiny pockets of electrons constituting the fourth zone surface for valence 3 lie at the corners of the regular hexagon given by displacing that hexagonal face in the [111] direction by half the distance to the face opposite it. (After W. Harrison, *Phys. Rev.* **118**, 1190 (1960).) Corresponding constructions for body-centered cubic metals can be found in the Harrison paper.

THE GEOMETRICAL STRUCTURE FACTOR IN MONATOMIC LATTICES WITH BASES

Nothing said up to now has exploited any properties of the potential $U(\mathbf{r})$ other than its periodicity, and, for convenience, inversion symmetry. If we pay somewhat closer attention to the form of U, recognizing that it will be made up of a sum of atomic potentials centered at the positions of the ions, then we can draw some further conclusions that are important in studying the electronic structure of monatomic lattices with a basis, such as the diamond and hexagonal close-packed (hcp) structures.

Suppose that the basis consists of identical ions at positions $\mathbf{d}_j$. Then the periodic potential $U(\mathbf{r})$ will have the form

$$U(\mathbf{r}) = \sum_{\mathbf{R}} \sum_{j} \phi(\mathbf{r} - \mathbf{R} - \mathbf{d}_j). \tag{9.31}$$

If we place this into Eq. (8.32) for $U_{\mathbf{K}}$, we find that

$$\begin{aligned} U_{\mathbf{K}} &= \frac{1}{v} \int_{\text{cell}} d\mathbf{r}\, e^{-i\mathbf{K}\cdot\mathbf{r}} \sum_{\mathbf{R},j} \phi(\mathbf{r} - \mathbf{R} - \mathbf{d}_j) \\ &= \frac{1}{v} \int_{\substack{\text{all}\\ \text{space}}} d\mathbf{r}\, e^{-i\mathbf{K}\cdot\mathbf{r}} \sum_{j} \phi(\mathbf{r} - \mathbf{d}_j), \end{aligned} \tag{9.32}$$

or

$$U_{\mathbf{K}} = \frac{1}{v} \phi(\mathbf{K}) S_{\mathbf{K}}^{*}, \tag{9.33}$$

where $\phi(\mathbf{K})$ is the Fourier transform of the atomic potential,

$$\phi(\mathbf{K}) = \int_{\substack{\text{all}\\ \text{space}}} d\mathbf{r}\, e^{-i\mathbf{K}\cdot\mathbf{r}} \phi(\mathbf{r}), \tag{9.34}$$

and $S_{\mathbf{K}}$ is the geometrical structure factor introduced in our discussion of X-ray diffraction (Chapter 6):

$$S_{\mathbf{K}} = \sum_{j} e^{i\mathbf{K}\cdot\mathbf{d}_j}. \tag{9.35}$$

Thus when the basis leads to a vanishing structure factor for some Bragg planes, i.e., when the X-ray diffraction peaks from these planes are missing, then the Fourier component of the periodic potential associated with such planes vanishes; i.e., the lowest-order splitting in the free electron levels disappears.

This result is of particular importance in the theory of metals with the hexagonal close-packed structure, of which there are over 25 (Table 4.4). The first Brillouin zone for the simple hexagonal lattice is a prism on a regular hexagon base. However, the structure factor associated with the hexagonal top and bottom of the prism vanishes (Problem 3, Chapter 6).

Therefore, according to nearly free electron theory, there is *no* first-order splitting of the free electron levels at these faces. It might appear that small splittings would still occur as a result of second-order (and higher-order) effects. However, if the one-electron Hamiltonian is independent of the spin, then in the hcp structure any Bloch level with wave vector $\mathbf{k}$ on the hexagonal face of the first Brillouin zone can be proved to be at least twofold degenerate. Accordingly, the splitting is rigorously zero. In a situation like this it is often more convenient to consider a representation of the zone structure in which those planes with zero gap are actually ignored. The regions that one is then led to consider are known as Jones zones or large zones.

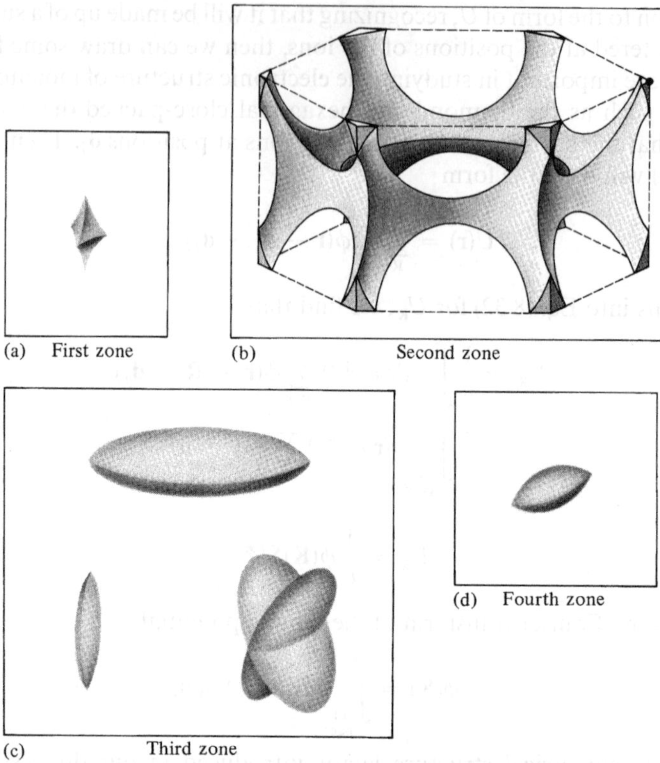

Figure 9.11
Free electron Fermi surface for a divalent hcp metal with ideal $c/a = 1.633$. Since the hcp structure is simple hexagonal with two atoms per primitive cell, there are four electrons per primitive cell to be accommodated. The resulting Fermi surface comes in many pieces, whose names reveal an interesting level of imagination and taste. (a) *The cap.* The *first zone* is almost entirely filled by the free electron sphere, but there are small unoccupied regions in the six upper and six lower corners. These can be assembled, by translations through reciprocal lattice vectors, into two of the objects shown. (b) *The monster.* Portions of the free electron sphere in the *second zone* can be translated back into the first zone to form one of the large structures shown in the second-zone picture. The monster encloses unoccupied levels. (c) Portions of the free electron sphere in the *third zone* can be reassembled into several electron-enclosing surfaces. There is one *lens*, two *cigars*, and three *butterflies*. (d) Those few occupied free electron levels in the *fourth zone* can be reassembled into three pockets of the type pictured.

These structures arise when there is significant splitting of the free electron levels on the hexagonal faces of the first zone as a result of spin-orbit coupling. When spin-orbit coupling is weak (as it is in the lighter elements), there is negligible splitting on these faces, and the appropriate structures are those shown in Figure 9.12. (From J. B. Ketterson and R. W. Stark, *Phys. Rev.* **156**, 751 (1967).)

IMPORTANCE OF SPIN-ORBIT COUPLING AT POINTS OF HIGH SYMMETRY

Until now we have regarded the electron spin as being completely inert dynamically. In fact, however, an electron moving through an electric field, such as that of the periodic potential $U(\mathbf{r})$, experiences a potential proportional to the scalar product of its spin magnetic moment with the vector product of its velocity and the electric field. This additional interaction is referred to as the *spin-orbit coupling*, and is of great importance in atomic physics (see Chapter 31). Spin-orbit coupling is important in calculating the almost free electron levels at points in k-space of high symmetry, since it often happens that levels that are rigorously degenerate when it is ignored are split by the spin-orbit coupling.

For example, the splitting of the electronic levels on the hexagonal faces of the first zone in hcp metals is entirely due to spin-orbit coupling. Since the strength of spin-orbit coupling increases with atomic number, this splitting is appreciable in the heavy hexagonal metals, but can be small enough to be ignored in the light ones. Correspondingly, there are two different schemes for constructing free electronlike Fermi surfaces in hexagonal metals. These are illustrated in Figures 9.11 and 9.12.

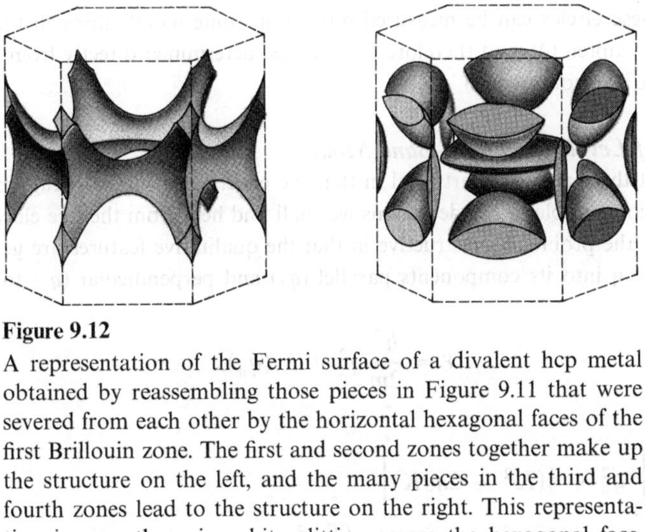

Figure 9.12
A representation of the Fermi surface of a divalent hcp metal obtained by reassembling those pieces in Figure 9.11 that were severed from each other by the horizontal hexagonal faces of the first Brillouin zone. The first and second zones together make up the structure on the left, and the many pieces in the third and fourth zones lead to the structure on the right. This representation ignores the spin orbit splitting across the hexagonal face. (After W. Harrison, *Phys. Rev.* **118**, 1190 (1960).)

PROBLEMS

1. *Nearly Free Electron Fermi Surface Near a Single Bragg Plane*

To investigate the nearly free electron band structure given by (9.26) near a Bragg plane, it is convenient to measure the wave vector $\mathbf{q}$ with respect to the point $\frac{1}{2}\mathbf{K}$ on the Bragg plane. If we write $\mathbf{q} = \frac{1}{2}\mathbf{K} + \mathbf{k}$, and resolve $\mathbf{k}$ into its components parallel ($k_{\parallel}$) and perpendicular ($k_{\perp}$)

to **K**, then (9.26) becomes

$$\mathcal{E} = \mathcal{E}^0_{\mathbf{K}/2} + \frac{\hbar^2}{2m}k^2 \pm \left(4\mathcal{E}^0_{\mathbf{K}/2}\frac{\hbar^2}{2m}{k_\parallel}^2 + |U_\mathbf{K}|^2\right)^{1/2}. \tag{9.36}$$

It is also convenient to measure the Fermi energy $\mathcal{E}_F$ with respect to the lowest value assumed by either of the bands given by (9.36) in the Bragg plane, writing:

$$\mathcal{E}_F = \mathcal{E}^0_{\mathbf{K}/2} - |U_K| + \Delta, \tag{9.37}$$

so that when $\Delta < 0$, no Fermi surface intersects the Bragg plane.

(a) Show that when $0 < \Delta < 2|U_\mathbf{K}|$, the Fermi surface lies entirely in the lower band and intersects the Bragg plane in a circle of radius

$$\rho = \sqrt{\frac{2m\Delta}{\hbar^2}}. \tag{9.38}$$

(b) Show that if $\Delta > |2U_\mathbf{K}|$, the Fermi surface lies in both bands, cutting the Bragg plane in two circles of radii ρ_1 and ρ_2 (Figure 9.6), and that the difference in the areas of the two circles is

$$\pi({\rho_2}^2 - {\rho_1}^2) = \frac{4m\pi}{\hbar^2}|U_\mathbf{K}|. \tag{9.39}$$

(The area of these circles can be measured directly in some metals through the de Haas–van Alphen effect (Chapter 14), and therefore $|U_\mathbf{K}|$ can be determined directly from experiment for such nearly free electron metals.)

2. *Density of Levels for a Two-Band Model*

To some extent this problem is artificial in that the effects of neglected Bragg planes can lead to corrections comparable to the deviations we shall find here from the free electron result. On the other hand, the problem is instructive in that the qualitative features are general.

If we resolve **q** into its components parallel ($q_\parallel$) and perpendicular ($q_\perp$) to **K**, then (9.26) becomes

$$\mathcal{E} = \frac{\hbar^2}{2m}q_\perp^2 + h_\pm(q_\parallel), \tag{9.40}$$

where

$$h_\pm(q_\parallel) = \frac{\hbar^2}{2m}\left[q_\parallel^2 + \tfrac{1}{4}(K^2 - 2q_\parallel K)\right] \pm \left\{\left[\frac{\hbar^2}{2m}\tfrac{1}{2}(K^2 - 2q_\parallel K)\right]^2 + |U_\mathbf{K}|^2\right\}^{1/2} \tag{9.41}$$

is only a function of $q_\parallel$. The density of levels can be evaluated from (8.57) by performing the integral in an appropriate primitive cell over wave vectors **q** in cylindrical coordinates with the z-axis along **K**.

(a) Show that when the integral over **q** is performed, the result for each band is

$$g(\mathcal{E}) = \frac{1}{4\pi^2}\left(\frac{2m}{\hbar^2}\right)(q_\parallel^{\max} - q_\parallel^{\min}), \tag{9.42}$$

where, for each band, $q_\parallel^{\max}$ and $q_\parallel^{\min}$ are the solutions to $\mathcal{E} = h_\pm(q_\parallel)$. Verify that the familiar free electron result is obtained in the limit $|U_\mathbf{K}| \to 0$.

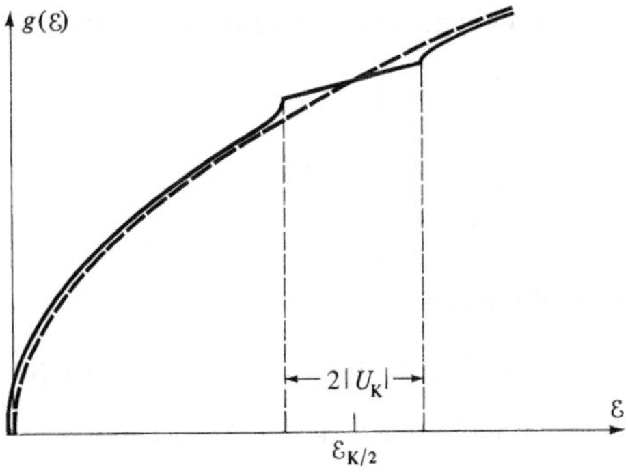

Figure 9.13
Density of levels in the two-band approximation. The dashed line is the free electron result Eq. (2.63). Note that in contrast to earlier figures in this chapter, this one explicitly shows second-order corrections to the free electron result far from the Bragg plane.

(b) Show that

$$q_{\|}^{\min} = -\sqrt{\frac{2m\mathcal{E}}{\hbar^2}} + O(U_{\mathbf{K}}^2), \quad (\mathcal{E} > 0), \qquad q_{\|}^{\max} = \tfrac{1}{2}\mathbf{K} \tag{9.43}$$

for the lower band, if the constant energy surface (at energy $\mathcal{E}$) cuts the zone plane (that is, $\mathcal{E}^0_{\mathbf{K}/2} - |U_{\mathbf{K}}| \leqslant \mathcal{E} \leqslant \mathcal{E}^0_{\mathbf{K}/2} + |U_{\mathbf{K}}|$).

(c) Show that for the *upper* band (9.42) should be interpreted as giving a density of levels

$$g_+(\mathcal{E}) = \frac{1}{4\pi^2}\left(\frac{2m}{\hbar^2}\right)(q_{\|}^{\max} - \tfrac{1}{2}K), \qquad \text{for } \mathcal{E} > \mathcal{E}_{\mathbf{K}/2} + |U_{\mathbf{K}}|. \tag{9.44}$$

(d) Show that $dg/d\mathcal{E}$ is singular at $\mathcal{E} = \mathcal{E}_{\mathbf{K}/2} \pm |U_{\mathbf{K}}|$, so that the density of levels has the form shown in Figure 9.13. (These singularities are not peculiar either to the weak potential or two-band approximations. See page 145.)

3. *Effect of Weak Periodic Potential at Places in k-Space Where Bragg Planes Meet*
Consider the point W ($\mathbf{k}_W = (2\pi/a)(1, \frac{1}{2}, 0)$) in the Brillouin zone of the fcc structure shown (see Figure 9.14). Here three Bragg planes ((200), (111), (11$\bar{1}$)) meet, and accordingly the free electron energies

$$\begin{aligned}
\mathcal{E}^0_1 &= \frac{\hbar^2}{2m}k^2, \\
\mathcal{E}^0_2 &= \frac{\hbar^2}{2m}\left(\mathbf{k} - \frac{2\pi}{a}(1, 1, 1)\right)^2, \\
\mathcal{E}^0_3 &= \frac{\hbar^2}{2m}\left(\mathbf{k} - \frac{2\pi}{a}(1, 1, \bar{1})\right)^2, \\
\mathcal{E}^0_4 &= \frac{\hbar^2}{2m}\left(\mathbf{k} - \frac{2\pi}{a}(2, 0, 0)\right)^2
\end{aligned} \tag{9.45}$$

are degenerate when $\mathbf{k} = \mathbf{k}_W$, and equal to $\mathcal{E}_W = \hbar^2 \mathbf{k}_W{}^2/2m$.

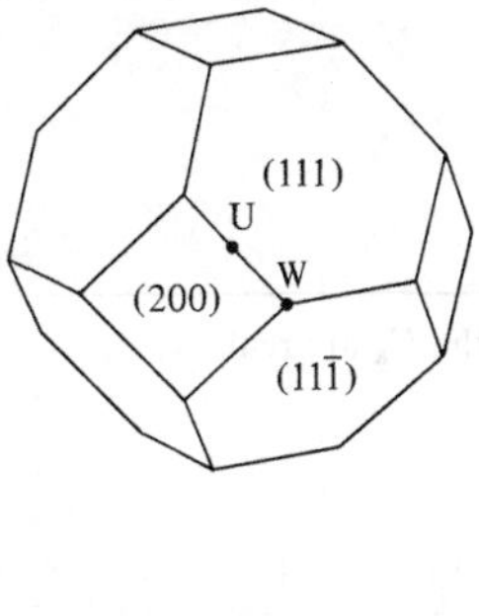

Figure 9.14
First Brillouin zone for a face-centered cubic crystal.

(a) Show that in a region of k-space near W, the first-order energies are given by solutions to[15]

$$\begin{vmatrix} \mathcal{E}_1^0 - \mathcal{E} & U_1 & U_1 & U_2 \\ U_1 & \mathcal{E}_2^0 - \mathcal{E} & U_2 & U_1 \\ U_1 & U_2 & \mathcal{E}_3^0 - \mathcal{E} & U_1 \\ U_2 & U_1 & U_1 & \mathcal{E}_4^0 - \mathcal{E} \end{vmatrix} = 0$$

where $U_2 = U_{200}$, $U_1 = U_{111} = U_{11\bar{1}}$, and that at W the roots are

$$\mathcal{E} = \mathcal{E}_W - U_2 \quad \text{(twice)}, \quad \mathcal{E} = \mathcal{E}_W + U_2 \pm 2U_1. \tag{9.46}$$

(b) Using a similar method, show that the energies at the point U ($\mathbf{k}_U = (2\pi/a)(1, \frac{1}{4}, \frac{1}{4})$) are

$$\mathcal{E} = \mathcal{E}_U - U_2, \quad \mathcal{E} = \mathcal{E}_U + \tfrac{1}{2}U_2 \pm \tfrac{1}{2}(U_2{}^2 + 8U_1{}^2)^{1/2}, \tag{9.47}$$

where $\mathcal{E}_U = \hbar^2 \mathbf{k}_U{}^2/2m$.

4. *Alternative Definition of Brillouin Zones*

Let $\mathbf{k}$ be a point in reciprocal space. Suppose spheres of radius k are drawn about every reciprocal lattice point $\mathbf{K}$ except the origin. Show that if $\mathbf{k}$ is in the interior of $n - 1$ spheres, and on the surface of none, then it lies in the interior of the nth Brillouin zone. Show that if $\mathbf{k}$ is in the interior of $n - 1$ spheres and on the surface of m additional spheres, then it is a point common to the boundaries of the nth, $(n + 1)$th, . . . , $(n + m)$th Brillouin zones.

5. *Brillouin Zones in a Two-Dimensional Square Lattice*

Consider a two-dimensional square lattice with lattice constant a.

(a) Write down, in units of $2\pi/a$, the radius of a circle that can accommodate m free electrons per primitive cell. Construct a table listing which of the first seven zones of the square lattice (Figure 9.15a) are completely full, which are partially empty, and which are completely empty for $m = 1, 2, \ldots, 12$. Verify that if $m \leqslant 12$, the occupied levels lie entirely within the first seven zones, and that when $m \geqslant 13$, levels in the eighth and higher zones become occupied.

(b) Draw pictures in suitable primitive cells of all branches of the Fermi surface for the cases $m = 1, 2, \ldots, 7$. The third zone surface for $m = 4$, for example, can be displayed as in Figure 9.15b.

15 Assume that the periodic potential U has inversion symmetry so that the $U_{\mathbf{K}}$ are real.

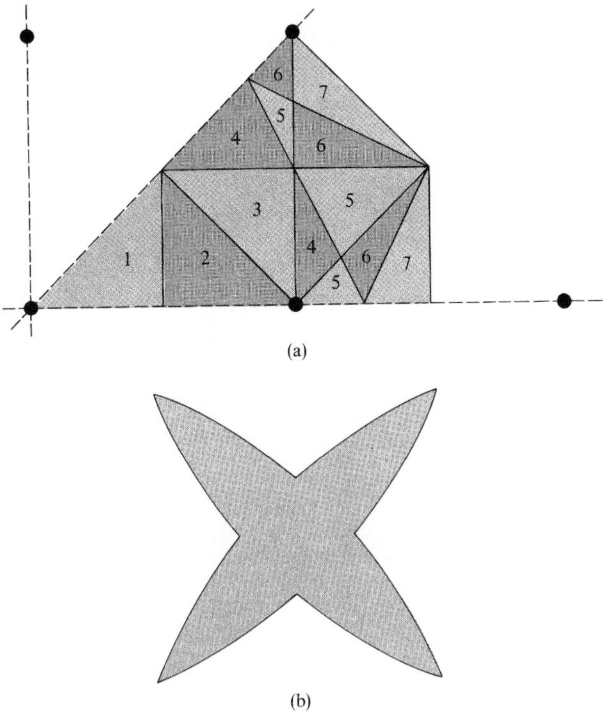

Figure 9.15

(a) The first seven Brillouin zones for the two-dimensional square lattice. Because of the lattice symmetry it is necessary to display only one eighth of the figure. The rest can be reconstructed by reflection in the dashed lines (which are not zone boundaries). (b) Fermi surface in the third zone for a square lattice with four electrons per unit cell. (The scale in (b) has been expanded considerably.)

10
The Tight-Binding Method

Linear Combinations of Atomic Orbitals

Application to Bands from *s*-Levels

General Features of Tight-Binding Levels

Wannier Functions

In Chapter 9 we calculated electronic levels in a metal by viewing it as a gas of nearly free conduction electrons, only weakly perturbed by the periodic potential of the ions. We can also take a very different point of view, regarding a solid (metal or insulator) as a collection of weakly interacting neutral atoms. As an extreme example of this, imagine assembling a group of sodium atoms into a body-centered cubic array with a lattice constant of the order of centimeters rather than angstroms. All electrons would then be in atomic levels localized at lattice sites, bearing no resemblance to the linear combinations of a few plane waves described in Chapter 9.

If we were to shrink the artificially large lattice constant of our array of sodium atoms, at some point before the actual lattice constant of metallic sodium was reached we would have to modify our identification of the electronic levels of the array with the atomic levels of isolated sodium atoms. This would become necessary for a particular atomic level, when the interatomic spacing became comparable to the spatial extent of its wave function, for an electron in that level would then feel the presence of the neighboring atoms.

The actual state of affairs for the 1*s*, 2*s*, 2*p* and 3*s* levels of atomic sodium is shown in Figure 10.1. The atomic wave functions for these levels are drawn about two nuclei separated by 3.7 Å, the nearest-neighbor distance in metallic sodium. The overlap of the 1*s* wave functions centered on the two sites is utterly negligible, indicating that these atomic levels are essentially unaltered in metallic sodium. The overlap of the 2*s*- and 2*p*-levels is exceedingly small, and one might hope to find levels in the metal very closely related to these. However, the overlap of the 3*s*-levels (which hold the atomic valence electrons) is substantial, and there is no reason to expect the actual electronic levels of the metal to resemble these atomic levels.

The *tight-binding approximation* deals with the case in which the overlap of atomic wave functions is enough to require corrections to the picture of isolated atoms, but not so much as to render the atomic description completely irrelevant. The approximation is most useful for describing the energy bands that arise from the partially filled *d*-shells of transition metal atoms and for describing the electronic structure of insulators.

Quite apart from its practical utility, the tight-binding approximation provides an instructive way of viewing Bloch levels complementary to that of the nearly free electron picture, permitting a reconciliation between the apparently contradictory features of localized atomic levels on the one hand, and free electron-like plane-wave levels on the other.

GENERAL FORMULATION

In developing the tight-binding approximation, we assume that in the vicinity of each lattice point the full periodic crystal Hamiltonian, H, can be approximated by the Hamiltonian, H_{at}, of a single atom located at the lattice point. We also assume that the bound levels of the atomic Hamiltonian are well localized; i.e., if ψ_n is a bound level of H_{at} for an atom at the origin,

$$H_{\text{at}}\psi_n = E_n\psi_n, \tag{10.1}$$

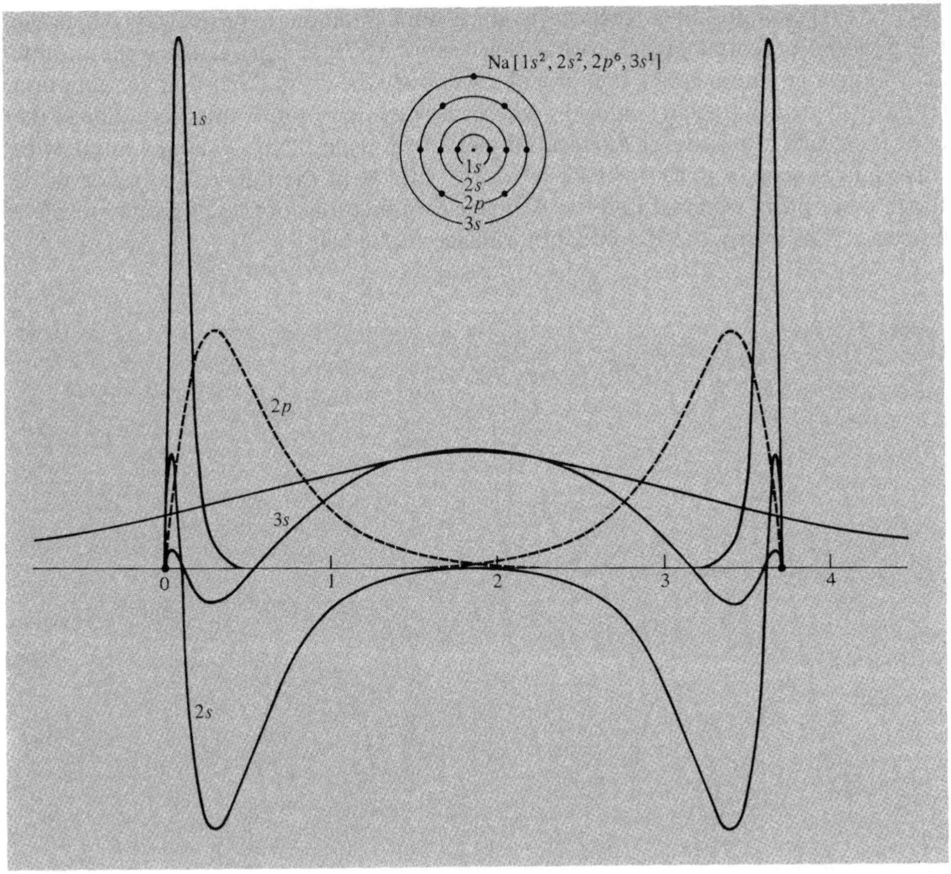

Figure 10.1
Calculated electron wave functions for the levels of atomic sodium, plotted about two nuclei separated by the nearest-neighbor distance in metallic sodium, 3.7 Å. The solid curves are $r\psi(r)$ for the 1*s*, 2*s*, and 3*s* levels. The dashed curve is *r* times the radial wave function for the 2*p* levels. Note how the 3*s* curves overlap extensively, the 2*s* and 2*p* curves overlap only a little, and the 1*s* curves have essentially no overlap. The curves are taken from calculations by D. R. Hartree and W. Hartree, *Proc. Roy. Soc.* **A193**, 299 (1948). The scale on the *r*-axis is in angstroms.

then we require that $\psi_n(\mathbf{r})$ be very small when r exceeds a distance of the order of the lattice constant, which we shall refer to as the "range" of ψ_n.

In the extreme case in which the crystal Hamiltonian begins to differ from H_{at} (for an atom whose lattice point we take as the origin) only at distances from $\mathbf{r} = \mathbf{0}$ that exceed the range of $\psi_n(\mathbf{r})$, the wave function $\psi_n(\mathbf{r})$ will be an excellent approximation to a stationary-state wave function for the full Hamiltonian, with eigenvalue E_n. So also will the wave functions $\psi_n(\mathbf{r} - \mathbf{R})$ for all $\mathbf{R}$ in the Bravais lattice, since H has the periodicity of the lattice.

To calculate corrections to this extreme case, we write the crystal Hamiltonian H as

$$H = H_{\text{at}} + \Delta U(\mathbf{r}), \tag{10.2}$$

where $\Delta U(\mathbf{r})$ contains all corrections to the atomic potential required to produce the full periodic potential of the crystal (see Figure 10.2). If $\psi_n(\mathbf{r})$ satisfies the atomic Schrödinger equation (10.1), then it will also satisfy the crystal Schrödinger equation (10.2), provided that $\Delta U(\mathbf{r})$ vanishes wherever $\psi_n(\mathbf{r})$ does not. If this were indeed the case, then each atomic level $\psi_n(\mathbf{r})$ would yield N levels in the periodic potential, with wave functions $\psi_n(\mathbf{r} - \mathbf{R})$, for each of the N sites $\mathbf{R}$ in the lattice. To preserve the Bloch description we must find the N linear combinations of these degenerate wave functions that satisfy the Bloch condition (see Eq. (8.6)):

$$\psi(\mathbf{r} + \mathbf{R}) = e^{i\mathbf{k}\cdot\mathbf{R}}\psi(\mathbf{r}). \tag{10.3}$$

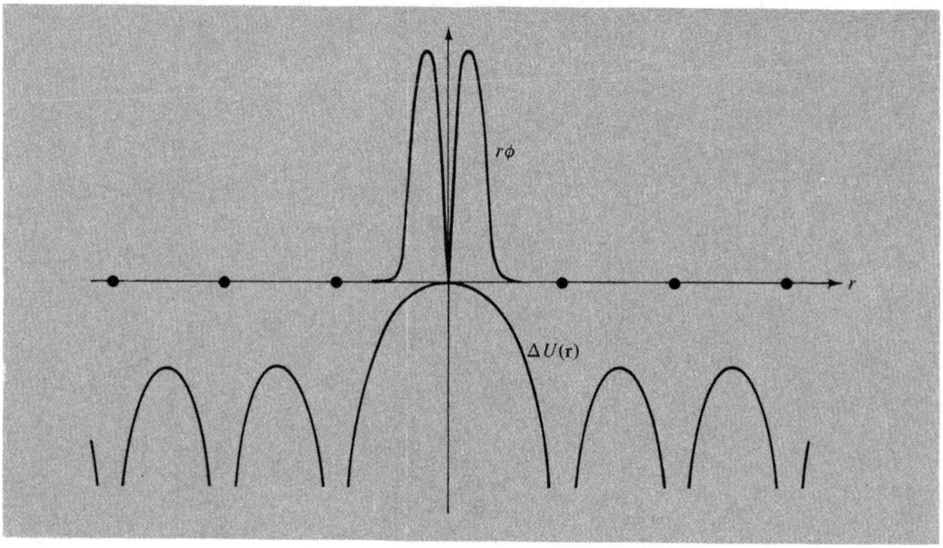

Figure 10.2
The lower curve depicts the function $\Delta U(\mathbf{r})$ drawn along a line of atomic sites. When $\Delta U(\mathbf{r})$ is added to a single atomic potential localized at the origin, the full periodic potential $U(\mathbf{r})$ is recovered. The upper curve represents r times an atomic wave function localized at the origin. When $r\phi(\mathbf{r})$ is large, $\Delta U(\mathbf{r})$ is small, and vice versa.

The N linear combinations we require are

$$\psi_{n\mathbf{k}}(\mathbf{r}) = \sum_{\mathbf{R}} e^{i\mathbf{k}\cdot\mathbf{R}}\psi_n(\mathbf{r} - \mathbf{R}), \tag{10.4}$$

where $\mathbf{k}$ ranges through the N values in the first Brillouin zone consistent with the Born-von Karman periodic boundary condition.[1] The Bloch condition (10.3) is verified for the wave functions (10.4) by noting that

[1] Except when explicitly studying surface effects, one should avoid the temptation to treat a finite crystal by restricting the summation on $\mathbf{R}$ in (10.4) to the sites of a finite portion of the Bravais lattice. It is far more convenient to sum over an infinite Bravais lattice (the sum converging rapidly because of the short range of the atomic wave function ψ_n) and to represent the finite crystal with the usual Born-von Karman boundary condition, which places the standard restriction (8.27) on $\mathbf{k}$, when the Bloch condition holds. With the sum taken over all sites, for example, it is permissible to make the crucial replacement of the summation variable $\mathbf{R}'$ by $\bar{\mathbf{R}} = \mathbf{R}' - \mathbf{R}$, in the second to last line of Eq. (10.5).

$$\begin{aligned}\psi(\mathbf{r}+\mathbf{R}) &= \sum_{\mathbf{R}'} e^{i\mathbf{k}\cdot\mathbf{R}'}\psi_n(\mathbf{r}+\mathbf{R}-\mathbf{R}') \\ &= e^{i\mathbf{k}\cdot\mathbf{R}}\left[\sum_{\mathbf{R}'} e^{i\mathbf{k}\cdot(\mathbf{R}'-\mathbf{R})}\psi_n(\mathbf{r}-(\mathbf{R}'-\mathbf{R}))\right] \\ &= e^{i\mathbf{k}\cdot\mathbf{R}}\left[\sum_{\bar{\mathbf{R}}} e^{i\mathbf{k}\cdot\bar{\mathbf{R}}}\psi_n(\mathbf{r}-\bar{\mathbf{R}})\right] \\ &= e^{i\mathbf{k}\cdot\mathbf{R}}\psi(\mathbf{r}). \end{aligned} \tag{10.5}$$

Thus the wave functions (10.4) satisfy the Bloch condition with wave vector **k**, while continuing to display the atomic character of the levels. The energy bands arrived at in this way, however, have little structure, $\mathcal{E}_n(\mathbf{k})$ being simply the energy of the atomic level, E_n, regardless of the value of **k**. To remedy this deficiency we must recognize that a more realistic assumption is that $\psi_n(\mathbf{r})$ becomes small, but not precisely zero, before $\Delta U(\mathbf{r})$ becomes appreciable (see Figure 10.2). This suggests that we seek a solution to the full crystal Schrödinger equation that retains the general form of (10.4):[2]

$$\psi(\mathbf{r}) = \sum_{\mathbf{R}} e^{i\mathbf{k}\cdot\mathbf{R}}\phi(\mathbf{r}-\mathbf{R}), \tag{10.6}$$

but with the function $\phi(\mathbf{r})$ not necessarily an exact atomic stationary-state wave function, but one to be determined by further calculation. If the product $\Delta U(\mathbf{r})\psi_n(\mathbf{r})$, though nonzero, is exceedingly small, we might expect the function $\phi(\mathbf{r})$ to be quite close to the atomic wave function $\psi_n(\mathbf{r})$ or to wave functions with which $\psi_n(\mathbf{r})$ is degenerate. Based on this expectation, one seeks a $\phi(\mathbf{r})$ that can be expanded in a relatively small number of localized atomic wave functions:[3,4]

$$\phi(\mathbf{r}) = \sum_{n} b_n\psi_n(\mathbf{r}). \tag{10.7}$$

If we multiply the crystal Schrödinger equation

$$H\psi(\mathbf{r}) = (H_{\text{at}} + \Delta U(\mathbf{r}))\psi(\mathbf{r}) = \mathcal{E}(\mathbf{k})\psi(\mathbf{r}) \tag{10.8}$$

by the atomic wave function $\psi_m{}^*(\mathbf{r})$, integrate over all **r**, and use the fact that

$$\int \psi_m{}^*(\mathbf{r})H_{\text{at}}\psi(\mathbf{r})\,d\mathbf{r} = \int (H_{\text{at}}\psi_m(\mathbf{r}))^*\psi(\mathbf{r})\,d\mathbf{r} = E_m \int \psi_m{}^*(\mathbf{r})\psi(\mathbf{r})\,d\mathbf{r}, \tag{10.9}$$

we find that

$$(\mathcal{E}(\mathbf{k}) - E_m)\int \psi_m{}^*(\mathbf{r})\psi(\mathbf{r})\,d\mathbf{r} = \int \psi_m{}^*(\mathbf{r})\,\Delta U(\mathbf{r})\psi(\mathbf{r})\,d\mathbf{r}. \tag{10.10}$$

---

2 It turns out (see p. 187) that any Bloch function can be written in the form (10.6), the function ϕ being known as a *Wannier function*, so no generality is lost in this assumption.

3 By including only localized (i.e., bound) atomic wave functions in (10.7) we make our first serious approximation. A complete set of atomic levels includes the ionized ones as well. This is the point at which the method ceases to be applicable to levels well described by the almost free electron approximation.

4 Because of this method of approximating ϕ, the tight-binding method is sometimes known as the method of the *linear combination of atomic orbitals* (or LCAO).

Placing (10.6) and (10.7) into (10.10) and using the orthonormality of the atomic wave functions,

$$\int \psi_m^*(\mathbf{r})\psi_n(\mathbf{r})\, d\mathbf{r} = \delta_{nm}, \tag{10.11}$$

we arrive at an eigenvalue equation that determines the coefficients $b_n(\mathbf{k})$ and the Bloch energies $\mathcal{E}(\mathbf{k})$:

$$\begin{aligned}(\mathcal{E}(\mathbf{k}) - E_m)b_m = & -(\mathcal{E}(\mathbf{k}) - E_m)\sum_n \left(\sum_{\mathbf{R}\neq 0} \int \psi_m^*(\mathbf{r})\psi_n(\mathbf{r} - \mathbf{R})e^{i\mathbf{k}\cdot\mathbf{R}}\, d\mathbf{r}\right) b_n \\ & + \sum_n \left(\int \psi_m^*(\mathbf{r})\, \Delta U(\mathbf{r})\psi_n(\mathbf{r})\, d\mathbf{r}\right) b_n \\ & + \sum_n \left(\sum_{\mathbf{R}\neq 0} \int \psi_m^*(\mathbf{r})\, \Delta U(\mathbf{r})\psi_n(\mathbf{r} - \mathbf{R})\, e^{i\mathbf{k}\cdot\mathbf{R}}\, d\mathbf{r}\right) b_n. \end{aligned} \tag{10.12}$$

The first term on the right of Eq. (10.12) contains integrals of the form[5]

$$\int d\mathbf{r}\, \psi_m^*(\mathbf{r})\psi_n(\mathbf{r} - \mathbf{R}). \tag{10.13}$$

We interpret our assumption of well-localized atomic levels to mean that (10.13) is small compared to unity. We assume that the integrals in the third term on the right of Eq. (10.12) are small, since they also contain the product of two atomic wave functions centered at different sites. Finally, we assume that the second term on the right of (10.12) is small because we expect the atomic wave functions to become small at distances large enough for the periodic potential to deviate appreciably from the atomic one.[6]

Consequently, the right-hand side of (10.13) (and therefore $(\mathcal{E}(\mathbf{k}) - E_m)b_m$) is always small. This is possible if $\mathcal{E}(\mathbf{k}) - E_m$ is small whenever b_m is not (and vice versa). Thus $\mathcal{E}(\mathbf{k})$ must be close to an atomic level, say E_0, and all the b_m except those going with that level and levels degenerate with (or close to) it in energy must be small:[7]

$$\mathcal{E}(\mathbf{k}) \approx E_0, \quad b_m \approx 0 \text{ unless } E_m \approx E_0. \tag{10.14}$$

If the estimates in (10.14) were strict equalities, we would be back to the extreme case in which the crystal levels were identical to the atomic ones. Now, however, we

---

[5] Integrals whose integrands contain a product of wave functions centered on different lattice sites are known as *overlap integrals*. The tight-binding approximation exploits the smallness of such overlap integrals. They also play an important role in the theory of magnetism (Chapter 32).

[6] This last assumption is on somewhat shakier ground than the others, since the ionic potentials need not fall off as rapidly as the atomic wave functions. However, it is also less critical in determining the conclusions we shall reach, since the term in question does not depend on **k**. In a sense this term simply plays the role of correcting the atomic potentials within each cell to include the fields of the ions outside the cell; it could be made as small as the other two terms by a judicious redefinition of the "atomic" Hamiltonian and levels.

[7] Note the similarity of this reasoning to that employed on pages 152 to 156. There, however, we concluded that the wave function was a linear combination of only a small number of plane waves, whose free electron energies were very close together. Here, we conclude that the wave function can be represented, through (10.7) and (10.6), by only a small number of atomic wave functions, whose atomic energies are very close together.

can determine the levels in the crystal more accurately, exploiting (10.14) to estimate the right-hand side of (10.12) by letting the sum over n run only through those levels with energies either degenerate with or very close to E_0. If the atomic level 0 is nondegenerate,[8] i.e., an s-level, then in this approximation (10.12) reduces to a single equation giving an explicit expression for the energy of the band arising from this s-level (generally referred to as an "s-band"). If we are interested in bands arising from an atomic p-level, which is triply degenerate, then (10.12) would give a set of three homogeneous equations, whose eigenvalues would give the $\mathcal{E}(\mathbf{k})$ for the three p-bands, and whose solutions $b(\mathbf{k})$ would give the appropriate linear combinations of atomic p-levels making up ϕ at the various $\mathbf{k}$'s in the Brillouin zone. To get a d-band from atomic d-levels, we should have to solve a 5×5 secular problem, etc.

Should the resulting $\mathcal{E}(\mathbf{k})$ stray sufficiently far from the atomic values at certain $\mathbf{k}$, it would be necessary to repeat the procedure, adding to the expansion (10.7) of ϕ those additional atomic levels whose energies the $\mathcal{E}(\mathbf{k})$ are approaching. In practice, for example, one generally solves a 6×6 secular problem that includes both d- and s-levels in computing the band structure of the transition metals, which have in the atomic state an outer s-shell and a partially filled d-shell. This procedure goes under the name of "s-d mixing" or "hybridization."

Often the atomic wave functions have so short a range that only nearest-neighbor terms in the sums over $\mathbf{R}$ in (10.12) need be retained, which very much simplifies subsequent analysis. We briefly illustrate the band structure that emerges in the simplest case.[9]

APPLICATION TO AN s-BAND ARISING FROM A SINGLE ATOMIC s-LEVEL

If all the coefficients b in (10.12) are zero except that for a single atomic s-level, then (10.12) gives directly the band structure of the corresponding s-band:

$$\mathcal{E}(\mathbf{k}) = E_s - \frac{\beta + \Sigma\gamma(\mathbf{R})e^{i\mathbf{k}\cdot\mathbf{R}}}{1 + \Sigma\alpha(\mathbf{R})e^{i\mathbf{k}\cdot\mathbf{R}}}, \tag{10.15}$$

where E_s is the energy of the atomic s-level, and

$$\beta = -\int d\mathbf{r}\, \Delta U(\mathbf{r})|\phi(\mathbf{r})|^2, \tag{10.16}$$

$$\alpha(\mathbf{R}) = \int d\mathbf{r}\, \phi^*(\mathbf{r})\phi(\mathbf{r} - \mathbf{R}), \tag{10.17}$$

and

$$\gamma(\mathbf{R}) = -\int d\mathbf{r}\, \phi^*(\mathbf{r})\, \Delta U(\mathbf{r})\phi(\mathbf{r} - \mathbf{R}). \tag{10.18}$$

---

[8] For the moment we ignore spin-orbit coupling. We can therefore concentrate entirely on the orbital parts of the levels. Spin can then be included by simply multiplying the orbital wave functions by the appropriate spinors, and doubling the degeneracy of each of the orbital levels.

[9] The simplest case is that of an s-band. The next most complicated case, a p-band, is discussed in Problem 2.

The coefficients (10.16) to (10.18) may be simplified by appealing to certain symmetries. Since ϕ is an s-level, $\phi(\mathbf{r})$ is real and depends only on the magnitude r. From this it follows that $\alpha(-\mathbf{R}) = \alpha(\mathbf{R})$. This and the inversion symmetry of the Bravais lattice, which requires that $\Delta U(-\mathbf{r}) = \Delta U(\mathbf{r})$, also imply that $\gamma(-\mathbf{R}) = \gamma(\mathbf{R})$. We ignore the terms in α in the denominator of (10.15), since they give small corrections to the numerator. A final simplification comes from assuming that only nearest-neighbor separations give appreciable overlap integrals.

Putting these observations together, we may simplify (10.15) to

$$\mathcal{E}(\mathbf{k}) = E_s - \beta - \sum_{\text{n.n.}} \gamma(\mathbf{R}) \cos \mathbf{k} \cdot \mathbf{R}, \tag{10.19}$$

where the sum runs only over those $\mathbf{R}$ in the Bravais lattice that connect the origin to its nearest neighbors.

To be explicit, let us apply (10.19) to a face-centered cubic crystal. The 12 nearest neighbors of the origin (see Figure 10.3) are at

$$\mathbf{R} = \frac{a}{2}(\pm 1, \pm 1, 0), \quad \frac{a}{2}(\pm 1, 0, \pm 1), \quad \frac{a}{2}(0, \pm 1, \pm 1). \tag{10.20}$$

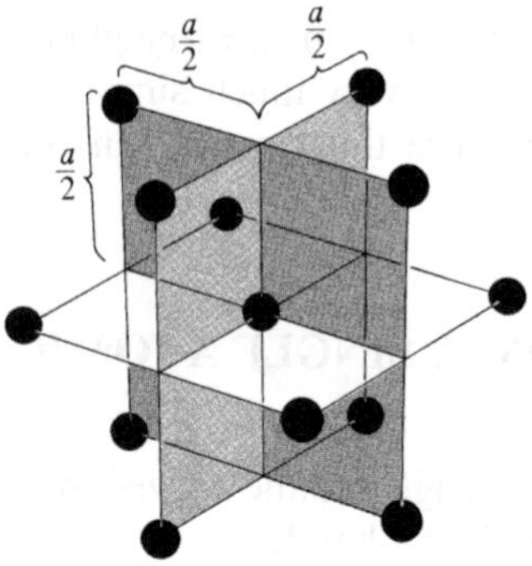

Figure 10.3
The 12 nearest neighbors of the origin in a face-centered cubic lattice with conventional cubic cell of side a.

If $\mathbf{k} = (k_x, k_y, k_z)$, then the corresponding 12 values of $\mathbf{k} \cdot \mathbf{R}$ are

$$\mathbf{k} \cdot \mathbf{R} = \frac{a}{2}(\pm k_i, \pm k_j), \qquad i, j = x, y; y, z; z, x. \tag{10.21}$$

Now $\Delta U(\mathbf{r}) = \Delta U(x, y, z)$ has the full cubic symmetry of the lattice, and is therefore unchanged by permutations of its arguments or changes in their signs. This, together with the fact that the s-level wave function $\phi(\mathbf{r})$ depends only on the magnitude of $\mathbf{r}$, implies that $\gamma(\mathbf{R})$ is the same constant γ for all 12 of the vectors (10.20). Consequently, the sum in (10.19) gives, with the aid of (10.21),

$$\mathcal{E}(\mathbf{k}) = E_s - \beta - 4\gamma(\cos \tfrac{1}{2}k_x a \cos \tfrac{1}{2}k_y a + \cos \tfrac{1}{2}k_y a \cos \tfrac{1}{2}k_z a + \cos \tfrac{1}{2}k_z a \cos \tfrac{1}{2}k_x a), \tag{10.22}$$

where

$$\gamma = -\int d\mathbf{r}\, \phi^*(x, y, z)\, \Delta U(x, y, z)\, \phi(x - \tfrac{1}{2}a, y - \tfrac{1}{2}a, z). \tag{10.23}$$

Equation (10.22) reveals the characteristic feature of tight-binding energy bands: The bandwidth—i.e., the spread between the minimum and maximum energies in the band—is proportional to the small overlap integral γ. Thus the tight-binding bands are narrow bands, and the smaller the overlap, the narrower the band. In the limit of vanishing overlap the bandwidth also vanishes, and the band becomes N-fold degenerate, corresponding to the extreme case in which the electron simply resides on any one of the N isolated atoms. The dependence of bandwidth on overlap integral is illustrated in Figure 10.4.

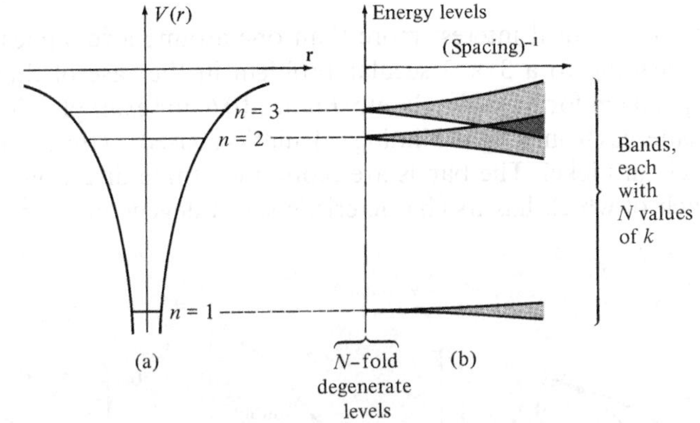

Figure 10.4
(a) Schematic representation of nondegenerate electronic levels in an atomic potential. (b) The energy levels for N such atoms in a periodic array, plotted as a function of mean inverse interatomic spacing. When the atoms are far apart (small overlap integrals) the levels are nearly degenerate, but when the atoms are closer together (larger overlap integrals), the levels broaden into bands.

In addition to displaying the effect of overlap on bandwidth, Eq. (10.22) illustrates several general features of the band structure of a face-centered cubic crystal that are not peculiar to the tight-binding case. Typical of these are the following:

1. In the limit of small ka, (10.22) reduces to:

$$\mathcal{E}(\mathbf{k}) = E_s - \beta - 12\gamma + \gamma k^2 a^2. \tag{10.24}$$

   This is independent of the direction of **k**—i.e., the constant-energy surfaces in the neighbourhood of $\mathbf{k} = \mathbf{0}$ are spherical.[10]
2. If $\mathcal{E}$ is plotted along any line perpendicular to one of the square faces of the first Brillouin zone (Figure 10.5), it will cross the square face with vanishing slope (Problem 1).

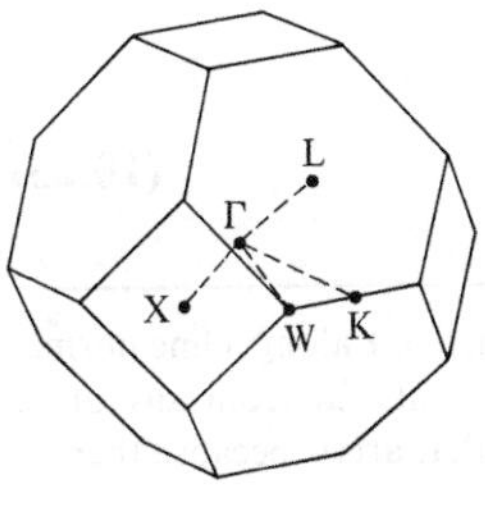

Figure 10.5
The first Brillouin zone for face-centered cubic crystals. The point Γ is at the center of the zone. The names K, L, W, and X are widely used for the points of high symmetry on the zone boundary.

10 This can be deduced quite generally for any nondegenerate band in a crystal with cubic symmetry.

3. If $\mathcal{E}$ is plotted along any line perpendicular to one of the hexagonal faces of the first Brillouin zone (Figure 10.5), it need not, in general, cross the face with vanishing slope (Problem 1).[11]

GENERAL REMARKS ON THE TIGHT-BINDING METHOD

1. In cases of practical interest more than one atomic level appears in the expansion (10.7), leading to a 3×3 secular problem in the case of three *p*-levels, a 5×5 secular problem for five *d*-levels, etc. Figure 10.6, for example, shows the band structure that emerges from a tight-binding calculation based on the 5-fold degenerate atomic 3-*d* levels in nickel. The bands are plotted for three directions of symmetry in the zone, each of which has its characteristic set of degeneracies.[12]

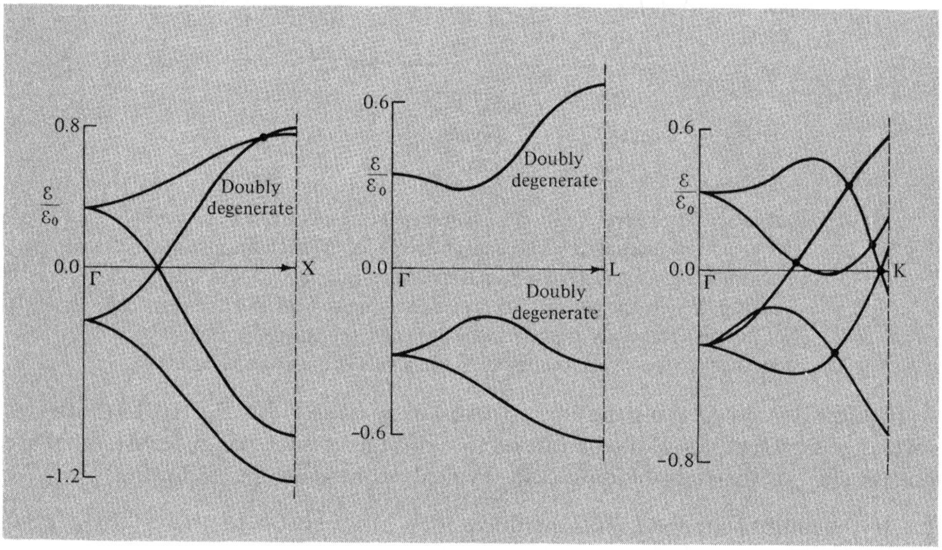

Figure 10.6
A tight-binding calculation of the 3*d* bands of nickel. (G. C. Fletcher, *Proc. Phys. Soc.* **A65**, 192 (1952).) Energies are given in units of $\mathcal{E}_0 = 1.349$ eV, so the bands are about 2.7 volts wide. The lines along which $\mathcal{E}$ is plotted are shown in Figure 10.5. Note the characteristic degeneracies along ΓX and ΓL, and the absence of degeneracy along ΓK. The great width of the bands indicates the inadequacy of so elementary a treatment.

2. A quite general feature of the tight-binding method is the relation between bandwidth and the overlap integrals

$$\gamma_{ij}(\mathbf{R}) = -\int d\mathbf{r}\, \phi_i^*(\mathbf{r})\, \Delta U(\mathbf{r}) \phi_j(\mathbf{r} - \mathbf{R}). \qquad \textbf{(10.25)}$$

---

[11] Compare the nearly free electron case (page 158), where the rate of change of $\mathcal{E}$ along a line normal to a Bragg plane was always found to vanish as the plane was crossed at points far from any other Bragg planes. The tight-binding result illustrates the more general possibility that arises because there is no plane of mirror symmetry parallel to the hexagonal face.

[12] The calculated bands are so wide as to cast doubt on the validity of the entire expansion. A more realistic calculation would have to include, at the very least, the effects of the 4*s*-level.

If the γ_{ij} are small, then the bandwidth is correspondingly small. As a rule of thumb, when the energy of a given atomic level increases (i.e., the binding energy decreases) so does the spatial extent of its wave function. Correspondingly, the low-lying bands in a solid are very narrow, but bandwidths increase with mean band energy. In metals the highest band (or bands) are very broad, since the spatial ranges of the highest atomic levels are comparable to a lattice constant, and the tight-binding approximation is then of doubtful validity.

3. Although the tight-binding wave function (10.6) is constructed out of localized atomic levels ϕ, an electron in a tight-binding level will be found, with equal probability, in any cell of the crystal, since its wave function (like any Bloch wave function) changes only by the phase factor $e^{i\mathbf{k}\cdot\mathbf{R}}$ as one moves from one cell to another a distance $\mathbf{R}$ away. Thus as $\mathbf{r}$ varies from cell to cell, there is superimposed on the atomic structure within each cell a sinusoidal variation in the amplitudes of Re ψ and Im ψ, as illustrated in Figure 10.7.

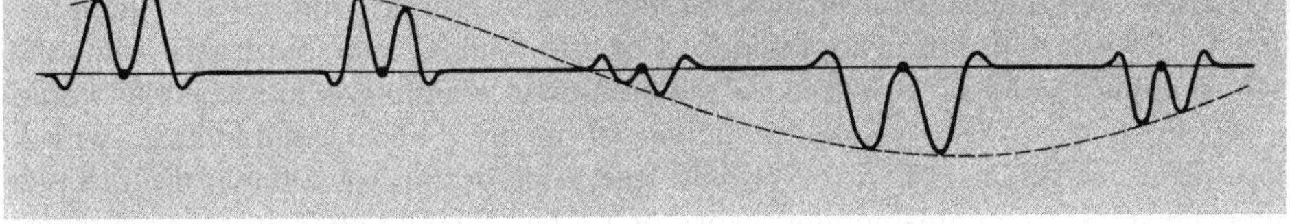

Figure 10.7
Characteristic spatial variation of the real (or imaginary) part of the tight-binding wave function (10.6).

A further indication that the tight-binding levels have a running wave or itinerant character comes from the theorem that the mean velocity of an electron in a Bloch level with wave vector $\mathbf{k}$ and energy $\mathcal{E}(\mathbf{k})$ is given by $\mathbf{v}(\mathbf{k}) = (1/\hbar)\,\partial\mathcal{E}/\partial\mathbf{k}$. (See Appendix E.) If $\mathcal{E}$ is independent of $\mathbf{k}$, $\partial\mathcal{E}/\partial\mathbf{k}$ is zero, which is consistent with the fact that in genuinely isolated atomic levels (which lead to zero bandwidth) the electrons are indeed tied to individual atoms. If, however, there is any nonzero overlap in the atomic wave functions, then $\mathcal{E}(\mathbf{k})$ will not be constant throughout the zone. Since a small variation in $\mathcal{E}$ implies a small nonzero value of $\partial\mathcal{E}/\partial\mathbf{k}$, and hence a small but nonzero mean velocity, as long as there is any overlap electrons will be able to move freely through the crystal! Decreasing the overlap only reduces the velocity; it does not eliminate the motion. One can view this motion as a quantum-mechanical tunneling from lattice site to lattice site. The less the overlap, the lower the tunneling probability, and hence the longer it takes to go a given distance.

4. In solids that are not monatomic Bravais lattices, the tight-binding approximation is more complicated. This problem arises in the hexagonal close-packed metals, which are simple hexagonal with a two-point basis. Formally, one can treat the two-point basis as a molecule, whose wave functions are assumed to be known, and proceed as above, using molecular instead of atomic wave functions. If the nearest-neighbor overlap remains small, then, in particular, it will be small in each "molecule," and an atomic *s*-level will give rise to two nearly degenerate molecular levels. Thus a single atomic *s*-level yields two tight-binding bands in the hexagonal close-packed structure.

Alternatively, one can proceed by continuing to construct linear combinations of atomic levels centered at the Bravais lattice points *and* at the basis points, generalizing (10.6) to

$$\psi(\mathbf{r}) = \sum_{\mathbf{R}} e^{i\mathbf{k}\cdot\mathbf{R}}(a\phi(\mathbf{r} - \mathbf{R}) + b\phi(\mathbf{r} - \mathbf{d} - \mathbf{R})), \tag{10.26}$$

(where **d** is the separation of the two basis atoms). This can be viewed as essentially the first approach, in which, however, approximate molecular wave functions are used, the approximation to the molecular levels being combined with the tight-binding approximation to the levels of the entire crystal.[13]

5. In the heavier elements spin-orbit coupling (see page 169) is of great importance in determining the atomic levels, and should therefore be included in a tight-binding treatment of the broadening of these levels into bands in the solid. In principle the extension is straightforward. We simply include in $\Delta U(\mathbf{r})$ the interaction between the electron's spin and the electric field of all ions except the one at the origin, incorporating that interaction into the atomic Hamiltonian. Once this is done we can no longer use spin-independent linear combinations of atomic orbital wave functions, but must work with linear combinations of both orbital and spin levels. Thus the tight-binding theory of an *s*-level, when spin-orbit coupling is appreciable, would approximate ϕ not by a single atomic *s*-level but by a linear combination (with **k** dependent coefficients) of two levels with the same orbital wave functions and two opposite spins. The tight-binding theory of a *d*-band would go from a 5×5 determinantal problem to a 10×10 one, etc. As mentioned in Chapter 9, effects of spin-orbit coupling, though often small, can frequently be quite crucial, as when they eliminate degeneracies that would rigorously be present if such coupling were ignored.[14]

6. All the analysis of electronic levels in a periodic potential in this chapter (and the preceding two) has been done within the independent electron approximation, which either neglects the interaction between electrons, or, at best, includes it in some average way through the effective periodic potential experienced by each single electron. We shall see in Chapter 32 that the independent electron approximation can fail when it gives at least one *partially* filled band that derives from well-localized atomic levels with small overlap integrals. In many cases of interest (notably in insulators and for the very low-lying bands in metals) this problem does not arise, since the tight-binding bands are so low in energy as to be completely filled. However, the possibility of such a failure of the independent electron approximation must be kept in mind when narrow tight-binding bands are derived from partially filled atomic shells—in metals, generally the *d*- and *f*-shells. One should be particularly aware of this possibility in solids with a magnetic structure.

This failure of the independent electron approximation obscures the simple picture the tight-binding approximation suggests: that of a continuous transition from the

---

[13] The "approximate molecular wave functions" will thus be **k**-dependent.

[14] The inclusion of spin-orbit coupling in the tight-binding method is discussed by J. Friedel, P. Lenghart, and G. Leman, *J. Phys. Chem. Solids* **25**, 781 (1964).

metallic to the atomic state as the interatomic distance is continuously increased.[15] If we took the tight-binding approximation at face value, then as the lattice constant in a metal increased, the overlap between all atomic levels would eventually become small, and all bands—even the partially filled conduction band (or bands)—would eventually become narrow tight-binding bands. As the conduction band narrowed, the velocity of the electrons in it would diminish and the conductivity of the metal would drop. Thus, we would expect a conductivity that dropped continuously to zero with the overlap integrals as the metal was expanded.

In fact, however, it is likely that a full calculation going beyond the independent electron approximation would predict that beyond a certain nearest-neighbor separation the conductivity should drop abruptly to zero, the material becoming an insulator (the so-called *Mott transition*).

The reason for this departure from the tight-binding prediction lies in the inability of the independent electron approximation to treat the very strong additional repulsion a second electron feels at a given atomic site when another electron is already there. We shall comment further on this in Chapter 32, but we mention the problem here because it is sometimes described as a failure of the tight-binding method.[16] This is somewhat misleading in that the failure occurs when the tight-binding approximation to the independent electron model is at its best; it is the independent electron approximation itself that fails.

WANNIER FUNCTIONS

We conclude this chapter with a demonstration that the Bloch functions for *any* band can always be written in the form (10.4) on which the tight-binding approximation is based. The functions ϕ that play the role of the atomic wave functions are known as *Wannier functions*. Such Wannier functions can be defined for any band, whether or not it is well described by the tight-binding approximation; but if the band is not a narrow tight-binding band, the Wannier functions will bear little resemblance to any of the electronic wave functions for the isolated atom.

To establish that any Bloch function $\psi_{n\mathbf{k}}(\mathbf{r})$ can be written in the form (10.4), we first note that considered as a function of $\mathbf{k}$ for fixed $\mathbf{r}$, $\psi_{n\mathbf{k}}(\mathbf{r})$ is periodic in the reciprocal lattice. It therefore has a Fourier series expansion in plane waves with wave vectors in the reciprocal of the reciprocal lattice, i.e., in the direct lattice. Thus for any fixed $\mathbf{r}$ we can write

$$\psi_{n\mathbf{k}}(\mathbf{r}) = \sum_{\mathbf{R}} f_n(\mathbf{R}, \mathbf{r})e^{i\mathbf{R}\cdot\mathbf{k}}, \tag{10.27}$$

where the coefficients in the sum depend on $\mathbf{r}$ as well as on the "wave vectors" $\mathbf{R}$, since for each $\mathbf{r}$ it is a different function of $\mathbf{k}$ that is being expanded.

---

15 A difficult procedure to realize in the laboratory, but a very tempting one to visualize theoretically, as an aid in understanding the nature of energy bands.

16 See, for example, H. Jones, *The Theory of Brillouin Zones and Electron States in Crystals*, North-Holland, Amsterdam, 1960, p. 229.

The Fourier coefficients in (10.27) are given by the inversion formula[17]

$$f_n(\mathbf{R}, \mathbf{r}) = \frac{1}{v_0} \int d\mathbf{k}\, e^{-i\mathbf{R}\cdot\mathbf{k}} \psi_{n\mathbf{k}}(\mathbf{r}). \tag{10.28}$$

Equation (10.27) is of the form (10.4), provided that the function $f_n(\mathbf{R}, \mathbf{r})$ depends on $\mathbf{r}$ and $\mathbf{R}$ only through their difference, $\mathbf{r} - \mathbf{R}$. But if $\mathbf{r}$ and $\mathbf{R}$ are both shifted by the Bravais lattice vector $\mathbf{R}_0$, then f is indeed unchanged as a direct consequence of (10.28) and Bloch's theorem, in the form (8.5). Thus $f_n(\mathbf{R}, \mathbf{r})$ has the form:

$$f_n(\mathbf{R}, \mathbf{r}) = \phi_n(\mathbf{r} - \mathbf{R}) \tag{10.29}$$

Unlike tight-binding atomic functions $\phi(\mathbf{r})$, the Wannier functions $\phi_n(\mathbf{r} - \mathbf{R})$ at different sites (or with different band indices) are orthogonal (see Problem 3, Eq. (10.35)). Since the complete set of Bloch functions can be written as linear combinations of the Wannier functions, the Wannier functions $\phi_n(\mathbf{r} - \mathbf{R})$ for all n and $\mathbf{R}$ form a complete orthogonal set. They therefore offer an alternative basis for an exact description of the independent electron levels in a crystal potential.

The similarity in form of the Wannier functions to the tight-binding functions leads one to hope that the Wannier functions will also be localized—i.e., that when $\mathbf{r}$ is very much larger than some length on the atomic scale, $\phi_n(\mathbf{r})$ will be negligibly small. To the extent that this can be established, the Wannier functions offer an ideal tool for discussing phenomena in which the spatial localization of electrons plays an important role. Perhaps the most important areas of application are these:

1. Attempts to derive a transport theory for Bloch electrons. The analog of free electron wave packets, electronic levels in a crystal that are localized in both $\mathbf{r}$ and $\mathbf{k}$, are conveniently constructed with the use of Wannier functions. The theory of Wannier functions is closely related to the theory of when and how the semiclassical theory of transport by Bloch electrons (Chapters 12 and 13) breaks down.
2. Phenomena involving localized electronic levels, due, for example, to attractive impurities that bind an electron. A very important example is the theory of donor and acceptor levels in semiconductors (Chapter 28).
3. Magnetic phenomena, in which localized magnetic moments are found to exist at suitable impurity sites.

Theoretical discussions of the range of Wannier functions are in general quite subtle.[18] Roughly speaking, the range of the Wannier function decreases as the band gap increases (as one might expect from the tight-binding approximation, in which the bands become narrower as the range of the atomic wave functions decreases). The various "breakdown" and "breakthrough" phenomena we shall mention in

---

[17] Here v_0 is the volume in k-space of the first Brillouin zone, and the integral is over the zone. Equations (10.27) and (10.28) (with $\mathbf{r}$ regarded as a fixed parameter) are just Eqs. (D.1) and (D.2) of Appendix D, with direct and reciprocal space interchanged.

[18] A relatively simple argument, but only in one dimension, is given by W. Kohn, *Phys. Rev.* **115**, 809 (1959). A more general discussion can be found in E. I. Blount, *Solid State Physics*, Vol. 13, Academic Press, New York, 1962, p. 305.

Chapter 12 that occur when the band gap is small find their reflection in the fact that theories based on the localization of the Wannier functions become less reliable in this limit.

PROBLEMS

1. (a) Show that along the principal symmetry directions shown in Figure 10.5 the tight-binding expression (10.22) for the energies of an s-band in a face-centered cubic crystal reduces to the following:

(i) Along ΓX $(k_y = k_z = 0,\ \ k_x = \mu\, 2\pi/a, \quad 0 \leqslant \mu \leqslant 1)$

$$\mathcal{E} = E_s - \beta - 4\gamma(1 + 2\cos\mu\pi).$$

(ii) Along ΓL $(k_x = k_y = k_z = \mu\, 2\pi/a, \quad 0 \leqslant \mu \leqslant \tfrac{1}{2})$

$$\mathcal{E} = E_s - \beta - 12\gamma\cos^2\mu\pi.$$

(iii) Along ΓK $(k_z = 0,\ \ k_x = k_y = \mu\, 2\pi/a, \quad 0 \leqslant \mu \leqslant \tfrac{3}{4})$

$$\mathcal{E} = E_s - \beta - 4\gamma(\cos^2\mu\pi + 2\cos\mu\pi).$$

(iv) Along ΓW $(k_z = 0,\ \ k_x = \mu\, 2\pi/a,\ \ k_y = \tfrac{1}{2}\mu\, 2\pi/a, \quad 0 \leqslant \mu \leqslant 1)$

$$\mathcal{E} = E_s - \beta - 4\gamma(\cos\mu\pi + \cos\tfrac{1}{2}\mu\pi + \cos\mu\pi\cos\tfrac{1}{2}\mu\pi).$$

(b) Show that on the square faces of the zone the normal derivative of $\mathcal{E}$ vanishes.

(c) Show that on the hexagonal faces of the zone, the normal derivative of $\mathcal{E}$ vanishes only along lines joining the center of the hexagon to its vertices.

2. ***Tight-Binding p-Bands in Cubic Crystals***

In dealing with cubic crystals, the most convenient linear combinations of three degenerate atomic p-levels have the form $x\phi(r)$, $y\phi(r)$, and $z\phi(r)$, where the function ϕ depends only on the magnitude of the vector $\mathbf{r}$. The energies of the three corresponding p-bands are found from (10.12) by setting to zero the determinant

$$|(\mathcal{E}(\mathbf{k}) - E_p)\,\delta_{ij} + \beta_{ij} + \tilde{\gamma}_{ij}(\mathbf{k})| = 0, \tag{10.30}$$

where

$$\tilde{\gamma}_{ij}(\mathbf{k}) = \sum_{\mathbf{R}} e^{i\mathbf{k}\cdot\mathbf{R}}\gamma_{ij}(\mathbf{R}),$$

$$\gamma_{ij}(\mathbf{R}) = -\int d\mathbf{r}\, \psi_i{}^*(\mathbf{r})\psi_j(\mathbf{r} - \mathbf{R})\,\Delta U(\mathbf{r}),$$

$$\beta_{ij} = \gamma_{ij}(\mathbf{R} = 0). \tag{10.31}$$

(A term multiplying $\mathcal{E}(\mathbf{k}) - E_p$, which gives rise to very small corrections analogous to those given by the denominator of (10.15) in the s-band case, has been omitted from (10.30).)

(a) As a consequence of cubic symmetry, show that

$$\beta_{xx} = \beta_{yy} = \beta_{zz} = \beta,$$
$$\beta_{xy} = 0. \tag{10.32}$$

(b) Assuming that the $\gamma_{ij}(\mathbf{R})$ are negligible except for nearest-neighbor $\mathbf{R}$, show that $\tilde{\gamma}_{ij}(\mathbf{k})$ is diagonal for a simple cubic Bravais lattice, so that $x\phi(r)$, $y\phi(r)$, and $z\phi(r)$ each generate independent bands. (Note that this ceases to be the case if the $\gamma_{ij}(\mathbf{R})$ for next nearest-neighbor $\mathbf{R}$ are also retained.)

(c) For a face-centered cubic Bravais lattice with only nearest-neighbor γ_{ij} appreciable, show that the energy bands are given by the roots of

$$0 = \begin{vmatrix} \begin{array}{l}\mathcal{E}(\mathbf{k}) - \mathcal{E}^0(\mathbf{k}) + \\ 4\gamma_0 \cos \tfrac{1}{2}k_y a \cos \tfrac{1}{2}k_z a\end{array} & -4\gamma_1 \sin \tfrac{1}{2}k_x a \sin \tfrac{1}{2}k_y a & -4\gamma_1 \sin \tfrac{1}{2}k_x a \sin \tfrac{1}{2}k_z a \\ -4\gamma_1 \sin \tfrac{1}{2}k_y a \sin \tfrac{1}{2}k_x a & \begin{array}{l}\mathcal{E}(\mathbf{k}) - \mathcal{E}^0(\mathbf{k}) + \\ 4\gamma_0 \cos \tfrac{1}{2}k_z a \cos \tfrac{1}{2}k_x a\end{array} & -4\gamma_1 \sin \tfrac{1}{2}k_y a \sin \tfrac{1}{2}k_z a \\ -4\gamma_1 \sin \tfrac{1}{2}k_z a \sin \tfrac{1}{2}k_x a & -4\gamma_1 \sin \tfrac{1}{2}k_z a \sin \tfrac{1}{2}k_y a & \begin{array}{l}\mathcal{E}(\mathbf{k}) - \mathcal{E}^0(\mathbf{k}) + \\ 4\gamma_0 \cos \tfrac{1}{2}k_x a \cos \tfrac{1}{2}k_y a\end{array} \end{vmatrix} \tag{10.33}$$

where

$$\begin{aligned} \mathcal{E}^0(\mathbf{k}) &= E_p - \beta \\ &\quad - 4\gamma_2(\cos \tfrac{1}{2}k_x a \cos \tfrac{1}{2}k_z a + \cos \tfrac{1}{2}k_x a \cos \tfrac{1}{2}k_y a + \cos \tfrac{1}{2}k_y a \cos \tfrac{1}{2}k_z a), \\ \gamma_0 &= -\int d\mathbf{r}\, [x^2 - y(y - \tfrac{1}{2}a)]\phi(r)\phi([x^2 + (y - \tfrac{1}{2}a)^2 + (z - \tfrac{1}{2}a)^2]^{1/2})\, \Delta U(\mathbf{r}), \\ \gamma_1 &= -\int d\mathbf{r}\, x(y - \tfrac{1}{2}a)\phi(r)\phi([(x - \tfrac{1}{2}a)^2 + (y - \tfrac{1}{2}a^2) + z^2]^{1/2})\, \Delta U(\mathbf{r}), \\ \gamma_2 &= -\int d\mathbf{r}\, x(x - \tfrac{1}{2}a)\phi(r)\phi([(x - \tfrac{1}{2}a)^2 + (y - \tfrac{1}{2}a)^2 + z^2]^{1/2})\, \Delta U(\mathbf{r}). \end{aligned} \tag{10.34}$$

(d) Show that all three bands are degenerate at $\mathbf{k} = \mathbf{0}$, and that when $\mathbf{k}$ is directed along either a cube axis (ΓX) or a cube diagonal (ΓL) there is a double degeneracy. Sketch the energy bands (in analogy to Figure 10.6) along these directions.

3. Prove that Wannier functions centered on different lattice sites are orthogonal,

$$\int \phi_n{}^*(\mathbf{r} - \mathbf{R})\phi_{n'}(\mathbf{r} - \mathbf{R}')\, d\mathbf{r} \propto \delta_{n,n'}\, \delta_{\mathbf{R},\mathbf{R}'}, \tag{10.35}$$

by appealing to the orthonormality of the Bloch functions and the identity (F.4) of Appendix F. Show also that

$$\int d\mathbf{r} \left| \phi_n(\mathbf{r}) \right|^2 = 1 \tag{10.36}$$

if the integral of the $|\psi_{n\mathbf{k}}(\mathbf{r})|^2$ over a primitive cell is normalized to unity.

11
Other Methods for Calculating Band Structure

In Chapters 9 and 10 we explored approximate solutions to the one-electron Schrödinger equation in the limiting cases of nearly free electrons, and tight binding. In most cases of interest the tight-binding approximation (at least in the simple form outlined in Chapter 10) is suitable only for the representation of bands arising from the ion core levels, while the nearly free electron approximation cannot be directly applied to any real solid.[1] The purpose of this chapter is therefore to describe some of the more common methods actually used in the calculation of real band structures.

We remarked in Chapter 8 that in merely writing down a separate Schrödinger equation[2]

$$\left(-\frac{\hbar^2}{2m}\nabla^2 + U(\mathbf{r})\right)\psi_{\mathbf{k}}(\mathbf{r}) = \varepsilon(\mathbf{k})\psi_{\mathbf{k}}(\mathbf{r}) \tag{11.1}$$

for each electron we are already enormously simplifying the actual problem of many *interacting* electrons in a periodic potential. In an exact treatment each electron cannot be described by a wave function determined by a single-particle Schrödinger equation, independent of all the others.

The independent electron approximation does not in fact entirely neglect electron-electron interactions. Rather it assumes that most of their important effects can be taken into account with a sufficiently clever choice for the periodic potential $U(\mathbf{r})$ appearing in the one-electron Schrödinger equation. Thus $U(\mathbf{r})$ contains not only the periodic potential due to the ions alone, but also periodic effects due to the interaction of the electron (whose wave function appears in (11.1)) with all the other electrons. The latter interaction depends on the configuration of the other electrons; i.e., it depends on *their* individual wave functions, which are also determined by a Schrödinger equation of the form (11.1). Thus to know the potential appearing in (11.1), one must first know all the solutions to (11.1). Since, however, to know the solutions one must know the potential, one is in for some difficult mathematical efforts.

The simplest (and often the most practical) procedure is to start with a shrewd guess, $U_0(\mathbf{r})$, for $U(\mathbf{r})$, calculate from (11.1) the wave functions for the occupied electronic levels, and from these recompute $U(\mathbf{r})$. If the new potential, $U_1(\mathbf{r})$ is the same as (or very close to) $U_0(\mathbf{r})$, one says that *self-consistency* has been achieved and takes $U = U_1$ for the actual potential. If U_1 differs from U_0, one repeats the procedure starting with U_1, taking U_2 as the actual potential if it is very close to U_1, and otherwise continuing on to the calculation of U_3. The hope is that this procedure will converge, eventually yielding a self-consistent potential that reproduces itself.[3]

We shall assume in this chapter (as in Chapters 8–10) that the potential $U(\mathbf{r})$ is a given function; i.e., that we are either engaged in the first step of this iterative procedure or, by a fortunate guess, are able to work with a reasonably self-consistent $U(\mathbf{r})$ from the start. The reliability of the methods we are about to describe is limited not only by the accuracy of the computed solutions to (11.1), which can be quite high, but also by the accuracy with which we have been able to estimate the potential $U(\mathbf{r})$.

---

[1] However, more sophisticated techniques often yield an analysis very much like the nearly free electron approximation in a suitably modified potential, known as the pseudopotential (see below).

[2] We continue to drop explicit reference to the band index n, except when this would lead to ambiguity.

[3] One must remember, however, that even the self-consistent solution is still only an approximate solution to the vastly more complex many-body problem.

The resulting $\mathcal{E}_n(\mathbf{k})$ display a disconcerting sensitivity to errors in the construction of the potential, and it is often the case that the final accuracy of the computed band structure is limited more by the problem of finding the potential than by the difficulties in solving the Schrödinger equation (11.1) for a given U. This is strikingly illustrated in Figure 11.1.

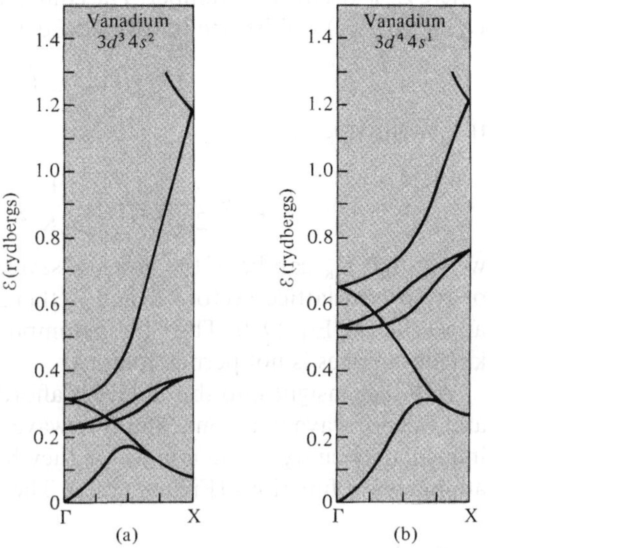

Figure 11.1
Energy bands for vanadium, calculated for two possible choices of crystal potential $U(\mathbf{r})$. Vanadium is body-centered cubic and the bands are plotted along the [100] direction from the origin to the Brillouin zone boundary. The atomic structure of vanadium is five electrons around a closed-shell argon configuration. The bands displayed are the $3d$ and $4s$ derived bands (and higher bands). (a) The bands are shown as calculated in a $U(\mathbf{r})$ derived from an assumed $3d^3 4s^2$ configuration for atomic vanadium. (b) The bands are shown based on an assumed $3d^4 4s^1$ atomic configuration. (From L. F. Matheiss, *Phys. Rev.* **A970** 134, (1964).)

Another point to emphasize at the start is that none of the methods we shall describe can be carried through analytically, except in the simplest one-dimensional examples. All require modern, high-speed computers for their execution. Progress in the theoretical calculation of energy bands has kept close pace with development of larger and faster computers, and the kinds of approximations one is likely to consider are influenced by available computational techniques.[4]

GENERAL FEATURES OF VALENCE-BAND WAVE FUNCTIONS

Since the low-lying core levels are well described by tight-binding wave functions, calculational methods aim at the higher-lying bands (which may be either filled, partially filled, or empty). These bands are referred to in this context, in contrast to the tight-binding core bands, as the *valence bands*.[5] The valence bands determine the electronic behavior of a solid in a variety of circumstances, electrons in the core levels being inert for many purposes.

The essential difficulty in practical calculations of the valence-band wave functions and energies is revealed when one asks why the nearly free electron approximation of Chapter 9 cannot be applied to the valence bands in an actual solid. A simple,

[4] See, for example, *Computational Methods in Band Theory*, P. M. Marcus, J. F. Janak, and A. R. Williams, eds., Plenum Press, New York, 1971; and *Methods in Computational Physics: Energy Bands in Solids*, Vol. 8, B. Alder, S. Fernbach, and M. Rotenburg, eds., Academic Press, New York, 1968.

[5] Unfortunately the same term, "valence band," is used in the theory of semiconductors with a rather more narrow meaning. See Chapter 28.

but superficial, reason is that the potential is not small. Very roughly we might estimate that, at least well within the ion core, $U(\mathbf{r})$ has the coulombic form

$$\frac{-Z_a e^2}{r}, \tag{11.2}$$

where Z_a is the atomic number. The contribution of (11.2) to the Fourier components $U_{\mathbf{K}}$ in Eq. (9.2) will be (see p. 167 and Eq. (17.73)):

$$U_{\mathbf{K}} \approx -\left(\frac{4\pi Z_a e^2}{K^2}\right)\frac{1}{v}. \tag{11.3}$$

If we write this as

$$|U_{\mathbf{K}}| \approx \frac{e^2}{2a_0}\left(\frac{a_0{}^3}{v}\right)\frac{1}{(a_0 K)^2}\,8\pi Z_a, \quad \frac{e^2}{2a_0} = 13.6\text{ eV}, \tag{11.4}$$

we see that $U_{\mathbf{K}}$ can be of the order of several electron volts for a very large number of reciprocal lattice vectors **K** and is therefore comparable to the kinetic energies appearing in Eq. (9.2). Thus the assumption that $U_{\mathbf{K}}$ is small compared to these kinetic energies is not permissible.

A deeper insight into this failure is afforded by considering the nature of the core and valence wave functions. The core wave functions are appreciable only within the immediate vicinity of the ion, where they have the characteristic oscillatory form of atomic wave functions (Figure 11.2a). These oscillations are a manifestation of the

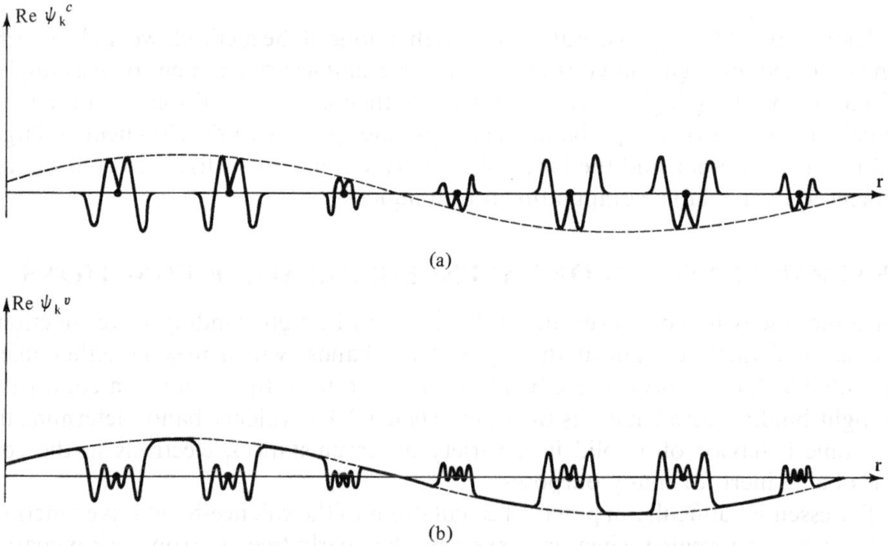

Figure 11.2
(a) Characteristic spatial dependence of a core wave function $\psi_{\mathbf{k}}{}^c(\mathbf{r})$. The curve shows Re ψ against position along a line of ions. Note the characteristic atomic oscillations in the vicinity of each ion. The dashed envelope of the atomic parts is sinusoidal, with wavelength $\lambda = 2\pi/k$. Between lattice sites the wave function is negligibly small. (b) Characteristic spatial dependence of a valence wave function $\psi_{\mathbf{k}}{}^v(\mathbf{r})$. The atomic oscillations are still present in the core region. The wave function need not be at all small between lattice sites, but it is likely to be slowly varying and plane-wavelike there.

high electronic kinetic energy within the core,[6] which, in combination with the high negative potential energy, produces the total energy of the core levels. Since valence levels have higher total energies than core levels, within the core region, where they experience the same large and negative potential energy as the core electrons, the valence electrons must have even higher kinetic energies. Thus within the core region the valence wave functions must be even more oscillatory than the core wave functions.

This conclusion can also be reached by an apparently different argument:

Eigenstates of the same Hamiltonian with different eigenvalues must be orthogonal. In particular any valence wave function $\psi_{\mathbf{k}}^{v}(\mathbf{r})$ and any core wave function $\psi_{\mathbf{k}}^{c}(\mathbf{r})$ must satisfy:

$$0 = \int d\mathbf{r}\, \psi_{\mathbf{k}}^{c}(\mathbf{r})^{*} \psi_{\mathbf{k}}^{v}(\mathbf{r}). \tag{11.5}$$

Core wave functions are appreciable only within the immediate vicinity of the ion, so the main contribution of this integral must come from the core region. It is enough to consider the contribution to (11.5) from the core region of a single ion, since Bloch's theorem ((8.3)) requires the integrand to be the same from cell to cell. Within this core region $\psi_{\mathbf{k}}^{v}(\mathbf{r})$ must have oscillations that carefully interlace with those of all the $\psi_{\mathbf{k}}^{c}(\mathbf{r})$ in order to cause the integrals (11.5) to vanish for all core levels.

Either of these arguments leads to the conclusion that a valence wave function should have the form pictured in Figure 11.2b. If, however, the valence wave functions have an oscillatory structure on the scale of the core region, a Fourier expansion such as (9.1) must contain many short wavelength plane waves, i.e., many terms with large wave vectors. Thus the nearly free electron method, which leads to an approximate wave function composed of a very small number of plane waves, must be untenable.

In one way or another, all of the calculational methods now in use are attempts to come to grips with the necessity for reproducing this detailed, atomic-like structure of the valence wave functions in the core region, while facing the fact that the valence levels are not of the tight-binding type, and therefore have appreciable wave functions in the interstitial regions.

THE CELLULAR METHOD

The first serious attempt to calculate band structure (aside from Bloch's original use of the tight-binding method) was the cellular method of Wigner and Seitz.[7] The method begins by observing that because of the Bloch relation (8.6):

$$\psi_{\mathbf{k}}(\mathbf{r} + \mathbf{R}) = e^{i\mathbf{k}\cdot\mathbf{R}}\psi_{\mathbf{k}}(\mathbf{r}), \tag{11.6}$$

it is enough to solve the Schrödinger equation (11.1) within a single primitive cell C_0. The wave function can then be determined via (11.6) in any other primitive cell from its values in C_0.

However, not every solution to (11.1) within C_0 leads in this way to an acceptable wave function for the entire crystal, since $\psi(\mathbf{r})$ and $\nabla\psi(\mathbf{r})$ must be continuous as $\mathbf{r}$

---

[6] The velocity operator is $(\hbar/mi)\nabla$, which means that the more rapidly a wave function varies in a region, the greater the electronic velocity must be in that region.

[7] E. P. Wigner and F. Seitz, *Phys. Rev.* **43**, 804 (1933); **46**, 509 (1934).

crosses the primitive cell boundary.[8] Because of (11.6), this condition can be phrased entirely in terms of the values of ψ within and on the surface of C_0. It is this boundary condition that introduces the wave vector **k** into the cellular solution, and eliminates all solutions except those for a discrete set of energies, which are just the band energies $\mathcal{E} = \mathcal{E}_n(\mathbf{k})$.

Boundary conditions within C_0 are

$$\psi(\mathbf{r}) = e^{-i\mathbf{k}\cdot\mathbf{R}}\psi(\mathbf{r} + \mathbf{R}), \tag{11.7}$$

and

$$\hat{\mathbf{n}}(\mathbf{r}) \cdot \nabla\psi(\mathbf{r}) = -e^{-i\mathbf{k}\cdot\mathbf{R}}\hat{\mathbf{n}}(\mathbf{r} + \mathbf{R}) \cdot \nabla\psi(\mathbf{r} + \mathbf{R}), \tag{11.8}$$

where **r** and **r** + **R** are both points on the surface of the cell and $\hat{\mathbf{n}}$ is an outward normal (see Problem 1).

The analytical problem is therefore to solve (11.1) within the primitive cell C_0 subject to these boundary conditions. To preserve the symmetry of the crystal, one takes the primitive cell C_0 to be the Wigner-Seitz primitive cell (Chapter 4) centered on the lattice point $\mathbf{R} = 0$.

The foregoing is an exact restatement of the problem. The first approximation of the cellular method is the replacement of the periodic potential $U(\mathbf{r})$ within the Wigner-Seitz primitive cell by a potential $V(r)$ with spherical symmetry about the origin (see Figure 11.3). One might, for example, choose $V(r)$ to be the potential of

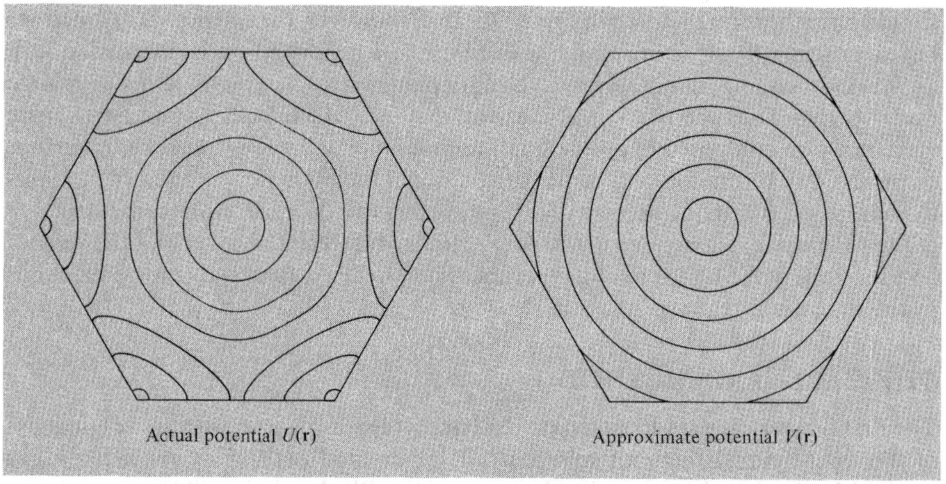

Figure 11.3
Equipotentials (i.e., curves of constant $U(\mathbf{r})$) within a primitive cell. For the actual crystal potential these will have spherical symmetry near the center of the cell where the potential is dominated by the contribution from the central ion. However near the boundary of the cell the potential will deviate substantially from spherical symmetry. The cellular method approximates the potential by a spherically symmetric one everywhere within the cell, with equipotentials as shown on the right.

[8] If ψ or $\nabla\psi$ were discontinuous at the cell boundary, then $\nabla^2\psi$ would have singularities (which are either δ-functions or derivatives of δ-functions) on the boundary. Since there are no such terms occurring in $U\psi$ on the boundary, the Schrödinger equation could not be satisfied.

a single ion at the origin, ignoring the fact that the neighbors of the origin will also contribute to $U(\mathbf{r})$ within C_0, especially near its boundaries. This approximation is made entirely for practical reasons, to render a difficult computational problem more manageable.

Once a potential has been chosen spherically symmetric inside C_0, then within the primitive cell a complete set of solutions to the Schrödinger equation (11.1) can be found of the form[9]

$$\psi_{lm}(\mathbf{r}) = Y_{lm}(\theta, \phi)\chi_l(r), \tag{11.9}$$

where $Y_{lm}(\theta, \phi)$ are spherical harmonics and $\chi_l(r)$ satisfies the ordinary differential equation

$$\chi_l''(r) + \frac{2}{r}\chi_l'(r) + \frac{2m}{\hbar^2}\left(\mathcal{E} - V(r) - \frac{\hbar^2}{2m}\frac{l(l+1)}{r^2}\right)\chi_l(r) = 0. \tag{11.10}$$

Given the potential $V(\mathrm{r})$ and given *any* value of $\mathcal{E}$, there is a unique $\chi_{l,\mathcal{E}}$ that solves (11.10) and is regular at the origin.[10] These $\chi_{l,\mathcal{E}}$ can be calculated numerically, ordinary differential equations being easy to handle on machines. Since any linear combination of solutions to Schrödinger's equation with the same energy is itself a solution,

$$\psi(\mathbf{r}, \mathcal{E}) = \sum_{lm} A_{lm} Y_{lm}(\theta, \phi)\chi_{l,\mathcal{E}}(r) \tag{11.11}$$

will solve (11.1) at energy $\mathcal{E}$ for arbitrary coefficients A_{lm}. However, (11.11) will only yield an acceptable wave function for the crystal if it satisfies the boundary conditions (11.7) and (11.8). It is in the imposition of these boundary conditions that the cellular method makes its next major approximation.

To begin with, one takes only as many terms in the expansion (11.11) as it is calculationally convenient to handle.[11] Since there is only a finite number of coefficients in the expansion, we can, for a general cell, fit the boundary condition only at a finite set of points on its surface. The imposition of this finite set of boundary conditions (chosen to be as many as there are unknown coefficients) leads to a set of **k**-dependent linear homogeneous equations for the A_{lm}, and the values of $\mathcal{E}$ for which the determinant of these equations vanishes are the required energies $\mathcal{E}_n(\mathbf{k})$.

---

[9] See, for example, D. Park, *Introduction to the Quantum Theory*, McGraw-Hill, New York, 1964, pp. 516–519, or any other book on quantum mechanics. There is, however, this important difference compared with the familiar atomic case: In atomic physics the boundary condition (that ψ vanishes at infinity) is also spherically symmetric, and consequently a single term of the form (11.9) gives a stationary state (i.e., the angular momentum is a good quantum number). In the present case (except for the spherical cellular model described below) the boundary condition does not have spherical symmetry. Therefore the stationary wave functions will be of the form (11.11) with nonvanishing coefficients for several distinct l and m values; i.e., angular momentum will not be a good quantum number.

[10] This assertion may be somewhat jarring to those who, from atomic physics, are used to the fact that only a discrete set of eigenvalues are found in any atomic problem, namely the energy levels of the atom for angular momentum l. This is because, in the atomic problem, we have the boundary condition that $\chi_l(r)$ vanishes as $r \to \infty$. Here we are only interested in χ_l within the Wigner-Seitz cell, and no such additional condition is required; ultimately the allowed values of $\mathcal{E}$ will be determined by the crystal boundary conditions (11.7) and (11.8). Imposing these does indeed lead back to a discrete set of energies: the $\mathcal{E}_n(\mathbf{k})$.

[11] Comforted by the assurance that eventually the expansion must converge, since for high enough angular momenta l the wave function will be very small everywhere within the cell.

In this way one can search for the eigenvalues $\mathcal{E}_n(\mathbf{k})$ for each fixed $\mathbf{k}$. Alternatively, one can fix $\mathcal{E}$, do a single numerical integration of (11.10), and then search for values of $\mathbf{k}$ for which the determinant vanishes. Provided that one has not been so unfortunate as to choose $\mathcal{E}$ in an energy gap, such values of $\mathbf{k}$ can always be found, and in this way the constant-energy surfaces can be mapped out.

Various ingenious techniques have been used to minimize the mismatch of the wave function at the boundaries due to the fact that the boundary conditions can only be imposed at a finite number of points; such cleverness, and the ability of computers to handle large determinants, have led to cellular calculations of very high accuracy,[12] producing band structures in substantial agreement with some of the other methods we shall describe.

The most famous application of the cellular method is the original calculation by Wigner and Seitz of the lowest energy level in the valence band of sodium metal. Since the bottom of the band is at $\mathbf{k} = \mathbf{0}$, the exponential factor disappears from the boundary conditions (11.7) and (11.8). Wigner and Seitz made the further approximation of replacing the Wigner-Seitz primitive cell by a sphere of radius r_0 with the same volume, thereby achieving a boundary condition with the same spherical symmetry as the potential $V(r)$. They could then consistently demand that the solution $\psi(\mathbf{r})$ itself have spherical symmetry, which requires that only the single term $l = 0$, $m = 0$ be retained in (11.11). Under these conditions the boundary conditions reduce to

$$\chi_0'(r_0) = 0. \tag{11.12}$$

Thus the solutions to the single equation (11.10) for $l = 0$, subject to the boundary condition (11.12), give the spherically symmetric cellular wave functions and energies.

Note that the problem has the same form as an atomic problem except that the atomic boundary condition—that the wave function vanish at infinity—is replaced by the cellular boundary condition—that the wave function have a vanishing radial derivative at r_0. The $3s^1$ atomic and cellular wave functions are plotted together in Figure 11.4. Note that the cellular wave function is larger than the atomic one in the interstitial region, but differs from it very little in the core region.

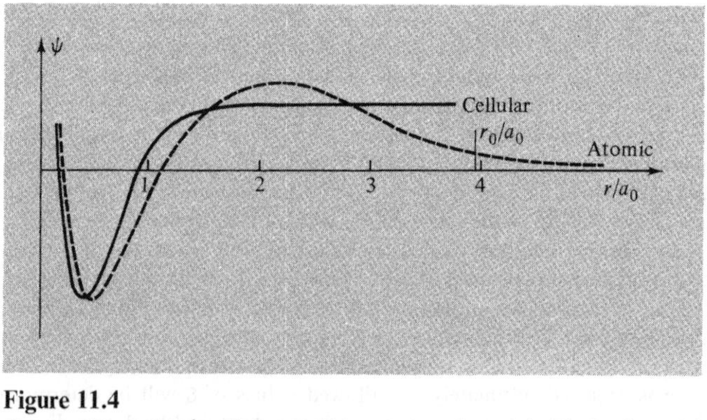

Figure 11.4
Comparison of $3s^1$ cellular (solid curve) and atomic (dashed curve) wave functions for sodium.

---

[12] Notably by S. L. Altmann and co-workers (see *Proc. Roy. Soc.* **A244**, 141, 153 (1958)).

There are perhaps two major difficulties with the cellular method:

1. The computational difficulties involved in numerically satisfying a boundary condition over the surface of the Wigner-Seitz primitive cell, a fairly complex polyhedral structure.
2. The physically questionable point of whether a potential representing an isolated ion is the best approximation to the correct potential within the entire Wigner–Seitz primitive cell. In particular, the potential used in the cellular calculations has a discontinuous derivative whenever the boundary between two cells is crossed (Figure 11.5), whereas in actual fact the potential is quite flat in such regions.

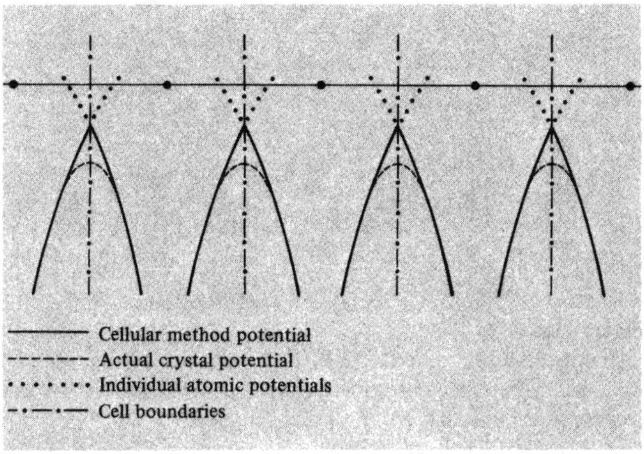

Figure 11.5
The cellular method potential has a discontinuous derivative midway between lattice points, but the actual potential is quite flat there.

A potential that overcomes both objections is the *muffin-tin potential*, which is taken to represent an isolated ion within a sphere of specified radius r_0 about each lattice point, and taken to be zero (i.e., constant) elsewhere (with r_0 chosen small enough that the spheres do not overlap). (See Figure 11.6.) The muffin-tin potential mitigates both problems, being flat in the interstitial regions, and leading to matching conditions on a spherical rather than a polyhedral surface.

Formally, the muffin-tin potential can be defined (for all $\mathbf{R}$) by:

$$\begin{aligned} U(\mathbf{r}) &= V(|\mathbf{r}-\mathbf{R}|), \quad \text{when } |\mathbf{r}-\mathbf{R}| < r_0 \quad \text{(the \textit{core} or \textit{atomic region})}, \\ &= V(r_0) = 0, \quad \text{when } |\mathbf{r}-\mathbf{R}| > r_0 \quad \text{(the \textit{interstitial region})}, \end{aligned} \tag{11.13}$$

where r_0 is less than half the nearest-neighbor distance.[13]

[13] Frequently r_0 is taken to be half the nearest-neighbor distance; i.e., the sphere is the inscribed sphere in the Wigner-Seitz cell. There are minor technical complications in the analysis in that case, which we avoid by requiring r to be less than that distance.

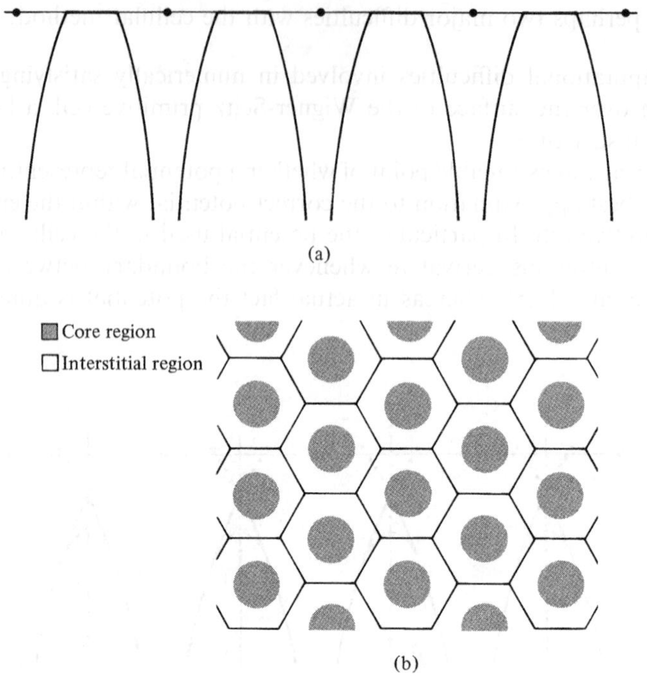

Figure 11.6
(a) The muffin-tin potential, plotted along a line of ions. (b) The muffin-tin potential is constant (zero) in the interstitial regions and represents an isolated ion in each core region.

If we agree that the function $V(r)$ is zero when its argument exceeds r_0, then we can write $U(\mathbf{r})$ very simply as

$$U(\mathbf{r}) = \sum_{\mathbf{R}} V(|\mathbf{r} - \mathbf{R}|). \tag{11.14}$$

Two methods are in wide use for computing the bands in a muffin-tin potential: the augmented plane-wave (APW) method and the method of Korringa, Kohn, and Rostoker (KKR).

THE AUGMENTED PLANE-WAVE METHOD (APW)

This approach, due to J. C. Slater,[14] represents $\psi_{\mathbf{k}}(\mathbf{r})$ as a superposition of a finite number of plane waves in the flat interstitial region, while forcing it to have a more rapid oscillatory atomic behavior in the core region. This is achieved by expanding $\psi_{\mathbf{k},\mathcal{E}}$ in a set of *augmented plane waves.*[15] The APW $\phi_{\mathbf{k},\mathcal{E}}$ is defined as follows:

1. $\phi_{\mathbf{k},\mathcal{E}} = e^{i\mathbf{k}\cdot\mathbf{r}}$ in the interstitial region. It is important to note that there is no constraint relating $\mathcal{E}$ and $\mathbf{k}$ (such as, for example, $\mathcal{E} = \hbar^2k^2/2m$). One can define

---

[14] *Phys. Rev.* **51**, 846 (1937).

[15] We append the energy of a level as an additional subscript when its explicit specification helps to avoid possible ambiguities.

an APW for any energy $\mathcal{E}$ and any wave vector **k**. *Thus any single APW does not satisfy the crystal Schrödinger equation for energy $\mathcal{E}$ in the interstitial region.*

2. $\phi_{\mathbf{k},\mathcal{E}}$ is *continuous* at the boundary between atomic and interstitial regions.
3. In the atomic region about **R**, $\phi_{\mathbf{k},\mathcal{E}}$ does satisfy the atomic Schrödinger equation:

$$-\frac{\hbar^2}{2m}\nabla^2\phi_{\mathbf{k},\mathcal{E}}(\mathbf{r}) + V(|\mathbf{r}-\mathbf{R}|)\phi_{\mathbf{k},\mathcal{E}}(\mathbf{r}) = \mathcal{E}\phi_{\mathbf{k},\mathcal{E}}(\mathbf{r}), \qquad |\mathbf{r}-\mathbf{R}| < r_0. \qquad \textbf{(11.15)}$$

Since **k** does not appear in this equation, $\phi_{\mathbf{k},\mathcal{E}}$ gets its **k** dependence only via the boundary condition (2) and the **k** dependence determined by (1) in the interstitial region.

It can be shown that these conditions determine a unique APW $\phi_{\mathbf{k},\mathcal{E}}$ for all **k** and $\mathcal{E}$. Note that in the interstitial region the APW satisfies not (11.15) but $H\phi_{\mathbf{k},\mathcal{E}} = (\hbar^2k^2/2m)\phi_{\mathbf{k},\mathcal{E}}$. Note also that, in general, $\phi_{\mathbf{k},\mathcal{E}}$ will have a discontinuous derivative on the boundary between interstitial and atomic regions, so that $\nabla^2\phi_{\mathbf{k},\mathcal{E}}$ will have delta-function singularities there.

The APW method tries to approximate the correct solution to the crystal Schrödinger equation (11.1) by a superposition of APW's, all with the same energy. For any reciprocal lattice vector **K** the APW $\phi_{\mathbf{k}+\mathbf{K},\mathcal{E}}$ satisfies the Bloch condition with wave vector **k** (Problem 2), and therefore the expansion of $\psi_{\mathbf{k}}(\mathbf{r})$ will be of the form

$$\psi_{\mathbf{k}}(\mathbf{r}) = \sum_{\mathbf{K}} c_{\mathbf{K}}\phi_{\mathbf{k}+\mathbf{K},\mathcal{E}(\mathbf{k})}(\mathbf{r}), \qquad \textbf{(11.16)}$$

where the sum is over reciprocal lattice vectors.

By taking the energy of the APW to be the actual energy of the Bloch level, we guarantee that $\psi_{\mathbf{k}}(\mathbf{r})$ satisfies the crystal Schrödinger equation in the atomic regions. The hope is that not too many augmented plane waves will suffice to approximate the solutions to the full Schrödinger equation in the interstitial region[16] and at the boundary. In practice, as many as a hundred APW's can be used; by the time this stage is reached, $\mathcal{E}(\mathbf{k})$ does not change appreciably when more APW's are added, and one feels with some confidence that good convergence has been achieved.

Because each APW has a discontinuous derivative at the boundary of the atomic and interstitial regions, it is best to work not with the Schrödinger equation but with an equivalent variational principle:

Given any *differentiable* (but not necessarily twice differentiable)[17] function $\psi(\mathbf{r})$, define the energy functional:

$$E[\psi] = \frac{\displaystyle\int\left(\frac{\hbar^2}{2m}|\nabla\psi(\mathbf{r})|^2 + U(\mathbf{r})|\psi(\mathbf{r})|^2\right)d\mathbf{r}}{\displaystyle\int|\psi(\mathbf{r})|^2\,d\mathbf{r}}. \qquad \textbf{(11.17)}$$

---

16 The reader is warned not to fall into the trap of thinking that the exact solutions to $-\hbar^2/2m\nabla^2\psi = \mathcal{E}\psi$ in the region of complex shape, where the muffin-tin potential is flat, must be linear combinations of plane waves $e^{i\mathbf{k}\cdot\mathbf{r}}$ with $\mathcal{E} = \hbar^2k^2/2m$.

17 The function ψ may have a kink where $\nabla\psi$ is discontinuous.

It can be shown[18] that a solution to the Schrödinger equation (11.1) satisfying the Bloch condition with wave vector **k** and energy $\mathcal{E}(\mathbf{k})$ makes (11.17) stationary with respect to differentiable functions $\psi(\mathbf{r})$ that satisfy the Bloch condition with wave vector **k**. The value of $E[\psi_{\mathbf{k}}]$ is just the energy $\mathcal{E}(\mathbf{k})$ of the level $\psi_{\mathbf{k}}$.

The variational principle is exploited by using the APW expansion (11.16) to calculate $E[\psi_{\mathbf{k}}]$. This leads to an approximation to $\mathcal{E}(\mathbf{k}) = E[\psi_{\mathbf{k}}]$ that depends on the coefficients $c_{\mathbf{K}}$. The demand that $E[\psi_{\mathbf{k}}]$ be stationary leads to the conditions $\partial E/\partial c_{\mathbf{K}} = 0$, which are a set of homogeneous equations in the $c_{\mathbf{K}}$. The coefficients in this set of equations depend on the sought for energy $\mathcal{E}(\mathbf{k})$, both through the $\mathcal{E}(\mathbf{k})$ dependence of the APW's and because the value of $E[\psi_{\mathbf{k}}]$ at the stationary point is $\mathcal{E}(\mathbf{k})$. Setting the determinant of these coefficients equal to zero gives an equation whose roots determine the $\mathcal{E}(\mathbf{k})$.

As in the cellular case, it is often preferable to work with a set of APW's of definite energy and search for the **k** at which the secular determinant vanishes, thereby mapping out the constant energy surfaces in **k**-space. With modern computing techniques it appears possible to include enough augmented plane waves to achieve excellent convergence,[19] and the APW method is one of the more successful schemes for calculating band structure.[20]

In Figure 11.7 we show portions of the energy bands for a few metallic elements, as calculated by L. F. Mattheiss using the APW method. One of the interesting results of this analysis is the extent to which the bands in zinc, which has a filled atomic *d*-shell, resemble the free electron bands. A comparison of Mattheiss' curves for titanium with the cellular calculations by Altmann (Figure 11.8) should, however, instill a healthy sense of caution: Although there are recognizable similarities, there are quite noticeable differences. These are probably due more to the differences in choice of potential than to the validity of the calculation methods, but they serve to indicate that one should be wary in using the results of first principles band-structure calculations.

THE GREEN'S FUNCTION METHOD OF KORRINGA, KOHN, AND ROSTOKER (KKR)

An alternative approach to the muffin-tin potential is provided by a method due to Korringa and to Kohn and Rostoker.[21] This starts from the integral form of the Schrödinger equation[22]

---

[18] For a simple proof (and a more detailed statement of the variational principle) see Appendix G.

[19] In some cases a very small number of APW's may suffice to give reasonable convergence for much the same reasons as in the case of the orthogonalized plane wave and pseudopotential methods, discussed below.

[20] Complete details on the method along with sample computer programs may even be found in textbook form: T. L. Loucks, *Augmented Plane Wave Method*, W. A. Benjamin, Menlo Park, California, 1967.

[21] J. Korringa, *Physica* **13**, 392 (1947); W. Kohn and N. Rostoker, *Phys. Rev.* **94**, 1111 (1954).

[22] Equation (11.18) is the starting point for the elementary theory of scattering. That it is equivalent to the ordinary Schrödinger equation (11.1) follows from the fact (Chapter 17, Problem 3) that G satisfies $(\mathcal{E} + \hbar^2\nabla^2/2m)G(\mathbf{r} - \mathbf{r}') = \delta(\mathbf{r} - \mathbf{r}')$. For an elementary discussion of these facts see, for example, D. S. Saxon, *Elementary Quantum Mechanics*, Holden-Day, San Francisco, 1968, p. 360 *et seq.* In scattering theory it is customary to include an inhomogeneous term $e^{i\mathbf{k}\cdot\mathbf{r}}$ in (11.18), where $\hbar k = \sqrt{2m\mathcal{E}}$, to satisfy the boundary condition appropriate to an incoming plane wave. Here, however, the boundary condition is the Bloch relation, which is satisfied by (11.18) without an inhomogeneous term.

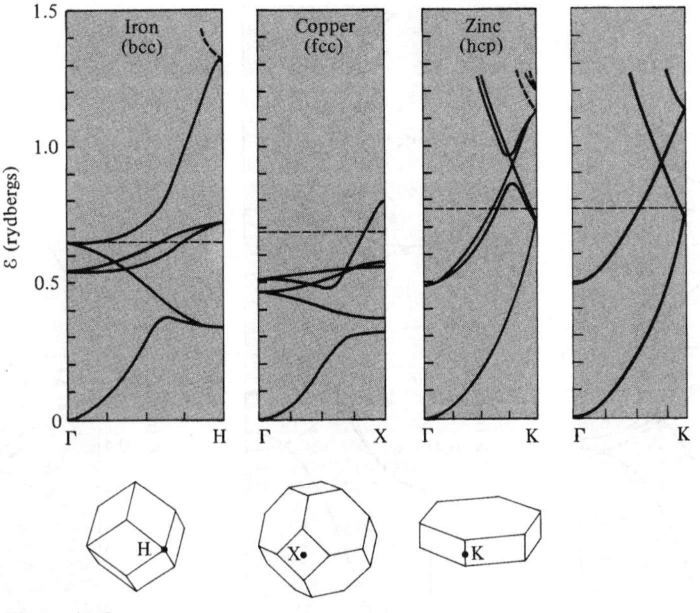

Figure 11.7
APW energy bands for iron, copper, and zinc, calculated by L. F. Mattheis, *Phys. Rev.* **134**, A970 (1964). The bands are plotted from the origin of **k**-space to the points indicated on the zone surfaces. Note the striking resemblance between the calculated bands of zinc and the free electron bands (pictured to the right). Zinc has two *s*-electrons outside of a closed-shell configuration. The horizontal dashed lines mark the Fermi energy.

$$\psi_{\mathbf{k}}(\mathbf{r}) = \int d\mathbf{r}'\, G_{\mathcal{E}(\mathbf{k})}(\mathbf{r} - \mathbf{r}')U(\mathbf{r}')\psi_{\mathbf{k}}(\mathbf{r}'), \tag{11.18}$$

where the integral is over all space and

$$\begin{aligned} G_{\mathcal{E}}(\mathbf{r} - \mathbf{r}') &= -\frac{2m}{\hbar^2}\frac{e^{iK|\mathbf{r}-\mathbf{r}'|}}{4\pi|\mathbf{r} - \mathbf{r}'|}, \\ K &= \sqrt{2m\mathcal{E}/\hbar^2}, \qquad \mathcal{E} > 0, \\ &= i\sqrt{2m(-\mathcal{E})/\hbar^2}, \qquad \mathcal{E} < 0. \end{aligned} \tag{11.19}$$

Substituting the form (11.14) for the muffin-tin potential into (11.18), and making the change of variables $\mathbf{r}'' = \mathbf{r}' - \mathbf{R}$ in each term of the resulting sum, we can rewrite (11.18) as

$$\psi_{\mathbf{k}}(\mathbf{r}) = \sum_{\mathbf{R}} \int d\mathbf{r}''\, G_{\mathcal{E}(\mathbf{k})}(\mathbf{r} - \mathbf{r}'' - \mathbf{R})V(r'')\psi_{\mathbf{k}}(\mathbf{r}'' + \mathbf{R}). \tag{11.20}$$

The Bloch condition gives $\psi_{\mathbf{k}}(\mathbf{r}'' + \mathbf{R}) = e^{i\mathbf{k}\cdot\mathbf{R}}\psi_{\mathbf{k}}(\mathbf{r}'')$, and we can therefore rewrite (11.20) (replacing $\mathbf{r}''$ by $\mathbf{r}'$):

$$\psi_{\mathbf{k}}(\mathbf{r}) = \int d\mathbf{r}'\, \mathcal{G}_{\mathbf{k},\mathcal{E}(\mathbf{k})}(\mathbf{r} - \mathbf{r}')V(r')\psi_{\mathbf{k}}(\mathbf{r}'), \tag{11.21}$$

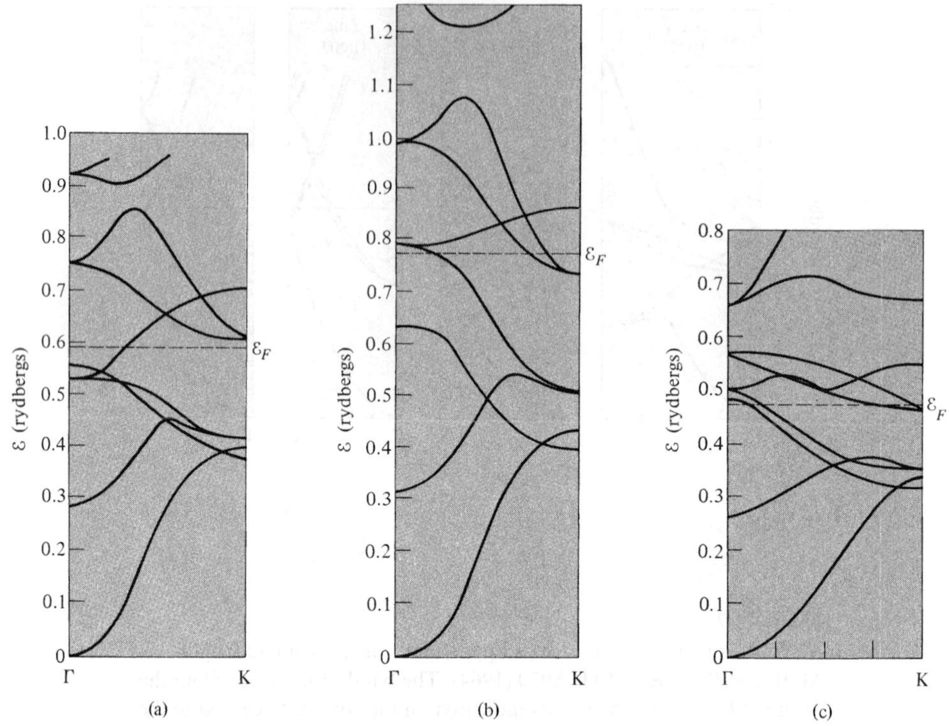

Figure 11.8
Three calculated band structures for titanium. Curves (a) and (b) were calculated by the cellular method for two possible potentials. They are taken from S. L. Altmann, in *Soft X-Ray Band Spectra*, D. Fabian (ed.), Academic Press—London, 1968. Curve (c) is from the APW calculation of Mattheis.

where

$$\mathcal{G}_{\mathbf{k},\mathcal{E}}(\mathbf{r} - \mathbf{r}') = \sum_{\mathbf{R}} G_{\mathcal{E}}(\mathbf{r} - \mathbf{r}' - \mathbf{R})e^{i\mathbf{k}\cdot\mathbf{R}}. \tag{11.22}$$

Equation (11.21) has the pleasing feature that *all* of the dependence on both wave vector **k** and crystal structure is contained in the function $\mathcal{G}_{\mathbf{k},\mathcal{E}}$, which can be calculated, once and for all, for a variety of crystal structures for specified values of $\mathcal{E}$ and **k**.[23] It is shown in Problem 3 that Eq. (11.21) implies that on the sphere of radius r_0, the values of $\psi_{\mathbf{k}}$ are constrained to satisfy the following integral equation:

$$0 = \int d\Omega' \left[\mathcal{G}_{\mathbf{k},\mathcal{E}(\mathbf{k})}(r_0\theta\phi, r_0\theta'\phi') \frac{\partial}{\partial r} \psi(r\theta'\phi') \bigg|_{r=r_0} - \psi(r_0\theta'\phi') \frac{\partial}{\partial r} \mathcal{G}_{\mathbf{k},\mathcal{E}(\mathbf{k})}(r_0\theta\phi, r\theta'\phi') \bigg|_{r=r_0} \right]. \tag{11.23}$$

[23] To do the **R**-sum one uses the same techniques as in calculations of the lattice energies of ionic crystals (Chapter 20).

Since the function $\psi_{\mathbf{k}}$ is continuous, it retains the form determined by the atomic problem (Eqs. (11.9) to (11.11)) at r_0. The approximation of the KKR method (which is exact for the muffin-tin potential up to this point) is to assume that $\psi_{\mathbf{k}}$ will be given to a reasonable degree of accuracy by keeping only a finite number (say N) of spherical harmonics in the expansion (11.11). By placing this truncated expansion in (11.23), multiplying by $Y_{lm}(\theta, \phi)$, and integrating the result over the solid angle $d\theta\, d\phi$ for all l and m appearing in the truncated expansion, we obtain a set of N linear equations for the A_{lm} appearing in the expansion (11.11). The coefficients in these equations depend on $\mathcal{E}(\mathbf{k})$ and $\mathbf{k}$ through $\mathcal{G}_{\mathbf{k},\mathcal{E}(\mathbf{k})}$ and through the radial wave function $\chi_{l,\mathcal{E}}$ and its derivative $\chi'_{l,\mathcal{E}}$. Setting the $N \times N$ determinant of the coefficients equal to zero once again gives an equation determining the relation between $\mathcal{E}$ and $\mathbf{k}$. As in the methods described earlier, one can either search for values of $\mathcal{E}$ giving a solution for fixed $\mathbf{k}$, or fix $\mathcal{E}$ and map out the surface in $\mathbf{k}$-space at which the determinant vanishes, which will then give the constant-energy surface $\mathcal{E}(\mathbf{k}) = \mathcal{E}$.

Both the KKR and APW methods can be regarded as techniques which, if carried out exactly for the muffin-tin potential, would lead to infinite-order determinantal conditions. These are then approximated by taking only a finite subdeterminant. In the APW method the truncation is in $\mathbf{K}$; the wave function is approximated in the interstitial region. In KKR, on the other hand, the sum over all $\mathbf{K}$ is effectively performed when $\mathcal{G}_{\mathbf{k},\mathcal{E}}$ is computed.[24] Instead, the approximation is in the form of the wave function in the atomic region. In both cases the procedure converges well if sufficiently many terms are retained; in practice the KKR method appears to require fewer terms in the spherical harmonic expansion than the APW technique requires in the $\mathbf{K}$ expansion. When the APW and KKR methods are applied to the same muffin-tin potential, they give results in substantial agreement.

The results of a KKR calculation for the $3s^2$ and $3p^1$ derived bands of aluminum are displayed in Figure 11.9. Note the extraordinary resemblance of the calculated

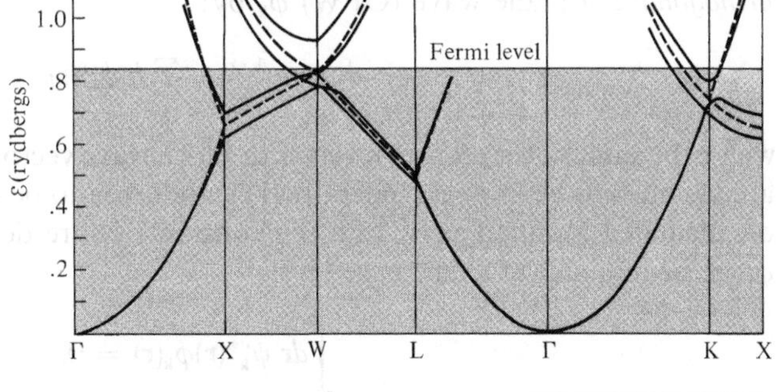

Figure 11.9
Calculated valence bands for aluminum (three electrons outside of a closed-shell neon configuration) compared with *free* electron bands (dashed lines). The bands are computed by the KKR method. (B. Segall, *Phys. Rev.* **124**, 1797 (1961).)

---

[24] It is not necessary to calculate $\mathcal{G}_{\mathbf{k}\mathcal{E}}$ for all values of $\mathbf{r}$, but only the integrals

$$\int d\Omega\, d\Omega'\ Y_{lm}{}^*(\theta\phi)\mathcal{G}_{\mathbf{k},\mathcal{E}}(r_0\theta\phi, r_0\theta'\phi')Y_{l'm'}(\theta'\phi') \text{ and}$$

$$\int d\Omega\, d\Omega'\ Y_{lm}{}^*(\theta\phi)\frac{\partial}{\partial r}\mathcal{G}_{\mathbf{k},\mathcal{E}}(r_0\theta\phi, r\theta'\phi')\Big|_{r=r_0} Y_{l'm'}(\theta'\phi').$$

These have been tabulated for various crystal structures over a range of $\mathcal{E}$ and $\mathbf{k}$ values, with r_0 usually taken as the radius of a sphere inscribed in a Wigner–Seitz cell.

bands to the free electron levels, plotted as dashed lines in the same figure. The only discernible effects of the interaction between electrons and ions are, as predicted by nearly free electron theory, to split the band degeneracies. This is a striking illustration of our observation (see page 152) that metals whose atomic configuration consists of a small number of s and p electrons outside of a rare-gas configuration have band structures that can be reproduced very well by the nearly free electron bands. The next two methods to be discussed attempt to shed some light on this remarkable fact.

THE ORTHOGONALIZED PLANE-WAVE METHOD (OPW)

An alternative method of combining rapid oscillations in the ion core region with plane-wavelike behavior interstitially, is the method of orthogonalized plane waves, due to Herring.[25] The OPW method does *not* require a muffin-tin potential to make calculations feasible, and is therefore of particular value if one insists on using an undoctored potential. In addition, the method affords some insight into why the nearly free electron approximation does so remarkably well in predicting the band structures of a variety of metals.

We begin by explicitly distinguishing between the core electrons and the valence electrons. The core wave functions are well localized about the lattice sites. The valence electrons, on the other hand, can be found with appreciable probability in the interstitial regions, where our hope is that their wave functions will be well approximated be a very small number of plane waves. Throughout this and the next section we shall affix superscripts c or v to wave functions to indicate whether they describe core or valence levels.

The difficulty with approximating a valence wave function by a few plane waves *everywhere* in space (as in the nearly free electron method) is that this hopelessly fails to produce the rapid oscillatory behavior required in the core region. Herring noted that this could be taken care of by using not simple plane waves, but plane waves orthogonalized to the core levels right from the start. Thus we define the *orthogonalized* plane wave (OPW) $\phi_{\mathbf{k}}$ by:

$$\phi_{\mathbf{k}} = e^{i\mathbf{k}\cdot\mathbf{r}} + \sum_c b_c \psi_{\mathbf{k}}{}^c(\mathbf{r}), \tag{11.24}$$

where the sum is over *all* core levels with Bloch wave vector **k**. The core wave functions are assumed to be known (generally they are taken to be tight-binding combinations of calculated atomic levels), and the constants b_c are determined by requiring that $\phi_{\mathbf{k}}$ be orthogonal to every core level:[26]

$$\int d\mathbf{r}\; \psi_{\mathbf{k}}^{c*}(\mathbf{r})\phi_{\mathbf{k}}(\mathbf{r}) = 0, \tag{11.25}$$

which implies that

$$b_c = -\int d\mathbf{r}\; \psi_{\mathbf{k}}^{c*}(\mathbf{r})e^{i\mathbf{k}\cdot\mathbf{r}}. \tag{11.26}$$

---

[25] C. Herring, *Phys. Rev.* **57**, 1169 (1940).

[26] We assume the normalization condition $\int d\mathbf{r}\, |\psi_{\mathbf{k}}{}^c|^2 = 1$. Note that $\phi_{\mathbf{k}'}$ is also orthogonal to $\psi_{\mathbf{k}}{}^c$, with $\mathbf{k}' \neq \mathbf{k}$ because of the Bloch condition.

The OPW $\phi_{\mathbf{k}}$ has the following properties characteristic of valence level wave functions:

1. By explicit construction it is orthogonal to all the core levels. It therefore also has the required rapid oscillations in the core region. This is particularly evident from (11.24), since the core wave functions $\psi_{\mathbf{k}}^{c}(\mathbf{r})$ appearing in $\phi_{\mathbf{k}}$ themselves oscillate in the core region.
2. Because the core levels are localized about lattice points, the second term in (11.24) is small in the interstitial region, where $\phi_{\mathbf{k}}$ is very close to the single plane wave $e^{i\mathbf{k}\cdot\mathbf{r}}$.

Since the plane wave $e^{i\mathbf{k}\cdot\mathbf{r}}$ and the core wave functions $\psi_{\mathbf{k}}^{c}(\mathbf{r})$ satisfy the Bloch condition with wave vector **k**, so will the OPW $\phi_{\mathbf{k}}$. We may therefore, as in the APW method, seek an expansion of the actual electronic eigenstates of the Schrödinger equation as linear combinations of OPW's:

$$\psi_{\mathbf{k}} = \sum_{\mathbf{K}} c_{\mathbf{K}} \phi_{\mathbf{k}+\mathbf{K}}. \tag{11.27}$$

As in the APW method, we can determine the coefficients $c_{\mathbf{K}}$ in (11.27) and the energies $\mathcal{E}(\mathbf{k})$ by inserting (11.27) into the variational principle (11.17), and requiring that the derivatives of the resulting expression with respect to all the $c_{\mathbf{K}}$'s vanish. The crystal potential $U(\mathbf{r})$ will enter into the resulting secular problem only through its OPW matrix elements:

$$\int \phi^{*}_{\mathbf{k}+\mathbf{K}}(\mathbf{r}) U(\mathbf{r}) \phi_{\mathbf{k}+\mathbf{K}'}(\mathbf{r})\, d\mathbf{r}. \tag{11.28}$$

The OPW method owes its success to the fact that although the plane-wave matrix elements of U are large, its OPW matrix elements turn out to be much smaller. Therefore, although it is hopeless to try to get convergence by expanding $\psi_{\mathbf{k}}$ in plane waves, the convergence of the expansion in OPW's is very much faster.

In practice the OPW method is used in two very different ways. On the one hand, one may carry out numerically a first principles OPW calculation, starting with an atomic potential, calculating its OPW matrix elements, and working with large enough secular problems (which may sometimes turn out to be remarkably small, but which can also require as many as a hundred OPW's) to ensure good convergence.

On the other hand, one frequently encounters "calculations" of band structure that appear to be nothing but the nearly free electron theory of Chapter 9, in which the Fourier components $U_{\mathbf{K}}$ of the potential are treated as adjustable parameters rather than known quantities. The $U_{\mathbf{K}}$ are determined by fitting the nearly free electron bands either to empirical data or to the bands calculated in detail by one of the more realistic methods. As an example of this, the KKR bands for aluminum, shown in Figure 11.9, can be reproduced with remarkable precision throughout the zone by a nearly free electron calculation that uses only four plane waves and requires only two parameters[27]: U_{111} and U_{200}.

[27] B. Segall, *Phys. Rev.* **124**, 1797 (1961). (A third parameter is used in the form for the free electron energy, which is written as $\alpha\hbar^2 k^2/2m$.) As it happens, these bands do not lead to a Fermi surface with the correct detailed structure (an illustration of the difficulty of getting precise potentials).

Since nearly free electron theory surely cannot work so well, it must be that the apparently nearly free electron secular problem is actually the final stage of a much more complicated analysis, such as that of the OPW method, the Fourier components $U_{\mathbf{K}}$ being OPW rather than plane-wave matrix elements of the potential. One therefore refers to such a calculation as an OPW calculation. In this context, however, such a designation is little more than a reminder that although the analysis is formally identical to nearly free electron theory, it can be placed on a more secure theoretical footing.

It is not at all clear, however, that the OPW approach is the best way to reduce the actual problem of an electron in a periodic potential to an effectively "nearly free" electron calculation. A more systematic way of studying this problem, as well as a variety of additional calculational approaches, is offered by the *pseudopotential methods.*

THE PSEUDOPOTENTIAL

The theory of the pseudopotential began as an extension of the OPW method. Aside from the possibility it offers of refining OPW calculations, it also provides at least a partial explanation for the success of nearly free electron calculations in fitting actual band structures.

We describe the pseudopotential method only in its earliest formulation,[28] which is basically a recasting of the OPW approach. Suppose that we write the exact wave function for a valence level as a linear combination of OPW's, as in (11.27). Let $\phi_{\mathbf{k}}{}^{v}$ be the plane-wave part of this expansion:

$$\phi_{\mathbf{k}}{}^{v}(\mathbf{r}) = \sum_{\mathbf{K}} c_{\mathbf{K}} e^{i(\mathbf{k}+\mathbf{K})\cdot\mathbf{r}}. \tag{11.29}$$

Then we can rewrite the expansions (11.27) and (11.24) as

$$\psi_{\mathbf{k}}{}^{v}(\mathbf{r}) = \phi_{\mathbf{k}}{}^{v}(\mathbf{r}) - \sum_{c} \left(\int d\mathbf{r}' \, \psi_{\mathbf{k}}^{c*}(\mathbf{r}')\phi_{\mathbf{k}}{}^{v}(\mathbf{r}') \right) \psi_{\mathbf{k}}{}^{c}(\mathbf{r}). \tag{11.30}$$

Since $\psi_{\mathbf{k}}{}^{v}$ is an exact valence wave function, it satisfies Schrödinger's equation with eigenvalue $\mathcal{E}_{\mathbf{k}}{}^{v}$:

$$H\psi_{\mathbf{k}}{}^{v} = \mathcal{E}_{\mathbf{k}}{}^{v}\psi_{\mathbf{k}}{}^{v}. \tag{11.31}$$

Substitution of (11.30) into (11.31) gives

$$H\phi_{\mathbf{k}}{}^{v} - \sum_{c} \left(\int d\mathbf{r}' \, \psi_{\mathbf{k}}^{c*}\phi_{\mathbf{k}}{}^{v} \right) H\psi_{\mathbf{k}}{}^{c} = \mathcal{E}_{\mathbf{k}}{}^{v} \left(\phi_{\mathbf{k}}{}^{v} - \sum_{c} \left(\int d\mathbf{r}' \, \psi_{\mathbf{k}}^{c*}\phi_{\mathbf{k}}{}^{v} \right) \psi_{\mathbf{k}}{}^{c} \right). \tag{11.32}$$

If we note that $H\psi_{\mathbf{k}}{}^{c} = \mathcal{E}_{\mathbf{k}}{}^{c}\psi_{\mathbf{k}}{}^{c}$ for the exact core levels, then we can rewrite (11.32) as

$$(H + V^{R})\phi_{\mathbf{k}}{}^{v} = \mathcal{E}_{\mathbf{k}}{}^{v}\phi_{\mathbf{k}}{}^{v}, \tag{11.33}$$

[28] E. Antoncik, *J. Phys, Chem. Solids* **10**, 314 (1959); J. C. Phillips and L. Kleinman, *Phys. Rev.* **116**, 287, 880 (1959).

where we have buried some rather cumbersome terms in the operator V^R, which is defined by

$$V^R\psi = \sum_c (\mathcal{E}_{\mathbf{k}}^v - \mathcal{E}_c)\left(\int d\mathbf{r}'\, \psi_{\mathbf{k}}^{c*}\psi\right)\psi_{\mathbf{k}}^c. \tag{11.34}$$

We have therefore arrived at an effective Schrödinger equation (11.33) satisfied by $\phi_{\mathbf{k}}^v$, the smooth part of the Bloch function. Since experience with the OPW method suggests that $\phi_{\mathbf{k}}^v$ can be approximated by a linear combination of a small number of plane waves, we might expect that the nearly free electron theory of Chapter 9 could be applied to finding the valence levels of $H + \text{V}_R$. This is the starting point for pseudopotential calculation and analysis.

The *pseudopotential* is defined to be the sum of the actual periodic potential U, and V^R:

$$H + V^R = -\frac{\hbar^2}{2m}\nabla^2 + V^{\text{pseudo}}. \tag{11.35}$$

The hope is that the pseudopotential is sufficiently small to justify a nearly free electron calculation of the valence levels. One can see a hint that this might be so from the fact that although the actual periodic potential is attractive near the ion cores, and thus $(\psi, U\psi) = \int d\mathbf{r}\, \psi^*(\mathbf{r})U(\mathbf{r})\psi(\mathbf{r})$ is negative, the corresponding matrix element of the potential V^R is, according to (11.34),

$$(\psi, V^R\psi) = \sum_c (\mathcal{E}_{\mathbf{k}}^v - \mathcal{E}_{\mathbf{k}}^c)\left|\int d\mathbf{r}\, \psi_{\mathbf{k}}^{c*}\psi\right|^2. \tag{11.36}$$

Since the valence energies lie above the core energies, this is always positive. Thus adding V^R to U provides at least a partial cancellation, and one might optimistically hope for it to lead to a potential weak enough to do nearly free electron calculations for $\phi_{\mathbf{k}}^v$ (the so-called pseudo wave function), treating the pseudopotential as a weak perturbation.

There are some peculiar features to the pseudopotential. Equation (11.34) implies that V^R (and hence the pseudopotential) is nonlocal; i.e., its effect on a wave function $\psi(\mathbf{r})$ is not merely to multiply it by some function of $\mathbf{r}$. In addition, the pseudopotential depends on the energy of the level being sought, $\mathcal{E}_{\mathbf{k}}^v$, which means that many of the basic theorems one is used to applying without further thought (such as the orthogonality of eigenfunctions belonging to different eigenvalues) are no longer applicable to H^{pseudo}.

The second difficulty can be removed by setting $\mathcal{E}_{\mathbf{k}}^v$ in (11.34) and in V^{pseudo} equal to the energy of the levels one is most interested in—generally the Fermi energy. Of course, once this replacement has been made, the eigenvalues of $H + V^R$ are no longer exactly those of the original Hamiltonian, except for the levels at the Fermi energy. Since these are frequently the levels of greatest interest, this need not be too great a price to pay. For example, one can, in this way, find the set of $\mathbf{k}$ for which $\mathcal{E}_{\mathbf{k}}^v = \mathcal{E}_F$, thereby mapping out the Fermi surface.

There turn out to be many ways other than (11.34) to define a V^R such that $H + V^R$ has the same *valence* eigenvalues as the actual crystal Hamiltonian H.

From such choices has arisen a wealth of pseudopotential lore, whose usefulness for anything other than justifying the nearly free electron Fermi surfaces has yet to be convincingly established.[29]

COMBINED METHODS

People have, of course, exercised considerable further ingenuity in combining the various techniques. Thus, for example, it may be useful to treat the *d*-bands of transition elements in a manner suggested by the tight-binding approximation and yet allow for *s-d* mixing, not by adding tight-binding functions for the *s*-band as well, but by combining one of the plane-wave methods we have described in some suitably self-consistent way. It goes without saying that we have only scratched the surface of several vast fields of endeavor in this survey of methods for computing the energy bands.

This and the preceding three chapters have been concerned with the abstract structural features of band structure. We now turn to some of the more direct observational manifestations of the electronic energy bands. Chapters 12 and 13 discuss the generalization of the transport theory of Drude and Sommerfeld to Bloch electrons; Chapter 14 discusses some of the techniques for direct observation of the Fermi surface; and Chapter 15 describes the band structures of some of the more familiar metals.

PROBLEMS

1. *Boundary Conditions on Electron Wave-Functions in Crystals*

Let $\mathbf{r}$ locate a point just within the boundary of a primitive cell C_0, and $\mathbf{r}'$ another point infinitesimally displaced from $\mathbf{r}$ just outside the same boundary. The continuity equations for $\psi(\mathbf{r})$ are

$$\lim_{\mathbf{r}\to\mathbf{r}'} [\psi(\mathbf{r}) - \psi(\mathbf{r}')] = 0,$$

$$\lim_{\mathbf{r}\to\mathbf{r}'} [\nabla\psi(\mathbf{r}) - \nabla\psi(\mathbf{r}')] = 0. \qquad \textbf{(11.37)}$$

(a) Verify that any point $\mathbf{r}$ on the surface of a primitive cell is separated by some Bravais lattice vector $\mathbf{R}$ from another surface point and that the normals to the cell at $\mathbf{r}$ and $\mathbf{r} + \mathbf{R}$ are oppositely directed.

(b) Using the fact that ψ can be chosen to have the Bloch form, show that the continuity conditions can equally well be written in terms of the values of ψ entirely within a primitive cell:

$$\psi(\mathbf{r}) = e^{-i\mathbf{k}\cdot\mathbf{R}}\psi(\mathbf{r} + \mathbf{R}),$$
$$\nabla\psi(\mathbf{r}) = e^{-i\mathbf{k}\cdot\mathbf{R}}\nabla\psi(\mathbf{r} + \mathbf{R}), \qquad \textbf{(11.38)}$$

for pairs of points on the surface separated by direct lattice vectors $\mathbf{R}$.

[29] A review of the pseudopotential and its applications can be found in *Solid State Physics*, Vol. 24, D. Turnbull and F. Seitz, eds., Academic, New York, 1970.

(c) Show that the only information in the second of equations (11.38) not already contained in the first is in the equation

$$\hat{\mathbf{n}}(\mathbf{r}) \cdot \nabla\psi(\mathbf{r}) = -e^{-i\mathbf{k}\cdot\mathbf{R}}\hat{\mathbf{n}}(\mathbf{r}+\mathbf{R}) \cdot \nabla\psi(\mathbf{r}+\mathbf{R}), \tag{11.39}$$

where the vector $\hat{\mathbf{n}}$ is normal to the surface of the cell.

2. Using the fact that the APW is continuous on the surfaces defining the muffin-tin potential, give an argument to show that the APW $\phi_{\mathbf{k}+\mathbf{K},\mathcal{E}}$ satisfies the Bloch condition with wave vector **k**.

3. The integral equation for a Bloch function in a periodic potential is given by Eq. (11.21) where, for potentials of the muffin-tin type, the region of integration is confined to $|\mathbf{r}'| < r_0$.

(a) From the definition (11.22) of $\mathcal{G}$ show that

$$\left(\frac{\hbar^2}{2m}\nabla'^2 + \mathcal{E}\right)\mathcal{G}_{\mathbf{k},\mathcal{E}}(\mathbf{r}-\mathbf{r}') = \delta(\mathbf{r}-\mathbf{r}'), \qquad r, r' < r_0. \tag{11.40}$$

(b) Show by writing

$$\mathcal{G}\nabla'^2\psi = \nabla' \cdot (\mathcal{G}\nabla'\psi - \nabla'\psi\mathcal{G}) + \psi\nabla'^2\mathcal{G},$$

that (11.21), (11.40), and the Schrödinger equation for $r' < r_0$ lead to

$$0 = \int_{r'<r_0} d\mathbf{r}'\, \nabla' \cdot \left[\mathcal{G}_{\mathbf{k},\mathcal{E}(\mathbf{k})}(\mathbf{r}-\mathbf{r}')\nabla'\psi_{\mathbf{k}}(\mathbf{r}') - \psi_{\mathbf{k}}(\mathbf{r}')\nabla'\mathcal{G}_{\mathbf{k},\mathcal{E}(\mathbf{k})}(\mathbf{r}-\mathbf{r}')\right]. \tag{11.41}$$

(c) Use Gauss' theorem to transform (11.41) to an integral over the surface of a sphere of radius $r' = r_0$ and show that when r is also set equal to r_0, Eq. (11.23) results.

12
The Semiclassical Model of Electron Dynamics

The Bloch theory (Chapter 8) extends the equilibrium free electron theory of Sommerfeld (Chapter 2) to the case in which a (nonconstant) periodic potential is present. In Table 12.1 we compare the major features of the two theories.

Table 12.1
COMPARISON OF SOMMERFELD AND BLOCH ONE-ELECTRON EQUILIBRIUM LEVELS

	SOMMERFELD	BLOCH
QUANTUM NUMBERS (EXCLUDING SPIN)	$\mathbf{k}$ ($\hbar\mathbf{k}$ is the momentum.)	$\mathbf{k}$, n ($\hbar\mathbf{k}$ is the crystal momentum and n is the band index.)
RANGE OF QUANTUM NUMBERS	$\mathbf{k}$ runs through all of k-space consistent with the Born-von Karman periodic boundary condition.	For each n, $\mathbf{k}$ runs through all wave vectors in a single primitive cell of the reciprocal lattice consistent with the Born-von Karman periodic boundary condition; n runs through an infinite set of discrete values.
ENERGY	$\mathcal{E}(\mathbf{k}) = \dfrac{\hbar^2 k^2}{2m}$.	For a given band index n, $\mathcal{E}_n(\mathbf{k})$ has no simple explicit form. The only general property is periodicity in the reciprocal lattice: $\mathcal{E}_n(\mathbf{k} + \mathbf{K}) = \mathcal{E}_n(\mathbf{k})$.
VELOCITY	The mean velocity of an electron in a level with wave vector $\mathbf{k}$ is: $\mathbf{v} = \dfrac{\hbar\mathbf{k}}{m} = \dfrac{1}{\hbar}\dfrac{\partial\mathcal{E}}{\partial\mathbf{k}}$.	The mean velocity of an electron in a level with band index n and wave vector $\mathbf{k}$ is: $\mathbf{v}_n(\mathbf{k}) = \dfrac{1}{\hbar}\dfrac{\partial\mathcal{E}_n(\mathbf{k})}{\partial\mathbf{k}}$.
WAVE FUNCTION	The wave function of an electron with wave vector $\mathbf{k}$ is: $\psi_{\mathbf{k}}(\mathbf{r}) = \dfrac{e^{i\mathbf{k}\cdot\mathbf{r}}}{V^{1/2}}$.	The wave function of an electron with band index n and wave vector $\mathbf{k}$ is: $\psi_{n\mathbf{k}}(\mathbf{r}) = e^{i\mathbf{k}\cdot\mathbf{r}}u_{n\mathbf{k}}(\mathbf{r})$ where the function $u_{n\mathbf{k}}$ has no simple explicit form. The only general property is periodicity in the direct lattice: $u_{n\mathbf{k}}(\mathbf{r} + \mathbf{R}) = u_{n\mathbf{k}}(\mathbf{r})$.

To discuss conduction we had to extend Sommerfeld's equilibrium theory to nonequilibrium cases. We argued in Chapter 2 that one could calculate the dynamic behavior of the free electron gas using ordinary classical mechanics, provided that there was no need to localize an electron on a scale comparable to the interelectronic distance. Thus the trajectory of each electron between collisions was calculated according to the usual classical equations of motion for a particle of momentum $\hbar\mathbf{k}$:

$$\dot{\mathbf{r}} = \frac{\hbar\mathbf{k}}{m},$$

$$\hbar\dot{\mathbf{k}} = -e\left(\mathbf{E} + \frac{1}{c}\mathbf{v} \times \mathbf{H}\right). \tag{12.1}$$

If pressed to justify this procedure from a quantum-mechanical point of view we would argue that (12.1) actually describes the behavior of a wave packet of free electron levels,

$$\psi(\mathbf{r}, t) = \sum_{\mathbf{k}'} g(\mathbf{k}') \exp\left[i\left(\mathbf{k}' \cdot \mathbf{r} - \frac{\hbar k'^2 t}{2m}\right)\right],$$
$$g(\mathbf{k}') \approx 0, \quad |\mathbf{k}' - \mathbf{k}| > \Delta k, \tag{12.2}$$

where $\mathbf{k}$ and $\mathbf{r}$ are the mean position and momentum about which the wave packet is localized (to within the limitation $\Delta x \, \Delta k > 1$ imposed by the uncertainty principle).

This approach has a simple and elegant generalization to electrons in a general periodic potential, which is known as the *semiclassical model.* Justifying the semiclassical model in detail is a formidable task, considerably more difficult than justifying the ordinary classical limit for free electrons. In this book we shall not offer a systematic derivation. Our emphasis instead will be on how the semiclassical model is used. We shall therefore simply describe the model, state the limitations on its validity, and extract some of its major physical consequences.[1]

The reader who is dissatisfied with the very incomplete and merely suggestive bases we shall offer for the semiclassical model is urged to examine the broad array of mysteries and anomalies of free electron theory that the model resolves. Perhaps a suitable attitude to take is this: If there were no underlying microscopic quantum theory of electrons in solids, one could still imagine a semiclassical mechanics (guessed by some late nineteenth-century Newton of crystalline spaces) that was brilliantly confirmed by its account of observed electronic behavior, just as classical mechanics was confirmed by its accounting for planetary motion, and only very much later given a more fundamental derivation as a limiting form of quantum mechanics.

As with free electrons, two questions arise in discussing conduction by Bloch electrons[2]: (a) What is the nature of the collisions? (b) How do Bloch electrons move between collisions? The semiclassical model deals entirely with the second question, but the Bloch theory also critically affects the first. Drude assumed that the electrons collided with the fixed heavy ions. This assumption cannot be reconciled with the very long mean free paths possible in metals, and fails to account for their observed temperature dependence.[3] The Bloch theory excludes it on theoretical grounds as well. Bloch levels are *stationary* solutions to the Schrödinger equation in the presence of the full periodic potential of the ions. If an electron in the level $\psi_{n\mathbf{k}}$ has a mean nonvanishing velocity (as it does unless $\partial\mathcal{E}_n(\mathbf{k})/\partial\mathbf{k}$ happens to vanish), then that velocity persists forever.[4] One cannot appeal to collisions with static ions as a mechanism to degrade the velocity, because the interaction of the electron with the fixed periodic array of ions has been *fully* taken into account *ab initio* in the Schrödinger equation solved by the Bloch wave function. Thus the conductivity of a perfect periodic crystal is infinite.

---

1 For one of the more recent efforts at a systematic derivation see J. Zak, *Phys. Rev.* **168**, 686 (1968). References to much of the earlier work are given therein. A very appealing treatment of Bloch electrons in a magnetic field (perhaps the most difficult area in which to derive the semiclassical model) is given by R. G. Chambers, *Proc. Phys. Soc.* **89**, 695 (1966), who explicitly constructs a time-dependent wave packet whose center moves along the orbit determined by the semiclassical equations of motion.

2 We shall use the term "Bloch electrons" to mean "electrons in a general periodic potential."

3 Page 9.

4 See page 141.

This result, so disconcerting to one's classical inclination to picture the electrons as suffering current degrading bumps with individual ions, can be understood as a simple manifestation of the wave nature of electrons. In a *periodic* array of scatterers a wave can propagate without attenuation because of the coherent constructive interference of the scattered waves.[5]

Metals have an electrical resistance because no real solid is a perfect crystal. There are always impurities, missing ions, or other imperfections that can scatter electrons, and at very low temperatures it is these that limit conduction. Even if imperfections could be entirely eliminated, however, the conductivity would remain finite because of thermal vibrations of the ions, which produce temperature-dependent distortions from perfect periodicity in the potential the electrons experience. These deviations from periodicity are capable of scattering electrons, and are the source of the temperature dependence of the electronic relaxation time that was noted in Chapter 1.

We defer a full discussion of the actual scattering mechanisms to Chapters 16 and 26. Here we only note that the Bloch theory now forces us to abandon Drude's naive picture of electron-ion scattering. We shall nevertheless continue to extract consequences that follow from the simple assumption that *some* scattering mechanism exists, irrespective of its detailed features.

Thus the main problem we face is how to describe the motion of Bloch electrons between collisions. The fact that the mean velocity of an electron in a definite Bloch level $\psi_{n\mathbf{k}}$ is[6]

$$\mathbf{v}_n(\mathbf{k}) = \frac{1}{\hbar}\frac{\partial \mathcal{E}_n(\mathbf{k})}{\partial \mathbf{k}} \tag{12.3}$$

is very suggestive. Consider a wave packet of Bloch levels from a given band, constructed in analogy to the free electron wave packet (12.2):

$$\psi_n(\mathbf{r}, t) = \sum_{\mathbf{k}'} g(\mathbf{k}')\psi_{n\mathbf{k}'}(\mathbf{r}) \exp\left[-\frac{i}{\hbar}\mathcal{E}_n(\mathbf{k}')t\right], \qquad g(\mathbf{k}') \approx 0, \quad |\mathbf{k}' - \mathbf{k}| > \Delta k \tag{12.4}$$

Let the spread in wave vector Δk be small compared with the dimensions of the Brillouin zone, so that $\mathcal{E}_n(\mathbf{k})$ varies little over all levels appearing in the wave packet. The formula for the velocity (12.3) can then be viewed as the familiar assertion that the group velocity of a wave packet is $\partial\omega/\partial\mathbf{k} = (\partial/\partial\mathbf{k})(\mathcal{E}/\hbar)$.

The semiclassical model describes such wave packets when it is unnecessary to specify the position of an electron on a scale comparable with the spread of the packet.

Let us estimate how broad the wave packet (12.4) must be when the spread in wave vector is small compared with the dimensions of the Brillouin zone. We examine the wave packet at points separated by a Bravais lattice vector. Setting $\mathbf{r} = \mathbf{r}_0 + \mathbf{R}$, and using the basic property (8.6) of the Bloch function, we can write (12.4) as

$$\psi_n(\mathbf{r}_0 + \mathbf{R}, t) = \sum_{\mathbf{k}'} [g(\mathbf{k}')\psi_{n\mathbf{k}'}(\mathbf{r}_0)] \exp\left[i\left(\mathbf{k}' \cdot \mathbf{R} - \frac{1}{\hbar}\mathcal{E}_n(\mathbf{k}')t\right)\right]. \tag{12.5}$$

[5] For a unified view of a variety of such phenomena, see L. Brillouin, *Wave Propagation in Periodic Structures*, Dover, New York, 1953.

[6] See page 141. The result is proved in Appendix E.

Viewed as a function of $\mathbf{R}$ for fixed $\mathbf{r}_0$, this is just a superposition of plane waves, of the form (12.2), with a weight function $\bar{g}(\mathbf{k}) = [g(\mathbf{k})\psi_{n\mathbf{k}}(\mathbf{r}_0)]$. Thus if Δk measures the region within which g (and hence $\bar{g}$) is appreciable,[7] then $\psi_n(\mathbf{r}_0 + \mathbf{R})$, in accordance with the usual rules for wave packets, should be appreciable within a region of dimensions $\Delta R \approx 1/\Delta k$. Since Δk is small compared with the zone dimensions, which are of the order of the inverse lattice constant $1/a$, it follows that ΔR must be large compared with a. This conclusion is independent of the particular value of $\mathbf{r}_0$, and we therefore conclude that *a wave packet of Bloch levels with a wave vector that is well defined on the scale of the Brillouin zone must be spread in real space over many primitive cells.*

The semiclassical model describes the response of the electrons to externally applied electric and magnetic fields that vary slowly over the dimensions of such a wave packet (Figure 12.1) and therefore exceedingly slowly over a few primitive cells.

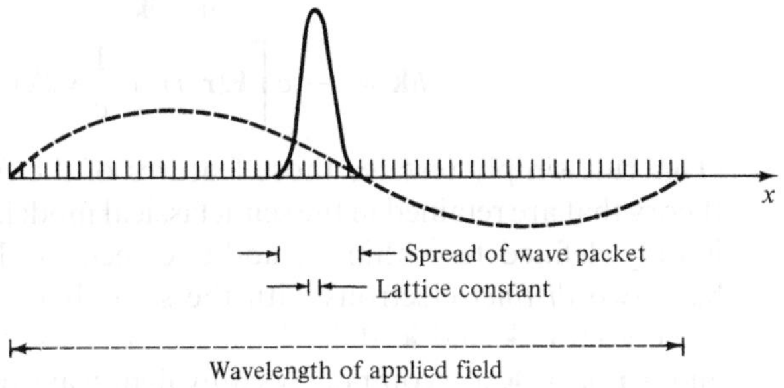

Figure 12.1
Schematic view of the situation described by the semiclassical model. The length over which the applied field (dashed line) varies is much greater than the spread in the wave packet of the electron (solid line), which in turn is much larger than the lattice constant.

In the semiclassical model such fields give rise to ordinary classical forces in an equation of motion describing the evolution of the position and wave vector of the packet. The subtlety of the semiclassical model that makes it more complicated than the ordinary classical limit of *free* electrons, is that the periodic potential of the lattice varies over dimensions that are *small* compared with the spread of the wave packet, and therefore cannot be treated classically. Thus the semiclassical model is a partial classical limit: The externally applied fields are treated classically, but the periodic field of the ions is not.

DESCRIPTION OF THE SEMICLASSICAL MODEL

The semiclassical model predicts how, in the absence of collisions, the position $\mathbf{r}$ and wave vector $\mathbf{k}$ of each electron[8] evolve in the presence of externally applied electric and magnetic fields. *This prediction is based entirely upon a knowledge of the band structure of the metal, i.e., upon the forms of the functions $\mathcal{E}_n(\mathbf{k})$, and upon no other explicit information about the periodic potential of the ions.* The model takes the $\mathcal{E}_n(\mathbf{k})$ as given functions, and says nothing about how to compute them. The aim of the model is to relate the band structure to the transport properties, i.e., the response

---

[7] If g is appreciable only in a neighborhood of $\mathbf{k}$ small compared with the dimensions of the zone, then $\psi_{n\mathbf{k}}(\mathbf{r}_0)$ will vary little over this range, and as a function of $\mathbf{k}$, $\bar{g}$ will differ little from a constant times g.

[8] Hereafter we shall speak of an electron as having both a position and a wave vector. What we are referring to, of course, is a wave packet, as described above.

of the electrons to applied fields or temperature gradients. One uses the model both to deduce transport properties from a given (calculated) band structure and to deduce features of the band structure from the observed transport properties.

Given the functions $\mathcal{E}_n(\mathbf{k})$, the semiclassical model associates with each electron a position $\mathbf{r}$, a wave vector $\mathbf{k}$, and a band index n. In the course of time and in the presence of external electric and magnetic fields $\mathbf{E}(\mathbf{r}, t)$ and $\mathbf{H}(\mathbf{r}, t)$ the position, wave vector, and band index are taken to evolve according to the following rules:

1. The band index n is a constant of the motion. The semiclassical model ignores the possibility of "interband transitions".
2. The time evolution of the position and wave vector of an electron with band index n are determined by the equations of motion:

$$\dot{\mathbf{r}} = \mathbf{v}_n(\mathbf{k}) = \frac{1}{\hbar}\frac{\partial \mathcal{E}_n(\mathbf{k})}{\partial \mathbf{k}}, \tag{12.6a}$$

$$\hbar\dot{\mathbf{k}} = -e\left[\mathbf{E}(\mathbf{r}, t) + \frac{1}{c}\,\mathbf{v}_n(\mathbf{k}) \times \mathbf{H}(\mathbf{r}, t)\right]. \tag{12.6b}$$

3. (This rule simply restates those features of the full quantum mechanical Bloch theory that are retained in the semiclassical model.) The wave vector of an electron is only defined to within an additive reciprocal lattice vector $\mathbf{K}$. One cannot have two *distinct* electrons with the same band index n and position $\mathbf{r}$, whose wave vectors $\mathbf{k}$ and $\mathbf{k}'$ differ by a reciprocal lattice vector $\mathbf{K}$; the labels n, $\mathbf{r}$, $\mathbf{k}$ and n, $\mathbf{r}$, $\mathbf{k} + \mathbf{K}$ are completely equivalent ways of describing the *same* electron.[9] All distinct wave vectors for a single band therefore lie in a single primitive cell of the reciprocal lattice. In thermal equilibrium the contribution to the electronic density from those electrons in the nth band with wave vectors in the infinitesimal volume element $d\mathbf{k}$ of k-space is given by the usual Fermi distribution (2.56):[10]

$$f(\mathcal{E}_n(\mathbf{k}))\frac{d\mathbf{k}}{4\pi^3} = \frac{d\mathbf{k}/4\pi^3}{e^{(\mathcal{E}_n(\mathbf{k})-\mu)/k_BT} + 1}. \tag{12.7}$$

COMMENTS AND RESTRICTIONS

A Many-Carrier Theory

Because the applied fields are assumed to cause no interband transitions, one can consider each band to contain a fixed number of electrons of a particular type. The properties of these types may differ considerably from band to band, since the kind of motion electrons with band index n can undergo depends on the particular form of $\mathcal{E}_n(\mathbf{k})$. In (or near) equilibrium, bands with all energies many k_BT above the Fermi energy $\mathcal{E}_F$ will be unoccupied. Thus one need not consider infinitely many carrier

---

[9] The semiclassical equations of motion (12.6) preserve this equivalence as time evolves. If $\mathbf{r}(t)$, $\mathbf{k}(t)$ give a solution for the nth band, then so will $\mathbf{r}(t)$, $\mathbf{k}(t) + \mathbf{K}$ for any reciprocal lattice vector $\mathbf{K}$, as a consequence of the periodicity of $\mathcal{E}_n(\mathbf{k})$.

[10] This assumes that interactions of the electron spin with any magnetic fields are of no consequence; if they are, then each spin population makes a contribution to n given by half (12.7) where $\mathcal{E}_n(\mathbf{k})$ must include the interaction energy of the given spin with the magnetic field.

types, but only those in bands with energies within a few k_BT of, or lower than, $\mathcal{E}_F$. Furthermore, we shall see below that bands in which all energies are many k_BT less than $\mathcal{E}_F$—i.e., bands that are completely filled in equilibrium—can also be ignored! As a result, only a small number of bands (or carrier types) need be considered in the description of a real metal or semiconductor.

Crystal Momentum Is Not Momentum

Note that within each band the equations of motion (12.6) are the same as the free electron equations (12.1) except that $\mathcal{E}_n(\mathbf{k})$ appears instead of the free electron energy $\hbar^2k^2/2m$. Nevertheless, the crystal momentum $\hbar\mathbf{k}$ is *not* the momentum of a Bloch electron, as emphasized in Chapter 8. The rate of change of an electron's momentum is given by the *total* force on the electron, but the rate of change of an electron's crystal momentum is given by Eq. (12.6), in which forces are exerted only by the external fields and not by the periodic field of the lattice.[11]

Limits of Validity

In the limit of zero periodic potential the semiclassical model must break down, for in that limit the electron will be a free electron. In a uniform electric field a free electron can continually increase its kinetic energy at the expense of electrostatic potential energy. However, the semiclassical model forbids interband transitions, and therefore requires that the energy of any electron remains confined within the limits of the band in which the electron originally found itself.[12] Thus there must be some minimum strength to a periodic potential before the semiclassical model can be applied. Such restrictions are not easy to derive, but have a very simple form, which we state here without proof.[13] At a given point in **k**-space the semiclassical equations will be valid for electrons in the nth band provided that the amplitudes of the slowly varying external electric and magnetic fields satisfy

$$eEa \ll \frac{[\mathcal{E}_{\text{gap}}(\mathbf{k})]^2}{\mathcal{E}_F}, \qquad (12.8)$$

$$\hbar\omega_c \ll \frac{[\mathcal{E}_{\text{gap}}(\mathbf{k})]^2}{\mathcal{E}_F}. \qquad (12.9)$$

In these inequalities the length a is of the order of a lattice constant, $\mathcal{E}_{\text{gap}}(\mathbf{k})$ is the difference between $\mathcal{E}_n(\mathbf{k})$ and the nearest energy $\mathcal{E}_{n'}(\mathbf{k})$ at the same point in k-space but in a different band, and ω_c is the angular cyclotron frequency (Eq. (1.18)).

Condition (12.8) is never close to being violated in a metal. Even with a current density as large as 10^2 amp/cm^2 and a resistivity as large as 100 μohm-cm, the field

---

11 Although the periodic lattice potential does play a crucial role in the semiclassical equations (through the structure of the function $\mathcal{E}_n(\mathbf{k})$ determined by that potential), the role cannot be that of a position-dependent force. To probe a force with the periodicity of the lattice one would have to localize an electron within a single primitive cell. Such a localization is inconsistent with the structure of the wave packets underlying the semiclassical model (see Figure 12.1), which are spread over many lattice sites.

12 This requirement is violated every time the free electron wave vector crosses a Bragg plane, since the electron then jumps from the lower free electron band to the higher one.

13 A rough justification is given in Appendix J.

in the metal will only be $E = \rho j = 10^{-2}$ volt/cm. Hence for a on the order of 10^{-8} cm, eEa is of order 10^{-10} eV. Since $\mathcal{E}_F$ is of the order of an electron volt or more, $\mathcal{E}_{\text{gap}}(\mathbf{k})$ must be as small as 10^{-5} eV before condition (12.8) is violated. In practice, gaps this small are never encountered except near points where two bands become degenerate, and then only in an exceedingly small region of **k**-space about such points. Typical small band gaps are of the order of 10^{-1} eV, and therefore (12.8) is satisfied with a factor of 10^{-8} to spare. The condition is of practical concern only in insulators and in homogeneous semiconductors, where it is possible to establish immense electric fields; when the condition is violated electrons can make an interband transition driven by the field, a phenomenon known as *electric breakdown*.

The condition (12.9) on the magnetic field strength is not as difficult to violate. The energy $\hbar\omega_c$ is of order 10^{-4} eV in a field of 10^4 gauss, in which case (12.9) fails for gaps as large as 10^{-2} eV. Although this is still a small energy gap, such gaps are not at all uncommon, especially when the gap is entirely due to a degeneracy split by spin-orbit coupling. When condition (12.9) fails to hold, electrons may not follow the orbits determined by the semiclassical equations of motion (12.6), a phenomenon known as *magnetic breakthrough* (or "breakdown"). The possibility of magnetic breakthrough must always be kept in mind in interpreting electronic properties in very strong magnetic fields.

In addition to the conditions (12.8) and (12.9) on the amplitude of the applied fields, one must add a low-frequency condition on the fields,

$$\hbar\omega \ll \mathcal{E}_{\text{gap}}, \tag{12.10}$$

or else a single photon could supply enough energy to produce an interband transition. There is also the condition on the wavelength of the applied fields,

$$\lambda \gg a, \tag{12.11}$$

that is necessary if wave packets can be meaningfully introduced at all.[14]

Basis for the Equations of Motion

As discussed above, Eq. (12.6a) is simply the statement that the velocity of a semiclassical electron is the group velocity of the underlying wave packet. Equation (12.6b) is considerably more difficult to justify. It is highly plausible in the presence of a static electric field as the simplest way to guarantee conservation of energy, for if the field is given by $\mathbf{E} = -\nabla\phi$, then we should expect each wave packet to move so that the energy

$$\mathcal{E}_n(\mathbf{k}(t)) - e\phi(\mathbf{r}(t)) \tag{12.12}$$

remains constant. The time derivative of this energy is

$$\frac{\partial \mathcal{E}_n}{\partial \mathbf{k}} \cdot \dot{\mathbf{k}} - e\nabla\phi \cdot \dot{\mathbf{r}}, \tag{12.13}$$

[14] It is also sometimes necessary to take into account further quantum effects due to the possibility of closed electronic k-space orbits in a magnetic field. This can be handled by an ingenious extension of the semiclassical model, and is therefore not a limitation in the sense of the restrictions described above. The problem arises in the theory of the de Haas–van Alphen effect and related phenomena, and is described in Chapter 14.

which Eq. (12.6a) permits us to write as

$$\mathbf{v}_n(\mathbf{k}) \cdot [\hbar\dot{\mathbf{k}} - e\nabla\phi]. \tag{12.14}$$

This will vanish if

$$\hbar\dot{\mathbf{k}} = e\nabla\phi = -e\mathbf{E}, \tag{12.15}$$

which is Eq. (12.6b) in the absence of a magnetic field. However, (12.15) is not necessary for energy to be conserved, since (12.14) vanishes if any term perpendicular to $\mathbf{v}_n(\mathbf{k})$ is added to (12.15). To justify with rigor that the only additional term should be $[\mathbf{v}_n(\mathbf{k})/c] \times \mathbf{H}$, and that the resulting equation should hold for time-dependent fields as well, is a most difficult matter, which we shall not pursue further. The dissatisfied reader is referred to Appendix H for a further way of rendering the semiclassical equations more plausible. There it is shown that they can be written in a very compact Hamiltonian form. To find a really compelling set of arguments, however, it is necessary to delve rather deeply into the (still growing) literature on the subject.[15]

CONSEQUENCES OF THE SEMICLASSICAL EQUATIONS OF MOTION

The rest of this chapter surveys some of the fundamental direct consequences of the semiclassical equations of motion. In Chapter 13 we shall turn to a more systematic way of extracting theories of conduction.

In most of the discussions that follow we shall consider a single band at a time, and shall therefore drop reference to the band index except when explicitly comparing the properties of two or more bands. For simplicity we shall also take the electronic equilibrium distribution function to be that appropriate to zero temperature. In metals finite temperature effects will have negligible influence on the properties discussed below. Thermoelectric effects in metals will be discussed in Chapter 13, and semiconductors will be treated in Chapter 28.

The spirit of the analysis that follows is quite similar to that in which we discussed transport properties in Chapters 1 and 2: We shall describe collisions in terms of a simple relaxation-time approximation, and focus most of our attention on the motion of electrons between collisions as determined (in contrast to Chapters 1 and 2) by the *semiclassical* equations of motion (12.6).

Filled Bands Are Inert

A filled band is one in which all the energies lie below[16] $\mathcal{E}_F$. Electrons in a filled band with wave vectors in a region of k-space of volume $d\mathbf{k}$ contribute $d\mathbf{k}/4\pi^3$ to the total electronic density (Eq. (12.7)). Thus the number of such electrons in a region of position space of volume $d\mathbf{r}$ will be $d\mathbf{r}\ d\mathbf{k}/4\pi^3$. One can therefore characterize a filled band semiclassically by the fact that the density of electrons in a six-dimensional rk-space (called phase space, in analogy to the rp-space of ordinary classical mechanics) is $1/4\pi^3$.

---

[15] See, for example, the references given in footnote 1.

[16] More generally, the energies should be so far below the chemical potential μ compared with k_BT that the Fermi function is indistinguishable from unity throughout the band.

The semiclassical equations (12.6) imply that a filled band remains a filled band at all times, even in the presence of space- and time-dependent electric and magnetic fields. This is a direct consequence of the semiclassical analogue of Liouville's theorem, which asserts the following:[17]

Given any region of six-dimensional phase space Ω_t, consider the point $\mathbf{r}'$, $\mathbf{k}'$ into which each point $\mathbf{r}$, $\mathbf{k}$ in Ω_t is taken by the semiclassical equations of motion between times[18] t and t'. The set of all such points $\mathbf{r}'$, $\mathbf{k}'$ constitutes a new region $\Omega_{t'}$, whose volume is the same as the volume of Ω_t (see Figure 12.2); i.e., phase space volumes are conserved by the semiclassical equations of motion.

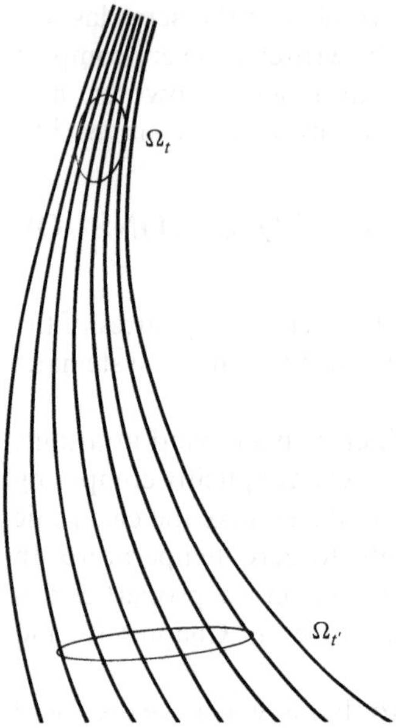

Figure 12.2
Semiclassical trajectories in rk-space. The region $\Omega_{t'}$ contains at time t' just those points that the semiclassical motion has carried from the region Ω_t at time t. Liouville's theorem asserts that Ω_t and $\Omega_{t'}$ have the same volume. (The illustration is for a two-dimensional rk-space lying in the plane of the page, i.e., for semiclassical motion in one dimension.)

This immediately implies that if the phase space density is $1/4\pi^3$ at time zero, it must remain so at all times, for consider any region Ω at time t. The electrons in Ω at time t are just those that were in some other region Ω_0 at time zero where, according to Liouville's theorem, Ω_0 has the same volume as Ω. Since the two regions also have the same number of electrons, they have the same phase space density of electrons. Because that density was $1/4\pi^3$, independent of the region at time 0, it must also be

[17] See Appendix H for a proof that the theorem applies to semiclassical motion. From a quantum mechanical point of view the inertness of filled bands is a simple consequence of the Pauli exclusion principle: The "phase space density" cannot increase if every level contains the maximum number of electrons allowed by the Pauli principle; furthermore, if interband transitions are prohibited, neither can it decrease, for the number of electrons in a level can only be reduced if there are some incompletely filled levels in the band for those electrons to move into. For logical consistency, however, it is necessary to demonstrate that this conclusion also follows directly from the semiclassical equations of motion, without reinvoking the underlying quantum mechanical theory that the model is meant to replace.

[18] The time t' need not be greater than t; i.e., the regions from which Ω_t evolved have the same volume as Ω_t, as well as the regions into which Ω_t will evolve.

$1/4\pi^3$, independent of the region at time t. Thus semiclassical motion between collisions cannot alter the configuration of a filled band, even in the presence of space- and time-dependent external fields.[19]

However, a band with a constant phase space density $1/4\pi^3$ cannot contribute to an electric or thermal current. To see this, note that an infinitesimal phase space volume element $d\mathbf{k}$ about the point $\mathbf{k}$ will contribute $d\mathbf{k}/4\pi^3$ electrons per unit volume, all with velocity $\mathbf{v}(\mathbf{k}) = (1/\hbar)\, \partial\mathcal{E}(\mathbf{k})/\partial\mathbf{k}$ to the current. Summing this over all $\mathbf{k}$ in the Brillouin zone, we find that the total contribution to the electric and energy current densities from a filled band is

$$\mathbf{j} = (-e) \int \frac{d\mathbf{k}}{4\pi^3} \frac{1}{\hbar} \frac{\partial \mathcal{E}}{\partial \mathbf{k}},$$

$$\mathbf{j}_{\mathcal{E}} = \int \frac{d\mathbf{k}}{4\pi^3} \mathcal{E}(\mathbf{k}) \frac{1}{\hbar} \frac{\partial \mathcal{E}}{\partial \mathbf{k}} = \frac{1}{2} \int \frac{d\mathbf{k}}{4\pi^3} \frac{1}{\hbar} \frac{\partial}{\partial \mathbf{k}} (\mathcal{E}(\mathbf{k}))^2. \qquad \textbf{(12.16)}$$

But both of these vanish as a consequence of the theorem[20] that the integral over any primitive cell of the gradient of a periodic function must vanish.

Thus only partially filled bands need be considered in calculating the electronic properties of a solid. This explains how that mysterious parameter of free electron theory, the number of conduction electrons, is to be arrived at: *Conduction is due only to those electrons that are found in partially filled bands.* The reason Drude's assignment to each atom of a number of conduction electrons equal to its valence is often successful is that in many cases those bands derived from the atomic valence electrons are the only ones that are partially filled.

Evidently a solid in which all bands are completely filled or empty will be an electrical and (at least as far as *electronic* transport of heat is concerned) thermal insulator. Since the number of levels in each band is just twice the number of primitive cells in the crystal, all bands can be filled or empty *only* in solids with an even number of electrons per primitive cell. Note that the converse is not true: Solids with an even number of electrons per primitive cell may be (and frequently are) conductors, since the overlap of band energies can lead to a ground state in which several bands are partially filled (see, for example, Figure 12.3). We have thus derived a necessary, but by no means sufficient, condition for a substance to be an insulator.

It is a reassuring exercise to go through the periodic table looking up the crystal structure of all insulating solid elements. They will all be found to have either even valence or (e.g., the halogens) a crystal structure that can be characterized as a lattice with a basis containing an even number of atoms, thereby confirming this very general rule.

---

[19] Collisions cannot alter this stability of filled bands either, provided that we retain our basic assumption (Chapter 1, page 6 and Chapter 13, page 245) that whatever else they do, the collisions cannot alter the distribution of electrons when it has its thermal equilibrium form. For a distribution function with the constant value $1/4\pi^3$ is precisely the zero temperature equilibrium form for any band all of whose energies lie below the Fermi energy.

[20] The theorem is proved in Appendix I. The periodic functions in this case are $\mathcal{E}(\mathbf{k})$ in the case of $\mathbf{j}$, and $\mathcal{E}(\mathbf{k})^2$ in the case of $\mathbf{j}_{\mathcal{E}}$.

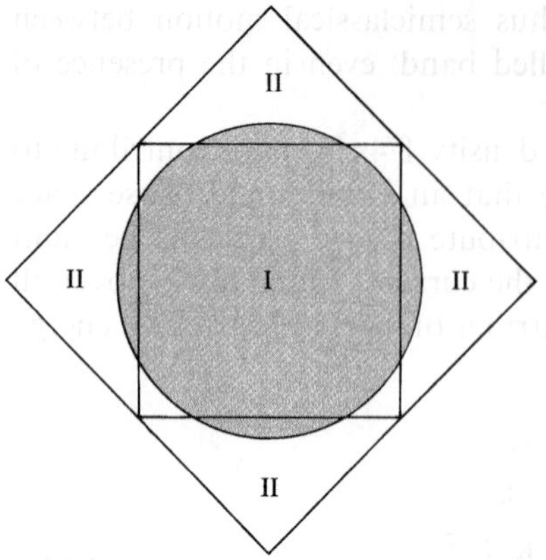

Figure 12.3
A two-dimensional illustration of why a divalent solid can be a conductor. A free electron circle, whose area equals that of the first Brillouin zone (I) of a square Bravais lattice, extends into the second zone (II), thus producing two partially filled bands. Under the influence of a sufficiently strong periodic potential the pockets of first-zone holes and second-zone electrons might shrink to zero. Quite generally, however, a weak periodic potential will always lead to this kind of overlap (except in one dimension).

Semiclassical Motion in an Applied DC Electric Field

In a uniform static electric field the semiclassical equation of motion for **k** (Eq. (12.6)) has the general solution

$$\mathbf{k}(t) = \mathbf{k}(0) - \frac{e\mathbf{E}t}{\hbar}. \tag{12.17}$$

Thus in a time t every electron changes its wave vector by the same amount. This is consistent with our observation that applied fields can have no effect on a filled band in the semiclassical model, for a uniform shift in the wave vector of *every* occupied level does not alter the phase space density of electrons when that density is constant, as it is for a filled band. However, it is somewhat jarring to one's classical intuition that by shifting the wave vector of every electron by the same amount we nevertheless fail to bring about a current-carrying configuration.

To understand this, one must remember that the current carried by an electron is proportional to its velocity, which is not proportional to **k** in the semiclassical model. The velocity of an electron at time t will be

$$\mathbf{v}(\mathbf{k}(t)) = \mathbf{v}\left(\mathbf{k}(0) - \frac{e\mathbf{E}t}{\hbar}\right). \tag{12.18}$$

Since $\mathbf{v}(\mathbf{k})$ is periodic in the reciprocal lattice, the velocity (12.18) is a bounded function of time and, when the field **E** is parallel to a reciprocal lattice vector, oscillatory! This is in striking contrast to the free electron case, where **v** is proportional to **k** and grows linearly in time.

The **k** dependence (and, to within a scale factor, the t dependence) of the velocity is illustrated in Figure 12.4, where both $\mathcal{E}(k)$ and $v(k)$ are plotted in one dimension. Although the velocity is linear in k near the band minimum, it reaches a maximum as the zone boundary is approached, and then drops back down, going to zero at the zone edge. In the region between the maximum of v and the zone edge the velocity actually decreases with increasing k, so that the acceleration of the electron is opposite to the externally applied electric force!

This extraordinary behavior is a consequence of the additional force exerted by the periodic potential, which, though no longer explicit in the semiclassical model, lies buried in it (through the functional form of $\mathcal{E}(\mathbf{k})$). As an electron approaches a

Figure 12.4
$\mathcal{E}(k)$ and $v(k)$ vs. k (or vs. time, via Eq. (12.17)) in one dimension (or three dimensions, in a direction parallel to a reciprocal lattice vector that determines one of the first-zone faces.)

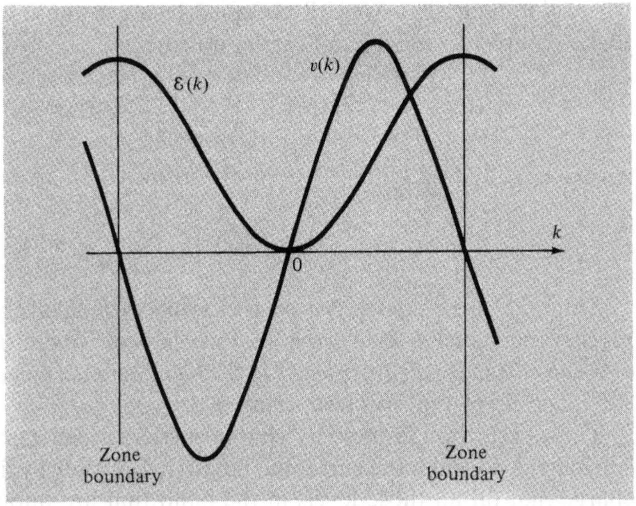

Bragg plane, the external electric field moves it toward levels in which it is increasingly likely to be Bragg-reflected back in the opposite direction.[21]

Thus if an electron could travel between collisions a distance in k-space larger than the dimensions of the zone, it would be possible for a DC field to induce an alternating current. Collisions, however, quite emphatically exclude this possibility. For reasonable values of the field and relaxation time the change in wave vector between two collisions, given by (12.17), is a minute fraction of the zone dimensions.[22]

But although the hypothetical effects of periodic motion in a DC field are inaccessible to observation, effects dominated by electrons that are close enough to the zone boundary to be decelerated by an applied field are readily observable through the curious behavior of "holes."

Holes

One of the most impressive achievements of the semiclassical model is its explanation for phenomena that free electron theory can account for only if the carriers have a positive charge. The most notable of these is the anomalous sign of the Hall coefficient in some metals (see page 58). There are three important points to grasp in understanding how the electrons in a band can contribute to currents in a manner suggestive of positively charged carriers:

1. Since electrons in a volume element $d\mathbf{k}$ about $\mathbf{k}$ contribute $-e\mathbf{v}(\mathbf{k})d\mathbf{k}/4\pi^3$ to the current density, the contribution of all the electrons in a given band to the current density will be

$$\mathbf{j} = (-e)\int_{\text{occupied}} \frac{d\mathbf{k}}{4\pi^3}\,\mathbf{v}(\mathbf{k}), \tag{12.19}$$

---

[21] For example, it is just in the vicinity of Bragg planes that plane-wave levels with different wave vectors are most strongly mixed in the nearly free electron approximation (Chapter 9).

[22] With an electric field of order 10^{-2} volt/cm and a relaxation time of order 10^{-14} sec, $eE\tau/\hbar$ is of order 10^{-1} cm^{-1}. Zone dimensions are of order $1/a \sim 10^8$ cm^{-1}.

where the integral is over all occupied levels in the band.[23] By exploiting the fact that a completely filled band carries no current,

$$0 = \int_{\text{zone}} \frac{d\mathbf{k}}{4\pi^3} \mathbf{v}(\mathbf{k}) = \int_{\text{occupied}} \frac{d\mathbf{k}}{4\pi^3} \mathbf{v}(\mathbf{k}) + \int_{\text{unoccupied}} \frac{d\mathbf{k}}{4\pi^3} \mathbf{v}(\mathbf{k}), \tag{12.20}$$

we can equally well write (12.19) in the form:

$$\mathbf{j} = (+e) \int_{\text{unoccupied}} \frac{d\mathbf{k}}{4\pi^3} \mathbf{v}(\mathbf{k}). \tag{12.21}$$

Thus *the current produced by occupying with electrons a specified set of levels is precisely the same as the current that would be produced if (a) the specified levels were unoccupied and (b) all other levels in the band were occupied but with particles of charge $+e$ (opposite to the electronic charge).*

Thus, even though the only charge carriers are electrons, we may, whenever it is convenient, consider the current to be carried entirely by fictitious particles of positive charge that fill all those levels in the band that are unoccupied by electrons.[24] The fictitious particles are called *holes.*

When one chooses to regard the current as being carried by positive holes rather than negative electrons, the electrons are best regarded as merely the absence of holes; i.e., levels occupied by electrons are to be considered unoccupied by holes. It must be emphasized that pictures cannot be mixed within a given band. If one wishes to regard electrons as carrying the current, then the unoccupied levels make no contribution; if one wishes to regard the holes as carrying the current, then the electrons make no contribution. One may, however, regard some bands in the electron picture and other bands in the hole picture, as suits one's convenience.

To complete the theory of holes we must consider the manner in which the set of unoccupied levels changes under the influence of applied fields:

2. *The unoccupied levels in a band evolve in time under the influence of applied fields precisely as they would if they were occupied by real electrons (of charge $-e$).*

This is because, given the values of **k** and **r** at $t = 0$, the semiclassical equations, being six first-order equations in six variables, uniquely determine **k** and **r** at all subsequent (and all prior) times, just as in ordinary classical mechanics the position and momentum of a particle at any instant determine its entire orbit in the presence of specified external fields. In Figure 12.5 we indicate schematically the orbits determined by the semiclassical equations, as lines in a seven-dimensional *rkt*-space. Because any point on an orbit uniquely specifies the entire orbit, two distinct orbits can have no points in common. We can therefore separate the orbits into occupied and unoccupied orbits, according to whether they contain occupied or unoccupied points at $t = 0$. At any time after $t = 0$ the unoccupied levels will lie on unoccupied orbits, and the occupied levels on occupied orbits. Thus the evolution of both occupied and unoccupied levels is completely determined by the structure of the

---

[23] This need not be the set of levels with energies less than $\mathcal{E}_F$, since we are interested in the non-equilibrium configurations brought about by applied fields.

[24] Note that this contains as a special case the fact that a filled band can carry no current, for a filled band has no unoccupied levels and therefore no fictitious positive carriers.

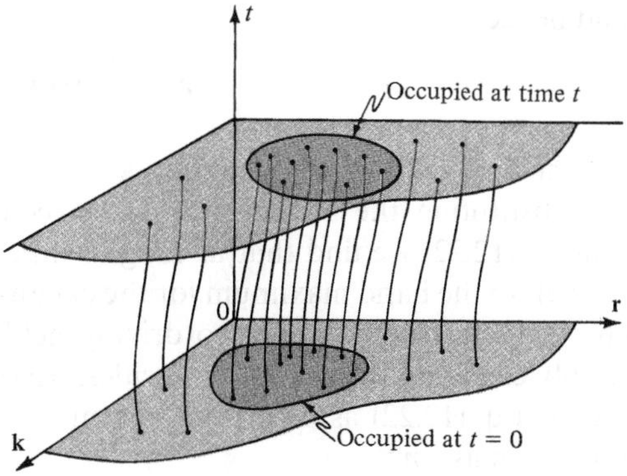

Figure 12.5
Schematic illustration of the time evolution of orbits in semiclassical phase space (here **r** and **k** are each indicated by a single coordinate). The occupied region at time t is determined by the orbits that lie in the occupied region at time $t = 0$.

orbits. But this structure depends only on the form of the semiclassical equations (12.6), and not on whether an electron happens actually to be following a particular orbit.

3. It therefore suffices to examine how electrons respond to applied fields to learn how holes do. The motion of an electron is determined by the semiclassical equation:

$$\hbar\dot{\mathbf{k}} = (-e)\left(\mathbf{E} + \frac{1}{c}\mathbf{v} \times \mathbf{H}\right). \qquad \textbf{(12.22)}$$

Whether or not the orbit of the electron resembles that of a free particle of negative charge depends on whether the acceleration, $d\mathbf{v}/dt$, is or is not parallel to $\dot{\mathbf{k}}$. Should the acceleration be opposite to $\dot{\mathbf{k}}$, then the electron would respond more like a positively charged free particle. As it happens, it is often the case that $d\mathbf{v}(\mathbf{k})/dt$ is indeed directed opposite to $\dot{\mathbf{k}}$ when $\mathbf{k}$ is the wave vector of an unoccupied level, for the following reason:

In equilibrium and in configurations that do not deviate substantially from equilibrium (as is generally the case for nonequilibrium electronic configurations of interest) the unoccupied levels often lie near the top of the band. If the band energy $\mathcal{E}(\mathbf{k})$ has its maximum value at $\mathbf{k}_0$, then if $\mathbf{k}$ is sufficiently close to $\mathbf{k}_0$, we may expand $\mathcal{E}(\mathbf{k})$ about $\mathbf{k}_0$. The linear term in $\mathbf{k} - \mathbf{k}_0$ vanishes at a maximum, and if we assume, for the moment, that $\mathbf{k}_0$ is a point of sufficiently high symmetry (e.g., cubic), then the quadratic term will be proportional to $(\mathbf{k} - \mathbf{k}_0)^2$. Thus

$$\mathcal{E}(\mathbf{k}) \approx \mathcal{E}(\mathbf{k}_0) - A(\mathbf{k} - \mathbf{k}_0)^2, \qquad \textbf{(12.23)}$$

where A is positive (since $\mathcal{E}$ is maximum at $\mathbf{k}_0$). It is conventional to define a positive quantity m^* with dimensions of mass by:

$$\frac{\hbar^2}{2m^*} = A. \qquad \textbf{(12.24)}$$

For levels with wave vectors near $\mathbf{k}_0$,

$$\mathbf{v}(\mathbf{k}) = \frac{1}{\hbar}\frac{\partial\mathcal{E}}{\partial\mathbf{k}} \approx -\frac{\hbar(\mathbf{k} - \mathbf{k}_0)}{m^*}, \qquad \textbf{(12.25)}$$

and hence

$$\mathbf{a} = \frac{d}{dt}\mathbf{v}(\mathbf{k}) = -\frac{\hbar}{m^*}\dot{\mathbf{k}}, \tag{12.26}$$

i.e., the acceleration is opposite to $\dot{\mathbf{k}}$.

Substituting the acceleration wave vector relation (12.26) into the equation of motion (12.22) we find that as long as an electron's orbit is confined to levels close enough to the band maximum for the expansion (12.23) to be accurate, the (negatively charged) electron responds to driving fields as if it had a negative mass $-m^*$. By simply changing the sign of both sides, we can equally well (and far more intuitively) regard Eq. (12.22) as describing the motion of a positively charged particle with a positive mass m^*.

Since the response of a hole is the same as the response an electron would have if the electron were in the unoccupied level (point 2 above), this completes the demonstration that holes behave in all respects like ordinary positively charged particles.

The requirement that the unoccupied levels lie sufficiently close to a highly symmetrical band maximum can be relaxed to a considerable degree.[25] We might expect dynamical behavior suggestive of positively or negatively charged particles, according to whether the angle between $\dot{\mathbf{k}}$ and the acceleration is greater than or less than 90° ($\dot{\mathbf{k}} \cdot \mathbf{a}$ negative or positive). Since

$$\dot{\mathbf{k}} \cdot \mathbf{a} = \dot{\mathbf{k}} \cdot \frac{d}{dt}\mathbf{v} = \dot{\mathbf{k}} \cdot \frac{d}{dt}\frac{1}{\hbar}\frac{\partial \mathcal{E}}{\partial \mathbf{k}} = \frac{1}{\hbar}\sum_{ij} \dot{k}_i \frac{\partial^2 \mathcal{E}}{\partial k_i\, \partial k_j} \dot{k}_j, \tag{12.27}$$

a sufficient condition that $\dot{\mathbf{k}} \cdot \mathbf{a}$ be negative is that

$$\sum_{ij} \Delta_i \frac{\partial^2 \mathcal{E}(\mathbf{k})}{\partial k_i\, \partial k_j} \Delta_j < 0 \qquad \text{(for any vector } \mathbf{\Delta}). \tag{12.28}$$

When $\mathbf{k}$ is at a local maximum of $\mathcal{E}(\mathbf{k})$, (12.28) must hold, for if the inequality was reversed for any vector $\mathbf{\Delta}_0$, then the energy would increase as $\mathbf{k}$ moved from the "maximum" in the direction of $\mathbf{\Delta}_0$. By continuity, (12.28) will therefore hold in some neighborhood of the maximum, and we can expect an electron to respond in a manner suggestive of positive charge, provided that its wave vector remains in that neighborhood.

The quantity m^* determining the dynamics of holes near band maxima of high symmetry is known as the "hole effective mass." More generally, one defines an "effective mass tensor":

$$[\mathbf{M}^{-1}(\mathbf{k})]_{ij} = \pm\frac{1}{\hbar^2}\frac{\partial^2 \mathcal{E}(\mathbf{k})}{\partial k_i\, \partial k_j} = \pm\frac{1}{\hbar}\frac{\partial v_i}{\partial k_j}, \tag{12.29}$$

where the sign is $-$ or $+$ according to whether $\mathbf{k}$ is near a band maximum (holes) or minimum (electrons). Since

$$\mathbf{a} = \frac{d\mathbf{v}}{dt} = \pm \mathbf{M}^{-1}(\mathbf{k})\hbar\dot{\mathbf{k}}, \tag{12.30}$$

---

[25] If the geometry of the unoccupied region of k-space becomes too complicated, however, the picture of holes becomes of limited usefulness.

the equation of motion (12.22) takes on the form

$$\mathbf{M}(\mathbf{k})\mathbf{a} = \mp e\left(\mathbf{E} + \frac{1}{c}\mathbf{v}(\mathbf{k}) \times \mathbf{H}\right). \tag{12.31}$$

The mass tensor plays an important role in determining the dynamics of holes located about anisotropic maxima (or electrons located about anisotropic minima). If the pocket of holes (or electrons) is small enough, one can replace the mass tensor by its value at the maximum (minimum), leading to a linear equation only slightly more complicated than that for free particles. Such equations describe quite accurately the dynamics of electrons and holes in semiconductors (Chapter 28).

Semiclassical Motion in a Uniform Magnetic Field

A wealth of important information on the electronic properties of metals and semiconductors comes from measurements of their response to various probes in the presence of a uniform magnetic field. In such a field the semiclassical equations are

$$\dot{\mathbf{r}} = \mathbf{v}(\mathbf{k}) = \frac{1}{\hbar}\frac{\partial \mathcal{E}(\mathbf{k})}{\partial \mathbf{k}}, \tag{12.32}$$

$$\hbar\dot{\mathbf{k}} = (-e)\frac{1}{c}\mathbf{v}(\mathbf{k}) \times \mathbf{H}. \tag{12.33}$$

It follows immediately from these equations that the component of **k** along the magnetic field and the electronic energy $\mathcal{E}(\mathbf{k})$ are both constants of the motion. These two conservation laws completely determine the electronic orbits in k-space: Electrons move along curves given by the intersection of surfaces of constant energy with planes perpendicular to the magnetic field (Figure 12.6).

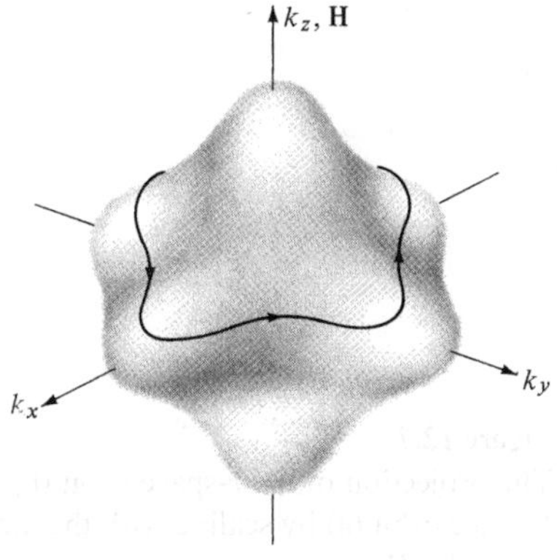

Figure 12.6
Intersection of a surface of constant energy with a plane perpendicular to the magnetic field. The arrow indicates the direction of motion along the orbit if the levels enclosed by the surface have lower energy than those outside.

The sense in which the orbit is traversed follows from observing that **v(k)**, being proportional to the k-gradient of $\mathcal{E}$, points in k-space from lower to higher energies. In conjunction with (12.33) this implies that if one were to imagine walking in k-space along the orbit in the direction of electronic motion with the magnetic field pointing from one's feet to one's head, then the high-energy side of the orbit would be on one's right. In particular, closed k-space orbits surrounding levels with energies higher than those on the orbit (hole orbits) are traversed in the opposite sense to closed orbits surrounding levels of lower energy (electron orbits). This is consistent with, but slightly more general than the conclusions reached in our discussion of holes.

The projection of the real space orbit in a plane perpendicular to the field, $\mathbf{r}_\perp = \mathbf{r} - \hat{\mathbf{H}}(\hat{\mathbf{H}} \cdot \mathbf{r})$, can be found by taking the vector product of both sides of (12.33) with a unit vector parallel to the field. This yields

$$\hat{\mathbf{H}} \times \hbar\dot{\mathbf{k}} = -\frac{eH}{c}(\dot{\mathbf{r}} - \hat{\mathbf{H}}(\hat{\mathbf{H}} \cdot \dot{\mathbf{r}})) = -\frac{eH}{c}\dot{\mathbf{r}}_\perp, \tag{12.34}$$

which integrates to

$$\mathbf{r}_\perp(t) - \mathbf{r}_\perp(0) = -\frac{\hbar c}{eH}\hat{\mathbf{H}} \times (\mathbf{k}(t) - \mathbf{k}(0)). \tag{12.35}$$

Since the cross product of a unit vector with a perpendicular vector is simply the second vector rotated through 90° about the unit vector, we conclude that the projection of the real space orbit in a plane perpendicular to the field is simply the k-space orbit, rotated through 90° about the field direction and scaled by the factor $\hbar c/eH$ (Figure 12.7).[26]

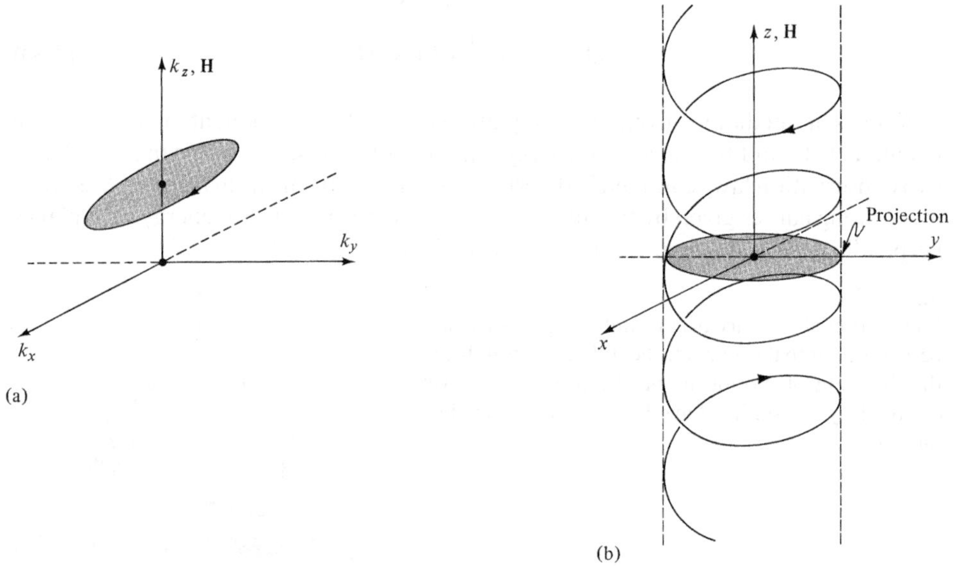

Figure 12.7
The projection of the r-space orbit (b) in a plane perpendicular to the field is obtained from the k-space orbit (a) by scaling with the factor $\hbar c/eH$ and rotation through 90° about the axis determined by H.

---

[26] The component of the real space orbit parallel to the field is not so simply described. Taking the field to be along the z-axis, we have

$$z(t) = z(0) + \int_0^t v_z(t)\,dt; \quad v_z = \frac{1}{\hbar}\frac{\partial \mathcal{E}}{\partial k_z}.$$

In contrast to the free electron case, v_z need not be constant (even though k_z is). Therefore the motion of the electron along the field need not be uniform.

Note that in the free electron case ($\mathcal{E} = \hbar^2k^2/2m$) the constant energy surfaces are spheres, whose intersections with planes are circles. A circle rotated through 90° remains a circle, and we recover the familiar result that a free electron moves along a circle when its motion is projected in a plane perpendicular to the field. In the semiclassical generalization the orbits need not be circular, and in many cases (Figure 12.8) they need not even be closed curves.

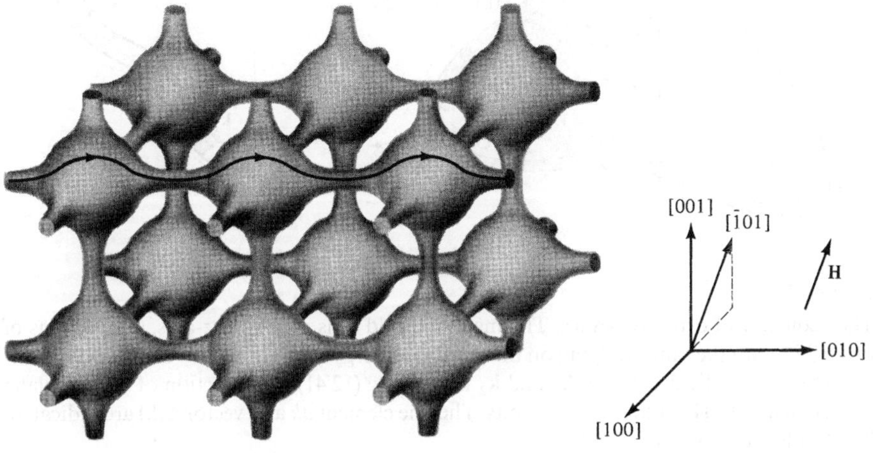

Figure 12.8
Representation in the repeated-zone scheme of a constant-energy surface with simple cubic symmetry, capable of giving rise to open orbits in suitably oriented magnetic fields. One such orbit is shown for a magnetic field parallel to [$\bar{1}$01]. For another example occurring in real metals, see page 292.

We can also express the rate at which the orbit is traversed in terms of certain geometrical features of the band structure. Consider an orbit of energy $\mathcal{E}$ in a particular plane perpendicular to the field (Figure 12.9a). The time taken to traverse that portion of the orbit lying between $\mathbf{k}_1$ and $\mathbf{k}_2$ is

$$t_2 - t_1 = \int_{t_1}^{t_2} dt = \int_{\mathbf{k}_1}^{\mathbf{k}_2} \frac{dk}{|\dot{\mathbf{k}}|}. \tag{12.36}$$

Eliminating $\dot{\mathbf{k}}$ through Eqs. (12.32) and (12.33) we have

$$t_2 - t_1 = \frac{\hbar^2 c}{eH} \int_{\mathbf{k}_1}^{\mathbf{k}_2} \frac{dk}{|(\partial\mathcal{E}/\partial\mathbf{k})_\perp|}. \tag{12.37}$$

where $(\partial\mathcal{E}/\partial\mathbf{k})_\perp$ is the component of $\partial\mathcal{E}/\partial\mathbf{k}$ perpendicular to the field, i.e., its projection in the plane of the orbit.

The quantity $|(\partial\mathcal{E}/\partial\mathbf{k})_\perp|$ has the following geometrical interpretation: Let $\mathbf{\Delta}(\mathbf{k})$ be a vector in the plane of the orbit that is perpendicular to the orbit at point $\mathbf{k}$ and that joins the point $\mathbf{k}$ to a neighboring orbit in the same plane of energy $\mathcal{E} + \Delta\mathcal{E}$ (Figure 12.9b). When $\Delta\mathcal{E}$ is very small we have

$$\Delta\mathcal{E} = \frac{\partial\mathcal{E}}{\partial\mathbf{k}} \cdot \mathbf{\Delta}(\mathbf{k}) = \left(\frac{\partial\mathcal{E}}{\partial\mathbf{k}}\right)_\perp \cdot \mathbf{\Delta}(\mathbf{k}). \tag{12.38}$$

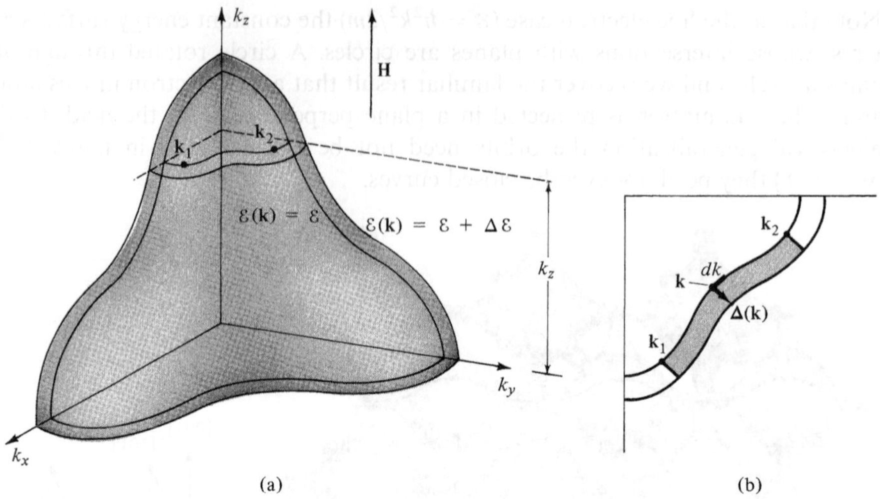

Figure 12.9
The geometry of orbit dynamics. The magnetic field **H** is along the z-axis. (a) Portions of two orbits with the same k_z, lying on the constant-energy surfaces $\mathcal{E}(\mathbf{k}) = \mathcal{E}$ and $\mathcal{E}(\mathbf{k}) = \mathcal{E} + \Delta\mathcal{E}$. The time of flight between $\mathbf{k}_1$ and $\mathbf{k}_2$ is given by (12.41). (b) A section of (a) in a plane perpendicular to **H** containing the orbits. The line element dk and vector $\boldsymbol{\Delta}(\mathbf{k})$ are indicated. The shaded area is $(\partial A_{1,2}/\partial\mathcal{E})\Delta\mathcal{E}$.

Furthermore, since $\partial\mathcal{E}/\partial\mathbf{k}$ is perpendicular to surfaces of constant energy, the vector $(\partial\mathcal{E}/\partial\mathbf{k})_\perp$ is perpendicular to the orbit, and hence parallel to $\boldsymbol{\Delta}(\mathbf{k})$. We can therefore replace (12.38) by

$$\Delta\mathcal{E} = \left|\left(\frac{\partial\mathcal{E}}{\partial\mathbf{k}}\right)_\perp\right| \Delta(\mathbf{k}) \qquad \textbf{(12.39)}$$

and rewrite (12.37):

$$t_2 - t_1 = \frac{\hbar^2 c}{eH}\frac{1}{\Delta\mathcal{E}}\int_{\mathbf{k}_1}^{\mathbf{k}_2} \Delta(\mathbf{k})\, dk. \qquad \textbf{(12.40)}$$

The integral in Eq. (12.40) just gives the area of the plane between the two neighboring orbits from $\mathbf{k}_1$ to $\mathbf{k}_2$ (Figure 12.9). Hence if we take the limit of (12.40) as $\Delta\mathcal{E} \to 0$, we have

$$t_2 - t_1 = \frac{\hbar^2 c}{eH}\frac{\partial A_{1,2}}{\partial\mathcal{E}}, \qquad \textbf{(12.41)}$$

where $\partial A_{1,2}/\partial\mathcal{E}$ is the rate at which the portion of the orbit between $\mathbf{k}_1$ and $\mathbf{k}_2$ starts to sweep out area in the given plane as $\mathcal{E}$ is increased.

The result (12.41) is most frequently encountered in the case in which the orbit is a simple closed curve, and $\mathbf{k}_1$ and $\mathbf{k}_2$ are chosen to give a single complete circuit ($\mathbf{k}_1 = \mathbf{k}_2$). The quantity $t_2 - t_1$ is then the period T of the orbit. If A is the k-space area enclosed by the orbit in its plane, then (12.41) gives[27]

[27] The quantities A and T depend on the energy $\mathcal{E}$ of the orbit and its plane, which is specified by k_z where the z-axis is taken to be along the field.

$$T(\mathcal{E}, k_z) = \frac{\hbar^2 c}{eH} \frac{\partial}{\partial \mathcal{E}} A(\mathcal{E}, k_z). \tag{12.42}$$

To make this resemble the free electron result[28]

$$T = \frac{2\pi}{\omega_c} = \frac{2\pi mc}{eH}, \tag{12.43}$$

it is customary to define a cyclotron effective mass $m^*(\mathcal{E}, k_z)$:

$$m^*(\mathcal{E}, k_z) = \frac{\hbar^2}{2\pi} \frac{\partial A(\mathcal{E}, k_z)}{\partial \mathcal{E}}. \tag{12.44}$$

We emphasize that this effective mass is not necessarily the same as other effective masses that it is convenient to define in other contexts, such as the specific heat effective mass. (See Problem 2.)

Semiclassical Motion in Perpendicular Uniform Electric and Magnetic Fields

When a uniform electric field **E** is present in addition to the static magnetic field **H**, the equation (12.35) for the projection of the real space orbit in a plane perpendicular to **H** acquires an additional term:

$$\mathbf{r}_\perp(t) - \mathbf{r}_\perp(0) = -\frac{\hbar c}{eH} \hat{\mathbf{H}} \times [\mathbf{k}(t) - \mathbf{k}(0)] + \mathbf{w}t, \tag{12.45}$$

where

$$\mathbf{w} = c\frac{E}{H}(\hat{\mathbf{E}} \times \hat{\mathbf{H}}). \tag{12.46}$$

Thus the motion in real space perpendicular to **H** is the superposition of (a) the k-space orbit rotated and scaled just as it would be if only the magnetic field were present and (b) a uniform drift with velocity **w**.[29]

To determine the k-space orbit we note that when **E** and **H** are perpendicular, the equation of motion (12.6b) can be written in the form

$$\hbar\dot{\mathbf{k}} = -\frac{e}{c}\frac{1}{\hbar}\frac{\partial \bar{\mathcal{E}}}{\partial \mathbf{k}} \times \mathbf{H}, \tag{12.47}$$

where[30]

$$\bar{\mathcal{E}}(\mathbf{k}) = \mathcal{E}(\mathbf{k}) - \hbar\mathbf{k} \cdot \mathbf{w}. \tag{12.48}$$

Equation (12.47) is the equation of motion an electron would have if only the magnetic field **H** were present, and if the band structure were given by $\bar{\mathcal{E}}(\mathbf{k})$ rather than $\mathcal{E}(\mathbf{k})$ (cf. Eq. (12.33)). We can therefore conclude from the analysis of that case that the k-space orbits are given by intersections of surfaces of constant $\bar{\mathcal{E}}$ with planes perpendicular to the magnetic field.

---

[28] See page 14 and Eq. (1.18). Equation (12.43) is derived from the general result (12.42) in Problem 1.

[29] The reader versed in electromagnetic theory will recognize **w** as the velocity of the frame of reference in which the electric field vanishes.

[30] For a free electron $\bar{\mathcal{E}}$ is simply the electron's energy in the frame moving with velocity **w** (to within a k-independent additive constant).

We have thus found an explicit geometrical construction for the semiclassical orbits in crossed electric and magnetic fields.

High-Field Hall Effect and Magnetoresistance[31]

We pursue further the analysis in crossed electric and magnetic fields when (a) the magnetic field is very large (typically 10^4 gauss or more; the pertinent physical condition depends on the band structure and will emerge below), and (b) $\overline{\mathcal{E}}(\mathbf{k})$ differs only slightly from $\mathcal{E}(\mathbf{k})$. The second assumption is almost certain to hold when the first does, since a typical wave vector $\mathbf{k}$ is at most of order $1/a_0$. Consequently

$$\hbar\mathbf{k}\cdot\mathbf{w} < \frac{\hbar}{a_0}c\frac{E}{H} = \left(\frac{e^2}{a_0}\right)\left(\frac{eEa_0}{\hbar\omega_c}\right). \tag{12.49}$$

Since eEa_0 is at most[32] of order 10^{-10} eV and $\hbar\omega_c$ is of order 10^{-4} eV in a field of 10^4 gauss, $\hbar\mathbf{k}\cdot\mathbf{w}$ is of order 10^{-6} Ry. Since $\mathcal{E}(\mathbf{k})$ is typically an appreciable fraction of a rydberg, $\overline{\mathcal{E}}(\mathbf{k})$ (Eq. (12.48)) is indeed close to $\mathcal{E}(\mathbf{k})$.

The limiting behavior of the current induced by the electric field in high magnetic fields is quite different depending on whether (a) all occupied (or all unoccupied) electronic levels lie on orbits that are closed curves or (b) some of the occupied and unoccupied levels lie on orbits that do not close on themselves, but are extended or "open" in k-space. Because $\overline{\mathcal{E}}(\mathbf{k})$ is very close to $\mathcal{E}(\mathbf{k})$, we shall assume that whichever of these criteria is satisfied by the orbits determined by $\overline{\mathcal{E}}$ is also satisfied by the orbits determined by $\mathcal{E}$.

***Case* 1** When all occupied (or all unoccupied) orbits are closed, we shall take the high magnetic field condition to mean that these orbits can be traversed many times between successive collisions. In the free electron case this reduces to the condition $\omega_c\tau \gg 1$, and we shall take this as indicating the order of magnitude required of the field in the general case. To satisfy the condition requires not only very large fields (10^4 gauss or more) but also very pure single crystals at very low temperatures, to guarantee long relaxation times. With some effort values of $\omega_c\tau$ as large as 100 or more have been achieved.

Suppose then that the period T is small compared with the relaxation time τ, for every orbit containing occupied levels.[33] To calculate the current density at time $t = 0$ we note[34] that $\mathbf{j} = -ne\mathbf{v}$, where $\mathbf{v}$ is the mean velocity acquired by an electron since its last collision, averaged over all occupied levels. Since the mean time back

---

[31] In applying the preceding analysis of crossed **E** and **H** fields to the theory of the Hall effect and magnetoresistance, we restrict ourselves to geometries symmetrically enough disposed with respect to the crystal axes that both Hall and applied electric fields are perpendicular to the magnetic field. However, similar conclusions can be reached by the more elaborate methods of Chapter 13, in the general case.

[32] See the paragraph following Eq. (12.9).

[33] For concreteness we assume the *occupied* levels all lie on closed orbits. If it is the *unoccupied* levels, we can appeal to the discussion of holes to justify essentially the same analysis, except that the current density will be given by $\mathbf{j} = +n_h e\mathbf{v}$, where n_h is the density of holes, and $\mathbf{v}$ is the mean velocity that an electron would acquire in time τ averaged over all unoccupied levels.

[34] See Eq. (1.4). The analysis that follows is quite similar in spirit to the Drude derivation of the DC conductivity in Chapter 1. We shall calculate the current carried by a single band, since the contributions of more than one band simply add together.

to the last collision is τ, we can conclude from (12.45) that the component of this velocity perpendicular to the magnetic field for a particular electron is just

$$\frac{\mathbf{r}_\perp(0) - \mathbf{r}_\perp(-\tau)}{\tau} = -\frac{\hbar c}{eH}\hat{\mathbf{H}} \times \frac{\mathbf{k}(0) - \mathbf{k}(-\tau)}{\tau} + \mathbf{w}. \tag{12.50}$$

Since all occupied orbits are closed, $\Delta\mathbf{k} = \mathbf{k}(0) - \mathbf{k}(-\tau)$ is bounded in time, so for sufficiently large τ the drift velocity $\mathbf{w}$ gives the dominant contribution to (12.50) and we have[35]

$$\lim_{\tau/T \to \infty} \mathbf{j}_\perp = -ne\mathbf{w} = -\frac{nec}{H}(\mathbf{E} \times \hat{\mathbf{H}}). \tag{12.51}$$

If it is the unoccupied levels that all lie on closed orbits, the corresponding result is[36]

$$\lim_{\tau/T \to \infty} \mathbf{j}_\perp = +\frac{n_h ec}{H}(\mathbf{E} \times \hat{\mathbf{H}}) \tag{12.52}$$

Equations (12.51) and (12.52) assert that when all relevant orbits are closed, the deflection of the Lorentz force is so effective in preventing electrons from acquiring from the electric field that the uniform drift velocity $\mathbf{w}$ perpendicular to $\mathbf{E}$ gives the dominant contribution to the current. The result embodied in Eqs. (12.51) and (12.52) is usually stated in terms of the Hall coefficient, defined (see Eq. (1.15) as the component of the electric field perpendicular to the current, divided by the product of the magnetic field and the current density. If the entire current is carried by electrons from a single band, for which (12.51) or (12.52) is valid, then the high-field Hall coefficient is simply[37]

$$R_\infty = -\frac{1}{nec}, \quad \text{electrons}; \qquad R_\infty = +\frac{1}{n_h ec}, \quad \text{holes}. \tag{12.53}$$

---

35 In writing the high-field limit in this form, we are interpreting the limit to mean large τ at fixed H, rather than large H at fixed τ. To demonstrate that the same leading term emerges in the latter case, or to estimate the value of $\omega_c\tau$ at which the leading term begins to dominate, requires a somewhat deeper analysis. We first note that if the electric field were zero, then the net contribution to the mean current of the term in (12.50) proportional to $\Delta\mathbf{k}$ would vanish when averaged over occupied orbits (since both $\mathbf{j}$ and $\mathbf{w}$ must vanish when $\mathbf{E} = 0$). When $\mathbf{E} \neq 0$, $\Delta\mathbf{k}$ no longer vanishes when averaged over orbits, because the replacement of $\mathcal{E}$ by $\bar{\mathcal{E}}$ (Eq. (12.48)) displaces all the k-space orbits in the same general direction. This is easily seen in the free electron case, since if $\mathcal{E}(\mathbf{k}) = \hbar^2k^2/2m$, then $\bar{\mathcal{E}}(\mathbf{k})$ is given (to within a dynamically irrelevant additive constant) by $\bar{\mathcal{E}}(\mathbf{k}) = \hbar^2(\mathbf{k} - m\mathbf{w}/\hbar)^2/2m$. Consequently, when averaged over all orbits, $\Delta\mathbf{k}$ will no longer give zero, but $m\mathbf{w}/\hbar$. It follows from (12.50) that the size of the contribution of $\Delta\mathbf{k}$ to the mean velocity $\mathbf{v}$, when averaged over orbits, will be $(mw/\hbar)(\hbar c/eH)(1/\tau) = w/(\omega_c\tau)$. This is smaller than the leading term w by $1/\omega_c\tau$. Thus the limiting form (12.51) does indeed become valid when the orbits can be traversed many times between collisions. For a general band structure the average of $\Delta\mathbf{k}$ will be more complicated (for example, it will depend on the particular orbit), but we can expect the free electron estimate to give the right order of magnitude if m is replaced by a suitably defined effective mass.

36 Since (12.51) and (12.52) are manifestly different, there can be no band in which all orbits (occupied and unoccupied) are closed curves. The topologically minded reader is invited to deduce this directly from the periodicity of $\mathcal{E}(\mathbf{k})$.

37 This remarkably general result is nothing but a compact way of expressing the dominance in the current of the drift velocity $\mathbf{w}$ in the high-field limit. It holds for quite general band structure precisely because the semiclassical equations preserve the fundamental role that $\mathbf{w}$ plays in free electron theory. It fails (see below) when some electron and hole orbits are open, because $\mathbf{w}$ then no longer dominates the high-field current.

This is just the elementary result (1.21) of free electron theory, reappearing under remarkably more general circumstances provided that (a) all occupied (or all unoccupied) orbits are closed, (b) the field is large enough that each orbit is traversed many times between collisions, and (c) the carriers are taken to be holes if it is the unoccupied orbits that are closed. Thus the semiclassical theory can account for the "anomalous" sign of some measured Hall coefficients,[38] as well as preserving, under fairly general conditions, the very valuable information about the carrier density that measured (high-field) Hall coefficients yield.

If several bands contribute to the current density, each of which has only closed electron (or hole) orbits, then (12.51) or (12.52) holds separately for each band, and the total current density in the high-field limit will be

$$\lim_{\tau/T\to\infty} \mathbf{j}_\perp = -\frac{n_{\text{eff}}ec}{H}(\mathbf{E}\times\hat{\mathbf{H}}), \tag{12.54}$$

where n_{eff} is the total density of electrons minus the total density of holes. The high-field Hall coefficient will then be

$$R_\infty = -\frac{1}{n_{\text{eff}}ec}. \tag{12.55}$$

Further aspects of the many-band case, including the question of how (12.55) must be modified when the density of electrons is equal to the density of holes (so-called compensated materials), are explored in Problem 4.

One can also verify (Problem 5) that since the corrections to the high-field current densities (12.51) and (12.52) are smaller by a factor of order $1/\omega_c\tau$, the transverse magnetoresistance[39] approaches a field-independent constant ("saturates") in the high-field limit[40] provided that the current is carried by a single band with closed electron (or hole) orbits. The many-band case is investigated in Problem 4, where it is shown that if all electron or hole orbits are closed in each band, the magnetoresistance continues to saturate unless the material is compensated, in which case it grows without limit with increasing magnetic field.

***Case* 2** The above conclusions change drastically if there is at least one band in which neither all occupied nor all unoccupied orbits are closed. This will be the case if at least some orbits at the Fermi energy are open, unbounded curves (Figure 12.8). Electrons in such orbits are no longer forced by the magnetic field to undergo a periodic motion along the direction of the electric field, as they are in closed orbits. Consequently the magnetic field is no longer effective in frustrating the ability of such electrons to acquire energy from the driving electric field. If the unbounded orbit stretches in a real space direction $\hat{\mathbf{n}}$, one would therefore expect to find a contribution to the current that does not vanish in the high-field limit, directed along $\hat{\mathbf{n}}$, and proportional to the projection of $\mathbf{E}$ along $\hat{\mathbf{n}}$:

$$\mathbf{j} = \sigma^{(0)}\hat{\mathbf{n}}(\hat{\mathbf{n}}\cdot\mathbf{E}) + \boldsymbol{\sigma}^{(1)}\cdot\mathbf{E}, \qquad \begin{aligned}&\sigma^{(0)} \to \text{constant as } H\to\infty,\\ &\boldsymbol{\sigma}^{(1)} \to 0 \text{ as } H\to\infty.\end{aligned} \tag{12.56}$$

---

[38] See Table 1.4, Figure 1.4, and p. 58.

[39] See page 12.

[40] In free electron theory the magnetoresistance is independent of magnetic field strength (page 14).

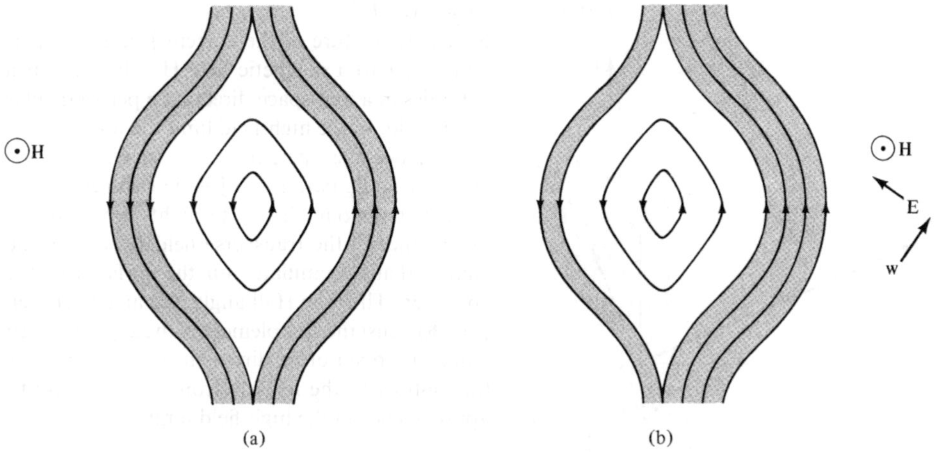

Figure 12.10
A section of constant-energy surfaces in a plane perpendicular to the magnetic field **H**, showing occupied open (shaded) and closed (unshaded) orbits. In (a) no electric field is present and the currents carried by open orbits in opposite directions cancel. In (b) an electric field **E** is present, leading in the steady state to an imbalance in oppositely directed populated open orbits, and hence a net current. (This follows from the fact that $\bar{\mathcal{E}}$ (Eq. (12.48)) is conserved by the semiclassical motion between collisions.)

This expectation is confirmed by the semiclassical equations, for the growth in wave vector ($\Delta\mathbf{k}$) of an electron emerging from a collision on an open orbit is not bounded in time, but grows at a rate[41] directly proportional to H, leading to a contribution to the average velocity (12.50) that is independent of magnetic field strength and directed along the real space direction of the open orbit in the high-field limit.[42]

The high-field form (12.56) is radically different from the forms (12.51) or (12.52) appropriate to carriers all of whose orbits are closed. As a result the Hall coefficient no longer has the simple high-field limit (12.53). Furthermore, the conclusion that the high-field magnetoresistance approaches a constant is no longer valid; indeed, the failure of the magnetoresistance to saturate is one of the characteristic signals that a Fermi surface can support open orbits.

To understand the implications of the limiting behavior (12.56) on the high-field magnetoresistance, consider an experiment (Figure 12.11) in which the direction of current flow (determined by the geometry of the specimen) does not lie along the direction $\hat{\mathbf{n}}$ of the open orbit in real space. Because of (12.56) this is possible in the

---

41 Since the rate at which the orbit is traversed is proportional to H (Eq. (12.41)).

42 Note that when $\mathbf{E} = 0$ this contribution must still give zero when averaged over all open orbits (since **j** and **w** are zero) so there must be oppositely directed open orbits whose contributions cancel. When an electric field is present, however, the orbits directed so as to extract energy from the field become more heavily populated at the expense of those directed so as to lose energy (Figure 12.10). This population difference is proportional to the projection of the drift velocity **w** along the k-space direction of the orbit, or, equivalently, to the projection of **E** along the real space direction of the orbit. This is the source of the $\hat{n} \cdot \mathbf{E}$ dependence in (12.56).

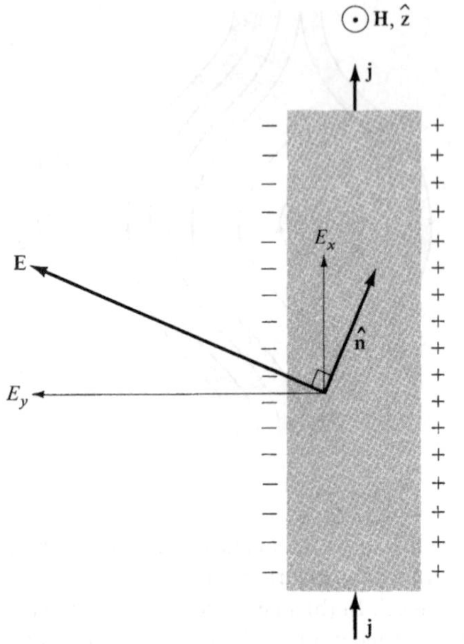

Figure 12.11
Schematic picture of the current **j** in a wire perpendicular to a magnetic field **H**, when an open orbit lies in a real-space direction **n̂** perpendicular to the field. In the high-field limit the total electric field **E** becomes perpendicular to **n̂**. Since the component E_x parallel to **j** is determined by the applied potential, this is brought about by the appearance of the transverse field E_y, due to the charge that accumulates on the surfaces of the specimen. Thus the Hall angle (the angle between **j** and **E**) is just the complement of the angle between **j** and the open-orbit direction. It therefore fails (in contrast to the free electron case, page 14) to approach 90° in the high-field limit.

high-field limit only if the projection of the electric field on **n̂**, $\mathbf{E} \cdot \hat{\mathbf{n}}$, vanishes.[43] The electric field therefore has the form (see Figure 12.11)

$$\mathbf{E} = E^{(0)}\hat{\mathbf{n}}' + E^{(1)}\hat{\mathbf{n}}, \tag{12.57}$$

where $\hat{\mathbf{n}}'$ is a unit vector perpendicular to both $\hat{\mathbf{n}}$ and $\hat{\mathbf{H}}$ ($\hat{\mathbf{n}}' = \hat{\mathbf{n}} \times \hat{\mathbf{H}}$), $E^{(0)}$ is independent of H in the high-field limit, and $E^{(1)}$ vanishes as $H \to \infty$.

The magnetoresistance is the ratio of the component of **E** along **j**, to j:

$$\rho = \frac{\mathbf{E} \cdot \hat{\mathbf{j}}}{j}. \tag{12.58}$$

When the current is not parallel to the direction $\hat{\mathbf{n}}$ of the open orbit, this gives in the high field-limit

$$\rho = \left(\frac{E^{(0)}}{j}\right) \hat{\mathbf{n}}' \cdot \hat{\mathbf{j}}. \tag{12.59}$$

To find $E^{(0)}/j$ we first substitute the electric field (12.57) into the field-current relation (12.56) to find, in the high-field limit, the leading behavior

$$\mathbf{j} = \sigma^{(0)}\hat{\mathbf{n}}E^{(1)} + \boldsymbol{\sigma}^{(1)} \cdot \hat{\mathbf{n}}'E^{(0)}. \tag{12.60}$$

Since $\hat{\mathbf{n}}'$ is perpendicular to $\hat{\mathbf{n}}$, this implies that

$$\hat{\mathbf{n}}' \cdot \mathbf{j} = E^{(0)}\hat{\mathbf{n}}' \cdot \boldsymbol{\sigma}^{(1)} \cdot \hat{\mathbf{n}}', \tag{12.61}$$

[43] In the experiment (Figure 12.11) the component of **E** along **j** is specified. However, in the steady state there will also be a Hall field perpendicular to **j** (see p. 12), which makes it possible for $\mathbf{E} \cdot \hat{\mathbf{n}}$ to vanish in the high-field limit.

or

$$\frac{E^{(0)}}{j} = \frac{\hat{\mathbf{n}}' \cdot \hat{\mathbf{j}}}{\hat{\mathbf{n}}' \cdot \boldsymbol{\sigma}^{(1)} \cdot \hat{\mathbf{n}}'}. \tag{12.62}$$

Substituting this result into Eq. (12.59), we find that in the high-field limit the leading term in the magnetoresistance is

$$\rho = \frac{(\hat{\mathbf{n}}' \cdot \hat{\mathbf{j}})^2}{\hat{\mathbf{n}}' \cdot \boldsymbol{\sigma}^{(1)} \cdot \hat{\mathbf{n}}'}. \tag{12.63}$$

Since $\boldsymbol{\sigma}^{(1)}$ vanishes in the high-field limit, this gives a magnetoresistance that grows without limit with increasing field, and is proportional to the square of the sine of the angle between the current and the real-space direction of the open orbit.

Thus the semiclassical model resolves another anomaly of free electron theory, providing two possible mechanisms[44] by which the magnetoresistance can grow without limit with increasing magnetic field.

We defer until Chapter 15 some illustrations of how these predictions are confirmed by the behavior of real metals, and turn next to a more systematic method of extracting transport coefficients from the semiclassical model.

PROBLEMS

1. For free electrons $\mathcal{E}(\mathbf{k}) = \hbar^2 k^2/2m$. Calculate $\partial A(\mathcal{E}, k_z)/\partial \mathcal{E}$ and show that the general expression (12.42) for the period in a magnetic field reduces to the free electron result (12.43).

2. For electrons near a band minimum (or maximum) $\mathcal{E}(\mathbf{k})$ has the form

$$\mathcal{E}(\mathbf{k}) = \text{constant} + \frac{\hbar^2}{2}(\mathbf{k} - \mathbf{k}_0) \cdot \mathbf{M}^{-1} \cdot (\mathbf{k} - \mathbf{k}_0), \tag{12.64}$$

where the matrix $\mathbf{M}$ is independent of $\mathbf{k}$. (Electrons in semiconductors are almost always treated in this approximation.)

(a) Calculate the cyclotron effective mass from (12.44) and show that it is independent of $\mathcal{E}$ and k_z, and given by

$$m^* = \left(\frac{|\mathbf{M}|}{\mathbf{M}_{zz}}\right)^{1/2}, \qquad \text{(cyclotron)} \tag{12.65}$$

where $|\mathbf{M}|$ is the determinant of the matrix $\mathbf{M}$.

(b) Calculate the electronic specific heat (2.80) resulting from the band structure (12.64), and by comparing it with the corresponding free electron result, show that the band structure contribution to the specific heat effective mass (page 48) is given by

$$m^* = |\mathbf{M}|^{1/3}. \qquad \text{(specific heat)} \tag{12.66}$$

3. When $\mathcal{E}(\mathbf{k})$ has the form (12.64), the semiclassical equations of motion are linear, and therefore easily solved.

---

[44] Open orbits or (Problem 4) compensation. In free electron theory the magnetoresistance is field-independent.

(a) Generalize the analysis in Chapter 1 to show that for such electrons the DC conductivity is given by

$$\boldsymbol{\sigma} = ne^2\tau\mathbf{M}^{-1}. \tag{12.67}$$

(b) Rederive the result (12.65) for the cyclotron effective mass by finding explicitly the time-dependent solutions to Eq. (12.31):

$$\mathbf{M}\cdot\frac{d\mathbf{v}}{dt} = -e\left(\mathbf{E} + \frac{\mathbf{v}}{c}\times\mathbf{H}\right), \tag{12.68}$$

and noting that the angular frequency is related to m^* by $\omega = eH/m^*c$.

4. The free electron result (1.19) for the current induced by an electric field perpendicular to a uniform magnetic field can be written in the form

$$\mathbf{E} = \boldsymbol{\rho}\cdot\mathbf{j} \tag{12.69}$$

where the resistivity tensor $\boldsymbol{\rho}$ has the form

$$\boldsymbol{\rho} = \begin{pmatrix} \rho & -RH \\ RH & \rho \end{pmatrix} \tag{12.70}$$

(It follows from the definitions (1.14) and (1.15) that ρ is the magnetoresistance and R the Hall coefficient.)

(a) Consider a metal with several partially filled bands, in each of which the induced current is related to the field by $\mathbf{E}_n = \boldsymbol{\rho}_n\mathbf{j}_n$, where the $\boldsymbol{\rho}_n$ have the form (12.70):[45]

$$\boldsymbol{\rho}_n = \begin{pmatrix} \rho_n & -R_nH \\ R_nH & \rho_n \end{pmatrix}. \tag{12.71}$$

Show that the total induced current is given by $\mathbf{E} = \boldsymbol{\rho}\cdot\mathbf{j}$, with

$$\boldsymbol{\rho} = (\Sigma\,\boldsymbol{\rho}_n{}^{-1})^{-1}. \tag{12.72}$$

(b) Show that if there are only two bands, then the Hall coefficient and magnetoresistance are given by:

$$R = \frac{R_1\rho_2{}^2 + R_2\rho_1{}^2 + R_1R_2(R_1 + R_2)H^2}{(\rho_1 + \rho_2)^2 + (R_1 + R_2)^2H^2}, \tag{12.73}$$

$$\rho = \frac{\rho_1\rho_2(\rho_1 + \rho_2) + (\rho_1R_2{}^2 + \rho_2R_1{}^2)H^2}{(\rho_1 + \rho_2)^2 + (R_1 + R_2)^2H^2}. \tag{12.74}$$

Note the explicit dependence on magnetic field strength even if the R_i and ρ_i are field-independent (as they are for free electron bands).

(c) Deduce directly from (12.73) the form (12.55) of the high-field Hall coefficient when both bands have closed orbits, and discuss the limiting high-field behavior in the case $n_{\text{eff}} = 0$ (i.e., for a compensated two-band metal). Show that in this case the magnetoresistance increases as H^2 with increasing field.

---

[45] The form (12.71) does not, in general, require that each band be free-electron-like, but only that the magnetic field lie along an axis of sufficient symmetry.

5. Since the correction to the high-field result (12.51) is of order H^2, the general form for the current induced in an electron band with closed orbits is

$$\mathbf{j} = -\frac{nec}{H}(\mathbf{E} \times \hat{\mathbf{H}}) + \boldsymbol{\sigma}^{(1)} \cdot \mathbf{E}, \tag{12.75}$$

where

$$\lim_{H \to \infty} H^2 \boldsymbol{\sigma}^{(1)} < \infty. \tag{12.76}$$

Show that the high-field magnetoresistance is given by

$$\rho = \frac{1}{(nec)^2} \lim_{H \to \infty} H^2 \sigma_{yy}^{(1)}, \tag{12.77}$$

where the y-axis is perpendicular to the magnetic field and the direction of current flow. Note that Eq. (12.76) requires the magnetoresistance to *saturate.*

6. The validity of the semiclassical result $\mathbf{k}(t) = \mathbf{k}(0) - e\mathbf{E}t/\hbar$ for an electron in a uniform electric field is strongly supported by the following theorem (which also provides a useful starting point for a rigorous theory of electric breakdown):

Consider the time-dependent Schrödinger equation for an electron in a periodic potential $U(\mathbf{r})$ and a uniform electric field:

$$i\hbar \frac{\partial \psi}{\partial t} = \left[-\frac{\hbar^2}{2m} \nabla^2 + U(\mathbf{r}) + e\mathbf{E} \cdot \mathbf{r} \right] \psi = H\psi. \tag{12.78}$$

Suppose that at time $t = 0$, $\psi(\mathbf{r}, 0)$ is a linear combination of Bloch levels, all of which have the same wave vector $\mathbf{k}$. Then at time t, $\psi(\mathbf{r}, t)$ will be a linear combination of Bloch levels,[46] all of which have the wave vector $\mathbf{k} - e\mathbf{E}t/\hbar$.

Prove this theorem by noting that the formal solution to the Schrödinger equation is

$$\psi(\mathbf{r}, t) = e^{-iHt/\hbar} \psi(\mathbf{r}, 0), \tag{12.79}$$

and by expressing the assumed property of the initial level and the property to be proved of the final level in terms of the effect on the wave function of translations through Bravais lattice vectors.

7. (a) Does the orbit in Figure 12.7 enclose occupied or unoccupied levels?
(b) Do the closed orbits in Figure 12.10 enclose occupied or unoccupied levels?

---

[46] The semiclassical theory of an electron in a uniform electric field is not exact in spite of this theorem, because the coefficients in the linear combination of Bloch levels will in general depend on time; thus interband transitions may occur.

13
The Semiclassical Theory of Conduction in Metals

Our discussion of electronic conduction in Chapters 1, 2, and 12 was often somewhat qualitative and frequently depended on simplifying features of the particular case being examined. In this chapter we describe a more systematic method of calculating conductivities, applicable to general semiclassical motion in the presence of space- and time-dependent perturbing fields and temperature gradients. The physical approximations underlying this analysis are no more rigorous or sophisticated than those used in Chapter 12, merely more precisely stated. However, the method by which the currents are calculated from the basic physical assumptions is more general and systematic, and of such a form that comparison with more accurate theories can easily be made (Chapter 16).

The description of conduction in this chapter will employ a nonequilibrium distribution function $g_n(\mathbf{r}, \mathbf{k}, t)$ defined so that $g_n(\mathbf{r}, \mathbf{k}, t)\, d\mathbf{r}\, d\mathbf{k}/4\pi^3$ is the number of electrons in the nth band at time t in the semiclassical phase space volume $d\mathbf{r}\, d\mathbf{k}$ about the point $\mathbf{r}$, $\mathbf{k}$. In equilibrium g reduces to the Fermi function,

$$g_n(\mathbf{r}, \mathbf{k}, t) \equiv f(\mathcal{E}_n(\mathbf{k})),$$
$$f(\mathcal{E}) = \frac{1}{e^{(\mathcal{E}-\mu)/k_B T} + 1}, \tag{13.1}$$

but in the presence of applied fields and/or temperature gradients it will differ from its equilibrium form.

In this chapter we shall derive a closed expression for g, based on (a) the assumption that between collisions the electronic motion is determined by the semiclassical equations (12.6), and (b) a particularly simple treatment of collisions, known as the relaxation-time approximation, that gives a precise content to the qualitative view of collisions we have put forth in earlier chapters. We shall then use the nonequilibrium distribution function to calculate the electric and thermal currents in several cases of interest beyond those considered in Chapter 12.

THE RELAXATION-TIME APPROXIMATION

Our fundamental picture of collisions retains the general features described, in Chapter 1, which we now formulate more precisely in a set of assumptions known as the relaxation-time approximation. We continue to assume that an electron experiences a collision in an infinitesimal time interval dt with probability dt/τ, but now allow for the possibility that the collision rate depends on the position, wave vector, and band index of the electron: $\tau = \tau_n(\mathbf{r}, \mathbf{k})$. We express the fact that collisions drive the electronic system toward local thermodynamic equilibrium in the following additional assumptions:

1. The distribution of electrons emerging from collisions at any time does not depend on the structure of the nonequilibrium distribution function $g_n(\mathbf{r}, \mathbf{k}, t)$ just prior to the collision.

2. If the electrons in a region about $\mathbf{r}$ have the equilibrium distribution appropriate to a local temperature[1] $T(\mathbf{r})$,

$$g_n(\mathbf{r}, \mathbf{k}, t) = g_n^0(\mathbf{r}, \mathbf{k}) = \frac{1}{e^{(\mathcal{E}_n(\mathbf{k}) - \mu(\mathbf{r}))/k_B T(\mathbf{r})} + 1}, \tag{13.2}$$

then collisions will not alter the form of the distribution function.

Assumption 1 asserts that collisions are completely effective in obliterating any information about the nonequilibrium configuration that the electrons may be carrying. This almost certainly overestimates the efficacy of collisions in restoring equilibrium (see Chapter 16).

Assumption 2 is a particularly simple way of representing quantitatively the fact that it is the role of collisions to maintain thermodynamic equilibrium at whatever local temperature is imposed by the conditions of the experiment.[2]

These two assumptions completely determine the form $dg_n(\mathbf{r}, \mathbf{k}, t)$ of the distribution function describing just those electrons that have emerged from a collision near point $\mathbf{r}$ in the time interval dt about t. According to assumption (1) dg cannot depend on the particular form of the full nonequilibrium distribution function $g_n(\mathbf{r}, \mathbf{k}, t)$. It therefore suffices to determine dg for any particular form of g. The simplest case is when g has the local equilibrium form (13.2), for according to assumption (2) the effect of collisions is then to leave this form unaltered. We know, however, that in the time interval dt a fraction $dt/\tau_n(\mathbf{r}, \mathbf{k})$ of the electrons in band n with wave vector $\mathbf{k}$ near position $\mathbf{r}$ will suffer a collision that *does* alter their band index and/or wave vector. If the form (13.2) of the distribution function is nevertheless to be unaltered, then the distribution of those electrons that *emerge* from collisions into band n with wave vector $\mathbf{k}$ during the same interval must precisely compensate for this loss. Thus:

$$dg_n(\mathbf{r}, \mathbf{k}, t) = \frac{dt}{\tau_n(\mathbf{r}, \mathbf{k})}\, g_n^0(\mathbf{r}, \mathbf{k}). \tag{13.3}$$

Equation (13.3) is the precise mathematical formulation of the relaxation-time approximation.[3]

---

[1] The only case we shall discuss in which the local equilibrium distribution is not the uniform equilibrium distribution (13.1) (with constant T and μ) is when the spatially varying temperature $T(\mathbf{r})$ is imposed by suitable application of sources and/or sinks of heat, as in a thermal conductivity measurement. In that case, since the electronic density n is constrained (electrostatically) to be constant, the local chemical potential will also depend on position, so that $\mu(\mathbf{r}) = \mu_{\text{equilib}}(n, T(\mathbf{r}))$. In the most general case the local temperature and chemical potential may depend on time as well as position. See, for example, Problem 4 at the end of this chapter, and Problem 1b in Chapter 16.

[2] A more fundamental theory would derive the fact that collisions play such a role, rather than assuming it.

[3] The relaxation-time approximation is critically reexamined in Chapter 16, where it is compared with a more accurate treatment of collisions.

Given these assumptions, we can compute the nonequilibrium distribution function in the presence of external fields and temperature gradients.[4]

CALCULATION OF THE NONEQUILIBRIUM DISTRIBUTION FUNCTION

Consider the group of electrons in the nth band that find themselves at time t in the volume element $d\mathbf{r}\,d\mathbf{k}$ about $\mathbf{r}$, $\mathbf{k}$. The number of electrons in this group is given in terms of the distribution function by

$$dN = g_n(\mathbf{r}, \mathbf{k}, t)\frac{d\mathbf{r}\,d\mathbf{k}}{4\pi^3}. \tag{13.4}$$

We can compute this number by grouping the electrons according to when they had their last collision. Let $\mathbf{r}_n(t')$, $\mathbf{k}_n(t')$ be the solution to the semiclassical equations of motion for the nth band that passes through the point $\mathbf{r}$, $\mathbf{k}$ when $t' = t$:

$$\mathbf{r}_n(t) = \mathbf{r}, \quad \mathbf{k}_n(t) = \mathbf{k}. \tag{13.5}$$

Electrons in the volume element $d\mathbf{r}\,d\mathbf{k}$ about $\mathbf{r}$, $\mathbf{k}$ at time t, whose *last* collision prior to t was in the interval dt' about t', must have emerged from that last collision in a phase space volume element $d\mathbf{r}'\,d\mathbf{k}'$ about $\mathbf{r}_n(t')$, $\mathbf{k}_n(t')$, since after t' their motion is determined entirely by the semiclassical equations, which must get them to $\mathbf{r}$, $\mathbf{k}$ at time t. According to the relaxation-time approximation (13.3), the total number of electrons emerging from collisions at $\mathbf{r}_n(t')$, $\mathbf{k}_n(t')$ into the volume element $d\mathbf{r}'\,d\mathbf{k}'$ in the interval dt' about t' is just:

$$\frac{dt'}{\tau_n(\mathbf{r}_n(t'), \mathbf{k}_n(t'))}\, g_n^0(\mathbf{r}_n(t'), \mathbf{k}_n(t'))\frac{d\mathbf{r}\,d\mathbf{k}}{4\pi^3}, \tag{13.6}$$

where we have exploited Liouville's theorem[5] to make the replacement

$$d\mathbf{r}'\,d\mathbf{k}' = d\mathbf{r}\,d\mathbf{k}. \tag{13.7}$$

Of this number, only a fraction $P_n(\mathbf{r}, \mathbf{k}, t; t')$ (which we shall compute below) actually survive from time t' to time t without suffering any further collisions. Therefore dN is given by multiplying (13.6) by this probability and summing over all possible times t' for the last collision prior to t:

$$dN = \frac{d\mathbf{r}\,d\mathbf{k}}{4\pi^3}\int_{-\infty}^{t}\frac{dt'\, g_n^0(\mathbf{r}_n(t'), \mathbf{k}_n(t'))P_n(\mathbf{r}, \mathbf{k}, t; t')}{\tau_n(\mathbf{r}_n(t'), \mathbf{k}_n(t'))}. \tag{13.8}$$

---

[4] Note the different roles played by the fields, which determine the electronic motion between collisions, and the temperature gradient, which determines the form (13.3) taken by the distribution of electrons emerging from collisions.

[5] The theorem states that volumes in rk-space are preserved by the semiclassical equations of motion. It is proved in Appendix H.

Comparing this with (13.4) gives

$$g_n(\mathbf{r}, \mathbf{k}, t) = \int_{-\infty}^{t} \frac{dt'\, g_n^0(\mathbf{r}_n(t'), \mathbf{k}_n(t'))P_n(\mathbf{r}, \mathbf{k}, t; t')}{\tau_n(\mathbf{r}_n(t'), \mathbf{k}_n(t'))}. \tag{13.9}$$

The structure of the result (13.9) is somewhat obscured by the notation, which reminds us explicitly that the distribution function is for the nth band evaluated at the point $\mathbf{r}$, $\mathbf{k}$, and that the t' dependence of the integrand is determined by evaluating g_n^0 and τ_n at the point $\mathbf{r}_n(t')$, $\mathbf{k}_n(t')$ of the semiclassical trajectory that passes through $\mathbf{r}$, $\mathbf{k}$ at time t. To avoid obscuring the simplicity of some further manipulations on (13.9) we temporarily adopt an abbreviated notation in which the band index n, the point $\mathbf{r}$, $\mathbf{k}$, and the trajectories $\mathbf{r}_n$, $\mathbf{k}_n$ are fixed and left implicit. Thus

$$\begin{gathered} g_n(\mathbf{r}, \mathbf{k}, t) \to g(t), \quad g_n^0(\mathbf{r}_n(t'), \mathbf{k}_n(t')) \to g^0(t'), \\ \tau_n(\mathbf{r}_n(t'), \mathbf{k}_n(t')) \to \tau(t'), \quad P_n(\mathbf{r}, \mathbf{k}, t; t') \to P(t, t'), \end{gathered} \tag{13.10}$$

so that (13.9) can be written:[6]

$$\boxed{g(t) = \int_{-\infty}^{t} \frac{dt'}{\tau(t')} g^0(t')P(t, t').} \tag{13.11}$$

We emphasize the simple structure of this formula. The electrons in a given phase space element at time t are grouped according to the time of their last collision. The number of electrons whose last collisions were in the time interval dt' about t' is the product of two factors:

1. The *total* number emerging from collisions in that time interval that are aimed so that if no further collision intervenes they will reach the given phase space element at time t. This number is determined by the relaxation-time approximation (13.3).
2. The fraction $P(t, t')$ of those electrons specified in (1) that actually do survive from t' to t without collisions.

It remains to compute $P(t, t')$, the fraction of electrons in band n that traverse the trajectory passing through $\mathbf{r}$, $\mathbf{k}$ at time t, without suffering collisions between t' and t. The fraction that survive from t' to t is less than the fraction that survive from $t' + dt'$ to t by the factor $[1 - dt'/\tau(t')]$, which gives the probability of an electron colliding between t' and $t' + dt'$. Thus

$$P(t, t') = P(t, t' + dt')\left[1 - \frac{dt'}{\tau(t')}\right]. \tag{13.12}$$

In the limit as $dt' \to 0$, this gives the differential equation

$$\frac{\partial}{\partial t'} P(t, t') = \frac{P(t, t')}{\tau(t')}, \tag{13.13}$$

---

[6] This result and the method by which it was constructed are associated with the name of R. G. Chambers, Proc. Phys. Soc. (London) **81**, 877 (1963).

whose solution, subject to the boundary condition

$$P(t, t) = 1, \tag{13.14}$$

is

$$P(t, t') = \exp\left(-\int_{t'}^{t} \frac{d\bar{t}}{\tau(\bar{t})}\right). \tag{13.15}$$

We may use Eq. (13.13) to rewrite the distribution function (13.11) in the form

$$g(t) = \int_{-\infty}^{t} dt'\, g^0(t') \frac{\partial}{\partial t'} P(t, t'). \tag{13.16}$$

It is convenient to integrate Eq. (13.16) by parts, using Eq. (13.14) and the physical condition that no electron can survive infinitely long without a collision: $P(t, -\infty) = 0$. The result is

$$g(t) = g^0(t) - \int_{-\infty}^{t} dt'\, P(t, t') \frac{d}{dt'} g^0(t'), \tag{13.17}$$

which expresses the distribution function as the local equilibrium distribution plus a correction.

To evaluate the time derivative of g^0, note (see Eqs. (13.10) and (13.2)) that it depends on time only through $\mathcal{E}_n(\mathbf{k}_n(t'))$, $T(\mathbf{r}_n(t'))$, and $\mu(\mathbf{r}_n(t'))$, so that[7]

$$\frac{dg^0(t')}{dt'} = \frac{\partial g^0}{\partial \mathcal{E}_n} \frac{\partial \mathcal{E}_n}{\partial \mathbf{k}} \cdot \frac{d\mathbf{k}_n}{dt'} + \frac{\partial g^0}{\partial T} \frac{\partial T}{\partial \mathbf{r}} \cdot \frac{d\mathbf{r}_n}{dt'} + \frac{\partial g^0}{\partial \mu} \frac{\partial \mu}{\partial \mathbf{r}} \cdot \frac{d\mathbf{r}_n}{dt'}. \tag{13.18}$$

If we use the semiclassical equations of motion (12.6) to eliminate $d\mathbf{r}_n/dt'$ and $d\mathbf{k}_n/dt'$ from Eq. (13.18), then Eq. (13.17) can be written as

$$g(t) = g^0 + \int_{-\infty}^{t} dt'\, P(t, t') \left[\left(-\frac{\partial f}{\partial \mathcal{E}}\right) \mathbf{v} \cdot \left(-e\mathbf{E} - \nabla\mu - \left(\frac{\mathcal{E} - \mu}{T}\right)\nabla T\right)\right], \tag{13.19}$$

where f is the Fermi function (13.1) (evaluated at the local temperature and chemical potential), and all the quantities in the brackets[8] depend on t' through their arguments $\mathbf{r}_n(t')$ and $\mathbf{k}_n(t')$.

SIMPLIFICATION OF THE NONEQUILIBRIUM DISTRIBUTION FUNCTION IN SPECIAL CASES

Equation (13.19) gives the semiclassical distribution function in the relaxation-time approximation under very general conditions, and is therefore applicable to a great variety of problems. In many cases, however, special circumstances permit considerably further simplification:

---

[7] If we were interested in applications in which the local temperature and chemical potential had an explicit time dependence, we should have to add terms in $\partial T/\partial t$ and $\partial \mu/\partial t$ to (13.18). An example is given in Problem 4.

[8] Note that a magnetic field $\mathbf{H}$ will not appear explicitly in (13.19) since the Lorentz force is perpendicular to $\mathbf{v}$. (It will, of course, appear implicitly through the time dependence of $\mathbf{r}_n(t')$ and $\mathbf{k}_n(t')$.)

1. *Weak Electric Fields and Temperature Gradients* As noted in Chapter 1, the electric fields and temperature gradients commonly applied to metals are almost invariably weak enough to permit calculation of the induced currents to linear order.[9] Since the second term in (13.19) is explicitly linear[10] in **E** and ∇T, the t' dependence of the integrand can be calculated at zero electric field and constant T.

2. *Spatially Uniform Electromagnetic Fields and Temperature Gradients, and Position-Independent Relaxation Times*[11] In this case the entire integrand in (13.19) will be independent of $\mathbf{r}_n(t')$. The only t' dependence (aside from possible explicit dependence of **E** and T on time) will be through $\mathbf{k}_n(t')$, which will be time-dependent if a magnetic field is present. Since the Fermi function f depends on **k** only through $\mathcal{E}_n(\mathbf{k})$, which is conserved in a magnetic field, the t' dependence of the integrand in (13.19) will be entirely contained in $P(t, t')$, $\mathbf{v}(\mathbf{k}_n(t'))$, and (if they are time-dependent) **E** and T.

3. *Energy-Dependent Relaxation Time* If τ depends on wave-vector only through $\mathcal{E}_n(\mathbf{k})$, then since $\mathcal{E}_n(\mathbf{k})$ is conserved in a magnetic field, $\tau(t')$ will not depend on t', and (13.15) reduces to

$$P(t, t') = e^{-(t-t')/\tau_n(\mathbf{k})}. \qquad \textbf{(13.20)}$$

There is no compelling reason why τ should depend on **k** only through $\mathcal{E}_n(\mathbf{k})$ in anisotropic systems, but when the nature of the electronic scattering does depend significantly on wave vector, the entire relaxation-time approximation is probably of questionable validity (see Chapter 16). Therefore most calculations in the relaxation-time approximation do make this additional simplification, and often even use a constant (energy-independent) τ. Since the distribution function (13.19) contains a factor $\partial f/\partial \mathcal{E}$, which is negligible except within $O(k_B T)$ of the Fermi energy, only the energy dependence of $\tau(\mathcal{E})$ in the neighborhood of $\mathcal{E}_F$ is significant in metals.

Under these conditions we can rewrite (13.19) as

$$g(\mathbf{k}, t) = g^0(\mathbf{k}) + \int_{-\infty}^{t} dt'\, e^{-(t-t')/\tau(\mathcal{E}(\mathbf{k}))} \left(-\frac{\partial f}{\partial \mathcal{E}}\right) \times \mathbf{v}(\mathbf{k}(t')) \cdot \left[-e\mathbf{E}(t') - \nabla\mu(t') - \frac{\mathcal{E}(\mathbf{k}) - \mu}{T}\nabla T(t')\right], \qquad \textbf{(13.21)}$$

---

[9] The linearization can be justified directly from (13.19) by first observing that the probability of an electron suffering no collisions in an interval of given length prior to t becomes negligibly small when the length of the interval is appreciably greater than τ. Consequently, only times t on the order of τ contribute appreciably to the integral in (13.19). However (see page 225), within such a time an electric field perturbs the electronic **k**-vector by an amount that is minute compared with the dimensions of the zone. This immediately implies that the **E** dependence of all terms in (13.19) is very weak. In a similar way, one can justify linearizing in the temperature gradient provided that the change in temperature over a mean free path is a small fraction of the prevailing temperature. However, one cannot linearize in the magnetic field, since it is entirely possible to produce magnetic fields in metals so strong that an electron can move distances in **k**-space comparable to the size of the zone within a relaxation time.

[10] The chemical potential varies in space only because the temperature does (see footnote 1) so $\nabla\mu$ is of order ∇T.

[11] In general, one might wish to allow τ to depend on position to take into account, for example, inhomogeneous distributions of impurities, special scattering effects associated with surfaces, etc.

where we have continued to suppress explicit reference to the band index n, but have reintroduced explicitly the $\mathbf{k}$ and t dependence.[12]

We conclude this chapter with applications of (13.21) to several cases of interest.

DC ELECTRICAL CONDUCTIVITY

If $\mathbf{H} = 0$, the $\mathbf{k}(t')$ appearing in (13.21) reduces to $\mathbf{k}$, and the time integration is elementary for static $\mathbf{E}$ and ∇T. If the temperature is uniform, we find:

$$g(\mathbf{k}) = g^0(\mathbf{k}) - e\mathbf{E} \cdot \mathbf{v}(\mathbf{k})\tau(\mathcal{E}(\mathbf{k}))\left(-\frac{\partial f}{\partial \mathcal{E}}\right). \tag{13.22}$$

Since the number of electrons per unit volume in the volume element $d\mathbf{k}$ is $g(\mathbf{k})\,d\mathbf{k}/4\pi^3$, the current density in a band is[13]

$$\mathbf{j} = -e\int \frac{d\mathbf{k}}{4\pi^3}\,\mathbf{v}(\mathbf{k})g. \tag{13.23}$$

Each partially filled band makes such a contribution to the current density; the total current density is the sum of these contributions over all bands. From (13.22) and (13.23) it can be written as $\mathbf{j} = \boldsymbol{\sigma}\mathbf{E}$, where the conductivity tensor $\boldsymbol{\sigma}$ is a sum of contributions from each band:[14]

$$\boldsymbol{\sigma} = \sum_n \boldsymbol{\sigma}^{(n)}, \tag{13.24}$$

$$\boldsymbol{\sigma}^{(n)} = e^2 \int \frac{d\mathbf{k}}{4\pi^3}\,\tau_n(\mathcal{E}_n(\mathbf{k}))\mathbf{v}_n(\mathbf{k})\mathbf{v}_n(\mathbf{k})\left(-\frac{\partial f}{\partial \mathcal{E}}\right)_{\mathcal{E}=\mathcal{E}_n(\mathbf{k})}. \tag{13.25}$$

The following properties of the conductivity are worthy of note:

1. ***Anisotropy*** In free electron theory $\mathbf{j}$ is parallel to $\mathbf{E}$, i.e. the tensor $\boldsymbol{\sigma}$ is diagonal: $\sigma_{\mu\nu} = \sigma\delta_{\mu\nu}$. In a general crystal structure $\mathbf{j}$ need not be parallel to $\mathbf{E}$, and the conductivity will be a tensor. In a crystal of cubic symmetry, however, $\mathbf{j}$ remains parallel to $\mathbf{E}$, for if the x-, y-, and z-axes are taken along the cubic axes then $\sigma_{xx} = \sigma_{yy} = \sigma_{zz}$. Furthermore, if a field in the x-direction induced any current in the y-direction, by exploiting the cubic symmetry one could equally well predict that the same current must arise in the $-y$-direction. The only consistent possibility is zero current, so that σ_{xy} must vanish (and, by symmetry, so must the other off-diagonal components). Hence $\sigma_{\mu\nu} = \sigma\delta_{\mu\nu}$ in crystals of cubic symmetry.

2. ***Irrelevance of Filled Bands*** The Fermi function has negligible derivative except when $\mathcal{E}$ is within k_BT of $\mathcal{E}_F$. Hence filled bands make no contribution to the conductivity, in accordance with the general discussion on pages 221–223.

---

[12] The quantity $\mathbf{k}(t')$ is the solution to the semiclassical equation of motion for band n in a uniform magnetic field $\mathbf{H}$, that is equal to $\mathbf{k}$ when $t' = t$.

[13] Here and elsewhere in this chapter integrations over $\mathbf{k}$ are assumed to be over a primitive cell unless otherwise specified.

[14] Since no current flows in equilibrium, the leading term in the distribution function, g^0, makes no contribution to (13.23). We are using a tensor notation in which $\mathbf{A} = \mathbf{bc}$ stands for $A_{\mu\nu} = b_\mu c_\nu$.

3. ***Equivalence of Particle and Hole Pictures in Metals*** In a metal, to an accuracy of order $(k_B T/\mathcal{E}_F)^2$ we can evaluate[15] (13.25) at $T = 0$. Since $(-\partial f/\partial \mathcal{E}) = \delta(\mathcal{E} - \mathcal{E}_F)$, the relaxation time can be evaluated at $\mathcal{E}_F$ and taken outside of the integral. Furthermore, since[16]

$$\mathbf{v}(\mathbf{k})\left(-\frac{\partial f}{\partial \mathcal{E}}\right)_{\mathcal{E}=\mathcal{E}_n(\mathbf{k})} = -\frac{1}{\hbar}\frac{\partial}{\partial \mathbf{k}} f(\mathcal{E}(\mathbf{k})), \tag{13.26}$$

we may integrate by parts[17] to find

$$\begin{aligned} \boldsymbol{\sigma} &= e^2 \tau(\mathcal{E}_F) \int \frac{d\mathbf{k}}{4\pi^3 \hbar} \frac{\partial}{\partial \mathbf{k}} \mathbf{v}(\mathbf{k}) f(\mathcal{E}(\mathbf{k})) \\ &= e^2 \tau(\mathcal{E}_F) \int_{\substack{\text{occupied} \\ \text{levels}}} \frac{d\mathbf{k}}{4\pi^3} \mathbf{M}^{-1}(\mathbf{k}). \end{aligned} \tag{13.27}$$

Since $\mathbf{M}^{-1}(\mathbf{k})$ is the derivative of a periodic function, its integral over the entire primitive cell must vanish,[18] and we may write (13.27) in the alternative form

$$\boldsymbol{\sigma} = e^2 \tau(\mathcal{E}_F) \int_{\substack{\text{unoccupied} \\ \text{levels}}} \frac{d\mathbf{k}}{4\pi^3} (-\mathbf{M}^{-1}(\mathbf{k})). \tag{13.28}$$

Comparing these two forms, we find that the contribution to the current can be regarded as coming from the unoccupied rather than the occupied levels, provided that the sign of the effective mass tensor is changed. This result was already implied by our discussion of holes (pages 225–229), but is repeated here to emphasize that it emerges from a more formal analysis as well.

4. ***Recovery of the Free Electron Result*** If $\mathbf{M}^{-1}{}_{\mu\nu} = (1/m^*)\,\delta_{\mu\nu}$ independent of $\mathbf{k}$ for all occupied levels in the band, then (13.27) reduces to the Drude form (Eq. (1.6)):

$$\sigma_{\mu\nu} = \frac{ne^2\tau}{m^*}, \tag{13.29}$$

with an effective mass. If $\mathbf{M}^{-1}{}_{\mu\nu} = -(1/m^*)\,\delta_{\mu\nu}$ independent of $\mathbf{k}$ for all unoccupied levels,[19] then (13.28) reduces to

$$\sigma_{\mu\nu} = \frac{n_h e^2 \tau}{m^*}, \tag{13.30}$$

where n_h is the number of unoccupied levels per unit volume; i.e., the conductivity of the band is of the Drude form, with m replaced by the effective mass m^* and the electronic density replaced by the density of holes.

---

15 See Appendix C.

16 We again drop explicit reference to the band index. The formulas below give the conductivity for a solid with a single band of carriers. If there is more than one band of carriers, one must sum on n to get the full conductivity.

17 See Appendix I.

18 This follows from the identity (I.1) of Appendix I, with one of the periodic functions taken to be unity. The mass tensor $\mathbf{M}^{-1}(\mathbf{k})$ is defined in Eq. (12.29).

19 Since the mass tensor is negative definite at the band maximum, m^* will be positive. For the relation between k-space volumes and particle densities see page 35.

AC ELECTRICAL CONDUCTIVITY

If the electric field is not static, but has the time dependence

$$\mathbf{E}(t) = \mathrm{Re}\,[\mathbf{E}(\omega)e^{-i\omega t}], \tag{13.31}$$

then the derivation of the conductivity from (13.21) proceeds just as in the DC case, except for an additional factor $e^{-i\omega t}$ in the integrand. One finds

$$\mathbf{j}(t) = \mathrm{Re}\,[\mathbf{j}(\omega)e^{-i\omega t}] \tag{13.32}$$

where

$$\mathbf{j}(\omega) = \boldsymbol{\sigma}(\omega)\cdot\mathbf{E}(\omega), \quad \boldsymbol{\sigma}(\omega) = \sum_n \boldsymbol{\sigma}^{(n)}(\omega), \tag{13.33}$$

and

$$\boldsymbol{\sigma}^{(n)}(\omega) = e^2 \int \frac{d\mathbf{k}}{4\pi^3} \frac{\mathbf{v}_n(\mathbf{k})\mathbf{v}_n(\mathbf{k})(-\partial f/\partial\mathcal{E})_{\mathcal{E}=\mathcal{E}_n(\mathbf{k})}}{[1/\tau_n(\mathcal{E}_n(\mathbf{k}))] - i\omega}. \tag{13.34}$$

Thus, as in the free electron case (Eq. (1.29)), the AC conductivity is just the DC conductivity divided by $1 - i\omega\tau$, except that we must now allow for the possibility that the relaxation time may differ from band to band.[20]

The form (13.34) permits a simple direct test of the validity of the semiclassical model in the limit $\omega\tau \gg 1$, where it reduces to

$$\boldsymbol{\sigma}^{(n)}(\omega) = -\frac{e^2}{i\omega}\int \frac{d\mathbf{k}}{4\pi^3}\,\mathbf{v}_n(\mathbf{k})\mathbf{v}_n(\mathbf{k})(-\partial f/\partial\mathcal{E})_{\mathcal{E}=\mathcal{E}_n(\mathbf{k})}, \tag{13.35}$$

or, equivalently (as derived in the DC case),

$$\sigma^{(n)}_{\mu\nu}(\omega) = -\frac{e^2}{i\omega}\int \frac{d\mathbf{k}}{4\pi^3} f(\mathcal{E}_n(\mathbf{k}))\,\frac{1}{\hbar^2}\frac{\partial^2\mathcal{E}_n(\mathbf{k})}{\partial k_\mu\,\partial k_\nu}. \tag{13.36}$$

Equation (13.36) determines the current induced to linear order in the AC electric field in the absence of collisions, since the high $\omega\tau$ limit can be interpreted to mean $\tau \to \infty$ for fixed ω. In the absence of collisions, however, it is an elementary quantum-mechanical calculation to compute exactly[21] the change to linear order in the Bloch wave functions induced by the electric field. Given these wave functions, one can calculate the expectation value of the current operator to linear order in the field, thereby arriving at a fully quantum-mechanical form for $\boldsymbol{\sigma}(\omega)$ that is not based on the approximations of the semiclassical model. Such a calculation is a straightforward exercise in first-order time-dependent perturbation theory. It is slightly too lengthy to include here, and we only quote the result:[22]

$$\sigma^{(n)}_{\mu\nu}(\omega) = -\frac{e^2}{i\omega}\int \frac{d\mathbf{k}}{4\pi^3} f(\mathcal{E}_n(\mathbf{k}))\,\frac{1}{\hbar^2}\Bigg[\frac{\hbar^2}{m}\delta_{\mu\nu} - \frac{\hbar^4}{m^2}\sum_{n'\neq n}\left(\frac{\langle n\mathbf{k}|\nabla_\mu|n'\mathbf{k}\rangle\,\langle n'\mathbf{k}|\nabla_\nu|n\mathbf{k}\rangle}{\hbar\omega + \mathcal{E}_n(\mathbf{k}) - \mathcal{E}_{n'}(\mathbf{k})} + \frac{\langle n\mathbf{k}|\nabla_\nu|n'\mathbf{k}\rangle\,\langle n'\mathbf{k}|\nabla_\mu|n\mathbf{k}\rangle}{-\hbar\omega + \mathcal{E}_n(\mathbf{k}) - \mathcal{E}_{n'}(\mathbf{k})}\right)\Bigg]. \tag{13.37}$$

---

[20] Within each band $\tau_n(\mathcal{E})$ can be replaced by $\tau_n(\mathcal{E}_F)$ with negligible error in metals.

[21] Within the independent electron approximation.

[22] The notation for the matrix elements of the gradient operator is as in Appendix E.

In general, this is quite different from (13.36). If, however, $\hbar\omega$ is small compared with the band gap for all occupied levels, then the frequencies in the denominators of (13.37) can be ignored, and the quantity in brackets reduces to the expression for $\partial^2\mathcal{E}_n(\mathbf{k})/\partial k_\mu\, \partial k_\nu$ derived in Appendix E (Eq. (E.11)). Equation (13.37) then reduces to the semiclassical result (13.36), confirming the assertion made in Chapter 12 that the semiclassical analysis should be valid provided that $\hbar\omega \ll \mathcal{E}_{\text{gap}}$ (Eq. (12.10)).[23]

THERMAL CONDUCTIVITY

In Chapters 1 and 2 we described the thermal current density as analogous to the electrical current density, with thermal energy being carried rather than electric charge. We can now give a more precise definition of the thermal current.

Consider a small fixed region of the solid within which the temperature is effectively constant. The rate at which heat appears in the region is just T times the rate at which the entropy of the electrons within the region changes ($dQ = T\,dS$). Thus[24] the thermal current density $\mathbf{j}^q$ is just the product of the temperature with the entropy current density, $\mathbf{j}^s$:

$$\mathbf{j}^q = T\mathbf{j}^s. \tag{13.38}$$

Since the volume of the region is fixed, changes in the entropy in the region are related to changes in the internal energy and number of electrons by the thermodynamic identity:

$$T\,dS = dU - \mu\,dN, \tag{13.39}$$

or, in terms of current densities,

$$T\mathbf{j}^s = \mathbf{j}^{\mathcal{E}} - \mu\mathbf{j}^n \tag{13.40}$$

where the energy and number current densities are given by[25]

$$\begin{Bmatrix} \mathbf{j}^{\mathcal{E}} \\ \mathbf{j}^n \end{Bmatrix} = \sum_n \int \frac{d\mathbf{k}}{4\pi^3} \begin{Bmatrix} \mathcal{E}_n(\mathbf{k}) \\ 1 \end{Bmatrix} \mathbf{v}_n(\mathbf{k})\, g_n(\mathbf{k}). \tag{13.41}$$

---

[23] As long as $\hbar\omega$ is small enough that no denominators in (13.37) vanish, the more general result simply provides quantitative corrections to the semiclassical approximation, which could, for example, be cast in the form of a power series in $\hbar\omega/\mathcal{E}_{\text{gap}}$. However, when $\hbar\omega$ becomes large enough for denominators to vanish (i.e., when the photon energy is large enough to cause interband transitions), then the semiclassical result fails qualitatively as well, for the detailed derivation of (13.37) includes the stipulation that when the denominator is singular, the result is to be interpreted in the limit as ω approaches the real axis of the complex frequency plane from above. (When no denominator vanishes, the result is independent of any infinitesimal imaginary part ω may have.) This introduces a real part into the conductivity, providing a mechanism for absorption in the absence of collisions, which the semiclassical model is incapable of producing. This additional real part is of critical importance in understanding the properties of metals at optical frequencies (see Chapter 15), in which interband transitions play a crucial role.

[24] This assumes that the entropy within the region changes only because electrons carry it in or out. Entropy can also be generated within the region through collisions. However, such entropy production can be shown to be a second-order effect in the applied temperature gradient and electric field (Joule heating—the "I^2R loss"—being the most familiar example) and can therefore be ignored in a linear theory.

[25] Note that these have the same form as the electrical current density except that the quantity carried by each electron is no longer its charge ($-e$) but its energy ($\mathcal{E}_n(\mathbf{k})$) or its number (unity). Note also that the number current is just the electric current divided by the charge: $j = -ej^n$. (Do not confuse the superscript n, indicating that j is the number current density, with the band index n.)

Combining (13.40) and (13.41) we find a thermal current density[26]

$$\mathbf{j}^q = \sum_n \int \frac{d\mathbf{k}}{4\pi^3} [\mathcal{E}_n(\mathbf{k}) - \mu]\mathbf{v}_n(\mathbf{k}) g_n(\mathbf{k}). \tag{13.42}$$

The distribution function appearing in (13.42) is given by (13.21), evaluated at $\mathbf{H} = 0$ in the presence of a uniform static electric field and temperature gradient[27]:

$$g(\mathbf{k}) = g^0(\mathbf{k}) + \tau(\mathcal{E}(\mathbf{k}))\left(-\frac{\partial f}{\partial \mathcal{E}}\right)\mathbf{v}(\mathbf{k}) \cdot \left[-e\boldsymbol{\mathcal{E}} + \frac{\mathcal{E}(\mathbf{k}) - \mu}{T}(-\nabla T)\right], \tag{13.43}$$

where

$$\boldsymbol{\mathcal{E}} = \mathbf{E} + \frac{\nabla \mu}{e}. \tag{13.44}$$

We can construct the electrical current density (13.23) and the thermal current density (13.42) from this distribution function:

$$\begin{aligned} \mathbf{j} &= \mathbf{L}^{11}\boldsymbol{\mathcal{E}} + \mathbf{L}^{12}(-\nabla T), \\ \mathbf{j}^q &= \mathbf{L}^{21}\boldsymbol{\mathcal{E}} + \mathbf{L}^{22}(-\nabla T), \end{aligned} \tag{13.45}$$

where the matrices $\mathbf{L}^{ij}$ are defined[28] in terms of

$$\boldsymbol{\mathfrak{L}}^{(\alpha)} = e^2 \int \frac{d\mathbf{k}}{4\pi^3}\left(-\frac{\partial f}{\partial \mathcal{E}}\right)\tau(\mathcal{E}(\mathbf{k}))\mathbf{v}(\mathbf{k})\mathbf{v}(\mathbf{k})(\mathcal{E}(\mathbf{k}) - \mu)^\alpha \tag{13.46}$$

by

$$\begin{aligned} \mathbf{L}^{11} &= \boldsymbol{\mathfrak{L}}^{(0)} \\ \mathbf{L}^{21} &= T\mathbf{L}^{12} = -\frac{1}{e}\boldsymbol{\mathfrak{L}}^{(1)}, \\ \mathbf{L}^{22} &= \frac{1}{e^2 T}\boldsymbol{\mathfrak{L}}^{(2)}. \end{aligned} \tag{13.47}$$

The structure of these results is simplified by defining[29]

$$\boldsymbol{\sigma}(\mathcal{E}) = e^2\tau(\mathcal{E}) \int \frac{d\mathbf{k}}{4\pi^3} \delta(\mathcal{E} - \mathcal{E}(\mathbf{k}))\mathbf{v}(\mathbf{k})\mathbf{v}(\mathbf{k}), \tag{13.48}$$

in terms of which

$$\boldsymbol{\mathfrak{L}}^{(\alpha)} = \int d\mathcal{E}\left(-\frac{\partial f}{\partial \mathcal{E}}\right)(\mathcal{E} - \mu)^\alpha \boldsymbol{\sigma}(\mathcal{E}). \tag{13.49}$$

---

[26] Since thermal conductivities are normally measured under conditions in which no electric current flows, it is often enough to identify the thermal current with the energy current (as we did in Chapter 1). When, however, heat and electric charge are carried simultaneously (as in the Peltier effect, described below), it is essential to use (13.42).

[27] See the discussion on pages 23 and 24 for why, in general, a temperature gradient will be accompanied by an electric field.

[28] To keep the notation as simple as possible the results that follow are given for the case in which all carriers lie in a single band, and the band index is suppressed. In the many-band case, each $\mathbf{L}$ must be replaced by the sum of the $\mathbf{L}$'s for all the partially filled bands. This generalization does not affect the validity of the Wiedemann-Franz law, but it can complicate the structure of the thermopower.

[29] Since $(-\partial f/\partial \mathcal{E}) = \delta(\mathcal{E} - \mathcal{E}_F)$ to an accuracy of order $(k_B T/\mathcal{E}_F)^2$ in metals, the notation is meant to remind one that the DC conductivity of a metal (13.25) is essentially $\boldsymbol{\sigma}(\mathcal{E}_F)$.

To evaluate (13.49) for metals we can exploit the fact that $(-\partial f/\partial \mathcal{E})$ is negligible except within $O(k_BT)$ of $\mu \approx \mathcal{E}_F$. Since the integrands in $\mathcal{L}^{(1)}$ and $\mathcal{L}^{(2)}$ have factors that vanish when $\mathcal{E} = \mu$, to evaluate them one must retain the first temperature correction in the Sommerfeld expansion.[30] When this is done, one finds with an accuracy of order $(k_BT/\mathcal{E}_F)^2$

$$\mathbf{L}^{11} = \boldsymbol{\sigma}(\mathcal{E}_F) = \boldsymbol{\sigma}, \tag{13.50}$$

$$\mathbf{L}^{21} = T\mathbf{L}^{12} = -\frac{\pi^2}{3e}(k_BT)^2\boldsymbol{\sigma}', \tag{13.51}$$

$$\mathbf{L}^{22} = \frac{\pi^2}{3}\frac{k_B{}^2T}{e^2}\boldsymbol{\sigma}, \tag{13.52}$$

where

$$\boldsymbol{\sigma}' = \frac{\partial}{\partial \mathcal{E}}\boldsymbol{\sigma}(\mathcal{E})\bigg|_{\mathcal{E}=\mathcal{E}_F}. \tag{13.53}$$

Equations (13.45) and (13.50) to (13.53) are the basic results of the theory of electronic contributions to the thermoelectric effects. They remain valid when more than one band is partially occupied, provided only that we interpret $\sigma_{ij}(\mathcal{E})$ to be the sum of (13.48) over *all* partially occupied bands.

To deduce the thermal conductivity from these results, we note that it relates the thermal current to the temperature gradient under conditions in which no electric current flows (as discussed in Chapter 1). The first of Eqs. (13.45) determines that if zero current flows, then

$$\boldsymbol{\mathcal{E}} = -(\mathbf{L}^{11})^{-1}\mathbf{L}^{12}(-\nabla T). \tag{13.54}$$

Substituting this into the second of Eqs. (13.45), we find that

$$\mathbf{j}^q = \mathbf{K}(-\nabla T), \tag{13.55}$$

where $\mathbf{K}$, the thermal conductivity tensor, is given by

$$\mathbf{K} = \mathbf{L}^{22} - \mathbf{L}^{21}(\mathbf{L}^{11})^{-1}\mathbf{L}^{12}. \tag{13.56}$$

It follows from Eqs. (13.50) to (13.52) and the fact that $\boldsymbol{\sigma}'$ is typically of order $\boldsymbol{\sigma}/\mathcal{E}_F$, that in metals the first term in (13.56) exceeds the second by a factor of order $(\mathcal{E}_F/k_BT)^2$. Thus

$$\mathbf{K} = \mathbf{L}^{22} + O(k_BT/\mathcal{E}_F)^2. \tag{13.57}$$

This is what one would have found if one had ignored the thermoelectric field from the start. We emphasize that its validity requires degenerate Fermi statistics. In semiconductors (13.57) is not a good approximation to the correct result (13.56).

If Eq. (13.57) is evaluated using (13.52), we find that

$$\mathbf{K} = \frac{\pi^2}{3}\left(\frac{k_B}{e}\right)^2 T\boldsymbol{\sigma}. \tag{13.58}$$

This is nothing but the Wiedemann-Franz law (see Eq. (2.93)) with a vastly more general range of validity. For arbitrary band structure, component by component, the thermal conductivity tensor is proportional to the electrical conductivity tensor

[30] See appendix C or Eq. (2.70), page 46.

with the universal constant of proportionality $\pi^2 k_B{}^2 T/3e^2$. Thus this remarkable experimental observation, made over a century ago, continues to reemerge from successively more refined theoretical models in an essentially unaltered form.

While rejoicing that the semiclassical model preserves this elegant result, one must not forget that deviations from the Wiedemann-Franz law are observed.[31] We shall see in Chapter 16 that this is a failure not of the semiclassical method, but of the relaxation-time approximation.

THE THERMOELECTRIC POWER

When a temperature gradient is maintained in a metal and no electric current is allowed to flow, there will be a steady-state electrostatic potential difference between the high- and low-temperature regions of the specimen.[32] Measuring this potential drop is not completely straightforward for several reasons:

1. To measure voltages accurately enough to detect a thermoelectric voltage, it is essential that the voltmeter connect points of the specimen at the same temperature. Otherwise, since the leads to the meter are in thermal equilibrium with the specimen at the contact points, there would be a temperature gradient within the circuitry of the meter itself, accompanied by an additional thermoelectric voltage Since no thermoelectric voltage develops between points of a single metal at the same temperature, one must use a circuit of two different metals (Figure 13.1), connected so that one junction is at a temperature T_1 and the other (bridged only by the voltmeter) at a temperature $T_0 \neq T_1$. Such a measurement yields the difference in the thermoelectric voltages developed in the two metals.

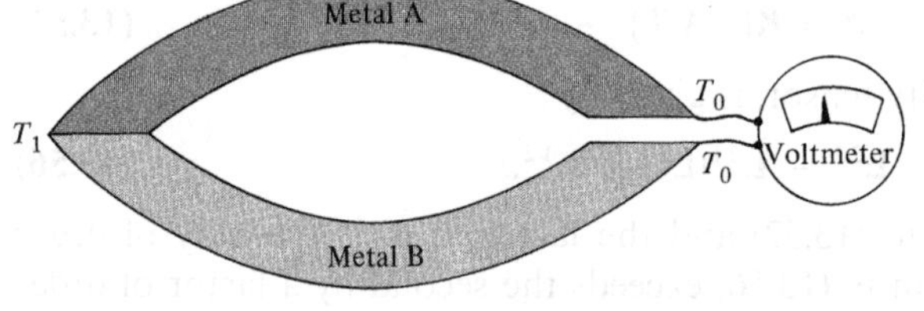

Figure 13.1
Circuit for measuring the difference in thermoelectric voltages developed in two different metals, in each of which the temperature varies from T_0 to T_1.

2. To measure the absolute thermoelectric voltage in a metal, one can exploit the fact that no thermoelectric voltage develops across a superconducting metal.[33] Hence when one of the metals in the bimetallic circuit is superconducting, the measurement yields directly the thermoelectric voltage across the other.[34]
3. The points in the circuit joined by the voltmeter have different electrostatic potentials *and* different chemical potentials.[35] If, as in most such devices, the

---

[31] See Chapter 3.

[32] The Seebeck effect. A crude but elementary discussion of the underlying physics is given on pages 23–25.

[33] See page 730.

[34] This makes it possible to measure the absolute thermoelectric voltage in a metal at temperatures up to 20 K (currently the highest temperature at which superconductivity has been observed). One can deduce it at higher temperatures from measurements of the Thomson effect (Problem 5).

[35] Although electrons will flow from one metal to the other so as to equalize the chemical potentials at the point of contact (see Chapter 18), there is still a chemical potential difference at the points joined by the voltmeter, because the temperature variation of the chemical potential differs in the two metals.

voltmeter reading is actually IR, where I is the small current flowing through a large resistance R, then it is essential to realize that the current is driven not just by the electric field $\mathbf{E}$, but by $\boldsymbol{\mathcal{E}} = \mathbf{E} + (1/e)\nabla\mu$. This is because the chemical potential gradient leads to a diffusion current, in addition to the current driven mechanically by the electric field.[36] As a result, the voltmeter reading will not be $-\int \mathbf{E} \cdot d\boldsymbol{\ell}$, but $-\int \boldsymbol{\mathcal{E}} \cdot d\boldsymbol{\ell}$.

The thermoelectric power (or thermopower) of a metal, Q, is defined as the proportionality constant between the contribution of the metal to the reading of such a voltmeter, and the temperature change:

$$-\int \boldsymbol{\mathcal{E}} \cdot d\boldsymbol{\ell} = Q\,\Delta T \tag{13.59}$$

or

$$\boldsymbol{\mathcal{E}} = Q\,\nabla T. \tag{13.60}$$

Since negligible current flows when the thermoelectric voltage is measured, Eq. (13.45) gives[37]

$$Q = \frac{L^{12}}{L^{11}}, \tag{13.61}$$

or, from Eqs. (13.50) and (13.51),

$$Q = -\frac{\pi^2}{3}\frac{k_B{}^2 T}{e}\frac{\sigma'}{\sigma}. \tag{13.62}$$

This has a considerably more complex structure than the free electron estimate (2.94), which is independent[38] of the relaxation time τ. We can cast σ' into a more useful form, by differentiating (13.48):

$$\frac{\partial}{\partial \mathcal{E}}\boldsymbol{\sigma}(\mathcal{E}) = \frac{\tau'(\mathcal{E})}{\tau(\mathcal{E})}\boldsymbol{\sigma}(\mathcal{E}) + e^2\tau(\mathcal{E})\int \frac{d\mathbf{k}}{4\pi^3}\,\delta'(\mathcal{E} - \mathcal{E}(\mathbf{k}))\mathbf{v}(\mathbf{k})\mathbf{v}(\mathbf{k}). \tag{13.63}$$

Since

$$\mathbf{v}(\mathbf{k})\delta'(\mathcal{E} - \mathcal{E}(\mathbf{k})) = -\frac{1}{\hbar}\frac{\partial}{\partial \mathbf{k}}\delta(\mathcal{E} - \mathcal{E}(\mathbf{k})), \tag{13.64}$$

an integration by parts gives[39]

$$\boldsymbol{\sigma}' = \frac{\tau'}{\tau}\boldsymbol{\sigma} + \frac{e^2\tau}{4\pi^3}\int d\mathbf{k}\,\delta(\mathcal{E}_F - \mathcal{E}(\mathbf{k}))\mathbf{M}^{-1}(\mathbf{k}). \tag{13.65}$$

---

[36] That it is this particular combination of field and chemical potential gradient that drives the electric current follows from Eq. (13.45). This phenomenon is often summarized in the assertion that a voltmeter measures not the electric potential but the "electrochemical potential".

[37] For simplicity we limit the discussion to cubic metals, for which the tensors $\mathbf{L}^{ij}$ are diagonal.

[38] If we take τ to be independent of energy, then in the free electron limit $\sigma'/\sigma = (3/2\mathcal{E}_F)$, and (13.62) reduces to $Q = -(\pi^2/2e)(k_B{}^2T/\mathcal{E}_F)$. This is a factor of 3 larger than the rough estimate (2.94). The disparity is due to the very crude way in which thermal averages of energies and velocities were treated in Chapters 1 and 2. It indicates that it is largely a lucky accident that the analogous derivation of the thermal conductivity gave the correct numerical factor.

[39] Although it is tempting to try to interpret $\sigma'(\mathcal{E}_F)$ as the variation of the physically measured DC conductivity with some suitably controlled parameters, this cannot be justified. The quantity $\sigma'(\mathcal{E}_F)$ means (within the relaxation-time approximation) no more (or less) than (13.65).

If the energy dependence of the relaxation time is unimportant, then the sign of the thermopower is determined by the sign of the effective mass, averaged over the Fermi surface, i.e., by whether the carriers are electrons or holes. This is consistent with the general theory of holes described in Chapter 12, and also provides a possible explanation for another of the anomalies of free electron theory.[40]

However, the thermopower is not a very valuable probe of the fundamental electronic properties of a metal; the energy dependence of τ is not well understood, the validity of the form (13.65) depends on the relaxation-time approximation, and, most important, vibrations of the lattice can affect the transport of thermal energy in a way that makes it very difficult to achieve an accurate theory of the thermopower.

OTHER THERMOELECTRIC EFFECTS

There is a variety of other thermoelectric effects. The Thomson effect is described in Problem 5, and we mention here only the Peltier effect.[41] If an electric current is driven in a bimetallic circuit that is maintained at a uniform temperature, then heat will be evolved at one junction and absorbed at the other (Figure 13.2). This is because an isothermal electric current in a metal is accompanied by a thermal current.

$$\mathbf{j}^q = \Pi \mathbf{j}, \tag{13.66}$$

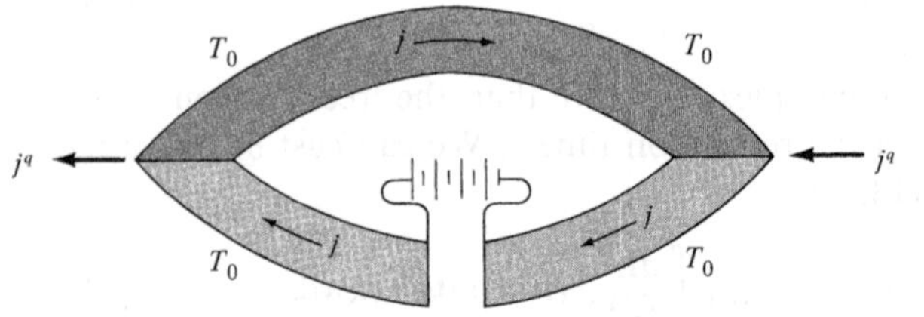

Figure 13.2
The Peltier effect. A current j is driven in a bimetallic circuit at uniform temperature T_0. To maintain the uniform temperature it is necessary to supply heat (via a thermal current j^q at one junction, and extract it at the other).

where Π is known as the Peltier coefficient. Because the electric current is uniform in the closed circuit and the Peltier coefficient differs from metal to metal, the thermal currents in the two metals will not be equal, and the difference must be evolved at one junction and supplied to the other if the uniform temperature is to be maintained.

If we set the temperature gradient in (13.45) equal to zero, then we find that the Peltier coefficient is given by

$$\Pi = \frac{L^{21}}{L^{11}}. \tag{13.67}$$

Because of the identity (13.51), the Peltier coefficient is simply related to the thermopower (13.61) by

$$\Pi = TQ, \tag{13.68}$$

a relation first deduced by Lord Kelvin.

---

[40] See Chapter 3.

[41] When a magnetic field, as well as a temperature gradient, is present, the number of possible measurements is further multiplied. The various thermomagnetic effects (Nernst, Ettingshausen, Righi-Leduc) are compactly summarized in H. B. Callen, *Thermodynamics*, Wiley, New York (1960), Chapter 17.

SEMICLASSICAL CONDUCTIVITY IN A UNIFORM MAGNETIC FIELD

The DC electrical conductivity at uniform temperature in a uniform magnetic field **H** can be cast in a form quite similar to the **H** = 0 result (13.25). In a magnetic field $\mathbf{v}(\mathbf{k}(t'))$ does depend on t', and the integral occurring in the nonequilibrium distribution function (13.21) can no longer be explicitly evaluated in the general case. Instead, the zero-field result (13.25) must be replaced by

$$\boldsymbol{\sigma}^{(n)} = e^2 \int \frac{d\mathbf{k}}{4\pi^3} \tau_n(\mathcal{E}_n(\mathbf{k}))\mathbf{v}_n(\mathbf{k})\overline{\mathbf{v}}_n(\mathbf{k}) \left(-\frac{\partial f}{\partial \mathcal{E}}\right)_{\mathcal{E}=\mathcal{E}_n(\mathbf{k})}, \tag{13.69}$$

where $\overline{\mathbf{v}}_n(\mathbf{k})$ is a weighted average of the velocity over the past history of the electron orbit[42] passing through **k**:

$$\overline{\mathbf{v}}_n(\mathbf{k}) = \int_{-\infty}^{0} \frac{dt}{\tau_n(\mathbf{k})} e^{t/\tau_n(\mathbf{k})} \mathbf{v}_n(\mathbf{k}_n(t)). \tag{13.70}$$

In the low-field limit the orbit is traversed very slowly, only points in the immediate vicinity of **k** contribute appreciably to the average in (13.70), and the zero field result is recovered. In the general case, and even in the high field limit, one must resort to some rather elaborate analysis, even to extract the information we arrived at in Chapter 12 from a direct examination of the semiclassical equations of motion. We shall not pursue such computations further here, but some of the applications of (13.70) are illustrated in Problem 6.

PROBLEMS

1. On page 250 we argued that in a metal with cubic symmetry the conductivity tensor is a constant times the unit matrix, i.e., that **j** is always parallel to **E**. Construct an analogous argument for a hexagonal close-packed metal, showing that the conductivity tensor is diagonal in a rectangular coordinate system with z taken along the c-axis, with $\sigma_{xx} = \sigma_{yy}$, so that the current induced by a field parallel or perpendicular to the c-axis is parallel to the field.

2. Deduce from (13.25) that at $T = 0$ (and hence to an excellent approximation at any $T \ll T_F$) the conductivity of a band with cubic symmetry is given by

$$\sigma = \frac{e^2}{12\pi^3\hbar} \tau(\mathcal{E}_F)\overline{v}S, \tag{13.71}$$

[42] Here $\mathbf{k}_n(t)$ is the solution to the semiclassical equations of motion (12.6) in a uniform magnetic field that passes through the point **k** at time zero ($\mathbf{k}_n(0) = \mathbf{k}$). (We have used the fact that the distribution function is independent of time when the fields are time-independent, and written the integral in (13.21) in the form it has at $t = 0$.)

where S is the area of Fermi surface in the band, and $\bar{v}$ is the electronic speed averaged over the Fermi surface:

$$\bar{v} = \frac{1}{S} \int dS \, |\mathbf{v}(\mathbf{k})|. \tag{13.72}$$

(Note that this contains, as a special case, the fact that filled or empty bands (neither of which have any Fermi surface) carry no current. It also provides an alternative way of viewing the fact that almost empty (few electrons) and almost filled (few holes) bands have low conductivity, since they will have very small amounts of Fermi surface.)

Verify that (13.71) reduces to the Drude result in the free electron limit.

3. Show that the equations describing the electric and thermal currents, (13.45) and (13.50) to (13.53), continue to hold in the presence of a uniform magnetic field, provided that Eq. (13.48) for $\boldsymbol{\sigma}(\mathcal{E})$ is generalized to include the effects of the magnetic field by replacing the second $\mathbf{v}(\mathbf{k})$ by $\bar{\mathbf{v}}(\mathbf{k})$, defined in Eq. (13.70).

4. The response of the conduction electrons to an electric field

$$\mathbf{E}(\mathbf{r}, t) = \text{Re}\left[\mathbf{E}(\mathbf{q}, \omega)e^{i(\mathbf{q}\cdot\mathbf{r}-\omega t)}\right], \tag{13.73}$$

which depends on position as well as time, requires some special consideration. Such a field will in general induce a spatially varying charge density

$$\begin{aligned} \rho(\mathbf{r}, t) &= -e\,\delta n(\mathbf{r}, t), \\ \delta n(\mathbf{r}, t) &= \text{Re}\left[\delta n(\mathbf{q}, \omega)e^{i(\mathbf{q}\cdot\mathbf{r}-\omega t)}\right]. \end{aligned} \tag{13.74}$$

Since electrons are conserved in collisions, the local equilibrium distribution appearing in the relaxation-time approximation (13.3) must correspond to a density equal to the actual instantaneous local density $n(\mathbf{r}, t)$. Thus even at uniform temperature one must allow for a local chemical potential of the form

$$\begin{aligned} \mu(\mathbf{r}, t) &= \mu + \delta\mu(\mathbf{r}, t), \\ \delta\mu(\mathbf{r}, t) &= \text{Re}\left[\delta\mu(\mathbf{q}, \omega)e^{i(\mathbf{q}\cdot\mathbf{r}-\omega t)}\right], \end{aligned} \tag{13.75}$$

where $\delta\mu(\mathbf{q}, \omega)$ is chosen to satisfy (to linear order in $\mathbf{E}$) the condition

$$\delta n(\mathbf{q}, \omega) = \frac{\partial n_{eq}(\mu)}{\partial \mu}\,\delta\mu(\mathbf{q}, \omega). \tag{13.76}$$

(a) Show, as a result, that at uniform temperature Eq. (13.22) must be replaced by[43]

$$\begin{aligned} g(\mathbf{r}, \mathbf{k}, t) &= f(\mathcal{E}(\mathbf{k})) + \text{Re}\left[\delta g(\mathbf{q}, \mathbf{k}, \omega)e^{i(\mathbf{q}\cdot\mathbf{r}-\omega t)}\right], \\ \delta g(\mathbf{q}, \omega) &= \left(-\frac{\partial f}{\partial \mathcal{E}}\right)\frac{(\delta\mu(\mathbf{q}, \omega)/\tau) - e\mathbf{v}(\mathbf{k})\cdot\mathbf{E}(\mathbf{q}, \omega)}{(1/\tau) - i[\omega - \mathbf{q}\cdot\mathbf{v}(\mathbf{k})]}. \end{aligned} \tag{13.77}$$

(b) By constructing the induced current and charge densities from the distribution function (13.77), show that the choice (13.75) of $\delta\mu(\mathbf{q}, \omega)$ is precisely what is required to insure that the equation of continuity (local charge conservation)

$$\mathbf{q}\cdot\mathbf{j}(\mathbf{q}, \omega) = \omega\rho(\mathbf{q}, \omega) \qquad \left(\nabla\cdot\mathbf{j} + \frac{\partial\rho}{\partial t} = 0\right) \tag{13.78}$$

is satisfied.

[43] See footnote 7.

(c) Show that if no charge density is induced, then the current is

$$\mathbf{j}(\mathbf{r}, t) = \mathrm{Re}\left[\boldsymbol{\sigma}(\mathbf{q}, \omega) \cdot \mathbf{E}(\mathbf{q}, \omega) e^{i(\mathbf{q}\cdot\mathbf{r}-\omega t)}\right],$$

$$\boldsymbol{\sigma}(\mathbf{q}, \omega) = e^2 \int \frac{d\mathbf{k}}{4\pi^3}\left(-\frac{\partial f}{\partial \mathcal{E}}\right)\frac{\mathbf{v}\mathbf{v}}{(1/\tau) - i[\omega - \mathbf{q}\cdot\mathbf{v}(\mathbf{k})]}. \tag{13.79}$$

Show that a sufficient condition for (13.79) to be valid is that the electric field $\mathbf{E}$ be perpendicular to a plane of mirror symmetry in which the wave vector $\mathbf{q}$ lies.

5. Consider a metal in which thermal and electric currents flow simultaneously. The rate at which heat is generated in a unit volume is related to the local energy and number densities by (cf. (13.39)):

$$\frac{dq}{dt} = \frac{du}{dt} - \mu\frac{dn}{dt}, \tag{13.80}$$

where μ is the local chemical potential. Using the equation of continuity,

$$\frac{dn}{dt} = -\nabla \cdot \mathbf{j}^n, \tag{13.81}$$

and the fact that the rate of change of the local energy density is determined by the rate at which electrons carry energy into the volume plus the rate at which the electric field does work,

$$\frac{du}{dt} = -\nabla \cdot \mathbf{j}^{\mathcal{E}} + \mathbf{E} \cdot \mathbf{j}, \tag{13.82}$$

show that (13.80) can be written in the form

$$\frac{dq}{dt} = -\nabla \cdot \mathbf{j}^q + \boldsymbol{\mathcal{E}} \cdot \mathbf{j}, \tag{13.83}$$

where $\mathbf{j}^q$ is the thermal current (given by (13.38) and (13.40)), and $\boldsymbol{\mathcal{E}} = \mathbf{E} + (1/e)\nabla\mu$. Assuming cubic symmetry, so that the tensors $\mathbf{L}^{ij}$ are diagonal, show that under conditions of uniform current flow ($\nabla \cdot \mathbf{j} = 0$) and uniform temperature gradient ($\nabla^2 T = 0$) that

$$\frac{dq}{dt} = \rho \mathbf{j}^2 + \frac{dK}{dT}(\nabla T)^2 - T\frac{dQ}{dT}(\nabla T)\cdot\mathbf{j} \tag{13.84}$$

where ρ is the resistivity, K is the thermal conductivity, and Q is the thermopower. By measuring the change in bulk heating as the current direction is reversed for fixed temperature gradient (known as the Thomson effect) one can therefore determine the temperature derivative of the thermopower, and thereby compute the value of Q at high temperatures, given its low-temperature value.

Compare the numerical coefficient of $\nabla T \cdot \mathbf{j}$ with that of the crude estimate in Problem 3, Chapter 1.

6. The average velocity $\bar{\mathbf{v}}$ (Eq. (13.70)) appearing in the expression (13.69) for the conductivity in a uniform magnetic field takes on a fairly simple form in the high-field limit.

(a) Show that for a closed orbit, the projection of $\bar{\mathbf{v}}$ in a plane perpendicular to $\mathbf{H}$ is

$$\bar{\mathbf{v}}_\perp = -\frac{\hbar c}{eH\tau}\hat{\mathbf{H}} \times [\mathbf{k} - \langle\mathbf{k}\rangle]_\perp + O\left(\frac{1}{H^2}\right), \tag{13.85}$$

where $\langle \mathbf{k} \rangle$ is the time average of the wave vector over the orbit:

$$\langle \mathbf{k} \rangle = \frac{1}{T} \oint \mathbf{k}\, dt. \tag{13.86}$$

(b) Show that for an open orbit the high-field limit of $\bar{\mathbf{v}}$ is just the average velocity of motion along the orbit (and hence parallel to the direction of the orbit).

(c) Show[44] that in the high-field limit, when $\mathbf{E} \cdot \mathbf{H} = 0$,

$$\mathbf{j}_\perp = -e \int \frac{d\mathbf{k}}{4\pi^3} \left(-\frac{\partial f}{\partial \mathbf{k}} \right) \mathbf{k} \cdot \mathbf{w}, \tag{13.87}$$

where $\mathbf{w} = c(\mathbf{E} \times \mathbf{H})/H^2$ is the drift velocity defined in (12.46). Deduce the forms (12.51) or (12.52) from (13.87), depending on whether the band is particle- or hole-like. (*Note:* Because $\mathbf{k}$ is *not* a periodic function in k-space, one cannot automatically integrate by parts in (13.87).)

(d) Deduce, from the result of (b), the limiting form (12.56) for the conductivity in the presence of open orbits. (*Hint:* Observe that $\bar{\mathbf{v}}$ is independent of the component of $\mathbf{k}$ parallel to the k-space direction of the open orbit.)

(e) Show from the general form of the semiclassical equation of motion in a magnetic field (12.6) that the conductivity tensor (13.69) for a given band in a uniform magnetic field has the functional dependence on H and τ of the form:

$$\boldsymbol{\sigma} = \tau \mathbf{F}(H\tau). \tag{13.88}$$

Deduce from (13.88) that when the current is carried by electrons in a single band (or if the relaxation time is the same for all bands), then

$$\frac{\rho_{xx}(H) - \rho_{xx}(0)}{\rho_{xx}(0)} \tag{13.89}$$

depends on H and τ only through the product $H\tau$ (Kohler's rule), for any diagonal component of the resistivity perpendicular to $\mathbf{H}$.

(f) Deduce from the properties of the semiclassical equations of motion in a magnetic field that

$$\sigma_{\mu\nu}(H) = \sigma_{\nu\mu}(-H). \tag{13.90}$$

This is known as an Onsager relation.[45] (*Hint:* Make the change of variables $\mathbf{k}(t) = \mathbf{k}'$, and appeal to Liouville's theorem in replacing the k-space integrals in (13.69) by integrals over $\mathbf{k}'$.)

---

[44] Argue that the term in $\langle \mathbf{k} \rangle$ in (13.85) makes no contribution because it depends only on $\mathcal{E}$ and k_z.

[45] Such relationships between transport coefficients were first formulated in very great generality by L. Onsager. The first equality in Eq. (13.51) is another example of an Onsager relation.

14
Measuring the Fermi Surface

There is a class of measurable quantities valued primarily because they contain detailed information about the geometric structure of the Fermi surface. Such quantities depend only on universal constants (e, h, c, or m), experimentally controlled variables (such as temperature, frequency, magnetic field strength, crystal orientation), and information about the electronic band structure that is entirely determined by the shape of the Fermi surface.

We have already encountered one such quantity, the high-field Hall constant, which (in uncompensated metals with no open orbits for the given field direction) is entirely determined by the k-space volume enclosed by the hole-like and particle-like branches of the Fermi surface.

Quantities yielding such Fermi surface information have a place of special importance in metals physics. Their measurement almost always requires single crystals of very pure substances at very low temperatures (to eliminate dependence on the relaxation time) and is frequently performed in very strong magnetic fields (to force the electrons to sample the Fermi surface geometry in the course of their semiclassical motion in k-space).

The importance of determining the Fermi surface of metals is clear: The shape of the Fermi surface is intimately involved in the transport coefficients of a metal (as discussed in Chapters 12 and 13) as well as in the equilibrium and optical properties (as will be illustrated in Chapter 15). An experimentally measured Fermi surface provides a target at which a first-principles band structure calculation can aim. It can also be used to provide data for fitting parameters in a phenomenological crystal potential, which can then be used to calculate other phenomena. If nothing else, Fermi surface measurements are of interest as a further test of the validity of the one-electron semiclassical theory, since there are now many independent ways of extracting Fermi surface information.

Of the techniques used to deduce the geometry of the Fermi surface, one has proved far and away the most powerful: the *de Haas–van Alphen effect* (and a group of closely related effects based on the same underlying physical mechanism). This phenomenon is almost entirely responsible for the vast and growing body of precise knowledge of the Fermi surfaces of a great number of metals. No other technique approaches it in power or simplicity. For this reason the bulk of this chapter is devoted to an exposition of the de Haas–van Alphen effect. We shall conclude with brief discussions of a selection of other effects which have been used to provide supplementary geometrical information.

THE de HAAS–van ALPHEN EFFECT

In Figure 14.1 are shown the results of a famous experiment by de Haas and van Alphen in 1930. They measured the magnetization M of a sample of bismuth as a function of magnetic field in high fields at 14.2 K, and found oscillations in M/H.

On the face of it this curious phenomenon, observed only at low temperatures and high fields, would not strike one as the extraordinary key to the electron structure of metals it has turned out to be. The full extent of its usefulness was only pointed out in 1952, by Onsager. Since the original experiment, and especially since around 1960,

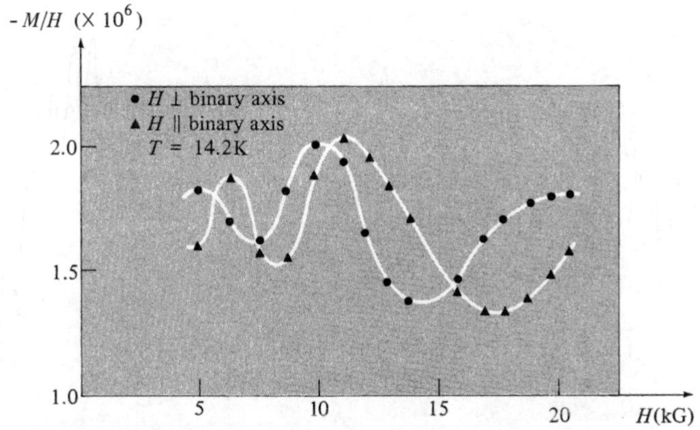

Figure 14.1
The data of de Haas and van Alphen. Magnetization per gram divided by field, plotted vs. field for two orientations of a crystal of bismuth, at 14.2 K. (W. J. de Haas and P. M. van Alphen, *Leiden Comm.* 208d, 212a (1930), and 220d (1932).)

careful observations have been made in many metals of this same oscillatory field dependence in the magnetic susceptibility,[1] $\chi = dM/dH$.

The oscillations display a remarkable regularity, if the susceptibility is plotted *not* against field, but against *inverse field.* It then becomes clear that χ has a periodic dependence on $1/H$, though frequently two or more periods are superposed. Some typical data are shown in Figure 14.2.

Similar oscillatory behavior has been observed not only in the susceptibility, but also in the conductivity (Shubnikov–de Haas effect), the magnetostriction (dependence of sample size on magnetic field strength), and, when measured with sufficient care, in almost all other quantities. Minute oscillations of this kind have even been observed in the high-field Hall "constant," a clear indication that the effect must lie in a failure of the semiclassical model. A variety of such effects are displayed in Figure 14.3.

The refinement of the de Haas–van Alphen effect into a powerful probe of the Fermi surface has been due largely to D. Shoenberg, whose history of the phenomenon[2] provides delightful and instructive reading. Two major techniques have been widely exploited to measure the oscillations. One, based on the fact that in a field a magnetized sample experiences a torque proportional to its magnetic moment,[3] simply measures the oscillations in angular position of a sample of the metal, attached to a

---

[1] When the magnetization varies linearly with the field one need not distinguish between M/H and $\partial M/\partial H$. Here, however, (and in the treatment of critical phenomena in Chapter 33) nonlinear effects are crucial. It is now generally agreed that in such cases the susceptibility should be defined as $\partial M/\partial H$.

[2] *Proc. 9th Internat. Conf. on Low Temperature Physics*, Daunt, Edwards, Milford, and Yaqub, ed. Plenum Press, New York, 1965, p. 665.

[3] The torque only exists when the magnetization is not parallel to the field. Because the effect is nonlinear, this is generally the case except when the field is in certain symmetry directions.

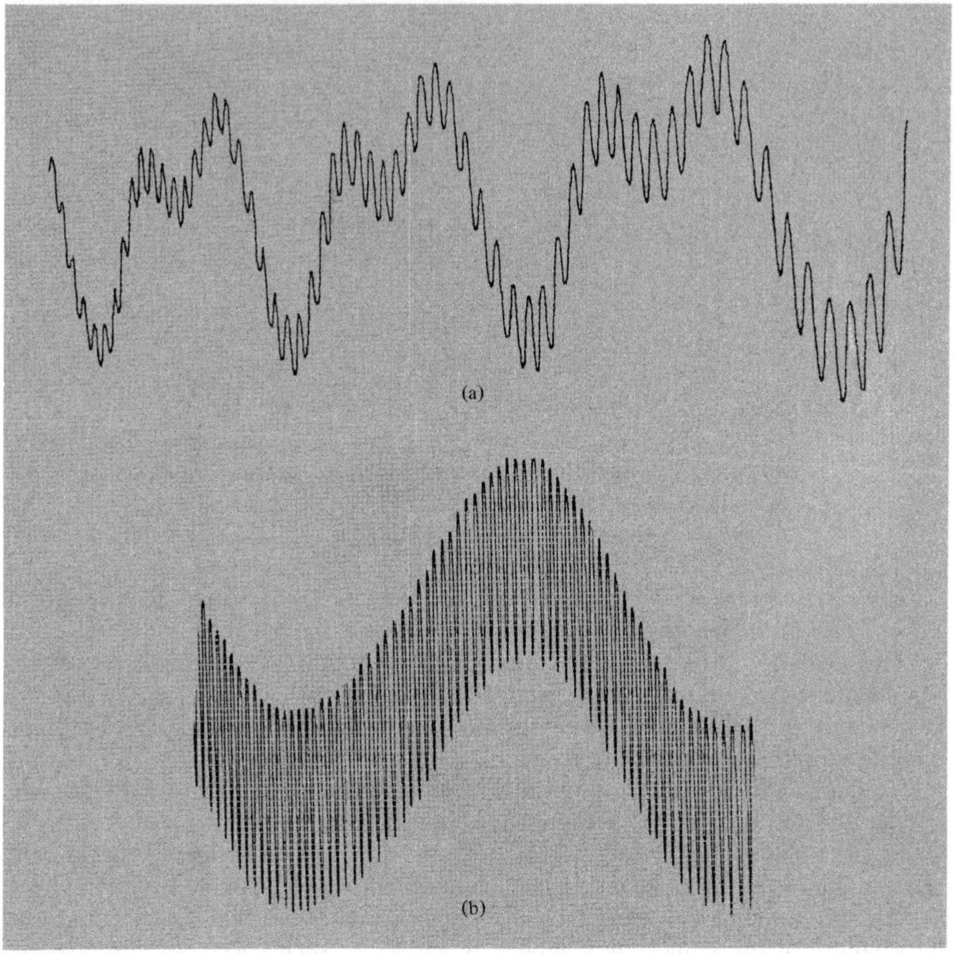

Figure 14.2
De Haas–van Alphen oscillations in (a) rhenium and (b) silver. (Courtesy of A. S. Joseph.)

filar suspension, as the magnetic field strength, and hence the magnetization $M(H)$ varies. The second technique, especially valuable if high fields are required, measures the voltage induced in a pickup coil surrounding the sample when a burst[4] of field is applied. Since this will be proportional to $dM/dt = (dM/dH)(dH/dt)$, one can measure the oscillations in the susceptibility as a function of field.

Even before the key to the theory of the de Haas–van Alphen effect for Bloch electrons was pointed out by Onsager, Landau[5] was able to account for the oscillations in free electron theory, as a direct consequence of the quantization of closed

[4] This "burst" of field is, of course, slowly varying on the scale of metallic relaxation times, so that the magnetization stays in equilibrium with the instantaneous value of the field.

[5] L. D. Landau, *Z. Phys.* **64**, 629 (1930). Note the date. Landau predicted the oscillations without knowing of de Hass and van Alphen's experiment, but thought that a magnetic field uniform enough to observe them could not be achieved (see Problem 3).

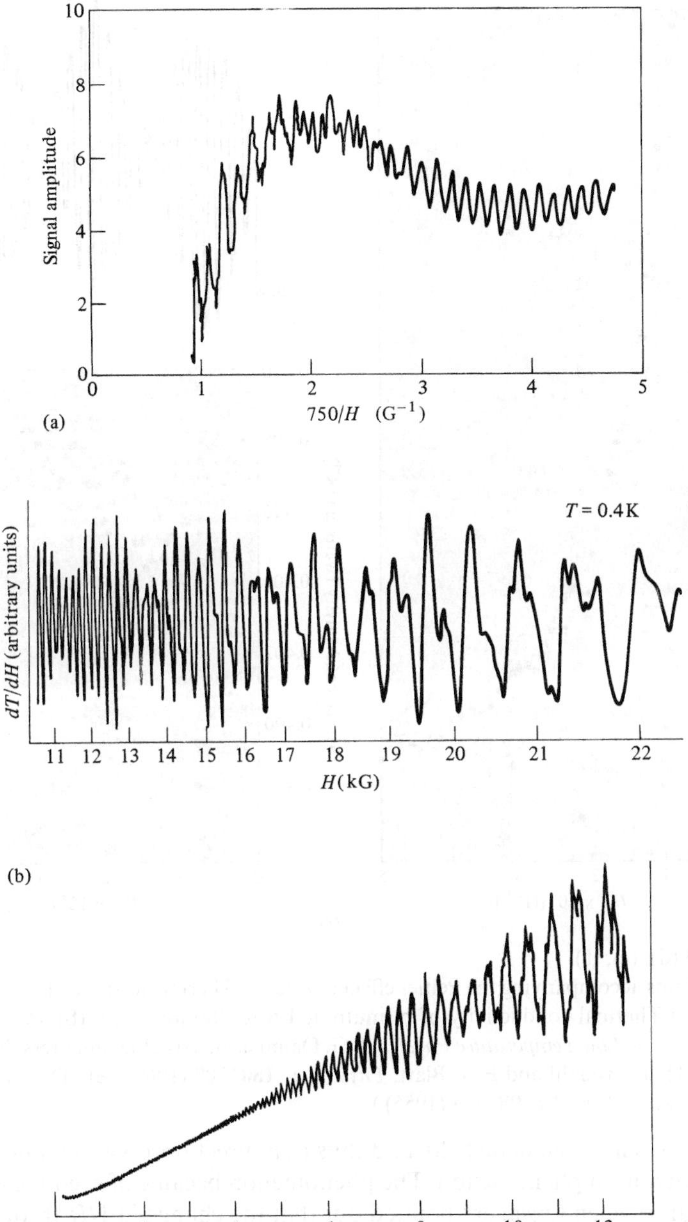

Figure 14.3 (a), (b), (c)
The ubiquity of the oscillations, of which the de Haas–van Alphen effect is the most celebrated example. (a) Sound attenuation in tungsten. (C. K. Jones and J. A. Rayne.) (b) dT/dH vs. field in antimony. (B. D. McCombe and G. Seidel.) (c) Magnetoresistance of gallium vs. field at 1.3 K.

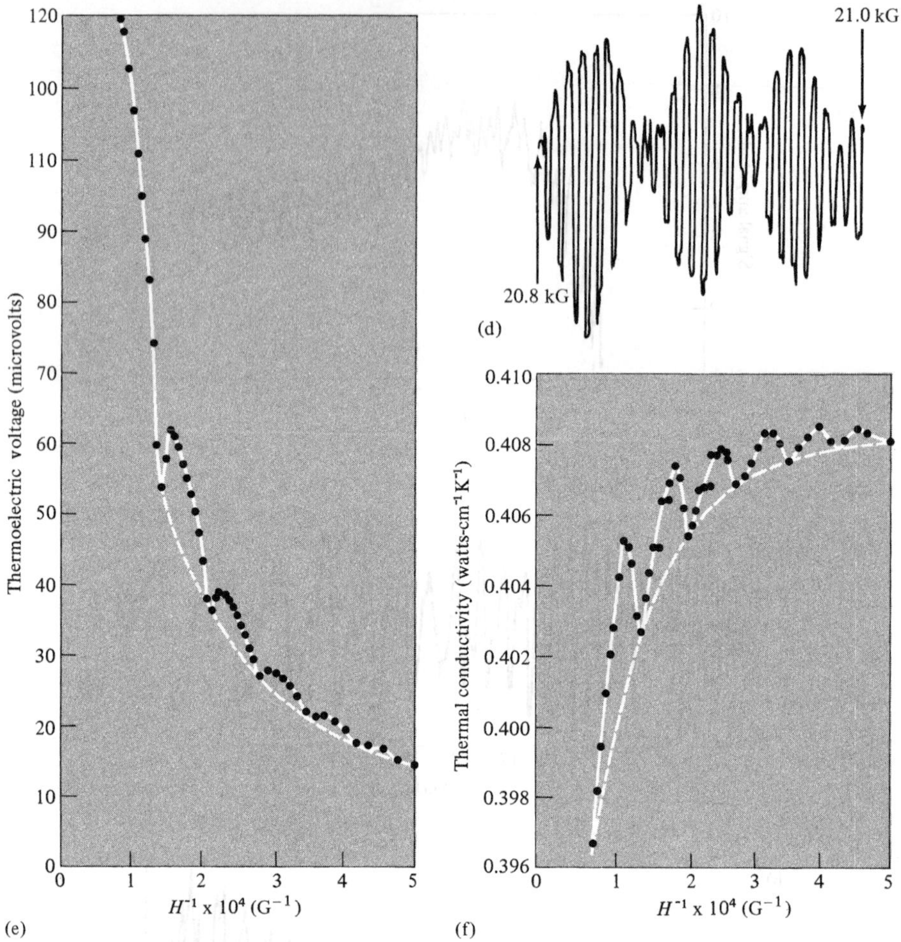

Figure 14.3 (d), (e), (f)
(d) Oscillations accompanying the Peltier effect in zinc. (e) Thermoelectric voltage of bismuth at 1.6 K. (f) Thermal conductivity of bismuth at 1.6 K. (Sources: (a), (b), (c)—*Proc. 9th Internat. Conf. on Low Temperature Physics*, J. G. Daunt et al., eds., Plenum Press, New York, 1965. (d)—H. J. Trodahl and F. J. Blatt, *Phys. Rev.* **180**, 709 (1969). (e), (f)—M. C. Steele and J. Babiskin, *Phys. Rev.* **98**, 359 (1955).)

electronic orbits in a magnetic field, and thus as a direct observational manifestation of a purely quantum phenomenon. The phenomenon became of even greater interest and importance when Onsager[6] pointed out that the change in $1/H$ through a single period of oscillation, $\Delta(1/H)$, was determined by the remarkably simple relation:

$$\Delta\left(\frac{1}{H}\right) = \frac{2\pi e}{\hbar c}\frac{1}{A_e} \tag{14.1}$$

where A_e is any extremal cross-sectional area of the Fermi surface in a plane normal to the magnetic field.

---

6 L. Onsager, *Phil. Mag.* **43**, 1006 (1952).

Some extremal areas are illustrated in Figure 14.4. If the z-axis is taken along the magnetic field, then the area of a Fermi surface cross section at height k_z is $A(k_z)$, and the extremal areas A_e are the values of $A(k_z)$ at the k_z where $dA/dk_z = 0$. (Thus maximum and minimum cross sections are among the extremal ones.)

Figure 14.4
Illustration of various extremal orbits. For **H** along the k_1-axis, (1) and (2) are maximum extremal orbits and (3) is a minimum extremal orbit. When the field is along the k_2-axis, only one extremal orbit, (4), is present.

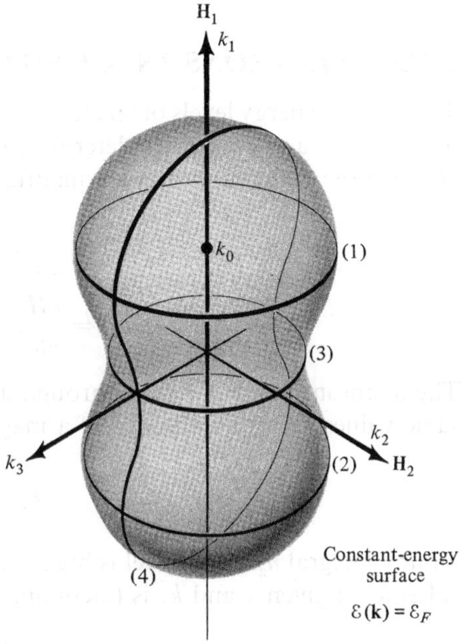

Since altering the magnetic field direction brings different extremal areas into play, all extremal areas of the Fermi surface can be mapped out. This frequently provides enough information to reconstruct the actual shape of the Fermi surface. In practice this may be a complex task, for if more than one extremal orbit is present in certain directions, or if more than one band is partially filled, several periods will be superimposed. Rather than directly disentangling the geometrical information from the data, it is often easier to guess at what the surface is (using, for example, an approximate calculation of the band structure), later refining the guess by testing it against the data.

The argument justifying (14.1) is simple but strikingly bold. The explanation cannot be classical, for a theorem of Bohr and van Leeuwen (see Chapter 31) asserts that no properties of a classical system in thermal equilibrium can depend in any way on the magnetic field. This powerful result applies to semiclassical systems (in the sense of Chapters 12 and 13) as well, so the de Haas–van Alphen effect is a definite failure of the semiclassical model. The failure arises whenever the semiclassical theory predicts closed orbits for the electronic motion projected on a plane perpendicular to the field. When this happens (as it generally does), the energies of motion perpendicular to **H** are quantized. To find these energy levels one must in principle return to the Schrödinger equation for an electron in the periodic crystalline potential in the presence of the magnetic field. The full solution of this problem is a formidable task, which has been accomplished only in the simple case of free electrons (i.e., zero periodic potential) in a magnetic field. We describe the results in the free electron case below, referring the reader to one of the standard texts for their derivation.[7] We shall not use the free electron results except to illustrate and test the validity of Onsager's far more general, but somewhat less rigorous, theory of the magnetic levels in a periodic potential.

FREE ELECTRONS IN A UNIFORM MAGNETIC FIELD

The orbital[8] energy levels of an electron in a cubical box with sides of length L parallel to the x-, y-, and z-axes are determined in the presence of a uniform magnetic field H along the z-direction by two quantum numbers, v and k_z:

$$\mathcal{E}_v(k_z) = \frac{\hbar^2}{2m} k_z^{\ 2} + \left(v + \frac{1}{2}\right) \hbar\omega_c,$$
$$\omega_c = \frac{eH}{mc}. \tag{14.2}$$

The quantum number v runs through all nonnegative integers, and k_z takes on the same values as in the absence of a magnetic field (Eq. (2.16)):

$$k_z = \frac{2\pi n_z}{L}, \tag{14.3}$$

for any integral n_z. Each level is highly degenerate. The number of levels with energy (14.2) for a given v and k_z is (including the factor of 2 for spin degeneracy):

$$\frac{2e}{hc} HL^2. \tag{14.4}$$

Since

$$\frac{hc}{2e} = 2.068 \times 10^{-7} \text{ G-cm}^2, \tag{14.5}$$

in a field of a kilogauss (a typical field for a de Haas–van Alphen experiment) and a sample 1 cm on a side, this degeneracy will be about 10^{10}. The degeneracy reflects the fact that a classical electron with a given energy and k_z spirals about a line parallel to the z-axis, which can have arbitrary x- and y-coordinates.[9]

Equation (14.2) is quite plausible: Since there is no component of Lorentz force along H, the energy of motion in the z-direction is unaffected by the field, and continues to be given by $\hbar^2 k_z^{\ 2}/2m$. However, the energy of motion perpendicular to the field, which would be $\hbar^2(k_x^{\ 2} + k_y^{\ 2})/2m$ if no field were present, is quantized in steps of $\hbar\omega_c$—Planck's constant times the frequency of the classical motion (page 14). This

---

[7] L. D. Landau and E. M. Lifshitz, *Quantum Mechanics*, (2nd ed.) Addison-Wesley, Reading, Mass., 1965, pp. 424–426, or R. E. Peierls, *Quantum Theory of Solids*, Oxford, New York, 1955, pp. 146–147. Peierls gives a better discussion of the rather subtle spatial boundary condition. The energy levels are found by reducing the problem, by a simple transformation, to that of a one-dimensional harmonic oscillator.

[8] Equation (14.2) does *not* include the interaction energy between the field and the electron spin. We consider the consequences of this additional term below, but for the moment we ignore it.

[9] This is why the degeneracy (14.4) is proportional to the cross-sectional area of the specimen.

phenomenon is called *orbit quantization*. The set of all levels with a given ν (and arbitrary k_z) is referred to collectively as the *νth Landau level*[10].

From this information a theory of the de Haas–van Alphen effect can be constructed for the free electron model. Rather than reproduce that analysis[11] we turn to a slightly modified version of Onsager's simple, but subtle, argument, which generalizes the free electron results to Bloch electrons and bears directly on the problem of Fermi surface determination.

LEVELS OF BLOCH ELECTRONS IN A UNIFORM MAGNETIC FIELD

Onsager's generalization of Landau's free electron results is only valid for magnetic levels with fairly high quantum numbers. However, we shall find that the de Haas–van Alphen effect is due to levels at the Fermi energy which almost always do have very high quantum numbers. In free electron theory, for example, unless almost all the electronic energy is in motion parallel to the field, a level of energy $\mathcal{E}_F$ must have a quantum number ν whose order of magnitude is $\mathcal{E}_F/\hbar\omega_c = \mathcal{E}_F/[(e\hbar/mc)H]$. Now

$$\frac{e\hbar}{mc} = \frac{\hbar}{m} \times 10^{-8}\ \text{eV/G} = 1.16 \times 10^{-8}\ \text{eV/G}. \tag{14.6}$$

Since $\mathcal{E}_F$ is typically several electron volts, even in fields as high as 10^4 G, the quantum number ν will be of order 10^4.

Energies of levels with very high quantum numbers can be accurately calculated with Bohr's correspondence principle, which asserts that the difference in energy of two adjacent levels is Planck's constant times the frequency of classical motion at the energy of the levels. Since k_z is a constant of the semiclassical motion, we apply this condition to levels with a specified k_z, and quantum numbers ν and $\nu + 1$.

Let $\mathcal{E}_\nu(k_z)$ be the energy of the νth allowed level[12] at the given k_z. The correspondence principle then gives

$$\mathcal{E}_{\nu+1}(k_z) - \mathcal{E}_\nu(k_z) = \frac{h}{T(\mathcal{E}_\nu(k_z), k_z)}, \tag{14.7}$$

---

10 It must be added that the above results are only valid when the radius of classical circular motion of an electron with energy $\mathcal{E}$ and momentum $\hbar k_z$ is not comparable with the cross-sectional dimensions of the box. For an electron with energy $\mathcal{E}_F$ and $k_z = 0$ the condition is most stringent:

$$L \gg r_c = \frac{v_F}{\omega_c} = \frac{\hbar k_F}{m\omega_c} = \left(\frac{\hbar c}{eH}\right) k_F.$$

At 10^3 gauss, $\hbar c/eH \approx 10^{-10}$ cm^2. Since k_F is typically about 10^8 cm^{-1}, the results are applicable to samples with dimensions on the order of centimeters, but fail when the sample is still as large as 0.1 mm.

11 It can be found in the book by Peierls cited in footnote 7.

12 Throughout the discussion that follows we consider a single band, and drop explicit reference to the band index. This is done primarily to avoid confusion between the band index n and the magnetic quantum number ν. Throughout this chapter $\mathcal{E}_\nu(k_z)$ is the νth allowed energy of an electron in the given band, with wave vector k_z. Should it be necessary to deal with more than one band, we would use the notation $\mathcal{E}_{n,\nu}(k_z)$.

where $T(\mathcal{E}, k_z)$ is the period of semiclassical motion on the orbit specified by $\mathcal{E}$ and k_z (Eq. (12.42)):

$$T(\mathcal{E}, k_z) = \frac{\hbar^2 c}{eH} \frac{\partial A(\mathcal{E}, k_z)}{\partial \mathcal{E}}, \tag{14.8}$$

and $A(\mathcal{E}, k_z)$ is the k-space area enclosed by the orbit. Combining (14.8) and (14.7), we may write (suppressing explicit reference to the variable k_z)

$$(\mathcal{E}_{\nu+1} - \mathcal{E}_\nu) \frac{\partial}{\partial \mathcal{E}} A(\mathcal{E}_\nu) = \frac{2\pi e H}{\hbar c}. \tag{14.9}$$

Because we are interested in $\mathcal{E}_\nu$ of the order of $\mathcal{E}_F$, we can greatly simplify (14.9). On the basis of the free electron results we expect that the energy difference between neighboring Landau levels will be of order $\hbar\omega_c$, which is at least 10^{-4} times smaller than the energies of the levels themselves. It is therefore an excellent approximation to take:

$$\frac{\partial}{\partial \mathcal{E}} A(\mathcal{E}_\nu) = \frac{A(\mathcal{E}_{\nu+1}) - A(\mathcal{E}_\nu)}{\mathcal{E}_{\nu+1} - \mathcal{E}_\nu}. \tag{14.10}$$

Placing this in (14.9) we find

$$A(\mathcal{E}_{\nu+1}) - A(\mathcal{E}_\nu) = \frac{2\pi e H}{\hbar c}, \tag{14.11}$$

which states that classical orbits at adjacent allowed energies (and the same k_z) enclose areas that differ by the fixed amount ΔA, where

$$\boxed{\Delta A = \frac{2\pi e H}{\hbar c}.} \tag{14.12}$$

Another way of stating this conclusion is that, at large ν, the area enclosed by the semiclassical orbit at an allowed energy and k_z must depend on ν according to:

$$\boxed{A(\mathcal{E}_\nu(k_z), k_z) = (\nu + \lambda)\,\Delta A,} \tag{14.13}$$

where λ is independent[13] of ν. This is Onsager's famous result (which he derived by an alternate route, using the Bohr-Sommerfeld quantization condition).

ORIGIN OF THE OSCILLATORY PHENOMENA

Underlying the de Haas–van Alphen and related oscillations is a sharp oscillatory structure in the electronic density of levels imposed by the quantization condition

[13] We shall follow the usual practice of assuming that λ is also independent of k_z and H. This is verified in Problem 1a for free electrons, and holds for any ellipsoidal band. Although it has not been proved in general, the reader is invited to show, as an exercise, that the conclusions reached below under the assumption of a constant λ are altered only if λ is an exceedingly rapidly varying function of either k_z or H. This is most unlikely.

(14.13). The level density will have a sharp peak[14] whenever $\mathcal{E}$ is equal to the energy of an extremal orbit[15] satisfying the quantization condition. The reason for this is shown in Figure 14.5. Figure 14.5a depicts the set of all orbits satisfying (14.13) for a given ν. These form a tubular structure (of cross-sectional area $(\nu + \lambda)\,\Delta A$) in k-space. The contribution to $g(\mathcal{E})\,d\mathcal{E}$ from the Landau levels associated with orbits on the νth such tube will be the number of such levels with energies between $\mathcal{E}$ and $\mathcal{E} + d\mathcal{E}$. This, in turn, is proportional to the area[16] of the portion of tube contained between the constant-energy surfaces of energies $\mathcal{E}$ and $\mathcal{E} + d\mathcal{E}$. Figure 14.5b shows this portion of tube when the orbits of energy $\mathcal{E}$ on the tube are *not* extremal, and Figure 14.5c shows the portion of tube when there *is* an extremal orbit of energy $\mathcal{E}$ on the tube. Evidently the area of the portion of tube is enormously enhanced in the latter case, as a result of the very slow energy variation of levels along the tube near the given orbit.

Most electronic properties of metals depend on the density of levels at the Fermi energy, $g(\mathcal{E}_F)$. It follows directly from the above argument[17] that $g(\mathcal{E}_F)$ will be singular

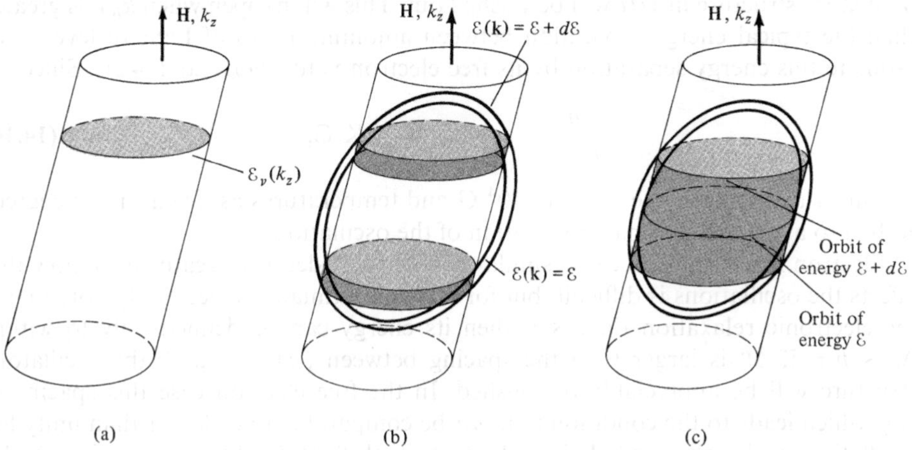

Figure 14.5
(a) A Landau tube. Its cross sections by planes perpendicular to **H** have the same area—$(\nu + \lambda)\,\Delta A$ for the νth tube—and are bounded by curves of constant energy $\mathcal{E}_\nu(k_z)$ at height k_z. (b) The portion of the tube containing orbits in the energy range from $\mathcal{E}$ to $\mathcal{E} + d\mathcal{E}$ when none of the orbits in that range occupy extremal positions on their constant-energy surfaces. (c) Same construction as in (b), except that $\mathcal{E}$ is now the energy of an extremal orbit. Note the great enhancement in the range of k_z for which the tube is contained between the constant-energy surfaces at $\mathcal{E}$ and $\mathcal{E} + d\mathcal{E}$.

---

[14] In fact, a more detailed analysis shows that the level density becomes singular as $(\mathcal{E} - \mathcal{E}_0)^{-1/2}$, when $\mathcal{E}$ is near the energy $\mathcal{E}_0$ of an extremal orbit satisfying the quantization condition.

[15] An extremal orbit of energy $\mathcal{E}$ is one that encloses an extremal cross-sectional area of the surface $\mathcal{E}(\mathbf{k}) = \mathcal{E}$.

[16] The density of levels contained in the tube is uniform along the field direction, allowed values of k_z being given by (14.3).

[17] Strictly speaking, the chemical potential (which is equal to $\mathcal{E}_F$ at zero temperature) also depends on magnetic field strength, complicating the argument, but this is a very small effect and can normally be ignored.

whenever the value of the magnetic field causes an extremal orbit on the Fermi surface to satisfy the quantization condition (14.13), i.e., whenever

$$(\nu + \lambda)\, \Delta A = A_e(\mathcal{E}_F). \tag{14.14}$$

Using the value (14.12) for ΔA, it follows that $g(\mathcal{E}_F)$ will be singular at regularly spaced intervals in $1/H$ given by

$$\Delta\left(\frac{1}{H}\right) = \frac{2\pi e}{\hbar c} \frac{1}{A_e(\mathcal{E}_F)}. \tag{14.15}$$

Thus oscillatory behavior as a function of $1/H$ with period (14.15) should appear in any quantity that depends on the level density at $\mathcal{E}_F$, which, at zero temperature, includes almost all characteristic metallic properties.

At nonzero temperatures typical metallic properties are determined by averages over a range of energies within $k_B T$ of $\mathcal{E}_F$. If this range is so broad that for *any* value of H extremal orbits satisfying (14.13) contribute appreciably to the average, then the oscillatory structure in $1/H$ will be washed out. This will happen when $k_B T$ is greater than the typical energy separation between adjoining tubes of Landau levels. We estimate this energy separation by its free electron value, $\hbar\omega_c$ (Eq. (14.2)). Since

$$\frac{e\hbar}{mck_B} = 1.34 \times 10^{-4}\ \mathrm{K/G}, \tag{14.16}$$

one must use fields of the order of 10^4 G and temperatures as low as a few degrees Kelvin to avoid the thermal obliteration of the oscillations.

Electron scattering can cause similar problems. A detailed treatment of how this affects the oscillations is difficult, but for a rough estimate we need only note that if the electronic relaxation time is τ, then its energy can be defined only to within $\Delta\mathcal{E} \sim \hbar/\tau$. If $\Delta\mathcal{E}$ is larger than the spacing between peaks in $g(\mathcal{E})$, the oscillatory structure will be appreciably diminished. In the free electron case this spacing is $\hbar\omega_c$, which leads to the condition that $\omega_c\tau$ be comparable to or larger than unity for oscillations to be observed. This is the same high-field condition that arises in the semiclassical theory of electronic transport (Chapters 12 and 13).

THE EFFECT OF ELECTRON SPIN ON THE OSCILLATORY PHENOMENA

Ignoring the effects of spin-orbit coupling,[18] the major complication introduced by electron spin is that the energy of each level will be increased or decreased by an amount

$$\frac{ge\hbar H}{4mc} \tag{14.17}$$

according to whether the spin is along or opposite to the field. The number g (not to be confused with the level density $g(\mathcal{E})$) is the "electron g-factor," which is very nearly equal to 2. If we denote the level density calculated by ignoring this additional energy

[18] Which are small in the lighter elements. See page 169.

by $g_0(\mathcal{E})$, then the result of these shifts is that the true level density $g(\mathcal{E})$ is given by

$$g(\mathcal{E}) = \frac{1}{2} g_0\left(\mathcal{E} + \frac{ge\hbar H}{4mc}\right) + \frac{1}{2} g_0\left(\mathcal{E} - \frac{ge\hbar H}{4mc}\right). \tag{14.18}$$

Note that the shift in the peaks is comparable to the separation between peaks (as estimated by the free electron value $e\hbar H/mc$). Cases have actually been observed in which for appropriate field directions this shift causes the oscillations in the two terms of (14.18) to fall 180° out of phase, leading to no net oscillation.

OTHER FERMI SURFACE PROBES

A variety of other experiments are used to probe the Fermi surface. In general, the information available from other techniques is geometrically less straightforward than the extremal areas supplied by the de Haas–van Alphen effect and related oscillations. Furthermore, this information is often more difficult to extract unambiguously from the data. We therefore confine ourselves to a brief survey of selected methods.

The Magnetoacoustic Effect

Fairly direct information about Fermi surface geometry can sometimes be extracted by measuring the attenuation of sound waves in a metal as they propagate perpendicular to a uniform magnetic field,[19] particularly if the wave is carried by displacements of the ions that are both perpendicular to its direction of propagation, and to the magnetic field (Figure 14.6). Since the ions are electrically charged, such a wave is accompanied by an electric field of the same frequency, wave vector, and polarization. The electrons in the metal can interact with the sound wave through this electric field, thereby supporting or hindering its propagation.

If conditions permit the electrons to complete many orbits in the magnetic field between collisions,[20] then the sound attenuation can depend on wavelength in a

Figure 14.6
The instantaneous displacement from equilibrium of the ions in a sound wave suitable for the magnetoacoustic effect. Only one row of ions is shown.

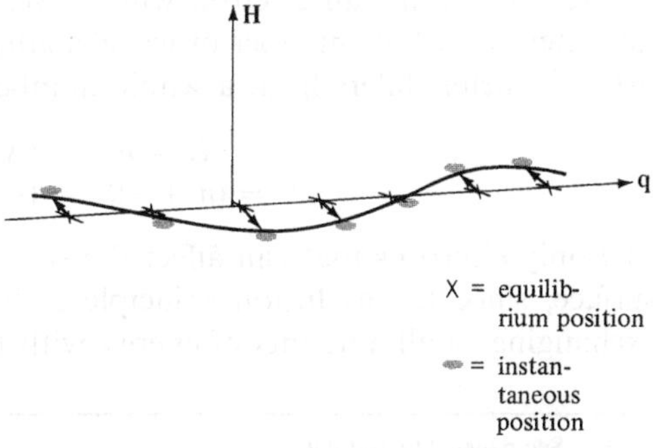

[19] A detailed theory of this phenomenon in the case of free electrons has been given by M. H. Cohen et al., *Phys. Rev.* **117**, 937 (1960).

[20] This requires $\omega_c\tau \gg 1$; i.e., the specimen must be a single crystal of high purity at low temperatures in a strong field.

manner reflecting the Fermi surface geometry. This is because[21] the electrons follow real space orbits whose projections in planes perpendicular to the field are simply cross sections of constant energy surfaces, scaled by the factor $\hbar c/eH$ (and rotated through 90°). When the wavelength of the sound is comparable to the dimensions of an electron's orbit[22] the extent to which the electric field of the wave perturbs the electron depends on how the wavelength l matches the maximum linear dimension l_c of the orbit along the direction of wave propagation (referred to in this context as the orbit's "diameter"). For example, electrons on orbits with diameters equal to half a wavelength (Figure 14.7a) can be accelerated (or decelerated) by the wave throughout their entire orbit, while electrons with orbit diameters equal to a whole wavelength (Figure 14.7b) must always be accelerated on parts of their orbit and decelerated on other parts.

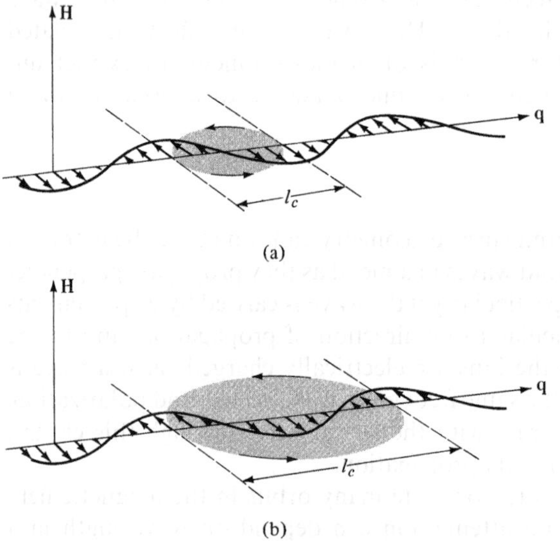

Figure 14.7
(a) An electron orbit with a diameter l_c equal to half a wavelength, positioned so as to be accelerated by the electric field accompanying the sound wave at all points of its orbit. (b) An electron orbit with a diameter equal to a whole wavelength. No matter where the orbit is positioned along the direction $\hat{\mathbf{q}}$, the kind of coherent acceleration (or deceleration) over the entire orbit possible in case (a) cannot occur.

More generally, an electron will be weakly coupled to the wave when its orbit diameter is a whole number of wavelengths, but can be strongly coupled when the orbit diameter differs from a whole number of wavelengths by half a wavelength:

$$\begin{aligned} l_c &= nl \qquad \text{(weakly coupled)}, \\ l_c &= (n + \tfrac{1}{2})l \qquad \text{(strongly coupled)}. \end{aligned} \tag{14.19}$$

The only electrons that can affect the sound attenuation are those near the Fermi surface, since the exclusion principle forbids electrons with lower energies from exchanging small amounts of energy with the wave. The Fermi surface has a con-

---

[21] See pages 229 and 230.

[22] A typical orbit diameter is of order v_F/ω_c. Since the angular frequency of the sound is of order v_s/l, when $l \approx l_c$ we have $\omega \approx \omega_c(v_s/v_F)$. Typical sound velocities are about 1 percent of the Fermi velocity, so electrons can complete many orbits during a single period of the waves of interest. In particular, during a single revolution of an electron, the electric field perturbing it can be regarded as static.

tinuous range of diameters, but the electrons on orbits with diameters near the extremal diameters play a dominant role, since there are many more of them.[23]

As a result the sound attenuation can display a periodic variation with inverse wavelength, in which the period (cf. Eq. (14.19)) is equal to the inverse of the extremal diameters of the Fermi surface along the direction of sound propagation:

$$\Delta\left(\frac{1}{l}\right) = \frac{1}{l_c}. \tag{14.20}$$

By varying the direction of propagation (to bring different extremal diameters into play) and by varying the direction of the magnetic field (to bring different Fermi surface cross sections into play), one can sometimes deduce the shape of the Fermi surface from this structure in the sound attenuation.

Ultrasonic Attenuation

Information about the Fermi surface can also be extracted from measurements of sound attenuation when no magnetic field is present. One no longer examines a resonant effect, but simply calculates the rate of attenuation assuming that it is entirely due to energy being lost to the electrons. It can be shown that if this is the case,[24] then the attenuation will be entirely determined by Fermi surface geometry. However, the geometrical information extracted in this way is, under the best of circumstances, nowhere near as simple as either the extremal areas furnished by the de Haas–van Alphen effect or the extremal diameters one can deduce from the magnetoacoustic effect.

Anomalous Skin Effect

One of the earliest Fermi surface determinations (in copper) was made by Pippard[25] from measurements of the reflection and absorption of microwave electromagnetic radiation (in the absence of a static magnetic field). If the frequency ω is not too high, such a field will penetrate into the metal a distance δ_0 (the "classical skin depth") given by[26]

$$\delta_0 = \frac{c}{\sqrt{2\pi\sigma\omega}}. \tag{14.21}$$

The derivation of (14.21) assumes that the field in the metal varies little over a mean free path: $\delta_0 \gg \ell$. When δ_0 is comparable to ℓ a much more complicated theory is required, and when $\delta_0 \ll \ell$ (the "extreme anomalous regime") the simple picture of an exponentially decaying field over a distance δ_0 breaks down completely. However, in the extreme anomalous case it can be shown that the field penetration and the

---

[23] This is quite analogous to the role played by cross sections of extremal area in the theory of the de Haas–van Alphen effect.

[24] In general, an unwarranted assumption. There are other mechanisms for sound attenuation. See, for example, Chapter 25.

[25] A. B. Pippard, *Phil. Trans. Roy. Soc.* **A250**, 325 (1957).

[26] See, for example, J. D. Jackson, *Classical Electrodynamics*, Wiley, New York, 1962, p. 225.

microwave reflectivity are now determined entirely by certain features of the Fermi surface geometry that depend only on the orientation of the Fermi surface with respect to the actual surface of the sample.

Cyclotron Resonance

This technique also exploits the attenuation of a microwave field as it penetrates a metal. Strictly speaking, the method does not measure Fermi surface geometry, but the "cyclotron mass" (12.44), determined by $\partial A/\partial \mathcal{E}$. This is done by observing the frequency at which an electric field resonates with the electronic motion in a uniform magnetic field. High $\omega_c\tau$ is required for the electrons to undergo periodic motion, and the resonance condition $\omega = \omega_c$ is satisfied at microwave frequencies.

Since the field does not penetrate far into the metal, electrons can absorb energy only when they are within a skin depth of the surface.[27] At microwave frequencies and large ω_c one is in the extreme anomalous regime, where the skin depth is quite small compared to the mean free path. Because the dimensions of the electron's real space orbit at the Fermi surface are comparable to the mean free path, the skin depth will also be small compared with the size of the orbit.

These considerations led Azbel' and Kaner[28] to suggest placing the magnetic field parallel to the surface, leading to the geometry shown in Figure 14.8. If the electron

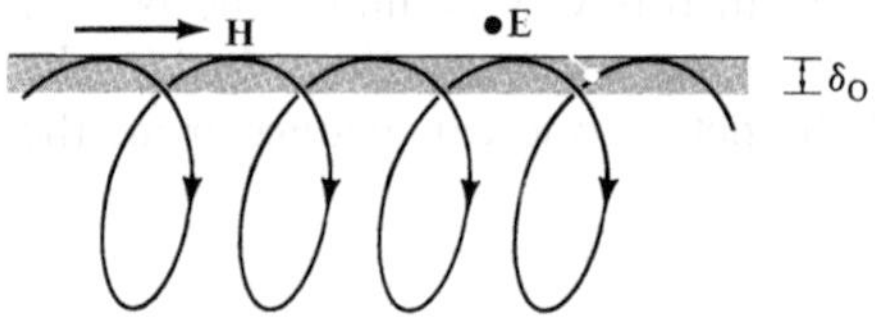

Figure 14.8
Parallel-field Azbel'-Kaner geometry.

experiences an electric field of the same phase each time it enters the skin depth, then it can resonantly absorb energy from the field. This will be the case if the applied field has completed an integral number of periods, T_E, each time the electron returns to the surface:

$$T = nT_E, \tag{14.22}$$

where T is the period of cyclotron motion and n is an integer. Since frequencies are inversely proportional to periods we can write (14.22) as

$$\omega = n\omega_c. \tag{14.23}$$

Usually one works at fixed frequency ω, and varies the strength of the magnetic field H, writing the resonant condition as

$$\frac{1}{H} = \frac{2\pi e}{\hbar^2 c\omega}\frac{1}{\partial A/\partial \mathcal{E}}n. \tag{14.24}$$

Thus if the absorption is plotted vs. $1/H$, resonant peaks due to a given cyclotron period will be uniformly spaced.

---

[27] In semiconductors the electron density is very much lower, a microwave field can penetrate much further, and the technique of cyclotron resonance is much more straightforward. (See Chapter 28.)

[28] M. I. Azbel' and E. A. Kaner, *Sov. Phys. JETP* **3**, 772 (1956).

Analysis of the data is complicated by the question of which orbits are providing the major contributions to the resonance. In the case of an ellipsoidal Fermi surface it can be shown that the cyclotron frequency depends only on the direction of the magnetic field, independent of the height, k_z, of the orbit. The method is therefore quite unambiguous in this case. However, when a continuum of periods is present for a given field direction, as happens whenever $T(\mathcal{E}_F, k_z)$ depends on k_z, some care must be exercised in interpreting the data. As usual, only orbits at the Fermi surface need be considered, for the exclusion principle prohibits electrons in lower-lying orbits from absorbing energy. A quantitative calculation indicates that the orbits at which the cyclotron period $T(\mathcal{E}_F, k_z)$ has its extremal value with respect to k_z are very likely to determine the resonant frequencies. However, the detailed frequency dependence of the energy loss can have quite a complicated structure, and one must be wary of the possibility that one may not always be measuring extremal values of $T(\mathcal{E}_F, k_z)$, but some rather complicated average of T over the Fermi surface. The situation is nowhere near as clearcut as it is in the de Haas–van Alphen effect.

Some typical cyclotron resonance data are shown in Figure 14.9. Note that several extremal periods are involved. The uniform spacing in $1/H$ of all the peaks produced by a single period is of great help in sorting out the rather complex structure.

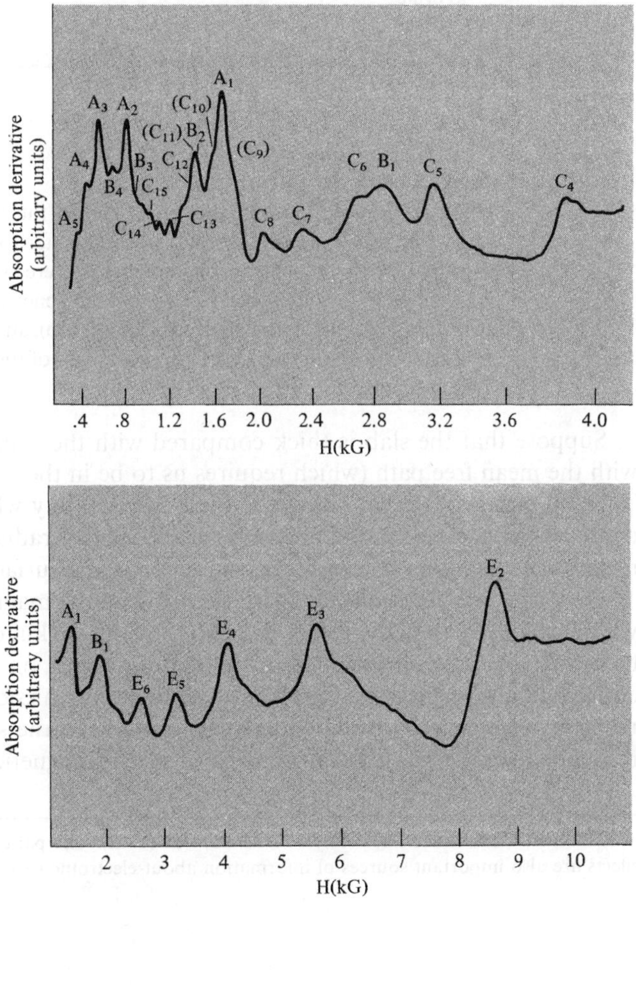

Figure 14.9
Typical cyclotron resonance peaks in aluminum at two different field orientations. Peaks in the field derivative of the absorbed power due to four distinct extremal cyclotron masses can be identified. (Peaks due to the same extremal mass are spaced uniformly in $1/H$, as can be verified by careful examination of the figure.) (T. W. Moore and F. W. Spong, *Phys. Rev.* **125**, 846 (1962).)

Size Effects

Another class of Fermi surface probes works with very thin specimens with parallel plane surfaces, looking for resonant effects produced by those electronic orbits that just fit between the two surfaces. The most straightforward of these is the parallel-field Gantmakher effect,[29] in which a thin plate of metal is placed in a magnetic field parallel to its surface, and exposed to microwave radiation polarized perpendicular to the field (Figure 14.10).

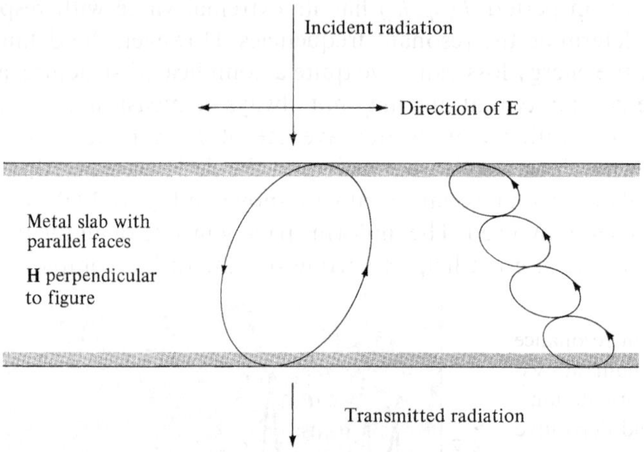

Figure 14.10
The parallel-field Gantmakher effect. When the thickness of the slab coincides with an extremal orbit diameter (or is an integral multiple of an extremal orbit diameter), there will be resonant transmission through the slab. The field can penetrate the slab only to within the skin depth (shaded region on top), and only electrons within a skin depth can reradiate energy out of the metal (shaded region on the bottom).

Suppose that the slab is thick compared with the skin depth, but not compared with the mean free path (which requires us to be in the extreme anomalous regime). Then an electric field can influence the electrons only when they are within a skin depth of the surface, and, conversely, electrons can radiate energy back out of the metal only when they are within a skin depth of the surface.

Now consider those electrons whose orbits in the magnetic field carry them from within a skin depth of the top of the slab to within a skin depth of the bottom. It can be shown that electrons in such orbits can reproduce, on the far side of the slab, the current induced by the driving electric field on the near side, thereby causing electromagnetic energy to be radiated from the far side of the slab. As a result, there is a resonant increase in the transmission of electromagnetic energy through the slab

[29] V. F. Gantmakher, *Sov. Phys. JETP* **15**, 982 (1962). The parallel- and tilted-field Gantmakher effects are also important sources of information about electronic relaxation times.

whenever the thickness and magnetic field are such that orbits can be so matched with the surfaces. Here, again, only electrons near the Fermi surface are effective, since only these are permitted by the exclusion principle to exchange energy with the field. Here, too, only orbits with extremal linear dimensions will contribute to the resonance.

Measurements of the Gantmakher effect are often made in the megahertz region to avoid the complex situation arising when the size resonances are superimposed on the cyclotron resonance frequencies, as might happen in the microwave regime. It is necessary, however, that the frequency be high enough to be in the anomalous regime.

The Fermi surface probes described above, together with a variety of related probes, have now been applied to a large number of metals. The information that has been extracted in this way is surveyed in Chapter 15.

PROBLEMS

1. (a) Show that the Onsager quantization condition (14.13) (with $\lambda = \frac{1}{2}$) applied to the orbits of a free electron leads directly to the free electron levels (14.2).

(b) Show that the degeneracy (14.4) of the free electron levels (14.2) is just the number of zero-field free electron levels with the given k_z, and with k_x and k_y within a planar region of area ΔA (Eq. (14.12)).

2. Using the fundamental relation (14.1), deduce the ratio of the areas of the two extremal orbits responsible for the oscillations in Figure 14.2b.

3. If there is any nonuniformity of the magnetic field over the sample of metal used in a de Haas–van Alphen experiment, then the structure in $g(\mathcal{E})$ will reflect this variation. Different regions will have maxima in $g(\mathcal{E})$ at different field strengths, and the susceptibility, which sums contributions from all regions, may lose its oscillatory structure. To avoid this, any spatial variation δH in the field must lead to a variation $\delta\mathcal{E}_\nu$ that is small compared with $\mathcal{E}_{\nu+1} - \mathcal{E}_\nu$ for the extremal orbits. Using the fact that $\partial A(\mathcal{E}, k_z)/\partial k_z$ vanishes for the extremal orbits, calculate $\partial\mathcal{E}_\nu(k_z)/\partial H$ from (14.13) for an extremal orbit. Deduce from this that to preserve the oscillatory structure the field inhomogeneity must satisfy

$$\frac{\delta H}{H} < \frac{\Delta A}{A}, \qquad \textbf{(14.25)}$$

where ΔA is given in (14.12).

4. (a) Show that in the range of microwave frequencies ($\omega \sim 10^{10}$ sec^{-1}) Eq. (1.33) for the propagation of an electromagnetic wave in a metal reduces to

$$-\nabla^2\mathbf{E} = \left(\frac{4\pi i\sigma\omega}{c^2}\right)\mathbf{E}. \qquad \textbf{(14.26)}$$

(b) Deduce from this the expression (14.21) for the classical skin depth.

(c) Why is this analysis incorrect when the field varies appreciably over a mean free path? (*Hint:* It is necessary to reexamine Drude's derivation of Ohm's law.)

15
Band Structure of Selected Metals

In this chapter we describe some of the better understood features of the band structures of specific metals, as deduced from experiment through techniques such as those described in Chapter 14. Our primary aim is simply to illustrate the great range and variety of band structures possessed by the metallic elements. When, however, a particular feature of its band structure is strikingly reflected in the physical properties of a metal, we shall point this out. In particular, we shall note examples of Fermi surfaces that afford clearcut illustrations of the influence of band structure on transport properties, as discussed in Chapters 12 and 13, as well as noting some of the more straightforward examples of how band structure can affect specific heats and optical properties.

THE MONOVALENT METALS

The monovalent metals have the simplest of all Fermi surfaces. They fall into two classes, the alkali metals and the noble metals, whose atomic structures and crystal structures are shown in Table 15.1.

Table 15.1
THE MONOVALENT METALS

ALKALI METALS (BODY-CENTERED CUBIC)[a]	NOBLE METALS (FACE-CENTERED CUBIC)
Li: $1s^2 2s^1$	—
Na: [Ne]$3s^1$	—
K: [Ar]$4s^1$	Cu: [Ar]$3d^{10}4s^1$
Rb: [Kr]$5s^1$	Ag: [Kr]$4d^{10}5s^1$
Cs: [Xe]$6s^1$	Au: [Xe]$4f^{14}5d^{10}6s^1$

[a] The Fermi surface of lithium is not well known because it has a so-called martensitic transformation to a mixture of crystalline phases at 77 K. Thus the bcc phase only exists at temperatures too high to observe the de Haas–van Alphen effect, and the low-temperature phase lacks the crystallinity necessary for a de Haas–van Alphen study. Sodium undergoes a similar martensitic transformation at 23 K, but with care the transformation can be partially inhibited, and good de Haas–van Alphen data on the bcc phase have been obtained. (We have also omitted the first and the last occupants of column IA of the periodic table from the list of alkali metals: Solid hydrogen is an insulator (and therefore not a monatomic Bravais lattice), though a metallic phase at very high pressures has been conjectured; and francium is radioactive, with a very short half-life.)

The Fermi surfaces of these metals are known with great precision (with the exception of lithium) and enclose a volume of k-space that accommodates just one electron per atom. All the bands are completely filled or empty except for a single half-filled conduction band. Of the two groups the noble metals are the more complicated. Their Fermi surfaces have a more complex topology and the influence on their properties of the filled d-band can be pronounced.

The Alkali Metals

The alkali metals have singly charged ions (whose core electrons form the tightly bound rare gas configuration and therefore give rise to very low-lying, very narrow, filled, tight-binding bands) outside of which a single conduction electron moves. If we treated the conduction electrons in the metal as completely free, then the Fermi surface would be a sphere of radius k_F, given by (see Eq. (2.21))

$$\frac{k_F{}^3}{3\pi^2} = n = \frac{2}{a^3}, \tag{15.1}$$

where a is the side of the conventional cubic cell (the bcc Bravais lattice has two atoms per conventional cell). In units of $2\pi/a$ (half the length of the side of the conventional cubic cell of the fcc reciprocal lattice) we can write:

$$k_F = \left(\frac{3}{4\pi}\right)^{1/3}\left(\frac{2\pi}{a}\right) = 0.620\left(\frac{2\pi}{a}\right). \tag{15.2}$$

The shortest distance from the center of the zone to a zone face (Figure 15.1) is

$$\Gamma\text{N} = \frac{2\pi}{a}\sqrt{(\tfrac{1}{2})^2 + (\tfrac{1}{2})^2 + 0^2} = 0.707\left(\frac{2\pi}{a}\right). \tag{15.3}$$

Therefore the free electron sphere is entirely contained within the first zone, approaching it most closely in the direction ΓN, where it reaches a fraction $k_F/\Gamma\text{N} = 0.877$ of the way to the zone face.

De Haas–van Alphen measurements of the Fermi surface confirm this free electron picture to a remarkable degree of precision, especially in Na and K, where deviations in k_F from the free electron value are at most a few parts in a thousand.[1] The deviations of these Fermi surfaces from perfect spheres are shown in Figure 15.1, which reveals both how small these deviations are, and how very precisely they are known.

Thus the alkalis furnish a spectacular example of the accuracy of the Sommerfeld free electron model. It would be wrong to conclude from this, however, that the effective crystalline potential is minute in the alkali metals. What it does suggest is that the weak pseudopotential method (Chapter 11) is well suited to describing the conduction electrons in the alkalis. Furthermore, even the pseudopotential need not be minute, for except near Bragg planes the deviation from free electron behavior occurs only to second order in the perturbing potential (Problem 5; see also Chapter 9). As a result, it can be shown that band gaps as large as an electron volt at the Bragg planes are still consistent with the nearly spherical Fermi surfaces (see Figure 15.3).

---

[1] The difficulty in observing such minute changes in the de Haas–van Alphen period as the orientation of the crystal is changed is neatly circumvented by working at constant magnetic field, and observing the change in susceptibility as the orientation of the crystal varies. Typical data are illustrated in Figure 15.2. The peak-to-peak distance is now associated with a change in extremal area of ΔA, which is typically $10^{-4}A$. Therefore, quite precise information can be extracted.

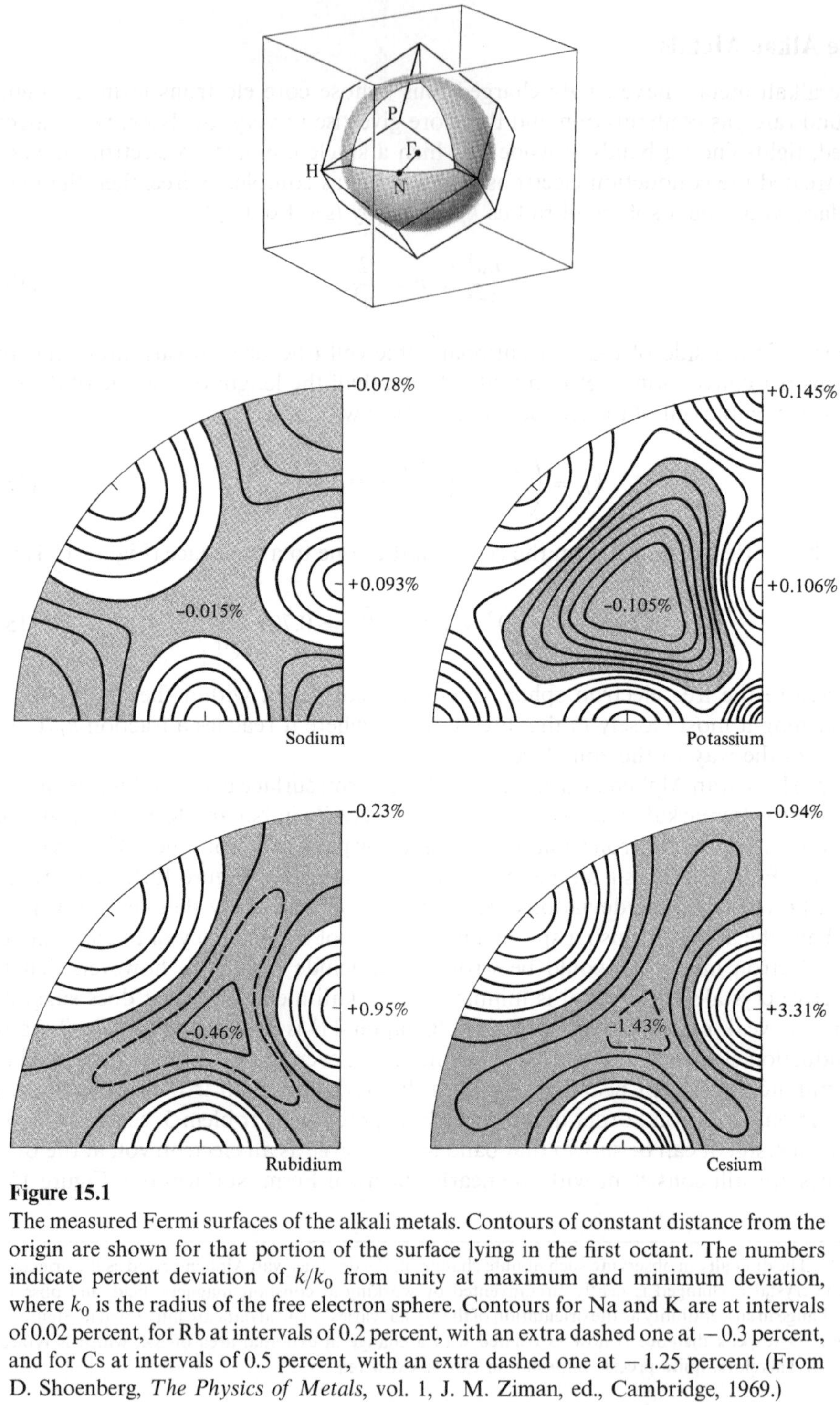

Figure 15.1
The measured Fermi surfaces of the alkali metals. Contours of constant distance from the origin are shown for that portion of the surface lying in the first octant. The numbers indicate percent deviation of k/k_0 from unity at maximum and minimum deviation, where k_0 is the radius of the free electron sphere. Contours for Na and K are at intervals of 0.02 percent, for Rb at intervals of 0.2 percent, with an extra dashed one at -0.3 percent, and for Cs at intervals of 0.5 percent, with an extra dashed one at -1.25 percent. (From D. Shoenberg, *The Physics of Metals*, vol. 1, J. M. Ziman, ed., Cambridge, 1969.)

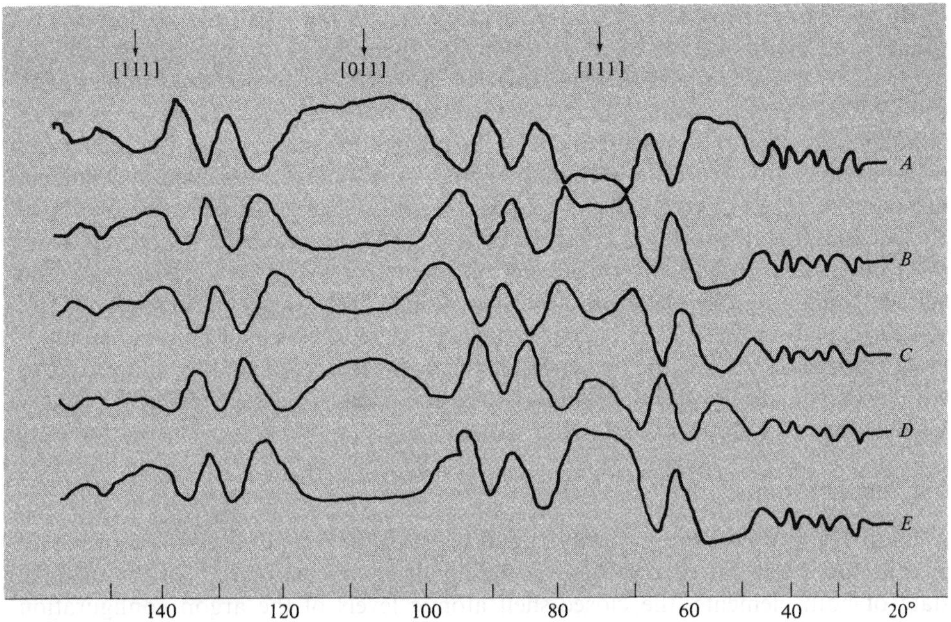

Figure 15.2
De Haas–van Alphen oscillations produced by rotating a potassium crystal in a fixed magnetic field. (From D. Shoenberg, *Low Temperature Physics LT9*, Plenum Press, New York, 1965.)

The alkali metals are unique in possessing nearly spherical Fermi surfaces lying entirely inside a single Brillouin zone. Because of this property the detailed semiclassical analysis of Chapter 12 reduces to the simple Sommerfeld free electron theory of Chapter 2 when applied to the transport properties of the alkalis. Since the analysis of free electrons is simpler than for general Bloch electrons, the alkalis provide a most valuable testing ground for studying various aspects of electronic behavior in metals, free of the formidable analytical complications imposed by band structure.

Figure 15.3
Illustrating that a fairly substantial energy gap at the Bragg plane (N) is possible, even though the bands are indistinguishable from free electron bands at $k_F = 0.877\ \Gamma\mathrm{N}$.

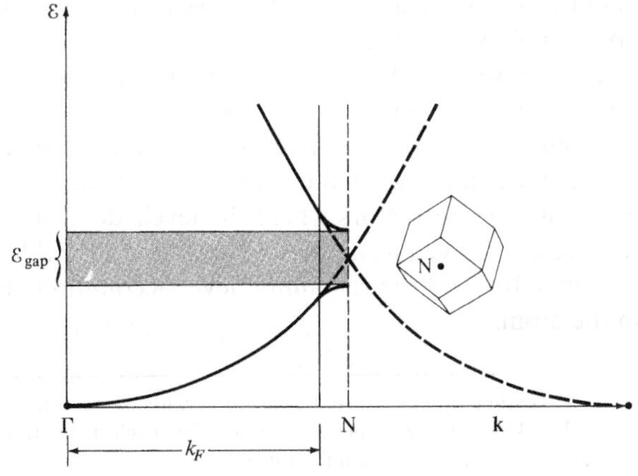

By and large the observed transport properties of the alkali metals[2] agree reasonably well with the observed sphericity of their Fermi surfaces—i.e., with the predictions of free electron theory. However, it is difficult to prepare samples sufficiently free of crystalline defects to test this with any stringency. Thus, for example, although it is clear that measurements of the magnetoresistance show it to be far less field-dependent in the alkalis than in many other metals, the field-independent behavior at large $\omega_c\tau$ required for spherical Fermi surfaces has yet to be observed, and measured Hall constants are still reported which deviate by a few percent from the value $-1/nec$ required by free electron theory (or by any simple closed Fermi surface containing one electronic level per atom). These discrepancies have led to some speculation that the electronic structure of the alkalis may be more complex than described here, but the evidence for this is far from compelling, and as of this writing the widely prevailing opinion is that the alkali metals do indeed have very nearly spherical Fermi surfaces.

The Noble Metals

A comparison of potassium ($[\mathrm{Ar}]4s^1$) and copper ($[\mathrm{Ar}]3d^{10}4s^1$) reveals the characteristic important differences between the alkali and noble metals. In the metallic state of both elements, the closed-shell atomic levels of the argon configuration ($1s^22s^22p^63s^23p^6$) give rise to very tightly bound bands, lying well below the energies of any of the remaining electronic levels in the metal. The electrons in these low-lying levels can be considered as part of the (for most purposes) inert ion cores, and the remaining bands can be constructed by considering either a bcc Bravais lattice of K^+ ion cores to which is added one electron per primitive cell, or an fcc Bravais lattice of Cu^{11+} ions to which are added eleven electrons ($3d^{10}4s^1$) per primitive cell.

In the case of potassium (and the other alkalis) the extra electron is accommodated by filling half of a band that is quite free electron-like, resulting in the very nearly spherical Fermi surfaces described above.

In the case of copper (and the other noble metals[3]) at least six bands are required (and six turn out to be enough) to accommodate the eleven additional electrons. Their structure is shown in Figure 15.4. For almost all wave vectors $\mathbf{k}$ the six bands can be seen to separate into five lying in a relatively narrow range of energies from about 2 to 5 eV below $\mathcal{E}_F$, and a sixth, with an energy anywhere from about 7 eV above to 9 eV below $\mathcal{E}_F$.

It is conventional to refer to the set of five narrow bands as the d-bands, and the remaining set of levels as the s-band. However, these designations must be used cautiously, since at some values of $\mathbf{k}$ all six levels are close together, and the distinction between d-band and s-band levels is not meaningful. The nomenclature reflects the fact that at wave vectors where the levels do clearly group into sets of five and one, the five are derived from the five orbital atomic d-levels, in the sense of tight binding (Chapter 10), and the remaining level accommodates what would be the $4s$ electron in the atom.

---

[2] Excepting lithium, whose Fermi surface, for the reasons mentioned in Table 15.1, is poorly known.

[3] The $4f$ band in gold lies low enough for its electrons to be considered part of the ion core, along with all those from the Xe configuration.

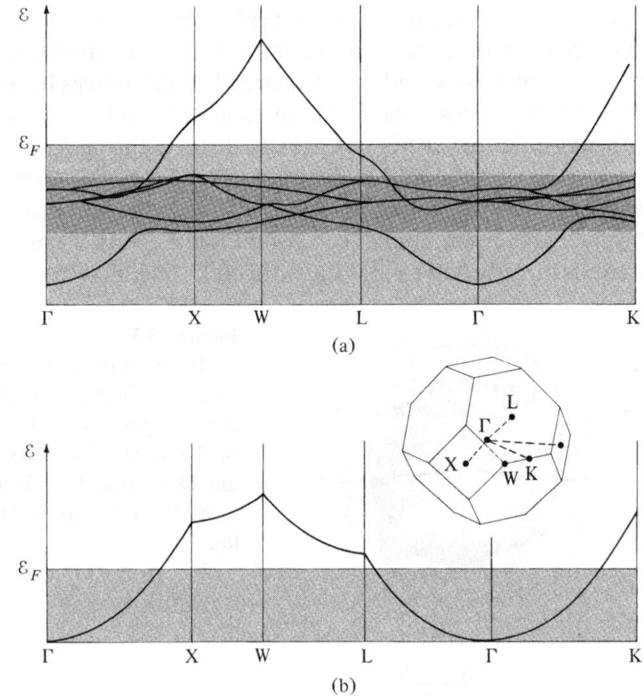

Figure 15.4
(a) Calculated energy bands in copper. (After G. A. Burdick, *Phys. Rev.* **129**, 138 (1963).) The ℰ vs. k curves are shown along several lines in the interior and on the surface of the first zone. (The point Γ is at the center of the zone.) The d-bands occupy the darkest region of the figure, whose width is about 3.5 eV. (b) The lowest-lying free electron energies along the same lines as in (a). (The energy scales in (a) and (b) are not the same.)

Note that the **k** dependence of the s-band levels, except where they approach the d-bands, bears a remarkable resemblance to the lowest free electron band for an fcc crystal (plotted in Figure 15.4b for comparison), especially if one allows for the expected modifications near the zone faces characteristic of a nearly free electron calculation (Chapter 9). Note also that the Fermi level lies far enough above the d-band for the s-band to intersect $ℰ_F$ at points where the resemblance to the free electron band is still quite recognizable.[4] Thus the calculated band structure indicates that for purposes of Fermi surface determination one might still hope for some success with a nearly free electron calculation. However, one must always keep in mind that not too far below the Fermi energy lurks a very complex set of d-bands, which can be expected to influence the metallic properties far more strongly than do any of the filled bands in the alkali metals.[5]

The Fermi surface for a single half-filled free electron band in an fcc Bravais lattice is a sphere entirely contained within the first Brillouin zone, approaching the surface of the zone most closely in the $\langle 111 \rangle$ directions, where it reaches 0.903 of the distance from the origin to the center of the hexagonal face. The de Haas–van Alphen

---

[4] However, the Fermi level is close enough to the d-band to make the s-band nomenclature somewhat dubious for conduction band levels on the Fermi surface. A more precise specification of how s-like or d-like a level is must be based on a detailed examination of its wave function. In this sense most, but by no means all, levels at the Fermi surface are s-like.

[5] The atomic ionization potentials provide a convenient reminder of the different roles played by filled bands in the alkali and noble metals. To remove the first (4s) and then the second (3p) electron from atomic potassium requires 4.34 and 31.81 eV, respectively. The corresponding figures for copper are 7.72 eV (4s) and 20.29 eV (3d).

effect in all three noble metals reveals that their Fermi surfaces are closely related to the free electron sphere; however, in the ⟨111⟩ directions contact is actually made with the zone faces, and the measured Fermi surfaces have the shape shown in Figure 15.5. Eight "necks" reach out to touch the eight hexagonal faces of the zone, but otherwise the surface is not grossly distorted from spherical. The existence of these necks is most strikingly evident in the de Haas–van Alphen oscillations for magnetic fields in the ⟨111⟩ directions, which contain two periods, determined by the extremal "belly" (maximum) and "neck" (minimum) orbits (Figure 15.6). The ratio of the two periods directly determines the ratio of the maximal to minimal ⟨111⟩ cross sections:[6]

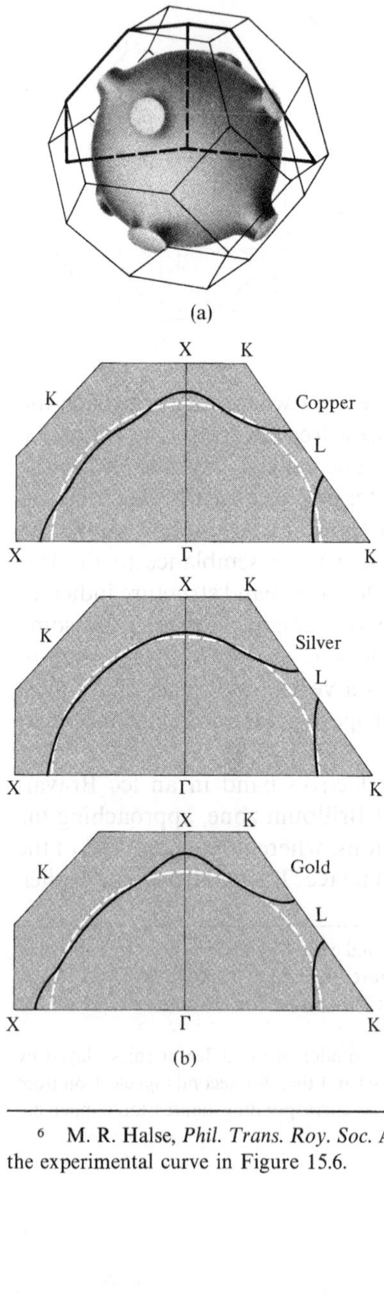

Figure 15.5
(a) In the three noble metals the free electron sphere bulges out in the ⟨111⟩ directions to make contact with the hexagonal zone faces. (b) Detailed cross sections of the surface for the separate metals. (D. Shoenberg and D. J. Roaf, *Phil. Trans. Roy. Soc.* **255**, 85 (1962).) The cross sections may be identified by a comparison with (a).

[6] M. R. Halse, *Phil. Trans. Roy. Soc.* **A265**, 507 (1969). The entry for silver can be read directly from the experimental curve in Figure 15.6.

METAL	A_{111} (BELLY)/A_{111} (NECK)
Cu	27
Ag	51
Au	29

Although a distorted sphere, bulging out to make contact with the hexagonal zone faces, is still a fairly simple structure, when viewed in the repeated-zone scheme the noble metal Fermi surface reveals a variety of exceedingly complex orbits. Some of the simplest are shown in Figure 15.7. The open orbits are responsible for the very dramatic behavior of the magnetoresistance of the noble metals (Figure 15.8), whose

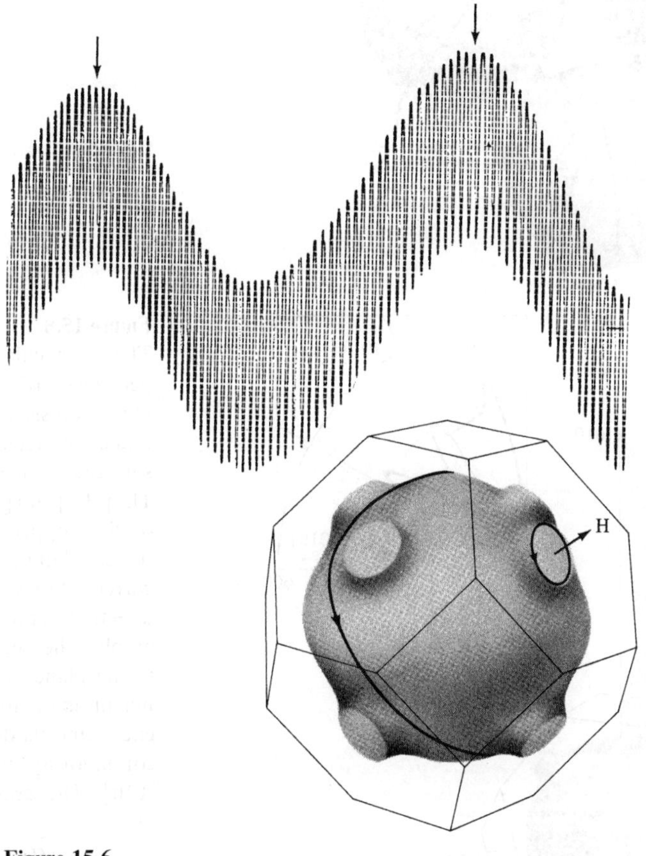

Figure 15.6
De Haas-van Alphen oscillations in silver. (Courtesy of A. S. Joseph.) The magnetic field is along a $\langle 111 \rangle$ direction. The two distinct periods are due to the neck and belly orbits indicated in the inset, the high-frequency oscillations coming from the larger belly orbit. By merely counting the number of high-frequency periods in a single low-frequency period (i.e., between the two arrows) one deduces directly that A_{111}(belly)/A_{111}(neck) = 51. (Note that it is not necessary to know either the vertical or horizontal scales of the graph to determine this fundamental piece of geometrical information!)

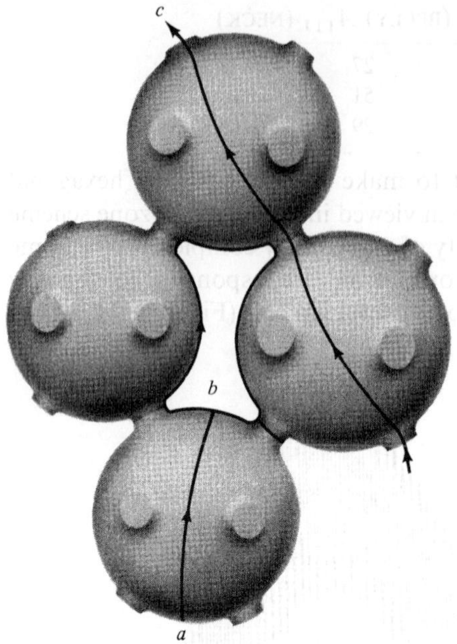

Figure 15.7
Indicating only a few of the surprisingly many types of orbits an electron can pursue in k-space when a uniform magnetic field is applied to a noble metal. (Recall that the orbits are given by slicing the Fermi surface with planes perpendicular to the field.) The figure displays (a) a closed particle orbit; (b) a closed hole orbit; (c) an open orbit, which continues in the same general direction indefinitely in the repeated-zone scheme.

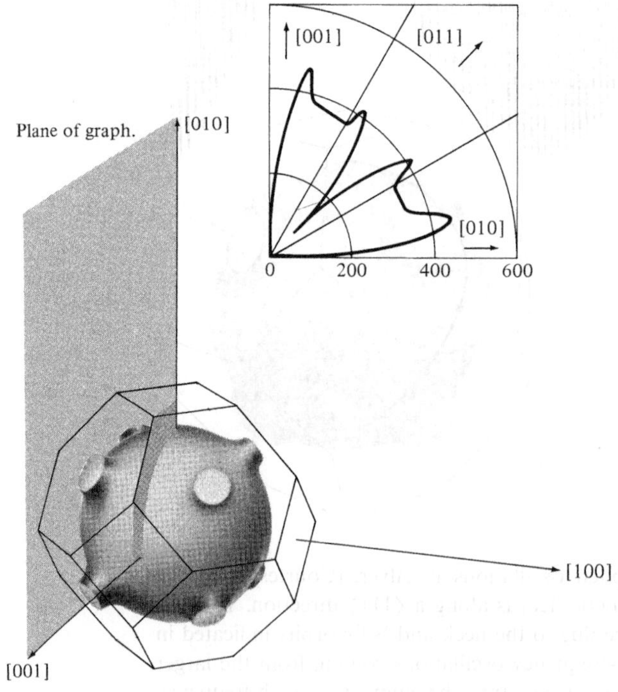

Figure 15.8
The spectacular direction dependence of the high-field magnetoresistance in copper that is characteristic of a Fermi surface supporting open orbits. The [001] and [010] directions of the copper crystal are as indicated in the figure, and the current flows in the [100] direction perpendicular to the graph. The magnetic field is in the plane of the graph. Its magnitude is fixed at 18 kilogauss, and its direction varied continuously from [001] to [010]. The graph is a polar plot of

$$\frac{\rho(H) - \rho(0)}{\rho(0)}$$

vs. orientation of the field. The sample is very pure and the temperature very low (4.2 K—the temperature of liquid helium) to insure the highest possible value for $\omega_c\tau$. (J. R. Klauder and J. E. Kunzler, *The Fermi Surface*, Harrison and Webb, eds., Wiley, New York, 1960.)

failure to saturate in certain directions is very neatly explained by the semiclassical theory (see pages 234–239).

Although the topology of the noble metal Fermi surfaces can lead to very complex transport properties, the Fermi surfaces have but a single branch, and therefore like the alkali metals the noble metals can be treated as one-band metals in analyses of their transport properties. All other known Fermi surfaces of metallic elements have more than one branch.

Because, however, of the very shallow *d*-bands, a one-band model is quite likely to be inadequate in explaining effects requiring more than a semiclassical analysis. The *d*-bands are revealed especially unambiguously in the optical properties of the noble metals.

Optical Properties of the Monovalent Metals

The color of a metal is determined by the frequency dependence of its reflectivity: Some frequencies are reflected more strongly than others. The very different colors of copper, gold, and aluminum indicate that this frequency dependence can vary strikingly from one metal to another.

In turn, the reflectivity of a metal is determined by its frequency-dependent conductivity, through one of the standard calculations of electromagnetic theory (Appendix K). Substituting the free electron form (1.29) into Eq. (K.6) yields a reflectivity in which properties of the specific metal appear only through the plasma frequency and the relaxation time. This free electron reflectivity lacks the structure necessary to account for the characteristic thresholds that appear in the reflectivities of real metals, as well as the striking variations from one metal to another.

Abrupt changes in the reflectivity are caused by the onset of new mechanisms for the absorption of energy. The free electron model gives a relatively structureless reflectivity because collisions provide the only mechanism for energy absorption. Incident radiation simply accelerates free electrons, and if there were no collisions the electrons would radiate back all the energy so acquired in the form of transmitted and reflected radiation. Since there is no transmission below the plasma frequency (see page 18 and also Problem 2), all radiation would be perfectly reflected in the absence of collisions. Above the plasma frequency, transmission is possible, and the reflectivity declines. The only effect of collisions on this is to round the sharp transition from perfect to partial reflection. Because of collisions some of the energy acquired by the electrons from the incident radiation is degraded into thermal energy (of the ions or impurities, for example), thereby diminishing the amount of reflected energy both above and below the plasma frequency. Because collisions have this effect at all frequencies, they introduce no striking frequency-dependent structure into the reflectivity.

For Bloch electrons the situation is quite different. A strongly frequency-dependent mechanism for absorbing incident energy is possible, which is most simply understood by regarding the incident radiation as a beam of photons of energy $\hbar\omega$ and momentum $\hbar\mathbf{q}$. A photon may lose energy by exciting an electron from a level with energy $\mathcal{E}$ to one with energy $\mathcal{E}' = \mathcal{E} + \hbar\omega$. In the free electron case momentum conservation imposes the additional constraint $\mathbf{p}' = \mathbf{p} + \hbar\mathbf{q}$, which proves impossible to satisfy (Problem 3), thereby prohibiting this type of energy loss. In the presence of a periodic potential, however, the translational symmetry of free space is broken, and momentum conservation does not hold. Nevertheless, a weaker conservation law is still in force because of the remaining translational symmetry of the periodic potential. This restricts the change in electron wave vector in a manner reminiscent of momentum conservation:

$$\mathbf{k}' = \mathbf{k} + \mathbf{q} + \mathbf{K}, \tag{15.4}$$

where $\mathbf{K}$ is a vector of the reciprocal lattice.

Equation (15.4) is a special case of "crystal momentum conservation," which is discussed in detail in Appendix M. Here, we only note that (15.4) is a highly plausible modification of the momentum conservation law satisfied in free space, since the electronic levels in a periodic potential, although not the single plane-wave levels of free space, can still be represented as superpositions of plane waves all of whose wave vectors differ only by vectors of the reciprocal lattice (see, for example, Eq. (8.42)).

Since a photon of visible light has a wavelength of order 5000 Å, the photon wave vector $\mathbf{q}$ is typically of order 10^5 cm^{-1}. Typical Brillouin zone dimensions, on the other hand, are of order $k_F \approx 10^8$ cm^{-1}. Thus the term $\mathbf{q}$ in (15.4) can shift the wave vector $\mathbf{k}$ by only a fraction of a percent of the dimensions of the Brillouin zone. Because two levels in the same band whose wave vectors differ by a reciprocal lattice vector are, in fact, identical, the shift by $\mathbf{K}$ can be ignored altogether, and we reach the important conclusion that the wave vector of a Bloch electron is essentially unchanged when it absorbs a photon.

For the electron's energy to change by $\hbar\omega$, typically a few electron volts, the electron must move from one band to another without appreciable change in wave vector. Such processes are known as interband transitions.[7] They can occur as soon as $\hbar\omega$ exceeds $\mathcal{E}_{n'}(\mathbf{k}) - \mathcal{E}_n(\mathbf{k})$ for some $\mathbf{k}$, and for two bands n and n', where $\mathcal{E}_n(\mathbf{k})$ is below the Fermi level (so that such an electron is available for excitation) and $\mathcal{E}_{n'}(\mathbf{k})$ is above the Fermi level (so that the final electronic level is not made unavailable by the Pauli principle). This critical energy or frequency is called the interband threshold.[8]

The interband threshold may be due either to the excitation of electrons from the conduction band (highest band containing some electrons) into higher unoccupied levels, or to the excitation of electrons from filled bands into unoccupied levels in the conduction band (lowest band containing some unoccupied levels).

In the alkali metals the filled bands lie far below the conduction band, and the excitation of conduction band electrons to higher levels gives the interband threshold. Since the Fermi surface in the alkali metals is so close to a free electron sphere, the bands above the conduction band are also quite close to free electron bands, especially

---

[7] More precisely, they are known as *direct interband transitions*. In general the analysis of optical data is complicated by the further possibility of *indirect interband transitions*, in which the electronic wave vector $\mathbf{k}$ is not conserved, the missing crystal momentum being carried away by a quantized lattice vibration or phonon. Since *phonon* energies are very much less than optical *photon* energies in the monovalent metals (Chapters 23 and 24), our general conclusions are not very sensitive to the possibility of indirect transitions, and we shall ignore them. They cannot be ignored, however, in a more precise quantitative theory.

[8] Interband transitions are explicitly prohibited in the semiclassical model of Chapters 12 and 13 by condition (12.10). When the frequency becomes comparable to the interband threshold, the semiclassical AC conductivity (13.34) must be used cautiously, if at all, since corrections coming from the more general form (13.37) may be quite important.

for values of $\mathbf{k}$ within the Fermi "sphere," which does not reach all the way to the zone faces. A free electron estimate of the threshold energy $\hbar\omega$ follows from observing that the occupied conduction band levels with energies closest to the next highest free electron levels at the same $\mathbf{k}$ occur at points on the Fermi sphere nearest to a Bragg plane; i.e., at points (Figure 15.1) where the Fermi sphere meets the lines ΓN. As a result, the interband threshold is

$$\hbar\omega = \frac{\hbar^2}{2m}(2k_0 - k_F)^2 - \frac{\hbar^2}{2m}k_F{}^2. \tag{15.5}$$

Here k_0 is the length of the line ΓN from the center of the zone to the midpoint of one of the zone faces (Figure 15.9), and satisfies (see page 285) $k_F = 0.877k_0$. If k_0 is expressed in terms of k_F, Eq. (15.5) gives

$$\hbar\omega = 0.64\mathcal{E}_F. \tag{15.6}$$

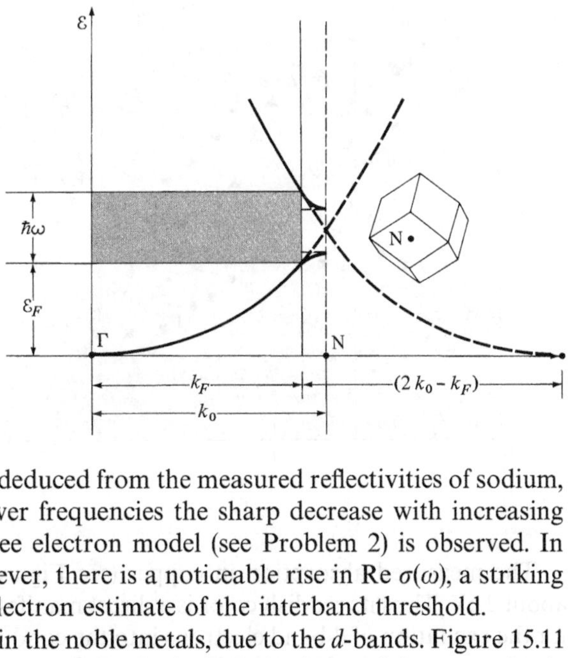

Figure 15.9
Free electron determination of the threshold energy for interband absorption in the alkali metals. Numerically, $\hbar\omega = 0.64\mathcal{E}_F$.

Figure 15.10 shows Re $\sigma(\omega)$ as deduced from the measured reflectivities of sodium, potassium, and rubidium. At lower frequencies the sharp decrease with increasing frequency characteristic of the free electron model (see Problem 2) is observed. In the neighborhood of $0.64\mathcal{E}_F$, however, there is a noticeable rise in Re $\sigma(\omega)$, a striking confirmation of the nearly free electron estimate of the interband threshold.

The situation is quite different in the noble metals, due to the d-bands. Figure 15.11 shows the computed band structure of Cu, including the lowest-lying completely empty bands. Note that these are also recognizable distortions of the free electron bands displayed below them. The threshold for exciting an electron up from the conduction band occurs at point b (which is where the Fermi surface "neck" meets the hexagonal zone face (Figure 15.5a)) with an energy proportional to the length of the upper vertical arrow—about 4 eV.

However, d-band electrons can be excited into unoccupied conduction band levels with considerably less energy than this. Such a transition occurs at the same point b, with an energy difference proportional to the length of the lower vertical arrow—about 2 eV. Another, somewhat lower, transition occurs at point a.

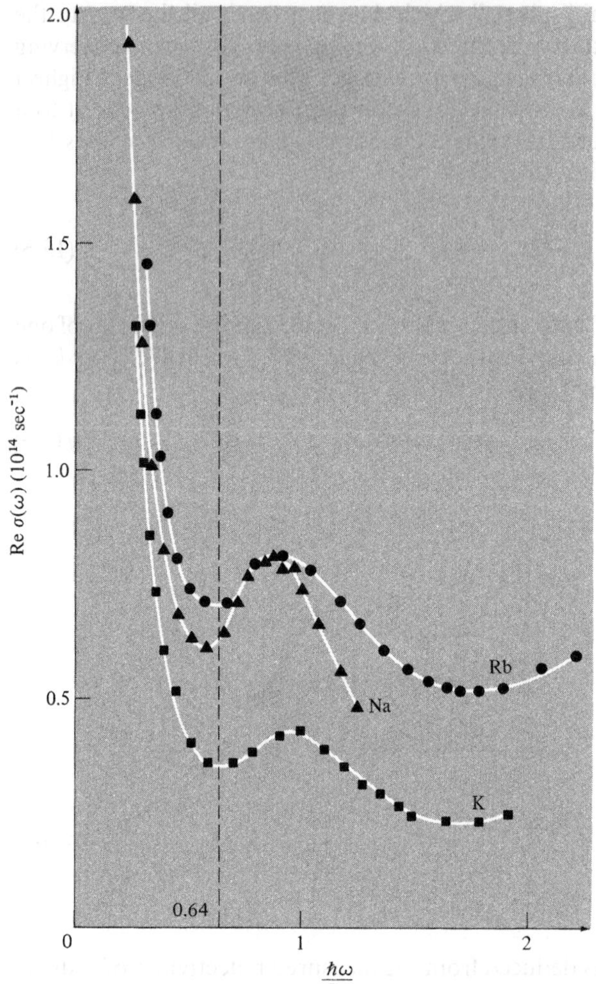

Figure 15.10
Re $\sigma(\omega)$, deduced from reflectivity measurements in three alkali metals. The interband threshold is evident, and occurs quite close to $0.64\mathcal{E}_F$, where $\mathcal{E}_F$ is the free electron Fermi energy given in Table 2.1, page 38. (Courtesy of N. Smith.)

The measured absorption in copper (see Figure 15.12) does increase sharply at about 2 eV. Thus its reddish color is a direct manifestation of the rather low threshold for the excitation of *d*-band electrons into the conduction band, 2eV lying somewhere in the orange part of the visible spectrum.[9]

The inference of band structure from optical properties remains simple in some of the polyvalent metals,[10] but it can be far from straightforward in others. Frequently, for example, there are points on the Fermi surface where the conduction band is degenerate with the next higher band, leading to interband transitions at arbitrarily low energies and the absence of any sharp interband threshold.

[9] A threshold at about the same energy also produces the yellowish color of gold. Silver, however, is more complicated: the threshold for *d*-band excitation and a plasmon-like threshold apparently merge at about 4 eV (Figure 15.12), resulting in a more uniform reflectivity throughout the visible range (about 2 to 4 eV).

[10] See, for example, the discussion of aluminum below.

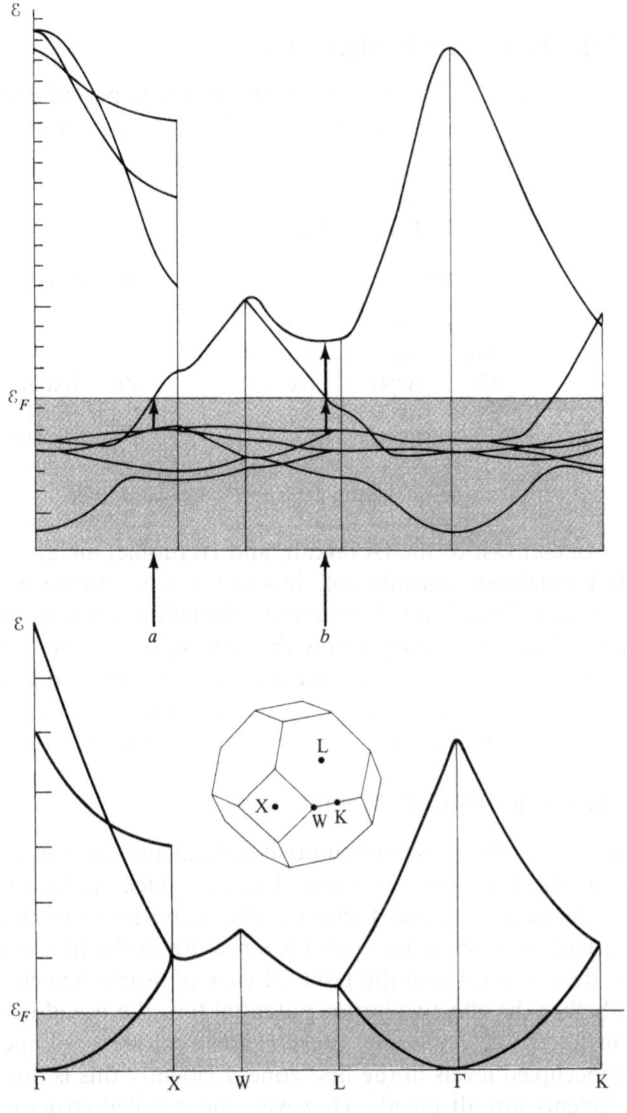

Figure 15.11
Burdick's calculated bands for copper, illustrating that the absorption threshold for transitions up from the conduction band is about 4 eV, while the threshold for transitions from the *d*-band to the conduction band is only about 2 eV. (The energy scale is in tenths of a rydberg (0.1 Ry = 1.36 eV).) Note the resemblance of the bands other than the *d*-bands to the free electron bands plotted below.

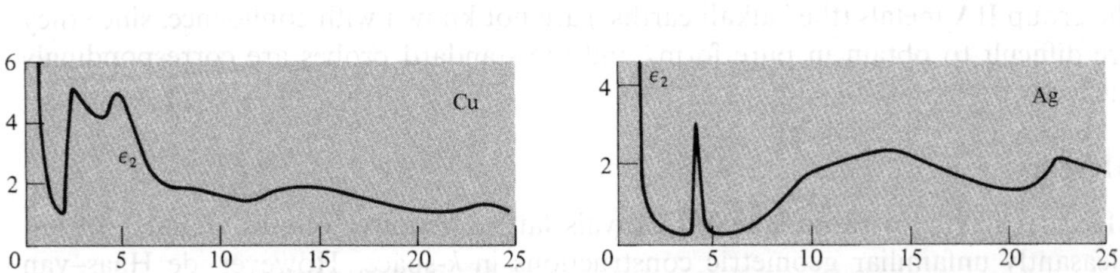

Figure 15.12
The imaginary part of the dielectric constant, $\epsilon_2(\omega) = \text{Im}\,\epsilon(\omega)$ vs. $\hbar\omega$, as deduced from reflectivity measurements. (H. Ehrenreich and H. R. Phillip, *Phys. Rev.* **128**, 1622 (1962).) Note the characteristic free electron behavior ($1/\omega^3$) below about 2 eV in copper and below about 4 eV in silver. The onset of interband absorption is quite apparent.

THE DIVALENT METALS

The divalent metals lie in the columns of the periodic table immediately to the right of the alkali and noble metals. Their electronic structure and crystal structures are given in Table 15.2.

Table 15.2
DIVALENT METALS

IIA METALS			IIB METALS		
Be:	$1s^2 2s^2$	hcp			
Mg:	[Ne]$3s^2$	hcp			
Ca:	[Ar]$4s^2$	fcc	Zn:	[Ar]$3d^{10}4s^2$	hcp
Sr:	[Kr]$5s^2$	fcc	Cd:	[Kr]$4d^{10}5s^2$	hcp
Ba:	[Xe]$6s^2$	bcc	Hg:	[Xe]$4f^{14}5d^{10}6s^2$	*

* Rhombohedral monatomic Bravais lattice.

In contrast to the IA (alkali) and IB (noble) metals, the properties of the IIA and IIB metals are considerably less drastically affected by the presence or absence of the filled *d*-band. Band structure calculations indicate that in zinc and cadmium the *d*-band lies completely below the bottom of the conduction band, while in mercury it overlaps the conduction band only in a narrow region quite near the bottom. As a result, the *d*-bands are relatively inert, and the variation in metallic properties with crystal structure is far more striking than the variation from column IIA to IIB.

The Cubic Divalent Metals

With two electrons per primitive cell, calcium, stronium, and barium could, in principle, be insulators. In the free electron model, the Fermi sphere has the same volume as the first zone and therefore intersects the zone faces. The free electron Fermi surface is thus a fairly complex structure in the first zone, and pockets of electrons in the second. From the point of view of nearly free electron theory, the question is whether the effective lattice potential (i.e., the pseudopotential) is strong enough to shrink the second-zone pockets down to zero volume, thereby filling up all the unoccupied levels in the first zone. Evidently this is not the case, since the group II elements are all metals. However, the detailed structures of the Fermi surfaces in the group IIA metals (the "alkali earths") are not known with confidence, since they are difficult to obtain in pure forms and the standard probes are correspondingly ineffective.

Mercury

Mercury, having a rhombohedral Bravais lattice, requires one to embark on unpleasantly unfamiliar geometric constructions in *k*-space. However, de Haas–van Alphen measurements have been made[11] indicating pockets of electrons in the second zone and a complex extended figure in the first.

---

[11] G. B. Brandt and J. A. Rayne, *Phys. Rev.* **148**, 644 (1966).

The Hexagonal Divalent Metals

Good de Haas–van Alphen data are available for beryllium, magnesium, zinc, and cadmium. The data suggest Fermi surfaces that are more or less recognizable distortions of the (extremely complex) structure found by simply drawing a free electron sphere containing four levels per primitive hexagonal cell (remember that the hcp structure has *two* atoms per primitive cell) and seeing how it is sliced up by the Bragg planes. This is illustrated in Figure 9.11 for the "ideal" ratio[12] $c/a = 1.633$.

A complication characteristic of all hcp metals arises from the vanishing of the structure factor on the hexagonal faces of the first zone, in the absence of spin-orbit coupling (page 169). It follows that a weak periodic potential (or pseudopotential) will not produce a first-order splitting in the free electron bands at these faces. This fact transcends the nearly free electron approximation: Quite generally, if spin-orbit coupling is neglected, there must be at least a twofold degeneracy on these faces. As a result, to the extent that spin-orbit coupling is small (as it is in the lighter elements) it is better to omit these Bragg planes in constructing the distorted free electron Fermi surface, leading to the rather simpler structures shown in Figure 9.12. Which picture is the more accurate depends on the size of the gaps induced by the spin-orbit coupling. It may happen that the gaps have such a size that the representation of Figure 9.11 is valid for the analysis of low-field galvanomagnetic data, while at high fields the probability of magnetic breakthrough at the gaps is large enough that the representation of Figure 9.12 is more appropriate.

This complication makes it rather difficult to disentangle de Haas–van Alphen data in hexagonal metals. Beryllium (with very weak spin-orbit coupling) has perhaps the simplest Fermi surface (Figure 15.13). The "coronet" encloses holes and the (two) "cigars" enclose electrons, so that beryllium furnishes a simple, if topologically grotesque, example of a compensated metal.

THE TRIVALENT METALS

Family resemblances diminish still further among the trivalent metals, and we consider only the simplest, aluminum.[13]

Aluminum

The Fermi surface of aluminum is very close to the free electron surface for a face-centered cubic monatomic Bravais lattice with three conduction electrons per atom,

---

[12] Be and Mg have c/a ratios close to the ideal value, but Zn and Cd have a c/a ratio about 15 percent larger.

[13] Boron is a semiconductor. The crystal structure of gallium (complex orthorhombic) leads to a free electron Fermi surface extending into the ninth zone. Indium has a centered tetragonal lattice that can be regarded as fcc, slightly stretched along one cube axis, and many of its electronic properties are recognizable distortions of those of aluminum. Thallium is the heaviest hcp metal, and therefore the one with strongest spin-orbit coupling. Its Fermi surface appears to resemble the free electron surface of Figure 9.11, in which the splittings on the hexagonal faces are retained (in contrast to beryllium, the lightest hcp metal).

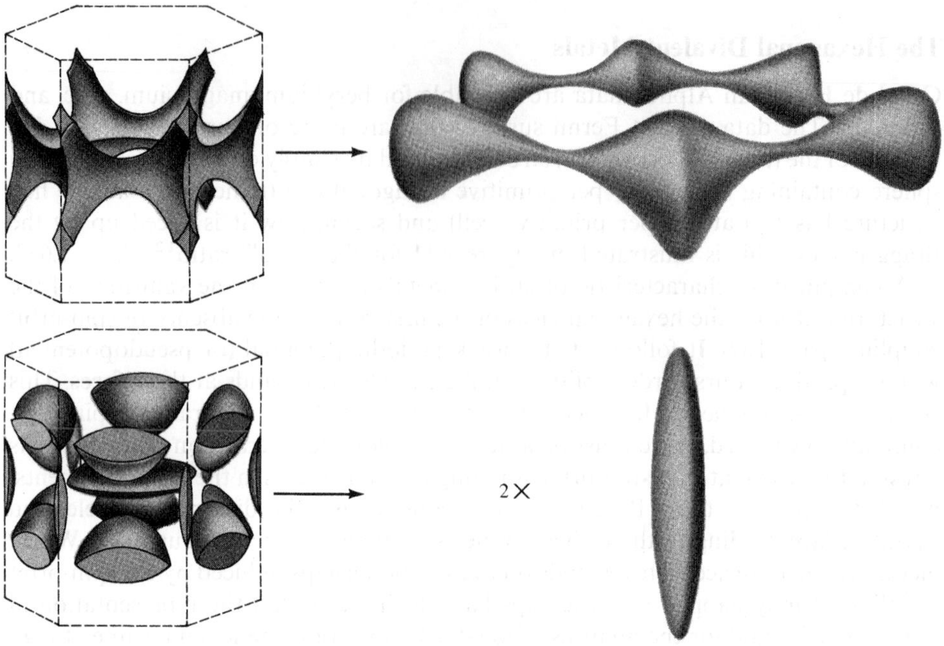

Figure 15.13
The measured Fermi surface of beryllium. (T. L. Loucks and P. H. Cutler, *Phys. Rev.* A **133**, 819 (1964).) The free electron "monster" (upper left) shrinks to a "coronet" (upper right) and all the other free electron pieces (lower left) vanish except for two "cigars" (lower right). The coronet encloses unoccupied levels, and the cigars contain electrons.

pictured in Figure 15.14. One can verify (Problem 4) that the free electron Fermi surface is entirely contained in the second, third, and fourth zones (Figure 15.14c). When displayed in a reduced-zone scheme the second-zone surface (Figure 15.14d) is a closed structure containing unoccupied levels, while the third-zone surface (Figure 15.14e) is a complex structure of narrow tubes. The amount of surface in the fourth zone is very small, enclosing tiny pockets of occupied levels.

The effect of a weak periodic potential is to eliminate the fourth-zone pockets of electrons, and reduce the third-zone surface to a set of disconnected "rings" (Figure 15.15). This is consistent with the de Haas–van Alphen data, which reveal no fourth-zone electron pockets and give the dimensions of the second- and third-zone surfaces quite precisely.

Aluminum provides a striking illustration of the semiclassical theory of Hall coefficients. The high-field Hall coefficient should be $R_H = -1/(n_e - n_h)ec$, where n_e and n_h are the number of levels per unit volume enclosed by the particle-like and hole-like branches of the Fermi surface. Since the first zone of aluminum is completely filled and accommodates two electrons per atom, one of the three valence electrons per atom remains to occupy second- and third-zone levels. Thus

$$n_e^{\mathrm{II}} + n_e^{\mathrm{III}} = \frac{n}{3}, \tag{15.7}$$

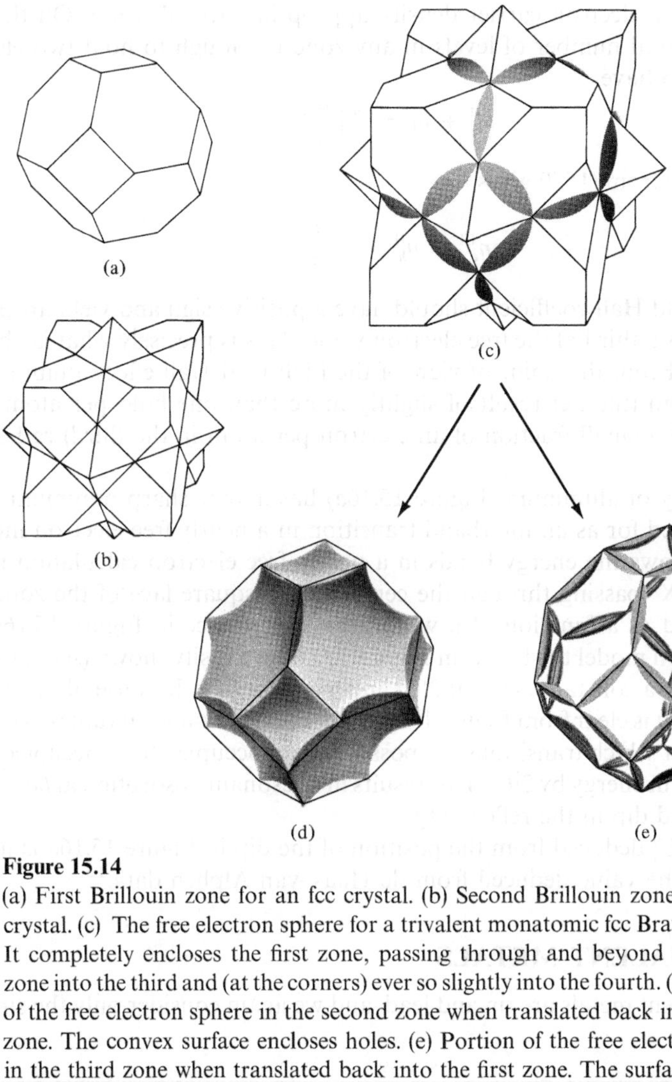

Figure 15.14
(a) First Brillouin zone for an fcc crystal. (b) Second Brillouin zone for an fcc crystal. (c) The free electron sphere for a trivalent monatomic fcc Bravais lattice. It completely encloses the first zone, passing through and beyond the second zone into the third and (at the corners) ever so slightly into the fourth. (d) Portion of the free electron sphere in the second zone when translated back into the first zone. The convex surface encloses holes. (e) Portion of the free electron sphere in the third zone when translated back into the first zone. The surface encloses particles. (The fourth-zone surface translates into microscopic pockets of electrons at all corner points.) (From R. Lück, doctoral dissertation, Technische Hochschule, Stuttgart, 1965.)

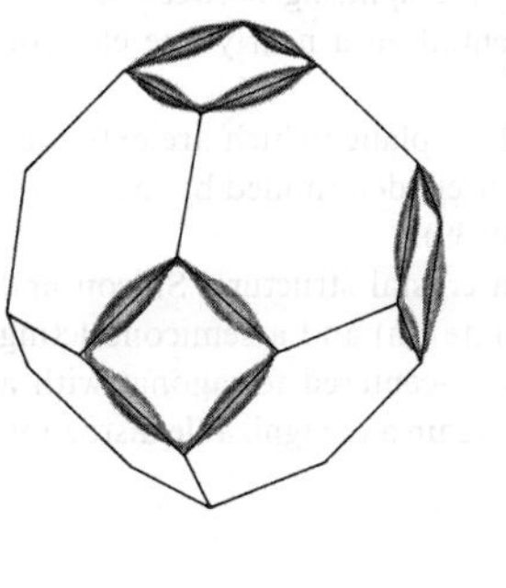

Figure 15.15
The third-zone surface of aluminum, in a reduced-zone scheme. (From N. W. Ashcroft, *Phil. Mag.* **8**, 2055 (1963).)

where n is the free electron carrier density appropriate to valence 3. On the other hand, since the total number of levels in any zone is enough to hold two electrons per atom, we also have

$$n_e^{\mathrm{II}} + n_h^{\mathrm{II}} = 2\left(\frac{n}{3}\right). \tag{15.8}$$

Subtracting (15.8) from (15.7) gives

$$n_e^{\mathrm{III}} - n_h^{\mathrm{II}} = -\frac{n}{3}. \tag{15.9}$$

Thus the high-field Hall coefficient should have a positive sign and yield an effective density of carriers a third of the free electron value. This is precisely what is observed (see Figure 1.4). From the point of view of the high-field Hall effect aluminum has one hole per atom (the net result of slightly more than one hole per atom in the second zone, and a small fraction of an electron per atom in the third) rather than three electrons.

The reflectivity of aluminum (Figure 15.16a) has a very sharp minimum, which is neatly accounted for as an interband transition in a nearly free electron model.[14] Figure 15.16b shows the energy bands in a nearly free electron calculation plotted along the line ΓX (passing through the center of the square face of the zone). The bands are plotted as a function of **k** within the square face in Figure 15.16c. In a nearly free electron model the bands in Figure 15.16c are easily shown (see Eq. (9.27)) to be displaced by a constant amount $2|U|$, independent of **k**. Because of the position of the Fermi level it is clear from Figure 15.16c that there is a range of values of **k** within the square face for which transitions are possible from occupied to unoccupied levels, all of which differ in energy by $2|U|$. This results in a resonant absorption at $\hbar\omega = 2|U|$, and a pronounced dip in the reflectivity.

The value of $|U|$ deduced from the position of the dip in Figure 15.16a is in good agreement with the value deduced from de Haas–van Alphen data.[15]

THE TETRAVALENT METALS

The only tetravalent metals are tin and lead, and we again consider only the simplest, lead.[16]

---

[14] The interband thresholds in the alkali metals were explained in terms of an essentially free electron model; i.e., it was unnecessary to take into account any of the distortions in the free electron bands produced by the lattice potential. The example discussed here is at the next level of complexity: the pertinent transition takes place between levels whose wave vectors lie in a Bragg plane, and the splitting between them is therefore entirely due to the first-order perturbation of the periodic potential, in a nearly free electron model.

[15] In the nearly free electron model the cross-sectional areas at a Bragg plane (which are extremal, and therefore directly accessible from the de Haas–van Alphen data) are entirely determined by the matrix element of the periodic potential $|U|$ associated with that plane. See Eq. (9.39).

[16] Carbon is an insulator or a semimetal (see below), depending on crystal structure. Silicon and germanium are semiconductors (Chapter 28). Tin has both a metallic (white tin) and a semiconducting (grey tin) phase. Grey tin has the diamond structure, but white tin is body-centered tetragonal with a two-atom basis. The Fermi surface has been measured and calculated and is again a recognizable distortion of the free electron surface.

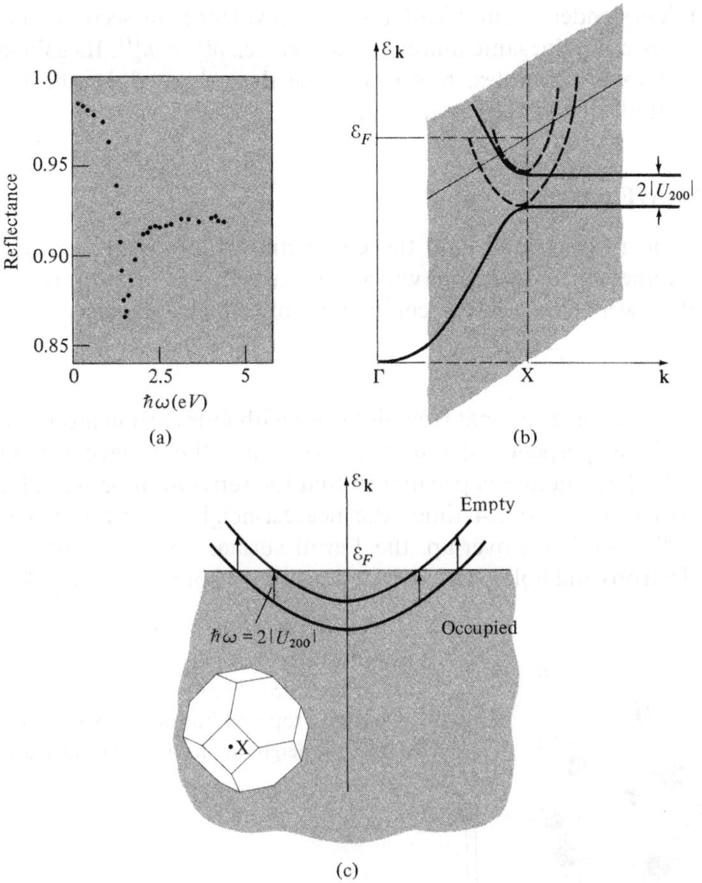

Figure 15.16
(a) Reflectivity of aluminum in the energy range $0 \leqslant \hbar\omega \leqslant 5$ eV. (H. E. Bennet, M. Silver, and E. J. Ashley, *J. Opt. Soc. Am.* **53**, 1089 (1963).) (b) Energy bands plotted along ΓX, and also in the square zone face perpendicular to ΓX (dashed). (c) A second view of the energy bands in the square zone face perpendicular to ΓX. For a weak pseudopotential these bands are nearly parallel and separated by an amount $2|U|$. When $\hbar\omega$ just exceeds this amount it becomes possible for electrons within $\hbar\omega$ of the Fermi energy to be excited from the lower to the upper band. This is the source of the structure in (a). (See N. W. Ashcroft and K. Sturm, *Phys. Rev. B* **3**, 1898 (1971).)

Lead

Like aluminum, lead has an fcc Bravais lattice and its free electron Fermi surface is rather similar, except that the sphere must have one third more volume and hence 10 percent greater radius, to accommodate four electrons per atom (see Figure 9.9). The fourth-zone electron pockets are therefore rather larger than in aluminum, but are still apparently eliminated by the crystal potential. The hole surface in the second zone is smaller than in aluminum, and the extended tubular particle surface in the third zone is less slender.[17] Since lead has an even valence, the second- and third-zone surfaces must contain the same number of levels; i.e., $n_h^{II} = n_e^{III}$. Its galvanomagnetic properties are rather complex, however, since the orbits of the third zone Fermi surface are not all of a single carrier type.

THE SEMIMETALS

The graphite form of carbon, and the conducting pentavalent elements are semimetals.[18] Semimetals are metals in which the carrier concentration is several orders of magnitude lower than the $10^{22}/\text{cm}^3$ typical of ordinary metals.

Graphite

Graphite has a simple hexagonal Bravais lattice with four carbon atoms per primitive cell. Lattice planes perpendicular to the c-axis have the honeycomb arrangement (Figure 15.17). The structure is peculiar in that the separation between lattice planes along the c-axis is almost 2.4 times the nearest-neighbor distance within planes. There is hardly any band overlap, the Fermi surface consisting primarily of tiny pockets of electrons and holes, with carrier densities of about $n_e = n_h = 3 \times 10^{18}/\text{cm}^3$.

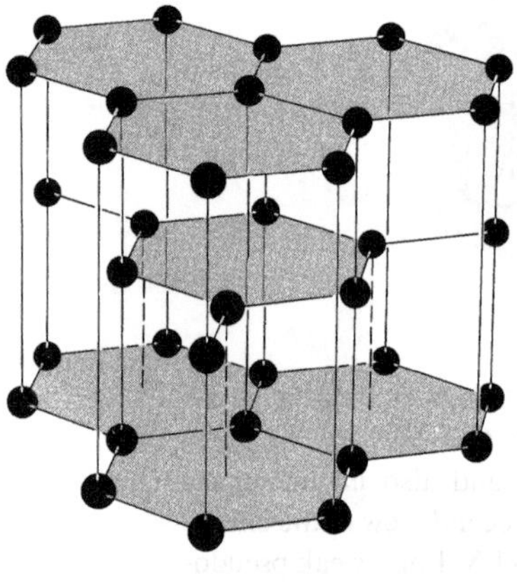

Figure 15.17
Crystal structure of graphite (not to scale). The distance between the top and bottom planes is almost 4.8 times the nearest-neighbor distance within planes.

The Pentavalent Semimetals

The noninsulating pentavalent elements, As ($[\text{Ar}]3d^{10}4s^2 4p^3$), Sb ($[\text{Kr}]4d^{10}5s^2 5p^3$), and Bi ($[\text{Xe}]4f^{14}5d^{10}6s^2 6p^3$) are also semimetals. All three have the same crystal structure: a rhombohedral Bravais lattice with a two-atom basis, as described in

[17] Since lead is a very heavy metal it is important to take into account spin-orbit coupling in computing the Fermi surface. See E. Fawcett, *Phys. Rev, Lett.* **6**, 534 (1961).

[18] Semimetals should not be confused with semiconductors. A pure semimetal at $T = 0$ is a conductor: there are partially filled electron and hole bands. A semiconductor, however, conducts only because carriers are either thermally excited or introduced by impurities. A pure semiconductor at $T = 0$ is an insulator (see Chapter 28).

Table 7.5. Having an even number of conduction electrons per primitive rhombohedral cell, they come very close to being insulators, but there is a slight band overlap, leading to a very small number of carriers. The Fermi surface of bismuth consists of several quite eccentric, ellipsoidal-shaped pockets of electrons and holes; the total density of electrons (and the total density of holes—these are compensated semimetals) is about $3 \times 10^{17}/\text{cm}^3$—down from typical metallic densities by about a factor of 10^5. Similar pockets have been observed in antimony, though apparently not so perfectly ellipsoidal, and with larger electron (and hole) densities—about $5 \times 10^{19}/\text{cm}^3$. In arsenic the common electron or hole density is $2 \times 10^{20}/\text{cm}^3$. The pockets are still less ellipsoidal, the hole pockets apparently being connected by thin "tubes" leading to an extended surface.[19]

These low carrier densities explain why the pentavalent metals provide such glaring exceptions to the data tabulated in Chapters 1 and 2 in rough support of free electron theory. Small pockets of carriers imply little Fermi surface area and hence a small density of levels at the Fermi energy. This is why[20] the linear term in the heat capacity of bismuth is only about 5 percent of the naive free electron value for a pentavalent element, and in antimony only about 35 percent (see Table 2.3). The resistivity of bismuth is typically 10 to 100 times larger than that of most metals, and in antimony it is about 3 to 30 times larger (see Table 1.2).

It is interesting to note that the crystal structure of bismuth (and the other two semimetals) is only a slight distortion of a simple cubic monatomic Bravais lattice, for it can be constructed as follows: Take a sodium chloride structure (Figure 4.24), stretch it out slightly along the (111) direction (so that the cube axes make equal angles of somewhat less than 90° with each other), and displace each chlorine site very slightly by the same amount in the (111) direction. The bismuth structure has one bismuth atom at each of the resulting sodium *and* chlorine sites.

As a result, the pentavalent semimetals provide a striking illustration of the crucial importance of crystal structure in determining metallic properties. Were they exactly simple cubic Bravais lattices, then, having an odd valence, they would be very good metals indeed. Thus the band gaps introduced by a very slight deviation from simple cubic change the effective number of carriers by a factor as large as 10^5!

THE TRANSITION METALS

The three rows of the periodic table extending from the alkali earths (calcium, strontium, and barium) to the noble metals (copper, silver, and gold) each contain nine transition elements, in which the *d*-shell that is empty in the alkali earths, and completely filled in the noble metals, is gradually filled in. The stable room temperature forms of the transition elements are either monatomic fcc or bcc Bravais lattices, or hcp structures. All are metals, but unlike the metals we have described up to now

---

19 See M. G. Priestley et al., *Phys. Rev.* **154**, 671 (1967).

20 In understanding these deviations from free electron theory it is also important to note that the effective masses in the pentavalent semimetals are generally substantially smaller than the free electron mass; as a result the disparity in the conductivity is not so great as the disparity in the number of carriers would suggest (since their velocities are higher for a given **k** than the velocity of a free electron).

(the noble metals and the so-called simple metals) their properties are to a considerable degree dominated by the *d*-electrons.

Calculated transition metal band structures indicate that the *d*-band not only lies high up in the conduction band (as in the noble metals), but in general (unlike the noble metals) extends through the Fermi energy. When levels on the Fermi surface are *d*-derived levels, the tight-binding approximation is probably a more sound conceptual starting point for estimating the Fermi surface than the nearly free electron (or OPW) constructions, and there is no longer any reason to expect the transition metal Fermi surfaces to resemble slightly distorted free electron spheres. A typical example, a suggested Fermi surface for bcc tungsten ($[Xe]4f^{14}5d^4 6s^2$), is shown in Figure 15.18.

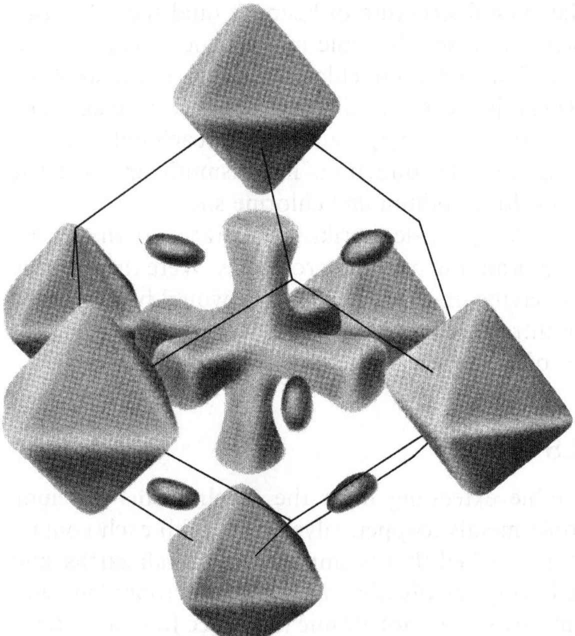

Figure 15.18
Proposed Fermi surface for bcc tungsten. The six octahedron-shaped pockets at the zone corners contain holes. They are all equivalent; i.e., any one can be taken into any other by a translation through a reciprocal lattice vector, so that all physically distinct levels in the group are contained in any one of them. The twelve smaller pockets in the centers of the zone faces (only five are visible) are also hole pockets. They are equivalent in pairs (from opposite faces). The structure in the center is an electron pocket. Tungsten has an even number of electrons, and is therefore a compensated metal. It therefore follows that the volume of a large hole pocket plus six times the volume of a small hole pocket is equal to the volume of the electron pocket at the center of the zone. Consistent with a Fermi surface composed entirely of closed pockets, the magnetoresistance has been observed to increase quadratically with H for all field directions, as predicted for a compensated metal without open orbits. Note that the surface, unlike those considered earlier, is not a distortion of the free electron surface. This is a consequence of the Fermi level lying within the *d*-band, and is characteristic of transition metals. (After A. V. Gold, as quoted in D. Shoenberg, *The Physics of Metals—*1. *Electrons*, J. M. Ziman, ed., Cambridge, 1969, p. 112.)

The d-bands are narrower than typical free electron conduction bands, and contain enough levels to accommodate ten electrons. Since the d-bands contain more levels in a narrower energy range, the density of levels is likely to be substantially higher than the free electron density of levels throughout the energy region where the d-band lies (see Figure 15.19). This effect can be observed in the electronic contribution to the low temperature specific heat. This was shown in Chapter 2 to be proportional to the density of levels at the Fermi energy (Eq. (2.80)).[21] An inspection of Table 2.3 reveals that the transition metal electronic specific heats are indeed significantly higher than those of the simple metals.[22,23]

Figure 15.19
Some qualitative features of the d-band and s-band contributions to the density of levels of a transition metal. The d-band is narrower and contains more levels than the s-band. Consequently when the Fermi level (separating the shaded and unshaded regions) lies within the d-band, the density of levels $g(\mathcal{E}_F)$ is very much larger than the free-electronlike contribution of the s-band alone. (An actual density of levels would have sharp kinks in it; see the description of van Hove singularities on page 145.) (From J. M. Ziman, *Electrons and Phonons*, Oxford, New York, 1960.)

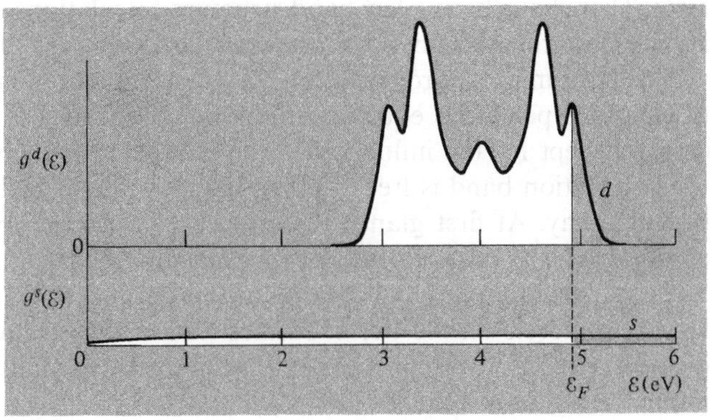

Studies of the transition metals are complicated by the fact that the partially filled d-bands can give rise to striking magnetic properties. Consequently, a far more subtle treatment of electron spin interactions is required than is found in any of the methods we have described. These points will be discussed further in Chapter 32.

The de Haas–van Alphen effect is also more difficult to measure in transition metals, since narrow bands lead to large values of $\partial A/\partial \mathcal{E}$ and therefore to low cyclotron frequencies (Eq. (12.42)). The high $\omega_c\tau$ regime is therefore more difficult to reach. Despite these complexities, de Haas–van Alphen data and the more conventional types of band structure calculations (for what they are worth) are now available for

---

[21] The derivation of Eq. (2.80) did not use specific properties of the free electron level density, and the result is equally valid for Bloch electrons.

[22] Note also that the specific heats of the semimetals are significantly lower, as their very low density of conduction electrons would lead one to expect.

[23] A detailed comparison of Eq. (2.80) with experiment is complicated by the fact that there are corrections due to electron-electron interactions (typically only a few percent) as well as corrections due to electron-phonon interactions (to be discussed in Chapter 26), which can be as large as 100 percent.

more than half of the transition metals, and data have been taken even in ferromagnetic samples.

THE RARE EARTH METALS

Between lanthanum and hafnium lie the rare earth metals. Their atomic configurations are characterized by partially filled 4*f* shells, which, like the partially filled *d*-shells of the transition metals, can lead to a variety of magnetic effects. The typical rare earth atomic configuration is $[Xe]4f^n5d^{(1 \text{ or } 0)}6s^2$. The solids can have many types of crystal structure, but by far the most common room temperature form is hexagonal close-packed.

There is currently very little Fermi surface data on the rare earth metals, for being chemically very similar to one another, they are difficult to obtain in sufficiently pure form. There have been a few band structure calculations, but in the absence of Fermi surface data their reliability is not easily assessed.

The customary approach is to treat the conduction band as containing a number of electrons per atom equal to the nominal chemical valence (three, in almost all cases). Except for the influence of the 5*d* atomic levels (which can be considerable) the conduction band is free electron-like; i.e., the 4*f* levels are not mixed in, in any essential way. At first glance this appears surprising, for one might expect the 4*f* atomic levels to broaden out into a partially filled[24] 4*f* band. Such a band would, like any partially filled band, contain the Fermi level, and therefore, conversely, at least some levels on the Fermi surface would have a strong 4*f* character. This is in obvious analogy to what indeed happens to the partially filled 3*d*-, 4*d*-, or 5*d* levels in the transition metals.

In spite of the analogy this does not occur in the rare earth metals, and levels on the Fermi surface have very little 4*f* character. The crucial difference is that the atomic 4*f* orbitals in the rare earth elements are much more localized than the highest occupied atomic *d*-levels in the transition metal elements. As a result, it seems quite likely that the independent electron approximation breaks down altogether for the 4*f* electrons, since they meet the requisite conditions (pages 186–187) of yielding narrow partially filled bands in a tight-binding analysis. Electron-electron interactions among the 4*f* electrons at each atomic site are in fact strong enough to yield local magnetic moments (Chapter 32).

It is sometimes asserted that the 4*f* band in the rare earths splits into two narrow parts: one fully occupied well below the Fermi level, and the other completely empty, well above (Figure 15.20). This picture has dubious validity, but is probably the best that can be had if one insists on applying the independent electron model to the 4*f* electrons. The gap between the two portions of the 4*f* band is then an attempt to represent the very stable spin configuration attained by the 4*f* electrons in the occupied portion of the band, in which any additional electrons are unable to participate.

However one chooses to describe the 4*f* electrons, it does appear to be the case that they can be considered as part of the ion cores in viewing the rare earth metal band structures, in spite of the fact that the atomic 4*f* shells are only partially filled.

---

[24] In most of the rare earths there are fewer than 14 electrons outside of the [Xe] configuration.

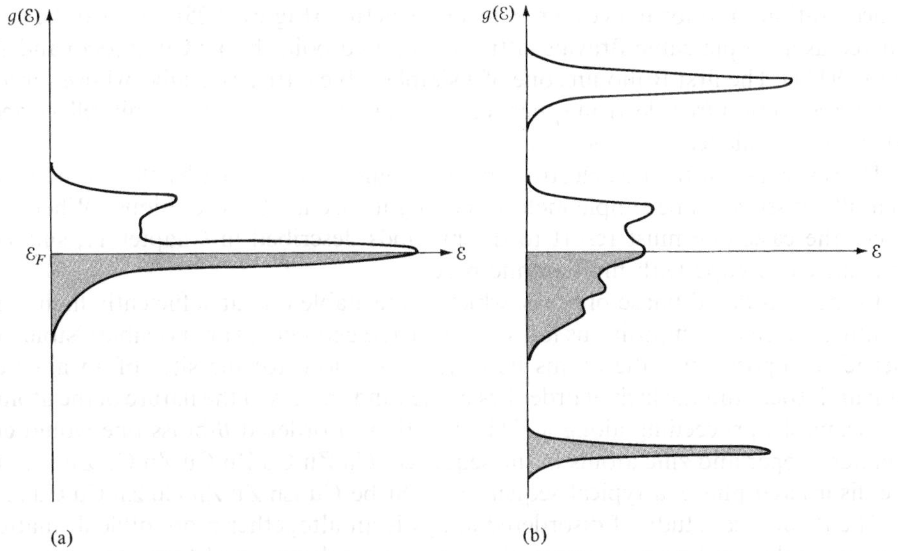

Figure 15.20
Two hypothetical level density curves for a rare earth metal. (a) The incorrect form, which naively superposes on a fairly broad *s-p-d* band a sharp *f*-band peak, at the Fermi energy. (b) The partially correct form, which has a fairly broad *s-p-d* form in the neighborhood of the Fermi energy, and two *f*-band peaks, one well below and one well above the Fermi energy. The most realistic point of view probably abandons the independent electron approximation (and hence the possibility of drawing simple one-electron densities of levels) for the 4*f* electrons.

ALLOYS

We conclude this chapter with the important reminder that the metallic state is not exhausted by a survey of the periodic table. The construction of alloys from the 70 or so elemental metals is an immense subject in its own right. Although there is no guarantee that any two metals will dissolve in each other (indium, for example, will not dissolve in gallium), most pairs do form so-called *binary alloys* over wide ranges of concentration. Ternary (three-component) alloys, tertiary (four-component) alloys, and so on, are also made and studied. Evidently we can construct in this way an immense number of different metals.[25]

Alloys are conveniently grouped into two broad classes: ordered and disordered. The ordered alloys, sometimes also called *stoichiometric*, have the translational symmetry of a Bravais lattice. Their structure is given by placing at each site of the Bravais lattice a multiatomic basis. As an example, the alloy known as β-brass has an ordered[26] phase in which the two components (copper and zinc) have equal

[25] Some alloys, like indium antimonide, are not metals, but semiconductors.

[26] There is also a disordered phase, and a sharp transition temperature, above which the ordering disappears. This order-disorder transition can be discussed in terms of the Ising model (in which copper at a site corresponds to "spin up" and zinc at a site, to "spin down"). See Chapter 33.

concentrations and form a cesium chloride structure (Figure 4.25). This can be considered as a simple cubic Bravais lattice with a two-point basis: Cu at (000) and Zn at $(a/2)(111)$. The first Brillouin zone of a simple cubic lattice is a cube, whose surface is intersected by a free electron sphere containing three electrons per unit cell (copper has nominal valence one, and zinc, two).[27]

However, use of the free electron model should be restricted by the requirement that all constituents be simple metals (making its use for brass dubious). When this is not the case, one must resort to the methods described in Chapter 11, suitably generalized to cope with multiatomic bases.

In the disordered phase of brass, which is the stable one at sufficiently high temperatures, even at compositions for which ordered geometries are possible ("stoichiometric" compositions), the atoms lie at (or very close to) the sites of an abstract Bravais lattice, and the lack of order lies in the randomness of the nature of the atoms. For example, proceeding along a (111) direction in ordered β-brass one would encounter copper and zinc atoms in the sequence: Cu Zn Cu Zn Cu Zn Cu Zn . . . . In the disordered phase, a typical sequence might be Cu Zn Zn Zn Cu Zn Cu Cu . . . .

The theoretical study of disordered alloys is an altogether more difficult matter. Because of the random assignment of atoms to sites there is no Bloch's theorem, and without the quantum number $\mathbf{k}$ one is at a loss to describe any electronic properties at all. On the other hand, such materials are clearly metals, frequently well described by simple Drude model calculations, and their heat capacities display the characteristic electronic contribution we have learned to expect from metals.

A striking difference, however, between disordered alloys and pure metals is that no matter how pure the disordered alloy is made, the substantial decrease in resistance with decreasing temperature characteristic of pure metals is absent. Thus the electrical resistance of the purest disordered brass at liquid-helium temperatures has only fallen to about half its room temperature value (as opposed to a drop of 10^{-4} in carefully prepared ordered metals). This phenomenon makes sense if we view one constituent of the alloy as a (highly concentrated) substitutional impurity in the lattice of the other, for the impurity scattering is then a major (temperature-independent) source of resistance at all temperatures. In very pure metals, in contrast, impurity scattering is only revealed at very low temperatures. Alternatively, one can simply note that in a disordered alloy the periodicity is destroyed, and the semiclassical analysis leading to undegraded currents in the absence of scattering mechanisms is no longer valid.

Thus the problem of the electronic structure of disordered alloys is a very difficult one, still largely unsolved, and of considerable current interest.

PROBLEMS

1. Verify that in a crystal with an fcc monatomic Bravais lattice, the free electron Fermi sphere for valence 1 reaches $(16/3\pi^2)^{1/6} = 0.903$ of the way from the origin to the zone face, in the [111] direction.

[27] See, for example, J. P. Jan, *Can. J. Phys.* **44**, 1787 (1966).

2. Using the general result given in Eq. (K. 6), examine the reflectivity implied by the free electron conductivity (1.29) when $\omega\tau \gg 1$. Show that the reflectivity is unity below the plasma frequency, and that $r = [{\omega_p}^2/4\omega^2]^2$ when $\omega \gg \omega_p$.

3. Prove that energy conservation and momentum conservation make it impossible for a free electron to absorb a photon. (*Note:* If you use the nonrelativistic form $\mathcal{E} = p^2/2m$ for the electronic energy, you will find that such absorption is possible, but only at an electronic energy so high (on the scale of mc^2) that the nonrelativistic approximation is not valid. It is therefore necessary to use the relativistic relation, $\mathcal{E} = (p^2c^2 + m^2c^4)^{1/2}$, to prove that the absorption is impossible at any electronic energy.)

4. The first Brillouin zone for a crystal with an fcc Bravais lattice extends furthest from the origin (Γ) at the point W where the square face and two hexagonal faces meet (Figure 15.4). Show that the free electron Fermi sphere for valence 3 extends beyond that point (specifically, $k_F/\Gamma\mathrm{W} = (1296/125\pi^2)^{1/6} = 1.008$), so that the first Brillouin zone is entirely filled.

5. In the alkali metals the free electron Fermi sphere is entirely contained in the first Brillouin zone, and the weak pseudopotential produces only slight deformations of this sphere, without altering the basic topology (in contrast to the noble metals). Various features of these deformations can be investigated using the methods of Chapter 9.

(a) In a weak periodic potential in the vicinity of a single Bragg plane, the two-plane-wave approximation (pages 156–159) may be used. Let the wave vector $\mathbf{k}$ have polar angles θ and ϕ, with respect to the reciprocal lattice vector $\mathbf{K}$ associated with the Bragg plane. If $\mathcal{E} < (\hbar^2/2m)(K/2)^2$ and $U_\mathbf{K}$ is sufficiently small, show that to order ${U_\mathbf{K}}^2$ the surface of energy $\mathcal{E}$ is given by

$$k(\theta, \phi) = \sqrt{\frac{2m\mathcal{E}}{\hbar^2}}\,(1 + \delta(\theta)), \tag{15.10}$$

where

$$\delta(\theta) = \frac{m|U_\mathbf{K}|^2/\mathcal{E}}{(\hbar K)^2 - 2\hbar K\cos\theta\,\sqrt{2m\mathcal{E}}}. \tag{15.11}$$

(b) Assuming that the result of the single Bragg plane approximation is valid through the zone, show that the shift in Fermi energy due to the weak periodic potential is given by $\mathcal{E}_F - \mathcal{E}_F^0 = \gamma$, where

$$\gamma = -\frac{1}{8}\frac{|U_\mathbf{K}|^2}{\mathcal{E}_F^0}\left(\frac{2k_F}{K}\right)\ln\left|\frac{1 + 2k_F/K}{1 - 2k_F/K}\right|. \tag{15.12}$$

(*Hint:* Note that the Fermi energy satisfies $n = \int (d\mathbf{k}/4\pi^3)\theta(\mathcal{E}_F - \mathcal{E}_\mathbf{k})$, where θ is the step function ($\theta(x) = 1$ for $x > 0$; $\theta(x) = 0$ for $x < 0$), and expand the θ function in γ.)

In the alkali metals the free electron Fermi sphere comes close to 12 Bragg planes, but since it is never close to more than one at a time, the shift in Fermi energy is just that calculated above, multiplied by 12.

16
Beyond the Relaxation-Time Approximation

Sources of Electronic Scattering

Scattering Probability and Relaxation Time

General Description of Collisions

The Boltzmann Equation

Impurity Scattering

Wiedemann-Franz Law

Matthiessen's Rule

Scattering in Isotropic Materials

The general semiclassical theory of conduction in Chapter 13 (as well as the arguments in Chapters 1 and 2) described electronic collisions as random, uncorrelated events that could be treated in a relaxation-time approximation. This approximation assumes that the form of the nonequilibrium electronic distribution function has no effect either on the rate at which a given electron experiences collisions or on the distribution of electrons emerging from collisions.[1]

No attempt was made to justify these assumptions. They are only used because they give the simplest representation of the fact that collisions do take place and are ultimately responsible for thermal equilibrium. Indeed, in detail the assumptions are certainly wrong. The rate at which an electron collides depends critically on the distribution of the other electrons, even in an independent electron approximation, because the Pauli exclusion principle allows an electron to be scattered only into empty electronic levels. Furthermore, the distribution of electrons that emerge from collisions does depend on the electronic distribution function, not only because the exclusion principle limits the available final levels, but also because the net output from collisions depends on the form of the input, which is determined by the distribution function.

Thus considerable restraint must be exercised in drawing conclusions from calculations based on the relaxation-time approximation. In general, such results can be used with confidence only when details of the collision process are clearly of little consequence. For example, the high-frequency ($\omega\tau \gg 1$) conductivity and the high-field ($\omega_c\tau \gg 1$) Hall coefficient are left unchanged by improvements in the relaxation-time approximation, since they describe limiting cases in which the number of collisions per cycle, or per period of revolution in the magnetic field, is vanishingly small.

Results whose validity transcends the relaxation-time approximation usually involve quantities that are independent of τ. The high-field Hall coefficient, for example, is $R_H = -1/nec$. However, one should not assume uncritically that all expressions in which τ fails to appear are valid beyond the relaxation-time approximation. A noteworthy counterexample is the Wiedemann-Franz law, which predicts the universal ratio $(\pi^2/3)(k_B/e)^2T$ for the ratio of thermal to electrical conductivities of a metal, independent[2] of the functional form of $\tau_n(\mathbf{k})$. Nevertheless, when the relaxation-time approximation is not made, we shall see that that the law holds only when the energy of each electron is conserved in each collision.

When specific features of the collision process are of importance, the relaxation-time approximation can still be highly informative, provided that the features are gross properties rather than fine details. Thus in describing semiconductors one often assigns different relaxation times to electrons and holes; i.e., one uses a τ that depends on the band index but not on the wave vector. If there is reason to believe that scattering processes are much more common in one band than the other, then such a simplification may be quite valuable in working out the general implications of such a disparity.

---

1 See pages 244 and 245.

2 See page 255.

However, results that are sensitive to detailed features of the functional form of $\tau_n(\mathbf{k})$ are to be viewed with suspicion. If, for example, one attempts to deduce $\tau(\mathbf{k})$ for a given band from a set of data and a theory based on the relaxation-time approximation, there is no reason to expect that different kinds of experimental determinations of $\tau(\mathbf{k})$ will not yield entirely different functions. The relaxation-time approximation overlooks the fact that the nature of the scattering does depend on the nonequilibrium electronic distribution function, which in general will differ from one experimental situation to another.

In this chapter we shall describe, and contrast to the relaxation-time approximation, the more accurate description of collisions used (except in the cases noted above) for anything more accurate than a rough, qualitative description of conduction.

SOURCES OF ELECTRONIC SCATTERING

How electronic collisions are to be described depends on the particular collision mechanisms of importance. We have already noted[3] the gross inadequacy of Drude's idea of collisions with individual ions. According to the Bloch theory, an electron in a perfectly periodic array of ions experiences no collisions at all. Within the independent electron approximation collisions can only arise from deviations from perfect periodicity. These fall into two broad categories:

1. **Impurities and Crystal Defects** Point defects (Chapter 30)—e.g., missing ions or an occasional ion in the wrong place—behave in much the same way as impurities, providing a localized scattering center. These are probably the simplest kinds of collision mechanisms to describe. In addition, however, there may also be more extended defects in which the lattice periodicity is violated along a line, or even over an entire plane.[4]
2. **Intrinsic Deviations from Periodicity in a Perfect Crystal, due to Thermal Vibrations of the Ions** Even in the absence of impurities or defects the ions do not remain rigidly fixed at the points of an ideal periodic array, for they possess some kinetic energy which increases with increasing temperature (Chapters 21–26). Below the melting temperature this energy is rarely enough to enable the ions to stray very far from their ideal equilibrium positions, and the primary effect of the thermal energy is to cause the ions to undergo small vibrations about these positions. The deviation of the ionic network from perfect periodicity due to these vibrations is the most important source of temperature dependence in the DC resistivity (Chapter 26), and is usually the dominant scattering mechanism at room temperature. As the temperature declines, the amplitude of the ionic vibrations drops, and eventually impurity and defect scattering dominates.

Aside from the scattering mechanisms due to deviations from perfect periodicity, another source of scattering, neglected in the independent electron approximation,

---

[3] See page 141.

[4] In this category we might include surface scattering, which becomes important, for example, in crystals whose dimensions are comparable to a mean free path.

arises from the interactions among the electrons. Electron-electron scattering[5] plays a relatively minor role in the theory of conduction in solids, for reasons to be described in Chapter 17. At high temperatures it is much less important than the scattering by thermal vibrations of the ions, and at low temperatures, except in crystals of exceptional purity and perfection, it is dominated by impurity or defect scattering.

SCATTERING PROBABILITY AND RELAXATION TIME

Instead of making the relaxation-time approximation, more realistic descriptions of collisions assume there is a probability per unit time (to be determined by suitable microscopic calculations) that an electron in band n with wave vector $\mathbf{k}$ will, as a result of a collision, be scattered into band n' with wave vector $\mathbf{k}'$. For simplicity we limit our discussion to a single band,[6] assuming that scattering occurs only within this band ($n' = n$). We also assume that the electron's spin is conserved in the scattering.[7] Finally, we assume that the collisions can be well localized in space and time, so that the collisions occurring at $\mathbf{r}$, t are determined by properties of the solid in the immediate vicinity of $\mathbf{r}$, t. Since all quantities affecting collisions at $\mathbf{r}$, t will then be evaluated at that point, to keep the notation simple we suppress explicit reference to these variables.

The scattering probability is written in terms of a quantity $W_{\mathbf{k},\mathbf{k}'}$, defined as follows: The probability in an infinitesimal time interval dt that an electron with wave vector $\mathbf{k}$ is scattered into any one of the group of levels (with the same spin) contained in the infinitesimal k-space volume element $d\mathbf{k}'$ about $\mathbf{k}'$, assuming that these levels are all unoccupied (and therefore not forbidden by the exclusion principle) is

$$\frac{W_{\mathbf{k},\mathbf{k}'}\,dt\,d\mathbf{k}'}{(2\pi)^3}. \tag{16.1}$$

The particular form taken by $W_{\mathbf{k},\mathbf{k}'}$ depends on the particular scattering mechanism being described. In general, W will have a quite complex structure, and may depend on the distribution function g. We shall consider a particularly simple form W may take below (Eq. (16.14)), but for the moment the points we wish to make depend only on the existence of W and not on its detailed structure.

Given the quantity $W_{\mathbf{k},\mathbf{k}'}$, and given the electronic distribution function g, we can explicitly construct the probability per unit time that an electron with wave vector $\mathbf{k}$ will suffer some collision. This quantity is, by definition, just $1/\tau(\mathbf{k})$, (page 244) and its structure reveals some of the limitations of the relaxation-time approximation. Since $W_{\mathbf{k},\mathbf{k}'}\,d\mathbf{k}'/(2\pi)^3$ is the probability per unit time that an electron with wave vector $\mathbf{k}$ will be scattered into the group of levels (with the same spin) contained in $d\mathbf{k}'$ about $\mathbf{k}'$, given that these levels are all unoccupied, the actual rate of transition must

---

[5] In the kinetic theory of gases, the analogue of electron-electron scattering is the sole source of scattering, aside from collisions with the walls of the container. In this sense the electron gas in a metal is quite unlike an ordinary gas.

[6] It is straightforward to generalize the discussion to cover the possibility of interband scattering. All of the points we wish to make, however, are fully illustrated by the one-band case.

[7] This is not the case when the scattering is due to magnetic impurities, and the lack of spin conservation can lead to some striking effects (pages 687–688).

be reduced by the fraction of these levels that actually are unoccupied (since transitions into occupied levels are forbidden by the exclusion principle). This fraction is just[8] $1 - g(\mathbf{k}')$. The total probability per unit time for a collision is given by summing over all final wave vectors $\mathbf{k}'$:

$$\frac{1}{\tau(\mathbf{k})} = \int \frac{d\mathbf{k}'}{(2\pi)^3} W_{\mathbf{k},\mathbf{k}'}[1 - g(\mathbf{k}')]. \tag{16.2}$$

It is evident from (16.2) that, in contrast to the relaxation time approximation, $\tau(\mathbf{k})$ is not a specified function of $\mathbf{k}$, but depends on the particular form assumed by the nonequilibrium distribution function g.

RATE OF CHANGE OF THE DISTRIBUTION FUNCTION DUE TO COLLISIONS

It is convenient to represent the information in (16.2) in a slightly different way. We define a quantity $(dg(\mathbf{k})/dt)_{\text{out}}$ so that the number of electrons per unit volume with wave vectors in the infinitesimal volume element $d\mathbf{k}$ about $\mathbf{k}$ that suffer a collision in the infinitesimal time interval dt is

$$-\left(\frac{dg(\mathbf{k})}{dt}\right)_{\text{out}} \frac{d\mathbf{k}}{(2\pi)^3}\, dt. \tag{16.3}$$

Since $d\mathbf{k}$ is infinitesimal, the effect of any collision on an electron in the volume element is to remove it from that volume element. Thus (16.3) can also be viewed as the number of electrons scattered out of the volume element $d\mathbf{k}$ about $\mathbf{k}$ in the time interval dt.

To evaluate $(dg(\mathbf{k})/dt)_{\text{out}}$, we simply note that because $dt/\tau(\mathbf{k})$ is the probability that any electron in the neighborhood of $\mathbf{k}$ is scattered in the time interval dt, the total number of electrons per unit volume in $d\mathbf{k}$ about $\mathbf{k}$ that suffer a collision is just $dt/\tau(\mathbf{k})$ times the number of electrons per unit volume in $d\mathbf{k}$ about $\mathbf{k}$, $g(\mathbf{k})\, d\mathbf{k}/(2\pi)^3$. Comparing this with (16.3) we find that

$$\begin{aligned}\left(\frac{dg(\mathbf{k})}{dt}\right)_{\text{out}} &= -\frac{g(\mathbf{k})}{\tau(\mathbf{k})} \\ &= -g(\mathbf{k}) \int \frac{d\mathbf{k}'}{(2\pi)^3} W_{\mathbf{k},\mathbf{k}'}[1 - g(\mathbf{k}')].\end{aligned} \tag{16.4}$$

This is not the only way in which the distribution function is affected by scattering: Electrons not only scatter out of the level $\mathbf{k}$, but are also scattered into it from other levels. We describe such processes in terms of a quantity $(dg(\mathbf{k})/dt)_{\text{in}}$, defined so that

$$\left(\frac{dg(\mathbf{k})}{dt}\right)_{\text{in}} \frac{d\mathbf{k}}{(2\pi)^3}\, dt \tag{16.5}$$

---

[8] The contribution to the electron density from electrons (with specified spin) in the volume element $d\mathbf{k}$ about $\mathbf{k}$ is (see page 244) $g(\mathbf{k})\, d\mathbf{k}/(2\pi)^3$. Since the maximum possible contribution to the density from electrons (with the specified spin) in this volume element occurs when all levels are occupied, and is equal to $d\mathbf{k}/(2\pi)^3$, $g(\mathbf{k})$ can also be interpreted as the fraction of levels in the volume element $d\mathbf{k}$ about $\mathbf{k}$ that are occupied. Hence $1 - g(\mathbf{k})$ is the fraction that are unoccupied.

is the number of electrons per unit volume that have arrived in the volume element $d\mathbf{k}$ about $\mathbf{k}$, as the result of a collision in the infinitesimal time interval dt.

To evaluate $(dg(\mathbf{k})/dt)_{\text{in}}$, consider the contribution from those electrons that, just prior to the collision, were in the volume element $d\mathbf{k}'$ about $\mathbf{k}'$. The total number of such electrons (with the specified spin) is $g(\mathbf{k}')d\mathbf{k}'/(2\pi)^3$. Of these, a fraction $W_{\mathbf{k}',\mathbf{k}}dt\, d\mathbf{k}/(2\pi)^3$ would be scattered into $d\mathbf{k}$ about $\mathbf{k}$, if all the levels in the volume element $d\mathbf{k}$ were empty; since, however, only a fraction $1 - g(\mathbf{k})$ are empty, the first fraction must be reduced by this factor. Therefore, the total number of electrons per unit volume arriving in the volume element $d\mathbf{k}$ about $\mathbf{k}$ from the volume element $d\mathbf{k}'$ about $\mathbf{k}'$ as the result of a collision in the time interval dt is

$$\left[g(\mathbf{k}')\frac{d\mathbf{k}'}{(2\pi)^3}\right]\left[W_{\mathbf{k}',\mathbf{k}}\frac{d\mathbf{k}}{(2\pi)^3}\,dt\right][1 - g(\mathbf{k})]. \tag{16.6}$$

Summing this over all $\mathbf{k}'$, and comparing the result with (16.5), we find that

$$\left(\frac{dg(\mathbf{k})}{dt}\right)_{\text{in}} = [1 - g(\mathbf{k})]\int\frac{d\mathbf{k}'}{(2\pi)^3}\,W_{\mathbf{k}',\mathbf{k}}\,g(\mathbf{k}'), \tag{16.7}$$

which has the same structure as (16.4) except for the interchange of $\mathbf{k}$ and $\mathbf{k}'$.

It is instructive to compare these expressions with the corresponding quantities in the relaxation-time approximation. The relaxation-time approximation for $(dg(\mathbf{k})/dt)_{\text{out}}$ differs from (16.4) only in that the collision rate, $1/\tau(\mathbf{k})$, is a definite specified function of $\mathbf{k}$ that does not (in contrast to (16.2)) depend on the distribution function g. The disparity between the expression (16.7) for $(dg(\mathbf{k})/dt)_{\text{in}}$ and the corresponding quantity in the relaxation-time approximation is more pronounced, for the relaxation-time approximation asserts that the distribution of electrons emerging from collisions in an interval dt is simply $dt/\tau(\mathbf{k})$ times the local equilibrium distribution function $g^0(\mathbf{k})$ (see Eq. (13.3)). These results are summarized in Table 16.1.

Table 16.1
A COMPARISON OF THE GENERAL TREATMENT OF COLLISIONS WITH THE SIMPLIFICATIONS OF THE RELAXATION-TIME APPROXIMATION

	RELAXATION-TIME APPROXIMATION	GENERAL
$\left(\frac{dg(\mathbf{k})}{dt}\right)_{\text{coll}}^{\text{out}}$	$-\frac{g(\mathbf{k})}{\tau(\mathbf{k})}$	$-\int\frac{d\mathbf{k}'}{(2\pi)^3}W_{\mathbf{k},\mathbf{k}'}[1 - g(\mathbf{k}')]g(\mathbf{k})$
$\left(\frac{dg(\mathbf{k})}{dt}\right)_{\text{coll}}^{\text{in}}$	$\frac{g^0(\mathbf{k})}{\tau(\mathbf{k})}$	$\int\frac{d\mathbf{k}'}{(2\pi)^3}W_{\mathbf{k}',\mathbf{k}}\,g(\mathbf{k}')[1 - g(\mathbf{k})]$
Comments	$\tau(\mathbf{k})$ is a specified function of $\mathbf{k}$ and does not depend on $g(\mathbf{k})$; $g^0(\mathbf{k})$ is the local equilibrium distribution function.	$W_{\mathbf{k},\mathbf{k}'}$ is a function of $\mathbf{k}$ and $\mathbf{k}'$, which in general may also depend on $g(\mathbf{k})$, or even on a second distribution function describing the local configuration of scatterers.

It is convenient to define $(dg/dt)_{\text{coll}}$ as the total rate at which the distribution function is changing due to collisions: $(dg(\mathbf{k})/dt)_{\text{coll}}\,dt\,d\mathbf{k}/(2\pi)^3$ is the change in the number of electrons per unit volume with wave vectors in the volume element $d\mathbf{k}$ about $\mathbf{k}$ in the time interval dt, due to all collisions. Since electrons can be scattered either into or out of $d\mathbf{k}$ by collisions, $(dg/dt)_{\text{coll}}$ is simply the sum of $(dg/dt)_{\text{in}}$ and $(dg/dt)_{\text{out}}$:

$$\left(\frac{dg(\mathbf{k})}{dt}\right)_{\text{coll}} = -\int \frac{d\mathbf{k}'}{(2\pi)^3}\{W_{\mathbf{k},\mathbf{k}'}g(\mathbf{k})[1 - g(\mathbf{k}')] - W_{\mathbf{k}',\mathbf{k}}g(\mathbf{k}')[1 - g(\mathbf{k})]\}. \quad \textbf{(16.8)}$$

In the relaxation-time approximation (see Table 16.1) this simplifies to

$$\left(\frac{dg(\mathbf{k})}{dt}\right)_{\text{coll}} = -\frac{[g(\mathbf{k}) - g^0(\mathbf{k})]}{\tau(\mathbf{k})}. \quad \begin{array}{c}\text{(relaxation-time}\\ \text{approximation)}\end{array} \quad \textbf{(16.9)}$$

DETERMINATION OF THE DISTRIBUTION FUNCTION: THE BOLTZMANN EQUATION

When the relaxation-time approximation is abandoned, one cannot construct an explicit representation of the nonequilibrium distribution function g in terms of the solutions to the semiclassical equations of motion, as we did in Chapter 13 by surveying all past times. One can, however, answer the more modest question of how g is to be constructed at time t from its value an infinitesimal time dt earlier.

To do this we first ignore the possibility of collisions taking place between $t - dt$ and t, correcting for this omission later. If no collisions occurred, then the $\mathbf{r}$ and $\mathbf{k}$ coordinates of every electron would evolve according to the semiclassical equations of motion (12.6):

$$\dot{\mathbf{r}} = \mathbf{v}(\mathbf{k}), \quad \hbar\dot{\mathbf{k}} = -e\left(\mathbf{E} + \frac{1}{c}\mathbf{v} \times \mathbf{H}\right) = \mathbf{F}(\mathbf{r}, \mathbf{k}). \quad \textbf{(16.10)}$$

Since dt is infinitesimal, we can find the explicit solution to these equations to linear order in dt: an electron at $\mathbf{r}, \mathbf{k}$ at time t must have been at $\mathbf{r} - \mathbf{v}(\mathbf{k})dt$, $\mathbf{k} - \mathbf{F}\,dt/\hbar$, at time $t - dt$. In the absence of collisions this is the only point electrons at $\mathbf{r}, \mathbf{k}$ can have come from, and every electron at this point will reach the point $\mathbf{r}, \mathbf{k}$. Consequently,[9]

$$g(\mathbf{r}, \mathbf{k}, t) = g(\mathbf{r} - \mathbf{v}(\mathbf{k})\,dt, \mathbf{k} - \mathbf{F}\,dt/\hbar, t - dt). \quad \textbf{(16.11)}$$

To take collisions into account we must add two correction terms to (16.11). The right-hand side is wrong because it assumes that all electrons get from $\mathbf{r} - \mathbf{v}\,dt$, $\mathbf{k} - \mathbf{F}\,dt/\hbar$ to $\mathbf{r}, \mathbf{k}$ in the time dt, ignoring the fact that some are deflected by collisions. It is also wrong because it fails to count those electrons found at $\mathbf{r}, \mathbf{k}$ at time t not as a result of their unimpeded semiclassical motion since time $t - dt$, but as the result

---

[9] In writing (16.11) we must appeal to Liouville's theorem (Appendix H) that volumes in phase space are preserved by the semiclassical equations of motion, for the argument only yields the result:

$$g(\mathbf{r}, \mathbf{k}, t)\,d\mathbf{r}(t)\,d\mathbf{k}(t) = g(\mathbf{r} - \mathbf{v}(\mathbf{k})t, \mathbf{k} - \mathbf{F}\,dt/\hbar, t - dt)\,d\mathbf{r}(t - dt)\,d\mathbf{k}(t - dt),$$

which expresses the equality of the numbers of electrons in $d\mathbf{r}(t)\,d\mathbf{k}(t)$ and $d\mathbf{r}(t - dt)\,d\mathbf{k}(t - dt)$. Liouville's theorem is required to permit cancellation of the phase-space volume elements from both sides.

of a collision between $t - dt$ and t. Adding these corrections, we find to leading order in dt:

$$\begin{aligned} g(\mathbf{r}, \mathbf{k}, t) = \\ & g(\mathbf{r} - \mathbf{v}(\mathbf{k})dt, \mathbf{k} - \mathbf{F}\, dt/\hbar, t - dt) && \text{(collisionless evolution)} \\ & + \left(\frac{\partial g(\mathbf{r}, \mathbf{k}, t)}{\partial t}\right)_{\text{out}} dt && \text{(correction: some fail to get there because of collisions)} \\ & + \left(\frac{\partial g(\mathbf{r}, \mathbf{k}, t)}{\partial t}\right)_{\text{in}} dt. && \text{(correction: some get there only because of collisions)} \end{aligned} \tag{16.12}$$

If we expand the left side to linear order in dt, then in the limit as $dt \to 0$, (16.12) reduces to

$$\boxed{\frac{\partial g}{\partial t} + \mathbf{v} \cdot \frac{\partial}{\partial \mathbf{r}} g + \mathbf{F} \cdot \frac{1}{\hbar} \frac{\partial}{\partial \mathbf{k}} g = \left(\frac{\partial g}{\partial t}\right)_{\text{coll}}.} \tag{16.13}$$

This is the celebrated Boltzmann equation. The terms on the left side are often referred to as the drift terms, and the term on the right side as the collision term. When the form (16.8) is used for the collision term, the Boltzmann equation in general becomes a nonlinear integrodifferential equation. This equation lies at the heart of the theory of transport in solids. Many subtle and ingenious techniques have been developed for extracting information about the distribution function, and hence the various conductivities.[10] We shall not pursue the subject further here, but shall refer to the Boltzmann equation only to the extent that it illuminates the limitations of the relaxation-time approximation.

If we replace the collision term in the Boltzmann equation by the relaxation-time approximation (16.9), the equation simplifies to a linear partial differential equation. It can be shown that the distribution function (13.17) that we constructed from the relaxation-time approximation is the solution to this equation (as must be the case, since identical assumptions underlie both derivations). We emphasize this equivalence because it is widespread practice to derive results such as those we found in Chapter 13 not directly from the explicit distribution function (13.17) given by the relaxation-time approximation, but by the apparently quite different route of solving the Boltzmann equation (16.13) with the collision term given by the relaxation-time approximation (16.9). The equivalence of the two approaches is illustrated in Problems 2 and 3, where some typical results of Chapter 13 are rederived from the Boltzmann equation in the relaxation-time approximation.

IMPURITY SCATTERING

We wish to compare some of the predictions of the relaxation-time approximation with those implied by the more accurate collision term (16.8). When we require a specific form for the collision probability $W_{\mathbf{k},\mathbf{k}'}$, we shall specialize to the case of

[10] A very thorough survey is given in J. M. Ziman, *Electrons and Phonons*, Oxford, 1960. There is also a remarkable series of review papers by I. M. Lifshitz and M. I. Kaganov, *Sov. Phys. Usp.* **2**, 831 (1960), **5**, 878 (1963), **8**, 805 (1966). [*Usp. Fiz. Nauk* **69**, 419 (1959), **78**, 411 (1962), **87**, 389 (1965).]

greatest analytic simplicity: elastic scattering by fixed substitutional impurities, located at random lattice sites throughout the crystal. This is not an artificial case, since scattering by the thermal vibrations of the ions (Chapter 26) and electron-electron scattering (Chapter 17) become increasingly weak as the temperature drops, whereas neither the concentration of impurities nor the electron-impurity interaction shows any striking temperature dependence. Thus at sufficiently low temperatures impurity scattering will be the dominant source of collisions in any real specimen. This scattering will be elastic provided that the energy gap between the impurity ground state and lowest excited state (typically of the order of electron volts) is large compared with $k_B T$. This will insure (a) that there are very few excited impurity ions to give up energy to electrons in collisions and (b) that there are very few empty electronic levels that lie low enough in energy to receive an electron after it has lost enough energy to excite the impurity ion out of its ground state.

If the impurities are sufficiently dilute[11] and the potential $U(\mathbf{r})$ describing the interaction between an electron and a single impurity at the origin is sufficiently weak, then it can be shown that

$$W_{\mathbf{k},\mathbf{k}'} = \frac{2\pi}{\hbar} n_i \delta(\mathcal{E}(\mathbf{k}) - \mathcal{E}(\mathbf{k}'))|\langle \mathbf{k}|U|\mathbf{k}'\rangle|^2, \qquad \textbf{(16.14)}$$

where n_i is the number of impurities per unit volume,

$$\langle \mathbf{k}|U|\mathbf{k}'\rangle = \int d\mathbf{r}\, \psi^*_{n\mathbf{k}'}(\mathbf{r})U(\mathbf{r})\psi_{n\mathbf{k}}(\mathbf{r}), \qquad \textbf{(16.15)}$$

and the Bloch functions are taken to be normalized so that

$$\int_{\text{cell}} d\mathbf{r}\, |\psi_{n\mathbf{k}}(\mathbf{r})|^2 = v_{\text{cell}}. \qquad \textbf{(16.16)}$$

Equation (16.14) can be derived by applying the "Golden Rule" of first-order time-dependent perturbation theory[12] to the scattering of a Bloch electron by each of the impurities. It is considerably harder to construct a more fundamental derivation by starting with the basic Hamiltonian for all the electrons and impurities, and deriving the entire Boltzmann equation with a collision term given by (16.8) and (16.14).[13]

We shall not pursue the derivation of (16.14) here, since we shall only exploit a few very general properties of the result:

1. Because of the delta function in (16.14), $W_{\mathbf{k},\mathbf{k}'} = 0$ unless $\mathcal{E}(\mathbf{k}) = \mathcal{E}(\mathbf{k}')$; the scattering is explicitly elastic.

2. $W_{\mathbf{k},\mathbf{k}'}$ is independent of the electronic distribution function g. This is a consequence of the independent electron approximation: The way in which an electron

---

[11] They must be dilute enough that one can treat the electrons as interacting with one impurity at a time.

[12] See, for example, L. D. Landau and E. M. Lifshitz, *Quantum Mechanics*, Addison-Wesley, Reading, Mass., 1965, Eq. (43.1).

[13] One of the earliest full analyses was given by J. M. Luttinger and W. Kohn, *Phys. Rev.* **108**, 590 (1957) and **109**, 1892 (1958).

interacts with an impurity is, aside from the restrictions imposed by the exclusion principle, independent of the other electrons. This is the major simplifying feature of impurity scattering. In electron-electron scattering, for example, $W_{\mathbf{k},\mathbf{k}'}$ depends on the distribution function g, since over and beyond the simple restrictions imposed by the exclusion principle, the probability of an electron scattering depends on what other electrons are available for it to interact with. The description of the scattering by the thermal vibrations of the ions is also more complicated since W will depend on the properties of the system of ions, which may be quite complex.

3. W has the symmetry

$$W_{\mathbf{k},\mathbf{k}'} = W_{\mathbf{k}',\mathbf{k}}, \tag{16.17}$$

which follows from the fact that U is Hermitian ($\langle \mathbf{k}|U|\mathbf{k}'\rangle = \langle \mathbf{k}'|U|\mathbf{k}\rangle^*$). It can be shown that this symmetry does not depend on the impurity-electron interaction being weak, but follows generally provided only that the crystal and impurity potentials are real and invariant under spatial inversions. The symmetry (16.17) is usually called "detailed balancing". For more general scattering mechanisms there are related, but more complex symmetries, which are of great importance in studies of the approach to thermodynamic equilibrium.

The symmetry (16.17) simplifies the collision term (16.8) to

$$\left(\frac{dg(\mathbf{k})}{dt}\right)_{\text{coll}} = -\int \frac{d\mathbf{k}'}{(2\pi)^3} W_{\mathbf{k},\mathbf{k}'}[g(\mathbf{k}) - g(\mathbf{k}')]. \tag{16.18}$$

Note that the quadratic terms in g appearing in (16.8) as a consequence of the exclusion principle cancel identically as a result of the symmetry (16.17).

In the remainder of this chapter we describe some typical problems whose proper formulation requires a more accurate description of collisions than the relaxation-time approximation can provide.

THE WIEDEMANN-FRANZ LAW

The derivation of the Wiedemann-Franz law in Chapter 13 appeared to be quite general. When the problem is reexamined without using the relaxation-time approximation, however, it can be shown that the law holds only when the energy of each electron is conserved in each collision. The corresponding mathematical requirement is that for any function $g(\mathbf{k})$ the scattering probability $W_{\mathbf{k},\mathbf{k}'}$ be such that

$$\int d\mathbf{k}'\, W_{\mathbf{k},\mathbf{k}'}\mathcal{E}(\mathbf{k}')g(\mathbf{k}') = \mathcal{E}(\mathbf{k}) \int d\mathbf{k}'\, W_{\mathbf{k},\mathbf{k}'}g(\mathbf{k}'). \tag{16.19}$$

This is clearly satisfied when W has the energy-conserving form, (16.14), but will not hold if $W_{\mathbf{k},\mathbf{k}'}$ can be nonzero for values of $\mathbf{k}$ and $\mathbf{k}'$ with $\mathcal{E}(\mathbf{k}) \neq \mathcal{E}(\mathbf{k}')$.

It would take us too far afield to demonstrate analytically that the condition of elastic scattering (16.19) is enough to insure the Wiedemann-Franz law. However, the physical reason for this is not hard to understand. Since the charge of each electron is permanently $-e$, the only way collisions can degrade an electric current is by changing the velocity of each electron. However, in a thermal current (Eq. (13.42)) the charge is replaced by $(\mathcal{E} - \mu)/T$. Therefore *if* energy is conserved in each collision (as charge certainly is), then thermal currents will be degraded in precisely the same manner and to the same extent as electric currents. If, however, collisions do not conserve the energy $\mathcal{E}$ of each electron, then a second mechanism becomes available for the degradation of a thermal current that has no electrical analogue: Collisions can alter the electron's energy $\mathcal{E}$ as well as its velocity. Since such inelastic collisions will have a substantially different effect on thermal and electric currents, there is no longer any reason to expect a simple relation to hold between electrical and thermal conductivities.[14]

Evidently the Wiedemann-Franz law will hold to a good approximation, if energy is conserved to a good approximation. The crucial requirement is that the change in energy of each electron in a collision should be small compared with $k_B T$. It turns out that the scattering by thermal vibrations of the ions can satisfy this condition at high temperatures. Since such scattering is the dominant high-temperature source of collisions, the Wiedemann-Franz law is generally well obeyed at both high and low[15] temperatures. However, in the intermediate temperature range (roughly ten to a few hundred degrees K), where inelastic collisions are both prevalent and capable of producing electronic energy losses of order $k_B T$, one expects and observes failures of the Wiedemann-Franz law.

MATTHIESSEN'S RULE

Suppose there are two physically distinguishable sources of scattering (for example, scattering by impurities and scattering by other electrons). If the presence of one mechanism does not alter the way in which the other mechanism functions, then the total collision rate W will be given by the sums of the collision rates due to the separate mechanisms:

$$W = W^{(1)} + W^{(2)}. \tag{16.20}$$

In the relaxation-time approximation this immediately implies that

$$\frac{1}{\tau} = \frac{1}{\tau^{(1)}} + \frac{1}{\tau^{(2)}}. \tag{16.21}$$

If, in addition, we assume a **k**-independent relaxation time for each mechanism, then, since the resistivity is proportional to $1/\tau$, we will have

$$\rho = \frac{m}{ne^2\tau} = \frac{m}{ne^2}\frac{1}{\tau^{(1)}} + \frac{m}{ne^2}\frac{1}{\tau^{(2)}} = \rho^{(1)} + \rho^{(2)}. \tag{16.22}$$

This asserts that the resistivity in the presence of several distinct scattering mechanisms is simply the sum of the resistivities one would have if each alone were present.

---

14 One sometimes encounters the assertion that the Wiedemann-Franz law fails because the relaxation time for thermal currents is different from the relaxation time for electric currents. This is, at best, a misleading oversimplification. The Wiedemann-Franz law fails, if inelastic scattering is present, because there are scattering processes that can degrade a thermal current without degrading an electric current. The failure is due not to the comparative *rates* at which electrons experience collisions, but to the comparative *effectiveness* of each single collision in degrading the two kinds of currents.

15 At low temperatures, as we have pointed out, the dominant source of collisions is elastic impurity scattering.

This proposition is known as Matthiessen's rule. At first glance its utility might seem questionable, since it is difficult to imagine how one might remove a source of scattering, keeping all other things constant. However, it does make certain general assertions of principle that are easily tested. For example, elastic impurity scattering should proceed at a temperature-independent rate (since neither the number of impurities nor their interaction with electrons is appreciably affected by the temperature), but the electron-electron scattering rate should go as T^2 (in the simplest theories: see Chapter 17). Thus Matthiessen's rule predicts a resistivity of the form $\rho = A + BT^2$ with temperature-independent coefficients A and B, if impurity and electron-electron scattering are the dominant mechanisms.

It is not hard to verify that Matthiessen's rule breaks down even in the relaxation-time approximation, if τ depends on $\mathbf{k}$. For the conductivity σ is then proportional to some average, $\bar{\tau}$, of the relaxation time (see, for example, Eq. (13.25)). Thus the resistivity, ρ, is proportional to $1/\bar{\tau}$, and Matthiessen's rule requires that

$$1/\bar{\tau} = 1/\overline{\tau^{(1)}} + 1/\overline{\tau^{(2)}}. \tag{16.23}$$

However, Eq. (16.21) gives only relations such as

$$\overline{(1/\tau)} = \overline{(1/\tau^{(1)})} + \overline{(1/\tau^{(2)})}, \tag{16.24}$$

which are not equivalent to (16.23) unless $\tau^{(1)}$ and $\tau^{(2)}$ are independent of $\mathbf{k}$.

A more realistic picture of collisions casts even graver doubts on the general validity of Matthiessen's rule, for the assumption that the scattering rate due to one mechanism is independent of the presence of the second becomes much less plausible as soon as the assumptions of the relaxation-time approximation are dropped. The actual rate at which an electron experiences collisions depends on the configuration of the other electrons, and this can be strongly affected by the presence of two competing scattering mechanisms unless, by some good fortune, it happens that the distribution function in the presence of each separate scattering mechanism is the same.

It can, however, be shown without making the relaxation-time approximation, that Matthiessen's rule holds as an inequality:[16]

$$\rho \geqslant \rho^{(1)} + \rho^{(2)}. \tag{16.25}$$

Quantitative analytic studies of the extent to which Matthiessen's rule fails are quite complex. The rule is certainly valuable as a crude guide to what to expect, but one must always bear in mind the possibility of gross failures—a possibility that is obscured by the naive relaxation-time approximation.

SCATTERING IN ISOTROPIC MATERIALS

It is sometimes asserted that the relaxation-time approximation can be justified in isotropic systems. This is an interesting and useful observation, but one must be aware

---

[16] See, for example, J. M. Ziman, *Electrons and Phonons*, Oxford, 1960, p. 286, and also Problem 4 below.

of its limitations. The point arises in the description of elastic impurity scattering in an isotropic metal. The two crucial conditions[17] are:

(*a*) The energy $\mathcal{E}(\mathbf{k})$ must depend only on the magnitude k of the vector $\mathbf{k}$.
(*b*) The probability of scattering between two levels $\mathbf{k}$ and $\mathbf{k}'$ must vanish unless $k = k'$ (i.e., it must be elastic), and must depend only on the common value of their energies and on the angle between $\mathbf{k}$ and $\mathbf{k}'$.

If condition (*a*) holds, then *in the relaxation-time approximation* the nonequilibrium distribution function in the presence of a static, spatially uniform electric field and temperature gradient, Eq. (13.43), has the general form[18]

$$g(\mathbf{k}) = g^0(\mathbf{k}) + \mathbf{a}(\mathcal{E}) \cdot \mathbf{k}, \tag{16.26}$$

where the vector function $\mathbf{a}$ depends on $\mathbf{k}$ only through its magnitude—i.e., only through $\mathcal{E}(\mathbf{k})$—and $g^0(\mathbf{k})$ is the local equilibrium distribution function. When the scattering is elastic impurity scattering and conditions (*a*) and (*b*) hold, one can show that if the solution to the Boltzmann equation in the relaxation-time approximation has the form (16.26),[19] *then it is also a solution to the full Boltzmann equation.*

To demonstrate this it suffices to show that the more correct Eq. (16.18) for $(dg/dt)_{\text{coll}}$ reduces to the form (16.9) assumed in the relaxation-time approximation, whenever the distribution function g has the form (16.26). We must therefore show that it is possible to find a function $\tau(\mathbf{k})$ that does not depend on the distribution function g, such that whenever g has the form (16.26) and the scattering is elastic isotropic impurity scattering, then

$$\int \frac{d\mathbf{k}'}{(2\pi)^3} W_{\mathbf{k},\mathbf{k}'}[g(\mathbf{k}) - g(\mathbf{k}')] = \frac{1}{\tau(\mathbf{k})}[g(\mathbf{k}) - g^0(\mathbf{k})]. \tag{16.27}$$

If we substitute the distribution function (16.26) into (16.27) and note that for elastic scattering $W_{\mathbf{k},\mathbf{k}'}$ vanishes unless $\mathcal{E}(\mathbf{k}) = \mathcal{E}(\mathbf{k}')$, then the vector $\mathbf{a}(\mathcal{E}')$ can be replaced by $\mathbf{a}(\mathcal{E})$ and removed from the integral, reducing the condition (16.27) to[20]

$$\mathbf{a}(\mathcal{E}) \cdot \int \frac{d\mathbf{k}'}{(2\pi)^3} W_{\mathbf{k},\mathbf{k}'}(\mathbf{k} - \mathbf{k}') = \frac{1}{\tau(\mathbf{k})} \mathbf{a}(\mathcal{E}) \cdot \mathbf{k}. \tag{16.28}$$

---

[17] A more detailed analysis reveals that these requirements need hold only for levels within $O(k_BT)$ of the Fermi surface. This is because the final form of the distribution function differs from the local equilibrium form only within this energy range; see, for example Eq. (13.43). As a result, the analysis that follows may apply not only to the ideal free electron gas, but also to the alkali metals, whose Fermi surfaces are remarkably spherical, provided that the scattering is sufficiently isotropic near the Fermi energy.

[18] This form also holds for time-dependent perturbations, and in the presence of spatially uniform static magnetic fields. If, however, any of the external fields or temperature gradients depend on position, then (16.26) does not hold, and the conclusion that the solution to the Boltzmann equation has the form given by the relaxation-time approximation can no longer be drawn.

[19] The distribution function constructed in Chapter 13 is a solution to the Boltzmann equation in the relaxation time approximation, as pointed out on page 320.

[20] Note that the equilibrium distribution g^0 depends on $\mathbf{k}$ only through $\mathcal{E}(\mathbf{k})$, and therefore drops out of (16.28) when the scattering is elastic.

We next resolve the vector $\mathbf{k}'$ into its components parallel and perpendicular to $\mathbf{k}$:

$$\mathbf{k}' = \mathbf{k}'_{\parallel} + \mathbf{k}'_{\perp} = (\hat{\mathbf{k}} \cdot \mathbf{k}')\hat{\mathbf{k}} + \mathbf{k}'_{\perp}. \tag{16.29}$$

Because the scattering is elastic and $W_{\mathbf{k},\mathbf{k}'}$ depends only on the angle between $\mathbf{k}$ and $\mathbf{k}'$, $W_{\mathbf{k},\mathbf{k}'}$ cannot depend on $\mathbf{k}'_{\perp}$, and therefore $\int d\mathbf{k}'\ W_{\mathbf{k},\mathbf{k}'}\mathbf{k}'_{\perp}$ must vanish. Consequently

$$\int d\mathbf{k}'\ W_{\mathbf{k},\mathbf{k}'}\mathbf{k}' = \int d\mathbf{k}'\ W_{\mathbf{k},\mathbf{k}'}\mathbf{k}'_{\parallel} = \hat{\mathbf{k}} \int d\mathbf{k}'\ W_{\mathbf{k},\mathbf{k}'}(\hat{\mathbf{k}} \cdot \hat{\mathbf{k}}')k'. \tag{16.30}$$

Finally, because $W_{\mathbf{k},\mathbf{k}'}$ vanishes unless the magnitudes of $\mathbf{k}$ and $\mathbf{k}'$ are equal, the factor k' in the last integrand in (16.30) can be replaced by k and brought outside the integral, where it can be combined with the unit vector $\hat{\mathbf{k}}$ to yield the vector $\mathbf{k}$:

$$\int d\mathbf{k}'\ W_{\mathbf{k},\mathbf{k}'}\mathbf{k}' = \mathbf{k} \int d\mathbf{k}'\ W_{\mathbf{k},\mathbf{k}'}(\hat{\mathbf{k}} \cdot \hat{\mathbf{k}}'). \tag{16.31}$$

It follows from the identity (16.31) that the left side of Eq. (16.28) does indeed have the same form as the right side, provided that $\tau(\mathbf{k})$ is defined by[21]

$$\frac{1}{\tau(\mathbf{k})} = \int \frac{d\mathbf{k}'}{(2\pi)^3} W_{\mathbf{k},\mathbf{k}'}(1 - \hat{\mathbf{k}} \cdot \hat{\mathbf{k}}'). \tag{16.32}$$

Thus the relaxation-time approximation (with a relaxation time given by (16.32) provides the same description as the full Boltzmann equation, when applied to spatially homogeneous disturbances in an isotropic metal with isotropic elastic impurity scattering.

Note that the relaxation-time given by (16.32) is a weighted average of the collision probability in which forward scattering ($\hat{\mathbf{k}} = \hat{\mathbf{k}}'$) receives very little weight. If θ is the angle between $\mathbf{k}$ and $\mathbf{k}'$, then $1 - \hat{\mathbf{k}} \cdot \hat{\mathbf{k}}' = 1 - \cos\theta \approx \theta^2/2$ for small angles. It is not unreasonable that small-angle scattering should make a very small contribution to the effective collision rate. If collisions were only in the forward direction (i.e., if $W_{\mathbf{k},\mathbf{k}'}$ vanished unless $\mathbf{k} = \mathbf{k}'$) then they would have no consequences at all. When the possible changes in wave vector are not zero, but very small, then the distribution of electronic wave vectors will be only slightly affected by collisions. A single collision, for example, would certainly not obliterate all traces of the fields that accelerated the electron, as maintained by the relaxation-time approximation, and therefore the effective inverse relaxation time (16.32) is very much smaller than the actual collision rate (16.2) when the scattering is predominantly forward scattering.

This point remains valid quite generally: Forward scattering will contribute less to effective "collision rates" than wide-angle scattering, unless a property is being measured that depends delicately on the precise direction in which certain electrons move. We shall see in Chapter 26 that this point is of importance in understanding the temperature dependence of the DC electrical resistivity.

---

21 Note that the relaxation time given by (16.32) can depend on the magnitude, but not the direction of $\mathbf{k}$.

PROBLEMS

1. Let $h(\mathbf{k})$ be any one-electron property whose total density is

$$H = \int \frac{d\mathbf{k}}{4\pi^3}\, h(\mathbf{k}) g(\mathbf{k}), \tag{16.33}$$

where g is the electronic distribution function. If, for example, $h(\mathbf{k})$ is the electronic energy, $\mathcal{E}(\mathbf{k})$, then H is the energy density u; if $h(\mathbf{k})$ is the electronic charge, $-e$, then H is the charge density, ρ. The value of the density H in the neighborhood of a point changes because electrons move into and out of the neighborhood, some as the result of the semiclassical equations of motion, and some as a result of collisions. The change in H due to collisions is

$$\left(\frac{dH}{dt}\right)_{\text{coll}} = \int \frac{d\mathbf{k}}{4\pi^3}\, h(\mathbf{k}) \left(\frac{\partial g}{\partial t}\right)_{\text{coll}}. \tag{16.34}$$

(a) Show from (16.8) that $(dH/dt)_{\text{coll}}$ vanishes provided that all collisions conserve h (i.e., provided that there is only scattering between levels $\mathbf{k}$ and $\mathbf{k}'$ with $h(\mathbf{k}) = h(\mathbf{k}')$).

(b) Show that if (16.8) is replaced by the relaxation-time approximation (16.9), then $(dH/dt)_{\text{coll}}$ vanishes only if the parameters $\mu(\mathbf{r}, t)$ and $T(\mathbf{r}, t)$, characterizing the local equilibrium distribution f, yield an equilibrium value of H equal to the actual value (16.33).

(c) Deduce the equation of continuity, $\nabla \cdot \mathbf{j} + \partial\rho/\partial t = 0$ from the Boltzmann equation (16.13).

(d) Deduce the equation of energy flow (13.83) from the Boltzmann equation (16.13), assuming that $(du/dt)_{\text{coll}} = 0$.

2. A metal is perturbed by a spatially uniform electric field and temperature gradient. Making the relaxation-time approximation (16.9) (where g_0 is the local equilibrium distribution appropriate to the imposed temperature gradient), solve the Boltzmann equation (16.13) to linear order in the field and temperature gradient, and verify that the solution is identical to (13.43).

3. A metal at constant temperature and in a uniform static magnetic field is perturbed by a uniform static electric field.

(a) Making the relaxation-time approximation (16.9), solve the Boltzmann equation (16.13) to linear order in the electric field (treating the magnetic field exactly), under the assumption

$$\mathcal{E}(\mathbf{k}) = \frac{\hbar^2 k^2}{2m^*}. \tag{16.35}$$

Verify that your solution is of the form (16.26).

(b) Construct the conductivity tensor from your solution, and verify that it agrees with what one finds by evaluating (13.69) and (13.70) for a single free electron band.

4. Consider the Boltzmann equation (16.13) for a metal in a static uniform electric field, with a collision term (16.18) appropriate to elastic impurity scattering.

(a) Assuming a nonequilibrium distribution function of the form

$$g(\mathbf{k}) = f(\mathbf{k}) + \delta g(\mathbf{k}), \tag{16.36}$$

where f is the equilibrium Fermi function and $\delta g(\mathbf{k})$ is of order E, derive to linear order in E an integral equation obeyed by δg, and show that the conductivity can be written in the form

$$\boldsymbol{\sigma} = e^2 \int \frac{d\mathbf{k}}{4\pi^3} \left(-\frac{\partial f}{\partial \mathcal{E}} \right) \mathbf{v}(\mathbf{k})\mathbf{u}(\mathbf{k}), \tag{16.37}$$

where $\mathbf{u}(\mathbf{k})$ is a solution to the integral equation

$$\mathbf{v}(\mathbf{k}) = \int \frac{d\mathbf{k}'}{(2\pi)^3} W_{\mathbf{k},\mathbf{k}'}[\mathbf{u}(\mathbf{k}) - \mathbf{u}(\mathbf{k}')]. \tag{16.38}$$

(b) Let $\alpha(\mathbf{k})$ and $\gamma(\mathbf{k})$ be any two functions of $\mathbf{k}$. Define

$$(\alpha, \gamma) = e^2 \int \frac{d\mathbf{k}}{4\pi^3} \left(-\frac{\partial f}{\partial \mathcal{E}} \right) \alpha(\mathbf{k})\gamma(\mathbf{k}) \tag{16.39}$$

so that (16.37) can be written compactly as

$$\sigma_{\mu\nu} = (v_\mu, u_\nu). \tag{16.40}$$

Define

$$\{\alpha, \gamma\} = e^2 \int \frac{d\mathbf{k}}{4\pi^3} \left(-\frac{\partial f}{\partial \mathcal{E}} \right) \alpha(\mathbf{k}) \int \frac{d\mathbf{k}'}{(2\pi)^3} W_{\mathbf{k},\mathbf{k}'}[\gamma(\mathbf{k}) - \gamma(\mathbf{k}')]. \tag{16.41}$$

Show that $\{\alpha, \gamma\} = \{\gamma, \alpha\}$, and that Eq. (16.38) implies

$$\{u_\mu, \gamma\} = (v_\mu, \gamma), \tag{16.42}$$

so that the conductivity can also be written in the form

$$\sigma_{\mu\nu} = \{u_\mu, u_\nu\}. \tag{16.43}$$

(c) Prove for arbitrary α and γ that

$$\{\alpha, \alpha\} \geqslant \frac{\{\alpha, \gamma\}^2}{\{\gamma, \gamma\}}. \tag{16.44}$$

(*Hint:* Prove that $\{\alpha + \lambda\gamma, \alpha + \lambda\gamma\} \geqslant 0$ for arbitrary λ, and choose a λ that minimizes the left-hand side of this inequality.)

(d) With the choice $\alpha = u_x$, deduce that σ_{xx} satisfies the inequality

$$\sigma_{xx} \geqslant \frac{e^2 \left[\int \frac{d\mathbf{k}}{4\pi^3} \left(-\frac{\partial f}{\partial \mathcal{E}} \right) v_x(\mathbf{k})\gamma(\mathbf{k}) \right]^2}{\int \frac{d\mathbf{k}}{4\pi^3} \left(-\frac{\partial f}{\partial \mathcal{E}} \right) \gamma(\mathbf{k}) \int \frac{d\mathbf{k}'}{(2\pi)^3} W_{\mathbf{k},\mathbf{k}'}[\gamma(\mathbf{k}) - \gamma(\mathbf{k}')]}. \tag{16.45}$$

for arbitrary functions γ.

(e) Suppose that $W = W^{(1)} + W^{(2)}$. Let γ be u_x, where $\mathbf{u}$ is the solution to (16.38). Let $\boldsymbol{\sigma}^{(1)}$ and $\boldsymbol{\sigma}^{(2)}$ be the conductivities one would have if only $W^{(1)}$ or $W^{(2)}$ were present. Deduce from (16.45), as it applies to $\boldsymbol{\sigma}^{(1)}$ and $\boldsymbol{\sigma}^{(2)}$, that

$$\frac{1}{\sigma_{xx}} \geqslant \frac{1}{\sigma_{xx}^{(1)}} + \frac{1}{\sigma_{xx}^{(2)}}. \tag{16.46}$$

17
Beyond the Independent Electron Approximation

The proper choice of the potential $U(\mathbf{r})$ appearing in the one-electron Schrödinger equation

$$-\frac{\hbar^2}{2m}\nabla^2\psi(\mathbf{r}) + U(\mathbf{r})\psi(\mathbf{r}) = \varepsilon\psi(\mathbf{r}) \tag{17.1}$$

is a subtle problem.[1] Underlying this problem is the question of how best to represent the effects of electron-electron interactions, a subject we have so far altogether ignored, by working in the independent electron approximation.

From a fundamental point of view it is impossible to describe electrons in a metal correctly[2] by so elementary an equation as (17.1), however ingenious the choice of $U(\mathbf{r})$, because of the enormously complicating effects of the interactions between electrons. A more accurate calculation of the electronic properties of a metal should start with the Schrödinger equation for the N-particle wave function of all N electrons in the metal,[3] $\Psi(\mathbf{r}_1 s_1, \mathbf{r}_2 s_2, \ldots, \mathbf{r}_N s_N)$:

$$H\Psi = \sum_{i=1}^{N}\left(-\frac{\hbar^2}{2m}\nabla_i^2\Psi - Ze^2\sum_{\mathbf{R}}\frac{1}{|\mathbf{r}_i - \mathbf{R}|}\Psi\right) + \frac{1}{2}\sum_{i\neq j}\frac{e^2}{|\mathbf{r}_i - \mathbf{r}_j|}\Psi = E\Psi. \tag{17.2}$$

Here the negative potential-energy term represents the attractive electrostatic potentials of the bare nuclei fixed at the points $\mathbf{R}$ of the Bravais lattice, and the last term represents the interactions of the electrons with each other.

One has no hope of solving an equation such as (17.2). Further progress requires some simplifying physical idea. One such idea is suggested by asking what choice of $U(\mathbf{r})$ would make the one-electron equation (17.1) least unreasonable. Evidently $U(\mathbf{r})$ should include the potentials of the ions:

$$U^{\text{ion}}(\mathbf{r}) = -Ze^2\sum_{\mathbf{R}}\frac{1}{|\mathbf{r} - \mathbf{R}|}. \tag{17.3}$$

In addition, however, we should like $U(\mathbf{r})$ to incorporate (at least approximately) the fact that the electron feels the electric fields of all the other electrons. If we treated the remaining electrons as a smooth distribution of negative charge with charge density ρ, the potential energy of the given electron in their field would be

$$U^{\text{el}}(\mathbf{r}) = -e\int d\mathbf{r}'\,\rho(\mathbf{r}')\frac{1}{|\mathbf{r} - \mathbf{r}'|}. \tag{17.4}$$

Furthermore, if we persisted in an independent electron picture, the contribution of an electron in the level[4] ψ_i to the charge density would be

$$\rho_i(\mathbf{r}) = -e|\psi_i(\mathbf{r})|^2. \tag{17.5}$$

---

[1] See the discussion at the beginning of Chapter 11.

[2] Even in the approximation of fixed immovable ions. We shall retain this assumption here, relaxing it in Chapters 21–26.

[3] We include explicitly the dependence of Ψ on the electron spin s, as well as the position $\mathbf{r}$.

[4] We let i stand for both the spin and orbital quantum numbers of the one-electron level.

The total electronic charge density would then be

$$\rho(\mathbf{r}) = -e \sum_i |\psi_i(\mathbf{r})|^2, \tag{17.6}$$

where the sum extends over all occupied one-electron levels in the metal.[5]

Placing (17.6) in (17.4) and letting $U = U^{\text{ion}} + U^{\text{el}}$ we arrive at the one-electron equation:

$$-\frac{\hbar^2}{2m}\nabla^2\psi_i(\mathbf{r}) + U^{\text{ion}}(\mathbf{r})\psi_i(\mathbf{r}) + \left[e^2 \sum_j \int d\mathbf{r}'\, |\psi_j(\mathbf{r}')|^2 \frac{1}{|\mathbf{r} - \mathbf{r}'|}\right]\psi_i(\mathbf{r}) = \varepsilon_i\psi_i(\mathbf{r}). \tag{17.7}$$

The set of equations (17.7) (there is one for each occupied one-electron level $\psi_i(\mathbf{r})$) is known as the *Hartree equations.* These nonlinear equations for the one-electron wave functions and energies are solved, in practice, by iteration: A form is guessed for U^{el} (the term in brackets in (17.7)) on the basis of which the equations are solved. A new U^{el} is then computed from the resulting wave functions, $\psi_i(\mathbf{r})$, and a new Schrödinger equation is solved. Ideally, the procedure is continued until further iterations do not materially alter the potential.[6]

The Hartree approximation fails to represent the way in which the particular (as opposed to the average) configuration of the other $N - 1$ electrons affects the electron under consideration, for Eq. (17.7) describes the electron as interacting only with the field obtained by averaging over the positions of the remaining electrons (with a weight determined by their wave functions). As crude an approximation as this is to the full Schrödinger equation (17.2), it still leads to a mathematical task of considerable numerical complexity. To improve upon the Hartree approximation is quite difficult.

There are, however, certain other important physical features of electron-electron interactions that cannot be treated in a simple self-consistent field approximation, but are nevertheless fairly well understood. In this chapter we shall survey the following:

1. The extension of the self-consistent field equations to include what is known as "exchange."
2. The phenomenon of "screening," which is of importance in developing a still more accurate theory of electron-electron interactions, and in accounting for the response of metallic electrons to charged particles such as ions, impurities, or other electrons.
3. The Fermi liquid theory of Landau, which provides a phenomenological way of predicting the qualitative effects of electron-electron interactions on the electronic properties of metals, as well as offering an explanation for the quite extraordinary success the independent electron approximation has had.

We shall not discuss any of the many attempts to develop a really systematic way of treating electron-electron interactions. Such efforts come under the general heading

---

[5] Although the electron does not interact with itself, it is not necessary to exclude its level from the sum in (17.6), for the inclusion of one extra spatially extended level among the 10^{22} or so occupied levels results in a negligible change in the density.

[6] For this reason the Hartree approximation is also known as the "self-consistent field approximation."

of "many-body problems," and have been dealt with, in recent years, through "field theoretic" or "Green's function" methods.

EXCHANGE: THE HARTREE-FOCK APPROXIMATION

The Hartree equations (17.7) have a fundamental inadequacy that is not at all evident from the derivation we gave. The defect emerges if we return to the exact N-electron Schrödinger equation and cast it into the equivalent variational form,[7] which asserts that a solution to $H\Psi = E\Psi$ is given by any state Ψ that makes stationary the quantity:

$$\langle H \rangle_\Psi = \frac{(\Psi, H\Psi)}{(\Psi, \Psi)}, \tag{17.8}$$

where

$$(\Psi, \Phi) = \sum_{s_1} \cdots \sum_{s_N} \int d\mathbf{r}_1 \ldots d\mathbf{r}_N \, \Psi^*(\mathbf{r}_1 s_1, \ldots, \mathbf{r}_N s_N) \, \Phi(\mathbf{r}_1 s_1, \ldots, \mathbf{r}_N s_N). \tag{17.9}$$

In particular, the ground-state wave function is that Ψ that minimizes (17.8). This property of the ground state is frequently exploited to construct approximate ground states by minimizing (17.8) not over all Ψ, but over a limited class of wave functions chosen to have a more tractable form.

It can be shown[8] that the Hartree equations (17.7) follow from minimizing (17.8) over all Ψ of the form:

$$\Psi(\mathbf{r}_1 s_1, \mathbf{r}_2 s_2, \ldots, \mathbf{r}_N s_N) = \psi_1(\mathbf{r}_1 s_1)\psi_2(\mathbf{r}_2 s_2) \ldots \psi_N(\mathbf{r}_N s_N), \tag{17.10}$$

where the ψ_i are a set of N orthonormal one-electron wave functions. Thus the Hartree equations give the best approximation to the full N-electron wave function that can be represented as a simple product of one-electron levels.

The wave function (17.10), however, is incompatible with the Pauli principle, which requires the sign of Ψ to change when any two of its arguments are interchanged:[9]

$$\Psi(\mathbf{r}_1 s_1, \ldots, \mathbf{r}_i s_i, \ldots, \mathbf{r}_j s_j, \ldots, \mathbf{r}_N s_N) = -\Psi(\mathbf{r}_1 s_1, \ldots, \mathbf{r}_j s_j, \ldots, \mathbf{r}_i s_i, \ldots, \mathbf{r}_N s_N). \tag{17.11}$$

Equation (17.11) cannot be satisfied for the product form (17.10) unless Ψ vanishes identically.

The simplest generalization of the Hartree approximation that incorporates the antisymmetry requirement (17.11) is to replace the trial wave function (17.10) by a

---

[7] See Appendix G. The discussion there is for the one-electron Schrödinger equation, but the general case is, if anything, simpler.

[8] We leave this as a straightforward exercise (Problem 1) for the reader.

[9] The antisymmetry of the N-electron wave function is the fundamental manifestation of the Pauli principle. The alternative statement of the principle, that no one-electron level can be multiply occupied, can only be formulated in an independent electron approximation. There it follows directly from the fact that (17.13) must vanish if any $\psi_i = \psi_j$. The Hartree state (17.10) *is* consistent (though not, like (17.13), automatically so) with the prohibition on multiple occupation, provided that no two ψ_i are the same. However, it fails the more fundamental test of antisymmetry.

Slater determinant of one-electron wave functions. This is a linear combination of the product (17.10) and all other products obtainable from it by permutation of the $\mathbf{r}_j s_j$ among themselves, added together with weights $+1$ or -1 so as to guarantee condition (17.11):

$$\Psi = \psi_1(\mathbf{r}_1 s_1)\psi_2(\mathbf{r}_2 s_2)\ldots\psi_N(\mathbf{r}_N s_N) - \psi_1(\mathbf{r}_2 s_2)\psi_2(\mathbf{r}_1 s_1)\ldots\psi_N(\mathbf{r}_N s_N) + \ldots. \quad \textbf{(17.12)}$$

This antisymmetrized product can be written compactly as the determinant of an $N \times N$ matrix:[10]

$$\Psi(\mathbf{r}_1 s_1, \mathbf{r}_2 s_2, \ldots, \mathbf{r}_N s_N) = \begin{vmatrix} \psi_1(\mathbf{r}_1 s_1)\psi_1(\mathbf{r}_2 s_2) & \ldots & \psi_1(\mathbf{r}_N s_N) \\ \psi_2(\mathbf{r}_1 s_1)\psi_2(\mathbf{r}_2 s_2) & \ldots & \psi_2(\mathbf{r}_N s_N) \\ \cdot & \cdot & \cdot \\ \cdot & \cdot & \cdot \\ \cdot & \cdot & \cdot \\ \psi_N(\mathbf{r}_1 s_1)\psi_N(\mathbf{r}_2 s_2) & \ldots & \psi_N(\mathbf{r}_N s_N) \end{vmatrix}. \quad \textbf{(17.13)}$$

With a little bookkeeping (Problem 2) it can be shown that if the energy (17.8) is evaluated in a state of the form (17.13), with orthonormal single electron wave functions $\psi_1 \ldots \psi_N$, then the result is:

$$\begin{aligned} \langle H \rangle_\Psi = &\sum_i \int d\mathbf{r}\, \psi_i^*(\mathbf{r}) \left(-\frac{\hbar^2}{2m}\nabla^2 + U^{\text{ion}}(\mathbf{r}) \right) \psi_i(\mathbf{r}) \\ &+ \frac{1}{2}\sum_{i,j} \int d\mathbf{r}\, d\mathbf{r}' \frac{e^2}{|\mathbf{r} - \mathbf{r}'|} |\psi_i(\mathbf{r})|^2 |\psi_j(\mathbf{r}')|^2 \\ &- \frac{1}{2}\sum_{i,j} \int d\mathbf{r}\, d\mathbf{r}' \frac{e^2}{|\mathbf{r} - \mathbf{r}'|} \delta_{s_i s_j} \psi_i^*(\mathbf{r})\psi_i(\mathbf{r}')\psi_j^*(\mathbf{r}')\psi_j(\mathbf{r}). \end{aligned} \quad \textbf{(17.14)}$$

Notice that the last term in (17.14) is negative and involves the product $\psi_i^*(\mathbf{r})\psi_i(\mathbf{r}')$ in place of the usual one-electron combination $|\psi_i(\mathbf{r})|^2$. Minimizing (17.14) with respect to the ψ_i^* (Problem 2) leads to a generalization of the Hartree equations known as the Hartree-Fock equations:

$$\begin{aligned} -\frac{\hbar^2}{2m}\nabla^2\psi_i(\mathbf{r}) &+ U^{\text{ion}}(\mathbf{r})\psi_i(\mathbf{r}) + U^{\text{el}}(\mathbf{r})\psi_i(\mathbf{r}) \\ &- \sum_j \int d\mathbf{r}' \frac{e^2}{|\mathbf{r} - \mathbf{r}'|} \psi_j^*(\mathbf{r}')\psi_i(\mathbf{r}')\psi_j(\mathbf{r})\delta_{s_i s_j} = \mathcal{E}_i\psi_i(\mathbf{r}), \end{aligned} \quad \textbf{(17.15)}$$

where U^{el} is defined in (17.4) and (17.6).

These equations differ from the Hartree equations (17.7) by an additional term on the left side, known as the *exchange term*. The complexity introduced by the exchange term is considerable. Like the self-consistent field U^{el} (often referred to as the *direct term*) it is nonlinear in ψ, but unlike the direct term it is not of the form $V(\mathbf{r})\psi(\mathbf{r})$. Instead, it has the structure $\int V(\mathbf{r}, \mathbf{r}')\psi(\mathbf{r}')\, d\mathbf{r}'$—i.e., it is an integral operator. As a result, the Hartree-Fock equations are in general quite intractable. The one exception is the free electron gas. When the periodic potential is zero (or constant)

---

[10] Since a determinant changes sign when any two columns are interchanged, this insures that the condition (17.11) holds.

the Hartree-Fock equations can be solved exactly by choosing the ψ_i to be a set of orthonormal plane waves.[11] Although the case of free electrons has dubious bearing on the problem of electrons in a real metal, the free electron solution suggests a further approximation that makes the Hartree-Fock equations in a periodic potential more manageable. We therefore comment briefly on the free electron case.

HARTREE-FOCK THEORY OF FREE ELECTRONS

The familiar set of free electron plane waves,

$$\psi_i(\mathbf{r}) = \left(\frac{e^{i\mathbf{k}_i \cdot \mathbf{r}}}{\sqrt{V}}\right) \times \text{spin function,} \tag{17.16}$$

in which each wave vector less than k_F occurs twice (once for each spin orientation) in the Slater determinant, gives a solution to the Hartree-Fock equation for free electrons. For if plane waves are indeed solutions, then the electronic charge density that determines U^{el} will be uniform. However, in the free electron gas the ions are represented by a uniform distribution of positive charge with the same density as the electronic charge. Hence the potential of the ions is precisely canceled by the direct term: $U^{\text{ion}} + U^{\text{el}} = 0$. Only the exchange term survives, which is easily evaluated by writing the Coulomb interaction in terms of its Fourier transform[12]:

$$\frac{e^2}{|\mathbf{r} - \mathbf{r}'|} = 4\pi e^2 \frac{1}{V} \sum_{\mathbf{q}} \frac{1}{q^2} e^{i\mathbf{q} \cdot (\mathbf{r}-\mathbf{r}')} \to 4\pi e^2 \int \frac{d\mathbf{q}}{(2\pi)^3} \frac{1}{q^2} e^{i\mathbf{q} \cdot (\mathbf{r}-\mathbf{r}')}. \tag{17.17}$$

If (17.17) is substituted into the exchange term in (17.15) and the ψ_i are all taken to be plane waves of the form (17.16), then the left side of (17.15) assumes the form

$$\mathcal{E}(\mathbf{k}_i)\psi_i, \tag{17.18}$$

where

$$\begin{aligned}\mathcal{E}(\mathbf{k}) &= \frac{\hbar^2 k^2}{2m} - \frac{1}{V} \sum_{k' < k_F} \frac{4\pi e^2}{|\mathbf{k} - \mathbf{k}'|^2} = \frac{\hbar^2 k^2}{2m} - \int_{k' < k_F} \frac{d\mathbf{k}'}{(2\pi)^3} \frac{4\pi e^2}{|\mathbf{k} - \mathbf{k}'|^2} \\ &= \frac{\hbar^2 k^2}{2m} - \frac{2e^2}{\pi} k_F F\left(\frac{k}{k_F}\right),\end{aligned} \tag{17.19}$$

and

$$F(x) = \frac{1}{2} + \frac{1 - x^2}{4x} \ln \left|\frac{1 + x}{1 - x}\right|. \tag{17.20}$$

This shows that plane waves do indeed solve (17.15), and that the energy of the one-electron level with wave vector **k** is given by (17.19). The function $F(x)$ is plotted in Figure 17.1a, and the energy $\mathcal{E}(\mathbf{k})$ in Figure 17.1b.

Several features of the energy (17.19) deserve comment:

1. Although the Hartree-Fock one-electron levels continue to be plane waves, the energy of an electron in the level $e^{i\mathbf{k} \cdot \mathbf{r}}$ is now given by $\hbar^2 k^2/2m$ plus a term describing

---

[11] More complicated solutions, known as spin density waves, are also possible (Chapter 32).

[12] See Problem 3.

Figure 17.1
(a) A plot of the function $F(x)$, defined by Eq. (17.20). Although the slope of this function diverges at $x = 1$, the divergence is logarithmic, and cannot be revealed by changing the scale of the plot. At large values of x the behavior is $F(x) \to 1/3x^2$. (b) The Hartree-Fock energy (17.19) may be written

$$\frac{\mathcal{E}_k}{\mathcal{E}_F^0} = \left[x^2 - 0.663\left(\frac{r_s}{a_0}\right)F(x)\right],$$

where $x = k/k_F$. This function is plotted here for $r_s/a_0 = 4$, and may be compared with the free electron energy (white line). Note that in addition to depressing the free electron energy substantially, the exchange term has led to a considerable increase in the bandwidth (in these units from 1 to 2.33), an effect not corroborated by experiments such as soft X-ray emission or photoelectron emission from metals, which purport to measure such bandwidths.

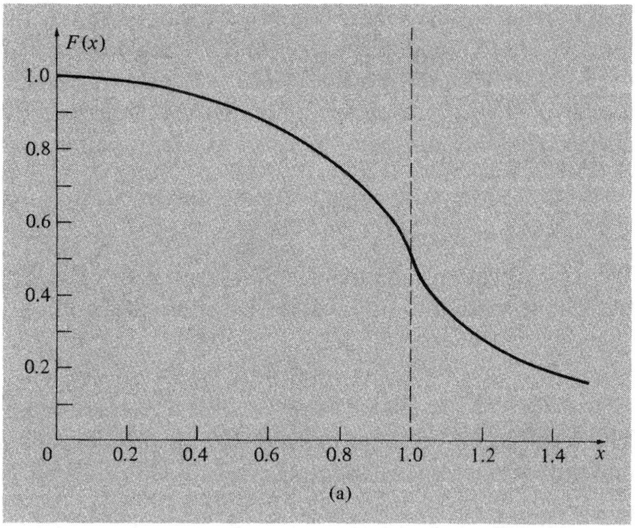

(a)

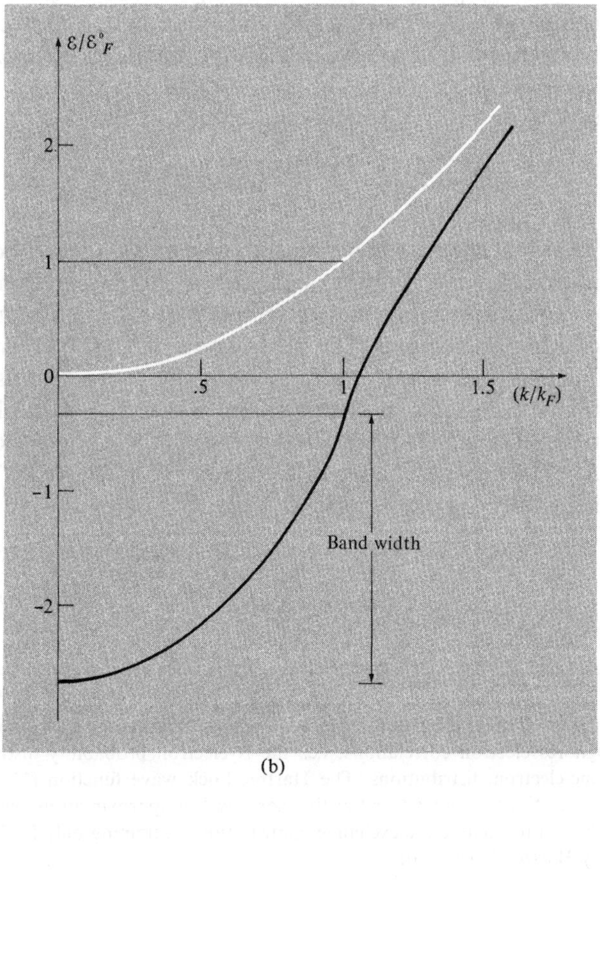

(b)

the effects of the electron-electron interaction. To compute the contribution of these interactions to the total energy of the N-electron system, we must sum this correction over all $k < k_F$, multiply by 2 (for the two spin levels that are occupied for each $\mathbf{k}$), and divide by 2 (because, in summing the interaction energy of a given electron over all electrons, we are counting each electron pair twice). In this way we find that

$$E = 2 \sum_{k<k_F} \frac{\hbar^2 k^2}{2m} - \frac{e^2 k_F}{\pi} \sum_{k<k_F} \left[1 + \frac{k_F{}^2 - k^2}{2kk_F} \ln \left| \frac{k_F + k}{k_F - k} \right| \right]. \qquad (17.21)$$

We have already evaluated the first term in Chapter 2 (Eq. (2.31)). If we transform the second term into an integral, it can be evaluated to give:

$$E = N \left[\frac{3}{5} \mathcal{E}_F - \frac{3}{4} \frac{e^2 k_F}{\pi} \right]. \qquad (17.22)$$

This result is conventionally written in terms of the rydberg ($e^2/2a_0 = 1\ \text{Ry} = 13.6\ \text{eV}$) and the parameter r_s/a_0 (page 4):

$$\frac{E}{N} = \frac{e^2}{2a_0} \left[\frac{3}{5} (k_F a_0)^2 - \frac{3}{2\pi} (k_F a_0) \right] = \left[\frac{2.21}{(r_s/a_0)^2} - \frac{0.916}{(r_s/a_0)} \right] \text{Ry}. \qquad (17.23)$$

Since r_s/a_0 in metals is in the range from 2 to 6, the second term in (17.23) is quite comparable to the first in size, and indicates that electron-electron interactions cannot be overlooked in any free electron estimate of the electronic energy of a metal.

2. With much labor the *exact* leading terms in a high-density (i.e., small r_s/a_0) expansion of the ground-state energy of the electron gas have been calculated:[13]

$$\frac{E}{N} = \left[\frac{2.21}{(r_s/a_0)^2} - \frac{0.916}{(r_s/a_0)} + 0.0622 \ln (r_s/a_0) - 0.096 + O(r_s/a_0) \right] \text{Ry}. \qquad (17.24)$$

Note that the first two terms are just the Hartree-Fock result (17.23). Since r_s/a_0 is not small in metals, this expansion is of doubtful relevance, but its derivation marked one of the first systematic attempts to work out a more accurate theory of electron-electron interactions. The next two terms in (17.24) and all other corrections to the Hartree-Fock result are conventionally referred to as the *correlation energy*. Note that the correlation energy is not a quantity with physical significance; it merely represents the error incurred in making a fairly crude first-order approximation.[14]

3. The *average* change in the energy of an electron from $\hbar^2 k^2/2m$ due to exchange is just the second term in E/N; i.e.,

$$\langle \mathcal{E}^{\text{exchg}} \rangle = -\frac{3}{4} \frac{e^2 k_F}{\pi} = -\frac{0.916}{(r_s/a_0)} \text{Ry}. \qquad (17.25)$$

---

[13] M. Gell-Mann and K. Brueckner, *Phys. Rev.* **106**, 364 (1957).

[14] Indeed, the name 'correlation energy" is something of a misnomer. The Hartree approximation ignores electron correlations; i.e., the N-electron probability distribution factors into a product of N one-electron distributions. The Hartree-Fock wave function (17.13) does not so factor; i.e., electron correlations are introduced at this next level of approximation. Nevertheless the "correlation energy" is defined to exclude the exchange contribution, containing only further corrections beyond that provided by Hartree-Fock theory.

This form led Slater[15] to suggest that in nonuniform systems and, in particular, in the presence of the periodic potential of the lattice, one could simplify the Hartree-Fock equations by replacing the exchange term in (17.15) by a local energy given by twice (17.25) with k_F evaluated at the local density; i.e., he proposed an equation in which the effect of exchange was taken into account by merely adding to the Hartree term $U^{\text{el}}(\mathbf{r})$ an additional potential $U^{\text{exchg}}(\mathbf{r})$, given by

$$U^{\text{exchg}}(\mathbf{r}) = -2.95(a_0{}^3 n(\mathbf{r}))^{1/3}\ \text{Ry}. \tag{17.26}$$

This procedure, gross and *ad hoc* though it is, is actually followed in many band structure calculations. There have been some controversies[16] over whether it is better to average the free electron exchange over all k or to evaluate it at $k = k_F$, but the crude nature of the approximation makes the dispute one of limited content. It is hard to say more for this simplification than that it approximates the effects of exchange by introducing a potential that favors regions of high density, in a way that roughly mimics the density dependence of the exchange term in the free electron energy density.

4. Equation (17.19) has one rather alarming feature: The derivative $\partial\mathcal{E}/\partial k$ becomes logarithmically infinite[17] at $k = k_F$. Since $(1/\hbar)\ \partial\mathcal{E}/\partial\mathbf{k}|_{k=k_F}$ is precisely the velocity of those electrons most important for metallic properties, this is an unsettling result. A singularity at $k = k_F$ in the one-electron energies makes the Sommerfeld expansion (2.70) invalid, and leads in this case to an electronic heat capacity at low temperatures going not as T, but as $T/|\ln T|$.

The singularity does not occur for a general noncoulombic potential, but can be traced back to the divergence of the Fourier transform $4\pi e^2/k^2$ of the interaction e^2/r, at $k = 0$. This in turn reflects the very long range of the inverse square force. If the Coulomb interaction were replaced, for example, by one of the form $e^2(e^{-k_0 r}/r)$, then its Fourier transform[18] would be $4\pi e^2/(k^2 + k_0{}^2)$, the $k = 0$ divergence would be eliminated, and the unphysical singularity of the Hartree-Fock energies removed. It can be argued (see below) that the potential appearing in the exchange term should be modified in just this way to take into account the fields of electrons other than the two at $\mathbf{r}$ and $\mathbf{r}'$, which rearrange themselves so as partially to cancel the fields the two electrons exert on one another. This effect, known as "screening," is of fundamental importance not only for its effects on the electron-electron interaction energy, but, more generally, in determining the behavior of any charge-carrying disturbance in a metal.[19]

SCREENING (GENERAL)

The phenomenon of screening is one of the simplest and most important manifestations of electron-electron interactions. We consider here only screening in a free

---

[15] J. C. Slater, *Phys. Rev.* **81**, 385 (1951); **82**, 538 (1951); **91**, 528 (1953).

[16] See, for example, W. Kohn and L. J. Sham, *Phys. Rev.*, **140**, A1193 (1965), and R. Gaspar, *Acta. Phys. Acad. Sci. Hung.* **3**, 263 (1954).

[17] See Figure 17.1.

[18] Problem 3.

[19] The ions in a metal are an important case and will be referred to in the context of dynamic screening in Chapter 26.

electron gas. The detailed theory of screening in the presence of a realistic periodic potential is much more complex, and one is often forced to use the free electron form of the theory even in discussions of real metals.

Suppose a positively charged particle is placed at a given position in the electron gas and rigidly held there. It will then attract electrons, creating a surplus of negative charge in its neighborhood, which reduces (or screens) its field. In treating this screening it is convenient to introduce two electrostatic potentials. The first, ϕ^{ext}, arises solely from the positively charged particle itself, and therefore satisfies Poisson's equation in the form:

$$-\nabla^2\phi^{\text{ext}}(\mathbf{r}) = 4\pi\rho^{\text{ext}}(\mathbf{r}), \tag{17.27}$$

where $\rho^{\text{ext}}(\mathbf{r})$ is the particle's charge density.[20] The second, ϕ, is the full physical potential, produced by both the positively charged particle and the cloud of screening electrons it induces. It therefore satisfies

$$-\nabla^2\phi(\mathbf{r}) = 4\pi\rho(\mathbf{r}), \tag{17.28}$$

where ρ is the full charge density,

$$\rho(\mathbf{r}) = \rho^{\text{ext}}(\mathbf{r}) + \rho^{\text{ind}}(\mathbf{r}), \tag{17.29}$$

and ρ^{ind} is the charge density induced in the electron gas by the presence of the external particle.

By analogy with the theory of dielectric media, one assumes that ϕ and ϕ^{ext} are linearly related by an equation of the form[21]

$$\phi^{\text{ext}}(\mathbf{r}) = \int d\mathbf{r}'\, \epsilon(\mathbf{r}, \mathbf{r}')\phi(\mathbf{r}'). \tag{17.30}$$

In a spatially uniform electron gas ϵ can depend only on the separation between the points $\mathbf{r}$ and $\mathbf{r}'$, but not on their absolute position:

$$\epsilon(\mathbf{r}, \mathbf{r}') = \epsilon(\mathbf{r} - \mathbf{r}'). \tag{17.31}$$

Thus (17.30) assumes the form

$$\phi^{\text{ext}}(\mathbf{r}) = \int d\mathbf{r}'\, \epsilon(\mathbf{r} - \mathbf{r}')\phi(\mathbf{r}'), \tag{17.32}$$

---

[20] The term "external" and superscript "ext" used to describe the applied charge are not meant to suggest that the charge is placed outside of the metal—it is in fact inside the metal—but refer only to its origin from some source of charge external to the system of electrons.

[21] The potential ϕ^{ext} is analogous to the electric displacement $\mathbf{D}$ (whose sources are the "free" charges extrinsic to the medium); the potential ϕ is analogous to the electric field $\mathbf{E}$, which arises from the total charge distribution, including both the "free" charges and the "bound" charges induced in the medium. The relation $\mathbf{D}(\mathbf{r}) = \int d\mathbf{r}'\epsilon(\mathbf{r} - \mathbf{r}')\mathbf{E}(\mathbf{r}')$ (or the corresponding relation (17.32)) reduces to the more familiar local relation $\mathbf{D}(\mathbf{r}) = \epsilon\mathbf{E}(\mathbf{r})$, with the dielectric constant ϵ being given by $\epsilon = \int d\mathbf{r}\epsilon(\mathbf{r})$, when $\mathbf{D}$ and $\mathbf{E}$ are spatially uniform fields (or, more generally, when the fields vary slowly on the scale of some r_0 for which $\epsilon(\mathbf{r}) \approx 0$, $r > r_0$.)

which implies[22] that the corresponding Fourier transforms satisfy

$$\phi^{\text{ext}}(\mathbf{q}) = \epsilon(\mathbf{q})\phi(\mathbf{q}), \tag{17.33}$$

where the Fourier transforms are defined by

$$\epsilon(\mathbf{q}) = \int d\mathbf{r}\, e^{-i\mathbf{q}\cdot\mathbf{r}}\epsilon(\mathbf{r}), \tag{17.34}$$

$$\epsilon(\mathbf{r}) = \int \frac{d\mathbf{q}}{(2\pi)^3}\, e^{i\mathbf{q}\cdot\mathbf{r}}\epsilon(\mathbf{q}), \tag{17.35}$$

with similar equations for ϕ and ϕ^{ext}.

The quantity $\epsilon(\mathbf{q})$ is called the (wave vector dependent) dielectric constant of the metal.[23] When written in the form

$$\boxed{\phi(\mathbf{q}) = \frac{1}{\epsilon(\mathbf{q})}\,\phi^{\text{ext}}(\mathbf{q})} \tag{17.36}$$

equation (17.33) asserts that the **q**th Fourier component of the total potential present in the electron gas is just the **q**th Fourier component of the external potential, reduced by the factor $1/\epsilon(\mathbf{q})$. This kind of relation is familiar in elementary discussions of dielectrics, where, however, the fields are generally uniform so that the dependence on wave vector does not come into play.

The quantity that turns out to be the most natural to calculate directly is not the dielectric constant $\epsilon(\mathbf{q})$ but the charge density $\rho^{\text{ind}}(\mathbf{r})$ induced in the electron gas by the total potential $\phi(\mathbf{r})$. We shall examine below how this can be calculated. When ρ^{ind} and ϕ are linearly related (as they will be for sufficiently weak ϕ), then their Fourier transforms will satisfy a relation of the form

$$\rho^{\text{ind}}(\mathbf{q}) = \chi(\mathbf{q})\phi(\mathbf{q}). \tag{17.37}$$

We can relate ϵ (the quantity of direct physical interest) to χ (the quantity that emerges naturally from a calculation) as follows:

The Fourier transforms of the Poisson equations (17.27) and (17.28) are

$$\begin{aligned} q^2\phi^{\text{ext}}(\mathbf{q}) &= 4\pi\rho^{\text{ext}}(\mathbf{q}), \\ q^2\phi(\mathbf{q}) &= 4\pi\rho(\mathbf{q}). \end{aligned} \tag{17.38}$$

Together with (17.29) and (17.37) these give

$$\frac{q^2}{4\pi}\left(\phi(\mathbf{q}) - \phi^{\text{ext}}(\mathbf{q})\right) = \chi(\mathbf{q})\phi(\mathbf{q}), \tag{17.39}$$

---

[22] Through the convolution theorem of Fourier analysis. We follow the usual physicists' practice of using the same symbol for a function and its Fourier transform, distinguishing the two by the symbol used for the argument.

[23] In elementary discussions of electrostatics it is sometimes asserted that the dielectric constant of a metal is infinite—i.e., that the charge is free to move and the medium is therefore infinitely polarizable. We shall find that the form of $\epsilon(\mathbf{q})$ is consistent with this, for in the limit of a spatially uniform applied field ($q \to 0$) $\epsilon(q)$ does indeed become infinite. (See Eq. (17.51).)

or

$$\phi(\mathbf{q}) = \phi^{\text{ext}}(\mathbf{q}) \Big/ \left(1 - \frac{4\pi}{q^2}\chi(\mathbf{q})\right). \tag{17.40}$$

Comparing this with (17.36) leads to the relation

$$\boxed{\epsilon(\mathbf{q}) = 1 - \frac{4\pi}{q^2}\chi(\mathbf{q}) = 1 - \frac{4\pi}{q^2}\frac{\rho^{\text{ind}}(\mathbf{q})}{\phi(\mathbf{q})}}. \tag{17.41}$$

Except for the assumption that the externally applied charge is weak enough to produce only a linear response in the electron gas, the analysis up to this point has been exact (though little more than a series of definitions). Serious approximations become necessary when one tries to calculate χ. Two widely prevalent theories of χ are employed, both simplifications of a general Hartree calculation of the charge induced by the impurity. The first—the Thomas-Fermi method—is basically the classical (more precisely, the semiclassical) limit of the Hartree theory; the second—the Lindhard method, also known as the random phase approximation (or RPA)—is basically an exact Hartree calculation of the charge density in the presence of the self-consistent field of the external charge plus electron gas, except that the Hartree calculation is simplified from the start by the fact that ρ^{ind} is only required to linear order in ϕ.

The Thomas-Fermi method has the advantage that it is applicable even when a linear relation between ρ^{ind} and ϕ does not hold; it has the disadvantage that it is reliable only for very slowly varying external potentials. When the Thomas-Fermi result is linearized, it is identical to the Lindhard result at small values of q, and less accurate than the Lindhard result when q is not small. We describe the two cases separately below.

THOMAS-FERMI THEORY OF SCREENING

In principle, to find the charge density in the presence of the total potential $\phi = \phi^{\text{ext}} + \phi^{\text{ind}}$ we must solve the one-electron Schrödinger equation,[24]

$$-\frac{\hbar^2}{2m}\nabla^2\psi_i(\mathbf{r}) - e\phi(\mathbf{r})\psi_i(\mathbf{r}) = \mathcal{E}_i\psi_i(\mathbf{r}), \tag{17.42}$$

and then construct the electronic density from the one-electron wave functions using (17.6). The Thomas-Fermi approach is based on a simplification in this procedure that can be made when the total potential $\phi(\mathbf{r})$ is a very slowly varying function of $\mathbf{r}$. Here we mean "slowly varying" in precisely the same sense as in Chapters 2 and

---

[24] Because ϕ is the total potential, arising from both the external charge and the charge density it induces in the electron gas, Eq. (17.42) is implicitly treating electron-electron interactions in the Hartree approximation. The self-consistency problem (at least in the linearized version of the theory) is contained in the stipulation that ϕ is related to the electronic charge density ρ^{ind} determined by the solutions to (17.42), through Eqs. (17.36) and (17.41).

12; i.e., we assume it is meaningful to specify the energy vs. wave vector relation of an electron at the position **r**, and we take this relation to be

$$\mathcal{E}(\mathbf{k}) = \frac{\hbar^2 k^2}{2m} - e\phi(\mathbf{r}). \tag{17.43}$$

Thus the energy is modified from its free electron value by the total local potential.

Evidently (17.43) makes sense only in terms of wave packets. These will have a typical spread in position at least of the order of $1/k_F$. We must therefore require that $\phi(\mathbf{r})$ vary slowly on the scale of a Fermi wavelength. In terms of Fourier components this means that the calculation will be reliable only for values of $\chi(\mathbf{q})$ with $q \ll k_F$. We shall verify this limitation explicitly when we turn to the more accurate Lindhard approach.

Thus we assume that the solutions to Eq. (17.42) describe a set of electrons with energies of the simple classical form (17.43). To calculate the charge density produced by these electrons we place their energies into the expression (2.58) for the electronic number density, to find (with $\beta = 1/k_B T$)

$$n(\mathbf{r}) = \int \frac{d\mathbf{k}}{4\pi^3} \frac{1}{\exp\left[\beta((\hbar^2 k^2/2m) - e\phi(\mathbf{r}) - \mu)\right] + 1}. \tag{17.44}$$

The induced charge density is just $-en(\mathbf{r}) + en_0$, where the second term is the charge density of the uniform positive background. The number density of the background is just the density of the electronic system when ϕ^{ext}, and hence ϕ, vanishes:[25]

$$n_0(\mu) = \int \frac{d\mathbf{k}}{4\pi^3} \frac{1}{\exp\left[\beta((\hbar^2 k^2/2m) - \mu)\right] + 1}. \tag{17.45}$$

We combine (17.44) and (17.45) to write,

$$\rho^{\text{ind}}(\mathbf{r}) = -e\left[n_0(\mu + e\phi(\mathbf{r})) - n_0(\mu)\right]. \tag{17.46}$$

This is the basic equation of nonlinear Thomas-Fermi theory.

In the present case we assume that ϕ is small enough for (17.46) to be expanded to give in leading order

$$\rho^{\text{ind}}(\mathbf{r}) = -e^2 \frac{\partial n_0}{\partial \mu} \phi(\mathbf{r}). \tag{17.47}$$

Comparing (17.47) with (17.37) we find that $\chi(\mathbf{q})$ is given by the constant

$$\chi(\mathbf{q}) = -e^2 \frac{\partial n_0}{\partial \mu}, \qquad \text{independent of } \mathbf{q}. \tag{17.48}$$

Substituting this in (17.41) gives the Thomas-Fermi dielectric constant[26]

$$\epsilon(\mathbf{q}) = 1 + \frac{4\pi e^2}{q^2} \frac{\partial n_0}{\partial \mu}. \tag{17.49}$$

---

[25] The values of the chemical potential μ appearing in (17.44) and (17.45) will be the same under the assumption that $\phi(\mathbf{r})$ is appreciable only in a finite region of the electron gas, outside of which the electron density is negligibly perturbed from its equilibrium value.

[26] As anticipated, this form for the dielectric constant does indeed become infinite as $q \to 0$. (See footnote 23.)

It is customary to define a Thomas-Fermi wave vector k_0 by:

$$k_0^2 = 4\pi e^2 \frac{\partial n_0}{\partial \mu}, \tag{17.50}$$

so that

$$\epsilon(\mathbf{q}) = 1 + \frac{k_0^2}{q^2}. \tag{17.51}$$

To illustrate the significance of k_0, consider the case where the external potential is that of a point charge:

$$\phi^{\text{ext}}(\mathbf{r}) = \frac{Q}{r}, \quad \phi^{\text{ext}}(\mathbf{q}) = \frac{4\pi Q}{q^2}. \tag{17.52}$$

The total potential in the metal will then be

$$\phi(\mathbf{q}) = \frac{1}{\epsilon(\mathbf{q})}\,\phi^{\text{ext}}(\mathbf{q}) = \frac{4\pi Q}{q^2 + k_0^2}. \tag{17.53}$$

The Fourier transform can be inverted to give (see Problem 3)

$$\phi(\mathbf{r}) = \int \frac{d\mathbf{q}}{(2\pi)^3}\, e^{i\mathbf{q}\cdot\mathbf{r}}\, \frac{4\pi Q}{q^2 + k_0^2} = \frac{Q}{r}\, e^{-k_0 r}. \tag{17.54}$$

Thus the total potential is of the coulombic form times an exponential damping factor that reduces it to a negligible size at distances greater than order $1/k_0$. This form is known as a *screened Coulomb potential*[27] or (from an analogous form in meson theory) a *Yukawa potential.*

We have thus extracted the anticipated result that the electrons screen out the field of the external charge. In addition, we have an explicit expression for the characteristic distance beyond which the disturbance is effectively screened. To estimate k_0 note that for a free electron gas, when $T \ll T_F$, $\partial n_0/\partial\mu$ is simply the density of levels at the Fermi energy, $g(\mathcal{E}_F) = mk_F/\hbar^2\pi^2$ (Eq. (2.64)). Therefore

$$\frac{k_0^2}{k_F^2} = \frac{4}{\pi}\frac{me^2}{\hbar^2 k_F} = \frac{4}{\pi}\frac{1}{k_F a_0} = \left(\frac{16}{3\pi^2}\right)^{2/3}\left(\frac{r_s}{a_0}\right);$$

$$k_0 = 0.815\, k_F \left(\frac{r_s}{a_0}\right)^{1/2} = \frac{2.95}{(r_s/a_0)^{1/2}}\,\text{Å}^{-1}. \tag{17.55}$$

Since r_s/a_0 is about 2 to 6 at metallic densities, k_0 is of the order of k_F; i.e., disturbances are screened in a distance which is similar to the interparticle spacing. Thus the electrons are highly effective in shielding external charges.

---

27 This form appears in the theory of electrolytes as originally given by P. Debye and E. Hückel, *Phys. Z.* **24**, 185, 305 (1923).

LINDHARD THEORY OF SCREENING

In the Lindhard approach[28] one returns to the Schrödinger equation (17.42) and does not make a semiclassical approximation, requiring a slowly varying ϕ. Instead, one exploits from the outset the fact that the induced density is required only to linear order in the total potential ϕ. It is then a routine matter to solve (17.42) to linear order by perturbation theory. Once one knows the electronic wave functions to linear order in ϕ one can also compute the linear change in electronic charge density via (17.6). The procedure is straightforward (Problem 5) and here we quote only the result. Equation (17.48) of linearized Thomas-Fermi theory must be generalized to

$$\chi(\mathbf{q}) = -e^2 \int \frac{d\mathbf{k}}{4\pi^3} \frac{f_{\mathbf{k}-\frac{1}{2}\mathbf{q}} - f_{\mathbf{k}+\frac{1}{2}\mathbf{q}}}{\hbar^2 \mathbf{k}\cdot\mathbf{q}/m}, \tag{17.56}$$

where $f_{\mathbf{k}}$ denotes the equilibrium Fermi function for a free electron with energy $\hbar^2k^2/2m$: $f_{\mathbf{k}} = 1/\{\exp[\beta(\hbar^2k^2/2m - \mu)] + 1\}$.

Note that when q is small compared to k_F the numerator of the integrand can be expanded about its value at $\mathbf{q} = 0$:

$$f_{\mathbf{k}\mp\frac{1}{2}\mathbf{q}} = f_{\mathbf{k}} \pm \frac{\hbar^2}{2}\frac{\mathbf{k}\cdot\mathbf{q}}{m}\frac{\partial}{\partial\mu} f_{\mathbf{k}} + O(q^2). \tag{17.57}$$

The term linear in $\mathbf{q}$ in this expansion gives the Thomas-Fermi result (17.48). Thus, as expected, in the limit of a slowly varying disturbance Lindhard theory reduces to Thomas-Fermi theory.[29] However, when q becomes comparable to k_F there is considerably more structure in the Lindhard dielectric constant. At $T = 0$ the integrals in (17.56) can be performed explicitly to give

$$\chi(\mathbf{q}) = -e^2\left(\frac{mk_F}{\hbar^2\pi^2}\right)\left[\frac{1}{2} + \frac{1-x^2}{4x}\ln\left|\frac{1+x}{1-x}\right|\right], \qquad x = \frac{q}{2k_F}. \tag{17.58}$$

The quantity in square brackets, which is 1 at $x = 0$, is the Lindhard correction[30] to the Thomas-Fermi result. Note that at $q = 2k_F$ the dielectric constant $\epsilon = 1 - 4\pi\chi/q^2$ is not analytic. As a result it can be shown that at large distances the screened potential ϕ of a point charge now has a term that goes (at $T = 0$) as:

$$\phi(\mathbf{r}) \sim \frac{1}{r^3}\cos 2k_F r. \tag{17.59}$$

Thus the screening at large distances has considerably more structure than the simple Yukawa potential predicted by the Thomas-Fermi theory, with a much more weakly decaying oscillatory term. Depending on the context these oscillations go under the name of Friedel oscillations or Ruderman-Kittel oscillations. We comment on them further in Chapter 26.

---

[28] J. Lindhard, *Kgl. Danske Videnskab. Selskab Mat.-Fys. Medd.* **28**, No. 8 (1954).

[29] Indeed, the Thomas-Fermi $\chi(\mathbf{q})$ can be characterized as the limit of the Lindhard $\chi(\mathbf{q})$ as $q \to 0$.

[30] The function in square brackets happens to be the function $F(x)$ appearing in the Hartree-Fock energy (17.19) for free electrons and plotted in Figure 17.1a.

FREQUENCY-DEPENDENT LINDHARD SCREENING

If the external charge density has time dependence $e^{-i\omega t}$, then the induced potential and charge density will also have such a time dependence, and the dielectric constant will depend on frequency as well as on wave vector. In the limiting case, where collisions can be ignored, the Lindhard argument can be straightforwardly generalized by using time-dependent rather than stationary perturbation theory. One finds that the static result (17.56) need be modified only by the addition of the quantity $\hbar\omega$ to the denominator of the integrand.[31] This more general form is of considerable importance in the theory of lattice vibrations in metals, as well as in the theory of superconductivity. Here we note only that when q approaches zero at fixed ω, the Lindhard dielectric constant,

$$\epsilon(\mathbf{q}, \omega) = 1 + \frac{4\pi e^2}{q^2} \int \frac{d\mathbf{k}}{4\pi^3} \frac{f_{\mathbf{k}-\frac{1}{2}\mathbf{q}} - f_{\mathbf{k}+\frac{1}{2}\mathbf{q}}}{\hbar^2 \mathbf{k} \cdot \mathbf{q}/m + \hbar\omega}, \tag{17.60}$$

reduces to the Drude result (1.37), which we derived under the assumption of a spatially uniform disturbance. Thus the more sophisticated Lindhard approach is consistent with more elementary investigations in the regimes in which they are applicable.

SCREENING THE HARTREE-FOCK APPROXIMATION

We have discussed screening by metallic electrons of an externally imposed charge distribution. However, screening will also affect the interaction of two electrons with each other, since from the point of view of the remaining electrons these two can be considered as external charges. If one returns to the Hartree-Fock equations and takes this point of view, one can make an important improvement. One cannot tamper with the Hartree self-consistent field term, since this is the term that gives rise to the screening in the first place. However, it is tempting[32] to replace the electron-electron interaction that occurs in the exchange term by its screened form, multiplying $1/(\mathbf{k} - \mathbf{k}')^2$ by the inverse dielectric constant $1/\epsilon(\mathbf{k} - \mathbf{k}')$ in (17.19). This eliminates the singularity responsible for the anomalous divergence in the one electron velocity $\mathbf{v}(\mathbf{k}) = (1/\hbar)(\partial\mathcal{E}(\mathbf{k})/\partial\mathbf{k})$ at $k = k_F$, for in the neighborhood of $q = 0$ the screened interaction approaches not e^2/q^2 but $e^2/k_0{}^2$. If one now calculates the value of $\mathbf{v}(\mathbf{k})$ at $k = k_F$, one finds that for values of r_s typical of metallic densities the velocity differs from its free electron value by only about 5 percent. Thus screening has characteristically reduced the importance of electron-electron interactions.[33]

---

[31] When the denominator vanishes, the integral is made unambiguous by the requirement that it be evaluated by giving ω a vanishingly small positive imaginary part.

[32] One of the successes of the Green's function approach is the justification in a systematic way of this apparently *ad hoc* introduction of screening into the exchange term.

[33] In the present case the reduction is rather dramatic, from a divergence to a minor correction of a few percent.

FERMI LIQUID THEORY

We conclude this chapter with a brief look at some deep and subtle arguments, primarily due to Landau,[34] that (a) explain the remarkable success of the independent electron approximation in spite of the strength of electron-electron interactions and (b) indicate how, in many cases, particularly in the calculation of transport properties, the consequences of electron-electron interactions can qualitatively be taken into account. Landau's approach is known as Fermi liquid theory. It was designed to deal with the liquid state of the isotope of helium of mass number 3, but is increasingly being applied to the theory of electron-electron interactions in metals.[35]

We first observe that up to this point our analysis of electron-electron interactions has led to a substantially modified energy vs. wave vector relation for the one-electron levels (e.g., Eq. (17.19)), but has not in any substantial way challenged the basic *structure* of the independent electron model, in which the electronic properties of a metal are viewed as arising from the occupation of a specified set of one-electron levels. Thus even in the Hartree-Fock approximation we continue to describe the stationary electronic states by specifying which one-electron levels ψ_i are present in the Slater determinant (17.13). The N-electron wave function therefore has exactly the same structure as it would have for noninteracting electrons, the only modification being that the form of the one-electron wave functions ψ_i may be affected by the interactions.[36] It is far from clear that this is a sensible way to describe the stationary states of the N-electron system. Suppose, for example, that the net electron-electron interaction were attractive, and so much so that pairs of electrons formed bound states.[37] Then the natural way to describe the electrons in a metal would be in terms of electron pairs. Such a metal could no more be adequately described by the stationary states of a set of independent single electrons than a gas of oxygen molecules could be described in terms of independent oxygen atoms.

Even if nothing as drastic as pair formation happens, it is still far from obvious that an independent electron description, with suitably modified energies, will be anywhere near the mark in describing the electrons in an actual metal. There is, however, reason to expect that this may be the case for electrons with energies near the Fermi energy.[38] The argument, due to Landau, can be divided into two stages. The first is fairly straightforward, but the second is very subtle indeed.

---

[34] L. D. Landau, *Sov. Phys. JETP* **3**, 920 (1957); **5**, 101 (1957); and **8**, 70 (1959).

[35] A thorough and fairly elementary survey of the theory of charged Fermi liquids up to 1966 can be found in *The Theory of Quantum Liquids I*, D. Pines and P. Nozieres, W. A. Benjamin, Menlo Park, California, 1966.

[36] For free electrons, not even this change is made. The wave functions continue to be plane waves.

[37] Something like this actually happens in a superconductor. See Chapter 34.

[38] There is no justification for the independent electron picture when the electronic energies are far from the Fermi energy, but happily, as we have seen in Chapters 2, 12, and 13, many of the most interesting electronic properties of a metal are almost completely determined by electrons within $k_B T$ of the Fermi energy. However, any property involving electronic levels well below or high above the Fermi energy (such as soft X-ray emission, photoelectric emission, or optical absorption) may very well be substantially affected by electron-electron interactions.

FERMI LIQUID THEORY: CONSEQUENCES OF THE EXCLUSION PRINCIPLE ON ELECTRON-ELECTRON SCATTERING NEAR THE FERMI ENERGY

Consider a set of noninteracting electrons. If we imagine gradually turning on the interactions between electrons, they will lead to two kinds of effects:

1. The energies of each one-electron level will be modified.[39] This is the kind of effect illustrated by the Hartree-Fock approximation and its refinements. We shall return to it below.
2. Electrons will be scattered in and out of the single electron levels, which are no longer stationary. This does not happen in the Hartree-Fock approximation, where one-electron levels continue to give valid stationary states of the interacting system. Whether this scattering is serious enough to invalidate the independent electron picture depends on how rapid the rate of scattering is. If it is sufficiently low, we could introduce a relaxation time and treat the scattering in the same way as the other scattering mechanisms we have discussed in our theories of transport processes. If it should happen (and we shall see that it usually does) that the electron-electron relaxation time is much larger than other relaxation times, then we can safely ignore it altogether and use the independent electron model with considerably more confidence, subject only to modifications required by the altered energy vs. **k** relation.[40]

One naively might expect the electron-electron scattering rate to be quite high, since the Coulomb interaction, even when screened, is rather strong. However, the exclusion principle comes dramatically to the rescue by reducing the scattering rate quite spectacularly in many cases of major interest. This reduction occurs when the electronic configuration differs only slightly from its thermal equilibrium form (as is the case in all of the transport processes we have investigated in Chapter 13). To illustrate the effect of the exclusion principle on the scattering rate, suppose, for example, that the N-electron state consists of a filled Fermi sphere (thermal equilibrium at $T = 0$) plus a single excited electron in a level with $\mathcal{E}_1 > \mathcal{E}_F$. In order for this electron to be scattered, it must interact with an electron of energy $\mathcal{E}_2$, which must be less than $\mathcal{E}_F$, since only electronic levels with energies less than $\mathcal{E}_F$ are occupied. The exclusion principle requires that these two electrons can only scatter into *unoccupied* levels, whose energies $\mathcal{E}_3$ and $\mathcal{E}_4$ must therefore be greater than $\mathcal{E}_F$. Thus we require that

$$\mathcal{E}_2 < \mathcal{E}_F, \quad \mathcal{E}_3 > \mathcal{E}_F, \quad \mathcal{E}_4 > \mathcal{E}_F. \tag{17.61}$$

In addition, energy conservation requires that

$$\mathcal{E}_1 + \mathcal{E}_2 = \mathcal{E}_3 + \mathcal{E}_4. \tag{17.62}$$

---

[39] We defer, for the moment, the question of whether it makes sense to speak of "one-electron levels" at all, when the interaction is turned on. (This is, of course, the central problem, which is why the argument is so subtle.)

[40] And subject to the delicate change in point of view associated with the introduction of "quasi-particles" (see below).

When $\mathcal{E}_1$ is exactly $\mathcal{E}_F$, conditions (17.61) and (17.62) can only be satisfied if $\mathcal{E}_2$, $\mathcal{E}_3$, and $\mathcal{E}_4$ are also all exactly $\mathcal{E}_F$. Thus the allowed wave vectors for electrons 2, 3, and 4 occupy a region of **k** space of *zero volume* (i.e., the Fermi surface), and therefore give a vanishingly small contribution to the integrals that make up the cross section for the process. In the language of scattering theory, one cay say that there is no phase space for the process. Consequently, *the lifetime of an electron at the Fermi surface at $T = 0$ is infinite.*

When $\mathcal{E}_1$ is a little different from $\mathcal{E}_F$, some phase space becomes available for the process, since the other three energies can now vary within a shell of thickness of order $|\mathcal{E}_1 - \mathcal{E}_F|$ about the Fermi surface, and remain consistent with (17.61) and (17.62). This leads to a scattering rate of order $(\mathcal{E}_1 - \mathcal{E}_F)^2$. The quantity appears squared rather than cubed, because once $\mathcal{E}_2$ and $\mathcal{E}_3$ have been chosen within the shell of allowed energies, energy conservation allows no further choice for $\mathcal{E}_4$.

If the excited electron is superimposed not on a filled Fermi sphere, but on a thermal equilibrium distribution of electrons at nonzero T, then there will be partially occupied levels in a shell of width k_BT about $\mathcal{E}_F$. This provides an additional range of choice of order k_BT in the energies satisfying (17.61) and (17.62), and therefore leads to a scattering rate going as $(k_BT)^2$, even when $\mathcal{E}_1 = \mathcal{E}_F$. Combining these considerations, we conclude that at temperature T, an electron of energy $\mathcal{E}_1$ near the Fermi surface has a scattering rate $1/\tau$ that depends on its energy and the temperature in the form

$$\frac{1}{\tau} = a(\mathcal{E}_1 - \mathcal{E}_F)^2 + b(k_BT)^2, \tag{17.63}$$

where the coefficients a and b are independent of $\mathcal{E}_1$ and T.

Thus the electronic lifetime due to electron-electron scattering can be made as large as one wishes by going to sufficiently low temperatures and considering electrons sufficiently close to the Fermi surface. Since it is only electrons within k_BT of the Fermi energy that significantly affect most low-energy metallic properties (those farther down are "frozen in" and there are negligibly few present farther up), the physically relevant relaxation time for such electrons goes as $1/T^2$.

To give a crude, but quantitative, estimate of this lifetime, we argue as follows: Assume that the temperature dependence of τ is completely taken into account by a factor $1/T^2$. We expect from lowest-order perturbation theory (Born approximation) that τ will depend on the electron-electron interaction through the square of the Fourier transform of the interaction potential. Our discussion of screening suggests that this can be estimated by the Thomas-Fermi screened potential, which is everywhere less than $4\pi e^2/k_0{}^2$. We therefore assume that the dependence of τ on temperature and electron-electron interaction is completely taken into account by the form:

$$\frac{1}{\tau} \propto \left(k_BT\right)^2 \left(\frac{4\pi e^2}{k_0{}^2}\right)^2. \tag{17.64}$$

Using the form (17.55) for k_0 we can write this as

$$\frac{1}{\tau} \propto \left(k_BT\right)^2 \left(\frac{\pi^2\hbar^2}{mk_F}\right)^2. \tag{17.65}$$

To establish the form of the proportionality constant we appeal to dimensional analysis. We have left at our disposal only the temperature-independent quantities characterizing a noninteracting electron gas: k_F, m, and $\hbar$. We can construct a quantity with dimensions of inverse time by multiplying (17.65) by $m^3/\hbar^7$, to get

$$\frac{1}{\tau} = A \frac{1}{\hbar} \frac{(k_B T)^2}{\mathcal{E}_F}. \tag{17.66}$$

Since no dimensionless factor can be constructed out of k_F, m, and $\hbar$, (17.66) is the only possible form. We take the dimensionless number A to be of order unity to within a power or two of ten.

At room temperature $k_B T$ is of the order of 10^{-2} eV, and $\mathcal{E}_F$ is of the order of electron volts. Therefore $(k_B T)^2/\mathcal{E}_F$ is of the order of 10^{-4} eV, which leads to a lifetime τ of the order of 10^{-10} second. In Chapter 1 we found that typical metallic relaxation times at room temperature were of the order of 10^{-14} second. We therefore conclude that at room temperature electron-electron scattering proceeds at a rate 10^4 times slower than the dominant scattering mechanism. This is a sufficiently large factor to allow for the power or two of ten error that might easily have crept into our crude dimensional analysis; there is no doubt that at room temperature electron-electron scattering is of little consequence in a metal. Since the electron-electron relaxation time increases as $1/T^2$ with falling temperature, it is quite possible that it can be of little consequence at all temperatures. It is certainly necessary to go to very low temperatures (to eliminate thermal scattering by the ionic vibrations) in very pure specimens (to eliminate impurity scattering) before one can hope to see effects of electron-electron scattering, and indications are only just emerging that it may be possible under these extreme conditions to see the characteristic T^2 dependence.

Therefore, at least for levels within $k_B T$ of the Fermi energy, electron-electron interactions do not appear to invalidate the independent electron picture. However, there is a serious gap in this argument, which brings us to the subtle part of Landau's theory.

FERMI LIQUID THEORY: QUASIPARTICLES

The above argument indicates that *if* the independent electron picture is a good first approximation, then at least for levels near the Fermi energy, electron-electron scattering will not invalidate that picture even if the interactions are strong. However, if the electron-electron interactions are strong it is not at all likely that the independent electron approximation will be a good first approximation, and it is therefore not clear that our argument has any relevance.

Landau cut this Gordian knot by acknowledging that the independent *electron* picture was not a valid starting point. He emphasized, however, that the argument described above remains applicable, provided that an independent *something* picture is still a good first approximation. He christened the "somethings" *quasiparticles* (or *quasielectrons*). If the quasiparticles obey the exclusion principle, then the argument we have given works as well for them as it does for independent electrons, acquiring thereby a much wider validity, provided that we can explain what a quasiparticle might be. Landau's definition of a quasiparticle is roughly this:

Suppose that as the electron-electron interactions are turned on, the states (at least the low-lying ones) of the strongly interacting N-electron system evolve in a continuous way from, and therefore remain in a one-to-one correspondence with, the states of the noninteracting N-electron system. We can specify the excited states of the noninteracting system by specifying how they differ from the ground state—i.e., by listing those wave vectors $\mathbf{k}_1, \mathbf{k}_2, \ldots, \mathbf{k}_n$ above k_F that describe occupied levels, and those, $\mathbf{k}_1', \mathbf{k}_2', \ldots, \mathbf{k}_m'$ below k_F, that describe unoccupied levels.[41] We then describe such a state by saying that m electrons have been excited out of the one-electron levels $\mathbf{k}_1', \ldots, \mathbf{k}_m'$, and n excited electrons are present in the one-electron levels $\mathbf{k}_1, \ldots, \mathbf{k}_n$. The energy of the excited state is just the ground state energy plus $\mathcal{E}(\mathbf{k}_1) + \ldots + \mathcal{E}(\mathbf{k}_n) - \mathcal{E}(\mathbf{k}_1') - \ldots - \mathcal{E}(\mathbf{k}_m')$, where, for free electrons, $\mathcal{E}(\mathbf{k}) = \hbar^2 k^2/2m$.

We now define quasiparticles implicitly, by asserting that the corresponding state of the interacting system is one in which m quasiparticles have been excited out of levels with wave vectors $\mathbf{k}_1' \ldots \mathbf{k}_m'$ and n excited quasiparticles are present in levels with wave vectors $\mathbf{k}_1 \ldots \mathbf{k}_n$. We say that the energy of the state is the ground-state energy plus $\mathcal{E}(\mathbf{k}_1) + \ldots + \mathcal{E}(\mathbf{k}_n) - \mathcal{E}(\mathbf{k}_1') - \ldots - \mathcal{E}(\mathbf{k}_m')$ where the quasiparticle $\mathcal{E}$ vs. $\mathbf{k}$ relation is, in general, quite difficult to determine.

It is not clear, of course, whether this is a consistent thing to do, for it implies that the excitation spectrum for the interacting system, though numerically different from that of the free system, nevertheless has a free electron type of structure. However we can now return to the argument of the preceding section and point out that this is at least a consistent possibility, for if the spectrum does have a structure like the free electron spectrum, then because of the exclusion principle quasiparticle-quasiparticle interactions will not drastically alter that structure, at least for quasiparticles near the Fermi surface.

This glimmering of an idea is a long way from a coherent theory. In particular, we must reexamine the rules for constructing quantities like electric and thermal currents from the distribution function, once we acknowledge that it is describing not electrons, but quasiparticles. Remarkably, these rules turn out to be very similar (but not identical) to what we would do if we were, in fact, dealing with electrons and not quasiparticles. We cannot hope to give an adequate account of this extraordinary subject here, and must refer the reader to the papers of Landau[34] and the book by Pines and Nozieres[35] for a fuller description.

The term "normal Fermi system" is used to refer to those systems of interacting particles obeying Fermi-Dirac statistics, for which the quasiparticle representation is valid. It can be shown by a difficult and ingenious argument of Landau's based on Green's function methods, that to all orders of perturbation theory (in the interaction) every interacting Fermi system is normal. This does not mean, however, that all electronic systems in metals are normal, for it is now well known that the superconducting ground state, as well as several kinds of magnetically ordered ground states, cannot be constructed in a perturbative way from the free electron ground

---

[41] Note that if we are comparing the N-electron excited state to an N-electron ground state, then n and m must be the same. They need not be the same if we are comparing the excited state of the N electron system to an N'-electron ground state. Note also that although we use a language appropriate to free electrons in describing the occupancy of levels, we could make the same points for a Fermi surface of general shape.

state. We can therefore only say that if a Fermi system is not normal, it is probably doing something else quite interesting and dramatic in its own right.

FERMI LIQUID THEORY: THE *f*-FUNCTION

Finally, assuming we are dealing with a normal Fermi system, we comment briefly on the remaining effects of electron-electron interactions on the electronic behavior. If a quasiparticle picture is valid, then the primary effect of electron-electron interactions is simply to alter the excitation energies $\mathcal{E}(\mathbf{k})$ from their free electron values. Landau pointed out that this has an important implication for the structure of transport theories. When electric or thermal currents are carried in a metal, the electronic distribution function $g(\mathbf{k})$ will differ from its equilibrium form $f(\mathbf{k})$. For truly independent electrons this has no bearing on the form of the $\mathcal{E}$ vs. $\mathbf{k}$ relation, but since the quasiparticle energy is a consequence of electron-electron interactions, it may well be altered when the configuration of the other electrons is changed. Landau noted that if the distribution function differed from its equilibrium form by $\delta n(\mathbf{k}) = g(\mathbf{k}) - f(\mathbf{k})$, then in a linearized theory[42] this would imply a change in the quasiparticle energy of the form[43]

$$\delta\mathcal{E}(\mathbf{k}) = \frac{1}{V}\sum_{\mathbf{k}'} f(\mathbf{k}, \mathbf{k}')\, \delta n(\mathbf{k}'). \tag{17.67}$$

This is precisely the state of affairs prevailing in Hartree-Fock theory, where $f(\mathbf{k}, \mathbf{k}')$ has the explicit form $4\pi e^2/(\mathbf{k} - \mathbf{k}')^2$. In a more accurate screened Hartree-Fock theory, f would have the form $4\pi e^2/[(\mathbf{k} - \mathbf{k}')^2 + k_0{}^2]$. In general neither of these approximate forms is correct, and the exact f-function is difficult to compute. Nevertheless, the existence of the relation (17.67) must be allowed for in a correct transport theory. It is beyond the scope of this book to carry out such a program. However one of its most important consequences is that for time independent processes the f-function drops completely out of the transport theory, and electron-electron interactions are of importance only insofar as they affect the scattering rate. This means, in particular, that stationary processes in a magnetic field at high $\omega_c\tau$ will be completely unaffected by electron-electron interactions and correctly given by the independent electron theory. These are precisely the processes that give valuable and extensive information about the Fermi surface, so that a major stumbling block to one's faith in the absolute validity of that information can be removed.

Although the f-function is beyond reliable computational techniques, one can try to deduce how its mere existence should affect various frequency-dependent transport properties. In most cases the effects appear to be small, and quite difficult to disentangle from band structure effects. However, attempts have recently been made to measure properties that do depend in a critical way on the f-function, in an effort to extract its values from experiments.[44]

---

[42] Such as almost all the transport theories used in practice.

[43] It is conventional to exclude from (17.67) the contribution to the change in energy associated with the macroscopic electromagnetic field produced by the currents or charge densities associated with the deviation from equilibrium; i.e., the f-function describes the exchange and correlation effects. Self-consistent field effects are explicitly dealt with separately, in the usual way.

[44] See, for example, P. M. Platzman, W. M. Walsh, Jr., and E-Ni Foo, *Phys. Rev.* **172**, 689 (1968).

FERMI LIQUID THEORY: CONCLUDING RULES OF THUMB

In summary, the independent electron picture is quite likely to be valid:

1. Provided that we are dealing only with electrons within $k_B T$ of $\mathcal{E}_F$.
2. Provided that we remember, when pressed, that we are not describing simple electrons anymore, but quasiparticles.
3. Provided that we allow for the effects of interaction on the $\mathcal{E}$ vs. **k** relation.
4. Provided that we allow for the possibility of an f-function in our transport theories.

PROBLEMS

1. *Derivation of the Hartree Equations from the Variational Principle*

(a) Show that the expectation value of the Hamiltonian (17.2) in a state of the form (17.10) is[45]

$$\langle H \rangle = \sum_i \int d\mathbf{r}\, \psi_i^*(\mathbf{r}) \left(-\frac{\hbar^2}{2m} \nabla^2 + U^{\text{ion}}(\mathbf{r}) \right) \psi_i(\mathbf{r}) + \frac{1}{2} \sum_{i \neq j} \int d\mathbf{r}\, d\mathbf{r}' \frac{e^2}{|\mathbf{r} - \mathbf{r}'|} |\psi_i(\mathbf{r})|^2 \, |\psi_j(\mathbf{r}')|^2, \quad \textbf{(17.68)}$$

provided that all the ψ_i satisfy the normalization condition $\int d\mathbf{r}\, |\psi_i|^2 = 1$.

(b) Expressing the constraint of normalization for each ψ_i with a Lagrange multiplier $\mathcal{E}_i$, and taking $\delta\psi_i$ and $\delta\psi_i^*$ as independent variations, show that the stationary condition

$$\delta_i \langle H \rangle = 0 \quad \textbf{(17.69)}$$

leads directly to the Hartree equations (17.7).

2. *Derivation of the Hartree-Fock Equations from the Variational Principle*

(a) Show that the expectation value of the Hamiltonian (17.2) in a state of the form (17.13) is given by (17.14).

(b) Show that when applied to Eq. (17.14) the procedure described in Problem 1(b) now leads to the Hartree-Fock equations (17.15).

3. *Properties of the Coulomb and Screened Coulomb Potentials*

(a) From the integral representation of the delta function,

$$\delta(\mathbf{r}) = \int \frac{d\mathbf{k}}{(2\pi)^3} e^{i\mathbf{k}\cdot\mathbf{r}} \quad \textbf{(17.70)}$$

and the fact that the Coulomb potential $\phi(\mathbf{r}) = -e/r$ satisfies Poisson's equation,

$$-\nabla^2 \phi(\mathbf{r}) = -4\pi e \delta(\mathbf{r}), \quad \textbf{(17.71)}$$

[45] Note the restriction (somewhat pedantic when large numbers of levels appear in the sum) $i \neq j$. Such a restriction is not present in the more general Hartree-Fock energy of Eq. (17.14) because there the direct and exchange terms for $i = j$ cancel identically.

argue that the electron-electron pair potential, $V(\mathbf{r}) = -e\phi(\mathbf{r}) = e^2/r$ can be written in the form

$$V(\mathbf{r}) = \int \frac{d\mathbf{k}}{(2\pi)^3} e^{i\mathbf{k}\cdot\mathbf{r}} V(\mathbf{k}), \tag{17.72}$$

where the Fourier transform $V(\mathbf{k})$ is given by

$$V(\mathbf{k}) = \frac{4\pi e^2}{k^2}. \tag{17.73}$$

(b) Show that the Fourier transform of the screened Coulomb interaction $V_s(\mathbf{r}) = (e^2/r)e^{-k_0 r}$ is

$$V_s(\mathbf{k}) = \frac{4\pi e^2}{k^2 + k_0{}^2}, \tag{17.74}$$

by substituting (17.74) into the Fourier integral

$$V_s(\mathbf{r}) = \int \frac{d\mathbf{k}}{(2\pi)^3} e^{i\mathbf{k}\cdot\mathbf{r}} V_s(\mathbf{k}) \tag{17.75}$$

and evaluating that integral in spherical coordinates. (The radial integral is best done as a contour integral.)

(c) Deduce from (17.74) that $V_s(r)$ satisfies

$$(-\nabla^2 + k_0{}^2)V_s(\mathbf{r}) = 4\pi e^2 \delta(\mathbf{r}). \tag{17.76}$$

4. ***Hartree-Fock Effective Mass near* k = 0**

Show that near the band minimum ($k = 0$) the Hartree-Fock one-electron energy (17.19) is parabolic in k:

$$\mathcal{E}(\mathbf{k}) \approx \frac{\hbar^2 k^2}{2m^*}, \tag{17.77}$$

where

$$\frac{m^*}{m} = \frac{1}{1 + 0.22(r_s/a_0)}. \tag{17.78}$$

5. ***Calculation of the Lindhard Response Function***

Using the formula of first-order stationary perturbation theory,

$$\psi_{\mathbf{k}} = \psi_{\mathbf{k}}^0 + \sum_{\mathbf{k}'} \frac{1}{\mathcal{E}_{\mathbf{k}} - \mathcal{E}_{\mathbf{k}'}} (\psi_{\mathbf{k}'}^0, V\psi_{\mathbf{k}}^0)\psi_{\mathbf{k}'}^0, \tag{17.79}$$

and expressing the charge density as

$$\rho(\mathbf{r}) = -e\sum f_{\mathbf{k}} |\psi_{\mathbf{k}}(\mathbf{r})|^2 = \rho^0(\mathbf{r}) + \rho^{\text{ind}}(\mathbf{r}), \tag{17.80}$$

(where $f_{\mathbf{k}}$ is the equilibrium Fermi distribution), show that the Fourier transform of the charge induced to first order in a total potential ϕ is given by

$$\rho^{\text{ind}}(\mathbf{q}) = -e^2 \int \frac{d\mathbf{k}}{4\pi^3} \frac{f_{\mathbf{k}-\frac{1}{2}\mathbf{q}} - f_{\mathbf{k}+\frac{1}{2}\mathbf{q}}}{\hbar^2(\mathbf{k}\cdot\mathbf{q}/m)} \phi(\mathbf{q}). \tag{17.81}$$

(Equation (17.56) then follows from the definition (17.37) of $\chi(\mathbf{q})$.)

18

Surface Effects

The Work Function

Contact Potentials

Thermionic Emission

Low-Energy Electron Diffraction

Field Ion Microscopy

Electronic Surface Levels

Because we have been interested primarily in bulk properties, we have ignored surfaces, working with the idealized model of an infinitely extended solid.[1] Our justification for this was that of the 10^{24} atoms in a macroscopic crystal of typically 10^8 atoms on a side, only about one in 10^8 reside near the surface.

In limiting ourselves to bulk properties we are neglecting the increasingly important field of surface physics, which deals with phenomena such as catalysis or crystal growth, which are entirely determined by the interaction of surface atoms with atoms impinging on the crystal. Since the microscopic structure of the surfaces of most specimens tends to be irregular and difficult to ascertain, the field of surface physics is quite complex, with nothing like the variety of simple and experimentally verifiable models available in the physics of bulk solids. We shall have nothing to say, even in this chapter, about such surface phenomena, limiting ourselves to a description of some of the important tools for determining surface structure.

However, even if we are interested only in bulk properties, we are still forced to deal with the surface whenever a measurement (e.g., the application of a voltmeter) removes an electron from the solid. The energy required to extract an electron, although it originates deep in the interior, is determined by surface as well as bulk conditions. This is because there are distortions in the electronic charge distribution near the surface that, because of the long range of the Coulomb potential, affect the energies of levels far inside. Such effects are crucial to the understanding of contact potentials (see below), thermionic emission (the boiling of electrons out of a metal at high temperatures), the photoelectric effect (the ejection of electrons by incident photons), or any other phenomena in which electrons are removed from a solid or pass from one solid to another.

In describing such phenomena a crucial role is played by the *work function*, defined as the minimum energy required to remove an electron from the interior of a solid to a position just outside. "Just outside" means a distance from the surface that is large on the atomic scale, but small compared with the linear dimensions of the crystal and will be given a more detailed specification in the discussion below.

EFFECT OF THE SURFACE ON THE BINDING ENERGY OF AN ELECTRON: THE WORK FUNCTION

To illustrate how the surface affects the energy required to remove an electron, let us compare the periodic potential $U^{\text{inf}}(\mathbf{r})$ of an idealized infinite crystal, with the potential $U^{\text{fin}}(\mathbf{r})$ appearing in the one-electron Schrödinger equation for a finite specimen of the same substance. For simplicity we consider only crystals from the cubic system that possess inversion symmetry. In the infinite (or periodically extended crystal) we can represent U^{inf} as a sum of contributions from primitive Wigner-Seitz cells about each lattice point:

$$U^{\text{inf}}(\mathbf{r}) = \sum_{\mathbf{R}} v(\mathbf{r} - \mathbf{R}), \tag{18.1}$$

[1] For mathematical convenience, we have also generally replaced the infinite solid with the periodically repeated solid portrayed by the Born–von Karman boundary conditions.

where

$$v(\mathbf{r}) = -e \int_C d\mathbf{r}' \, \rho(\mathbf{r}') \frac{1}{|\mathbf{r} - \mathbf{r}'|}. \tag{18.2}$$

The integration in (18.2) is over a Wigner-Seitz cell C centered at the origin; $\rho(\mathbf{r})$ is the total charge density, electronic and ionic.[2]

At distances from the cell that are large compared with its dimensions we can make the multipole expansion of electrostatics, writing

$$\begin{aligned} \frac{1}{|\mathbf{r} - \mathbf{r}'|} &= \frac{1}{r} - (\mathbf{r}' \cdot \nabla) \frac{1}{r} + \frac{1}{2} (\mathbf{r}' \cdot \nabla)^2 \frac{1}{r} + \cdots \\ &= \frac{1}{r} + \frac{\mathbf{r}' \cdot \hat{\mathbf{r}}}{r^2} + \frac{3(\mathbf{r}' \cdot \hat{\mathbf{r}})^2 - r'^2}{r^3} + \frac{1}{r} O\left(\frac{r'}{r}\right)^3, \end{aligned} \tag{18.3}$$

to find that

$$v(\mathbf{r}) = -e \frac{Q}{r} - e \frac{\mathbf{p} \cdot \hat{\mathbf{r}}}{r^2} + O\left(\frac{1}{r^3}\right), \tag{18.4}$$

where

$$Q = \int_C d\mathbf{r}' \, \rho(\mathbf{r}') \tag{18.5}$$

is the total charge of the cell, and

$$\mathbf{p} = \int_C d\mathbf{r}' \, \mathbf{r}' \, \rho(\mathbf{r}') \tag{18.6}$$

is its total dipole moment.

Since the crystal is electrically neutral and $\rho(\mathbf{r})$ has the periodicity of the Bravais lattice, each primitive cell must be electrically neutral: $Q = 0$. Furthermore, in a crystal with inversion symmetry the contribution of the Wigner-Seitz cell to the dipole moment vanishes. As a consequence of cubic symmetry the coefficient of the $1/r^3$ term (the quadrupole potential) will also vanish,[3] and since inversion symmetry also requires the coefficient of $1/r^4$ to vanish, we can conclude that the contribution of a Wigner-Seitz cell to $v(\mathbf{r})$ decreases as $1/r^5$—i.e., quite rapidly—at large distances from the cell.

---

[2] The one-electron Schrödinger equation we have in mind is therefore the self-consistent Hartree equation discussed in Chapters 11 and 17.

[3] This follows from the fact that $\int_c d\mathbf{r}' \, r_i' r_j' \rho(\mathbf{r}')$ must vanish when $i \neq j$, and must equal its average value, $\frac{1}{3} \int d\mathbf{r}' \, r'^2 \rho(\mathbf{r}')$, when $i = j$. Consequently the first term in

$$\int_C d\mathbf{r}' \left[\frac{3(\mathbf{r}' \cdot \hat{\mathbf{r}})^2}{r^3} - \frac{r'^2}{r^3} \right] \rho(\mathbf{r}')$$

must cancel the second.

If the crystal does not have cubic symmetry our general conclusions are unaffected, but considerable care must be exercised in dealing with the quadrupole term. A $1/r^3$ dependence by itself does not diminish quite rapidly enough with distance to insure that remote cells are without mutual influence, and the angular dependence of the quadrupole potential must also be taken into account. This makes the discussion much more technical, and is not, for our purposes, worth the trouble.

Thus the contribution to $U^{\text{inf}}(\mathbf{r})$ from cells far (on the atomic scale) from the point $\mathbf{r}$ is negligibly small, and $U^{\text{inf}}(\mathbf{r})$ is very well approximated by the contribution from cells within a few lattice constants of $\mathbf{r}$.

Now consider a finite crystal. Suppose that we could represent the ionic configuration by simply occupying some finite region V of the Bravais lattice occupied in the infinite crystal. Suppose, furthermore, that the electronic charge density in the Wigner-Seitz cell about each ion remained completely undistorted from the form it takes in the infinite crystal, even in cells near the surface (Figure 18.1a). Under these assumptions each occupied cell would continue to contribute $v(\mathbf{r} - \mathbf{R})$ to the potential, and we would have:

$$U^{\text{fin}}(\mathbf{r}) = \sum_{\mathbf{R} \text{ in } V} v(\mathbf{r} - \mathbf{R}). \tag{18.7}$$

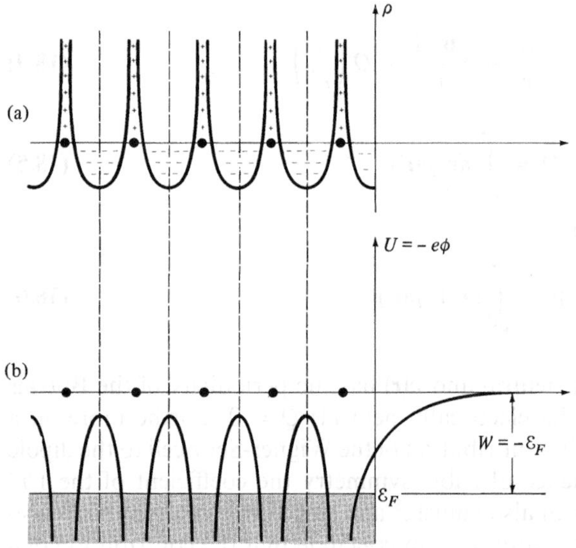

Figure 18.1
(a) The electric charge density near the surface of a finite crystal if there were no distortion in cells near the surface. The density is plotted along a line of ions. Vertical dashed lines indicate cell boundaries. (b) The form of the crystal potential U (or the electrostatic potential $\phi = -U/e$) determined by the charge density in (a), along the same line. Far from the crystal U and ϕ drop to zero. The (negative) Fermi energy is indicated on the vertical axis. The shading below the Fermi energy is meant to suggest the filled electronic levels in the metal. Since the lowest electronic levels outside the metal have zero energy, an energy $W = -\mathcal{E}_F$ must be supplied to remove an electron.

If (18.7) were correct, then at points $\mathbf{r}$ far inside the crystal on the atomic scale, U^{fin} would differ from U^{inf} only because the finite crystal lacked cells at sites $\mathbf{R}$ far from $\mathbf{r}$. Since such cells make a negligible contribution to the potential at $\mathbf{r}$, $U^{\text{fin}}(\mathbf{r})$ would be indistinguishable from $U^{\text{inf}}(\mathbf{r})$ when $\mathbf{r}$ was more than a few lattice constants inside the crystal. Furthermore, when $\mathbf{r}$ was more than a few lattice constants outside of the crystal, $U^{\text{fin}}(\mathbf{r})$ would be vanishingly small, because of the rapidly diminishing inverse fifth-power dependence of the contribution to U^{fin} from each occupied cell of the cubic crystal (Figure 18.1b).

As a result, the energy of the highest occupied electronic level well within the crystal would still be $\mathcal{E}_F$, where $\mathcal{E}_F$ is the Fermi energy calculated for the ideal infinite crystal with the periodic potential U^{inf}. Furthermore the lowest energy of an electronic level outside of the crystal would be zero (since U^{fin} would approach zero outside of the crystal, and the kinetic energy of a free electron can be made arbitrarily small).

Therefore if there were no distortion in the charge distribution in surface cells, the minimum energy required to remove an electron from the interior of the crystal to a point just outside the crystal would be[4]

$$W = 0 - \mathcal{E}_F = -\mathcal{E}_F. \tag{18.8}$$

This conclusion is incorrect. The actual charge distribution in cells near the surface of a finite crystal does differ from the charge distribution of cells in the interior. For one thing the positions of the surface ions will, in general, be slightly displaced from their ideal Bravais lattice positions. Furthermore the electronic charge distribution in cells near the surface need not have the symmetry of the Bravais lattice (Figure 18.2a). Such cells will, in general, have a nonvanishing electric dipole moment and may even yield a nonvanishing net electrical surface charge.

The particular way in which the charge distribution in cells near the surface differs from that in the bulk depends on such details as whether the surface is plane or rough, and, if plane, on the orientation of the plane with respect to the crystallographic axes. Determining this distorted charge distribution for various types of surfaces is a difficult problem in surface physics that we shall not explore. Our primary interest is in the consequences of such a distortion.

We first consider the case in which the distortion of the surface cells does not result in a net macroscopic charge per unit area of metallic surface. If we require the metal as a whole to be electrically neutral, this will be the case if all its surfaces have equivalent structures, either because they are crystallographically equivalent planes or, if rough, because they have been prepared by identical processes. At distances far (on the atomic scale) from such an electrically neutral surface, the charge distributions of the individual distorted surface cells will continue to yield no net macroscopic electric field.[5] However, within the surface layer in which the cells are distorted, this distortion will give rise to appreciable electric fields, against which an amount of work $W_s = \int e\mathbf{E} \cdot d\boldsymbol{\ell}$ must be performed, in moving an electron through the layer.

The value of W_s is determined by the manner in which the charge distribution in the surface cells differs from that of the bulk, which, in turn, depends on the type of surface being considered. In some models[5] the distortion in the charge of surface cells is represented as a uniform macroscopic surface density of dipoles, and with this model

---

[4] Since electrons are bound to the metal, work must be done to extract them, so $\mathcal{E}_F$ must be negative. This is not hard to reconcile with the convention of free electron theory that $\mathcal{E}_F = \hbar^2 k_F^2/2m$. The point is that in theories of bulk properties that use the model of an infinite metal, there is no reason to consider any particular value for the arbitrary additive constant in the electronic energy; we implicitly committed ourselves to a particular choice for that constant by assigning the energy zero to the lowest electronic level. With that convention, the potential energy for an electron outside of the crystal must be large and positive (larger, in fact, than $\mathcal{E}_F$) if electrons are indeed to be bound to the metal. In the present discussion, however, we adopted the familiar electrostatic convention of taking the potential to be zero at large distances from a finite metallic specimen. To be consistent with this alternative convention, it is necessary to add a large negative constant to the energy of each electronic level within the metal. One can regard that negative constant as providing a crude representation of the attractive potential of the lattice of ions. The value of that constant has no effect on the determination of bulk properties, but in comparing energies inside and outside of the metal one must either introduce such a term explicitly, or abandon the convention that the potential is zero far from the metal.

[5] See, for example, Problem 1a.

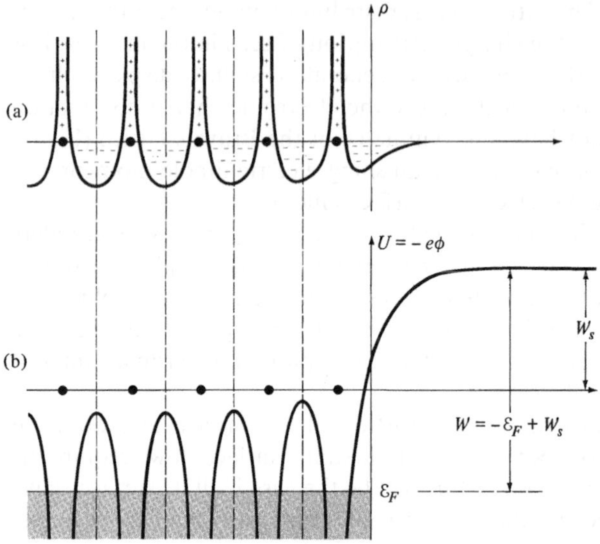

Figure 18.2
(a) The actual form of the electric charge density near the surface of a crystal (neglecting possible slight displacements of the ions near the surface from their sites in the infinite crystal). Note the electron deficiency in the two cells nearest the surface and the presence of electronic charge in the first "cell" on the vacuum side of the surface. It is this kind of distortion that produces the "double layer" described below. (b) The form of the crystal potential U determined by the charge density in (a). If the additive constant is chosen so that U resembles the potential of Fig. 18.1b far inside the crystal, then outside of the crystal U will not approach zero, but the value W_s equal to the work that must be done to carry an electron through the electric field in the double layer. The lowest levels outside the crystal now have an energy W_s, and therefore an energy $W = -\mathcal{E}_F + W_s$ must be supplied to remove an electron.

in mind, the surface layer (quite generally) is often referred to as the "double layer."

The work W_s done by the field prevailing in the double layer must be added to the expression (18.8) that gives the work function when the distortion of the surface cells is ignored. The correct work function is therefore given by[6]

$$W = -\mathcal{E}_F + W_s. \tag{18.9}$$

The corresponding form for the crystal potential $U(\mathbf{r})$ is shown in Fig. 18.2b.

If the crystal faces are not equivalent, then there is nothing to prohibit a net macroscopic surface charge developing in each face in addition to the double layers, provided the total charge on all surfaces of the crystal vanishes. Indeed, it is easy to see that small but nonvanishing surface charges must arise for the following reason:

Consider a crystal with two inequivalent faces F and F'. Their work functions W and W' need not be the same, since the contributions W_s and W_s' to the work functions

---

[6] Like Eq. (18.8), Eq. (18.9) assumes that $\mathcal{E}_F$ has been calculated for the infinite crystal with a particular choice for the additive constant in the periodic potential, namely the one that insures that for a finite crystal where distortion of the surface cell charge distributions is not taken into account, U vanishes at large distances from the crystal.

will arise from double layers of different internal structures. Consider now extracting an electron at the Fermi level through face F, and then bringing it back in through face F' to an interior level that is again at the Fermi energy (Figure 18.3). The total work done in such a cycle must vanish, if energy is to be conserved. However, the work done in extracting and reintroducing the electron is $W - W'$, which need not vanish if the surfaces are not equivalent. There must therefore be an electric field outside of the metal against which a compensating amount of work is done as the electron is carried from face F to face F'; i.e., the two faces of the crystals must be at different electrostatic potentials ϕ and ϕ' satisfying

$$-e(\phi - \phi') = W - W'. \tag{18.10}$$

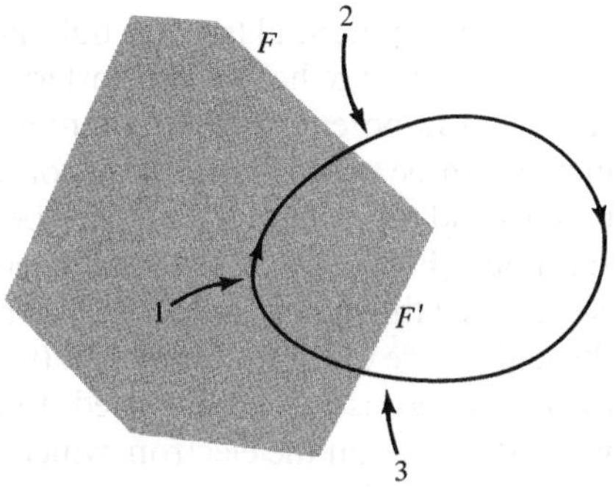

Figure 18.3
Zero total work is done in taking an electron from an interior level at the Fermi energy over the path shown, returning it at the end to an interior level at the Fermi energy. That work, however, is the sum of three contributions: W (in going from 1 to 2), $e(\phi - \phi')$ (in going from 2 to 3, where ϕ and ϕ' are the electrostatic potentials just outside faces F and and F'), and $-W'$ (in going from 3 back to 1).

Since the double layers cannot yield macroscopic fields outside of the metal these fields must arise from net macroscopic distributions of electric charge on the surfaces.[7] The quantity of charge that must be redistributed among the surfaces to produce such exterior fields is minute, compared with the amount of charge that is redistributed among neighboring surface cells in establishing the double layer.[8] Correspondingly, the electric field prevailing inside the double layer is immense compared with the exterior electric field arising from the net surface charge.[9]

When all surfaces of the solid are not equivalent, the work function is defined for a particular surface to include only the work that must be done against the field in its double layer (a feature intrinsic to that surface), but not the additional work that must be done against exterior fields that may be present due to the redistribution of surfaces charges (a quantity that depends on what other surfaces happen to be exposed). Because such exterior fields are minute compared with the fields in the double layers, one insures that only the latter fields contribute to the work function of a surface by defining it to be the minimum work required to remove an electron through that surface to a point that is far away on the atomic scale (so that the electron

---

[7] The condition that the crystal as a whole be neutral requires only that the sum over all faces of the macroscopic surface charge on each should vanish.

[8] See Problem 1b.

[9] The potential drop from face to face is comparable to the potential drop across the double layers (see Eq. (18.10)). The former drop, however, takes place over macroscopic distances (of the order of the dimensions of the crystal faces) while the latter occurs over microscopic distances (of the order of the thickness of the double layer—i.e., a few lattice constants).

has passed through the entire double layer) but not far on the scale of the dimensions of the macroscopic crystal faces (so that fields existing outside of the crystal do negligible work on the electron[10]).

CONTACT POTENTIALS

Suppose two metals are connected in a way that permits electrons to flow freely from one to the other. When equilibrium is attained, the electrons in each metal must be at the same chemical potential. This is achieved by a momentary flow of charge from the surface of one to the surface of the other. The surface charge on each metal gives rise to a potential in the interior that uniformly shifts all of the bulk levels together with the chemical potential (so that bulk properties in the interior remain unaltered).

Because charge has been transferred, the two metals will no longer be at the same electrostatic potential. The potential difference between any two faces of the two metals can be expressed in terms of their work functions by the same argument we used to find the potential difference between two inequivalent faces of a single metallic specimen (Figure 18.3). One again observes that if an electron at the Fermi level[11] is extracted through a face of the first metal (with work function W), and reintroduced through a face of the second metal (with work function W') at the (same) Fermi level, then if energy is to be conserved, there must be an exterior electric field that does work $W - W'$ on the electron, which requires, in turn, a potential difference between the two faces given by

$$-e(\phi - \phi') = W - W'. \tag{18.11}$$

The two metals (before and after their electrons have come to equilibrium) are depicted schematically in Figure 18.4. Since some contact must be made between the metals to permit their electrons to come to equilibrium, the potential difference (18.11) is known as a contact potential.

THE MEASUREMENT OF WORK FUNCTIONS VIA THE MEASUREMENT OF CONTACT POTENTIALS

Equation (18.11) suggests that a simple way to measure the work function of a metal[12] would be to measure the contact potential between it and a metal of known work function. This cannot be done simply by connecting a galvanometer across two faces, for if it could one would have produced a flow of current in a circuit without a sustaining source of energy.

---

[10] Even if all faces are equivalent, the interaction of the removed electron with those remaining in the metal will induce macroscopic surface charges (giving rise to the "image charge" of electrostatics) whose contribution to the work function is also made negligible by this definition.

[11] In metals at room temperature and below, the chemical potential differs negligibly from the Fermi energy. See Eq. (2.77).

[12] When speaking of the work function of a metal without reference to a particular crystal face one has in mind the value appropriate to some rough (on the microscopic scale) face, which therefore represents some average value of the work functions for crystallographically well-defined faces.

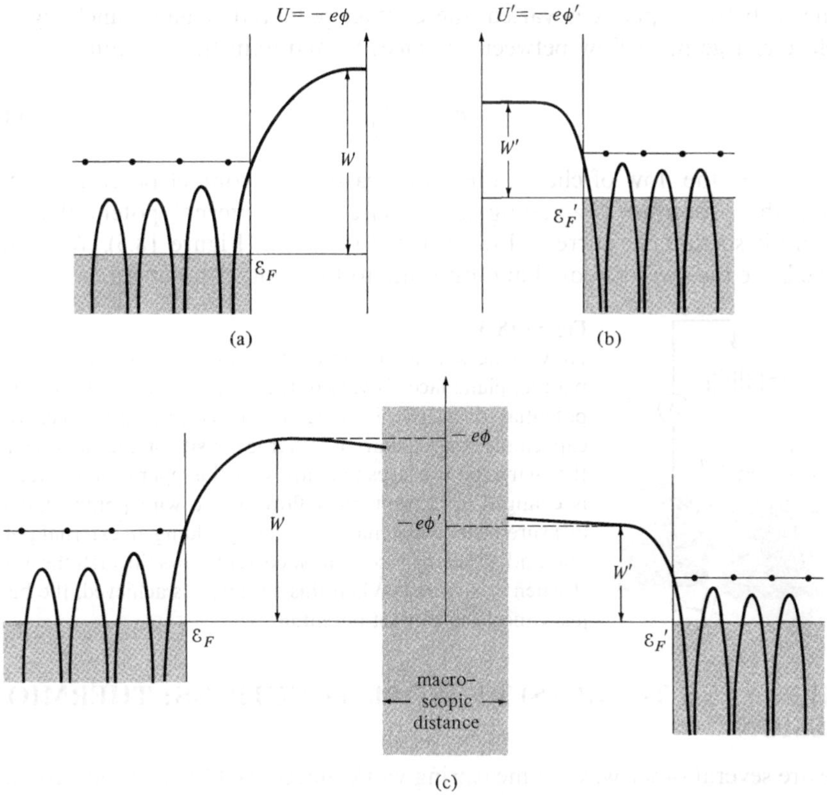

Figure 18.4
(a) The crystal potential U (or electrostatic potential $\phi = U/(-e)$) for a metal with work function W and Fermi energy $\mathcal{E}_F$. (The figure is essentially that of Figure 18.2b.) (b) A similar plot for a second metal with work function W' and Fermi energy $\mathcal{E}_F'$, electrically insulated from the first. In (c) the two metals have been joined together by a conducting wire, so that charge can pass freely from one to the other. The only result is to introduce small amounts of net surface charge in each metal, sufficient to shift uniformly the level structures in (a) and (b) so as to bring the two Fermi levels into coincidence. Because of the slight surface charges on the metals the potentials outside them are no longer strictly constant, and there is a potential drop from one metal to the other given by $-e(\phi - \phi') = W - W'$.

However, a simple method due to Kelvin can be used to measure contact potentials. Suppose the two samples are arranged so that the two crystal faces form a plane parallel capacitor. If there is a potential difference V between the faces, then there will be a charge per unit area given by

$$\sigma = \frac{E}{4\pi} = \frac{V}{4\pi d}, \qquad \textbf{(18.12)}$$

where d is the distance between the faces. If the plates are connected and no external voltage imposed, the potential difference will just be the contact potential V_c. As the distance d between plates is varied, the contact potential remains unchanged, and therefore charge must flow between the faces to maintain the relation

$$\sigma = \frac{V_c}{4\pi d}. \tag{18.13}$$

By measuring the flow of charge one can measure the contact potential. One can simplify the procedure by adding to the circuit an external potential bias and adjusting it so that no current flows when d is varied (Figure 18.5). When this is achieved, the bias will be equal and opposite to the contact potential.

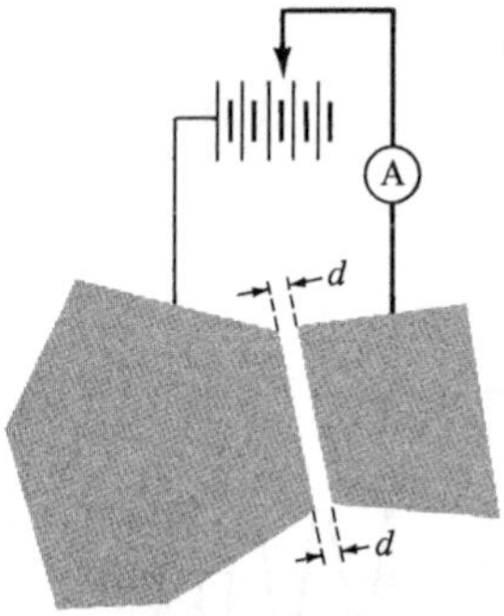

Figure 18.5
How to measure a contact potential. As the distance between two parallel plane faces is varied, the capacitance will vary. Since the potential difference is fixed at the contact potential, variable capacitance implies variable charge density on the faces. To permit the charge on the faces to readjust as the separation between them is changed, a current must flow in the wire joining them. The measurement can be made simpler by adding an external potential bias and adjusting it so that *no* current flows through the ammeter A when d is varied. When this situation is achieved, the bias will just cancel the contact potential.

OTHER WAYS TO MEASURE WORK FUNCTIONS: THERMIONIC EMISSION

There are several other ways of measuring work functions. One method exploits the photoelectric effect, measuring the minimum photon energy necessary to eject an electron through the crystal face, and setting that equal to W.

Another method, of some importance in designing cathode-ray tubes, measures the temperature dependence of the electron current flowing from the face of a hot metal. To understand such *thermionic emission*, first consider the idealized case in which a metal surface is in thermal equilibrium with a dilute electron gas outside of the metal (Figure 18.6). At temperature T the electronic distribution function is

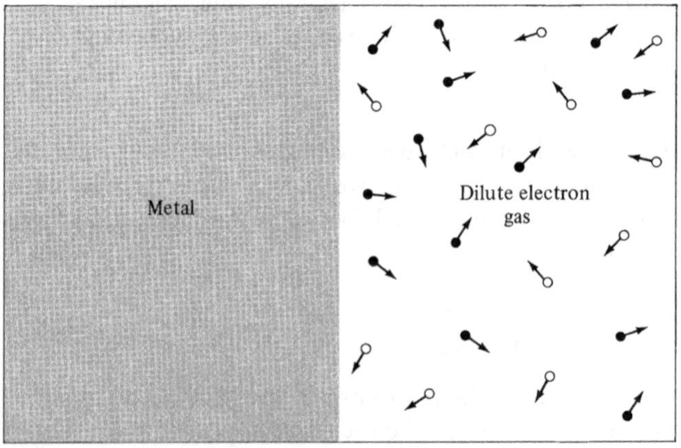

Figure 18.6
A simple model of thermionic emission. The current flowing out of a metal when electrons that escape are continually swept away is calculated by assuming that the metal is in thermal equilibrium with a dilute gas of free electrons, and identifying the thermionic current with that carried by those electrons moving away from the surface (solid circles).

$$f(\mathbf{k}) = \frac{1}{\exp\left[(\varepsilon_n(\mathbf{k}) - \mu)/k_B T\right] + 1}. \tag{18.14}$$

Inside the metal $\varepsilon_n(\mathbf{k})$ is determined by the band structure.[13] Outside of the metal, $\varepsilon_n(\mathbf{k})$ should be taken to have the free particle form

$$\frac{\hbar^2 k^2}{2m} - e\phi, \tag{18.15}$$

where ϕ is the local value of the electrostatic potential.[14] If the additive constant in the periodic potential is defined according to the convention of Figure 18.2b (so that the expression (18.9) gives the work function), then outside of the double layer we have (see Figure 18.2b):

$$-e\phi = W_s. \tag{18.16}$$

Therefore the distribution of external electrons beyond the double layer will be

$$f(\mathbf{k}) = \frac{1}{\exp\left[(\hbar^2 k^2/2m + W_s - \mu)/k_B T\right] + 1}. \tag{18.17}$$

However, Eq. (18.9) permits this to be written in the form[15]

$$f(\mathbf{k}) = \frac{1}{\exp\left[(\hbar^2 k^2/2m + W)/k_B T\right] + 1}, \tag{18.18}$$

where W is the work function of the surface.

Since work functions are typically a few electron volts in size (see Table 18.1), W/k_B is of order 10^4K. Therefore at temperatures below several thousand degrees (18.18) reduces to:[16]

$$f(\mathbf{k}) = \exp\left[-(\hbar^2 k^2/2m + W)/k_B T\right]. \tag{18.19}$$

The electronic current density flowing away from the surface is given by adding the contributions from all electrons with positive $v_x = \hbar k_x/m$, where the positive x-direction is taken to be the direction of the normal out of the surface:

$$j = -e \int_{k_x > 0} \frac{d\mathbf{k}}{4\pi^3} v_x f(\mathbf{k}) = e^{-W/k_B T}(-e) \int_{k_x > 0} \frac{d\mathbf{k}}{4\pi^3} \frac{\hbar k_x}{m} e^{-\hbar^2 k^2/2mk_B T}. \tag{18.20}$$

The integration is elementary, giving for the current per unit area emitted by the surface,

$$\begin{aligned} j &= -\frac{em}{2\pi^2 \hbar^3}(k_B T)^2 e^{-W/k_B T} \\ &= 120\ \text{amp-cm}^{-2}\text{-K}^{-2}(T^2 e^{-W/k_B T}). \end{aligned} \tag{18.21}$$

---

13 The derivation given here does not require the bands to be free electron-like. The result (18.21) is independent of the details of the band structure.

14 Ignore for the moment the contribution of the dilute electron gas itself to this potential. The current calculated by ignoring such complications (known as space-charge effects) is called the saturation current.

15 See footnote 11.

16 The experimental confirmation of the Maxwell–Boltzman form (18.19) of this distribution function was a considerable obstruction on the road to the discovery that metallic electrons obey Fermi-Dirac statistics.

This result, known as the *Richardson-Dushman equation*, asserts that if $\ln(j/T^2)$ is plotted against $1/k_BT$, the resulting curve will be a straight line with slope $-W$. In this way the absolute work function can be determined.

In practice, the neglected space-charge effects are made unimportant by the application of a small electric field that sweeps away the electrons as they are emitted. In addition, for our model to be applicable, the flow of electrons away from the metal must be dominated by electrons that have originated within the metal, and not by electrons from the gas outside, which have been reflected back after striking the surface. If surface scattering is important, the current will be less than predicted by (18.21).

MEASURED WORK FUNCTIONS OF SELECTED METALS

In Table 18.1 we list the work functions for some typical metals as determined by these three methods. Generally speaking, the various methods all give results that differ by about 5 percent. Since the variation in work function over different crystallographic faces can easily be of this size, it is not worth quoting the separate results. Nor should a number for the metal as a whole be relied on to more than a few percent.

Table 18.1
WORK FUNCTIONS OF TYPICAL METALS

METAL	W (eV)	METAL	W (eV)	METAL	W (eV)
Li	2.38	Ca	2.80	In	3.8
Na	2.35	Sr	2.35	Ga	3.96
K	2.22	Ba	2.49	Tl	3.7
Rb	2.16	Nb	3.99	Sn	4.38
Cs	1.81	Fe	4.31	Pb	4.0
Cu	4.4	Mn	3.83	Bi	4.4
Ag	4.3	Zn	4.24	Sb	4.08
Au	4.3	Cd	4.1	W	4.5
Be	3.92	Hg	4.52		
Mg	3.64	Al	4.25		

Source: V. S. Fomenko, *Handbook of Thermionic Properties*, G. V. Samsanov, ed., Plenum Press Data Division, New York, 1966. (Values given are the author's distillation of many different experimental determinations.)

We conclude out discussion of surfaces with brief descriptions of two of the major techniques used to probe surface structure.

LOW-ENERGY ELECTRON DIFFRACTION

The structure of the surface of a crystal specimen with a good plane surface (on the microscopic level) can be studied by the technique of low-energy electron diffraction (LEED). The basis for the method is very much the same as in the theory of X-ray diffraction, modified to take into account the fact that the diffracting surface is only periodic in two dimensions (i.e., in its own plane). Elastically scattered electrons are more suitable for the study of surfaces than X rays, because they penetrate only a very short distance into the solid, so that the diffraction pattern is determined almost entirely by the surface atoms.

The energies of the electrons required for such a study are easily estimated. A free electron with wave vector **k** has energy

$$E = \frac{\hbar^2 k^2}{2m} = (ka_0)^2 \mathrm{Ry} = 13.6(ka_0)^2 \text{ eV}, \tag{18.22}$$

from which it follows that the de Broglie wavelength of an electron is related to its energy in electron volts by

$$\lambda = \frac{12.3}{(E_{\mathrm{eV}})^{1/2}} \text{ Å}. \tag{18.23}$$

Since the electron wavelengths must be of the order of a lattice constant or less, energies of some tens of electron volts or higher are necessary.

To understand qualitatively the pattern produced in an electron diffraction measurement, suppose the scattering is elastic,[17] and that the incident and scattered electron wave vectors are **k** and **k**′. Suppose, furthermore, that the crystal surface is a lattice plane perpendicular to the reciprocal lattice vector $\mathbf{b}_3$ (see page 90). Choose a set of primitive vectors including $\mathbf{b}_3$ for the reciprocal lattice, and primitive vectors $\mathbf{a}_i$ for the direct lattice, satisfying

$$\mathbf{a}_i \cdot \mathbf{b}_j = 2\pi\delta_{ij}. \tag{18.24}$$

If the electron beam penetrates so little that only scattering from the surface plane is significant, then the condition for constructive interference is that the change **q** in the wave vector of the scattered electron satisfy

$$\mathbf{q} \cdot \mathbf{d} = 2\pi \times \text{integer}, \qquad \mathbf{q} = \mathbf{k}' - \mathbf{k}, \tag{18.25}$$

for all vectors **d** joining lattice points in the plane of the surface (cf. Eq. (6.5)).

Since such **d** are perpendicular to $\mathbf{b}_3$, they can be written as:

$$\mathbf{d} = n_1\mathbf{a}_1 + n_2\mathbf{a}_2. \tag{18.26}$$

Writing **q** in the general form

$$\mathbf{q} = \sum_{i=1}^{3} q_i \mathbf{b}_i, \tag{18.27}$$

we find that conditions (18.25) and (18.26) require

$$\begin{aligned} q_1 &= 2\pi \times \text{integer}, \\ q_2 &= 2\pi \times \text{integer}, \\ q_3 &= \text{arbitrary}. \end{aligned} \tag{18.28}$$

Since $\mathbf{b}_3$ is normal to the surface, these conditions will be satisfied by discrete *lines*[18] in **q**-space perpendicular to the crystal surface (as opposed to discrete *points* in the case of diffraction by a three-dimensional lattice). Consequently, even when the

---

[17] In fact, however, the elastically scattered component is generally a very small fraction of the total reflected flux of electrons.

[18] Sometimes referred to as "rods" in the literature.

energy conservation condition $k = k'$ is added, there will always be nontrivial solutions unless the incident wave vector is too small.

Equation (18.28) (together with an experimental arrangement that selects out the elastically scattered component) enables one to deduce the surface Bravais lattice from the structure of the reflected pattern. If scattering from more than one surface layer is important, the general structure of the pattern will be unchanged, for lower-lying planes will produce a weaker version of the same pattern (attenuated because only a smaller fraction of the beam can penetrate to the next layer).

Much more information is contained in the detailed distribution of scattered electrons than the mere arrangement of atoms in the surface plane, but to extract such information is a difficult problem, whose solution remains elusive.

THE FIELD ION MICROSCOPE

Elastic low-energy electron diffraction reveals the structure of the Fourier transform of the surface charge density—i.e., its shape in k-space. The structure in real space can be seen by the technique of the field ion microscope. The surface must be that of a sharply pointed specimen (Figure 18.7), with a point that is more or less hemispherical on the atomic scale, and with a radius of order a few thousand angstroms. The specimen is placed in a high vacuum facing an electrode. A large voltage is applied between the

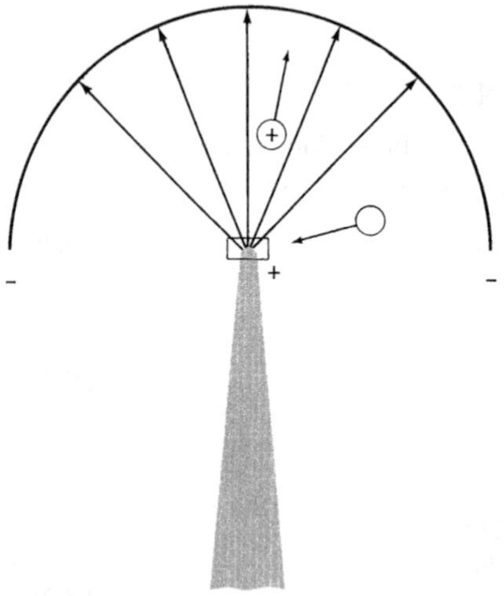

Figure 18.7
Schematic representation of a field ion microscope. The sample (pointed cone) is at a positive potential with respect to the plate, so that field lines point radially outward. A neutral helium atom (empty circle) will be attracted to the region of high field by the induced dipole interaction. A helium ion (circle with plus) will be strongly repelled along field lines. The field will only be strong enough to ionize helium atoms in the immediate vicinity of the tip. The basic assumption is that most of the helium atoms are ionized in the immediate neighborhood of surface atoms, where the field is at its strongest. Since the variation of the field near the surface reflects the variation of the atomic structure, the pattern of ions impinging on the plate should give a representation of the atomic structure of the tip.

specimen and the electrode, with a polarity chosen to make the specimen positive. Neutral helium atoms are then introduced into the vacuum chamber, which are polarized by the field. The interaction of the field with this induced dipole moment draws the helium atoms into the region of strongest field—i.e., toward the tip of the specimen. At a few atomic spacings from the tip the field becomes so strong that an electron can be torn from the helium atom. When this happens, the atom becomes a positively charged helium ion, and is energetically repelled from the tip toward the electrode. If the field strength is adjusted so that ionization occurs only very close to the surface, one might hope that the angular distribution of ions fleeing from the tip would reflect the microscopic structure of the surface—via the structure of the field in the immediate vicinity of the surface—magnified by the ratio of the radius of the collecting electrode to the radius of the hemispherical sample.

In fact, the pictures so obtained not only reflect the detailed symmetry of the crystal, but actually indicate the positions of individual atoms (Figure 18.8). The technique can be used to study the behavior of individual atomic impurities.

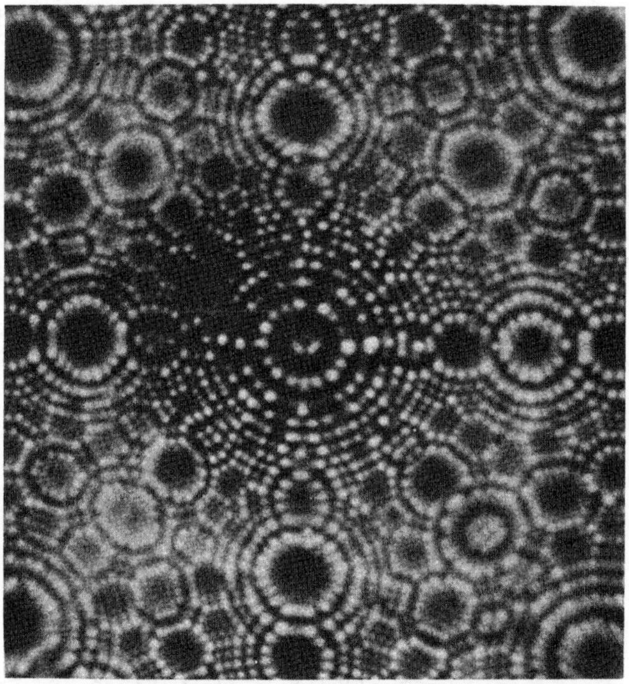

Figure 18.8
A field ion micrograph of a gold tip. In this example the imaging gas is neon rather than helium. (From R. S. Averbach and D. N. Seidman, *Surface Science* **40**, 249 (1973). We thank Professor Seidman for giving us the original micrograph.)

ELECTRONIC SURFACE LEVELS

Any attempt to describe a solid surface in detail must use the fact that in addition to the Bloch solutions to the one-electron Schrödinger equation for the familiar periodically extended crystal, there are other solutions with complex wave vectors describing electronic levels that are localized in the neighborhood of the surface of a real crystal. We have properly neglected such levels in our earlier discussions of bulk properties. The number of surface levels compared with the number of Bloch levels turns out to be at most of the order of the number of surface atoms compared with the number of atoms in the entire crystal, i.e., about one in 10^8 for a macroscopic specimen. As a result the surface levels make negligible contribution to bulk properties except for exceedingly small specimens. However, they are of considerable importance in determining the structure of the crystal surface. For example, they must play a role in any genuinely microscopic calculation of the structure of the surface dipole layer.

To understand qualitatively how these surface levels arise, we reexamine our derivation of Bloch's theorem in Chapter 8.

The argument leading to the Bloch form

$$\psi(\mathbf{r}) = e^{i\mathbf{k}\cdot\mathbf{r}}u(\mathbf{r}), \qquad u(\mathbf{r} + \mathbf{R}) = u(\mathbf{r}), \tag{18.29}$$

did not require the wave vector **k** to be real. This further restriction emerged from an application of the Born-von Karman periodic boundary condition. This boundary condition, however, is an artifact of the infinite crystal. If it is abandoned, we can find many more solutions to the infinite crystal Schrödinger equation, having the form

$$\psi(\mathbf{r}) = [e^{i\mathbf{k}\cdot\mathbf{r}}u(\mathbf{r})]e^{-\boldsymbol{\kappa}\cdot\mathbf{r}} \tag{18.30}$$

where **k** is now the real part of the Bloch wave vector, which may also have an imaginary part $\boldsymbol{\kappa}$.

The wave function (18.30) grows without bound in the direction opposite to $\boldsymbol{\kappa}$, and decays exponentially in the direction of $\boldsymbol{\kappa}$. Since the electronic density is everywhere finite, such levels have no relevance for an infinite crystal. If, however, there is a plane surface perpendicular to $\boldsymbol{\kappa}$, then one might attempt to join a solution of the form (18.30) within the crystal, that grows exponentially as the surface is approached, with one that is exponentially damped outside of the crystal (Figure 18.9).

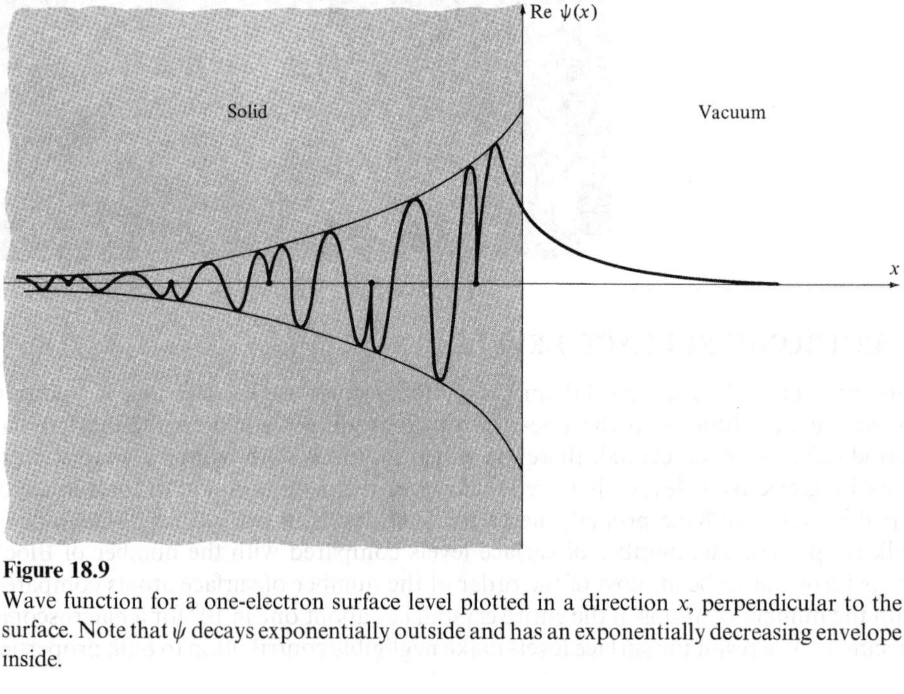

Figure 18.9
Wave function for a one-electron surface level plotted in a direction x, perpendicular to the surface. Note that ψ decays exponentially outside and has an exponentially decreasing envelope inside.

In general, for a fixed component of **k** parallel to the surface this matching will be possible only for a discrete set of **κ** (as is the case for any problem concerning localized levels).

To explore this problem further would take us well beyond the scope of this book, for we should first have to reexamine the entire discussion of Bloch functions without the restriction that the wave vector **k** be real, and then discuss the problem of how such Bloch functions with complex wave vectors can be matched onto exponentially decaying levels in empty space. Features of such solutions in the nearly free electron approximation are explored in Problem 2.

PROBLEMS

1. ***Some Problems in Electrostatics Bearing on Contact Potentials and the Double Layer***

(a) Consider a plane surface of a metal perpendicular to the x-axis. Perhaps the simplest model of the distortion in the charge densities of the cells near the surface is one that ignores any variation in the plane of the surface, describing the deviation in the charge density from its form in the bulk by a function of the single variable x, $\delta\rho(x)$. The condition that there is no net macroscopic charge density on the surface is

$$0 = \int dx\, \delta\rho(x). \qquad \textbf{(18.31)}$$

The charge density $\delta\rho(x)$ will give rise to an electric field $E(x)$ also normal to the surface. Deduce directly from Gauss's law ($\nabla \cdot \mathbf{E} = 4\pi\, \delta\rho$) that if the field vanishes on one side of the double layer (as it does inside the metal) then it must also vanish on the other side. Deduce also that the work that must be done to move an electron through the double layer is just

$$W_s = -4\pi e P, \qquad \textbf{(18.32)}$$

where P is the dipole moment per unit surface area produced by the double layer. (*Hint:* Write the work as an integral and introduce a judicious integration by parts.)

(b) Show that the charge density that must be supplied to a conducting sphere of radius 1 cm to raise its potential from zero to 1 volt is of order 10^{-10} electrons per square angstrom.

2. ***Electron Surface Levels for a Weak Periodic Potential***[19]

The method of Chapter 9 can be used to investigate electronic surface levels. Consider a semi-infinite crystal with a plane surface, perpendicular to a reciprocal lattice vector **K** (crystal surfaces are parallel to lattice planes). Taking the x-axis to lie along **K** and the origin to be a Bravais lattice point, to a rough approximation we can represent the potential of the semi-infinite crystal by $V(\mathbf{r}) = U(\mathbf{r})$, $x < a$; $V(\mathbf{r}) = 0$, $x > a$. Here $U(\mathbf{r})$ is the periodic potential of the infinite crystal. The distance a lies between zero and the interplanar distance in the family of planes parallel to the surface, and should be chosen in any particular problem to give that $U(\mathbf{r})$ which most resembles the actual potential at the surface.

[19] See E. T. Goodwin, *Proc. Camb. Phil. Soc.* **35**, 205 (1935).

We continue to assume that the Fourier components $U_{\mathbf{K}}$ are real. However, if we wish the lowest level outside of the crystal to have energy 0, we can no longer neglect the zeroth Fourier component U_0 within the crystal, as we did in Chapter 9. The result of retaining U_0 is simply to shift the formulas of Chapter 9 for the levels within the crystal by that amount. Note that U_0 need not be small for the method of Chapter 9 to work, in contrast to the Fourier components $U_{\mathbf{K}}$, with $\mathbf{K} \neq \mathbf{0}$.

We examine a Bloch level for the infinite crystal with a wave vector **k** that is near the Bragg plane determined by **K**, but near no other Bragg plane, so that in a weak periodic potential the wave function of the level is a linear combination of plane waves with wave vectors **k** and **k** − **K**. In Chapter 9 we required **k** to be real, to satisfy the Born-von Karman boundary condition. In a semi-infinite crystal, however, the component of **k** normal to the crystal surface need not be real, provided that it yields a decaying wave in the negative x-direction (into the metal). Outside of the metal, the Bloch function must be joined onto a solution of the free-space Schrödinger equation that decays in the positive x-direction (away from the metal). Thus, outside the metal we take

$$\psi(\mathbf{r}) = e^{-px + i\mathbf{k}_{\parallel}\cdot\mathbf{r}}, \qquad x > a, \tag{18.33}$$

and inside,

$$\psi(\mathbf{r}) = e^{qx + ik_0x + i\mathbf{k}_{\parallel}\cdot\mathbf{r}}(c_{\mathbf{k}} + c_{\mathbf{k}-\mathbf{K}}e^{-iKx}), \qquad x < a, \tag{18.34}$$

where $\mathbf{k}_{\parallel}$ is the part of **k** parallel to the surface, and the coefficients in (18.34) are determined by the secular equation (9.24) (with the energy $\mathcal{E}$ shifted by the constant U_0):

$$\begin{aligned}(\mathcal{E} - \mathcal{E}^0_{\mathbf{k}} - U_0)c_{\mathbf{k}} - U_{\mathbf{K}}c_{\mathbf{k}-\mathbf{K}} &= 0,\\ -U_{\mathbf{K}}c_{\mathbf{k}} + (\mathcal{E} - \mathcal{E}^0_{\mathbf{k}-\mathbf{K}} - U_0)c_{\mathbf{k}-\mathbf{K}} &= 0.\end{aligned} \tag{18.35}$$

(a) Verify that for (18.35) to yield real energies for $\mathbf{q} \neq 0$ it is necessary that $k_0 = K/2$.
(b) Show that when $k_0 = K/2$, the resulting energies are

$$\mathcal{E} = \frac{\hbar^2}{2m}\left(k_{\parallel}{}^2 + \frac{1}{4}K^2 - q^2\right) + U_0 \pm \sqrt{U_{\mathbf{K}}{}^2 - \left(\frac{\hbar^2}{2m}qK\right)^2}. \tag{18.36}$$

(c) Show that continuity of ψ and $\nabla\psi$ at the surface leads to the condition

$$p + q = \frac{1}{2}K\tan\left(\frac{K}{2}a + \delta\right), \tag{18.37}$$

where

$$\frac{c_{\mathbf{k}}}{c_{\mathbf{k}-\mathbf{K}}} = e^{2i\delta}. \tag{18.38}$$

(d) Taking the case $a = 0$, and using the fact that outside of the metal

$$\mathcal{E} = \frac{\hbar^2}{2m}(k_{\parallel}{}^2 - p^2), \tag{18.39}$$

show that (18.35) to (18.39) have a solution:

$$q = -\frac{1}{4}K\frac{U_{\mathbf{K}}}{\mathcal{E}_0}\sin 2\delta,$$

where

$$\sec^2 \delta = -\frac{(U_0 + U_{\mathbf{K}})}{\mathcal{E}_0}, \qquad \mathcal{E}_0 = \frac{\hbar^2}{2m}\left(\frac{K}{2}\right)^2. \tag{18.41}$$

(Note that this solution exists only when U_0 and $U_{\mathbf{K}}$ are negative, and $|U_0| + |U_{\mathbf{K}}| > \mathcal{E}_0$.)

19
Classification of Solids

In Chapter 7 we discussed the classification of solids on the basis of the symmetry of their crystal structures. The categories described there are very important, but are based entirely on a single aspect of the solid: its geometrical symmetries. Such a classification scheme is blind to important structural aspects of a solid that affect its physical (even if not its purely geometrical) properties. Thus within each of the seven crystal systems one can find solids exhibiting the full range of electrical, mechanical, and optical properties.

In this chapter we describe another, less rigorous, classification scheme, which is not based on symmetry, but emphasizes physical properties. The scheme is based on the configuration of the valence electrons.[1]

The most important distinction determined by the valence electrons is that between metals and insulators. As we have seen (Chapter 8) the difference between metals and insulators depends on whether there are (metals) or are not (insulators) any partially filled energy bands.[2] In perfect crystals at zero temperature, provided that the independent electron approximation is valid, this is a completely rigorous criterion, leading to two unambiguous categories.[3]

The basis for these two categories is the distribution of electrons, not in real space, but in wave vector space. Nowhere near as rigorous a criterion can be found to distinguish metals from insulators based on the distribution of electrons in real space. One can only make the qualitative observation that the electronic distribution in metals is generally not nearly so concentrated in the vicinity of the ion cores as it is in insulators. This is illustrated in Figure 19.1, where the wave functions of the occupied electronic levels in atomic sodium and atomic neon are plotted about two centers whose separation is equal to the nearest-neighbor separation in the solid. The electronic density in sodium remains appreciable even midway between the atoms, but it is quite small in neon. If one tried to construct from this an argument that solid neon should be insulating and solid sodium conducting, the chain of thought would be something like this: Appreciable overlap of atomic wave functions suggests—from the point of view of tight-binding theory (Chapter 10)—the presence of broad bands, which in turn leads to the possibility of considerable band overlap, and, hence, metallic properties. One is thus led rather quickly back to k-space, where the only really satisfactory criterion can be given.

---

1 As elsewhere in this book, we take the view that solids are composed of ion cores (i.e., nuclei and those electrons so strongly bound as to be negligibly perturbed from their atomic configurations by their environment in the solid) and valence electrons (i.e., those electrons whose configuration in the solid may differ significantly from that in the isolated atom.) As emphasized earlier, the distinction between core and valence electrons is one of convenience. In metals—especially in simple metals—it is frequently enough to consider only the conduction electrons as valence electrons, including all others in the rigid ion core. In transition metals, however, it may be of considerable importance to consider the electrons in the highest d-shells as valence rather than core electrons. In saying that the classification scheme is based on the valence electrons, we mean only that it depends on those aspects of the atomic electronic configuration that are significantly altered when the atoms are assembled together into the solid.

2 The distinction also depends on the validity of the independent electron approximation (or, less stringently, on the validity of the quasiparticle picture (Chapter 17)).

3 At nonzero temperatures the distinction may be blurred in insulators with small energy gaps, because of thermal excitation of electrons into the conduction band. Such solids are known as intrinsic semiconductors. Impurities in an otherwise insulating solid may also contribute electrons that are easily thermally excited into the conduction band, leading to extrinsic semiconductors. The characteristic properties of semiconductors are discussed in Chapter 28. From the point of view of this chapter (which is concerned only with perfect crystals at $T = 0$), all semiconductors are insulators.

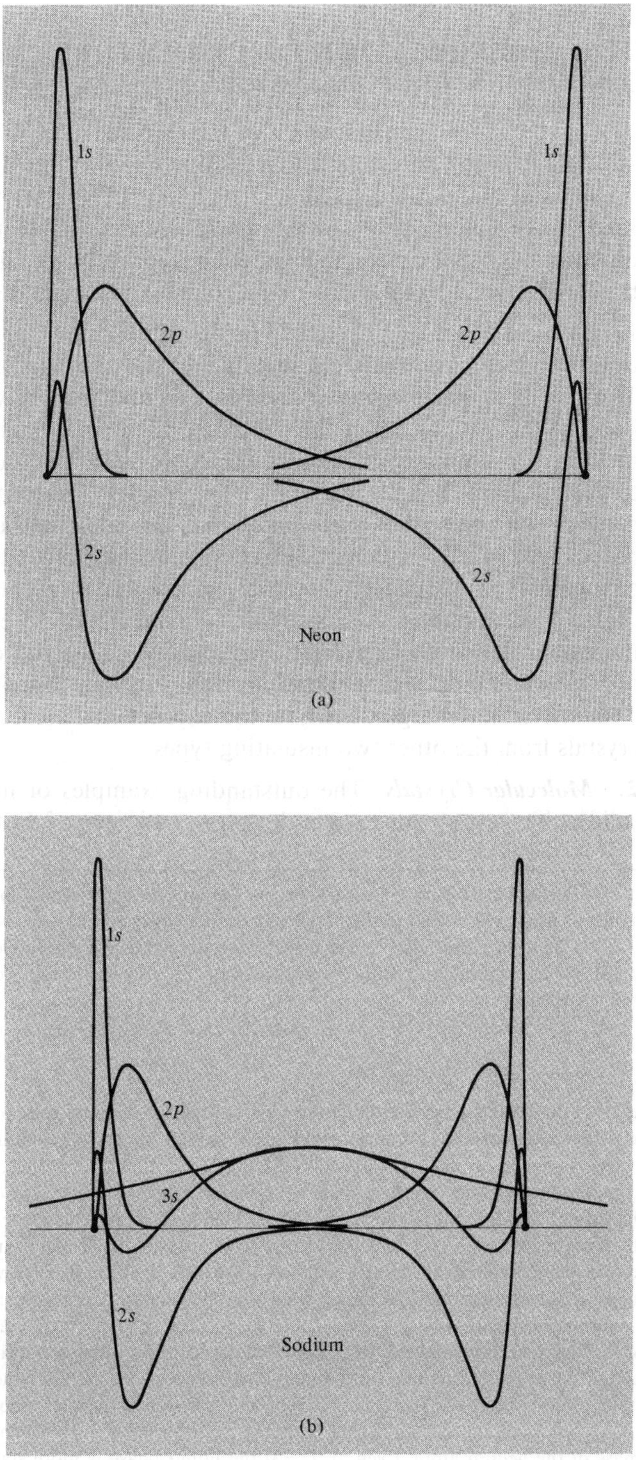

Figure 19.1
The calculated radial atomic wave functions $r\psi(r)$ for (a) neon $[1s^2 2s^2 2p^6]$ and (b) sodium $[1s^2 2s^2 2p^6 3s^1]$. The wave functions are drawn about two centers whose separation is taken to be the observed nearest-enighbor distance in the solid (neon, 3.1Å; sodium, 3.7Å.) There is a very small overlap of the $2s$ and $2p$ orbitals in neon. The $2s$ and $2p$ orbitals overlap considerably less in sodium, but there is enormous overlap of the $3s$ wave functions. (Curves are from calculations by D. R. Hartree and W. Hartree, *Proc. Roy. Soc.* **A**193, 299 (1948).)

THE CLASSIFICATION OF INSULATORS

The distinction between metals and insulators is based on the electronic distribution in k-space, which specifies which of the possible **k**-levels are occupied. However, it is of great utility to draw further distinctions, within the family of insulators, on the basis of the spatial electronic distribution. There are three broadly recognizable kinds of insulating solids with clearly distinguishable types of spatial electronic distributions.[4] The categories cannot be rigorously specified, and we shall encounter borderline cases, but the prototypes of each class are easily delineated.

1. ***Covalent Crystals*** These can be described as having spatial electronic distributions not that dissimilar from metals, but with no partially filled bands in k-space. Thus the electrons in covalent crystals need not be sharply localized in the neighborhood of the ion cores. On the other hand, covalent crystals are not likely to have the nearly uniform distribution of electronic density in the interstitial region characteristic of simple metals, whose one-electron wave functions are nearly plane waves between ion cores. It is more likely that the interstitial electronic distribution will be localized in certain preferred directions, leading to what are known in the language of chemistry as "bonds". An example of a covalent crystal (also known as a valence crystal) is diamond, an insulator with a band gap of 5.5 eV. There is an appreciable interstitial density of electrons in diamond, which is highly concentrated in the vicinity of the lines joining each carbon ion core to its four nearest neighbors (Figure 19.2).[5] This interstitial charge density is the characteristic feature distinguishing covalent crystals from the other two insulating types.

2. ***Molecular Crystals*** The outstanding examples of molecular crystals[6] are the solid noble gases, neon, argon, krypton, and xenon.[7] In the atomic state they have

---

[4] Because hydrogen is in several respects unique among atoms, a fourth type—the hydrogen-bonded solid—is frequently added to the list. We shall briefly describe this type of solid at the end of the chapter.

[5] Chemists would refer to this distribution of charge as the four electronic carbon bonds. From the point of view of Bloch theory, however, it is simply a property of the occupied electronic levels that leads to a charge density

$$\rho(\mathbf{r}) = -e \sum_{\substack{\text{all valence}\\ \text{band levels}}} |\psi(\mathbf{r})|^2,$$

that is appreciable in certain directions far from the ion cores, in spite of there being a substantial energy gap between occupied and unoccupied levels, so that the crystal is an insulator.

[6] The name "molecular crystal" reflects the fact that the entities out of which such substances are composed differ little from the isolated individual molecules. In the case of the noble gases the "molecules" happen to be identical to the atoms. Nevertheless, these structures are called molecular crystals, rather than atomic crystals, to permit the inclusion of such additional substances as solid hydrogen, nitrogen, etc. In these materials the constituent entities are H_2 or N_2 molecules, which are weakly perturbed from their free form in the solid. Although solid hydrogen and nitrogen are perhaps more deserving of the name "molecular crystal" than the solid noble gases, they provide considerably less clearcut examples, since within each molecule the electronic distribution is not localized about the two ion cores. Therefore if one should focus one's attention on the individual ion cores as the fundamental building blocks, one would have to describe solids such as hydrogen and nitrogen as partially molecular and partially covalent.

[7] Solid helium is a somewhat pathological example of a molecular solid, because of the very light mass of the helium atom. Even at $T = 0$ the liquid phase is more stable, unless considerable external pressure is applied.

Figure 19.2
The electronic charge distribution on a plane section of the conventional cubic cell of diamond, as suggested by X-ray diffraction data. The curves in (a) are curves of constant electronic density. The numbers along the curves indicate electronic density in electrons per cubic angstrom. The plane section of the cell that (a) describes is displayed in (b). Note that the electron density is quite high (5.02 electrons per cubic angstrom, as compared with 0.034 in the regions of lowest density) at the points where the plane intersects the nearest neighbor bonds. This is characteristic of covalent crystals. (Based on a figure from Y. K. Syrkin and M. E. Dyatkina, *Structure of Molecules and the Chemical Bond*, translated and revised by M. A. Partridge and D. O. Jordan, Interscience, New York, 1950.)

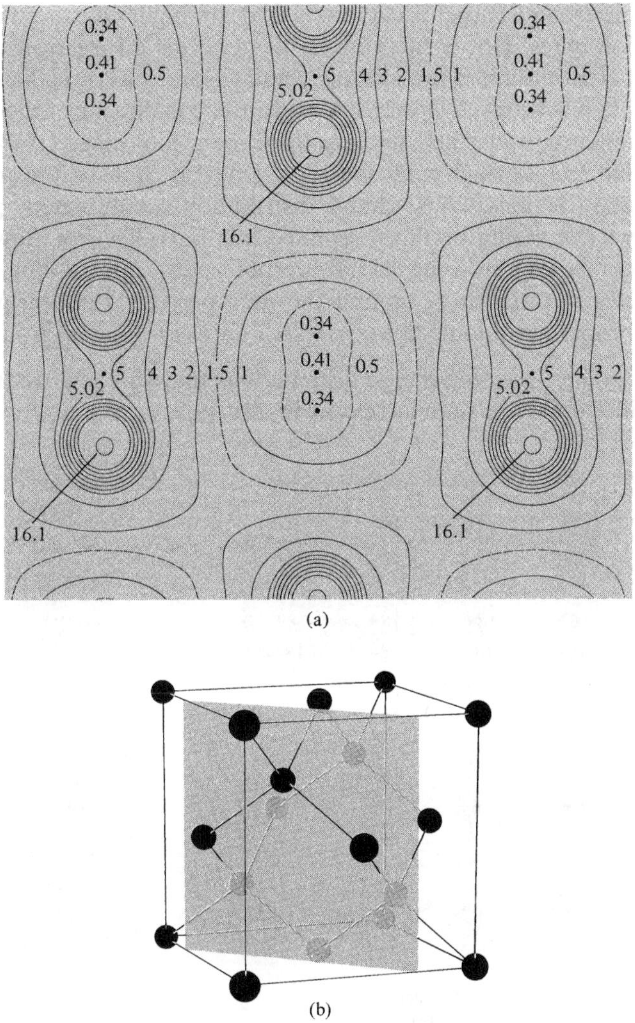

completely filled electronic shells—a highly stable configuration which is only weakly perturbed in the solid. From the point of view of band structure the noble gases are fine examples of extreme tight-binding solids; i.e., there is very little electronic density between ion cores, all the electrons remaining well localized in the neighborhood of their parent ions. For many purposes the whole theory of band structure is somewhat beside the point for such solids, since all electrons can be considered to be core electrons.[8] A discussion of molecular crystals must start with a consideration of those weak perturbations on the atoms that do occur when the solid is formed.

3. *Ionic Crystals* Ionic crystals, such as sodium chloride are compounds composed of a metallic and nonmetallic element. Like molecular crystals, ionic crystals have

[8] The conduction band levels have been calculated and, as expected, lie several electron volts above the filled band that accommodates the eight atomic valence electrons.

electronic charge distributions that are highly localized in the neighborhood of the ion cores. However, in molecular crystals all electrons remain very close to their parent atoms, while in ionic crystals some electrons have strayed so far from their parents as to be closely bound to the constituent of the opposite type. Indeed, one might consider an ionic crystal to be a molecular crystal in which the constituent molecules (which come in two varieties) are not sodium and chlorine atoms, but Na^+ and Cl^- ions, whose charge distribution is only weakly perturbed in the solid from what it would be in the isolated free ions. Because, however, the localized entities composing an ionic crystal are not neutral atoms, but charged ions, the immense electrostatic forces among the ions play an overbearing role in determining the properties of ionic crystals, which are very different from those of molecular solids.

The electronic charge distributions that characterize the three basic categories of insulators are summarized in Figure 19.3.

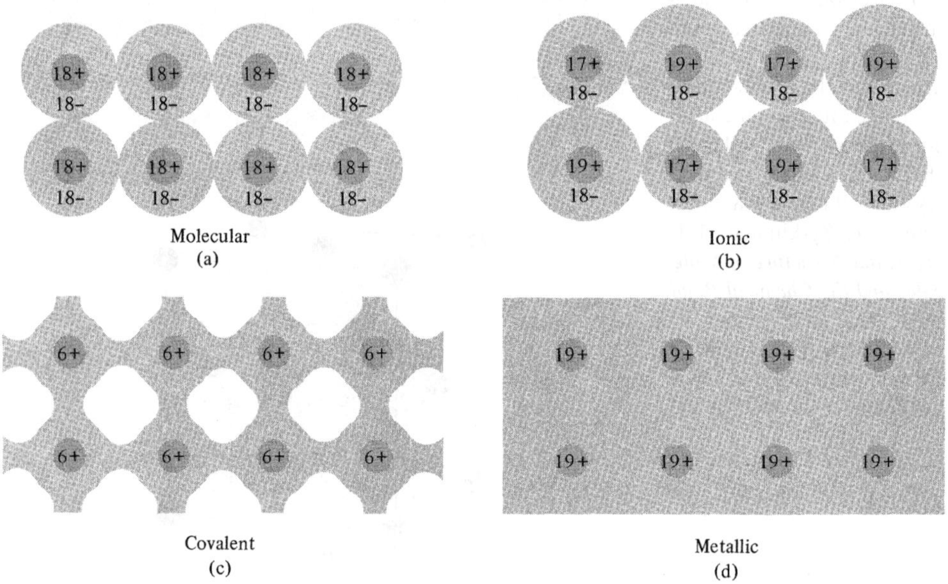

Figure 19.3
Highly schematic two-dimensional representation of the electronic charge distribution in the basic solid types. The small circles represent the positively charged nuclei, and the shaded parts, regions in which the electronic density is appreciable (though by no means uniform). We have (a) molecular (represented by two-dimensional "argon"): (b) ionic ("potassium chloride"); (c) covalent ("carbon"); (d) metallic ("potassium").

One frequently finds these distinctions drawn with emphasis not so much on the spatial configuration of the electron as on the so-called bonding. This is a point of view particularly dear to chemists, for whom the question of overwhelming importance in classifying a solid is what holds it together. The two points of view are closely related, since the Coulomb attraction between the electrons and atomic nuclei is the ultimate "glue" binding any solid together. Thus the nature of the binding depends critically on the spatial arrangement of the electrons. However, from the point of view of contemporary physics, and especially in the study of macroscopic solids, the energy required to assemble an object is not nearly as fundamental a property as it can be for the chemist. We have therefore chosen to emphasize spatial electronic structure in describing the categories, rather than (as has become traditional) bonding. To the physicist, bonding is only one of many properties strongly affected by this spatial distribution.

One should nevertheless be aware of the nomenclatural implications of this chemical point of view: One speaks of "the metallic bond," "the ionic bond," "the covalent bond," and "the hydrogen bond" when referring to the characteristic way in which the electrostatic forces conspire to hold solids of the corresponding types together. We shall have more to say about the binding energies (also known as cohesive energies) of the various solid types in Chapter 20.

In the remainder of this chapter we wish to amplify the distinctions among the basic categories of solids, emphasizing both the very different kinds of models used in describing the extreme types, as well as the continuity that prevails between the different categories. The discussion that follows vastly oversimplifies each type. What we are offering is a series of models on the level of sophistication of the Drude model of a metal, from which to begin the analysis of the several categories. More quantitative aspects of these models will be found in Chapter 20.

IONIC CRYSTALS

The most naive model of an ionic crystal treats all ions as impenetrable charged spheres. The crystal is held together by the electrostatic attraction between positively and negatively charged spheres, and is prevented from collapsing by their impenetrability.

The impenetrability is a consequence of the Pauli exclusion principle and the stable closed-shell electronic configurations of the ions. When two ions are brought so close together that their electronic charge distributions start to overlap, the exclusion principle requires that the excess charge introduced in the neighborhood of each ion by the other be accommodated in unoccupied levels. However, the electronic configuration of both positive and negative ions is of the stable closed-shell ns^2np^6 variety, which means that a large energy gap exists between the lowest unoccupied levels and the occupied ones. As a result it costs much energy to force the charge distributions to overlap; i.e., a strongly repulsive force exists between ions, whenever they are so close together that their electronic charge distributions interpenetrate.

For the qualitative points we wish to make in this chapter it is enough to consider the ions as impenetrable spheres—i.e., to take the potential representing this repulsive force as infinite within a certain distance, and zero beyond that distance. We stress, though, that the ions are not rigorously impenetrable. In detailed calculations on ionic crystals one must assume a less simplistic form for the dependence of the repulsive potential on interionic separation. (We shall give an elementary illustration of this in Chapter 20.) Furthermore, in a more realistic picture the ions cannot be taken to be rigorously spherical, since they are distorted from their (rigorously spherical) shape in free space by their neighbors in the crystal.

ALKALI HALIDES (I–VII IONIC CRYSTALS)

The ideal ionic crystal of spherical, charged billiard balls is most nearly realized by the alkali halides. These crystals are all cubic at normal pressures. The positive ion (cation) is one of the alkali metals (Li^+, Na^+, K^+, Rb^+, or Cs^+) and the negative ion (anion) one of the halogens (F^-, Cl^-, Br^-, or I^-). They all crystallize under normal conditions in the sodium chloride structure (Figure 19.4a) except for CsCl, CsBr, and CsI, which are most stable in the cesium chloride structure (Figure 19.4b).

To understand why the basic entities in these structures are ions rather than atoms, consider, for example, RbBr. An isolated bromine atom can actually attract an additional electron to form the stable anion Br^-, with the closed-shell electronic configuration of krypton. The additional electron has a binding energy[9] of about

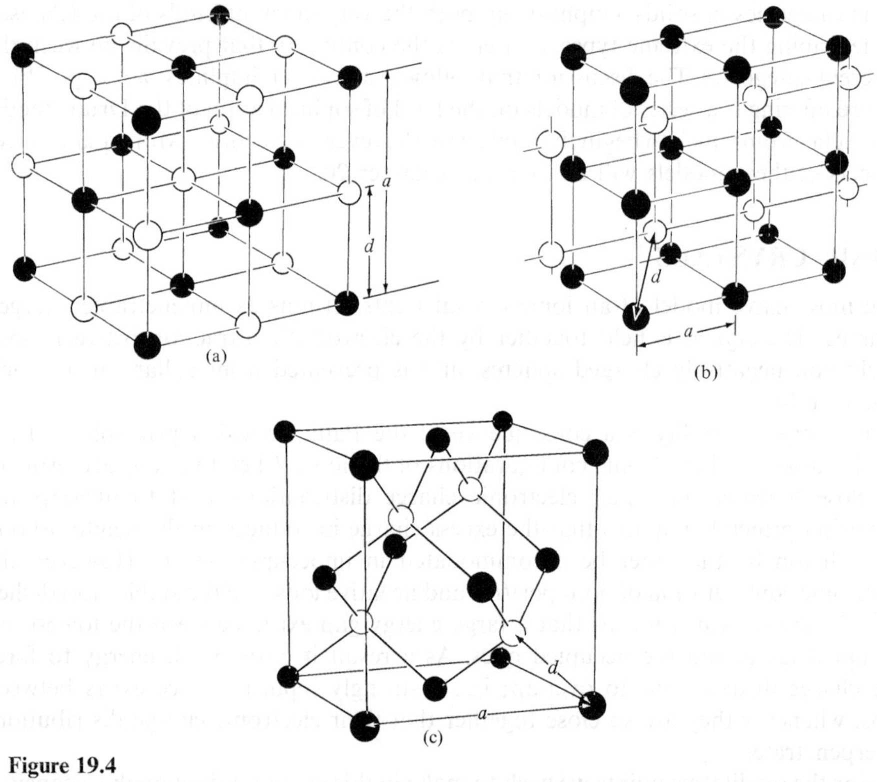

Figure 19.4
(a) The sodium chloride structure; (b) the cesium chloride structure; (c) the zincblende (sphalerite) structure. The side of the conventional cubic cell, a, and the nearest-neighbor distance, d, are indicated in each case. Inspection of the figure reveals that they are related by: (a) sodium chloride: $d = a/2$; (b) cesium chloride: $d = \sqrt{3}a/2$; (c) zincblende: $d = \sqrt{3}a/4$. For detailed descriptions of these structures see Chapter 4.

---

[9] This is usually put by saying that bromine has an "electron affinity" of 3.5 eV. It may at first appear surprising that a neutral atom can bind an extra electron. The binding is possible because the cloud of atomic electrons surrounding the nucleus is not completely effective in screening its field over the sixth and last p-orbital (the one holding the extra electron) which penetrates rather far into the cloud.

3.5 eV. However, to produce the cation Rb^+ (also with the closed-shell krypton configuration) it costs 4.2 eV to liberate the extra electron. It might therefore appear that a rubidium atom and a bromine atom together would have an energy 0.5 eV lower than the corresponding ions. This is indeed the case, provided that the ions are very far apart. If, however, we bring the ions together, the energy of the pair is lowered by their attractive electrostatic interaction. In crystalline RbBr the interionic distance is about $r = 3.4$ Å. A pair of ions at this distance has an additional Coulomb energy of $-e^2/r = -4.2$ eV, which more than compensates for the 0.5 eV favoring the atoms over the ions at large separation.

The picture of an alkali halide as a set of spherical ions packed together is confirmed by the electronic charge distributions inferred from X-ray diffraction data. Figure 19.5 shows the charge distribution these experiments suggest for sodium chloride.

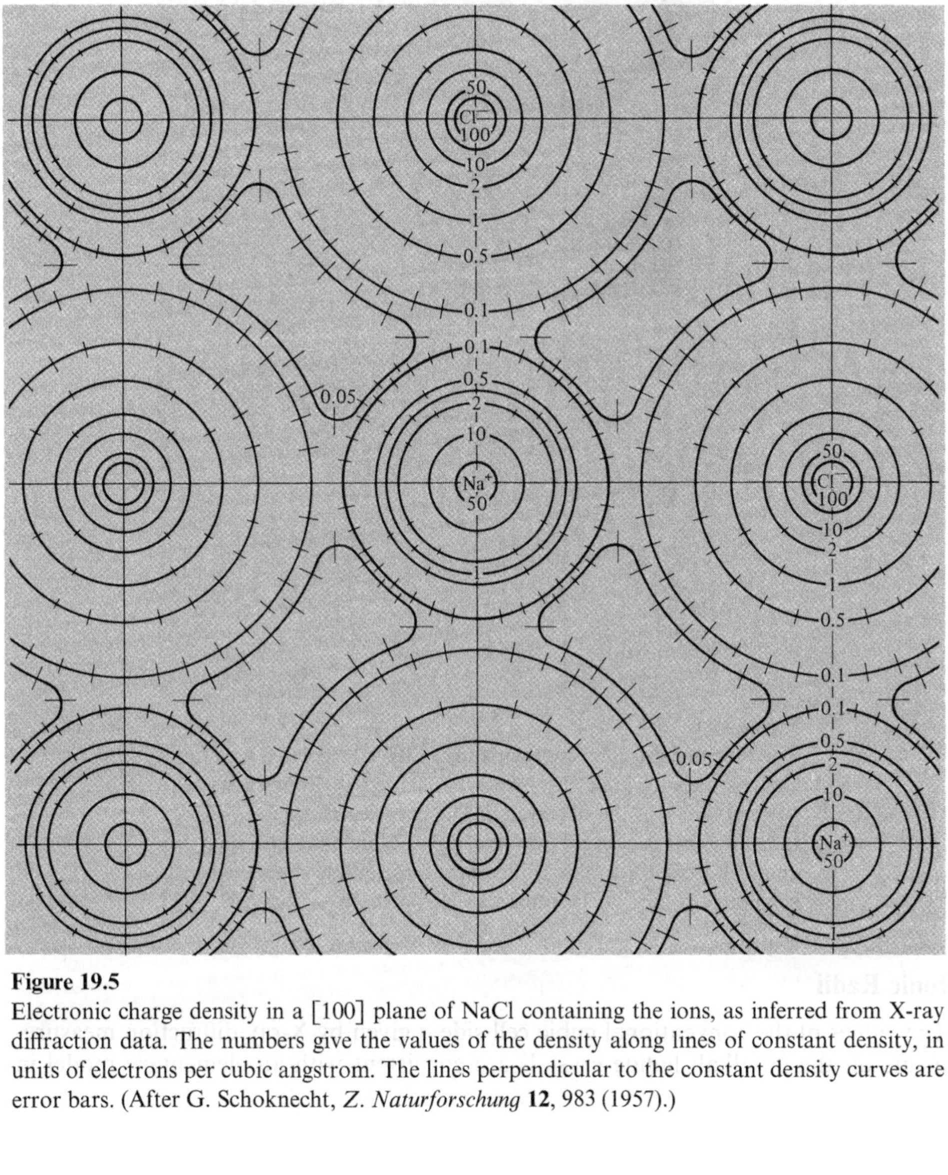

Figure 19.5
Electronic charge density in a [100] plane of NaCl containing the ions, as inferred from X-ray diffraction data. The numbers give the values of the density along lines of constant density, in units of electrons per cubic angstrom. The lines perpendicular to the constant density curves are error bars. (After G. Schoknecht, *Z. Naturforschung* **12**, 983 (1957).)

The idea that the alkali halides are composed of slightly distorted localized ions is also confirmed by band structure calculations. Figure 19.6 shows the energy bands calculated for KCl as a function of (externally imposed) lattice constant, compared with the corresponding levels of the free ions. The band energies can differ by as much as half a rydberg from the energies of the levels of the isolated ions even at large separations, because of the interionic Coulomb interactions. However, the band *widths* at the observed lattice constant are all exceedingly narrow, indicating that there is little overlap of the ionic charge distributions.

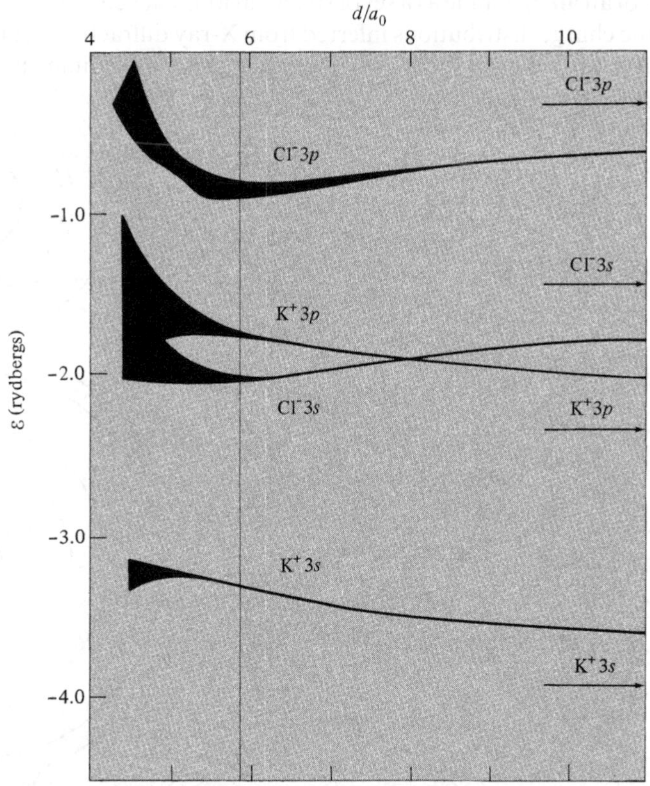

Figure 19.6
The four highest filled energy bands of KCl, calculated as a function of the interionic distance d (measured in Bohr radii). The vertical line is at the observed d. The energies of the free ions are indicated by the arrows on the right. Note that although the energies are considerably shifted in the crystalline state, the bands remain very narrow. (From L. P. Howard, *Phys. Rev.* **109**, 1927 (1958).)

Ionic Radii

The values of the conventional cubic cell side a given by X-ray diffraction measurements on the 20 alkali halide crystals are consistent with an elementary model in which the ions are regarded as impenetrable spheres of a definite radius r, known as the *ionic radius.* Let d be the distance between the centers of neighboring positive and negative ions, so that $d = a/2$ in the sodium chloride structure, and $a\sqrt{3}/2$ in the cesium chloride structure (see Figure 19.4). Table 19.1 lists the values of d for the alkali halide crystals.[10] If we take each of the nine ions to be a sphere of definite radius, then in most cases we can fit the nearest-neighbor distance d_{XY} for the alkali halide XY to an accuracy of around 2 percent by taking $d_{XY} = r_X + r_Y$. The only exceptions are LiCl, LiBr, and LiI, where the radius sum falls short of d by 6, 7, and 8 percent, respectively, and NaBr and NaI, where the radius sum falls short by 3 and 4 percent.

Leaving aside, for the moment, these exceptions, this shows that the observed lattice constants can be reproduced to a few percent accuracy by assuming the ions to be hard spheres of the specified radii, close packed into a sodium chloride (or cesium chloride) structure. However, the choice of ionic radii is not unique, for by adding a fixed amount Δr to all the alkali radii and subtracting the same Δr from the halogen radii ($r_X \to r + \Delta r, r_Y \to r - \Delta r$) the value of $r_X + r_Y$ will be unaffected.[11] This ambiguity can be resolved and the anomalous behavior of the lithium halides can be explained by the following further observation:

Our assertion that the nearest-neighbor distance d is given by the sum of the radii of the ions whose centers are separated by d assumes that such ions are in fact in contact (Figure 19.7). This will be the case provided the radius $r^>$ of the larger ion is not too much greater than the radius $r^<$ of the smaller. Should the disparity be too great, however, the smaller ions may not touch the larger ones at all (Figure 19.8). In that case d will be independent of the size of the smaller ion altogether, and be determined entirely by the size of the larger. The relation $r^+ + r^- = d$ appropriate to the sodium chloride structure must then be replaced by $\sqrt{2}r^> = d$ (Figure 19.7 and 19.8). The critical radius ratio, at which the smaller ion loses contact, occurs when each large ion touches both the small ion centered on the nearest-neighbor site and the large ion centered on the next nearest-neighbor site (Figure 19.9). The critical ratio satisfies (see Figure 19.9)

$$\frac{r^>}{r^<} = \frac{1}{\sqrt{2} - 1} = \sqrt{2} + 1 = 2.41 \qquad \text{(sodium chloride structure).} \qquad \textbf{(19.1)}$$

From the values of $r^>/r^<$ given in Table 19.1, one finds that the critical value 2.41 is exceeded only in LiCl, LiBr, and LiI. It is thus to be expected that the value observed for d exceeds the radial sum in these lithium halides, for in these cases d should be compared not with $r^+ + r^-$, but with $\sqrt{2}r^>$. This latter quantity is listed in square brackets after the value of $r^>/r^<$ for the three lithium halides. It fits the observed d to the same 2 percent accuracy that the values of $r^+ + r^-$ yield in the cases in which

---

[10] By "nearest-neighbor distance" we always mean the minimum distance between ionic *centers.* Thus (for example) in Figure 19.8 the nearest-neighbor distance is d, even though the large circles touch each other, but not the little circles. The distance between the center of a big circle and the center of a little circle is less than the distance between centers of neighboring big circles.

[11] This leads to rival schemes of ionic radii being put forth, but those quoted in Table 19.1 are still the most widely used.

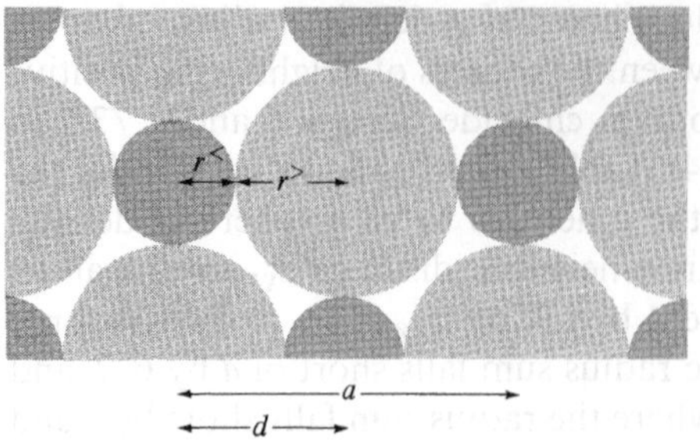

Figure 19.7
A [100] plane of the sodium chloride structure containing the centers of ions. Each large ion makes contact only with the neighboring small ions. Therefore the nearest-neighbor distance d is equal to the sum of the ionic radii, $r^> + r^<$. This is the normal state of affairs.

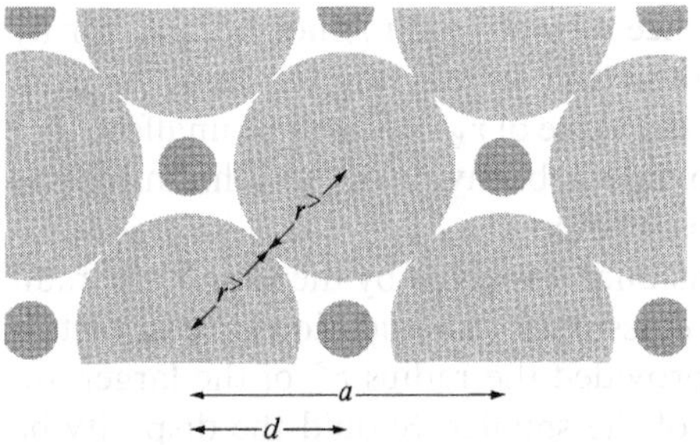

Figure 19.8
Same plane as in Figure 19.7, but now the disparity in ionic radii is so great that each large ion makes contact only with the nearest large ion. In this case the nearest-neighbor distance d (defined to be the shortest distance between ionic centers) is related only to the larger ionic radius $r^>$ by $d = \sqrt{2}r^>$.

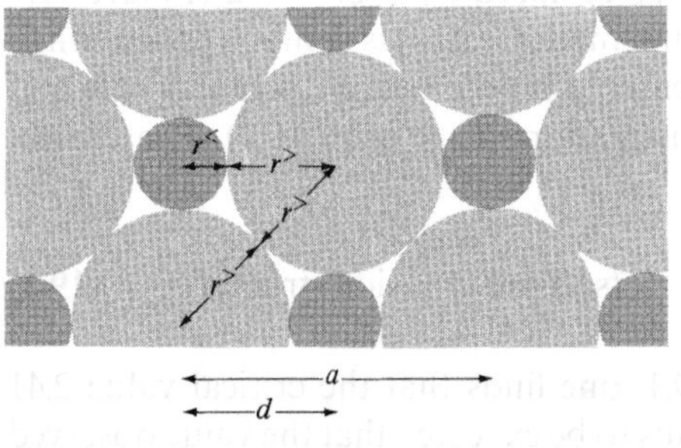

Figure 19.9
The state of affairs when the radius ratio has its critical value $r^>/r^< = \sqrt{2} + 1$. For smaller ratios Figure 19.7 holds; for larger ratios, Figure 19.8 holds. The value for the critical ratio follows from the observation that when the ratio is critical the relations $d = \sqrt{2}\,r^>$ and $d = r^+ + r^- = r^> + r^<$ must *both* hold.

they are applicable. It is not significant that *one* of the three errant lithium halides can be brought into line in this way, since the free variable Δr can be chosen to bring this about. However, the fact that the other two are also thereby made to agree lends great credence to the picture of the ions as impenetrable spheres, with the radii given in Table 19.1.

A similar calculation for the cesium chloride structure yields the smaller critical value

$$\frac{r_>}{r_<} = \frac{1}{2}(\sqrt{3} + 1) = 1.37 \qquad \text{(cesium chloride).} \qquad \textbf{(19.2)}$$

This value is not exceeded by the radius ratios of the three alkali halides with this

Table 19.1
PROPOSED IONIC RADII FOR THE ALKALI HALIDES[a]

	Li^+ (0.60)	Na^+ (0.95)	K^+ (1.33)	Rb^+ (1.48)	Cs^+ (1.69)
F^- (1.36)					
d	2.01	2.31	2.67	2.82	3.00
$r^- + r^+$	1.96	2.31	2.69	2.84	3.05
$r^>/r^<$	2.27	1.43	1.02	1.09	1.24
Cl^- (1.81)					
d	2.57	2.82	3.15	3.29	3.57
$r^- + r^+$	2.41	2.76	3.14	3.29	3.50
$r^>/r^<$	3.02 [2.56]	1.91	1.36	1.22	1.07
Br^- (1.95)					
d	2.75	2.99	3.30	3.43	3.71
$r^- + r^+$	2.55	2.90	3.28	3.43	3.64
$r^>/r^<$	3.25 [2.76]	2.05	1.47	1.32	1.15
I^- (2.16)					
d	3.00	3.24	3.53	3.67	3.95
$r^- + r^+$	2.76	3.11	3.49	3.64	3.85
$r^>/r^<$	3.60[3.05]	2.27	1.62	1.46	1.28

[a] The ionic radius (in angstroms) is given in parentheses immediately following the name of each ion. The following additional information (all in angstroms) is listed in the box corresponding to each alkali halide:

1. The nearest-neighbor distance d.[10] In the sodium chloride structure, $d = a/2$, where a is the side of the conventional cubic cell; in the cesium chloride structure (CsCl, CsBr, and CsI), $d = \sqrt{3}a/2$. (See Figure 19.4.)
2. The sum of the ionic radii, $r^- + r^+$.
3. The ratio of the ionic radii, $r^>/r^<$. In the three cases where this ratio is so large that d is not given by the radius sum, the new theoretical value ($\sqrt{2}r^>$) is listed in square brackets immediately after the radius ratio.

Source: L. Pauling, *The Nature of the Chemical Bond*, 3rd ed., Cornell University Press, Ithaca, New York, 1960, p. 514.

structure, and the radial sums are in good agreement with the observed lattice constants.[12]

II–VI Ionic Crystals

Doubly ionized elements from columns II and VI of the periodic table may also form ionic crystals. Except for the beryllium compounds and MgTe, these also assume the sodium chloride structure. It can be seen from Table 19.2 that $d = a/2$ again

[12] Agreement can be somewhat improved by increasing the ionic radii in the cesium chloride structure to account for the fact that each ion has eight nearest neighbors rather than the six it is surrounded by in the sodium chloride structure. Consequently, the Pauli principle repulsion is stronger, and the ions will not be squeezed so closely together.

Table 19.2
PROPOSED IONIC RADII FOR THE DOUBLY IONIZED ELEMENTS FROM COLUMNS II AND VI OF THE PERIODIC TABLE[a]

	Be^{++} (0.31)	Mg^{++} (0.65)	Ca^{++} (0.99)	Sr^{++} (1.13)	Ba^{++} (1.35)
O^{--} (1.40)					
d	1.64	2.10	2.40	2.54	2.76
$r^- + r^+$	1.71	2.05	2.39	2.53	2.75
$r^>/r^<$	4.52	2.15	1.41	1.24	1.04
S^{--} (1.84)					
d	2.10	2.60	2.85	3.01	3.19
$r^- + r^+$	2.15	2.49	2.83	2.97	3.19
$r^>/r^<$	5.94[2.25]	2.83[2.60]	1.86	1.63	1.36
Se^{--} (1.98)					
d	2.20	2.72	2.96	3.11	3.30
$r^- + r^+$	2.29	2.63	2.97	3.11	3.33
$r^>/r^<$	6.39[2.42]	3.05 [2.80]	2.00	1.75	1.47
Te^{--} (2.21)					
d	2.41	2.75	3.17	3.33	3.50
$r^- + r^+$	2.52	2.86	3.20	3.34	3.56
$r^>/r^<$	7.13[2.71]	3.40	2.23	1.96	1.64

[a] The ionic radius (in angstroms) is given in parentheses immediately following the name of each ion. In addition, the following information is given in the box corresponding to each compound:

1. The nearest-neighbor distance d.
2. The sum of the ionic radii, $r^+ + r^-$.
3. The ratio of the ionic radii, $r^>/r^<$.

All compounds have the sodium chloride structure except for BeS, BeSe, and BeTe (zincblende) and BeO, MgTe (wurtzite.) In the two magnesium compounds for which the radius ratio exceeds the critical value 2.42 for the sodium chloride structure, the corrected theoretical value $d = \sqrt{2}r^>$ is given in brackets. In the zincblende structures, the critical ratio 4.45 is exceeded in all cases, and the corrected value $d = \sqrt{6}r^>/2$ is given in brackets. These (and crystals with the wurtzite structure) are better treated as covalent.

Source: L. Pauling, *The Nature of the Chemical Bond*, 3rd ed., Cornell University Press, Ithaca, N.Y., 1960, p. 514.

agrees to within a few percent with $r^+ + r^-$ for the calcium, strontium, and barium salts, and MgO. In MgS and MgSe the critical ratio $r^>/r^< = 2.42$ is exceeded; here $d = a/2$ agrees with $\sqrt{2}r^>$ to within about 3 percent.

However, agreement with the ionic radii is not nearly as satisfactory for MgTe and the beryllium compounds. BeS, BeSe, and BeTe crystallize in the zincblende (sphalerite) structure (see Chapter 4 and Figure 19.4), and the other two in the wurtzite structure.[13] The critical ratio[14] $r^>/r^<$ is $2 + \sqrt{6} = 4.45$, which is exceeded in all three beryllium compounds with the zincblende structure.[15] The quoted value of d (taken to be $\sqrt{3}/4$ times the measured side a of the conventional cubic cell) should then be compared not with $r^+ + r^-$, but with $\sqrt{6}r^>/2$, which is listed in square brackets in Table 19.2 after $r^>/r^<$. The agreement is relatively poor compared with the impressive agreement we found in the crystals with the sodium chloride structure.

One reason for this is that beryllium (and to some extent magnesium) is substantially more difficult to ionize than the other elements in column II. (First ionization potentials, in electron volts: Be, 9.32; Mg, 7.64; Ca, 6.11; Sr, 5.69; Ba, 5.21.) Thus the cost in energy to produce widely separated ions, as opposed to atoms, is quite high in the beryllium compounds. In addition, because it is so small, the beryllium ion cannot take full advantage of the crystal structures with high coordination number to compensate for this by maximizing the interionic Coulomb interaction: The anions would be repelled by the overlap of their own charge distributions, before they came close enough to the beryllium ion (as, for example, in Figure 19.8). These considerations suggest that in the beryllium compounds we are already moving away from the realm of purely ionic crystals.

As it turns out, tetrahedrally coordinated structures (such as the zincblende and wurtzite structures) tend to be primarily covalently bonded. The tetrahedrally coordinated II–VI compounds are more covalent than ionic in character.[16]

III–V CRYSTALS (MIXED IONIC AND COVALENT)

The crystals that pair elements from columns III and V of the periodic table are still less ionic in character. They almost all assume the zincblende structure characteristic of covalent crystals. Some examples are given in Table 19.3. Most behave as semiconductors rather than insulators; i.e., their band gaps are relatively small. This is another indication that their ionic nature is quite weak, and the electrons not strongly localized. The III–V compounds are thus good examples of substances that are partially ionic and partially covalent. They are conventionally described as primarily covalent, with some residual concentration of excess charge about the ion cores.

COVALENT CRYSTALS

The striking difference in interstitial charge distributions in ionic and covalent crystals is seen by comparing Figures 19.2 and 19.5; the electron density along a nearest-

---

[13] In the zincblende structure, a view along a [111] direction reveals that the atoms of a given kind are stacked in the sequence . . . *ABCABC* . . . while maintaining tetrahedral bonds with those of the other kind. The underlying Bravais lattice is cubic. There is another arrangement that preserves the tetrahedral bonding but stacks the atoms of a given kind in the sequence . . . *ABABAB* . . . . This is the wurtzite structure, whose underlying Bravais lattice is hexagonal.

[14] The reader is invited to verify these numerological assertions.

[15] This is also the critical ratio for the wurtzite structure, provided the c/a of the underlying hcp structure is close to ideal, as it almost always is.

[16] It is possible to specify *covalent radii*, for use in tetrahedrally coordinated structures, which are almost as successful in fitting their lattice constants as are the ionic radii in ionic crystals. (See L. Pauling, *The Nature of the Chemical Bond*, 3rd ed., Cornell University Press, Ithaca, New York, 1960.

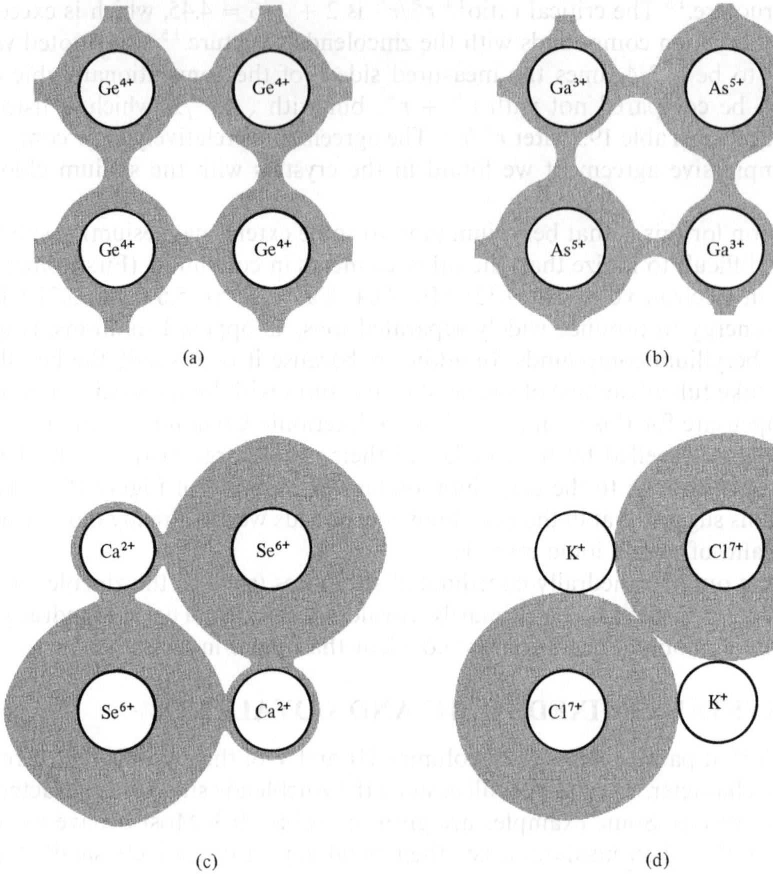

Figure 19.10

A highly schematic representation of the continuity from perfect covalent to perfect ionic crystals. (a) *Perfectly covalent germanium.* Four electrons per unit cell are identically distributed about the Ge^{4+} ion cores. The electronic density is large in certain directions in the interstitial region. (b) *Covalent gallium arsenide.* The interstitial electron density has diminished somewhat and there is a slight tendency for the cloud of electrons around each As^{5+} ion core to have somewhat more charge than is necessary to compensate the positive charge, while the electron cloud around each Ga^{3+} ion core has somewhat less. The crystal thus has a very slight ionic character as well. (c) *Ionic calcium selenide.* The Ca^{2+} ion is almost denuded of valence electrons and the cloud of electrons around the Se^{6+} ion core is almost the full eight necessary to produce Se^{--}. (It would be more conventional, in fact, to represent the selenium as an Se^{--} ion core lacking a small fraction of an electron.) The crystal is weakly covalent, to the extent that the Ca^{2+} is slightly shielded by electrons in its immediate neighborhood, and the Se^{6+} does not have quite enough electrons to filled the outer eight shells completely, making Se^{--}. The covalent character is also seen in the slight distortion of the charge distribution outward along nearest-neighbor lines. (d) *Perfectly ionic potassium chloride.* The K^+ ion is bare of excess electrons, and all eight electrons cluster around the Cl^{7+} to make Cl^-. (It would be more conventional to show no electrons at all, and simply draw a Cl^- ion core for the chlorine.)

neighbor line drops below 0.1 electron per cubic angstrom in NaCl, while it does not get less than 5 electrons per cubic angstrom, in the covalent crystal *par excellence*, diamond.

Diamond is typical of the crystal structures formed by the elements from column IV of the period table: carbon, silicon, germanium, and (grey) tin (see Table 4.3). These elements all crystallize in the tetrahedrally coordinated diamond structure. In chemical terminology, each atom participates in four covalent bonds, by sharing an electron with each of its four neighbors. Although the ultimate basis for the bonding remains electrostatic, the reason why the crystal remains bound is considerably more complex than the simple picture of oppositely charged billiard balls that suits the ionic crystals so well. We shall have more to say about this in Chapter 20.

Table 19.3
SOME COVALENT III–V COMPOUNDS[a]

	Al	Ga	In
P	5.45	5.45	5.87
As	5.62	5.65	6.04
Sb	6.13	6.12	6.48

[a] All have the zincblende structure. The side of the conventional cubic cell (in angstroms) is given.

The continuous variation in the nature of the charge distribution as one progresses from the extreme ionic I–VII compounds through the progressively more ambiguous II–VI, and III–V compounds over to the extreme covalent elements from column IV, is indicated schematically in Figure 19.10.

Covalent crystals are not as good insulators as ionic crystals; this is consistent with the delocalization of charge in the covalent bond. The semiconductors are all covalent crystals, sometimes (as in the III–V compounds) with a small touch of ionic bonding.

MOLECULAR CRYSTALS

As one moves to the right in the periodic table from any of the elements in the upper part of column IV, the solids one encounters become successively more insulating (or, if one starts lower down in column IV, less metallic) and more weakly bound (with, for example, lower and lower melting points). At the extreme right of the table are the column VIII elements, which afford the best example of molecular solids. The solid noble gases (except for helium) all crystallize in monatomic fcc Bravais lattices. The electronic configuration of each atom is of the stable closed-shell type, which is little deformed in the solid. The solid is held together by very weak forces, the so-called *van der Waals* or *fluctuating dipole* forces. The qualitative physical origin of this force is easily explained:[17]

[17] Note that, again, the ultimate source of the bonding must be electrostatic. However, the way in which the electrostatic attraction now manifests itself is so different (from, for example, the ionic crystals) that this type of binding is given a separate name of its own. A more rigorous quantum-mechanical derivation of the van der Waals attraction is given in Problem 1.

Consider two atoms (1 and 2) separated by a distance r. Although the average charge distribution in a single rare gas atom is spherically symmetric, at any instant there may be a net dipole moment (whose time-averaged value must vanish). If the instantaneous dipole moment of atom 1 is $\mathbf{p}_1$, then there will be an electric field proportional to p_1/r^3 at a distance r from the atom.[18] This will induce a dipole moment in atom 2 proportional to the field:

$$p_2 = \alpha E \sim \frac{\alpha p_1}{r^3}, \tag{19.3}$$

where α is the polarizability[19] of the atom. Since two dipoles have an energy of interaction proportional to the product of their moments divided by the cube of the distance between them, there will be a lowering of energy of order

$$\frac{p_2 p_1}{r^3} \sim \frac{\alpha {p_1}^2}{r^6}, \tag{19.4}$$

associated with the induced moment. Since this quantity depends on ${p_1}^2$, its time average does not vanish, even though the average value of $\mathbf{p}_1$ is zero. Because this force falls off rapidly with distance, it is exceedingly weak, whence the low melting and boiling points of the condensed noble gases.

For a more precise treatment of molecular solids one must also consider the fluctuating dipolar interaction between groups of three or more atoms that cannot be represented as a sum of pair interactions. These fall off more rapidly than $1/r^6$, but are also important in arriving at an accurate theory of the solid state.[20]

The elements in groups V, VI, and VII (with the exception of metallic polonium and the semimetals antimony and bismuth) partake to varying degrees of both molecular and covalent character. Solid oxygen and nitrogen, as we have mentioned, are molecular crystals in which the weakly perturbed entities are not free atoms, but O_2 or N_2 molecules. Within these molecules the binding is covalent, so that the electronic distribution in the crystal as a whole has a mixed molecular and covalent structure. There are also substances (sulfur and selenium are examples) that possess elaborate crystal structures and for which no simple categorization of their bonding is possible.

METALS

As one moves to the left in the periodic table from column IV, one enters the family of metals; i.e., the covalent bond expands until there is an appreciable density of electrons throughout the interstitial regions, and, in k-space, appreciable band overlap. The prime examples of metallic crystals are the alkali metals of column I, which for many purposes can be accurately described by the Sommerfeld free electron model, in

---

[18] We have in mind a distance rather farther from the atom than its linear dimensions. When one gets too close to the atom the dipole approximation is no longer valid, but, more importantly, the strong core-core repulsion begins to dominate the attractive fluctuating dipole interaction.

[19] See Chapter 27.

[20] B. M. Axilrod and E. Teller, *J. Chem. Phys.* **22**, 1619 (1943); B. M. Axilrod, *J. Chem. Phys.* **29**, 719, 724 (1951).

which the valence electrons are completely separated from their ion cores and form a nearly uniform gas.

More generally, one can find aspects of covalent and molecular bonding even in metals, particularly in the noble metals in which the filled atomic *d*-shells are not very tightly bound, and as a consequence suffer considerable distortion in the metal.

It is instructive to compare the ionic radii of the metallic elements (as calculated from the structure of the ionic crystals they participate in) with the nearest-neighbor distance in the metal (Table 19.4). It is evident that the concept of ionic radius is completely irrelevant in determining the alkali metal lattice constants. This is consistent with the fact that such quantities as the alkali metal compressibilities are of the order of their electron gas values; the ions are genuinely small objects, embedded in a sea of electrons. In the noble metals, on the other hand, as noted in Chapter 15, the closed *d*-shell plays a far more important role in determining metallic properties than do the ion cores of the alkali metals. This is reflected in the fact that in Cu, Ag, and Au, the nearest-neighbor distances in the metal are not that much larger than the ionic radii in ionic crystals. In both the ionic crystals and (to only a slightly lesser degree) the metal, size is determined by the *d*-shells.

Table 19.4
IONIC RADII COMPARED WITH HALF THE NEAREST-NEIGHBOR DISTANCES IN METALS

METAL	SINGLY IONIZED IONIC RADIUS, r_{ion} (Å)	NEAREST-NEIGHBOR HALF DISTANCE IN METAL, r_{met} (Å)	r_{met}/r_{ion}
Li	0.60	1.51	2.52
Na	0.95	1.83	1.93
K	1.33	2.26	1.70
Rb	1.48	2.42	1.64
Cs	1.69	2.62	1.55
Cu	0.96	1.28	1.33
Ag	1.26	1.45	1.15
Au	1.37	1.44	1.05

HYDROGEN-BONDED CRYSTALS

Some classifications list hydrogen-bonded crystals as a fourth category of insulator. This is in recognition of the fact that hydrogen is unique in three important ways:

1. The ion core of a hydrogen atom is a bare proton of order 10^{-13} cm in radius, a factor of 10^5 smaller than any other ion core.
2. Hydrogen is but one electron shy of the stable helium configuration, which, uniquely among stable configurations, has not eight but only two electrons in the outermost shell.
3. The first ionization potential of atomic hydrogen is unusually high (H, 13.59 eV; Li, 5.39 eV; Na, 5.14 eV; K, 4.34 eV; Rb, 4.18 eV; Cs, 3.89 eV).

As a result of these properties hydrogen can play a role unlike any other element in crystalline structures. Because of its large ionization potential, it is much more difficult to remove an electron completely from hydrogen, and it therefore does not behave as an alkali metal ion (of minute radius) in forming ionic crystals. On the other hand, it cannot behave as the atoms in typical covalent crystals, for lacking only one electron from a closed-shell configuration it can, in chemical terms, form only one covalent bond through electron sharing.[21] Finally, because the proton has, for all practical purposes, no size at all, it can essentially sit on the surface of the large negative ions, resulting in a type of structure unattainable with any other positive ion.

One manifestation of these peculiar properties is illustrated in Figure 19.11 for the case of ice. The electron from the hydrogen atom is, like the proton, fairly well localized in the neighborhood of the oxygen ions. The positive proton resides close to a single oxygen ion, along the line joining it to one of its neighbors, thereby helping to bind the two oxygen ions together. (Note the lack of regularity in the positions of the protons. This can be observed thermodynamically in the large "residual entropy" possessed by ice at low temperatures, corresponding to the large number of ways of assigning a proton to either end of each bond, consistent with two protons being close to each oxygen atom.)

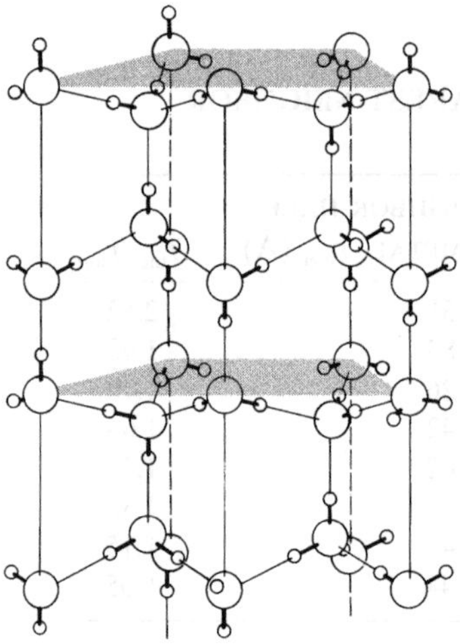

Figure 19.11
The crystal structure of one of the many phases of ice. The large circles are oxygen ions; the small circles are protons. Ice is an example in which hydrogen bonding plays a crucial role. (After L. Pauling, *The Nature of The Chemical Bond*, 3rd. ed., Cornell University Press, Ithaca, New York, 1960.)

This completes our descriptive survey of some of the various solid types. We now turn to some of the elementary quantitative implications of the various structures on bulk solid properties—particularly the binding, or cohesive, energy.

PROBLEMS

1. *Origin of the van der Waals Force*

Consider two noble gas atoms a distance R apart, represented by fixed nuclei of charge Ze located at $\mathbf{0}$ and $\mathbf{R}$, each surrounded by Z electrons. The electrons bound to the nucleus at $\mathbf{0}$ have co-

[21] This contrasts with the four bonds that are present in tetrahedrally coordinated covalent crystals, as the result of forming two closed shells from eight electrons.

ordinates $\mathbf{r}_i^{(1)}$ and those bound to the nucleus at $\mathbf{R}$ have coordinates $\mathbf{r}_i^{(2)}$, $i = 1, \ldots, Z$. We suppose $\mathbf{R}$ to be so large that there is negligible overlap between the electronic charge distributions about the two nuclei.[22] Let H_1 and H_2 be the Hamiltonians for atoms 1 and 2 alone. The Hamiltonian for the two-atom system will be $H = H_1 + H_2 + U$, where U gives the Coulomb interaction between all pairs of charged particles, one from atom 1, and one from atom 2:

$$U = e^2\left[\frac{Z^2}{R} - \sum_{i=1}^{Z}\left(\frac{Z}{|\mathbf{R} - \mathbf{r}_i^{(1)}|} + \frac{Z}{r_i^{(2)}}\right) + \sum_{i,j=1}^{Z}\frac{1}{|\mathbf{r}_i^{(1)} - \mathbf{r}_j^{(2)}|}\right]. \tag{19.5}$$

To second order in perturbation theory, the interaction energy between the two atoms will be given by:

$$\Delta E = \langle 0|U|0\rangle + \sum_n \frac{|\langle 0|U|n\rangle|^2}{E_0 - E_n}, \tag{19.6}$$

where $|0\rangle$ is the ground state of the unperturbed two-atom system, and $|n\rangle$ its excited states.

(a) Show that the first-order term in (19.6) is just the electrostatic interaction energy between two distributions of charge density $\rho^{(1)}(\mathbf{r})$ and $\rho^{(2)}(\mathbf{r})$, where $\rho^{(1)}$ and $\rho^{(2)}$ are the ground-state charge distributions of atoms 1 and 2.

(b) Prove that if the charge distributions have zero overlap and are spherically symmetric, this interaction energy is identically zero.[23]

(c) The assumption that there is negligible overlap between the electronic states on the two atoms also means that the wave functions appearing in the second order term in (19.6) are negligibly small unless $|\mathbf{r}_i^{(1)}|$ and $|\mathbf{r}_i^{(2)} - \mathbf{R}|$ are small compared with R. Show that if (19.5) is expanded in these quantities, the leading nonvanishing term is

$$-\frac{e^2}{R^3}\sum_{i,j}\left[3(\mathbf{r}_i^{(1)} \cdot \hat{\mathbf{R}})([\mathbf{r}_j^{(2)} - \mathbf{R}] \cdot \hat{\mathbf{R}}) - \mathbf{r}_i^{(1)} \cdot (\mathbf{r}_j^{(2)} - \mathbf{R})\right]. \tag{19.7}$$

(d) Show, as a result, that the leading term in (19.6) varies as $1/R^6$ and is negative.

2. *Geometrical Relations in Diatomic Crystals*

Verify that the critical ratios $r^{>}/r^{<}$ are $(\sqrt{3} + 1)/2$ for the cesium chloride structure and $2 + \sqrt{6}$ for the zincblende structure, as asserted in the text.

---

[22] Because of this we can ignore the Pauli principle as it affects the interchange of electrons between atoms, and regard the electrons on atom 1 as distinguishable from those on atom 2. In particular, we need not antisymmetrize the states appearing in (19.6).

[23] If the atoms are too close together, overlap cannot be ignored, and it leads to a strong (short-range) repulsion. The very slight overlap when the atoms are far apart yields corrections to the interaction that fall off exponentially with separation.

20
Cohesive Energy

The cohesive energy of a solid is the energy required to disassemble it into its constituent parts—i.e., its binding energy.[1] This energy depends, of course, on what the constituent parts are considered to be. They are generally taken to be the individual atoms of the chemical elements out of which the solid is composed, but other conventions are sometimes used. For example, it might be convenient to define the cohesive energy of solid nitrogen as that required to separate it into a set of isolated nitrogen molecules, rather than atoms. Knowing the binding energy of an isolated nitrogen molecule, one can easily convert from one definition to the other. Similarly, in the alkali halide crystals we shall discuss the energy necessary to separate the solid into isolated ions, rather than atoms. The link between the two energies is provided by the first ionization potential of the alkali metal atom and the electron affinity of the halogen atom.

In the early days of solid state physics much effort was devoted to computing cohesive energies, and the subject pervaded the theory of solids much more than it does today. Older discussions of the classification of solids, for example, rely heavily on the nature of the cohesion, rather than emphasizing (as in Chapter 19) the (closely related) spatial electronic arrangement. The importance of the cohesive energy is that it is the ground-state energy of the solid; its sign, for example, determines whether the solid will be stable at all. Indeed, its generalization to nonzero temperatures, the Helmholtz free energy, if known as a function of volume and temperature, contains all equilibrium thermodynamic information about the solid. However, solid state physics has come to focus more and more on nonequilibrium properties (e.g., transport properties and optical properties) and the study of cohesion no longer plays the dominant role it once did.

In this chapter we shall discuss some elementary facts about cohesive energies at zero temperature. We shall calculate these energies for an externally imposed lattice constant, and shall therefore be considering solids under external pressure. By calculating the rate of change of cohesive energy with lattice constant we can find the pressure necessary to maintain a given volume, and therefore determine the equilibrium lattice constant as that requiring zero pressure[2] for its maintenance. In the same way we can calculate the compressibility of the solid—i.e., the change in volume produced by a given change in pressure. This is more accessible to physical measurement than the cohesive energy itself, since it does not require dismantling of the solid into its constituents.

Throughout this chapter we shall treat the ion cores as classical particles, which can be perfectly localized with zero kinetic energy at the sites of a lattice. This is incorrect, for it violates the uncertainty principle. If an ion core is confined to a region of linear dimensions Δx, the uncertainty in its momentum will be of order $\hbar/\Delta x$. It will therefore have a kinetic energy of order $\hbar^2/M(\Delta x)^2$, known as the zero-point kinetic energy, whose contribution to the energy of the solid must be taken

---

[1] It is frequently given in kilocalories per mole. A useful conversion factor is 23.05 kcal/mole = 1 eV/molecule.

[2] More correctly, atmospheric pressure. However, the difference in size between a solid at atmospheric pressure and one in vacuum is negligible at the level of accuracy of our analysis.

into account. Furthermore, since the ions are not perfectly localized (for that would imply infinite zero-point kinetic energy), the deviations in their potential energy from that of classical particles fixed at lattice sites must be allowed for. We shall not be able to do this in any but the crudest way (Problem 1) until Chapter 23, where we describe the theory of lattice vibrations. For now we only note that the lighter the ionic mass, the larger is the zero-point kinetic energy, and the more suspect the approximation of perfectly localized ions. We shall see some simple evidence for the importance of zero-point motion in the lighter noble gases in our discussion below.[3] In most other cases the errors introduced by the neglect of zero-point motion are of the order of 1 percent or less.

Having noted this oversimplification, we turn to the other, generally more important, factors contributing to the binding energy of the various types of solids.[4] We start with the molecular solids (the crude theory of which is particularly simple) treating them as atoms held together by the short-range fluctuating dipole interaction, and prevented from coming close together by the still shorter-range core-core repulsion.[5] On a similar level of sophistication, ionic crystals are somewhat more subtle, because the basic building blocks are now electrically charged ions, and problems arise connected with the very long range of the interionic force. On the other hand, the electrostatic interaction energy of the ions is so large that it completely dominates all other sources of attraction.[6] In this respect the crude theory of ionic crystals is the simplest of all.

However, when we turn to covalent crystals and metals we find that even a crude theory is difficult to construct. The basic problem is that the arrangement of the valence electrons, whether in the well-localized bonds of good covalent insulators or the electron gas of the alkali metals, is vastly different from what it is in either the isolated atoms or the isolated ions. Our discussion in these cases will be highly qualitative.

For simplicity we shall discuss only cubic crystals in this chapter, and consider the energy of the solid as a function of the cubic cell side, a. In so doing we are ignoring crystals whose energy may depend on more than one geometrical parameter (e.g., on c and a in hcp structures). We shall also ignore deformations of cubic crystals from their equilibrium size and shape more general than a mere uniform compression (which preserves their cubic symmetry). The physics of more complex deformations is no different, but the geometrical aspects of the more general deformations can be complicated. We shall limit our discussion of such deformations to the less fundamental description given in the discussion of elastic constants in Chapter 22.

---

[3] Only in solid helium do considerations determined by zero-point motion become of really crucial importance. The mass of helium is so light that quantum effects prevent it from solidifying at all, unless an external pressure is imposed.

[4] We emphasize again that the only attractive forces at work are electrostatic, but the way in which they manifest themselves varies so dramatically from category to category as to require in each case separate discussions, and even separate nomenclature.

[5] Recall (page 379) that this is simply a crude way of representing classically some effects of the Pauli exclusion principle, when applied to filled atomic shells.

[6] Such as the fluctuating dipole interactions between the ions.

MOLECULAR CRYSTALS: THE NOBLE GASES

We consider only the simplest molecular crystals, in which the constituent entities are noble gas atoms. We omit solid helium, because of the crucial role played by quantum effects.[7] As described in Chapter 19, the atoms in a solid noble gas are only slightly distorted from the stable closed-shell configuration they possess in the free state. Such small distortion as occurs can be described by the fluctuating dipole interaction and represented as a weak attractive potential which varies as the inverse sixth power of the interatomic separation. It is this weak attraction that holds the solid together.

When the atoms approach one another too closely, the repulsion of the ion core comes into play and is crucial in determining the equilibrium size of the solid. At short distances this repulsion must be stronger than the attraction, and it has become conventional to represent it in the form of a power law as well. The power generally chosen is 12, and the resulting potential then has the form

$$\phi(r) = -\frac{A}{r^6} + \frac{B}{r^{12}}, \tag{20.1}$$

where A and B are positive constants, and r is the distance between atoms. The potential is usually written in the dimensionally more appealing form

$$\phi(r) = 4\epsilon\left[\left(\frac{\sigma}{r}\right)^{12} - \left(\frac{\sigma}{r}\right)^{6}\right], \qquad \begin{aligned}\sigma &= (B/A)^{1/6},\\ \epsilon &= A^2/4B,\end{aligned} \tag{20.2}$$

and is known as the Lennard-Jones 6–12 potential. There is no reason for choosing the exponent in the repulsive term to be 12, other than the resulting analytic simplicity and the requirement that the number be larger than 6. With this choice, however, the thermodynamic properties of gaseous neon, argon, krypton, and xenon at low densities can be well reproduced by suitable choices of the parameters ϵ and σ for each. The values so obtained are displayed[8] in Table 20.1.

Table 20.1
VALUES OF THE LENNARD-JONES PARAMETERS FOR THE NOBLE GASES[a]

	Ne	Ar	Kr	Xe
$\epsilon(10^{-13}$ erg)	0.050	0.167	0.225	0.320
ϵ(eV)	0.0031	0.0104	0.0140	0.0200
σ(Å)	2.74	3.40	3.65	3.98

[a] As deduced from properties of the low-density gases (second virial coefficient).
Source: N. Bernardes, *Phys. Rev.* **112**, 1534 (1958).

[7] And also because solid helium (of either isotope) does not exist at zero pressure; 25 atmospheres are required for ^{4}He, and 33 for ^{3}He.

[8] These values must be used with caution in the solid state, for at high densities the interaction cannot be represented as a sum of pair potentials (see page 390). If one nevertheless insists on fitting data in the solid with a sum of pair potentials of the form (20.2), the best choice of ϵ and σ need not be the same as that determined by properties in the gaseous state.

We emphasize that the precise form of the potential (20.2) should not be taken too seriously. It is nothing more than a simple way of taking into account the following:

1. The potential is attractive and varies as $1/r^6$ at large separations.
2. The potential is strongly repulsive at small separations.
3. The parameters ϵ and σ measure the strength of the attraction and the radius of the repulsive core, as determined by fitting data in the gaseous state.

Note that ϵ is only of order 0.01 eV, consistent with the very weak binding of the solidified noble gases. The Lennard-Jones potential is shown in Figure 20.1.

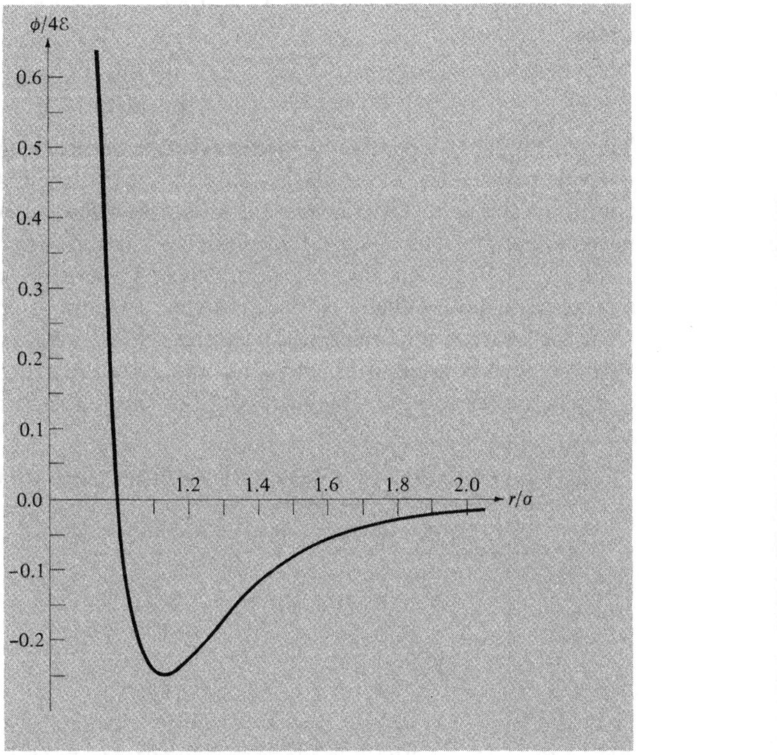

Figure 20.1
The Lennard Jones 6–12 potential (Eq. (20.2)).

Let us try to fit some of the observed properties of the solid noble gases using only the data from the gaseous state contained in Table 20.1, and the potential (20.2). We treat the rare gas solid as a set of classical particles, localized with negligible kinetic energy at the points of the observed face-centered cubic Bravais lattice. To compute the total potential energy of the solid first note that the energy of interaction of the atom at the origin with all the others is

$$\sum_{\mathbf{R} \neq \mathbf{0}} \phi(\mathbf{R}). \tag{20.3}$$

If we multiply this by N, the total number of atoms in the crystal, we get twice the total potential energy of the crystal, because we have thereby counted the interaction energy of each atom pair twice. Thus the energy per particle, u, is just:

$$u = \frac{1}{2} \sum_{\mathbf{R} \neq \mathbf{0}} \phi(\mathbf{R}), \tag{20.4}$$

where the sum is over all nonzero vectors in the fcc Bravais lattice.

It is convenient to write the length of the Bravais lattice vector **R** as a dimensionless number, $\alpha(\mathbf{R})$, times the nearest-neighbor separation, r. Equations (20.2) and (20.4) then give

$$u = 2\epsilon \left[A_{12} \left(\frac{\sigma}{r} \right)^{12} - A_6 \left(\frac{\sigma}{r} \right)^6 \right], \tag{20.5}$$

where

$$A_n = \sum_{\mathbf{R} \neq \mathbf{0}} \frac{1}{\alpha(\mathbf{R})^n}. \tag{20.6}$$

The constants A_n depend only on the crystal structure (in this case, fcc) and the number n. Evidently when n is very large, only the nearest neighbors of the origin will contribute to the sum (20.6). Since by definition $\alpha(\mathbf{R}) = 1$, when **R** is a vector joining nearest neighbors, as $n \to \infty$, A_n approaches the number of nearest neighbors, which is 12 for the fcc Bravais lattice. As n diminishes, A_n increases, since next-nearest neighbors start to contribute. When n is 12, A_n is given to a tenth of a percent by the contributions from the nearest, next-nearest, and third-nearest neighbors of the origin. Values of A_n have been computed for most common crystal structures and a range of values of n. The values of A_n for the most common cubic structures are shown in Table 20.2.

Table 20.2
THE LATTICE SUMS A_n FOR THE THREE CUBIC BRAVAIS LATTICES[a]

n	SIMPLE CUBIC	BODY-CENTERED CUBIC	FACE-CENTERED CUBIC
$\leqslant 3$	∞	∞	∞
4	16.53	22.64	25.34
5	10.38	14.76	16.97
6	8.40	12.25	14.45
7	7.47	11.05	13.36
8	6.95	10.36	12.80
9	6.63	9.89	12.49
10	6.43	9.56	12.31
11	6.29	9.31	12.20
12	6.20	9.11	12.13
13	6.14	8.95	12.09
14	6.10	8.82	12.06
15	6.07	8.70	12.04
16	6.05	8.61	12.03
$n \geqslant 17$	$6 + 12(1/2)^{n/2}$	$8 + 6(3/4)^{n/2}$	$12 + 6(1/2)^{n/2}$

[a] A_n is the sum of the inverse nth powers of the distances from a given Bravais lattice point to all others, where the unit of distance is taken to be the distance between nearest neighbors (Eq. (20.6)). To the accuracy of the table only nearest- and next-nearest neighbors contribute when $n \geqslant 17$, and the given formulas may be used.

Source: J. E. Jones and A. E. Ingham, *Proc. Roy. Soc.* (*London*) **A107**, 636 (1925).

Equilibrium Density of the Solid Noble Gases

To find the nearest-neighbor separation in equilibrium, r_0, and hence the density, we need only minimize (20.5) with respect to r, to find that $\partial u/\partial r = 0$ at

$$r_0^{\text{th}} = \left(\frac{2A_{12}}{A_6}\right)^{1/6} \sigma = 1.09\sigma. \qquad \textbf{(20.7)}$$

In Table 20.3 the theoretical value $r_0^{\text{th}} = 1.09\sigma$ is compared with the measured value, r_0^{exp}. Agreement is quite good, although r_0^{exp} becomes progressively bigger than r_0^{th} as the atomic mass becomes lighter. This can be understood as an effect of the zero-point kinetic energy we have neglected. This energy becomes greater, the smaller the volume into which the atoms are squeezed. It should therefore behave as an effectively repulsive force, increasing the lattice constant over the value given by (20.7). Since the zero-point energy becomes more important with decreasing atomic mass, we should expect (20.7) to fall short of r_0^{exp} most for the lightest masses.

Table 20.3
NEAREST-NEIGHBOR DISTANCE r_0, COHESIVE ENERGY u_0, AND BULK MODULUS B_0 AT ZERO PRESSURE FOR THE SOLID NOBLE GASES[a]

		Ne	Ar	Kr	Xe
r_0 (angstroms)	(Experiment)	3.13	3.75	3.99	4.33
$r_0 = 1.09\sigma$	(Theory)	2.99	3.71	3.98	4.34
u_0 (eV/atom)	(Experiment)	−0.02	−0.08	−0.11	−0.17
$u_0 = -8.6\epsilon$	(Theory)	−0.027	−0.089	−0.120	−0.172
B_0 (10^{10} dyne/cm^2)[b]	(Experiment)	1.1	2.7	3.5	3.6
$B_0 = 75\epsilon/\sigma^3$	(Theory)	1.81	3.18	3.46	3.81

[a] The theoretical values are those calculated from the elementary classical theory.
[b] One atmosphere of pressure = 1.01×10^6 dynes/cm^2; 1 bar of pressure = 10^6 dynes/cm^2.
Source: Data quoted by M. L. Klein, G. K. Horton, and J. L. Feldman, *Phys. Rev.* **184**, 968 (1969); D. N. Batchelder, et al., *Phys. Rev.* **162**, 767 (1967); E. R. Dobbs and G. O. Jones, *Rep. Prog. Phys.* **xx**, 516 (1957).

Equilibrium Cohesive Energy of the Solid Noble Gases

If we substitute the equilibrium nearest-neighbor separation (20.7) into the energy per particle (20.5) we find the equilibrium cohesive energy:

$$u_0^{\text{th}} = -\frac{\epsilon {A_6}^2}{2A_{12}} = -8.6\epsilon. \qquad \textbf{(20.8)}$$

If u_0^{th} is compared with the measured value u_0^{exp} (Table 20.3), the agreement is again found to be good, although $|u_0^{\text{th}}|$ increasingly exceeds $|u_0^{\text{exp}}|$ as the atomic mass decreases. This again makes sense as an effect of the neglected zero-point motion. We have ignored a positive term in the energy (kinetic energy is always positive) that decreases the binding, and becomes more important with decreasing atomic mass.

Equilibrium Bulk Modulus of the Solid Noble Gases

The bulk modulus $B = -V(\partial P/\partial V)_T$ can also be calculated in terms of ϵ and σ. Since the pressure at $T = 0$ is given by $P = -dU/dV$, where U is the total energy, we can write B in terms of the energy per particle $u = U/N$ and volume per particle $v = V/N$ as

$$B = v\frac{\partial}{\partial v}\left(\frac{\partial u}{\partial v}\right). \tag{20.9}$$

The volume per particle v in a fcc lattice is $v = a^3/4$, where the side a of the conventional cubic cell is related to the nearest-neighbor separation r by $a = \sqrt{2}r$. We may therefore write

$$v = \frac{r^3}{\sqrt{2}}, \qquad \frac{\partial}{\partial v} = \frac{\sqrt{2}}{3r^2}\frac{\partial}{\partial r}, \tag{20.10}$$

and rewrite the bulk modulus as

$$B = \frac{\sqrt{2}}{9}r\frac{\partial}{\partial r}\frac{1}{r^2}\frac{\partial}{\partial r}u. \tag{20.11}$$

The equilibrium separation r_0 is that which minimizes the energy per particle u. Therefore $\partial u/\partial r$ vanishes in equilibrium and (20.11) reduces to

$$B_0^{\text{th}} = \frac{\sqrt{2}}{9r_0}\frac{\partial^2 u}{\partial r^2}\bigg|_{r=r_0} = \frac{4\epsilon}{\sigma^3}A_{12}\left(\frac{A_6}{A_{12}}\right)^{5/2} = \frac{75\epsilon}{\sigma^3}. \tag{20.12}$$

Comparing B_0^{th} with the measured B_0^{exp} (Table 20.3) we find good agreement in xenon and krypton, but the experimental bulk modulus is about 20 percent larger in argon and 60 percent larger in neon. The mass dependence again suggests that these discrepancies are due to the neglected zero-point motion.

IONIC CRYSTALS

The simplest theory of cohesion in the ionic crystals makes the same physical simplifications as the theory of cohesion in molecular crystals: One assumes that the cohesive energy is entirely given by the potential energy of classical particles localized at the equilibrium positions.[9] Because the particles in ionic crystals are electrically charged ions, by far the largest term in the interaction energy is the interionic Coulomb interaction. This varies as the inverse first power of the interionic distance, and completely overwhelms the inverse sixth-power fluctuating dipole interaction[10] and can be taken as the sole source of binding in rough calculations.

---

[9] We shall define the cohesive energy of an ionic crystal to be the energy required to disassemble it into isolated ions, rather than atoms. If one wishes the cohesive energy with respect to isolated atoms, one must supplement our analysis with calculations or measurements of ionization potentials and electron affinities.

[10] Such a term, however, is present in ionic crystals as well, and must be allowed for in more precise calculations.

In determining the equilibrium lattice parameters one must still take into account the strong short-range core-core repulsion due to the Pauli principle, without which the crystal would collapse. We therefore represent the total cohesive energy per ion pair[11] in the form

$$u(r) = u^{\text{core}}(r) + u^{\text{coul}}(r), \tag{20.13}$$

where r is the nearest-neighbor distance.[12]

The calculation of $u^{\text{coul}}(r)$ is not as straightforward as the calculation of the attractive energy in molecular crystals, because of the very long range of the Coulomb potential. Consider, for example, the sodium chloride structure (Figure 19.4a), which we can represent as a fcc Bravais lattice of negative anions at sites $\mathbf{R}$, and a second Bravais lattice of positive cations displaced by $\mathbf{d}$ from the first, where $\mathbf{d}$ is a translation through $a/2$ along a cube side. We again measure all interionic distances in terms of the nearest-neighbor distance $r = a/2$:

$$\begin{aligned} |\mathbf{R}| &= \alpha(\mathbf{R})r, \\ |\mathbf{R} + \mathbf{d}| &= \alpha(\mathbf{R} + \mathbf{d})r. \end{aligned} \tag{20.14}$$

It is then tempting to proceed as in the earlier case, writing the total potential energy of a single cation (or a single anion) as

$$-\frac{e^2}{r}\left\{\frac{1}{\alpha(\mathbf{d})} + \sum_{\mathbf{R} \neq \mathbf{0}}\left(\frac{1}{\alpha(\mathbf{R} + \mathbf{d})} - \frac{1}{\alpha(\mathbf{R})}\right)\right\}. \tag{20.15}$$

If there are N ions in the crystal, the total potential energy will be half of N times (20.15):

$$U = -\frac{N}{2}\frac{e^2}{r}\left\{\frac{1}{\alpha(\mathbf{d})} + \sum_{\mathbf{R} \neq \mathbf{0}}\left(\frac{1}{\alpha(\mathbf{R} + \mathbf{d})} - \frac{1}{\alpha(\mathbf{R})}\right)\right\}. \tag{20.16}$$

The energy per ion pair is this divided by the number $N/2$ of ion pairs:

$$u^{\text{coul}}(r) = -\frac{e^2}{r}\left\{\frac{1}{\alpha(\mathbf{d})} + \sum_{\mathbf{R} \neq \mathbf{0}}\left(\frac{1}{\alpha(\mathbf{R} + \mathbf{d})} - \frac{1}{\alpha(\mathbf{R})}\right)\right\}. \tag{20.17}$$

However, $1/r$ falls off so slowly with distance that Eq. (20.17) is not a well-defined sum. Mathematically speaking, it is only a conditionally convergent series, and can therefore be summed to any value whatsoever, depending on the order in which the summation is performed!

This is not just a mathematical nuisance. It reflects the physical fact that Coulomb interactions are of such long range that the energy of a collection of charged particles can depend crucially on the configuration of a negligibly small fraction of them on the surface. We have already encountered this problem in Chapter 18. In the present case we can put the point as follows:

---

[11] It is customary to calculate the cohesive energy per ion pair, rather than per ion. If there are N ions, then there will be $N/2$ ion pairs.

[12] We called the nearest-neighbor distance d in Chapter 19 to avoid confusion with ionic radii. Here we call it r because it is esthetically distressing to take derivatives with respect to d. And we are esthetes, at heart.

If we included only a finite set of ions in the summation there would be no ambiguity, and the sum would give the electrostatic energy of that finite crystal. Summing the infinite series in a particular order corresponds to constructing the infinite crystal as a particular limiting form of successively larger and larger finite crystals. If the interionic interactions were of short enough range one could prove that the limiting energy per ion pair would not depend on how the infinite crystal was built up (provided that the surface of successive finite constructions was not wildly irregular). However, with the long-range Coulomb interaction one can construct the infinite crystal in such a way that arbitrary distributions of surface charge and/or dipolar layers are present at all stages. By judiciously choosing the form of these surface charges one can arrange things so that the energy per ion pair u approaches any desired value in the limit of an infinite crystal. This is the physics underlying the mathematical ambiguity in Eq. (20.17).

The disease being thus diagnosed, the cure is obvious: The series must be summed in such a way that at all stages of the summation there are no appreciable contributions to the energy from charges at the surface. There are many ways this can be guaranteed. For example, one can break up the crystal into electrically neutral cells whose charge distributions have the full cubic symmetry (see Figure 20.2). The energy of a finite subcrystal composed of n such cells will then be just n times the energy of a single cell, plus the cell-cell interaction energy. The internal energy of a cell is easily calculated since the cell contains only a small number of charges. But the interaction energy between cells will fall off as the inverse *fifth* power of the distance between the cells,[13] and thus the cell-cell interaction energy will be a rapidly converging summation, which, in the limit of an infinite crystal, will not depend on the order of summation.

There are numerically more powerful, but more complex, ways of computing such

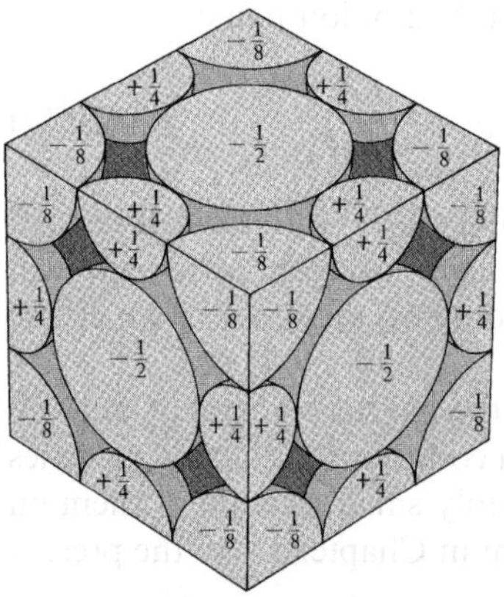

Figure 20.2
One possible way of dividing up the sodium chloride structure into cubical cells, whose electrostatic interaction energy falls off rapidly (as the inverse fifth power) with intercellular distance. Each cell contains four units of positive charge, made up of a whole unit at the center and twelve quarter units on the edges, and four units of negative charge, made up of six half units on the faces and eight eighth units at the corners. For computation each sphere can be represented as a point charge at its center. (The energies of interaction of the surface point charges of two adjacent cubes must not be counted.)

---

[13] This is because the charge distribution within each cell has the full cubic symmetry. See page 355. Note also that a minor problem arises if some ions lie on the boundary between cells. Their charge must then be divided between cells so as to maintain the full symmetry of each cell. Having done this, one must be careful not to include the self-energy of the divided ion in the interaction energy between the cells sharing it.

Coulomb lattice sums, which are, however, all guided by the same physical criterion. The most famous is due to Ewald.[14]

The result of all such calculations is that the electrostatic interaction per ion pair has the form:

$$u^{\text{coul}}(r) = -\alpha \frac{e^2}{r}, \tag{20.18}$$

where α, known as the Madelung constant, depends only on the crystal structure. Values of α for the most important cubic structures are given in Table 20.4. Note that α is an increasing function of coordination number; i.e., the more nearest neighbors (of opposite charge), the lower the electrostatic energy. Since the Coulomb interaction has so long a range this is not an obvious result. Indeed, the amount by which the electrostatic energy of the cesium chloride structure (coordination number 8) is lower than that of a sodium chloride structure with the same nearest-neighbor distance r (coordination number 6) is less than 1 percent, although the nearest-neighbor contribution is lower by 33 percent.

Table 20.4
THE MADELUNG CONSTANT α FOR SOME CUBIC CRYSTAL STRUCTURES

CRYSTAL STRUCTURE	MADELUNG CONSTANT α
Cesium chloride	1.7627
Sodium chloride	1.7476
Zincblende	1.6381

The dominant contribution of the Coulomb energy to the cohesive energy of the alkali halides is demonstrated in Table 20.5, where $u^{\text{coul}}(r)$ is evaluated at the experimentally observed nearest-neighbor separations, and compared with the experimentally determined cohesive energies. It can be seen that the u^{coul} alone accounts for the bulk of the observed binding, being in all cases about 10 percent lower than the measured cohesive energy.

It is to be expected that the electrostatic energy alone overestimates the strength of the binding, for Eq. (20.18) omits any contribution from the positive potential representing the short-range core-core repulsion. This weakens the binding. We can see that the resulting correction will be small by noting that the potential representing the core-core repulsion is a very rapidly varying function of ionic separation. If we were to represent the core as infinitely repulsive hard spheres, we should find a cohesive energy exactly given by the electrostatic energy at minimum separation (Figure 20.3). Evidently this is too extreme. We acquire more latitude by letting the repulsion vary with an inverse power law, writing the total energy per ion pair as

$$u(r) = -\frac{\alpha e^2}{r} + \frac{C}{r^m}. \tag{20.19}$$

[14] P. P. Ewald, *Ann. Physik* **64**, 253 (1921). A particularly nice discussion can be found in J. C. Slater, *Insulators Semiconductors and Metals*, McGraw-Hill, New York, 1967, pp. 215–220.

Table 20.5
MEASURED COHESIVE ENERGY AND ELECTROSTATIC ENERGY FOR THE ALKALI HALIDES WITH THE SODIUM CHLORIDE STRUCTURE

	Li	Na	K	Rb	Cs
F	−1.68[a] −2.01[b]	−1.49 −1.75	−1.32 −1.51	−1.26 −1.43	−1.20 −1.34
Cl	−1.38 −1.57	−1.27 −1.43	−1.15 −1.28	−1.11 −1.23	
Br	−1.32 −1.47	−1.21 −1.35	−1.10 −1.22	−1.06 −1.18	
I	−1.23 −1.34	−1.13 −1.24	−1.04 −1.14	−1.01 −1.10	

[a] The upper figure in each box is the measured cohesive energy (compared with separated ions) in units of 10^{-11} erg per ion pair. Source: M. P. Tosi, *Solid State Physics*, vol. 16, F. Seitz and D. Turnbull, eds., Academic Press, New York, 1964, p. 54.

[b] The lower figure in each box is the electrostatic energy as given by Eq. (20.18), evaluated at the observed nearest-neighbor separation r.

The equilibrium separation r_0 is then determined by minimizing u. Setting $u'(r_0)$ equal to zero gives

$$r_0^{m-1} = \frac{mC}{e^2\alpha}. \tag{20.20}$$

In the noble gases we used the corresponding equation to determine r_0 (Eq. (20.7)), but now, lacking an independent measurement of C, we may use it to determine

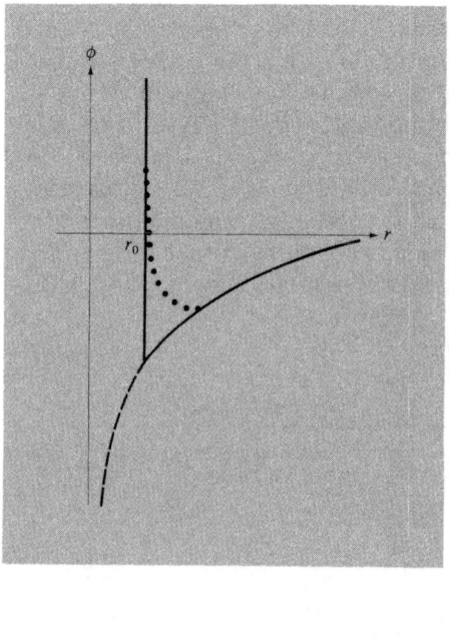

Figure 20.3
Graph of the pair potential, which is infinitely repulsive when $r < r_0$ and Coulombic when $r > r_0$. The dashed curve is the extension of the Coulomb potential. The dotted curve is how the potential would be affected if the repulsion were a power law, rather than being infinitely strong.

C in terms of the experimentally measured r_0:

$$C = \frac{\alpha e^2 r_0^{m-1}}{m}. \tag{20.21}$$

We can then substitute this back into (20.19) to find that the theoretical cohesive energy per ion pair is

$$u_0^{\text{th}} = u(r_0) = -\frac{\alpha e^2}{r_0}\frac{m-1}{m}. \tag{20.22}$$

As expected, this is only slightly smaller than (20.18) for large m.

In the noble gases we chose $m = 12$, for reasons of calculational convenience, noting that this led to reasonable agreement with the data. The motivation for $m = 12$ is lacking in the alkali halides,[15] and if a power law is used to represent the repulsion, we might just as well determine the exponent by fitting the data as closely as possible. It is not advisable to fix m by setting (20.22) equal to the observed cohesive energy, for (20.22) is so slowly varying a function of m that small errors in the experimental measurement will cause large alterations in m. A better procedure is to find an independent measurement determining m. We can then use that m in (20.22) to see whether the agreement with the experimental cohesive energies is thereby improved over the 10 percent agreement in Table 20.5.

Such an independent determination of m is provided by the experimentally measured bulk moduli. If B_0 and r_0 are, respectively, the equilibrium bulk modulus and nearest-neighbor separation, then (see Problem 2) m has the value

$$m = 1 + \frac{18 B_0 r_0^3}{|u^{\text{coul}}(r_0)|}. \tag{20.23}$$

The values of m obtained from the measured values of B_0 and r_0 are listed in Table 20.6. They vary from about 6 to 10. When the purely electrostatic contributions to the cohesive energy are corrected by the factor $(m - 1)/m$, agreement with the observed cohesive energies is considerably improved, being 3 percent or better, except from the troublesome[16] lithium halides and sodium iodide.

This is as much as (if not more than) one can expect from so crude a theory. A better analysis would make several improvements:

1. The core-core repulsion is probably better represented in an exponential form (the so-called Born-Mayer potential being a popular choice) than as a power law.
2. The inverse sixth-power fluctuating dipole force between ion cores should be taken into account.
3. The zero-point vibrations of the lattice should be allowed for.

However, these improvements will not alter our main conclusion, that the major part (90 percent) of the cohesive energy in the ionic crystals is due simply to the electrostatic Coulomb interactions among the ions, considered as fixed-point charges.

---

[15] One might expect m to be rather less than 12 simply because the halogen ions owing to their excess negative charge should have a rather lower density of electrons at their surface than the corresponding noble gas atoms.

[16] See page 383.

Table 20.6
MEASURED DATA[a] AND DERIVED QUANTITIES[b] FOR THE ALKALI HALIDES WITH THE SODIUM CHLORIDE STRUCTURE

COMPOUND	(1) r (Å)	(2) B $\left(10^{11}\ \frac{\text{dynes}}{\text{cm}^2}\right)$	(3) u $\left(10^{-11}\ \frac{\text{ergs}}{\text{ion pair}}\right)$	(4) u^{coul} $\left(= -\frac{Ae^2}{r}\right)$	(5) m $\left(= 1 + \frac{18Br^3}{\lvert u^{\text{coul}}\rvert}\right)$	(6) u^{th} $\left(= \frac{m-1}{m} u^{\text{coul}}\right)$
LiF	2.01	6.71	−1.68	−2.01	5.88	−1.67
LiCl	2.56	2.98	−1.38	−1.57	6.73	−1.34
LiBr	2.75	2.38	−1.32	−1.47	7.06	−1.26
LiI	3.00	1.72	−1.23	−1.34	7.24	−1.15
NaF	2.31	4.65	−1.49	−1.75	6.90	−1.50
NaCl	2.82	2.40	−1.27	−1.43	7.77	−1.25
NaBr	2.99	1.99	−1.21	−1.35	8.09	−1.18
NaI	3.24	1.51	−1.13	−1.24	8.46	−1.09
KF	2.67	3.05	−1.32	−1.51	7.92	−1.32
KCl	3.15	1.75	−1.15	−1.28	8.69	−1.13
KBr	3.30	1.48	−1.10	−1.22	8.85	−1.08
KI	3.53	1.17	−1.04	−1.14	9.13	−1.02
RbF	2.82	2.62	−1.26	−1.43	8.40	−1.26
RbCl	3.29	1.56	−1.11	−1.23	9.13	−1.10
RbBr	3.43	1.30	−1.06	−1.18	9.00	−1.05
RbI	3.67	1.05	−1.01	−1.10	9.49	−0.98
CsF	3.00	2.35	−1.20	−1.34	9.52	−1.20

[a] The first three columns give measured data. (1) Nearest-neighbor distance r (from R. W. G. Wyckoff, *Crystal Structures*, 2nd ed., Interscience, New York, 1963). (2) Bulk modulus (from M. P. Tosi, *Solid State Physics*, vol. 16, F. Seitz and D. Turnbull, eds., Academic Press, New York, 1964, p. 44). (3) Cohesive energy (*Ibid.*, p. 54).

[b] The last three columns give derived quantities. (4) Coulomb contribution (20.18) to the cohesive energy, $u^{\text{coul}} = 4.03/r(\text{Å}) \times 10^{-11}$ erg/ion pair. (5) Repulsive exponent m in terms of measured bulk modulus and nearest-neighbor distance, according to (20.23). (6) Corrected theoretical cohesive energy, obtained by multiplying u^{coul} by $(m - 1)/m$; it should be compared with the measured cohesive energy in column (3).

COHESION IN COVALENT CRYSTALS AND METALS

The crude theories we have made of the cohesive energies of molecular and ionic crystals are as accurate as they are, primarily because in these solids the valence electron configuration is not appreciably distorted from what it is in the isolated atoms (molecular crystals) or ions (ionic crystals). This ceases to be so in covalent crystals and metals, which are characterized by distributions of valence electrons that differ substantially from anything found either in isolated atoms or in isolated ions of the underlying material. Consequently, to calculate the cohesive energy of such solids one cannot merely calculate the classical potential energy of a set of weakly or negligibly deformed atoms or ions, arranged in the appropriate crystal structure. Instead, even the simplest calculations must include a computation of the energy levels of the valence electrons in the presence of the periodic potential of the ion cores.

Thus a theory of the cohesive energy of covalent crystals and metals must contain a computation of their band structure.[17] For this reason there is no model of cohesion in such solids of remotely comparable simplicity to those we have described for molecular and ionic crystals. The basis for calculations of comparable accuracy must be provided by the techniques we have described in Chapters 10, 11, and 17. We shall limit ourselves here to a few qualitative remarks about covalent crystals, and some crude and highly inaccurate estimates for metals, based on a free electron picture.

Cohesion in Covalent Crystals

The theory of cohesion in good covalent insulators is quite similar to the theory of chemical bonding in molecules,[18] a subject beyond the scope of this book.[19] The way in which the electrostatic forces conspire to hold covalent crystals together is considerably more subtle than the simple electrostatic attraction between point ions, which works so well in describing ionic crystals, or even the fluctuating dipole interaction we used in describing the solid noble gases. Consider, for concreteness, the case of diamond (carbon). Suppose that a group of carbon atoms are placed onto the sites of a diamond lattice, but with so large a lattice constant that the energy of the collection is just the sum of the energies of the isolated atoms (i.e., the cohesive energy is zero). Cohesion will result if the energy of the collection can be lowered by reducing the lattice constant to its observed value. As the lattice constant is reduced, there will eventually be some overlap in the atomic wave functions centered on different sites (cf. the discussion in Chapter 10). If the outermost atomic shells were filled (as they are in the noble gas atoms or in the ions forming an ionic crystal) this overlap would result in the short-range core-core repulsion, and would raise the

---

[17] Indeed, cohesive-energy calculations provided the earliest motivations for accurate band structure calculations. It was only later that it became generally recognized that the band structure itself was of fundamental interest, independent of the problem of cohesion.

[18] The basic text is L. Pauling, *The Nature of the Chemical Bond*, 3rd ed., Cornell University Press, Ithaca, New York, 1960.

[19] However, an elementary discussion of the hydrogen molecule is given in Chapter 32.

energy over that of isolated atoms. However, the core-core repulsion of filled atomic shells is a consequence of the Pauli exclusion principle, together with the fact that the only available electronic levels, if the outer shells are filled, lie much higher in energy. If the outermost electronic shells are only partially filled (as in carbon), then the electrons in the outer shells can rearrange themselves with considerably more flexibility when the wave functions of neighboring atoms start to overlap, since other levels with comparable energy in the same shell are available.

It turns out that under these circumstances the overlap of the outermost shells generally leads to a lowering of the total electronic energy, with the electrons forming levels that are not localized about a single ion core. There is no simple reason why this is so. The less localized the electronic wave function, the smaller need be the maximum electronic momentum required by the uncertainty principle, and hence the lower the electronic kinetic energy. To this must be added estimates of the change in potential energy in the less localized levels. The net result is generally a lowering of the energy.[20]

Cohesion in Free Electron Metals

Turning to the other extreme, we can compare a solid not with a set of atoms but with a free electron gas. In Chapter 2 we noted that the pressure of a free electron gas at the density of the alkali metals gave their observed compressibilities to within a factor of two or less. To go from this to a crude theory of cohesion in the alkali metals, we must add to the electron gas kinetic energy the total electrostatic potential energy. This contains, among other things, the energy of attraction between the positively charged ions and the negatively charged electron gas, without which the metal would not be bound at all.

We treat the ions in an alkali metal as point charges localized at the sites of the body-centered cubic Bravais lattice. We treat the electrons as a uniform compensating background of negative charge. The total electrostatic energy per atom of such a configuration can be computed by techniques similar to those used in the elementary theory of ionic crystals. The result for a bcc lattice is[21]

$$u^{\text{coul}} = -\frac{24.35}{(r_s/a_0)} \text{ eV/atom}, \tag{20.24}$$

where r_s is the radius of the Wigner-Seitz sphere (the volume per electron is $4\pi r_s^3/3$) and a_0 is the Bohr radius. As expected, this term favors high densities (i.e., low r_s).

The attractive electrostatic energy (20.24) must be balanced against the electronic kinetic energy per atom. Since there is one free electron per atom in the alkali metals, we have (see Chapter 2, page 37):

$$u^{\text{kin}} = \frac{3}{5}\,\mathcal{E}_F = \frac{30.1}{(r_s/a_0)^2} \text{ eV/atom}. \tag{20.25}$$

---

[20] The discussion of the hydrogen molecule in Chapter 32 provides an illustration of this in an especially simple case.

[21] See, for example, C. A. Sholl, *Proc. Phys. Soc.* **92**, 434 (1967).

If we wished to be more accurate, we should have to replace (20.25) by the complete ground-state energy per electron of a uniform electron gas[22] at the density $3/4\pi r_s^{\,3}$. The computation of this is quite difficult (see Chapter 17) and, considering the crudeness of the electron gas model, of questionable utility for an estimate of real cohesive energies. Here we shall include only the exchange correction to (20.25) (see Eq. (17.25)):

$$u^{\text{ex}} = -\frac{0.916}{(r_s/a_0)}\,\text{Ry/atom} = -\frac{12.5}{(r_s/a_0)}\,\text{eV/atom}. \qquad \textbf{(20.26)}$$

Note that the exchange correction to the electron gas energy has the same density dependence as the average electrostatic energy (20.24) and is about half its size. This indicates the importance of electron-electron interactions in metallic cohesion and the consequent difficulties that any adequate theory of cohesion must cope with.

Adding these three contributions, we find that

$$u = \frac{30.1}{(r_s/a_0)^2} - \frac{36.8}{(r_s/a_0)}\,\text{eV/atom}. \qquad \textbf{(20.27)}$$

Minimizing this with respect to r_s gives:

$$\frac{r_s}{a_0} = 1.6. \qquad \textbf{(20.28)}$$

The observed values of r_s/a_0 range from 2 to 6 in the alkali metals.[23] The failure of (20.28) even to come close is in (perhaps healthy) contrast to our earlier successes, and indicates the difficulty in coming to terms with metallic cohesion with any simple picture. A particularly striking qualitative failure of (20.28) is its prediction of the same r_s for all alkali metals. This result would not be affected by a more accurate determination of the total electron gas energy, for that would still have the form $E(r_s)$, and, minimizing $E(r_s) - 24.35\,(a_0/r_s)$, would still lead to a unique equilibrium value of r_s, independent of the alkali metal.

Evidently some other scale of length must be introduced to distinguish among the alkali metals, and it is not hard to see what that must be. Our treatment has pictured the ions as points, even though the real ion cores have nonnegligible radii. The approximation of point ions is not as absurd in metals as it would be in molecular or ionic crystals, since the fraction of the total volume occupied by the ions is considerably smaller in metals. However, in making that approximation we have ignored at least two important effects. If the ion core has a nonzero radius, then the conduction electron gas is largely prevented from entering that fraction of the metallic volume occupied by the ion cores. Even in a very crude theory, this means that the density of the electron gas is greater than we have estimated, and therefore its kinetic energy is also greater. Furthermore, because the conduction electrons are excluded from the ion core regions, they cannot get as close to the positively charged ions as the picture

---

[22] Excluding the average electrostatic energy of the electrons and ions, which is already accounted for in (20.24). This average electrostatic energy is just the Hartree energy (Chapter 17), which vanishes when the ions are treated as a *uniform* positive background of compensating charge, rather than as the localized point charges on which the computation of (20.24) is based.

[23] Table 1.1.

underlying (20.24) assumes. We should therefore expect the electrostatic energy to be less negative than we have estimated.

Both these effects should cause the equilibrium value of r_s/a_0 to increase with increasing ion core radius (Problem 4). This is consistent with the observed alkali metal densities. Evidently, any even moderately accurate computation of this crucial effect must be quite subtle, requiring good estimates of both the conduction electron wave functions and the crystal potential appearing in the one-electron Schrödinger equation.

PROBLEMS

1. A measure of the importance of quantum effects in the noble gases is the de Boer parameter. We calculated the energy per atom $u(r)$ of a noble gas (Eq. (20.5)) under the assumption that it was entirely potential energy. In a quantum theory, however, there will be zero-point vibrations even at $T = 0$, leading to a correction to Eq. (20.5) proportional to $\hbar$.

(a) Show, on purely dimensional grounds, that if the correction is strictly linear in h, then the correction to the energy must have the form

$$\Delta u = \epsilon \Lambda f(r/\sigma), \tag{20.29}$$

where f depends on the particular noble gas in question only through the ratio r/σ, and

$$\Lambda = \frac{h}{\sigma\sqrt{M\epsilon}}. \tag{20.30}$$

The quantity Λ, known as the de Boer parameter, is listed in Table 20.7. Since h/σ is the uncertainty in the momentum of a particle localized to within a distance σ, Λ^2 is roughly the ratio of the kinetic energy of zero-point motion of an atom to the magnitude of the attractive interaction. The size of Λ is thus a measure of the importance of quantum effects (and a glance at Table 20.7 immediately indicates why our purely classical discussion cannot hope to cope with solid helium).

Table 20.7
THE DE BOER PARAMETER FOR THE NOBLE GASES, INCLUDING THE TWO HELIUM ISOTOPES

^{3}He	^{4}He	Ne	Ar	Kr	Xe
3.1	2.6	0.59	0.19	0.10	0.064

(b) Let r_c be the equilibrium interparticle distance computed by minimizing the classical energy (20.5) and $r_c + \Delta r$ be the value obtained by minimizing the classical energy plus the quantum correction (20.29). Show, under the assumption $\Delta r \ll r_c$, that the ratio of the values of $\Delta r/r_c$ for any two noble gases is equal to the ratio of their de Boer parameters.

(c) Show that the result of (b) also holds for the fractional changes in internal energy and bulk moduli, due to quantum corrections.

These conclusions are compared with the data for neon and argon in Table 20.8. (In the cases of krypton and xenon the deviations from the classical values are too small to be reliably extracted from the data; in the case of the helium isotopes the de Boer parameter is too large for this analysis to be reliable.) Chapter 25 describes how the effects of zero-point vibrations can be more accurately taken into account.

Table 20.8
COMPARATIVE SIZE OF QUANTUM CORRECTIONS TO THE EQUILIBRIUM PROPERTIES OF NEON AND ARGON

X	X_{Ne}	X_{Ar}	X_{Ne}/X_{Ar}
Λ	0.59	0.19	3.1
$\Delta r/r^c$	0.047	0.011	4.3
$\Delta u/u^c$	0.26	0.10	2.6
$\Delta B/B^c$	0.39	0.15	2.6

2. Show that the bulk modulus for an ionic crystal with the NaCl structure is given by

$$B_0 = \frac{1}{18r_0} \left.\frac{d^2u}{dr^2}\right|_{r=r_0}, \tag{20.31}$$

where r_0 is the equilibrium nearest-neighbor separation. Show that the form (20.19) for the total energy per ion pair gives

$$B_0 = \frac{(m-1)}{18} \frac{\alpha e^2}{r_0^{\ 4}}, \tag{20.32}$$

and hence that

$$m = 1 + \frac{18B_0 r_0^{\ 3}}{|u_{\text{coul}}(r_0)|}, \tag{20.33}$$

where $u_{\text{coul}}(r)$ is the energy per ion pair of a crystal of point charges at nearest-neighbor separation r.

3. We can use the form (20.19) of the cohesive energy per ion pair to investigate the stability of the possible crystal structure an ionic crystal may assume. Assuming that the coupling constant C characterizing the contribution of the short-range repulsion is proportional to the coordination number Z, show that the equilibrium cohesive energy for different lattice types varies as $(\alpha^m/Z)^{1/(m-1)}$, and use the values of α in Table 20.4 to construct a table of relative stability according to the value of m. (*Hint:* First examine the cases of large or small m.)

4. (a) As a very crude model of an alkali metal, suppose that the charge of each valence electron is uniformly distributed through a sphere of radius r_s about each ion. Show that the electrostatic energy per electron is then

$$u^{\text{coul}} = -\frac{9a_0}{5r_s}\ \text{Ry/electron} = -\frac{24.49}{(r_s/a_0)}\ \text{eV/electron}. \tag{20.34}$$

(This is remarkably close to the result (20.24) for a bcc lattice of ions immersed in a completely uniform distribution of compensating negative charge.)

(b) In a real metal the valence electrons are largely excluded from the ion core. If we take this into account by uniformly distributing the charge of each electron in the region between spheres of radius r_c and r_s about each ion, and then replacing the potential of each ion by the pseudopotential,

$$\begin{aligned} V_{ps}(r) &= -\frac{e^2}{r}, \qquad r > r_c \\ &= 0, \qquad r < r_c, \end{aligned} \tag{20.35}$$

show that (20.34) must be replaced to leading order in r_c/r_s by

$$-\frac{9a_0}{5r_s} + \frac{3(r_c/a_0)^2}{(r_s/a_0)^3} \text{ Ry/electron.} \qquad \textbf{(20.36)}$$

(c) Taking the energy per particle to be the sum of the kinetic (20.25), exchange (20.26), and potential (20.36) energies, show that the equilibrium value of r_s/a_0 is given by

$$r_s/a_0 = 0.82 + 1.82(r_c/a_0)[1 + O(a_0/r_c)^2], \qquad \textbf{(20.37)}$$

and compare this with the values given in Table 1.1 and Table 19.4.

21

Failures of the Static Lattice Model

In Chapter 3 we reviewed the limitations of the free electron theory of metals, citing a variety of phenomena that could only be explained by the presence of the periodic potential arising from the lattice of ions.[1] In subsequent chapters the periodic array of ions has played a critical role in our treatment of metals and insulators. In all such discussions we have taken the ions to constitute a fixed, rigid, immobile periodic array.[2] This, however, is only an approximation to the actual ionic configuration,[3] for the ions are not infinitely massive, nor are they held in place by infinitely strong forces. Consequently, in a classical theory the static lattice model can be valid only at zero temperature. At nonzero temperatures each ion must have some thermal energy, and therefore a certain amount of motion in the vicinity of its equilibrium position. Furthermore, in a quantum theory even at zero temperature the static lattice model is incorrect, because the uncertainty principle ($\Delta x \, \Delta p \gtrsim h$) requires localized ions to possess some nonvanishing mean square momentum.[4]

The oversimplified model of immobile ions has been impressively successful in accounting for a wealth of detailed metallic equilibrium and transport properties dominated by the behavior of the conduction electrons, provided that one does not inquire into the source of the electronic collisions. We have also had some success with the static lattice model in accounting for the equilibrium properties of ionic and molecular insulators.

However, it is necessary to go beyond the static lattice model to fill in the many gaps in our understanding of metals (some of which—for example, the theory of the temperature dependence of the DC conductivity—are substantial) and to achieve anything beyond the most rudimentary theory of insulators. The limitations of a static lattice theory are particularly severe in the theory of insulators, since the electronic system in an insulator is comparatively passive, all electrons residing in filled bands. Except in phenomena that supply enough energy to excite electrons across the energy gap E_g between the top of the highest filled band and the lowest empty levels, insulators are electronically quiescent. If we adhered to the static lattice approximation in insulators, we would have no degrees of freedom left to account for their rich and varied properties.

In this chapter we shall summarize some of the ways in which the static lattice model fails to cope with experimental facts. In subsequent chapters we shall turn to the dynamical theory of lattice vibrations, which in one form or another will be the primary subject of Chapters 22 to 27.

We have grouped the major shortcomings of the static lattice model into three broad categories:

---

[1] Recall that, when used in this general way, the word "ion" means ions in ionic crystals, ion cores in metals and covalent crystals, and atoms in a rare gas solid.

[2] Except in Chapter 20, where we considered a uniform expansion of the array, and briefly considered atomic zero-point motion in the solid noble gases.

[3] We do not have in mind the fact that any real crystal has imperfections—i.e., static deviations from perfect periodicity (see Chapter 30). These can still be described in terms of a static lattice. Our concern here is with dynamic deviations from periodicity associated with the vibrations of the ions about their equilibrium positions. These always occur, even in an otherwise perfect crystal.

[4] We have discussed this consequence of the uncertainty principle on page 396, and have seen some simple evidence for it in our estimates of the cohesive energies of the rare gas solids (Problem 1, Chapter 20).

1. Failures to explain equilibrium properties.
2. Failures to explain transport properties.
3. Failures to explain the interaction of various types of radiation with the solid.

EQUILIBRIUM PROPERTIES

All equilibrium properties are affected in varying degrees by the lattice vibrations. We list below some of the most important.

Specific Heat

The static lattice model attributes the specific heat of a metal to the electronic degrees of freedom. It predicts a linear temperature dependence at temperatures well below the Fermi temperature—i.e., all the way to the melting point. Such linear behavior is indeed observed (Chapter 2), but only up to temperatures of order 10K. At higher temperatures the specific heat increases much more rapidly (as T^3) and at still higher temperatures (typically between 10^2 and 10^3K) approaches a roughly constant value. This additional (and, above 10K, dominant) contribution to the specific heat is entirely due to the hitherto neglected degrees of freedom of the lattice of ions.

Insulators provide further evidence that the ions contribute to the specific heat. If static lattice theory were literally correct, the thermal energy of an insulator would differ from that at $T = 0$ only to the extent that electrons were thermally excited across the energy gap E_g. The number of electrons so excited can be shown (Chapter 28) to have a temperature dependence dominated by $e^{-E_g/2k_BT}$, at temperatures below E_g/k_B (i.e., at all temperatures of interest, if E_g is as big as an electron volt). This exponential also dominates the behavior of $c_v = du/dT$. However, the observed low-temperature specific heat of insulators is not exponential, but varies, as T^3. Both in insulators and in metals this T^3 contribution to c_v can be explained by introducing the motion of the lattice into the theory in a quantum-mechanical way.

Equilibrium Density and Cohesive Energies

We have mentioned in Chapter 20 that zero-point vibrations must be included in a calculation of the ground-state energy of a solid, and hence in a calculation of its equilibrium density and cohesive energy. The contribution of the zero-point vibrations of the ions is considerably smaller than the potential energy terms in most crystals, but, as we have seen, leads to easily observable effects in neon and argon.[5]

Thermal Expansion

The equilibrium density of a solid depends on the temperature. In the static lattice model the only effect of temperature is the excitation of electrons. In insulators such excitations are of negligible importance at temperatures below E_g/k_B. The thermal

---

[5] In solid helium the zero-point vibrations are so substantial that they cannot be ignored even in a first approximation. For this reason, the solid forms of the two helium isotopes (of mass number 3 and 4) are often referred to as *quantum solids*.

expansion of insulators (and, as it turns out, of metals, too) is critically related to the ionic degrees of freedom. In a sense this is merely the $T \neq 0$ version of the point made in the preceding paragraph, but the lattice vibrations generally provide only a small correction to the $T = 0$ equilibrium size, whereas they play a central role in determining the thermal expansion.

Melting

At sufficiently high temperatures solids melt; i.e., the ions leave their equilibrium positions and wander great distances through the resulting liquid. Here the static lattice hypothesis fails with a vengeance; however, even below the melting point, when the ions remain in the vicinity of their equilibrium positions, it is clear that any adequate theory of the melting process (and only very crude ones exist) must take into account the increasing amplitude of the lattice vibrations with increasing temperature.

TRANSPORT PROPERTIES

In Chapters 1, 2, 12, and 13 we examined those transport properties of a metal that depend almost entirely on its electronic structure. However, many aspects of transport in metals, and all aspects of transport in insulators, can only be understood when the vibrations of the lattice are taken into account.

Temperature Dependence of the Electronic Relaxation Time

In a perfect periodic potential an electron would suffer no collisions and the electrical and thermal conductivities of such a metal would be infinite. We have on occasion referred to the fact that one of the major sources of scattering in a metal is the deviation of the lattice from perfect periodicity due to the thermal vibrations of the ions about their equilibrium sites. This is responsible for the characteristic T^5 term in the electrical resistivity at low temperatures, as well as its linear growth with T at high temperatures (Chapter 26). The static lattice model cannot explain these facts.

Failure of the Wiedemann-Franz Law

The failure of the Wiedemann-Franz law at intermediate temperatures (page 59) finds a simple explanation in the theory of how electrons are scattered by the lattice vibrations.

Superconductivity

Below a certain temperature (20 K or substantially less) the resistivity of certain metals (known as superconductors) drops abruptly to zero. A full explanation for this was not given until 1957. It is now understood, and one of the crucial parts of the explanation is the influence of lattice vibrations on the effective interaction between two electrons in a metal (Chapter 34). If the lattice were rigorously static, there would be no superconductors.[6]

Thermal Conductivity of Insulators

Most metallic transport properties have no analogues in insulators. However, electrical insulators do conduct heat. To be sure, they do not conduct it as well as metals: The far end of a silver spoon dipped in coffee becomes hot much faster than the handle of the ceramic cup. However, from the point of view of a static lattice model there is no mechanism for insulators to conduct appreciable amounts of heat at all; there are simply too few electrons in partially filled bands to do the job. The thermal conductivity of insulators is due predominantly to the lattice degrees of freedom.

Transmission of Sound

Insulators transmit not only heat; they transmit sound as well, in the form of vibrational waves in the lattice of ions. In the static lattice model electrical insulators would also be acoustic insulators.

INTERACTION WITH RADIATION

We have discussed the interaction of radiation with solids in Chapter 6 (X radiation) and parts of Chapters 1 and 15 (optical properties of metals). There is a wealth of additional data on the response of solids to radiation that cannot be explained in terms of the response of electrons in a fixed array of ions. Some important examples are the following:

Reflectivity of Ionic Crystals

Ionic crystals exhibit a sharp maximum in their reflectivity at frequencies in the infrared, corresponding to values of $\hbar\omega$ far below their electronic energy gap. The phenomenon therefore cannot be due to an electronic excitation. It arises from the fact that the electric field in the radiation exerts oppositely directed forces on the positive and negative ions, thereby displacing them with respect to one another. The proper explanation of this phenomenon requires a theory of lattice vibrations.

Inelastic Scattering of Light

When laser light is scattered from crystals some components of the reflected beam have small shifts in frequency (Brillouin and Raman scattering). The explanation of this phenomenon requires a quantum theory of lattice vibrations.

---

[6] More precisely, there might still be superconductors, but they would be quite different from the ones we know today. Alternative mechanisms for superconductivity have been proposed that do not rely on the effect of lattice vibrations on the electron-electron interaction, but examples of superconductivity based on such mechanisms have yet to be found.

Scattering of X Rays

The intensity of X radiation in the Bragg peaks predicted by the static lattice model is incorrect. The thermal vibrations of the ions about their equilibrium positions (and even the zero-point vibrations at $T = 0$) diminish the amplitude of the Bragg peaks. In addition, because the lattice is not static, there is a background of scattered X radiation in directions that do not satisfy the Bragg condition.

Scattering of Neutrons

When neutrons[7] are scattered off crystalline solids, they are found to lose energy only in definite discrete amounts, which depend on the change of momentum suffered in the scattering. The quantum theory of lattice vibrations gives a very simple explanation of this phenomenon, and neutrons are one of the most valuable probes of a solid we possess.

The foregoing is in no sense an exhaustive list of the ways in which lattice vibrations make themselves felt. However, it does illustrate most of the important functions the lattice vibrations perform:

1. The ability of the ions to vibrate about their equilibrium positions is essential in determining any equilibrium property of a solid that is not dominated by a very much larger contribution from the electrons.
2. The lattice vibrations provide a mechanism for transporting energy through a solid.
3. The lattice vibrations are a crucial source of electronic scattering in metals, and can profoundly affect the interaction between electrons.
4. The lattice vibrations play a role in the response of a solid to any probe that couples to the ions, such as visible light, X rays, or neutrons.

We shall examine these and other aspects of the lattice vibrations in Chapters 22 to 27.

---

[7] From the quantum-mechanical point of view a beam of neutrons of energy E and momentum $\mathbf{p}$ can be regarded as a beam of radiation of angular frequency $\omega = E/\hbar$ and wave vector $\mathbf{k} = \mathbf{p}/\hbar$.

22

Classical Theory of the Harmonic Crystal

The Harmonic Approximation

The Adiabatic Approximation

Specific Heat of a Classical Crystal

One-Dimensional Monatomic Bravais Lattice

One-Dimensional Lattice with a Basis

Three-Dimensional Monatomic Bravais Lattice

Three-Dimensional Lattice with a Basis

Relation to Theory of Elasticity

In relaxing the artificial assumption that the ions sit without moving at the sites **R** of a Bravais lattice, we shall rely extensively on two weaker assumptions:

1. We shall assume that the mean equilibrium position of each ion is a Bravais lattice site. We can then continue to associate with each ion a particular Bravais lattice site **R**, about which the ion oscillates, but the site **R** is now only the mean position of the ion, and not its fixed instantaneous position.
2. We shall assume that the typical excursions of each ion from its equilibrium position are small compared with the interionic spacing (in a sense to be made more precise below).

Assumption 1 accounts for the observed crystalline structure of solids by asserting that the Bravais lattice is still there in spite of ionic motion, but that it describes the average ionic configuration rather than the instantaneous one. Note that although this assumption permits a wide range of possible ionic motion, it does not allow for ionic diffusion: The oscillations of each ion are assumed to be forever about a particular Bravais lattice site **R**. Except when the possibility of ions interchanging equilibrium positions becomes important (as it might, for example, near the melting point) this assumption is not terribly restrictive.

Assumption 2, however, is not made out of any strong conviction in its general validity, but on grounds of analytical necessity. It leads to a simple theory—the *harmonic approximation*—from which precise quantitative results can be extracted. These results are often in excellent agreement with observed solid properties. Some properties, however, cannot be accounted for by a harmonic theory, and to explain them it is necessary to go to an *anharmonic* theory (Chapter 25). Even in such cases, however, the method of computation implicitly continues to rely on assumption 2, although it is exploited in a more sophisticated way. When assumption 2 genuinely fails (as appears to be the case in solid helium), one must deal from the start with a theory of formidable analytical complexity, and it is only quite recently that any progress has been made along these lines.[1]

Because of assumption 1 we may unambiguously label each ion with the Bravais lattice site **R** about which it oscillates.[2] We denote the position of the ion whose mean position is **R** by $\mathbf{r}(\mathbf{R})$ (see Figure 22.1). If the static lattice approximation were valid, i.e., if each ion were stationary at its Bravais lattice site, we would have $\mathbf{r}(\mathbf{R}) = \mathbf{R}$. In the more realistic case, however, $\mathbf{r}(\mathbf{R})$ will deviate from its average value **R**, and we may write at any given time:[3]

$$\mathbf{r}(\mathbf{R}) = \mathbf{R} + \mathbf{u}(\mathbf{R}), \tag{22.1}$$

---

[1] Under the romantic name of "the theory of quantum solids." The nomenclature refers to the fact that according to classical theory, assumption 2 holds in any solid at low enough T. It is only the uncertainty principle that requires some deviations from equilibrium in the ionic positions no matter how low the temperature.

[2] In much of this chapter we shall deal explicitly only with monatomic Bravais lattices; i.e., solids whose crystal structures consist of a single ion per primitive cell located at the sites $\mathbf{R} = n_1\mathbf{a}_1 + n_2\mathbf{a}_2 + n_3\mathbf{a}_3$ of a Bravais lattice. The generalization of the discussion to lattices with a basis of n atoms per primitive cell, located at $\mathbf{R} + \mathbf{d}_1, \mathbf{R} + \mathbf{d}_2, \ldots, \mathbf{R} + \mathbf{d}_n$, is straightforward, but the notation can be cumbersome.

[3] More generally, in describing a lattice with a basis, we would let $\mathbf{r}_j(\mathbf{R})$ describe the position of the jth basis atom in the primitive cell about **R**, and write $\mathbf{r}_j(\mathbf{R}) = \mathbf{R} + \mathbf{d}_j + \mathbf{u}_j(\mathbf{R})$.

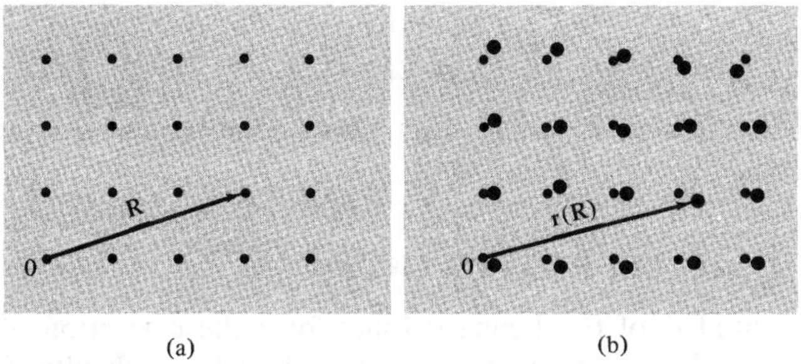

Figure 22.1
(a) The Bravais lattice of points, specified by vectors **R**. (b) A particular instantaneous configuration of ions. The ion whose mean position is **R** is found at **r**(**R**).

where **u**(**R**) is the deviation from equilibrium of the ion whose equilibrium site is **R**. (See Figure 22.2.)

To make the discussion concrete, let us reexamine our treatment of the cohesive energy of the noble gases (Chapter 20) within this broader framework. We continue to assume that a pair of atoms separated by **r** contributes an amount $\phi(\mathbf{r})$ to the potential energy of the crystal, where ϕ is, for example, the Lennard-Jones potential (20.2). If the static lattice model were correct, and each atom remained fixed at its Bravais lattice site, the total potential energy of the crystal would just be the sum of the contributions of all distinct pairs:

$$U = \frac{1}{2}\sum_{\mathbf{RR}'} \phi(\mathbf{R} - \mathbf{R}') = \frac{N}{2}\sum_{\mathbf{R}\neq 0} \phi(\mathbf{R}). \tag{22.2}$$

If, however, we allow for the fact that the atom whose average position is **R** will in general be found at a position $\mathbf{r}(\mathbf{R}) \neq \mathbf{R}$, then we must replace (22.2) by

$$U = \frac{1}{2}\sum_{\mathbf{RR}'} \phi(\mathbf{r}(\mathbf{R}) - \mathbf{r}(\mathbf{R}')) = \frac{1}{2}\sum_{\mathbf{RR}'} \phi(\mathbf{R} - \mathbf{R}' + \mathbf{u}(\mathbf{R}) - \mathbf{u}(\mathbf{R}')). \tag{22.3}$$

Thus the potential energy now depends on the dynamical variables **u**(**R**), and we must come to grips with the dynamical (or statistical mechanical) problem governed by the Hamiltonian:[4]

Figure 22.2
The relation between the Bravais lattice vector **R**, the instantaneous position **r**(**R**) of the ion that oscillates about **R**, and the ionic displacement, $\mathbf{u}(\mathbf{R}) = \mathbf{r}(\mathbf{R}) - \mathbf{R}$.

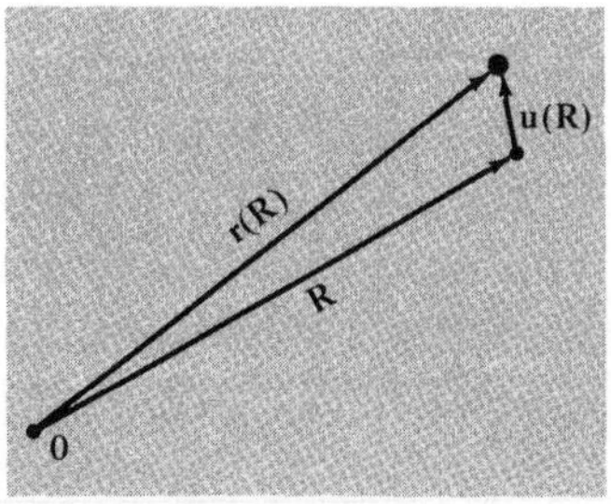

[4] In lattice dynamics one generally takes the canonical coordinates to be the **u**(**R**) rather than the **r**(**R**); i.e., each ion is referred to a different origin.

$$H = \sum_{\mathbf{R}} \frac{\mathbf{P}(\mathbf{R})^2}{2M} + U, \tag{22.4}$$

where $\mathbf{P}(\mathbf{R})$ is the momentum of the atom whose equilibrium position is $\mathbf{R}$, and M is the atomic mass.

THE HARMONIC APPROXIMATION

For a pair potential ϕ of the Lennard-Jones form the extraction of useful exact information from this Hamiltonian is a task of hopeless difficulty. One therefore resorts to an approximation based on the expectation that the atoms will not deviate substantially from their equilibrium positions. If all the $\mathbf{u}(\mathbf{R})$ are small,[5] then we can expand the potential energy U about its equilibrium value, using the three-dimensional form of Taylor's theorem:

$$f(\mathbf{r} + \mathbf{a}) = f(\mathbf{r}) + \mathbf{a} \cdot \nabla f(\mathbf{r}) + \frac{1}{2}(\mathbf{a} \cdot \nabla)^2 f(\mathbf{r}) + \frac{1}{3!}(\mathbf{a} \cdot \nabla)^3 f(\mathbf{r}) + \cdots. \tag{22.5}$$

Applying this to each term of (22.3), with $\mathbf{r} = \mathbf{R} - \mathbf{R}'$ and $\mathbf{a} = \mathbf{u}(\mathbf{R}) - \mathbf{u}(\mathbf{R}')$, we find:

$$\begin{aligned} U = \frac{N}{2}\sum \phi(\mathbf{R}) &+ \frac{1}{2}\sum_{\mathbf{RR}'} (\mathbf{u}(\mathbf{R}) - \mathbf{u}(\mathbf{R}')) \cdot \nabla\phi(\mathbf{R} - \mathbf{R}') \\ &+ \frac{1}{4}\sum_{\mathbf{RR}'} [(\mathbf{u}(\mathbf{R}) - \mathbf{u}(\mathbf{R}')) \cdot \nabla]^2 \phi(\mathbf{R} - \mathbf{R}') + O(u^3). \end{aligned} \tag{22.6}$$

The coefficient of $\mathbf{u}(\mathbf{R})$ in the linear term is just

$$\sum_{\mathbf{R}'} \nabla\phi(\mathbf{R} - \mathbf{R}'). \tag{22.7}$$

This, however, is just minus the force exerted on the atom at $\mathbf{R}$ by all the other atoms, when each is placed at its equilibrium position. It must therefore vanish, since there is no net force on any atom in equilibrium.

Since the linear term in (22.6) vanishes, the first nonvanishing correction to the equilibrium potential energy is given by the quadratic term. In the *harmonic approximation* only this term is retained, and the potential energy is written as

$$U = U^{\text{eq}} + U^{\text{harm}}, \tag{22.8}$$

where U^{eq} is the equilibrium potential energy (22.2) and

$$\begin{aligned} U^{\text{harm}} &= \frac{1}{4} \sum_{\substack{\mathbf{RR}' \\ \mu,\nu = x,y,z}} [u_\mu(\mathbf{R}) - u_\mu(\mathbf{R}')]\phi_{\mu\nu}(\mathbf{R} - \mathbf{R}')[u_\nu(\mathbf{R}) - u_\nu(\mathbf{R}')], \\ \phi_{\mu\nu}(\mathbf{r}) &= \frac{\partial^2 \phi(\mathbf{r})}{\partial r_\mu \, \partial r_\nu}. \end{aligned} \tag{22.9}$$

---

[5] More precisely, if $\mathbf{u}(\mathbf{R}) - \mathbf{u}(\mathbf{R}')$ is small for all atom pairs with appreciable $\phi(\mathbf{R} - \mathbf{R}')$. The absolute displacement of an atom can be large. What is important is that its displacement be small with respect to those atoms with which it has any appreciable interaction.

Since U^{eq} is just a constant (i.e., independent of the $\mathbf{u}$'s and $\mathbf{P}$'s), it can be ignored in many dynamical problems,[6] and one frequently acts as if the total potential energy were just U^{harm}, dropping the superscript altogether when no ambiguity is likely to result.

The harmonic approximation is the starting point for all theories of lattice dynamics (except, perhaps, in solid helium). Further corrections to U, especially those of third and fourth order in the u's, are known as anharmonic terms, and are of considerable importance in understanding many physical phenomena, as we shall see in Chapter 25. They are generally treated as small perturbations on the dominant harmonic term.

The harmonic potential energy is usually written not in the form (22.9), but in the more general form

$$U^{\text{harm}} = \frac{1}{2} \sum_{\substack{\mathbf{R}\mathbf{R}' \\ \mu\nu}} u_\mu(\mathbf{R}) D_{\mu\nu}(\mathbf{R} - \mathbf{R}') u_\nu(\mathbf{R}'). \tag{22.10}$$

Evidently (22.9) has this structure, with

$$D_{\mu\nu}(\mathbf{R} - \mathbf{R}') = \delta_{\mathbf{R},\mathbf{R}'} \sum_{\mathbf{R}''} \phi_{\mu\nu}(\mathbf{R} - \mathbf{R}'') - \phi_{\mu\nu}(\mathbf{R} - \mathbf{R}'). \tag{22.11}$$

THE ADIABATIC APPROXIMATION

Aside from the fact that it is more compact, we use the form (22.10) rather than (22.9) because in general the ion-ion interaction cannot be represented as a simple sum of pair interactions of the form (22.3). Indeed, except in particularly simple cases (such as the noble gases) the quantities D appearing in (22.10) are quite difficult to calculate. In ionic crystals the difficulty comes from the long-range nature of the Coulomb interaction between ions. In covalent crystals and metals the difficulty is a more profound one, for there the ionic motion is inextricably coupled to the motion of the valence electrons. This is because in covalent crystals and metals the electronic arrangement, and hence the contribution of the valence electrons to the total energy of the solid, depends in detail on the particular arrangement of the ion cores. Thus when the solid is deformed by displacing the ion cores from their equilibrium positions, the electronic wave functions will be deformed as well, in a way that can be quite difficult to deduce with any precision.[7]

One copes with this problem by making the so-called adiabatic approximation. The adiabatic approximation is based on the fact that typical electronic velocities are much greater than typical ionic velocities. As we saw in Chapter 2, the significant electronic velocity is $v_F \approx 10^8$ cm/sec. On the other hand, typical ionic velocities

---

[6] Of course, it cannot always be ignored. As we have seen in Chapter 20, it is of crucial importance in determining the absolute energy of the crystal, its equilibrium size, or its equilibrium compressibility.

[7] This can be a problem even in ionic crystals. The outermost core electrons can be sufficiently weakly bound for the ions to undergo significant polarization when displaced from their equilibrium positions. A theory that takes this into account is known as the shell model (not to be confused with the shell model of nuclear physics). See Chapter 27.

are at most of order 10^5 cm/sec.[8] One therefore assumes that because the ions move so slowly on the scale of velocities of relevance to the electrons, at any moment the electrons will be in their ground state for that particular instantaneous ionic configuration. In computing the force constants appearing in (22.10), one must then supplement the interaction between ion cores with terms representing the dependence of the additional electronic energy on the instantaneous ionic configuration specified by the $\mathbf{u}(\mathbf{R})$. In practice, this can be quite difficult to do, and a more practical approach is to consider the quantities D as empirical parameters, to be measured directly by experiment (Chapter 24).[9]

SPECIFIC HEAT OF A CLASSICAL CRYSTAL: THE LAW OF DULONG AND PETIT

Having abandoned the static lattice approximation, we can no longer evaluate equilibrium properties (as we did in Chapter 20) by simply assuming that each ion sits quietly at its Bravais lattice site $\mathbf{R}$. We must now average over all possible ionic configurations, giving each configuration or state a weight proportional to e^{-E/k_BT}, where E is the energy of the configuration.[10] Thus if we treat the crystal classically, its thermal energy density is given by

$$u = \frac{1}{V}\frac{\int d\Gamma\, e^{-\beta H}H}{\int d\Gamma\, e^{-\beta H}}, \qquad \beta = \frac{1}{k_BT}, \tag{22.12}$$

where we have used a compact notation in which $d\Gamma$ stands for the volume element in crystal phase space:

$$d\Gamma = \prod_{\mathbf{R}} d\mathbf{u}(\mathbf{R})\, d\mathbf{P}(\mathbf{R}) = \prod_{\mathbf{R},\mu} du_\mu(\mathbf{R})\, dp_\mu(\mathbf{R}). \tag{22.13}$$

We can also write (22.12) in the more useful form:

$$u = -\frac{1}{V}\frac{\partial}{\partial\beta}\ln\int d\Gamma\, e^{-\beta H}, \tag{22.14}$$

as can be verified by explicitly differentiating the logarithm in (22.14).

---

[8] This will emerge in the subsequent analysis. We shall find that typical ionic vibration frequencies are at most of the order of $0.01\mathcal{E}_F/\hbar$. Since the amplitude of the ionic vibrations is small compared with the size of the unit cell, $a = O(1/k_F)$, the ionic velocity is of order less than $0.01\mathcal{E}_F/\hbar k_F \approx 0.01 v_F$.

[9] There is, however, a highly developed theory of how to compute D for metals. See Chapter 26.

[10] This is the fundamental rule of equilibrium statistical mechanics. It applies whether the system under consideration is treated classically or quantum-mechanically, provided that the states under discussion are states of the *full N-particle system* (i.e., not single particle levels). By a classical state we mean a specified set of values for the $3N$ canonical coordinates $\mathbf{u}(\mathbf{R})$ and $3N$ canonical momenta $\mathbf{P}(\mathbf{R})$, i.e., a point in phase space. By a quantum state we mean a stationary-state solution to the N-particle Schrödinger equation: $H\Psi = E\Psi$.

In the harmonic approximation the temperature dependence of the integral appearing in (22.14) is easily extracted by making the change of variables:

$$\begin{aligned} \mathbf{u}(\mathbf{R}) &= \beta^{-1/2}\bar{\mathbf{u}}(\mathbf{R}), \quad d\mathbf{u}(\mathbf{R}) = \beta^{-3/2}\, d\bar{\mathbf{u}}(\mathbf{R}), \\ \mathbf{P}(\mathbf{R}) &= \beta^{-1/2}\bar{\mathbf{P}}(\mathbf{R}), \quad d\mathbf{P}(\mathbf{R}) = \beta^{-3/2}\, d\bar{\mathbf{P}}(\mathbf{R}). \end{aligned} \tag{22.15}$$

The integral in (22.14) can then be written as

$$\begin{aligned} \int d\Gamma\, e^{-\beta H} &= \int d\Gamma \exp\left[-\beta\left(\sum \frac{\mathbf{P}(\mathbf{R})^2}{2M} + U^{\text{eq}} + U^{\text{harm}}\right)\right] \\ &= e^{-\beta U^{\text{eq}}}\beta^{-3N}\left\{\int \prod_{\mathbf{R}} d\bar{\mathbf{u}}(\mathbf{R})\, d\bar{\mathbf{P}}(\mathbf{R})\right. \\ &\quad \left.\times \exp\left[-\sum \frac{1}{2M}\bar{\mathbf{P}}(\mathbf{R})^2 - \frac{1}{2}\sum \bar{u}_\mu(\mathbf{R})D_{\mu\nu}(\mathbf{R}-\mathbf{R}')\bar{u}_\nu(\mathbf{R}')\right]\right\}. \end{aligned} \tag{22.16}$$

The entire integral appearing in braces in (22.16) is independent of temperature, and therefore makes no contribution to the β-derivative when (22.16) is substituted into (22.14). The thermal energy therefore reduces simply to:

$$u = -\frac{1}{V}\frac{\partial}{\partial\beta}\ln\left(e^{-\beta U^{\text{eq}}}\beta^{-3N} \times \text{const}\right) = \frac{U^{\text{eq}}}{V} + \frac{3N}{V}k_B T \tag{22.17}$$

or[11]

$$u = u^{\text{eq}} + 3nk_B T. \tag{22.18}$$

Note that this reduces to the result $u = u^{\text{eq}}$ of static lattice theory at $T = 0$ (as we expect from classical theory, which ignores zero-point motion). At nonzero temperatures the static result is corrected by the simple additive term $3nk_BT$. Since k_BT is only a few hundredths of an electron volt, even at room temperature, this is generally a small correction. It is far more useful to consider the specific heat, $c_v = (\partial u/\partial T)_v$, (which is also much more easily measured than the internal energy). The static lattice contribution to u drops out of c_v, which is determined entirely by the temperature-dependent correction:[12]

$$c_v = \frac{\partial u}{\partial T} = 3nk_B. \tag{22.19}$$

---

[11] When it is necessary to distinguish the number of ions per unit volume from the number of conduction electrons per unit volume we shall use subscripts (n_i or n_e). In simple metals $n_e = Zn_i$, where Z is the valence.

[12] Experiments measure the specific heat at constant pressure, c_p, but we have calculated the specific heat at constant volume, c_v. In a gas these differ substantially, but in a solid they are almost identical. This is seen most intuitively from the thermodynamic identity: $c_p/c_v = (\partial P/\partial V)_S/(\partial P/\partial V)_T$. The two specific heats differ to the extent that the adiabatic and isothermal compressibilities differ. Since u^{eq} is the dominant term in the internal energy of a solid, thermal considerations are of little consequence in determining the compressibility. Thus the work necessary to compress a solid by a definite amount depends very little on whether the solid is thermally insulated (adiabatic) or in contact with a heat bath at temperature T (isothermal) during the compression. Ordinarily the two specific heats differ by less than a percent at room temperature, and substantially less than a percent, at lower temperatures.

This result, that the specific heat due to the lattice vibrations (i.e., the entire specific heat of an insulator) is just $3k_B$ per ion, is known as the law of Dulong and Petit. In a monatomic solid in which there are 6.022×10^{23} ions per mole, it is most commonly encountered in the form:[13]

$$c_v^{\text{molar}} = 5.96 \text{ cal/mole-K}. \qquad \textbf{(22.20)}$$

In Figure 22.3 we display the measured specific heats of solid argon, krypton, and xenon. At temperatures of the order of 100 K and above, the measured specific heats are quite close to the Dulong and Petit value. However:

1. As the temperature drops, the specific heat falls well below the Dulong and Petit value, heading toward zero at zero temperature.
2. Even as the temperature gets large, it seems fairly clear that the curve is not approaching the precise Dulong and Petit value.

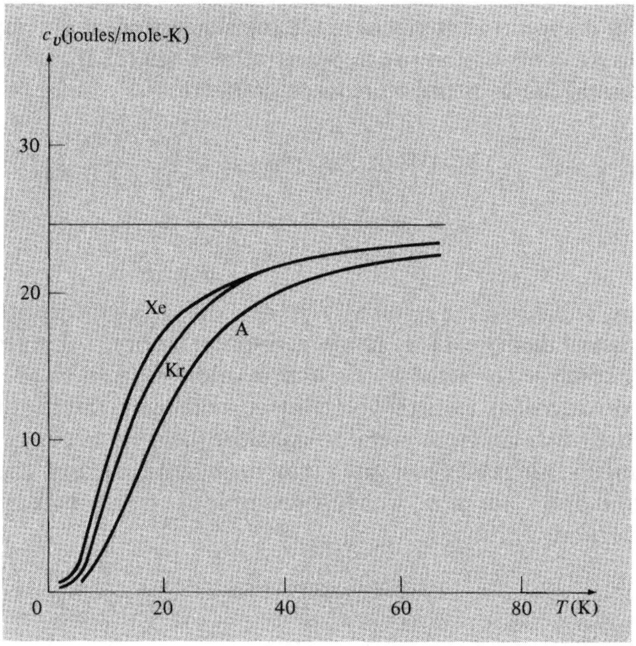

Figure 22.3
Measured specific heats of argon, xenon, and krypton. The horizontal line is the classical Dulong and Petit value. (Quoted in M. L. Klein, G. K. Horton, and J. L. Feldman, *Phys. Rev.* **184**, 68 (1969).)

Point 2 can be explained on purely classical grounds as a failure of the harmonic approximation. According to classical theory, at very low temperatures the thermal energies are simply insufficient to permit an ion to wander any appreciable distance from its equilibrium site, and the harmonic approximation becomes very good as the

---

[13] $k_B = 1.38 \times 10^{-16}$ erg/K; 4.184×10^7 ergs $= 1$ cal.

temperature drops.[14] However, at higher temperatures the ions possess sufficient energy to stray far enough from their equilibrium positions for the neglected anharmonic terms (i.e., terms beyond the quadratic ones in the expansion of U in powers of the ionic displacements $\mathbf{u}$) to become important. Thus classical statistical mechanics indicates that the law of Dulong and Petit need not be perfectly obeyed at high temperatures, but should become more and more closely followed as the temperature drops.

Thus the low-temperature behavior (point 1) is completely inexplicable classically. The quantum theory is required to account for the low-temperature specific heat of the lattice and, except at rather high temperatures (of order 10^2 K, judging from Figure 22.3), we cannot expect to get very far in a theory of lattice dynamics that adheres to a purely classical picture.[15] We must therefore turn to a quantum theory of lattice dynamics to explain physical phenomena governed by the lattice vibrations.

In spite of this blatant failure of classical mechanics, however, it is essential to understand the classical theory of lattice vibrations before attempting to construct a quantum theory. The reason for this lies in the quadratic structure of the harmonic Hamiltonian. Because it is quadratic in the displacements $\mathbf{u}(\mathbf{R})$ and momenta $\mathbf{P}(\mathbf{R})$ it represents a special case of the general classical problem of small oscillations, which can be solved exactly.[16] In the solution a general motion of the N ions is represented as a superposition (or linear combination) of $3N$ normal modes of vibration, each with its own characteristic frequency ν. But it is a basic result of the quantum theory that the allowed energies of an oscillator with frequency ν are given by

$$(n + \tfrac{1}{2})h\nu, \qquad n = 0, 1, 2, \ldots. \tag{22.21}$$

The generalization of this result to $3N$ independent oscillators is the obvious one: The allowed energies of the $3N$ oscillator system are given by assigning to each oscillator a half integral multiple of its frequency times h, and adding up the contributions from each oscillator. In the case of the harmonic crystal the frequencies of the $3N$ normal modes provide such a set of frequencies, out of which all the energy levels of the crystal can be constructed.[17]

An analysis of the classical normal modes of a lattice of ions is therefore of great utility, even though a purely classical theory of lattice vibrations is clearly inadequate. We must examine the classical normal modes of the crystal before we can correct the

---

[14] In fact, the harmonic approximation becomes asymptotically exact as $T \to 0$ in a classical theory, for at $T = 0$ (infinite β) only values of the $\mathbf{u}$ that give an absolute minimum to the energy (i.e., $\mathbf{u}(\mathbf{R}) \to 0$) contribute to the exact integral (22.12). At sufficiently small T only $\mathbf{u}(\mathbf{R})$ in the immediate vicinity of 0 will give any appreciable contribution. Thus at sufficiently small T the exact Hamiltonian is equal to its harmonic approximation at all values of $\mathbf{u}$ that contribute appreciably to the integral. On the other hand, it is also true that at very low temperatures only very small values of the $\mathbf{u}(\mathbf{R})$ give any appreciable contribution to the integral (22.16) in which the Hamiltonian has been replaced by its harmonic approximation. Thus in both the exact integral (22.12) and its harmonic approximation (22.16) the integrands are appreciable at low temperatures only where they agree.

[15] A similar problem arose in connection with the electronic contribution to the specific heat of a metal, where the classical result $(3/2)k_B$ per electron fails at temperatures below the Fermi temperature.

[16] See any text on classical mechanics.

[17] We shall state this more precisely at the beginning of Chapter 23. A summary of the detailed quantum-mechanical proof is given in Appendix L.

law of Dulong and Petit and go on to describe a variety of other properties of the dynamical lattice. The rest of this chapter is therefore devoted to a study of the classical harmonic crystal. We approach the problem in the following stages:

1. Normal modes of a one-dimensional monatomic Bravais lattice.
2. Normal modes of a one-dimensional lattice with a basis.
3. Normal modes of a three-dimensional monatomic Bravais lattice.
4. Normal modes of a three-dimensional lattice with a basis.

In principle the analysis is the same in all four cases, but the purely notational complexities of the most general case (4) tend to obscure important physical aspects, which are revealed more clearly in the simpler cases.

We conclude the chapter by relating this analysis to the classical theory of an elastic continuous medium.

NORMAL MODES OF A ONE-DIMENSIONAL MONATOMIC BRAVAIS LATTICE

Consider a set of ions of mass M distributed along a line at points separated by a distance a, so that the one-dimensional Bravais lattice vectors are just $\mathbf{R} = na$, for integral n. Let $u(na)$ be the displacement along the line from its equilibrium position, of the ion that oscillates about na (Figure 22.4). For simplicity we assume that only

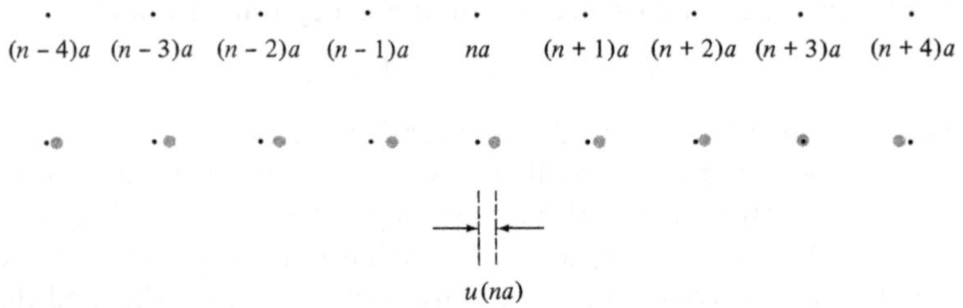

Figure 22.4
At any instant, the ion whose equilibrium position is na is displaced from equilibrium by an amount $u(na)$.

neighboring ions interact, so we may take the harmonic potential energy (22.9) to have the form

$$U^{\text{harm}} = \frac{1}{2} K \sum_n [u(na) - u([n+1]a)]^2, \tag{22.22}$$

(where $K = \phi''(a)$, $\phi(x)$ being the interaction energy of two ions a distance x apart along the line). The equations of motion are then

$$M\ddot{u}(na) = -\frac{\partial U^{\text{harm}}}{\partial u(na)} = -K[2u(na) - u([n-1]a) - u([n+1]a)]. \tag{22.23}$$

These are precisely the equations that would be obeyed if each ion were connected to its two neighbors by perfect massless springs with spring constant K (and equilibrium length a, although the equations are in fact independent of the equilibrium length of the spring). The resulting motion is most easily visualized in terms of such a model (Figure 22.5).

Figure 22.5
If only nearest-neighbor forces are kept, the harmonic approximation for the one-dimensional Bravais lattice describes a model in which each ion is tied to its neighbors by perfect springs.

If the chain of ions has only a finite number, N, then we must specify how the ions at the two ends are to be described. We could take them to interact only with their neighbors on the interior side, but this would complicate the analysis without materially altering the final results. For if N is large, and if we are not interested in end effects, then the precise way in which the ions at the ends are treated is immaterial, and we may choose the approach on grounds of mathematical convenience. As in the case of the electron gas (Chapter 2) by far the most convenient choice is the Born-von Karman periodic boundary condition. In the linear chain of ions this boundary condition is easily specified: We simply join the two remote ends of the chain back together by one more of the same springs that connect internal ions (Figure 22.6). If we take the ions to occupy sites $a, 2a, \ldots, Na$, then we can use Eq. (22.22) to describe each of the N ions ($n = 1, 2, \ldots, N$), provided that we interpret the quantities $u([N+1]a)$ and $u(0)$ that occur in the equations of motion for $u(Na)$ and $u(a)$, respectively, as[18]

$$u([N+1]a) = u(a); \quad u(0) = u(Na). \tag{22.24}$$

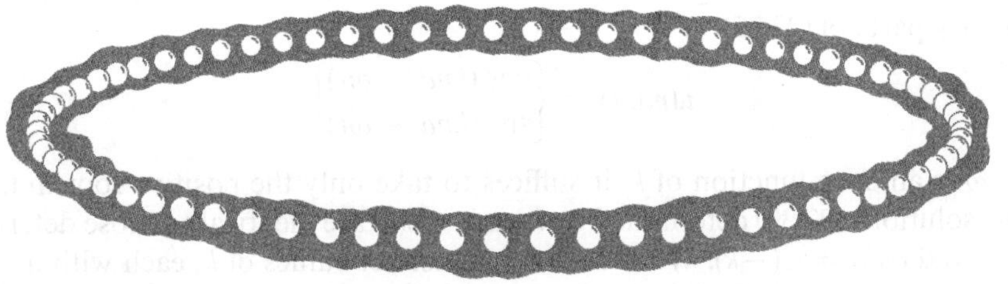

Figure 22.6
The Born-von Karman or periodic boundary condition for the linear chain.

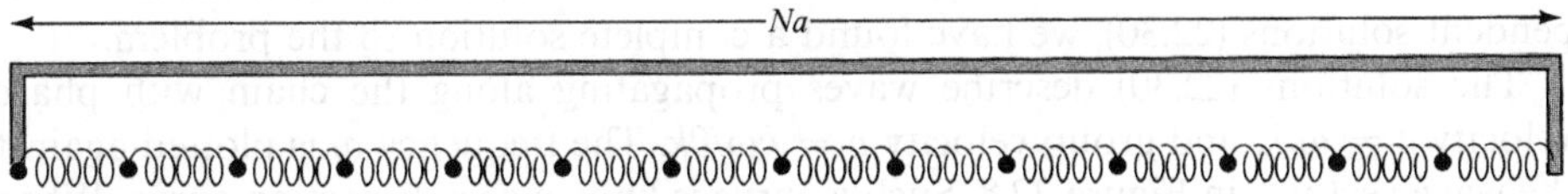

Figure 22.7
An alternative representation of the Born-von Karman boundary condition. The object connecting the ion on the extreme left with the spring on the extreme right is a massless rigid rod of length $L = Na$.

[18] An alternative interpretation of the Born-von Karman boundary condition is to consider not the deformation of the chain into a loop, but an explicit mechanical constraint forcing ion N to interact with ion 1 via a spring of spring constant K (Figure 22.7). This picture is probably more helpful in interpreting the boundary condition in three dimensions, and is especially useful to keep in mind when considering questions involving the total momentum of a finite crystal, or the question of why a crystal assumes the equilibrium size it does.

We seek solutions to (22.23) of the form:

$$u(na, t) \propto e^{i(kna - \omega t)}. \tag{22.25}$$

The periodic boundary condition (22.24) requires that

$$e^{ikNa} = 1, \tag{22.26}$$

which in turn requires k to have the form:

$$k = \frac{2\pi}{a}\frac{n}{N}, \qquad n \text{ an integer.} \tag{22.27}$$

Note that if k is changed by $2\pi/a$, the displacement $u(na)$ defined by (22.25) is unaffected. Consequently, there are just N values of k consistent with (22.27) that yield distinct solutions. We take them to be the values lying between $-\pi/a$ and π/a.[19]

Upon substituting (22.25) into (22.23) we find that

$$\begin{aligned} -M\omega^2 e^{i(kna - \omega t)} &= -K[2 - e^{-ika} - e^{ika}]e^{i(kna - \omega t)} \\ &= -2K(1 - \cos ka)e^{i(kna - \omega t)}, \end{aligned} \tag{22.28}$$

and thus we have a solution for a given k, provided that $\omega = \omega(k)$, where

$$\omega(k) = \sqrt{\frac{2K(1 - \cos ka)}{M}} = 2\sqrt{\frac{K}{M}}\,|\sin \tfrac{1}{2}ka|. \tag{22.29}$$

The solutions describing the actual ionic displacements are given by the real or imaginary parts of (22.25):

$$u(na, t) \propto \begin{Bmatrix} \cos (kna - \omega t) \\ \sin (kna - \omega t) \end{Bmatrix}. \tag{22.30}$$

Since ω is an even function of k, it suffices to take only the positive root in (22.29), for the solutions (22.30) determined by k and $-\omega(k)$ are identical to those determined by $-k$ and $\omega(k) = \omega(-k)$. We thus have N distinct values of k, each with a unique frequency $\omega(k)$, so Eq. (22.30) yields $2N$ independent solutions.[20] An arbitrary motion of the chain is determined by specifying the N initial positions and N initial velocities of the ions. Since these can always be fit with a linear combination of the $2N$ independent solutions (22.30), we have found a complete solution to the problem.

The solutions (22.30) describe waves propagating along the chain with phase velocity $c = \omega/k$, and group velocity $v = \partial\omega/\partial k$. The frequency ω is plotted against the wave vector k in Figure 22.8. Such a curve is known as a dispersion curve. When k is small compared with π/a (i.e., when the wavelength is large compared with the interparticle spacing), ω is linear in k:

$$\omega = \left(a\sqrt{\frac{K}{M}}\right)|k|. \tag{22.31}$$

This is the type of behavior we are accustomed to in the cases of light waves and ordinary sound waves. If ω is linear in k, then the group velocity is the same as the

---

[19] This is just the one-dimensional version of requiring **k** to lie in the first Brillouin zone (Chapter 8).

[20] Although there are $2N$ solutions, there are only N "normal modes", for the sine solution is simply the cosine solution, shifted in time by $\pi/2\omega$.

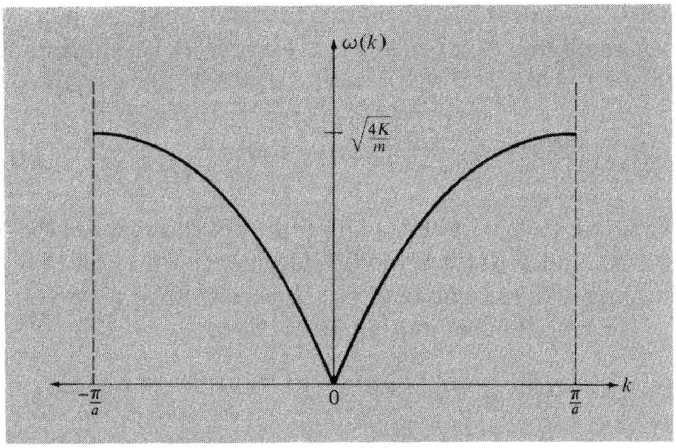

Figure 22.8
Dispersion curve for a monatomic linear chain with only nearest-neighbor interactions. Note that ω is linear for small k, and that $\partial\omega/\partial k$ vanishes at the boundaries of the zone ($k = \pm\pi/a$).

phase velocity, and both are independent of frequency. One of the characteristic features of waves in discrete media, however, is that the linearity ceases to hold at wavelengths short enough to be comparable to the interparticle spacing. In the present case ω falls below ck as k increases, and the dispersion curve actually becomes flat (i.e., the group velocity drops to zero) when k reaches $\pm\pi/a$.

If we drop the assumption that only nearest neighbors interact, very little changes in these results. The functional dependence of ω on k becomes more complex, but we continue to find N normal modes of the form (22.25) for the N allowed values of k. Furthermore, the angular frequency $\omega(k)$ remains linear in k for k small compared with π/a, and satisfies $\partial\omega/\partial k = 0$ at $k = \pm\pi/a$.[21]

NORMAL MODES OF A ONE-DIMENSIONAL LATTICE WITH A BASIS

We consider next a one-dimensional Bravais lattice with *two* ions per primitive cell, with equilibrium positions na and $na + d$. We take the two ions to be identical, but take $d \leqslant a/2$, so the force between neighboring ions depends on whether their separation is d or $a - d$ (Figure 22.9).[22] For simplicity we again assume that only nearest

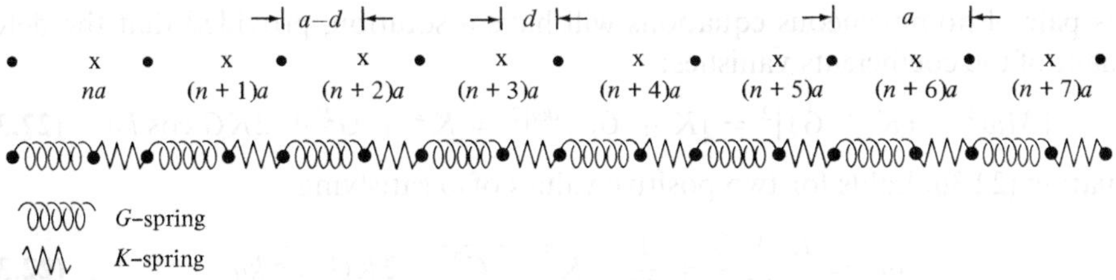

Figure 22.9
The diatomic linear chain of identical atoms, connected by springs of alternating strengths.

---

21 See Problem 1. These conclusions are correct provided that the interaction is of finite range—i.e., provided that an ion interacts only with its first through mth nearest neighbors, where m is a fixed integer (independent of N). If the interaction has infinitely long range, then it must fall off faster than the inverse cube of the interionic distance (in one dimension) if the frequencies are to be linear in k for small k.

22 An equally instructive problem arises when the forces between all neighboring ionic pairs are identical, but the ionic mass alternates between M_1 and M_2 along the chain. See Problem 2.

neighbors interact, with a force that is stronger for pairs separated by d than for pairs separated by $a - d$ (since $a - d$ exceeds d). The harmonic potential energy (22.9) can then be written:

$$U^{\text{harm}} = \frac{K}{2}\sum_n [u_1(na) - u_2(na)]^2 + \frac{G}{2}\sum_n [u_2(na) - u_1([n+1]a)]^2, \quad \textbf{(22.32)}$$

where we have written $u_1(na)$ for the displacement of the ion that oscillates about the site na, and $u_2(na)$ for the displacement of the ion that oscillates about $na + d$. In keeping with our choice $d \leqslant a/2$, we also take $K \geqslant G$.

The equations of motion are

$$\begin{aligned} M\ddot{u}_1(na) &= -\frac{\partial U^{\text{harm}}}{\partial u_1(na)} \\ &= -K[u_1(na) - u_2(na)] - G[u_1(na) - u_2([n-1]a)], \\ M\ddot{u}_2(na) &= -\frac{\partial U^{\text{harm}}}{\partial u_2(na)} \\ &= -K[u_2(na) - u_1(na)] - G[u_2(na) - u_1([n+1]a)]. \end{aligned} \quad \textbf{(22.33)}$$

We again seek a solution representing a wave with angular frequency ω and wave vector k:

$$\begin{aligned} u_1(na) &= \epsilon_1 e^{i(kna - \omega t)}, \\ u_2(na) &= \epsilon_2 e^{i(kna - \omega t)}. \end{aligned} \quad \textbf{(22.34)}$$

Here ϵ_1 and ϵ_2 are constants to be determined, whose ratio will specify the relative amplitude and phase of the vibration of the ions within each primitive cell. As in the monatomic case, the Born-von Karman periodic boundary condition again leads to the N nonequivalent values of k given by (22.27).

If we substitute (22.34) into (22.33) and cancel a common factor of $e^{i(kna - \omega t)}$ from both equations, we are left with two coupled equations:

$$\begin{aligned} [M\omega^2 - (K+G)]\epsilon_1 + (K + Ge^{-ika})\epsilon_2 &= 0, \\ (K + Ge^{ika})\epsilon_1 + [M\omega^2 - (K+G)]\epsilon_2 &= 0. \end{aligned} \quad \textbf{(22.35)}$$

This pair of homogeneous equations will have a solution, provided that the determinant of the coefficients vanishes:

$$[M\omega^2 - (K+G)]^2 = |K + Ge^{-ika}|^2 = K^2 + G^2 + 2KG\cos ka. \quad \textbf{(22.36)}$$

Equation (22.36) holds for two positive values of ω satisfying

$$\omega^2 = \frac{K+G}{M} \pm \frac{1}{M}\sqrt{K^2 + G^2 + 2KG\cos ka}, \quad \textbf{(22.37)}$$

with

$$\frac{\epsilon_2}{\epsilon_1} = \mp \frac{K + Ge^{ika}}{|K + Ge^{ika}|}. \quad \textbf{(22.38)}$$

For each of the N values of k there are thus *two* solutions, leading to a total of $2N$ normal modes, as is appropriate to the $2N$ degrees of freedom (two ions in each of N primitive cells). The two ω vs. k curves are referred to as the two *branches* of the dispersion relation, and are plotted in Figure 22.10. The lower branch has the same structure as the single branch we found in the monatomic Bravais lattice: ω vanishes linearly in k for small k, and the curve becomes flat at the edges of the Brillouin zone. This branch is known as the *acoustic branch* because its dispersion relation is of the form $\omega = ck$ characteristic of sound waves, at small k. The second branch starts at $\omega = \sqrt{2(K + G)/M}$ at $k = 0$ and decreases with increasing k down to $\sqrt{2K/M}$ at the zone edge. This branch is known as the *optical branch* because the long wavelength optical modes in ionic crystals can interact with electromagnetic radiation, and are responsible for much of the characteristic optical behavior of such crystals (Chapter 27).

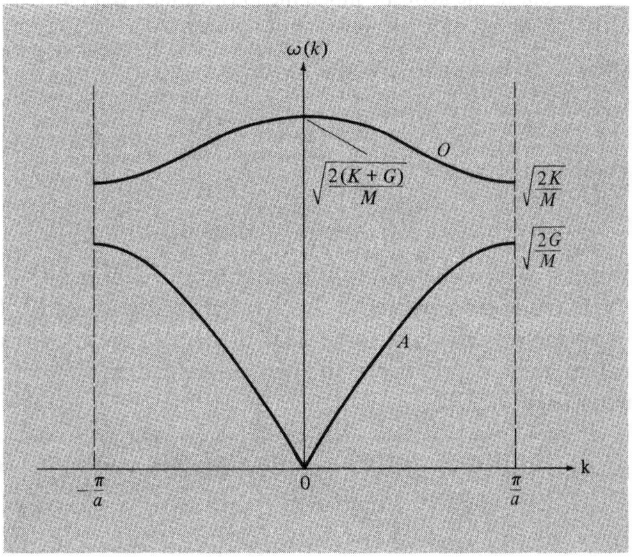

Figure 22.10
Dispersion relation for the diatomic linear chain. The lower branch is the acoustic branch and has the same structure as the single branch present in the monatomic case (Figure 22.8). In addition, there is now an optical branch (upper branch.)

We can gain some insight into the nature of the two branches by considering some special cases in more detail:

***Case* 1** $k \ll \pi/a$ Here $\cos ka \approx 1 - (ka)^2/2$, and to leading order in k the roots become:

$$\omega = \sqrt{\frac{2(K + G)}{M}} - O(ka)^2, \tag{22.39}$$

$$\omega = \sqrt{\frac{KG}{2M(K + G)}}(ka). \tag{22.40}$$

When k is very small, (22.38) reduces to $\epsilon_2 = \mp \epsilon_1$. The lower sign belongs to the acoustic mode, and describes a motion in which the two ions in the cell move in phase with one another (Figure 22.11). The upper sign belongs to the high-frequency optical mode, and describes a motion in which the two atoms in the cell are 180° out of phase.

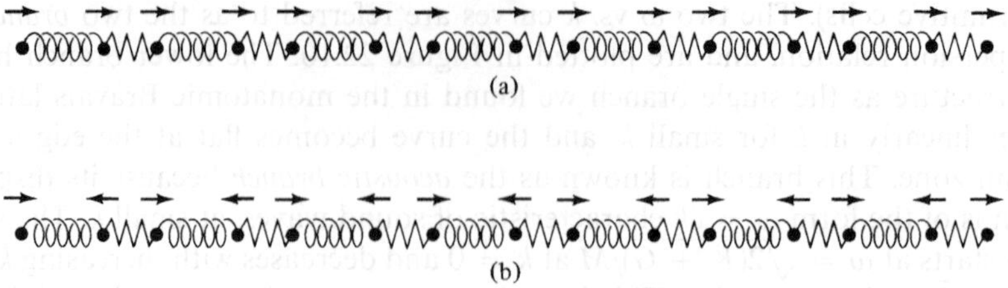

Figure 22.11
The long wavelength acoustic (a) and optical (b) modes in the diatomic linear chain. The primitive cell contains the two ions joined by the K-spring, represented by a jagged line. In both cases the motion of every primitive cell is identical, but in the acoustic mode the ions within a cell move together, while they move 180° out of phase in the optical mode.

***Case* 2** $k = \pi/a$ Now the roots are

$$\omega = \sqrt{\frac{2K}{M}}, \qquad \epsilon_1 = -\epsilon_2; \tag{22.41}$$

$$\omega = \sqrt{\frac{2G}{M}}, \qquad \epsilon_1 = \epsilon_2. \tag{22.42}$$

When $k = \pi/a$, the motions in neighboring cells are 180° out of phase, and therefore the two solutions must be as pictured in Figure 22.12. In each case, only one type of spring is stretched. Note that if the two spring constants were the same, there would be no gap between the two frequencies at $k = \pi/a$. The reason for this is clear from Figure 22.12.

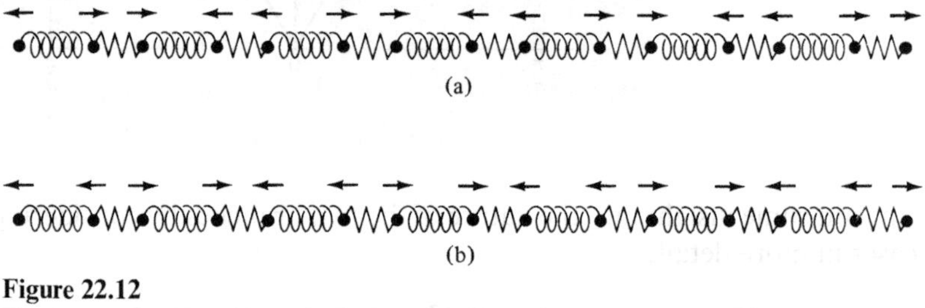

Figure 22.12
The acoustic (a) and optical (b) modes of the diatomic linear chain, when $k = \pm\pi/a$, at the edges of the Brillouin zone. Now the motion changes by 180° from cell to cell. However, as in Figure 22.11, the ions within each cell move in phase in the acoustic mode, and 180° out of phase in the optical mode. Note that if the K- and G-springs were identical the motion would be the same in both cases. This is why the two branches become degenerate at the edges of the zone when $K = G$.

***Case* 3** $K \gg G$ To leading order in G/K we have:

$$\omega = \sqrt{\frac{2K}{M}}\left[1 + O\left(\frac{G}{K}\right)\right], \qquad \epsilon_1 \approx -\epsilon_2; \tag{22.43}$$

$$\omega = \sqrt{\frac{2G}{M}}\left|\sin \tfrac{1}{2}ka\right|\left[1 + O\left(\frac{G}{K}\right)\right], \qquad \epsilon_1 \approx \epsilon_2. \tag{22.44}$$

The optical branch now has a frequency that is independent of k, to leading order in G/K, and equal to the vibrational frequency of a single diatomic molecule composed of two mass M ions connected by a spring K. Consistent with this picture of independent molecular vibrations in each primitive cell, the atomic motions within each cell are 180° out of phase (to leading order in G/K) whatever the wavelength of the normal mode. Because G/K is not zero, these molecular vibrations are very weakly coupled, and the result is a small spread of order G/K in the optical band frequencies, as k varies through the Brillouin zone.[23]

The acoustic branch (22.44) is (to leading order in G/K) just that for a linear chain of atoms of mass $2M$ coupled by the weak spring G (compare (22.44) with (22.29)). This is consistent with the fact that $\epsilon_1 = \epsilon_2$; i.e., within each cell the atoms move in phase, and the strong K-spring is hardly stretched at all.

This case suggests the following characterization of the difference between optical and acoustic branches:[24] An acoustic mode is one in which all ions within a primitive cell move essentially in phase, as a unit, and the dynamics are dominated by the interaction between cells; an optical mode, on the other hand, is one in which the ions within each primitive cell are executing what is essentially a molecular vibratory mode, which is broadened out into a band of frequencies by virtue of the intercellular interactions.

***Case* 4** $K = G$ In this case we are really dealing with a monatomic Bravais lattice of lattice constant $a/2$, and the analysis of the preceding section is applicable. It is nevertheless instructive to see how that analysis emerges in the limit $K \to G$. This is the subject of Problem 3.

NORMAL MODES OF A MONATOMIC THREE-DIMENSIONAL BRAVAIS LATTICE

We now consider a general three-dimensional harmonic potential (Eq. (22.10)). To keep from being blinded by indices it is often convenient to adopt a matrix notation, writing a quantity like

$$\sum_{\mu\nu} u_\mu(\mathbf{R}) D_{\mu\nu}(\mathbf{R} - \mathbf{R}') u_\nu(\mathbf{R}') \tag{22.45}$$

as the vector product of $\mathbf{u}(\mathbf{R})$ with the vector obtained by operating on the vector $\mathbf{u}(\mathbf{R}')$ with the matrix $\mathbf{D}(\mathbf{R} - \mathbf{R}')$. With this convention the harmonic potential (22.10) can be written:

$$U^{\text{harm}} = \frac{1}{2} \sum_{\mathbf{RR}'} \mathbf{u}(\mathbf{R}) \mathbf{D}(\mathbf{R} - \mathbf{R}') \mathbf{u}(\mathbf{R}'). \tag{22.46}$$

In discussing the normal modes of the crystal it is helpful to exploit some general symmetries that must be obeyed by the matrices $\mathbf{D}(\mathbf{R} - \mathbf{R}')$, regardless of the particular forms of the interionic forces.

---

[23] Note the similarity of this case to the tight-binding theory of electronic energy levels (Chapter 10), in which weakly coupled atomic energy levels broaden into a narrow band. In the present case weakly coupled molecular vibrational levels broaden into a narrow band.

[24] This simple physical interpretation does not hold in the general case.

***Symmetry* 1**

$$D_{\mu\nu}(\mathbf{R} - \mathbf{R}') = D_{\nu\mu}(\mathbf{R}' - \mathbf{R}). \tag{22.47}$$

Since the D's are coefficients in the quadratic form (22.10), they can always be chosen to have this symmetry. Alternatively, it follows from the general definition of $D_{\mu\nu}(\mathbf{R} - \mathbf{R}')$ as a second derivative of the exact interaction potential,

$$D_{\mu\nu}(\mathbf{R} - \mathbf{R}') = \left.\frac{\partial^2 U}{\partial u_\mu(\mathbf{R})\, \partial u_\nu(\mathbf{R}')}\right|_{\mathbf{u}\equiv 0}, \tag{22.48}$$

because of the independence of order of differentiation.

***Symmetry* 2**

$$D_{\mu\nu}(\mathbf{R} - \mathbf{R}') = D_{\mu\nu}(\mathbf{R}' - \mathbf{R}) \quad \text{or} \quad \mathsf{D}(\mathbf{R}) = \mathsf{D}(-\mathbf{R}), \tag{22.49}$$

or, by virtue of (22.47),

$$D_{\mu\nu}(\mathbf{R} - \mathbf{R}') = D_{\nu\mu}(\mathbf{R} - \mathbf{R}'). \tag{22.50}$$

This symmetry follows from the fact that every Bravais lattice has inversion symmetry. This implies that the energy of a configuration in which the ion associated with site **R** has a displacement $\mathbf{u}(\mathbf{R})$ must be the same as the energy of the configuration in which the ion associated with site **R** has a displacement $-\mathbf{u}(-\mathbf{R})$.[25] Equation (22.49) is just the condition that (22.45) be unchanged by this replacement ($\mathbf{u}(\mathbf{R}) \to -\mathbf{u}(-\mathbf{R})$) for arbitrary values of the $\mathbf{u}(\mathbf{R})$.

***Symmetry* 3**

$$\sum_{\mathbf{R}} D_{\mu\nu}(\mathbf{R}) = 0 \quad \text{or} \quad \sum_{\mathbf{R}} \mathsf{D}(\mathbf{R}) = 0. \tag{22.51}$$

This follows from the fact that if every ion is given the *same* displacement **d** from equilibrium (i.e., $\mathbf{u}(\mathbf{R}) \equiv \mathbf{d}$), then the entire crystal will simply be displaced without internal distortion, and U^{harm} will have the same value as it has when all the $\mathbf{u}(\mathbf{R})$ vanish, namely zero:

$$0 = \sum_{\substack{\mathbf{R}\mathbf{R}' \\ \mu\nu}} d_\mu D_{\mu\nu}(\mathbf{R} - \mathbf{R}') d_\nu = \sum_{\mu\nu} N d_\mu d_\nu \Big(\sum_{\mathbf{R}} D_{\mu\nu}(\mathbf{R})\Big). \tag{22.52}$$

The relation (22.51) is simply the condition that (22.52) vanish for arbitrary choices of the vector **d**.

Armed with these symmetries, we may proceed as follows:

We have $3N$ equations of motion (one for each of the three components of the displacements of the N ions):

$$M\ddot{u}_\mu(\mathbf{R}) = -\frac{\partial U^{\text{harm}}}{\partial u_\mu(\mathbf{R})} = -\sum_{\mathbf{R}'\nu} D_{\mu\nu}(\mathbf{R} - \mathbf{R}') u_\nu(\mathbf{R}'), \tag{22.53}$$

---

[25] That is, $\mathbf{r}(\mathbf{R}) \to -\mathbf{r}(-\mathbf{R})$.

or, in matrix notation,

$$M\ddot{\mathbf{u}}(\mathbf{R}) = -\sum_{\mathbf{R}'} \mathbf{D}(\mathbf{R} - \mathbf{R}')\mathbf{u}(\mathbf{R}'). \tag{22.54}$$

As in the one-dimensional cases we seek solutions to the equations of motion in the form of simple plane waves:

$$\mathbf{u}(\mathbf{R}, t) = \boldsymbol{\epsilon} e^{i(\mathbf{k}\cdot\mathbf{R} - \omega t)}. \tag{22.55}$$

Here $\boldsymbol{\epsilon}$ is a vector, to be determined, that describes the direction in which the ions move. It is known as the *polarization vector* of the normal mode.

We continue to use the Born-von Karman periodic boundary condition, requiring that $\mathbf{u}(\mathbf{R} + N_i\mathbf{a}_i) = \mathbf{u}(\mathbf{R})$ for each of the three primitive vectors $\mathbf{a}_i$, where the N_i are large integers satisfying $N = N_1N_2N_3$. This restricts the allowed wave vectors $\mathbf{k}$ to those of the form:[26]

$$\mathbf{k} = \frac{n_1}{N_1}\mathbf{b}_1 + \frac{n_2}{N_2}\mathbf{b}_2 + \frac{n_3}{N_3}\mathbf{b}_3, \qquad n_i \text{ integral}, \tag{22.56}$$

where the $\mathbf{b}_i$ are the reciprocal lattice vectors satisfying $\mathbf{b}_i \cdot \mathbf{a}_j = 2\pi\delta_{ij}$. As in our discussion of the one-dimensional case, only $\mathbf{k}$ within a single primitive cell of the reciprocal lattice will yield distinct solutions; i.e., if one adds a reciprocal lattice vector $\mathbf{K}$ to the $\mathbf{k}$ appearing in (22.55) the displacements of all ions are left completely unaltered, because of the basic property $e^{i\mathbf{K}\cdot\mathbf{R}} \equiv 1$, of the reciprocal lattice vectors. As a result there will be just N nonequivalent values of $\mathbf{k}$ of the form (22.56) which can be chosen to lie in any primitive cell of the reciprocal lattice. It is generally convenient to take that cell to be the first Brillouin zone.

If we substitute (22.55) into (22.54) we find a solution whenever $\boldsymbol{\epsilon}$ is an eigenvector of the three-dimensional eigenvalue problem:

$$M\omega^2\boldsymbol{\epsilon} = \mathbf{D}(\mathbf{k})\boldsymbol{\epsilon}. \tag{22.57}$$

Here $\mathbf{D}(\mathbf{k})$, known as the *dynamical matrix*, is given by

$$\mathbf{D}(\mathbf{k}) = \sum_{\mathbf{R}} \mathbf{D}(\mathbf{R})e^{-i\mathbf{k}\cdot\mathbf{R}}. \tag{22.58}$$

The three solutions to (22.57) for each of the N allowed values of $\mathbf{k}$, give us $3N$ normal modes. In discussing these solutions it is useful to translate the symmetries of $\mathbf{D}(\mathbf{R})$ into corresponding symmetries of $\mathbf{D}(\mathbf{k})$. It follows from (22.49) and (22.51) that $\mathbf{D}(\mathbf{k})$ can be written in the form:

$$\begin{aligned}\mathbf{D}(\mathbf{k}) &= \frac{1}{2}\sum_{\mathbf{R}} \mathbf{D}(\mathbf{R})[e^{-i\mathbf{k}\cdot\mathbf{R}} + e^{i\mathbf{k}\cdot\mathbf{R}} - 2] \\ &= \sum_{\mathbf{R}} \mathbf{D}(\mathbf{R})[\cos(\mathbf{k}\cdot\mathbf{R}) - 1] \\ &= -2\sum_{\mathbf{R}} \mathbf{D}(\mathbf{R})\sin^2(\tfrac{1}{2}\mathbf{k}\cdot\mathbf{R}).\end{aligned} \tag{22.59}$$

---

[26] Compare the discussion on page 136, where identical restrictions were imposed on the allowed wave vectors for an electronic wave function in a periodic potential.

Equation (22.59) explicitly demonstrates that $\mathbf{D}(\mathbf{k})$ is an even function of $\mathbf{k}$, and a real matrix. In addition, Eq. (22.50) implies that $\mathbf{D}(\mathbf{k})$ is a symmetric matrix. It is a theorem of matrix algebra that every real symmetric three-dimensional matrix has three real eigenvectors, $\boldsymbol{\epsilon}_1$, $\boldsymbol{\epsilon}_2$, $\boldsymbol{\epsilon}_3$, which satisfy

$$\mathbf{D}(\mathbf{k})\boldsymbol{\epsilon}_s(\mathbf{k}) = \lambda_s(\mathbf{k})\boldsymbol{\epsilon}_s(\mathbf{k}), \tag{22.60}$$

and can be normalized so that

$$\boldsymbol{\epsilon}_s(\mathbf{k}) \cdot \boldsymbol{\epsilon}_{s'}(\mathbf{k}) = \delta_{ss'}, \qquad s, s' = 1, 2, 3. \tag{22.61}$$

Evidently the three normal modes with wave vector $\mathbf{k}$ will have polarization vectors $\boldsymbol{\epsilon}_s(\mathbf{k})$ and frequencies $\omega_s(\mathbf{k})$, given by[27]

$$\omega_s(\mathbf{k}) = \sqrt{\frac{\lambda_s(\mathbf{k})}{M}}. \tag{22.62}$$

In the one-dimensional monatomic Bravais lattice we found that $\omega(\mathbf{k})$ vanished linearly with k at small k. In the three-dimensional monatomic Bravais lattice this remains the case for each of the three branches. This follows from Eq. (22.59), for when $\mathbf{k} \cdot \mathbf{R}$ is small for all $\mathbf{R}$ connecting sites whose ions have any appreciable interaction, then we can approximate the sine by[28]

$$\sin^2 (\tfrac{1}{2}\mathbf{k} \cdot \mathbf{R}) \approx (\tfrac{1}{2}\mathbf{k} \cdot \mathbf{R})^2 \tag{22.63}$$

and therefore

$$\mathbf{D}(\mathbf{k}) \approx -\frac{k^2}{2} \sum_{\mathbf{R}} (\hat{\mathbf{k}} \cdot \mathbf{R})^2 \mathbf{D}(\mathbf{R}), \qquad \hat{\mathbf{k}} = \frac{\mathbf{k}}{k}. \tag{22.64}$$

Consequently, in the long-wavelength (small k) limit we can write

$$\omega_s(\mathbf{k}) = c_s(\hat{\mathbf{k}})k, \tag{22.65}$$

where the $c_s(\hat{\mathbf{k}})$ are the square roots of the eigenvalues of the matrix

$$-\frac{1}{2M} \sum_{\mathbf{R}} (\hat{\mathbf{k}} \cdot \mathbf{R})^2 \mathbf{D}(\mathbf{R}). \tag{22.66}$$

Note that, in general, the c_s will depend on the direction $\hat{\mathbf{k}}$ of propagation of the wave as well as on the branch index s.

Typical dispersion curves for a three-dimensional monatomic Bravais lattice are shown in Figure 22.13.

---

[27] It can be shown that if $\mathbf{D}(\mathbf{k})$ has any negative eigenvalue, then there will be an ionic configuration for which U^{harm} is negative, contradicting the assumption that U^{eq} is the minimum energy. Consequently, the frequencies $\omega_s(\mathbf{k})$ are real. As in the one dimensional case, it suffices to take only the positive square root in (22.62).

[28] If the interaction does not decrease rapidly enough with distance, this procedure may be impermissible. A sufficient condition for its validity is that

$$\sum_{\mathbf{R}} R^2 \mathbf{D}(\mathbf{R})$$

converge, which is guaranteed provided that $\mathbf{D}(\mathbf{R})$ decreases faster than $\frac{1}{R^5}$ in three dimensions (cf. footnote 21).

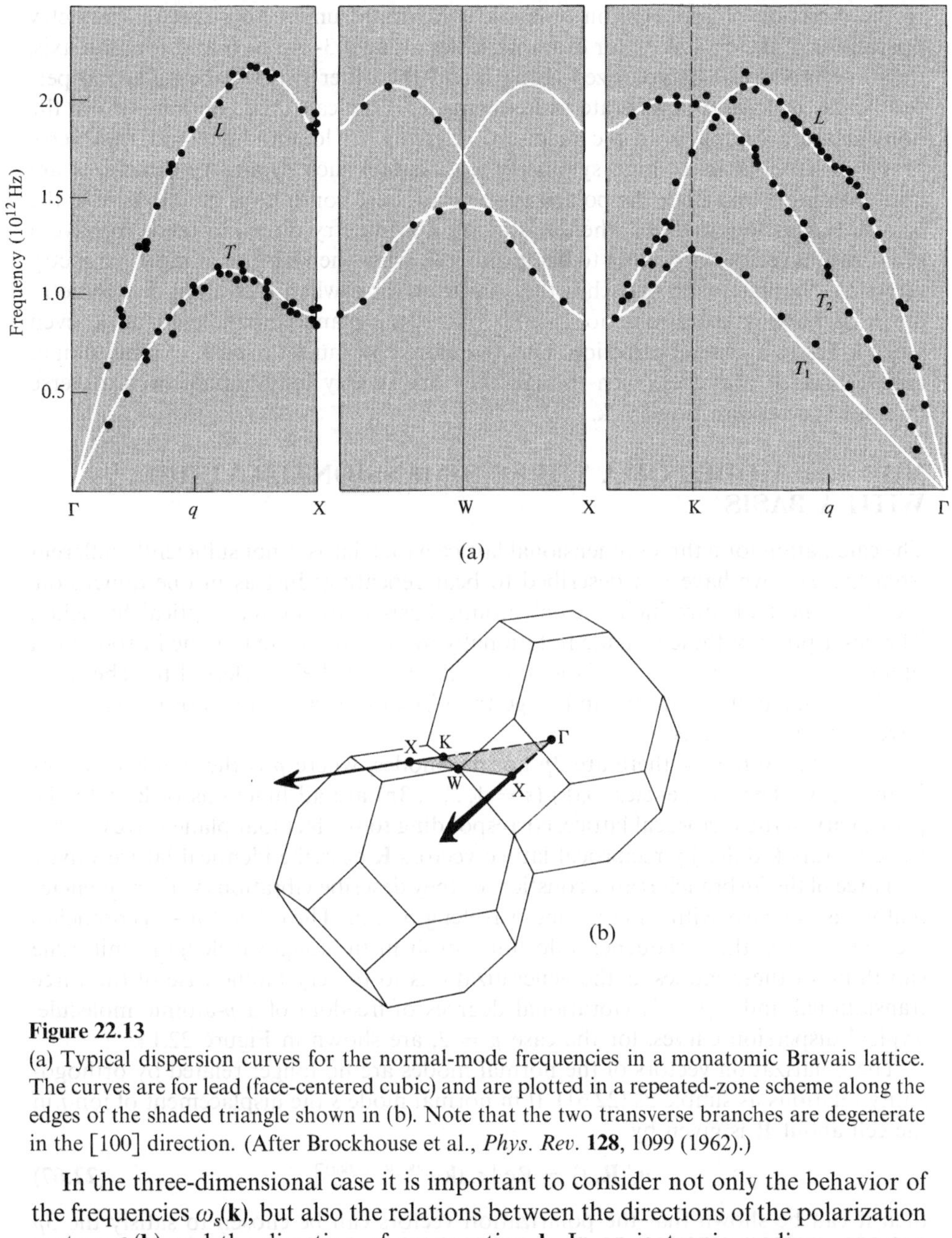

Figure 22.13
(a) Typical dispersion curves for the normal-mode frequencies in a monatomic Bravais lattice. The curves are for lead (face-centered cubic) and are plotted in a repeated-zone scheme along the edges of the shaded triangle shown in (b). Note that the two transverse branches are degenerate in the [100] direction. (After Brockhouse et al., *Phys. Rev.* **128**, 1099 (1962).)

In the three-dimensional case it is important to consider not only the behavior of the frequencies $\omega_s(\mathbf{k})$, but also the relations between the directions of the polarization vectors $\boldsymbol{\epsilon}_s(\mathbf{k})$ and the direction of propagation $\mathbf{k}$. In an isotropic medium, one can always choose the three solutions for a given $\mathbf{k}$ so that one branch (the longitudinal branch) is polarized along the direction of propagation ($\boldsymbol{\epsilon} \parallel \mathbf{k}$), and the other two (the transverse branches) are polarized perpendicular to the direction of propagation ($\boldsymbol{\epsilon} \perp \mathbf{k}$).

In an anisotropic crystal the polarization vectors need not be so simply related to the direction of propagation unless **k** is invariant under appropriate symmetry operations of the crystal. If, for example, **k** lies along a 3-, 4-, or 6-fold rotation axis, then one mode will be polarized along **k** and the other two will be polarized perpendicular to **k** (and degenerate in frequency).[29] One can then continue to use the nomenclature of the isotropic medium, referring to longitudinal and transverse branches. In crystals of high symmetry (e.g., cubic) such symmetry directions are quite common, and since the polarization vectors are continuous functions of **k**, the branch that is longitudinal when **k** is along a symmetry direction tends to have a polarization vector fairly close to being along **k**, even when **k** is not along a symmetry direction. Similarly, the branches that are transverse when **k** is along a symmetry direction have polarizations not too far out of the plane perpendicular to **k**, even when **k** lies in a general direction. One therefore continues to speak of longitudinal and transverse branches, even though they are strictly longitudinal or transverse only for special directions of **k**.

NORMAL MODES OF A THREE-DIMENSIONAL LATTICE WITH A BASIS

The calculation for a three-dimensional lattice with a basis is not sufficiently different from the case we have just described to bear repeating. Just as in one dimension, the chief effect of introducing a polyatomic basis is to produce optical branches. The description of these is made notationally more complicated by the introduction of an index to specify which of the ions in the basis is being referred to. The main results of the analysis are by and large the obvious extrapolations of the cases we have already considered:

For each value of **k** there are $3p$ normal modes, where p is the number of ions in the basis. The frequencies $\omega_s(\mathbf{k})$ $(s = 1, \ldots, 3p)$ are all functions of **k**, with the periodicity of the reciprocal lattice, corresponding to the fact that plane waves whose wave vectors **k** differ by reciprocal lattice vectors **K** describe identical lattice waves.

Three of the $3p$ branches are acoustic; i.e., they describe vibrations with frequencies that vanish linearly with k in the long-wavelength limit. The other $3(p - 1)$ branches are optical; i.e., their frequencies do not vanish in the long-wavelength limit. One can think of these modes as the generalizations to the crystalline case of the three translational and $3(p - 1)$ vibrational degrees of freedom of a p-atomic molecule. Typical dispersion curves, for the case $p = 2$, are shown in Figure 22.14.

The polarization vectors of the normal modes are no longer related by orthogonality relations as simple as (22.61). If in normal mode s the displacement of ion i in the cell about **R** is given by

$$\mathbf{u}_s{}^i(\mathbf{R}, t) = \text{Re}\left[\boldsymbol{\epsilon}_s{}^i(\mathbf{k})e^{i(\mathbf{k}\cdot\mathbf{R}-\omega_s(\mathbf{k})t)}\right], \tag{22.67}$$

then it can be shown that the polarization vectors can be chosen to satisfy the $3p$ generalized orthogonality relations:

$$\sum_{i=1}^{p} \boldsymbol{\epsilon}_s{}^{i*}(\mathbf{k}) \cdot \boldsymbol{\epsilon}_{s'}{}^{i}(\mathbf{k}) M_i = \delta_{ss'}, \tag{22.68}$$

---

[29] See Problem 4. Note, however, that the three polarization vectors are orthogonal for *general* directions of **k** (Eq. 22.61).

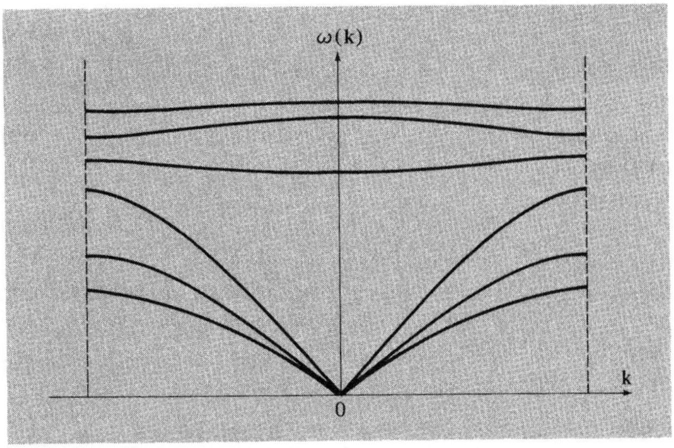

Figure 22.14
Typical dispersion curves along a general direction in k-space for a lattice with a two-ion basis. The three lower curves (acoustic branches) are linear in k for small k. The three upper curves (optical branches) will be quite flat if the intracellular interactions are much stronger than those between cells. Note that the direction of **k** is not one of high symmetry, since there is no degeneracy.

where M_i is the mass of the ith type of basis ion. In general the polarization vectors need not be real,[30] nor does the orthogonality relation (22.68) lend itself to a simple general geometrical interpretation.

CONNECTION WITH THE THEORY OF ELASTICITY

The classical theory of elasticity ignores the microscopic atomic structure of a solid, and treats it as a continuum. A general deformation of the solid is described in terms of a continuous displacement field $\mathbf{u}(\mathbf{r})$, specifying the vector displacement of the part of the solid that in equilibrium occupies position $\mathbf{r}$. The fundamental assumption of the theory is that the contribution to the energy density of the solid at point $\mathbf{r}$ depends only on the value of $\mathbf{u}(\mathbf{r})$ in the immediate vicinity of $\mathbf{r}$, or, more precisely, only on the first derivatives of $\mathbf{u}(\mathbf{r})$ at the point $\mathbf{r}$.

We can derive the continuum theory of elasticity from the theory of lattice vibrations, by considering only lattice deformations that vary slowly on a scale determined by the range of the interionic forces. We must also assume that one can specify the deformation of the basis ions within each primitive cell entirely in terms of the vector field $\mathbf{u}(\mathbf{r})$, specifying the displacement of the entire cell. For simplicity we restrict our discussion to monatomic Bravais lattices, where this assumption is trivially valid.

To derive the classical theory of elasticity we first note that the symmetries (22.49) and (22.51) permit us to write the harmonic potential energy (22.10) in the form:

$$U^{\text{harm}} = -\frac{1}{4}\sum_{\mathbf{RR}'} \{\mathbf{u}(\mathbf{R}') - \mathbf{u}(\mathbf{R})\}\mathsf{D}(\mathbf{R} - \mathbf{R}')\{\mathbf{u}(\mathbf{R}') - \mathbf{u}(\mathbf{R})\}. \tag{22.69}$$

We only consider displacements $\mathbf{u}(\mathbf{R})$ that have a very slight variation from cell to cell. We can then consider a smooth continuous function $\mathbf{u}(\mathbf{r})$, which is equal to $\mathbf{u}(\mathbf{R})$ when $\mathbf{r}$ is a Bravais lattice site. If $\mathbf{u}(\mathbf{r})$ varies little over the range of $\mathsf{D}(\mathbf{R} - \mathbf{R}')$, then to an excellent approximation (which becomes exact in the limit of very long-wavelength disturbances) we can make the replacement

$$\mathbf{u}(\mathbf{R}') = \mathbf{u}(\mathbf{R}) + (\mathbf{R}' - \mathbf{R}) \cdot \nabla \mathbf{u}(\mathbf{r})\big|_{\mathbf{r}=\mathbf{R}} \tag{22.70}$$

[30] That is, perpendicular components of the displacement in the normal mode will not be in phase, and the mode will have an elliptical polarization.

in (22.69), to find that

$$U^{\text{harm}} = \frac{1}{2} \sum_{\mathbf{R},\mu\nu\sigma\tau} \left(\frac{\partial}{\partial x_\sigma} u_\mu(\mathbf{R}) \right) \left(\frac{\partial}{\partial x_\tau} u_\nu(\mathbf{R}) \right) E_{\sigma\mu\tau\nu} . \tag{22.71}$$

The quantities $E_{\sigma\mu\tau\nu}$, which constitute a tensor of the fourth rank, are given in terms of **D** by[31]

$$E_{\sigma\mu\tau\nu} = -\frac{1}{2} \sum_{\mathbf{R}} R_\sigma D_{\mu\nu}(\mathbf{R}) R_\tau . \tag{22.72}$$

Since the $\mathbf{u}(\mathbf{r})$ are slowly varying, we can equally well write (22.71) as an integral,

$$U^{\text{harm}} = \frac{1}{2} \sum_{\substack{\sigma\tau \\ \mu\nu}} \int d\mathbf{r} \left(\frac{\partial}{\partial x_\sigma} u_\mu(\mathbf{r}) \right) \left(\frac{\partial}{\partial x_\tau} u_\nu(\mathbf{r}) \right) \bar{E}_{\sigma\mu\tau\nu} , \tag{22.73}$$

where

$$\bar{E}_{\sigma\mu\tau\nu} = \frac{1}{v} E_{\sigma\mu\tau\nu} , \tag{22.74}$$

and v is the volume of the primitive cell.

Equation (22.73) is the starting point in the analysis of classical elasticity theory. We pursue the subject further, to extract the symmetries of the tensor $E_{\sigma\mu\tau\nu}$ that the theory exploits.

First note that it follows directly from (22.72) and (22.50) that $E_{\sigma\mu\tau\nu}$ is unaltered by the interchange $(\mu \leftrightarrow \nu)$ or the interchange $(\sigma \leftrightarrow \tau)$. Thus it is enough to specify the value of $E_{\sigma\mu\tau\nu}$ for the six values:

$$xx, \quad yy, \quad zz, \quad yz, \quad zx, \quad xy \tag{22.75}$$

of the pair $\mu\nu$, and the same six values of the pair $\sigma\tau$. This indicates that $6 \times 6 = 36$ independent numbers are required to specify the energy for a given deformation. A further general argument reduces their number to 21; it can be reduced still further by exploiting the symmetry of the particular crystal at hand.

Further Reduction in the Number of Independent Elastic Constants

The energy of a crystal is unaffected by a rigid rotation. However, under a rotation through the infinitesimal angle $\delta\omega$ about an axis $\hat{\mathbf{n}}$ passing through the origin, each Bravais lattice vector will be shifted by

$$\mathbf{u}(\mathbf{R}) = \delta\boldsymbol{\omega} \times \mathbf{R}, \qquad \delta\boldsymbol{\omega} = \delta\omega\hat{\mathbf{n}}. \tag{22.76}$$

If we substitute (22.76) into (22.71), we must find that $U^{\text{harm}} = 0$ for arbitrary $\delta\boldsymbol{\omega}$. It is not difficult to show that this implies that U^{harm} can depend on the derivatives $(\partial/\partial x_\sigma)u_\mu$ only in the symmetrical combination (the strain tensor):

$$\epsilon_{\sigma\mu} = \frac{1}{2} \left(\frac{\partial}{\partial x_\sigma} u_\mu + \frac{\partial}{\partial x_\mu} u_\sigma \right). \tag{22.77}$$

---

[31] Evidently our theory will make sense only if **D**(**R**) vanishes rapidly enough at large R for the sum in (22.72) to converge. This will be satisfied trivially if **D**(**R**) vanishes for R greater than some R_0, and also for **D**(**R**) of infinitely long range, provided it vanishes faster than $1/R^5$.

Consequently, we can rewrite (22.73) as

$$U^{\text{harm}} = \frac{1}{2}\int d\mathbf{r}\left[\sum_{\substack{\sigma\mu\\ \tau\nu}} \mathcal{E}_{\sigma\mu}c_{\sigma\mu\tau\nu}\mathcal{E}_{\tau\nu}\right], \tag{22.78}$$

where

$$c_{\sigma\mu\tau\nu} = -\frac{1}{8v}\sum_{\mathbf{R}}\left[R_\sigma D_{\mu\nu}R_\tau + R_\mu D_{\sigma\nu}R_\tau + R_\sigma D_{\mu\tau}R_\nu + R_\mu D_{\sigma\tau}R_\nu\right]. \tag{22.79}$$

It is clear from (22.79) and the symmetry (22.50) of **D** that $c_{\sigma\mu\tau\nu}$ is invariant under the transposition $\sigma\mu \leftrightarrow \tau\nu$. In addition, it follows directly from (22.79) that $c_{\sigma\mu\tau\nu}$ is invariant under the transpositions $\sigma \leftrightarrow \mu$ or $\tau \leftrightarrow \nu$. As a result, the number of independent components of $c_{\sigma\mu\tau\nu}$ is reduced to 21.

Crystal Symmetries

Depending on the crystal system, one can further reduce the number of independent elastic constants.[32] The maximum number required for each of the seven crystal systems is shown in Table 22.1. For example, in the cubic case the only three independent components are

Table 22.1
NUMBER OF INDEPENDENT ELASTIC CONSTANTS

CRYSTAL SYSTEM	POINT GROUPS	ELASTIC CONSTANTS
Triclinic	all	21
Monoclinic	all	13
Orthorhombic	all	9
Tetragonal	C_4, C_{4h}, S_4	7
	$C_{4v}, D_4, D_{4h}, D_{2d}$	6
Rhombohedral	C_3, S_6	7
	C_{3v}, D_3, D_{3d}	6
Hexagonal	all	5
Cubic	all	3

$$\begin{aligned} c_{11} &= c_{xxxx} = c_{yyyy} = c_{zzzz}, \\ c_{12} &= c_{xxyy} = c_{yyzz} = c_{zzxx}, \\ c_{44} &= c_{xyxy} = c_{yzyz} = c_{zxzx}. \end{aligned}$$

All the other components (in which x, y, or z must appear as an index an odd number of times) vanish because the energy of a cubic crystal cannot change when the sign of a single component of the displacement field along any of the cubic axes is reversed.

Unfortunately, the language in which elasticity theory is conventionally couched fails to take full advantage of the simple tensor notation. In particular, the displacement field is usually described not by (22.77), but by the strain components

$$\begin{aligned} e_{\mu\nu} &= \mathcal{E}_{\mu\nu}, \qquad \mu = \nu \\ &= 2\mathcal{E}_{\mu\nu}, \qquad \mu \neq \nu, \end{aligned} \tag{22.80}$$

which are, in turn, simplified to e_α, $\alpha = 1, \ldots, 6$ according to the rule:

$$xx \to 1, \quad yy \to 2, \quad zz \to 3, \quad yz \to 4, \quad zx \to 5, \quad xy \to 6. \tag{22.81}$$

[32] See, for example, A. E. H. Love, *A Treatise on the Mathematical Theory of Elasticity*, Dover, New York, 1944, page 159.

Instead of Eq. (22.78), one writes:

$$U = \frac{1}{2}\sum_{\alpha\beta}\int d\mathbf{r}\; e_\alpha C_{\alpha\beta} e_\beta, \tag{22.82}$$

where the elements of the 6 × 6 matrix of C's are related to the components of the tensor $c_{\sigma\mu\tau\nu}$ by

$$\begin{gathered} C_{\alpha\beta} = c_{\sigma\mu\tau\nu}, \\ \text{where } \alpha \leftrightarrow \sigma\mu \\ \text{and } \beta \leftrightarrow \tau\nu, \\ \text{as specified in (22.81).} \end{gathered} \tag{22.83}$$

The quantities $C_{\alpha\beta}$ are called the elastic stiffness constants (or the elastic moduli). The elements of the 6 × 6 matrix S that is inverse to C are called the elastic compliance constants (or simply the elastic constants).

Given the potential-energy density (22.78), the macroscopic theory of elasticity goes on to construct a wave equation for $\mathbf{u}(\mathbf{r}, t)$. The neatest way to do this is to note that the kinetic energy associated with a given deformation field $\mathbf{u}(\mathbf{r})$ can be written in the form

$$T = \rho \int d\mathbf{r}\, \frac{1}{2}\dot{\mathbf{u}}(\mathbf{r}, t)^2, \tag{22.84}$$

where ρ is the mass density of the lattice: $\rho = MN/V$. We can then write a Lagrangian for the medium in the form:

$$\begin{aligned} L = T - U = \frac{1}{2}\int d\mathbf{r}\Bigg[&\rho\dot{\mathbf{u}}(\mathbf{r})^2 \\ &- \frac{1}{4}\sum_{\substack{\mu\nu \\ \sigma\tau}} c_{\sigma\mu\nu\tau}\left(\frac{\partial}{\partial x_\sigma}u_\mu(\mathbf{r}) + \frac{\partial}{\partial x_\mu}u_\sigma(\mathbf{r})\right)\left(\frac{\partial}{\partial x_\tau}u_\nu(\mathbf{r}) + \frac{\partial}{\partial x_\nu}u_\tau(\mathbf{r})\right)\Bigg]. \end{aligned} \tag{22.85}$$

Hamilton's principle,

$$\delta \int dt\, L = 0,$$

then implies the equations of motion:[33]

$$\rho\ddot{u}_\mu = \sum_{\sigma\nu\tau} c_{\mu\sigma\nu\tau}\frac{\partial^2 u_\tau}{\partial x_\sigma\,\partial x_\nu}. \tag{22.86}$$

If a solution is sought of the form

$$\mathbf{u}(\mathbf{r}, t) = \boldsymbol{\epsilon} e^{i(\mathbf{k}\cdot\mathbf{r} - \omega t)} \tag{22.87}$$

---

[33] These can also, of course, be derived in a more elementary and picturesque way by considering the forces acting on small volume elements. The superiority of the Lagrangian derivation is only evident when tensor notation is used.

then ω will have to be related to $\mathbf{k}$ through the eigenvalue equation

$$\rho\omega^2\epsilon_\mu = \sum_\tau \left(\sum_{\sigma\nu} c_{\mu\sigma\nu\tau} k_\sigma k_\nu \right) \epsilon_\tau . \qquad \textbf{(22.88)}$$

This has the same structure as (and through (22.79) can be shown to be identical to) the results (22.65) and (22.66) derived in the long wavelength limit of the general harmonic theory. Thus in the limit of long wavelengths the normal modes of the discrete crystal reduce to the sound waves of the elastic continuum. Conversely, by measuring the velocities of sound in the solid, one can extract information about the force constants via (22.88) and the microscopic definition of the $c_{\sigma\mu\nu\tau}$ in (22.79).

In Table 22.2 we list the elastic constants for some representative cubic solids.

Table 22.2
ELASTIC CONSTANTS FOR SOME CUBIC CRYSTALS[a]

SUBSTANCE	C_{11}	C_{12}	C_{44}	REFERENCE[b]
Li (78 K)	0.148	0.125	0.108	1
Na	0.070	0.061	0.045	2
Cu	1.68	1.21	0.75	3
Ag	1.24	0.93	0.46	3
Au	1.86	1.57	0.42	3
Al	1.07	0.61	0.28	4
Pb	0.46	0.39	0.144	5
Ge	1.29	0.48	0.67	1
Si	1.66	0.64	0.80	3
V	2.29	1.19	0.43	6
Ta	2.67	1.61	0.82	6
Nb	2.47	1.35	0.287	6
Fe	2.34	1.36	1.18	7
Ni	2.45	1.40	1.25	8
LiCl	0.494	0.228	0.246	9
NaCl	0.487	0.124	0.126	9
KF	0.656	0.146	0.125	9
RbCl	0.361	0.062	0.047	10

[a] Elastic constants in 10^{12} dynes-cm^{-2} at 300 K.
[b] References are as follows:

1. H. B. Huntington, *Solid State Phys.* **7**, 214 (1958).
2. P. Ho and A. L. Ruoff, *J. Phys. Chem. Solids* **29**, 2101 (1968).
3. J. deLaunay, *Solid State Phys.* **2**, 220 (1956).
4. P. Ho and A. L. Ruoff, *J. Appl. Phys.* **40**, 3 (1969).
5. P. Ho and A. L. Ruoff, *J. Appl. Phys.* **40**, 51 (1969).
6. D. I. Bolef, *J. Appl. Phys.* **32**, 100 (1961).
7. J. A. Rayne and B. S. Chandrasekhar, *Phys. Rev.* **122**, 1714 (1961).
8. G. A. Alers et al., *J. Phys. Chem. Solids* **13**, 40 (1960).
9. J. T. Lewis et al., *Phys. Rev.* **161**, 877 (1969).
10. M. Ghafelebashi et al., *J. Appl. Phys.* **41**, 652, 2268 (1970).

PROBLEMS

1. *Linear Chain with mth Nearest-Neighbor Interactions*

Reexamine the theory of the linear chain, without making the assumption that only nearest neighbors interact, replacing Eq. (22.22) by

$$U^{\text{harm}} = \sum_n \sum_{m>0} \frac{1}{2} K_m [u(na) - u([n+m]a)]^2. \tag{22.89}$$

(a) Show that the dispersion relation (22.29) must be generalized to

$$\omega = 2\sqrt{\sum_{m>0} K_m \frac{(\sin^2 \frac{1}{2}mka)}{M}}. \tag{22.90}$$

(b) Show that the long-wavelength limit of the dispersion relation, (22.31), must be generalized to:

$$\omega = a\left(\sum_{m>0} m^2 K_m/M\right)^{1/2} |k|, \tag{22.91}$$

provided that $\Sigma m^2 K_m$ converges.

(c) Show that if $K_m = 1/m^p$ $(1 < p < 3)$, so that the sum does not converge, then in the long-wavelength limit

$$\omega \propto k^{(p-1)/2}. \tag{22.92}$$

(*Hint:* It is no longer permissible to use the small-k expansion of the sine in (22.90), but one can replace the sum by an integral in the limit of small k.)

(d) Show that in the special case $p = 3$,

$$\omega \sim k\sqrt{|\ln k|}. \tag{22.93}$$

2. *Diatomic Linear Chain*

Consider a linear chain in which alternate ions have mass M_1 and M_2, and only nearest neighbors interact.

(a) Show that the dispersion relation for the normal modes is

$$\omega^2 = \frac{K}{M_1 M_2}\left(M_1 + M_2 \pm \sqrt{M_1{}^2 + M_2{}^2 + 2M_1 M_2 \cos ka}\right). \tag{22.94}$$

(b) Discuss the form of the dispersion relation and the nature of the normal modes when $M_1 \gg M_2$.

(c) Compare the dispersion relation with that of the monatomic linear chain (22.29) when $M_1 \approx M_2$.

3. *Lattice with a Basis Viewed as a Weakly Perturbed Monatomic Bravais Lattice*

It is instructive to examine the dispersion relation (22.37) for the one-dimensional lattice with a basis, in the limit in which the coupling constants K and G become very close:

$$K = K_0 + \Delta, \quad G = K_0 - \Delta, \qquad \Delta \ll K_0. \tag{22.95}$$

(a) Show that when $\Delta = 0$, the dispersion relation (22.37) reduces to that for a monatomic linear chain with nearest-neighbor coupling. (*Warning:* If the length of the unit cell in the diatomic chain is a, then when $K = G$ it will reduce to a monatomic chain with lattice constant $a/2$; furthermore, the Brillouin zone $(-\pi/a < k < \pi/a)$ for the diatomic chain will be only half the Brillouin zone $(-\pi/(a/2) < k < \pi/(a/2))$ of the monatomic chain. You must therefore explain how two branches (acoustic and optical) in half the zone reduce back to one branch in the full zone. To demonstrate the reduction convincingly you should examine the behavior of the amplitude ratio, Eq. (22.38), when $\Delta = 0$.)

(b) Show that when $\Delta \neq 0$, but $\Delta \ll K_0$, then the dispersion relation differs from that of the monatomic chain only by terms of order $(\Delta/K_0)^2$, except when $|\pi - ka|$ is of order Δ/K_0. Show that when this happens the distortion of the dispersion relation for the monatomic chain is linear in Δ/K_0.[34]

4. *Polarization of the Normal Modes of a Monatomic Bravais Lattice*

(a) Show that if $\mathbf{k}$ lies along a 3-, 4-, or 6-fold axis, then one normal mode is polarized along $\mathbf{k}$, and the other two are degenerate and polarized perpendicular to $\mathbf{k}$.

(b) Show that if $\mathbf{k}$ lies in a plane of mirror symmetry, then one normal mode has a polarization perpendicular to $\mathbf{k}$, and the other two have polarization vectors lying in the mirror plane.

(c) Show that if the point $\mathbf{k}$ lies in a Bragg plane that is parallel to a plane of mirror symmetry, then one normal mode is polarized perpendicular to the Bragg plane, while the other two have polarizations lying in the plane. (Note that in this case the modes cannot be strictly longitudinal and transverse unless $\mathbf{k}$ is perpendicular to the Bragg plane.)

To answer these questions, one must note that any operation that leaves both $\mathbf{k}$ and the crystal invariant must transform one normal mode with wave vector $\mathbf{k}$ into another. In particular, the set of three (orthogonal) polarization vectors must be invariant under such operations. In applying this fact one must remember that if two normal modes are degenerate, then any vector in the plane spanned by their polarization vectors is also a possible polarization vector.

5. *Normal Modes of a Three-Dimensional Crystal*

Consider a face-centered cubic monatomic Bravais lattice in which each ion interacts only with its (twelve) nearest neighbors. Assume that the interaction between a pair of neighboring ions is described by a pair potential ϕ that depends only on the distance r between the pair of ions.

(a) Show that the frequencies of the three normal modes with wave vector $\mathbf{k}$ are given by

$$\omega = \sqrt{\lambda/M} \tag{22.96}$$

where the λ are the eigenvalues of the 3×3 matrix:

$$\mathbf{D} = \sum_{\mathbf{R}} \sin^2(\tfrac{1}{2}\mathbf{k}\cdot\mathbf{R})[A\mathbf{1} + B\hat{\mathbf{R}}\hat{\mathbf{R}}]. \tag{22.97}$$

Here the sum is over the twelve nearest neighbors of $\mathbf{R} = 0$:

$$\frac{a}{2}(\pm\hat{\mathbf{x}} \pm \hat{\mathbf{y}}), \quad \frac{a}{2}(\pm\hat{\mathbf{y}} \pm \hat{\mathbf{z}}), \quad \frac{a}{2}(\pm\hat{\mathbf{z}} \pm \hat{\mathbf{x}}); \tag{22.98}$$

$\mathbf{1}$ is the unit matrix $((\mathbf{1})_{\mu\nu} = \delta_{\mu\nu})$, and $\hat{\mathbf{R}}\hat{\mathbf{R}}$ is the diadic formed from the unit vectors $\hat{\mathbf{R}} = \mathbf{R}/R$ (i.e., $(\hat{\mathbf{R}}\hat{\mathbf{R}})_{\mu\nu} = \hat{\mathbf{R}}_\mu\hat{\mathbf{R}}_\nu$). The constants A and B are: $A = 2\phi'(d)/d$, $B = 2[\phi''(d) - \phi'(d)/d]$, where

---

[34] Note the analogy to the nearly free electron model of Chapter 9: The free electron gas corresponds to the monatomic linear chain; the weak periodic potential corresponds to the small change in coupling between alternate nearest-neighbor pairs.

d is the equilibrium nearest-neighbor distance. (This follows from Equations (22.59) and (22.11).)

(b) Show that when $\mathbf{k}$ is in the (100) direction ($\mathbf{k} = (k, 0, 0)$ in rectangular coordinates), then one normal mode is strictly longitudinal, with frequency

$$\omega_L = \sqrt{\frac{8A + 4B}{M}} \sin \tfrac{1}{4}ka, \tag{22.99}$$

and the other two are strictly transverse and degenerate, with frequency

$$\omega_T = \sqrt{\frac{8A + 2B}{M}} \sin \tfrac{1}{4}ka. \tag{22.100}$$

(c) What are the frequencies and polarizations of the normal modes when $\mathbf{k}$ is along a [111] direction ($\mathbf{k} = (k, k, k)/\sqrt{3}$)?

(d) Show that when $\mathbf{k}$ is along the [110] direction ($\mathbf{k} = (k, k, 0)/\sqrt{2}$), then one mode is strictly longitudinal, with frequency

$$\omega_L = \sqrt{\frac{8A + 2B}{M} \sin^2\left(\frac{1}{4}\frac{ka}{\sqrt{2}}\right) + \frac{2A + 2B}{M} \sin^2\left(\frac{1}{2}\frac{ka}{\sqrt{2}}\right)}, \tag{22.101}$$

one is strictly transverse and polarized along the z-axis ($\boldsymbol{\epsilon} = (0, 0, 1)$), with frequency

$$\omega_T^{(1)} = \sqrt{\frac{8A + 4B}{M} \sin^2\left(\frac{1}{4}\frac{ka}{\sqrt{2}}\right) + \frac{2A}{M} \sin^2\left(\frac{1}{2}\frac{ka}{\sqrt{2}}\right)}, \tag{22.102}$$

and the third is strictly transverse and perpendicular to the z-axis, with frequency

$$\omega_T^{(2)} = \sqrt{\frac{8A + 2B}{M} \sin^2\left(\frac{1}{4}\frac{ka}{\sqrt{2}}\right) + \frac{2A}{M} \sin^2\left(\frac{1}{2}\frac{ka}{\sqrt{2}}\right)}. \tag{22.103}$$

(e) Sketch the dispersion curves along the lines ΓX and ΓKX, (Figure 22.13), assuming that $A = 0$. (*Note:* The length of ΓX is $2\pi/a$.)

23

Quantum Theory of the Harmonic Crystal

Normal Modes and Phonons

High-Temperature Specific Heat

Low-Temperature Specific Heat

Models of Debye and Einstein

Comparison of Lattice and Electronic Specific Heats

Density of Normal Modes (Phonon Level Density)

Analogy with Theory of Blackbody Radiation

In Chapter 22 we found that the contribution of the lattice vibrations to the specific heat of a classical harmonic crystal was independent of the temperature (the law of Dulong and Petit). However, as the temperature drops below room temperature, the specific heat of all solids starts to decline below the classical value, and eventually is observed to vanish as T^3 (in insulators) or $AT + BT^3$ (in metals). The explanation of this behavior was one of the earliest triumphs of the quantum theory of solids.

In a quantum theory of the specific heat of a harmonic crystal, the classical expression (22.12) for the thermal energy density u must be replaced by the general quantum-mechanical result

$$u = \frac{1}{V}\sum_i E_i e^{-\beta E_i} \Big/ \sum_i e^{-\beta E_i}, \qquad \beta = 1/k_B T, \tag{23.1}$$

where E_i is the energy of the ith stationary state of the crystal, and the sum is over all stationary states.

The energies of these stationary states are given by the eigenvalues of the harmonic Hamiltonian:[1]

$$H^{\text{harm}} = \sum_{\mathbf{R}} \frac{1}{2M} P(\mathbf{R})^2 + \frac{1}{2}\sum_{\mathbf{RR'}} u_\mu(\mathbf{R}) D_{\mu\nu}(\mathbf{R} - \mathbf{R'}) u_\nu(\mathbf{R'}). \tag{23.2}$$

The detailed procedure by which these eigenvalues are extracted is summarized in Appendix L. The result of this calculation is so simple and intuitively plausible that we simply state it here, unencumbered by its straightforward, but rather lengthy, derivation.

To specify the energy levels of an N-ion harmonic crystal, one regards it as $3N$ independent oscillators, whose frequencies are those of the $3N$ classical normal modes described in Chapter 22. The contribution to the total energy of a particular normal mode with angular frequency $\omega_s(\mathbf{k})$ can have only the discrete set of values

$$(n_{\mathbf{k}s} + \tfrac{1}{2})\hbar\omega_s(\mathbf{k}), \tag{23.3}$$

where $n_{\mathbf{k}s}$, the excitation number of the normal mode, is restricted to the values 0, 1, 2, ... A state of the entire crystal is specified by giving the excitation numbers for each of the $3N$ normal modes. The total energy is just the sum of the energies of the individual normal modes:

$$E = \sum_{\mathbf{k}s} (n_{\mathbf{k}s} + \tfrac{1}{2})\hbar\omega_s(\mathbf{k}). \tag{23.4}$$

The thermal energy (23.1) can be evaluated from (23.4) straightforwardly. Before doing so, however, we digress to describe the language in which the excited states of the harmonic crystal are customarily discussed.

---

[1] See Eqs. (22.8) and (22.10). We display only the form appropriate to a monatomic Bravais lattice, but the discussion that follows is quite general. We have added the kinetic energy, which no longer disappears from the problem at an early stage, (as it does in classical statistical mechanics). We drop, for the moment, the additive constant U^{eq}. This has the effect of subtracting U^{eq}/V from the energy density (23.1). Since U^{eq} does not depend on temperature, this has no effect on the specific heat. Should we require, however, the volume dependence of the internal energy, it would be necessary to retain U^{eq}.

NORMAL MODES vs. PHONONS

We have described the result (23.4) in terms of the excitation number $n_{\mathbf{k}s}$ of the normal mode with wave vector **k** in branch s. This nomenclature can be quite clumsy, especially when describing processes in which energy is exchanged among the normal modes, or between the normal modes and other systems, such as electrons, incident neutrons, or incident X rays. Usually the language of normal modes is replaced by an equivalent corpuscular description, which is analogous to the terminology used in the quantum theory of the electromagnetic field. In that theory the allowed energies of a normal mode of the radiation field in a cavity are given by $(n + \frac{1}{2})\hbar\omega$, where ω is the angular frequency of the mode. It is the universal practice, however, to speak not of the quantum number of excitation of the mode, n, but of the number, n, of *photons* of that type that are present. In precisely the same way, instead of saying that the normal mode of branch s with wave vector **k** is in its $n_{\mathbf{k}s}$th excited state, one says that there are $n_{\mathbf{k}s}$ *phonons* of type s with wave vector **k** present in the crystal.

The term "phonon" stresses this analogy with photons. The latter are the quanta of the radiation field that (in the appropriate frequency range) describes classical light; the former are the quanta of the ionic displacement field that (in the appropriate frequency range) describes classical sound. Although the language of phonons is more convenient than that of normal modes, the two nomenclatures are completely equivalent.

GENERAL FORM OF THE LATTICE SPECIFIC HEAT

To calculate the contribution of the lattice vibrations to the internal energy, we substitute the explicit form (23.4) of the energy levels into the general formula (23.1). To expedite the calculation we introduce the quantity

$$f = \frac{1}{V}\ln\left(\sum_i e^{-\beta E_i}\right). \tag{23.5}$$

The identity

$$u = -\frac{\partial f}{\partial \beta} \tag{23.6}$$

can be verified by explicit differentiation of (23.5). To evaluate f, note that $e^{-\beta E}$ occurs exactly once for every energy E of the form (23.4) in the expansion of the product

$$\prod_{\mathbf{k}s}\left(e^{-\beta\hbar\omega_s(\mathbf{k})/2} + e^{-3\beta\hbar\omega_s(\mathbf{k})/2} + e^{-5\beta\hbar\omega_s(\mathbf{k})/2} + \cdots\right). \tag{23.7}$$

The individual terms in this product are just convergent geometric series, that can be explicitly summed, to give

$$f = \frac{1}{V}\ln\prod_{\mathbf{k}s}\frac{e^{-\beta\hbar\omega_s(\mathbf{k})/2}}{1 - e^{-\beta\hbar\omega_s(\mathbf{k})}}. \tag{23.8}$$

Differentiating f, as required by (23.6), we find that the internal energy density is given by

$$\frac{1}{V}\sum_{\mathbf{k}s} \hbar\omega_s(\mathbf{k})[n_s(\mathbf{k}) + \tfrac{1}{2}], \tag{23.9}$$

where

$$n_s(\mathbf{k}) = \frac{1}{e^{\beta\hbar\omega_s(\mathbf{k})} - 1}. \tag{23.10}$$

Comparing Eq. (23.9), for the mean thermal energy density of the crystal at temperature T, with Eq. (23.4), for the energy in a particular stationary state, leads one to conclude that $n_s(\mathbf{k})$ is simply the mean excitation number of the normal mode $\mathbf{k}s$ at temperature T. In the phonon language, $n_s(\mathbf{k})$ is the mean number of phonons of type $\mathbf{k}s$ present in thermal equilibrium[2] at temperature T.

Thus the simple classical expression for the energy density of a harmonic crystal at temperature T, Eq. (22.18), must be generalized to[3]

$$u = u^{\text{eq}} + \frac{1}{V}\sum_{\mathbf{k}s} \tfrac{1}{2}\hbar\omega_s(\mathbf{k}) + \frac{1}{V}\sum_{\mathbf{k}s} \frac{\hbar\omega_s(\mathbf{k})}{e^{\beta\hbar\omega_s(\mathbf{k})} - 1}. \tag{23.11}$$

As $T \to 0$ the third term vanishes, but, in contrast with the classical result (22.18), there remains not only the energy u^{eq} of the equilibrium configuration, but also a second term, giving the energy of the zero-point vibrations of the normal modes. All the temperature dependence of u (and hence the entire contribution to the specific heat) comes from the third term, whose variation with temperature is far more complex than the simple linear form of the classical result. In the quantum theory of the harmonic solid the specific heat is no longer constant, but is given by

$$c_v = \frac{1}{V}\sum_{\mathbf{k}s} \frac{\partial}{\partial T} \frac{\hbar\omega_s(\mathbf{k})}{e^{\beta\hbar\omega_s(\mathbf{k})} - 1}, \tag{23.12}$$

which now depends in a detailed way on the frequency spectrum of the normal modes.

Certain general features of the specific heat (23.12) emerge in limiting cases, which we now examine.

HIGH-TEMPERATURE SPECIFIC HEAT

When $k_B T/\hbar$ is large compared with all the phonon frequencies (i.e., when every normal mode is in a highly excited state), then the argument of the exponential will be small in each term of (23.12) and we can expand:

---

[2] Those familiar with the ideal Bose gas will recognize Eq. (23.10) as a special case of the Bose-Einstein distribution function, giving the number of bosons with energy $\hbar\omega_s(\mathbf{k})$ in thermal equilibrium at temperature T, when the chemical potential μ is taken to be zero. The lack of freedom in selecting μ comes from the fact that the total number of bosons in thermal equilibrium is not an independent variable at our disposal in the case of phonons (as it is, for example, in the case of ^{4}He atoms) but is entirely determined by the temperature.

[3] For comparison with Eq. (22.18) we reintroduce the constant giving the potential energy of the static equilibrium distribution.

$$\frac{1}{e^x - 1} = \frac{1}{x + \frac{1}{2}x^2 + \frac{1}{6}x^3 + \cdots} = \frac{1}{x}\left[1 - \frac{x}{2} + \frac{x^2}{12} + O(x^3)\right],$$
$$x = \frac{\hbar\omega}{k_B T} \ll 1. \tag{23.13}$$

If we keep only the leading term in this expansion, then the summand in (23.12) reduces to the constant $k_B T$, and the specific heat reduces to k_B times the density of normal modes, $3N/V$. This is just the classical law of Dulong and Petit (22.19).

Additional terms in the expansion (23.13) yield the high-temperature quantum corrections to the Dulong and Petit law. The linear term in x (in the square brackets) gives a temperature-independent term in the thermal energy (which is precisely equal to minus the zero-point energy) and therefore does not affect the specific heat. The leading correction is thus given by the term in square brackets quadratic in x. When this is substituted into (23.12) it gives a correction to the Dulong and Petit specific heat c_v^0 of the form:

$$c_v = c_v^0 + \Delta c_v, \qquad \frac{\Delta c_v}{c_v^0} = -\frac{\hbar^2}{12(k_B T)^2}\frac{1}{3N}\sum \omega_s(\mathbf{k})^2. \tag{23.14}$$

At temperatures high enough for this expansion to be valid, the anharmonic corrections to the classical specific heat, which are not included in the Dulong and Petit value,[4] are likely to be of significance, and will tend to mask the quantum correction (23.14).[5]

LOW-TEMPERATURE SPECIFIC HEAT

To discuss the specific heat more generally, we first observe that in the limit of a large crystal the set of discrete wave vectors summed over in Eq. (23.12) becomes dense on the scale over which the summand has appreciable variation. We can therefore replace the sum by an integral according to the general prescription (2.29) for any set of wave vectors satisfying the Born-von Karman boundary conditions, rewriting (23.12) as

$$c_v = \frac{\partial}{\partial T}\sum_s \int \frac{d\mathbf{k}}{(2\pi)^3} \frac{\hbar\omega_s(\mathbf{k})}{e^{\hbar\omega_s(\mathbf{k})/k_B T} - 1}, \tag{23.15}$$

where the integral may be taken over the first Brillouin zone.

At very low temperatures, modes with $\hbar\omega_s(\mathbf{k}) \gg k_B T$ will contribute negligibly to (23.15), since the integrand will vanish exponentially. However, because $\omega_s(\mathbf{k}) \to 0$ as $k \to 0$ in the three acoustic branches, this condition will fail to be satisfied by acoustic modes of sufficiently long-wavelength, no matter how low the temperature. These modes (and only these) will continue to contribute appreciably to the specific heat. Bearing this in mind, we can make the following simplifications in (23.15), all of which result in a vanishingly small fractional error, in the zero-temperature limit:

---

[4] See point 2, page 428, and the discussion following it.

[5] Indeed, at such high temperatures real crystals are likely to have melted—a rather extreme form of anharmonic behavior.

1. Even if the crystal has a polyatomic basis, we may ignore the optical modes in the sum over s, since their frequencies are bounded below.[6]
2. We may replace the dispersion relation $\omega = \omega_s(\mathbf{k})$ for the three acoustic branches by its long-wavelength form (22.65) $\omega = c_s(\hat{\mathbf{k}})k$. This will be valid provided that $k_B T/\hbar$ is substantially less than those frequencies at which the acoustic dispersion curves begin to differ appreciably from their long-wavelength linear forms.
3. We may replace the k-space integration over the first Brillouin zone by an integration over all of k-space. This is because the integrand is negligibly small unless $\hbar c_s(\hat{\mathbf{k}})k$ is of order $k_B T$, which happens only in the immediate vicinity of $\mathbf{k} = \mathbf{0}$ at low temperatures.

These three simplifications are illustrated in Figure 23.1.

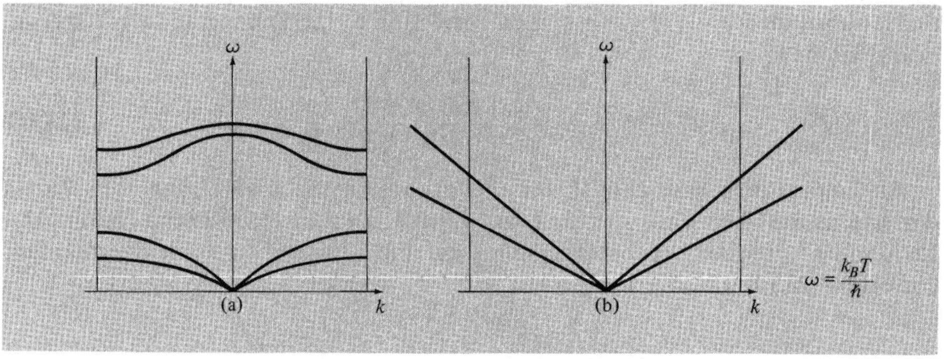

Figure 23.1
The simplifications that can be made in evaluating the low-temperature specific heat of a harmonic crystal. (a) Typical normal-mode dispersion relations for a diatomic crystal along a particular direction in k-space (taken to be a direction of high enough symmetry for two of the acoustic and two of the optical branches to be degenerate). (b) The spectrum that replaces (a) in evaluating the integral (23.15). The acoustic branches are replaced by linear branches, extending over all k (i.e., the integral is extended from the first zone to all of k-space) and the optical branches are ignored. This is justified because frequencies large compared with $k_B T/\hbar$ (those parts of the dispersion curves in (a) and (b) above the horizontal dashed line) make negligible contributions to (23.15), and because the parts of the dispersion curves describing modes that do contribute (the parts below the horizontal dashed line) are identical in (a) and in (b).

Thus at very low temperatures (23.15) may be simplified to

$$c_v = \frac{\partial}{\partial T} \sum_s \int \frac{d\mathbf{k}}{(2\pi)^3} \frac{\hbar c_s(\hat{\mathbf{k}})k}{e^{\hbar c_s(\hat{\mathbf{k}})k/k_B T} - 1}, \tag{23.16}$$

where the integral is over all of k-space. We evaluate the integral in spherical coor-

---

[6] Under certain special conditions (usually associated with an imminent change in crystal structure) an optical branch may dip down almost to zero frequency (acquiring what is known as a "soft mode"). When this happens, there will be an additional contribution to the low-temperature specific heat from that optical branch.

dinates, writing $d\mathbf{k} = k^2 dk\, d\Omega$. If we make the change of variables $\beta\hbar c_s(\hat{\mathbf{k}})k = x$ in the k integration, then (23.16) becomes

$$c_v = \frac{\partial}{\partial T} \frac{(k_B T)^4}{(\hbar c)^3} \frac{3}{2\pi^2} \int_0^\infty \frac{x^3\, dx}{e^x - 1}, \tag{23.17}$$

where $1/c^3$ is the average of the inverse third power of the long-wavelength phase velocities of the three acoustic modes:

$$\frac{1}{c^3} = \frac{1}{3} \sum_s \int \frac{d\Omega}{4\pi} \frac{1}{c_s(\hat{\mathbf{k}})^3}. \tag{23.18}$$

The definite integral in (23.17) can be evaluated by writing[7]

$$\int_0^\infty \frac{x^3\, dx}{e^x - 1} = \sum_{n=1}^\infty \int_0^\infty x^3 e^{-nx}\, dx = 6 \sum_{n=1}^\infty \frac{1}{n^4} = \frac{\pi^4}{15}. \tag{23.19}$$

Therefore, at very low temperatures[8]

$$c_v \approx \frac{\partial}{\partial T} \frac{\pi^2}{10} \frac{(k_B T)^4}{(\hbar c)^3} = \frac{2\pi^2}{5} k_B \left(\frac{k_B T}{\hbar c}\right)^3. \tag{23.20}$$

This relation can be checked by comparing the measured low-temperature specific heat with the measured elastic constants, which are directly related to the phase velocities appearing in the definition (23.18) of c. In the alkali halides, for example, the disagreement has been found to be less than the experimental errors in the measurements (typically about a percent).[9]

Since (23.20) remains valid only as long as $k_B T/\hbar$ is small compared with all phonon frequencies that are not in the linear part of the spectrum, we might expect that this would require $k_B T/\hbar$ to be a small fraction of the frequencies near the zone edge. This requires T to be well below room temperature. Since the Dulong and Petit law starts to fail as the temperature drops below room temperature, there is a considerable temperature range over which neither the high- nor low-temperature evaluations of the specific heat are valid, and one must work with the general form (23.15). However, it is quite common to use an approximate interpolation scheme for this intermediate temperature range.

INTERMEDIATE-TEMPERATURE SPECIFIC HEAT: THE MODELS OF DEBYE AND EINSTEIN

The earliest quantum theories of lattice specific heats, due to Einstein and Debye, did not use phonon spectra of the general form we have been considering, but assumed

---

7 See also Appendix C, Eqs. (C. 11) to (C. 13).

8 We reemphasize that this result becomes asymptotically exact (within the harmonic approximation) as $T \to 0$; i.e., it can be written as an equality:

$$\lim_{T\to 0} \frac{c_v}{T^3} = \frac{2\pi^2}{5} \frac{k_B{}^4}{\hbar^3 c^3}.$$

9 J. T. Lewis et al., *Phys. Rev.* **161**, 877 (1967).

normal-mode dispersion relations of an especially simple structure. Their results, based on crude approximations to the normal-mode dispersion relations, are still of use as interpolation formulas. Debye's theory has had a considerable impact on the nomenclature of the subject, and even on the way in which data are presented.

The Debye Interpolation Scheme

The Debye model replaces all branches of the vibrational spectrum with three branches, each with the same linear dispersion relation:[10]

$$\omega = ck. \tag{23.21}$$

In addition, the integral in (23.15) over the first Brillouin zone is replaced by an integral over a sphere of radius k_D, chosen to contain precisely N allowed wave vectors, where N is the number of ions in the crystal. Since the volume of k-space per wave vector is $(2\pi)^3/V$ (see page 37), this requires $(2\pi)^3 N/V$ to equal $4\pi k_D{}^3/3$, so that k_D is determined by the relation[11]

$$\boxed{n = \frac{k_D{}^3}{6\pi^2}.} \tag{23.22}$$

As a result of these simplifications Eq. (23.15) reduces to

$$c_v = \frac{\partial}{\partial T}\frac{3\hbar c}{2\pi^2}\int_0^{k_D}\frac{k^3\,dk}{e^{\beta\hbar ck}-1}. \tag{23.23}$$

In evaluating the integral (23.23) it is convenient to define a Debye frequency by

$$\boxed{\omega_D = k_D c} \tag{23.24}$$

and a Debye temperature by

$$\boxed{k_B\Theta_D = \hbar\omega_D = \hbar ck_D.} \tag{23.25}$$

Evidently k_D is a measure of the inverse interparticle spacing, ω_D is a measure of the maximum phonon frequency, and Θ_D is a measure of the temperature above which all modes begin to be excited, and below which modes begin to be "frozen out."[12]

---

[10] In the case of a lattice with a polyatomic basis, the representation of the $3p$ branches of the phonon spectrum by only three is compensated by the volume of the Debye sphere being p times the volume of the first Brillouin zone. This point is elaborated in the discussion of the Einstein model.

[11] When, in applications to metals, there is a danger of confusing the density of ions with the density of conduction electrons, we shall denote the former by n_i and the latter by n_e. The two are related by $n_e = Zn_i$, where Z is the nominal valence. Since the free electron Fermi wave vector k_F satisfies $k_F{}^3/3\pi^2 = n_e$, k_D is related to k_F in a metal by $k_D = (2/Z)^{1/3}k_F$.

[12] One can also regard Θ_D and ω_D as measures of the "stiffness" of the crystal.

If we make the change of variables $\hbar ck/k_BT = x$, then (23.23) can be written in terms of the Debye temperature:

$$c_v = 9nk_B\left(\frac{T}{\Theta_D}\right)^3 \int_0^{\Theta_D/T} \frac{x^4 e^x\,dx}{(e^x - 1)^2}. \qquad \textbf{(23.26)}$$

This formula expresses the specific heat at all temperatures in terms of a single empirical parameter, Θ_D. One reasonable way to pick Θ_D (though it is by no means the only way used) is by making (23.26) agree with the observed specific heat at low temperatures. This will be assured (at least in the harmonic approximation) if the speed c in (23.21) or (23.25) is related to the exact phonon spectrum through (23.18). The resulting form for the low-temperature specific heat is[13]

$$c_v = \frac{12\pi^4}{5}nk_B\left(\frac{T}{\Theta_D}\right)^3 = 234\left(\frac{T}{\Theta_D}\right)^3 nk_B. \qquad \textbf{(23.27)}$$

Values of Θ_D for some of the alkali halides, determined by fitting the T^3 term in their low-temperature specific heats, are quoted in Table 23.1.

Table 23.1
DEBYE TEMPERATURES FOR THE ALKALI HALIDE CRYSTALS[a]

	F	Cl	Br	I
Li	730	422	—	—
Na	492	321	224	164
K	336	231	173	131
Rb	—	165	131	103

[a] Given in degrees kelvin. All values were obtained by comparing the constant in the T^3 fit to the low-temperature specific heat to Eq. (23.27), except for NaF, KF, and NaBr, where Θ_D was deduced from the measured elastic constants through (23.18), and (23.25). (In cases where values were obtained by both methods, they agree to within a percent or two, which is about the size of the experimental uncertainty in the numbers.)
Source: J. T. Lewis et al., *Phys. Rev.* **161**, 877 (1967).

Unfortunately Θ_D is not always chosen by this convention. Partly because the Debye result (23.26) was viewed by some as far more general than a rough interpolation formula, the practice arose of fitting observed heat capacities with (23.26) by allowing Θ_D to depend on temperature. There is no good reason for doing this,

[13] This can be derived directly from (23.26) by noting that for $T < \Theta_D$, the upper limit of the integral can be extended to infinity with exponentially small error. It is also equivalent to the exact result (23.20), provided c is eliminated in favor of Θ_D and the ionic density via (23.22) and (23.25).

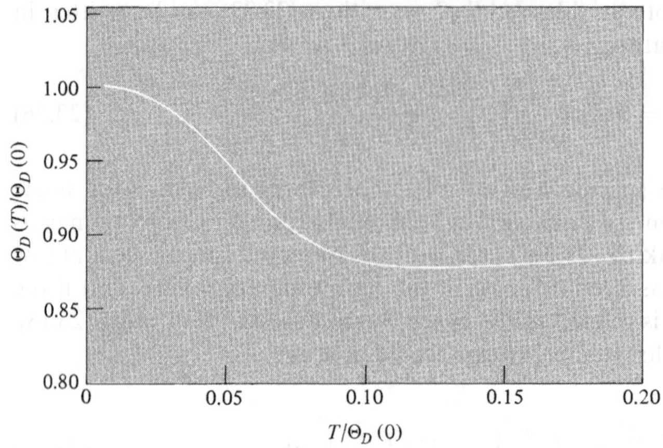

Figure 23.2
Debye temperature as a function of temperature for argon and krypton. This is a widely used way of presenting specific heat data. (L. Finegold and N. Phillips, *Phys. Rev.* **177**, 1383 (1969).)

but the practice has persisted down to the present day—to the extent that results of heat-capacity measurements are sometimes reported in terms of $\Theta_D(T)$ rather than the data themselves.[14] It is helpful, in converting such information back to specific heats, to have a graph of the Debye c_v as a function of Θ_D/T. This is shown in Figure 23.3, and some numerical values of the function are given in Table 23.2. In Table 23.3 we give some Debye temperatures for selected elements that were determined by fitting the observed heat capacity with the Debye formula (23.26) at the point where the heat capacity is about half the Dulong and Petit value.

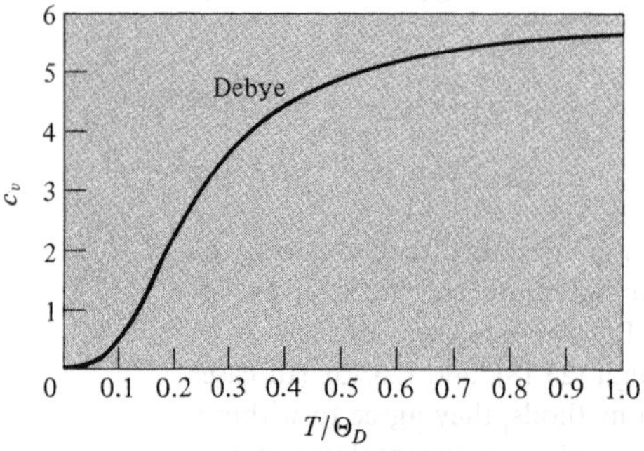

Figure 23.3
Specific heat in the Debye approximation (in cal/mole-K) vs. T/Θ_D. (From J. de Launay, *op. cit.*; see Table 23.2.)

Note that at temperatures well above Θ_D we can replace the integrand in (23.26) by its form for small x, and the Dulong and Petit result emerges. (This is to be expected, since it was built into the formula through the definition of k_D.) Thus the Debye temperature plays the same role in the theory of lattice vibrations as the Fermi temperature plays in the theory of electrons in metals: Both are a measure of the temperature separating the low-temperature region where quantum statistics must be used from the high-temperature region where classical statistical mechanics is valid. However, in the electronic case actual temperatures are always well below T_F, whereas Θ_D (see Table 23.3) is typically of the order of 10^2 K, so both classical and quantum regimes can be encountered.

[14] See, for example, Figure 23.2.

Table 23.2
TEMPERATURE DEPENDENCE OF THE DEBYE SPECIFIC HEAT[a]

T/Θ_D	$c_v/3nk_B$	T/Θ_D	$c_v/3nk_B$	T/Θ_D	$c_v/3nk_B$
0.00	0	0.35	0.687	0.70	0.905
0.05	0.00974	0.40	0.746	0.75	0.917
0.10	0.0758	0.45	0.791	0.80	0.926
0.15	0.213	0.50	0.825	0.85	0.934
0.20	0.369	0.55	0.852	0.90	0.941
0.25	0.503	0.60	0.874	0.95	0.947
0.30	0.608	0.65	0.891	1.00	0.952

[a] The table entries are the ratios of the Debye to the Dulong-Petit specific heats, that is, $c_v/3nk_B$, with c_v given by (23.26).
Source: J. de Launay, *Solid State Physics*, vol. 2, F. Seitz and D. Turnbull, eds., Academic Press, New York, 1956.

Table 23.3
DEBYE TEMPERATURES FOR SELECTED ELEMENTS[a]

ELEMENT	Θ_D (K)	ELEMENT	Θ_D (K)
Li	400	A	85
Na	150	Ne	63
K	100		
		Cu	315
Be	1000	Ag	215
Mg	318	Au	170
Ca	230		
		Zn	234
B	1250	Cd	120
Al	394	Hg	100
Ga	240		
In	129	Cr	460
Tl	96	Mo	380
		W	310
C (diamond)	1860	Mn	400
Si	625	Fe	420
Ge	360	Co	385
Sn (grey)	260	Ni	375
Sn (white)	170	Pd	275
Pb	88	Pt	230
As	285	La	132
Sb	200	Gd	152
Bi	120	Pr	74

[a] The temperatures were determined by fitting the observed specific heats c_v to the Debye formula (23.26) at the point where $c_v = 3nk_B/2$.
Source: J. de Launay, *Solid State Physics*, vol. 2, F. Seitz and D. Turnbull, eds., Academic Press, New York, 1956.

The Einstein Model

In the Debye model of a crystal with a polyatomic basis, the optical branches of the spectrum are represented by the high k values of the same linear expression (23.21) whose low k values give the acoustic branch (Figure 23.4a). An alternative scheme is to apply the Debye model only to the three acoustic branches of the spectrum. The optical branches are represented by the "Einstein approximation," which replaces the frequency of each optical branch by a frequency ω_E that does not depend on $\mathbf{k}$ (see Figure 23.4b). The density n in (23.22), (23.26), and (23.27) must then be taken as the number of primitive cells per unit volume of crystal, and (23.26) will give only

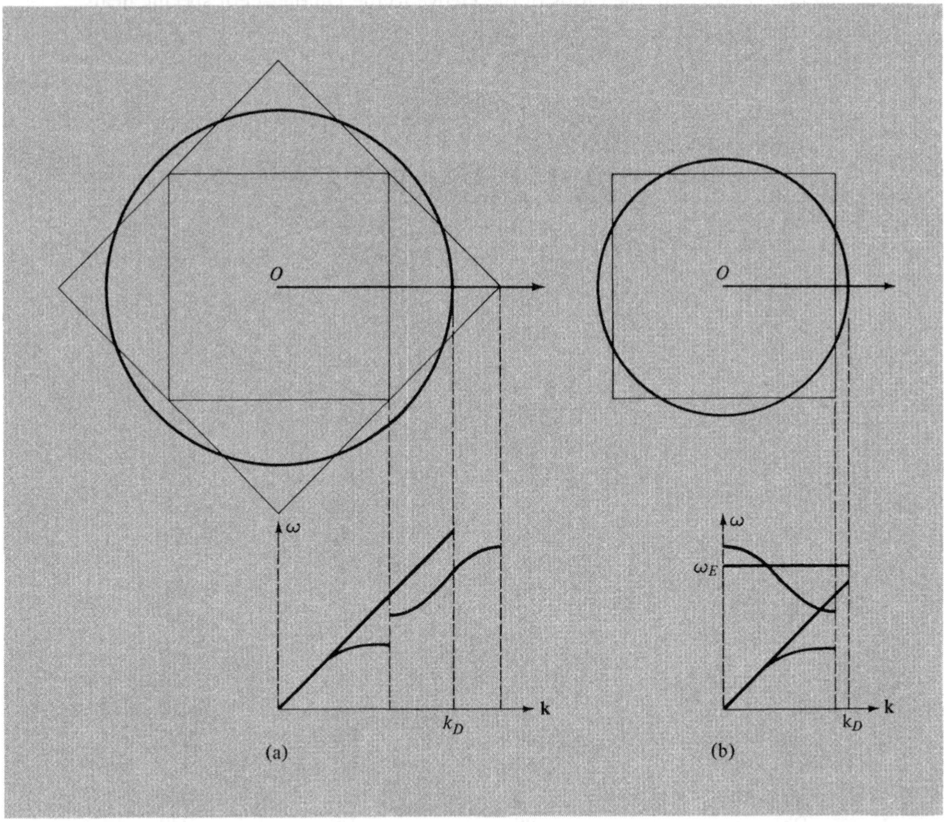

Figure 23.4
Two different ways of approximating the acoustic and optical branches of a diatomic crystal (illustrated in two dimensions along a line of symmetry). (a) *The Debye approximation.* The first two zones of the square lattice are replaced by a circle with the same total area, and the entire spectrum is replaced by a linear one within the circle. (b) *Debye approximation for the acoustic branch and Einstein approximation for the optical branch.* The first zone is replaced by a circle with the same area, the acoustic branch is replaced by a linear branch within the circle, and the optical branch is replaced by a constant branch within the circle.

the contribution of the acoustic branches to the specific heat.[15] Each optical branch will contribute

$$\frac{n\hbar\omega_E}{e^{\hbar\omega_E/k_BT} - 1} \tag{23.28}$$

to the thermal energy density in the Einstein approximation, so if there are p such branches there will be an additional term

$$c_v^{\text{optical}} = pnk_B \frac{(\hbar\omega_E/k_BT)^2 e^{\hbar\omega_E/k_BT}}{(e^{\hbar\omega_E/k_BT} - 1)^2} \tag{23.29}$$

in the specific heat.[16]

The characteristic features of the Einstein term (23.29) are that (a) well above the Einstein temperature $\Theta_E = \hbar\omega_E/k_B$ each optical mode simply contributes the constant k_B/V to the specific heat, as required by the classical law of Dulong and Petit, and (b) at temperatures well below the Einstein temperature the contribution of the optical modes to the specific heat drops exponentially, reflecting the difficulty in thermally exciting any optical modes at low temperatures.

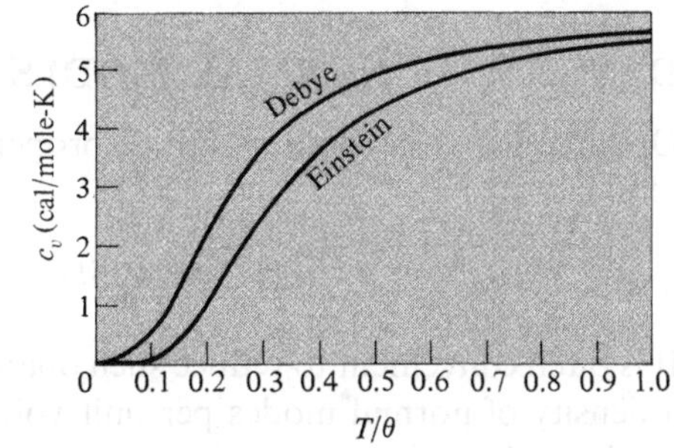

Figure 23.5
A comparison of the Debye and Einstein approximations to the specific heat of an insulating crystal. Θ is either the Debye or the Einstein temperature, depending on which curve is being examined. Both curves are normalized to approach the Dulong and Petit value of 5.96 cal/mole-K at high temperatures. In fitting to a solid with an m-ion basis, the Einstein curve should be given $m - 1$ times the weight of the Debye one. (From J. de Launay, *op. cit*; see Table 23.2.)

COMPARISON OF LATTICE AND ELECTRONIC SPECIFIC HEATS

It is useful to have a measure of the temperature at which the specific heat of a metal ceases to be dominated by the electronic contribution (linear in T) rather than the contribution from the lattice vibrations (cubic in T). If we divide the electronic

[15] Note that in Eq. (23.27) for the low-temperature specific heat this redefinition of n is precisely compensated by the redefinition of Θ_D, so that the coefficient of T^3 is unchanged. This reflects the fact that the optical branches do not contribute to the low-temperature specific heat, the form of which must therefore be independent of how they are treated.

[16] The first application of quantum mechanics to the theory of the specific heats of solids was made by Einstein, who proposed a total specific heat of the form (23.29). Although this form did produce the observed decline from the high-temperature form of Dulong and Petit, it fell far too rapidly to zero at very low temperatures (see Figure 23.5). Debye subsequently observed that because a solid could support elastic waves of very long-wavelength, and hence very low frequency, the picture of a solid as a set of identical oscillators, on which Einstein's formula was based, could not be correct. Nevertheless, the Einstein model does apply fairly well to the contribution of a relatively narrow optical branch to the specific heat, and in that role the model is still used.

contribution to the specific heat (Eq. (2.81)) by the low-temperature form (23.27) of the phonon contribution, and note that the electronic density is Z times the density of ions, where Z is the nominal valence, we find

$$\frac{c_v^{\mathrm{el}}}{c_v^{\mathrm{ph}}} = \frac{5}{24\pi^2} Z \frac{\Theta_D{}^3}{T^2 T_F}. \tag{23.30}$$

Thus the phonon contribution begins to exceed the electronic contribution at a temperature T_0, given by

$$T_0 = 0.145 \left(\frac{Z\Theta_D}{T_F}\right)^{1/2} \Theta_D. \tag{23.31}$$

Since Debye temperatures are of the order of room temperature, while Fermi temperatures are several tens of thousands of degrees Kelvin, the temperature T_0 is typically a few percent of the Debye temperature—i.e., a few degrees Kelvin. This explains why the linear term in the heat capacity of metals is only observed at low temperatures.

DENSITY OF NORMAL MODES (PHONON-LEVEL DENSITY)

One frequently encounters lattice properties that, like the specific heat (23.15), are of the form

$$\frac{1}{V} \sum_{\mathbf{k}s} Q(\omega_s(\mathbf{k})) = \sum_s \int \frac{d\mathbf{k}}{(2\pi)^3} Q(\omega_s(\mathbf{k})). \tag{23.32}$$

It is often convenient to reduce such quantities to frequency integrals, by introducing a density of normal modes per unit volume,[17] $g(\omega)$, defined so that $g(\omega)\,d\omega$ is the total number of modes with frequencies in the infinitesimal range between ω and $\omega + d\omega$, divided by the total volume of the crystal. In terms of g, the sum or integral in (23.32) takes on the form

$$\int d\omega\, g(\omega) Q(\omega). \tag{23.33}$$

By comparing (23.33) with (23.32) it is clear that the density of normal modes can be represented in the form

$$g(\omega) = \sum_s \int \frac{d\mathbf{k}}{(2\pi)^3} \delta(\omega - \omega_s(\mathbf{k})). \tag{23.34}$$

The density of normal modes is also called the phonon density of levels, for if we describe the lattice in the phonon rather than the normal-mode language, each normal mode corresponds to a possible level for a single phonon.

By following precisely the same steps that led to the representation (8.63) for the

[17] Compare the quite analogous discussion of the electronic density of levels on pages 143–145. Generally $g(\omega)$ is taken to give the contributions of all branches of the phonon spectrum, but one can also define separate $g_s(\omega)$ for each branch.

electron density of levels, one can represent the phonon density of levels in the alternative form

$$g(\omega) = \sum_s \int \frac{dS}{(2\pi)^3} \frac{1}{|\nabla\omega_s(\mathbf{k})|}, \tag{23.35}$$

where the integral is over that surface in the first zone on which $\omega_s(\mathbf{k}) \equiv \omega$. Just as in the electronic case, because $\omega_s(\mathbf{k})$ is periodic there will be a structure of singularities in $g(\omega)$, reflecting the fact that the group velocity appearing in the denominator of (23.35) must vanish at some frequencies. As in the electronic case, the singularities are known as van Hove singularities.[18] A typical density of levels displaying these singularities is shown in Figure 23.6, and a concrete illustration of how the singularities arise in the linear chain is given in Problem 3.

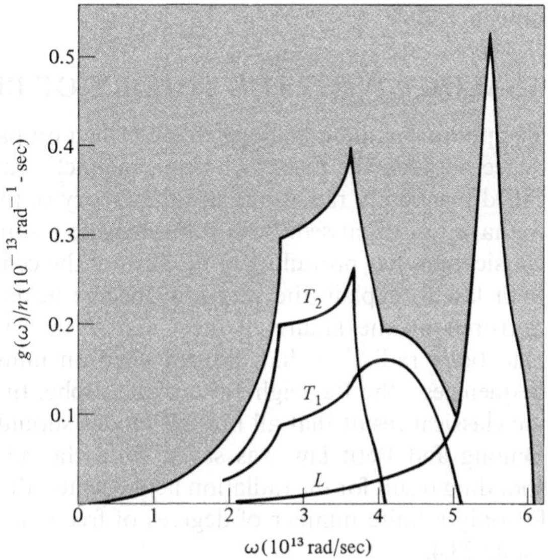

Figure 23.6
Phonon density of levels in aluminum, as deduced from neutron scattering data (Chapter 24). The highest curve is the full density of levels. Separate level densities for the three branches are also shown. (After R. Stedman, L. Almqvist, and G. Nilsson, *Phys. Rev.* **162**, 549 (1967).)

The Debye approximation and its limitations are quite compactly expressed in terms of the density of levels. If all three branches of the spectrum have the linear dispersion relation (23.21), and if the wave vectors of the normal modes are assumed to lie within a sphere of radius k_D rather than the first Brillouin zone, then (23.34) becomes simply:

$$\begin{aligned} g_D(\omega) &= 3\int_{k<k_D} \frac{d\mathbf{k}}{(2\pi)^3}\,\delta(\omega - ck) = \frac{3}{2\pi^2}\int_0^{k_D} k^2\,dk\,\delta(\omega - ck) \\ &= \begin{cases} \dfrac{3}{2\pi^2}\dfrac{\omega^2}{c^3}, & \omega < \omega_D = k_D c; \\ 0, & \omega > \omega_D. \end{cases} \end{aligned} \tag{23.36}$$

This simple parabolic behavior is clearly a rather gross approximation to the form characteristic of real solids (Figure 23.6). The choice of k_D does guarantee that the

[18] The singularities were in fact first noted in the context of the theory of lattice vibrations.

area under the curve $g_D(\omega)$ will be the same as under the correct curve, and if, in addition, the velocity c is chosen according to (23.18), then the curves will be in agreement in the neighborhood of $\omega = 0$. The former property is sufficient to produce the Dulong and Petit law at high temperatures, and the latter guarantees the correct specific heat at low temperatures.[19]

In a similar way, an Einstein model for an optical branch amounts to the approximation:

$$g_E(\omega) = \int_{\text{zone}} \frac{d\mathbf{k}}{(2\pi)^3}\, \delta(\omega - \omega_E) = n\delta(\omega - \omega_E), \tag{23.37}$$

which can be expected to give reasonable results provided that the frequency variation of the property Q being calculated is not appreciable over the width of the actual optical branch.

ANALOGY WITH THE THEORY OF BLACKBODY RADIATION

The photon-phonon analogy we described on page 453 carries over to a correspondence between the theory of electromagnetic radiation in thermal equilibrium (so-called blackbody radiation) and the theory of the vibrational energy of a solid that we have just discussed. Both subjects were a source of mystery in the context of the classical physics prevailing at the turn of the century. The failure of the Dulong and Petit law to explain the very low specific heats of solids at low temperatures was mirrored by the failure of the classical theory to predict an energy density for blackbody radiation that did not yield an infinite answer when summed over all frequencies—the Rayleigh-Jeans catastrophe. In both cases the problem came from the classical result that all normal modes should contribute $k_B T$ to the energy. The Dulong and Petit law was saved from the self-contradiction afflicting the corresponding result for the radiation field because the discrete nature of the solid allowed for only a finite number of degrees of freedom. The two theories are compared in Table 23.4.

Table 23.4
A COMPARISON OF PHONONS AND PHOTONS

	PHONONS	PHOTONS
Number of normal modes	$3p$ modes for each $\mathbf{k}$, $\omega = \omega_s(\mathbf{k})$	Two modes for each $\mathbf{k}$, $\omega = ck$ ($c \approx 3 \times 10^{10}$ cm/sec)
Restriction on wave vector	$\mathbf{k}$ confined to first Brillouin zone	$\mathbf{k}$ arbitrary
Thermal energy density	$\sum_s \int \frac{d\mathbf{k}}{(2\pi)^3} \frac{\hbar\omega_s(\mathbf{k})}{e^{\beta\hbar\omega_s(\mathbf{k})} - 1}$ (integral over first Brillouin zone)	$2\int \frac{d\mathbf{k}}{(2\pi)^3} \frac{\hbar ck}{e^{\beta\hbar ck} - 1}$ (integral over all $\mathbf{k}$)

19 One can somewhat improve the overall fit with a refinement of the Debye model that uses three different sound velocities for the three branches.

Because of the simple general form for the photon dispersion relation, the exact expression for the thermal energy of blackbody radiation is very similar to the Debye approximation to the thermal energy of a harmonic crystal. The differences are:

1. The sound velocity is replaced by the velocity of light.
2. The formula for blackbody radiation has an extra factor of $\frac{2}{3}$, corresponding to the fact that there are only two branches to the photon spectrum (electromagnetic radiation must be transverse: there is no longitudinal branch).
3. The upper limit of the integral is not k_D but ∞, since there is no restriction on the maximum allowed photon wave vector.

Point 3 means that the formulas for blackbody radiation always assume the form appropriate to the extreme low-temperature limit in the crystalline case. This is reasonable, since no matter what the temperature the vast majority (infinitely many) of the normal modes of the radiation field will have $\hbar ck$ greater than k_BT. Together with the exact linearity in k of the photon dispersion relation, this means that we are always in the region where the heat capacity is rigorously cubic. As a result, we can read off the exact thermal energy density for blackbody radiation from Eq. (23.20) for the low-temperature specific heat $c_v = \partial u/\partial T$ due to lattice vibrations, by interpreting c to be the velocity of light and multiplying by $\frac{2}{3}$ (to remove the contribution of the longitudinal acoustic branch). The result is the Stefan-Boltzmann law:

$$u = \frac{\pi^2}{15} \frac{(k_BT)^4}{(\hbar c)^3}. \tag{23.38}$$

Similarly, the thermal energy density in the frequency range from ω to $\omega + d\omega$ is

$$\frac{\hbar\omega g(\omega)\, d\omega}{e^{\beta\hbar\omega} - 1}. \tag{23.39}$$

The appropriate level density is just two thirds of the Debye form (23.36), without the cutoff at ω_D. This yields

$$\frac{\hbar}{\pi^2} \frac{\omega^3}{c^3} \frac{d\omega}{e^{\beta\hbar\omega} - 1}, \tag{23.40}$$

which is the Planck radiation law.

PROBLEMS

1. *High Temperature Specific Heat of a Harmonic Crystal*

(a) Show that Eq. (23.14), for the leading high-temperature quantum corrections to the law of Dulong and Petit, can also be written in the form:

$$\frac{\Delta c_v}{c_v^0} = -\frac{1}{12} \int d\omega\, g(\omega) \left(\frac{\hbar\omega}{k_BT}\right)^2 \Bigg/ \int d\omega\, g(\omega) \tag{23.41}$$

where $g(\omega)$ is the density of normal modes.

(b) Show that the next term in the high-temperature expansion of c_v/c_v^0 is

$$\frac{1}{240}\int d\omega\, g(\omega)\left(\frac{\hbar\omega}{k_B T}\right)^4 \Bigg/ \int d\omega\, g(\omega). \tag{23.42}$$

(c) Show that if the crystal is a monatomic Bravais lattice of ions acting only through pair potentials $\phi(\mathbf{r})$, then (within the harmonic approximation) the second moment of the frequency distribution that appears in (23.41) is given by

$$\int d\omega\, \omega^2 g(\omega) = \frac{n}{M}\sum_{\mathbf{R}\neq 0} \nabla^2\phi(\mathbf{R}). \tag{23.43}$$

2. *Low Temperature Specific Heat in d-Dimensions, and for Nonlinear Dispersion Laws*

(a) Show that Eq. (23.36), for the density of normal modes in the Debye approximation, gives the exact (within the harmonic approximation) leading *low-frequency* behavior of $g(\omega)$, provided that the velocity c is taken to be that given in Eq. (23.18).

(b) Show that in a d-dimensional harmonic crystal, the low-frequency density of normal modes varies as ω^{d-1}.

(c) Deduce from this that the low-temperature specific heat of a harmonic crystal vanishes as T^d in d dimensions.

(d) Show that if it should happen that the normal mode frequencies did not vanish linearly with k, but as k^ν, then the low-temperature specific heat would vanish as $T^{d/\nu}$, in d dimensions.

3. *van Hove Singularities*

(a) In a linear harmonic chain with only nearest-neighbor interactions, the normal-mode dispersion relation has the form (cf. Eq. (22.29)) $\omega(k) = \omega_0|\sin(ka/2)|$, where the constant ω_0 is the maximum frequency (assumed when k is on the zone boundary). Show that the density of normal modes in this case is given by

$$g(\omega) = \frac{2}{\pi a\sqrt{\omega_0^2 - \omega^2}}. \tag{23.44}$$

The singularity at $\omega = \omega_0$ is a van Hove singularity.

(b) In three dimensions the van Hove singularities are infinities not in the normal mode density itself, but in its derivative. Show that the normal modes in the neighborhood of a maximum of $\omega(\mathbf{k})$, for example, lead to a term in the normal-mode density that varies as $(\omega_0 - \omega)^{1/2}$.

24
Measuring Phonon Dispersion Relations

It is possible to extract the detailed form of the normal-mode dispersion relations $\omega_s(\mathbf{k})$ from experiments in which the lattice vibrations exchange energy with an external probe. The most informative of such probes is a beam of neutrons. One can view the energy lost (or gained) by a neutron while interacting with a crystal as being due to the emission (or absorption) of phonons, and by measuring the emergent angles and energies of the scattered neutrons one can extract direct information about the phonon spectrum. Similar information can be obtained when the probe is electromagnetic radiation, the two most important cases being X rays and visible light.

The broad, general principles underlying these experiments are much the same, whether the incident particles are neutrons or photons, but the information one extracts from electromagnetic probes is generally more limited or more difficult to interpret. On the other hand, electromagnetic probes—particularly X-ray analyses—are of crucial importance for those solids that are not susceptible to analysis through neutron scattering. One example[1] is solid helium-3, in which neutron spectroscopy is made impossible by the enormous cross section for a helium-3 nucleus capturing a neutron.

Neutrons and photons probe the phonon spectrum in different ways primarily because of their very different energy-momentum relations:

$$\text{Neutrons:} \qquad E_n = \frac{p^2}{2M_n}, \qquad M_n = 1838.65m_e = 1.67 \times 10^{-24}\ \text{gm}, \tag{24.1}$$

$$\text{Photons:} \qquad E_\gamma = pc, \qquad c = 2.99792 \times 10^{10}\ \text{cm/sec}. \tag{24.2}$$

Throughout the energy ranges of interest for the measurement of phonon dispersion relations, these two energy-momentum relations are spectacularly different (see Figure 24.1). However, the part of the general analysis that does not exploit the particular form of the E vs. p relation of the probe is much the same in either case. Therefore, although we shall start with a discussion of neutron scattering, we shall be able to apply to photons those aspects of the discussion that do not depend on the particular form (24.1) of the neutron energy-momentum relation.

NEUTRON SCATTERING BY A CRYSTAL

Consider a neutron, of momentum $\mathbf{p}$ and energy $E = p^2/2M_n$, that is incident upon a crystal. Since the neutron interacts strongly only with the atomic nuclei in the crystal,[2] it will pass without difficulty into the crystal[3] and subsequently emerge with momentum $\mathbf{p}'$ and energy $E' = p'^2/2M_n$.

---

[1] A more subtle example is vanadium, in which the natural relative abundances of the vanadium isotopes conspire numerically with the isotopic variation in neutron scattering amplitude, in such a way as to cancel out almost completely the informative (so-called *coherent*) part of the scattering. The combination of scattering amplitudes can be changed by isotope enrichment.

[2] The neutron has no electric charge, so it interacts with electrons only through the relatively weak coupling of its magnetic moment to the magnetic moment of the electrons. This is of considerable importance in the study of magnetically ordered solids (Chapter 33), but is of little consequence in determining phonon spectra.

[3] Typical nuclear radii are of order 10^{-13} cm, and typical internuclear distances in the solid are of order 10^{-8} cm. Hence the nuclei occupy only 10^{-15} of the total volume of the solid.

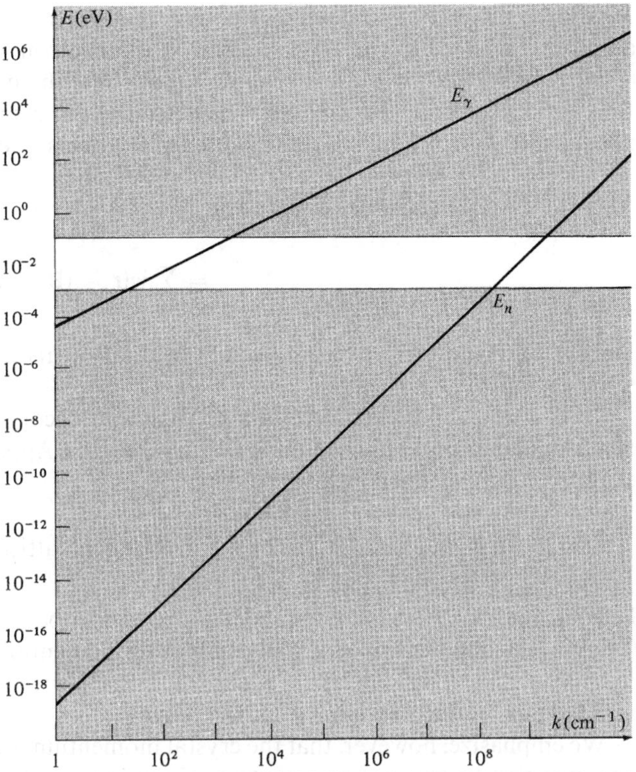

Figure 24.1
Neutron (n) and photon (γ) energy-momentum relations. When $k = 10^n$ cm^{-1}, $E_n = 2.07 \times 10^{2n-19}$ eV and $E_\gamma = 1.97 \times 10^{n-5}$ eV. Typical thermal energies lie in or near the white band.

We assume that the ions in the crystal are well described by the harmonic approximation. Later we shall indicate how our conclusions must be modified by the inevitable anharmonic terms in the ion-ion interaction. Suppose that at the start of the experiment the crystal is in a state with phonon occupation numbers[4] $n_{\mathbf{k}s}$, and after the experiment, as a result of its interaction with the neutron, the crystal is in a state with phonon numbers $n'_{\mathbf{k}s}$. Conservation of energy requires that

$$E' - E = -\sum_{\mathbf{k}s} \hbar\omega_{\mathbf{k}s}\,\Delta n_{\mathbf{k}s}, \qquad \Delta n_{\mathbf{k}s} = n'_{\mathbf{k}s} - n_{\mathbf{k}s}; \tag{24.3}$$

i.e., the change in the energy of the neutron is equal to the energy of the phonons it has absorbed during its passage through the crystal, minus the energy of the phonons it has emitted.[5]

Thus the change in neutron energy upon passage through the crystal contains information about the phonon frequencies. A second conservation law is required to disentangle this information from the scattering data. The second law is known as the *conservation of crystal momentum.* It is a very general consequence of a symmetry of the neutron-ion interaction,

$$H_{n\text{-}i} = \sum_{\mathbf{R}} w(\mathbf{r} - \mathbf{R} - \mathbf{u}(\mathbf{R})). \tag{24.4}$$

[4] By a state with phonon occupation numbers $n_{\mathbf{k}s}$ we mean one in which $n_{\mathbf{k}s}$ phonons of type $\mathbf{k}s$ are present—i.e., one in which the $\mathbf{k}s$-th normal mode is in its $n_{\mathbf{k}s}$th excited state.

[5] A neutron may lose energy or gain it, depending on the balance between the energies of the emitted and absorbed phonons.

Here w is the (very short-ranged) potential of interaction between a neutron and an atomic nucleus of the crystal, and $\mathbf{r}$ is the neutron coordinate. The interaction (24.4) is unaffected by a transformation that shifts the neutron coordinate $\mathbf{r}$ by any Bravais lattice vector $\mathbf{R}_0$ and also permutes the ion displacement variables $\mathbf{u}(\mathbf{R})$ by $\mathbf{u}(\mathbf{R}) \to \mathbf{u}(\mathbf{R} - \mathbf{R}_0)$, for if we make both replacements, then (24.4) becomes:

$$H_{n\text{-}i} \to \sum_{\mathbf{R}} w(\mathbf{r} + \mathbf{R}_0 - \mathbf{R} - \mathbf{u}(\mathbf{R} - \mathbf{R}_0))$$

$$= \sum_{\mathbf{R}} w(\mathbf{r} - (\mathbf{R} - \mathbf{R}_0) - \mathbf{u}(\mathbf{R} - \mathbf{R}_0)). \quad \textbf{(24.5)}$$

Since we are summing over all vectors in the Bravais lattice, (24.5) is precisely the same as (24.4).[6]

One of the fundamental results of quantum theory is that symmetries of the Hamiltonian imply conservation laws. It is shown in Appendix M that this particular symmetry implies the conservation law:

$$\mathbf{p}' - \mathbf{p} = -\sum_{\mathbf{k}s} \hbar\mathbf{k}\, \Delta n_{\mathbf{k}s} + (\text{reciprocal lattice vector} \times \hbar). \quad \textbf{(24.6)}$$

If we define the *crystal momentum* of a phonon to be $\hbar$ times its wave vector, then (24.6) makes an assertion strikingly similar to momentum conservation: *The change in neutron momentum is just the negative of the change in total phonon crystal momentum, to within an additive reciprocal lattice vector.*

We emphasize, however, that the crystal momentum of a phonon is not, in general, accompanied by any real momentum of the ionic system. "Crystal momentum" is simply a name for $\hbar$ times the phonon wave vector.[7] The name is intended to suggest that $\hbar\mathbf{k}$ frequently plays a role quite similar to that of a momentum, as it evidently does in Eq. (24.6). Since a crystal possesses translational symmetry, it is not surprising that there should be a conservation law rather like momentum conservation;[8] but because this symmetry is only that of a Bravais lattice (as opposed to the full translational symmetry of empty space), it is also not surprising that the conservation law is weaker than momentum conservation (i.e., crystal momentum is only conserved to within an additive reciprocal lattice vector).

Because there are two conservation laws, it turns out to be possible to extract the explicit forms of the $\omega_s(\mathbf{k})$ from the neutron scattering data in a simple way. To show this we examine the distribution of scattered neutrons emerging from the crystal, classifying the types of scattering that can take place according to the total number

---

[6] This is strictly the case only for a neutron interacting with an infinite crystal. To the extent that surface scattering is important (and in neutron scattering it is not), crystal momentum will not be conserved.

[7] The nomenclature is quite analogous to that used in Chapter 8, where we defined the crystal momentum of a Bloch electron with wave vector $\mathbf{k}$, to be $\hbar\mathbf{k}$. The identical terminology is deliberate, for in processes in which phonon and electronic transitions occur together, the total crystal momentum of the electron-phonon system is conserved (to within a reciprocal lattice vector $\times\ \hbar$). (See Appendix M and Chapter 26.)

[8] The law of momentum conservation follows from the complete translational invariance of empty space.

of phonons that a neutron has exchanged energy with while passing through the crystal.

Zero-Phonon Scattering

In this case the final state of the crystal is identical to its initial state. Energy conservation (Eq. (24.3)) implies that the energy of the neutron is unchanged (i.e., the scattering is *elastic*) and crystal momentum conservation (Eq. (24.6)) implies that the neutron momentum can only change by $\hbar\mathbf{K}$, where $\mathbf{K}$ is a reciprocal lattice vector. If we write the incident and scattered neutron momenta as:

$$\mathbf{p} = \hbar\mathbf{q}, \quad \mathbf{p}' = \hbar\mathbf{q}', \tag{24.7}$$

then these restrictions become:

$$q' = q, \quad \mathbf{q}' = \mathbf{q} + \mathbf{K}. \tag{24.8}$$

Equations (24.8) are precisely the Laue conditions that the incident and scattered X-ray wave vectors must satisfy in order for elastically scattered X rays to produce a Bragg peak (see page 99). Since a neutron with momentum $\mathbf{p} = \hbar\mathbf{q}$ can be viewed as a plane wave with wave vector $\mathbf{q}$, this emergence of the Laue condition is to be expected. We conclude that elastically scattered neutrons, which create or destroy *no* phonons, are found only in directions that satisfy the Bragg condition, and give precisely the same structural information about the crystal as described in the discussion of elastic X-ray scattering in Chapter 6.

One-Phonon Scattering

It is the neutrons that absorb or emit precisely one phonon that convey the most important information. In the case of absorption (which is generally the more important) conservation of energy and crystal momentum implies that

$$\begin{aligned} E' &= E + \hbar\omega_s(\mathbf{k}), \\ \mathbf{p}' &= \mathbf{p} + \hbar\mathbf{k} + \hbar\mathbf{K}, \end{aligned} \tag{24.9}$$

where $\mathbf{k}$ and s are the wave vector and branch index of the absorbed phonon. In the case of emission we have:

$$\begin{aligned} E' &= E - \hbar\omega_s(\mathbf{k}), \\ \mathbf{p}' &= \mathbf{p} - \hbar\mathbf{k} + \hbar\mathbf{K}, \end{aligned} \tag{24.10}$$

where the phonon has been emitted into branch s, with wave vector $\mathbf{k}$.

In either case we can use the crystal momentum law to represent $\mathbf{k}$ in terms of the neutron momentum transfer, $\mathbf{p}' - \mathbf{p}$. Furthermore, the additive reciprocal lattice vector that appears in this relation can be ignored, when the resulting expression for $\mathbf{k}$ is substituted into the energy conservation law, for each $\omega_s(\mathbf{k})$ is a periodic function in the reciprocal lattice:

$$\omega_s(\mathbf{k} \pm \mathbf{K}) = \omega_s(\mathbf{k}). \tag{24.11}$$

As a result, the two conservation laws yield one equation:

$$\frac{p'^2}{2M_n} = \frac{p^2}{2M_n} + \hbar\omega_s\left(\frac{\mathbf{p}' - \mathbf{p}}{\hbar}\right), \qquad \text{phonon absorbed,} \tag{24.12}$$

or

$$\frac{p'^2}{2M_n} = \frac{p^2}{2M_n} - \hbar\omega_s\left(\frac{\mathbf{p} - \mathbf{p}'}{\hbar}\right), \qquad \text{phonon emitted.} \tag{24.13}$$

In a given experiment, the incident neutron momentum and energy are usually specified. Thus for a given phonon dispersion relation $\omega_s(\mathbf{k})$ the only unknowns in (24.12) and (24.13) are the three components of the final neutron momentum $\mathbf{p}'$. Quite generally, a single equation relating the three components of a vector $\mathbf{p}'$ will (if it has any solutions at all) specify a surface (or surfaces) in three-dimensional $\mathbf{p}'$-space. If we only examine neutrons emerging in a definite direction we will specify the direction of $\mathbf{p}'$, and can therefore expect to find solutions at only a single point on the surface (or a finite number of points on the surfaces).[9]

If we select a general direction we will see neutrons scattered by one-phonon processes only at a few discrete values of p', and correspondingly only at a few discrete energies $E' = p'^2/2M_n$. Knowing the energy and the direction in which the scattered neutron emerges, we can construct $\mathbf{p}' - \mathbf{p}$ and $E' - E$, and can therefore conclude that the crystal has a normal mode whose frequency is $(E' - E)/\hbar$ and wave vector is $\pm(\mathbf{p}' - \mathbf{p})/\hbar$. We have therefore measured a point in the crystal's phonon spectrum. By varying all the parameters at our disposal (incident energy, orientation of the crystal, and direction of detection) we can collect a large number of such points, and quite effectively map out the entire phonon spectrum (Figure 24.2). This can be accomplished, however, only if it is possible to distinguish the neutrons scattered in one-phonon processes from the others. We consider explicitly the case of two-phonon processes.

Two-Phonon Scattering

In a two-phonon process a neutron can absorb or emit two phonons, or emit one and absorb another (which can also be described as the scattering of a single phonon). To be concrete we discuss the case of two-phonon absorption. The conservation laws will then have the form:

$$\begin{aligned} E' &= E + \hbar\omega_s(\mathbf{k}) + \hbar\omega_{s'}(\mathbf{k}'), \\ \mathbf{p}' &= \mathbf{p} + \hbar\mathbf{k} + \hbar\mathbf{k}' + \hbar\mathbf{K}. \end{aligned} \tag{24.14}$$

If we eliminate $\mathbf{k}'$ through the crystal momentum conservation law, we arrive at a single restriction:

$$E' = E + \hbar\omega_s(\mathbf{k}) + \hbar\omega_{s'}\left(\frac{\mathbf{p}' - \mathbf{p}}{\hbar} - \mathbf{k}\right). \tag{24.15}$$

[9] Alternatively, by specifying the direction of $\mathbf{p}'$ we leave a single unknown variable (the magnitude p') in (24.12) or (24.13) and therefore expect at most a finite number of solutions.

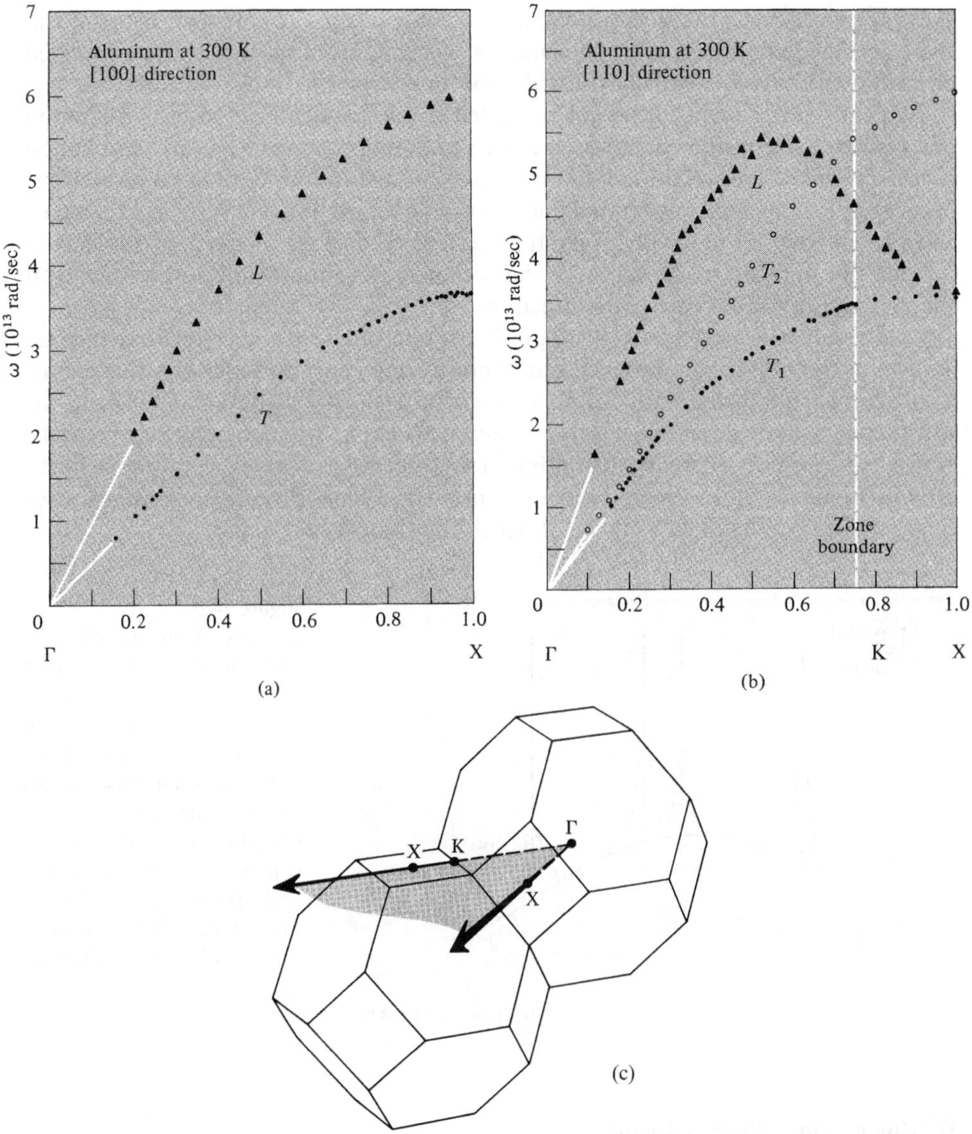

Figure 24.2
Phonon dispersion relations in aluminum, measured along the *k*-space lines ΓX and ΓKX by neutron scattering. The estimated error in frequency is 1 to 2 percent. Each point represents an observed neutron group. (After J. Yarnell et al., *Lattice Dynamics*, R. F. Wallis, ed., Pergamon, New York, 1965.) Note that the two transverse branches are degenerate along ΓX (4-fold axis), but not along ΓK (2-fold axis). See Chapter 22.

For each fixed value of **k**, the discussion we gave in analyzing the one-phonon case can be repeated: For a given direction of detection, scattered neutrons will occur only at a small set of discrete energies. However, **k** can now be varied continuously throughout the first Brillouin zone, since the wave vector of the absorbed phonons is *not* at our disposal. As **k** varies, the discrete energies of the emergent neutron will also vary. Therefore the totality of neutrons emerging from the process in a specified direction will have a *continuous* distribution of energies.

Evidently this conclusion is not restricted to the particular type of two-phonon process we have considered, nor even to two-phonon processes. Only in one-phonon processes are the conservation laws restrictive enough to forbid all but a discrete set of energies for the neutrons scattered in a given direction. If a neutron has exchanged energy with two or more phonons, the number of degrees of freedom sufficiently exceeds the number of conservation laws that a continuum of scattered neutron energies can be observed in any direction.

As a result it is possible to distinguish the one-phonon processes from the rest (known as the multiphonon background) not by any characteristics of a single scattered neutron, but by the statistical structure of the energy distribution of the neutrons scattered in a given direction. The one-phonon processes will contribute sharp peaks at isolated energies, whereas the multiphonon processes will give a continuous background (Figure 24.3). The energy and momentum transfer of the one-phonon processes can therefore be identified as those at which the sharp peaks occur.

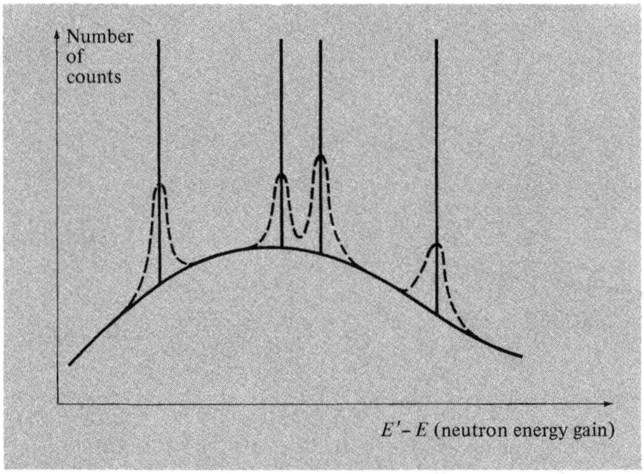

Figure 24.3
Relative numbers of scattered neutrons in a given direction as a function of neutron energy. The smooth curve is the background due to multiphonon processes. In an ideal harmonic crystal the one-phonon processes would contribute sharp peaks. In a real crystal these peaks are broadened (dashed curves) due to phonon lifetime effects.

Widths of One-Phonon Peaks

Some typical neutron distributions are shown in Figure 24.4. Note that although the one-phonon peaks are generally quite unambiguous, they are not perfectly sharp, as our analysis has suggested. This is because real crystals are not perfectly harmonic. The stationary states of the harmonic approximation are only approximate stationary states: Even if the real crystal is in such a state (characterized by a specific set of phonon occupation numbers) at one time, it will eventually evolve into a superposition of other such states (characterized by different phonon occupation numbers). If, however, the harmonic stationary states are reasonably good approximations to the exact ones, this decay may be slow enough to permit one to continue to describe processes occurring within the crystal in terms of phonons, provided that one assigns

Figure 24.4
Some typical experimental neutron groups. In all cases the number of neutrons emerging in a fixed direction for a fixed incident energy is plotted against a variable that distinguishes scattered neutron energies. (a) Copper. (G. Gobert and B. Jacrot, *J. Phys. Radium* **19** (1959).) (b) Germanium. (I. Pelah et al., *Phys. Rev.* **108**, 1091 (1957).)

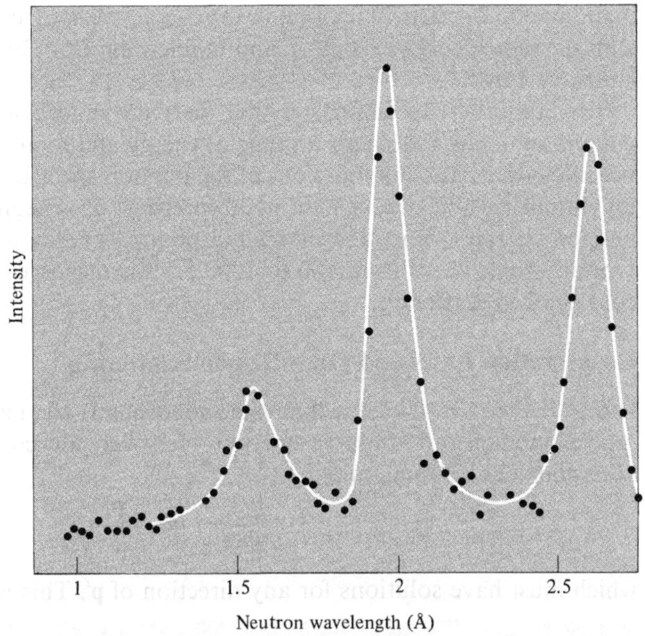

(a)

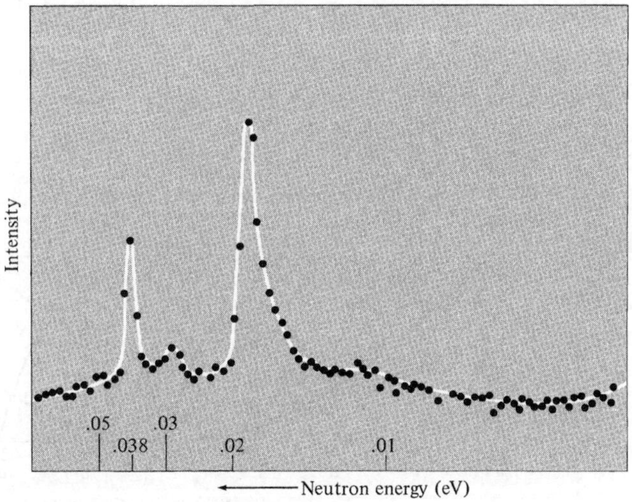

(b)

to the phonons finite lifetimes, reflecting the eventual decay of the approximate harmonic stationary state. Associated with a phonon of lifetime τ, there will be an uncertainty $\hbar/\tau$ in the phonon energy. The energy conservation law determining the one-phonon peaks will then be correspondingly weakened.

These points will be taken up in more detail in Chapter 25. Here we merely note that the one-phonon peaks, though broadened, are still clearly identifiable. The fact that they are indeed due to one-phonon processes is strikingly confirmed by the consistency of the $\omega_s(\mathbf{k})$ curves inferred from their positions, for there is considerable redundancy in the data furnished by the one-phonon peaks. One can extract information about a particular phonon in several ways, by considering scattering events with the same energy exchange, and momentum transfers that differ by a reciprocal lattice vector.[10]

It is important to emphasize that there are indeed solutions to the one-phonon conservation law (24.12) for a range of energy and momentum transfers sufficient to permit a systematic mapping out of the phonon spectrum. To see this, suppose first, for simplicity, that the incident neutron energy E is negligibly small on the scale of phonon energies. Since the maximum phonon energy is of order $k_B\Theta_D$, and Θ_D is typically somewhere from 100 to 1000 K, this means we are dealing with so-called cold-neutron scattering.

Conservation Laws and One-Phonon Scattering

If $E = 0$, then Eq. (24.13) will have no solutions at all (a zero-energy neutron cannot *emit* a phonon and conserve energy). However, the conservation law for phonon *absorption* (24.12) reduces to

$$\frac{p'^2}{2m} = \hbar\omega_s\left(\frac{\mathbf{p}'}{\hbar}\right), \tag{24.16}$$

which must have solutions for any direction of $\mathbf{p}'$. This is evident from Figure 24.5.

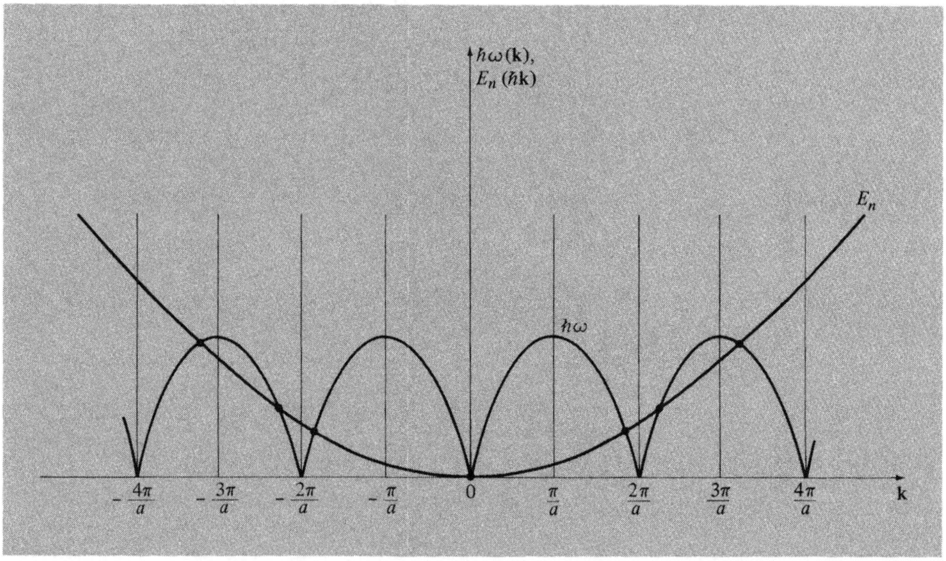

Figure 24.5
One-dimensional demonstration of the fact that the conservation laws for one-phonon absorption can always be satisfied for zero-energy incident neutrons. The equation $\hbar^2k^2/2M_n = \hbar\omega(\mathbf{k})$ is satisfied wherever the two curves intersect.

---

[10] It is also possible to extract information about the polarization vectors. This follows from the fact (derived in Appendix N) that the cross section for a given one-phonon process is proportional to

$$|\boldsymbol{\epsilon}_s(\mathbf{k}) \cdot (\mathbf{p} - \mathbf{p}')|^2,$$

where $\boldsymbol{\epsilon}_s(\mathbf{k})$ is the polarization vector for the phonon involved, and $\mathbf{p}' - \mathbf{p}$ is the neutron momentum transfer.

It follows analytically from the fact that the neutron energy vanishes quadratically for small p', while $\hbar\omega_s(\mathbf{p}'/\hbar)$ either vanishes linearly (acoustic branch) or approaches a constant (optical branch). Hence for small enough p' the neutron energy is always less than that of the phonon for any direction of $\mathbf{p}'$. However, as p' increases, the neutron energy can increase without bound, while $\hbar\omega_s(\mathbf{p}'/\hbar)$ is bounded above by the maximum phonon energy in the branch. By continuity there is therefore at least one value of p' for each direction of $\mathbf{p}'$ for which the left and right sides of (24.16) are equal. There must be at least one such solution for each branch s of the phonon spectrum. Usually there is more than one solution (Figure 24.5). This is because the final neutron energy is comparatively small (even when $\mathbf{p}'/\hbar$ is at the surface of the Brillouin zone), for a neutron of wave vector $\mathbf{q}$ (measured in inverse angstroms) has an energy

$$E_N = 2.1(q[\text{Å}^{-1}])^2 \times 10^{-3}\ \text{eV},$$
$$\frac{E_N}{k_B} = 24(q[\text{Å}^{-1}])^2\ \text{K}. \tag{24.17}$$

Thus E_n/k_B is small compared with typical Θ_D, even when q is at a zone boundary.

When the incident neutron energy is not zero, there will continue to be solutions corresponding to the absorption of a phonon in each branch (Figure 24.6). If a

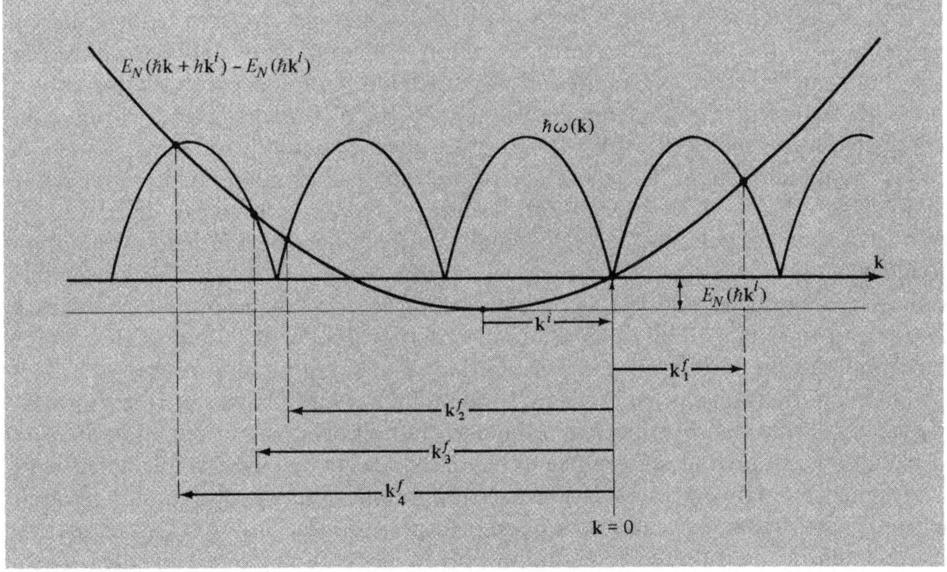

Figure 24.6
Graphical solution to the one-phonon conservation laws when the incident neutron has wave vector $\mathbf{k}^i$. The conservation law for phonon absorption can be written

$$E_N(\hbar\mathbf{k} + \hbar\mathbf{k}^i) - E_N(\hbar\mathbf{k}^i) = \hbar\omega(\mathbf{k}),$$

where $\hbar\mathbf{k}$ is the momentum of the scattered neutron, and $E_N(\mathbf{p}) = p^2/2M_N$. To draw the left-hand side of this equation, one displaces the neutron energy-momentum curve horizontally so that it is centered at $\mathbf{k} = -\mathbf{k}^i$ rather than $\mathbf{k} = 0$, and displaces it downward by an amount $E_N(\hbar\mathbf{k}^i)$. Solutions occur wherever this displaced curve intersects the phonon dispersion curve $\hbar\omega(\mathbf{k})$. In the present case there are solutions for four different scattered neutron wave vectors, $\mathbf{k}_1{}^f \cdots \mathbf{k}_4{}^f$.

certain threshold energy is exceeded, additional solutions will become possible corresponding to the emission of a phonon. Therefore there is no lack of one-phonon peaks, and ingenious techniques have been developed for mapping out the phonon spectrum of a crystal along various directions in k-space with considerable accuracy (a few percent) and at a large number of points.

ELECTROMAGNETIC SCATTERING BY A CRYSTAL

Precisely the same conservation laws (energy and crystal momentum) apply to the scattering of photons by the ions in a crystal, but because of the very different quantitative form of the photon energy-momentum relation, simple direct information about the entire phonon spectrum is much more difficult to extract than from neutron scattering data. The two most commonly employed electromagnetic techniques, each with its limitations, are the inelastic scattering of X rays and of visible light.

X-Ray Measurements of Phonon Spectra

Our discussion of X-ray scattering in Chapter 6 was based on the model of a static lattice (which is why it is equivalent to the zero-phonon elastic scattering described in our discussion of neutron scattering). When the assumption of a rigid static lattice of ions is relaxed, it is possible for X-ray photons, like neutrons, to be inelastically scattered with the emission and/or absorption of one or more phonons. However, the change in the energy of an inelastically scattered photon is extremely difficult to measure. A typical X-ray energy is several keV (10^3 eV), whereas a typical phonon energy is several meV (10^{-3} eV), and at most a few hundredths of an eV, for Θ_D of order room temperature. In general the resolution of such minute photon frequency shifts is so difficult that one can only measure the *total* scattered radiation of all frequencies, as a function of scattering angle, in the diffuse background of radiation found at angles away from those satisfying the Bragg condition. Because of this difficulty in energy resolution, the characteristic structure of the one-phonon processes is lost and their contribution to the total radiation scattered at any angle, cannot be simply distinguished from the contribution of the multiphonon processes.

Some information can be extracted along different lines, however. It is shown in Appendix N that the contribution of the one-phonon processes to the total intensity of radiation scattered at a given angle is entirely determined by a simple function of the frequencies and polarizations of those few phonons taking part in the one-phonon events. Therefore one can extract the phonon dispersion relations from a measurement of the intensity of scattered X radiation as a function of angle and incident X-ray frequency, provided that one can find some way of subtracting from this intensity the contribution from the multiphonon processes. Generally one attempts to do this by a theoretical calculation of the multiphonon contribution. In addition, however, one must allow for the fact that X rays, unlike neutrons, interact strongly with electrons. There will be a contribution to the intensity due to inelastically scattered electrons (the so-called Compton background), which must also be corrected for.

As a result of these considerations X-ray scattering is a far less powerful probe of the phonon spectrum than neutron scattering. The great virtue of neutrons is that good energy resolution is possible, and once the scattered energies have been resolved, the highly informative one-phonon processes are clearly identifiable.

Optical Measurements of Phonon Spectra

If photons of visible light (usually from a high intensity laser beam) are scattered with the emission or absorption of phonons, the energy (or frequency) shifts are still very small, but they can be measured, generally by interferometric techniques. Therefore one can isolate the one-phonon contribution to the light scattering, and extract the values of $\omega_s(\mathbf{k})$ for the phonons participating in the process. Because, however, the photon wave vectors (of order 10^5 cm^{-1}) are small compared with the Brillouin zone dimensions (of order 10^8 cm^{-1}), information is provided only about phonons in the immediate neighborhood of $\mathbf{k} = \mathbf{0}$. The process is referred to as *Brillouin scattering*, when the phonon emitted or absorbed is acoustic, and *Raman scattering*, when the phonon is optical.

In examining the conservation laws for these processes, one must note that the photon wave vectors inside the crystal will differ from their free space values by a factor of the index of refraction of the crystal n (since the frequencies in the crystal are unchanged, and the velocity is c/n). Therefore if the free space wave vectors of the incident and scattered photons are $\mathbf{q}$ and $\mathbf{q}'$, and the corresponding angular frequencies are ω and ω', conservation of energy and crystal momentum in a one-phonon process requires

$$\hbar\omega' = \hbar\omega \pm \hbar\omega_s(\mathbf{k}) \tag{24.18}$$

and

$$\hbar n\mathbf{q}' = \hbar n\mathbf{q} \pm \hbar\mathbf{k} + \hbar\mathbf{K}. \tag{24.19}$$

Here the upper sign refers to processes in which a phonon is absorbed (known as the *anti-Stokes* component of the scattered radiation) and the lower sign refers to processes in which a phonon is emitted (the *Stokes* component). Since the photon wave vectors $\mathbf{q}$ and $\mathbf{q}'$ are small in magnitude compared with the dimensions of the Brillouin zone, for phonon wave vectors $\mathbf{k}$ in the first zone the crystal momentum conservation law (24.19) can be obeyed only if the reciprocal lattice vector $\mathbf{K}$ is zero.

The two types of process are shown in Figure 24.7, and the constraint imposed

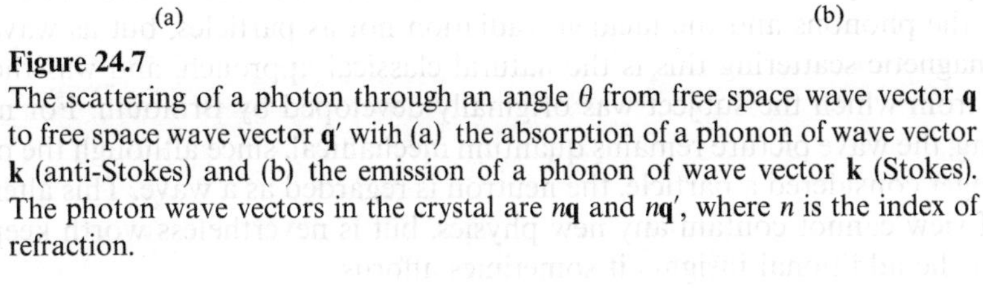

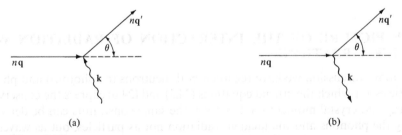

Figure 24.7
The scattering of a photon through an angle θ from free space wave vector $\mathbf{q}$ to free space wave vector $\mathbf{q}'$ with (a) the absorption of a phonon of wave vector $\mathbf{k}$ (anti-Stokes) and (b) the emission of a phonon of wave vector $\mathbf{k}$ (Stokes). The photon wave vectors in the crystal are $n\mathbf{q}$ and $n\mathbf{q}'$, where n is the index of refraction.

by crystal momentum conservation in Figure 24.8. Since the energy of any phonon is at most of order $\hbar\omega_D \approx 10^{-2}$ eV, the photon energy (typically a few eV), and hence the magnitude of the photon wave vector, is changed very little—i.e., the triangle in Figure 24.8 is very nearly isosceles. It follows immediately that the magnitude k of the phonon wave vector is related to the angular frequency of the light and the scattering angle θ by

$$k = 2nq \sin \tfrac{1}{2}\theta = (2\omega n/c) \sin \tfrac{1}{2}\theta. \tag{24.20}$$

The direction of **k** is determined by the construction in Figure 24.8, and the frequency, $\omega_s(\mathbf{k})$ by the measured (small) change in photon frequency.

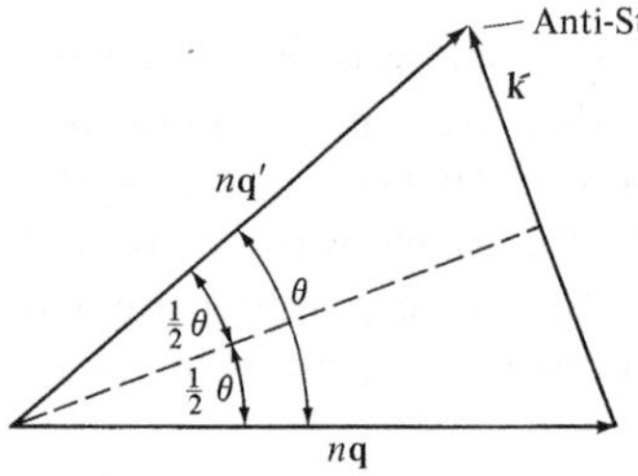

Figure 24.8
Geometrical derivation of Eq. (24.20). Because the photon energy is virtually unchanged, the triangle is isosceles. Because the process takes place within the crystal, the photon wave vectors are $n\mathbf{q}$ and $n\mathbf{q}'$, where n is the index of refraction of the crystal. The figure is drawn for the case of phonon absorption (anti-Stokes). It also describes the case of phonon emission (Stokes) if the direction of **k** is reversed.

In the case of Brillouin scattering, the phonon is an acoustic phonon near the origin of k-space, and $\omega_s(\mathbf{k})$ has the form $\omega_s(\mathbf{k}) = c_s(\hat{\mathbf{k}})k$ (Eq. (22.65)). Equation (24.20) then relates the sound velocity $c_s(\hat{\mathbf{k}})$ to the scattering angle and shift in photon energy $\Delta\omega$ by

$$c_s(\hat{\mathbf{k}}) = \frac{\Delta\omega}{2\omega}\frac{c}{n}(csc \tfrac{1}{2}\theta). \tag{24.21}$$

Some typical data are shown in Figure 24.9.

WAVE PICTURE OF THE INTERACTION OF RADIATION WITH LATTICE VIBRATIONS

In the above discussion we have regarded both neutrons (or photons) and phonons as particles, for which the crucial equations (24.3) and (24.6) express the conservation of energy and crystal momentum. However, the same constraints can be derived by viewing the phonons and the incident radiation not as particles, but as waves. For electromagnetic scattering this is the natural classical approach, and was the point of view from which the subject was originally developed by Brillouin. For neutron scattering, the wave picture remains quantum mechanical, since although the phonon is no longer considered a particle, the neutron is regarded as a wave. This alternative point of view cannot contain any new physics, but is nevertheless worth keeping in mind for the additional insights it sometimes affords.

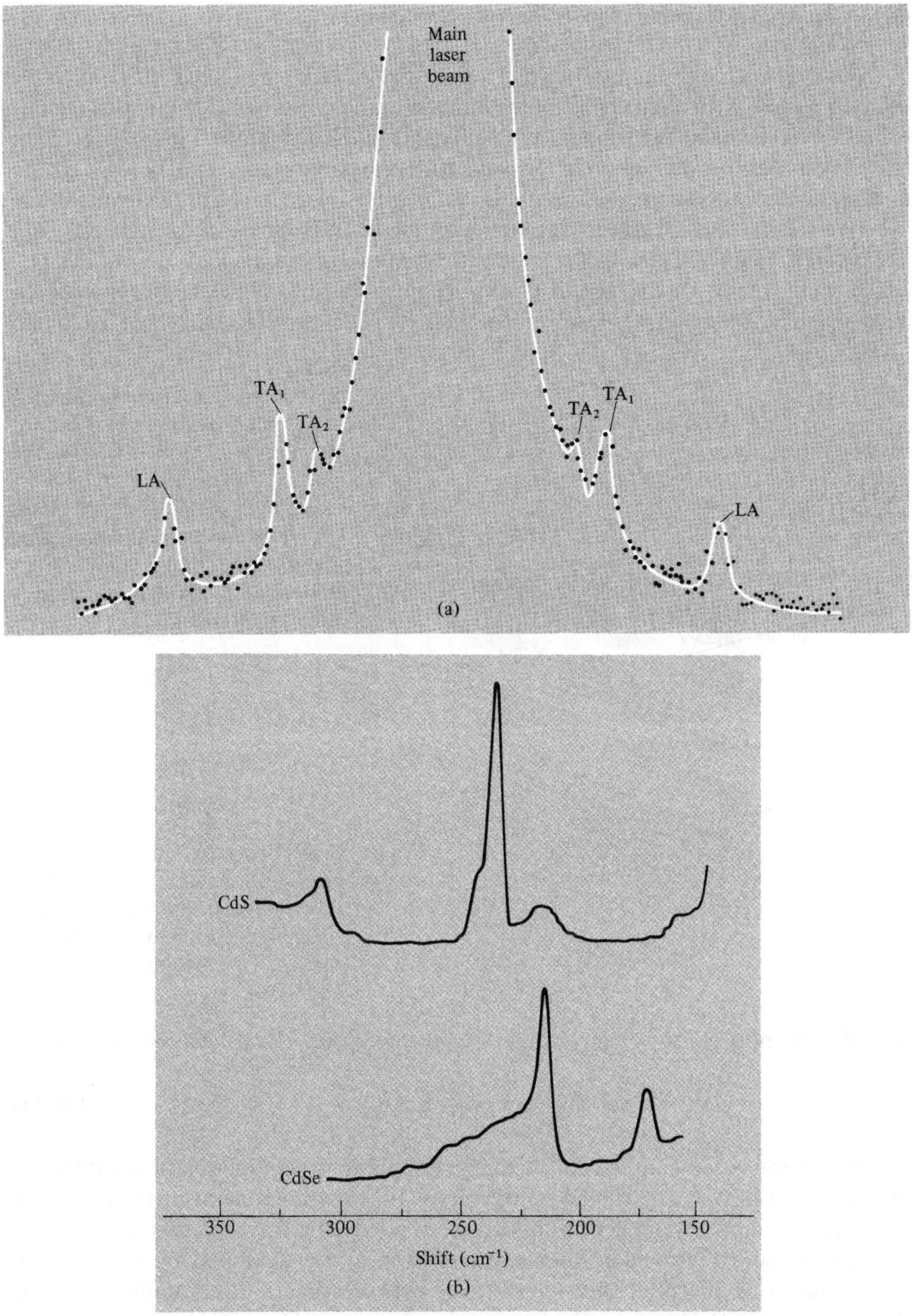

Figure 24.9
(a) Characteristic structure of a Brillouin spectrum. Intensity is plotted vs. frequency. There are clearly identifiable peaks at frequencies above and below the frequency of the main laser beam, corresponding to one longitudinal and two transverse acoustic branches. (S. Fray et al., *Light Scattering Spectra of Solids*, G. B. Wright, ed., Springer, New York, 1969.) (b) The Raman spectra of CdS and CdSe, revealing peaks determined by the longitudinal and transverse optical phonons. (R. K. Chang et al., *ibid.*)

Consider, then, the interaction of a wave with angular frequency $E/\hbar$ and wave vector $\mathbf{q} = \mathbf{p}/\hbar$, with a particular normal mode of the crystal with angular frequency ω and wave vector $\mathbf{k}$. We suppose that only this particular normal mode is excited; i.e., we consider the interaction of the wave with one phonon at a time. We also ignore for the moment the microscopic structure of the crystal, regarding the normal mode of interest as a wavelike disturbance in a continuous medium. If the disturbance did not move it would present to the incident radiation a periodic variation in density, which would act as a diffraction grating (Figure 24.10), the scattered wave being determined by Bragg's law. However, the disturbance is not stationary, but is moving

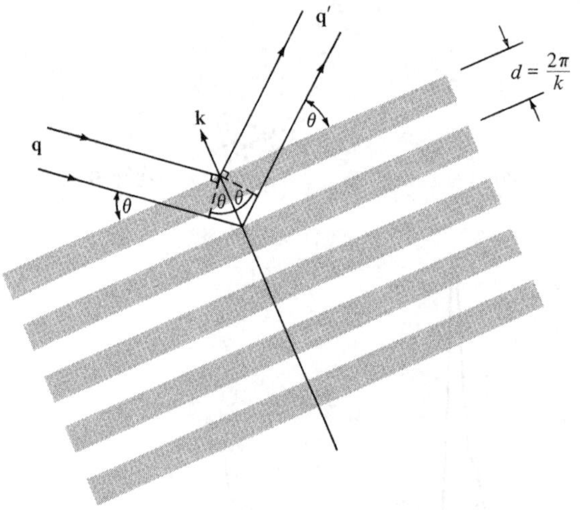

Figure 24.10
Scattering of a neutron by a phonon in a frame of reference in which the phonon phase velocity is zero. The phonon appears as a static diffraction grating; i.e., it results in regions of alternating high and low ionic density. The Bragg condition (p. 97), $m\lambda = 2d \sin\theta$ (m an integer), can be written as

$$\frac{2\pi m}{q} = \frac{4\pi}{k} \sin\theta$$

or

$$mk = 2q \sin\theta$$

or

$$mk = (\mathbf{q}' - \mathbf{q}) \cdot \hat{\mathbf{k}}.$$

Since Bragg reflection is specular (angle of incidence equals angle of reflection) and since the magnitude q' equals the magnitude q, it follows that $\mathbf{q}' - \mathbf{q}$ must be parallel to $\mathbf{k}$, and therefore $\mathbf{q}' - \mathbf{q} = m\mathbf{k}$.

with the phonon phase velocity which is directed along $\mathbf{k}$ and has magnitude ω/k:

$$\mathbf{v} = \frac{\omega}{k} \hat{\mathbf{k}}. \tag{24.22}$$

This complication can be dealt with by describing the diffraction in the frame of reference that moves with the phase velocity $\mathbf{v}$. In that frame the disturbance will be stationary, and the Bragg condition can be applied. The wave vectors (of both the lattice wave and the incident and scattered waves) are unaltered by a change of frame, since this affects neither the distance between planes of constant phase nor their orientation.[11] However, the frequencies undergo a Doppler shift:

---

[11] More precisely, the change in wave vector is a relativistic effect, which we ignore because the phase velocity $\mathbf{v}$ is small compared with c. The Doppler shift formulas (24.23) are also used in their nonrelativistic forms.

$$\bar{\omega} = \omega - \mathbf{k} \cdot \mathbf{v},$$
$$\frac{\bar{E}}{\hbar} = \frac{E}{\hbar} - \mathbf{q} \cdot \mathbf{v},$$
$$\frac{\bar{E}'}{\hbar} = \frac{E'}{\hbar} - \mathbf{q}' \cdot \mathbf{v}. \qquad (24.23)$$

Since Bragg reflection by a stationary grating leaves the frequency of the incident wave unaltered, $\bar{E}'$ must be equal to $\bar{E}$. The transformation law (24.23) then implies that in the original frame the frequency of the scattered wave must be shifted:

$$\frac{E'}{\hbar} = \frac{E}{\hbar} + (\mathbf{q}' - \mathbf{q}) \cdot \mathbf{v}. \qquad (24.24)$$

The change in wave vector under a Bragg reflection has the form:

$$\mathbf{q}' = \mathbf{q} + m\mathbf{k}, \qquad (24.25)$$

where the integer m is the order of the Bragg reflection (as demonstrated in Figure 24.10).[12] This relation holds in either frame, since wave vectors are invariant under change of frame.

Substituting (24.25) into (24.24), we find that the frequency shift in the original frame is given by

$$\frac{E'}{\hbar} = \frac{E}{\hbar} + m\mathbf{k} \cdot \mathbf{v}. \qquad (24.26)$$

If we substitute into (24.26) the explicit form (24.22) for the phase velocity $\mathbf{v}$, we find that

$$E' = E + m\hbar\omega. \qquad (24.27)$$

Equations (24.25) and (24.27) reveal that an mth-order Bragg reflection in the moving frame corresponds to a process that we would describe in the laboratory frame as the absorption or emission of m phonons of a given type. Multiphonon processes involving several normal modes evidently will correspond to successive Bragg reflections from the corresponding moving diffraction gratings.

The wave vector condition (24.25) might appear to lack the arbitrary additive reciprocal lattice vector that is present in the crystal momentum conservation law (24.6). Actually, it is implicit in (24.25) as well, as soon as we acknowledge that the crystal is not a continuum, but a discrete system. Only in a continuum is it possible to assign a unique wave vector $\mathbf{k}$ to each normal mode. In a discrete lattice the normal mode wave vector is defined only to within an additive reciprocal lattice vector (see page 439).

Thus from the wave point of view the law of energy conservation is simply a statement of the Doppler shift for a wave reflected from a moving diffraction grating; the law of crystal momentum conservation is the Bragg condition for that grating, the additive reciprocal lattice vector expressing the variety of orientations that the grating

---

[12] Note that m can have either sign, depending on the side of the grating from which the wave is incident.

may be considered to have, due to the discrete periodic nature of the underlying crystal.

PROBLEMS

1. (a) Draw diagrams showing some possible two-phonon processes in which a neutron enters with momentum $\mathbf{p}$ and leaves with momentum $\mathbf{p}'$. In labeling the diagrams take due account of the conservation laws.

(b) Repeat (a) for three-phonon processes.

2. (a) Repeat the method of graphical solution given in Figure 24.6 for the case of phonon emission.

(b) Verify that when the incident neutron energy is zero, no solutions are possible.

(c) In qualitative terms, how does the number of solutions depend on the incident wave vector $\mathbf{k}_i$?

3. This problem is based on Appendices L and N.

(a) Using the definition of W given in Eq. (N.17) and the expansion (L.14) for $\mathbf{u}(\mathbf{R})$, show that the Debye-Waller factor has the form:

$$e^{-2W} = \exp\left\{-v\int \frac{d\mathbf{k}}{(2\pi)^d}\sum_s \frac{\hbar}{2M\omega_s(\mathbf{k})}(\mathbf{q}\cdot\boldsymbol{\epsilon}_s(\mathbf{k}))^2 \coth \tfrac{1}{2}\beta\hbar\omega_s(\mathbf{k})\right\} \tag{24.28}$$

where v is the appropriate cell volume.

(b) Show that $e^{-2W} = 0$ in one and two dimensions. (Consider the behavior of the integrand for small k.) What are the implications of this for the possible existence of one- or two-dimensional crystalline ordering?

(c) Estimate the size of the Debye-Waller factor for a three-dimensional crystal.

25
Anharmonic Effects in Crystals

Fundamental Inadequacy of Harmonic Models

General Aspects of Anharmonic Theories

Equation of State and
Thermal Expansion of a Crystal

The Grüneisen Parameter

Thermal Expansion of Metals

Phonon Collisions

Lattice Thermal Conductivity

Umklapp Processes

Second Sound

In Chapter 21 we reviewed the evidence forcing us to abandon the model of a static lattice of ions[1], and in subsequent chapters we have cautiously relaxed this oversimplification. We have, however, relied on two less restrictive simplifying assumptions:

1. **Assumption of Small Oscillations** We have assumed that although the ions are not rigidly confined to their equilibrium sites, their displacements from those equilibrium sites are small.
2. **Harmonic Approximation** We have assumed that we can calculate accurately the properties of solids by retaining only the leading nonvanishing term in the expansion of the ionic interaction energy about its equilibrium value.

The assumption of small oscillations appears to be reasonable in most solids (with the important exception of solid helium) for temperatures well below the melting point. In any event, it is forced on us by computational necessity. When it fails, one must resort to very complex approximation schemes whose validity is far from clear.

When the assumption of small oscillations does hold, one might be tempted to conclude that corrections to the harmonic approximation are of interest only in calculations of high precision. This is incorrect. There are many important physical phenomena that cannot be explained in a purely harmonic theory because they are *entirely* due to the neglected higher-order terms in the expansion of the ionic interaction energy about its equilibrium value.

In this chapter we shall survey some of those phenomena whose explanation requires the presence of such *anharmonic terms.* We have already encountered two examples:

1. The quantum theory of the harmonic crystal predicts that the specific heat should obey the classical law of Dulong and Petit at high temperatures ($T \gg \Theta_D$). The failure of the high-temperature specific heat to approach this value is an anharmonic effect (see page 428 and also page 455).
2. In our discussion of neutron scattering (Chapter 24) we argued that the inelastic neutron scattering cross section should have sharp peaks at energies allowed by the conservation laws governing one-phonon processes. However, the observed peaks, though quite clearly peaks, have a measurable width (see Figure 24.4). We interpreted this broadening to be a consequence of the fact that the eigenstates of the harmonic Hamiltonian were not true stationary states of the crystal—i.e., that anharmonic corrections to the harmonic approximation were significant. The width of the one-phonon peaks is a direct measure of the strength of the anharmonic part of the ionic interaction energy.

Additional phenomena dominated by the anharmonic terms can be grouped into equilibrium and transport properties:

1. **Equilibrium Properties** There is a large class of equilibrium properties in crystals, observable at any temperature, whose consistent explanation requires

---

[1] We continue to use the word "ion" in the broadest sense, including in the term, for example, the neutral atoms or molecules making up a molecular solid.

the presence of anharmonic terms in the ionic interaction energy. The most important of these is thermal expansion. In a rigorously harmonic crystal the equilibrium size would not depend on temperature. The existence of anharmonic terms is also implied by the fact that the elastic constants depend on volume and temperature, and by the fact that the adiabatic and isothermal elastic constants are not the same.

2. **Transport Properties** The thermal conductivity of an insulating solid is limited in a perfect crystal only by the anharmonic terms in the ionic interaction energy. A rigorously harmonic crystal would have an infinite thermal conductivity. This is probably the most important transport property determined by the anharmonic terms, but they also play essential roles in almost any of the processes by which the lattice vibrations transmit energy.

GENERAL ASPECTS OF ANHARMONIC THEORIES

The standard description of anharmonic terms is simple in principle, though plagued in practice by notational complexities. The small-oscillation assumption is retained which permits one to keep only the leading corrections to the harmonic terms in the expansion of the ionic interaction energy, U, in powers of the ionic displacements $\mathbf{u}$. Thus one replaces (22.8) and (22.10) by:

$$U = U^{\text{eq}} + U^{\text{harm}} + U^{\text{anh}}, \tag{25.1}$$

where (see (22.10)) the anharmonic correction terms have the form:

$$U^{\text{anh}} = \sum_{n=3}^{\infty} \frac{1}{n!} \sum_{\mathbf{R}_1 \cdots \mathbf{R}_n} D^{(n)}_{\mu_1 \cdots \mu_n}(\mathbf{R}_1 \cdots \mathbf{R}_n) u_{\mu_1}(\mathbf{R}_1) \cdots u_{\mu_n}(\mathbf{R}_n), \tag{25.2}$$

where

$$D^{(n)}_{\mu_1 \cdots \mu_n}(\mathbf{R}_1 \cdots \mathbf{R}_n) = \partial^n U / \partial u_{\mu_1}(\mathbf{R}_1) \cdots \partial u_{\mu_n}(\mathbf{R}_n)\big|_{\mathbf{u}\equiv 0}. \tag{25.3}$$

In the spirit of the assumption of small oscillations one might be tempted to retain only the leading terms (cubic in the $\mathbf{u}$) in U^{anh}, and this is frequently done. However, there are two reasons for retaining the quartic terms as well:

1. The Hamiltonian that retains only the cubic anharmonic terms is unstable: The potential energy can be made as large and negative as desired by assigning suitable values to the $\mathbf{u}$'s (see Problem 1). This implies that the cubic Hamiltonian has no ground state,[2] so by replacing the full Hamiltonian with one truncated at the cubic anharmonic terms, we have replaced a well-defined physical problem by one with spectacular, but artificial, mathematical pathologies. Nevertheless, the additional cubic terms are frequently treated as a small perturbation, and physically sensible results are found in spite of the formal absurdity of the procedure. If one insists, however, on dealing with a well-defined problem, one must retain the quartic terms as well.

---

[2] See, for example, G. Baym, *Phys. Rev.* **117**, 886 (1960).

2. Contributions from the cubic terms often behave anomalously not for the reason given above but because of the rather stringent requirements the conservation laws impose on processes mediated by these terms. When this happens quartic terms can be of comparable importance even when the small oscillations assumption is a good one.

It is the almost universal practice to retain no terms beyond the quartic anharmonic terms in detailed calculations, unless one is engaged in proving very general types of results, or dealing with crystals (notably solid helium) for which the entire assumption of small oscillations is of doubtful validity. Furthermore, there is a tendency in practice to keep only the cubic anharmonic terms, though one must always keep in mind the pitfalls described above.

EQUATION OF STATE AND THERMAL EXPANSION OF A CRYSTAL

To calculate the equation of state, we write the pressure as $P = -(\partial F/\partial V)_T$, where F, the Helmholtz free energy, is given by $F = U - TS$. Since the entropy S and internal energy U are related by

$$T\left(\frac{\partial S}{\partial T}\right)_V = \left(\frac{\partial U}{\partial T}\right)_V, \tag{25.4}$$

we can express the pressure entirely in terms of the internal energy, in the form:[3]

$$P = -\frac{\partial}{\partial V}\left[U - T\int_0^T \frac{dT'}{T'}\frac{\partial}{\partial T'}U(T', V)\right]. \tag{25.5}$$

If the small-oscillations assumption is valid, then the internal energy of an insulating crystal should be accurately given by the result (23.11) of the harmonic approximation:

$$U = U^{\text{eq}} + \tfrac{1}{2}\sum_{\mathbf{k}s}\hbar\omega_s(\mathbf{k}) + \sum_{\mathbf{k}s}\frac{\hbar\omega_s(\mathbf{k})}{e^{\beta\hbar\omega_s(\mathbf{k})} - 1}. \tag{25.6}$$

Substituting this into the general form, (25.5), one finds that[4]

$$P = -\frac{\partial}{\partial V}\left[U^{\text{eq}} + \sum\tfrac{1}{2}\hbar\omega_s(\mathbf{k})\right] + \sum_{\mathbf{k}s}\left(-\frac{\partial}{\partial V}(\hbar\omega_s(\mathbf{k}))\right)\frac{1}{e^{\beta\hbar\omega_s(\mathbf{k})} - 1}. \tag{25.7}$$

This result has a very simple structure: The first term (which is all that survives at $T = 0$) is the negative volume derivative of the ground-state energy. At nonzero temperatures this must be supplemented with the negative volume derivative of the phonon energies, the contribution of each phonon level being weighted with its mean occupation number.

According to (25.7), the equilibrium pressure depends on temperature only because the normal-mode frequencies depend on the equilibrium volume of the crystal. If,

---

[3] We use the fact that the entropy density vanishes at $T = 0$ (third law of thermodynamics) to dispose of an integration constant.

[4] See Problem 2.

however, the potential energy of the crystal were rigorously of the harmonic form (Eqs. (22.46) and (22.8))

$$U^{eq} + \frac{1}{2} \sum_{\mathbf{RR}'} \mathbf{u}(\mathbf{R})\mathbf{D}(\mathbf{R} - \mathbf{R}')\mathbf{u}(\mathbf{R}'), \tag{25.8}$$

with force constants $\mathbf{D}$ that were independent of the $\mathbf{u}(\mathbf{R})$, then the normal mode frequencies could have no volume dependence at all.[5]

To see this, note that to determine the volume dependence of the normal-mode frequencies, we must examine the small-oscillations problem not only for the original Bravais lattice given by vectors $\mathbf{R}$, but also for the expanded (or contracted) lattices given by the vectors[6] $\bar{\mathbf{R}} = (1 + \epsilon)\mathbf{R}$, whose volumes differ by the factor $(1 + \epsilon)^3$ from the volume of the original lattice. If the potential energy is rigorously of the form (25.8) even when the $\mathbf{u}(\mathbf{R})$ are not small, then the new small-oscillations problem is easily reduced back to the old one. For the ionic positions $\mathbf{r}(\mathbf{R}) = \bar{\mathbf{R}} + \bar{\mathbf{u}}(\bar{\mathbf{R}})$ can also be written as $\mathbf{r}(\mathbf{R}) = \mathbf{R} + \mathbf{u}(\mathbf{R})$, provided that the displacements $\mathbf{u}$ with respect to the original lattice are related to the displacements $\bar{\mathbf{u}}$ with respect to the expanded (or contracted) lattice by

$$\mathbf{u}(\mathbf{R}) = \epsilon\mathbf{R} + \bar{\mathbf{u}}(\bar{\mathbf{R}}). \tag{25.9}$$

If the potential energy is rigorously given by (25.8), then to evaluate the energy of the configuration given by $\mathbf{r}(\mathbf{R}) = \bar{\mathbf{R}} + \bar{\mathbf{u}}(\bar{\mathbf{R}})$ we need not perform a new expansion of U about the new equilibrium positions $\bar{\mathbf{R}}$, but may simply substitute the equivalent displacements $\mathbf{u}$ given by (25.9) into (25.8). The resulting expression for the potential energy of the configuration in which the ions are displaced by $\bar{\mathbf{u}}(\bar{\mathbf{R}})$ from equilibrium positions at $\bar{\mathbf{R}}$ is[7]

$$U^{eq} + \frac{1}{2}\epsilon^2 \sum_{\mathbf{RR}'} \mathbf{R}\mathbf{D}(\mathbf{R} - \mathbf{R}')\mathbf{R}' + \frac{1}{2} \sum_{\mathbf{RR}'} \bar{\mathbf{u}}(\mathbf{R})\mathbf{D}(\mathbf{R} - \mathbf{R}')\bar{\mathbf{u}}(\mathbf{R}'). \tag{25.10}$$

The first two terms in (25.10) are independent of the new displacements $\bar{\mathbf{u}}$ and give the potential energy of the new equilibrium configuration. The dynamics are determined by the term quadratic in the $\bar{\mathbf{u}}$. Since the coefficients in this term are identical to the coefficients of the corresponding term in (25.8), the dynamics of the oscillations about the new equilibrium positions will be identical to those of the old. The normal-mode frequencies are therefore unaffected by the change in equilibrium volume.

Because the normal-mode frequencies of a rigorously harmonic crystal are unaffected by a change in volume, the pressure given by (25.7) depends only on volume, but not on temperature. Thus in a rigorously harmonic crystal the pressure required to maintain a given volume does not vary with temperature. Since

$$\left(\frac{\partial V}{\partial T}\right)_P = -\frac{(\partial P/\partial T)_V}{(\partial P/\partial V)_T}, \tag{25.11}$$

---

[5] This is a generalization of the familiar observation that the frequency of a harmonic oscillator does not depend on the amplitude of vibration.

[6] For simplicity we consider only monatomic Bravais lattices whose symmetry is such that a uniform isotropic expansion (or contraction) yields a new *equilibrium* configuration (as opposed, for example, to a crystal of orthorhombic symmetry, where the scale factor $(1 + \epsilon)$ would differ along different crystal axes). The final result, however, is quite general.

[7] Terms linear in the $\bar{\mathbf{u}}$ must vanish if the new sites $\bar{\mathbf{R}}$ do indeed give an equilibrium configuration of the crystal.

it also follows that the equilibrium volume cannot vary with temperature at fixed pressure. Thus the coefficient of thermal expansion,[8]

$$\alpha = \frac{1}{l}\left(\frac{\partial l}{\partial T}\right)_P = \frac{1}{3V}\left(\frac{\partial V}{\partial T}\right)_P = \frac{1}{3B}\left(\frac{\partial P}{\partial T}\right)_V, \tag{25.12}$$

must vanish.

The absence of thermal expansion in a rigorously harmonic lattice implies thermodynamically several other anomalies. The constant-volume and constant-pressure specific heats are related by:

$$c_p = c_v - \frac{T(\partial P/\partial T)_V^2}{V(\partial P/\partial V)_T} \tag{25.13}$$

and must therefore be identical in such a solid. So must the adiabatic and isothermal compressibilities, since

$$\frac{c_p}{c_v} = \frac{(\partial P/\partial V)_S}{(\partial P/\partial V)_T}. \tag{25.14}$$

Results such as these are anomalous, because in real crystals the force constants **D** in the harmonic approximation to the potential energy do depend on the equilibrium lattice about which the harmonic expansion is made. Implicit in this dependence is the fact that in real crystals the harmonic approximation is not exact. Indeed, it is possible to express the amount by which the normal-mode frequencies change when the equilibrium lattice vectors are changed from **R** to $(1 + \epsilon)\mathbf{R}$, in terms of the coefficients of the anharmonic terms appearing in the expansion of the potential energy[9] about the equilibrium positions **R**. In this way, measurements of the thermal expansion coefficient can be made to yield information about the magnitude of the anharmonic corrections to the energy.

THERMAL EXPANSION; THE GRÜNEISEN PARAMETER

Having recognized that the phonon frequencies of a real crystal do depend on the equilibrium volume, we may continue with the analysis of the equation of state (25.7). Substituting this form of the pressure into (25.12), we find that the coefficient of thermal expansion may be written as

$$\alpha = \frac{1}{3B}\sum_{\mathbf{k}s}\left(-\frac{\partial}{\partial V}\hbar\omega_{\mathbf{k}s}\right)\frac{\partial}{\partial T}n_s(\mathbf{k}), \tag{25.15}$$

where $n_s(\mathbf{k}) = [e^{\beta\hbar\omega_s(\mathbf{k})} - 1]^{-1}$. If we compare this with the formula (23.12) for the specific heat, which can be written in the form

$$c_v = \sum_{\mathbf{k}s}\frac{\hbar\omega_s(\mathbf{k})}{V}\frac{\partial}{\partial T}n_s(\mathbf{k}), \tag{25.16}$$

the following representation is suggested for the thermal expansion coefficient α:

---

[8] We continue to assume a symmetric enough crystal that all linear dimensions scale in the same way with temperature. Crystals of noncubic symmetry have direction-dependent expansion coefficients. We have introduced the bulk modulus B, defined (Eq. (2.35)) by $B = -V(\partial P/\partial V)_T$.

[9] See Problem 4.

First, define a quantity

$$c_{vs}(\mathbf{k}) = \frac{\hbar\omega_s(\mathbf{k})}{V}\frac{\partial}{\partial T} n_s(\mathbf{k}), \tag{25.17}$$

which is the contribution of the normal-mode **k**, s to the specific heat. Next, define a quantity $\gamma_{\mathbf{k}s}$, known as the *Grüneisen parameter* for the mode **k**s, as the negative logarithmic derivative of the frequency of the mode with respect to volume; i.e.,

$$\gamma_{\mathbf{k}s} = -\frac{V}{\omega_s(\mathbf{k})}\frac{\partial\omega_s(\mathbf{k})}{\partial V} = -\frac{\partial(\ln\omega_s(\mathbf{k}))}{\partial(\ln V)}. \tag{25.18}$$

Finally, define an overall *Grüneisen parameter*

$$\gamma = \frac{\sum_{\mathbf{k},s}\gamma_{\mathbf{k}s}c_{vs}(\mathbf{k})}{\sum_{\mathbf{k},s}c_{vs}(\mathbf{k})}, \tag{25.19}$$

as the weighted average of the $\gamma_{\mathbf{k}s}$, in which the contribution of each normal mode is weighted by its contribution to the specific heat. Using these definitions, we may write (25.15) in the simple form:

$$\alpha = \frac{\gamma c_v}{3B}. \tag{25.20}$$

The coefficient of thermal expansion is represented in this rather peculiar way because in the simplest models the volume dependence of the normal-mode frequencies is contained in a universal multiplicative factor, and therefore the $\gamma_{\mathbf{k}s}$ are the same for all normal modes. Under these circumstances, (25.15) reduces directly to (25.20) without need for the intervening definitions. In a Debye model, for example, all the normal-mode frequencies scale linearly with the cutoff frequency ω_D, and therefore

$$\gamma_{\mathbf{k}s} \equiv -\frac{\partial(\ln\omega_D)}{\partial(\ln V)}. \tag{25.21}$$

Since the bulk modulus appearing in the denominator of (25.20) is only weakly temperature-dependent,[10] theories with constant $\gamma_{\mathbf{k}s}$ predict that the coefficient of thermal expansion should have the same temperature dependence as the specific heat. In particular, it should approach a constant at temperatures large compared with Θ_D, and should vanish as T^3 as $T \to 0$.

The representation (25.20) preserves these two limiting forms. In any real solid the $\gamma_{\mathbf{k}s}$ will not be the same for all normal modes, and γ will therefore depend on temperature. Nevertheless, Eq. (25.19) implies that γ approaches a constant value as $T \to 0$, and a (different) constant value at temperatures large compared with Θ_D. Consequently, the limiting temperature dependence of the thermal expansion coefficient, even in the general case, will be

$$\begin{aligned} \alpha &\sim T^3, & T &\to 0; \\ \alpha &\sim \text{constant}, & T &\gg \Theta_D. \end{aligned} \tag{25.22}$$

Some Grüneisen parameters and their variation with temperature are given in Table 25.1 and Figure 25.1.

---

10 In any event, B is directly measurable, so its slight temperature dependence can be allowed for.

Table 25.1
LINEAR EXPANSION COEFFICIENTS AND GRÜNEISEN PARAMETERS FOR SOME ALKALI HALIDES

T (K)		LiF	NaCl	NaI	KCl	KBr	KI	RbI	CsBr
0	α	0	0	0	0	0	0	0	0
	γ	1.70	0.90	1.04	0.32	0.29	0.28	−0.18	2.0
20	α	0.063	0.62	5.1	0.74	2.23	4.5	6.0	10.0
	γ	1.60	0.96	1.22	0.53	0.74	0.79	0.85	—
65	α	3.6	15.8	27.3	17.5	22.5	26.0	28.0	35.2
	γ	1.59	1.39	1.64	1.30	1.42	1.35	1.35	—
283	α	32.9	39.5	45.1	36.9	38.5	40.0	39.2	47.1
	γ	1.58	1.57	1.71	1.45	1.49	1.47	—	2.0

[a] The units of the linear expansion coefficient α are in 10^{-6} K^{-1}.
Source: G. K. White, *Proc. Roy Soc. London* **A286**, 204 (1965).

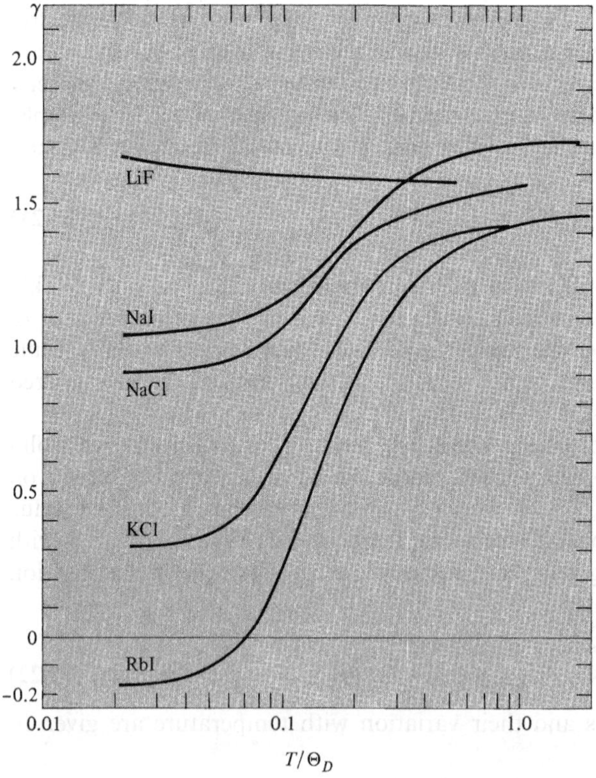

Figure 25.1
Grüneisen parameter vs. T/Θ_D for some alkali halide crystals. (From G. K. White, *Proc. Roy Soc. London* **A286**, 204 (1965).)

THERMAL EXPANSION OF METALS

The above discussion assumes that the only degrees of freedom are ionic—i.e., that the solid is an insulator. In the case of a metal, we can estimate the effects of the additional electronic degrees of freedom through Eq. (25.12). Again, the bulk modulus is very weakly temperature-dependent and can be replaced by its value at $T = 0$. For a rough estimate of the electronic contribution to $(\partial P/\partial T)_V$ we simply add to the contribution from the lattice vibrations that of a free electron gas. Since the free electron gas equation of state is (see Eq. (2.101))

$$P = \frac{2}{3}\frac{U}{V} \tag{25.23}$$

it follows that

$$\left(\frac{\partial P^{\text{el}}}{\partial T}\right)_V = \frac{2}{3}c_v^{\text{el}} \tag{25.24}$$

and therefore the coefficient of thermal expansion becomes

$$\alpha = \frac{1}{3B}\left(\gamma c_v^{\text{ion}} + \frac{2}{3}c_v^{\text{el}}\right). \tag{25.25}$$

Since the Grüneisen parameter is typically of order unity, the electronic contribution to the temperature dependence of the expansion coefficient will be appreciable only at temperatures at which the electronic contribution to the specific heat is comparable to that of the ions—i.e., at temperatures of the order of 10 K or less (see Eq. (23.30)).[11] Thus the most striking predicted difference between the expansion coefficient of metals and insulators is that at very low temperatures α should vanish linearly in T in metals but as T^3 in insulators. This behavior is confirmed by experiment.[12]

Some characteristic thermal expansion coefficients for metals are given in Table 25.2.

LATTICE THERMAL CONDUCTIVITY: THE GENERAL APPROACH

As we have noted in Chapters 22 and 23, thermal energy can be stored in the vibrational normal modes of the crystal. Since these modes are elastic waves, one can also transport thermal energy through the lattice of ions by setting up suitable wave packets of normal modes, just as one can send impulses down a stretched elastic string by plucking one end. At low temperatures the fact that the allowed energies of a normal mode are quantized is of critical importance, and it is much more convenient to describe this transport of energy in the phonon language.

---

[11] The electrons will, of course, make a substantial contribution to the (roughly temperature independent) bulk modulus (see pp. 39, 40).

[12] See G. K. White, *Proc. Roy Soc. London*, **A286**, 204 (1965), and K. Andres, *Phys. Kondens. Mater.* **2**, 294 (1964).

Table 25.2
LINEAR EXPANSION COEFFICIENTS FOR SELECTED METALS AT ROOM TEMPERATURE

METAL	COEFFICIENT[a]	METAL	COEFFICIENT[a]
Li	45	Ca	22.5
Na	71	Ba	18
K	83	Nb	7.1
Rb	66	Fe	11.7
Cs	97	Zn	61 (‖)
Cu	17.0		14 (⊥)
Ag	18.9	Al	23.6
Au	13.9	In	−7.5 (‖)
Be	9.4 (‖)		50 (⊥)
	11.7 (⊥)	Pb	28.8
Mg	25.7 (‖)	Ir	6.5
	24.3 (⊥)		

[a] The units are $10^{-6}\ K^{-1}$. In the noncubic cases separate coefficients are listed for expansion parallel and perpendicular to the axis of highest symmetry.
Source: W. B. Pearson, *A Handbook of Lattice Spacings and Structures of Metals and Alloys*, Pergamon, New York, 1958.

In using the phonon picture to describe energy transport one considers phonons whose positions are localized within a definite region, which is small on the scale of the macroscopic dimensions of the crystal, though it must be large on the scale of the interionic spacing. Since a single normal mode with a definite wave vector **k** involves the motion of ions throughout the entire crystal, a state consisting of a single phonon with wave vector **k** cannot describe a localized disturbance of the crystal. However, by superposing states of the crystal in each of which a normal mode with wave vector in some small neighborhood $\Delta\mathbf{k}$ of **k** is excited, one can construct localized phonon-like disturbances. The justification for the change from wave to particle language relies on the properties of wave packets. Rather than exploring the detailed and relatively unilluminating mathematics of wave packets, we simply stress the analogy with the electronic case,[13] and allow ourselves the same liberty with phonons: By sacrificing some precision in the specification of phonon wave vector we can construct phonon wave functions[14] that are localized on a scale $\Delta x \approx 1/\Delta k$.

In a perfectly harmonic crystal the phonon states are stationary states. Therefore, if a distribution of phonons is established that carries a thermal current (for example, by having an excess of phonons with similarly directed group velocities) that dis-

[13] See pages 50 and 216. The point at issue here is completely parallel to that we discussed in replacing the wave-mechanical description of electrons with the classical picture of localized particles. The transport of energy by thermal currents can be viewed in the same way as the transport of charge by electrical currents. The carriers are now phonons and the quantity carried by each phonon is its energy $\hbar\omega_s(\mathbf{k})$.

[14] Note that the uncertainty in wave vector must be small compared with the dimensions of the Brillouin zone if we are to assign a wave vector with the conventional properties to the phonon. Since the zone dimensions are on the order of the inverse lattice constant, Δx must be large compared with the interionic spacing, and as one might have guessed the phonons are not localized on the microscopic scale.

tribution will remain unaltered in the course of time, and the thermal current will remain forever undegraded. *A perfectly harmonic crystal would have an infinite thermal conductivity.*[15]

The thermal conductivity of real insulators[16] is not infinite for several reasons:

1. The inevitable lattice imperfections, impurities, isotopic inhomogeneities, and the like (Chapter 30) that afflict real crystals act as scattering centers for the phonons, and help to degrade any thermal current.
2. Even in a perfect, pure crystal, the phonons would eventually be scattered by the surface of the sample, and this would limit the thermal current.
3. Even in a perfect, pure, and infinite crystal, the stationary states of the harmonic Hamiltonian are only approximate stationary states of the full anharmonic Hamiltonian, and therefore a state with a definite set of phonon occupation numbers will not remain unchanged in the course of time.

In the present discussion we are primarily interested in the last point, which is the only intrinsic source of thermal resistance that cannot in principle be systematically reduced by making bigger and better crystals.

The point of view generally taken in discussing this effect of anharmonicity is to view the anharmonic corrections to the harmonic Hamiltonian H_0, as perturbations which cause transitions from one harmonic eigenstate to another—i.e., which lead to the creation, destruction, or scattering of phonons. Thus the anharmonic part of the ionic interaction plays the same role in the theory of heat transport in an insulator as impurities or the electron-phonon interaction do in the theory of the transport of charge in a metal.

If the anharmonic terms are small[17] compared with the harmonic part of the Hamiltonian, then it should suffice to calculate their effects in perturbation theory, and this is what is generally done. It can be shown (Appendix O) that in lowest-order perturbation theory an anharmonic term of degree n in the ionic displacements $\mathbf{u}$ can cause transitions between two eigenstates of the harmonic Hamiltonian, precisely n of whose phonon occupation numbers differ. Thus the cubic anharmonic term can cause the following kinds of transitions:

1. All phonon occupation numbers in the initial and final states are unchanged except that $n_{\mathbf{k}s} \rightarrow n_{\mathbf{k}s} - 1$, $n_{\mathbf{k}'s'} \rightarrow n_{\mathbf{k}'s'} + 1$, and $n_{\mathbf{k}''s''} \rightarrow n_{\mathbf{k}''s''} + 1$. Evidently such a transition can be thought of as an event in which a phonon from branch s with wave vector $\mathbf{k}$ decays into two with wave vectors and branch indices $\mathbf{k}'s'$ and $\mathbf{k}''s''$.
2. All phonon occupation numbers in the initial and final states are unchanged except that $n_{\mathbf{k}s} \rightarrow n_{\mathbf{k}s} - 1$, $n_{\mathbf{k}'s'} \rightarrow n_{\mathbf{k}'s'} - 1$, and $n_{\mathbf{k}''s''} \rightarrow n_{\mathbf{k}''s''} + 1$. Such a transition can be viewed as an event in which two phonons with wave vectors and branch

[15] This is analogous to the fact (see page 141) that electrons in a perfect periodic potential (without defects or lattice vibrations) would be perfect electrical conductors.

[16] We speak of insulators, though our remarks also apply to the ionic contribution to the thermal conductivity of metals. The latter, however, is generally masked by an electronic contribution that is one or two orders of magnitude larger.

[17] The fact that one-phonon peaks are clearly identifiable in the scattering of neutrons from crystals is some indication of this (see Chapter 24).

indices $\mathbf{k}s$ and $\mathbf{k}'s'$ merge to form a single phonon in branch s'' with wave vector $\mathbf{k}''$.

Such processes are often depicted schematically as in Figure 25.2.

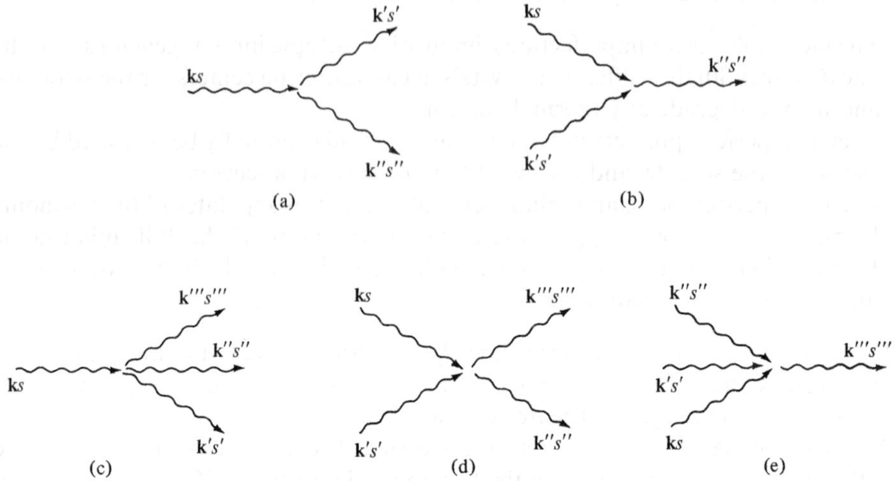

Figure 25.2
Processes produced in lowest-order perturbation theory by cubic and quartic anharmonic terms. (a) Cubic: one phonon decays into two. (b) Cubic: two phonons merge into one. (c) Quartic: one phonon decays into three. (d) Quartic: two phonons turn into two others (phonon-phonon scattering). (e) Quartic: three phonons merge into one.

The two other cubic processes that come to mind (three phonons disappearing or three new phonons being created) are forbidden by energy conservation. Since the total energy of the three phonons must be positive, energy would be lost in a three-phonon annihilation, and found from nowhere in a three-phonon creation.

In a similar vein, the quartic terms can yield transitions which are simply characterized as ones in which one phonon decays into three, three merge into one, or two of a given kind are replaced by two others (see Figure 25.2).

Higher-order anharmonic terms contribute transitions as well, but, in the spirit of the assumption of small oscillations, the cubic and quartic terms are expected to be the most important. Frequently, only transitions caused by the cubic terms are considered. However, as we have mentioned, the conservation laws often impose very stringent restrictions on the processes produced by the cubic terms. As a result, even though the quartic terms are smaller than the cubic, so few cubic processes may be allowed that the both types of process may yield comparable transition rates.

In the discussion that follows we shall not appeal to any detailed features of the anharmonic terms, other than those embodied in the conservation laws of energy and crystal momentum.[18] If the phonon occupation numbers are $n_{\mathbf{k}s}$ before the transition and $n'_{\mathbf{k}s}$ after, then energy conservation requires

$$\sum \hbar\omega_s(\mathbf{k})n_{\mathbf{k}s} = \sum \hbar\omega_s(\mathbf{k})n'_{\mathbf{k}s}, \tag{25.26}$$

[18] Equation (M.18). A full discussion of crystal momentum conservation is given in Appendix M. See also page 471.

and crystal momentum conservation requires

$$\sum \mathbf{k} n_{\mathbf{k}s} = \sum \mathbf{k} n'_{\mathbf{k}s} + \mathbf{K}, \tag{25.27}$$

where **K** is some reciprocal lattice vector.

These transitions are usually referred to by the term "collisions" to emphasize the analogy with electronic transport. Here, however, the term includes processes in which a single phonon decays into several, several merge into one, and similar generalized "collisions" that a theory without number conservation permits. To the extent that the small oscillations assumption is valid and high-order anharmonic terms are unimportant, only a small number of phonons will participate in any given "collision." One can then treat the transport of energy by phonons with a Boltzmann equation (Chapter 16) containing collision terms describing those processes in which phonons can be scattered with appreciable probability. For a more elementary qualitative theory, one may even introduce a single phonon relaxation time τ, which specifies the probability per unit time of a phonon suffering any of the various types of collision.[19]

LATTICE THERMAL CONDUCTIVITY: ELEMENTARY KINETIC THEORY

We shall not explore here the detailed approach to lattice thermal conductivity embodied in the phonon Boltzmann equation, but shall, instead, illustrate some of the important physical features of the problem using an elementary relaxation-time approximation, analogous to that used in our discussion of electronic transport in metals in Chapters 1 and 2.

For simplicity, we continue to treat only monatomic Bravais lattices, where the phonon spectrum has only acoustic branches. Since we are concerned primarily with qualitative features of thermal conduction rather than precise results, we shall also make the Debye approximation, when it is convenient, letting the phonon dispersion relation be $\omega = ck$ for all three acoustic branches.

Suppose a small temperature gradient is imposed along the x-direction in an insulating crystal (Figure 25.3). As in the Drude model (cf. page 6) we assume that collisions maintain local thermodynamic equilibrium in a particularly simple way. Those phonons emerging from collisions at position x are taken to contribute to the nonequilibrium energy density an amount proportional to the equilibrium energy density at temperature $T(x)$: $u(x) = u^{\text{eq}}[T(x)]$. Each phonon at a given point will contribute to the thermal current density in the x-direction an amount equal to the product of the x-component of its velocity with its contribution to the energy density.[20] However, the average contribution of a phonon to the energy density depends on the position of its last collision. There is thus a correlation between where a phonon

[19] In precise analogy with the electron relaxation time introduced in the discussion of the Drude model. The argument that follows is also analogous to the electronic case except that the phonons are uncharged (no thermoelectric field), the phonon number density depends on temperature, and phonons need not be conserved, particularly at the ends of specimens.

[20] For a more detailed description of this procedure see the quite analogous discussion of the electronic contribution to the thermal conductivity of a metal on pages 20–22.

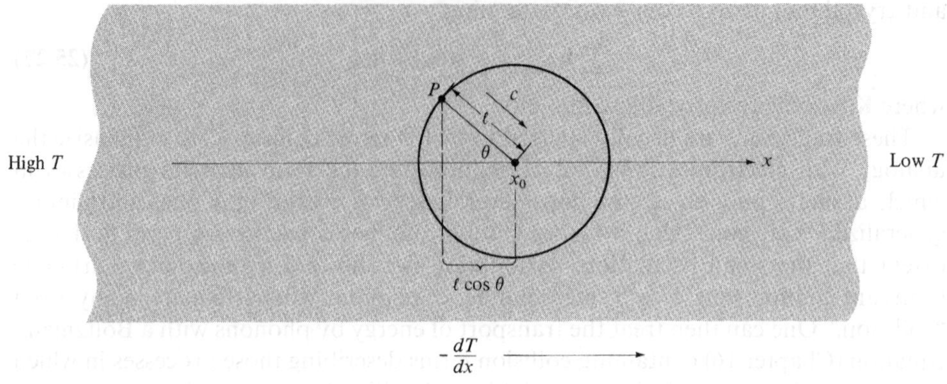

Figure 25.3
Heat propagation by phonons in the presence of a uniform temperature gradient along the x-axis. The thermal current at x_0 is carried by phonons whose last collision was, on the average, a distance $\ell = c\tau$ away from x_0. Phonons with velocities making an angle θ with the x-axis at x_0 collided last at a point P a distance $\ell \cos \theta$ up the temperature gradient, and therefore carry an energy density $u(x_0 - \ell \cos \theta)$ with x-velocity $c \cos \theta$. The net thermal current is proportional to the product of these quantities averaged over all solid angles.

has come from (i.e., the direction of its velocity) and its contribution to the average energy density, which results in a net thermal current.

To estimate this thermal current we average the product of the energy density and the x-velocity over all the places where the phonon's last collision might have occurred. Assuming, in the spirit of the Drude model, that the collision occurred a distance $\ell = c\tau$ from the point x_0 (at which the current is to be calculated), in a direction making an angle θ to the x-axis (Figure 25.3) we have

$$j = \langle c_x u(x_0 - \ell \cos \theta)\rangle_\theta = \int_0^\pi c \cos \theta \, u(x_0 - \ell \cos \theta) \frac{2\pi \, d\theta}{4\pi} \sin \theta,$$
$$= \frac{1}{2}\int_{-1}^{1} \mu d\mu \, cu(x_0 - \ell\mu). \tag{25.28}$$

To linear order in the temperature gradient we then have

$$j = -c\ell \frac{\partial u}{\partial x} \cdot \frac{1}{2}\int_{-1}^{1} \mu^2 \, d\mu = \frac{1}{3} c\ell \frac{\partial u}{\partial T}\left(-\frac{\partial T}{\partial x}\right), \tag{25.29}$$

or

$$j = \kappa\left(-\frac{\partial T}{\partial x}\right), \tag{25.30}$$

where the thermal conductivity κ is given by

$$\kappa = \tfrac{1}{3} c_v c \ell = \tfrac{1}{3} c_v c^2 \tau. \tag{25.31}$$

Here c_v is the specific heat of the phonons and is one of the quantities that determines the temperature dependence of κ. The other[21] is the phonon collision rate, τ^{-1}. We

---

[21] In the Debye model the phonon speed c is a temperature-independent constant. Even in a more accurate model, in which c^2 should be replaced by some suitable average, it will not make a strong contribution to the temperature dependence of κ, in contrast to a classical gas, where $c^2 \sim k_B T$.

have discussed $c_v(T)$ in Chapter 23, but the question of the temperature dependence of τ^{-1} is one of great subtlety and complexity, which took many years to understand fully. The problems that arise depend on whether the high-temperature ($T \gg \Theta_D$) or low-temperature ($T \ll \Theta_D$) regime is considered.

***Case* 1** ($T \gg \Theta_D$) At high temperatures the total number of phonons present in the crystal is proportional to T because the thermal equilibrium phonon occupation numbers reduce to:

$$n_s(\mathbf{k}) = \frac{1}{e^{\hbar\omega_s(\mathbf{k})/k_BT} - 1} \approx \frac{k_BT}{\hbar\omega_s(\mathbf{k})}. \tag{25.32}$$

Since a given phonon that contributes to the thermal current is more likely to be scattered the more other phonons there are present to do the scattering, we should expect the relaxation time to decline with increasing temperature. Furthermore, since at high temperatures the phonon specific heat obeys the law of Dulong and Petit and is temperature-independent, we should expect the thermal conductivity itself to decline with increasing temperature, in the high-temperature regime.

This is confirmed by experiment. The rate of decline is generally given by

$$\kappa \sim \frac{1}{T^x}, \tag{25.33}$$

where x is somewhere between 1 and 2. The precise theory of the power law is quite complex, having to do with competition between scattering processes produced by cubic and quartic anharmonic terms.[22] This is a case where processes governed by the cubic terms are so stringently limited by the conservation laws that the quartic terms, even when very much smaller, can produce enough processes to even up the balance.

***Case* 2** ($T \ll \Theta_D$) At any temperature T, only phonons with energies comparable to or less than k_BT will be present in appreciable numbers. In particular, when $T \ll \Theta_D$, the phonons present will have $\omega_s(\mathbf{k}) \ll \omega_D$, and $k \ll k_D$. With this in mind, consider a phonon collision mediated by the cubic or quartic anharmonic terms. Since only a small number of phonons are involved, the total energy and total crystal momentum of those phonons about to participate in the collision must be small compared with $\hbar\omega_D$ and k_D. Since energy is conserved in the collision, the total energy of the phonons emerging from the collision must continue to be small compared with $\hbar\omega_D$. This is only possible if the wave vector of each, and hence their total wave vector, is small compared with k_D. However, both initial and final total wave vectors can be small compared with k_D (which is comparable in size to a reciprocal lattice vector) only if the additive reciprocal lattice vector $\mathbf{K}$ appearing in the crystal momentum conservation law is zero. Thus, at very low temperatures, the only collisions occurring with appreciable probability are those that conserve the total crystal momentum exactly, and not just to within an additive reciprocal lattice vector.

This very important conclusion is sometimes stated in terms of a distinction between so-called *normal* and *umklapp* processes: A normal process is a phonon collision in which the total initial and final crystal momenta are strictly equal; in an

22 See, for example, C. Herring, *Phys. Rev.* **95**, 954 (1954), and references cited therein.

umklapp process they differ by a nonzero reciprocal lattice vector. Evidently this distinction depends on the primitive cell in which one chooses to specify the phonon wave vector (Figure 25.4). That cell is almost always taken to be the first Brillouin zone.[23] The effect of low temperatures on crystal momentum conservation is sometimes summarized in the assertion that *at sufficiently low temperatures the only scattering processes that can occur at an appreciable rate are normal processes: umklapp processes are "frozen out."*

The freezing out of umklapp processes is of critical significance for the low-

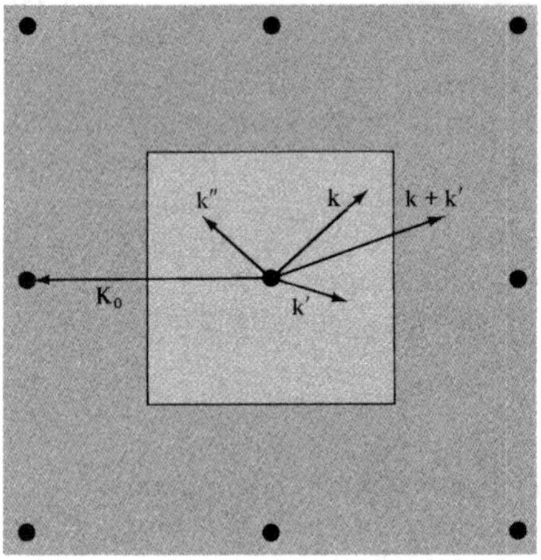

(a)

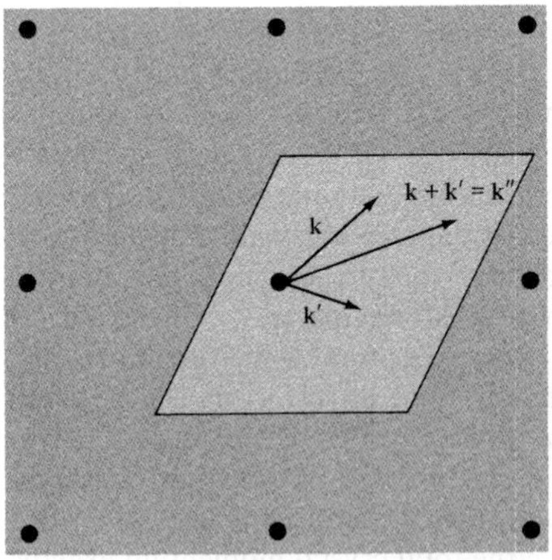

(b)

Figure 25.4

An umklapp process (illustrated in the two-dimensional square lattice). The points are reciprocal lattice points, the square region in (a) is the first Brillouin zone, and the parallelogram in (b) is an alternative primitive cell. Two phonons with wave vectors **k** and **k**′ are allowed by crystal momentum conservation to merge into a single phonon of wave vector **k**″. If all phonon wave vectors are specified in the first Brillouin zone, then **k**″ differs from **k** + **k**′ by the nonzero reciprocal lattice vector $\mathbf{K}_0$, and the process is an umklapp process. If, however, all phonon wave vectors are specified in the primitive cell shown in (b), then **k**″ = **k** + **k**′, and the process is a normal process. Given the primitive cell in which phonon wave vectors are to be specified, the distinction between umklapp and normal processes is unambiguous, for every phonon level has a unique wave vector in that cell, and the sums of the initial and final wave vectors are uniquely determined. The process is normal if the two sums agree, and umklapp if they differ by a nonzero reciprocal lattice vector. However, by changing to another primitive cell, one can make some umklapp processes normal, and vice versa. (Note that the vectors **k** and **k**′ are the same in (a) and (b).)

---

[23] The rarity of collisions that add a reciprocal lattice vector to the total crystal momentum at low temperatures can be unambiguously described in terms of the freezing out of umklapp processes, provided that the primitive cell contains a neighborhood of the point $\mathbf{k} = 0$ that is large enough to include all phonon wave vectors $\mathbf{k}$ with $\hbar\omega_s(\mathbf{k})$ large compared with $k_B T$. The first Brillouin zone is obviously such a choice.

temperature thermal conductivity. If only normal processes occur, then the total phonon wave vector

$$\sum_{s} \sum_{\text{1st Bz}} \mathbf{k} n_s(\mathbf{k}) \tag{25.34}$$

will be conserved. However, in the thermal equilibrium state, with mean phonon occupation numbers:

$$n_s(\mathbf{k}) = \frac{1}{e^{\beta\hbar\omega_s(\mathbf{k})} - 1}, \tag{25.35}$$

the total phonon wave vector (25.34) vanishes, since $\omega_s(-\mathbf{k}) = \omega_s(\mathbf{k})$. Thus if one should start with a phonon distribution with nonvanishing total crystal momentum, then normal collisions alone cannot bring about full thermodynamic equilibrium, *even if there is no temperature gradient present.* In fact, it can be shown[24] that in the absence of a temperature gradient, if all collisions conserve crystal momentum, then the phonon distribution function will relax to the steady-state form:

$$n_s^{\mathbf{w}}(\mathbf{k}) = \frac{1}{e^{\beta(\hbar\omega_s(\mathbf{k}) - \mathbf{w}\cdot\mathbf{k})} - 1}, \tag{25.36}$$

where $\mathbf{w}$ is a constant determined by the condition that

$$\sum \mathbf{k} n_s^{\mathbf{w}}(\mathbf{k}) \tag{25.37}$$

be equal to the initial total crystal momentum.

Evidently the distribution function (25.36) is not symmetric in $\mathbf{k}$ and will, in general, support a nonvanishing thermal current density:[25]

$$j^{\text{th}} = \frac{1}{V} \sum_{\mathbf{k}s} \hbar\omega_s(\mathbf{k}) \frac{\partial\omega_s(\mathbf{k})}{\partial\mathbf{k}} n_s^{\mathbf{w}}(\mathbf{k}) \neq 0. \tag{25.38}$$

This is tantamount to the assertion that *in the absence of umklapp processes the thermal conductivity of an insulating crystal is infinite.*[26]

---

[24] Equation (25.36) can be derived with the aid of a phonon Boltzmann equation; see, for example, J. M. Ziman, *Electrons and Phonons*, Oxford, 1960, Chapter VIII. We shall not derive it here, since the only point we wish to draw from it is intuitively quite plausible: If the steady-state distribution function leads to a nonzero total crystal momentum, then it must violate the symmetry leading to vanishing thermal current. The thermal current will therefore also be nonzero, barring fortuitous cancellations. A similar point arises in the theory of the low temperature electrical resistivity of a metal. See pps. 527, 528.

[25] That current will be parallel to the total crystal momentum, in crystals of cubic symmetry, and, more generally, will flow in a direction not too far from the direction of the total crystal momentum.

[26] Implicit in this discussion is the fact that phonons can appear and disappear at the ends of the sample. This becomes clear when one tries to apply the same argument to a dilute classical gas, in which collisions conserve *real* momentum. Such a gas, contained in a long cylindrical vessel, does not have infinite thermal conductivity. Our argument fails in this case because the gas cannot penetrate the ends of the vessel, and the resulting accumulation of molecules at the ends leads to diffusion currents which reduce the total momentum back to zero. Phonons, however, though they can be reflected, can also be absorbed at the ends of a cylindrical crystalline sample by transferring their energy to the heat baths at either end. There is therefore no inconsistency in assuming a steady-state distribution with net nonvanishing crystal momentum throughout the specimen. The thermal current in a crystal in the absence of umklapp processes is more analogous to the transport of heat by convection in a gas flowing through an open-ended cylinder.

A perfect infinite anharmonic crystal has a finite thermal conductivity at low temperatures only because there will still be some small probability of crystal-momentum-destroying umklapp processes that do degrade the thermal current. Since the change in total crystal momentum in an umklapp process is equal to a non-vanishing reciprocal lattice vector (of a size comparable to k_D), this means that at least one of the phonons involved in a cubic or quartic umklapp collision must itself have a crystal momentum that is not small compared with k_D. That same phonon must also have an energy that is not small compared with $\hbar\omega_D$, and energy conservation therefore requires at least one phonon with an energy not small compared with $\hbar\omega_D$ to be present before the collision. When T is small compared with Θ_D, the mean number of such phonons is

$$n_s(\mathbf{k}) = \frac{1}{e^{\hbar\omega_s(\mathbf{k})/k_BT} - 1} \approx \frac{1}{e^{\Theta_D/T} - 1} \approx e^{-\Theta_D/T}. \tag{25.39}$$

Thus, as the temperature drops, the number of phonons that can participate in umklapp processes drops exponentially. Without the umklapp processes the thermal conductivity would be infinite, and we therefore expect that the effective relaxation time appearing in the thermal conductivity must vary as

$$\tau \sim e^{T_0/T} \tag{25.40}$$

at temperatures well below Θ_D, where T_0 is a temperature of order Θ_D. Determining T_0 precisely requires analysis of great complexity, an analysis which also leads to powers of T multiplying the exponential. But these are small corrections to the dominant exponential behavior, the qualitative form of which is a straightforward consequence of the freezing out of umklapp processes.

When the temperature reaches a point where the exponential increase in the thermal conductivity sets in, the conductivity increases so rapidly with decreasing temperature that the phonon mean free path soon becomes comparable to the mean free path due to the scattering of phonons by lattice imperfections or impurities, or even to the mean free path describing the scattering of phonons by the sides of the finite specimen. Once this happens, the mean free path in (25.31) ceases to be the intrinsic one due to the anharmonic terms, and must be replaced by a temperature-independent length determined by the spatial distribution of imperfections or the size[27] of the specimen. The temperature dependence of κ then becomes that of the specific heat, which declines as T^3 at temperatures well below Θ_D.[28]

Surveying the full temperature range, we expect the very low temperature thermal conductivity to be limited by temperature-independent scattering processes determined by the geometry and purity of the sample. It will therefore rise as T^3 with the phonon specific heat. This rise continues until a temperature is reached at which umklapp processes become frequent enough to yield a mean free path shorter than

---

[27] This regime is known as the Casimir limit. See H. B. G. Casimir, *Physica* **5**, 595 (1938).

[28] The degree to which this result is dependent on *crystallinity* is indicated by experiments on glasses and amorphous materials, where, for $T \leqslant 1$ K, the thermal conductivity rises roughly as T^2. See R. C. Zeller and R. O. Pohl, *Phys. Rev.* **B4**, 2029 (1971).

the temperature-independent one. At this point the thermal conductivity will reach a maximum, beyond which it will decline rapidly due to the factor $e^{T_0/T}$, reflecting the exponential increase in the frequency of umklapp processes with rising temperature. The decline in κ continues up to temperatures well above Θ_D, but the drastic exponential decline is quickly replaced by a slow power law, reflecting simply the increasing number of phonons available to participate in scattering at high temperatures.

Some typical measured thermal conductivities that illustrate these general trends are displayed in Figure 25.5.

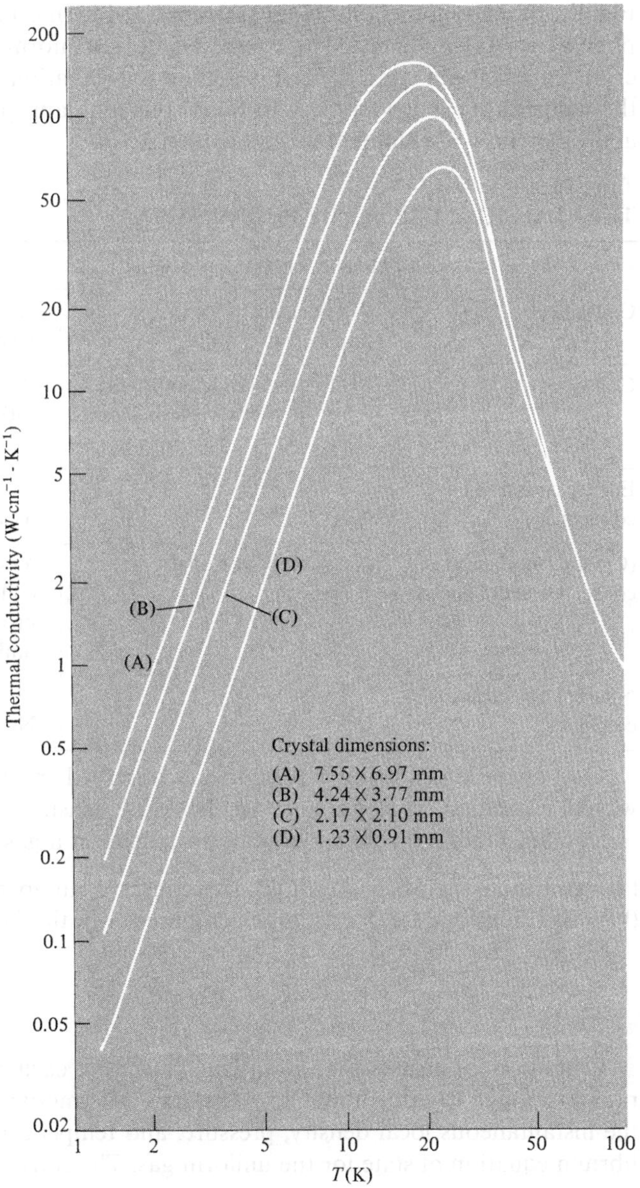

Figure 25.5
Thermal conductivity of isotopically pure crystals of LiF. Below about 10 K the conductivity is limited by surface scattering. Therefore the temperature dependence comes entirely from the T^3 dependence of the specific heat, and the larger the cross-sectional area of the sample, the larger the conductivity. As the temperature rises, umklapp processes become less rare, and the conductivity reaches a maximum when the mean free path due to phonon-phonon scattering is comparable to that due to surface scattering. At still higher temperatures the conductivity falls because the phonon-phonon scattering rate is rapidly increasing, while the phonon specific heat is starting to level off. (After P. D. Thatcher, *Phys. Rev.* **156**, 975 (1967).)

SECOND SOUND

As we have had occasion to note, there is an analogy between the phonons in an insulator and the molecules of an ordinary classical gas. Like the molecules of a gas, phonons can exchange energy and (crystal) momentum in collisions, and can transport thermal energy from one region to another. Unlike the molecules in a gas, however, the number of phonons need not be conserved in a collision or at the surfaces of the containing vessel (the container, in the case of phonons, being the crystal itself). Finally, though momentum is always conserved in intermolecular collisions in a gas, phonon crystal momentum is conserved only in normal collisions, and therefore crystal momentum conservation is a good conservation law only to the extent that the temperature is low enough to freeze out umklapp processes. These similarities and differences are summarized in Table 25.3.

Table 25.3
THE CLASSICAL GAS vs. THE PHONON GAS

	CLASSICAL GAS OF MOLECULES	PHONON GAS
Container	A vessel with impenetrable walls	A crystal, which is the medium that sustains the phonons
Collisions	Molecules collide with each other and with walls of vessel	Phonons collide with each other, with surface of crystal, and with impurities
Energy conserved in collisions	Yes	Yes
(Crystal) momentum conserved in collisions	Yes (except at walls)	Yes (except at surfaces and in collisions with impurities), provided that $T \ll \Theta_D$, so that umklapp processes are frozen out
Number conserved in collisions	Yes	No

One of the most striking phenomena observed in an ordinary gas is sound, a wavelike oscillatory disturbance in the local density of molecules. According to elementary kinetic theory sound can propagate in a gas provided that:

(a) Collisions between the molecules conserve number, energy, and momentum.
(b) The collision rate, $1/\tau$, is large compared with the frequency of the sound wave, $\nu = \omega/2\pi$:

$$\omega \ll \frac{1}{\tau}. \tag{25.41}$$

Condition (b) insures that at any instant in the oscillatory cycle collisions occur rapidly enough to bring about a local state of thermodynamic equilibrium, in which the instantaneous local density, pressure, and temperature are related by the equilibrium equation of state for the uniform gas. The conservation laws (condition (a)) are essential for the establishment of this equilibrium. The momentum conservation law is of critical importance, in requiring the instantaneous local equilibrium configuration to have a net nonvanishing momentum (sometimes referred to as "local equilibrium in a moving frame of reference"), which is the kinematical basis for the oscillation.

In considering whether sound has an analogue in the phonon gas, one must recognize that the phonon gas differs in two relevant ways from an ordinary gas:

1. Number is not conserved in collisions.
2. Crystal momentum is not conserved exactly, although it becomes conserved to increasingly greater accuracy as the temperature drops and umklapp processes are "frozen out."

The first point presents no serious problems: The loss of one of the conservation laws reflects the fact that the equilibrium phonon distribution function

$$\frac{1}{e^{\hbar\omega_s(\mathbf{k})/k_BT} - 1} \tag{25.42}$$

is determined entirely by the temperature, while the equilibrium distribution function in an ideal gas depends on both temperature and density. Since local equilibrium is specified by one less variable in the phonon gas, one less conservation law is required to maintain it.

However, momentum conservation is quite essential for the propagation of sound, and this means that the rate of umklapp collisions, which destroy crystal momentum, must be small compared with the frequency of the oscillation:

$$\frac{1}{\tau_u} \ll \omega. \tag{25.43}$$

This condition has no analogue in the theory of sound in an ordinary gas. However, the analogue of Eq. (25.41) must continue to hold, where the relevant relaxation time is that describing the momentum-conserving normal collisions, τ_N,

$$\omega \ll \frac{1}{\tau_N}, \tag{25.44}$$

for it is still essential that local thermodynamic equilibrium be maintained on a time scale short compared to the period of an oscillation. Combining conditions (25.43) and (25.44), we find that the frequency must lie in the "window"

$$\frac{1}{\tau_u} \ll \omega \ll \frac{1}{\tau_N}. \tag{25.45}$$

Therefore the analogue of sound in the phonon gas will exist at temperatures low enough for the rate of normal collisions to be substantially greater than the rate of umklapps and at frequencies intermediate between the two collision rates. The phenomenon, known as *second sound*, can be regarded as an oscillation in the local phonon number density (just as ordinary sound is an oscillation in the local density of molecules) or, perhaps more pertinently for phonons (since their chief attribute is that they carry energy), as an oscillation in the local energy density. Since the local equilibrium number and energy densities of the phonon gas in a crystal are uniquely determined by the local temperature, second sound will manifest itself as a wavelike oscillation in the temperature. Its observation is most favorable in solids of very high isotopic purity (since any deviation from a perfect Bravais lattice, including the presence of occasional ions of different isotopic mass, will lead to collisions in which crystal momentum is not conserved) and with fairly large anharmonic terms (since a high rate of normal phonon collisions is required to maintain local thermodynamic equilibrium). These considerations make solid helium and the ionic crystal sodium fluoride promising media for the observation of second sound. In both crystals heat impulses have indeed been observed to propagate at the velocity predicted by the second-sound wave equation, rather than in the diffusional manner associated with ordinary thermal conduction.[29] The prediction and detection of second sound is one of the great triumphs of the theory of lattice vibrations.

PROBLEMS

1. *Instability of a Theory with Only Cubic Anharmonicity*

Show that if only cubic corrections to the harmonic potential energy are retained, then that potential energy can be made negative and arbitrarily large in magnitude by a suitable choice of the ionic displacements $\mathbf{u}(\mathbf{R})$. (*Hint:* Take an arbitrary set of displacements and consider the effect on the total potential energy of multiplying them all by a scale factor and changing all of their signs.)

2. *Equation of State of the Harmonic Crystal*

Derive the form (25.7) for the pressure in the harmonic approximation, by substituting the harmonic form (25.6) of the internal energy U into the general thermodynamic relation (25.5). (*Hint:* Change the integration variable from T' to $x = \hbar\omega_s(\mathbf{k})/T'$, and integrate by parts with respect to x, taking due care of the integrated terms.)

3. *Grüneisen Parameters in One Dimension*

Consider a one-dimensional array of N atoms interacting through pair potentials $\phi(r)$ and constrained to have a length $L = Na$ (i.e., the equilibrium lattice constant is constrained to be a).

(a) Show that if only nearest-neighbor interactions are appreciable, then the k-dependent Grüneisen parameters are in fact independent of k and given by

$$\gamma = -\frac{a}{2}\frac{\phi'''(a)}{\phi''(a)}. \tag{25.46}$$

(b) Show that if next-nearest-neighbor interactions are retained, the Grüneisen parameters for the individual normal modes will, in general, depend on the wave vector.

---

29 The observation of second sound in solid ^{4}He is reported by Ackerman et al., *Phys. Rev. Lett.* **16**, 789 (1966), and in solid ^{3}He by C. C. Ackerman and W. C. Overton, Jr., *Phys. Rev. Lett.* **22**, 764 (1969). The onset of second sound in NaF is reported by T. F. McNelly et al., *Phys. Rev. Lett.* **24**, 100 (1970). The subject is reviewed by C. C. Ackerman and R. A. Guyer, *Annals of Physics* **50**, 128 (1968).

4. *General Form of the Grüneisen Parameters*

If the harmonic approximation is not made, the full ionic potential energy of a monatomic Bravais lattice will have the form

$$U^{\mathrm{eq}} + \frac{1}{2}\sum_{\substack{\mu\nu \\ \mathbf{R}\mathbf{R}'}} u_\mu(\mathbf{R})u_\nu(\mathbf{R}')D_{\mu\nu}(\mathbf{R} - \mathbf{R}')$$

$$+ \frac{1}{6}\sum_{\substack{\mu\nu\lambda \\ \mathbf{R}\mathbf{R}'\mathbf{R}''}} u_\mu(\mathbf{R})u_\nu(\mathbf{R}')u_\lambda(\mathbf{R}'')D_{\mu\nu\lambda}(\mathbf{R}, \mathbf{R}', \mathbf{R}'') + \cdots, \tag{25.47}$$

where $\mathbf{u}(\mathbf{R})$ gives the displacement from the equilibrium position $\mathbf{R}$.

(a) Show that if the expansion is made not about the equilibrium positions $\mathbf{R}$, but about sites $\bar{\mathbf{R}} = (1 + \eta)\mathbf{R}$, then the coefficients of the quadratic term in the new expansion are given to linear order in η by

$$\bar{D}_{\mu\nu}(\bar{\mathbf{R}} - \bar{\mathbf{R}}') = D_{\mu\nu}(\mathbf{R} - \mathbf{R}') + \eta\, \delta D_{\mu\nu}(\mathbf{R} - \mathbf{R}'), \tag{25.48}$$

where

$$\delta D_{\mu\nu}(\mathbf{R} - \mathbf{R}') = \sum_{\lambda\mathbf{R}''} D_{\mu\nu\lambda}(\mathbf{R}, \mathbf{R}', \mathbf{R}'')R_\lambda''. \tag{25.49}$$

Note that only the cubic term in (25.47) contributes to this order in η.

(b) Show that the Grüneisen parameter for the normal-mode $\mathbf{k}s$ is given by

$$\gamma_{\mathbf{k}s} = \frac{\boldsymbol{\epsilon}(\mathbf{k}s)\delta\mathbf{D}(\mathbf{k})\boldsymbol{\epsilon}(\mathbf{k}s)}{6M\omega_s(\mathbf{k})^2}. \tag{25.50}$$

5. *Three-Phonon Processes in One Dimension*

Consider a process in which two phonons combine to give a third (or one phonon decays into two others). Let all phonons be acoustic, assume that the two transverse branches lie below the longitudinal branch, and assume that $d^2\omega/dk^2 \leqslant 0$ for all three branches.

(a) Interpreting the conservation laws graphically (as, for example, in Figure 24.5), show that there can be no process in which all three phonons belong to the same branch.

(b) Show that the only possible processes are those in which the single phonon is in a branch higher than at least one of the members of the pair; i.e.,

$$\text{Transverse} + \text{transverse} \leftrightarrow \text{longitudinal}$$

or

$$\text{Transverse} + \text{longitudinal} \leftrightarrow \text{longitudinal}.$$

26
Phonons in Metals

The general theory of lattice vibrations expounded in Chapters 22 and 23 is applicable to both metals and insulators. However, its detailed application to metals is complicated by two aspects of the metallic state:

1. **The Ions Are Charged** This leads to difficulties associated with the very long range of the direct electrostatic interaction between ions.[1]
2. **Conduction Electrons Are Present** Even the simplest theory of lattice vibrations in a metal must acknowledge the presence of a set of electrons that cannot be considered rigidly bound up in the ion cores. The conduction electrons interact with the ions through electrostatic forces which are just as strong as the direct Coulomb forces between ions, and it is therefore essential to know what they too are doing in the course of a lattice vibration.

As it turns out, these mobile conduction electrons provide just the mechanism required to remove the problems connected with the long range of the direct electrostatic interaction between ions.

ELEMENTARY THEORY OF THE PHONON DISPERSION RELATION

Suppose that we were to ignore the forces the conduction electrons exert on the ions. Then the theory of metallic lattice vibrations would be just the theory of the normal modes of a set of N charged particles of charge Ze and mass M in a volume V. In the limit of long wavelengths, except for the difference in particle mass and charge,[2] this is just the problem we analyzed in Chapter 1 (pages 19, 20), where we found that an electron gas could sustain density oscillations at the plasma frequency ω_p, given by

$$\omega_p^2 = \frac{4\pi n_e e^2}{m}. \tag{26.1}$$

Making the substitutions $e \to Ze$, $m \to M$, $n_e \to n_i = n_e/Z$, we can conclude in the same way that a set of charged point ions should undergo long-wavelength vibrations at an ionic plasma frequency Ω_p, given by

$$\begin{aligned}\Omega_p^2 &= \frac{4\pi n_i (Ze)^2}{M} \\ &= \left(\frac{Zm}{M}\right)\omega_p^2.\end{aligned} \tag{26.2}$$

This contradicts the conclusion in Chapter 22 that the long-wavelength normal-mode frequencies of a monatomic Bravais lattice should vanish linearly with k. That result is inapplicable because the approximation (22.64) leading to the linear form

---

[1] This problem also arises in ionic crystals, as we have seen in Chapter 20. We will discuss the theory of their lattice vibrations in Chapter 27.

[2] We also ignore the fact that, unlike an electron gas, the ions are distributed at lattice sites in equilibrium, and have short-range repulsive interactions. A more careful analysis is given in Problem 1.

for $\omega(\mathbf{k})$ at small k is only valid if the forces between ions separated by R are negligibly small for R of order $1/k$. But the inverse square force falls off so slowly with distance that no matter how small k is, interactions of ions separated by $R \gtrsim 1/k$ can contribute substantially to the dynamical matrix (22.59).[3] Nevertheless, the phonon spectra of metals clearly possess branches in which ω vanishes linearly with k. This can be seen directly from neutron scattering, and also from the T^3 term in the specific heat,[4] characteristic of such a linear dependence.[5]

To understand why the phonon dispersion is linear at small k it is essential, when considering ionic motion, to take the conduction electrons into account.

The response of the electrons is treated in the adiabatic approximation (see Chapter 22, p. 425) which assumes that at any time the electrons take on the configurations they would have if the ions were frozen into their instantaneous positions. Furthermore we have seen in Chapter 17 that in the presence of an external charge distribution (in this case the instantaneous distribution of the ions) the electron gas distributes itself so as to shield or screen the fields produced by that distribution. Thus as the ions execute their comparatively sluggish vibrations, the nimble conduction electrons are continually redistributing themselves so as to cancel out the long-range part of the ionic field, yielding an effective ionic field that is short-ranged, and therefore capable of leading to a phonon dispersion relation that is linear in k at long wavelengths.

One often refers to the original direct Coulomb interaction between ions as the "bare" ion-ion interaction, and the short-range effective interaction produced by conduction electron screening as the "dressed" interaction.

To estimate the actual phonon frequencies from this picture we consider the ionic configuration in a phonon of wave vector $\mathbf{k}$, to constitute, as far as the conduction electrons are concerned, an external charge density[6] with wave vector $\mathbf{k}$. According to (17.36) the field associated with such a distribution is reduced (by virtue of the screening action of the electrons) by $1/\epsilon(\mathbf{k})$, where $\epsilon(\mathbf{k})$ is the electron gas dielectric constant. Since the square of the phonon frequency, $\omega(\mathbf{k})^2$, is proportional to the restoring force, and hence to the field, we must reduce (26.2) by $1/\epsilon(\mathbf{k})$, for it was derived without taking screening into account. This yields a "dressed" phonon frequency given by

$$\omega(\mathbf{k})^2 = \frac{\Omega_p{}^2}{\epsilon(\mathbf{k})}. \tag{26.3}$$

As $k \to 0$ the dielectric constant is given by the Thomas-Fermi form (17.51):

$$\epsilon(\mathbf{k}) = 1 + \frac{k_0{}^2}{k^2}, \tag{26.4}$$

---

[3] Consider, for example, a plane sheet of charged particles, which gives rise to an electric field independent of the distance from the sheet. See also Chapter 22, footnote 28.

[4] See Problem 2.

[5] See Problem 2, Chapter 23.

[6] This ignores complications stemming from the discreteness of the lattice (and hence the ambiguity of $\mathbf{k}$ to within an additive reciprocal lattice vector).

and therefore, as $k \to 0$,

$$\omega(\mathbf{k}) \approx ck, \quad c^2 = \frac{\Omega_p^{\,2}}{k_0^{\,2}} = \frac{Zm}{M}\frac{\omega_p^{\,2}}{k_0^{\,2}}. \tag{26.5}$$

To see that this gives a reasonable value for the phonon velocity, we estimate k_0 by its free electron value[7] (17.55) given by

$$\frac{4\pi e^2}{k_0^{\,2}} = \frac{\hbar^2\pi^2}{mk_F}, \tag{26.6}$$

and evaluate the electron plasma frequency using (2.21),

$$n_e = \frac{k_F^{\,3}}{3\pi^2}. \tag{26.7}$$

Thus the velocity of sound is given by

$$\boxed{c^2 = \frac{1}{3}Z\frac{m}{M}v_F^{\,2},} \tag{26.8}$$

which is known as the Bohm-Staver relation.[8]

Since the electron-ion mass ratio is typically of order 10^{-4} or 10^{-5}, this predicts a sound velocity about a hundredth of the Fermi velocity, or of order 10^6 cm/sec, in agreement with observed orders of magnitude. Alternatively, since

$$\frac{\Theta_D}{T_F} = \frac{\hbar ck_D/k_B}{\frac{1}{2}\hbar k_F v_F/k_B} = \frac{2k_D}{k_F}\frac{c}{v_F} \approx \frac{c}{v_F}, \tag{26.9}$$

---

[7] Alternatively, one can use the exact long-wavelength relation (17.50)

$$\frac{4\pi e^2}{k_0^{\,2}} = \frac{1}{\partial n_e/\partial\mu},$$

together with the thermodynamic identity

$$\frac{n}{\partial n/\partial\mu} = \frac{\partial P}{\partial n},$$

to write (26.5) in the form

$$c^2 = \frac{\partial P_{\text{el}}}{\partial\rho_{\text{ion}}}, \qquad \rho_{\text{ion}} = \frac{Mn_e}{Z}.$$

Since continuum mechanics predicts (ignoring anisotropy) that the sound velocity of any medium is given by the square root of the derivative of pressure with respect to mass density, (26.5) is as good as the approximation that the compressibility of a metal is dominated by the electronic contribution (the mass density, of course, *is* dominated by the ionic contribution) and (26.8) is as good as the approximation that the compressibility is dominated by the *free* electron contribution. By coincidence, this is very nearly the case in the alkali metals (see page 39). However, it is evident that (26.8) overlooks, at the least, both electron-electron interactions and ionic core-core repulsion.

[8] D. Bohm and T. Staver, *Phys. Rev.* **84**, 836 (1950).

Eq. (26.8) accounts for the fact that the Debye temperature in a metal is typically of order room temperature, while the Fermi temperature is several tens of thousands of degrees Kelvin.

KOHN ANOMALIES

The assumption that the Coulombic part of the effective ion-ion interaction is reduced by the electronic dielectric constant also has implications for normal modes of short wavelength. At wave vectors that are not small compared with k_F, the Thomas-Fermi dielectric constant (26.4) must be replaced by the more accurate Lindhard result,[9] which is singular[10] when the wave vector $\mathbf{q}$ of the disturbance has magnitude $2k_F$. W. Kohn pointed out[11] that this singularity should be conveyed, via the screened ion-ion interaction, into the phonon spectrum itself, resulting in weak but discernible "kinks" (infinities in $\partial\omega/\partial\mathbf{q}$) at values of $\mathbf{q}$ corresponding to extremal diameters of the Fermi surface.

Highly accurate neutron measurements of $\omega(\mathbf{q})$ are required to reveal these anomalies. Such measurements have been performed[12] and indicate a structure of singularities that is consistent with the Fermi surface geometry deduced from quite unrelated data.

DIELECTRIC CONSTANT OF A METAL

Our discussion of screening in Chapter 17 was based on the electron gas model, which treats the ions as an inert uniform background of positive charge. This overlooks the fact that an external source of charge can induce fields in a metal by distorting the charge distribution of the ions, as well as that of the electrons. There are circumstances in which one is legitimately interested in the screening action of the electrons alone.[13] However, one also often wishes to consider the screening of an external source by all the charged particles in the metal—ions and electrons. We are now in a position to consider an elementary treatment of this additional ionic source of screening.

We define, in the usual way (see Chapter 17), the total dielectric constant of the metal as the proportionality constant relating the Fourier transform of the total potential in the metal to the Fourier transform of the potential of the external charge:

$$\epsilon\phi^{\text{total}} = \phi^{\text{ext}}. \tag{26.10}$$

It is instructive to relate the total dielectric constant ϵ to the dielectric constant ϵ^{el} of the electron gas alone, the dielectric constant $\epsilon^{\text{ion}}_{\text{bare}}$ of the bare ions alone, and

---

[9] See Chapter 17, page 343.

[10] Its derivative has a logarithmic divergence.

[11] W. Kohn, *Phys. Rev. Lett.* **2**, 393 (1959).

[12] R. Stedman et al., *Phys. Rev.* **162**, 545 (1967).

[13] For example, in our derivation of the Bohm-Staver relation (26.8). In that argument the ions were treated as a source of charge external to the electron gas, and not as additional constituents of the screening medium.

the dielectric constant $\epsilon^{\text{ion}}_{\text{dressed}}$ appropriate to the dressed ions—i.e., describing a set of "particles" (ions with their clouds of screening electrons) interacting through a screened interaction V^{eff}.

If we were to consider the medium to be the electrons alone, including the ions in with the explicitly external sources through a total "external" potential $\phi^{\text{ext}} + \phi^{\text{ion}}$, then we could write

$$\epsilon^{\text{el}}\phi^{\text{total}} = \phi^{\text{ext}} + \phi^{\text{ion}}. \tag{26.11}$$

On the other hand, we could also consider the medium to consist of the bare ions alone, regarding the electrons as an additional source of external potential. We should then have

$$\epsilon^{\text{ion}}_{\text{bare}}\phi^{\text{total}} = \phi^{\text{ext}} + \phi^{\text{el}}. \tag{26.12}$$

Adding these last two equations together and subtracting the definition (26.10) of ϵ we find

$$(\epsilon^{\text{el}} + \epsilon^{\text{ion}}_{\text{bare}} - \epsilon)\phi^{\text{total}} = \phi^{\text{ext}} + \phi^{\text{el}} + \phi^{\text{ion}}; \tag{26.13}$$

but since $\phi^{\text{tot}} = \phi^{\text{ext}} + \phi^{\text{el}} + \phi^{\text{ion}}$, we have deduced that[14]

$$\boxed{\epsilon = \epsilon^{\text{el}} + \epsilon^{\text{ion}}_{\text{bare}} - 1.} \tag{26.14}$$

Equation (26.14) gives the dielectric constant of the metal in terms of those of the electrons and the bare ions. However, it is often more convenient to deal not with bare, but with the dressed ions. By dressed ions one means the ions together with their cloud of screening electrons; i.e., particles that give rise to an effective potential that is the bare ionic potential screened by electrons. The dielectric constant $\epsilon^{\text{ion}}_{\text{dressed}}$ describes the total potential that would be established by such a set of particles in the presence of a given external potential. To describe the response of a metal (as opposed to a set of dressed ions) to an external potential we must note that in addition to "dressing" the ions, the electrons also screen the external potential; i.e., the "external potential" to be screened by the dressed ions is not the bare external potential, but the electronically screened one.

Thus the response of a metal to a potential ϕ^{ext} can be viewed as the response of a set of dressed ions to a potential $(1/\epsilon^{\text{el}})\phi^{\text{ext}}$, and we have

$$\phi^{\text{total}} = \frac{1}{\epsilon^{\text{ion}}_{\text{dressed}}}\frac{1}{\epsilon^{\text{el}}}\phi^{\text{ext}}. \tag{26.15}$$

Comparing this with the definition (26.10) of ϵ, we have:

$$\boxed{\frac{1}{\epsilon} = \frac{1}{\epsilon^{\text{ion}}_{\text{dressed}}}\frac{1}{\epsilon^{\text{el}}}} \tag{26.16}$$

---

[14] In terms of the polarizability $\alpha = (\epsilon - 1)/4\pi$, this is simply a special case of the assertion that the polarizability of a medium composed of several types of carriers is the sum of the polarizabilities of the individual carriers.

which should be used in place of (26.14) if we wish a description based on dressed, rather than bare, ions. The two formulations must, of course, be equivalent. Writing (26.14) in the form:

$$\frac{1}{\epsilon} = \frac{1}{\epsilon^{\text{el}}} \frac{1}{1 + (\epsilon^{\text{ion}}_{\text{bare}} - 1)/\epsilon^{\text{el}}} \tag{26.17}$$

we see that consistency with (26.16) requires that[15]

$$\epsilon^{\text{ion}}_{\text{dressed}} = 1 + \frac{1}{\epsilon^{\text{el}}} (\epsilon^{\text{ion}}_{\text{bare}} - 1). \tag{26.18}$$

To study the rough quantitative significance of the ionic contribution to the dielectric constant, we use the simplest expressions available for ϵ^{el} and $\epsilon^{\text{ion}}_{\text{bare}}$. For the former we use the Thomas-Fermi result (26.4).[16] For the latter we can simply transcribe the result (1.37) for the dielectric constant of a gas of charged particles, provided that we replace the electronic plasma frequency (26.1) by that of the ions (26.2).[17] Thus with

$$\epsilon^{\text{ion}}_{\text{bare}} = 1 - \frac{\Omega_p^{\,2}}{\omega^2}, \tag{26.19}$$

the total dielectric constant (26.14) becomes

$$\boxed{\epsilon = 1 + \frac{k_0^{\,2}}{q^2} - \frac{\Omega_p^{\,2}}{\omega^2},} \tag{26.20}$$

and the dielectric constant (26.18) of the dressed ions becomes

$$\boxed{\epsilon^{\text{ion}}_{\text{dressed}} = 1 - \frac{\Omega_p^{\,2}/\epsilon^{\text{el}}}{\omega^2} = 1 - \frac{\omega(\mathbf{q})^2}{\omega^2},} \tag{26.21}$$

where we have used the screening relation (26.3) to introduce the dressed phonon frequency $\omega(\mathbf{q})$. Note that $\epsilon^{\text{ion}}_{\text{dressed}}$ has the same form as $\epsilon^{\text{ion}}_{\text{bare}}$, with the bare phonon frequency Ω_p replaced by the dressed frequency.

---

15 In terms of polarizabilities this is the reasonable assertion that

$$\alpha^{\text{ion}}_{\text{dressed}} = \frac{\alpha^{\text{ion}}_{\text{bare}}}{\epsilon^{\text{el}}}.$$

16 By using the static electronic dielectric constant we are restricting our attention to disturbances of wave vector $\mathbf{q}$ whose frequencies are low enough to satisfy $\omega \ll qv_F$.

17 Equation (26.19) (like (1.37), from which it is taken), ignores the dependence on wave vector $\mathbf{q}$. This is valid if the characteristic particle velocity carries one a distance small compared with a wavelength of the disturbance, in a period of the disturbance: $v/\omega \ll 1/q$, or $\omega \gg qv$. Since a typical ionic velocity is very much smaller than v_F, there is a large range of frequencies and wave vectors for which one may consistently use $\epsilon(\mathbf{q}, \omega) \approx \epsilon(\mathbf{q}, \omega = 0)$ for the electrons (cf. footnote 16) and $\epsilon(\mathbf{q}, \omega) \approx \epsilon(\mathbf{q} = 0, \omega)$ for the bare ions.

We may substitute (26.21) into the form (26.16) for the total dielectric constant, to find:

$$\frac{1}{\epsilon} = \left(\frac{1}{1 + k_0^2/q^2}\right)\left(\frac{\omega^2}{\omega^2 - \omega(\mathbf{q})^2}\right), \tag{26.22}$$

which is, of course, equivalent to (26.20).

The most important consequences of the form (26.22) of the total dielectric constant arise in treating the effective electron-electron interaction in a metal. We therefore continue our discussion from this point of view.

EFFECTIVE ELECTRON-ELECTRON INTERACTION

In Chapter 17 we argued that for many purposes the Fourier transform of the electron-electron Coulomb interaction should be screened by the electronic dielectric constant,

$$\frac{4\pi e^2}{k^2} \to \frac{4\pi e^2}{k^2 \epsilon^{\mathrm{el}}} = \frac{4\pi e^2}{k^2 + k_0^2}, \tag{26.23}$$

to represent the effect of the other electrons in screening the interaction between a given pair. However, the ions also screen interactions, and we should have used not ϵ^{el} but the full dielectric constant. Using the form (26.22) we find that (26.23) should be replaced by

$$\frac{4\pi e^2}{k^2} \to \frac{4\pi e^2}{k^2 \epsilon} = \frac{4\pi e^2}{k^2 + k_0^2}\left(1 + \frac{\omega(\mathbf{k})^2}{\omega^2 - \omega(\mathbf{k})^2}\right). \tag{26.24}$$

Thus the effect of the ions is to multiply (26.23) by a correction factor that depends on frequency as well as wave vector. The frequency dependence reflects the fact that the screening action of the ions is not instantaneous, but limited by the (small on the scale of v_F) velocity of propagation of elastic waves in the lattice. As a result the part of the effective electron-electron interaction mediated by the ions is retarded.

In using (26.24) as an effective interaction between a pair of electrons, one needs to know how ω and $\mathbf{k}$ depend on the quantum numbers of the pair. From the analysis of Chapter 17 we know that when the effective interaction is taken to have the frequency-independent form (26.23), then $\mathbf{k}$ is to be taken as the difference in the wave vectors of the two electronic levels. By analogy, when the effective interaction is frequency-dependent, we shall take ω as the difference in the angular frequencies (i.e., the energies divided by $\hbar$) of the levels. Thus, given two electrons with wave vectors $\mathbf{k}$ and $\mathbf{k}'$ and energies $\mathcal{E}_{\mathbf{k}}$ and $\mathcal{E}_{\mathbf{k}'}$, we take their effective interaction to be[18]

$$v_{\mathbf{k},\mathbf{k}'}^{\mathrm{eff}} = \frac{4\pi e^2}{q^2 + k_0^2}\left[1 + \frac{\omega(\mathbf{q})^2}{\omega^2 - \omega(\mathbf{q})^2}\right]; \qquad \mathbf{q} = \mathbf{k} - \mathbf{k}', \quad \omega = \frac{\mathcal{E}_{\mathbf{k}} - \mathcal{E}_{\mathbf{k}'}}{\hbar}. \tag{26.25}$$

[18] This form was arrived at through the work of Fröhlich and of Bardeen and Pines (H. Fröhlich, *Phys. Rev.* **79**, 845 (1950); J. Bardeen and D. Pines, *Phys. Rev.* **99**, 1140 (1955)). The argument we have given should be viewed more as an indication of the plausibility of (26.25) than as a derivation. A systematic derivation of (26.25), together with a specification of the circumstances under which it can be used as an effective interaction, requires the use of field theoretic ("Green's function") methods.

There are two important qualitative features[19] of v^{eff}:

1. The dressed phonon frequency $\omega(\mathbf{q})$ is of order ω_D or less. Thus when the energies of the two electrons differ by much more than $\hbar\omega_D$, the phonon correction to their effective interaction is negligibly small. Since the range of variation of electronic energies, $\mathcal{E}_F$, is typically 10^2 to 10^3 times $\hbar\omega_D$, only electrons with energies quite close together have an interaction appreciably affected by the phonons.
2. When, however, the electronic energy difference is less than $\hbar\omega_D$, the phonon contribution has the opposite sign from the electronically screened interaction, and is larger in magnitude; i.e., the sign of the effective electron-electron interaction is reversed. This phenomenon, known as "overscreening" is a crucial ingredient in the modern theory of superconductivity. We shall return to it in Chapter 34.

PHONON CONTRIBUTION TO THE ELECTRONIC ENERGY-WAVE VECTOR RELATION

Aside from its crucial role in the theory of superconductivity, the effective interaction (26.25) has important implications for the less dramatic properties of conduction electrons. In Chapter 17 we noted that the simplest correction to the electronic energy $\mathcal{E}_{\mathbf{k}}$ due to electron-electron interactions was the exchange term of Hartree-Fock theory:

$$\Delta\mathcal{E}_{\mathbf{k}} = -\int \frac{d\mathbf{k}'}{(2\pi)^3} \frac{4\pi e^2}{|\mathbf{k}-\mathbf{k}'|^2} f(\mathbf{k}'). \tag{26.26}$$

We observed that this correction led to a spurious singularity in $\partial\mathcal{E}/\partial\mathbf{k}$ at $k = k_F$, which we removed by screening the interaction in (26.26) with the electronic dielectric constant. In a more accurate treatment the screening should be described not by the electronic dielectric constant alone, but by the full dielectric constant of the metal. This suggests in conjunction with (26.25) that the screened modification of the Hartree-Fock exchange term should be taken to be

$$\Delta\mathcal{E}_{\mathbf{k}} = -\int \frac{d\mathbf{k}'}{(2\pi)^3} \frac{4\pi e^2}{|\mathbf{k}-\mathbf{k}'|^2 + k_0{}^2} \times \left\{1 + \frac{\omega(\mathbf{k}-\mathbf{k}')^2}{[(\mathcal{E}_{\mathbf{k}} - \mathcal{E}_{\mathbf{k}'})/\hbar]^2 - \omega(\mathbf{k}-\mathbf{k}')^2}\right\} f(\mathbf{k}'). \tag{26.27}$$

Because the ionic contribution to the screening depends on the electronic energy, this is a rather complicated integral equation for $\mathcal{E}_{\mathbf{k}}$. However, by exploiting the fact that the phonon energy $\hbar\omega(\mathbf{k}-\mathbf{k}')$ is very small compared with $\mathcal{E}_F$, it is possible to

[19] The vanishing of (26.25) at $\omega = 0$ should not be taken seriously. It would mean that for very slowly varying disturbances the ions have time to adjust themselves so as to cancel perfectly the field of the electrons. This cannot be so, if only because electrons are point particles, while ions have impenetrable cores. This was ignored in the derivation of the bare ionic dielectric constant, which took into account only the Coulomb interactions between ions. More accurate calculations that take finite core effects into account eliminate the perfect cancellation.

extract the most important information from (26.27) without solving the full integral equation (see Problem 3). The most important conclusions are:

1. The value of $\mathcal{E}_F$ and the shape of the Fermi surface are unaffected by the ionic correction to the screening; i.e., they are correctly given by ignoring the second term within the braces in (26.27).
2. When $\mathcal{E}_{\mathbf{k}}$ is close to $\mathcal{E}_F$ on the scale of $\hbar\omega_D$, one finds that

$$\mathcal{E}_{\mathbf{k}} - \mathcal{E}_F = \frac{\mathcal{E}_{\mathbf{k}}^{TF} - \mathcal{E}_F}{1 + \lambda}, \tag{26.28}$$

where $\mathcal{E}_{\mathbf{k}}^{TF}$ is the energy calculated in the absence of the ionic correction to the screening, and λ is given by an integral over the Fermi surface:

$$\lambda = \int \frac{dS'}{8\pi^3 \hbar v(\mathbf{k}')} \frac{4\pi e^2}{(\mathbf{k} - \mathbf{k}')^2 + k_0{}^2}. \tag{26.29}$$

In particular, this means that the phonon correction to the electronic velocity and density of levels at the Fermi surface are given by[20]

$$\begin{aligned} \mathbf{v}(\mathbf{k}) &= \frac{1}{\hbar}\frac{\partial \mathcal{E}}{\partial \mathbf{k}} = \frac{1}{(1 + \lambda)}\mathbf{v}^0(\mathbf{k}), \\ g(\mathcal{E}_F) &= (1 + \lambda) g^0(\mathcal{E}_F). \end{aligned} \tag{26.30}$$

These corrections apply only to one-electron energy levels well within $\hbar\omega_D$ of $\mathcal{E}_F$. However, at temperatures well below room temperature ($k_B T \ll \hbar\omega_D$) these are precisely the electronic levels that determine the great bulk of metallic properties, and therefore corrections due to ionic screening must be taken into account. This becomes particularly clear when we estimate the size of λ.

Since k_0 is of order k_F (see (17.55)), we have that

$$\lambda \lesssim \frac{4\pi e^2}{k_0{}^2} \int \frac{dS'}{8\pi^3 \hbar v(\mathbf{k}')}. \tag{26.31}$$

However, from (17.50) and (8.63), we find:

$$\frac{4\pi e^2}{k_0{}^2} = \frac{\partial n}{\partial \mu} = \frac{1}{g(\mathcal{E}_F)} = \left[\int \frac{dS'}{4\pi^3 \hbar v(\mathbf{k}')}\right]^{-1}. \tag{26.32}$$

Thus λ in this simple model is less than, but of order, unity. As a result, in many metals the correction due to ionic screening of the electron-electron interaction (more commonly known as the phonon correction) is the major reason for deviations of the density of levels from its free electron value, being more important than either band structure effects or corrections due to direct electron-electron interactions.[21]

---

[20] We assume a spherical Fermi surface, so λ is constant. The superscript 0 indicates the Thomas-Fermi value.

[21] In determining the effect of the electron-phonon interaction on various one-electron properties it is not enough simply to replace the uncorrected density of levels by Eq. (26.30). One must, in general, reexamine the full derivation in the presence of the effective interaction (26.27). One finds, for example, that the specific heat (Eq. (2.80)) should be corrected by the factor $(1 + \lambda)$, but that the Pauli susceptibility (Eq. (31.69)) should not (see Chapter 31, page 663, footnote 29).

3. When $\mathcal{E}_{\mathbf{k}}$ is several times $\hbar\omega_D$ from $\mathcal{E}_F$, then

$$\mathcal{E}_{\mathbf{k}} - \mathcal{E}_F = (\mathcal{E}_{\mathbf{k}}^{TF} - \mathcal{E}_F)\left[1 - O\left(\frac{\hbar\omega_D}{\mathcal{E}_{\mathbf{k}} - \mathcal{E}_F}\right)^2\right], \tag{26.33}$$

and the phonon correction rapidly becomes insignificant.

These results are summarized in Figure 26.1.

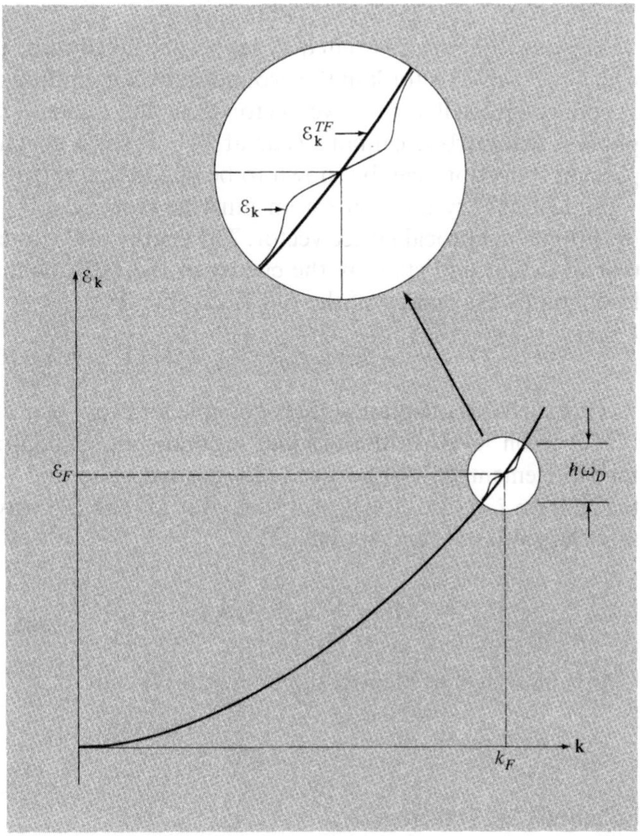

Figure 26.1
Correction to the electronic $\mathcal{E}$ vs. **k** relation due to screening by the ions (electron-phonon correction). The correction (lighter curve) is appreciable only within $\hbar\omega_D$ of $\mathcal{E}_F$, where the slope of the uncorrected curve can be considerably reduced.

THE ELECTRON-PHONON INTERACTION

According to (26.27) the ionic screening adds to the energy of an electron with wave vector **k** an amount

$$v_{\mathbf{k},\mathbf{k}'}^{\text{eff}} = \frac{1}{V}\left(\frac{4\pi e^2}{(\mathbf{k} - \mathbf{k}')^2 + k_0{}^2}\right)\left(\frac{[\hbar\omega(\mathbf{k} - \mathbf{k}')]^2}{[\hbar\omega(\mathbf{k} - \mathbf{k}')]^2 - (\mathcal{E}_{\mathbf{k}} - \mathcal{E}_{\mathbf{k}'})^2}\right) \tag{26.34}$$

for each occupied electronic level $\mathbf{k}'$ (with the same spin). There is, however, an alternative way of deriving the effect of lattice deformations on the electronic energy. Without making any explicit reference to screening, one can simply calculate the change in the total energy of the metal due to the fact that the electrons (charged particles) can interact with the phonons (density waves in the charged lattice of ions). If this interaction is described by some interaction Hamiltonian V^{ep}, then the change in the energy of the metal due to the interaction will be given in second-order perturbation theory by an expression of the form:

$$\Delta E = \sum_i \frac{|\langle 0|V^{ep}|i\rangle|^2}{E_0 - E_i}. \tag{26.35}$$

The most important excited states $|i\rangle$ are taken to be those in which an electron, which had wave vector $\mathbf{k}$ in the ground state, has emitted a phonon with wave vector $\mathbf{q}$, ending up itself with wave vector $\mathbf{k}'$ in the excited state. (Processes in which a phonon is absorbed cannot occur at $T = 0$ since no phonons are present. Multi-phonon transitions can be shown to be of considerably less importance.)

Since total crystal momentum must be conserved,[22] it follows that $\mathbf{k}' + \mathbf{q} = \mathbf{k}$, to within a reciprocal lattice vector. The energy of the intermediate state differs from that of the ground state by the energy of the extra phonon and the new electronic level, minus the energy of the old electronic level:

$$E_i - E_0 = \mathcal{E}_{\mathbf{k}'} + \hbar\omega(\mathbf{k} - \mathbf{k}') - \mathcal{E}_{\mathbf{k}}. \tag{26.36}$$

One such intermediate state is possible for every pair of occupied and unoccupied one-electron levels in the original ground-state configuration. If we let $g_{\mathbf{k},\mathbf{k}'}$ be the matrix element of V^{ep} between the ground state and such an excited state, then the sum on i in (26.35) is just a sum over all pairs of wave vectors of occupied and unoccupied levels, and we have:

$$\Delta E = \sum_{\mathbf{k}\mathbf{k}'} n_{\mathbf{k}}(1 - n_{\mathbf{k}'}) \frac{|g_{\mathbf{k},\mathbf{k}'}|^2}{\mathcal{E}_{\mathbf{k}} - \mathcal{E}_{\mathbf{k}'} - \hbar\omega(\mathbf{k} - \mathbf{k}')}. \tag{26.37}$$

It is natural[23] to identify the v^{eff} in (26.34) with

$$v^{\text{eff}}_{\mathbf{k}\mathbf{k}'} = \frac{\partial^2 \Delta E}{\partial n_{\mathbf{k}}\, \partial n_{\mathbf{k}'}}. \tag{26.38}$$

Equation (26.37) then gives:

$$\begin{aligned} v^{\text{eff}}_{\mathbf{k},\mathbf{k}'} &= -|g_{\mathbf{k},\mathbf{k}'}|^2 \left(\frac{1}{\mathcal{E}_{\mathbf{k}} - \mathcal{E}_{\mathbf{k}'} - \hbar\omega(\mathbf{k} - \mathbf{k}')} + \frac{1}{\mathcal{E}_{\mathbf{k}'} - \mathcal{E}_{\mathbf{k}} - \hbar\omega(\mathbf{k}' - \mathbf{k})} \right) \\ &= |g_{\mathbf{k},\mathbf{k}'}|^2 \left[\frac{2\hbar\omega(\mathbf{k} - \mathbf{k}')}{[\hbar\omega(\mathbf{k} - \mathbf{k}')]^2 - (\mathcal{E}_{\mathbf{k}} - \mathcal{E}_{\mathbf{k}'})^2} \right]. \end{aligned} \tag{26.39}$$

---

22 See Appendix M, page 788.

23 This argument (which is not rigorous) is in the spirit of the Landau approach (Chapter 17), which holds that the appropriate quasiparticle energies follow from the first derivative (with respect to occupation number) of (26.37) and the modification of those energies due to the occupation of those levels, from the second derivative.

By requiring this form of the effective interaction to agree with the result (26.34), we can deduce the electron-phonon coupling constant:

$$|g_{\mathbf{k},\mathbf{k}'}|^2 = \frac{1}{V}\frac{4\pi e^2}{|\mathbf{k}-\mathbf{k}'|^2 + k_0{}^2}\frac{1}{2}\hbar\omega_{\mathbf{k}-\mathbf{k}'}. \tag{26.40}$$

The most important feature of this result is that, in the limit of small $|\mathbf{k}-\mathbf{k}'|$, g^2 vanishes linearly with $|\mathbf{k}-\mathbf{k}'|$. If we write the free electron estimate (26.6) for k_0 in the form:

$$\frac{4\pi e^2}{k_0{}^2} = \frac{2\mathcal{E}_F}{3n_e} \tag{26.41}$$

we find that

$$|g_{\mathbf{k},\mathbf{k}'}|^2 \approx \frac{\hbar\omega(\mathbf{k}-\mathbf{k}')\mathcal{E}_F}{3n_eV} = \frac{\hbar\omega(\mathbf{k}-\mathbf{k}')\mathcal{E}_F}{3NZ}, \qquad |\mathbf{k}-\mathbf{k}'| \ll k_0. \tag{26.42}$$

The fact that the square of the electron-phonon coupling constant vanishes linearly with the wave vector of the phonon has important consequences for the theory of the electrical resistivity of a metal.

THE TEMPERATURE-DEPENDENT ELECTRICAL RESISTIVITY OF METALS

We have noted[24] that Bloch electrons in a perfect periodic potential can sustain an electric current even in the absence of any driving electric field; i.e., their conductivity is infinite. The finite conductivity of metals is entirely due to deviations in the lattice of ions from perfect periodicity. The most important such deviation is that associated with the thermal vibrations of the ions about their equilibrium positions, for it is an intrinsic source of resistivity, present even in a perfect sample free from such crystal imperfections as impurities, defects, and boundaries.

The quantitative theory of the temperature dependence of the resistivity provided by the lattice vibrations starts from the observation that the periodic potential of a set of rigid ions,

$$U^{\text{per}}(\mathbf{r}) = \sum_{\mathbf{R}} V(\mathbf{r}-\mathbf{R}), \tag{26.43}$$

is only an approximation to the true, aperiodic potential:

$$U(\mathbf{r}) = \sum_{\mathbf{R}} V[\mathbf{r}-\mathbf{R}-\mathbf{u}(\mathbf{R})] = U^{\text{per}}(\mathbf{r}) - \sum_{\mathbf{R}} \mathbf{u}(\mathbf{R})\cdot\nabla V(\mathbf{r}-\mathbf{R}) + \cdots. \tag{26.44}$$

The difference between these two forms can be considered as a perturbation that acts on the stationary one-electron levels of the periodic Hamiltonian, causing transitions among Bloch levels that lead to the degradation of currents.

[24] See, for example, Chapter 12, page 215.

As is generally the case with transitions caused by lattice vibrations, they can be considered here as processes in which an electron absorbs or emits a phonon (or phonons), changing its energy by the phonon energy and its wave vector (to within a reciprocal lattice vector) by the phonon wave vector. Indeed, this picture of the scattering of electrons by lattice vibrations is very similar to the picture in Chapter 24 of the scattering of neutrons by lattice vibrations.

The simplest theories of the lattice contribution to the resistivity of metals assume that the scattering is dominated by processes in which an electron emits (or absorbs) a single phonon. If the electronic transition is from a level with wave vector $\mathbf{k}$ and energy $\mathcal{E}_{\mathbf{k}}$ to one with wave vector $\mathbf{k}'$ and energy $\mathcal{E}_{\mathbf{k}'}$, then energy and crystal momentum conservation[25] require that the energy of the phonon involved satisfy

$$\mathcal{E}_{\mathbf{k}} = \mathcal{E}_{\mathbf{k}'} \pm \hbar\omega(\mathbf{k} - \mathbf{k}'), \tag{26.45}$$

where the plus (minus) sign is appropriate to phonon emission (absorption) (and where we assume that $\omega(-\mathbf{q}) = \omega(\mathbf{q})$). This equation can be viewed as a constraint on the wave vectors $\mathbf{q}$ of phonons capable of participating in a one-phonon process with an electron with wave vector $\mathbf{k}$, namely

$$\omega(\mathbf{q}) = \pm \frac{1}{\hbar}\left[\mathcal{E}_{\mathbf{k}+\mathbf{q}} - \mathcal{E}_{\mathbf{k}}\right]. \tag{26.46}$$

As in the case of neutron scattering, this constraint, being a single restriction, determines a two-dimensional surface of allowed wave vectors in the three-dimensional phonon wave vector space. Indeed, since $\hbar\omega(\mathbf{q})$ is a minute energy on the electronic energy scale, the surface of allowed $\mathbf{q}$ for a given $\mathbf{k}$ is very close to the set of vectors connecting $\mathbf{k}$ to all other points on the constant energy surface $\mathcal{E}_{\mathbf{k}'} = \mathcal{E}_{\mathbf{k}}$ (Figure 26.2).

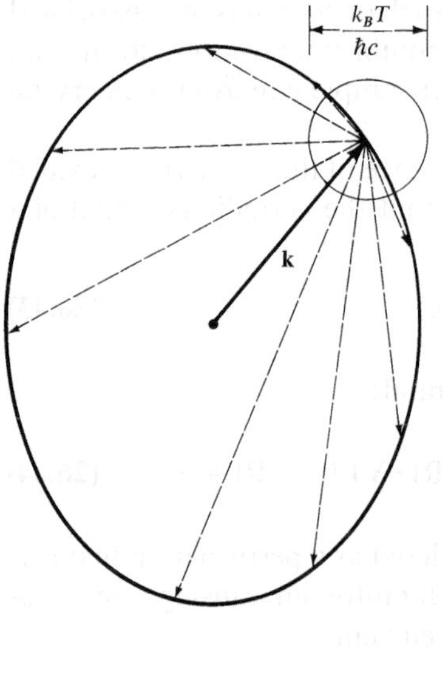

Figure 26.2
Construction of the wave vectors of those phonons allowed by the conservation laws to participate in a one-phonon scattering event with an electron with wave vector $\mathbf{k}$. Because the phonon energy is at most $\hbar\omega_D \ll \mathcal{E}_F$, the surface containing the tips of the phonon wave vectors originating from $\mathbf{k}$ differs only slightly from the Fermi surface. At temperatures well below Θ_D the only phonons that can actually participate in scattering events have wave vectors whose tips lie within the small sphere of size $k_B T/\hbar c$ about the tip of the wave vector $\mathbf{k}$.

---

[25] See Appendix M.

At high temperatures ($T \gg \Theta_D$) the number of phonons in any normal mode is given by

$$n(\mathbf{q}) = \frac{1}{e^{\beta\hbar\omega(\mathbf{q})} - 1} \approx \frac{k_B T}{\hbar\omega(\mathbf{q})}. \qquad \textbf{(26.47)}$$

Thus the total number of phonons on the surface of allowed wave vectors for the scattering of a given electron is directly proportional to T. Since the number of scatterers increases linearly with T, so will the resistivity:

$$\boxed{\rho \sim T, \qquad T \gg \Theta_D.} \qquad \textbf{(26.48)}$$

At low temperatures ($T \ll \Theta_D$) things are rather more complicated. We first note that only phonons with $\hbar\omega(\mathbf{q})$ comparable to or less than $k_B T$ can be absorbed or emitted by electrons. In the case of absorption this is immediately obvious, since these are the only phonons present in appreciable numbers. It is also true in the case of emission, for in order to emit a phonon an electron must be far enough above the Fermi level for the final electronic level (whose energy is lower by $\hbar\omega(\mathbf{q})$) to be unoccupied; since levels are occupied only to within order $k_B T$ above $\mathcal{E}_F$, and unoccupied only to within order $k_B T$ below, only phonons with energies $\hbar\omega(\mathbf{q})$ of order $k_B T$ can be emitted.

Well below the Debye temperature, the condition $\hbar\omega(\mathbf{q}) \lesssim k_B T$ requires q to be small compared with k_D. In this regime ω is of order cq so the wave vectors q of the phonons are of order $k_B T/\hbar c$ or less. Thus within the surface of phonons that the conservation laws permit to be absorbed or emitted, only a subsurface of linear dimensions proportional to T, and hence of area proportional to T^2, can actually participate.

We conclude that the number of phonons that can scatter an electron declines as T^2 well below the Debye temperature. However, the electronic scattering rate declines even faster, for when q is small the square of the electron-phonon coupling constant (26.42) vanishes linearly with q. Well below Θ_D the physically relevant phonons have wave vectors q of order $k_B T/\hbar c$, and therefore the scattering rate (which is proportional to the square of the coupling constant) for those processes that can take place, declines linearly with T.

Combining these two features, we conclude that for T well below Θ_D the net electron-phonon scattering rate declines as T^3:

$$\boxed{\frac{1}{\tau^{\text{el-ph}}} \sim T^3, \qquad T \ll \Theta_D.} \qquad \textbf{(26.49)}$$

However, low-temperature electron-phonon scattering is one of those cases in which the rate at which the current degrades is not simply proportional to the scattering rate. This is because well below Θ_D any given one-phonon process can change the electronic wave vector by only a very small amount (namely the wave vector of the participating phonon, which is small compared with k_D or k_F). Provided that the electronic velocity $\mathbf{v}(\mathbf{k})$ does not undergo large variations between Fermi surface points separated by very small $\mathbf{q}$, the velocity will also not change very much in a single scattering event. Thus as the temperature declines, the scattering becomes more concentrated in the forward direction and is therefore less effective in degrading a current.

The quantitative consequences of this for the low-temperature phonon resistivity are suggested by the analysis of Chapter 16 (pages 324–326). There we showed, in the case of elastic scattering in an isotropic metal, that the effective scattering rate appearing in the resistivity is proportional to an angular average of the actual scattering rate, weighted with the factor $1 - \cos\theta$, where θ is the scattering angle (Figure 26.3). At very low temperatures phonon scattering is very nearly elastic (the energy change being small compared with $\hbar\omega_D$), and we can apply this result with some confidence, at least in metals with isotropic Fermi surfaces. Since $\sin(\theta/2) = q/2k_F$ (Figure 26.3), $1 - \cos\theta = 2\sin^2(\theta/2) = \frac{1}{2}(q/k_F)^2$. But $q = O(k_BT/\hbar c)$ for T well below Θ_D, and this introduces a final factor of T^2 into the low-temperature resistivity.

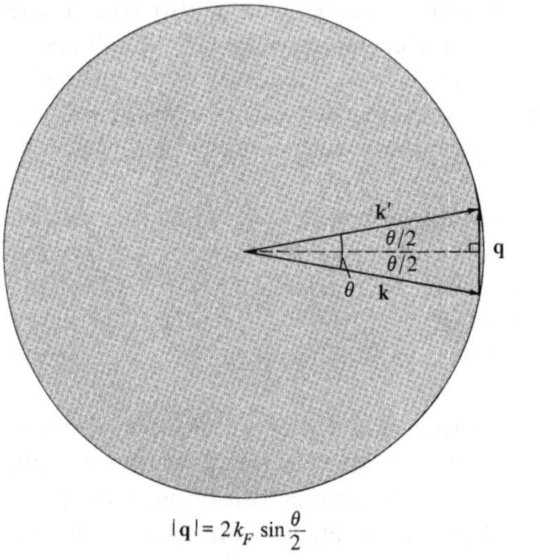

Figure 26.3
Small-angle scattering on a spherical Fermi surface. Since the scattering is nearly elastic, $k \approx k' \approx k_F$. When the phonon wave vector $\mathbf{q}$ (and hence θ) are small, we have $\theta/2 \approx q/2k_F$.

The additional factor of T^2, expressing the growing predominance of forward scattering with declining temperature, arises even in anisotropic metals (with certain exceptions noted in the following section). When combined with the T^3 dependence of the scattering rate, it leads to the "Bloch T^5 law":

$$\rho \sim T^5, \qquad T \ll \Theta_D. \tag{26.50}$$

MODIFICATION OF THE T⁵ LAW BY UMKLAPP PROCESSES

The factor of T^2 in the low-temperature resistivity due to the dominance of forward scattering depends on the assumption that electronic levels with nearly the same wave vectors have nearly the same velocities. For sufficiently complex Fermi surfaces, or when interband scattering is possible this need not be so. In such cases the current can still be efficiently degraded even though the change in wave vector (but not velocity) is small in each scattering event, and the low-temperature resistivity need not decline as rapidly as T^5.

One of the most important examples of a small change in wave vector causing a large change in velocity arises when a nearly free electron Fermi surface approaches a Bragg plane (Figure 26.4). A small wave vector **q** can then join Fermi surface levels on opposite sides of the plane, with nearly oppositely directed velocities. Such events are called "umklapp processes."[26] Within the nearly free electron point of view, one can regard the large change in velocity as due to a phonon-induced Bragg reflection.[27]

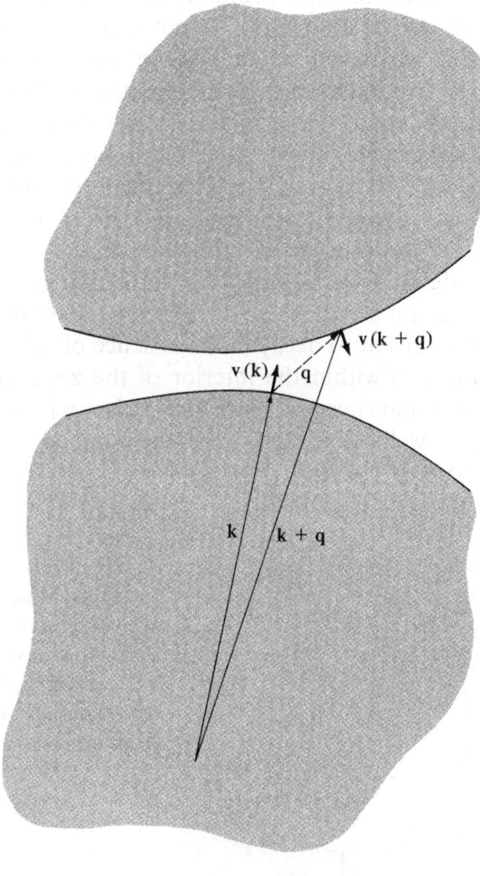

Figure 26.4
A simple umklapp process. The wave vectors **k** and **k** + **q** differ by an amount small compared with k_F(or k_D), but the velocities **v**(**k**) and **v**(**k** + **q**) are not close together.

PHONON DRAG

Peierls[28] pointed out a way in which the low temperature resistivity might decline more rapidly than T^5. This behavior has yet to be observed, owing, presumably, to

---

[26] Compare the discussion in Chapter 25, pages 501–502.

[27] If the Fermi surface is displayed in the first zone rather than in an extended-zone scheme, then the change in wave vector is only small modulo a reciprocal lattice vector. Umklapp processes are sometimes said to be those in which the additive reciprocal lattice vector appearing in the crystal momentum conservation law is nonzero. As emphasized in Chapter 25, this distinction is not independent of the choice of primitive cell. The crucial point for electron-phonon scattering is whether small changes in electron crystal momentum (to within a possible additive reciprocal lattice vector) can result in large changes in electron velocity. When put this way, the criterion is independent of primitive cell.

[28] R. E. Peierls, *Ann. Phys.* (5) **12**, 154 (1932).

its being masked by temperature-independent scattering by defects (which eventually dominates the resistivity at low enough temperatures).

The derivation of the T^5 law assumes that the phonons are in thermal equilibrium, whereas in fact the nonequilibrium nature of the current-carrying electronic distribution should lead, through electron-phonon scattering, to a phonon distribution that is also out of equilibrium. Suppose (to take a simple case) that the Fermi surface lies within the first Brillouin zone. We define umklapp processes as those in which total crystal momentum is not conserved, under the convention that the primitive cell in which individual electron and phonon wave vectors are specified is the first zone. If the total crystal momentum of the combined electron-phonon system were initially nonzero then in the absence of umklapp processes, it would remain nonzero at all subsequent times, even in the absence of an electric field,[29] and the electron-phonon system could not come to complete thermal equilibrium. Instead, the electrons and phonons would drift along together, maintaining their nonzero crystal momentum and also a nonzero electric current.

Metals (free from defects) have finite conductivities only because umklapp processes can occur. These do degrade the total crystal momentum and make it possible for a current to decay in the absence of a driving electric field. If however, the Fermi surface is within the interior of the zone, then there is a minimum phonon wave vector and energy (Figure 26.5) below which umklapp processes cannot occur. When $k_B T$ is well below this energy, the number of phonons available for such events should become proportional to $\exp(-\hbar\omega_{\text{min}}/k_B T)$, and therefore the resistivity should drop exponentially in $1/T$.

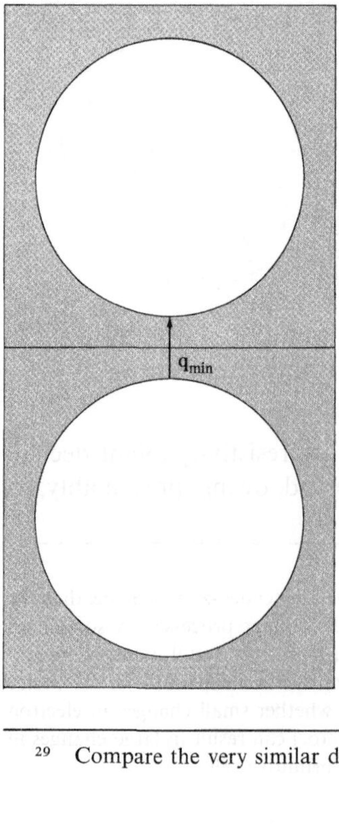

Figure 26.5
Extended-zone picture of a metal whose Fermi surface is completely contained in the first zone. Here q_{min} is the minimum wave vector for a phonon that can participate in an umklapp process. At temperatures below those corresponding to the energy of this phonon, the contribution from umklapp scattering should drop exponentially.

---

[29] Compare the very similar discussion of the thermal conductivity of an insulator in Chapter 25.

PROBLEMS

1. *A More Detailed Treatment of the Phonon Dispersion Relation in Metals*

In deriving the Bohm-Staver relation (26.8) we regarded the ions as point particles, interacting only through Coulomb forces. A more realistic model would take the ions as extended distributions of charge, and allow for the impenetrability of the ion cores by an effective ion-ion interaction in additional to the Coulomb interaction. Since the core-core repulsion is short-ranged, it leads to no difficulties in the usual treatment of lattice vibrations, and can be described by a dynamical matrix $\mathbf{D}^c$ in the manner described in Chapter 22. We can therefore treat lattice vibrations in metals by the methods of Chapter 22, provided that we take the full dynamical matrix $\mathbf{D}$ to be $\mathbf{D}^c$ plus a term arising from the Coulomb interactions between the ionic charge distributions, as screened by the electrons.

Let us take the ion at position $\mathbf{R} + \mathbf{u}(\mathbf{R})$ to have a charge distribution $\rho[\mathbf{r} - \mathbf{R} - \mathbf{u}(\mathbf{R})]$, so that the electrostatic force on such an ion is given by $\int d\mathbf{r}\, \mathbf{E}(\mathbf{r})\rho[\mathbf{r} - \mathbf{R} - \mathbf{u}(\mathbf{R})]$, where $\mathbf{E}(\mathbf{r})$ is the electric field reduced by electronic screening,[30] due to all the other ions (whose charge density is $\sum_{\mathbf{R}' \neq \mathbf{R}} \rho[\mathbf{r} - \mathbf{R}' - \mathbf{u}(\mathbf{R}')]$).

(a) Expand this additional electrostatic interaction to linear order in the ionic displacements $\mathbf{u}$, and, assuming that the electronic screening is described by the static dielectric constant[31] $\epsilon(\mathbf{q})$, show that the dynamical matrix appearing in Eq. (22.57) must now be taken to be

$$D_{\mu\nu}(\mathbf{k}) = D^c_{\mu\nu}(\mathbf{k}) + V_{\mu\nu}(\mathbf{k}) + \sum_{\mathbf{K}\neq 0} [V_{\mu\nu}(\mathbf{k} + \mathbf{K}) - V_{\mu\nu}(\mathbf{K})],$$

$$V_{\mu\nu}(\mathbf{q}) = \frac{4\pi n q_\mu q_\nu |\rho(\mathbf{q})|^2}{q^2 \epsilon(q)}. \qquad (26.51)$$

(b) Show that if the electronic screening is not taken into account ($\epsilon \equiv 1$), Eq. (26.51) predicts in the long-wavelength limit a longitudinal normal mode at the ionic plasma frequency (26.2).

(c) Show that when screening is taken into account via the Thomas-Fermi dielectric function (26.4), all phonon frequencies vanish linearly with k at long wavelengths, though the dispersion relation is not of the simple Bohm-Staver form (26.5).

2. *Electronic vs. Ionic Contributions to the Specific Heat of Metals*

(a) Estimating the sound velocity in a metal by the Bohm-Staver relation (26.8), show that

$$\frac{\hbar\omega_D}{\mathcal{E}_F} = \left(\frac{2^{8/3}}{3}\frac{Z^{1/3}m}{M}\right)^{1/2}. \qquad (26.52)$$

---

[30] The screening theory of Chapter 17 was based on the assumption that the external potential was a weak perturbation on the electron gas. Since this is not the case for the potential of the ions, a relation of the form $\phi^{\text{total}}(\mathbf{q}) = (1/\epsilon)\phi^{\text{ion}}(\mathbf{q})$ is not strictly valid. One *can* find a linear relation between the *deviations* of the total and ionic potentials from their equilibrium values. To derive it, however, one must take the system perturbed by the ions to be not a free electron gas, but a gas of electrons in the presence of the full equilibrium periodic potential. The formula describing the screening is then more complex. These extra difficulties are referred to as band structure effects. We ignore them in this problem.

[31] More precisely, we should use the frequency-dependent dielectric function $\epsilon(\mathbf{q}, \omega)$, where ω is the frequency of the normal mode being examined. When, however, ω is less than ω_D, the frequency dependence of the dielectric function (17.60) is quite negligible. This observation is the analytic justification of the adiabatic approximation.

(b) Using this result and Eq. (23.30), show that the low temperature electronic and ionic contributions to the specific heat are related by

$$\frac{c_v^{\text{el}}}{c_v^{\text{ion}}} = \left(\frac{5}{12\pi^2}\right) Z \left(\frac{4Z^{1/3}m}{3M}\right)^{3/2} \left(\frac{\mathcal{E}_F}{k_B T}\right)^2. \tag{26.53}$$

(c) Estimating the mass of the ion by AM_p, where A is the mass number and M_p is the proton mass ($M_p = 1836$ m) show that the electronic specific heat exceeds the ionic contribution when the temperature falls below

$$T_0 = 5.3Z^{1/2}\left(\frac{Z}{A}\right)^{3/4}\left(\frac{a_0}{r_s}\right)^2 \times 10^2 \text{ K}. \tag{26.54}$$

(d) Evaluate T_0 for sodium, aluminum, and lead.

(e) Show that the magnitude of the leading (cubic) term in the lattice specific heat is bigger than the magnitude of the cubic correction to the electronic specific heat (Eq. (2.102), evaluated in the free electron approximation) by a factor

$$\frac{1}{Z}\left(\frac{3M}{Z^{1/3}m}\right)^{3/2}. \tag{26.55}$$

3. *Phonon Corrections to the Electronic Energy*

In the limit $\omega_D \to 0$, the correction (26.27) to the electronic energy reduces to that of the Hartree-Fock approximation as modified by Thomas-Fermi screening (Chapter 17, page 344):

$$\mathcal{E}_{\mathbf{k}}^{TF} = \mathcal{E}_{\mathbf{k}}^{0} - \int_{k'<k_F} \frac{d\mathbf{k}'}{(2\pi)^3} \frac{4\pi e^2}{|\mathbf{k} - \mathbf{k}'|^2 + k_0^{\,2}}. \tag{26.56}$$

When the phonon frequencies are not considered vanishingly small, (26.27) differs appreciably from (26.56) only for those values of the integration variable $\mathbf{k}'$ for which $\mathcal{E}_{\mathbf{k}'}$ is within $O(\hbar\omega_D)$ of $\mathcal{E}_{\mathbf{k}}$. Since $\hbar\omega_D$ is small compared with $\mathcal{E}_F$, the region of $\mathbf{k}'$ for which the correction is appreciable is a shell about the surface $\mathcal{E}_{\mathbf{k}'} = \mathcal{E}_{\mathbf{k}}$ that is thin (on the scale of the dimensions of the zone). We may exploit this fact to simplify the correction term by writing the integral over $\mathbf{k}'$ as an integral over energy $\mathcal{E}'$, and an integral over the constant-energy surfaces $\mathcal{E}_{\mathbf{k}'} = \mathcal{E}'$. As $\mathcal{E}'$ varies, the variation of the term in $(\mathcal{E}_{\mathbf{k}} - \mathcal{E}')^2$ in the denominator of (26.27) is very important, since the denominator vanishes within this range. However, the remaining $\mathcal{E}'$ dependence of the integrand (due to the fact that $\mathbf{k}'$ is constrained to a surface of energy $\mathcal{E}'$) leads to very little variation as $\mathcal{E}'$ varies within $O(\hbar\omega_D)$ of $\mathcal{E}_F$. It is thus a good approximation to replace the $\mathbf{k}'$ integrations over the surfaces $\mathcal{E}_{\mathbf{k}'} = \mathcal{E}'$ by integrations over the single surface $\mathcal{E}_{\mathbf{k}'} = \mathcal{E}_{\mathbf{k}}$. After this replacement, the only $\mathcal{E}'$ dependence left comes from the explicit term in $\mathcal{E}'$ in the denominator. The integral over $\mathcal{E}'$ is then easily performed.

(a) Show that under this approximation,

$$\mathcal{E}_{\mathbf{k}} = \mathcal{E}_{\mathbf{k}}^{TF} - \int_{\mathcal{E}_{\mathbf{k}'}=\mathcal{E}_{\mathbf{k}}} \frac{dS'}{8\pi^3|\partial\mathcal{E}/\partial\mathbf{k}|} \frac{4\pi e^2}{|\mathbf{k} - \mathbf{k}'|^2 + k_0^{\,2}} \times \frac{1}{2}\hbar\omega(\mathbf{k} - \mathbf{k}') \ln\left|\frac{\mathcal{E}_F - \mathcal{E}_{\mathbf{k}} - \hbar\omega(\mathbf{k} - \mathbf{k}')}{\mathcal{E}_F - \mathcal{E}_{\mathbf{k}} + \hbar\omega(\mathbf{k} - \mathbf{k}')}\right|. \tag{26.57}$$

(b) Show that (26.57) immediately implies that the phonon-corrected Fermi surface $\mathcal{E}_{\mathbf{k}} = \mathcal{E}_F$ is identical to the uncorrected Fermi surface, $\mathcal{E}_{\mathbf{k}}^{TF} = \mathcal{E}_F$.

(c) Show that when $\mathcal{E}_{\mathbf{k}}$ is several times $\hbar\omega_D$ from $\mathcal{E}_F$, the phonon correction is smaller than the Thomas-Fermi correction by $O(\hbar\omega_D/\mathcal{E}_F)(\hbar\omega_D/[\mathcal{E}_{\mathbf{k}} - \mathcal{E}_F])$.

(d) Show that when $\mathcal{E}_{\mathbf{k}} - \mathcal{E}_F$ is small compared with $\hbar\omega_D$, (26.57) reduces to (26.28) and (26.29).

27
Dielectric Properties of Insulators

Macroscopic Electrostatic Maxwell Equations

Theory of the Local Field

Clausius-Mossotti Relation

Theory of the Polarizability

Long-Wavelength Optical Modes in Ionic Crystals

Optical Properties of Ionic Crystals

Reststrahlen

Covalent Insulators

Pyroelectric and Ferroelectric Crystals

Because charge cannot flow freely in insulators, applied electric fields of substantial amplitude can penetrate into their interiors. There are at least three broad contexts in which it is important to know how the internal structure of an insulator, both electronic and ionic, readjusts when an additional electric field is superimposed on the electric field associated with the periodic lattice potential:

1. We may place a sample of the insulator in a static electric field such as that existing between the plates of a capacitor. Many important consequences of the resulting internal distortion can be deduced if one knows the static dielectric constant ϵ_0 of the crystal, whose calculation is therefore an important aim of any microscopic theory of insulators.
2. We may be interested in the optical properties of the insulator—i.e., in its response to the AC electric field associated with electromagnetic radiation. In this case the important quantity to calculate is the frequency-dependent dielectric constant $\epsilon(\omega)$, or, equivalently, the index of refraction, $n = \sqrt{\epsilon}$.
3. In an ionic crystal, even in the absence of externally applied fields, there may be long-range electrostatic forces between the ions in addition to the periodic lattice potential, when the lattice is deformed from its equilibrium configuration (as, for example, in the course of executing a normal mode). Such forces are often best dealt with by considering the additional electric field giving rise to them, whose sources are intrinsic to the crystal.

In dealing with any of these phenomena the theory of the macroscopic Maxwell equations in a medium is a most valuable tool. We begin with a review of the electrostatic aspects of this theory.

MACROSCOPIC MAXWELL EQUATIONS OF ELECTROSTATICS

When viewed on the atomic scale, the charge density $\rho^{\text{micro}}(\mathbf{r})$ of any insulator is a very rapidly varying function of position, reflecting the microscopic atomic structure of the insulator. On the same atomic scale the electrostatic potential $\phi^{\text{micro}}(\mathbf{r})$ and the electric field $\mathbf{E}^{\text{micro}}(\mathbf{r}) = -\nabla\phi^{\text{micro}}(\mathbf{r})$ also have strong and rapid variations since they are related to $\rho^{\text{micro}}(\mathbf{r})$ by

$$\nabla \cdot \mathbf{E}^{\text{micro}}(\mathbf{r}) = 4\pi\rho^{\text{micro}}(\mathbf{r}). \tag{27.1}$$

On the other hand, in the conventional *macroscopic* electromagnetic theory of an insulator the charge density $\rho(\mathbf{r})$, potential $\phi(\mathbf{r})$, and electric field $\mathbf{E}(\mathbf{r})$ show no such rapid variation.[1] Specifically, in the case of an insulator bearing no excess charge beyond that of its component ions (or atoms or molecules), the macroscopic electrostatic field is determined by the macroscopic Maxwell equation:[2]

$$\nabla \cdot \mathbf{D}(\mathbf{r}) = 0, \tag{27.2}$$

[1] Indeed, in an insulating medium in the absence of any externally applied fields, $\phi(\mathbf{r})$ is zero (or constant).

[2] More generally, one writes $\nabla \cdot \mathbf{D} = 4\pi\rho$, where ρ is the so-called free charge—i.e., that part of the macroscopic charge density due to excess charges not intrinsic to the medium. Throughout the following discussion we assume that there is no free charge, so that our macroscopic charge density is always the so-called bound charge of macroscopic electrostatics. The inclusion of free charge is straightforward, but not relevant to any of the applications we wish to make here.

in conjunction with the equation giving the macroscopic electric field **E** in terms of the electric displacement **D** and polarization density **P**,

$$\mathbf{D}(\mathbf{r}) = \mathbf{E}(\mathbf{r}) + 4\pi\mathbf{P}(\mathbf{r}). \tag{27.3}$$

These imply (in the absence of free charge) that the macroscopic electric field satisfies

$$\nabla \cdot \mathbf{E}(\mathbf{r}) = -4\pi\nabla \cdot \mathbf{P}(\mathbf{r}), \tag{27.4}$$

where **P** (to be defined in detail below) is generally a very slowly varying function of position inside an insulator.

Although it is very convenient to work with the macroscopic Maxwell equations, it is also essential to deal with the microscopic field acting on individual ions.[3] One must therefore keep the relation between macroscopic and microscopic quantities clearly in mind. The connection, first derived by Lorentz, can be made as follows:[4]

Suppose we have an insulator (not necessarily in its equilibrium configuration) that is described (at an instant) by a microscopic charge density $\rho^{\text{micro}}(\mathbf{r})$, which reflects the detailed atomic arrangement of electrons and nuclei and which gives rise to the rapidly varying microscopic field, $\mathbf{E}^{\text{micro}}(\mathbf{r})$. The macroscopic field $\mathbf{E}(\mathbf{r})$ is defined to be an average of $\mathbf{E}^{\text{micro}}$ over a region about **r** that is small on the macroscopic scale, but large compared with characteristic atomic dimensions a (Figure 27.1). We make the averaging procedure explicit by using a positive normalized weight function f, satisfying:

$$f(\mathbf{r}) \geqslant 0; \quad f(\mathbf{r}) = 0, \quad r > r_0; \quad \int d\mathbf{r}\, f(\mathbf{r}) = 1; \quad f(-\mathbf{r}) = f(\mathbf{r}). \tag{27.5}$$

The distance r_0 beyond which f vanishes is large compared with atomic dimensions

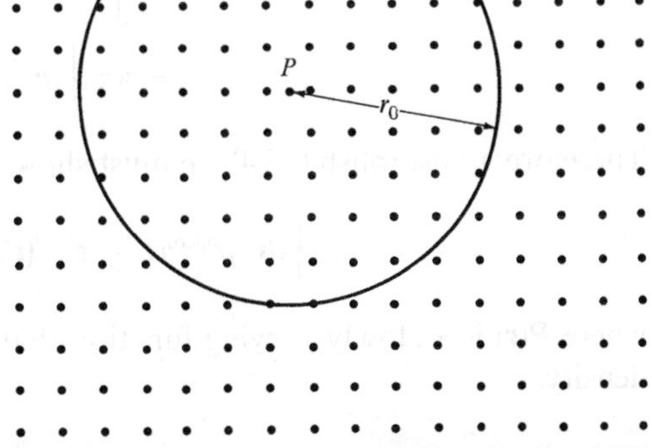

Figure 27.1
The value of a macroscopic quantity at a point P is an average of the microscopic quantity over a region of dimensions r_0 in the neighborhood of P, where r_0 is large compared to the interparticle spacing a.

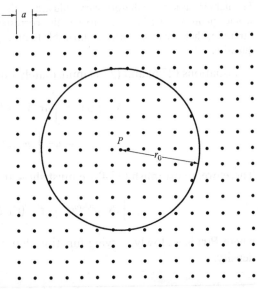

[3] We continue with our convention of using the single term "ion" to refer to the ions in ionic crystals, but also the atoms or molecules making up molecular crystals.

[4] The following discussion is very similar to a derivation of all the macroscopic Maxwell equations by G. Russakoff, *Am. J. Phys.* **10**, 1188 (1970).

a, but small on the scale over which macroscopically defined quantities vary.[5] We also require that f vary slowly; i.e., $|\nabla f|/f$ should not be appreciably greater than the minimum value, of order $1/r_0$, required by Eqs. (27.5). Beyond these assumptions, the form of the macroscopic theory is independent of the properties of the weight function f.

We can now give a precise definition of the macroscopic electric field $\mathbf{E}(\mathbf{r})$ at the point $\mathbf{r}$: it is the average of the microscopic field in a region of radius r_0 about $\mathbf{r}$, with points displaced by $-\mathbf{r}'$ from $\mathbf{r}$ receiving a weight proportional to $f(\mathbf{r}')$; i.e.,

$$\mathbf{E}(\mathbf{r}) = \int d\mathbf{r}'\, \mathbf{E}^{\text{micro}}(\mathbf{r} - \mathbf{r}')f(\mathbf{r}'). \tag{27.6}$$

Loosely speaking, the operation specified by (27.6) washes out those features of the microscopic field that vary rapidly on the scale of r_0, and preserves those features that vary slowly on the scale of r_0 (Figure 27.2). Note, for example, that if $\mathbf{E}^{\text{micro}}$ should happen to vary slowly on the scale of r_0 (as would be the case if the point $\mathbf{r}$ were in empty space, far from the insulator), then $\mathbf{E}(\mathbf{r})$ would equal $\mathbf{E}^{\text{micro}}(\mathbf{r})$.

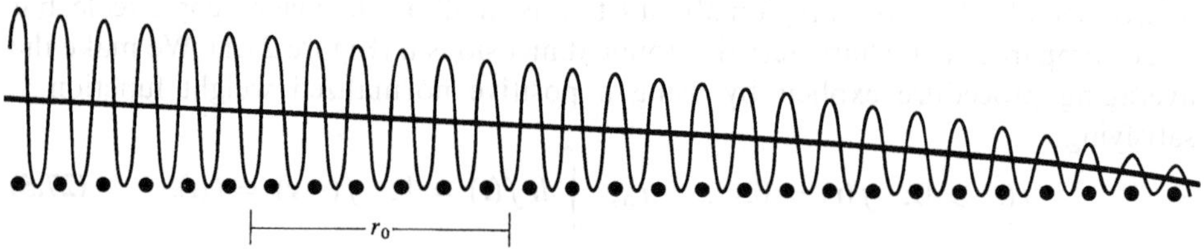

Figure 27.2
The lighter rapidly oscillatory curve illustrates the characteristic spatial variation of a microscopic quantity. The heavier curve is the corresponding macroscopic quantity. Only spatial variations that occur on a scale comparable to or larger than r_0 are preserved in the macroscopic quantity.

Equations (27.6) and (27.1) immediately imply that

$$\begin{aligned}\nabla \cdot \mathbf{E}(\mathbf{r}) &= \int d\mathbf{r}'\, \nabla \cdot \mathbf{E}^{\text{micro}}(\mathbf{r} - \mathbf{r}')f(\mathbf{r}') \\ &= 4\pi \int d\mathbf{r}'\, \rho^{\text{micro}}(\mathbf{r} - \mathbf{r}')f(\mathbf{r}'). \end{aligned} \tag{27.7}$$

Therefore, to establish (27.4) we must show that

$$\int d\mathbf{r}'\, \rho^{\text{micro}}(\mathbf{r} - \mathbf{r}')f(\mathbf{r}')\, d\mathbf{r}' = -\nabla \cdot \mathbf{P}(\mathbf{r}), \tag{27.8}$$

where $\mathbf{P}(\mathbf{r})$ is a slowly varying function that can be interpreted as a dipole moment density.

---

[5] More precisely, the macroscopic Maxwell equations are valid only when the variation in the macroscopic fields is sufficiently slow that their minimum characteristic wavelength allows a choice of r_0 satisfying $\lambda \gg r_0 \gg a$. This condition can be satisfied by the field associated with visible light ($\lambda \sim 10^4 a$), but not by the field associated with X rays ($\lambda \sim a$).

We shall discuss only the case in which the microscopic charge density can be resolved into a sum of contributions from ions (or atoms or molecules) located at positions $\mathbf{r}_j$ characterized by individual charge distributions $\rho_j(\mathbf{r} - \mathbf{r}_j)$:

$$\rho^{\text{micro}}(\mathbf{r}) = \sum_j \rho_j(\mathbf{r} - \mathbf{r}_j). \tag{27.9}$$

Such a resolution is quite natural in ionic or molecular solids, but is rather more difficult to achieve in covalent crystals, where important parts of the electronic charge distribution are not readily associated with a particular site in the crystal. Our discussion is therefore primarily applicable to the former two insulating categories. Quite a different approach is required to calculate the dielectric properties of covalent crystals. We shall return to this point below.

We are interested in nonequilibrium configurations of the insulator in which the ions are displaced from their equilibrium positions $\mathbf{r}_j^0$ and are deformed from their equilibrium shapes,[6] which are described by charge densities ρ_j^0. Thus $\rho^{\text{micro}}(\mathbf{r})$ will not, in general, be equal to the equilibrium microscopic charge density,

$$\rho_0^{\text{micro}}(\mathbf{r}) = \sum_j \rho_j^0(\mathbf{r} - \mathbf{r}_j^0). \tag{27.10}$$

Using (27.9) we can write (27.7) as:

$$\begin{aligned}\nabla \cdot \mathbf{E}(\mathbf{r}) &= 4\pi \sum_j \int d\mathbf{r}'\, \rho_j(\mathbf{r} - \mathbf{r}_j - \mathbf{r}')f(\mathbf{r}') \\ &= 4\pi \sum_j \int d\bar{\mathbf{r}}\, \rho_j(\bar{\mathbf{r}})f(\mathbf{r} - \mathbf{r}_j^0 - (\bar{\mathbf{r}} + \Delta_j)),\end{aligned} \tag{27.11}$$

where $\Delta_j = \mathbf{r}_j - \mathbf{r}_j^0$. The displacement Δ_j of the jth ion from its equilibrium position is a microscopic distance of order a or less. Furthermore the charge density $\rho_j(\bar{\mathbf{r}})$ vanishes when $\bar{r}$ exceeds a microscopic distance of order a. Since the variation in the weight function f is very small over distances of order a, we can expand (27.11) in what is effectively a series in powers of a/r_0 by using the Taylor expansion:

$$f(\mathbf{r} - \mathbf{r}_j^0 - (\bar{\mathbf{r}} + \Delta_j)) = \sum_{n=0}^{\infty} \frac{1}{n!}\left[-(\bar{\mathbf{r}} + \Delta_j)\cdot\nabla\right]^n f(\mathbf{r} - \mathbf{r}_j^0). \tag{27.12}$$

If we substitute the first two terms[7] from (27.12) into (27.11) we find that

$$\nabla \cdot \mathbf{E}(\mathbf{r}) = 4\pi\left[\sum_j e_j f(\mathbf{r} - \mathbf{r}_j^0) - \sum_j (\mathbf{p}_j + e_j\Delta_j)\cdot\nabla f(\mathbf{r} - \mathbf{r}_j^0)\right], \tag{27.13}$$

where

$$e_j = \int d\bar{\mathbf{r}}\, \rho_j(\bar{\mathbf{r}}), \quad \mathbf{p}_j = \int d\bar{\mathbf{r}}\, \rho_j(\bar{\mathbf{r}})\bar{\mathbf{r}}. \tag{27.14}$$

---

[6] We have in mind applications (a) to monatomic Bravais lattices (in which the $\mathbf{r}_j^0$ are just the Bravais lattice vectors $\mathbf{R}$ and all of the functions ρ_j^0 are identical, and (b) to lattices with a basis, in which the $\mathbf{r}_j^0$ run through all vectors $\mathbf{R}$, $\mathbf{R} + \mathbf{d}$, etc., and there are as many distinct functional forms for the ρ_j^0 as there are distinct types of ions in the basis.

[7] We shall find that the first ($n = 0$) term makes no contribution to (27.11), and we must therefore retain the next ($n = 1$) term to get the leading contribution.

The quantities e_j and $\mathbf{p}_j$ are simply the total charge and dipole moment of the jth ion.

In the case of a monatomic Bravais lattice the charge of each "ion" must be zero (since the crystal is neutral and all "ions" are identical). In addition the equilibrium positions $\mathbf{r}_j^0$ are the Bravais lattice sites $\mathbf{R}$, so (27.13) reduces to

$$\nabla \cdot \mathbf{E}(\mathbf{r}) = -4\pi\nabla \cdot \sum_{\mathbf{R}} f(\mathbf{r} - \mathbf{R})\mathbf{p}(\mathbf{R}), \tag{27.15}$$

where $\mathbf{p}(\mathbf{R})$ is the dipole moment of the atom at site $\mathbf{R}$.

With a straightforward generalization of the definition of $\mathbf{p}(\mathbf{R})$, this result remains valid (to leading order in a/r_0) even when we allow for ionic charge and a polyatomic basis. To see this, suppose that the $\mathbf{r}_j^0$ now run through the sites $\mathbf{R} + \mathbf{d}$ of a lattice with a basis. We can then label p_j and e_j by the Bravais lattice vector $\mathbf{R}$ and basis vector $\mathbf{d}$ specifying the equilibrium position of the jth ion:[8]

$$\mathbf{p}_j \to \mathbf{p}(\mathbf{R}, \mathbf{d}), \quad e_j \to e(\mathbf{d}), \quad \mathbf{r}_j^0 \to \mathbf{R} + \mathbf{d}, \quad \boldsymbol{\Delta}_j \to \mathbf{u}(\mathbf{R}, \mathbf{d}). \tag{27.16}$$

Since d is a microscopic length of order a, we can perform the further expansion:

$$f(\mathbf{r} - \mathbf{R} - \mathbf{d}) \approx f(\mathbf{r} - \mathbf{R}) - \mathbf{d} \cdot \nabla f(\mathbf{r} - \mathbf{R}). \tag{27.17}$$

Substituting this into (27.13) and dropping terms of higher than linear order in a/r_0, we again recover (27.15), where $\mathbf{p}(\mathbf{R})$ is now the dipole moment of the entire primitive cell[9] associated with $\mathbf{R}$:

$$\mathbf{p}(\mathbf{R}) = \sum_{\mathbf{d}} [e(\mathbf{d})\mathbf{u}(\mathbf{R}, \mathbf{d}) + \mathbf{p}(\mathbf{R}, \mathbf{d})]. \tag{27.18}$$

Comparing (27.15) with the macroscopic Maxwell equation (27.4), we find that the two are consistent if the polarization density is defined by

$$\mathbf{P}(\mathbf{r}) = \sum_{\mathbf{R}} f(\mathbf{r} - \mathbf{R})\mathbf{p}(\mathbf{R}). \tag{27.19}$$

If we are dealing with distortions from equilibrium whose form does not vary much from cell to cell on the microscopic scale, then $\mathbf{p}(\mathbf{R})$ will vary only slowly from cell to cell, and we can evaluate (27.19) as an integral:

$$\mathbf{P}(\mathbf{r}) = \frac{1}{v}\sum_{\mathbf{R}} v f(\mathbf{r} - \mathbf{R})\mathbf{p}(\mathbf{R}) \approx \frac{1}{v}\int d\bar{\mathbf{r}}\, f(\mathbf{r} - \bar{\mathbf{r}})\mathbf{p}(\bar{\mathbf{r}}), \tag{27.20}$$

where $\mathbf{p}(\bar{\mathbf{r}})$ is a smooth, slowly varying continuous function equal to the polarization of the cells in the immediate vicinity of $\bar{\mathbf{r}}$, and v is the volume of the equilibrium primitive cell.

---

[8] Ions separated by Bravais lattice vectors have the same total charge, so e_j depends only on $\mathbf{d}$, and not on $\mathbf{R}$.

[9] In deriving (27.18) we have used the fact that the total charge of the primitive cell, $\Sigma e(\mathbf{d})$, vanishes. We have also neglected an additional term, $\Sigma\, \mathbf{d}e(\mathbf{d})$, which is the dipole moment of the primitive cell in the undistorted equilibrium crystal. In most crystals this term vanishes for the most natural choices of primitive cell. If it did not vanish, the crystal would have a polarization density in equilibrium in the absence of distorting forces or external electric fields. Such crystals do exist, and are known as pyroelectrics. We shall discuss them later in this chapter, where we shall also make clearer what is meant by "most natural choices of primitive cell" (see page 554).

We shall restrict our use of the macroscopic Maxwell equations to situations in which the variation in cellular polarization is appreciable only over distances large compared with the dimensions r_0 of the averaging region; this is certainly the case for fields whose wavelengths are in the visible part of the spectrum or longer. Since the integrand in (27.20) vanishes when $\bar{\mathbf{r}}$ is more than r_0 from $\mathbf{r}$, then if $\mathbf{p}(\bar{\mathbf{r}})$ varies negligibly over a distance r_0 from $\bar{\mathbf{r}}$, we can replace $\mathbf{p}(\bar{\mathbf{r}})$ by $\mathbf{p}(\mathbf{r})$, and bring it outside the integral to obtain:

$$\mathbf{P}(\mathbf{r}) = \frac{\mathbf{p}(\mathbf{r})}{v} \int d\bar{\mathbf{r}}\, f(\mathbf{r} - \bar{\mathbf{r}}). \qquad (27.21)$$

Since $\int d\mathbf{r}'\, f(\mathbf{r}') = 1$, we finally have

$$\mathbf{P}(\mathbf{r}) = \frac{1}{v}\, \mathbf{p}(\mathbf{r}); \qquad (27.22)$$

i.e., provided the dipole moment of each cell varies appreciably only on the macroscopic scale, then the macroscopic Maxwell equation (27.4) holds with the polarization density $\mathbf{P}(\mathbf{r})$ defined to be the dipole moment of a primitive cell in the neighborhood of $\mathbf{r}$, divided by its equilibrium volume.[10]

THEORY OF THE LOCAL FIELD

To exploit macroscopic electrostatics, a theory is required relating the polarization density $\mathbf{P}$ back to the macroscopic electric field $\mathbf{E}$. Since each ion has microscopic dimensions, its displacement and distortion will be determined by the force due to the *microscopic* field at the position of the ion, diminished by the contribution to the microscopic field from the ion itself. This field is frequently called the local (or effective) field, $\mathbf{E}^{\text{loc}}(\mathbf{r})$.

We can exploit macroscopic electrostatics to simplify the evaluation of $\mathbf{E}^{\text{loc}}(\mathbf{r})$ by dividing space into regions near to and far from $\mathbf{r}$. The far region is to contain all external sources of field, all points outside the crystal, and only points inside the crystal that are far from $\mathbf{r}$ compared with the dimensions r_0 of the averaging region used in (Eq. (27.6)). All other points are said to be in the near region (Figure 27.3). The reason for this division is that the contribution to $\mathbf{E}^{\text{loc}}(\mathbf{r})$ of all charge in the far region will vary negligibly over a distance r_0 about $\mathbf{r}$, and would be unaffected if we were to apply the averaging procedure specified in (27.6). Therefore the contribution to $\mathbf{E}^{\text{loc}}(\mathbf{r})$ of all charge in the far region is just the macroscopic field, $\mathbf{E}^{\text{macro}}_{\text{far}}(\mathbf{r})$, that would exist at $\mathbf{r}$ if only the charge in the far region were present:

$$\mathbf{E}^{\text{loc}}(\mathbf{r}) = \mathbf{E}^{\text{loc}}_{\text{near}}(\mathbf{r}) + \mathbf{E}^{\text{micro}}_{\text{far}}(\mathbf{r}) = \mathbf{E}^{\text{loc}}_{\text{near}}(\mathbf{r}) + \mathbf{E}^{\text{macro}}_{\text{far}}(\mathbf{r}). \qquad (27.23)$$

Now $\mathbf{E}(\mathbf{r})$, the full macroscopic field at $\mathbf{r}$, is constructed by averaging the microscopic field within r_0 of $\mathbf{r}$ due to all charges, in both the near and the far regions; i.e.,

$$\mathbf{E}(\mathbf{r}) = \mathbf{E}^{\text{macro}}_{\text{far}}(\mathbf{r}) + \mathbf{E}^{\text{macro}}_{\text{near}}(\mathbf{r}), \qquad (27.24)$$

---

[10] The derivation of this intuitive result permits us to estimate corrections when required.

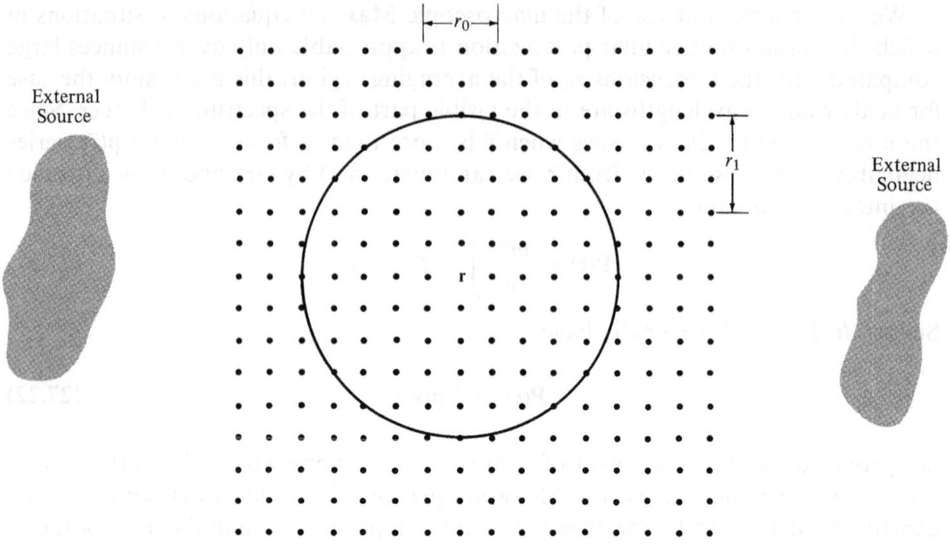

Figure 27.3
In calculating the local field at a point r it is convenient to consider separately contributions from the *far region* (i.e., all the crystal outside the sphere of radius r_1 about $\mathbf{r}$ and all external sources of field) and from the *near region* (i.e., all points within the sphere about $\mathbf{r}$). The far region is taken to be far from $\mathbf{r}$ on the scale of the averaging length r_0, to insure that the microscopic field due to charges in the far region is equal to its macroscopic average.

where $\mathbf{E}_{\text{near}}^{\text{macro}}(\mathbf{r})$ is the macroscopic field that would exist at $\mathbf{r}$ if only the charges in the near region[11] were present. We can therefore rewrite Eq. (27.23) as:

$$\boxed{\mathbf{E}^{\text{loc}}(\mathbf{r}) = \mathbf{E}(\mathbf{r}) + \mathbf{E}_{\text{near}}^{\text{loc}}(\mathbf{r}) - \mathbf{E}_{\text{near}}^{\text{macro}}(\mathbf{r}).} \qquad \textbf{(27.25)}$$

Thus we have related the unknown local field at $\mathbf{r}$ to the macroscopic electric field[12] at $\mathbf{r}$ and additional terms that depend only on the configuration of charges in the near region.

We shall apply (27.25) only to nonequilibrium configurations of the crystal with negligible spatial variation from cell to cell over distances of order r_1, the size of the near region.[13] In such cases $E_{\text{near}}^{\text{macro}}(\mathbf{r})$ will be the macroscopic field due to a *uniformly* polarized medium, whose shape is that of the near region. If we choose the near region

---

11 Including, of course, the ion on which we are calculating the force.

12 A further complication of a purely macroscopic nature is peripheral to the argument here, in which $\mathbf{E}(\mathbf{r})$ is assumed to be given. If the internal field and polarization are produced by placing the sample in a specified field $\mathbf{E}^{\text{ext}}$, then an additional problem in macroscopic electrostatics must be solved to determine the macroscopic field $\mathbf{E}$ in the interior of the sample, since the discontinuity in the polarization density $\mathbf{P}$ at the surface of the sample acts as a bound surface charge, and contributes an additional term to the macroscopic field in the interior. For certain simply shaped samples in uniform external fields the induced polarization $\mathbf{P}$ and the macroscopic field $\mathbf{E}$ in the interior will both be constant and parallel to $\mathbf{E}^{\text{ext}}$, and one can write: $\mathbf{E} = \mathbf{E}^{\text{ext}} - N\mathbf{P}$, where N, the "depolarization factor," depends on the geometry of the sample. The most important elementary case is the sphere, for which $N = 4\pi/3$. For a general ellipsoid (in which $\mathbf{P}$ need not be parallel to $\mathbf{E}$) see E. C. Stoner, *Phil. Mag.* **36**, 803 (1945).

13 Note that we are now very macroscopic indeed, requiring that $\lambda \gg r_1 \gg r_0 \gg a$.

to be a sphere, then this field is given by the following elementary result of electrostatics (see Problem 1): The macroscopic field anywhere inside a uniformly polarized sphere is just $\mathbf{E} = -4\pi\mathbf{P}/3$, where $\mathbf{P}$ is the polarization density. Therefore if the near region is a sphere over which $\mathbf{P}$ has negligible spatial variation, then Eq. (27.25) becomes

$$\mathbf{E}^{\text{loc}}(\mathbf{r}) = \mathbf{E}(\mathbf{r}) + \mathbf{E}^{\text{loc}}_{\text{near}}(\mathbf{r}) + \frac{4\pi\mathbf{P}(\mathbf{r})}{3}. \tag{27.26}$$

We are thus left with the problem of calculating the microscopic local field $\mathbf{E}^{\text{loc}}_{\text{near}}(\mathbf{r})$ appropriate to a spherical region whose center is taken to be the ion on which the field acts. Inside this region the charge density is the same in every cell (except for the removal of the ion at the center on which we are calculating the force). In most applications this calculation is done under the following simplifying assumptions:

1. The spatial dimensions and the displacement from equilibrium of each ion are considered to be so small that the polarizing field acting on it can be taken to be uniform over the whole ion and equal to the value of $\mathbf{E}^{\text{loc}}$ at the equilibrium position of the ion.
2. The spatial dimensions and the displacement from equilibrium of each ion are considered to be so small that the contribution to the local field at the equilibrium position of the given ion, from the ion whose equilibrium position is $\mathbf{R} + \mathbf{d}$, is accurately given by the field of a dipole of moment $e(\mathbf{d})\mathbf{u}(\mathbf{R} + \mathbf{d}) + \mathbf{p}(\mathbf{R} + \mathbf{d})$.

Since the dipole moments of ions at equivalent sites (displaced from each other by Bravais lattice vectors $\mathbf{R}$) are identical within the near region over which $\mathbf{P}$ has negligible variation, the calculation of $\mathbf{E}^{\text{loc}}_{\text{near}}$ at an equilibrium site reduces to the type of lattice sum we described in Chapter 20. Furthermore, in the special case in which every equilibrium site in the equilibrium crystal is a center of cubic symmetry, it is easily shown (Problem 2) that this lattice sum must vanish; i.e., $\mathbf{E}^{\text{loc}}_{\text{near}}(\mathbf{r}) = 0$ at every equilibrium site. Since this case includes both the solid noble gases and the alkali halides, it is the only one we consider. For these crystals we may assume that the field polarizing each ion in the neighborhood of $\mathbf{r}$ is[14]

$$\boxed{\mathbf{E}^{\text{loc}}(\mathbf{r}) = \mathbf{E}(\mathbf{r}) + \frac{4\pi\mathbf{P}(\mathbf{r})}{3}.} \tag{27.27}$$

This result, sometimes known as the Lorentz relation, is widely used in theories of dielectrics. It is very important to remember the assumptions underlying it, particularly that of cubic symmetry about every equilibrium site.

Sometimes (27.27) is written in terms of the dielectric constant ϵ of the medium, using the constitutive relation[15]

$$\mathbf{D}(\mathbf{r}) = \epsilon\mathbf{E}(\mathbf{r}), \tag{27.28}$$

---

[14] Note that implicit in this relation is the fact that the local field acting on an ion depends only on the general location of the ion but not (in a lattice with a basis) on the type of ion (i.e., it depends on $\mathbf{R}$ but not on $\mathbf{d}$). This convenient simplification is a consequence of our assumption that every ion occupies a position of cubic symmetry.

[15] In noncubic crystals $\mathbf{P}$, and therefore $\mathbf{D}$, need not be parallel to $\mathbf{E}$, so ϵ is a tensor.

together with the relation (27.3) between **D**, **E**, and **P**, to express **P**(**r**) in terms of **E**(**r**):

$$\mathbf{P}(\mathbf{r}) = \frac{\epsilon - 1}{4\pi}\mathbf{E}(\mathbf{r}). \tag{27.29}$$

Using this to eliminate **P**(**r**) from (27.27), one finds that

$$\boxed{\mathbf{E}^{\mathrm{loc}}(\mathbf{r}) = \frac{\epsilon + 2}{3}\mathbf{E}(\mathbf{r}).} \tag{27.30}$$

Yet another way of expressing the same result is in terms of the *polarizability*, α, of the medium. The polarizability $\alpha(\mathbf{d})$ of the type of ion at position **d** in the basis is defined to be the ratio of its induced dipole moment to the field actually acting on it. Thus

$$\mathbf{p}(\mathbf{R} + \mathbf{d}) + e\mathbf{u}(\mathbf{R} + \mathbf{d}) = \alpha(\mathbf{d})\, \mathbf{E}^{\mathrm{loc}}(\mathbf{r})\big|_{\mathbf{r}\approx\mathbf{R}}. \tag{27.31}$$

The polarizability α of the medium is defined as the sum of the polarizabilities of the ions in a primitive cell:

$$\alpha = \sum_{\mathbf{d}} \alpha(\mathbf{d}). \tag{27.32}$$

Since (cf. (27.18) and (27.22)).

$$\mathbf{P}(\mathbf{r}) = \frac{1}{v}\sum_{\mathbf{d}}\Big[\mathbf{p}(\mathbf{R}, \mathbf{d}) + e(\mathbf{d})\mathbf{u}(\mathbf{R}, \mathbf{d})\Big]_{\mathbf{R}\approx\mathbf{r}}, \tag{27.33}$$

it follows that

$$\mathbf{P}(\mathbf{r}) = \frac{\alpha}{v}\mathbf{E}^{\mathrm{loc}}(\mathbf{r}). \tag{27.34}$$

Using (27.29) and (27.30) to express both **P** and $\mathbf{E}^{\mathrm{loc}}$ in terms of **E**, we find that (27.34) implies that

$$\boxed{\frac{\epsilon - 1}{\epsilon + 2} = \frac{4\pi\alpha}{3v}.} \tag{27.35}$$

This equation, known as the Clausius-Mossotti relation,[16] provides a valuable link between macroscopic and microscopic theories. A microscopic theory is required to calculate α, which gives the response of the ions to the actual field $\mathbf{E}^{\mathrm{loc}}$ acting on them. The resulting ε can then be used, in conjunction with the macroscopic Maxwell equations, to predict the optical properties of the insulator.

THEORY OF THE POLARIZABILITY

Two terms contribute to the polarizability α. The contribution from **p** (see Eq. (27.31)), the "atomic polarizability," arises from the distortion of the ionic charge distribution.

---

[16] When written in terms of the index of refraction, $n = \sqrt{\epsilon}$, the Clausius-Mossotti relation is known as the Lorentz-Lorenz relation. (In the recent physics and chemistry literature of England and the United States it has become the widespread practice to misspell the last name of O. F. Mossotti with a single "s," and/or to interchange his initials.)

The contribution from $e\mathbf{u}$, the "displacement polarizability," arises from ionic displacements. There is no displacement polarizability in molecular crystals where the "ions" are uncharged, but in ionic crystals it is comparable to the atomic polarizability.

Atomic Polarizability

We allow the local field acting on the ion in question to be frequency-dependent, writing

$$\mathbf{E}^{\text{loc}} = \text{Re}\,(\mathbf{E}_0 e^{-i\omega t}), \tag{27.36}$$

where $\mathbf{E}_0$ is independent of position (assumption 1, page 541). The simplest classical theory of atomic polarizability treats the ion as an electronic shell of charge $Z_i e$ and mass $Z_i m$ tied to a heavy, immobile, undeformable ion core, by a harmonic spring, of spring constant $K = Z_i m\omega_0^2$ (Figure 27.4). If the displacement of the shell

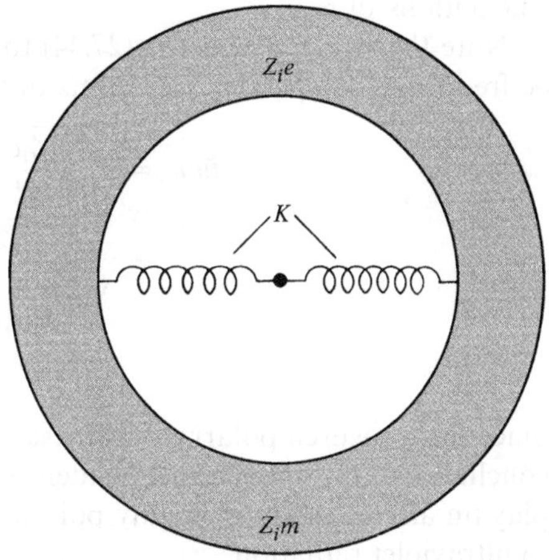

Figure 27.4
Crude classical model of atomic polarizability. The ion is represented as a charged shell of charge $Z_i e$ and mass $Z_i m$ tied to an immobile nucleus by a spring of force constant $K = Z_i m\omega_0^2$.

from its equilibrium position is given by

$$\mathbf{r} = \text{Re}\,(\mathbf{r}_0 e^{-i\omega t}), \tag{27.37}$$

then the equation of motion of the shell,

$$Z_i m\ddot{\mathbf{r}} = -K\mathbf{r} - Z_i e\mathbf{E}^{\text{loc}}, \tag{27.38}$$

implies that

$$\mathbf{r}_0 = -\frac{e\mathbf{E}_0}{m(\omega_0^2 - \omega^2)}. \tag{27.39}$$

Since the induced dipole moment is $\mathbf{p} = -Z_i e\mathbf{r}$, we have

$$\mathbf{p} = \text{Re}\,(\mathbf{p}_0 e^{-i\omega t}), \tag{27.40}$$

with

$$\mathbf{p}_0 = \frac{Z_i e^2}{m(\omega_0^2 - \omega^2)}\mathbf{E}_0. \tag{27.41}$$

Defining the frequency-dependent atomic polarizability by

$$\mathbf{p}_0 = \alpha^{\text{at}}(\omega)\mathbf{E}_0, \tag{27.42}$$

we have

$$\alpha^{\text{at}}(\omega) = \frac{Z_i e^2}{m(\omega_0{}^2 - \omega^2)}. \tag{27.43}$$

The model leading to (27.43) is, of course, very crude. However, for our purposes the most important feature of the result is that if ω is small compared with ω_0, the polarizability will be independent of frequency and equal to its static value:

$$\alpha^{\text{at}} = \frac{Z_i e^2}{m\omega_0{}^2}. \tag{27.44}$$

We would expect ω_0, the frequency of vibration of the electronic shell, to be of the order of an atomic excitation energy divided by $\hbar$. This suggests that, unless $\hbar\omega$ is of the order of several electron volts, we can take the atomic polarizability to be independent of frequency. This is confirmed by more accurate quantum-mechanical calculations of α.

Note that we can also use (27.44) to estimate the frequency below which α^{at} will be frequency-independent, in terms of the observed static polarizabilities:

$$\begin{aligned}\hbar\omega_0 &= \sqrt{\frac{\hbar^2 Z_i e^2}{m\alpha^{\text{at}}}} \\ &= \sqrt{\frac{4a_0{}^3 Z_i}{\alpha^{\text{at}}}\,\frac{e^2}{2a_0}}, \qquad a_0 = \frac{\hbar^2}{me^2}, \\ &= \sqrt{Z_i\left(\frac{10^{-24}\ \text{cm}^3}{\alpha^{\text{at}}}\right)} \times 10.5\ \text{eV}. \end{aligned} \tag{27.45}$$

Since the measured polarizabilities (see Table 27.1) are of the order of 10^{-24} cm^3, we conclude that the frequency dependence of the atomic polarizability will not come into play (in all but the most highly polarizable of ions) until frequencies corresponding to ultraviolet radiation.

Table 27.1
ATOMIC POLARIZABILITIES OF THE HALOGEN IONS, NOBLE GAS ATOMS, AND ALKALI METAL IONS[a]

HALOGENS		NOBLE GASES		ALKALI METALS	
		He	0.2	Li^+	0.03
F^-	1.2	Ne	0.4	Na^+	0.2
Cl^-	3	Ar	1.6	K^+	0.9
Br^-	4.5	Kr	2.5	Rb^+	1.7
I^-	7	Xe	4.0	Cs^+	2.5

[a] In units of 10^{-24} cm^3. Note that entries in the same row have the same electronic shell structure, but increasing nuclear charge.
Source: A. Dalgarno, *Advances Phys.* **11**, 281 (1962).

Displacement Polarizability

In ionic crystals we must consider the dipole moment due to the displacement of the charged ions by the electric field, in addition to the atomic polarization resulting from the deformation of their electronic shells by the field. We begin by ignoring the atomic polarization (*rigid-ion approximation*). To simplify the discussion we also consider only crystals with two ions per primitive cell, of charges e and $-e$. If the ions are undeformable, then the dipole moment of the primitive cell is just

$$\mathbf{p} = e\mathbf{w}, \qquad \mathbf{w} = \mathbf{u}^+ - \mathbf{u}^-, \tag{27.46}$$

where $\mathbf{u}^\pm$ is the displacement of the positive or negative ion from its equilibrium position.

To determine $\mathbf{w}(\mathbf{r})$ we note that the long-range electrostatic forces between ions are already contained in the field $\mathbf{E}^{\text{loc}}$. The remaining short-range interionic forces (e.g., higher-order electrostatic multipole moments and core-core repulsion) will fall off rapidly with distance, and we may assume that they produce a restoring force for an ion at $\mathbf{r}$ that depends only on the displacement of the ions in its vicinity. Since we are considering disturbances that vary slowly on the atomic scale, in the vicinity of $\mathbf{r}$ all ions of the same charge move as a whole with the same displacement, $\mathbf{u}^+(\mathbf{r})$ or $\mathbf{u}^-(\mathbf{r})$. Thus the short-range part of the restoring force acting on an ion at $\mathbf{r}$ will simply be proportional to[17] the relative displacement $\mathbf{w}(\mathbf{r}) = \mathbf{u}^+(\mathbf{r}) - \mathbf{u}^-(\mathbf{r})$ of the two oppositely charged sublattices in the neighborhood of $\mathbf{r}$.

Consequently in a distortion of the crystal with slow spatial variation on the microscopic scale, the displacements of the positive and negative ions satisfy equations of the form:

$$\begin{aligned} M_+\ddot{\mathbf{u}}^+ &= -k(\mathbf{u}^+ - \mathbf{u}^-) + e\mathbf{E}^{\text{loc}}, \\ M_-\ddot{\mathbf{u}}^- &= -k(\mathbf{u}^- - \mathbf{u}^+) - e\mathbf{E}^{\text{loc}}, \end{aligned} \tag{27.47}$$

which can be written

$$\ddot{\mathbf{w}} = \frac{e}{M}\mathbf{E}^{\text{loc}} - \frac{k}{M}\mathbf{w}, \tag{27.48}$$

where M is the ionic reduced mass, $M^{-1} = (M_+)^{-1} + (M_-)^{-1}$. Letting $\mathbf{E}^{\text{loc}}$ be an AC field of the form (27.36), we find that

$$\mathbf{w} = \text{Re}\,(\mathbf{w}_0 e^{-i\omega t}), \quad \mathbf{w}_0 = \frac{e\mathbf{E}_0/M}{\bar{\omega}^2 - \omega^2}, \tag{27.49}$$

where

$$\bar{\omega}^2 = \frac{k}{M}. \tag{27.50}$$

Accordingly,

$$\alpha^{\text{dis}} = \frac{p_0}{E_0} = \frac{ew_0}{E_0} = \frac{e^2}{M(\bar{\omega}^2 - \omega^2)}. \tag{27.51}$$

Note that the displacement polarizability (27.51) has the same form as the atomic polarizability (27.43). However the resonant frequency $\bar{\omega}$ is characteristic of lattice vibrational frequencies, and therefore $\hbar\bar{\omega} \approx \hbar\omega_D \approx 10^{-1}$ to 10^{-2} eV. This can be 10^2 to 10^3 times smaller than the atomic frequency ω_0, and therefore, in contrast to the atomic polarizability, the displacement polarizability has a significant frequency dependence in the infrared and optical range.

---

[17] The proportionality constant in general will be a tensor, but it reduces to a constant in a crystal of cubic symmetry, the only case we consider here.

Note also that because the ionic mass M is about 10^4 times the electronic mass m, the static ($\omega = 0$) ionic and displacement polarizabilities may well be of the same size. This means that the rigid-ion model we have used is unjustifiable, and (27.51) must be corrected to take into account the atomic polarizability of the ions as well. The most naive way to do this is simply to add the two types of contribution to the polarizability:

$$\alpha = (\alpha^+ + \alpha^-) + \frac{e^2}{M(\bar{\omega}^2 - \omega^2)}, \tag{27.52}$$

where α^+ and α^- are the atomic polarizabilities of the positive and negative ions. There is no real justification for this, since the first term in (27.52) was calculated on the assumption that the ions were immobile but polarizable, while the second was calculated for ions that could be moved, but not deformed. Evidently a more reasonable approach would combine the models that lead on the one hand to (27.43) and, on the other, to (27.51), calculating in one step the response to the local field of ions that can be both displaced and deformed. Such theories are known as *shell model* theories. They generally lead to results that differ considerably in numerical detail from those predicted by the more naive (27.52), but have many of the same basic structural features. We therefore explore the consequences of (27.52) further, indicating later how it should be modified in a more reasonable model.

In conjunction with the Clausius-Mossotti relation (27.35), the approximation expressed by (27.52) leads to a dielectric constant $\epsilon(\omega)$ for an ionic crystal, given by

$$\frac{\epsilon(\omega) - 1}{\epsilon(\omega) + 2} = \frac{4\pi}{3v}\left(\alpha^+ + \alpha^- + \frac{e^2}{M(\bar{\omega}^2 - \omega^2)}\right). \tag{27.53}$$

In particular, the static dielectric constant is given by

$$\frac{\epsilon_0 - 1}{\epsilon_0 + 2} = \frac{4\pi}{3v}\left(\alpha^+ + \alpha^- + \frac{e^2}{M\bar{\omega}^2}\right), \qquad (\omega \ll \bar{\omega}), \tag{27.54}$$

while the high-frequency[18] dielectric constant satisfies

$$\frac{\epsilon_\infty - 1}{\epsilon_\infty + 2} = \frac{4\pi}{3v}(\alpha^+ + \alpha^-), \qquad (\bar{\omega} \ll \omega \ll \omega_0). \tag{27.55}$$

It is convenient to write $\epsilon(\omega)$ in terms of ϵ_0 and ϵ_∞, since the two limiting forms are readily measured: ϵ_0 is the static dielectric constant of the crystal, while ϵ_∞ is the dielectric constant at optical frequencies, and is therefore related to the index of refraction, n, by $n^2 = \epsilon_\infty$. We have

$$\frac{\epsilon(\omega) - 1}{\epsilon(\omega) + 2} = \frac{\epsilon_\infty - 1}{\epsilon_\infty + 2} + \frac{1}{1 - (\omega^2/\bar{\omega}^2)}\left(\frac{\epsilon_0 - 1}{\epsilon_0 + 2} - \frac{\epsilon_\infty - 1}{\epsilon_\infty + 2}\right), \tag{27.56}$$

---

[18] In this context, by "high frequencies" we shall always mean frequencies high compared with lattice vibrational frequencies, but low compared with atomic excitation frequencies. The frequency of visible light generally satisfies this condition.

which can be solved for $\epsilon(\omega)$:

$$\epsilon(\omega) = \epsilon_\infty + \frac{\epsilon_\infty - \epsilon_0}{(\omega^2/\omega_T{}^2) - 1}, \tag{27.57}$$

where

$$\omega_T{}^2 = \bar{\omega}^2 \left(\frac{\epsilon_\infty + 2}{\epsilon_0 + 2}\right) = \bar{\omega}^2 \left(1 - \frac{\epsilon_0 - \epsilon_\infty}{\epsilon_0 + 2}\right). \tag{27.58}$$

Application to Long-Wavelength Optical Modes of Ionic Crystals

To calculate the normal mode dispersion relations in an ionic crystal we could proceed by the general techniques described in Chapter 22. However, we would encounter severe computational difficulties because of the very long range of the interionic electrostatic interactions. Techniques have been developed for dealing with this problem, similar to those exploited in calculating the cohesive energy of an ionic crystal (Chapter 20). However, for long-wavelength optical modes one can avoid such calculations by stating the problem as one in macroscopic electrostatics:

In a long-wavelength ($\mathbf{k} \approx \mathbf{0}$) optical mode the oppositely charged ions in each primitive cell undergo oppositely directed displacements, giving rise to a nonvanishing polarization density $\mathbf{P}$. Associated with this polarization density there will in general be a macroscopic electric field $\mathbf{E}$ and an electric displacement $\mathbf{D}$, related by

$$\mathbf{D} = \epsilon\mathbf{E} = \mathbf{E} + 4\pi\mathbf{P}. \tag{27.59}$$

In the absence of free charge, we have

$$\nabla \cdot \mathbf{D} = 0. \tag{27.60}$$

Furthermore, $\mathbf{E}^{\text{micro}}$ is the gradient of a potential.[19] It follows from (27.6) that $\mathbf{E}$ is also, so that

$$\nabla \times \mathbf{E} = \nabla \times (-\nabla\phi) = 0. \tag{27.61}$$

In a cubic crystal $\mathbf{D}$ is parallel to $\mathbf{E}$ (i.e., ϵ is not a tensor) and therefore, from (27.59), both are parallel to $\mathbf{P}$. If all three have the spatial dependence,

$$\begin{Bmatrix} \mathbf{D} \\ \mathbf{E} \\ \mathbf{P} \end{Bmatrix} = \text{Re} \begin{Bmatrix} \mathbf{D}_0 \\ \mathbf{E}_0 \\ \mathbf{P}_0 \end{Bmatrix} e^{i\mathbf{k}\cdot\mathbf{r}}, \tag{27.62}$$

then (27.60) reduces to $\mathbf{k} \cdot \mathbf{D}_0 = 0$, which requires that

$$\mathbf{D} = 0 \quad \text{or} \quad \mathbf{D}, \mathbf{E}, \text{and } \mathbf{P} \perp \mathbf{k}, \tag{27.63}$$

while (27.61) reduces to $\mathbf{k} \times \mathbf{E}_0 = 0$, which requires that

$$\mathbf{E} = 0 \quad \text{or} \quad \mathbf{E}, \mathbf{D}, \text{and } \mathbf{P} \parallel \mathbf{k}. \tag{27.64}$$

In a longitudinal optical mode the (nonzero) polarization density $\mathbf{P}$ is parallel to $\mathbf{k}$,

---

19 At optical frequencies one might worry about keeping only electrostatic fields, since the right side of the full Maxwell equation, $\nabla \times \mathbf{E} = -(1/c)\, \partial\mathbf{B}/\partial t$ need not be negligible. We shall see shortly, however, that a full electrodynamic treatment leads to very much the same conclusions.

and Eq. (27.63) therefore requires that **D** must vanish. This is consistent with (27.59) only if

$$\mathbf{E} = -4\pi\mathbf{P}, \quad \epsilon = 0 \qquad \text{(longitudinal mode).} \tag{27.65}$$

On the other hand, in a transverse optical mode the (nonzero) polarization density **P** is perpendicular to **k**, which is consistent with (27.64) only if **E** vanishes. This, however, is consistent with (27.59) only if

$$\mathbf{E} = 0, \quad \epsilon = \infty \qquad \text{(transverse mode).} \tag{27.66}$$

According to (27.57), $\epsilon = \infty$ when $\omega^2 = \omega_T{}^2$, and therefore the result (27.66) identifies ω_T as the frequency of the long-wavelength ($\mathbf{k} \to 0$) transverse optical mode. The frequency ω_L of the longitudinal optical mode is determined by the condition $\epsilon = 0$ (Eq. (27.65)), and (27.57) therefore gives

$$\boxed{\omega_L{}^2 = \frac{\epsilon_0}{\epsilon_\infty}\,\omega_T{}^2.} \tag{27.67}$$

This equation, relating the longitudinal and transverse optical-mode frequencies to the static dielectric constant and index of refraction, is known as the *Lyddane-Sachs-Teller relation.* Note that it follows entirely from the interpretation that (27.65) and (27.66) lend to the zeros and poles of $\epsilon(\omega)$, together with the functional form of (27.57)—i.e., the fact that in the frequency range of interest ϵ as a function of ω^2 is a constant plus a simple pole. As a result, the relation has a validity going well beyond the crude approximation (27.52) of additive polarizabilities, and also applies to the far more sophisticated shell model theories of diatomic ionic crystals.

Since the crystal is more polarizable at low frequencies[20] than at high, ω_L exceeds ω_T. It may seem surprising that ω_L should differ at all from ω_T in the limit of long wavelengths, since in this limit the ionic displacements in any region of finite extent are indistinguishable. However, because of the long range of electrostatic forces, their influence can always persist over distances comparable to the wavelength, no matter how long that wavelength may be; thus longitudinal and transverse optical modes will always experience different electrostatic restoring forces.[21] Indeed, if we use the Lorentz relation (27.27) we find from (27.65) that the electrostatic restoring force in a long-wavelength longitudinal optical mode is given by the local field

$$(\mathbf{E}^{\text{loc}})_L = \mathbf{E} + \frac{4\pi\mathbf{P}}{3} = -\frac{8\pi\mathbf{P}}{3} \qquad \text{(longitudinal),} \tag{27.68}$$

while (from (27.66)) it is given in a long-wavelength transverse optical mode by

$$(\mathbf{E}^{\text{loc}})_T = \frac{4\pi\mathbf{P}}{3} \qquad \text{(transverse).} \tag{27.69}$$

[20] At frequencies well above the natural vibrational frequencies of the ions, they fail to respond to an oscillatory force, and one has only atomic polarizability. At low frequencies both mechanisms can contribute.

[21] This argument, based on instantaneous action at a distance, must be reexamined when the electrostatic approximation (27.61) is dropped (see footnotes 19 and 25).

Thus in a longitudinal mode the local field acts to reduce the polarization (i.e., it adds to the short-range restoring force proportional to $k = M\omega^2$) while in a transverse mode it acts to support the polarization (i.e., it reduces the short-range restoring force). This is consistent with (27.58), which predicts that ω_T is less than $\bar{\omega}$ (since $\epsilon_0 - \epsilon_\infty$ is positive). It is also consistent with (27.67), which, with the aid of (27.58), can be written:

$$\omega_L{}^2 = \bar{\omega}^2\left(1 + 2\frac{\epsilon_0 - \epsilon_\infty}{\epsilon_0 + 2}\frac{1}{\epsilon_\infty}\right), \tag{27.70}$$

which indicates that ω_L exceeds $\bar{\omega}$.

The Lyddane-Sachs-Teller relation (27.67) has been confirmed by comparing measurements of ω_L and ω_T from neutron scattering, with measured values of the dielectric constant and index of refraction. In two alkali halides (NaI and KBr), ω_L/ω_T and $(\epsilon_0/\epsilon_\infty)^{1/2}$ were found to agree to within the experimental uncertainty of the measurements (a few percent).[22]

However, because it is merely a consequence of the analytic form of $\epsilon(\omega)$, the validity of the Lyddane-Sachs-Teller relation does not provide a very stringent test of a theory. A more specific prediction can be constructed from Eqs. (27.54), (27.55), and (27.58), which combine together to give

$$\frac{9}{4\pi}\frac{(\epsilon_0 - \epsilon_\infty)}{(\epsilon_\infty + 2)^2}\,\omega_T{}^2 = \frac{e^2}{Mv}. \tag{27.71}$$

Since e^2/Mv is determined entirely by the ionic charge, the ionic reduced mass, and the lattice constant, the right side of (27.71) is known. However, measured values of ϵ_0, ϵ_∞, and ω_T in the alkali halides lead to a value for the left side of (27.71) that can be expressed in the form $(e^*)^2/Mv$, where e^* (known as the *Szigeti charge*) ranges between about $0.7e$ and $0.9e$. This should *not* be taken as evidence that the ions are not fully charged, but as a telling sign of the failure of the crude assumption (27.52) that atomic and displacement polarizabilities simply add to give the total polarizability.

To remedy this defect, one must turn to a shell model theory in which atomic and displacement polarizations are calculated together, by allowing the electronic shell to move relative to the ion core (as done above in calculating the atomic polarizability) at the same time as the ion cores are themselves displaced.[23] The general structural form (27.57) of $\epsilon(\omega)$ is preserved in such a theory, but the specific forms for the constants ϵ_0, ϵ_∞, and ω_T can be quite different.

Application to the Optical Properties of Ionic Crystals

The above discussion of the transverse optical mode is not completely accurate, based as it is on the electrostatic approximation (27.61) to the Maxwell equation:[24]

$$\nabla \times \mathbf{E} = -\frac{1}{c}\frac{\partial \mathbf{B}}{\partial t}. \tag{27.72}$$

---

[22] A. D. B. Woods et al., *Phys. Rev.* **131**, 1025 (1963).

[23] An early and particularly simple model is given by S. Roberts, *Phys. Rev.* **77**, 258 (1950).

[24] Our discussion of the longitudinal optical mode is founded entirely on the Maxwell equation $\nabla \cdot \mathbf{D} = 0$ and remains valid in a fully electrodynamic analysis.

When (27.61) is replaced by the more general (27.72), the conclusion (27.66) that the transverse optical mode frequency is determined by the condition $\epsilon(\omega) = \infty$ must be replaced by the more general result (Eq. (1.34)) that transverse fields with angular frequency ω and wave vector $\mathbf{k}$ can propagate only if

$$\epsilon(\omega) = \frac{k^2c^2}{\omega^2}. \tag{27.73}$$

Thus for optical modes with wave vectors satisfying $kc \gg \omega$, the approximation $\epsilon = \infty$ is reasonable. The frequencies of optical phonons are of order $\omega_D = k_D s$, where s is the speed of sound in the crystal, so this requires that

$$\frac{k}{k_D} \gg \frac{s}{c}. \tag{27.74}$$

Since k_D is comparable to the dimensions of the Brillouin zone, while s/c is of order 10^{-4} to 10^{-5}, the electrostatic approximation is fully justified except for optical modes whose wave vectors are only a small fraction of a percent of the dimensions of the zone from $\mathbf{k} = 0$.

We can describe the structure of the transverse modes all the way down to $\mathbf{k} = 0$, by plotting ϵ vs. ω (Eq. (27.57)) (Figure 27.5). Note that ϵ is negative between ω_T and ω_L, so Eq. (27.73) requires kc to be imaginary. Thus no radiation can propagate in the crystal between the transverse and longitudinal optical frequencies. Outside this forbidden range ω is plotted vs. k in Figure 27.6. The dispersion relation has two branches, lying entirely below ω_T and entirely above ω_L. The lower branch has the form $\omega \equiv \omega_T$ except when k is so small as to be comparable to ω_T/c. It describes the electric field accompanying a transverse optical mode in the constant-frequency region. However, when k is of order ω_T/c the frequency falls below ω_T, vanishing as $kc/\sqrt{\epsilon_0}$, a relation characteristic of ordinary electromagnetic radiation in a medium with dielectric constant ϵ_0.

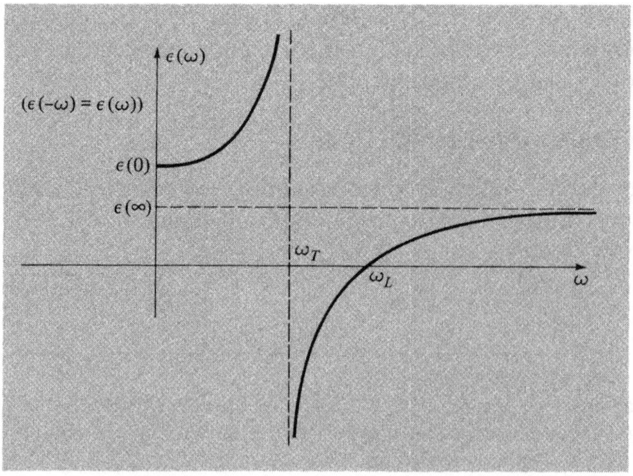

Figure 27.5
Frequency-dependent dielectric constant for a diatomic ionic crystal.

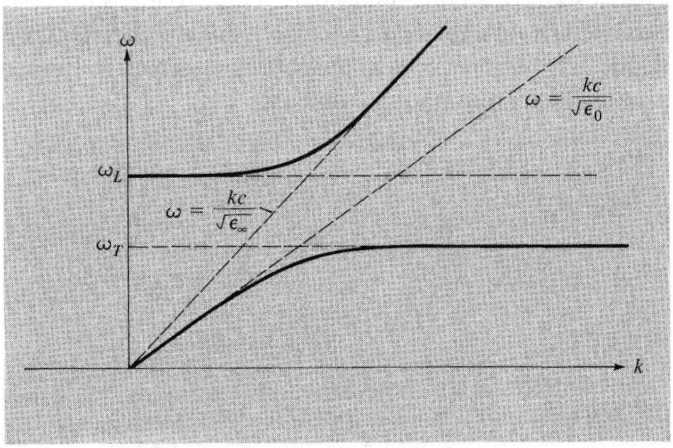

Figure 27.6
Solutions to the dispersion relation $\omega = kc/\sqrt{\varepsilon(\omega)}$ for *transverse* electromagnetic modes propagating in a diatomic ionic crystal. (The relation to Figure 27.5 is most readily seen by rotating the figure through 90° and considering it to be a plot of $k = \omega\sqrt{\varepsilon(\omega)}/c$, vs. ω.) In the linear regions one mode is clearly photonlike and one clearly optical phononlike. In the curved regions both modes have a mixed nature, and are sometimes referred to as "polaritons."

The upper branch, on the other hand, assumes the linear form $\omega = kc/\sqrt{\epsilon_\infty}$, characteristic of electromagnetic radiation in a medium with dielectric constant ϵ_∞, when k is large compared with ω_T/c, but as k approaches zero, the frequency does not vanish linearly, but levels off to ω_L.[25]

Finally, note that if the dielectric constant is a real number, then the reflectivity of the crystal is given by (see Eq. (K.6) in Appendix K)

$$r = \left(\frac{\sqrt{\epsilon} - 1}{\sqrt{\epsilon} + 1}\right)^2. \tag{27.75}$$

As $\epsilon \to \infty$, the reflectivity approaches unity. Thus all incident radiation should be perfectly reflected at the frequency of the transverse optical mode. This effect can be amplified by repeated reflections of a ray from crystal faces. Since n reflections will diminish the intensity by r^n, after very many reflections only the component of radiation with frequencies very close to ω_T will survive. This surviving radiation is known as the *reststrahl* (residual ray). Such repeated reflections provide a very precise way of measuring ω_T, as well as a method for producing very monochromatic radiation in the infrared.

---

[25] Thus as $k \to 0$ a transverse mode *does* occur at the same frequency as the longitudinal mode (see page 548). The reason this behavior emerges in an electrodynamic, but not an electrostatic, analysis is basically the finite velocity of signal propagation in an electrodynamic theory. Electromagnetic signals can only propagate with the speed of light, and therefore no matter how long their spatial range, they can be effective in distinguishing longitudinal from transverse modes only if they can travel a distance comparable to a wavelength in a time small compared with a period (i.e., $kc \gg \omega$). The argument on page 548, which explains why ω_L and ω_T differ, implicitly assumes that Coulomb forces act instantaneously at a distance, and becomes invalid when this assumption fails.

To the extent that the lattice vibrations are anharmonic (and therefore damped) ϵ will also have an imaginary part. This broadens the reststrahl resonance. The typical behavior of observed frequency-dependent dielectric constants in ionic crystals, as deduced from their optical properties, is shown in Figure 27.7. Alkali halide dielectric properties are summarized in Table 27.2.

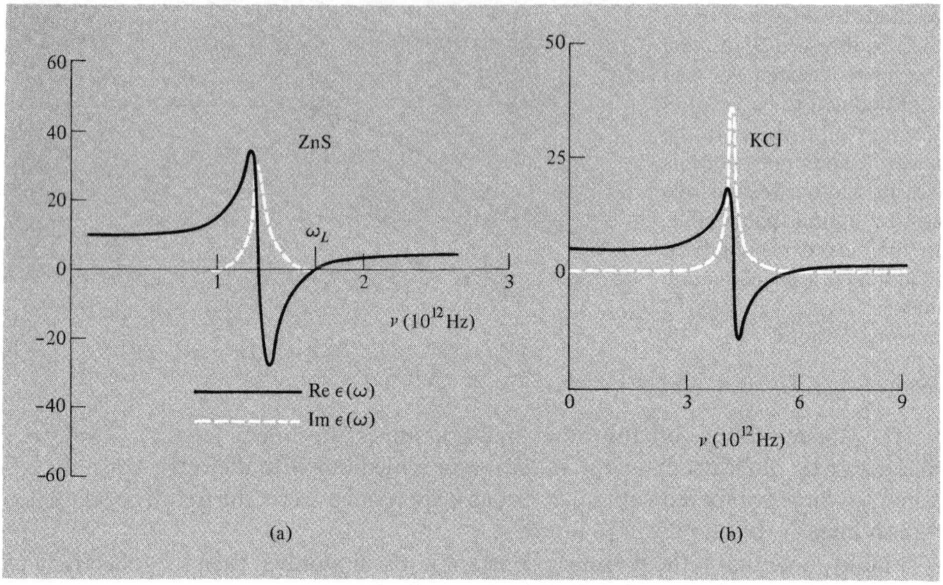

Figure 27.7
(a) Real (solid line) and imaginary (dashed line) parts of the dielectric constant of zinc sulfide. (After F. Abeles and J. P. Mathieu, *Annales de Physique* **3**, 5 (1958); quoted by E. Burstein, *Phonons and Phonon Interactions*, T. A. Bak, ed., W. A. Benjamin, Menlo Park, California, 1964.) (b) Real (solid line) and imaginary (dashed line) parts of the dielectric constant of potassium chloride. (After G. R. Wilkinson and C. Smart; quoted by D. H. Martin, *Advances Phys.* **14**, 39 (1965).)

COVALENT INSULATORS

The above analysis of ionic and molecular crystals has relied on the possibility of resolving the charge distribution of the crystal into contributions from identifiable ions (atoms, molecules) as in (27.9). In covalent crystals, however, appreciable electronic charge density resides between ions (forming the so-called covalent bonds). This part of the total charge distribution is uniquely a property of the condensed state of matter, bearing no resemblance to the charge distribution of single isolated ions (atoms, molecules). Furthermore, since it comes from the most loosely bound atomic electrons, it makes a very important contribution to the polarizability of the crystal. Therefore, in calculating dielectric properties of covalent crystals one must deal with the polarizability of the crystal as a whole, either invoking band theory from the start or developing a phenomenology of "bond polarizabilities."

We shall not pursue this subject here, except to point out that covalent crystals can have quite large dielectric constants, reflecting the relatively delocalized structure of their electronic charge distributions. Static dielectric constants for selected covalent crystals are listed in Table 27.3. As we shall see (Chapter 28), the fact that the dielectric

Table 27.2
STATIC DIELECTRIC CONSTANT, OPTICAL DIELECTRIC CONSTANT, AND TRANSVERSE OPTICAL PHONON FREQUENCY FOR ALKALI HALIDE CRYSTALS

COMPOUND	ϵ_0	ϵ_∞	$\hbar\omega_T/k_B$[a]
LiF	9.01	1.96	442
NaF	5.05	1.74	354
KF	5.46	1.85	274
RbF	6.48	1.96	224
CsF	—	2.16	125
LiCl	11.95	2.78	276
NaCl	5.90	2.34	245
KCl	4.84	2.19	215
RbCl	4.92	2.19	183
CsCl	7.20	2.62	151
LiBr	13.25	3.17	229
NaBr	6.28	2.59	195
KBr	4.90	2.34	166
RbBr	4.86	2.34	139
CsBr	6.67	2.42	114
LiI	16.85	3.80	—
NaI	7.28	2.93	167
KI	5.10	2.62	156
RbI	4.91	2.59	117.5
CsI	6.59	2.62	94.6

[a] From the reststrahl peak; in degrees Kelvin.
Source: R. S. Knox and K. J. Teegarden, *Physics of Color Centers*, W. B. Fowler, ed., Academic Press, New York, 1968, page 625.

Table 27.3
STATIC DIELECTRIC CONSTANTS FOR SELECTED COVALENT AND COVALENT-IONIC CRYSTALS OF THE DIAMOND, ZINCBLENDE, AND WURTZITE STRUCTURES[a]

CRYSTAL	STRUCTURE	ϵ_0	CRYSTAL	STRUCTURE	ϵ_0
C	*d*	5.7	ZnO	*w*	4.6
Si	*d*	12.0	ZnS	*w*	5.1
Ge	*d*	16.0	ZnSe	*z*	5.8
Sn	*d*	23.8	ZnTe	*z*	8.3
SiC	*z*	6.7	CdS	*w*	5.2
GaP	*z*	8.4	CdSe	*w*	7.0
GaAs	*z*	10.9	CdTe	*z*	7.1
GaSb	*z*	14.4	BeO	*w*	3.0
InP	*z*	9.6	MgO	*z*	3.0
InAs	*z*	12.2			
InSb	*z*	15.7			

[a] Quoted by J. C. Phillips, *Phys. Rev. Lett.* **20**, 550 (1968).

constants can be quite substantial is a point of considerable importance in the theory of impurity levels in semiconductors.

PYROELECTRICITY

In deriving the macroscopic equation

$$\nabla \cdot \mathbf{E} = -4\pi \nabla \cdot \mathbf{P} \tag{27.76}$$

for ionic crystals, we assumed (see footnote 9) that the equilibrium dipole moment of the primitive cell,

$$\mathbf{p}_0 = \sum_{\mathbf{d}} \mathbf{d} e(\mathbf{d}), \tag{27.77}$$

vanished, and therefore ignored a term

$$\Delta \mathbf{P} = \frac{\mathbf{p}_0}{v} \tag{27.78}$$

in the polarization density $\mathbf{P}$. As Figure 27.8 demonstrates, the value of the dipole moment $\mathbf{p}_0$ is not independent of the choice of primitive cell. However, since only the divergence of $\mathbf{P}$ has physical significance, an additive constant vector $\Delta\mathbf{P}$ does not affect the physics implied by the macroscopic Maxwell equations.

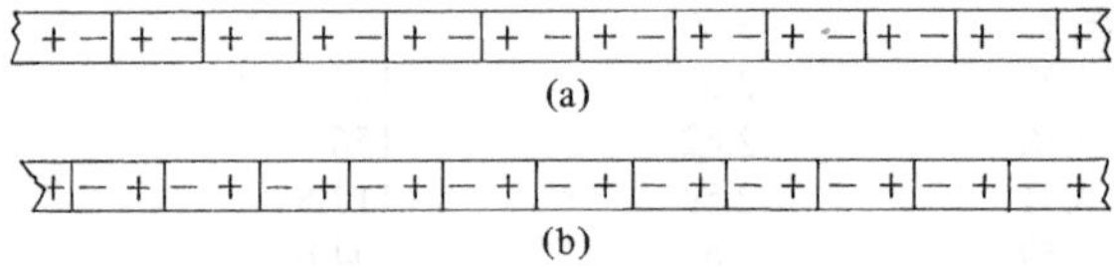

Figure 27.8
The dipole moment of the primitive cell depends on the choice of primitive cell. This is illustrated for a one-dimensional ionic crystal.

There would be nothing more to say if all crystals were infinite in extent. However, real crystals have surfaces, at which the macroscopic polarization density $\mathbf{P}$ drops discontinuously to zero, thereby contributing a singular term on the right side of (27.76). This term is conventionally interpreted as a bound surface charge per unit area, whose magnitude is the normal component of $\mathbf{P}$ at the surface, P_n. Thus an additive constant in $\mathbf{P}$ is far from inconsequential in a finite crystal.

In a finite crystal, however, we must reexamine our assumption that each primitive cell has zero total charge:

$$\sum_{\mathbf{d}} e(\mathbf{d}) = 0. \tag{27.79}$$

In an infinite crystal of identical cells this is merely the statement that the crystal as a whole is neutral, but in a crystal with surfaces, only the interior cells are identically occupied, and charge neutrality is perfectly consistent with partially filled, and therefore charged, surface cells (Figure 27.9). Should one's choice of cell lead to surface cells containing net charge, an additional term would have to be added to (27.76) to represent this bound surface charge, ρ_s. When the choice of cell is changed, both P_n and ρ_s will change, in such a way that the total net macroscopic surface charge density, $P_n + \rho_s$, is unchanged.

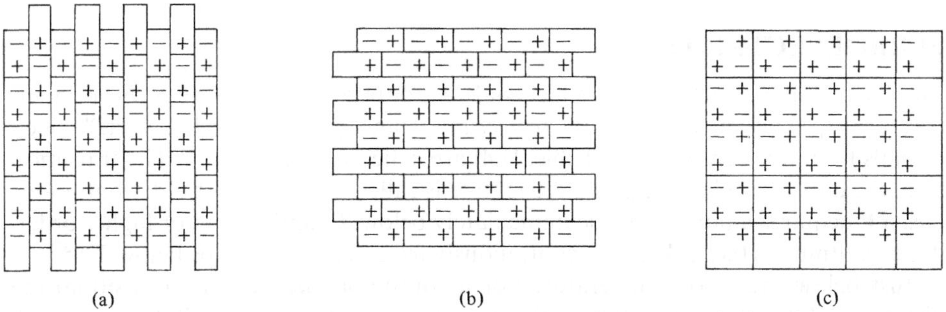
(a) (b) (c)

Figure 27.9
The "natural" choice of primitive cell is one that leads to uncharged cells at the surface. The cells chosen in (a) and (b) violate this criterion, and their contribution to the polarization density is cancelled by the contribution from the charged surface cells. The cell in (c) (which is non-primitive) leads to uncharged surface cells and has no dipole moment.

Thus the "natural" choice of cell for which (27.76) is valid without an additional term representing the unbalanced charge in surface cells, is a cell whose neutrality is maintained even at the surfaces of actual physical specimens.[26]

Crystals whose natural primitive cells have a nonvanishing dipole moment $\mathbf{p}_0$ are called *pyroelectric*.[27] In equilibrium a perfect specimen of a pyroelectric crystal has a total dipole moment of $\mathbf{p}_0$ times the number of cells in the crystal,[28] and thus a polarization density $\mathbf{P} = \mathbf{p}_0/v$ throughout the crystal, even in the absence of an external field. This immediately implies some severe restrictions on the point-group symmetries of a pyroelectric crystal, for a symmetry operation must preserve all crystalline properties and, in particular, the direction of $\mathbf{P}$. Thus the only possible rotation axis is one parallel to $\mathbf{P}$, and furthermore there cannot be mirror planes perpendicular to that axis. This excludes all point groups except (Table 7.3) C_n and C_{nv} ($n = 2, 3, 4, 6$) and C_1 and C_{1h}. A glance at Table 7.3 reveals that these are the only point groups compatible with the location of a directed object (an arrow, for example) at each site.[29]

---

[26] This often requires a cell that is not primitive (see Figure 27.9), but it is easily verified that the earlier analysis in this chapter is in no way affected by using a larger microscopic cell.

[27] The name (pyro = fire) reflects the fact that under ordinary conditions the moment of a pyroelectric crystal will be masked by neutralizing layers of ions from the atmosphere that collect on the faces of the crystal. If, however, the crystal is heated, then the masking will no longer be complete, since the polarization will change due to thermal expansion of the crystal, neutralizing ions will be evaporated, etc. Thus the effect was first thought to be the production of an electric moment by heat. (Sometimes the term "polar crystal" is used instead of "pyroelectric crystal." However, "polar crystal" is also widely used as a synonym for "ionic crystal" (whether pyroelectric or not), and the term is therefore best avoided.) The net polarization can also be masked by a domain structure, as in ferromagnets (see Chapter 33).

[28] The dipole moment of the surface cells need not be $\mathbf{p}_0$, but in the limit of a large crystal this will have a negligible effect on the total dipole moment, since the overwhelming majority of cells will be in the interior.

[29] Some crystals, though nonpyroelectric in the absence of external stresses can develop a spontaneous dipole moment when mechanically strained; i.e., by suitable squeezing, their crystal structures can be distorted to ones that can sustain a dipole moment. Such crystals are called *piezoelectric*. The point group of a piezoelectric crystal (when unstrained) cannot contain the inversion.

FERROELECTRICITY

The most stable structure of some crystals is nonpyroelectric above a certain temperature T_c (known as the *Curie temperature*) and pyroelectric below it.[30] Such crystals (examples are given in Table 27.4) are called ferroelectrics.[31] The transition from the unpolarized to the pyroelectric state is called first order if it is discontinuous (i.e., if **P** acquires a nonzero value immediately below T_c) and second or higher order, if it is continuous (i.e., if **P** grows continuously from zero as T drops below T_c).[32]

Just below the Curie temperature (for a continuous ferroelectric transition) the distortion of the primitive cell from the unpolarized configuration will be very small, and it is therefore possible, by applying an electric field opposite to this small polarization, to diminish and even reverse it. As T drops farther below T_c, the distortion of the cell increases, and very much stronger fields are required to reverse the direction of **P**. This is sometimes taken as the essential attribute of ferroelectrics, which are then defined as pyroelectric crystals whose polarization can be reversed by applying a strong electric field. This is done to include those crystals one feels would satisfy the first definition (existence of a Curie temperature), except that they melt before the conjectured Curie temperature can be reached. Well below the Curie temperature, however, the reversal of polarization may require so drastic a restructuring of the crystal as to be impossible even in the strongest attainable fields.

Immediately below the Curie temperature of a continuous ferroelectric transition, the crystal spontaneously and continuously distorts to a polarized state. One would therefore expect the dielectric constant to be anomalously large in the neighborhood of T_c, reflecting the fact that it requires very little applied field to alter substantially the displacement polarization of the crystal. Dielectric constants as large as 10^5 have been observed near ferroelectric transition points. In an ideal experiment the dielectric constant should actually become infinite precisely at T_c. For a continuous transition this simply expresses the fact that as T_c is approached from above, the net restoring force opposing a lattice distortion from the unpolarized to the polarized phase vanishes.

If the restoring force opposing a particular lattice distortion vanishes, there should be a zero-frequency normal mode whose polarization vectors describe precisely this distortion. Since the distortion leads to a net dipole moment and therefore involves a relative displacement between ions of opposite charge, the mode will be an optical mode. In the vicinity of the transition, relative displacements will be large, anharmonic terms will be substantial, and this "soft" mode should be rather strongly damped.

These two observations (infinite static dielectric constant and a zero-frequency optical mode) are not independent. One implies the other by the Lyddane-Sachs-Teller relation (27.67), which requires the transverse optical-mode frequency to vanish whenever the static dielectric constant is infinite.

---

[30] Transitions back and forth are also known: e.g., there can be a range of temperatures for the pyroelectric phase, above and below which the crystal is unpolarized.

[31] The name stresses the analogy with ferromagnetic materials, which have a net *magnetic* moment. It is not meant to suggest that iron has any special relation to the phenomenon.

[32] Sometimes the term "ferroelectric" is reserved for crystals in which the transition is second order.

Perhaps the simplest type of ferroelectric crystal (and the one most widely studied) is the perovskite structure, shown in Figure 27.10. Other ferroelectrics tend to be substantially more complex. Some characteristic examples are given in Table 27.4.

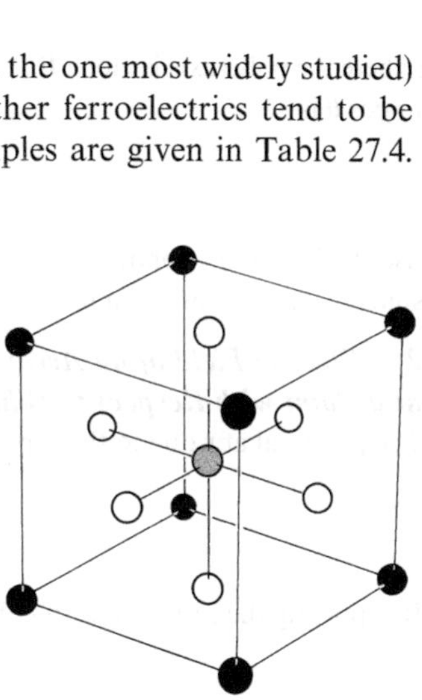

Figure 27.10
The perovskite structure, characteristic of the barium titanate ($BaTiO_3$) class of ferroelectrics in the unpolarized phase. The crystal is cubic, with Ba^{++} ions at the cube corners, O^{--} ions at the centers of the cube faces, and Ti^{4+} ions at the cube centers. The first transition is to a tetragonal structure, the positive ions being displaced relative to the negative ones, along a [100] direction. The perovskite structure is an example of a cubic crystal in which every ion is *not* at a point of full cubic symmetry. (The Ba^{++} and Ti^{4+} are, but the O^{--} ions are not.) Therefore the local field acting on the oxygen ions is more complicated than that given by the simple Lorentz formula. This is important in understanding the mechanism for the ferroelectricity.

Table 27.4
SELECTED FERROELECTRIC CRYSTALS

NAME	FORMULA	T_c (K)	P (μC/cm^2)	at T (K)
Potassium dihydrogen phosphate	KH_2PO_4	123	4.75	96
Potassium dideuterium phosphate	KD_2PO_4	213	4.83	180
Rubidium dihydrogen phosphate	RbH_2PO_4	147	5.6	90
Rubidium dideuterium phosphate	RbD_2PO_4	218	—	—
Barium titanate	$BaTiO_3$	393	26.0	300
Lead titanate	$PbTiO_3$	763	>50	300
Cadmium titanate	$CdTiO_3$	55	—	—
Potassium niobate	$KNbO_3$	708	30.0	523
Rochelle salt	$NaKC_4H_4O_6 \cdot 4D_2O$	$\left\{\begin{matrix}297\\255\end{matrix}\right\}^a$	0.25	278
Deuterated Rochelle salt	$NaKC_4H_2D_2O_6 \cdot 4D_2O$	$\left\{\begin{matrix}308\\251\end{matrix}\right\}^a$	0.35	279

[a] Has upper and lower T_c.
Source: F. Jona and G. Shirane, *Ferroelectric Crystals*, Pergamon, New York, 1962, p. 389.

PROBLEMS

1. *Electric Field of a Neutral Uniformly Polarized Sphere of Radius* a
Far from the sphere, the potential ϕ will be that of a point dipole of moment $p = 4\pi Pa^3/3$:

$$\phi = \frac{P \cos \theta}{r^2}, \qquad \textbf{(27.80)}$$

(where the polar axis is along **P**). Using the fact that the general solution to $\nabla^2\phi = 0$ proportional to $\cos\theta$ is

$$\frac{A\cos\theta}{r^2} + Br\cos\theta, \tag{27.81}$$

use the boundary conditions at the surface of the sphere to show that the potential inside the sphere leads to a uniform field $\mathbf{E} = -4\pi\mathbf{P}/3$.

2. *Electric Field of an Array of Identical Dipoles with Identical Orientations, at a Point with Respect to Which the Array Has Cubic Symmetry*

The potential at **r** due to the dipole at **r**′ is

$$\phi = -\mathbf{p}\cdot\nabla\frac{1}{|\mathbf{r}-\mathbf{r}'|}. \tag{27.82}$$

By applying the restrictions of cubic symmetry to the tensor

$$\sum_{\mathbf{r}'}\nabla_\mu\nabla_\nu\frac{1}{|\mathbf{r}-\mathbf{r}'|}, \tag{27.83}$$

and noting that $\nabla^2(1/r) = 0$, $\mathbf{r} \neq 0$, show that $\mathbf{E}(\mathbf{r})$ must vanish, when the positions $\mathbf{r}'$ of the dipoles have cubic symmetry about **r**.

3. *Polarizability of a Single Hydrogen Atom*

Suppose an electric field **E** is applied (along the x-axis) to a hydrogen atom in its ground state with wave function

$$\psi_0 \propto e^{-r/a_0}. \tag{27.84}$$

(a) Assume a trial function for the atom in the field of the form

$$\psi \propto \psi_0(1+\gamma x) = \psi_0 + \delta\psi, \tag{27.85}$$

and determine γ by minimizing the total energy.

(b) Calculate the polarization

$$p = \int d\mathbf{r}\,(-e)\,x\,(\psi_0\,\delta\psi^* + \psi_0^*\,\delta\psi), \tag{27.86}$$

using the best trial function, and show that this leads to a polarizability $\alpha = 4a_0{}^3$. (The exact answer is $4.5a_0{}^3$.)

4. *Orientational Polarization*

The following situation sometimes arises in pure solids and liquids whose molecules have permanent dipole moments (such as water or ammonia) and also in solids such as ionic crystals with some ions replaced by others with permanent moments (such as OH^- in KCl).

(a) An electric field tends to align such molecules; thermal disorder favors misalignment. Using equilibrium statistical mechanics, write down the probability that the dipole makes an angle in the range from θ to $\theta + d\theta$ with the applied field. If there are N such dipoles of moment p, show that their total dipole moment in thermal equilibrium is

$$Np\langle\cos\theta\rangle = NpL\left(\frac{pE}{k_BT}\right), \tag{27.87}$$

where $L(x)$, the "Langevin function," is given by

$$L(x) = \coth x - \left(\frac{1}{x}\right). \tag{27.88}$$

(b) Typical dipole moments are of order 1 Debye unit (10^{-18} in esu). Show that for an electric field of order 10^4 volts/cm the polarizability at room temperature can be written as

$$\alpha = \frac{p^2}{3k_BT}. \tag{27.89}$$

5. *Generalized Lyddane-Sachs-Teller Relation*

Suppose that the dielectric constant $\epsilon(\omega)$ does not have a single pole as a function of ω^2 (as in (27.57)) but has the more general structure:

$$\epsilon(\omega) = A + \sum_{i=1}^{n} \frac{B_i}{\omega^2 - \omega_i^2}. \tag{27.90}$$

Show directly from (27.90) that the Lyddane-Sachs-Teller relation (27.67) is generalized to

$$\frac{\epsilon_0}{\epsilon_\infty} = \prod \left(\frac{\omega_i^0}{\omega_i}\right)^2, \tag{27.91}$$

where the ω_i^0 are the frequencies at which ϵ vanishes. (*Hint:* Write the condition $\epsilon = 0$ as an nth-degree polynomial in ω^2, and note that the product of the roots is simply related to the value of the polynomial at $\omega = 0$.) What is the significance of the frequencies ω_i and ω_i^0?

28
Homogeneous Semiconductors

General Properties of Semiconductors

Examples of Semiconductor Band Structure

Cyclotron Resonance

Carrier Statistics in Thermal Equilibrium

Intrinsic and Extrinsic Semiconductors

Statistics of Impurity Levels in Thermal Equilibrium

Thermal Equilibrium Carrier Densities of Impure Semiconductors

Impurity Band Conduction

Transport in Nondegenerate Semiconductors

In Chapter 12 we observed that electrons in a completely filled band can carry no current. Within the independent electron model this result is the basis for the distinction between insulators and metals: In the ground state of an insulator all bands are either completely filled or completely empty; in the ground state of a metal at least one band is partially filled.

We can characterize insulators by the *energy gap*, E_g, between the top of the highest filled band(s) and the bottom of the lowest empty band(s) (see Figure 28.1). A solid with an energy gap will be nonconducting at $T = 0$ (unless the DC electric field is so strong and the energy gap so minute that electric breakdown can occur (Eq. (12.8)) or unless the AC field is of such high frequency that $\hbar\omega$ exceeds the energy gap).

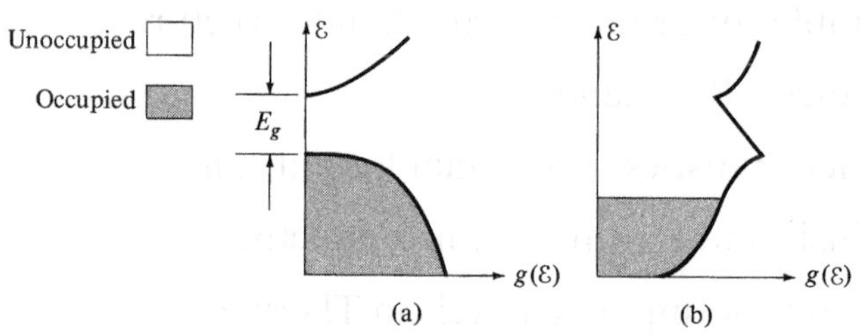

Figure 28.1
(a) In an insulator there is a region of forbidden energies separating the highest occupied and lowest unoccupied levels. (b) In a metal the boundary occurs in a region of allowed levels. This is indicated schematically by plotting the density of levels (horizontally) vs. energy (vertically).

However, when the temperature is not zero there is a nonvanishing probability that some electrons will be thermally excited across the energy gap into the lowest unoccupied bands, which are called, in this context, the *conduction bands*, leaving behind unoccupied levels in the highest occupied bands, called *valence bands*. The thermally excited electrons are capable of conducting, and hole-type conduction can occur in the band out of which they have been excited.

Whether such thermal excitation leads to appreciable conductivity depends critically on the size of the energy gap, for the fraction of electrons excited across the gap at temperature T is, as we shall see, roughly of order $e^{-E_g/2k_BT}$. With an energy gap of 4 eV at room temperature ($k_BT \approx 0.025$ eV) this factor is $e^{-80} \approx 10^{-35}$, and essentially no electrons are excited across the gap. If, however, E_g is 0.25 eV, then the factor at room temperature is $e^{-5} \approx 10^{-2}$, and observable conduction will occur.

Solids that are insulators at $T = 0$, but whose energy gaps are of such a size that thermal excitation can lead to observable conductivity at temperatures below the melting point, are known as *semiconductors*. Evidently the distinction between a semiconductor and an insulator is not a sharp one, but roughly speaking the energy gap in most important semiconductors is less than 2 eV and frequently as low as a few tenths of an electron volt. Typical room temperature resistivities of semiconductors are between 10^{-3} and 10^{9} ohm-cm (in contrast to metals, where $\rho \approx 10^{-6}$ ohm-cm, and good insulators, where ρ can be as large as 10^{22} ohm-cm).

Since the number of electrons excited thermally into the conduction band (and therefore the number of holes they leave behind in the valence band) varies exponentially with $1/T$, the electrical conductivity of a semiconductor should be a very rapidly *increasing* function of temperature. This is in striking contrast to the case of metals. The conductivity of a metal (Eq. (1.6)),

$$\sigma = \frac{ne^2\tau}{m}, \tag{28.1}$$

declines with increasing temperature, for the density of carriers n is independent of temperature, and all temperature dependence comes from the relaxation time τ, which generally decreases with increasing temperature because of the increase in electron-phonon scattering. The relaxation time in a semiconductor will also decrease with increasing temperature, but this effect (typically described by a power law) is quite overwhelmed by the very much more rapid increase in the density of carriers with increasing temperature.[1]

Thus the most striking feature of semiconductors is that, unlike metals, their electrical resistance declines with rising temperature; i.e., they have a "negative coefficient of resistance." It was this property that first brought them to the attention of physicists in the early nineteenth century.[2] By the end of the nineteenth century a considerable body of semiconducting lore had been amassed; it was observed that the thermopowers of semiconductors were anomalously large compared with those of metals (by a factor of 100 or so), that semiconductors exhibited the phenomenon of photoconductivity, and that rectifying effects could be obtained at the junction of two unlike semiconductors. Early in the twentieth century, measurements of the Hall effect[3] were made confirming the fact that the temperature dependence of the conductivity was dominated by that of the number of carriers, and indicating that in many substances the sign of the dominant carrier was positive rather than negative.

Phenomena such as these were a source of considerable mystery until the full development of band theory many years later. Within the band theory they find simple explanations. For example, photoconductivity (the increase in conductivity produced by shining light on a material) is a consequence of the fact that if the band

---

[1] Thus the conductivity of a semiconductor is not a good measure of the collision rate, as it is in a metal. It is often advantageous to separate from the conductivity a term whose temperature dependence reflects only that of the collision rate. This is done by defining the *mobility*, μ, of a carrier, as being the ratio of the drift velocity it achieves in a field E, to the field strength: $v_d = \mu E$. If the carriers have density n and charge q, the current density will be $j = nqv_d$, and therefore the conductivity is related to the mobility by $\sigma = nq\mu$. The concept of mobility has little independent use in discussions of metals, since it is related to the conductivity by a temperature-independent constant. However, it plays an important role in descriptions of semiconductors (and any other conductors where the carrier density can vary, such as ionic solutions), enabling one to disentangle two distinct sources of temperature dependence in the conductivity. The usefulness of the mobility will be illustrated in our discussion of inhomogeneous semiconductors in Chapter 29.

[2] M. Faraday, *Experimental Researches on Electricity*, 1839, Facsimile Reprint by Taylor and Francis, London. R. A. Smith, *Semiconductors*, Cambridge University Press, 1964, provides one of the most pleasant introductions to the subject available. Most of the information in our brief historical survey is drawn from it.

[3] One might expect that the number of excited electrons would equal the number of holes left behind, so that the Hall effect would yield little direct information on the number of carriers. However, as we shall see, the number of electrons need not equal the number of holes in an impure semiconductor, and these were the only ones available at the time of the early experiments.

gap is small, then visible light can excite electrons across the gap into the conduction band, resulting in conduction by those electrons and by the holes left behind. The thermopower, to take another example, is roughly a hundred times larger in a semiconductor than in a metal. This is because the density of carriers is so low in a semiconductor that they are properly described by Maxwell-Boltzmann statistics (as we shall see below). Thus the factor of 100 is the same factor by which the early theories of metals (prior to Sommerfeld's introduction of Fermi-Dirac statistics) overestimated the thermopower (page 25).

The band theoretic explanations of these and other characteristic semiconducting properties will be the subject of this chapter and the next.

Compilation of reliable information on semiconductors in the early days was substantially impeded by the fact that data are enormously sensitive to the purity of the sample. An example of this is shown in Figure 28.2, where the resistivity of germanium is plotted vs. T for a variety of impurity concentrations. Note that concentrations as low as parts in 10^8 can lead to observable effects, and that the resistivity can vary at a given temperature by a factor of 10^{12}, as the impurity concentration changes by only a factor of 10^3. Note also that, for a given impurity concentration, the resistivity eventually falls onto a common curve as the temperature increases. This latter resistivity which is evidently the resistivity of an ideal perfectly pure sample, is known as the *intrinsic* resistivity, while the data for the various samples, except at temperatures so high that they agree with the intrinsic curve, are referred to as *extrinsic properties*. Quite generally, a semiconductor is intrinsic if its electronic properties are dominated by electrons thermally excited from the valence to the conduction band, and extrinsic if its electronic properties are dominated by electrons contributed to the conduction band by impurities (or captured from the valence band by impurities) in a manner to be described below. We shall return shortly to the question of why semiconducting properties are so very sensitive to the purity of the specimen.

EXAMPLES OF SEMICONDUCTORS

Semiconducting crystals come primarily from the covalent class of insulators.[4] The simple semiconducting elements are from column IV of the periodic table, silicon and germanium being the two most important elemental semiconductors. Carbon, in the form of diamond, is more properly classified as an insulator, since its energy gap is of order 5.5 eV. Tin, in the allotropic form of grey tin, is semiconducting, with a very small energy gap. (Lead, of course, is metallic.) The other semiconducting elements, red phosphorus, boron, selenium, and tellurium, tend to have highly complex crystal structures, characterized, however, by covalent bonding.

In addition to the semiconducting elements there is a variety of semiconducting compounds. One broad class, the III–V semiconductors, consists of crystals of the zincblende structure (page 81) composed of elements from columns III and V of the

---

[4] Among the various categories of insulating crystals, the covalent crystals have a spatial distribution of electronic charge most similar to metals. (See Chapter 19.)

Specimen	Donor concentration (cm^{-3})
1	5.3×10^{14}
2	9.3×10^{14}
5	1.6×10^{15}
7	2.3×10^{15}
8	3.0×10^{15}
10	5.2×10^{15}
12	8.5×10^{15}
15	1.3×10^{16}
17	2.4×10^{16}
18	3.5×10^{16}
20	4.5×10^{16}
21	5.5×10^{16}
22	6.4×10^{16}
23	7.4×10^{16}
24	8.4×10^{16}
25	1.2×10^{17}
26	1.3×10^{17}
27	2.7×10^{17}
29	9.5×10^{17}

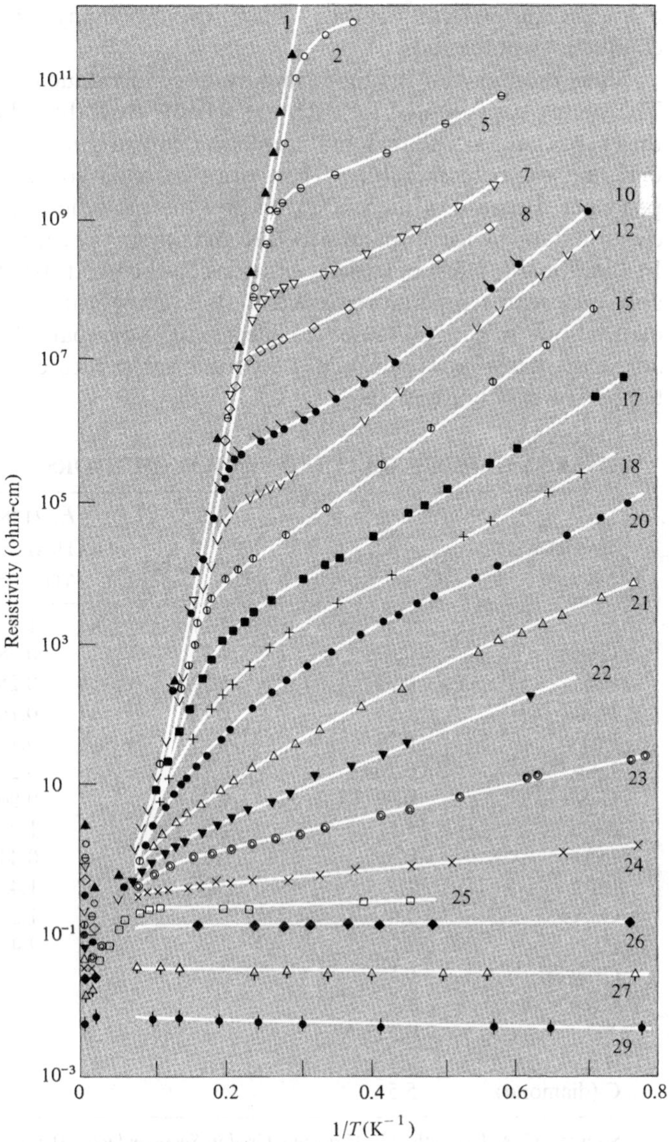

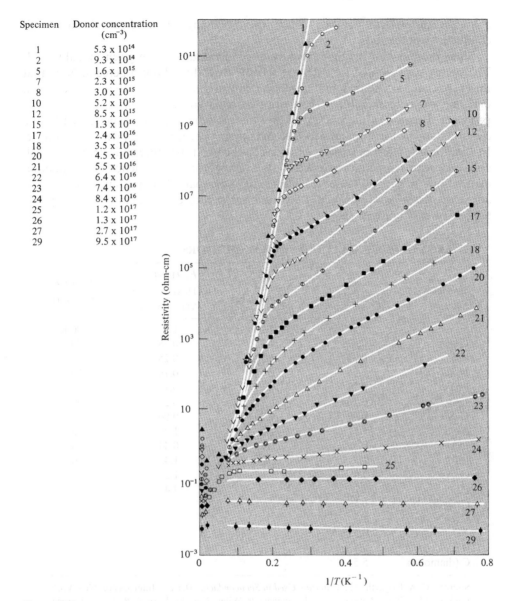

Figure 28.2
The resistivity of antimony-doped germanium as a function of $1/T$ for several impurity concentrations. (From H. J. Fritzsche, *J. Phys. Chem. Solids* **6**, 69 (1958).)

periodic table. As described in Chapter 19, the bonding in such compounds is also predominantly covalent. Semiconducting crystals made up of elements from columns II and VI begin to have a strong ionic as well as a covalent character. These are known as *polar semiconductors*, and can have either the zincblende structure or, as in the case of lead selenide, telluride, or sulfide, the sodium chloride structure more characteristic of ionic bonding. There are also many far more complicated semiconducting compounds.

Some examples of the more important semiconductors are given in Table 28.1. The energy gaps quoted for each are reliable to within about 5 percent. Note that the energy gaps are all temperature-dependent, varying by about 10 percent between 0 K and room temperature. There are two main sources of this temperature dependence. Because of thermal expansion the periodic potential experienced by the electrons (and hence the band structure and the energy gap) can vary with temperature. In addition, the effect of lattice vibrations on the band structure and energy gap[5] will also vary with temperature, reflecting the temperature dependence of the phonon distribution. In general these two effects are of comparable importance, and lead to an energy gap that is linear in T at room temperature and quadratic at very low temperatures (Figure 28.3).

Table 28.1
ENERGY GAPS OF SELECTED SEMICONDUCTORS

MATERIAL	E_g $(T = 300\text{ K})$	E_g $(T = 0\text{ K})$	E_0 (LINEAR EXTRAPOLATION TO $T = 0$)	LINEAR DOWN TO
Si	1.12 eV	1.17	1.2	200 K
Ge	0.67	0.75	0.78	150
PbS	0.37	0.29	0.25	
PbSe	0.26	0.17	0.14	20
PbTe	0.29	0.19	0.17	
InSb	0.16	0.23	0.25	100
GaSb	0.69	0.79	0.80	75
AlSb	1.5	1.6	1.7	80
InAs	0.35	0.43	0.44	80
InP	1.3		1.4	80
GaAs	1.4		1.5	
GaP	2.2		2.4	
Grey Sn	0.1			
Grey Se	1.8			
Te	0.35			
B	1.5			
C (diamond)	5.5			

Sources: C. A. Hogarth, ed., *Materials Used in Semiconductor Devices*, Interscience, New York, 1965; O. Madelung, *Physics of III–V Compounds*, Wiley, New York, 1964; R. A. Smith, *Semiconductors*, Cambridge University Press, 1964.

The energy gap can be measured in several ways. The optical properties of the crystal are one of the most important sources of information. When the frequency of an incident photon becomes large enough for $\hbar\omega$ to exceed the energy gap, then, just as in metals (see pages 293, 294) there will be an abrupt increase in the absorption

[5] Via, for example, the kinds of effects described in Chapter 26.

Figure 28.3
Typical temperature dependence of the energy gap of a semiconductor. Values of E_0, $E_g(0)$, and $E_g(300\ \text{K})$ for several materials are listed in Table 28.1.

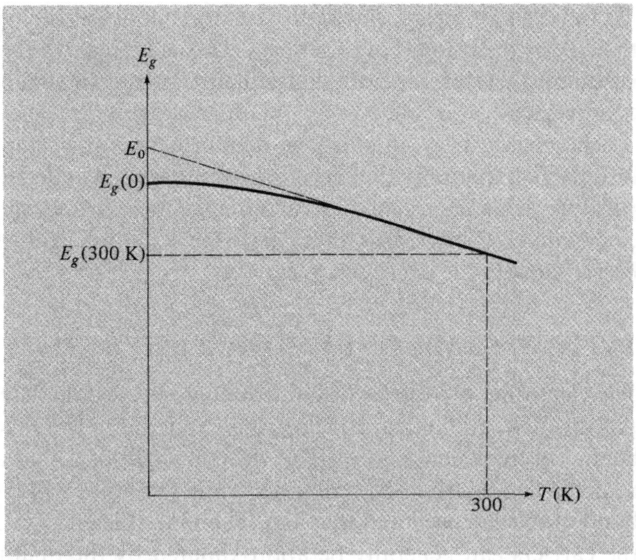

of incident radiation. If the conduction band minimum occurs at the same point in k-space as the valence band maximum, then the energy gap can be directly determined from the optical threshold. If, as is often the case, the minima and maxima occur at different points in k-space, then for crystal momentum to be conserved a phonon must also participate in the process,[6] which is then known as an "indirect transition" (Figure 28.4). Since the phonon will supply not only the missing crystal momentum

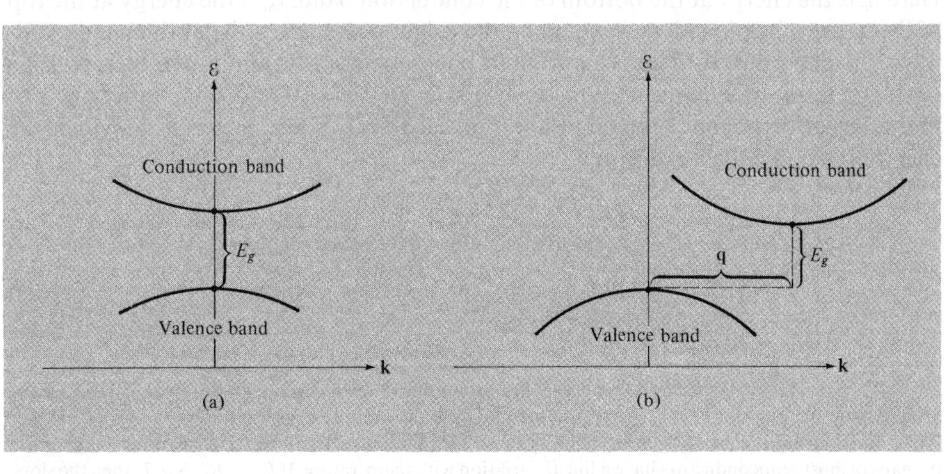

Figure 28.4
Photon absorption via (a) direct and (b) indirect transitions. In (a) the optical threshold is at $\omega = E_g/\hbar$; in (b) it occurs at $E_g/\hbar - \omega(\mathbf{q})$, since the phonon of wave vector $\mathbf{q}$ that must be absorbed to supply the missing crystal momentum also supplies an energy $\hbar\omega(\mathbf{q})$.

---

[6] At optical frequencies the crystal momentum supplied by the photon itself is negligibly small.

$\hbar\mathbf{k}$, but also an energy $\hbar\omega(\mathbf{k})$, the photon energy at the optical threshold will be less than E_g by an amount of order $\hbar\omega_D$. This is typically a few hundredths of an electron volt, and therefore of little consequence except in semiconductors with very small energy gaps.[7]

The energy gap may also be deduced from the temperature dependence of the intrinsic conductivity, which is predominantly a reflection of the very strong temperature dependence of the carrier densities. These vary (as we shall see below) essentially as $e^{-E_g/2k_BT}$, so that if $-\ln(\sigma)$ is plotted against $1/2k_BT$, the slope[8] should be very nearly the energy gap, E_g.

TYPICAL SEMICONDUCTOR BAND STRUCTURES

The electronic properties of semiconductors are completely determined by the comparatively small numbers of electrons excited into the conduction band and holes left behind in the valence band. The electrons will be found almost exclusively in levels near the conduction band minima, while the holes will be confined to the neighborhood of the valence band maxima. Therefore the energy vs. wave vector relations for the carriers can generally be approximated by the quadratic forms they assume in the neighborhood of such extrema:[9]

$$\begin{aligned} \mathcal{E}(\mathbf{k}) &= \mathcal{E}_c + \frac{\hbar^2}{2}\sum_{\mu\nu} k_\mu(\mathbf{M}^{-1})_{\mu\nu}k_\nu \qquad \text{(electrons)}, \\ \mathcal{E}(\mathbf{k}) &= \mathcal{E}_v - \frac{\hbar^2}{2}\sum_{\mu\nu} k_\mu(\mathbf{M}^{-1})_{\mu\nu}k_\nu \qquad \text{(holes)}. \end{aligned} \tag{28.2}$$

Here $\mathcal{E}_c$ is the energy at the bottom of the conduction band, $\mathcal{E}_v$ is the energy at the top of the valence band, and we have taken the origin of k-space to lie at the band maximum or minimum. If there is more than one maximum or minimum, there will be one such term for each point. Since the tensor $\mathbf{M}^{-1}$ is real and symmetric, one can find a set of orthogonal principal axes for each such point, in terms of which the energies have the diagonal forms

$$\begin{aligned} \mathcal{E}(\mathbf{k}) &= \mathcal{E}_c + \hbar^2\left(\frac{k_1^2}{2m_1} + \frac{k_2^2}{2m_2} + \frac{k_3^2}{2m_3}\right) \qquad \text{(electrons)}, \\ \mathcal{E}(\mathbf{k}) &= \mathcal{E}_v - \hbar^2\left(\frac{k_1^2}{2m_1} + \frac{k_2^2}{2m_2} + \frac{k_3^2}{2m_3}\right) \qquad \text{(holes)}. \end{aligned} \tag{28.3}$$

---

[7] To extract a really accurate band gap from the optical absorption data, however, it is necessary to determine the phonon spectrum and use it to analyze the indirect transitions.

[8] In deducing the energy gap in this way, however, one must remember that at room temperature the gaps of most semiconductors have a linear variation with temperature. If $E_g = E_0 - AT$, then the slope of the graph will be not E_g but E_0, the linear extrapolation of the room temperature gap to zero temperature (Figure 28.3). Values of E_0 extracted from this linear extrapolation procedure are also given in Table 28.1.

[9] The inverse of the matrix of coefficients in (28.2) is called $\mathbf{M}$ because it is a special case of the general effective mass tensor introduced on page 228. The electron mass tensor will not, of course, be the same as the hole mass tensor, but to avoid a multiplicity of subscripts we use the single generic symbol $\mathbf{M}$ for both.

Thus the constant energy surfaces about the extrema are ellipsoidal in shape, and are generally specified by giving the principal axes of the ellipsoids, the three "effective masses," and the location in k-space of the ellipsoids. Some important examples are:

Silicon The crystal has the diamond structure, so the first Brillouin zone is the truncated octahedron appropriate to a face-centered cubic Bravais lattice. The conduction band has six symmetry-related minima at points in the ⟨100⟩ directions, about 80 percent of the way to the zone boundary (Figure 28.5). By symmetry each

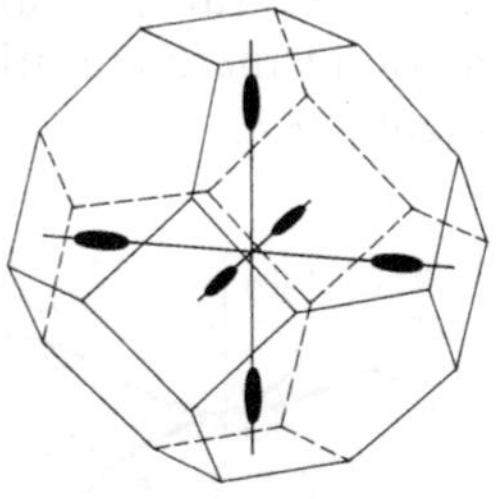

Figure 28.5
Constant-energy surfaces near the conduction band minima in silicon. There are six symmetry-related ellipsoidal pockets. The long axes are directed along ⟨100⟩ directions.

of the six ellipsoids must be an ellipsoid of revolution about a cube axis. They are quite cigar-shaped, being elongated along the cube axis. In terms of the free electron mass m, the effective mass along the axis (the longitudinal effective mass) is $m_L \approx 1.0m$ while the effective masses perpendicular to the axis (the transverse effective mass) are $m_T \approx 0.2m$. There are two degenerate valence band maxima, both located at $\mathbf{k} = \mathbf{0}$, which are spherically symmetric to the extent that the ellipsoidal expansion is valid, with masses of $0.49m$ and $0.16m$ (Figure 28.6).

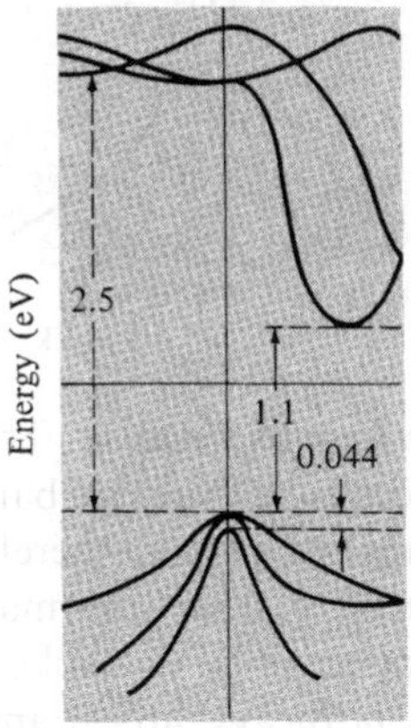

Figure 28.6
Energy bands in silicon. Note the conduction band minimum along [100] that gives rise to the ellipsoids of Figure 28.5. The valence band maximum occurs at $\mathbf{k} = 0$, where two degenerate bands with different curvatures meet, giving rise to "light holes" and "heavy holes." Note also, the third band, only 0.044 eV below the valence band maximum. This band is separated from the other two only by spin-orbit coupling. At temperatures on the order of room temperature ($k_BT = 0.025$ eV) it too may be a significant source of carriers. (From C. A. Hogarth, ed., *Materials Used in Semiconductor Devices*, Interscience, New York, 1965.)

Germanium The crystal structure and Brillouin zone are as in silicon. However, the conduction band minima now occur at the zone boundaries in the ⟨111⟩ directions. Minima on parallel hexagonal faces of the zone represent the same physical levels, so there are four symmetry-related conduction band minima. The ellipsoidal constant energy surfaces are ellipsoids of revolution elongated along the ⟨111⟩ directions, with effective masses $m_L \approx 1.6m$, and $m_T \approx 0.08m$ (Figure 28.7). There are again two

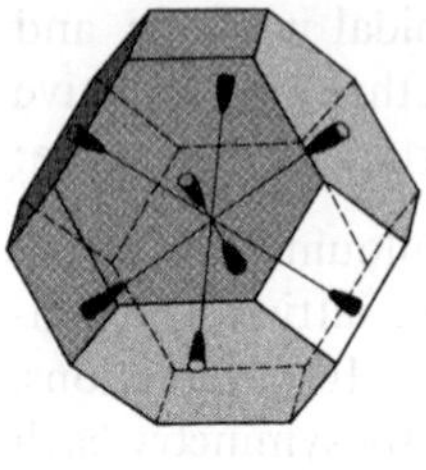

Figure 28.7
Constant-energy surfaces near the conduction band minima in germanium. There are eight symmetry-related half ellipsoids with long axes along ⟨111⟩ directions centered on the midpoints of the hexagonal zone faces. With a suitable choice of primitive cell in k-space these can be represented as four ellipsoids, the half ellipsoids on opposite faces being joined together by translations through suitable reciprocal lattice vectors.

degenerate valence bands, both with maxima at $\mathbf{k} = \mathbf{0}$, which are spherically symmetric in the quadratic approximation with effective masses of $0.28m$ and $0.044m$ (Figure 28.8).

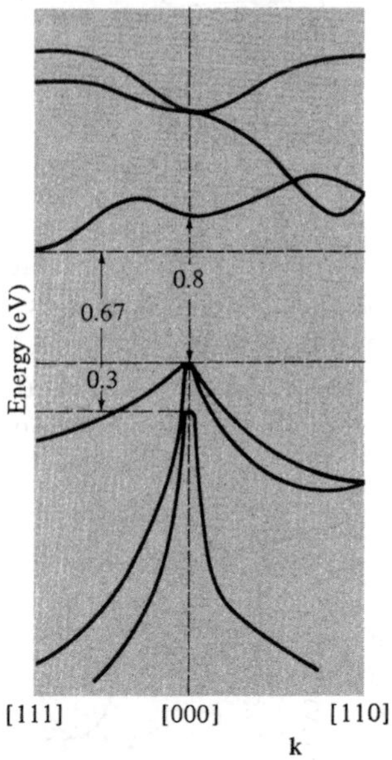

Figure 28.8
Energy bands in germanium. Note the conduction band minimum along [111] at the zone boundary that gives rise to the four ellipsoidal pockets of Figure 28.7. The valence band maximum, as in silicon, is at $\mathbf{k} = \mathbf{0}$, where two degenerate bands with different curvatures meet, giving rise to two pockets of holes with distinct effective masses. (From C. A. Hogarth, ed., *Materials Used in Semiconductor Devices*, Interscience, New York, 1965.)

Indium antimonide This compound, which has the zincblende structure, is interesting because all valence band maxima and conduction band minima are at $\mathbf{k} = 0$. The energy surfaces are therefore spherical. The conduction band effective mass is very small, $m^* \approx 0.015m$. Information on the valence band masses is less unambiguous, but there appear to be two spherical pockets about $\mathbf{k} = 0$, one with an effective mass of about $0.2m$ (heavy holes) and another with effective mass of about $0.015m$ (light holes).

CYCLOTRON RESONANCE

The effective masses discussed above are measured by the technique of cyclotron resonance. Consider an electron close enough to the bottom of the conduction band (or top of the valence band) for the quadratic expansion (28.2) to be valid. In the presence of a magnetic field **H** the semiclassical equations of motion (12.32) and (12.33) imply that the velocity **v**(**k**) obeys the single set of equations

$$\mathsf{M}\frac{d\mathbf{v}}{dt} = \mp\frac{e}{c}\mathbf{v}\times\mathbf{H}. \tag{28.4}$$

In a constant uniform field (taken along the z-axis) it is not difficult to show (Problem 1) that (28.4) has an oscillatory solution

$$\mathbf{v} = \mathrm{Re}\,\mathbf{v}_0 e^{-i\omega t}, \tag{28.5}$$

provided that

$$\omega = \frac{eH}{m^*c}, \tag{28.6}$$

where m^*, the "cyclotron effective mass," is given by

$$m^* = \left(\frac{\det \mathsf{M}}{M_{zz}}\right)^{1/2}. \tag{28.7}$$

This result can also be written in terms of the eigenvalues and principal axes of the mass tensor as (Problem 1):

$$m^* = \sqrt{\frac{m_1m_2m_3}{\hat{H}_1{}^2m_1 + \hat{H}_2{}^2m_2 + \hat{H}_3{}^2m_3}}, \tag{28.8}$$

where the $\hat{H}_i$ are the components along the three principal axes of a unit vector parallel to the field.

Note that the cyclotron frequency depends, for a given ellipsoid, on the orientation of the magnetic field with respect to that ellipsoid, but not on the initial wave vector or energy of the electron. Thus for a given orientation of the crystal with respect to the field, all electrons in a given ellipsoidal pocket of conduction band electrons (and, by the same token, all holes in a given ellipsoidal pocket of valence band holes) precess at a frequency entirely determined by the effective mass tensor describing that pocket. There will therefore be a small number of distinct cyclotron frequencies. By noting how these resonant frequencies shift as the orientation of the magnetic field is varied, one can extract from (28.8) the kind of information we quoted above.

To observe cyclotron resonance it is essential that the cyclotron frequency (28.6) be larger than or comparable to the collision frequency. As in the case of metals, this generally requires working with very pure samples at very low temperatures, to reduce both impurity scattering and phonon scattering to a minimum. Under such conditions the electrical conductivity of a semiconductor will be so small that (in contrast to the case of a metal (page 278)) the driving electromagnetic field can penetrate far enough into the sample to excite the resonance without any difficulties associated with a skin depth. On the other hand, under such conditions of low temperatures and purity the number of carriers available in thermal equilibrium to participate in the resonance may well be so small that carriers will have to be created by other means—such as photoexcitation. Some typical cyclotron resonance data are shown in Figure 28.9.

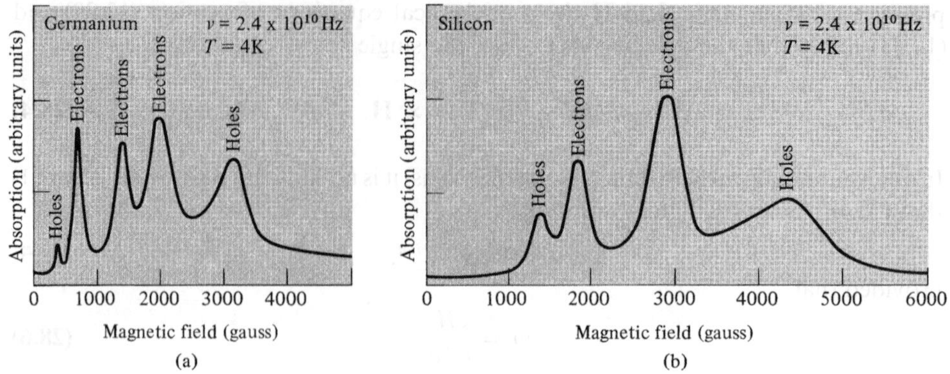

Figure 28.9
Typical cyclotron resonance signals in (a) germanium and (b) silicon. The field lies in a (110) plane and makes an angle with the [001] axis of 60° (Ge) and 30° (Si). (From G. Dresselhaus et al., *Phys. Rev.* **98**, 368 (1955).)

NUMBER OF CARRIERS IN THERMAL EQUILIBRIUM

The most important property of any semiconductor at temperature T is the number of electrons per unit volume in the conduction band, n_c, and the number of holes[10] per unit volume in the valence band, p_v. The determination of these as a function of temperature is a straightforward, though sometimes algebraically complicated, exercise in the application of Fermi-Dirac statistics to the appropriate set of one-electron levels.

The values of $n_c(T)$ and $p_v(T)$ depend critically, as we shall see, on the presence of impurities. However, there are certain general relations that hold regardless of the purity of the sample, and we consider these first. Suppose the density of levels (page 143) is $g_c(\mathcal{E})$ in the conduction band and $g_v(\mathcal{E})$ in the valence band. The effect of impurities, as we shall see below, is to introduce additional levels at energies between the top of the valence band, $\mathcal{E}_v$, and the bottom of the conduction band, $\mathcal{E}_c$, without, however, appreciably altering the form of $g_c(\mathcal{E})$ and $g_v(\mathcal{E})$. Since conduction is entirely due to electrons in conduction band levels or holes in valence band levels, regardless of the concentration of impurities the numbers of carriers present at temperature T will be given by

$$n_c(T) = \int_{\mathcal{E}_c}^{\infty} d\mathcal{E}\, g_c(\mathcal{E}) \frac{1}{e^{(\mathcal{E}-\mu)/k_BT} + 1},$$

$$p_v(T) = \int_{-\infty}^{\mathcal{E}_v} d\mathcal{E}\, g_v(\mathcal{E}) \left(1 - \frac{1}{e^{(\mathcal{E}-\mu)/k_BT} + 1}\right)$$

$$= \int_{-\infty}^{\mathcal{E}_v} d\mathcal{E}\, g_v(\mathcal{E}) \frac{1}{e^{(\mu-\mathcal{E})/k_BT} + 1}. \tag{28.9}$$

[10] Hole densities are conventionally denoted by the letter p (for positive). This widely used notation exploits the coincidence that the n denoting the *number* density of electrons can also be regarded as standing for "negative."

Impurities affect the determination of n_c and p_v only through the value of the chemical potential[11] μ to be used in Eq. (28.9). To determine μ one must know something about the impurity levels. However, one can extract some useful information from (28.9) which is independent of the precise value of the chemical potential, provided only that it satisfies the conditions:

$$\begin{aligned} \mathcal{E}_c - \mu &\gg k_B T, \\ \mu - \mathcal{E}_v &\gg k_B T. \end{aligned} \tag{28.10}$$

There will be a range of values of μ for which (28.10) holds even for energy gaps $E_g = \mathcal{E}_c - \mathcal{E}_v$ as small as a few tenths of an electron volt and temperatures as high as room temperature. Our procedure will be to assume the validity of (28.10), use it to simplify (28.9), and then, from the values of n_c and p_v so obtained and the appropriate information about possible impurity levels, compute the actual value of the chemical potential to check whether it does indeed lie in the range given by (28.10). If it does, the semiconductor is described as "nondegenerate," and the procedure is a valid one. If it does not, one is dealing with a "degenerate semiconductor" and must work directly with Eq. (28.9) without making the simplifications implied by (28.10).

Given Eq. (28.10), then since every conduction band level exceeds $\mathcal{E}_c$ and every valence band level is less than $\mathcal{E}_v$, we may simplify the statistical factors in (28.9):

$$\begin{aligned} \frac{1}{e^{(\mathcal{E}-\mu)/k_B T} + 1} &\approx e^{-(\mathcal{E}-\mu)/k_B T}, \qquad \mathcal{E} > \mathcal{E}_c; \\ \frac{1}{e^{(\mu-\mathcal{E})/k_B T} + 1} &\approx e^{-(\mu-\mathcal{E})/k_B T}, \qquad \mathcal{E} < \mathcal{E}_v. \end{aligned} \tag{28.11}$$

Equations (28.9) thereby reduce to

$$\boxed{\begin{aligned} n_c(T) &= N_c(T) e^{-(\mathcal{E}_c - \mu)/k_B T}, \\ p_v(T) &= P_v(T) e^{-(\mu - \mathcal{E}_v)/k_B T}, \end{aligned}} \tag{28.12}$$

where

$$\begin{aligned} N_c(T) &= \int_{\mathcal{E}_c}^{\infty} d\mathcal{E}\, g_c(\mathcal{E}) e^{-(\mathcal{E}-\mathcal{E}_c)/k_B T}, \\ P_v(T) &= \int_{-\infty}^{\mathcal{E}_v} d\mathcal{E}\, g_v(\mathcal{E}) e^{-(\mathcal{E}_v-\mathcal{E})/k_B T}. \end{aligned} \tag{28.13}$$

Because the ranges of integration in (28.13) include the points where the arguments of the exponentials vanish, $N_c(T)$ and $P_v(T)$ are relatively slowly varying functions

---

[11] It is the widespread practice to refer to the chemical potential of a semiconductor as "the Fermi level," a somewhat unfortunate terminology. Since the chemical potential almost always lies in the energy gap, there is no one-electron level whose energy is actually at "the Fermi level" (in contrast to the case of a metal). Thus the usual definition of the Fermi level (that energy below which the one-electron levels are occupied and above which they are unoccupied in the ground state of a metal) does not specify a unique energy in the case of a semiconductor: Any energy in the gap separates occupied from unoccupied levels at $T = 0$. The term "Fermi level" should be regarded as nothing more than a synonym for "chemical potential," in the context of semiconductors.

of temperature, compared with the exponential factors they multiply in (28.12). This is their most important feature. Usually, however, one can evaluate them explicitly. Because of the exponential factors in the integrands of (28.13) only energies within $k_B T$ of the band edges contribute appreciably, and in this range the quadratic approximation, (28.2) or (28.3), is generally excellent. The level densities can then be taken to be (Problem 3):

$$g_{c,v}(\mathcal{E}) = \sqrt{2|\mathcal{E} - \mathcal{E}_{c,v}|}\,\frac{m_{c,v}^{3/2}}{\hbar^3\pi^2}, \tag{28.14}$$

and the integrals (28.13) then give

$$N_c(T) = \frac{1}{4}\left(\frac{2m_c k_B T}{\pi\hbar^2}\right)^{3/2},$$
$$P_v(T) = \frac{1}{4}\left(\frac{2m_v k_B T}{\pi\hbar^2}\right)^{3/2}. \tag{28.15}$$

Here m_c^3 is the product of the principal values of the conduction band effective mass tensor (i.e., its determinant),[12] and m_v^3 is similarly defined.

Equation (28.15) can be cast in the numerically convenient forms:

$$N_c(T) = 2.5\left(\frac{m_c}{m}\right)^{3/2}\left(\frac{T}{300\text{ K}}\right)^{3/2} \times 10^{19}/\text{cm}^3,$$
$$P_v(T) = 2.5\left(\frac{m_v}{m}\right)^{3/2}\left(\frac{T}{300\text{ K}}\right)^{3/2} \times 10^{19}/\text{cm}^3, \tag{28.16}$$

where T is to be measured in degrees Kelvin. Since the exponential factors in (28.12) are less than unity by at least an order of magnitude, and since m_c/m and m_v/m are typically of the order of unity, Eq. (28.16) indicates that 10^{18} or 10^{19} carriers/cm^3 is an absolute upper limit to the carrier concentration in a nondegenerate semiconductor.

We still cannot infer $n_c(T)$ and $p_v(T)$ from (28.12) until we know the value of the chemical potential μ. However, the μ dependence disappears from the product of the two densities:

$$\begin{aligned} n_c p_v &= N_c P_v e^{-(\mathcal{E}_c - \mathcal{E}_v)/k_B T} \\ &= N_c P_v e^{-E_g/k_B T}. \end{aligned} \tag{28.17}$$

This result (sometimes called the "law of mass action"[13]) means that at a given temperature it suffices to know the density of one carrier type to determine that of the other. How this determination is made depends on how important the impurities are as a source of carriers.

---

12 If there is more than one conduction band minimum one must add together terms of the form (28.14) and (28.15) for each minimum. These sums will continue to have the same forms as (28.14) and (28.15), provided that the definition of m_c is altered to $m_c^{3/2} \to \sum m_c^{3/2}$.

13 The analogy with chemical reactions is quite precise: A carrier is provided by the dissociation of a combined electron and hole.

Intrinsic Case

If the crystal is so pure that impurities contribute negligibly to the carrier densities, one speaks of an "intrinsic semiconductor." In the intrinsic case, conduction band electrons can only have come from formerly occupied valence band levels, leaving holes behind them. The number of conduction band electrons is therefore equal to the number of valence band holes:

$$n_c(T) = p_v(T) \equiv n_i(T). \tag{28.18}$$

Since $n_c = p_v$, we may write their common value n_i as $(n_c p_v)^{1/2}$. Equation (28.17) then gives

$$n_i(T) = [N_c(T)P_v(T)]^{1/2} e^{-E_g/2k_BT}, \tag{28.19}$$

or, from (28.15) and (28.16):

$$\boxed{\begin{aligned} n_i(T) &= \frac{1}{4}\left(\frac{2k_BT}{\pi\hbar^2}\right)^{3/2} (m_c m_v)^{3/4} e^{-E_g/2k_BT} \\ &= 2.5\left(\frac{m_c}{m}\right)^{3/4}\left(\frac{m_v}{m}\right)^{3/4}\left(\frac{T}{300\ \text{K}}\right)^{3/2} e^{-E_g/2k_BT} \times 10^{19}/\text{cm}^3. \end{aligned}} \tag{28.20}$$

We may now establish in the intrinsic case the condition for the validity of assumption (28.10) on which our analysis has been based. Defining μ_i to be the value of the chemical potential in the intrinsic case, we find that Eqs. (28.12) give values of n_c and p_v equal to n_i (Eq. (28.19)), provided that

$$\mu = \mu_i = \mathcal{E}_v + \tfrac{1}{2}E_g + \tfrac{1}{2}k_BT \ln\left(\frac{P_v}{N_c}\right), \tag{28.21}$$

or, from Eq. (28.15),

$$\boxed{\mu_i = \mathcal{E}_v + \tfrac{1}{2}E_g + \tfrac{3}{4}k_BT \ln\left(\frac{m_v}{m_c}\right).} \tag{28.22}$$

This asserts that as $T \to 0$, the chemical potential μ_i lies precisely in the middle of the energy gap. Furthermore, since $\ln(m_v/m_c)$ is a number of order unity, μ_i will not wander from the center of the energy gap by more than order k_BT. Consequently, at temperatures k_BT small compared with E_g, the chemical potential will be found far from the boundaries of the forbidden region, $\mathcal{E}_c$ and $\mathcal{E}_v$, compared with k_BT (Figure 28.10), and the condition for nondegeneracy (28.10) will be satisfied. Therefore (28.20) is a valid evaluation of the common value of n_c and p_v in the intrinsic case, provided only that E_g is large compared with k_BT, a condition that is satisfied in almost all semiconductors at room temperature and below.

Extrinsic Case: Some General Features

If impurities contribute a significant fraction of the conduction band electrons and/or valence band holes, one speaks of an "extrinsic semiconductor." Because of these

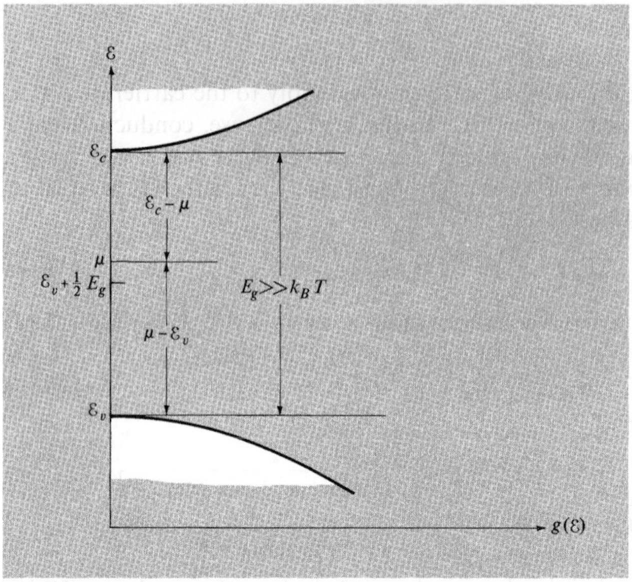

Figure 28.10
In an intrinsic semiconductor with an energy gap E_g large compared with k_BT, the chemical potential μ lies within order k_BT of the center of the energy gap, and is therefore far compared with k_BT from both boundaries of the gap at $\mathcal{E}_c$ and $\mathcal{E}_v$.

added sources of carriers the density of conduction band electrons need no longer be equal to the density of valence band holes:

$$n_c - p_v = \Delta n \neq 0. \tag{28.23}$$

Since the law of mass action Eq. (28.17) holds regardless of the importance of impurities, we can use the definition (28.19) of $n_i(T)$ to write quite generally,

$$n_c p_v = n_i^2. \tag{28.24}$$

Equations (28.24) and (28.23) permit one to express the carrier densities in the extrinsic case in terms of their intrinsic values n_i and the deviation Δn from intrinsic behavior:

$$\begin{Bmatrix} n_c \\ p_v \end{Bmatrix} = \frac{1}{2}\left[(\Delta n)^2 + 4n_i^2\right]^{1/2} \pm \tfrac{1}{2}\Delta n. \tag{28.25}$$

The quantity $\Delta n/n_i$, which measures the importance of the impurities as a source of carriers, can be given a particularly simple expression as a function of chemical potential μ, if we note that Eqs. (28.12) have the form[14]

$$n_c = e^{\beta(\mu - \mu_i)}n_i; \quad p_v = e^{-\beta(\mu - \mu_i)}n_i. \tag{28.26}$$

Therefore

$$\frac{\Delta n}{n_i} = 2 \sinh \beta(\mu - \mu_i). \tag{28.27}$$

[14] To verify these relations one need not substitute the explicit definitions of n_i and μ_i; it is enough to note that n_c and p_v are proportional to exp $(\beta\mu)$ and exp $(-\beta\mu)$, respectively, and that both reduce to n_i when $\mu = \mu_i$.

We have noted that if the energy gap E_g is large compared with $k_B T$, then the intrinsic chemical potential μ_i will satisfy the assumption (28.10) of nondegeneracy. But Eq. (28.27) requires that if μ_i is far from $\mathcal{E}_c$ or $\mathcal{E}_v$ on the scale of $k_B T$, then μ must be as well, unless Δn is many orders of magnitude larger than the intrinsic carrier density n_i. Thus the nondegeneracy assumption underlying the derivation of (28.27) is valid when $E_g \gg k_B T$, unless we are in a region of extreme extrinsic behavior.

Note also that when Δn is large compared with n_i, then Eq. (28.25) asserts that the density of one carrier type is essentially equal to Δn, while that of the other type is smaller by a factor of order $(n_i/\Delta n)^2$. Thus when impurities do provide the major source of carriers, one of the two carrier types will be dominant. An extrinsic semiconductor is called "*n*-type" or "*p*-type" according to whether the dominant carriers are electrons or holes.

To complete the specification of the carrier densities in extrinsic semiconductors one must determine Δn or μ. To do this we must examine the nature of the electronic levels introduced by the impurities and the statistical mechanics of the occupation of these levels in thermal equilibrium.

IMPURITY LEVELS

Impurities that contribute to the carrier density of a semiconductor are called *donors* if they supply additional electrons to the conduction band, and *acceptors* if they supply additional holes to (i.e., capture electrons from) the valence band. Donor impurities are atoms that have a higher chemical valence than the atoms making up the pure (host) material, while acceptors have a lower chemical valence.

Consider, for example, the case of substitutional impurities in a group IV semiconductor. Suppose that we take a crystal of pure germanium, and replace an occasional germanium atom by its neighbor to the right in the periodic table, arsenic (Figure 28.11). The germanium ion has charge $4e$ and contributes four valence electrons, while the arsenic ion has charge $5e$ and contributes five valence electrons. If, to a first approximation, we ignore the difference in structure between the arsenic and germanium ion cores, we can represent the substitution of an arsenic atom for

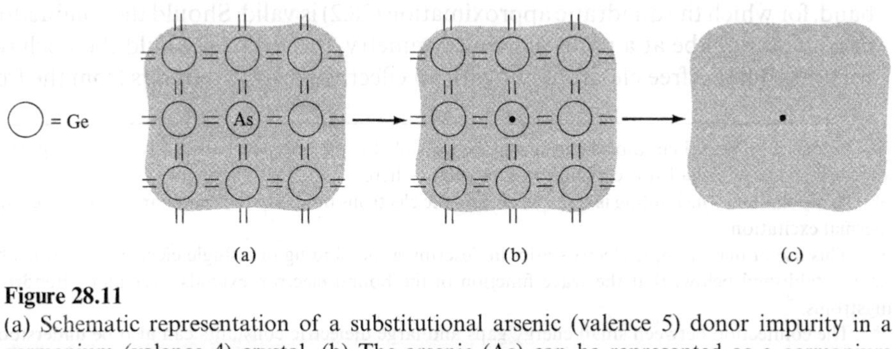

Figure 28.11
(a) Schematic representation of a substitutional arsenic (valence 5) donor impurity in a germanium (valence 4) crystal. (b) The arsenic (As) can be represented as a germanium atom *plus* an additional unit of positive charge fixed at the site of the atom (circled dot). (c) In the semiclassical approximation, in which the pure semiconductor is treated as a homogeneous medium, the arsenic impurity is represented as a fixed point charge $+e$ (dot).

a germanium atom by a slightly less drastic modification, in which the germanium atom is not removed, but an additional fixed positive charge of e is placed at its site, along with an additional electron.

This is the general model for a semiconductor doped with donor impurities. Distributed irregularly[15] throughout the perfect pure crystal are N_D fixed attractive centers of charge $+e$, per unit volume, along with the same number of additional electrons. As expected, each such center of charge $+e$ can bind[16] one of the additional electrons of charge $-e$. If the impurity were not embedded in the semiconductor, but in empty space, the binding energy of the electron would just be the first ionization potential of the impurity atom, 9.81 eV for arsenic. However (*and this is of crucial importance in the theory of semiconductors*), since the impurity is embedded in the medium of the pure semiconductor, this binding energy is enormously reduced (to 0.013 eV for the case of arsenic in germanium). This happens for two reasons:

1. The field of the charge representing the impurity must be reduced by the static dielectric constant ϵ of the semiconductor.[17] These are quite large ($\epsilon \approx 16$ in germanium), being typically between about 10 and 20 but ranging in some cases as high as 100 or more. The large dielectric constants are consequences of the small energy gaps. If there were no overall energy gap, the crystal would be a metal instead of a semiconductor, and the static dielectric constant would be infinite, reflecting the fact that a static electric field can induce a current in which electrons move arbitrarily far from their original positions. If the energy gap is not zero, but small, then the dielectric constant will not be infinite, but can be quite large, reflecting the relative ease with which the spatial distribution of electrons can be deformed.[18]
2. An electron moving in the medium of the semiconductor should be described not by the free space energy-momentum relation, but by the semiclassical relation (Chapter 12) $\mathcal{E}(\mathbf{k}) = \mathcal{E}_c(\mathbf{k})$, where $\hbar\mathbf{k}$ is the electronic crystal momentum, and $\mathcal{E}_c(\mathbf{k})$ is the conduction band energy-momentum relation; i.e., the additional electron introduced by the impurity should be thought of as being in a superposition of conduction band levels of the pure host material, which is appropriately altered by the additional localized charge $+e$ representing the impurity. The electron can minimize its energy by using only levels near the bottom of the conduction band, for which the quadratic approximation (28.2) is valid. Should the conduction band minimum be at a point of cubic symmetry, the electron would then behave very much like a free electron, but with an effective mass that differs from the free

---

[15] Under very special circumstances it may be possible for the impurities themselves to be regularly arranged in space. We shall not consider this possibility here.

[16] As we shall see, the binding is quite weak, and the electrons bound to the center are readily liberated by thermal excitation.

[17] This use of macroscopic electrostatics in describing the binding of a single electron is justified by the fact (established below) that the wave function of the bound electron extends over many hundreds of angstroms.

[18] The connection between small energy gaps and large dielectric constants can also be understood from the point of view of perturbation theory: The size of the dielectric constant is a measure of the extent to which a weak electric field distorts the electronic wave function. But a small energy gap means there will be small energy denominators, and hence large changes, in the first-order wave functions.

electron mass m. More generally, the energy wave vector relation will be some anisotropic quadratic function of k. In either case, however, to a first approximation, we may represent the electron as moving in free space but with a mass given by some appropriately defined effective mass m^*, rather than the free electron mass. In general, this mass will be smaller than the free electron mass, often by a factor of 0.1 or even less.

These two observations suggest that we may represent an electron in the presence of a donor impurity of charge e within the medium of the semiconductor, as a particle of charge $-e$ and mass m^*, moving in free space in the presence of an attractive center of charge e/ϵ. This is precisely the problem of a hydrogen atom, except that the product $-e^2$ of the nuclear and electronic charges must be replaced by $-e^2/\epsilon$, and the free electron mass m, by m^*. Thus the radius of the first Bohr orbit, $a_0 = \hbar^2/me^2$, becomes

$$r_0 = \frac{m}{m^*}\epsilon a_0, \tag{28.28}$$

and the ground-state binding energy, $me^4/2\hbar^2 = 13.6$ eV becomes

$$\mathcal{E} = \frac{m^*}{m}\frac{1}{\epsilon^2} \times 13.6 \text{ eV}. \tag{28.29}$$

For reasonable values of m^*/m and ϵ, the radius r_0 can be 100 Å or more. This is very important for the consistency of the entire argument, for both the use of the semiclassical model and the use of the macroscopic dielectric constant are predicated on the assumption that the fields being described vary slowly on the scale of a lattice constant.

Furthermore, typical values of m^*/m and ϵ can lead to a binding energy $\mathcal{E}$ smaller than 13.6 eV by a factor of a thousand or more. Indeed, since small energy gaps are generally associated with large dielectric constants, it is almost always the case that *the binding energy of an electron to a donor impurity is small compared with the energy gap of the semiconductor*. Since this binding energy is measured relative to the energy of the conduction band levels from which the bound impurity level is formed, we conclude that donor impurities introduce additional electronic levels at energies $\mathcal{E}_d$ which are lower than the energy $\mathcal{E}_c$ at the bottom of the conduction band by an amount that is small compared with the energy gap E_g (Figure 28.12).

Figure 28.12
Level density for a semiconductor containing both donor and acceptor impurities. The donor levels $\mathcal{E}_d$ are generally close to the bottom of the conduction band, $\mathcal{E}_c$ compared with E_g, and the acceptor levels, $\mathcal{E}_a$, are generally close to the top of the valence band, $\mathcal{E}_v$.

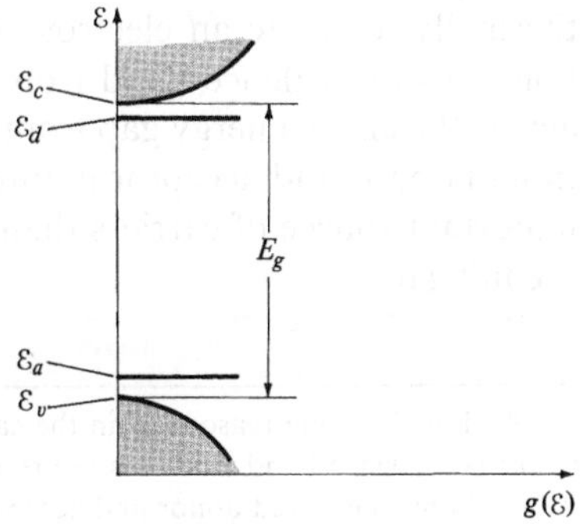

A similar argument can be applied to acceptor impurities, whose valence is one less than that of the host atoms (e.g., gallium in germanium). Such an impurity can be represented by the superimposition of a fixed charge $-e$ on top of a host atom, along with the presence of one less electron in the crystal. The missing electron can be represented as a bound hole, attracted by the excess negative charge representing the impurity, with a binding energy that is again small[19] on the scale of the energy gap, E_g. In terms of the electron picture this bound hole will be manifested as an additional electronic level at an energy $\mathcal{E}_a$ lying slightly above the top of the valence band (Figure 28.12). The hole is bound when the level is empty. The binding energy of the hole is just the energy $\mathcal{E}_a - \mathcal{E}_v$ necessary to excite an electron from the top of the valence band into the acceptor level, thereby filling the hole in the vicinity of the acceptor and creating a free hole in the valence band.

Table 28.2

LEVELS OF GROUP V (DONORS) AND GROUP III (ACCEPTORS) IMPURITIES IN SILICON AND GERMANIUM

GROUP III ACCEPTORS (TABLE ENTRY IS $\mathcal{E}_a - \mathcal{E}_v$)					
	B	Al	Ga	In	Tl
Si	0.046 eV	0.057	0.065	0.16	0.26
Ge	0.0104	0.0102	0.0108	0.0112	0.01

GROUP V DONORS (TABLE ENTRY IS $\mathcal{E}_c - \mathcal{E}_d$)				
	P	As	Sb	Bi
Si	0.044 eV	0.049	0.039	0.069
Ge	0.0120	0.0127	0.0096	—

ROOM TEMPERATURE ENERGY GAPS ($E_g = \mathcal{E}_c - \mathcal{E}_v$)	
Si	1.12 eV
Ge	0.67 eV

Source: P. Aigrain and M. Balkanski, *Selected Constants Relative to Semiconductors*, Pergamon, New York, 1961.

The single most important fact about these donor and acceptor levels is that they lie very close to the boundaries of the forbidden energy region.[20] It is far easier thermally to excite an electron into the conduction band from a donor level, or a hole into the valence band from an acceptor level, than it is to excite an electron across the entire energy gap from valence to conduction band. Unless the concentration of donor and acceptor impurities is very small, they will therefore be a far more important source of carriers than the intrinsic mechanism of exciting carriers across the full gap.

[19] For the same reasons as in the case of donor impurities, the binding energy of the hole is quite weak; i.e., valence band electrons are readily lifted into the acceptor level by thermal excitation.

[20] Some measured donor and acceptor levels are given in Table 28.2.

POPULATION OF IMPURITY LEVELS IN THERMAL EQUILIBRIUM

To assess the extent to which carriers can be thermally excited from impurity levels, we must compute the mean number of electrons in the levels at a given temperature and chemical potential. We assume that the density of impurities is low enough that the interaction of electrons (or holes) bound at different impurity sites is negligible. We may then calculate the number density of electrons n_d (or holes p_a) bound to donor (or acceptor) sites by simply multiplying by the density of donors N_d (or acceptors N_a) the mean number of electrons (or holes) there would be if there were only a single impurity. For simplicity we assume that the impurity introduces only a single one-electron orbital level.[21] We calculate its mean occupancy as follows:

Donor Level If we ignored electron-electron interactions the level could either be empty, could contain one electron of either spin, or two electrons of opposite spins. However, the Coulomb repulsion of two localized electrons raises the energy of the doubly occupied level so high that double occupation is essentially prohibited. Quite generally, the mean number of electrons in a system in thermal equilibrium is given by:

$$\langle n \rangle = \frac{\sum N_j e^{-\beta(E_j - \mu N_j)}}{\sum e^{-\beta(E_j - \mu N_j)}}, \tag{28.30}$$

where the sum is over all states of the system, E_j and N_j, are the energy and number of electrons in state j, and μ is the chemical potential. In the present case the system is a single impurity with just three states: one with no electrons present which makes no contribution to the energy, and two with a single electron present of energy $\mathcal{E}_d$. Therefore (28.30) gives

$$\langle n \rangle = \frac{2e^{-\beta(\mathcal{E}_d - \mu)}}{1 + 2e^{-\beta(\mathcal{E}_d - \mu)}} = \frac{1}{\frac{1}{2}e^{\beta(\mathcal{E}_d - \mu)} + 1}, \tag{28.31}$$

so that[22]

$$\boxed{n_d = \frac{N_d}{\frac{1}{2}e^{\beta(\mathcal{E}_d - \mu)} + 1}.} \tag{28.32}$$

Acceptor Level In contrast to a donor level, an acceptor level, when viewed as an electronic level, can be singly or doubly occupied, but not empty. This is easily seen from the hole point of view. An acceptor impurity can be regarded as a fixed, negatively charged attractive center superimposed on an unaltered host atom. This additional charge $-e$ can weakly bind one hole (corresponding to one electron being in the

---

[21] There is no general reason why a donor site cannot have more than one bound level, and we assume a single one only to simplify our discussion. Our qualitative conclusions, however, are quite general (see Problem 4c).

[22] Some insight into the curious factor of $\frac{1}{2}$ that emerges in (28.32) in contrast to the more familiar distribution function of Fermi-Dirac statistics can be gained by examining what happens as the energy of the doubly occupied level drops from $+\infty$ down to $2\mathcal{E}_d$. See Problem 4.

acceptor level). The binding energy of the hole is $\mathcal{E}_a - \mathcal{E}_v$, and when the hole is "ionized" an additional electron moves into the acceptor level. However, the configuration in which no electrons are in the acceptor level corresponds to two holes being localized in the presence of the acceptor impurity, which has a very high energy due to the mutual Coulomb repulsion of the holes.[23]

Bearing this in mind, we can calculate the mean number of electrons at an acceptor level from (28.30) by noting that the state with no electrons is now prohibited, while the two-electron state has an energy that is $\mathcal{E}_a$ higher than the two one-electron states. Therefore

$$\langle n \rangle = \frac{2e^{\beta\mu} + 2e^{-\beta(\mathcal{E}_a - 2\mu)}}{2e^{\beta\mu} + e^{-\beta(\mathcal{E}_a - 2\mu)}} = \frac{e^{\beta(\mu - \mathcal{E}_a)} + 1}{\frac{1}{2}e^{\beta(\mu - \mathcal{E}_a)} + 1}. \tag{28.33}$$

The mean number of holes in the acceptor level is the difference between the maximum number of electrons the level can hold (two) and the actual mean number of electrons in the level ($\langle n \rangle$): $\langle p \rangle = 2 - \langle n \rangle$, and therefore $p_a = N_a \langle p \rangle$ is given by

$$p_a = \frac{N_a}{\frac{1}{2}e^{\beta(\mu - \mathcal{E}_a)} + 1}. \tag{28.34}$$

THERMAL EQUILIBRIUM CARRIER DENSITIES OF IMPURE SEMICONDUCTORS

Consider a semiconductor doped with N_d donor impurities and N_a acceptor impurities per unit volume. To determine the carrier densities we must generalize the constraint $n_c = p_v$ (Eq. (28.18)) that enabled us to find these densities in the intrinsic (pure) case. We can do this by first considering the electronic configuration at $T = 0$. Suppose $N_d \geqslant N_a$. (The case $N_d < N_a$ is equally straightforward and leads to the same result (28.35).) Then in a unit volume of semiconductor N_a of the N_d electrons supplied by the donor impurities can drop from the donor levels into the acceptor levels.[24] This gives a ground-state electronic configuration in which the valence band and acceptor levels are filled, $N_d - N_a$ of the donor levels are filled, and the conduction band levels are empty. In thermal equilibrium at temperature T the electrons will be redistributed among these levels, but since their total number remains the same, the number of electrons in conduction band or donor levels, $n_c + n_d$, must exceed its value at $T = 0$, $N_d - N_a$, by precisely the number of empty levels (i.e., holes), $p_v + p_a$, in the valence band and acceptor levels:

$$n_c + n_d = N_d - N_a + p_v + p_a. \tag{28.35}$$

[23] When describing acceptor levels as electronic levels one usually ignores the electron that *must* be in the level, considering only the presence or absence of the second electron. One describes the level as empty or filled according to whether the second electron is absent or present.

[24] Since $\mathcal{E}_d$ is just below the conduction band minimum $\mathcal{E}_c$, and $\mathcal{E}_a$ is just above the valence band maximum, $\mathcal{E}_v$, we have $\mathcal{E}_d > \mathcal{E}_a$ (see Figure 28.12).

This equation, together with the explicit forms we have found for n_c, p_v, n_d, and n_a as functions of μ and T, permits one to find μ as a function of T, and therefore to find the thermal equilibrium carrier densities at any temperature. A general analysis is rather complicated, and we consider here only a particularly simple and important case:

Suppose that

$$\begin{aligned} \mathcal{E}_d - \mu &\gg k_B T, \\ \mu - \mathcal{E}_a &\gg k_B T. \end{aligned} \tag{28.36}$$

Since $\mathcal{E}_d$ and $\mathcal{E}_a$ are close to the edges of the gap, this is only slightly more restrictive than the nondegeneracy assumption (28.10). Condition (28.36) and the expressions (28.32) and (28.34) for n_d and p_a insure that thermal excitation fully "ionizes" the impurities, leaving only a negligible fraction with bound electrons or holes: $n_d \ll N_d$, $p_a \ll N_a$. Equation (28.35) therefore becomes

$$\Delta n = n_c - p_v = N_d - N_a, \tag{28.37}$$

so Eqs. (28.25) and (28.27) now give the carrier densities and chemical potential as explicit functions of the temperature alone:

$$\left\{\begin{matrix} n_c \\ p_v \end{matrix}\right\} = \frac{1}{2}[(N_d - N_a)^2 + 4n_i^2]^{1/2} \pm \frac{1}{2}[N_d - N_a] \tag{28.38}$$

$$\frac{N_d - N_a}{n_i} = 2 \sinh \beta(\mu - \mu_i). \tag{28.39}$$

If the gap is large compared with $k_B T$, the assumption (28.36) we began with should remain valid unless μ is quite far from μ_i on the scale of $k_B T$. According to Eq. (28.39), this will only happen when $|N_d - N_a|$ is several orders of magnitude greater than the intrinsic carrier density n_i. Therefore Eq. (28.38) correctly describes the transition from predominantly intrinsic behavior ($n_i \gg |N_d - N_a|$) well into the region of predominantly extrinsic behavior ($n_i \ll |N_d - N_a|$). Expanding (28.38), we find that at low impurity concentrations the corrections to the purely intrinsic carrier densities are

$$\left\{\begin{matrix} n_c \\ p_v \end{matrix}\right\} \approx n_i \pm \tfrac{1}{2}(N_d - N_a), \tag{28.40}$$

while for a considerable range of carrier concentrations in the extrinsic regime,

$$\left.\begin{aligned} n_c &\approx N_d - N_a \\ p_v &\approx \frac{n_i^2}{N_d - N_a} \end{aligned}\right\} \quad N_d > N_a;$$
$$\left.\begin{aligned} n_c &\approx \frac{n_i^2}{N_a - N_d} \\ p_v &\approx N_a - N_d \end{aligned}\right\} \quad N_a > N_d. \tag{28.41}$$

Equation (28.41) is quite important in the theory of semiconducting devices (Chapter 29). It asserts that the net excess of electrons (or holes) $N_d - N_a$ introduced by the impurities is almost entirely donated to the conduction (or valence) band; the other band has the very much smaller carrier density $n_i^2/(N_d - N_a)$, as required by the law of mass action, (28.24).

If the temperature is too low (or the impurity concentration too high), condition (28.36) eventually fails to hold, and either n_d/N_d or p_a/N_a (but not both) ceases to be negligible, i.e., one of the impurity types is no longer fully ionized by thermal excitation. As a result, the dominant carrier density declines with decreasing temperature (Figure 28.13).[25]

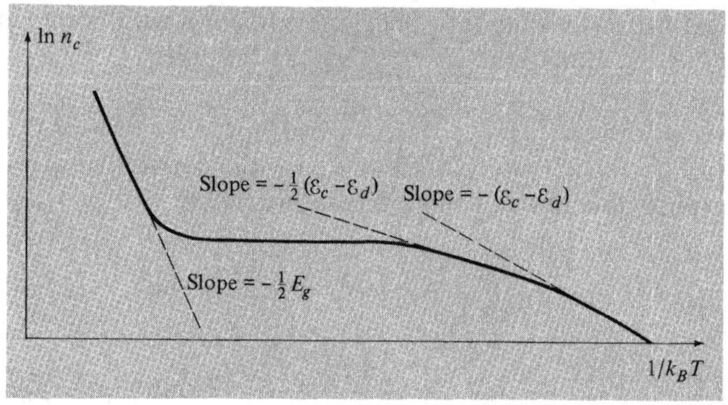

Figure 28.13
Temperature dependence of the majority carrier density (for the case $N_d > N_a$). The two high-temperature regimes are discussed in the text; the very low-temperature behavior is described in Problem 6.

IMPURITY BAND CONDUCTION

As the temperature approaches zero, so does the fraction of ionized impurities, and therefore also the density of carriers in the conduction or valence bands. Nevertheless, some small residual conductivity is observed even at the lowest temperatures. This is because the wave function of an electron (or hole) bound to an impurity site has considerable spatial extent, and therefore the overlap of wave functions at different impurity sites is possible even at fairly low concentrations. When this overlap is not negligible, it is possible for an electron to tunnel from one site to another. The resulting transport of charge is known as "impurity band conduction."

The use of the term "band" in this context is based on an analogy with the tight-binding method (Chapter 10), which shows that a set of atomic levels with a single energy can broaden into a band of energies, when wave function overlap is taken into account. The impurities, however, are usually not situated at the sites of a Bravais lattice, and one must therefore be cautious in attributing to the impurity "bands" features associated with electronic bands in *periodic* potentials.[26]

---

[25] This behavior is described more quantitatively in Problem 6.

[26] The problem of electronic behavior in aperiodic potentials (which arises not only in connection with impurity bands, but also, for example, in the case of disordered alloys) is still in its infancy, and is one of the very lively areas of current research in solid state physics.

THE THEORY OF TRANSPORT IN NONDEGENERATE SEMICONDUCTORS

It is a straightforward consequence (Problem 7) of Fermi-Dirac statistics and the nondegeneracy assumption (28.10) that the thermal equilibrium velocity distribution for electrons near a particular conduction band minimum (or holes near a particular valence band maximum) has the form:

$$f(\mathbf{v}) = n \frac{|\det \mathbf{M}|^{1/2}}{(2\pi k_B T)^{3/2}} \exp\left\{-\frac{\beta}{2}\sum_{\mu\nu} v_\mu \mathbf{M}_{\mu\nu} v_\nu\right\}, \tag{28.42}$$

where n is their contribution to the total carrier density.

This is just the form assumed by the thermal equilibrium molecular velocity distribution in a classical gas, with two exceptions:

1. In a classical gas, the density of molecules n is specified; in a semiconductor, n is an extremely sensitive function of temperature.
2. In a classical gas the mass tensor $\mathbf{M}$ is diagonal.

As a result, the theory of transport in a nondegenerate semiconductor is similar to the theory of transport in a classical gas of several charged components,[27] and many results of the classical theory can be applied directly to semiconductors, when allowance is made for the temperature dependence of the carrier densities and tensor character of the mass. For example, the anomalously high thermopower of a semiconductor (page 563) is only anomalous in comparison with the thermopower of metals; it is quite in accord with the properties of a classical charged gas. Indeed, the thermopower of metals was considered anomalously low in the early days of electron theory, before it was realized that metallic electrons must be described by Fermi-Dirac, rather than classical, statistics.

PROBLEMS

1. *Cyclotron Resonance in Semiconductors*

(a) Show that the formulas (28.6) and (28.7) for the cyclotron resonance frequency follow from substituting the oscillatory velocity (28.5) into the semiclassical equation of motion (28.4), and requiring that the resulting homogeneous equation have a nonzero solution.

(b) Show that (28.7) and (28.8) are equivalent representations of the cyclotron effective mass by evaluating (28.7) in the coordinate system in which the mass tensor $\mathbf{M}$ is diagonal.

2. *Interpretation of Cyclotron Resonance Data*

(a) Compare the cyclotron resonance signal from silicon in Figure 28.9b with the geometry of the conduction band ellipsoids shown in Figure 28.5, and explain why there are only two electron peaks although there are six pockets of electrons.

---

[27] Such a theory was extensively developed by Lorentz, as an attempt at refining the Drude model of metals. Although Lorentz's theory requires substantial modification to be applicable to metals (i.e., the introduction of degenerate Fermi-Dirac statistics and band structure), many of his results can be applied to the description of nondegenerate semiconductors with very little alteration.

(b) Verify that the positions of the electron resonances in Figure 28.9b are consistent with the electron effective masses given for silicon on page 569 and the formulas, (28.6) and (28.8), for the resonance frequency.

(c) Repeat (a) for the resonance in germanium (Figure 28.9a), noting that Figure 28.7 shows four electron pockets.

(d) Verify that the positions of the electron resonances in Figure 28.9a are consistent with the electron effective masses given for germanium on page 569.

3. *Level Density for Ellipsoidal Pockets*

(a) Show that the contribution of an ellipsoidal pocket of electrons to the conduction band density of levels $g_c(\mathcal{E})$, is given by $(d/d\mathcal{E})h(\mathcal{E})$, where $h(\mathcal{E})$ is the number of levels per unit volume in the pocket with energies less than $\mathcal{E}$.

(b) Show, similarly, that the contribution of an ellipsoidal pocket of holes to the valence band density of levels $g_v(\mathcal{E})$ is given by $(d/d\mathcal{E})h(\mathcal{E})$, where $h(\mathcal{E})$ is the number of electronic levels per unit volume in the pocket with energies greater than $\mathcal{E}$.

(c) Using the fact that a volume Ω of k-space contains $\Omega/4\pi^3$ electronic levels per cubic centimeter and the formula $V = (4\pi/3)abc$ for the volume of the ellipsoid $x^2/a^2 + y^2/b^2 + z^2/c^2 = 1$, show that formulas (28.14) follow directly from (a) and (b), when the conduction (or valence) band has a single ellipsoidal pocket.

4. *Statistics of Donor Levels*

(a) Show that if the energy of a doubly occupied donor level is taken to be $2\mathcal{E}_d + \Delta$, then Eq. (28.32) must be replaced by

$$n_d = N_d \frac{1 + e^{-\beta(\mathcal{E}_d - \mu + \Delta)}}{\frac{1}{2}e^{\beta(\mathcal{E}_d - \mu)} + 1 + \frac{1}{2}e^{-\beta(\mathcal{E}_d - \mu + \Delta)}}. \tag{28.43}$$

(b) Verify that Eq. (28.43) reduces to (28.32) as $\Delta \to \infty$, and that it reduces to the expected result for independent electrons as $\Delta \to 0$.

(c) Consider a donor impurity with many bound electronic orbital levels, with energies $\mathcal{E}_i$. Assuming that the electron-electron Coulomb repulsion prohibits more than a single electron from being bound to the impurity, show that the appropriate generalization of (28.32) is

$$\frac{N_d}{1 + \frac{1}{2}(\sum e^{-\beta(\mathcal{E}_i - \mu)})^{-1}} \tag{28.44}$$

Indicate how (if at all) this alters the results described on pages 582–584.

5. *Constraint on Carrier Densities in p-Type Semiconductors*

Describe the electronic configuration of a doped semiconductor as $T \to 0$, when $N_a > N_d$. Explain why (28.35) (derived in the text when $N_d \geqslant N_a$) continues to give a correct constraint on the electron and hole densities at nonzero temperatures, when $N_a > N_d$.

6. *Carrier Statistics in Doped Semiconductors at Low Temperatures*

Consider a doped semiconductor with $N_d > N_a$. Assume that the nondegeneracy condition (28.10) holds, but that $(N_d - N_a)/n_i$ is so large that (28.39) does not necessarily yield a value of μ compatible with (28.36).

(a) Show under these conditions that p_v is negligible compared with n_c, and p_a is negligible compared with N_a, so that the chemical potential is given by the quadratic equation

$$N_c e^{-\beta(\mathcal{E}_c - \mu)} = N_d - N_a - \frac{N_d}{\frac{1}{2}e^{\beta(\mathcal{E}_d - \mu)} + 1}. \tag{28.45}$$

(b) Deduce from this that if the temperature drops so low that n_c ceases to be given by $N_d - N_a$ (Eq. (28.41)), then there is a transition to a regime in which

$$n_c = \sqrt{\frac{N_c(N_d - N_a)}{2}}\, e^{-\beta(\mathcal{E}_c - \mathcal{E}_d)/2}. \tag{28.46}$$

(c) Show that as the temperature drops still lower, there is another transition to a regime in which

$$n_c = \frac{N_c(N_d - N_a)}{N_a}\, e^{-\beta(\mathcal{E}_c - \mathcal{E}_d)}. \tag{28.47}$$

(d) Derive the results analogous to (28.45)–(28.47) when $N_a > N_d$.

7. *Velocity Distribution for Carriers in an Ellipsoidal Pocket*
Derive the velocity distribution (28.42) from the **k**-space distribution function

$$f(\mathbf{k}) \propto \frac{1}{e^{\beta(\mathcal{E}(\mathbf{k}) - \mu)} + 1}, \tag{28.48}$$

by assuming the nondegeneracy condition (28.10), changing from the variable **k** to the variable **v**, and noting that the contribution of the pocket to the carrier density is just $n = \int d\mathbf{v}\, f(\mathbf{v})$.

29

Inhomogeneous Semiconductors

As used by music lovers, the term "solid state physics" refers only to the subject of inhomogeneous semiconductors, and it would be more accurate if it were this latter phrase that festooned the brows of countless tuners and amplifiers. The prevailing usage reflects the fact that modern solid state physics has had its most dramatic and extensive technological consequences through the electronic properties of semiconducting devices. These devices use semiconducting crystals in which the concentrations of donor and acceptor impurities have been made nonuniform in a carefully controlled manner. We shall not attempt to survey here the great variety of semiconducting devices, but will only describe the broad physical principles that underlie their operation. These principles come into play in determining how the densities and currents of electrons and holes are distributed in an inhomogeneous semiconductor, both in the absence and in the presence of an applied electrostatic potential.

The inhomogeneous semiconductors of interest are, ideally, single crystals in which the local concentration of donor and acceptor impurities varies with position. One way to make such crystals is to vary the concentration of impurities in the "melt" as the growing crystal is slowly extracted, thus producing a variation in impurity concentration along one spatial direction. Delicate methods of fabrication are needed because it is generally important, for efficient operation, that there be no great increase in electronic scattering associated with the variation in impurity concentration.

We shall illustrate the physics of inhomogeneous semiconductors by considering the simplest example, the *p-n* junction. This is a semiconducting crystal in which the impurity concentration varies only along a given direction (taken to be the x-axis) and only in a small region (taken to be around $x = 0$). For negative x the crystal has a preponderance of acceptor impurities (i.e., it is *p*-type) while for positive x it has a preponderance of donor impurities (i.e., it is *n*-type) (Figure 29.1). The manner in which the densities of donors and acceptors $N_d(x)$ and $N_a(x)$ vary with position is

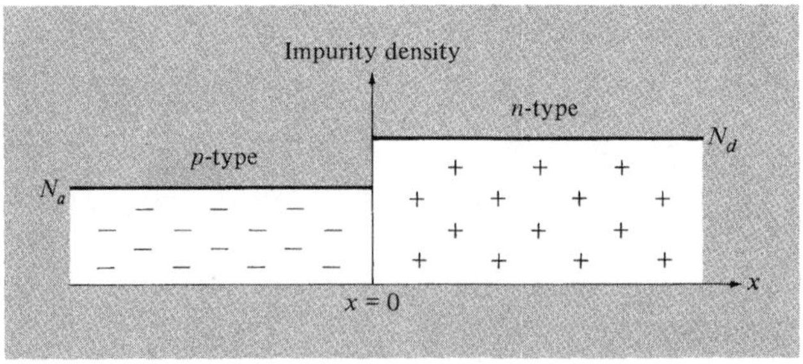

Figure 29.1
The impurity densities along a *p-n* junction in the case of an "abrupt junction," for which donor impurities dominate at positive x, and acceptor impurities at negative x. The donors are represented by (+) to indicate their charge when ionized, and the acceptors by (−). For a junction to be abrupt, the region about $x = 0$ where the impurity concentrations change must be narrow compared with the "depletion layer" in which the carrier densities are nonuniform. (Typical plots of the carrier densities are superimposed on this figure in Figure 29.3.)

called the "doping profile." The term "junction" is used to refer both to the device as a whole and, more specifically, to the transition region about $x = 0$ in which the doping profile is nonuniform.

As we shall see below, the nonuniformity in impurity concentrations induces a nonuniformity in the densities $n_c(x)$ and $p_v(x)$ of conduction band electrons and valence band holes, which in turn gives rise to a potential $\phi(x)$. The region in which these carrier densities are nonuniform is known as the "depletion layer" (or "space-charge region"). The depletion layer can extend for a range of about 10^2 to 10^4 Å around the (generally more narrow) transition region in which the doping profile varies, as we shall see below. Within the depletion layer, except near its boundaries, the total density of carriers is very much less than it is in the homogeneous regions farther away from the transition region. The existence of a depletion layer is one of the crucial properties of the *p-n* junction. One of our main concerns will be to explain why such a layer is induced by the variation in impurity concentrations, and how its structure changes with the application of an external potential V.

For simplicity we shall consider here only "abrupt junctions," in which the transition region is so sharp that variation in impurity concentrations[1] can be represented by a single discontinuous change at $x = 0$:

$$N_d(x) = \begin{Bmatrix} N_d, & x > 0 \\ 0, & x < 0 \end{Bmatrix},$$

$$N_a(x) = \begin{Bmatrix} 0, & x > 0 \\ N_a, & x < 0 \end{Bmatrix}. \tag{29.1}$$

Abrupt junctions are not only conceptually the simplest, but also the type of greatest practical interest. How sharp the actual transition region must be made for (29.1) to give a reasonable model of a physical junction will emerge in the analysis below. We shall find that a junction may be regarded as abrupt if the transition region in the actual doping profile is small in extent compared with the depletion layer. In most cases this permits the transition region to extend for 100 Å or more. A junction that cannot be treated as abrupt is called a "graded junction."

THE SEMICLASSICAL MODEL

To calculate the response of an inhomogeneous semiconductor to an applied electrostatic potential, or even to compute the distribution of electric charge in the absence of an applied potential, one almost always uses the semiclassical model of Chapter 12. When a potential $\phi(x)$ is superimposed on the periodic potential of the crystal, the semiclassical model treats the electrons in the nth band as classical particles (i.e., as wave packets) governed by the Hamiltonian

$$H_n = \mathcal{E}_n\left(\frac{\mathbf{p}}{\hbar}\right) - e\phi(x). \tag{29.2}$$

---

[1] It is not essential that there be only donor impurities in the n-type region and only acceptor impurities in the p-type region. It suffices for each impurity type to be the dominant one in its own region. In what follows N_d may be viewed as the excess density of donors over acceptors and N_a as the excess density of acceptors over donors.

Such a treatment is valid provided that the potential $\phi(x)$ varies sufficiently slowly. How slow this variation must be is, in general, a very difficult question to answer. At the very least, one requires that the change in electrostatic energy $e\Delta\phi$ over a distance of the order of the lattice constant be small compared with the band gap E_g, but the condition may well be even more stringent than this.[2] In the case of the *p-n* junction the potential ϕ has almost all of its spatial variation within the depletion layer. There, as we shall see, the energy $e\phi$ changes by about E_g, over a distance that is typically a few hundred angstroms or more (so that the field in the depletion layer can be as large as 10^6 volts per meter). Although this satisfies the minimum necessary condition for the validity of the semiclassical model (the change in $e\phi$ over a lattice constant is no more than a fraction of a percent of E_g), the variation is strong enough that one cannot exclude the possibility that the semiclassical description may break down in the depletion layer. Thus one should bear in mind the possibility that the field in the depletion layer may be strong enough to induce tunneling of electrons from valence band to conduction band levels, leading to a conductivity considerably in excess of the semiclassical prediction.

Having issued this warning, however, we shall follow the general practice of assuming the validity of the semiclassical description, so that we may explore its consequences. Before describing the semiclassical theory of the currents that flow in a *p-n* junction in the presence of an applied potential, we first examine the case of the *p-n* junction in thermal equilibrium, in the absence of applied potentials and current flow.

THE *p-n* JUNCTION IN EQUILIBRIUM

We wish to determine the carrier densities and the electrostatic potential $\phi(x)$ induced by the nonuniform doping. We assume that nondegenerate conditions hold throughout the material, so that the carrier densities at each position x have the "Maxwellian" forms analogous to the densities (28.12) we found in the uniform case. In the non-uniform case, to derive the carrier densities at position x along the junction in the presence of a potential $\phi(x)$, the semiclassical procedure is simply to repeat the analysis for the uniform case, but using the semiclassical one-electron energy (29.2), in which each level is shifted by $-e\phi(x)$. Using the forms (28.3) of the $\mathcal{E}(\mathbf{k})$ appropriate to levels near the conduction band minimum or valence band maximum, we see that the effect of this is simply to shift the constants $\mathcal{E}_c$ and $\mathcal{E}_v$ by $-e\phi(x)$. Thus Eq. (28.12) for the equilibrium carrier densities is generalized to

$$n_c(x) = N_c(T)\exp\left\{-\frac{[\mathcal{E}_c - e\phi(x) - \mu]}{k_BT}\right\},$$

$$p_v(x) = P_v(T)\exp\left\{-\frac{[\mu - \mathcal{E}_v + e\phi(x)]}{k_BT}\right\}. \tag{29.3}$$

The potential $\phi(x)$ must be determined self-consistently, as that potential arising (via Poisson's equation) when the carrier densities have the forms (29.3). We examine

---

[2] A crude argument appropriate to metals is given in Appendix J. Analogous arguments (of comparable crudity) can be developed for semiconductors.

this problem in the special case (again, the case of major practical interest) in which far from the transition region on either side extrinsic conditions prevail, in which the impurities are fully "ionized" (see pages 583–584). Thus far away on the n-side the density of conduction band electrons is very nearly equal to the density N_d of donors, while far away on the p-side the density of valence band holes is very nearly equal to the density N_a of acceptors:

$$N_d = n_c(\infty) = N_c(T) \exp\left\{-\frac{[\mathcal{E}_c - e\phi(\infty) - \mu]}{k_B T}\right\},$$
$$N_a = p_v(-\infty) = P_v(T) \exp\left\{-\frac{[\mu - \mathcal{E}_v + e\phi(-\infty)]}{k_B T}\right\}. \qquad \textbf{(29.4)}$$

Since the entire crystal is in thermal equilibrium, the chemical potential does not vary with position. In particular, the same value of μ appears in either of Eqs. (29.4); this immediately requires that the total potential drop across the junction be given by[3]

$$e\phi(\infty) - e\phi(-\infty) = \mathcal{E}_c - \mathcal{E}_v + k_B T \ln\left[\frac{N_d N_a}{N_c P_v}\right], \qquad \textbf{(29.5)}$$

or

$$\boxed{e\,\Delta\phi = E_g + k_B T \ln\left[\frac{N_d N_a}{N_c P_v}\right].} \qquad \textbf{(29.6)}$$

An alternative way of representing the information in (29.3) and (29.6) is sometimes helpful. If we define a position-dependent "electrochemical potential" $\mu_e(x)$ by

$$\mu_e(x) = \mu + e\phi(x), \qquad \textbf{(29.7)}$$

then we may write the carrier densities (29.3) as

$$n_c(x) = N_c(T) \exp\left\{-\frac{[\mathcal{E}_c - \mu_e(x)]}{k_B T}\right\},$$
$$p_v(x) = P_v(T) \exp\left\{-\frac{[\mu_e(x) - \mathcal{E}_v]}{k_B T}\right\}. \qquad \textbf{(29.8)}$$

These have precisely the form of the relations (28.12) for a homogeneous semiconductor, except that the constant chemical potential μ is replaced by the electrochemical potential $\mu_e(x)$. Thus $\mu_e(\infty)$ is the chemical potential of a homogeneous n-type crystal whose properties are identical to those of the inhomogeneous crystal far on the n-side of the transition region, while $\mu_e(-\infty)$ is the chemical potential of a homogeneous p-type crystal identical to the inhomogeneous crystal far on the p-side. The relation (29.6) can equally well be written as[4]

$$e\,\Delta\phi = \mu_e(\infty) - \mu_e(-\infty). \qquad \textbf{(29.9)}$$

---

[3] The derivation of (29.5) requires the validity of (29.3) only far from the depletion layer, where ϕ is indeed slowly varying. It therefore holds even when the semiclassical model fails in the transition region.

[4] This follows directly from (29.7). Equation (29.9) is sometimes summarized in the rule that the total potential drop is such as to bring the "Fermi levels at the two ends of the junction" into coincidence. This point of view is evidently inspired by the representation of Figure 29.2b.

Figure 29.2a shows the electrochemical potential plotted as a function of position along the *p-n* junction. We have assumed (as will be demonstrated below) that ϕ varies monotonically from one end to the other. Figure 29.2b shows an alternative representation of the same information in which the potential ϕ giving the position dependence in (29.3) is regarded as shifting $\mathcal{E}_c$ (or $\mathcal{E}_v$) rather than μ. In either case, the significance of the diagrams is that at any particular position x along the junction, the carrier densities are those that would be found in a piece of homogeneous material with the impurity concentrations prevailing at x, and with a chemical potential that is positioned with respect to the band edges as shown in the vertical section of the diagrams at x.

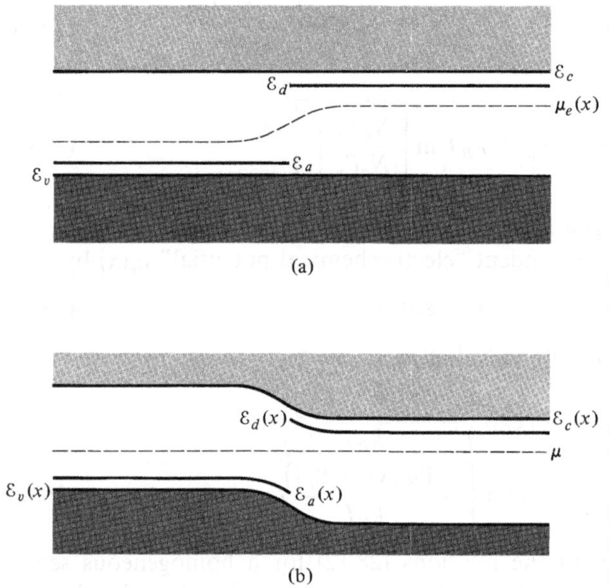

Figure 29.2
Two equivalent ways of representing the effect of the internal potential $\phi(x)$ on the electron and hole densities of a *p-n* junction. (a) The electrochemical potential $\mu_e(x) = \mu + e\phi(x)$ is plotted along the *p-n* junction. The carrier densities at any point x are those that would be found in a uniform semiconductor characterized by the fixed band and impurity energies $\mathcal{E}_c$, $\mathcal{E}_v$, $\mathcal{E}_d$, and $\mathcal{E}_a$, at a chemical potential equal to $\mu_e(x)$. (b) Here $\mathcal{E}_c(x) = \mathcal{E}_c - e\phi(x)$ is the energy of an electron wave packet localized about x formed from levels very near the conduction band minimum, and similarly for $\mathcal{E}_v(x)$. The energies of the local impurity levels are $\mathcal{E}_d(x) = \mathcal{E}_d - e\phi(x)$ and $\mathcal{E}_a(x) = \mathcal{E}_a - e\phi(x)$. The (constant) chemical potential is also shown. The carrier densities at any point x are those that would be found in a uniform semiconductor characterized by band and impurity energies equal to $\mathcal{E}_c(x)$, $\mathcal{E}_d(x)$, $\mathcal{E}_a(x)$, and $\mathcal{E}_v(x)$ at the fixed chemical potential μ.

Equation (29.6) (or its equivalent form, (29.9)) serves as the boundary condition in a differential equation determining the potential $\phi(x)$. The differential equation is simply Poisson's equation,[5]

$$-\nabla^2\phi = -\frac{d^2\phi}{dx^2} = \frac{4\pi\rho(x)}{\epsilon}, \tag{29.10}$$

[5] Here ϵ is the static dielectric constant of the semiconductor. The use of the macroscopic equation is possible because ϕ varies over the depletion layer, which is large on the interatomic scale.

relating the potential $\phi(x)$ to the charge distribution $\rho(x)$ giving rise to it. To express $\rho(x)$ in terms of ϕ and get a closed equation, we first note that if (as we have assumed) the impurities are fully ionized far from the junction, then they will remain fully ionized[6] at all x. Consequently the charge density due to the impurities and the carriers is[7]

$$\rho(x) = e[N_d(x) - N_a(x) - n_c(x) + p_v(x)]. \tag{29.11}$$

When the carrier and impurity densities (29.3) and (29.1) are substituted into the form (29.11) for the charge density, and the result is substituted into Poisson's equation (29.10), one finds a nonlinear differential equation for $\phi(x)$ whose exact solution usually requires numerical techniques.[8] However, a quite reasonable description of $\phi(x)$ may be had by exploiting the fact that the total change in $e\phi$ is of order $E_g \gg k_BT$. The relevance of this fact emerges when we combine (29.3) and (29.4) to write

$$\begin{aligned} n_c(x) &= N_d e^{-e[\phi(\infty) - \phi(x)]/k_BT}, \\ p_v(x) &= N_a e^{-e[\phi(x) - \phi(-\infty)]/k_BT}. \end{aligned} \tag{29.12}$$

Suppose that the change in ϕ occurs within a region $-d_p \leqslant x \leqslant d_n$. Outside of this region, ϕ has its asymptotic value, and therefore $n_c = N_d$ on the n side, $p_v = N_a$ on the p side, and $\rho = 0$. Within the region, except quite near the boundaries, $e\phi$ differs by many k_BT from its asymptotic value, so $n_c \ll N_d$, $p_v \ll N_a$. Thus, except in the vicinity of $x = -d_p$ and $x = d_n$, the charge density (29.11) between $-d_p$ and d_n is quite accurately given by $\rho(x) = e[N_d(x) - N_a(x)]$, there being no appreciable carrier charge to cancel the charges of the "ionized" impurities. The points $x = -d_p$ and $x = d_n$ therefore mark the boundaries of the depletion layer.

Combining these observations, and using the form (29.1) for the impurity densities, we find that except for x just greater than $-d_p$ or just less than d_n, Poisson's equation is well approximated by

$$\phi''(x) = \begin{cases} 0, & x > d_n, \\ \dfrac{-4\pi e N_d}{\epsilon}, & d_n > x > 0, \\ \dfrac{4\pi e N_a}{\epsilon}, & 0 > x > -d_p, \\ 0, & -d_p > x. \end{cases} \tag{29.13}$$

---

6 If ϕ is monotonic (as we shall find below) this follows from the fact that the degree of ionization of an impurity increases, the farther the chemical potential is from the impurity level. See Figure 29.2 and Eqs. (28.32) and (28.34).

7 The density of holes on the far n-side has the very small value $p_v(\infty) = n_i^2/N_d$ required by the law of mass action. However, the density of electrons on the far n-side actually exceeds N_d by this same small amount so as to insure that $n_c(\infty) - p_v(\infty) = N_d$. In computing the total charge density, if we ignore this small correction in n_c (as we have done in writing (29.4)), then we should also ignore the small compensating density of holes on the far n-side. Similar remarks apply to the small concentration of electrons on the far p-side. These "minority carrier densities" have negligible effect on the total balance of charge. We shall see below, however, that they play an important role in determining the flow of currents in the presence of an applied potential.

8 Some aspects of that equation are investigated in Problem 1.

This immediately integrates to give

$$\phi(x) = \begin{cases} \phi(\infty), & x > d_n, \\ \phi(\infty) - \left(\dfrac{2\pi e N_d}{\epsilon}\right)(x - d_n)^2, & d_n > x > 0, \\ \phi(-\infty) + \left(\dfrac{2\pi e N_a}{\epsilon}\right)(x + d_p)^2, & 0 > x > -d_p, \\ \phi(-\infty), & x < -d_p. \end{cases} \tag{29.14}$$

The boundary conditions (continuity of ϕ and its first derivative) are explicitly obeyed by the solution (29.14) at $x = -d_p$ and $x = d_n$. Requiring them to hold at $x = 0$ gives two additional equations that determine the lengths d_n and d_p. Continuity of ϕ' at $x = 0$ implies that

$$N_d d_n = N_a d_p, \tag{29.15}$$

which is just the condition that the excess of positive charge on the n-side of the junction be equal to the excess of negative charge on the p-side. Continuity of ϕ at $x = 0$ requires that

$$\left(\frac{2\pi e}{\epsilon}\right)(N_d d_n^2 + N_a d_p^2) = \phi(\infty) - \phi(-\infty) = \Delta\phi. \tag{29.16}$$

Together with (29.15) this determines the lengths d_n and d_p:

$$d_{n,p} = \left\{\frac{(N_a/N_d)^{\pm 1}}{(N_d + N_a)} \frac{\epsilon\Delta\phi}{2\pi e}\right\}^{1/2}. \tag{29.17}$$

To estimate the sizes of these lengths we may write Eq. (29.17) in the numerically more convenient form

$$d_{n,p} = 105\left\{\frac{(N_a/N_d)^{\pm 1}}{10^{-18}(N_d + N_a)}[\epsilon e\,\Delta\phi]_{\text{eV}}\right\}^{1/2} \text{Å}. \tag{29.18}$$

The quantity $\epsilon e\,\Delta\phi$ is typically of order 1 eV, and since typical impurity concentrations are in the range from 10^{14} to 10^{18} per cubic centimeter, the lengths d_n and d_p, which give the extent of the depletion layer, will generally be from 10^4 to 10^2 Å. The field within the depletion layer is of order $\Delta\phi/(d_n + d_p)$, and for d's of this size is therefore in the range from 10^5 to 10^7 volts per meter, for an energy gap of 0.1 eV.

The resulting picture of the depletion layer is shown in Figure 29.3. The potential ϕ varies monotonically through the layer, as asserted above. Except at the boundaries of the layer, the carrier concentrations are negligible compared with the impurity concentrations, so the charge density is that of the ionized impurities. Outside of the depletion layer the carrier concentrations balance the impurity concentrations, and the charge density is zero.

The mechanism establishing such a region of sharply reduced carrier densities is relatively simple. Suppose that one initially were able to impose carrier concentrations that gave charge neutrality at every point in the crystal. Such a configuration could not be maintained, for electrons would begin to diffuse from the n-side (where their concentration was high) to the p-side (where their concentration was very low), and

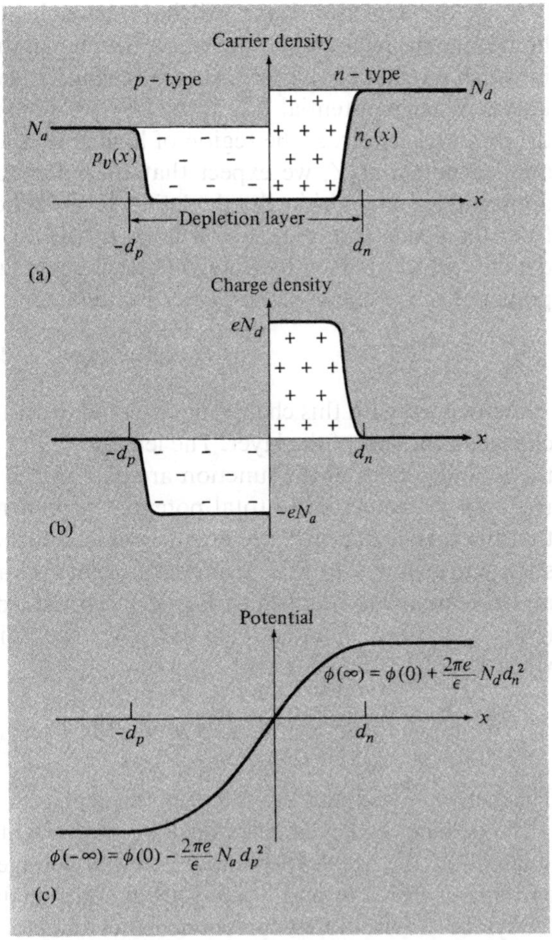

Figure 29.3
(a) Carrier densities, (b) charge density, and (c) potential $\phi(x)$ plotted vs. position across an abrupt *p-n* junction. In the analysis in the text the approximation was made that the carrier densities and charge density are constants except for discontinuous changes at $x = -d_p$ and $x = d_n$. More precisely (see Problem 1), these quantities undergo rapid change over regions just within the depletion layer whose extent is a fraction of order $(k_B T/E_g)^{1/2}$ of the total extent of the depletion layer. The extent of the depletion layer is typically from 10^2 to 10^4 Å.

holes would diffuse in the opposite direction. As this diffusion continued, the resulting transfer of charge would build up an electric field opposing further diffusive currents, until an equilibrium configuration was reached in which the effect of the field on the currents precisely canceled the effect of diffusion. Because the carriers are highly mobile, in this equilibrium configuration the carrier densities are very low wherever the field has an appreciable value. This is precisely the state of affairs depicted in Figure 29.3.

ELEMENTARY PICTURE OF RECTIFICATION BY A *p-n* JUNCTION

We now consider the behavior of a *p-n* junction when an external voltage V is applied. We shall take V to be positive if its application raises the potential of the *p*-side with respect to the *n*-side. When $V = 0$ we found above that there is a depletion layer some 10^2 to 10^4 Å in extent about the transition point where the doping changes from *p*-type to *n*-type, in which the density of carriers is reduced greatly below its value in the homogeneous regions farther away. Because of its greatly reduced carrier density, the depletion layer will have a much higher electrical resistance than the homogeneous regions, and the whole device can therefore be viewed as a series circuit in which a relatively high resistance is sandwiched between two relatively low resistances. When a potential V is applied across such a circuit, almost all of the potential drop will occur across the region of high resistance. Thus even in the presence of an applied potential V, we expect that the potential $\phi(x)$ along the device will vary appreciably only within the depletion layer. When $V = 0$ we found that $\phi(x)$ rose from the p-side of the depletion layer to the n-side by the amount (which we now denote by $(\Delta\phi)_0$) given by Eq. (29.6), so we conclude that when $V \neq 0$ the change in potential across the depletion layer is modified to

$$\Delta\phi = (\Delta\phi)_0 - V. \tag{29.19}$$

Associated with this change in potential drop across the depletion layer, there is a change in the size of the layer. The lengths d_n and d_p giving the extent of the layer on the n- and p-sides of the junction are determined by Eqs. (29.15) and (29.16), which use only the value of the total potential drop across the layer, and the assumption that the carrier densities are greatly reduced throughout almost all of the layer. We shall find below that this assumption remains valid when $V \neq 0$, and therefore d_n and d_p continue to be given by Eq. (29.17) provided that we take the value of $\Delta\phi$ to be $(\Delta\phi)_0 - V$. Since d_n and d_p vary as $(\Delta\phi)^{1/2}$ according to Eq. (29.17), we conclude that when $V \neq 0$,

$$d_{n,p}(V) = d_{n,p}(0)\left[1 - \frac{V}{(\Delta\phi)_0}\right]^{1/2}. \tag{29.20}$$

This behavior of ϕ and the extent of the depletion layer are illustrated in Figure 29.4.

To deduce the dependence on V of the current that flows when a p-n junction is "biased" by the application of an external voltage, we must consider separately the currents of electrons and holes. Throughout the discussion that follows we shall use the symbol J for number current densities and j for electrical current densities, so that

$$j_e = -eJ_e, \quad j_h = eJ_h. \tag{29.21}$$

When $V = 0$, both J_e and J_h vanish. This does not, of course, mean that no individual carriers flow across the junction, but only that as many electrons (or holes) flow in one direction as in the other. When $V \neq 0$, this balance is disrupted. Consider, for example, the current of holes across the depletion layer. This has two components:

1. A hole current flows from the n- to the p-side of the junction, known as the hole *generation current*. As the name indicates, this current arises from holes that are generated just on the n-side of the depletion layer by the thermal excitation of electrons out of valence band levels. Although the density of such holes on the n-side ("minority carriers") is minute compared with the density of electrons ("majority carriers"), they play an important role in carrying current across the junction. This is because any such hole that wanders into the depletion layer is immediately swept over to the p-side of the junction by the strong electric field

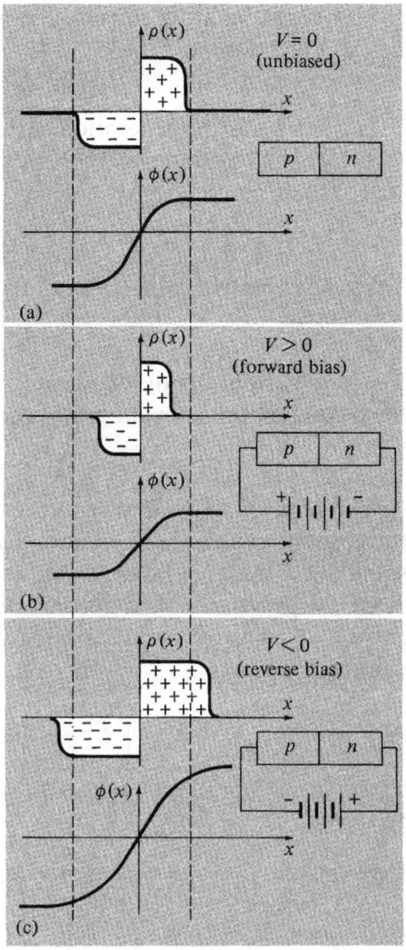

Figure 29.4
The charge density ρ and potential ϕ in the depletion layer (a) for the unbiased junction, (b) for the junction with $V > 0$ (forward bias), and (c) for the junction with $V < 0$ (reverse bias). The positions $x = d_n$ and $x = -d_p$ that mark the boundaries of the depletion layer when $V = 0$ are given by the dashed lines. The depletion layer and change in ϕ are reduced by a forward bias and increased by a reverse bias.

that prevails within the layer. The resulting generation current is insensitive to the size of the potential drop across the depletion layer, since any hole, having entered the layer from the *n*-side, will be swept through to the *p*-side.[9]

2. A hole current flows from the *p*- to the *n*-side of the junction, known as the hole *recombination current*.[10] The electric field in the depletion layer acts to oppose such a current, and only holes that arrive at the edge of the depletion layer with a thermal energy sufficient to surmount the potential barrier will contribute to

[9] The density of holes giving rise to the hole generation current will also be insensitive to the size of V, provided that eV is small compared with E_g, for this density is entirely determined by the law of mass action and the density of electrons. The latter density differs only slightly from the value N_c outside of the depletion layer when eV is small compared with E_g, as will emerge from the more detailed analysis below.

[10] So named because of the fate suffered by such holes upon arriving on the *n*-side of the junction, where one of the abundant electrons will eventually drop into the empty level that constitutes the hole.

the recombination current. The number of such holes is proportional to $e^{-e\Delta\phi/k_BT}$, and therefore[11]

$$J_h^{\text{rec}} \propto e^{-e[(\Delta\phi)_0 - V]/k_BT}. \tag{29.22}$$

In contrast to the generation current, the recombination current is highly sensitive to the applied voltage V. We can compare their magnitudes by noting that when $V = 0$ there can be no net hole current across the junction:

$$J_h^{\text{rec}}\big|_{V=0} = J_h^{\text{gen}}. \tag{29.23}$$

Taken together with Eq. (29.22), this requires that

$$J_h^{\text{rec}} = J_h^{\text{gen}} e^{eV/k_BT}. \tag{29.24}$$

The total current of holes flowing from the p- to the n-side of the junction is given by the recombination current minus the generation current:

$$J_h = J_h^{\text{rec}} - J_h^{\text{gen}} = J_h^{\text{gen}}(e^{eV/k_BT} - 1). \tag{29.25}$$

The same analysis applies to the components of the electron current, except that the generation and recombination currents of electrons flow oppositely to the corresponding currents of holes. Since, however, the electrons are oppositely charged, the electrical generation and recombination currents of electrons are parallel to the electrical generation and recombination currents of holes. The total electrical current density is thus:

$$\boxed{j = e(J_h^{\text{gen}} + J_e^{\text{gen}})(e^{eV/k_BT} - 1).} \tag{29.26}$$

This has the highly asymmetric form characteristic of rectifiers, as shown in Figure 29.5.

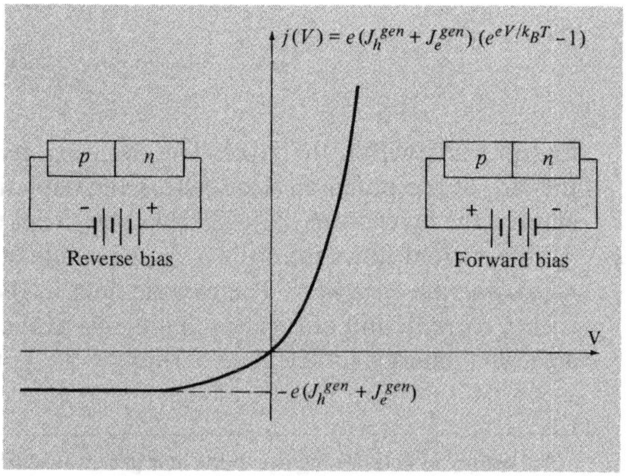

Figure 29.5
Current vs. applied voltage V for a p-n junction. The relation is valid for eV small compared with the energy gap, E_g. The saturation current $(eJ_h^{\text{gen}} + eJ_e^{\text{gen}})$ varies with temperature as e^{-E_g/k_BT}, as established below.

[11] In assuming that (29.22) gives the dominant dependence of the hole recombination current on V, we are assuming that the density of holes just on the p-side of the depletion layer differs only slightly from N_a. We shall find that this is also the case provided that eV is small compared with the energy gap E_g.

GENERAL PHYSICAL ASPECTS OF THE NONEQUILIBRIUM CASE

The foregoing discussion provides no estimate of the size of the prefactor $e(J_h^{\text{gen}} + J_e^{\text{gen}})$ appearing in (29.26). In addition, in the nonequilibrium ($V \neq 0$) case the local carrier densities will not in general be determined by the local potential ϕ through the simple equilibrium Maxwellian relations (29.3). In the nonequilibrium case it requires further analysis to construct a picture of the carrier densities in the neighborhood of the transition region that is comparable in detail to the picture we gave for the equilibrium case.

In this more detailed approach it is not especially helpful to resolve the electron and hole currents across the junction into generation and recombination currents. Instead, at each point x (both inside and outside the depletion layer) we shall write equations relating the total electron and hole currents, $J_e(x)$ and $J_h(x)$, the electron and hole densities, $n_c(x)$ and $p_v(x)$, and the potential $\phi(x)$ (or, equivalently, the electric field, $E(x) = -d\phi(x)/dx$). We shall find five such equations, which will enable us, in principle, to find these five quantities. This method is a direct generalization of the approach we followed in our analysis of the equilibrium ($V = 0$) case. In equilibrium the electron and hole currents vanish, there are only three unknowns, and the three equations we used were Poisson's equation, and the two equations (29.3) that relate $n_c(x)$ and $p_v(x)$ to $\phi(x)$ in thermal equilibrium. Thus the nonequilibrium problem can be viewed as that of finding the appropriate equations to replace the equilibrium relation (29.3), when $V \neq 0$ and currents flow.

We first observe that in the presence of both an electric field and a carrier density gradient, the carrier current density can be written as the sum of a term proportional to the field (the *drift current*) and a term proportional to the density gradient (the *diffusion current*):

$$\boxed{\begin{aligned} J_e &= -\mu_n n_c E - D_n \frac{dn_c}{dx}, \\ J_h &= \mu_p p_v E - D_p \frac{dp_v}{dx}. \end{aligned}} \qquad \textbf{(29.27)}$$

The positive[12] proportionality constants μ_n and μ_p appearing in Eq. (29.27) are known as the electron and hole *mobilities*. We have introduced the mobilities, rather than writing the drift current in terms of conductivities, to make explicit the manner in which the drift current depends on the carrier densities. If only electrons at uniform density are present, then $\sigma E = j = -eJ_e = e\mu_n nE$. Using the Drude form $\sigma = ne^2\tau/m$ for the conductivity (Eq. (1.6)) we find that

$$\mu_n = \frac{e\tau_n^{\text{coll}}}{m_n}, \qquad \textbf{(29.28)}$$

[12] The signs in Eq. (29.27) have been chosen to make the mobilities positive; the hole drift current is along the field, and the electron drift current is opposite to the field.

and, similarly,

$$\mu_p = \frac{e\tau_p^{\text{coll}}}{m_p}, \tag{29.29}$$

where m_n and m_p are the appropriate effective masses, and τ_n^{coll} and τ_p^{coll} are the carrier collision times.[13]

The positive[14] proportionality constants D_n and D_p appearing in Eq. (29.27) are known as the electron and hole *diffusion constants.* They are related to the mobilities by the *Einstein relations*:[15]

$$\boxed{\mu_n = \frac{eD_n}{k_B T}, \quad \mu_p = \frac{eD_p}{k_B T}.} \tag{29.30}$$

The Einstein relations follow directly from the fact that the electron and hole currents must vanish in thermal equilibrium: Only if the mobilities and diffusion constants are related by (29.30) will the currents given by (29.27) be zero when the carrier densities have the equilibrium form (29.3)[16] (as is easily verified by direct substitution of (29.3) into (29.27)).

The relation (29.27) giving the currents in terms of the density gradients and field, together with the forms (29.28)–(29.30) for the mobilities and diffusion constants, can also be derived directly from the kind of simple kinetic argument used in Chapter 1 (see Problem 2).

Note that in thermal equilibrium, Eq. (29.27) and the conditions $J_e = J_h = 0$ contain all information necessary to determine the carrier densities, for when the currents vanish we may integrate Eq. (29.27) to rederive (with the aid of the Einstein relations (29.30)) the thermal equilibrium densities (29.3). When $V \neq 0$ and currents flow, we require a further equation, which can be viewed as the generalization to the nonequilibrium case of the equilibrium conditions of vanishing currents. If the numbers of carriers were conserved, the required generalization would simply be the equations of continuity,

$$\frac{\partial n_c}{\partial t} = -\frac{\partial J_e}{\partial x},$$
$$\frac{\partial p_v}{\partial t} = -\frac{\partial J_h}{\partial x}, \tag{29.31}$$

which express the fact that the change in the number of carriers in a region is entirely determined by the rate at which carriers flow into and out of the region. However, carrier numbers are not conserved. A conduction band electron and a valence band

---

[13] In semiconductors there is another lifetime of fundamental importance (see below), the *recombination time.* The superscript "coll" has been affixed to the collision mean free times to distinguish them from the recombination times.

[14] They are positive because the diffusion current flows from high- to low-density regions. In zero field, Eq. (29.27) is sometimes known as Fick's law.

[15] The Einstein relations are very general, arising in any treatment of charged particles that obey Maxwell-Boltzmann statistics, such as the ions in an electrolytic solution.

[16] The generalization of (29.30) to the degenerate case is described in Problem 3.

hole can be *generated* by the thermal excitation of an electron out of a valence band level. Furthermore a conduction band electron and a valence band hole can *recombine* (i.e., the electron can drop into the empty level that is the hole), resulting in the disappearance of one carrier of each type. Terms must be added to the continuity equations describing these other ways in which the number of carriers in a region can change:

$$\frac{\partial n_c}{\partial t} = \left(\frac{dn_c}{dt}\right)_{g\text{-}r} - \frac{\partial J_e}{\partial x},$$
$$\frac{\partial p_v}{\partial t} = \left(\frac{dp_v}{dt}\right)_{g\text{-}r} - \frac{\partial J_h}{\partial x}. \tag{29.32}$$

To determine the forms of $(dn_c/dt)_{g\text{-}r}$ and $(dp_v/dt)_{g\text{-}r}$ we note that generation and recombination act to restore thermal equilibrium when the carrier densities deviate from their equilibrium values. In regions where n_c and p_v exceed their equilibrium values, recombination occurs faster than generation, leading to a decrease in the carrier densities, while in regions where they fall short of their equilibrium values, generation occurs faster than recombination, leading to an increase in the carrier densities. In the simplest models these processes are described by electron and hole lifetimes,[17] τ_n and τ_p. The rate at which each carrier density changes due to recombination and generation is set proportional to its deviation from the form determined by the other carrier density and the law of mass action (28.24):

$$\left(\frac{dn_c}{dt}\right)_{g\text{-}r} = -\frac{(n_c - n_c^0)}{\tau_n},$$
$$\left(\frac{dp_v}{dt}\right)_{g\text{-}r} = -\frac{(p_v - p_v^0)}{\tau_p}, \tag{29.33}$$

where $n_c^0 = n_i^2/p_v$, and $p_v^0 = n_i^2/n_c$.

To interpret these equations, note that the first, for example, expresses the change in electron carrier density due to generation and recombination in an infinitesimal time dt as

$$n_c(t + dt) = \left(1 - \frac{dt}{\tau_n}\right) n_c(t) + \left(\frac{dt}{\tau_n}\right) n_c^0. \tag{29.34}$$

The first term on the right of Eq. (29.34) expresses the destruction, through recombination, of a fraction dt/τ_n of the electron carriers; i.e., τ_n is the average electronic lifetime before recombination occurs. The second term on the right expresses the creation through thermal generation of n^0/τ_n electron carriers per unit volume, per unit time. Note that, as required, Eqs. (29.33) give carrier densities that decrease when they exceed their equilibrium values, increase when they are less than their equilibrium values, and do not change when they are equal to their equilibrium values.

The lifetimes τ_n and τ_p are generally much longer than overall electron or hole collision times, τ_n^{coll} and τ_p^{coll}, for the recombination (or generation) of an electron

[17] Also known as "recombination times." Conservation of total electric charge requires that the recombination rates be proportional to the densities of the other carrier type: $(1/\tau_n)/(1/\tau_p) = p_v/n_c$.

and a hole is an interband transition (electron goes from valence to conduction band (generation) or conduction band to valence band (recombination)). Ordinary collisions, which conserve the number of carriers, are intraband transitions. Reflecting this, typical lifetimes range between 10^{-3} and 10^{-8} second, while the collision times are similar to those found in metals—i.e., 10^{-12} or 10^{-13} second.

In the presence of a static external potential the *p-n* junction, though not in thermal equilibrium, is in a steady state; i.e., the carrier densities will be constant in time: $dn_c/dt = dp_v/dt = 0$. Using this fact and the forms (29.33) for the rates at which recombination and generation change the carrier densities, we find that the continuity equation (29.32) requires

$$\frac{dJ_e}{dx} + \frac{n_c - n_c^0}{\tau_n} = 0,$$
$$\frac{dJ_h}{dx} + \frac{p_v - p_v^0}{\tau_p} = 0. \qquad (29.35)$$

These are the equations that replace the equilibrium conditions $J_e = J_h = 0$, when $V \neq 0$.

One very important application of Eqs. (29.35) and (29.27) is in regions where the electric field E is negligibly small and the majority carrier density is constant. In that case the minority carrier drift current can be ignored compared with the minority carrier diffusion current, and Eqs. (29.27) and (29.35) reduce to a single equation for the minority carrier density with a constant recombination time:

$$D_n \frac{d^2 n_c}{dx^2} = \frac{n_c - n_c^0}{\tau_n},$$
$$D_p \frac{d^2 p_v}{dx^2} = \frac{p_v - p_v^0}{\tau_p}. \qquad (E \approx 0) \qquad (29.36)$$

The solutions to these equations vary exponentially in x/L, where the lengths

$$L_n = (D_n \tau_n)^{1/2}, \quad L_p = (D_p \tau_p)^{1/2}, \qquad (29.37)$$

are known as the electron and hole *diffusion lengths*. Suppose, for example, (to take a case that will be of some importance below) that we are in the region of uniform potential on the *n*-side of the depletion layer, so that the equilibrium density p_v^0 has the constant value $p_v(\infty) = n_i^2/N_d$. If the density of holes is constrained to have the value $p_v(x_0) \neq p_v(\infty)$ at a point x_0, then the solution to Eq. (29.36) for $x \geqslant x_0$ is

$$p_v(x) = p_v(\infty) + [p_v(x_0) - p_v(\infty)]e^{-(x-x_0)/L_p}. \qquad (29.38)$$

Thus the diffusion length is a measure of the distance it takes for the density to relax back to its equilibrium value.

One would expect that the distance L over which a deviation from equilibrium density can be maintained would be roughly the distance a carrier can travel before undergoing recombination. This is not immediately obvious from the forms (29.37) for the diffusion lengths L_n and L_p, but it is revealed when one rewrites (29.37) using (a) the Einstein relations (29.30) between the diffusion constant and the mobility,

(b) the Drude form (29.28), or (29.29), for the mobility, (c) the relation $\frac{1}{2}mv_{\text{th}}^2 = \frac{3}{2}k_BT$ between the mean square carrier velocity and the temperature under nondegenerate conditions, and (d) the definition $\ell = v_{\text{th}}\tau^{\text{coll}}$ of the carrier mean free path between collisions. Making these substitutions one finds:

$$L_n = \left(\frac{\tau_n}{3\tau_n^{\text{coll}}}\right)^{1/2} \ell_n,$$
$$L_p = \left(\frac{\tau_p}{3\tau_p^{\text{coll}}}\right)^{1/2} \ell_p. \tag{29.39}$$

Assuming that the direction of a carrier is random after each collision, a series of N collisions can be viewed as a random walk of step length ℓ. It is easily shown[18] that in such a walk the total displacement is $N^{1/2}\ell$. Since the number of collisions a carrier can undergo in a recombination time is the ratio of the recombination time to the collision time, Eq. (29.39) does indeed show that the diffusion length measures the distance a carrier can go before undergoing recombination.

Using the typical values given on page 604 for the collision time and the (very much longer) recombination time, we find that (29.39) gives a diffusion length that can be between 10^2 and 10^5 mean free paths.

We can estimate the size of the generation currents that appear in the I-V relation (29.26), in terms of the diffusion lengths and carrier lifetimes. We first note that, by definition of the lifetime, holes are created by thermal generation at a rate p_v^0/τ_p per unit volume. Such a hole stands an appreciable chance of entering the depletion layer (and then being swiftly swept across to the n-side) before undergoing recombination, provided that it is created within a diffusion length L_p of the boundary of the depletion layer. Therefore the flow of thermally generated holes per unit area into the depletion layer per second will be of order $L_p p_v^0/\tau_p$. Since $p_v^0 = n_i^2/N_d$, we have

$$J_h^{\text{gen}} = \left(\frac{n_i^2}{N_d}\right)\frac{L_p}{\tau_p}, \tag{29.40}$$

and, similarly,

$$J_e^{\text{gen}} = \left(\frac{n_i^2}{N_a}\right)\frac{L_n}{\tau_n}. \tag{29.41}$$

The sum of the currents appearing in (29.40) and (29.41) is known as the *saturation current*, since it is the maximum current that can flow through the junction when V is negative ("reversed bias"). Because the temperature dependence of n_i^2 is dominated by the factor e^{-E_g/k_BT} (Eq. (28.19)), the saturation current is strongly temperature-dependent.

A MORE DETAILED THEORY OF THE NONEQUILIBRIUM *p-n* JUNCTION

Using the concepts of drift and diffusion currents we can give a more detailed description of the behavior of the p-n junction when $V \neq 0$. The equilibrium p-n

[18] See for example, F. Reif, *Fundamentals of Statistical and Thermal Physics*, McGraw-Hill, New York, 1965, p. 16.

junction has two characteristic regions: the depletion layer, in which the electric field, space charge, and carrier density gradients are large, and the homogeneous regions outside of the depletion layer, in which they are quite small. In the non-equilibrium case the position beyond which the electric field and space charge are small differs from the position beyond which the carrier density gradients are small. Thus when $V \neq 0$, the *p-n* junction is characterized not by two, but by three different regions (described compactly in Table 29.1):

1. **The Depletion Layer** As in the equilibrium case, this is a region in which the electric field, space charge, and carrier density gradients are all large. When $V \neq 0$, according to Eq. (29.20) the depletion layer is narrower than or wider than is the case for $V = 0$ depending on whether V is positive (forward bias) or negative (reverse bias).
2. **The Diffusion Regions** These are regions (extending a distance of the order of a diffusion length out from the boundaries of the depletion layer) in which the electric field and space charge are small, but the carrier density gradients remain appreciable (though not as large as in the depletion layer).
3. **The Homogeneous Regions** Beyond the diffusion regions the electric field, space charge, and carrier density gradients are all very small, as in the equilibrium homogeneous regions.

The diffusion region (2) is not present in the equilibrium case. It arises when $V \neq 0$ for the following reason:

In equilibrium ($V = 0$) the change in carrier densities across the depletion layer is just enough to join the homogeneous equilibrium values on the high-density sides ($n_c(\infty) = N_d$, $p_v(-\infty) = N_a$), to the homogeneous equilibrium values on the low-density sides[19] ($n_c(-\infty) = n_i^2/N_a$, $p_v(+\infty) = n_i^2/N_d$). When $V \neq 0$, however, we have noted that the extent of the depletion layer and the size of the potential drop across the layer differ from their equilibrium values. Consequently (as we shall see explicitly below) the change in carrier densities across the layer can no longer fit the difference in the homogeneous equilibrium values appropriate to the two sides, and a further region must appear in which the carrier densities relax from their values at the boundaries of the depletion layer to the values appropriate to the more remote homogeneous region (Figure 29.6).

Table 29.1 summarizes these properties, and also indicates the characteristic behavior of the electron and hole drift and diffusion currents in each of the three regions, when a current j flows in the junction[20]:

1. *In the depletion layer* there are both drift and diffusion currents. In the equilibrium case they are equal and opposite for each carrier type, yielding no net currents of

---

19 In discussing the equilibrium case above, we approximated the minority carrier densities (i.e., the homogeneous equilibrium values on the low-density side) by zero (cf. note 7). This was appropriate because we were describing only the space-charge density, to which the minority carriers make a minute contribution compared with the majority carriers. The contribution of the minority carriers to the current, however, is not negligible, and it is necessary for us to use the values quoted here (determined by the values of the majority carrier densities and the law of mass action).

20 In the steady state the total electric current is uniform along the junction: j cannot depend on x.

Table 29.1
THE THREE CHARACTERISTIC REGIONS IN A BIASED *p-n* JUNCTION [a]

	HOMOGENEOUS p-TYPE	DIFFUSION REGION ← $O(L_p)$ →	DEPLETION LAYER ← d_p →← d_n →	DIFFUSION REGION ← $O(L_n)$ →	HOMOGENEOUS n-TYPE
ELECTRIC FIELD OR SPACE CHARGE	SMALL		LARGE	SMALL	
∇p, ∇n	SMALL	LARGE			SMALL
p	LARGE		SMALL		
n	SMALL			LARGE	
j_h^{drift}	$\approx j$	$O(j)$	$\gg j$	≈ 0	≈ 0
$j_h^{\text{diffusion}}$	≈ 0	$O(j)$	$\gg j$	$O(j)$	≈ 0
$j_e^{\text{diffusion}}$	≈ 0	$O(j)$	$\gg j$	$O(j)$	≈ 0
j_e^{drift}	≈ 0	≈ 0	$\gg j$	$O(j)$	$\approx j$

[a] The positions and extent of the regions are indicated at the top of the table. The column beneath each region gives the orders of magnitude of the important physical quantities.

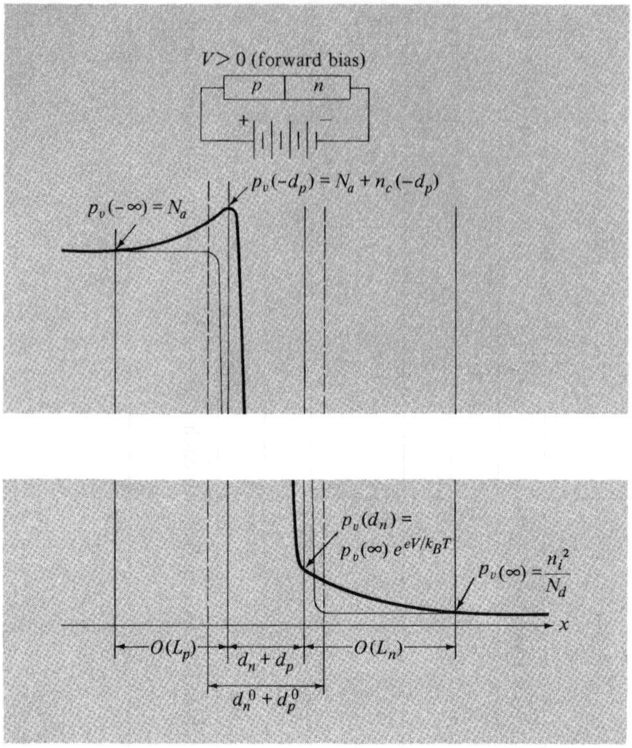

Figure 29.6
The density of holes (heavy curve) along a *p-n* junction with $V > 0$ (forward bias). The vertical solid lines give the boundaries of the depletion layer and diffusion regions. Note the break in vertical scale. For comparison, the density of holes when $V = 0$ (unbiased junction) is shown as a light curve, together with the boundaries of the depletion layer in the unbiased case (dashed vertical lines). The density of electrons behaves in a similar manner. When V is negative (reverse bias), the density of holes falls below its asymptotic value in the diffusion region. Note that although the excess density in the biased case over the unbiased case has the same size in both diffusion regions, on the *p*-side it represents a minute percentage change in the carrier density, while on the *n*-side it is a very large percentage change.

electrons or holes. In the nonequilibrium case the net current flowing across the depletion layer results from a slight imbalance between the drift and diffusion currents of each carrier type; i.e., the drift and diffusion currents are separately quite large compared with the total current. Once we have constructed a complete picture of the currents flowing in the junction it is easy to verify this explicitly (Problem 4). It is a consequence of the very large electric field and density gradients in the diffusion layer (which more than compensate for the very low carrier densities).

2. *In the diffusion regions*, the carrier densities are nearer to the values they have in the homogeneous regions. The majority carrier density has become so large that its drift current is appreciable, even though the field is now very small; the minority carrier drift current is quite negligible in comparison. Because the carrier densities continue to vary in the diffusion regions, both diffusion currents (proportional not to the density, but its gradient) are appreciable. Typically, all currents in the diffusion region except the negligible minority carrier drift current are of order j.
3. *In the homogeneous regions* the diffusion currents are negligible, and the entire current is carried by the majority carrier drift current.

Given this picture of the individual carrier drift and diffusion currents, we can readily calculate the total current j flowing in the junction for a given value of V. To simplify the analysis we make one further assumption:[21] We assume that the passage of carriers across the depletion layer is so swift that negligible generation and recombination occur within the layer. If this is so, then the total currents of electrons and holes, J_e and J_h, will be constant across the depletion layer in the steady state. Consequently, in the expression $j = -eJ_e + eJ_h$ for the total current, we may separately evaluate J_e and J_h at whatever points along the depletion layer it is most convenient to do so. The most convenient point for the electron current is at the boundary between the depletion layer and the diffusion region on the *p*-side, and the most convenient point for the hole current is at the other boundary.[22] We therefore write

$$j = -eJ_e(-d_p) + eJ_h(d_n). \tag{29.42}$$

This representation is useful because at the boundaries between the depletion layer and the diffusion regions the minority carrier currents are purely diffusive (see Table 29.1). Consequently, if we could calculate the position dependence of the minority carrier densities within the diffusion regions, we could immediately compute their currents, using Eq. (29.27) (with $E = 0$):

$$\begin{aligned} J_e(-d_p) &= -D_n \left.\frac{dn_c}{dx}\right|_{x=-d_p}, \\ J_h(d_n) &= -D_p \left.\frac{dp_v}{dx}\right|_{x=d_n}. \end{aligned} \tag{29.43}$$

Because, however, the minority carrier drift currents are negligible in the diffusion regions, the minority carrier densities satisfy the diffusion equation (29.36). If we let $p_v(d_n)$ be the density of holes at the boundary of the depletion layer on the *n*-side, and if we note that far from that boundary on the *n*-side p_v approaches the value $p_v(\infty) = n_i^2/N_d$, then the solution (29.38) to the diffusion equation (29.36) is

$$p_v(x) = \frac{n_i^2}{N_d} + \left[p_v(d_n) - \frac{n_i^2}{N_d}\right] e^{-(x-d_n)/L_p}, \qquad x \geqslant d_n. \tag{29.44}$$

Similarly, the electron density within the diffusion region on the *p*-side is given by

$$n_c(x) = \frac{n_i^2}{N_a} + \left[n_c(-d_p) - \frac{n_i^2}{N_a}\right] e^{(x+d_p)/L_n}, \qquad x \leqslant -d_p. \tag{29.45}$$

Substituting these densities into (29.43) we find that the minority carrier currents at the boundaries of the depletion layer are

$$\begin{aligned} J_e(-d_p) &= -\frac{D_n}{L_n}\left[n_c(-d_p) - \frac{n_i^2}{N_a}\right], \\ J_h(d_n) &= \frac{D_p}{L_p}\left[p_v(d_n) - \frac{n_i^2}{N_d}\right], \end{aligned} \tag{29.46}$$

---

[21] Also generally the case. When the assumption fails, the full set of equations must be integrated across the depletion layer.

[22] The more elementary picture of rectification given above also focused on the electron current originating on the hole side of the junction, and vice versa.

so that the total current, (29.42), is

$$j = \frac{eD_n}{L_n}\left[n_c(-d_p) - \frac{n_i^2}{N_a}\right] + \frac{eD_p}{L_p}\left[p_v(d_n) - \frac{n_i^2}{N_d}\right]. \qquad (29.47)$$

It only remains for us to find the amounts by which the minority carrier densities differ from their homogeneous equilibrium values at the boundaries of the depletion layer. In equilibrium we found the variation in carrier densities across the depletion layer by using the equilibrium expression (29.3) for the variation of the carrier densities in a potential $\phi(x)$. We have noted above that this expression follows from the fact that in equilibrium the drift currents are equal and opposite to the diffusion currents. In the general nonequilibrium case (e.g., in the diffusion region) the drift and diffusion currents are not in balance, and Eq. (29.3) does not hold. However, in the depletion layer there is a near balance between drift and diffusion currents,[23] and therefore to a reasonable approximation the carrier densities do obey Eq. (29.3), changing by a factor $e^{-e\Delta\phi/k_BT}$ as the depletion layer is crossed:

$$\begin{aligned} n_c(-d_p) &= n_c(d_n)e^{-e\Delta\phi/k_BT} = [n_c(d_n)e^{-e(\Delta\phi)_0/k_BT}]e^{eV/k_BT}, \\ p_v(d_n) &= p_v(-d_p)e^{-e\Delta\phi/k_BT} = [p_v(-d_p)e^{-e(\Delta\phi)_0/k_BT}]e^{eV/k_BT}. \end{aligned} \qquad (29.48)$$

When $eV \ll E_g$, then V will be small compared with $(\Delta\phi)_0$, and the carrier densities on the minority side $[n_c(-d_p)$ and $p_v(d_n)]$ will continue to be very small compared with their values of the majority side $[n_c(d_n)$ and $p_v(-d_p)]$, just as they are when $V = 0$. Consequently, the conditions that the space charge vanish at the boundaries of the depletion layer,

$$\begin{aligned} n_c(d_n) - p_v(d_n) &= N_d, \\ p_v(-d_p) - n_c(-d_p) &= N_a, \end{aligned} \qquad (29.49)$$

give values for the majority carrier densities $n_c(d_n)$ and $p_v(-d_p)$ that differ from their equilibrium values N_d and N_a only by factors that are very close to unity. Thus, to an excellent approximation, when $eV \ll E_g$,

$$\begin{aligned} n_c(-d_p) &= [N_d e^{-e(\Delta\phi)_0/k_BT}]e^{eV/k_BT}, \\ p_v(d_n) &= [N_a e^{-e(\Delta\phi)_0/k_BT}]e^{eV/k_BT}, \end{aligned} \qquad (29.50)$$

or, equivalently,[24]

$$\begin{aligned} n_c(-d_p) &= \frac{n_i^2}{N_a}\, e^{eV/k_BT}, \\ p_v(d_n) &= \frac{n_i^2}{N_d}\, e^{eV/k_BT}. \end{aligned} \qquad (29.51)$$

Substituting these results into the expression (29.47) for the total current, we find that

$$j = en_i^2\left(\frac{D_n}{L_nN_a} + \frac{D_p}{L_pN_d}\right)(e^{eV/k_BT} - 1). \qquad (29.52)$$

---

23 This is verified in Problem 4.

24 This follows from the form (29.6) of $(\Delta\phi)_0$ and the form (28.19) of n_i. It also follows directly from (29.50) by requiring that it yield the correct equilibrium values $n_c(-d_p) = n_i^2/N_a$ and $p_v(d_n) = n_i^2/N_c$, when $V = 0$.

This has the form (29.26) with the generation currents given explicitly by

$$J_e^{\text{gen}} = \left(\frac{n_i^2}{N_a}\right)\frac{D_n}{L_n},$$
$$J_h^{\text{gen}} = \left(\frac{n_i^2}{N_d}\right)\frac{D_p}{L_p}. \tag{29.53}$$

If we eliminate the diffusion constants appearing in (29.53) with the aid of (29.37), the expressions for the generation currents agree with the rough estimates (29.40) and (29.41).

PROBLEMS

1. *The Depletion Layer in Thermal Equilibrium*

(a) Show that if the exact (nondegenerate) form (29.3) is retained for the carrier densities, then Poisson's equation (which we approximated by Eq. (29.13) in the text) becomes the following differential equation for the variable $\psi = (e\phi + \mu - \mu_i)/k_BT$:

$$\frac{d^2\psi}{dx^2} = K^2\left(\sinh\psi - \frac{\Delta N(x)}{2n_i}\right), \tag{29.54}$$

where $K^2 = 8\pi n_i e^2/k_BT\epsilon$, $\Delta N(x)$ is the doping profile, $\Delta N(x) = N_d(x) - N_a(x)$, and n_i and μ_i are the carrier density and chemical potential for an impurity-free sample at the same temperature.

(b) The text discussed the case of a *p-n* junction made of highly extrinsic material, with N_d, $N_a \gg n_i$. In the opposite case of a lightly doped nearly intrinsic semiconductor, with

$$n_i \gg N_d, N_a, \tag{29.55}$$

we can find the electrostatic potential to high precision for an arbitrary doping profile, as follows:

(i) Assume that $\psi \ll 1$, so that $\sinh\psi \approx \psi$. Show that the solution to (29.54) is then given by:

$$\psi(x) = \frac{1}{2}K\int_{-\infty}^{\infty} dx'\, e^{-K|x-x'|}\,\frac{\Delta N(x')}{2n_i}. \tag{29.56}$$

(ii) Show that this solution and (29.55) imply that ψ is indeed much less than 1, justifying the initial ansatz.

(iii) Show that if ΔN varies along more than one dimension, then in the nearly intrinsic case:

$$\phi(\mathbf{r}) = e\int d\mathbf{r}'\,\Delta N(\mathbf{r}')\,\frac{e^{-K|\mathbf{r}-\mathbf{r}'|}}{|\mathbf{r}-\mathbf{r}'|}. \tag{29.57}$$

(iv) The above result is identical in form to the screened Thomas-Fermi potential produced by impurities in a metal [Eq. (17.54)]. Show that the Thomas-Fermi wave vector [Eq. (17.50)] for a free electron gas has precisely the form of K, except that v_F must be replaced by the thermal velocity appropriate to Boltzmann statistics, and the carrier density must be taken as $2n_i$. (Why this last factor of 2?) The quantity K is the screening length of Debye-Hückel theory.

(c) Some insight into the general solution to (29.54) may be gained by the simple expedient of changing the names of the variables:

$$\psi \to u, \quad x \to t, \quad K^2 \to \frac{1}{m}. \tag{29.58}$$

The equation then describes the displacement u of a particle of mass m moving under the influence of a force that depends upon position (u) and time (t). In the case of an abrupt junction this force is time-independent before and after $t = 0$. Sketch the "potential energy" before and after $t = 0$, and deduce from your sketch a qualitative argument that the solution to (29.54) that becomes asymptotically constant as $x \to \pm\infty$ can vary appreciably only in the neighborhood of $x = 0$.

(d) Show that conservation of "energy" before and after $t = 0$ in the mechanical model for the abrupt junction described above permits one to show that the exact potential at $x = 0$ is that given by the approximate solution (29.14) plus a correction $\Delta\phi$, given by:

$$\Delta\phi = -\frac{kT}{e}\left(\frac{\sqrt{N_d^{\,2} + 4n_i^{\,2}} - \sqrt{N_a^{\,2} + 4n_i^{\,2}}}{N_d + N_a}\right). \tag{29.59}$$

Comment on how important a correction to ϕ this is, and how reliable the carrier densities (29.12) given by the approximate solution (29.14) are likely to be in the depletion region.

(e) As in (d), find and discuss the approximate and exact electric fields at $x = 0$.

2. *Derivation of the Einstein Relations from Kinetic Theory*

Show that the phenomenological equations (29.27) relating the carrier currents to the electric field and carrier density gradients, follow from elementary kinetic arguments such as were used in Chapter 1, with mobilities of the form (29.28) and (29.29), and diffusion constants of the form

$$D = \tfrac{1}{3}\langle v^2\rangle \tau^{\text{coll}}. \tag{29.60}$$

Show that the Einstein relations (29.30) are satisfied provided the mean square thermal velocity $\langle v^2\rangle$ is given by Maxwell-Boltzmann statistics.

3. *Einstein Relations in the Degenerate Case*

When dealing with degenerate inhomogeneous semiconductors one must generalize the equilibrium carrier densities (29.3) to

$$\begin{aligned} n_c(x) &= n_c^0\,(\mu + e\phi(x)), \\ p_v(x) &= p_v^0\,(\mu + e\phi(x)), \end{aligned} \tag{29.61}$$

where $n_c^0(\mu)$, and $p_v^0(\mu)$ are the carrier densities of the homogeneous semiconductor as a function of chemical potential.[25]

(a) Show that the expression (29.9) for $\Delta\phi$ *and the interpretation that precedes it* continue to follow directly from (29.61).

(b) Show by a slight generalization of the argument on page 602 that

$$\mu_n = eD_n\frac{1}{n}\frac{\partial n}{\partial \mu}, \quad \mu_p = -eD_p\frac{1}{p}\frac{\partial p}{\partial \mu}. \tag{29.62}$$

---

[25] Note that the functional forms of $n_c^0\,(\mu)$ and $p_v^0\,(\mu)$ do *not* depend on the doping (though of course the value of μ does).

(c) In an inhomogeneous semiconductor, not in equilibrium, with carrier densities $n_c(x)$ and and $p_v(x)$, one sometimes defines electron and hole *quasichemical potentials*[26] $\tilde{\mu}_e(x)$ and $\tilde{\mu}_h(x)$ by requiring the carrier densities to have the equilibrium form (29.61):

$$n_c(x) = n_c^0\,(\tilde{\mu}_e(x) + e\phi(x)), \quad p_v(x) = p_v^0\,(\tilde{\mu}_h(x) + e\phi(x)). \tag{29.63}$$

Show that, as a consequence of the Einstein relations (29.62), the total drift plus diffusion currents are just

$$\begin{aligned} J_e &= -\mu_n n_c \frac{d}{dx}\frac{1}{e}\,\tilde{\mu}_e(x), \\ J_h &= \mu_p p_v \frac{d}{dx}\frac{1}{e}\,\tilde{\mu}_h(x). \end{aligned} \tag{29.64}$$

Note that these have the form of pure drift currents in an electrostatic potential $\phi = (-1/e)\tilde{\mu}$.

4. *Drift and Diffusion Currents in the Depletion Layer*

Noting that the electric field in the depletion layer is of order $\Delta\phi/d$, $d = d_n + d_p$, and that the carrier densities there exceed their minority values substantially (except at the edges of the layer), show that the assumption that the drift (and hence diffusion) currents in the depletion layer greatly exceed the total current is very well satisfied.

5. *Fields in the Diffusion Region*

Verify the assumption that the potential ϕ undergoes negligible variation in the diffusion region, by estimating its change across the diffusion region as follows:

(a) Find the electron drift current at d_n by noting that the total electron current is continuous across the depletion layer and calculating explicitly the electron diffusion current at d_n.

(b) Noting that the electron density is very close to N_d at d_n, find an expression for the electric field at d_n necessary to produce the drift current calculated in (a).

(c) Assuming that the field found in (b) sets the scale for the electric field in the diffusion region, show that the change in ϕ across the diffusion region is of order $(k_BT/e)(n_i/N_d)^2$.

(d) Why is this indeed negligible?

6. *Saturation Current*

Estimate the size of the saturation electric current in a *p-n* junction at room temperature, if the band gap is 0.5 eV, the donor (or acceptor) concentrations $10^{18}/\text{cm}^3$, the recombination times 10^{-5} second, and the diffusion lengths 10^{-4} cm.

---

26 Since we are not in equilibrium, $\tilde{\mu}_e$ need not equal $\tilde{\mu}_h$.

30
Defects in Crystals

Thermodynamics of Point Defects

Schottky and Frenkel Defects

Annealing

Electrical Conductivity of Ionic Crystals

Color Centers

Polarons and Excitons

Dislocations

Strength of Crystals

Crystal Growth

Stacking Faults and Grain Boundaries

By a crystalline defect one generally means any region where the microscopic arrangement of ions differs drastically from that of a perfect crystal. Defects are called surface, line, or point defects, according to whether the imperfect region is bounded on the atomic scale in one, two, or three dimensions.

Like human defects, those of crystals come in a seemingly endless variety, many dreary and depressing, and a few fascinating. In this chapter we shall describe a few of those imperfections whose presence has a profound effect on at least one major physical property of the solid. One could argue that almost any defect meets this test; for example, isotopic inhomogeneity can alter both the phonon spectrum and the character of the neutron scattering. The examples we will consider, however, are somewhat more dramatic.[1] The two most important kinds of defects we shall mention are:

1. **Vacancies and Interstitials** These are point defects, consisting of the absence of ions (or presence of extra ions). Such defects are entirely responsible for the observed electrical conductivity of ionic crystals, and can profoundly alter their optical properties (and, in particular, their color). Furthermore, their presence is a normal thermal equilibrium phenomenon, so they can be an intrinsic feature of real crystals.
2. **Dislocations** These are line defects which, though probably absent from the ideal crystal in thermal equilibrium, are almost invariably present in any real specimen. Dislocations are essential in explaining the observed strength (or rather, the lack of shear strength) of real crystals, and the observed rates of crystal growth.

POINT DEFECTS: GENERAL THERMODYNAMIC FEATURES

Point defects are present even in the thermal equilibrium crystal, as we can illustrate by considering only the simplest kind: a *vacancy* or *Schottky defect* in a monatomic Bravais lattice. A vacancy occurs whenever a Bravais lattice site that would normally be occupied by an ion in the perfect crystal has no ion associated with it (Figure 30.1). If the number, n, of such vacancies at temperature T is an extensive thermodynamic variable (i.e., if it is proportional to the total number of ions, N, when N is very large),

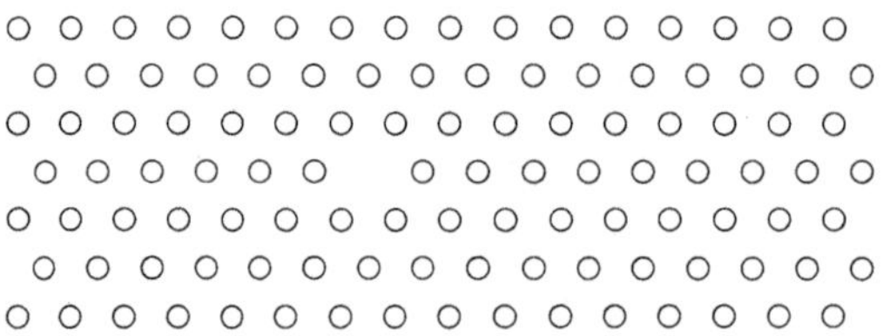

Figure 30.1
A portion of a monatomic Bravais lattice containing a vacancy or Schottky defect.

[1] Not only is our choice of defects highly selective, but we have also included in the chapter some phenomena (polarons and excitons) that are not generally regarded as defects at all. This is because they bear strong resemblances to other phenomena that are regarded as defects, and therefore arise rather naturally in such a discussion.

then we can estimate its size by minimizing the appropriate thermodynamic potential. If the crystal is at constant pressure P, this will be the Gibbs free energy,

$$G = U - TS + PV.$$

To see how G depends on n, it is simplest to think of an N-ion crystal containing n vacant sites, as an $(N + n)$-ion perfect crystal, from which n ions have been removed. Thus the volume, $V(n)$, will to a first approximation be just $(N + n)v_0$, where v_0 is the volume per ion in the perfect crystal.

For any particular choice of the n sites to be deprived of their ions, we can, in principle, calculate $F_0(n) = U - TS$ for that particular imperfect crystal. If n is very small[2] compared with N, then we might expect this to depend only on the number of vacancies but not on their spatial arrangement.[3] We must add to the entropy S for a fixed configuration of vacancies, a further contribution S^{config} expressing the disorder arising from the $(N + n)!/N!n!$ ways of choosing the n vacant sites out of the $N + n$:

$$S^{\text{config}} = k_B \ln \frac{(N + n)!}{N!n!}. \tag{30.1}$$

Thus the full Gibbs free energy is

$$G(n) = F_0(n) - TS^{\text{config}}(n) + P(N + n)v_0. \tag{30.2}$$

Using Stirling's formula, valid for large X,

$$\ln X! \approx X(\ln X - 1) \tag{30.3}$$

we can evaluate

$$\frac{\partial S^{\text{config}}}{\partial n} = k_B \ln\left(\frac{N + n}{n}\right) \approx k_B \ln\left(\frac{N}{n}\right); \qquad n \ll N, \tag{30.4}$$

and therefore

$$\frac{\partial G}{\partial n} = \frac{\partial F_0}{\partial n} + Pv_0 - k_B T \ln\left(\frac{N}{n}\right). \tag{30.5}$$

When $n \ll N$, we can write

$$\frac{\partial F_0}{\partial n} \approx \left.\frac{\partial F_0}{\partial n}\right|_{n=0} = \mathcal{E}, \tag{30.6}$$

where $\mathcal{E}$ is independent of n. Therefore (30.5) tells us that G is minimized by

$$n = Ne^{-(\mathcal{E} + Pv_0)/k_B T}. \tag{30.7}$$

---

[2] This is not inconsistent with our assertion that n is an extensive variable of order N. Extensivity requires $\lim_{N\to\infty}(n/N) \neq 0$. However, n being small compared with N requires not that the limit vanish, but only that it be much less than unity. This is always the case for point defects in crystals. Indeed, if the number of defects were close to the number of ions, we would have no business speaking of a crystal at all.

[3] This will certainly not be true for configurations in which appreciable numbers of vacancies lie close together, since the presence of one can then affect the energy required to form another. However, when $n \ll N$, such configurations will be quite rare.

To calculate $\mathcal{E}$ we could (as in Chapter 22) write the total potential energy of a typical $(N + n)$ ion lattice with n vacancies as $U = U^{\text{eq}} + U^{\text{harm}}$ (see (22.8)). We could then calculate F_0 from the partition function

$$e^{-\beta F_0} = \sum_E e^{-\beta E} = e^{-\beta U^{\text{eq}}} \sum_{E_{\text{harm}}} e^{-\beta E_{\text{harm}}}, \qquad \beta = \frac{1}{k_B T}, \tag{30.8}$$

where E^{harm} runs through the eigenvalues of the harmonic part of the Hamiltonian. Evidently this will yield an F_0, which is the equilibrium potential energy of the lattice with vacancies plus the free energy of the phonons

$$F_0 = U^{\text{eq}} + F^{\text{ph}}. \tag{30.9}$$

The second term is generally small compared with the first, so that to a first approximation $\mathcal{E}$ is just

$$\mathcal{E}_0 = \left.\frac{\partial U^{\text{eq}}}{\partial n}\right|_{n=0}, \tag{30.10}$$

the temperature-independent potential energy required to remove one ion. At normal (e.g., atmospheric) pressures Pv_0 is negligible in comparison, and therefore

$$n = Ne^{-\beta \mathcal{E}_0}. \tag{30.11}$$

Since $\mathcal{E}_0$ can be expected to be of the order of electron volts,[4] n/N will indeed be small, but not zero.

The phonon correction to (30.11) from the second term in (30.9) generally increases n somewhat. This is because the introduction of vacancies tends to lower some of the normal-mode frequencies (and hence the associated phonon energies), thereby leading to a negative $\partial F^{ph}/\partial n$. A simple model of this effect is discussed in Problem 1.

The analysis above has assumed that only one kind of point defect can occur: a vacancy at a Bravais lattice site. In general, of course, there can be more than one kind of vacancy (in polyatomic lattices). There is also the possibility of extra ions occupying regions not occupied in the perfect crystal, a type of point defect known as an *interstitial*. Thus we should generalize our analysis to allow for n_j point defects of the jth type. If all n_j are small compared with N, each type of defect will occur in numbers given by the obvious generalization of (30.7) (ignoring the small correction Pv_0):

$$n_j = N_j e^{-\beta \mathcal{E}_j}, \quad \mathcal{E}_j = \left.\frac{\partial F_0}{\partial n_j}\right|_{n_j=0}, \tag{30.12}$$

where N_j is the number of sites where a defect of type j can occur.

The $\mathcal{E}_j$ are generally quite large compared with $k_B T$ and if, in addition, the smallest two values of $\mathcal{E}_j$ (say $\mathcal{E}_1$ and $\mathcal{E}_2$) are themselves far apart compared with $k_B T$, then n_1 will be very much greater than all the other n_j; i.e., the defect with the smallest $\mathcal{E}_j$ will be overwhelmingly the most abundant type.

However, (30.12) is only correct if the numbers of each kind of defect are independent, since it follows from minimizing the free energy independently with respect

---

[4] We would expect it to be roughly the size of the cohesive energy per particle. See Chapter 20.

to all the n_j. Should there be constraints among the n_j, we must reexamine the problem. The most important such constraint is charge neutrality. We cannot have a set of defects consisting entirely of positive ion vacancies in an ionic crystal, without creating unbalanced positive charge, with its prohibitively large Coulomb energy. This excess charge must be balanced either by positive ion interstitials, negative ion vacancies, or some combination of these.[5] Thus the free energy must be minimized subject to the constraint:

$$0 = \Sigma q_j n_j, \tag{30.13}$$

where q_j is the charge of the jth defect type ($q_j = +e$ for a negative ion vacancy or positive ion interstitial, and $q_j = -e$ for a positive ion vacancy or negative ion interstitial). If we introduce a Lagrange multiplier λ, then we can take the constraint into account by minimizing not G, but $G + (\lambda \Sigma q_j n_j)$. This leads to the replacement of (30.12) by

$$n_j = N_j e^{-\beta(\mathcal{E}_j + \lambda q_j)}, \tag{30.14}$$

where the unknown λ is determined by requiring that (30.14) satisfy the constraint (30.13).

Usually the lowest $\mathcal{E}_j$ for each charge type is separated in energy by many $k_B T$ from the next lowest.[6] Consequently there will be one dominant type of defect of each charge, whose numbers are given by

$$\begin{aligned} n_+ &= N_+ e^{-\beta(\mathcal{E}_+ + \lambda e)} \\ n_- &= N_- e^{-\beta(\mathcal{E}_- - \lambda e)}, \qquad \mathcal{E}_\pm = \min_{(q_j = \pm e)} (\mathcal{E}_j) \end{aligned} \tag{30.15}$$

Since the densities of all the other defect types satisfy

$$\begin{aligned} n_j &\ll n_+, \quad q_j = +e, \\ n_j &\ll n_-, \quad q_j = -e, \end{aligned} \tag{30.16}$$

charge neutrality requires to high precision that

$$n_+ = n_-. \tag{30.17}$$

Since (30.15) also requires

$$n_+ n_- = N_+ N_- e^{-\beta(\mathcal{E}_+ + \mathcal{E}_-)}, \tag{30.18}$$

we find that

$$n_+ = n_- = \sqrt{N_+ N_-}\, e^{-\beta(\mathcal{E}_+ + \mathcal{E}_-)/2}. \tag{30.19}$$

Thus the constraint of charge neutrality reduces the concentration of the most abundant defect type and raises the concentration of the most abundant type of the opposite charge. It does so in a manner that changes the values they would have in the absence of the constraint to the geometric mean of those values.

Even in simple diatomic ionic crystals, there are several ways in which charge neutrality can thus be achieved (Figure 30.2). There can be essentially equal numbers

---

5 We ignore the possibility of color center formation. See below.

6 When this is not the case, then the distinction developed below between Schottky and Frenkel defects cannot be made. See Problem 2.

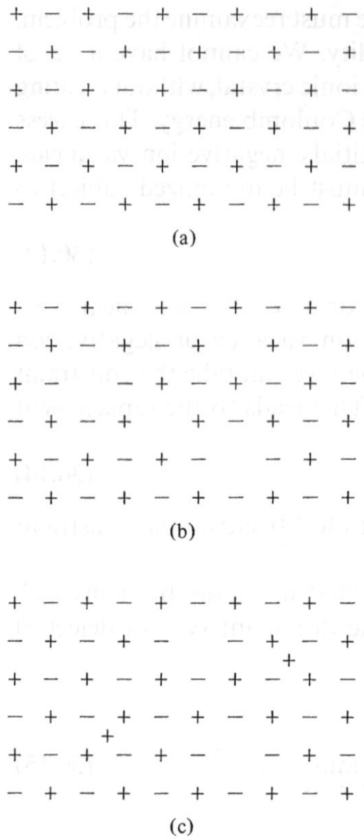

Figure 30.2
(a) A perfect ionic crystal. (b) An ionic crystal with point defects of the Schottky type (equal numbers of positive and negative ion vacancies). (c) An ionic crystal with defects of the Frenkel type (equal numbers of positive ion vacancies and interstitials).

of positive and negative ion vacancies, known, in this context, also as *Schottky defects*. On the other hand, there can be essentially equal numbers of vacancies and interstitials of the same ion, known as *Frenkel defects*. Alkali halides have defects of the Schottky type; silver halides, of the Frenkel type. (The third possibility, equal numbers of positive and negative ion interstitials, does not seem to occur, interstitials being generally more costly in energy than vacancies of the same ion.)

DEFECTS AND THERMODYNAMIC EQUILIBRIUM

It is most unlikely that *line* or *surface* defects can, like point defects, have a non-vanishing concentration in thermal equilibrium. The energy of formation of one of these more extended defects will be proportional to the linear dimensions ($N^{1/3}$) or cross-sectional area ($N^{2/3}$) of the crystal. However, the number of ways of introducing one (provided that it is not excessively "wiggly" (lines) or "wobbly" (surfaces)) does not appear to be more than logarithmic in N, as for point defects. Thus although the cost in energy of a single point defect (independent of N) is more than compensated for by the gain in entropy (of order ln N), this is probably not the case for the line and surface defects.

Line and surface defects are, in all likelihood, metastable configurations of the crystal. However, thermal equilibrium may well be approached so slowly that for practical purposes the defects may be considered to be frozen in. It is also easy to arrange nonequilibrium concentrations of point defects, which can have considerable permanence (for example, by quickly cooling a crystal that had been in equilibrium). The equilibrium concentration of point defects can be brought back to the Maxwell-Boltzmann form, and the density of line and surface defects correspondingly reduced toward zero by the slow application and removal of heat. The restoration of equilibrium defect concentrations in this way is known as *annealing*.

POINT DEFECTS: THE ELECTRICAL CONDUCTIVITY OF IONIC CRYSTALS

Ionic crystals, those electronic insulators *par excellence*, have a nonvanishing electrical conductivity. Typical resistivities depend sensitively on temperature and purity of the specimen, and can range, in alkali halide crystals, from 10^2 to 10^8 ohm-cm (which should be compared with typical metallic resistivities, which are of order microhm centimeters). Conduction cannot be due to the thermal excitation of electrons from the valence to the conduction bands, as in semiconductors (eq. (28.20)), for the band gap is so large that few if any of the 10^{23} electrons can be so aroused. There is direct evidence that the charge is carried not by electrons, but by the ions themselves: After the passage of a current, atoms corresponding to the appropriate ions are found deposited at the electrodes in numbers proportional to the total charge carried by the current.

The ability of the ions so to conduct is enormously enhanced by the presence of vacancies. It requires far less work to move a vacancy through a crystal than to force an ion through the dense ionic array of a perfect crystal (Figure 30.3).

There is abundant evidence that ionic conduction depends on the movement of vacancies. The conductivity is observed to increase with temperature exponentially in $1/T$, reflecting the temperature dependence of the thermal equilibrium vacancy concentration (30.14).[7] Furthermore, at low temperatures, the conductivity of a monovalent ionic crystal doped with divalent impurities (e.g., Ca in NaCl) is found to be proportional to the concentration of divalent impurities, in spite of the fact that the material deposited at the cathode continues to be the monovalent atom. As shown in Figure 30.4, the important function of the impurity is to force, via charge neutrality, the creation of a Na^+ vacancy for each Ca^{++} ion incorporated substitutionally into the lattice at an Na^+ site.[8] Thus the more Ca introduced, the greater the number of Na^+ vacancies, and the greater the conduction.[9]

---

[7] This by itself is not completely convincing, since the ionic diffusion constant, which depends on the probability of the ion having enough thermal energy to cross a potential barrier will also vary exponentially in $1/T$. Consequently, so will the mobility and conductivity. (See Chapter 29 for definitions of diffusion constant and mobility.)

[8] Direct evidence for this is that the density of the doped crystal is less than that of the pure crystal, even though a calcium atom has a greater mass than one of sodium.

[9] This phenomenon is quite analogous to the impurity doping of semiconductors. See Chapter 28, and Problem 3.

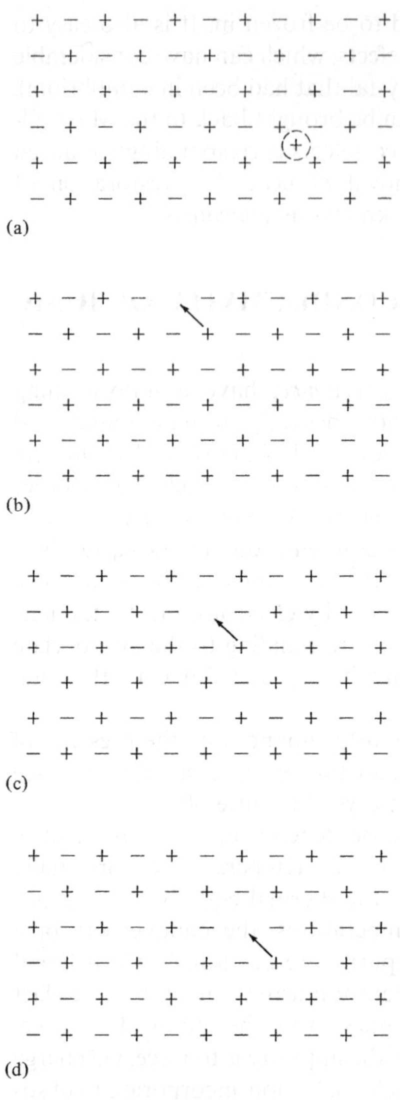

Figure 30.3
(a) It is very difficult to pass an extra positively charged ion through a perfect crystal. However, by successive motion of positively charged ions into neighboring vacancies, (b)–(d), a unit of positive charge can be moved with relative ease across the whole crystal.

COLOR CENTERS

We have indicated that charge neutrality requires vacancies of one constituent of a diatomic ionic crystal to be balanced, either by an equal number of interstitials of the same constituent (Frenkel) or by an equal number of vacancies of the other constituent (Schottky). It is also, however, possible to balance the missing charge of a negative ion vacancy with an electron localized in the vicinity of the point defect whose charge it is replacing.

Such an electron can be regarded as bound to an effectively positively charged center, and will, in general, have a spectrum of energy levels.[10] Excitations between

---

[10] See Problem 5.

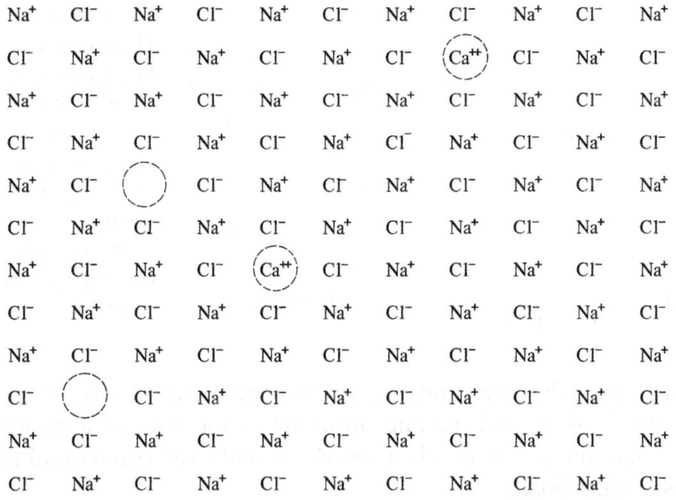

Figure 30.4
The introduction of n Ca^{++} ions into NaCl results in n Na^+ ions being replaced by Ca^{++} and the creation of an additional n Na^+ vacancies to preserve charge neutrality.

these levels produce a series of optical absorption lines quite analogous to those of single isolated atoms. These excitation energies occur in the optically forbidden band between $\hbar\omega_T$ and $\hbar\omega_L$, for the perfect crystal (see Chapter 27), and therefore stand out as quite striking peaks in the optical absorption spectrum (Figure 30.5). These and other such defect-electron structures are known as color centers, since their presence imparts a strong color to the otherwise transparent perfect crystal.

Color centers have been extensively studied in the alkali halides, which can be colored by exposure to X- or γ-radiation (with the ensuing production of defects by the very high-energy photons) or, more instructively, by heating an alkali halide crystal in a vapor of the alkali metal. In this last case excess alkali atoms (whose numbers can range from one in 10^7 to as high as one in 10^3) are incorporated into the crystal, as subsequent chemical analysis demonstrates. However, the mass density of the colored crystal diminishes in proportion to the concentration of excess alkali

Figure 30.5
The absorption spectrum of KCl, showing peaks associated with the various combinations of F-centers; e.g., the F-center itself, the M-center, and the R-center. R. H. Silsbee, *Phys. Rev.* **A180**, 138 (1965).)

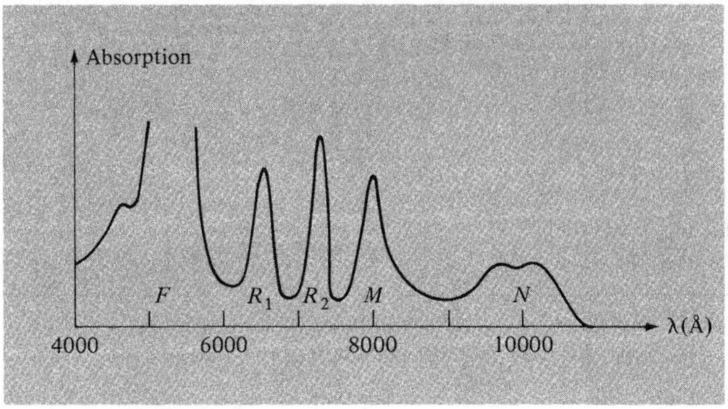

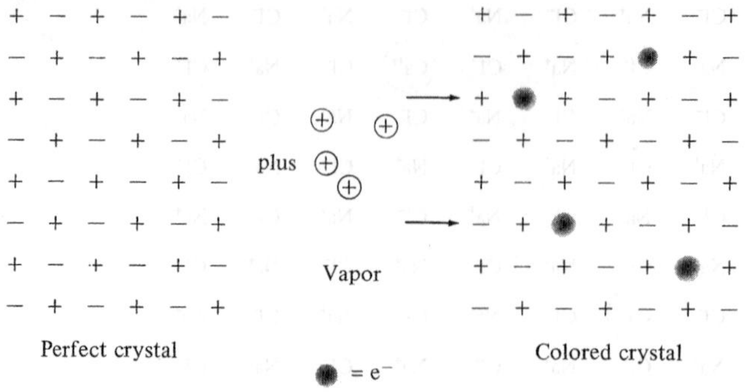

Figure 30.6
Heating a perfect alkali halide crystal in the vapor of the alkali metal can produce a crystal with an excess of alkali ions. There is a corresponding concentration of negative ion vacancies whose sites are now occupied by the (highly localized) excess electrons.

atoms, demonstrating that the atoms are not absorbed interstitially. Instead, the alkali metal atoms are ionized and assume positions in the sites of a perfect positively charged sublattice, and the excess electrons are bound to an equal number of negative ion vacancies (Figure 30.6).

Striking evidence for the validity of this picture is provided by the fact that the absorption spectrum so produced is not substantially altered if, for example, we heat potassium chloride in a vapor of sodium, rather than potassium metal. This confirms the fact that the primary role of the vapor of metal atoms is to introduce negative ion vacancies and to supply the neutralizing electron, whose energy levels produce the absorption spectrum.

An electron bound to a negative ion vacancy (known as an *F*-center[11]) is capable of reproducing many of the qualitative features of ordinary atomic spectra, with the added complication that it moves in a field of cubic, rather than spherical, symmetry. This permits one to exercise one's knowledge of group theory (e.g., how angular momentum multiplets are split by cubic fields). Indeed, by stressing the crystal, one can reduce the cubic symmetry, producing diagnostic perturbations, which are helpful in disentangling a wealth of additional structure in the absorption spectrum. The extra structure is present because the simple *F*-center is not the only way the electrons and vacancies can conspire to color the crystal.[12] Two other possibilities are: (a) The *M*-center (Figure 30.7a), in which two neighboring negative ion vacancies in a (100) plane bind two electrons. (b) The *R*-center (Figure 30.7b), in which three neighboring negative ion vacancies in a (111) plane bind three electrons.

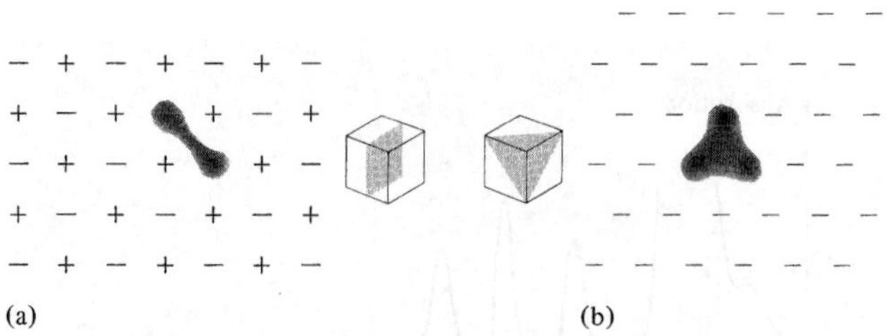

Figure 30.7
(a) The *M*-center, in which two neighboring negative ion vacancies in a (100) plane bind two electrons. (b) The *R*-center, in which three negative ion vacancies in a (111) plane bind three electrons.

---

[11] An abbreviation for the German *Farbzentrum*.

[12] It is, however, the most abundant of the centers.

It took considerable ingenuity to demonstrate that these various categories of defects were indeed responsible for the observed spectra. Identification is made possible by noting that each has a characteristic response to the effects of stress or electric fields on its level structure.

The resonances in the optical absorption produced by color centers are not nearly so sharp as those produced by the excitation of isolated atoms. This is because the linewidth is inversely proportional to the excited-state lifetime. Isolated atoms can only decay radiatively, a relatively slow process, but the "atom" represented by an *F*-center is strongly coupled to the rest of the solid, and can lose its energy by the emission of phonons.

One might think that by heating an alkali halide crystal in *halogen* gas, one could also introduce alkali metal vacancies to which holes could be bound, but these *antimorphs* to the *F*-center and its cousins have not been observed. Holes can be bound to point imperfections, but the imperfections have not been observed to be positive ion vacancies. Indeed, the most studied hole center, the V_K-center is based on no vacancies at all, but only on the possibility of a hole binding two neighboring negative ions (e.g. chlorine) into something which has a spectrum rather like Cl_2^-

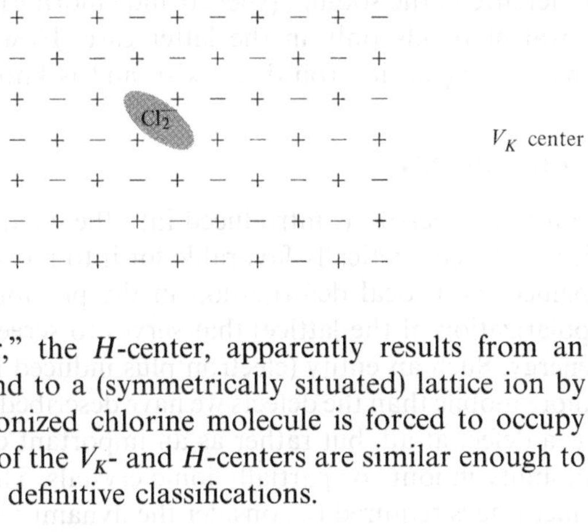

Figure 30.8
Color centers involving the binding of holes do not involve positive ion vacancies. The V_K-center is based on the possibility of a hole binding two neighboring negative ions.

(Figure 30.8). A similar "hole center," the *H*-center, apparently results from an interstitial chlorine ion being so bound to a (symmetrically situated) lattice ion by a hole (Figure 30.9); i.e., the singly ionized chlorine molecule is forced to occupy a single negative ion site. The spectra of the V_K- and *H*-centers are similar enough to have greatly impeded efforts to make definitive classifications.

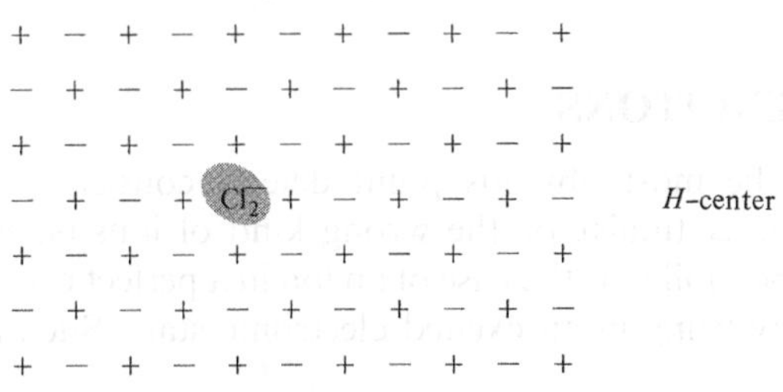

Figure 30.9
The *H*-center, which is thought to result from the binding (by a hole) of an interstitial chloride ion to a symmetrically situated lattice ion. The result is a singly ionized chlorine molecule that is forced to occupy a single negative ion site.

Having started upon the diagnosis and construction of color centers, one can continue at some length. For example, one can look for, or make, a simple *F*-center in which one of the six nearest neighboring positive ions has been replaced by an

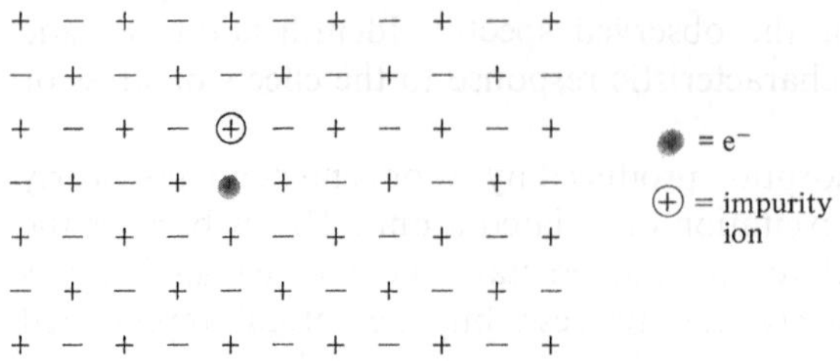

Figure 30.10
The F_A-center, in which one of the six nearest-neighbor positive ions surrounding a negative ion vacancy is replaced by an impurity ion, thereby lowering the symmetry of the levels of the bound electron. This type of conjunction of impurity and vacancy is often energetically favorable.

impurity (Figure 30.10). One then has an F_A-center, whose reduced symmetry delights spectroscopists.

Finally, pursuing the search for opposites we might inquire whether the antimorph of the V_K-center has been observed: a localized electron, serving to bind two neighboring positively charged ions together. Because (for example) Cl_2 molecules exist (covalently bonded) and Na_2 molecules in general do not, the answer is no. Indeed, the asymmetry between electron and hole centers is precisely due to the crucial difference in the sodium (*s*-level) and chlorine (*p*-level) valence electrons, which form covalent bonds only in the latter case. However, something rather less localized than the V_K antimorph does exist, and is known as the *polaron.*

POLARONS

When an electron is introduced into the conduction band of a perfect ionic crystal, it may be energetically favorable for it to move in a spatially localized level, accompanied by a local deformation in the previously perfect ionic arrangement (i.e., a polarization of the lattice) that serves to screen its field and reduce its electrostatic energy. Such an entity (electron plus induced lattice polarization) happens to be far more mobile than the defects we have described up to now, and is generally not viewed as a defect at all, but rather as an important complication in the theory of electron mobility in ionic or partially ionic crystals. Theories of polarons are quite intricate, since one is required to consider the dynamics of an electron that is strongly coupled to the ionic degrees of freedom.[13]

EXCITONS

The most obvious point defects consist of missing ions (vacancies), excess ions (interstitials), or the wrong kind of ions (substitutional impurities). A more subtle possibility is the case of an ion in a perfect crystal, that differs from its colleagues only by being in an excited electronic state. Such a "defect" is called a *Frenkel exciton.*

[13] Two general references on polarons are C. G. Kuper and G. D. Whitfield, eds., *Polarons and Excitons*, Plenum Press, New York, 1963, and the article by J. Appel, in *Solid State Physics*, vol. 21, Academic Press, New York, 1968, page 193. (The reader should be warned that our introduction of the polaron as a mobile antimorph of the V_K center reflects only our own desperate efforts at literary continuity. It is not the orthodox point of view.)

Since any ion is capable of being so excited, and since the coupling between the ions' outer electronic shells is strong, the excitation energy can actually be transferred from ion to ion. Thus the Frenkel exciton can move through the crystal without the ions themselves having to change places, as a result of which it is (like the polaron) far more mobile than vacancies, interstitials, or substitutional impurities. Indeed, for most purposes it is better not to think of the exciton as being localized at all. It is more accurate to describe the electronic structure of a crystal containing an exciton, as a quantum-mechanical superposition of states, in which it is equally probable that the excitation is associated with any ion in the crystal. This latter view bears the same relation to specific excited ions, as the Bloch tight-binding levels (Chapter 10) bear to the individual atomic levels, in the theory of band structures.

Thus the exciton is probably better regarded as one of the more complex manifestations of electronic band structure than as a crystal defect. Indeed, once one recognizes that the proper description of an exciton is really a problem in electronic band structure, one can adopt a very different view of the same phenomenon:

Suppose we have calculated the electronic ground state of an insulator in the independent electron approximation. The lowest excited state of the insulator will evidently be given by removing one electron from the highest level in the highest occupied band (the valence band) and placing it into the lowest-lying level of the lowest unoccupied band (conduction band).[14] Such a rearrangement of the distribution of electrons does not alter the self-consistent periodic potential in which they move (Eqs. (17.7) or (17.15)). This is because the Bloch electrons are not localized (since $|\psi_{nk}(r)|^2$ is periodic), and therefore the change in local charge density produced by changing the level of a single electron will be of order $1/N$ (since only an Nth of the electron's charge will be in any given cell), i.e. negligibly small. Thus the electronic energy levels do not have to be recomputed for the excited configuration and the first excited state will lie an energy $\mathcal{E}_c - \mathcal{E}_v$ above the energy of the ground state, where $\mathcal{E}_c$ is the conduction band minimum and $\mathcal{E}_v$ the valence band maximum.

However, there is another way to make an excited state. Suppose we form a one-electron level by superposing enough levels near the conduction band minimum to form a well-localized wave packet. Because we need levels in the neighborhood of the minimum to produce the wave packet, the energy $\overline{\mathcal{E}}_c$ of the wave packet will be somewhat greater than $\mathcal{E}_c$. Suppose, in addition, that the valence band level we depopulate is also a wave packet, formed of levels in the neighborhood of the valence band maximum (so that its energy $\overline{\mathcal{E}}_v$ is somewhat less than $\mathcal{E}_v$), and chosen so that the center of the wave packet is spatially very near the center of the conduction band wave packet. If we ignored electron-electron interactions, the energy required to move an electron from valence to conduction band wave packets would be $\overline{\mathcal{E}}_c - \overline{\mathcal{E}}_v > \mathcal{E}_c - \mathcal{E}_v$, but because the levels are localized, there will, in addition, be a non-negligible amount of negative Coulomb energy due to the electrostatic attraction of the (localized) conduction band electron and (localized) valence band hole.

This additional negative electrostatic energy can reduce the total excitation energy to an amount that is less than $\mathcal{E}_c - \mathcal{E}_v$, so the more complicated type of excited state, in which the conduction band electron is spatially correlated with the valence

---

[14] We use the nomenclature introduced on page 562.

band hole it left behind, is the true lowest excited state of the crystal. Evidence for this is (a) the onset of optical absorption at energies below the interband continuum threshold (Figure 30.11) and (b) the following elementary theoretical argument, indicating that one always does better by exploiting the electron-hole attraction:

Let us consider the case in which the localized electron and hole levels extend over many lattice constants. We may then make the same type of semiclassical argument that we used to deduce the form of the impurity levels in semiconductors (Chapter 28). We regard the electron and hole as particles of masses m_c and m_v (the conduction and valence band effective masses (see (28.3)), which we take, for simplicity, to be isotropic). They interact through an attractive Coulomb interaction screened by the dielectric constant ϵ of the crystal. Evidently this is just the hydrogen atom problem, with the hydrogen atom reduced mass μ ($1/\mu = 1/M_{\text{proton}} + 1/m_{\text{electron}} \approx 1/m_{\text{electron}}$) replaced by the reduced effective mass m^* ($1/m^* = 1/m_c + 1/m_v$), and the electronic charge replaced by e^2/ϵ. Thus there will be bound states, the lowest of which extends over a Bohr radius given by:

$$a_{ex} = \frac{\hbar^2}{m^*(e^2/\epsilon)} = \epsilon \frac{m}{m^*} a_0. \tag{30.20}$$

The energy of the bound state will be lower than the energy ($\mathcal{E}_c - \mathcal{E}_v$) of the non-interacting electron and hole by

$$\begin{aligned} E_{ex} &= \frac{(e^2/\epsilon)}{2a_0^*} = \frac{m^*}{m}\frac{1}{\epsilon^2}\frac{e^2}{2a_0} \\ &= \frac{m^*}{m}\frac{1}{\epsilon^2}(13.6)\text{ eV}. \end{aligned} \tag{30.21}$$

The validity of this model requires that a_{ex} be large on the scale of the lattice (i.e., $a_{ex} \gg a_0$), but since insulators with small energy gaps tend to have small effective masses and large dielectric constants, that is not difficult to achieve, particularly in semiconductors. Such hydrogenic spectra have in fact been observed in the optical absorption that occurs below the interband threshold.

The exciton described by this model is known as the *Mott-Wannier exciton.* Evidently as the atomic levels out of which the band levels are formed become more tightly bound, ϵ will decrease, m^* will increase, $a_0{}^*$ will decrease, the exciton will become more localized, and the Mott-Wannier picture will eventually break down. The Mott-Wannier exciton and the Frenkel exciton are opposite extremes of the same phenomenon. In the Frenkel case, based as it is on a single excited ionic level, the electron and hole are sharply localized on the atomic scale. The exciton spectra of the solid rare gases fall in this class.[15]

LINE DEFECTS: DISLOCATIONS

One of the most spectacular failures of the model of a solid as a perfect crystal is its inability to account for the order of magnitude of the force necessary to deform a

[15] For a general reference on excitons, see R. S. Knox, *Excitons*, Academic Press, New York, 1963.

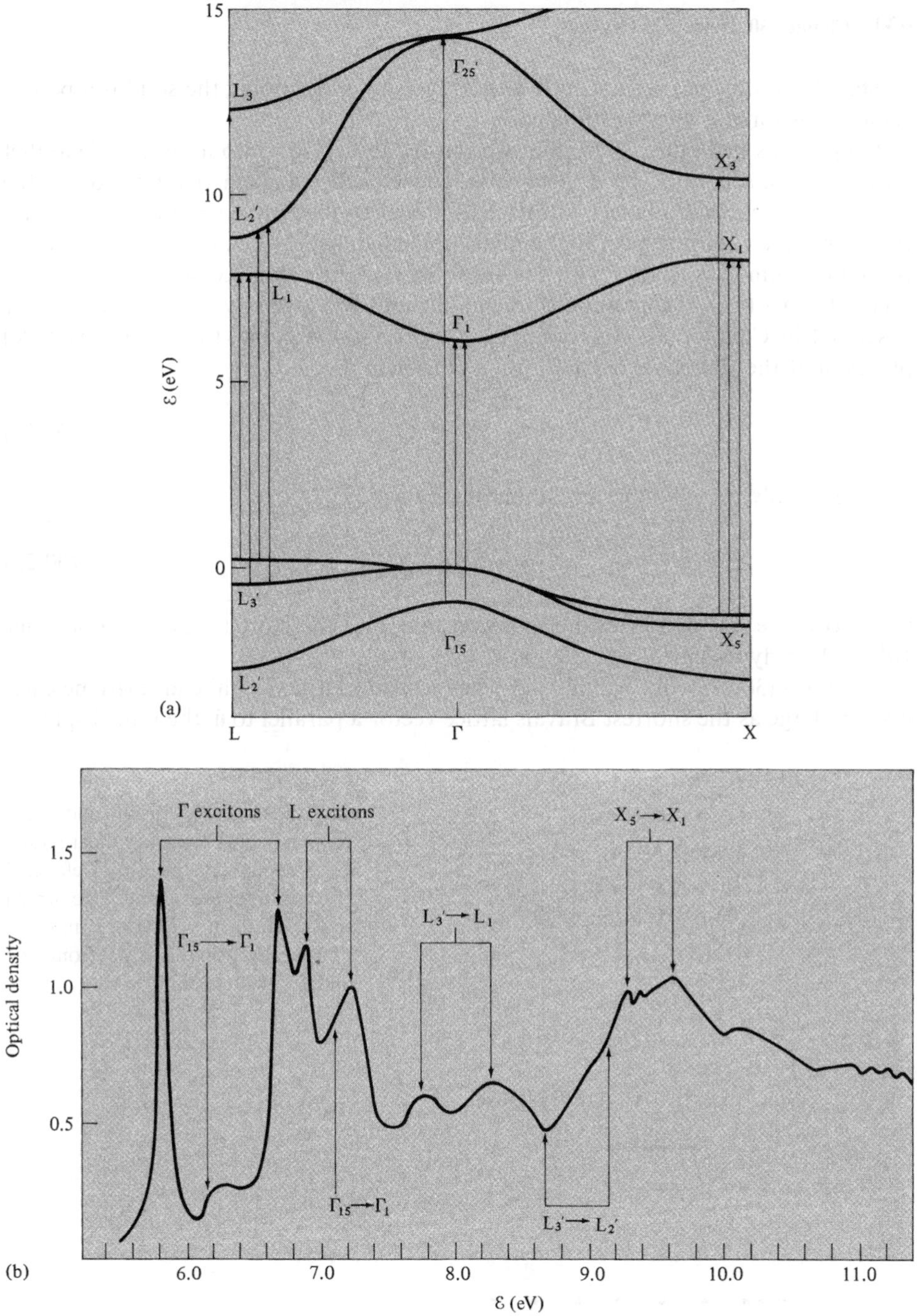

Figure 30.11
(a) The band structure of KI as inferred by J. C. Phillips (*Phys. Rev.* **136**, A1705 (1964) from its optical absorption spectrum. (b) The exciton spectrum associated with the various valence and conduction band maxima and minima. (After J. E. Eby, K. J. Teegarden, and D. B. Dutton, *Phys. Rev.* **116**, 1099 (1959), as summarized by J. C. Phillips, "Fundamental Optical Spectra of Solids," in *Solid State Physics*, vol. 18, Academic Press, New York, 1966.)

crystal plastically (i.e., permanently and irreversibly). Assuming the solid is a perfect crystal, this force is easily estimated:

Suppose, as in Figure 30.12, that we resolve the crystal into a family of parallel lattice planes, separated by a distance d, and consider a shear deformation of the crystal in which each plane is displaced parallel to itself in a specified direction $\hat{\mathbf{n}}$ by an amount x, with respect to the plane immediately below it. Let the extra energy per unit volume associated with the shear be $u(x)$. For small x, we expect u to be quadratic in x ($x = 0$ corresponds to equilibrium) and given by the elasticity theory described in Chapter 22. For example, if the crystal is cubic, the planes are (100) planes, and the direction is [010], then (Problem 4)

$$u = 2\left(\frac{x}{d}\right)^2 C_{44}. \tag{30.22}$$

More generally we will have a relation of the form:

$$u = \frac{1}{2}\left(\frac{x}{d}\right)^2 G, \tag{30.23}$$

where G is the size of a typical elastic constant, and therefore (Table 22.2) of order 10^{11} to 10^{12} dynes/cm^2.

The form (30.23) will certainly fail when x is too large. To take an extreme case, if x is as large as the shortest Bravais lattice vector $\mathbf{a}$ parallel to $\hat{\mathbf{n}}$, then the displaced

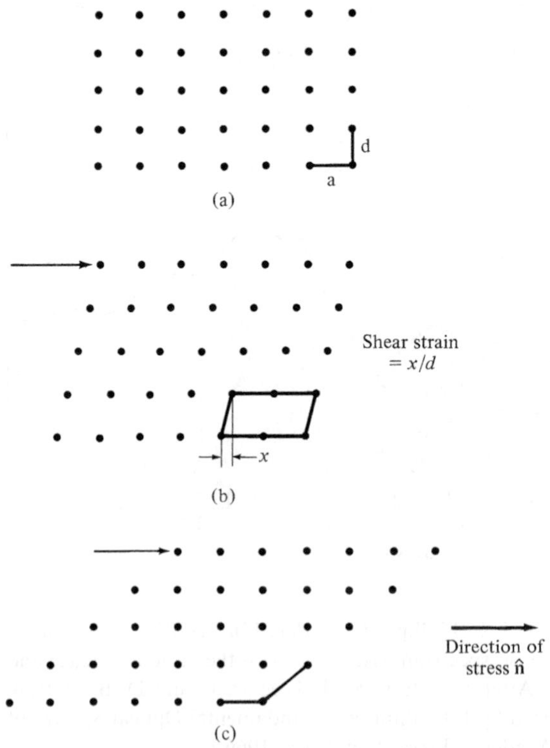

Figure 30.12
An undeformed crystal undergoes progressively increasing shear strain. (a) Perfect crystal. (b) Deformed crystal. In (c) the crystal is deformed so far that the new interior configuration is indistinguishable from the undeformed crystal.

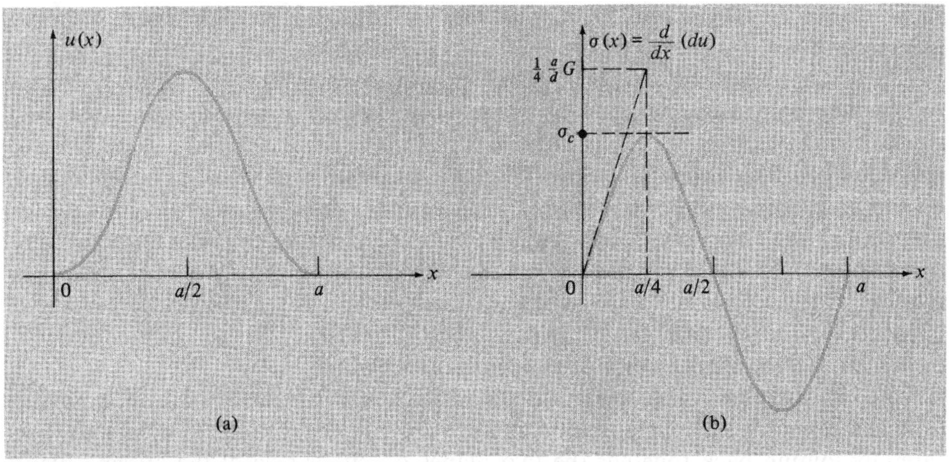

Figure 30.13
(a) The behavior of the additional energy per unit volume, $u(x)$, due to a shear strain x. Note that $u(x + a) = u(x)$. (b) A plot of the force per unit area per plane necessary to maintain the strain x. In this simple model the order of magnitude of the maximum or critical stress σ_c can be estimated by taking the value of σ at $x = a/4$, or alternatively by extrapolating the linear region of $\sigma(x)$ to this value of x.

configuration (ignoring small surface effects) will be indistinguishable from the undeformed crystal, and $u(a)$ will be 0. Indeed, as a function of x, u will be periodic with period a: $u(x + a) = u(x)$, reducing to the form (30.23) only when $x \ll a$ (Figure 30.13a). As a result, starting from the perfect crystal the force $\sigma(x)$ per unit area of plane (per plane) necessary to maintain the displacement x, known as the shear stress, will not increase indefinitely with x. We estimate its maximum size as follows:

If the crystal is made up of N planes of area A, then the volume is $V = ANd$, and the shear stress is given by

$$\sigma = \frac{1}{NA}\frac{d}{dx}(Vu) = d\left(\frac{du}{dx}\right). \tag{30.24}$$

This will be maximum at some displacement x_0 between 0 and $a/2$ (Figure 30.13b). If we roughly estimate the value at maximum by extrapolating the linear region of $\sigma(x)$ (valid for small x) out to $x = a/4$, then we find that the critical shear stress is of order:

$$\sigma_c \approx \frac{d}{dx}\frac{1}{2}G\frac{x^2}{d}\bigg|_{x=a/4} = \frac{1}{4}\frac{a}{d}G \approx 10^{11}\ \text{dynes/cm}^2. \tag{30.25}$$

If a shear greater than σ_c is applied, there is nothing to prevent one plane from sliding over another; i.e., the crystal undergoes *slip*. It is evident from Figure 30.13b that (30.25) gives only a rough estimate of the critical shear stress. However, the observed critical shear stress in apparently well-prepared "single crystals" can be less than the estimate (30.25) by as much as a factor of 10^4! An error of this magnitude suggests that the description of slip, on which the estimate (30.25) is based, is simply incorrect.

The actual process by which slip occurs in most cases is far more subtle. A crucial role is played by a special kind of linear defect known as a *dislocation.* The two simplest kinds, *screw dislocations* and *edge dislocations*, are illustrated in Figure 30.14, and described in more detail below. Dislocation densities in actual crystals depend on the preparation of the specimen,[16] but can range from 10^2 to $10^{12}/\text{cm}^2$. Along a linear dislocation the crystal is in so high a state of local distortion that the additional distortion required to move the dislocation sideways by one lattice constant requires relatively little additional applied stress. Furthermore, the net effect of moving a dislocation through many lattice constants is a displacement by a lattice constant[17] of the two halves of the crystal separated by the plane of motion.[18]

One can imagine constructing an edge dislocation (Figure 30.14a) by removing from the crystal a half plane of atoms terminating in the dislocation line, and then carefully joining back together the two planes on either side of the missing plane

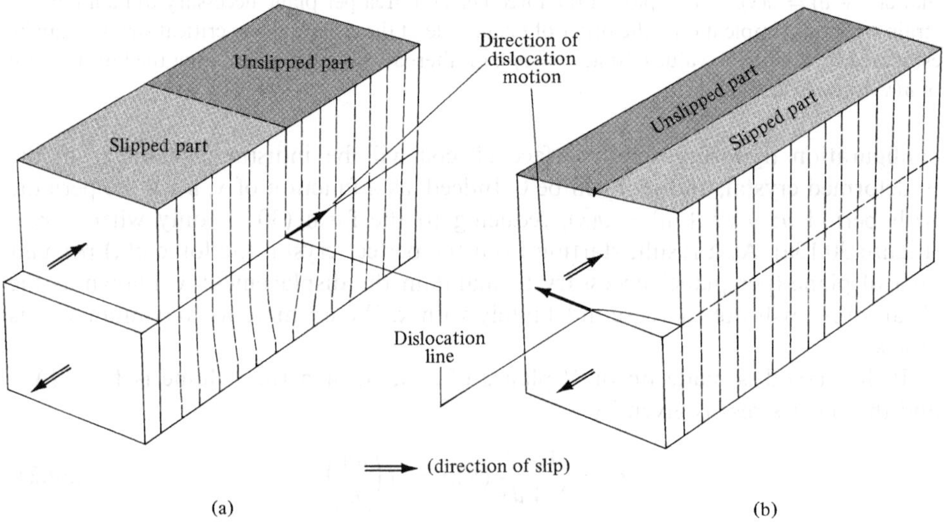

Figure 30.14
(a) Slip in a crystal via the motion of an *edge* dislocation. (b) Slip in a crystal via the motion of a *screw* dislocation.

---

[16] As mentioned above linear defects are not a thermodynamic equilibrium phenomenon. There is therefore no intrinsic value to the dislocation density (which can be considerably reduced by annealing).

[17] There is another kind of slip, mediated by dislocations, in which the slipped portion of the crystal bears a more complex relation to the unslipped part. See the description of "twinning," below.

[18] The analogy is often made to the passage of a linear ripple across a carpet. The effect is a slight displacement of the carpet, made far more easily than by sliding the entire undeformed carpet the same distance.

in a way that restores the basic order of the perfect crystal everywhere except in the vicinity of the dislocation line.[19]

Similarly, a screw dislocation (Figure 30.14b) can be "constructed" by imagining a plane terminating at the dislocation line, above which the crystal has been displaced by a lattice vector parallel to the line, and then rejoined to the part of the crystal below in a way that preserves the basic crystalline order everywhere except near the line itself.

More generally, dislocations need not be rectilinear. One can describe a general

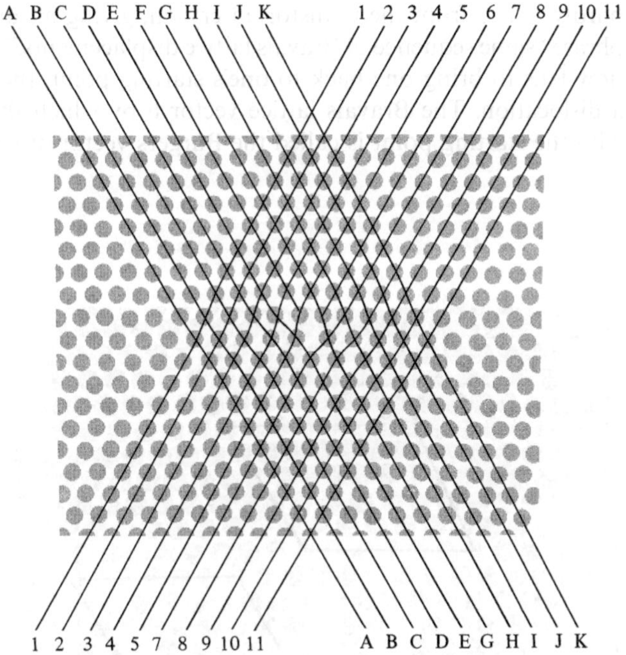

Figure 30.15
Ambiguities in the "constructive" definition of a dislocation. One plane of a crystal is shown, perpendicular to a single edge dislocation. (The point where the dislocation intersects the plane is most easily perceived by viewing the figure at a low angle along either of the families of parallel lines.) One can describe the dislocation as being produced by the insertion of the extra plane of atoms that intersects the upper half of the figure in the line 6, or, equally well, by the insertion of the extra plane of atoms that intersects the upper half of the figure in the line *F*. Alternatively, the dislocation can be viewed as produced by the removal of a plane from the lower half of the figure, and that plane can either have been the one that was between 5 and 7 or between *E* and *G*. The figure is based on the "bubble raft" photographs of Bragg and Nye. (*Proc. Roy. Soc.*, **A190**, 474 (1947).)

[19] Only the dislocation line itself has an absolute significance, however. Given an edge dislocation, there are any number of places from which the "removed plane" might have been taken. Indeed, one may also think of the dislocation as having been constructed by the insertion of an extra plane (Figure 30.15). The same is true of screw dislocations.

dislocation as any linear region in the crystal (either a closed curve or a curve that terminates on the surface) with the following properties:

1. Away from the region the crystal is locally only negligibly different from the perfect crystal.
2. In the neighborhood of the region the atomic positions are substantially different from the original crystalline sites.
3. There exists a nonvanishing *Burgers vector.*

The Burgers vector is defined as follows: Consider a closed curve in the perfect crystal passing through a succession of lattice sites, which therefore can be traversed by a series of displacements by Bravais lattice vectors (Figure 30.16, lower curve). Now traverse the same sequence of Bravais lattice displacements in the putatively dislocated crystal (upper curve in Figure 30.16). The test path should be far enough away from the dislocation that the configuration of the crystal in its neighborhood hardly differs from the undistorted crystal, giving an unambiguous meaning to the phrase "same sequence of Bravais lattice displacements." If the series of displacements now fails to bring one back to one's starting point, then the curve has surrounded a dislocation. The Bravais lattice vector **b** by which the endpoint fails to coincide with the starting point is called the Burgers vector of the dislocation.[20]

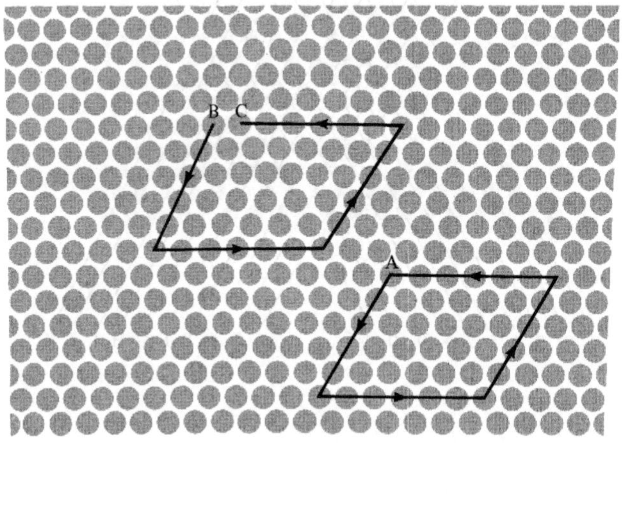

Figure 30.16
Two paths in a lattice plane. The lower path is in a dislocation free region. If one starts at A and moves five steps down, six to the right, five up, and six to the left, one returns to A. The upper path surrounds a dislocation. (The dislocation line is perpendicular to the lattice plane.) If one starts at B and moves through the same sequence of steps (five down, six to the right, five up, and six to the left) one does not return to the starting point B, but to C. The vector from B to C is the Burgers vector **b**. (The dislocation that the second path surrounds is seen most readily by viewing the page at a very low angle.)

[20] If $\mathbf{b} = 0$, the linear defect is not a dislocation (unless it happens that the path has surrounded two dislocations with Burgers vectors of equal magnitudes and opposite directions). A linear array of vacancies, for example, satisfies criteria 1 and 2, but is not a dislocation. (If an ion in the interior of the lower curve in Figure 30.16 is withdrawn, the path still closes.)

Some thought should convince one that the Burgers vector of a given dislocation does not depend on the path around the dislocation chosen for the test. The Burgers vector is perpendicular to an edge dislocation, and parallel to a screw dislocation. Dislocations more complex than edge or screw dislocations can still be described by a single path-independent Burgers vector, though the relation between the direction of the Burgers vector and the geometry of the dislocated region will not be as simple as it is for edge and screw dislocations.[21]

CRYSTALLINE STRENGTH

The weakness of good crystals was a mystery for many years, in part, no doubt, because the observed data easily led one to the wrong conclusion. Relatively poorly prepared crystals were found to have yield strengths close to the high value we first estimated for the perfect crystal. However, as the crystals were improved (for example, by annealing) the yield strengths were found to drop drastically, falling by several orders of magnitude in very well prepared crystals. It was natural to assume that the yield strength was approaching that of a perfect crystal as specimens were improved, but, in fact, quite the opposite was happening.

Three people independently came up with the explanation in 1934,[22] inventing[23] the dislocation to account for the data. They suggested that almost all real crystals contain dislocations, and that plastic slip occurs through their motion as described above. There are then two ways of making a strong crystal. One is to make an essentially perfect crystal, free of all dislocations. This is extremely difficult to achieve.[24] Another way is to arrange to impede the flow of dislocations, for although dislocations move with relative ease in a perfect crystal, if they encounter interstitials, impurities, or even other dislocations crossing their paths, the work required to move them can increase considerably.

Thus the poorly prepared crystal is hard because it is infested with dislocations and defects, and these interfere so seriously with each other's motion that slip can occur only by the more drastic means described earlier. However, as the crystal is purified and improved, dislocations largely move out of the crystal, vacancies and interstitials are reduced to their (low) thermal equilibrium concentrations, and the unimpeded motion of those dislocations that remain makes it possible for the crystal to deform with ease. At this point the crystal is very soft. If one could continue the process of refinement to the point where all dislocations were removed, the crystal

---

[21] If one imagines making a closed dislocation with scalpel and glue, cutting a surface bounding a circuit in the crystal, displacing the surfaces on either side of the cut, and then regluing them together after removing or adding any atoms that arenow required to restore perfect order, then the Burgers vector is the amount by which the surfaces have been displaced. The topological definition (which is equivalent) is perhaps more intuitive since it does not require contemplation of these abstruse operations.

[22] G. I. Taylor, E. Orowan, and G. Polyani. (G. I. Taylor, *Proc. Roy. Soc.* **A145**, 362 (1934); E. Orowan, *Z. Phys.* **89**, 614 (1934); G. Polyani, *Z. Phys.* **98**, 660 (1934).) Dislocations were introduced into the continuum theory of elasticity, some 30 years earlier, by V. Volterra.

[23] Dislocations were not directly observed for almost another 10 years.

[24] See, however, the description of "whiskers" below.

would again become hard. In certain cases this has actually been observed, as we shall see below.

WORK HARDENING

It is a familiar fact that a bar of soft metal, after repeated bendings back and forth, eventually refuses to be bent, and breaks. This is an example of *work hardening*. With every bending, more and more dislocations flow into the metal, until there are so many that they impede each other's flow. Then the crystal is incapable of further plastic deformation, and breaks under subsequent stress.

DISLOCATIONS AND CRYSTAL GROWTH

The problem of plastic (i.e., irreversible) flow was unraveled by attributing it to the motion of dislocations. An equally perplexing problem was that of crystal growth, which was solved by invoking the existence of screw dislocations. Suppose we grow a large crystal by exposing a small piece of crystal to a vapor of the same atoms. Atoms from the vapor will condense into lattice positions more readily if neighbors surrounding the site are already in place. Thus an atom is relatively weakly attracted to a perfect crystal plane, more strongly attracted to a step between two planes, and most strongly attracted to a corner (Figure 30.17). If one assumes that grown crystals are perfect, and that growth takes place plane by plane, then whenever a new plane is required, atoms must condense onto the plane below as in Figure 30.17a. Because of the relatively weak binding in this case, such processes (known as "nucleating the next layer") occur far too slowly to account for the observed rate of crystal growth. If, however, the crystal contains a screw dislocation, it is never necessary to nucleate a new plane, for the local planar structure can wind endlessly about the screw dislocation like a spiral ramp (as in Figure 30.17d).

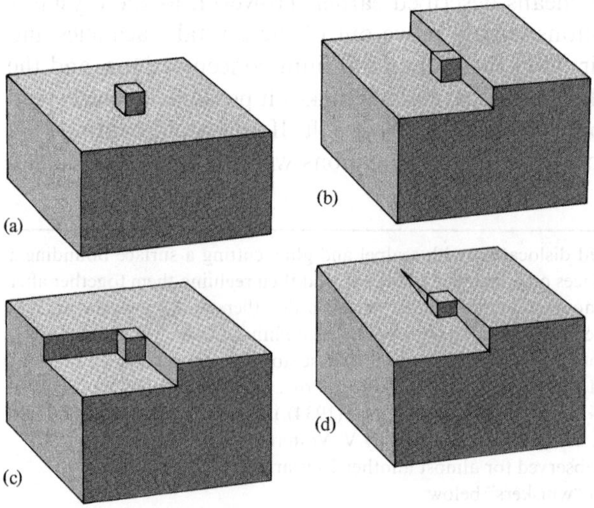

Figure 30.17
Atoms are relatively weakly attracted to perfect crystal planes (a), are more strongly attracted to a step between two planes (b), and are most strongly attracted to a corner (c). If the crystal contains a screw dislocation, as in (d), then by adding atoms as shown, the local planar structure can spiral endlessly around the dislocation. Crystals can grow much more rapidly in this way, since the nucleation of new planes by the process shown in (a) is never required.

WHISKERS

The type of crystal growth described above can lead to a very long, thin, whisker-shaped crystal winding about and extending a single screw dislocation. Such whiskers may contain only a single dislocation (the nucleating screw dislocation itself) and are observed to have yield strengths comparable to those predicted by the perfect crystal model.

OBSERVATION OF DISLOCATIONS AND OTHER DEFECTS

One of the earliest confirmations that dislocations (and other varieties of defects) can indeed exist in naturally formed crystals was provided by the observations of Bragg and Nye[25] on rafts of identical bubbles floating on the surface of soap solutions. The bubbles are held together by surface tension and a two-dimensional array of them approximate very well a section of crystal. Point defects, dislocations, and grain boundaries were all found to occur in the bubble arrays.

Direct observations in solids have since been made by the technique of transmission electron microscopy. Chemical etching also reveals the intersection of dislocations with solid surfaces. At such points the solid is in a state of considerable strain, and the nearby atoms may be preferentially dislodged by chemical action.

SURFACE IMPERFECTIONS: STACKING FAULTS

There is a more complex kind of slip, mediated by dislocations, in which the applied stress causes the coherent formation of dislocations in successive crystal planes. As each dislocation moves through the crystal it leaves in its wake a lattice plane displaced by a non-Bravais lattice vector, and the result of the passage of the family of dislocations is a region in which the crystalline ordering is the mirror image (in the plane of slip) of the original crystal. Such processes are known as "twinning," and the inverted region is known as a "deformation twin."

For example, in a perfect face-centered cubic, successive (111) planes are arranged in the pattern

$$\ldots\widehat{ABC}\widehat{ABC}\widehat{ABC}\widehat{ABC}\ldots, \tag{30.26}$$

as in Figure 4.21. After slip giving rise to a deformation twin, the pattern will be

$$\ldots ABCABCABCAB\overset{\Downarrow}{C}BACBACBA\ldots, \tag{30.27}$$

where the double arrow indicates the boundary of the slipped region.

Misplaced planes of atoms such as these are known as "stacking faults." Another example is the arrangement

$$\ldots ABCABCAB\overset{\Downarrow}{A}BCABCABC\ldots, \tag{30.28}$$

[25] W. L. Bragg and J. F. Nye, *Proc. Roy. Soc.*, **A190**, 474 (1947).

in which a given plane (indicated by the double arrow) is out of step, falling into the hexagonal close-packed sequence rather than that appropriate to face-centered cubic, after which the regular (unmirrored) fcc arrangement is resumed.

LOW-ANGLE GRAIN BOUNDARY

A *grain boundary* is formed by the junction of two single crystals of different orientation along a common planar surface. When the difference in orientations is *small*, the boundary is referred to as a *low-angle* grain boundary. An example known as a *tilt* boundary is shown in Figure 30.18. It is formed from a linear sequence of edge dislocations. There is also a *twist* boundary, which is formed from a sequence of screw dislocations. In general, low-angle grain boundaries are composed of a mixture of both types.

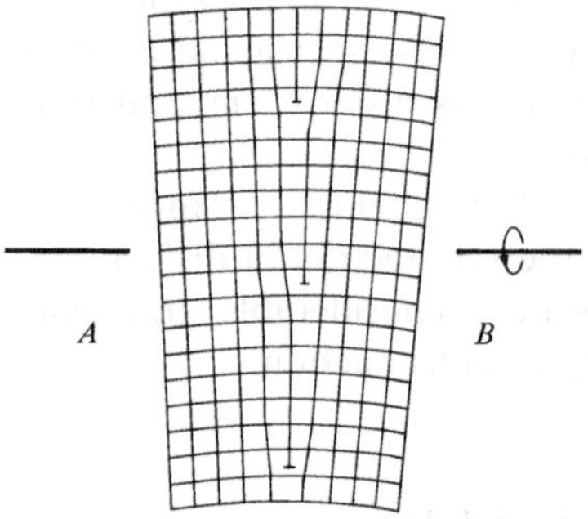

Figure 30.18
A low-angle tilt boundary (a type of low angle grain boundary) can be regarded as formed from a sequence of edge dislocations. If the section B of the crystal is twisted relative to A about the axis shown by a small amount, we may generate (in addition) a twist component in the boundary. A twist boundary, if of small angle, may be viewed as composed of a sequence of screw dislocations.

Unless very carefully prepared, most real crystals consist of many slightly misaligned grains, separated by low-angle grain boundaries. The misalignment is little enough for X-ray diffraction to reveal sharp Bragg peaks, but the existence of the grains has an important effect on the intensity of the peaks.

PROBLEMS

1. *Phonon Correction to the Density of Vacancies*

A more accurate treatment of the equilibrium number of vacancies in a monatomic Bravais lattice would multiply (30.11) by a phonon correction (see (30.9)):

$$n = Ne^{-\beta\varepsilon_0}e^{-\beta(\partial F^{ph}/\partial n)}. \tag{30.29}$$

Make a simple Einstein theory of the normal modes of the crystal with vacancies, i.e., treat each ion as an independent oscillator, but let the oscillator frequency be either ω_E or $\bar{\omega}_E$, according to whether the ion has or does not have one of its (z) nearest neighboring sites vacant. Show that in this model (30.29) becomes:

$$n = Ne^{-\beta\varepsilon_0}\left[\frac{1 - e^{-\beta\hbar\omega_E}}{1 - e^{-\beta\hbar\bar{\omega}_E}}\right]^{3z} \tag{30.30}$$

Since $\bar{\omega}_E < \omega_E$ (why?) the phonon correction favors vacancy formation. Discuss its form when $T \gg \Theta_E$ and when $T \ll \Theta_E$.

2. *Mixed Schottky and Frenkel Defects*

Consider a diatomic ionic crystal in which the energies of formation for positive and negative ion vacancies and interstitials are given by $\mathcal{E}^v_+, \mathcal{E}^v_-, \mathcal{E}^i_+, \mathcal{E}^i_-$. If negative ion interstitials are prohibited (i.e., if $\mathcal{E}^i_-$ is much larger than the others on the scale of k_BT), then positive ion vacancies will be the only possible negatively charged defects. Their charge can be balanced either by negative ion vacancies (Schottky) or positive ion interstitials (Frenkel) depending on whether $\mathcal{E}^i_+ - \mathcal{E}^v_- \gg k_BT$ or $\mathcal{E}^v_- - \mathcal{E}^i_+ \gg k_BT$, respectively. In the Schottky case Eq. (30.19) gives

$$(n^v_+)_s = (n^v_-)_s = [N^v_+N^v_-e^{-\beta(\mathcal{E}^v_+ + \mathcal{E}^v_-)}]^{1/2}, \tag{30.31}$$

and in the Frenkel case,

$$(n^v_+)_f = (n^i_+)_f = [N^v_+N^i_+e^{-\beta(\mathcal{E}^v_+ + \mathcal{E}^i_+)}]^{1/2}. \tag{30.32}$$

Show that if neither case applies (i.e., if $\mathcal{E}^i_+ - \mathcal{E}^v_- = O(k_BT)$, then the concentrations of the three defect types will be given by

$$\begin{aligned} n^v_+ &= [(n^v_+)_s{}^2 + (n^v_+)_f{}^2]^{1/2}, \\ n^i_+ &= \frac{(n^i_+)_f{}^2}{n^v_+}, \\ n^v_- &= \frac{(n^v_-)_s{}^2}{n^v_+}. \end{aligned} \tag{30.33}$$

Verify that these reduce back to (30.31) and (30.32) in the appropriate limits.

3. *Point Defects in Calcium-Doped Sodium Chloride*

Consider a crystal of Ca-doped NaCl, with n_{Ca} calcium atoms per cubic centimeter. Noting that pure NaCl has defects of the Schottky type with concentrations

$$n^v_+ = n^v_- = n_i = (N_+N_-)^{1/2}e^{-\beta(\mathcal{E}_+ + \mathcal{E}_-)/2}, \tag{30.34}$$

show that the defect densities in the doped crystal are given by

$$\begin{aligned} n^v_+ &= \frac{1}{2}\left[\sqrt{4n_i{}^2 + n_{\text{Ca}}{}^2} + n_{\text{Ca}}\right], \\ n^v_- &= \frac{1}{2}\left[\sqrt{4n_i{}^2 + n_{\text{Ca}}{}^2} - n_{\text{Ca}}\right]. \end{aligned} \tag{30.35}$$

(Note the similarity to the theory of doped semiconductors; see Eq. (28.38).)

4. *Shear Stress of a Perfect Crystal*

Show from (22.82) that (30.22) is valid for a cubic crystal.

5. *Simple Model of an F-Center*

Figure 30.19b shows the positions of the maxima of F-center bands (illustrated for the chlorides in Figure 30.19a) as a function of lattice constant a. Take as a model of the F-center an electron

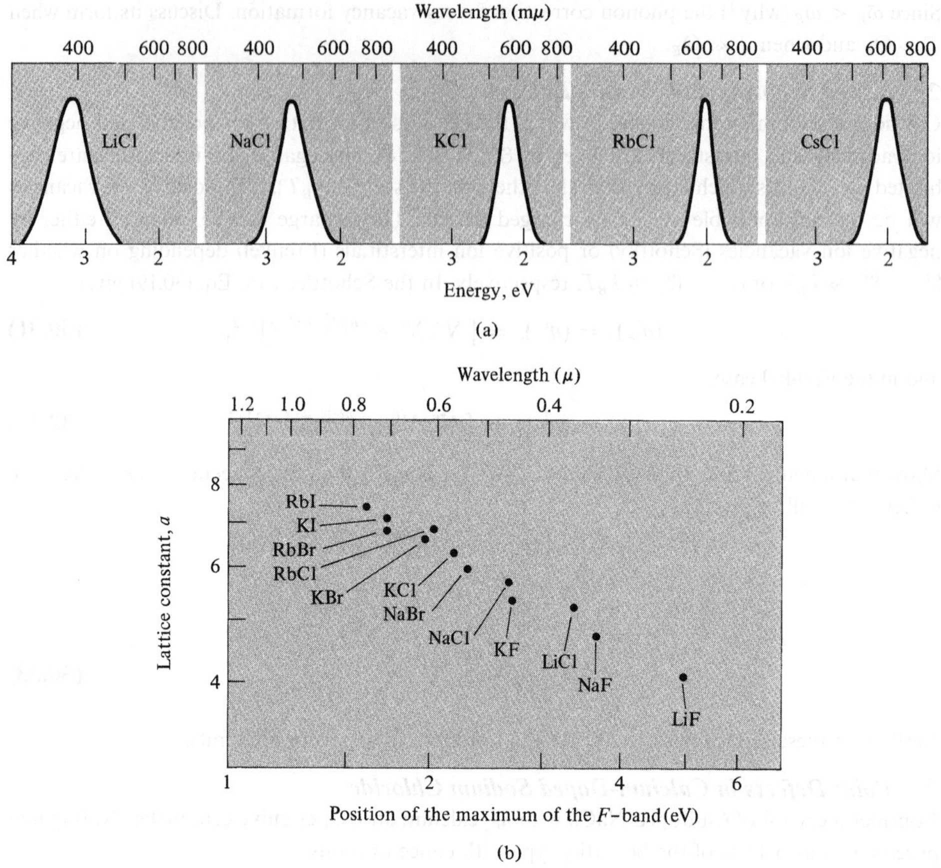

Figure 30.19
(a) Absorption bands of *F*-centers in some alkali chlorides. (b) The dependence on lattice constant of the maximum in the *F*-center absorption band. (From Schulman and Compton, *Color Centers in Solids*, Pergamon, New York, 1962.

trapped by a vacancy potential of the form $V(r) = 0$, $r < d$; $V(r) = \infty$, $r > d$, where d is proportional to the lattice constant a. Show that the spectrum scales as $1/d^2$, so that if the peaks are associated with the same types of excitation,

$$\lambda_{\max} \propto a^2 \tag{30.36}$$

where $\lambda_{\max}$ is the wavelength corresponding to the observed maximum in the F-band absorption. (Equation (30.36) is known as the Mollwo relation.)

6. *Burgers Vector*
What is the smallest Burgers vector parallel to a [111] direction that a dislocation may have in an fcc crystal?

7. *Elastic Energy of a Screw Dislocation*
Consider a region of crystal of radius r about a screw dislocation with Burgers vector **b** (Figure 30.20). Provided r is sufficiently large, the shear strain is $b/2\pi r$. (What happens close to the dis-

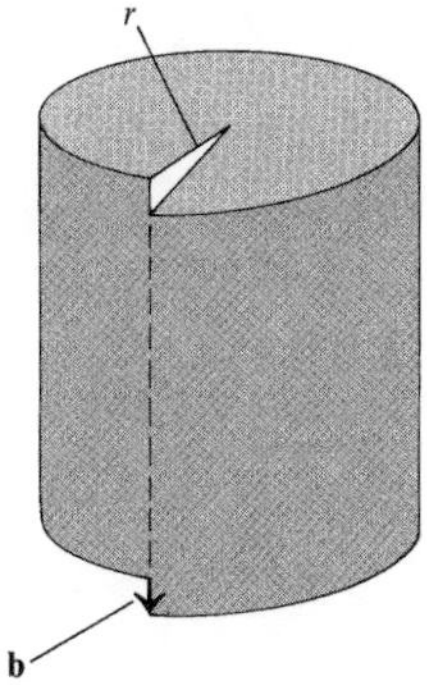

Figure 30.20
A screw dislocation and its Burgers vector **b**.

location?) Assuming that stress and strain are related by Eq. (30.23), show that the total elastic energy per unit length of the screw dislocation is

$$G\frac{b^2}{4\pi}\ln\frac{R}{r_0}, \tag{30.36}$$

where R and r_0 are upper and lower limits on r. What physical considerations determine reasonable values for these quantities?

31
Diamagnetism and Paramagnetism

In the preceding chapters we have considered the effect of a magnetic field only on metals, and only insofar as the motion of conduction electrons in the field revealed the metal's Fermi surface. In the next three chapters we turn our attention to some of the more intrinsically magnetic properties of solids: the magnetic moments they exhibit in the presence (and sometimes even in the absence) of applied magnetic fields.

In this chapter we shall first review the theory of atomic magnetism. We shall then consider those magnetic properties of insulating solids that can be understood in terms of the properties of their individual atoms or ions with, if necessary, suitable modifications to take into account effects of the crystalline environment. We shall also consider those magnetic properties of metals that can be at least qualitatively understood in the independent electron approximation.

In none of the applications of this chapter shall we discuss at any length electron-electron interactions. This is because in the case of insulators we shall base our analysis on results of atomic physics (whose derivation depends critically, of course, on such interactions), and because in the case of metals the phenomena we shall describe here are at least roughly accounted for in an independent electron model. In Chapter 32 we shall turn to an examination of the physics underlying those electron-electron interactions that can profoundly affect the characteristically magnetic properties of metals and insulators. In Chapter 33 we shall describe the further magnetic phenomena (such as ferromagnetism and antiferromagnetism) that can result from these interactions.

MAGNETIZATION DENSITY AND SUSCEPTIBILITY

At $T = 0$ the *magnetization density* $M(H)$ of a quantum-mechanical system of volume V in a uniform magnetic field[1] H is defined to be[2]

$$M(H) = -\frac{1}{V}\frac{\partial E_0(H)}{\partial H}, \tag{31.1}$$

where $E_0(H)$ is the ground-state energy in the presence of the field H. If the system is in thermal equilibrium at temperature T, then one defines the magnetization density as the thermal equilibrium average of the magnetization density of each excited state of energy $E_n(H)$:

$$M(H, T) = \frac{\sum\limits_n M_n(H)e^{-E_n/k_BT}}{\sum\limits_n e^{-E_n/k_BT}}, \tag{31.2}$$

[1] We shall take H to be the field that acts on the individual microscopic magnetic moments within the solid. As in the case of a dielectric solid (cf. Chapter 27), this need not be the same as the applied field. However, for the paramagnetic and diamagnetic substances to be discussed in this chapter the local field corrections are very small, and will be ignored.

[2] For simplicity we assume that **M** is parallel to **H**. More generally one should write a vector equation: $M_\mu = -(1/V)\partial E_0/\partial H_\mu$, and the susceptibility (defined below) will be a tensor. In Problem 1 it is shown that this definition is equivalent to the more familiar one of Ampère encountered in the conventional formulations of classical macroscopic electrodynamics.

where

$$M_n(H) = -\frac{1}{V}\frac{\partial E_n(H)}{\partial H}. \tag{31.3}$$

This can also be written in the thermodynamic form

$$M = -\frac{1}{V}\frac{\partial F}{\partial H}, \tag{31.4}$$

where F, the magnetic Helmholtz free energy, is defined by the fundamental statistical mechanical rule:

$$e^{-F/k_BT} = \sum_n e^{-E_n(H)/k_BT}. \tag{31.5}$$

The *susceptibility* is defined as[3]

$$\chi = \frac{\partial M}{\partial H} = -\frac{1}{V}\frac{\partial^2 F}{\partial H^2}. \tag{31.6}$$

The magnetization can be measured by recording the force exerted on a specimen by an inhomogeneous field that varies slowly over the sample, for the change in free energy on moving the specimen[4] from x to $x + dx$ will be:[5]

$$dF = F(H(x + dx)) - F(H(x)) = \frac{\partial F}{\partial H}\frac{\partial H}{\partial x}\,dx = -VM\frac{\partial H}{\partial x}\,dx, \tag{31.7}$$

and therefore the force per unit volume f exerted on the specimen by the field is

$$f = -\frac{1}{V}\frac{dF}{dx} = M\frac{\partial H}{\partial x}. \tag{31.8}$$

CALCULATION OF ATOMIC SUSCEPTIBILITIES: GENERAL FORMULATION

In the presence of a uniform magnetic field the Hamiltonian of an ion (or atom) is modified in the following major ways:[6]

---

[3] Often, as we shall see, M is very accurately linear in H for attainable field strengths, in which case the definition reduces to $\chi = M/H$. Note also that χ is dimensionless (in CGS units) since H^2 has the dimensions of energy per unit volume.

[4] Equal to the mechanical work done on the specimen if the temperature is held constant.

[5] We take the field to be along the z-direction and move the sample in the x-direction.

[6] Note the following ways in which, generally, one does *not* bother to modify it. One almost always neglects the effects of the magnetic field on the translational motion of the *ion*; i.e., one does not make the replacement (31.9) for the momentum operators describing the ionic nuclei. Furthermore, unless one is explicitly interested in nuclear spin effects (as in the case of magnetic resonance experiments) one ignores the analogue of (31.12) for the nuclear spins. In both cases these simplifications are justified by the very much greater nuclear mass, which makes the nuclear contribution to the magnetic moment of the solid some 10^6 to 10^8 times smaller than the electronic contribution. Finally, the replacement (31.9) in the electron momentum operators that appear in the terms describing spin-orbit coupling leads to very small corrections compared with the direct coupling of the electron spin to the magnetic field, and is also generally ignored.

1. In the total electronic kinetic energy, $T_0 = \sum p_i^2/2m$, the momentum of each electron (of charge $-e$) is replaced by[7]

$$\mathbf{p}_i \rightarrow \mathbf{p}_i + \frac{e}{c}\mathbf{A}(\mathbf{r}_i), \tag{31.9}$$

where **A** is the vector potential. In this chapter we shall take **A** in a uniform field **H** to have the form:

$$\mathbf{A} = -\tfrac{1}{2}\mathbf{r} \times \mathbf{H}, \tag{31.10}$$

so that the conditions

$$\mathbf{H} = \nabla \times \mathbf{A} \quad \text{and} \quad \nabla \cdot \mathbf{A} = 0 \tag{31.11}$$

are both satisfied.

2. The interaction energy of the field with each electron spin $\mathbf{s}^i = \frac{1}{2}\boldsymbol{\sigma}_i$ must be added to the Hamiltonian:[8]

$$\Delta\mathcal{H} = g_0\mu_B H\mathbf{S}_z, \qquad \left(\mathbf{S}_z = \sum_i \mathbf{s}_z{}^i\right). \tag{31.12}$$

Here μ_B, the *Bohr magneton*, is given by

$$\mu_B = \frac{e\hbar}{2mc} = 0.927 \times 10^{-20}\ \text{erg/G}$$
$$= 0.579 \times 10^{-8}\ \text{eV/G}, \tag{31.13}$$

and g_0, the electronic *g-factor*, is given by

$$g_0 = 2\left[1 + \frac{\alpha}{2\pi} + O(\alpha^2) + \cdots\right], \qquad \alpha = \frac{e^2}{\hbar c} \approx \frac{1}{137}$$
$$= 2.0023, \tag{31.14}$$

which, to the accuracy of most measurements of interest in solids, can be taken to be precisely 2.

As a result of (31.9) the total electronic kinetic energy operator must be replaced by

$$T = \frac{1}{2m}\sum_i \left[\mathbf{p}_i + \frac{e}{c}\mathbf{A}(\mathbf{r}_i)\right]^2 = \frac{1}{2m}\sum_i \left(\mathbf{p}_i - \frac{e}{2c}\mathbf{r}_i \times \mathbf{H}\right)^2, \tag{31.15}$$

---

[7] In a purely classical theory (taking the electron's spin to be a quantum phenomenon) this would be the only effect of the field. It can then easily be demonstrated from classical statistical mechanics that the thermal equilibrium magnetization must always vanish (Bohr–von Leeuwen theorem) for the sum defining the free energy becomes an integral over a $6N$-dimensional N-electron phase space:

$$e^{-\beta F} = \int \prod_{i=1}^{N} d\mathbf{p}_i\, d\mathbf{r}_i \exp\left[-\beta H(\mathbf{r}_1, \ldots, \mathbf{r}_N; \mathbf{p}_1, \ldots, \mathbf{p}_N)\right].$$

Since the magnetic field enters *only* in the form $\mathbf{p}_i + e\mathbf{A}(\mathbf{r}_i)/c$ it can be eliminated entirely by a simple shift of origin in the momentum integrations (the limits of which run from $-\infty$ to ∞ and are therefore unaffected by the shift). But if F does not depend on H, then the magnetization, being proportional to $\partial F/\partial H$, must vanish. Thus a quantum theory is required from the start to explain any magnetic phenomena.

[8] In magnetic problems we use script $\mathcal{H}$ for the Hamiltonian to avoid confusion with the magnetic field strength H. We also use dimensionless spins (with integral or half-integral values) so that the angular momentum is $\hbar$ times the spin.

which can be expanded to give

$$T = T_0 + \mu_B \mathsf{L} \cdot \mathbf{H} + \frac{e^2}{8mc^2} H^2 \sum_i (x_i^2 + y_i^2), \tag{31.16}$$

where L is the total electronic orbital angular momentum:[9]

$$\hbar \mathsf{L} = \sum_i \mathbf{r}_i \times \mathbf{p}_i. \tag{31.17}$$

The spin term (31.12) combines with (31.16) to give the following field-dependent terms in the Hamiltonian:

$$\Delta\mathcal{H} = \mu_B(\mathsf{L} + g_0\mathsf{S}) \cdot \mathbf{H} + \frac{e^2}{8mc^2} H^2 \sum_i (x_i^2 + y_i^2). \tag{31.18}$$

We shall see below that the energy shifts produced by (31.18) are generally quite small on the scale of atomic excitation energies, even for the highest presently attainable laboratory field strengths. Therefore one can compute the changes in the energy levels induced by the field with ordinary perturbation theory. To compute the susceptibility, a second derivative with respect to the field, one must retain terms up to the second order in H, and must therefore use the famous result of second-order perturbation theory:[10]

$$E_n \to E_n + \Delta E_n; \quad \Delta E_n = \langle n|\Delta\mathcal{H}|n\rangle + \sum_{n' \neq n} \frac{|\langle n|\Delta\mathcal{H}|n'\rangle|^2}{E_n - E_{n'}}. \tag{31.19}$$

Inserting (31.18) into (31.19) and retaining terms through those quadratic in H, we find that to second order,[11]

$$\Delta E_n = \mu_B \mathbf{H} \cdot \langle n|\mathsf{L} + g_0\mathsf{S}|n\rangle + \sum_{n' \neq n} \frac{|\langle n|\mu_B\mathbf{H} \cdot (\mathsf{L} + g_0\mathsf{S})|n'\rangle|^2}{E_n - E_{n'}} + \frac{e^2}{8mc^2} H^2 \langle n|\sum_i (x_i^2 + y_i^2)|n\rangle. \tag{31.20}$$

Equation (31.20) is the basis for theories of the magnetic susceptibility of individual atoms, ions, or molecules. It also underlies theories of the susceptibilities of those solids that can be represented as a collection of only slightly deformed individual ions, i.e., ionic and molecular solids. In such cases the susceptibility is computed ion by ion.

---

[9] We measure L in the same dimensionless units as the spin, so that each component of L has integral eigenvalues, and the orbital angular momentum in conventional units is $\hbar\mathsf{L}$. We also use sans serif boldface type to denote angular momentum operators. Note also that by L we shall mean the vector operator whose components are L_x, L_y, and L_z. (Similar remarks apply to the spin operator S, and the total angular momentum operator J.)

[10] D. Park, *Introduction to the Quantum Theory*. McGraw-Hill, New York, 1964, Chapter 8. Note that if the nth level is degenerate, as is often the case, the states n must be chosen so as to diagonalize $\Delta\mathcal{H}$ in the degenerate subspace. This is not hard to arrange, as we shall see.

[11] The quantity e^2/mc^2 may be written as $\alpha^2 a_0$.

Before applying (31.20) to particular cases, we first observe that unless the term that is linear in H vanishes identically (as it sometimes does), it will almost always be the dominant term even when the field is very strong ($\sim 10^4$ gauss). If it does not vanish, $\langle n|(L_z + g_0 S_z)|n\rangle$ will be of order unity so that

$$\mu_B \mathbf{H} \cdot \langle n|(\mathbf{L} + g_0 \mathbf{S})|n\rangle = O(\mu_B H) \sim \frac{\hbar e H}{mc} \sim \hbar\omega_c. \tag{31.21}$$

This is of order 10^{-4} eV when H is of order 10^4 gauss (thereby substantiating our earlier assertion that the energy shifts are small). To estimate the size of the last term in $\Delta\mathcal{H}$, we note that $\langle n|(x_i^2 + y_i^2)|n\rangle$ will be of the order of the square of a typical atomic dimension, so that

$$\frac{e^2}{8mc^2} H^2 \langle n|\sum_i (x_i^2 + y_i^2)|n\rangle = O\left[\left(\frac{eH}{mc}\right)^2 m a_0^2\right] \approx (\hbar\omega_c)\left(\frac{\hbar\omega_c}{e^2/a_0}\right). \tag{31.22}$$

Since e^2/a_0 is about 27 eV, this term is smaller than the linear term (31.21) by a factor of about 10^{-5}, even at fields as high as 10^4 gauss. The second term in (31.20) can also be shown to be smaller than the first by a factor of order $\hbar\omega_c/\Delta$, where $\Delta = \min |E_n - E_{n'}|$ is a typical atomic excitation energy. In most cases Δ will be large enough to make this factor quite small.

SUSCEPTIBILITY OF INSULATORS WITH ALL SHELLS FILLED: LARMOR DIAMAGNETISM

The simplest application of these results is to a solid composed of ions[12] with all electronic shells filled. Such an ion has zero spin and orbital angular momentum in its ground state[13] $|0\rangle$

$$\mathbf{J}|0\rangle = \mathbf{L}|0\rangle = \mathbf{S}|0\rangle = 0. \tag{31.23}$$

Consequently only the third term in (31.20) contributes to the field-induced shift in the ground-state energy:[14]

$$\Delta E_0 = \frac{e^2}{8mc^2} H^2 \langle 0|\sum_i (x_i^2 + y_i^2)|0\rangle = \frac{e^2}{12mc^2} H^2 \langle 0|\sum r_i^2|0\rangle. \tag{31.24}$$

If (as is the case at all but very high temperatures) there is negligible probability of the ion being in any but its ground state in thermal equilibrium, then the susceptibility of a solid composed of N such ions is given by

$$\chi = -\frac{N}{V}\frac{\partial^2 \Delta E_0}{\partial H^2} = -\frac{e^2}{6mc^2}\frac{N}{V}\langle 0|\sum_i r_i^2|0\rangle. \tag{31.25}$$

---

[12] As in earlier chapters we continue to use the term "ion" to mean ion or atom, the latter being an ion of charge 0.

[13] This is because the ground state of a closed-shell ion is spherically symmetric. It is also an especially simple consequence of Hund's rules (see below).

[14] The last form follows from the spherical symmetry of the closed-shell ion:

$$\langle 0|\Sigma x_i^2|0\rangle = \langle 0|\Sigma y_i^2|0\rangle = \langle 0|\Sigma z_i^2|0\rangle = \tfrac{1}{3}\langle 0|\Sigma r_i^2|0\rangle$$

This is known as the *Larmor diamagnetic susceptibility*.[15] The term *diamagnetism* is applied to cases of negative susceptibility—i.e., cases in which the induced moment is opposite to the applied field.

Equation (31.25) should describe the magnetic response of the solid noble gases and of simple ionic crystals such as the alkali halides, since in these solids the ions are only slightly distorted by their crystalline environment. Indeed, in the alkali halides the susceptibilities can be represented, to within a few percent, as a sum of independent susceptibilities for the positive and negative ions. These ionic susceptibilities also give accurately the contribution of the alkali halides to the susceptibility of solutions in which they are dissolved.

Susceptibilities are usually quoted as molar susceptibilities, based on the magnetization per mole, rather than per cubic centimeter. Thus χ^{molar} is given by multiplying χ by the volume of a mole, $N_A/[N/V]$, where N_A is Avogadro's number. It is also conventional to define a mean square ionic radius by

$$\langle r^2 \rangle = \frac{1}{Z_i} \sum \langle 0 | r_i{}^2 | 0 \rangle, \tag{31.26}$$

where Z_i is the *total* number of electrons in the ion. Thus the molar susceptibility is written:

$$\chi^{\text{molar}} = -Z_i N_A \frac{e^2}{6mc^2} \langle r^2 \rangle = -Z_i \left(\frac{e^2}{\hbar c} \right)^2 \frac{N_A a_0{}^3}{6} \langle (r/a_0)^2 \rangle. \tag{31.27}$$

Since $a_0 = 0.529$ Å, $e^2/\hbar c = 1/137$, and $N_A = 0.6022 \times 10^{24}$,

$$\chi^{\text{molar}} = -0.79 Z_i \times 10^{-6} \langle (r/a_0)^2 \rangle \text{ cm}^3/\text{mole}. \tag{31.28}$$

The quantity $\langle (r/a_0)^2 \rangle$ is of order unity, as is the number of moles per cubic centimeter (by which the molar susceptibility must be multiplied to get the dimensionless susceptibility defined in (31.6)). We conclude that diamagnetic susceptibilities are typically of order 10^{-5}; i.e., M is minute compared with H.

Molar susceptibilities for the noble gases and the alkali halide ions are given in Table 31.1.

Table 31.1

MOLAR SUSCEPTIBILITIES OF NOBLE GAS ATOMS AND ALKALI HALIDE IONS[a]

ELEMENT	SUSCEPTIBILITY	ELEMENT	SUSCEPTIBILITY	ELEMENT	SUSCEPTIBILITY
		He	−1.9	Li^+	−0.7
F^-	−9.4	Ne	−7.2	Na^+	−6.1
Cl^-	−24.2	A	−19.4	K^+	−14.6
Br^-	−34.5	Kr	−28	Rb^+	−22.0
I^-	−50.6	Xe	−43	Cs^+	−35.1

[a] In units of 10^{-6} cm^3/mole. Ions in each row have the same electronic configuration.

Source: R. Kubo and T. Nagamiya, eds., *Solid State Physics*, McGraw-Hill, New York, 1969, p. 439.

[15] It is also frequently referred to as the Langevin susceptibility.

When a solid contains some ions with partially filled electronic shells, its magnetic behavior is very different. Before we can apply the general result (31.20) to this case, we must review the basic facts about the low-lying states of such ions.

GROUND STATE OF IONS WITH A PARTIALLY FILLED SHELL: HUND'S RULES

Suppose we have a free[16] atom or ion in which all electronic shells are filled or empty except for one, whose one-electron levels are characterized by orbital angular momentum l. Since for given l there are $2l + 1$ values l_z can have ($l, l - 1, l - 2, \ldots, -l$) and two possible spin directions for each l_z, such a shell will contain $2(2l + 1)$ one-electron levels. Let n be the number of electrons in the shell, with $0 < n < 2(2l + 1)$. If the electrons did not interact with one another, the ionic ground state would be degenerate, reflecting the large number of ways of putting n electrons into more than n levels. However, this degeneracy is considerably (though in general not completely) lifted by electron-electron Coulomb interactions as well as by the electron spin-orbit interaction. Except for the very heaviest ions (where spin-orbit coupling is very strong) the lowest-lying levels after the degeneracy is lifted can be described by a simple set of rules, justified both by complex calculations and by the analysis of atomic spectra. Here we shall simply state the rules, since we are more interested in their implications for the magnetic properties of solids than in their underlying justification.[17]

1. ***Russel-Saunders Coupling*** To a good approximation[18] the Hamiltonian of the atom or ion can be taken to commute with the total electronic spin and orbital angular momenta, $\mathbf{S}$ and $\mathbf{L}$, as well as with the total electronic angular momentum $\mathbf{J} = \mathbf{L} + \mathbf{S}$. Therefore the states of the ion can be described by quantum numbers L, L_z, S, S_z, J, and J_z, indicating that they are eigenstates of the operators $\mathbf{L}^2$, $\mathbf{L}_z$, $\mathbf{S}^2$, $\mathbf{S}_z$, $\mathbf{J}^2$, and $\mathbf{J}_z$ with eigenvalues $L(L + 1)$, L_z, $S(S + 1)$, S_z, $J(J + 1)$, and J_z, respectively. Since filled shells have zero orbital, spin, and total angular momentum, these quantum numbers describe the electronic configuration of the partially filled shell, as well as the ion as a whole.

2. ***Hund's First Rule*** Out of the many states one can form by placing n electrons into the $2(2l + 1)$ levels of the partially filled shell, those that lie lowest in energy have the largest total spin S that is consistent with the exclusion principle. To see what that value is, one notes that the largest value S can have is equal to the largest magnitude that S_z can have. If $n \leqslant 2l + 1$, all electrons can have parallel spins without multiple occupation of any one-electron level in the shell, by assigning them levels with different values of l_z. Hence $S = \frac{1}{2}n$, when $n \leqslant 2l + 1$. When $n = 2l + 1$,

---

[16] We shall discuss how the behavior of the free atom or ion is modified by the crystalline environment on pages 656–659.

[17] The rules are discussed in most quantum mechanics texts. See, for example, L. D. Landau and E. M. Lifshitz, *Quantum Mechanics*, Addison Wesley, Reading, Mass., 1965.

[18] The total angular momentum J is always a good quantum number for an atom or ion, but L and S are good quantum numbers only to the extent that spin-orbit coupling is unimportant.

S has its maximum value, $l + \frac{1}{2}$. Since electrons after the $(2l + 1)$th are required by the exclusion principle to have their spins opposite to the spins of the first $2l + 1$, S is reduced from its maximum value by half a unit for each electron after the $(2l + 1)$th.

3. ***Hund's Second Rule*** The total orbital angular momentum L of the lowest-lying states has the largest value that is consistent with Hund's first rule, and with the exclusion principle. To determine that value, one notes that it is equal to the largest magnitude that L_z can have. Thus the first electron in the shell will go into a level with $|l_z|$ equal to its maximum value l. The second, according to rule 2, must have the same spin as the first, and is therefore forbidden by the exclusion principle from having the same value of l_z. The best it can do is to have $|l_z| = l - 1$, leading to a total L of $l + (l - 1) = 2l - 1$. Continuing in this way, if the shell is less than half filled, we will have $L = l + (l - 1) + \cdots + [l - (n - 1)]$. When the shell is precisely half filled, all values of l_z must be assumed, and therefore $L = 0$. The second half of the shell is filled with electrons with spin opposite to those in the first half, and therefore the exclusion principle allows us again to go through the same series of values for L we traversed in filling the first half.

4. ***Hund's Third Rule*** The first two rules determine the values of L and S assumed by the states of lowest energy. This still leaves $(2L + 1)(2S + 1)$ possible states. These can be further classified according to their total angular momentum J, which, according to the basic rules of angular momentum composition, can take on all integral values between $|L - S|$ and $L + S$. The degeneracy of the set of $(2L + 1)(2S + 1)$ states is lifted by the spin-orbit coupling, which, within this set of states, can be represented by a term in the Hamiltonian of the simple form $\lambda(\mathbf{L} \cdot \mathbf{S})$. Spin-orbit coupling will favor maximum J (parallel orbital and spin angular momenta) if λ is negative, and minimum J (antiparallel orbital and spin angular momenta) if λ is positive. As it turns out, λ is positive for shells that are less than half filled and negative for shells that are more than half filled. As a result, the value J assumes in the states of lowest energy is:

$$
\begin{aligned}
J &= |L - S|, \qquad && n \leqslant (2l + 1), \\
J &= L + S, \qquad && n \geqslant (2l + 1).
\end{aligned} \tag{31.29}
$$

In magnetic problems one usually deals only with the set of $(2L + 1)(2S + 1)$ states determined by Hund's first two rules, all others lying so much higher in energy as to be of no interest. Furthermore, it is often enough to consider only the $2J + 1$ lowest lying of these specified by the third rule.

The rules are easier to apply than their description might suggest; indeed, in determining the lowest lying J-multiplet (known as a *term*) for ions in a solid, one really encounters only 22 cases of interest: 1 to 9 electrons in a d-shell ($l = 2$) or 1 to 13 electrons in an f-shell ($l = 3$).[19] For unfortunate historical reasons the ground-state multiplet in these cases is not described by the simple triad of numbers

[19] Partially filled p-shells contain valence electrons, and invariably broaden into bands in the solid. Thus the configuration of the electrons they contain in the solid is in no sense a slight distortion of the configuration in the free atom, and the analysis of this chapter is inapplicable.

SLJ. Instead, the orbital angular momentum L is given by a letter, according to the hallowed spectroscopic code:

$$\begin{aligned} L &= 0\ 1\ 2\ 3\ 4\ 5\ 6 \\ X &= S\ P\ D\ F\ G\ H\ I \end{aligned} \tag{31.30}$$

The spin is specified by affixing the number $2S + 1$ (known as the multiplicity) to the letter as a superprefix, and only J is given as the number J, affixed as a right subscript. Thus the lowest lying J-multiplet is described by the symbol: $^{(2S+1)}X_J$.

The cases of major interest for the study of magnetism in solids are given in Table 31.2.

Table 31.2
GROUND STATES OF IONS WITH PARTIALLY FILLED *d*- OR *f*-SHELLS, AS CONSTRUCTED FROM HUND'S RULES[a]

d-shell ($l = 2$)

n	$l_z = 2$,	1,	0,	−1,	−2	S	$L = \lvert\Sigma l_z\rvert$	J		SYMBOL
1	↓					1/2	2	3/2		$^2D_{3/2}$
2	↓	↓				1	3	2	$J = \lvert L - S\rvert$	$^3F_2$
3	↓	↓	↓			3/2	3	3/2		$^4F_{3/2}$
4	↓	↓	↓	↓		2	2	0		$^5D_0$
5	↓	↓	↓	↓	↓	5/2	0	5/2		$^6S_{5/2}$
6	⇅	↑	↑	↑	↑	2	2	4		$^5D_4$
7	⇅	⇅	↑	↑	↑	3/2	3	9/2	$J = L + S$	$^4F_{9/2}$
8	⇅	⇅	⇅	↑	↑	1	3	4		$^3F_4$
9	⇅	⇅	⇅	⇅	↑	1/2	2	5/2		$^2D_{5/2}$
10	⇅	⇅	⇅	⇅	⇅	0	0	0		$^1S_0$

f-shell ($l = 3$)

n	$l_z = 3$,	2,	1,	0,	−1,	−2,	−3	S	$L = \lvert\Sigma l_z\rvert$	J		
1	↓							1/2	3	5/2		$^2F_{5/2}$
2	↓	↓						1	5	4		$^3H_4$
3	↓	↓	↓					3/2	6	9/2		$^4I_{9/2}$
4	↓	↓	↓	↓				2	6	4	$J = \lvert L - S\rvert$	$^5I_4$
5	↓	↓	↓	↓	↓			5/2	5	5/2		$^6H_{5/2}$
6	↓	↓	↓	↓	↓	↓		3	3	0		$^7F_0$
7	↓	↓	↓	↓	↓	↓	↓	7/2	0	7/2		$^8S_{7/2}$
8	⇅	↑	↑	↑	↑	↑	↑	3	3	6		$^7F_6$
9	⇅	⇅	↑	↑	↑	↑	↑	5/2	5	15/2		$^6H_{15/2}$
10	⇅	⇅	⇅	↑	↑	↑	↑	2	6	8	$J = L + S$	$^5I_8$
11	⇅	⇅	⇅	⇅	↑	↑	↑	3/2	6	15/2		$^4I_{15/2}$
12	⇅	⇅	⇅	⇅	⇅	↑	↑	1	5	6		$^3H_6$
13	⇅	⇅	⇅	⇅	⇅	⇅	↑	1/2	3	7/2		$^2F_{7/2}$
14	⇅	⇅	⇅	⇅	⇅	⇅	⇅	0	0	0		$^1S_0$

[a] ↑ = spin $\frac{1}{2}$; ↓ = spin $-\frac{1}{2}$.

SUSCEPTIBILITY OF INSULATORS CONTAINING IONS WITH A PARTIALLY FILLED SHELL: PARAMAGNETISM

There are two cases to distinguish:

1. If the shell has $J = 0$ (as is the case for shells that are one electron short of being half filled) then the ground state is nondegenerate (as in the case of a filled shell) and the linear term in the energy shift (31.20) vanishes.[20] However (in contrast to the case of a filled shell), the second term in (31.20) need not vanish, and the shift in the ground-state energy due to the field will be given by

$$\Delta E_0 = \frac{e^2}{8mc^2} H^2 \langle 0 | \sum_i (x_i^2 + y_i^2) | 0 \rangle - \sum_n \frac{|\langle 0 | \mu_B \mathbf{H} \cdot (\mathbf{L} + g_0 \mathbf{S}) | n \rangle|^2}{E_n - E_0}. \quad \textbf{(31.31)}$$

When the solid contains N/V such ions per unit volume, the susceptibility is

$$\begin{aligned} \chi &= -\frac{N}{V}\frac{\partial^2 E_0}{\partial H^2} \\ &= -\frac{N}{V}\left[\frac{e^2}{4mc^2}\langle 0 | \sum_i (x_i^2 + y_i^2) | 0 \rangle - 2\mu_B{}^2 \sum_n \frac{|\langle 0 | (\mathbf{L}_z + g_0 \mathbf{S}_z | n \rangle|^2}{E_n - E_0}\right]. \end{aligned} \quad \textbf{(31.32)}$$

The first term is just the Larmor diamagnetic susceptibility discussed above. The second term has a sign opposite to that of the first (since the energies of excited states necessarily exceed that of the ground state). It therefore favors alignment of the moment parallel to the field, a behavior known as *paramagnetism*. This paramagnetic correction to the Larmor diamagnetic susceptibility is known as *Van Vleck paramagnetism*.[21] The magnetic behavior of ions with a shell one electron short of being half filled is determined by a balance between Larmor diamagnetism and Van Vleck paramagnetism, *provided* that only the ground state is occupied with appreciable probability in thermal equilibrium, so the free energy is just the ground-state energy. In many such cases, however, the next lowest J-multiplet is close enough to the $J = 0$ ground state that its contribution to the free energy (and hence to the susceptibility) is appreciable, and a more complicated formula than (31.32) is required.

2. If the shell does not have $J = 0$ (i.e., in all cases except for closed shells and shells one electron short of being half filled), then the first term in the energy shift (31.20) will not vanish and, as we indicated above, will almost always be so much larger than the other two that they can safely be ignored. In this case the ground state is $(2J + 1)$-fold degenerate in zero field, and we are faced with the problem of evaluating and diagonalizing the $(2J + 1)$-dimensional square matrix[22]

$$\langle JLSJ_z | (\mathbf{L}_z + g_0 \mathbf{S}_z) | JLSJ_z' \rangle; \qquad J_z, J_z' = -J, \ldots, J. \quad \textbf{(31.33)}$$

---

[20] This follows from the symmetry of states with $J = 0$, as shown in Problem 4.

[21] Van Vleck paramagnetism also arises in the susceptibilities of molecules of a more complex structure than the single ions we consider here.

[22] See the remark in footnote 10.

This task is made simple by a theorem (Wigner-Eckart),[23] which states that the matrix elements of any vector operator in the $(2J + 1)$-dimensional space of eigenstates of $\mathbf{J}^2$ and $\mathbf{J}_z$ with a given value of J, are proportional to the matrix elements of $\mathbf{J}$ itself:

$$\langle JLSJ_z|(\mathbf{L} + g_0\mathbf{S})|JLSJ_z'\rangle = g(JLS)\langle JLSJ_z|\mathbf{J}|JLSJ_z'\rangle. \tag{31.34}$$

The significant feature of this result is that the proportionality constant $g(JLS)$ does not depend on the values of J_z or J_z'.

In particular, since the matrix elements of $\mathbf{J}_z$ are

$$\langle JLSJ_z|\mathbf{J}_z|JLSJ_z'\rangle = J_z\delta_{J_z,J_z'} \tag{31.35}$$

it follows that

$$\langle JLSJ_z|(\mathbf{L}_z + g_0\mathbf{S}_z)|JLSJ_z'\rangle = g(JLS)J_z\delta_{J_zJ_z'}. \tag{31.36}$$

This solves the secular problem: i.e., the matrix is already diagonal in the states of definite J_z, and the $(2J + 1)$-fold degenerate ground state is therefore split into states with definite values of J_z whose energies are uniformly separated by $g(JLS)\mu_B H$.

The value of $g(JLS)$ (known as the Landé g-factor) is easily computed (Appendix P):

$$g(JLS) = \tfrac{1}{2}(g_0 + 1) - \tfrac{1}{2}(g_0 - 1)\frac{L(L+1) - S(S+1)}{J(J+1)}, \tag{31.37}$$

or, taking the electron g-factor g_0 to be exactly 2,

$$g(JLS) = \frac{3}{2} + \frac{1}{2}\left[\frac{S(S+1) - L(L+1)}{J(J+1)}\right]. \tag{31.38}$$

One sometimes encounters the result (31.34), which can be written in the equivalent form,

$$\langle JLSJ_z|(\mathbf{L} + g_0\mathbf{S})|JLSJ_z'\rangle = \langle JLSJ_z|g(JLS)\mathbf{J}|JLSJ_z'\rangle, \tag{31.39}$$

written without the surrounding state vectors:

$$\mathbf{L} + g_0\mathbf{S} = g(JLS)\mathbf{J}. \tag{31.40}$$

We stress that this relation is valid only within the $(2J + 1)$ dimensional set of states that make up the degenerate atomic ground state in zero field; i.e., (31.40) is obeyed only for matrix elements taken between states that are diagonal in J, L, and S. If the splitting between the zero-field atomic ground-state multiplet and the first excited multiplet is large compared with k_BT (as is frequently the case), then only the $(2J + 1)$ states in the ground-state multiplet will contribute appreciably to the free energy. In that case (and only in that case) Eq. (31.40) permits one to interpret the first term in the energy shift (31.20) as expressing the interaction $(-\boldsymbol{\mu} \cdot \mathbf{H})$ of the field with a magnetic moment that is proportional to the total angular momentum of the ion,[24]

$$\boldsymbol{\mu} = -g(JLS)\mu_B\mathbf{J}. \tag{31.41}$$

---

[23] For a proof see, for example, K. Gottfried, *Quantum Mechanics*, vol. 1, W. A. Benjamin, Menlo Park, Calif., 1966, pp. 302–304.

[24] Within the ground-state multiplet the energy of the ion in a field $\mathbf{H}$ is then given by the operator $-\boldsymbol{\mu} \cdot \mathbf{H}$. This is a very simple example of a "spin Hamiltonian" (see pages 679–681).

Because the zero-field ground state is degenerate, it is never permissible to calculate the susceptibility by equating the free energy to the ground-state energy (as we did in the case of the nondegenerate shells with $J = 0$), for as the field goes to zero, the splitting of the $(2J + 1)$ lowest-lying states will be small compared with k_BT. To obtain the susceptibility we must therefore do an additional statistical mechanical calculation.

Magnetization of a Set of Identical Ions of Angular Momentum J: Curie's Law

If only the lowest $2J + 1$ states are thermally excited with appreciable probability, then the free energy (31.5) is given by:

$$e^{-\beta F} = \sum_{J_z=-J}^{J} e^{-\beta\gamma H J_z}, \qquad \gamma = g(JLS)\mu_B, \qquad \beta = \frac{1}{k_B T}. \tag{31.42}$$

This geometric series is easily summed to give:

$$e^{-\beta F} = \frac{e^{\beta\gamma H(J+1/2)} - e^{-\beta\gamma H(J+1/2)}}{e^{\beta\gamma H/2} - e^{-\beta\gamma H/2}}. \tag{31.43}$$

The expression (31.4) for the magnetization of N such ions in a volume V then gives

$$M = -\frac{N}{V}\frac{\partial F}{\partial H} = \frac{N}{V}\gamma J B_J(\beta\gamma J H), \tag{31.44}$$

where the *Brillouin function* $B_J(x)$ is defined by

$$B_J(x) = \frac{2J+1}{2J}\coth\frac{2J+1}{2J}x - \frac{1}{2J}\coth\frac{1}{2J}x. \tag{31.45}$$

This is plotted in Figure 31.1 for several values of J.

Note that as $T \to 0$ for fixed H, $M \to (N/V)\gamma J$; i.e., each ion is completely aligned by the field, $|J_z|$ having its maximum (or "saturation") value J. However, this case

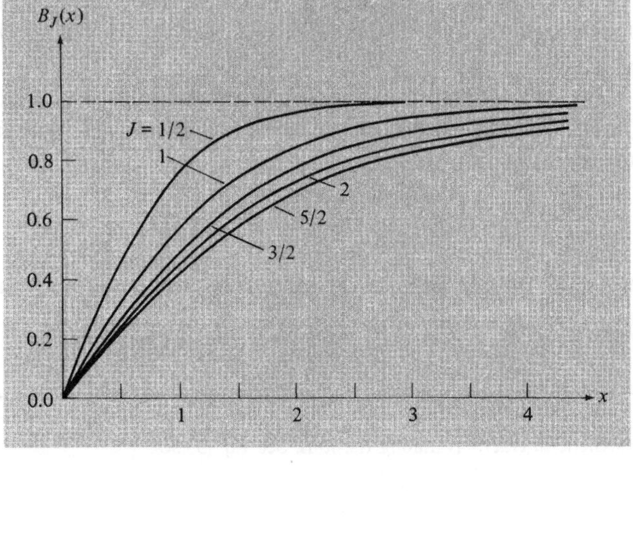

Figure 31.1
Plot of the Brillouin function $B_J(x)$ for various values of the spin J.

arises only when $k_BT \ll \gamma H$; since $\gamma H/k_B \approx \hbar\omega_c/k_B \approx 1$ K in a field of 10^4 gauss, one normally encounters the opposite limit except at the lowest temperatures and highest fields.

When $\gamma H \ll k_BT$ the small-x expansion,

$$\coth x \approx \frac{1}{x} + \tfrac{1}{3}x + O(x^3), \quad B_J(x) \approx \frac{J+1}{3J}x + O(x^3), \tag{31.46}$$

gives

$$\boxed{\chi = \frac{N}{V}\frac{(g\mu_B)^2}{3}\frac{J(J+1)}{k_BT}, \qquad (k_BT \gg g\mu_BH),} \tag{31.47}$$

or

$$\chi^{\text{molar}} = N_A \frac{(g\mu_B)^2}{3}\frac{J(J+1)}{k_BT}. \tag{31.48}$$

This variation of the susceptibility inversely with the temperature is known as Curie's law. It characterizes paramagnetic systems with "permanent moments" whose alignment is favored by the field, and opposed by thermal disorder. Although the condition $k_BT \gg g\mu_BH$ for the validity of Curie's law is satisfied for an enormous range of fields and temperatures, it is important to remember that the "law" is subject to this restriction.[25]

The paramagnetic susceptibility (31.47) is larger than the temperature-independent Larmor diamagnetic susceptibility (31.25) by a factor of order 500 at room temperature (Problem 7) and, therefore, when an ion is present that has a partially filled shell with nonzero J, the contribution of the shell to the total susceptibility of the solid completely dominates the diamagnetic contribution from the other (filled) shells. From our estimate that diamagnetic susceptibilities are of order 10^{-5} (see page 649) we conclude that room temperature paramagnetic susceptibilities should be of order 10^{-2} to 10^{-3}.

Curie's Law in Solids

We now examine the extent to which the above theory of the paramagnetism of free ions continues to describe the behavior of the ions when they are part of the structure of a solid.

Insulating crystals containing rare earth ions (which have partially filled electronic f-shells) are found to obey Curie's law quite well. One frequently writes the law in the form:

$$\chi = \frac{1}{3}\frac{N}{V}\frac{{\mu_B}^2 p^2}{k_BT}, \tag{31.49}$$

where p, the "effective Bohr magneton number," is given by

$$p = g(JLS)[J(J+1)]^{1/2}. \tag{31.50}$$

---

[25] On the other hand the law does hold at very high temperatures even when there are appreciable magnetic *interactions* among the ions. See Eq. (33.50).

In Table 31.3 the value of p determined by the coefficient of $1/T$ in the measured susceptibility, is compared with that given by (31.50) and the Landé g-factor (31.38).

Table 31.3
CALCULATED AND MEASURED EFFECTIVE MAGNETON NUMBERS p FOR RARE EARTH IONS[a]

ELEMENT (TRIPLY IONIZED)	BASIC ELECTRON CONFIGURATION	GROUND-STATE TERM	CALCULATED[b] p	MEASURED[c] p
La	$4f^0$	^1S	0.00	diamagnetic
Ce	$4f^1$	$^2F_{5/2}$	2.54	2.4
Pr	$4f^2$	$^3H_4$	3.58	3.5
Nd	$4f^3$	$^4I_{9/2}$	3.62	3.5
Pm	$4f^4$	$^5I_4$	2.68	—
Sm	$4f^5$	$^6H_{5/2}$	0.84	1.5
Eu	$4f^6$	$^7F_0$	0.00	3.4
Gd	$4f^7$	$^8S_{7/2}$	7.94	8.0
Tb	$4f^8$	$^7F_6$	9.72	9.5
Dy	$4f^9$	$^6H_{15/2}$	10.63	10.6
Ho	$4f^{10}$	$^5I_8$	10.60	10.4
Er	$4f^{11}$	$^4I_{15/2}$	9.59	9.5
Tm	$4f^{12}$	$^3H_6$	7.57	7.3
Yb	$4f^{13}$	$^2F_{7/2}$	4.54	4.5
Lu	$4f^{14}$	^1S	0.00	diamagnetic

[a] Note the discrepancy in Sm and Eu having its origin in low-lying J-multiplets assumed absent in the theory.
[b] Equation (31.50).
[c] Equation (31.49).
Source: J. H. Van Vleck, *The Theory of Electric and Magnetic Susceptibilities*, Oxford, 1952, p. 243; see also R. Kubo and T. Nagamiya, eds., *Solid State Physics*, McGraw-Hill, New York, 1969, p. 451.

The agreement is excellent, except for samarium and europium. In the latter case we have $J = 0$, and our analysis is clearly not applicable. In both cases, however, the discrepancy has been accounted for by recognizing that the J-multiplet lying just above the ground state is so close in energy that (a) the energy denominators in the second term in the energy (31.20) (neglected in the derivation of Curie's law) are small enough for it to be important, and (b) the probability of some ions being thermally excited out of the state(s) of lowest J (also neglected in deriving Curie's law) can be appreciable.

Thus in all cases the magnetism of rare earth ions in an insulating solid is well described by treating them as isolated ions. This is not the case, however, for *transition metal* ions in an insulating solid. Indeed, for transition metal ions from the iron group one finds that although Curie's law is obeyed, the value of p determined from it is given by (31.50) only if one assumes that although S is still given by Hund's rules, L is zero and hence J is equal to S. (See Table 31.4.) This phenomenon is known as the *quenching* of the orbital angular momentum, and is a particular example of a general phenomenon known as *crystal field splitting*.

Table 31.4
CALCULATED AND MEASURED EFFECTIVE MAGNETON NUMBERS p FOR THE IRON ($3d$) GROUP IONS[a]

ELEMENT (AND IONIZATION)	BASIC ELECTRON CONFIGURATION	GROUND-STATE TERM	CALCULATED[b] p ($J = S$)	CALCULATED[b] p ($J = \|L \pm S\|$)	MEASURED[c] p
Ti^{3+}	$3d^1$	$^2D_{3/2}$	1.73	1.55	—
V^{4+}	$3d^1$	$^2D_{3/2}$	1.73	1.55	1.8
V^{3+}	$3d^2$	$^3F_2$	2.83	1.63	2.8
V^{2+}	$3d^3$	$^4F_{3/2}$	3.87	0.77	3.8
Cr^{3+}	$3d^3$	$^4F_{3/2}$	3.87	0.77	3.7
Mn^{4+}	$3d^3$	$^4F_{3/2}$	3.87	0.77	4.0
Cr^{2+}	$3d^4$	$^5D_0$	4.90	0	4.8
Mn^{3+}	$3d^4$	$^5D_0$	4.90	0	5.0
Mn^{2+}	$3d^5$	$^6S_{5/2}$	5.92	5.92	5.9
Fe^{3+}	$3d^5$	$^6S_{5/2}$	5.92	5.92	5.9
Fe^{2+}	$3d^6$	$^5D_4$	4.90	6.70	5.4
Co^{2+}	$3d^7$	$^4F_{9/2}$	3.87	6.54	4.8
Ni^{2+}	$3d^8$	$^3F_4$	2.83	5.59	3.2
Cu^{2+}	$3d^9$	$^2D_{5/2}$	1.73	3.55	1.9

[a] Because of quenching, much better theoretical values are obtained by taking J equal to S, the total spin, than by taking the value $J = |L \pm S|$ appropriate to the free ion.
[b] Equation (31.50). In the case $J = S$, one takes $L = 0$.
[c] Equation (31.49).
Source: J. H. Van Vleck, *The Theory of Electric and Magnetic Susceptibilities*, Oxford, 1952, p. 285; R. Kubo and T. Nagamiya, eds., *Solid State Physics*, McGraw-Hill, New York, 1969, p. 453.

Crystal field splitting is unimportant for rare earth ions, because their partially filled $4f$ shells lie deep inside the ion (beneath filled $5s$ and $5p$ shells). In contrast to this, the partially filled d-shells of transition metal ions are the outermost electronic shells, and are therefore far more strongly influenced by their crystalline environment. The electrons in the partially filled d-shells are subject to nonnegligible electric fields that do not have spherical symmetry, but only the symmetry of the crystalline site at which the ion is located. As a result, the basis for Hund's rules is partially invalidated.

As it turns out, the first two of Hund's rules can be retained, even in the crystalline environment. The crystal field must, however, be introduced as a perturbation on the $(2S + 1)(2L + 1)$-fold set of states determined by the first two rules. This perturbation acts in addition to the spin-orbit coupling. Therefore Hund's third rule (which resulted from the action of the spin-orbit coupling alone) must be modified.

In the case of the transition metal ions from the iron group (partially filled $3d$ shells) the crystal field is very much larger than the spin-orbit coupling, so that to a first approximation a new version of Hund's third rule can be constructed in which the perturbation of spin-orbit coupling is ignored altogether, in favor of the crystal field perturbation. This latter perturbation will *not* split the spin degeneracy, since it depends only on spatial variables and therefore commutes with $\mathbf{S}$, but it can completely lift the degeneracy of the orbital L-multiplet, if it is sufficiently asymmetric.[26] The result will then be a ground-state multiplet in which the mean value of every component of $\mathbf{L}$ vanishes (even though $\mathbf{L}^2$ still has the mean value $L(L + 1)$). One can interpret this classically as arising from a precession of the orbital angular momentum in the crystal field, so that although its magnitude is unchanged, all its components average to zero.

The situation for the higher transition metal series (partially filled $4d$ or $5d$ shells) is more complex, since in the heavier elements the spin-orbit coupling is stronger. The multiplet splitting due to spin-orbit coupling may be comparable to (or greater than) the crystal field splitting. In general cases like these, considerations of how the crystal fields can rearrange the levels into structures different from those implied by Hund's third rule, are based on fairly subtle applications of group theory. We shall not explore them here, but mention two important principles that come into play:

1. The less symmetric the crystal field, the lower the degeneracy one expects the exact ionic ground state to have. There is, however, an important theorem (due to Kramers) asserting that no matter how unsymmetric the crystal field, an ion possessing an odd number of electrons must have a ground state that is at least doubly degenerate, even in the presence of crystal fields and spin-orbit interactions.
2. One might expect that the crystal field would often have such high symmetry (as at sites of cubic symmetry) that it would produce less than the maximum lifting of degeneracy allowed by the theorem of Kramers. However, another theorem, due to Jahn and Teller, asserts that if a magnetic ion is at a crystal site of such high symmetry that its ground-state degeneracy is not the Kramers minimum, then it will be energetically favorable for the crystal to distort (e.g., for the equilibrium position of the ion to be displaced) in such a way as to lower the symmetry enough to remove the degeneracy. Whether this lifting of degeneracy is large enough to be important (i.e., comparable to k_BT or to the splitting in applied magnetic fields) is not guaranteed by the theorem. If it is not large enough, the Jahn-Teller effect will not be observable.

THERMAL PROPERTIES OF PARAMAGNETIC INSULATORS: ADIABATIC DEMAGNETIZATION

Since the Helmholtz free energy is $F = U - TS$, where U is the internal energy, the magnetic entropy $S(H, T)$ is given by

$$S = k_B\beta^2 \frac{\partial F}{\partial \beta}, \quad \beta = \frac{1}{k_BT}, \tag{31.51}$$

[26] If one adds the spin-orbit coupling to the Hamiltonian, as an additional perturbation on the crystal field, even the remaining $(2S + 1)$-fold degeneracy of the ground state will be split. However this additional splitting may well be small compared with both k_BT and the splitting in an applied magnetic field, in which case it can be ignored. Evidently this is the case for the transition metal ions from the iron group.

(since $U = (\partial/\partial\beta)\beta F$). The expression (31.42) for the free energy of a set of non-interacting paramagnetic ions reveals that βF depends on β and H only through their product; i.e., F has the form

$$F = \frac{1}{\beta}\Phi(\beta H). \tag{31.52}$$

Consequently the entropy has the form

$$S = k_B[-\Phi(\beta H) + \beta H \Phi'(\beta H)], \tag{31.53}$$

which depends only on the product $\beta H = H/k_BT$. As a result, by adiabatically (i.e., at fixed S) lowering the field acting on a spin system (slowly enough so that thermal equilibrium is always maintained) we will lower the temperature of the spin system by a proportionate amount, for if S is unchanged then H/T cannot change, and therefore

$$T_{\text{final}} = T_{\text{initial}}\left(\frac{H_{\text{final}}}{H_{\text{initial}}}\right). \tag{31.54}$$

This can be used as a practical method for achieving low temperatures only in a temperature range where the specific heat of the spin system is the dominant contribution to the specific heat of the entire solid. In practice this restricts one to temperatures far below the Debye temperature (see Problem 10), and the technique has proved useful for cooling from a few degrees Kelvin down to a few hundredths (or, if one is skillful, thousandths) of a degree.

The limit on the temperatures one can reach by adiabatic demagnetization is set by the limits on the validity of the conclusion that the entropy depends only on H/T. If this were rigorously correct one could cool all the way to zero temperature by completely removing the field. But this assumption must fail at small fields, for otherwise the zero-field entropy would not depend on temperature. In reality the entropy in zero field must depend on temperature, so that the entropy density can drop to zero with decreasing temperature, as required by the third law of thermodynamics. The temperature dependence of the zero-field entropy is brought about by the existence of magnetic interactions between the paramagnetic ions, the increased importance of crystal field splitting at low temperatures, and other such effects that are left out of the analysis leading to (31.53). When these are taken into account, the result (31.54) for the final temperature must be replaced by the general result $S(H_{\text{initial}}, T_{\text{initial}}) = S(0, T_{\text{final}})$ and one must have a detailed knowledge of the temperature dependence of the zero-field entropy, to compute the final temperature (see Figure 31.2).

Evidently the most effective materials are those where the inevitable decline with temperature of the zero-field entropy sets in at the lowest possible temperature. One therefore uses paramagnetic salts with well-sheltered (to minimize crystal field splitting), well-separated (to minimize magnetic interactions) magnetic ions. Countering this, of course, is the lower magnetic specific heat resulting from a lower density of magnetic ions. The most popular substances used at present are of the type $Ce_2Mg_3(NO_3)_{12} \cdot (H_20)_{24}$.

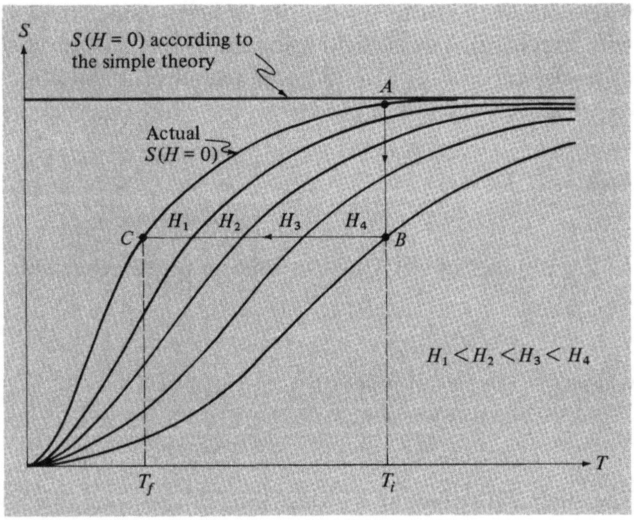

Figure 31.2
Plots of the entropy of a system of interacting spins for various values of external magnetic field, H. (The dashed line represents the constant $Nk_B \ln(2J + 1)$ for independent spins in zero field.) The cooling cycle is this: Starting at A (T_i, $H = 0$), we proceed isothermally to B, raising the field in the process from zero to H_4. The next step is to remove the field adiabatically (constant S), thereby moving to C and achieving a temperature T_f.

SUSCEPTIBILITY OF METALS: PAULI PARAMAGNETISM

None of the above discussions bear on the problem of the contribution of conduction electrons to the magnetic moment of a metal. The conduction electrons are not spatially localized like electrons in partially filled ionic shells, nor, because of the stringent constraints of the exclusion principle, do they respond independently like electrons localized on different ions.

However, within the independent electron approximation the problem of conduction electron magnetism can be solved. The solution is quite complicated, owing to the intricate way in which the electron orbital motion responds to the field. If we neglect the orbital response (i.e., consider the electron to have only a spin magnetic moment, but no charge), then we may proceed as follows:

Each electron will contribute $-\mu_B/V$ (taking $g_0 = 2$) to the magnetization density if its spin is parallel to the field H, and μ_B/V, if antiparallel. Hence if $n_\pm$ is the number of electrons per unit volume with spin parallel (+) or antiparallel (−) to H, the magnetization density will be

$$M = -\mu_B(n_+ - n_-). \qquad \textbf{(31.55)}$$

If the electrons interact with the field only through their magnetic moments, then the only effect of the field is to shift the energy of each electronic level by $\pm\mu_B H$, according to whether the spin is parallel (+) or antiparallel (−) to H. We can express this simply in terms of the density of levels for a given spin. Let $g_\pm(\mathcal{E})\,d\mathcal{E}$ be the number of electrons of the specified spin per unit volume in the energy range $\mathcal{E}$ to $\mathcal{E} + d\mathcal{E}$.[27] In the absence of the field we would have

$$g_\pm(\mathcal{E}) = \tfrac{1}{2}g(\mathcal{E}), \qquad (H = 0), \qquad \textbf{(31.56)}$$

[27] To avoid confusing the density of levels with the g-factor, we shall always make the energy argument of the level density explicit. A subscript distinguishes the Bohr magneton μ_B from the chemical potential μ.

where $g(\mathcal{E})$ is the ordinary density of levels. Since the energy of each electronic level with spin parallel to the field is shifted up from its zero field value by $\mu_B H$, the number of levels with energy $\mathcal{E}$ in the presence of H is the same as the number with energy $\mathcal{E} - \mu_B H$ in the absence of H:

$$g_+(\mathcal{E}) = \tfrac{1}{2}g(\mathcal{E} - \mu_B H). \tag{31.57}$$

Similarly,

$$g_-(\mathcal{E}) = \tfrac{1}{2}g(\mathcal{E} + \mu_B H). \tag{31.58}$$

The number of electrons per unit volume of each spin species is given by

$$n_\pm = \int d\mathcal{E}\; g_\pm(\mathcal{E})\, f(\mathcal{E}), \tag{31.59}$$

where f is the Fermi function

$$f(\mathcal{E}) = \frac{1}{e^{\beta(\mathcal{E}-\mu)} + 1}. \tag{31.60}$$

The chemical potential μ is determined by noting that the total electronic density is given by

$$n = n_+ + n_-. \tag{31.61}$$

Eliminating μ through this relation we can use (31.59) and (31.55) to find the magnetization density as a function of the electronic density n. In the nondegenerate case $(f \approx e^{-\beta(\mathcal{E}-\mu)})$ this leads back to our earlier theory of paramagnetism, giving precisely (31.44) with $J = 1/2$. (See Problem 8.)

However, in metals one is very much in the degenerate case. The important variation in the density of levels $g(\mathcal{E})$ is on the scale of $\mathcal{E}_F$, and since $\mu_B H$ is only of order $10^{-4}\mathcal{E}_F$ even at 10^4 gauss, we can, with negligible error, expand the density of levels:

$$g_\pm(\mathcal{E}) = \tfrac{1}{2}g(\mathcal{E} \pm \mu_B H) = \tfrac{1}{2}g(\mathcal{E}) \pm \tfrac{1}{2}\mu_B H g'(\mathcal{E}). \tag{31.62}$$

In conjunction with (31.59) this gives

$$n_\pm = \tfrac{1}{2}\int g(\mathcal{E})\, f(\mathcal{E})\, d\mathcal{E} \mp \tfrac{1}{2}\,\mu_B H \int d\mathcal{E}\; g'(\mathcal{E}) f(\mathcal{E}), \tag{31.63}$$

so that, from (31.61),

$$n = \int g(\mathcal{E})\, f(\mathcal{E})\, d\mathcal{E}. \tag{31.64}$$

This is precisely the formula for the electronic density in the absence of the field, and thus the chemical potential μ can be taken to have its zero field value, Eq. (2.77):

$$\mu = \mathcal{E}_F\left[1 + O\left(\frac{k_B T}{\mathcal{E}_F}\right)^2\right]. \tag{31.65}$$

In conjunction with Eq. (31.55), Eq. (31.63) gives a magnetization density

$$M = {\mu_B}^2 H \int g'(\mathcal{E})\, f(\mathcal{E})\, d\mathcal{E}, \tag{31.66}$$

or, integrating by parts,

$$M = \mu_B{}^2 H \int g(\mathcal{E}) \left(-\frac{\partial f}{\partial \mathcal{E}} \right) d\mathcal{E}. \qquad \textbf{(31.67)}$$

At zero temperature, $-\partial f/\partial \mathcal{E} = \delta(\mathcal{E} - \mathcal{E}_F)$, so that

$$M = \mu_B{}^2 H g(\mathcal{E}_F). \qquad \textbf{(31.68)}$$

Since (see Chapter 2) the $T \neq 0$ corrections to $-\partial f/\partial \mathcal{E}$ are of order $(k_B T/\mathcal{E}_F)^2$, Eq. (31.68) is also valid at all but the very highest temperatures ($T \approx 10^4$ K).

It follows from (31.68) that the susceptibility is

$$\boxed{\chi = \mu_B{}^2 g(\mathcal{E}_F).} \qquad \textbf{(31.69)}$$

This is known as the *Pauli paramagnetic susceptibility*. In contrast to the susceptibility of paramagnetic ions given by Curie's law, the Pauli susceptibility of conduction electrons is essentially independent of temperature. In the free electron case the density of levels has the form $g(\mathcal{E}_F) = mk_F/\hbar^2\pi^2$, and the Pauli susceptibility takes on the simple form

$$\chi_{\text{Pauli}} = \left(\frac{\alpha}{2\pi} \right)^2 (a_0 k_F), \qquad \textbf{(31.70)}$$

where $\alpha = e^2/\hbar c = 1/137$. An alternative form is

$$\chi_{\text{Pauli}} = \left(\frac{2.59}{r_s/a_0} \right) \times 10^{-6}. \qquad \textbf{(31.71)}$$

These expressions reveal that χ_{Pauli} has the minute size characteristic of diamagnetic susceptibilities, in contrast to the strikingly larger paramagnetic susceptibilities of magnetic ions. This is because the exclusion principle is far more effective than thermal disorder in suppressing the tendency of the spin magnetic moments to align with the field. Another way of comparing Pauli paramagnetism with the paramagnetism of magnetic ions is to note that the Pauli susceptibility can be cast into the Curie's law form (Eq. (31.47), but with a fixed temperature of order T_F playing the role of T. Thus the Pauli susceptibility is hundreds of times smaller, even at room temperatures.[28]

Values of the Pauli susceptibility, both measured and theoretical (from Eq. (31.71)) are given in Table 31.5 for the alkali metals. The rather significant discrepancy between the two sets of figures is mainly a result of the neglected electron-electron interactions (see Problem 12).[29]

---

[28] Until Pauli's theory, the absence of a strong Curie's law paramagnetism in metals was another of the outstanding anomalies in the free electron theory of metals; as in the case of the specific heat, the anomaly was removed by observing that electrons obey Fermi-Dirac, rather than classical, statistics.

[29] The reader who recalls the large correction to the electronic density of levels appearing in the electronic specific heat, which arises from the electron-phonon interaction, may be surprised to learn that a similarly large correction does not arise in the Pauli susceptibility. There is an important difference between the two cases. When the specific heat is computed, one calculates a fixed temperature-independent correction to the electronic density of levels, and then inserts that *fixed* density of levels into formulas (such as (2.79)) telling how the energy changes as the temperature varies. When a magnetic field is varied, however, the density of levels itself changes. We have already noted, for example (ignoring phonon

Table 31.5
COMPARISON OF FREE ELECTRON AND MEASURED PAULI SUSCEPTIBILITIES

METAL	r_s/a_0	$10^6\chi_{\text{Pauli}}$ (from Eq. (31.71))	$10^6\chi_{\text{Pauli}}$ (measured)[a]
Li	3.25	0.80	2.0
Na	3.93	0.66	1.1
K	4.86	0.53	0.8_5
Rb	5.20	0.50	0.8
Cs	5.62	0.46	0.8

[a] The measured values are taken from the following sources: Li: R. T. Schumacher and C. P. Slichter, *Phys. Rev.* **101**, 58 (1956); Na: R. T. Schumacher and W. E. Vehse, *J. Phys. Chem. Solids* **24**, 297 (1965); K: S. Schultz and G. Dunifer, *Phys. Rev. Lett.* **18**, 283 (1967); Rb, Cs: J. A. Kaeck, *Phys. Rev.* **175**, 897 (1968).

CONDUCTION ELECTRON DIAMAGNETISM

In the foregoing discussion of conduction electron magnetism we considered only the paramagnetic effects arising from the coupling of the intrinsic spin of the electrons with the applied field H. There are also diamagnetic effects arising from the coupling of the field to the orbital motion of the electrons. We discussed this at some length in Chapter 14, where we found that at very low temperatures, high fields, and high purities ($\omega_c\tau = eH\tau/mc \gg 1$) there was a complicated oscillatory structure to the dependence of M on H. In ordinary specimens the condition of high $\omega_c\tau$ is not met, and the oscillatory structure is not perceptible. However, the dependence of M on H does not average out to zero: There is a net nonvanishing magnetization antiparallel to H, known as the *Landau diamagnetism*, that is due to the orbital electronic motion induced by the field. For *free* electrons it can be shown[30] that

$$\chi_{\text{Landau}} = -\tfrac{1}{3}\chi_{\text{Pauli}}. \tag{31.72}$$

If the electrons move in a periodic potential but are otherwise independent, the analysis becomes quite complicated, but again results in a diamagnetic susceptibility of the same order of magnitude as the paramagnetic susceptibility. In practice, of

corrections), that the density of levels for each spin population is shifted up or down in energy with changing field. The phonon correction to this result occurs in a neighborhood (of width $\hbar\omega_D$, which is large compared with the shift $\hbar\omega_c$ due to the field) of the Fermi level. But the Fermi level does not shift with field, in contrast to the uncorrected density of levels. Consequently, one cannot simply substitute a phonon-corrected density of levels into (31.68) as one can in (2.79) because the corrected density of levels varies with field in an intrinsically different way from the uncorrected one. A careful analysis reveals that because the phonon correction is tied to the Fermi level, it has very little effect on the magnetization as the field varies, leading to a correction factor in the susceptibility that is only of order $(m/M)^{1/2}$ (in contrast to the correction of order unity in the specific heat).

[30] See, for example, R. E. Peierls, *Quantum Theory of Solids*, Oxford, 1955, pp. 144–149. An analysis that takes band structure into account can be found in P. K. Misra and L. M. Roth, *Phys. Rev.* **177**, 1089 (1969) and references therein.

course, it is the *total* susceptibility that is revealed by a measurement of the bulk moment induced by a field, and this is a combination of the Pauli paramagnetic susceptibility, the Landau diamagnetic susceptibility, and the Larmor diamagnetic susceptibility (of the closed-shell ion cores). As a result, it is not at all a straightforward matter to isolate experimentally a particular term in the susceptibility. Thus the Pauli susceptibilities quoted in Table 31.5 were obtained by quite indirect methods, one of which we now describe.

MEASUREMENT OF PAULI PARAMAGNETISM BY NUCLEAR MAGNETIC RESONANCE

To distinguish the electron spin paramagnetic contribution to the susceptibility of a metal from the other sources of magnetization, one requires a probe that couples much more strongly to the spin magnetic moments of the conduction electrons than it does to fields arising from electronic translational motion. The magnetic moments of the ionic *nuclei* provide such a probe.

A nucleus of angular momentum I possesses a magnetic moment $\mathbf{m}_N = \gamma_N \mathbf{I}$ (which is typically smaller in order of magnitude than the electronic magnetic moment by the electron to nuclear mass ratio). In an applied magnetic field the $(2I + 1)$ degenerate nuclear spin levels are split by an amount $\gamma_N H$. This splitting can be detected by observing the resonant absorption of energy at the angular frequency $\gamma_N H/\hbar$.[31,32]

The field determining the frequency of the nuclear magnetic resonance is, of course, the field acting directly on the nucleus. In nonparamagnetic substances the field at the nucleus differs by small diamagnetic corrections (known as the chemical shift) from the externally applied field. In metals, however, there is a more important[33] source of field at the nucleus:

The conduction electrons (whose wave functions are generally derived, at least in part, from atomic *s*-shells) have nonvanishing wave functions at the ionic nuclei. When an electron actually overlaps the nucleus, however, there is a direct magnetic coupling of their moments proportional to $\mathbf{m}_e \cdot \mathbf{m}_N$.[34] If the conduction electron gas had no net paramagnetic moment, this coupling would yield no net shift in the nuclear resonance since electrons of all spin orientations would be found, with equal probability, at the nuclear position.[35] However, the same field in which the nuclei precess also produces the Pauli paramagnetic imbalance in the electronic spin populations.

---

[31] In practical nuclear magnetic resonance experiments the resonance is observed by fixing the frequency of the perturbing radio-frequency (RF) field, and varying the applied magnetic field.

[32] An excellent introduction to nuclear magnetic resonance is given by C. Slichter, *Principles of Magnetic Resonance*, Harper Row, New York, 1963.

[33] Since isolated nuclei are hard to work with, one generally deals only with relative shifts. One knows that the shift in metals is more important because it differs from the shift in salts of the same metal by far more than the shift from one salt to another.

[34] Known, variously, as the hyperfine interaction, the Fermi interaction, or the contact interaction.

[35] This picture assumes, of course, that the nucleus experiences the average electron field—i.e., that each nucleus interacts with many electron spins in the time it takes to complete a single period of precession. Since this time is typically 10^{-6} second in a strong field, the condition is very well satisfied, for a conduction electron moves with velocity v_F (of order 10^8 cm/sec) and therefore takes about 10^{-21} second to cross an atomic nucleus (whose radius is of order 10^{-13} cm).

There is therefore a net electronic moment leading to an effective field at the nucleus that is proportional to the conduction electron spin susceptibility.

The shift produced by this field, known as the Knight shift, is measured by noting the difference in resonant frequency between the metallic element in (for example) a nonparamagnetic salt, and in the metallic state. Unfortunately the Knight shift is proportional not only to the Pauli susceptibility, but also to the magnitude squared of the conduction electron wave function at the ionic nucleus. It is therefore necessary to have some estimate of this (usually from a calculation) to extract the Pauli susceptibility from the measured Knight shift.

ELECTRON DIAMAGNETISM IN DOPED SEMICONDUCTORS

Doped semiconductors provide an example of a conducting material in which the conduction electron diamagnetism can be substantially larger than the paramagnetism. One first measures the susceptibility of the intrinsic material at very low temperatures, where it is almost entirely due to the diamagnetism of the ion cores. This contribution will also be present in the doped material, and by subtracting it from the total susceptibility one can extract the contribution to the susceptibility of the doped material due to the carriers introduced by the doping.[36]

Consider the case in which the carriers go into bands with spherical symmetry, so that $\mathcal{E}(\mathbf{k}) = \hbar^2 k^2/m^*$. (For concreteness we consider donor impurities, and measure $\mathbf{k}$ with respect to the conduction band minimum.) According to (31.69), the paramagnetic susceptibility is proportional to the density of levels.[37] Since this is proportional to m for free electrons, the Pauli susceptibility of the carriers will be reduced[38] by a factor m^*/m. On the other hand, the Landau susceptibility is enhanced by a factor m/m^*, since the coupling of the electron orbital motion to the fields is proportional to $e(\mathbf{v}/c) \times \mathbf{H}$, which is inversely proportional to m^*. As a result,

$$\frac{\chi_{\text{Landau}}}{\chi_{\text{Pauli}}} \sim \left(\frac{m}{m^*}\right)^2. \tag{31.73}$$

Thus there are semiconductors in which the electron Landau diamagnetism completely dominates the Pauli spin paramagnetism, and can therefore be directly extracted, by measuring the amount by which the susceptibility exceeds that of the undoped material.[39]

This completes our survey of those magnetic properties of solids that can be understood without taking into explicit account the interactions between the sources of magnetic moment. In Chapter 32 we shall turn to the theory underlying such interactions, and in Chapter 33 we shall return to the examination of further magnetic properties of solids, which depend on them crucially.

---

[36] The change in the diamagnetic susceptibility due to the different closed-shell structure of the donor ions gives a very small correction.

[37] These considerations can be shown to remain valid even if the conduction band electrons are not degenerate.

[38] The ratio m^*/m is typically 0.1 or smaller.

[39] For a review, see R. Bowers, *J. Phys. Chem. Solids* **8**, 206 (1959).

PROBLEMS

1. The classical definition of the magnetic moment **m** of a particle of charge $-e$, due to its orbital motion, was given by Ampère as the average over the orbit of

$$-\frac{e}{2c}(\mathbf{r} \times \mathbf{v}). \tag{31.74}$$

Show that our definition, $\mathbf{m} = -\partial E/\partial \mathbf{H}$, reduces to this form by showing from (31.15) that

$$\mathbf{m} = -\frac{e}{2mc}\sum_i \mathbf{r}_i \times \left(\mathbf{p}_i - \frac{e}{2c}\mathbf{r}_i \times \mathbf{H}\right) \tag{31.75}$$

and

$$\mathbf{v}_i = \frac{\partial H}{\partial \mathbf{p}_i} = \frac{1}{m}\left(\mathbf{p}_i - \frac{e}{2c}\mathbf{r}_i \times \mathbf{H}\right). \tag{31.76}$$

2. The Pauli spin matrices satisfy the simple identity

$$(\mathbf{a} \cdot \boldsymbol{\sigma})(\mathbf{b} \cdot \boldsymbol{\sigma}) = \mathbf{a} \cdot \mathbf{b} + i(\mathbf{a} \times \mathbf{b}) \cdot \boldsymbol{\sigma} \tag{31.77}$$

provided all the components of **a** and **b** commute with those of $\boldsymbol{\sigma}$. If the components of **a** commute among themselves, $\mathbf{a} \times \mathbf{a} = 0$. Since the components of **p** so commute, in the absence of a magnetic field we could equally well write the kinetic energy of a spin-$\frac{1}{2}$ particle in the form $(\boldsymbol{\sigma} \cdot \mathbf{p})^2/2m$. However when a field is present, the components of $\mathbf{p} + e\mathbf{A}/c$ no longer commute among themselves. Show, as a result, that (31.77) implies:

$$\frac{1}{2m}\left[\boldsymbol{\sigma} \cdot \left(\mathbf{p} + \frac{e\mathbf{A}}{c}\right)\right]^2 = \frac{1}{2m}\left(\mathbf{p} + \frac{e\mathbf{A}}{c}\right)^2 + \frac{e\hbar}{mc}\frac{1}{2}\boldsymbol{\sigma} \cdot \mathbf{H}, \tag{31.78}$$

which thus gives both the spin and orbital contributions to the magnetic part of the Hamiltonian in one compact formula (provided that $g_0 = 2$).

3. (a) Show that Hund's rules for a shell of angular momentum l containing n electrons can be summarized in the formulas:

$$\begin{aligned} S &= \tfrac{1}{2}[(2l + 1) - |2l + 1 - n|], \\ L &= S|2l + 1 - n|, \\ J &= |2l - n|S. \end{aligned} \tag{31.79}$$

(b) Verify that that the two ways of counting the degeneracy of a given LS-multiplet give the same answer; i.e., verify that

$$(2L + 1)(2S + 1) = \sum_{|L-S|}^{L+S} (2J + 1). \tag{31.80}$$

(c) Show that the total splitting of an LS-multiplet due to the spin orbit interaction $\lambda(\mathbf{L} \cdot \mathbf{S})$ is

$$\begin{aligned} E_{J_{\max}} - E_{J_{\min}} &= \lambda S(2L + 1), \qquad L > S, \\ &= \lambda L(2S + 1), \qquad S > L, \end{aligned} \tag{31.81}$$

and that the splittings between successive J-multiplets within the LS-multiplet is

$$E_{J+1} - E_J = \lambda(J + 1). \tag{31.82}$$

4. (a) The angular momentum commutation relations are summarized in the vector operator identities

$$\mathbf{L} \times \mathbf{L} = i\mathbf{L}, \quad \mathbf{S} \times \mathbf{S} = i\mathbf{S}. \tag{31.83}$$

Deduce from these identities and the fact that all components of $\mathbf{L}$ commute with all components of $\mathbf{S}$ that

$$[\mathbf{L} + g_0\mathbf{S}, \hat{\mathbf{n}} \cdot \mathbf{J}] = i\hat{\mathbf{n}} \times (\mathbf{L} + g_0\mathbf{S}), \tag{31.84}$$

for any (c-number) unit vector $\hat{\mathbf{n}}$.

(b) A state $|0\rangle$ with zero total angular momentum satisfies

$$\mathbf{J}_x|0\rangle = \mathbf{J}_y|0\rangle = \mathbf{J}_z|0\rangle = 0. \tag{31.85}$$

Deduce from (31.84) that

$$\langle 0|(\mathbf{L} + g_0\mathbf{S})|0\rangle = 0, \tag{31.86}$$

even though $\mathbf{L}^2$ and $\mathbf{S}^2$ need not vanish in the state $|0\rangle$, and $(\mathbf{L} + g_0\mathbf{S})|0\rangle$ need not be zero.

(c) Deduce the Wigner-Eckart theorem (Eq. (31.34) in the special case $J = 1/2$, from the commutation relations (31.84).

5. Suppose that within the set of $(2L + 1)(2S + 1)$ lowest-lying ionic states the crystal field can be represented in the form $a\mathbf{L}_x{}^2 + b\mathbf{L}_y{}^2 + c\mathbf{L}_z{}^2$, with a, b, and c all different. Show in the special case $L = 1$ that if the crystal field is the dominant perturbation (compared with spin-orbit coupling), then it will yield a $(2S + 1)$-fold degenerate set of ground states in which every matrix element of every component of $\mathbf{L}$ vanishes.

6. The susceptibility of a simple metal has a contribution $\chi_{\text{c.e.}}$ from the conduction electrons and a contribution χ_{ion} from the diamagnetic response of the closed-shell core electrons. Taking the conduction electron susceptibility to be given by the free electron values of the Pauli paramagnetic and Landau diamagnetic susceptibilities, show that

$$\frac{\chi_{\text{ion}}}{\chi_{\text{c.e.}}} = -\frac{1}{3}\frac{Z_c}{Z_v}\langle (k_F r)^2\rangle, \tag{31.87}$$

where Z_v is the valence, Z_c is the number of core electrons, and $\langle r^2\rangle$ is the mean square ionic radius defined in (31.26).

7. Consider an ion with a partially filled shell of angular momentum J, and Z additional electrons in filled shells. Show that the ratio of the Curie's law paramagnetic susceptibility to the Larmor diamagnetic susceptibility is

$$\frac{\chi_{\text{par}}}{\chi_{\text{dia}}} = -\frac{2J(J + 1)}{Zk_BT}\frac{\hbar^2}{m\langle r^2\rangle}, \tag{31.88}$$

and deduce from this the numerical estimate on page 656.

8. Show that the magnetization of a *nondegenerate* electron gas is precisely given by the result (31.44) for independent moments (with J set equal to 1/2), by inserting the low-density expansion of the Fermi function, $f \approx e^{-\beta(\mathcal{E}-\mu)}$ into (31.59).

9. If one writes the free energy (31.5) in the form

$$e^{-\beta F} = \sum_n e^{-\beta E_n} = \sum_n \langle n|e^{-\beta\mathcal{H}}|n\rangle = \text{Tr}\, e^{-\beta\mathcal{H}}, \tag{31.89}$$

then it is easy to deduce Curie's law directly at high temperature without going through the algebra of Brillouin functions, for when $\mathcal{H} \ll k_BT$, we can expand $e^{-\beta\mathcal{H}} = 1 - \beta\mathcal{H} + (\beta\mathcal{H})^2/2 - \cdots$. Evaluate the free energy to second order in the field, using the fact that

$$\text{Tr}(\mathbf{J}_\mu \mathbf{J}_\nu) = \tfrac{1}{3}\delta_{\mu\nu}\, \text{Tr}\, \mathbf{J}^2,$$

and extract the high-temperature susceptibility (31.47).

10. Show that for an ideal paramagnet whose free energy has the form (31.52), the specific heat at constant field is simply related to the susceptibility by

$$c_H = T\left(\frac{\partial s}{\partial T}\right)_H = \frac{H^2\chi}{T}, \tag{31.90}$$

or, in the regime where Curie's law is valid,

$$c_H = \frac{1}{3}\frac{N}{V} k_B J(J+1)\left(\frac{g\mu_B H}{k_B T}\right)^2. \tag{31.91}$$

Estimating the lattice vibrational contribution to the specific heat by Eq. (23.27), show that the lattice contribution drops below that of the spins at a temperature T_0 of order

$$T_0 \approx \left(\frac{N}{N_i}\right)^{1/5}\left(\frac{g\mu_B H}{k_B\Theta_D}\right)^{2/5}\Theta_D. \tag{31.92}$$

(Here N_i is the total number of ions, and N is the number of paramagnetic ions.) What is a typical size for $g\mu_B H/k_B\Theta_D$ in a field of 10^4 gauss?

11. Show that if T is small compared with the Fermi temperature, the temperature-dependent correction to the Pauli susceptibility (31.69) is given by

$$\chi(T) = \chi(0)\left(1 - \frac{\pi^2}{6}(k_BT)^2\left[\left(\frac{g'}{g}\right)^2 - \frac{g''}{g}\right]\right) \tag{31.93}$$

where g, g', and g'' are the density of levels and its derivatives at the Fermi energy. Show that for free electrons this reduces to

$$\chi(T) = \chi(0)\left(1 - \frac{\pi^2}{12}\left(\frac{k_BT}{\mathcal{E}_F}\right)^2\right). \tag{31.94}$$

12. Because of electron-electron interactions, the shift in an electron's energy due to the interaction of its spin magnetic moment with a field H will have an additional term expressing the change in the distribution of electrons with which a given electron interacts. In the Hartree-Fock approximation (see, for example, Eq. (17.19)) this will be of the form

$$\mathcal{E}_{\pm}(\mathbf{k}) = \mathcal{E}_0(\mathbf{k}) \pm \mu_B H - \int \frac{d\mathbf{k}'}{(2\pi)^3} v(|\mathbf{k} - \mathbf{k}'|) f(\mathcal{E}_{\pm}(\mathbf{k})). \tag{31.95}$$

Show that when $k_B T \ll \mathcal{E}_F$ this leads to an integral equation for $\mathcal{E}_+ - \mathcal{E}_-$ that has the solution:

$$[\mathcal{E}_+(\mathbf{k}) - \mathcal{E}_-(\mathbf{k})]_{k=k_F} = \frac{2\mu_B H}{1 - v_0 g(\mathcal{E}_F)}, \tag{31.96}$$

where v_0 is an average of v over *all* solid angles,

$$v_0 = \frac{1}{2} \int_{-1}^{1} dx\, v(\sqrt{2k_F{}^2(1 - x)}). \tag{31.97}$$

By what factor is the Pauli susceptibility now modified?

32 Electron Interactions and Magnetic Structure

Electrostatic Origins of Magnetic Interactions

Magnetic Properties of a Two-Electron System

Failure of the Independent Electron Approximation

Spin Hamiltonians

Direct, Super, Indirect, and Itinerant Exchange

Magnetic Interactions in the Free Electron Gas

The Hubbard Model

Local Moments

The Kondo Theory of the Resistance Minimum

The simple theory of paramagnetism in solids, described in Chapter 31, assumes that the discrete sources of magnetic moment (e.g., the ionic shells of nonzero angular momentum in insulators, or the conduction electrons in simple metals) do not interact with one another. We have seen that this assumption must be dropped in order, for example, to predict the lowest temperatures attainable by adiabatic demagnetization, or to estimate accurately the Pauli spin paramagnetism of metallic conduction electrons.

There are, however, more spectacular consequences of magnetic interactions.[1] Some solids, known as ferromagnets, have a nonvanishing magnetic moment, or "spontaneous magnetization," even in the absence of a magnetic field.[2] If there were no magnetic interactions, in the absence of a field the individual magnetic moments would be thermally disordered, would point in random directions, and could not sum to a net moment for the solid as a whole (Figure 32.1a). The net parallel orientation of the moments in a ferromagnetic (Figure 32.1b) must be due to interactions between them. In other solids, known as antiferromagnets, although there is no net total moment in the absence of a field, there is a far from random spatial pattern of the individual magnetic moments, due to magnetic interactions favoring antiparallel orientations of neighboring moments (Figure 32.1c).

The theory of the origin of magnetic interactions is one of the less well developed of the fundamental areas of solid state physics. The problem is best understood in

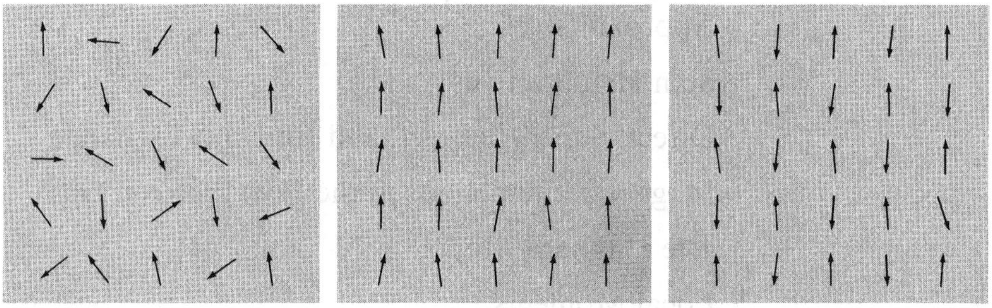

(a) (b) (c)

Figure 32.1
Typical distribution of directions for the local magnetic moments when no magnetic field is present (a) in a solid with inconsequential magnetic interactions, (b) in a ferromagnetic solid below its critical temperature, and (c) in an antiferromagnetic solid below its critical temperature. Cases (b) and (c) illustrate *magnetically ordered* states.

---

[1] We use the term "magnetic interaction" to describe any dependence of the energy of two or more magnetic moments on their relative directions. We shall see that the most important contribution to this energy dependence is usually electrostatic rather than magnetic in origin. The nomenclature can be misleading if one forgets that "magnetic" refers only to the effects of the interactions, and not necessarily their sources.

[2] The spontaneous magnetization decreases with increasing temperature, and vanishes above a certain critical temperature (see Chapter 33). When we use the term "ferromagnet" we shall mean a ferromagnetic material below its critical temperature. We use the term "antiferromagnet" with the same understanding.

insulators, where the magnetic ions are well separated, though even here the theory is quite complex. To keep things as simple as possible we shall illustrate some of the basic physics applicable to insulators (and, with considerable modifications and embellishments, to metals as well) only in the simple case of a single hydrogen molecule, which the charitable reader will be asked to regard as a solid with $N = 2$, rather than $O(10^{23})$. We shall then indicate how the ideas suggested by the hydrogen molecule are generalized to real solids containing large numbers of atoms. Finally, we shall illustrate some of the further complexities one encounters in the theory of magnetic moments and their interactions, in metals.

Readers who wish to be spared the rather uncomfortably compromised exposition forced on us by the difficulty and incompleteness of the subject might merely want to note the following two major points, which the remainder of the chapter will serve to illustrate in rather more detail:

1. One's first expectation might be that the magnetic interactions between discrete moments arise from their magnetic fields, either directly through magnetic dipole-dipole interactions, or, less directly, through spin-orbit coupling. These, however, are not often the dominant magnetic interaction. By far the most important source of magnetic interaction is the ordinary *electrostatic* electron-electron interaction. Indeed, to a first approximation many theories of magnetism ignore dipole-dipole and spin-orbit coupling altogether, retaining only Coulomb interactions.
2. To explain magnetic ordering in solids, in the great majority of cases it is necessary to go well beyond the independent electron approximation, upon which band theory, with its impressive successes in accounting for nonmagnetic properties, is based. It is rarely enough merely to introduce electron-electron interactions into band theory in the form of self-consistent fields. Indeed, the development of a tractable model of a magnetic metal, capable of describing both the characteristic electron spin correlations as well as the electronic transport properties predicted by simple band theory, remains one of the major unsolved problems of modern solid state theory.

ESTIMATE OF MAGNETIC DIPOLAR INTERACTION ENERGIES

Before explaining how magnetic interactions can arise from purely electrostatic couplings, we estimate the direct dipolar interaction energy of two magnetic dipoles $\mathbf{m}_1$ and $\mathbf{m}_2$, separated by $\mathbf{r}$:

$$U = \frac{1}{r^3}[\mathbf{m}_1 \cdot \mathbf{m}_2 - 3(\mathbf{m}_1 \cdot \hat{\mathbf{r}})(\mathbf{m}_2 \cdot \hat{\mathbf{r}})]. \tag{32.1}$$

Atomic magnetic dipole moments have a magnitude $m_1 \approx m_2 \approx g\mu_B \approx e\hbar/mc$ (pages 646 and 654). Hence the size of U (ignoring its angular dependence) will be

$$U \approx \frac{(g\mu_B)^2}{r^3} \approx \left(\frac{e^2}{\hbar c}\right)^2 \left(\frac{a_0}{r}\right)^3 \frac{e^2}{a_0} \approx \frac{1}{(137)^2}\left(\frac{a_0}{r}\right)^3 \text{Ry}. \tag{32.2}$$

In a magnetic solid, moments are typically about 2 Å apart, and hence U is no more than 10^{-4} eV. This is minute compared with electrostatic energy differences between atomic states, which are typically a fraction of an electron volt. Therefore, if we can find a reason why the electrostatic energy of a pair of magnetic ions (or electrons) depends on the direction of their moments (and such a reason is supplied by the Pauli exclusion principle, as we shall see below), then we should expect that source of magnetic interaction to be far more important than the dipolar interaction.[3]

One can also generally rule out spin-orbit coupling as a major source of magnetic interaction. To be sure, it is quite important in determining the total magnetic moment of individual atoms, and is therefore a significant intraatomic source of magnetic interaction. Even in this case, however, the first two Hund rules (page 650) are determined purely by electrostatic energy considerations. Only the third rule, giving the final splitting within the LS-multiplet is based on spin-orbit coupling. In those paramagnetic insulators in which crystal field splitting quenches the orbital angular momentum (page 657) even this consequence of spin-orbit coupling is superseded by purely electrostatic effects.

MAGNETIC PROPERTIES OF A TWO-ELECTRON SYSTEM: SINGLET AND TRIPLET STATES

To illustrate how the Pauli principle can lead to magnetic effects even when there are no spin-dependent terms in the Hamiltonian, we consider a two-electron system with a *spin-independent* Hamiltonian. Because H does not depend on spin, the general stationary state Ψ will be the product of a purely orbital stationary state whose wave function $\psi(\mathbf{r}_1, \mathbf{r}_2)$ satisfies the orbital Schrödinger equation,

$$H\psi = -\frac{\hbar^2}{2m}(\nabla_1{}^2 + \nabla_2{}^2)\psi + V(\mathbf{r}_1, \mathbf{r}_2)\psi = E\psi, \qquad (32.3)$$

with any linear combination of the four spin states[4]

$$|\uparrow\uparrow\rangle, \quad |\uparrow\downarrow\rangle, \quad |\downarrow\uparrow\rangle, \quad |\downarrow\downarrow\rangle. \qquad (32.4)$$

We can choose these linear combinations to have definite values of the total spin S, and its component S_z along an axis. The appropriate linear combinations can be tabulated as follows:[5]

---

[3] A strong hint that the dipolar interaction is far too weak is provided by the ferromagnetic critical temperatures in iron, cobalt, and nickel, which are many hundreds of degrees K. If the spins were aligned by magnetic dipolar interactions, one would expect ferromagnetic alignment to be thermally obliterated above a few degrees K (1 K $\sim 10^{-4}$ eV). On the other hand, in solids with widely separated magnetic moments the dipolar interactions may dominate those of electrostatic origin. Dipolar interactions are also of crucial importance in accounting for the phenomenon of ferromagnetic domains (page 718).

[4] These symbols denote spin states with both electrons in levels of definite s_z. In the state $|\uparrow\downarrow\rangle$, for example, electron 1 has $s_z = 1/2$, and electron 2, $s_z = -1/2$.

[5] See, for example, D. Park, *Introduction to the Quantum Theory*, McGraw-Hill, New York, 1964, pp. 154–156.

STATE	S	S_z
$\frac{1}{\sqrt{2}}(\lvert\uparrow\downarrow\rangle - \lvert\downarrow\uparrow\rangle)$	0	0
$\lvert\uparrow\uparrow\rangle$	1	1
$\frac{1}{\sqrt{2}}(\lvert\uparrow\downarrow\rangle + \lvert\downarrow\uparrow\rangle)$	1	0
$\lvert\downarrow\downarrow\rangle$	1	-1

Note that the one state with $S = 0$ (known as the singlet state) changes sign when the spins of the two electrons are interchanged, while the three states with $S = 1$ (known as the triplet states) do not. The Pauli exclusion principle requires that the *total* wave function Ψ change sign under the simultaneous interchange of both space and spin coordinates. Since the total wave function is the product of its spin and orbital parts, it follows that solutions to the orbital Schrödinger equation (32.3) that do not change sign under interchange of $\mathbf{r}_1$ and $\mathbf{r}_2$ (symmetric solutions) must describe states with $S = 0$, while solutions that do change sign (antisymmetric solutions) must go with $S = 1$.[6] There is thus a strict correlation between the spatial symmetry of the solution to the (spin-independent) orbital Schrödinger equation and the total spin: Symmetric solutions require singlet spin states; and antisymmetric, triplets.

If E_s and E_t are the lowest eigenvalues of (32.3) associated with the singlet (symmetric) and triplet (antisymmetric) solutions, then the ground state will have spin zero or one, depending only on whether E_s is less than or greater than E_t, a question, we stress again, that is settled completely by an examination of the *spin-independent* Schrödinger equation (32.3).

As it happens, for two-electron systems there is an elementary theorem that the ground-state wave function for (32.3) must be symmetric.[7] Thus E_s must be less than E_t, and the ground state must have zero total spin. However, the theorem holds only for two-electron systems,[8] and it is therefore important to find a way to estimate $E_s - E_t$ that can be generalized to the analogous problem for an N-atom solid. We continue to use the two-electron system to illustrate this method (in spite of the theorem assuring us that the singlet state has the lowest energy) because it reveals most simply the inadequacy of the independent electron approximation in magnetic problems.

---

[6] All solutions to (32.3) can be taken to be symmetric or antisymmetric, because of the symmetry of V (which contains all electrostatic interactions among the two electrons and two protons fixed at $\mathbf{R}_1$ and $\mathbf{R}_2$). See Problem 1.

[7] See Problem 2.

[8] In *one* spatial dimension it has been proved that the ground state of *any* number of electrons with arbitrary spin-independent interactions must have zero total spin (E. Lieb and D. Mattis, *Phys. Rev.* **125**, 164 (1962)). The theorem cannot be generalized to three dimensions (where, for example, Hund's rules (see Chapter 31) provide many counterexamples).

CALCULATION OF THE SINGLET-TRIPLET SPLITTING: FAILURE OF THE INDEPENDENT ELECTRON APPROXIMATION

The singlet-triplet energy splitting measures the extent to which the antiparallel ($S = 0$) spin alignment of two electrons is more favorable than the parallel ($S = 1$). Since $E_s - E_t$ is the difference between eigenvalues of a Hamiltonian containing only electrostatic interactions, this energy should be of the order of electrostatic energy differences, and therefore quite capable of being the dominant source of magnetic interaction, even when explicitly spin-dependent interactions are added to the Hamiltonian. We shall describe some approximate methods for calculating $E_s - E_t$. Our aim is not to extract numerical results (though the methods are used for that purpose), but to illustrate in the very simple two-electron case the very subtle (especially when N is large) inadequacies of the independent electron approximation, in dealing with electron spin correlations.

Suppose, then, we begin by trying to solve the two-electron problem (32.3) in the independent electron approximation; i.e., we ignore the electron-electron Coulomb interaction in $V(\mathbf{r}_1, \mathbf{r}_2)$, retaining only the interaction of each electron with the two ions (which we take to be fixed at $\mathbf{R}_1$ and $\mathbf{R}_2$). The two-electron Schrödinger equation (32.3) then assumes the form

$$(h_1 + h_2)\psi(\mathbf{r}_1, \mathbf{r}_2) = E\psi(\mathbf{r}_1, \mathbf{r}_2), \tag{32.5}$$

where[9]

$$h_i = -\frac{\hbar^2}{2m}\nabla_i^2 - \frac{e^2}{|\mathbf{r}_i - \mathbf{R}_1|} - \frac{e^2}{|\mathbf{r}_i - \mathbf{R}_2|}, \qquad i = 1, 2. \tag{32.6}$$

Because the Hamiltonian in (32.5) is a sum of one-electron Hamiltonians, the solution can be constructed from solutions of the one-electron Schrödinger equation:

$$h\psi(\mathbf{r}) = \mathcal{E}\psi(\mathbf{r}). \tag{32.7}$$

If $\psi_0(\mathbf{r})$ and $\psi_1(\mathbf{r})$ are the two solutions to (32.7) of lowest energy, with energies $\mathcal{E}_0 < \mathcal{E}_1$, then the symmetric solution of lowest energy to the approximate two-electron Schrödinger equation (32.5) is

$$\psi_s(\mathbf{r}_1, \mathbf{r}_2) = \psi_0(\mathbf{r}_1)\psi_0(\mathbf{r}_2), \quad E_s = 2\mathcal{E}_0, \tag{32.8}$$

and the lowest antisymmetric solution is

$$\psi_t(\mathbf{r}_1, \mathbf{r}_2) = \psi_0(\mathbf{r}_1)\psi_1(\mathbf{r}_2) - \psi_0(\mathbf{r}_2)\psi_1(\mathbf{r}_1), \quad E_t = \mathcal{E}_0 + \mathcal{E}_1. \tag{32.9}$$

The singlet-triplet energy splitting is then

$$E_s - E_t = \mathcal{E}_0 - \mathcal{E}_1, \tag{32.10}$$

which is consistent with the general theorem $E_s < E_t$ for two-electron systems.

In arriving at the ground-state energy $2\mathcal{E}_0$, we have merely followed the steps of

[9] The following discussion would apply unchanged if one were to approximate the electron-electron interaction by a self-consistent field which modified the bare electron-ion Coulomb interaction. (See page 192.)

band theory, specialized to the case of an $N = 2$ "solid," first solving the one-electron problem (32.7), and then filling the lowest $N/2$ one-electron levels with two electrons (of opposite spin) per level. In spite of this pleasing familiarity, the wave function (32.8) is manifestly a very bad approximation to the ground state of the exact Schrödinger equation (32.3) when the protons are very far apart, for in that case it fails quite disastrously to deal with the electron-electron Coulomb interaction. This becomes evident when one examines the structure of the one-electron wave functions $\psi_0(\mathbf{r})$ and $\psi_1(\mathbf{r})$. For well-separated protons these solutions to (32.7) are given to an excellent approximation by the tight-binding method (Chapter 10) specialized to the case in which $N = 2$. The tight-binding method takes the one-electron stationary-state wave functions of the solid to be linear combinations of atomic stationary-state wave functions centered at the lattice points $\mathbf{R}$. When $N = 2$ the correct linear combinations are[10]

$$\begin{aligned} \psi_0(\mathbf{r}) &= \phi_1(\mathbf{r}) + \phi_2(\mathbf{r}), \\ \psi_1(\mathbf{r}) &= \phi_1(\mathbf{r}) - \phi_2(\mathbf{r}), \end{aligned} \qquad \textbf{(32.11)}$$

where $\phi_i(\mathbf{r})$ is the ground-state electronic wave function for a single hydrogen atom whose proton is fixed at $\mathbf{R}_i$. If the one-electron levels have this form (which is essentially exact for well-separated protons), then the two-electron wave functions (32.8) and (32.9) (given by the independent electron approximation) become

$$\begin{aligned} \psi_s(\mathbf{r}_1, \mathbf{r}_2) = \phi_1(\mathbf{r}_1)\phi_2(\mathbf{r}_2) + \phi_2(\mathbf{r}_1)\phi_1(\mathbf{r}_2) \\ + \phi_1(\mathbf{r}_1)\phi_1(\mathbf{r}_2) + \phi_2(\mathbf{r}_1)\phi_2(\mathbf{r}_2), \end{aligned} \qquad \textbf{(32.12)}$$

and

$$\psi_t(\mathbf{r}_1, \mathbf{r}_2) = 2[\phi_2(\mathbf{r}_1)\phi_1(\mathbf{r}_2) - \phi_1(\mathbf{r}_1)\phi_2(\mathbf{r}_2)]. \qquad \textbf{(32.13)}$$

Equation (32.12) gives an excellent approximation to the ground state of the Schrödinger equation (32.5), in which electron-electron interactions are ignored. However, it gives a very bad approximation to the original Schrödinger equation (32.3), in which electron-electron interactions are retained. To see this, note that the first and second terms in (32.12) are quite different from the third and fourth. In the first two terms each electron is localized in a hydrogenic orbit in the neighborhood of a different nucleus. When the two protons are far apart the interaction energy of the two electrons is small, and the description of the molecule as two slightly perturbed atoms (implied by the first two terms in (32.12)) is quite good. However, in each of the last two terms in (32.12) both electrons are localized in hydrogenic orbits about the *same* proton. Their interaction energy is therefore considerable no matter how far apart the protons are. Thus the last two terms in (32.12) describe the hydrogen molecule as a H^- ion and a bare proton—a highly inaccurate picture when electron-electron interactions are allowed for.[11]

---

[10] See Problem 3. If we choose phases so that the ϕ_i are real and positive (which can be done for the hydrogen atom ground state), then the linear combination with the positive sign will have the lower energy since it has no nodes.

[11] This failure of the independent electron approximation to describe accurately the hydrogen molecule is analogous to the failure of the independent electron approximation treated in our discussion of the tight-binding method in Chapter 10. The problem does not arise in the case of a filled band (or, in the molecular analogue, two nearby helium atoms) because a more accurate wave function also must place two electrons in each localized orbital.

The ground state (32.12) of the independent electron approximation therefore gives a 50 percent probability of both electrons being together on the same ion. The independent electron triplet state (32.13) does not suffer from this defect. Consequently, when we introduce electron-electron interactions into the Hamiltonian, the triplet (32.13) will surely give a lower mean energy than the singlet (32.12), when the protons are far enough apart.

This does not mean, however, that the true ground state is a triplet. A symmetric state which never places two electrons on the same proton, and is therefore of much lower energy than the independent electron ground state, is given by taking only the first two terms in (32.12):

$$\bar{\psi}_s(\mathbf{r}_1, \mathbf{r}_2) = \phi_1(\mathbf{r}_1)\phi_2(\mathbf{r}_2) + \phi_2(\mathbf{r}_1)\phi_1(\mathbf{r}_2). \quad \textbf{(32.14)}$$

The theory that takes its approximations to the singlet and triplet ground states of the full Hamiltonian (32.3) to be proportional to (32.14) and (32.13) is known as the Heitler-London approximation.[12] Evidently the Heitler-London singlet state (32.14) is far more accurate for widely separated protons than the independent electron singlet state (32.12). When appropriately generalized, it should be more suitable for discussing magnetic ions in an insulating crystal.

On the other hand, when the protons are very close together the independent electron approximation (32.8) is closer to the true ground state than the Heitler-London approximation (32.14), as is easily seen in the extreme case in which the two protons actually coincide. The independent electron approximation starts with two one-electron wave functions appropriate to a single doubly charged nucleus, whereas the Heitler-London approximation works with one-electron wave functions for a singly charged nucleus. These are far too extended in space to form a good starting point for the description of what is now not a hydrogen molecule, but a helium atom.

The foregoing analysis was intended primarily to emphasize, through the simple example of a two-electron system, that one cannot apply the concepts of band theory, based as it is on the independent electron approximation, to account for magnetic interactions in insulating crystals. As for the Heitler-London method itself, it too has shortcomings, for although it gives very accurate singlet and triplet energies for large spatial separations,[13] its prediction for the very small singlet-triplet energy splitting is considerably less certain when the ions are far apart. The method is therefore quite treacherous to use uncritically.[14] We nevertheless give below the form of the

---

[12] In the context of molecular physics, the description based on the independent electron ground state (32.12) is known as the Hund-Mulliken approximation, or the method of molecular orbitals. Other terminology is associated with the fact that the Heitler-London approximation to the ground state can be written as a linear combination of *two* independent electron approximation two-electron states:

$$\bar{\psi}(\mathbf{r}_1, \mathbf{r}_2) = \psi_0(\mathbf{r}_1)\psi_0(\mathbf{r}_2) - \psi_1(\mathbf{r}_1)\psi_1(\mathbf{r}_2),$$

a simple example of a state of affairs referred to as "configuration mixing." The Heitler-London states $\bar{\psi}_s$ and ψ_t are known as the "bonding" and "anti-bonding" states.

[13] Unlike the independent electron approximation.

[14] A thorough critique of the Heitler-London method has been given by C. Herring, "*Direct Exchange Between Well Separated Atoms,*" in *Magnetism*, vol. 2B, G. T. Rado and H. Suhl, eds., Academic Press, New York, 1965.

Heitler-London result for $E_s - E_t$, because it is both a starting point for more refined treatments and the source of a nomenclature that pervades much of the subject of magnetism:

The Heitler-London approximation uses the singlet and triplet wave functions (32.14) and (32.13) to estimate the singlet-triplet splitting as

$$E_s - E_t = \frac{(\bar{\psi}_s, H\bar{\psi}_s)}{(\bar{\psi}_s, \bar{\psi}_s)} - \frac{(\psi_t, H\psi_t)}{(\psi_t, \psi_t)}, \tag{32.15}$$

where H is the full Hamiltonian (32.3). In the limit of large spatial separations this splitting can be shown (Problem 4) to reduce simply to

$$\frac{1}{2}(E_s - E_t) = \int d\mathbf{r}_1\, d\mathbf{r}_2\, [\phi_1(\mathbf{r}_1)\phi_2(\mathbf{r}_2)] \left(\frac{e^2}{|\mathbf{r}_1 - \mathbf{r}_2|} + \frac{e^2}{|\mathbf{R}_1 - \mathbf{R}_2|} - \frac{e^2}{|\mathbf{r}_1 - \mathbf{R}_1|} - \frac{e^2}{|\mathbf{r}_2 - \mathbf{R}_2|} \right) [\phi_2(\mathbf{r}_1)\phi_1(\mathbf{r}_2)]. \tag{32.16}$$

Because this is a matrix element between two states that differ solely through the exchange of the coordinates of the two electrons, the singlet-triplet energy difference is referred to as an *exchange splitting* or, when viewed as a source of magnetic interaction, an *exchange interaction.*[15]

Since the atomic orbital $\phi_i(\mathbf{r})$ is strongly localized in the neighborhood of $\mathbf{r} = \mathbf{R}_i$, the factors $\phi_1(\mathbf{r}_1)\phi_2(\mathbf{r}_1)$ and $\phi_1(\mathbf{r}_2)\phi_2(\mathbf{r}_2)$ in the integrand of (32.16) insure that the singlet-triplet energy splitting will fall off rapidly with the distance $|\mathbf{R}_1 - \mathbf{R}_2|$ between protons.

THE SPIN HAMILTONIAN AND THE HEISENBERG MODEL

There is a way to express the dependence of the spin of a two-electron system on the singlet-triplet energy splitting, which, though unnecessarily complicated in this simple case, is of fundamental importance in analyzing the energetics of the spin configurations of real insulating solids. One first notes that when the two protons are far apart, the ground state describes two independent hydrogen atoms and is therefore fourfold degenerate (since each electron can have two spin orientations). One next considers the protons to be brought a bit closer together, so that there is a splitting ($E_s \neq E_t$) of the fourfold degeneracy due to interactions between the atoms, which is, however, small compared with all other excitation energies of the two-electron system. Under such conditions these four states will play the dominant role in determining many important properties of the molecule,[16] and one often simplifies the analysis by ignoring the higher states altogether, representing the molecule as a simple four-state system. If we do represent a general state of the molecule as a linear combination of the four lowest states, it is convenient to have an operator, known as the spin Hamiltonian,

---

[15] One should not, however, be lulled by this nomenclature into forgetting that underlying the exchange interaction are nothing but electrostatic interaction energies, and the Pauli exclusion principle.

[16] For example, the thermal equilibrium properties when $k_B T$ is comparable to $E_s - E_t$, but small enough that no states other than the four are thermally excited.

whose eigenvalues are the same as those of the original Hamiltonian within the four-state manifold, and whose eigenfunctions give the spin of the corresponding states.

To construct the spin Hamiltonian for a two-electron system, note that each individual electron spin operator satisfies $\mathbf{S}_i^2 = \frac{1}{2}(\frac{1}{2} + 1) = \frac{3}{4}$, so that the total spin $\mathbf{S}$ satisfies

$$\mathbf{S}^2 = (\mathbf{S}_1 + \mathbf{S}_2)^2 = \frac{3}{2} + 2\mathbf{S}_1 \cdot \mathbf{S}_2. \tag{32.17}$$

Since $\mathbf{S}^2$ has the eigenvalue $S(S + 1)$ in states of spin S, it follows from (32.17) that the operator $\mathbf{S}_1 \cdot \mathbf{S}_2$ has eigenvalue $-\frac{3}{4}$ in the singlet ($S = 0$) state and $+\frac{1}{4}$ in the triplet ($S = 1$) states. Consequently the operator

$$\mathcal{H}^{\text{spin}} = \tfrac{1}{4}(E_s + 3E_t) - (E_s - E_t)\mathbf{S}_1 \cdot \mathbf{S}_2 \tag{32.18}$$

has eigenvalue E_s in the singlet state, and E_t in each of the three triplet states, and is the desired spin Hamiltonian.

By redefining the zero of energy we may omit the constant $(E_s + 3E_t)/4$ common to all four states, and write the spin Hamiltonian as

$$\mathcal{H}^{\text{spin}} = -J\mathbf{S}_1 \cdot \mathbf{S}_2, \quad J = E_s - E_t. \tag{32.19}$$

Since $\mathcal{H}^{\text{spin}}$ is the scalar product of the vector spin operators $\mathbf{S}_1$ and $\mathbf{S}_2$, it will favor parallel spins if J is positive and antiparallel if J is negative.[17] Note that in contrast to the magnetic dipolar interaction (32.1), the coupling in the spin Hamiltonian depends only on the relative orientation of the two spins, but not on their directions with respect to $\mathbf{R}_1 - \mathbf{R}_2$. This is a general consequence of the spin independence of the original Hamiltonian and (one should note) holds without any assumption about its spatial symmetry. One must include terms that break rotational symmetry in spin space (such as dipolar interactions or spin-orbit coupling) in the original Hamiltonian, in order to produce a spin Hamiltonian with anisotropic coupling.[18]

When N is large, rather than merely reexpressing some known results (as when $N = 2$), the spin Hamiltonian contains in highly compact form some exceedingly complex information about the low-lying levels.[19] When N ions of spin S are widely separated,[20] the ground state will be $(2S + 1)^N$-fold degenerate. The spin Hamiltonian describes the splitting of this vastly degenerate ground state when the ions are somewhat closer together, but still far enough apart that the splittings are small compared with any other excitation energies. One can (in a variety of ways) construct an operator function of the $\mathbf{S}_i$ whose eigenvalues give the split levels. What is remarkable, however, is that for many cases of interest the form of the spin Hamiltonian is simply that

---

[17] Since J is positive or negative depending on whether E_t or E_s is lower, this simply restates the fact that the spins are parallel in the triplet state and antiparallel in the singlet.

[18] Such anisotropy is of critical importance in understanding the existence of easy and hard directions of magnetization and plays a role in the theory of domain formation. (See page 720.)

[19] In general, this information is not easily extracted, even from a spin Hamiltonian. In contrast to the case $N = 2$ one does not know the low-lying levels from the start, and finding a spin Hamiltonian is only half the problem. There remains the highly nontrivial problem of finding the eigenvalues of that spin Hamiltonian. (See, for example, pages 701–709).

[20] For simplicity we assume that for each ion $J = S$ (i.e., $L = 0$). This restriction is not essential to the development of a spin Hamiltonian.

for the two-spin case, summed over all pairs of ions:

$$H^{\text{spin}} = -\sum J_{ij}\mathbf{S}_i \cdot \mathbf{S}_j. \tag{32.20}$$

We shall not go into the question of when (32.20) can be justified, which is quite a complex matter.[21] However one should note:

1. For only products of pairs of spin operators to appear in (33.20) it is necessary that all magnetic ions be far enough apart that the overlap of their electronic wave functions is very small.
2. When the angular momentum of each ion contains an orbital as well as a spin part, the coupling in the spin Hamiltonian may depend on the absolute as well as the relative spin orientations.

The spin Hamiltonian (32.20) is known as the Heisenberg[22] Hamiltonian, and the J_{ij} are known as the exchange coupling constants (or parameters or coefficients). Extracting information even from the Heisenberg Hamiltonian is, in general, so difficult a task, that it by itself is taken as the starting point for many quite profound investigations of magnetism in solids. One must remember, however, that much subtle physics and quite complex approximations must be delved into before one can even arrive at a Heisenberg Hamiltonian.

DIRECT EXCHANGE, SUPEREXCHANGE, INDIRECT EXCHANGE, AND ITINERANT EXCHANGE

The magnetic interaction we have just described is known as *direct exchange*, because it arises from the direct Coulomb interaction among electrons from the two ions. It often happens that two magnetic ions are separated by a nonmagnetic ion (i.e., one with all electronic shells closed). It is then possible for the magnetic ions to have a magnetic interaction mediated by the electrons in their common nonmagnetic neighbors, which is more important than their direct exchange interaction. This type of magnetic interaction is called *superexchange*. (See Figure 32.2.)

Yet another source of magnetic interaction can occur between electrons in the partially filled f-shells in the rare earth metals. In addition to their direct exchange coupling, the f-electrons are coupled through their interactions with the conduction electrons. This mechanism (in a sense the metallic analogue of superexchange in insulators) is known as *indirect exchange*. It can be stronger than the direct exchange coupling, since the f-shells generally overlap very little.

There are also important exchange interactions in metals among the conduction electrons themselves, often referred to as *itinerant exchange*.[23] To emphasize the great generality of exchange interactions, we give below a brief discussion of itinerant exchange in the case farthest away from the well-localized electrons for which the Heitler-London theory of direct exchange was devised: the free electron gas.

---

[21] A very thorough discussion is given by C. Herring (see footnote 14).

[22] In the older literature it is known as the Heisenberg-Dirac Hamiltonian.

[23] A discussion of itinerant exchange is given by C. Herring in *Magnetism*, vol. 4, G. T. Rado and H. Suhl. eds., Academic Press, New York, 1966.

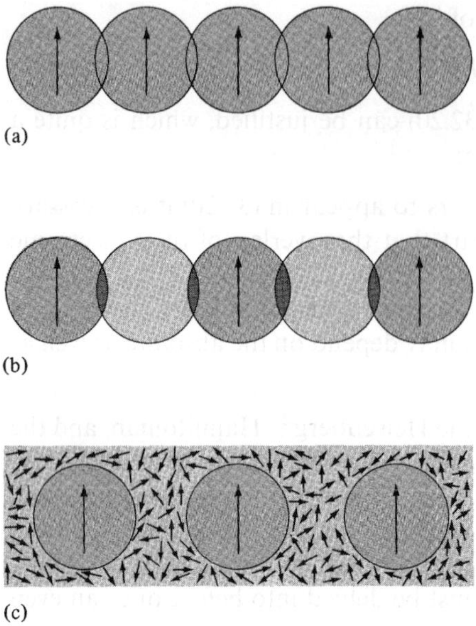

Figure 32.2
Schematic illustrations of (a) direct exchange, in which the magnetic ions interact because their charge distributions overlap; (b) superexchange, in which magnetic ions with nonoverlapping charge distributions interact because both have overlap with the same nonmagnetic ion; and (c) indirect exchange, in which in the absence of overlap a magnetic interaction is mediated by interactions with the conduction electrons.

MAGNETIC INTERACTIONS IN THE FREE ELECTRON GAS

The theory of magnetism in a free electron gas is hopelessly inadequate as an approach to the problem of magnetism in real metals. However, the subject is not without interest because (a) it provides another simple model in which magnetic structure is implied in the absence of explicitly spin-dependent interactions, (b) its complexity indicates the magnitude of the problem one faces in real metals, and (c) a correct (currently nonexistent) general theory of magnetism in metals will undoubtedly have to find a way of dealing simultaneously with both the localized (as described above) *and* itinerant (as described below) aspects of exchange.

F. Bloch[24] first pointed out that the Hartree-Fock approximation can predict ferromagnetism in a gas of electrons interacting only through their mutual Coulomb interactions. Within this approximation we showed in Chapter 17 that if every one-electron level with wave vector less than k_F is occupied by two electrons of opposite spin, then the ground-state energy of N free electrons is (Eq. (17.23)):

$$E = N\left[\frac{3}{5}(k_F a_0)^2 - \frac{3}{2\pi}(k_F a_0)\right]\text{Ry} \qquad \left(1\ \text{Ry} = \frac{e^2}{2a_0}\right). \tag{32.21}$$

The first term in (32.21) is the total kinetic energy, and the second, known as the exchange energy, is the Hartree-Fock approximation to the effect of electron-electron Coulomb interactions.

In deriving (32.21), however, we assumed that every occupied one-electron orbital level would be occupied by two electrons of opposite spin. A more general possibility, leading to a net spin imbalance, would be to fill each one-electron level with k less than some $k_\uparrow$ with spin-up electrons, and each with $k < k_\downarrow$, with spin-down electrons.

---

24 *Z. Physik* **57**, 545 (1929).

Since (see Eq. (17.15)) the exchange interaction in Hartree-Fock theory is only between electrons of the same spin, we will have an equation of the form (32.21) for each spin population:

$$E_\uparrow = N_\uparrow\left[\frac{3}{5}(k_\uparrow a_0)^2 - \frac{3}{2\pi}(k_\uparrow a_0)\right]\text{Ry},$$
$$E_\downarrow = N_\downarrow\left[\frac{3}{5}(k_\downarrow a_0)^2 - \frac{3}{2\pi}(k_\downarrow a_0)\right]\text{Ry}, \tag{32.22}$$

where the total energy and total number of electrons are

$$E = E_\uparrow + E_\downarrow,$$
$$\frac{N}{V} = \frac{N_\uparrow}{V} + \frac{N_\downarrow}{V} = \frac{k_\uparrow{}^3}{6\pi^2} + \frac{k_\downarrow{}^3}{6\pi^2} = \frac{k_F{}^3}{3\pi^2}. \tag{32.23}$$

Equation (32.21) is the form E takes if $N_\uparrow = N_\downarrow = N/2$, but we may now ask whether one cannot get a lower energy by dropping this assumption. If so, then the ground state will have a nonvanishing magnetization density,

$$M = -g\mu_B\frac{N_\uparrow - N_\downarrow}{V}, \tag{32.24}$$

and the electron gas will be ferromagnetic.

For simplicity, we consider only the opposite extreme,[25] taking $N_\downarrow = N$ and $N_\uparrow = 0$. Then E will be $E_\downarrow$, and $k_\downarrow$ will be $2^{1/3}k_F$ (according to (32.23), and therefore

$$E = N\left[\frac{3}{5}2^{2/3}(k_F a_0)^2 - \frac{3}{2\pi}2^{1/3}(k_F a_0)\right]. \tag{32.25}$$

Compared with the nonmagnetic case (32.21), the positive kinetic energy in (32.25) is larger by $2^{2/3}$, and the magnitude of the negative exchange energy is larger by $2^{1/3}$. Hence the energy of the fully magnetized state is lower than that of the unmagnetized state when exchange energy dominates kinetic energy. This happens for small k_F—i.e., at low densities. As the density decreases, the transition from nonmagnetic to fully magnetic ground states occurs when the energies (32.21) and (32.25) become equal—i.e., when

$$k_F a_0 = \frac{5}{2\pi}\frac{1}{2^{1/3}+1}, \tag{32.26}$$

or (see Eq. (2.22)) when

$$\frac{r_s}{a_0} = \frac{2\pi}{5}(2^{1/3}+1)\left(\frac{9\pi}{4}\right)^{1/3} = 5.45. \tag{32.27}$$

Cesium is the only metallic element with so low a conduction electron density that r_s exceeds this value, but there are metallic compounds[26] with $r_s/a_0 > 5.45$. In none

---

[25] It can be shown that values of $N_\uparrow$ and $N_\downarrow$ between the extremes $N_\uparrow = N_\downarrow$, and $N_\uparrow$ (or $N_\downarrow$) $= N$, $N_\downarrow$ (or $N_\uparrow$) $= 0$, give a higher energy than one or the other limiting case.

[26] For example, the metal amines. See J. J. Lagowski and M. J. Sienko, eds., *Metal Ammonia Solutions*, Butterworth, London, 1970.

of these cases, however, are the materials ferromagnetic, even though their band structures are reasonably well described by the free electron model.

The simple criterion (32.26) for low-density ferromagnetism is invalidated by further theoretical considerations as well:

1. Even within the Hartree-Fock approximation there are still more complicated choices of one-electron levels that lead to a lower energy than either the fully magnetized or nonmagnetic solutions. These solutions, called spin density waves, were discovered by Overhauser[27] and give an antiferromagnetic ground state, at densities in the neighborhood of that given by (32.27).
2. The Hartree-Fock approximation is improved if one lets the electrons screen the exchange interaction (page 344), and thereby reduce its spatial range. This improvement alters the Hartree-Fock prediction quite drastically. For example, in the extreme short-range case of a delta function potential, it predicts ferromagnetism at *high* densities and a nonmagnetic state at *low* densities.
3. At very low densities the true ground state of the *free* electron gas bears no resemblance to any of the forms described above. In the low density limit the free electron gas can be shown to crystallize, taking on a configuration (the *Wigner crystal*) whose description is quite outside the reach of the independent electron approximation.[28]

Thus the best Hartree-Fock ground state is by no means obvious and, worse still, simple attempts to improve Hartree-Fock theory can drastically alter its predictions. It is currently thought that the free electron gas is probably not ferromagnetic at any density, but a rigorous proof of this is lacking. Certainly ferromagnetism has only been observed experimentally in metals whose free ions contain partially filled *d*- or *f*-shells, a state of affairs hopelessly beyond the range of competence of the free electron model. To account for magnetic ordering in metals, itinerant exchange interactions must be combined with specific features of the band structure[29] and/or the kinds of atomic considerations leading to Hund's rules.

---

[27] A. W. Overhauser, *Phys. Rev. Lett.* **4**, 462 (1960); Phys. Rev. **128**, 1437 (1962); see also C. Herring (footnote 23). Introducing screening eliminates the spin density wave. However, certain special features of the band structure of chromium make it possible to resurrect the spin density wave by introducing band structure into the theory in a fairly simple way, and it is currently believed that such a theory accounts for the antiferromagnetism of chromium. See, for example, T. M. Rice, *Phys. Rev.* **B2**, 3619 (1970) and references cited therein.

[28] E. Wigner, *Trans. Farad. Soc.* **34**, 678 (1938).

[29] A widely held view of the ferromagnetism of nickel, for example, simply combines the free electron picture of itinerant ferromagnetism with the band theory. Thus the energy bands of nickel are calculated in the usual way except that one allows for a self-consistent exchange field (frequently taken to be simply a constant) that can differ for electrons of opposite spins when the two spin populations differ. By suitably choosing the exchange field one can construct a ground state for nickel (which has a total of ten electrons in 3*d* and 4*s* levels in the atomic state) in which one *d*-band is filled (five electrons per atom) with spin-up electrons. The second *d*-band contains spin-down electrons, but is displaced (relative to the first) upward in energy across the Fermi level, so that it is somewhat short of being filled (containing only 4.4 electrons per atom). The missing 0.6 electron per atom resides in a free electron band, with randomly oriented spins. Since the population of the spin-up band exceeds that of the spin-down band by 0.6 electron per atom, the solid has a net magnetization. See E. C. Stoner, *Rept. Prog. Phys.* **11**, 43 (1947) for a review of early work along these lines, and C. Herring (footnote 23) for a more up-to-date survey.

THE HUBBARD MODEL

J. Hubbard[30] has proposed a highly oversimplified model that attempts to come to grips with these problems, containing the bare minimum of features necessary to yield both bandlike and localized behavior in suitable limits. In the Hubbard model the vast set of bound and continuum electron levels of each ion is reduced to a single localized orbital level. The states of the model are given by specifying the four possible configurations of each ion (its level can either be empty, contain one electron with either of two spins, or two electrons of opposite spin). The Hamiltonian for the Hubbard model contains two types of terms: (a) a term diagonal in these states, that is just a positive energy U times the number of doubly occupied ionic levels (plus an (unimportant) energy $\mathcal{E}$ times the number of electrons); and (b) a term off-diagonal in these states that has nonvanishing matrix elements t between just those pairs of states that differ only by a single electron having been moved (without change in spin) from a given ion to one of its neighbors. The first set of terms, in the absence of the second, would favor local magnetic moments, since it would suppress the possibility of a second electron (with oppositely directed spin) at singly occupied sites. The second set of terms in the absence of the first, can be shown to lead to a conventional band spectrum and one-electron Bloch levels in which each electron is distributed throughout the entire crystal. When both sets of terms are present, even this simple model has proved too difficult for exact analysis, although much interesting information has been extracted in special cases. If, for example, the total number of electrons is equal to the total number of sites, then in the limit of negligible intrasite repulsion ($t \gg U$) one has an ordinary half-filled metallic band. In the opposite limit ($U \gg t$), however, one can derive an antiferromagnetic Heisenberg spin Hamiltonian (having an exchange constant $|J| = 4t^2/U$) to describe the low-lying excitations. No one, however, has given a rigorous solution to how the model changes from a nonmagnetic metal to an antiferromagnetic insulator as t/U is varied.

LOCALIZED MOMENTS IN ALLOYS

We have pointed out the perils in treating magnetism in conductors from a purely itinerant point of view. On the other hand, in metals for which the magnetic ions have a partially filled d-shell in the free atom, one's natural inclination might be to go to the other extreme, treating the d-electrons with the same techniques used in magnetic insulators—i.e., with the concept of direct exchange (supplemented by the indirect exchange that can also arise in metals).[31] This approach is also perilous. An ion that in isolation has a magnetic electronic shell may retain only a fraction of its magnetic moment, or none at all, when placed in a metallic environment. This is very well illustrated by the properties of dilute magnetic alloys.

When small amounts of transition metal elements are dissolved in a nonmagnetic (often free electronlike) metal, the resulting alloy may or may not display a localized

---

[30] J. Hubbard, *Proc. Roy. Soc.* **A276**, 238 (1963); **A277**, 237 (1964), **A281**, 401 (1964). The model is applied to the hydrogen molecule in Problem 5.

[31] With the f-shells in rare earth metals this is probably a reasonable procedure.

moment (Table 32.1).[32] The moment of the free ion is determined by Hund's rules (pages 650–652), which in turn are based on considerations of the intraionic Coulomb (and, to a lesser degree, spin-orbit) interactions. A theory of localized moments in dilute magnetic alloys must determine how these considerations are modified when the ion is not free, but embedded in a metal.[33]

Table 32.1
PRESENCE OR ABSENCE OF LOCALIZED MOMENTS WHEN TRANSITION METAL IMPURITIES ARE DISSOLVED IN NONMAGNETIC HOSTS[a]

	HOST			
IMPURITY	Au	Cu	Ag	Al
Ti	No	—	—	No
V	?	—	—	No
Cr	Yes	Yes	Yes	No
Mn	Yes	Yes	Yes	?
Fe	Yes	Yes	—	No
Co	?	?	—	No
Ni	No	No	—	No

[a] Presence is indicated by a "Yes" entry, absence, by "No." A question mark indicates that the situation is uncertain. For some combinations of host and impurity there are metallurgical difficulties in achieving reproducible dilute alloys, due chiefly to problems of insolubility. These account for the majority of blank entries.
Source: A. J. Heeger, *Solid State Physics* **23**, F. Seitz and D. Turnbull, eds., Academic Press, New York, 1969.

Even if we were to ignore all interactions between the electrons on the magnetic ion, and the electrons and ions of the host metal, there would still be a simple mechanism by which the net moment of the ion could be altered. Depending on the position of the levels of the ion relative to the metallic Fermi level, electrons might leave the ion for the conduction band of the host, or drop from the conduction band into lower-lying ionic levels, thereby altering or, in some cases, even eliminating the moment of the ion. In addition to this, because levels of the ion are degenerate with the continuum of conduction band levels in the host, there will be a mixing of levels, in which the ionic ones become spatially less localized, while nearby conduction band levels have their charge distributions altered in the neighborhood of the magnetic

[32] The criterion for a localized moment is a term in the magnetic susceptibility inversely proportional to the temperature, with a coefficient proportional to the density of magnetic ions as predicted by Curie's law, Eq. (31.47).

[33] In Chapter 31 we considered the analogous problem arising when the ion was embedded in an insulator, and found that effects of the crystal field could easily be more important than the Hund's rules couplings, with a corresponding alteration in the net magnetic moment.

ion. This, in turn, will radically alter the intra-ionic energetics that determine the net spin configuration of those electrons that are localized near the ion. For example, as the ionic levels become less and less localized, the intra-ionic Coulomb repulsion will become of progressively less importance.

In short, the problem is one of considerable complexity. Some insight has been provided by a model due to P. W. Anderson,[34] in which all the levels of the magnetic ion are replaced by a single localized level (as the ions are treated in the Hubbard model), and the coupling between localized and band levels is reduced to a bare minimum. It is a measure of the complexity of problems in which both localized and band features play an important role that even the highly oversimplified Anderson model has failed to yield an exact solution, in spite of its having been subject to an onslaught of theoretical analysis.

THE KONDO THEORY OF THE RESISTANCE MINIMUM

The existence of localized moments in dilute alloys that couple to the conduction electrons has important consequences for the electrical conductivity. The magnetic impurities act as scattering centers, and if they are the predominant type of impurity or lattice imperfection, then at low enough temperatures the scattering they cause will be the primary source of electrical resistance.[35] In Chapter 16 we found that nonmagnetic scatterers lead to a temperature independent term toward which the resistivity monotonically drops with decreasing temperature (the so-called residual resistivity). In magnetic alloys, however, it has been known since 1930[36] that instead of dropping monotonically the resistivity has a rather shallow minimum occurring at a low ($O(10\text{ K})$) temperature that depends weakly on the concentration of magnetic impurities (Figure 32.3).

It was not until 1963 that this minimum was shown by J. Kondo[37] to arise from some unexpected features of the scattering of conduction electrons that arise only when the scattering center has a magnetic moment. In such a case the exchange interaction between the conduction electrons and the local moment leads to scattering events in which the electronic spin is flipped (with a compensating change of spin on the local moment). Prior to Kondo's analysis, this scattering was treated only to leading order in perturbation theory and was found not to differ qualitatively from nonmagnetic scattering, such as we described in Chapter 16. Kondo discovered, however, that in all higher orders of perturbation theory the magnetic scattering cross section is divergent, yielding an infinite resistivity.

The divergence depends critically on there being a sharp cutoff in the conduction electron wave vector distribution, as there is at $T = 0$. Subsequent analysis by Kondo and others has shown that the thermal rounding of the electron distribution removes

---

34 P. W. Anderson, *Phys. Rev.* **124**, 41 (1961). See also A. J. Heeger, *Solid State Physics*, vol. 23, F. Seitz and D. Turnbull, eds., Academic Press, New York, 1969, p. 293.

35 Recall that the contribution due to phonon scattering decreases as T^5.

36 W. Meissner and B. Voigt, *Ann. Phys.* **7**, 761, 892 (1930).

37 J. Kondo, *Prog. Theoret. Phys.* **32**, 37 (1964); *Solid State Physics*, vol. 23, F. Seitz and D. Turnbull, eds., Academic Press, New York, 1969, p. 183.

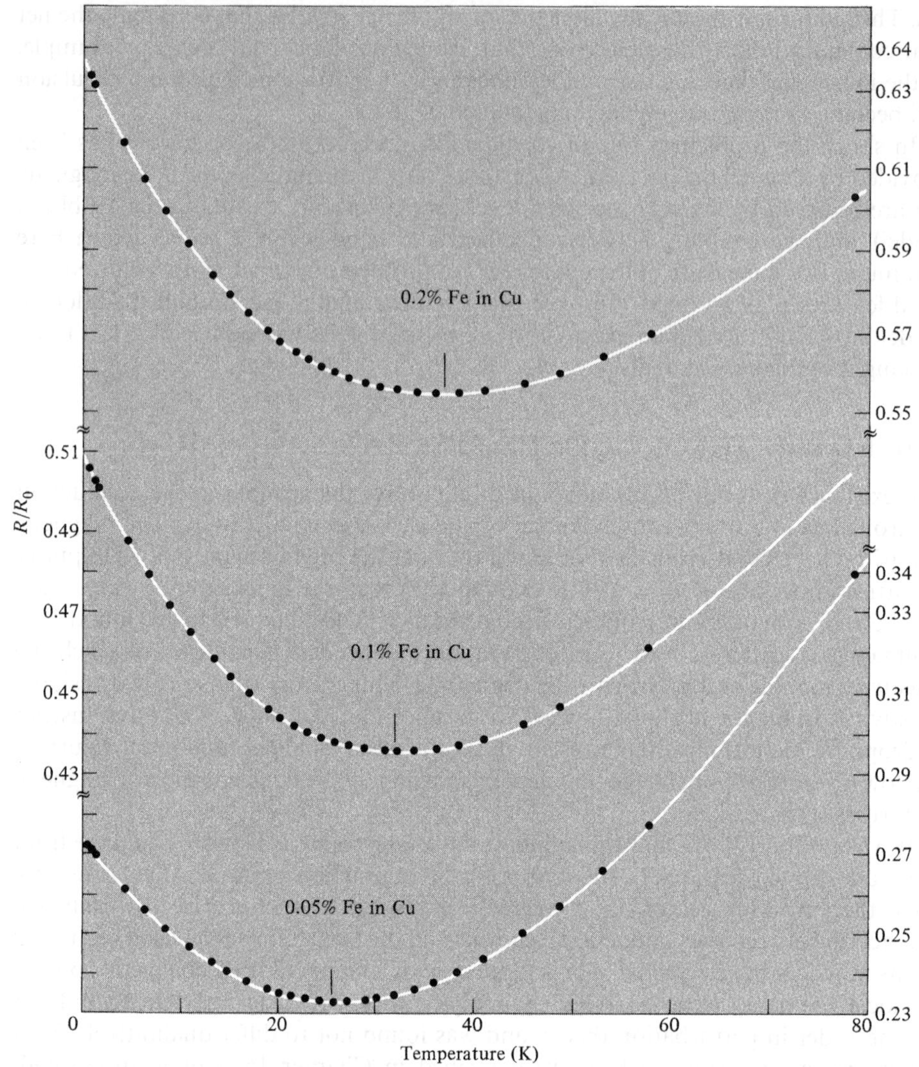

Figure 32.3
The resistance minimum for various dilute alloys of iron in copper. (R_0 is the resistivity at 0°C.) The position of the minimum depends on the concentration of iron. (From J. P. Franck et al., *Proc. Roy, Soc.* **A263**, 494 (1961).)

the divergence, yielding instead a term in the impurity contribution to the resistivity that *increases* with decreasing temperature. It is the balancing of this term against the phonon contribution, which decreases as the temperature falls, that results in the resistance minimum.

The foregoing has offered little more than a glimpse into the difficult, subtle, often fascinating questions that one encounters in almost any attempt to account for magnetic interactions. It is, however, only half the problem. Even if one is *given* a suitably simplified model of the important magnetic interactions (for example, the

Heisenberg Hamiltonian (32.20)), one must still face the task of extracting information of physical interest from the model. In general this turns out to be as difficult, subtle, and fascinating as the problem of deriving the model in the first place. A view of this aspect of magnetism is given in Chapter 33.

PROBLEMS

1. *Symmetry of Two-Electron Orbital Wave Functions*
Prove that the stationary states of the orbital Schrödinger equation for a two-electron system with a symmetric potential—i.e., Eq. (32.3) with $V(\mathbf{r}_1, \mathbf{r}_2) = V(\mathbf{r}_2, \mathbf{r}_1)$—can be chosen to be either symmetric or antisymmetric. (The proof is quite analogous to the first proof of Bloch's theorem in Chapter 8.)

2. *Proof That the Two-Electron Ground State of a Spin-Independent Hamiltonian Is a Singlet*

(a) The mean energy of a two-electron system with Hamiltonian (32.3) in the state ψ can be written (after an integration by parts in the kinetic energy term) in the form:

$$E = \int d\mathbf{r}_1\, d\mathbf{r}_2 \left[\frac{\hbar^2}{2m}\{|\nabla_1\psi|^2 + |\nabla_2\psi|^2\} + V(\mathbf{r}_1, \mathbf{r}_2)|\psi|^2\right]. \qquad \textbf{(32.28)}$$

Show that the lowest value (32.28) assumes over all normalized antisymmetric differentiable wave functions ψ that vanish at infinity is the triplet ground-state energy E_t, and that when symmetric functions are used the lowest value is the singlet ground-state energy E_s.

(b) Using (i) the result of (a), (ii) the fact that the triplet ground state ψ_t can be taken to be real when V is real, and (iii) the fact that $|\psi_t|$ is symmetric, deduce that $E_s \leqslant E_t$.

3. *Symmetry of One-Electron Orbital Wave Functions for the Hydrogen Molecule*
Prove (in much the same way as in Problem 1) that if a one-electron potential has a plane of mirror symmetry, then the stationary one-electron levels can be chosen so that they are either invariant or change sign under reflection in that plane. (This establishes that Eq. (32.11) gives the correct linear combinations of atomic orbitals for the two-proton potential.)

4. *Heitler-London Singlet-Triplet Splitting*
Derive the Heitler-London estimate (32.16) for the difference in singlet and triplet ground-state energies for the hydrogen molecule. (In showing that (32.15) reduces to (32.16) for well-separated protons, it is essential to take into account the following points: (a) The one-electron wave functions ϕ_1 and ϕ_2 out of which (32:13) and (32.14) are constructed are exact ground-state wave functions for a single electron in a hydrogen atom at $\mathbf{R}_1$ and $\mathbf{R}_2$, respectively. (b) The criterion for well-separated protons is that they be far apart compared with the range of a one-electron hydrogenic wave function. (c) The electrostatic field outside of a spherically symmetric distribution of charge is precisely the field one would have if all the charge were concentrated in a single point charge at the center of the sphere. It is also convenient to include in the Hamiltonian the (constant) interaction energy $e^2/|\mathbf{R}_1 - \mathbf{R}_2|$ of the two protons.)

5. *Hubbard Model of the Hydrogen Molecule*
The Hubbard model represents an atom at $\mathbf{R}$ by a single orbital electronic level $|\mathbf{R}\rangle$. If the level is empty (no electron on the atom) the energy is zero, if one electron (of either spin) is in the level the energy is $\mathcal{E}$, and if two electrons (of necessarily opposite spins) are in the level the energy

is $2\mathcal{E} + U$, the additional positive energy U representing the intra-atomic Coulomb repulsion between the two localized electrons. (The exclusion principle prevents more than two electrons from occupying the level.)

The Hubbard model for a two-atom molecule consists of two such orbital levels, $|\mathbf{R}\rangle$ and $|\mathbf{R}'\rangle$, representing electrons localized at $\mathbf{R}$ and $\mathbf{R}'$, respectively. For simplicity, one takes the two levels to be orthogonal:

$$\langle \mathbf{R}|\mathbf{R}'\rangle = 0. \tag{32.29}$$

We first consider the problem of two "protons" and one electron (i.e., $H_2{}^+$). If the one-electron Hamiltonian h were diagonal in $|\mathbf{R}\rangle$ and $|\mathbf{R}'\rangle$, the stationary levels would describe a hydrogen atom and a proton. We know, however, that if the protons are not too far apart, there will be a probability for the electron tunneling from one to the other, which leads to an ionized hydrogen molecule. We represent this amplitude for tunneling by an off-diagonal term in the one-electron Hamiltonian:

$$\langle \mathbf{R}|h|\mathbf{R}'\rangle = \langle \mathbf{R}'|h|\mathbf{R}\rangle = -t, \tag{32.30}$$

where we may choose the phases of $|\mathbf{R}\rangle$ and $|\mathbf{R}'\rangle$ to make the number t real and positive. This, in conjunction with the diagonal terms

$$\langle \mathbf{R}|h|\mathbf{R}\rangle = \langle \mathbf{R}'|h|\mathbf{R}'\rangle = \mathcal{E}, \tag{32.31}$$

defines the one-electron problem.

(a) Show that the one-electron stationary levels are

$$\frac{1}{\sqrt{2}}(|\mathbf{R}\rangle \mp |\mathbf{R}'\rangle) \tag{32.32}$$

with corresponding eigenvalues

$$\mathcal{E} \pm t. \tag{32.33}$$

As a first approach to the two-electron problem (the hydrogen molecule) we make the independent electron approximation for the singlet (spatially symmetric) ground state, putting both electrons into the one-electron level of lowest energy, to get a total energy of $2(\mathcal{E} - t)$. This ignores entirely the interaction energy U arising when two electrons are found on the same proton. The crudest way to improve upon the estimate $2(\mathcal{E} - t)$ is to add the energy U, multiplied by the probability of actually finding two electrons on the same proton, when the molecule is in the ground state of the independent electron approximation.

(b) Show that this probability is $\frac{1}{2}$, so that the improved independent electron estimate of the ground-state energy is

$$E_{ie} = 2(\mathcal{E} - t) + \tfrac{1}{2}U. \tag{32.34}$$

(This result is just the Hartree (or self-consistent field) approximation, applied to the Hubbard model. See Chapters 11 and 17.)

The full set of singlet (spatially symmetric) states of the two-electron problem are:

$$\Phi_0 = \frac{1}{\sqrt{2}}(|\mathbf{R}\rangle|\mathbf{R}'\rangle + |\mathbf{R}'\rangle|\mathbf{R}\rangle),$$
$$\Phi_1 = |\mathbf{R}\rangle|\mathbf{R}\rangle, \quad \Phi_2 = |\mathbf{R}'\rangle|\mathbf{R}'\rangle, \tag{32.35}$$

where $|\mathbf{R}\rangle|\mathbf{R}'\rangle$ has electron 1 on the ion at $\mathbf{R}$, and electron 2 on the ion at $\mathbf{R}'$, etc.

(c) Show that the approximate ground state wave function in the independent electron approximation can be written in terms of the states (32.35) as

$$\Phi_{ie} = \frac{1}{\sqrt{2}}\Phi_0 + \frac{1}{2}(\Phi_1 + \Phi_2) \tag{32.36}$$

The matrix elements of the *full* two-electron Hamiltonian,

$$H = h_1 + h_2 + V_{12}, \tag{32.37}$$

in the space of singlet states are $H_{ij} = (\Phi_i, H\Phi_j)$, where

$$\begin{pmatrix} H_{00} & H_{01} & H_{02} \\ H_{10} & H_{11} & H_{12} \\ H_{20} & H_{21} & H_{22} \end{pmatrix} = \begin{pmatrix} 2\mathcal{E} & -\sqrt{2}\,t & -\sqrt{2}\,t \\ -\sqrt{2}\,t & 2\mathcal{E} + U & 0 \\ -\sqrt{2}\,t & 0 & 2\mathcal{E} + U \end{pmatrix} \tag{32.38}$$

Note that the diagonal elements in the states Φ_1 and Φ_2 that place two electrons on the same proton contain the extra Coulomb repulsion U; the Coulomb repulsion is not present in the diagonal element in the state Φ_0, since in Φ_0 the electrons are on different protons. This appearance of U is the only effect of the electron-electron interaction V_{12}. Note also that the one-electron tunneling amplitude t connects only states in which a single electron has been moved from one proton to the other (it would take a further two-body interaction to give a nonvanishing matrix element between states in which the positions of two electrons are changed). Convince yourself that the factor of $\sqrt{2}$ in (32.38) is correct.

The Heitler-London approximation to the singlet ground state is just Φ_0, and therefore the Heitler-London estimate of the ground-state energy is just H_{00}, so

$$E_{\text{HL}} = 2\mathcal{E}. \tag{32.39}$$

(d) Show that the exact ground-state energy of the Hamiltonian (32.38) is

$$E = 2\mathcal{E} + \tfrac{1}{2}U - \sqrt{4t^2 + \tfrac{1}{4}U^2}. \tag{32.40}$$

Plot this energy, the independent electron approximation (32.34) to the ground-state energy, and the Heitler-London approximation to the ground-state energy as functions of U (for fixed $\mathcal{E}$ and t). Comment on the behavior for large and small U/t and why it is physically reasonable. How do these three energies compare when $U = 2t$?

(e) Show that the exact ground state of the Hamiltonian (32.38) is (to within a normalization constant)

$$\Phi = \frac{1}{\sqrt{2}}\Phi_0 + \left(\sqrt{1 + \left(\frac{U}{4t}\right)^2} - \frac{U}{4t}\right)\frac{1}{2}(\Phi_1 + \Phi_2) \tag{32.41}$$

What is the probability in this state of finding two electrons on the same ion? Plot your answer as a function of U (for fixed $\mathcal{E}$ and t) and comment on its behavior for small and large U/t.

33

Magnetic Ordering

Types of Magnetic Structure

Observation of Magnetic Structure

Thermodynamic Properties at the Onset of Magnetic Ordering

Ground State of the Heisenberg Ferro- and Antiferro-magnet

Low-Temperature Properties: Spin Waves

High-Temperature Properties: Corrections to Curie's Law

Analysis of the Critical Point

Mean Field Theory

Effects of Dipolar Interactions: Domains, Demagnetization Factors

In the preceding two chapters we were primarily concerned with how magnetic moments can be established in solids, and how electron-electron Coulomb interactions, together with the Pauli exclusion principle, can give rise to effective interactions among moments. In this chapter we shall start from the existence of such interacting moments, without further inquiry into the theory of their origin. We shall survey the types of magnetic structure such moments are observed to yield, and shall examine some typical problems encountered in trying to deduce the behavior of these structures, even when one's starting point is not the fundamental Hamiltonian of the solid, but a "simple" phenomenological model of interacting moments.

In describing observed properties of magnetic structures we shall avoid committing ourselves to a particular model of the underlying magnetic interactions. In most of our theoretical analysis, however, we shall use the spin Hamiltonian (32.20) of Heisenberg. As it turns out, even starting from the Heisenberg model, it is an exceedingly difficult task to deduce the magnetic properties of the solid as the temperature and applied field vary. No systematic solution has been found even to this simplified model problem, although much partial information has been extracted in a variety of important cases.

We shall discuss the following representative subjects:

1. The types of magnetic ordering that have been observed.
2. The theory of the very low-lying states of magnetically ordered systems.
3. The theory of high-temperature magnetic properties.
4. The critical region of temperatures in which magnetic ordering disappears.
5. A very crude phenomenological theory of magnetic ordering (mean field theory).
6. Some important consequences of magnetic dipolar interactions in ferromagnetically ordered solids.

TYPES OF MAGNETIC STRUCTURE

We shall use the language appropriate to solids in which the magnetic ions are localized at lattice sites, indicating below how the discussion may be generalized to itinerant electron magnetism.

If there were no magnetic interactions, individual magnetic moments would, in the absence of a field, be thermally disordered at any temperature, and the vector moment of each magnetic ion would average to zero.[1] In some solids, however, individual magnetic ions have nonvanishing average vector moments below a critical temperature T_c. Such solids are called *magnetically ordered.*

The individual localized moments in a magnetically ordered solid may or may not add up to a net magnetization density for the solid as a whole. If they do, the microscopic magnetic ordering is revealed by the existence of a macroscopic bulk magnetization density (even in the absence of an applied field) known as the *spontaneous magnetization*, and the ordered state is described as *ferromagnetic.*

More common is the case in which the individual local moments sum to zero total moment, and no spontaneous magnetization is present to reveal the microscopic ordering. Such magnetically ordered states are called *antiferromagnetic.*

---

[1] This was demonstrated in Chapter 31. Equation (31.44) gives $M = 0$ when $H = 0$ at any T.

In the simplest ferromagnets all the local moments have the same magnitude and average direction. The simplest antiferromagnetic state arises when the local moments fall on two interpenetrating *sublattices* of identical structure.[2] Within each sublattice the moments have the same magnitude and average direction, but the net moments of the two sublattices are oppositely directed, summing to zero total moment (see Figure 33.1).

Figure 33.1
Some simple antiferromagnetic spin arrangements. (a) Antiferromagnetic ordering on a body-centered cubic lattice. Spins of the same kind form two interpenetrating simple cubic lattices. (b) Antiferromagnetic ordering on a simple cubic lattice. Spins of the same kind form two interpenetrating face-centered cubic lattices.

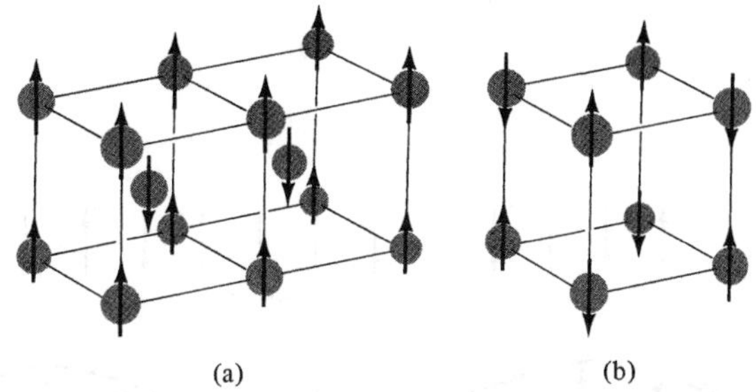

The term "ferromagnetic" is also used in a more restrictive sense, when one distinguishes among the varieties of ferromagnetic states that can occur when there are many (not necessarily identical) magnetic ions per primitive cell. In such contexts the term "ferromagnetic" is often reserved for those magnetic structures in which *all* the local moments have a positive component along the direction of spontaneous magnetization. Solids possessing a spontaneous magnetization that fail to satisfy this criterion are called *ferrimagnets*.[3] In a simple ferrimagnet the exchange coupling between nearest neighbors may favor antiparallel alignment, but because neighboring magnetic ions are not identical their moments will not cancel, leaving a net moment for the solid as a whole.

Some of the many types of magnetic ordering are illustrated schematically in Figure 33.2. Many magnetic structures are so complex that it is better to describe them explicitly rather than using one of the above three categories.

Similar distinctions can be made for magnetically ordered metals, even though the concept of a localized magnetic ion may be inapplicable. One specifies the ordering in terms of the spin density, which is defined so that at any point $\mathbf{r}$ and along any direction $\hat{\mathbf{z}}$, $s_z(\mathbf{r}) = \frac{1}{2}[n_\uparrow(\mathbf{r}) - n_\downarrow(\mathbf{r})]$, where $n_\uparrow(\mathbf{r})$ and $n_\downarrow(\mathbf{r})$ are the contributions of the two spin populations to the electronic density when the spins are resolved into components along the z-axis. In a magnetically ordered metal the local spin density fails to vanish. In a ferromagnetic metal $\int d\mathbf{r}\, s_z(\mathbf{r})$ is also nonvanishing for some direction $\hat{\mathbf{z}}$, while it is zero for any choice of $\hat{\mathbf{z}}$ in an antiferromagnetic metal, even though $s_z(\mathbf{r})$ itself does not vanish.

The magnetic structures observed in metals can also be complex. Antiferromagnetic chromium, for example, has a nonvanishing periodic spin density, whose period is

---

2 For example, a simple cubic lattice can be viewed as two interpenetrating face-centered cubic lattices; a body-centered cubic lattice can be viewed as two interpenetrating simple cubic lattices. A face-centered cubic lattice, however, cannot be so represented.

3 After the ferrites. See W. P. Wolf, *Repts. Prog. Phys.* **24**, 212 (1961) for a review.

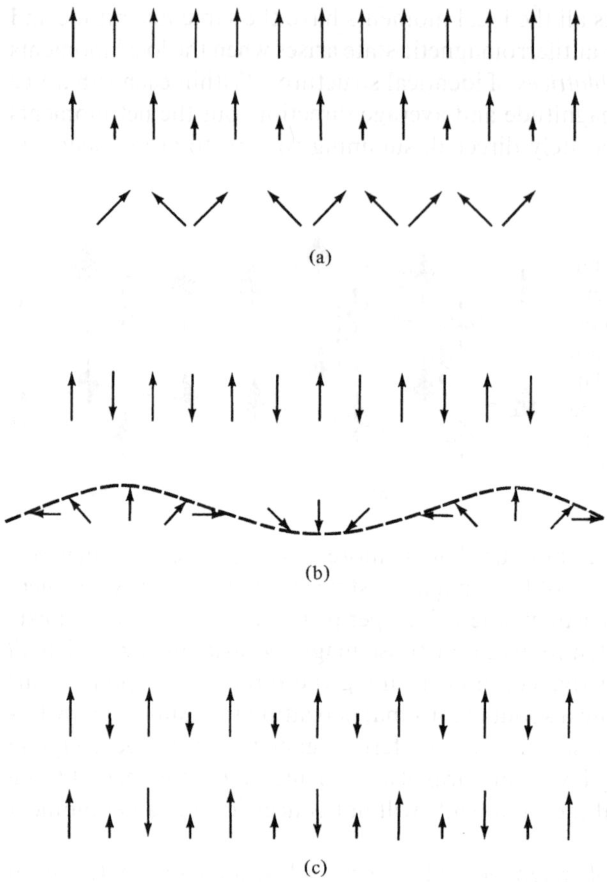

Figure 33.2
Linear arrays of spins illustrating possible (a) ferromagnetic, (b) antiferromagnetic, and (c) ferrimagnetic orderings.

unrelated to the periodicity of the lattice under normal conditions, being determined instead by the Fermi surface geometry.

Some examples of magnetically ordered solids are listed in Tables 33.1 to 33.3.

OBSERVATION OF MAGNETIC STRUCTURES

The magnetic ordering of a solid with a spontaneous magnetization is clearly revealed by the resulting macroscopic magnetic field.[4] However the magnetic ordering in antiferromagnetic solids yields no macroscopic field, and must be diagnosed by more subtle means. Low-energy neutrons are an excellent probe of the local moments, for the neutron has a magnetic moment that couples to the electronic spin in the solid. This leads to peaks in the elastic neutron scattering cross section in addition to those due to the nonmagnetic Bragg reflection of the neutrons by the ionic nuclei (see page 473). The magnetic reflections can be distinguished from the nonmagnetic ones because they weaken and disappear as the temperature rises through the critical

[4] A *caveat*: This is often masked by a domain structure. See pages 718–722.

Table 33.1
SELECTED FERROMAGNETS, WITH CRITICAL TEMPERATURES T_c AND SATURATION MAGNETIZATION M_0

MATERIAL	T_c (K)	M_0 (gauss)[a]
Fe	1043	1752
Co	1388	1446
Ni	627	510
Gd	293	1980
Dy	85	3000
$CrBr_3$	37	270
Au_2MnAl	200	323
Cu_2MnAl	630	726
Cu_2MnIn	500	613
EuO	77	1910
EuS	16.5	1184
MnAs	318	870
MnBi	670	675
$GdCl_3$	2.2	550

[a] At $T = 0(K)$.
Source: F. Keffer, *Handbuch der Physik*, vol. 18, pt. 2, Springer, New York, 1966; P. Heller, *Rep. Progr. Phys.*, **30**, (pt. II), 731 (1967).

Table 33.2
SELECTED ANTIFERROMAGNETS, WITH CRITICAL TEMPERATURES T_c

MATERIAL	T_c (K)	MATERIAL	T_c (K)
MnO	122	$KCoF_3$	125
FeO	198	MnF_2	67.34
CoO	291	FeF_2	78.4
NiO	600	CoF_2	37.7
$RbMnF_3$	54.5	$MnCl_2$	2
$KFeF_3$	115	VS	1040
$KMnF_3$	88.3	Cr	311

Source: F. Keffer, *Handbuch der Physik*, vol. 18, pt. 2, Springer, New York, 1966.

temperature at which the ordering vanishes, and also by the way they vary with applied magnetic fields.[5] (See Figure 33.3.)

Nuclear magnetic resonance[6] offers another way to probe the microscopic spin structure. The ionic nuclei feel the dipolar magnetic fields of nearby electrons. Nuclear magnetic resonance in magnetically ordered solids can therefore be observed even

[5] A comprehensive survey of both theoretical and experimental aspects of neutron scattering from magnetically ordered solids can be found in Y. A. Izyumov and R. P. Ozerov, *Magnetic Neutron Diffraction*, Plenum Press, New York, 1970.

[6] See page 665.

Table 33.3
SELECTED FERRIMAGNETS, WITH CRITICAL TEMPERATURES T_c AND SATURATION MAGNETIZATION M_0

MATERIAL	T_c (K)	M_0 (gauss)[a]
Fe_3O_4 (magnetite)	858	510
$CoFe_2O_4$	793	475
$NiFe_2O_4$	858	300
$CuFe_2O_4$	728	160
$MnFe_2O_4$	573	560
$Y_3Fe_5O_{12}$ (YIG)	560	195

[a] At $T = 0$(K).
Source: F. Keffer, *Handbuch der Physik*, vol. 18, pt. 2, Springer, New York, 1966.

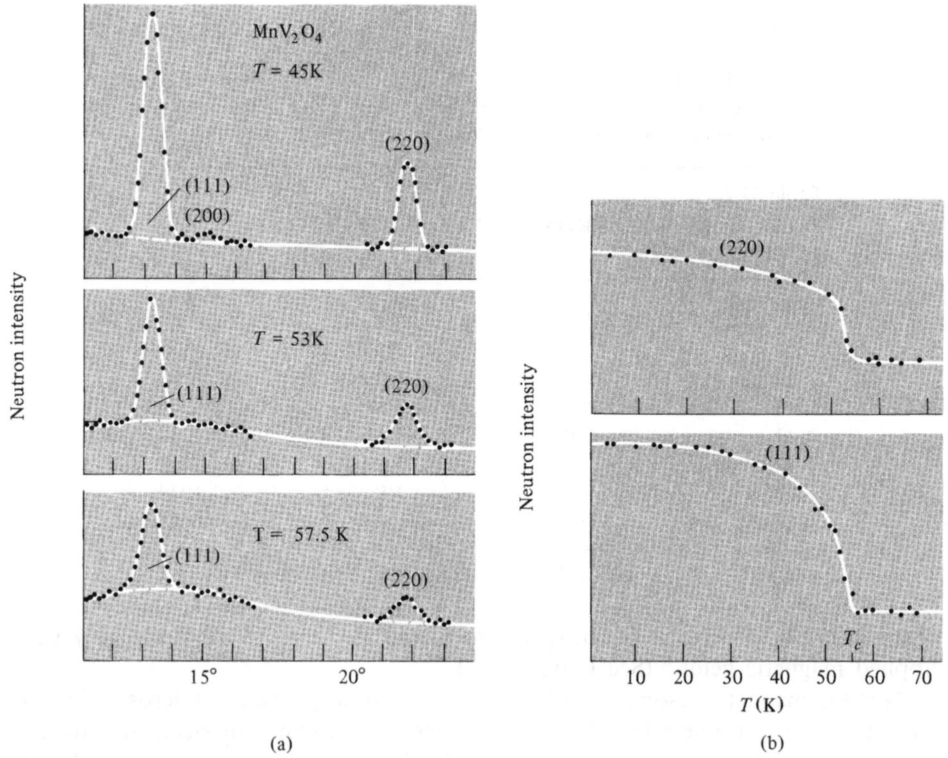

Figure 33.3
(a) Neutron Bragg peaks in manganese vanadite (MnV_2O_4), an antiferromagnet with $T_c = 56$ K. The intensity of the peaks decreases as T rises to T_c. (b) Intensity of the (220) and (111) peaks vs. temperature. Above T_c the temperature dependence is very slight. (From R. Plumier, *Proceedings of the International Conference on Magnetism*, Nottingham, 1964.)

in the absence of applied fields, the field at the nucleus (and hence the resonance frequency) being entirely due to the ordered moments. Thus nuclear magnetic resonance can be used, for example, to measure the macroscopically inaccessible net magnetization of each antiferromagnetic sublattice (see, for example, Figure 33.4).

THERMODYNAMIC PROPERTIES AT THE ONSET OF MAGNETIC ORDERING

The critical temperature T_c above which magnetic ordering vanishes is known as the Curie temperature in ferromagnets (or ferrimagnets) and the Néel temperature (often written T_N) in antiferromagnets. As the critical temperature is approached from below, the spontaneous magnetization (or, in antiferromagnets, the sublattice magnetization) drops continuously to zero. The observed magnetization just below T_c is well described by a power law.

$$M(T) \sim (T_c - T)^{\beta}, \qquad \textbf{(33.1)}$$

where β is typically between 0.33 and 0.37 (see Figure 33.4).

The onset of ordering is also signaled as the temperature drops to T_c from above, most notably by the zero-field susceptibility. In the absence of magnetic interactions the susceptibility varies inversely with T at all temperatures (Curie's law, page 656). In a ferromagnet, however, the susceptibility is observed to diverge as T drops to T_c, following the power law:

$$\chi(T) \sim (T - T_c)^{-\gamma}, \qquad \textbf{(33.2)}$$

where γ is typically between 1.3 and 1.4 (see Figure 33.5). In an antiferromagnet the susceptibility rises to a maximum a little above T_c, and then drops toward T_c, with a large maximum in its slope at the critical point (see Figure 33.6).

There is also a characteristic singularity in the zero-field specific heat at a magnetic critical point:

$$c(T) \sim (T - T_c)^{-\alpha}. \qquad \textbf{(33.3)}$$

The singularity is not nearly so strong as in the susceptibility, the exponent α being of order 0.1 or less.[7]

The critical region of temperatures is probably the most difficult to handle theoretically. We shall comment further on the theory of the critical region below, but shall first turn our attention to the regimes of low ($T \ll T_c$) and high ($T \gg T_c$) temperatures, which are more amenable to analysis.

---

[7] A variety of other critical exponents can also be defined and measured. Excellent reviews of magnetic and other critical points have been given by M. E. Fisher, *Rep. Progr. Phys.* **30** (pt. II), 615 (1967); P. Heller, *Rep. Progr. Phys.* **30** (pt. II), 731 (1967); and L. P. Kadanoff et al., *Rev. Mod. Phys.* **39**, 395 (1967). A theory of the critical point that permits the numerical calculation of critical exponents has been developed by K. G. Wilson. It is based on the methods of the renormalization group and has been given an elementary review by S. Ma, *Rev. Mod. Phys.* **45**, 589 (1973). See also M. E. Fisher, *Rev. Mod. Phys.* **46**, 597 (1974).

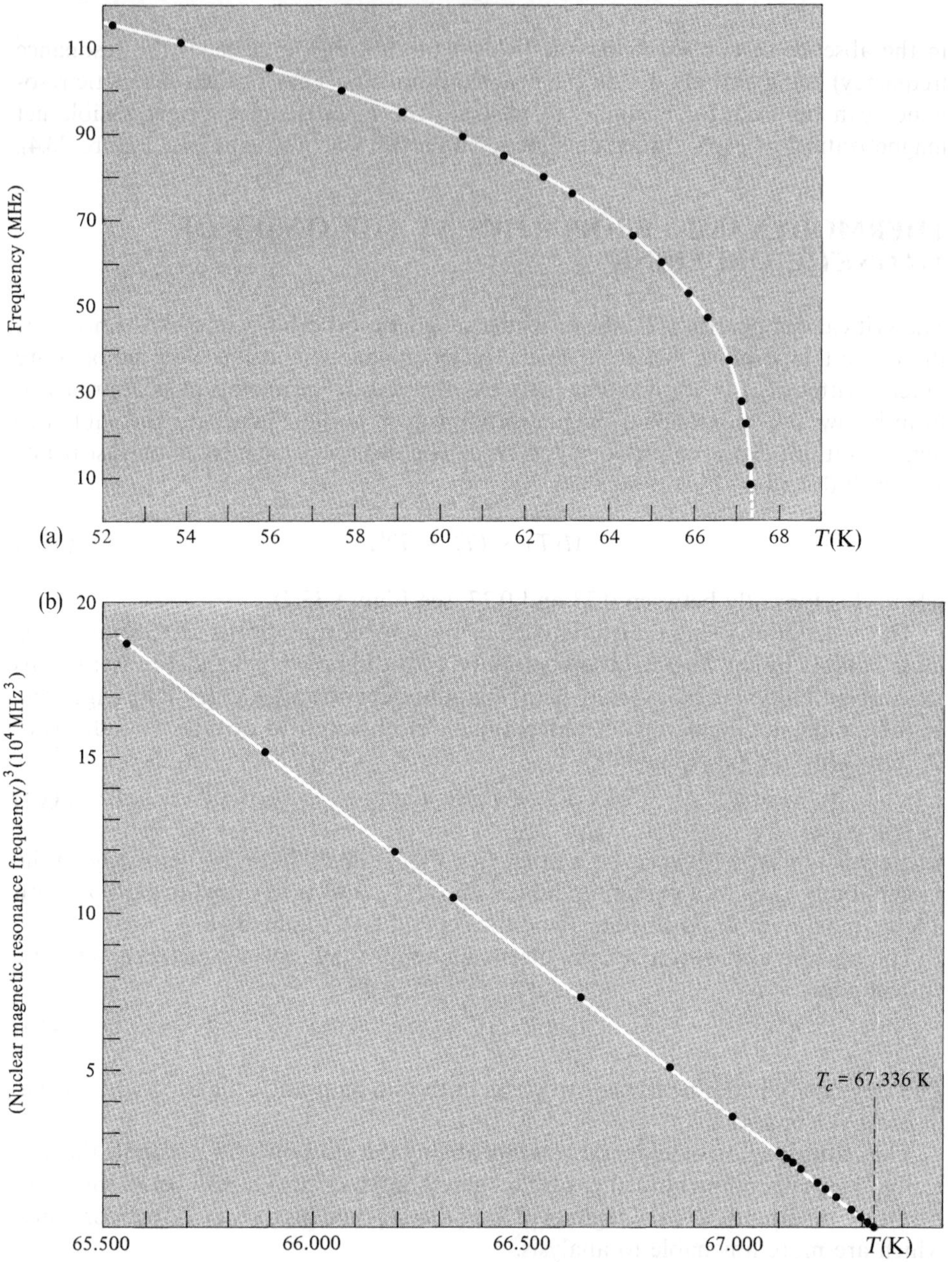

Figure 33.4
(a) The temperature dependence of the zero field ^{19}F nuclear magnetic resonance frequency in the antiferromagnet MnF_2. The resonance frequency vanishes at the antiferromagnetic critical temperature $T_c = 67.336$ K. (From P. Heller and G. B. Benedek, *Phys. Rev. Lett.* **8**, 428 (1962).) (b) Temperature dependence of the cube of the zero-field ^{19}F nuclear magnetic resonance frequency in MnF_2 in the immediate vicinity of T_c (note that the temperature scale is much expanded compared with that in (a)). If the frequency is proportional to the sublattice magnetization, this demonstrates that the magnetization vanishes as $(T_c - T)^{1/3}$ to very high accuracy.

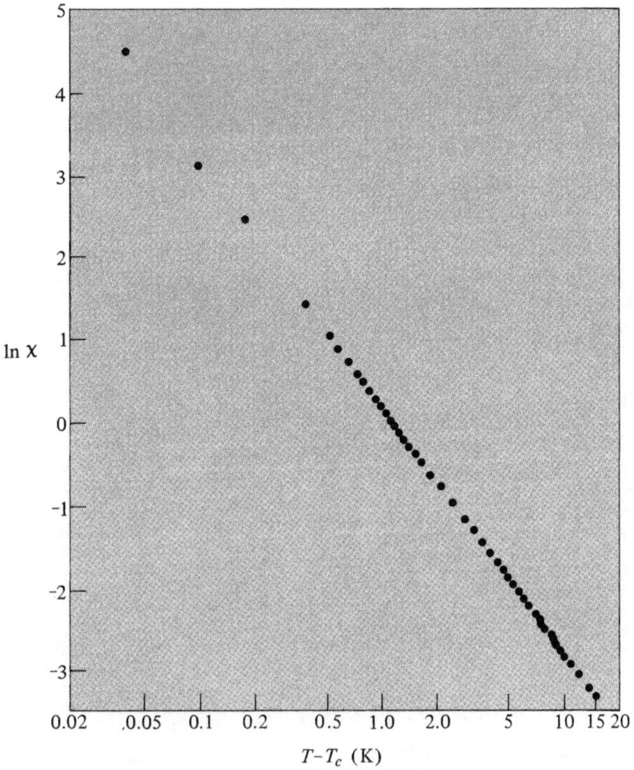

Figure 33.5
The susceptibility of iron (with a small amount of dissolved tungsten) above the critical temperature $T_c =$ 1043 K. A power law is very closely obeyed in this range, and the slope gives $\chi \sim (T - T_c)^{-1.33}$. (From J. E. Noakes et al., *J. Appl. Phys.* **37**, 1264 (1966). Note that $\log_{10} \chi = 0.4343 \ln \chi$.)

ZERO-TEMPERATURE PROPERTIES: GROUND STATE OF THE HEISENBERG FERROMAGNET

We consider a set of magnetic ions at Bravais lattice sites **R**, whose low lying excitations can be described by a ferromagnetic Heisenberg Hamiltonian (Eq. (32.20)),[8]

$$\mathcal{H} = -\frac{1}{2}\sum_{\mathbf{RR}'} \mathbf{S}(\mathbf{R}) \cdot \mathbf{S}(\mathbf{R}')J(\mathbf{R} - \mathbf{R}') - g\mu_B H \sum_{\mathbf{R}} S_z(\mathbf{R}),$$
$$J(\mathbf{R} - \mathbf{R}') = J(\mathbf{R}' - \mathbf{R}) \geqslant 0. \qquad \textbf{(33.4)}$$

We describe this Hamiltonian as ferromagnetic because a positive exchange interaction J favors parallel spin alignment. Effects due to magnetic dipolar coupling between the moments are not included in the interaction J, but can be taken into account by suitable definition of the field **H** (whose direction we take to define the z-axis) that acts on the local spins. We shall elaborate on this below (p. 722), but

[8] It is the widespread practice to refer to the operators in the Heisenberg Hamiltonian as spin operators, even though the spin operator for each ion here represents its *total* angular momentum which, in general, has both a spin and an orbital part. It is also the usual practice to take these fictitious spins to be parallel to the magnetic moment of the ion, rather than to its total angular momentum; i.e., the term in H in (33.4) is preceded by a minus sign (for positive $g\mu_B$), when **H** is along the positive z-axis.

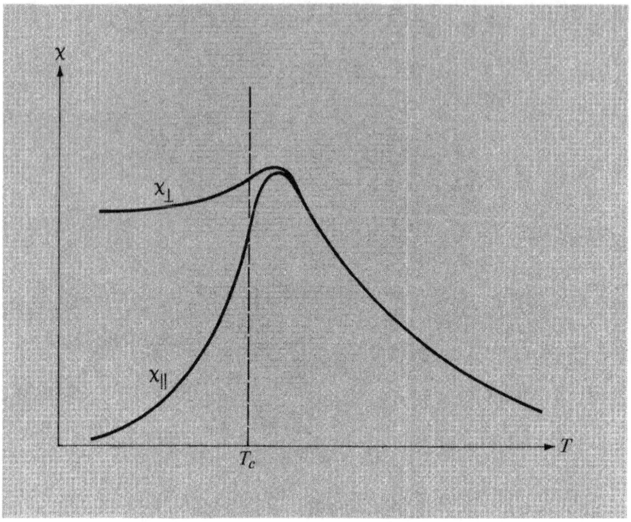

Figure 33.6
Characteristic temperature dependence of the susceptibility of an antiferromagnet near the critical temperature. Below T_c the susceptibility depends very much on whether the field is applied parallel or perpendicular to the direction of sublattice magnetization. Note that if the antiferromagnet were perfectly isotropic this would not be the case: Whatever the direction of the applied field, the sublattice magnetization would rotate into the energetically most favorable orientation with respect to the field (presumably perpendicular) and there would be only one susceptibility ($\chi_\perp$). The orientation dependence below T_c is due to crystalline anisotropy. The anisotropy is also responsible for the slight difference in $\chi_\parallel$ and $\chi_\perp$ above T_c, where parallel and perpendicular now refer to the axis along which (because of anisotropy) the sublattice magnetization below T_c prefers to lie. (See M. E. Fisher, *Phil. Mag.* **7**, 1731 (1962).)

merely note for now that **H** is the local field (in the sense of Chapter 27) acting on each magnetic ion, which is not necessarily equal to the externally applied field.

If we were to regard the spins appearing in the Hamiltonian (33.4) as classical vectors, we should expect the state of lowest energy to be the one with all spins aligned along the z-axis, parallel to the magnetic field and to each other. This suggests, as a candidate for the quantum-mechanical ground state $|0\rangle$, one that is an eigenstate of $\mathbf{S}_z(\mathbf{R})$ for every $\mathbf{R}$ with the maximum eigenvalue, S:

$$|0\rangle = \prod_{\mathbf{R}} |S\rangle_{\mathbf{R}}, \tag{33.5}$$

where

$$\mathbf{S}_z(\mathbf{R})|S\rangle_{\mathbf{R}} = S|S\rangle_{\mathbf{R}}. \tag{33.6}$$

To verify that $|0\rangle$ is indeed an eigenstate of $\mathcal{H}$, we rewrite the Hamiltonian (33.4) in terms of the operators

$$\mathbf{S}_{\pm}(\mathbf{R}) = \mathbf{S}_x(\mathbf{R}) \pm i\mathbf{S}_y(\mathbf{R}), \tag{33.7}$$

which have the property[9]

$$\mathbf{S}_{\pm}(\mathbf{R})|S_z\rangle_{\mathbf{R}} = \sqrt{(S \mp S_z)(S + 1 \pm S_z)}\,|S_z \pm 1\rangle_{\mathbf{R}}. \tag{33.8}$$

Separating the terms in $\mathbf{S}_z$ from those containing $\mathbf{S}_+$ or $\mathbf{S}_-$, we may write

$$\begin{aligned}\mathcal{H} = &-\frac{1}{2}\sum_{\mathbf{R},\mathbf{R}'} J(\mathbf{R} - \mathbf{R}')\mathbf{S}_z(\mathbf{R})\mathbf{S}_z(\mathbf{R}') - g\mu_B H \sum_{\mathbf{R}} \mathbf{S}_z(\mathbf{R}) \\ &-\frac{1}{2}\sum_{\mathbf{R},\mathbf{R}'} J(\mathbf{R} - \mathbf{R}')\mathbf{S}_-(\mathbf{R}')\mathbf{S}_+(\mathbf{R}).\end{aligned} \tag{33.9}$$

Since $\mathbf{S}_+(\mathbf{R})|S_z\rangle_{\mathbf{R}} = 0$ when $S_z = S$, it follows that when $\mathcal{H}$ acts on $|0\rangle$, only the terms in $\mathbf{S}_z$ contribute to the result. But $|0\rangle$ is constructed to be an eigenstate of each $\mathbf{S}_z(\mathbf{R})$ with eigenvalue S, and therefore

$$\mathcal{H}|0\rangle = E_0|0\rangle, \tag{33.10}$$

where

$$E_0 = -\frac{1}{2}S^2 \sum_{\mathbf{R},\mathbf{R}'} J(\mathbf{R} - \mathbf{R}') - Ng\mu_B HS. \tag{33.11}$$

Thus $|0\rangle$ is indeed an eigenstate of $\mathcal{H}$. To prove that E_0 is the energy of the *ground* state, we consider any other eigenstate $|0'\rangle$ of $\mathcal{H}$ with eigenvalue E_0'. Since

$$E_0' = \langle 0'|\mathcal{H}|0'\rangle, \tag{33.12}$$

it follows that when all the $J(\mathbf{R} - \mathbf{R}')$ are positive, E_0' has the lower bound

$$-\frac{1}{2}\sum_{\mathbf{R}\mathbf{R}'} J(\mathbf{R} - \mathbf{R}') \max \langle \mathbf{S}(\mathbf{R}) \cdot \mathbf{S}(\mathbf{R}')\rangle - g\mu_B H \sum_{\mathbf{R}} \max \langle \mathbf{S}_z(\mathbf{R})\rangle \tag{33.13}$$

where max $\langle X\rangle$ is the largest diagonal matrix element that the operator X can assume (in any state whatsoever). In Problem 1 it is shown that[10]

$$\begin{aligned}\langle \mathbf{S}(\mathbf{R}) \cdot \mathbf{S}(\mathbf{R}')\rangle &\leqslant S^2, \qquad \mathbf{R} \neq \mathbf{R}', \\ \langle \mathbf{S}_z(\mathbf{R})\rangle &\leqslant S.\end{aligned} \tag{33.14}$$

Combining these inequalities with the bound (33.13) for E_0' and comparing the resulting inequality with the form (33.11) of E_0, we conclude that E_0' cannot be less than E_0, and therefore E_0 must be the energy of the ground state.

---

9 See, for example, A. Messiah, *Quantum Mechanics*, Wiley, New York, p. 512, 1962.

10 These results may seem "obvious" to one's classical intuition, but one should reflect on the fact that min $\langle \mathbf{S}(\mathbf{R}) \cdot \mathbf{S}(\mathbf{R}')\rangle$ is not $-S^2$, but $-S(S + 1)$ (Problem 1). And, of course, if $\mathbf{R} = \mathbf{R}'$, then max $\langle \mathbf{S}(\mathbf{R}) \cdot \mathbf{S}(\mathbf{R}')\rangle$ is not S^2, but $S(S + 1)$.

ZERO-TEMPERATURE PROPERTIES: GROUND STATE OF THE HEISENBERG ANTIFERROMAGNET

Finding the ground state of the Heisenberg antiferromagnet is an unsolved problem except in the special case of a one-dimensional array of spin 1/2 ions with coupling only between nearest neighbors.[11] The difficulty is illustrated by the case in which the spins reside on two sublattices, each spin interacting only with those on the other sublattice. In the absence of an applied field the Hamiltonian is

$$\mathcal{H} = \frac{1}{2}\sum_{\mathbf{R},\mathbf{R}'} |J(\mathbf{R} - \mathbf{R}')|\mathbf{S}(\mathbf{R}) \cdot \mathbf{S}(\mathbf{R}'). \tag{33.15}$$

An obvious guess for the ground state is to put each sublattice into a ferromagnetic ground state of the form (33.5), with oppositely directed sublattice magnetizations. If the spins were classical vectors this would take maximum advantage of the antiferromagnetic coupling between sublattices, and would yield a ground-state energy

$$E_0 = -\frac{1}{2}\sum_{\mathbf{R},\mathbf{R}'} |J(\mathbf{R} - \mathbf{R}')|S^2. \tag{33.16}$$

In contrast to the ferromagnetic case, however, the terms $\mathbf{S}_-(\mathbf{R})\mathbf{S}_+(\mathbf{R}')$ in the Hamiltonian (33.9), when acting on such a state do not always give zero, but also yield a state in which a spin in the "up" sublattice has had its z-component reduced by unity and a spin in the "down" sublattice is correspondingly raised. Thus the state is *not* an eigenstate.

All that can easily be established is that (33.16) is an upper bound to the true ground-state energy (Problem 2). A lower bound can also be found (Problem 2) leading to the inequality:

$$-\frac{1}{2}S(S+1)\sum_{\mathbf{R},\mathbf{R}'} |J(\mathbf{R} - \mathbf{R}')| \leqslant E_0 \leqslant -\frac{1}{2}S^2\sum_{\mathbf{R},\mathbf{R}'} |J(\mathbf{R} - \mathbf{R}')|. \tag{33.17}$$

In the limit of large spin (when the spins become, in effect, classical vectors) the ratio of these bounds approaches unity. However, they are far from restrictive when S is small. In the one-dimensional nearest-neighbor spin 1/2 chain, for example, the bounds give $-0.25NJ \geqslant E_0 \geqslant -0.75NJ$, while Bethe's exact result is $E_0 = -NJ[\ln 2 - (1/4)] = -0.443NJ$. Thus more elaborate analysis is needed to estimate accurately the antiferromagnetic ground-state energy.

LOW-TEMPERATURE BEHAVIOR OF THE HEISENBERG FERROMAGNET: SPIN WAVES

We can not only exhibit the exact ground state of the Heisenberg ferromagnet, but can also find some of its low-lying excited states. The knowledge of these states underlies the theory of the low-temperature properties of the Heisenberg ferromagnet.

[11] H. A. Bethe, *Z. Physik*, **71**, 205 (1931).

At zero temperature the ferromagnet is in its ground state (33.5), the mean "spin" of each ion is S, and the magnetization density (known as the *saturation magnetization*) is

$$M = g\mu_B \frac{N}{V} S. \tag{33.18}$$

When $T \neq 0$, we must weight the mean magnetization of all states with the Boltzmann factor e^{-E/k_BT}. Very near $T = 0$, only low-lying states will have appreciable weight. To construct some of these low-lying states we examine a state[12] $|\mathbf{R}\rangle$ differing from the ground state $|0\rangle$ only in that the spin at site $\mathbf{R}$ has had its z-component reduced from S to $S - 1$:

$$|\mathbf{R}\rangle = \frac{1}{\sqrt{2S}} \mathbf{S}_-(\mathbf{R})|0\rangle. \tag{33.19}$$

The state $|\mathbf{R}\rangle$ remains an eigenstate of the terms containing $\mathbf{S}_z$ in the Hamiltonian (33.9). Because, however, the spin at $\mathbf{R}$ does not assume its maximum z-component, $\mathbf{S}_+(\mathbf{R})|\mathbf{R}\rangle$ will not vanish, and $\mathbf{S}_-(\mathbf{R}')\mathbf{S}_+(\mathbf{R})$ simply shifts the site at which the spin is reduced from $\mathbf{R}$ to $\mathbf{R}'$. Thus[13]

$$\mathbf{S}_-(\mathbf{R}')\mathbf{S}_+(\mathbf{R})|\mathbf{R}\rangle = 2S|\mathbf{R}'\rangle. \tag{33.20}$$

If in addition we note that

$$\begin{aligned} \mathbf{S}_z(\mathbf{R}')|\mathbf{R}\rangle &= S|\mathbf{R}\rangle, & \mathbf{R}' &\neq \mathbf{R}, \\ &= (S-1)|\mathbf{R}\rangle, & \mathbf{R}' &= \mathbf{R}, \end{aligned} \tag{33.21}$$

then it follows that

$$\mathcal{H}|\mathbf{R}\rangle = E_0|\mathbf{R}\rangle + g\mu_B H|\mathbf{R}\rangle + S \sum_{\mathbf{R}'} J(\mathbf{R} - \mathbf{R}')[|\mathbf{R}\rangle - |\mathbf{R}'\rangle], \tag{33.22}$$

where E_0 is the ground-state energy (33.11).

Although $|\mathbf{R}\rangle$ is therefore not an eigenstate of $\mathcal{H}$, $\mathcal{H}|\mathbf{R}\rangle$ *is* a linear combination of $|\mathbf{R}\rangle$ and other states with only a single lowered spin. Because J depends on $\mathbf{R}$ and $\mathbf{R}'$ only in the translationally invariant combination $\mathbf{R} - \mathbf{R}'$, it is straightforward to find linear combinations of these states that *are* eigenstates.[14] Let

$$|\mathbf{k}\rangle = \frac{1}{\sqrt{N}} \sum_{\mathbf{R}} e^{i\mathbf{k}\cdot\mathbf{R}}|\mathbf{R}\rangle. \tag{33.23}$$

Equation (33.22) implies that

$$\begin{aligned} \mathcal{H}|\mathbf{k}\rangle &= E_{\mathbf{k}}|\mathbf{k}\rangle, \\ E_{\mathbf{k}} &= E_0 + g\mu_B H + S \sum_{\mathbf{R}} J(\mathbf{R})(1 - e^{i\mathbf{k}\cdot\mathbf{R}}). \end{aligned} \tag{33.24}$$

---

12 Exercise: Verify that $|\mathbf{R}\rangle$ is normalized to unity.

13 Exercise: Verify that the numerical factor $2S$ is correct.

14 The analysis that follows closely parallels the discussion in Chapter 22 of normal modes in a harmonic crystal. In particular, the state $|\mathbf{k}\rangle$ can be formed for just N distinct wave vectors lying in the first Brillouin zone, if we invoke the Born-von Karman boundary condition. Since values of $\mathbf{k}$ that differ by a reciprocal lattice vector lead to identical states, it suffices to consider only these N values. The reader should also verify, using the appropriate identities from Appendix F, that the states $|\mathbf{k}\rangle$ are orthonormal: $\langle\mathbf{k}|\mathbf{k}'\rangle = \delta_{\mathbf{k}\mathbf{k}'}$.

Taking advantage of the symmetry, $J(-\mathbf{R}) = J(\mathbf{R})$, we can write the excitation energy $\mathcal{E}(\mathbf{k})$ of the state $|\mathbf{k}\rangle$ (i.e., the amount by which its energy exceeds that of the ground state) as

$$\mathcal{E}(\mathbf{k}) = E_{\mathbf{k}} - E_0 = 2S \sum_{\mathbf{R}} J(\mathbf{R}) \sin^2 (\tfrac{1}{2}\mathbf{k} \cdot \mathbf{R}) + g\mu_B H. \tag{33.25}$$

To give a physical interpretation of the state $|\mathbf{k}\rangle$ we note the following:

1. Since $|\mathbf{k}\rangle$ is a superposition of states in each of which the total spin is diminished from its saturation value NS by one unit, the total spin in the state $|\mathbf{k}\rangle$ itself has the value $NS - 1$.
2. The probability of the lowered spin being found at a particular site $\mathbf{R}$ in the state $|\mathbf{k}\rangle$ is $|\langle\mathbf{k}|\mathbf{R}\rangle|^2 = 1/N$; i.e., the lowered spin is distributed with equal probability among all the magnetic ions.
3. We define the transverse spin correlation function in the state $|\mathbf{k}\rangle$ to be the expectation value of

$$\mathbf{S}_\perp(\mathbf{R}) \cdot \mathbf{S}_\perp(\mathbf{R}') = \mathbf{S}_x(\mathbf{R})\mathbf{S}_x(\mathbf{R}') + \mathbf{S}_y(\mathbf{R})\mathbf{S}_y(\mathbf{R}'). \tag{33.26}$$

A straightforward evaluation (Problem 4) gives

$$\langle\mathbf{k}|\mathbf{S}_\perp(\mathbf{R}) \cdot \mathbf{S}_\perp(\mathbf{R}')|\mathbf{k}\rangle = \frac{2S}{N} \cos\left[\mathbf{k} \cdot (\mathbf{R} - \mathbf{R}')\right], \qquad \mathbf{R} \neq \mathbf{R}'. \tag{33.27}$$

Thus on the average each spin has a small transverse component, perpendicular to the direction of magnetization, of size $(2S/N)^{1/2}$; the orientations of the transverse components of two spins separated by $\mathbf{R} - \mathbf{R}'$ differ by an angle $\mathbf{k} \cdot (\mathbf{R} - \mathbf{R}')$.

The microscopic magnetization in the state $|\mathbf{k}\rangle$ suggested by these facts is pictured in Figure 33.7. One describes the state $|\mathbf{k}\rangle$ as a state containing a spin wave (or "magnon") of wave vector $\mathbf{k}$ and energy $\mathcal{E}(\mathbf{k})$ (Eq. (33.25)).

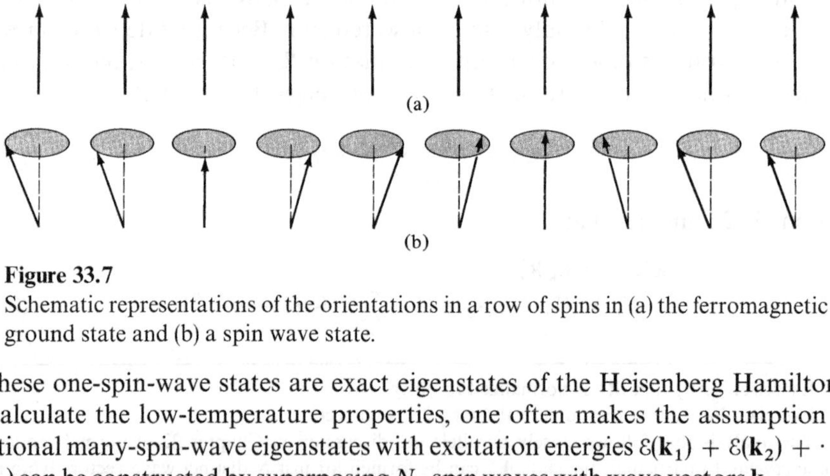

Figure 33.7
Schematic representations of the orientations in a row of spins in (a) the ferromagnetic ground state and (b) a spin wave state.

These one-spin-wave states are exact eigenstates of the Heisenberg Hamiltonian. To calculate the low-temperature properties, one often makes the assumption that additional many-spin-wave eigenstates with excitation energies $\mathcal{E}(\mathbf{k}_1) + \mathcal{E}(\mathbf{k}_2) + \cdots + \mathcal{E}(\mathbf{k}_{N_0})$ can be constructed by superposing N_0 spin waves with wave vectors $\mathbf{k}_1, \ldots, \mathbf{k}_{N_0}$. On the basis of the analogy with phonons in a harmonic crystal (where many-phonon states are exact stationary states, just as one-phonon states are) this appears to be a reasonable assumption. However, in the case of spin waves it is only an approximation. Spin waves do *not* rigorously obey the superposition principle. Nevertheless, it has been shown that this approximation correctly reproduces the dominant term in the low-temperature spontaneous magnetization. We therefore pursue the approximation further, using it to calculate $M(T)$, with the warning that if one wishes to go beyond the leading correction to the $T = 0$ result, one must resort to a far more sophisticated analysis.

If the low-lying excited states of the ferromagnet have excitation energies of the form

$$\sum \mathcal{E}(\mathbf{k})n_{\mathbf{k}}, \qquad n_{\mathbf{k}} = 0, 1, 2, \ldots, \tag{33.28}$$

then the mean number of spin waves with wave vector $\mathbf{k}$ at temperature T will be given by[15]

$$n(\mathbf{k}) = \langle n_{\mathbf{k}} \rangle = \frac{1}{(e^{\mathcal{E}(\mathbf{k})/k_B T} - 1)}. \tag{33.29}$$

Since the total spin is reduced from its saturation value NS by one unit per spin wave, the magnetization at temperature T satisfies

$$M(T) = M(0)\left[1 - \frac{1}{NS}\sum_{\mathbf{k}} n(\mathbf{k})\right], \tag{33.30}$$

or

$$M(T) = M(0)\left[1 - \frac{V}{NS}\int \frac{d\mathbf{k}}{(2\pi)^3}\frac{1}{(e^{\mathcal{E}(\mathbf{k})/k_B T} - 1)}\right]. \tag{33.31}$$

The *spontaneous* magnetization is evaluated from (33.31) using the form (33.25) assumed by the $\mathcal{E}(\mathbf{k})$, in the limit of vanishing magnetic field:

$$\mathcal{E}(\mathbf{k}) = 2S \sum_{\mathbf{R}} J(\mathbf{R}) \sin^2 (\tfrac{1}{2}\mathbf{k} \cdot \mathbf{R}). \tag{33.32}$$

At very low temperatures we can evaluate (33.31) in the same way as we extracted the low-temperature lattice specific heat in Chapter 23. As $T \to 0$, only spin waves with vanishingly small excitation energies will contribute appreciably to the integral. Since we have taken all the exchange constants $J(\mathbf{R})$ to be positive, the energy of a spin wave is vanishingly small only in the limit as $k \to 0$, when it becomes:

$$\mathcal{E}(\mathbf{k}) \approx \frac{S}{2}\sum_{\mathbf{R}} J(\mathbf{R})(\mathbf{k} \cdot \mathbf{R})^2. \tag{33.33}$$

We may insert this form into (33.31) for all $\mathbf{k}$, for when the approximation (33.33) ceases to be valid, both the exact and approximate $\mathcal{E}(\mathbf{k})$ will be so large that they will give a negligibly small contribution to the integral as $T \to 0$. For the same reason we may extend the integration over the first Brillouin zone to all of k-space with negligible error at low temperatures. If, finally, we make the change of variables $\mathbf{k} = (k_B T)^{1/2}\mathbf{q}$, we arrive at the result:

$$M(T) = M(0)\left[1 - \frac{V}{NS}(k_B T)^{3/2}\int \frac{d\mathbf{q}}{(2\pi)^3}\left\{\exp\left[S\sum_{\mathbf{R}} J(\mathbf{R})\frac{(\mathbf{q} \cdot \mathbf{R})^2}{2}\right] - 1\right\}^{-1}\right]. \tag{33.34}$$

---

[15] See the analogous discussion for phonons on pages 453–454.

This asserts that as the temperature rises from $T = 0$, the spontaneous magnetization should deviate from its saturation value by an amount proportional to $T^{3/2}$, a result known as the Bloch $T^{3/2}$ law. The $T^{3/2}$ law is well confirmed by experiment[16] (Figure 33.8). The result (33.34) has also been shown[17] to be the exact leading term in the low-temperature expansion of the deviation of the spontaneous magnetization from saturation.

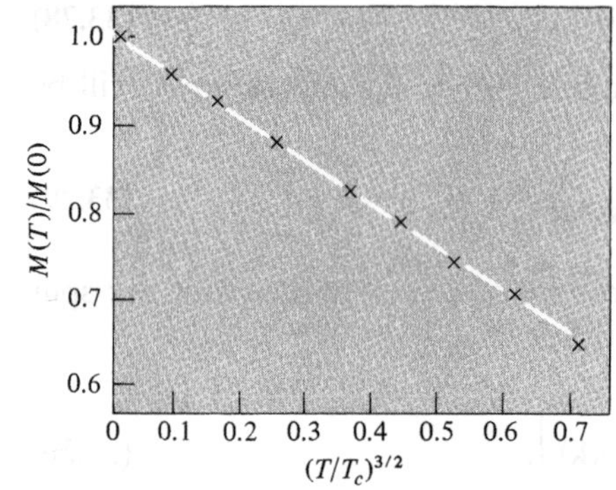

Figure 33.8
The ratio of the spontaneous magnetization at temperature T to its saturation ($T = 0$) value as a function of $(T/T_c)^{3/2}$ for ferromagnetic gadolinium ($T_c = 293$ K). The linearity of the curve accords with the Bloch $T^{3/2}$ law. (After F. Holtzberg et al., *J. Appl. Phys.* **35**, 1033 (1964).)

Another implication of (33.34) has also been rigorously verified. In one and two spatial dimensions the integral in (33.34) diverges at small q. One generally interprets this to mean that at any nonzero temperature so many spin waves are excited that the magnetization is completely eliminated. This conclusion, that there can be no spontaneous magnetization in the one- and two-dimensional isotropic Heisenberg models, has been proved directly without making the spin wave approximation.[18]

Spin waves are not peculiar to the isotropic Heisenberg ferromagnet. There is a spin wave theory of the low-lying excitations of the antiferromagnet which is rather more complex, as one might guess from the fact that even the antiferromagnetic ground state is unknown. The theory predicts a spin wave excitation energy that, in contrast to the ferromagnetic case, is linear in k at long wavelengths.[19]

Spin wave theories have also been constructed for itinerant models of magnetism. Quite generally, one expects spin waves to exist whenever there is a direction associated with the local order that can vary spatially in a continuous manner at a cost in energy that becomes small as the wavelength of the variation becomes very long.

We noted above that elastic magnetic neutron scattering can reveal magnetic

---

[16] In isotropic ferromagnets; if there is significant anisotropy in the exchange coupling, the spin wave excitation energy will not vanish for small k, and the $T^{3/2}$ law fails. See Problem 5.

[17] F. Dyson, *Phys. Rev.* **102**, 1230 (1956). Dyson also calculated several of the higher-order corrections. That his calculation was something of a *tour de force* is indicated by the fact that prior to his work there were almost as many disagreeing "corrections" to the $T^{3/2}$ term as there were published papers on the subject.

[18] The proof (N. D. Mermin and H. Wagner, *Phys. Rev. Lett.* **17**, 1133 (1966)) is based on an argument of P. C. Hohenberg. See N. D. Mermin, *J. Phys. Soc. Japan* **26**. Supplement, 203 (1969) for a review and for other applications of the method in solids.

[19] An elementary phenomenological analysis is given by F. Keffer et al., *Am. J. Phys.* **21**, 250 (1953).

structure, just as elastic nonmagnetic neutron scattering can reveal the spatial arrangement of the ions. The analogy extends to inelastic scattering: Inelastic magnetic neutron scattering reveals the spin wave spectrum in the same way as inelastic nonmagnetic neutron scattering reveals the phonon spectrum. Thus there are "one-spin-wave" peaks in the magnetic part of the inelastic scattering cross section, in which the change in neutron energy and wave vector are given by the excitation energy and wave vector of a spin wave. Observation of these peaks confirms the k^2 dependence of the spin wave excitation energy in ferromagnets (and also the linear dependence on k in antiferromagnets). (See Figure 33.9.)

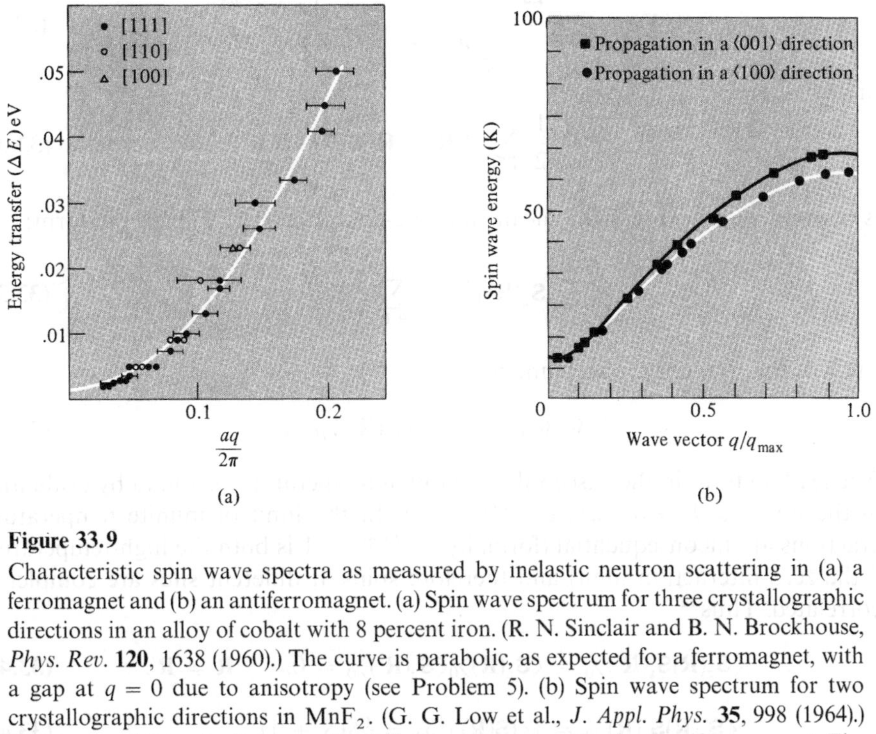

Figure 33.9
Characteristic spin wave spectra as measured by inelastic neutron scattering in (a) a ferromagnet and (b) an antiferromagnet. (a) Spin wave spectrum for three crystallographic directions in an alloy of cobalt with 8 percent iron. (R. N. Sinclair and B. N. Brockhouse, *Phys. Rev.* **120**, 1638 (1960).) The curve is parabolic, as expected for a ferromagnet, with a gap at $q = 0$ due to anisotropy (see Problem 5). (b) Spin wave spectrum for two crystallographic directions in MnF_2. (G. G. Low et al., *J. Appl. Phys.* **35**, 998 (1964).) The curve exhibits the linear small-q behavior characteristic of an antiferromagnet. The gap at $q = 0$ is again due to anisotropy.

HIGH-TEMPERATURE SUSCEPTIBILITY

Except in artificially simplified models, no one has succeeded in calculating the zero-field susceptibility $\chi(T)$ of the Heisenberg model in closed form, when magnetic interactions are present. It has, however, been possible to compute many terms in the expansion of the susceptibility in inverse powers of the temperature. The leading term is inversely proportional to T, independent of the exchange constants, and by itself gives the Curie's law susceptibility (page 656) characteristic of noninteracting moments. Subsequent terms give corrections to Curie's law.

The high-temperature expansion starts from the exact identity[20]

$$\chi(T) = \frac{g\mu_B}{V}\frac{\partial}{\partial H}\langle\sum_{\mathbf{R}} S_z(\mathbf{R})\rangle\Big|_{H=0}$$
$$= \frac{1}{V}\frac{1}{k_BT}(g\mu_B)^2\langle[\sum_{\mathbf{R}} S_z(\mathbf{R})]^2\rangle_{H=0}. \tag{33.35}$$

Here the angular brackets denote an equilibrium average in the absence of an applied field:

$$\langle X\rangle_{H=0} = \frac{\sum_\alpha \langle\alpha|X|\alpha\rangle e^{-\beta E_\alpha}}{\sum_\alpha e^{-\beta E_\alpha}} = \frac{\text{Tr}\, Xe^{-\beta\mathcal{H}_0}}{\text{Tr}\, e^{-\beta\mathcal{H}_0}}, \tag{33.36}$$

where

$$\mathcal{H}_0 = -\frac{1}{2}\sum_{\mathbf{R}\neq\mathbf{R}'} J(\mathbf{R}-\mathbf{R}')\mathbf{S}(\mathbf{R})\cdot\mathbf{S}(\mathbf{R}'). \tag{33.37}$$

It is convenient to express the mean square z-component of spin in the form:

$$\langle[\sum_{\mathbf{R}} S_z(\mathbf{R})]^2\rangle = \sum_{\mathbf{R}'\mathbf{R}} \Gamma(\mathbf{R},\mathbf{R}'), \tag{33.38}$$

where Γ is the spin correlation function,

$$\Gamma(\mathbf{R},\mathbf{R}') = \langle S_z(\mathbf{R})S_z(\mathbf{R}')\rangle_{H=0}. \tag{33.39}$$

The leading term in the susceptibility at high temperatures is found by evaluating Γ in the limit as $T\to\infty$ (i.e., $e^{-\mathcal{H}_0/k_BT}\to 1$). In the limit of infinite temperature, interactions are inconsequential (formally, $e^{-J/k_BT}\to 1$ is both the high-temperature and the zero-interaction limit) and therefore spins at different sites are completely uncorrelated. Thus[21]

$$\langle S_z(\mathbf{R})S_z(\mathbf{R}')\rangle_0 = \langle S_z(\mathbf{R})\rangle_0\langle S_z(\mathbf{R}')\rangle_0 = 0, \qquad \mathbf{R}\neq\mathbf{R}', \tag{33.40}$$

but

$$\langle S_z(\mathbf{R})S_z(\mathbf{R})\rangle_0 = \tfrac{1}{3}\langle(\mathbf{S}(\mathbf{R}))^2\rangle_0 = \tfrac{1}{3}S(S+1). \tag{33.41}$$

Combining these, we have

$$\langle S_z(\mathbf{R})S_z(\mathbf{R}')\rangle_0 = \tfrac{1}{3}S(S+1)\,\delta_{\mathbf{R},\mathbf{R}'}. \tag{33.42}$$

The leading correction to the behavior of Γ for $T\to\infty$ is given by retaining the first term in the expansion of the statistical weight:

$$e^{-\beta\mathcal{H}_0} = 1 - \beta\mathcal{H}_0 + O(\beta\mathcal{H}_0)^2. \tag{33.43}$$

---

[20] If the spins in the Hamiltonian (33.4) were classical vectors, this would follow directly from the definition (31.6). The fact that they are operators does not invalidate the derivation, provided that the component of the total spin along the field commutes with the Hamiltonian.

[21] We introduce the notation $\langle X\rangle_0 = \lim_{T\to\infty}\langle X\rangle$. Note that $\langle X\rangle_0 = \text{Tr}\,X/\text{Tr}\mathbf{1}$.

Inserting this into (33.39) we find that

$$\Gamma(\mathbf{R}, \mathbf{R}') \approx \frac{\frac{1}{3}S(S+1)\,\delta_{\mathbf{R},\mathbf{R}'} - \beta\langle \mathbf{S}_z(\mathbf{R})\mathbf{S}_z(\mathbf{R}')\mathcal{H}_0\rangle_0}{1 - \beta\langle\mathcal{H}_0\rangle_0}. \tag{33.44}$$

At infinite T (i.e., in the absence of interactions) we have

$$\langle \mathbf{S}(\mathbf{R})\cdot\mathbf{S}(\mathbf{R}')\rangle_0 = 0, \qquad \mathbf{R} \neq \mathbf{R}',$$
$$\langle\mathcal{H}_0\rangle_0 = 0, \tag{33.45}$$

so the denominator in (33.44) remains unity. The correction to the leading term in the numerator, however, is

$$\beta\frac{1}{2}\sum_{\mathbf{R}_1,\mathbf{R}_2} J(\mathbf{R}_1 - \mathbf{R}_2)\langle \mathbf{S}_z(\mathbf{R})\mathbf{S}_z(\mathbf{R}')\ \mathbf{S}(\mathbf{R}_1)\cdot\mathbf{S}(\mathbf{R}_2)\rangle_0. \tag{33.46}$$

Because the spins at different sites are independent in the $T \to \infty$ limit, (33.46) fails to vanish only when $\mathbf{R}_1 = \mathbf{R}$, $\mathbf{R}_2 = \mathbf{R}'$, or vice versa. It therefore reduces to

$$\beta J(\mathbf{R} - \mathbf{R}')\sum_{\mu=x,y,z}\langle \mathbf{S}_z(\mathbf{R})\mathbf{S}_\mu(\mathbf{R})\rangle_0\langle \mathbf{S}_z(\mathbf{R}')\mathbf{S}_\mu(\mathbf{R}')\rangle_0. \tag{33.47}$$

Since different components of a given spin are uncorrelated, this further simplifies to

$$\beta J(\mathbf{R} - \mathbf{R}')\langle \mathbf{S}_z{}^2(\mathbf{R})\rangle_0\langle \mathbf{S}_z{}^2(\mathbf{R}')\rangle_0 = \beta J(\mathbf{R} - \mathbf{R}')\left(\frac{S(S+1)}{3}\right)^2. \tag{33.48}$$

Collecting these results, we find that the high-temperature expansion (33.44) yields

$$\Gamma(\mathbf{R}, \mathbf{R}') = \frac{S(S+1)}{3}\left[\delta_{\mathbf{R},\mathbf{R}'} + \frac{S(S+1)}{3}\beta J(\mathbf{R} - \mathbf{R}') + O(\beta J)^2\right]. \tag{33.49}$$

Thus at high temperatures the correlation function for two distinct spins is simply proportional to the exchange interaction itself. This is reasonable, since one expects a positive (i.e., ferromagnetic) exchange coupling between two spins to favor their parallel alignment (and hence lead to a positive value of their inner product), while a negative (i.e., antiferromagnetic) coupling should favor antiparallel alignment. The result neglects, however, the possibility that two distinct spins may be more strongly correlated by their common coupling to other spins than by their direct coupling. Terms that can be so interpreted do appear when one carries the high-temperature expansion to still higher orders in J/k_BT.

Inserting the correlation function (33.49) into the susceptibility (33.35), using Eq. (33.38), we find the high-temperature susceptibility:

$$\chi(T) = \frac{N}{V}\frac{(g\mu_B)^2}{3k_BT}S(S+1)\left[1 + \frac{\theta}{T} + O\left(\frac{\theta}{T}\right)^2\right], \tag{33.50}$$

where

$$\theta = \frac{S(S+1)}{3}\frac{J_0}{k_B}, \qquad J_0 = \sum_{\mathbf{R}} J(\mathbf{R}). \tag{33.51}$$

The susceptibility (33.50) has the form of Curie's law (Eq. (31.47)) multiplied by the correction factor $(1 + \theta/T)$, which is greater than or less than unity, according to whether the coupling is predominantly ferromagnetic or antiferromagnetic.[22] Thus even well above the critical temperature one can obtain a clue as to the nature of the ordering that sets in below T_c from the temperature dependence of the susceptibility.[23]

ANALYSIS OF THE CRITICAL POINT

Quantitative theories of magnetic ordering have proved most difficult to construct near the critical temperature T_c at which the ordering disappears. The difficulty is not peculiar to the problem of magnetism. The critical points of liquid-vapor transitions, superconducting transitions (Chapter 34), the superfluid transition in liquid He_4, and order-disorder transitions in alloys, to name just a few, present quite strong analogies and give rise to quite similar theoretical difficulties.

One computational approach[24] has been to calculate as many terms as possible in the high-temperature expansion of (for example) the susceptibility, and extrapolate down in T to the singularity, thereby obtaining the critical temperature as well as the exponent γ (Eq. (33.2)). Highly sophisticated extrapolation techniques have been developed[25] and the value of γ thereby obtained is quite compatible with the observed divergence. Unfortunately, a similar approach cannot easily be used for the spontaneous magnetization of the Heisenberg model. If the series expansion for $M(T)$ were known about $T = 0$, one could extrapolate upward to the singularity, obtaining both a check on the T_c evaluated by extrapolating the susceptibility downward, as well as the critical exponent β (Eq. (33.1)). Unfortunately, however, low-temperature expansions of $M(T)$ require calculating corrections to the spin wave approximation. Although this is possible to a limited extent, it cannot be carried to anywhere near the level of the systematic procedure available for calculating high-temperature expansions.

Another approach is to simplify the Hamiltonian still further. The price that one pays is a model that bears only a generic resemblance to the original physical problem, except occasionally in peculiar cases that happen (generally *ex post facto*) to resemble the new models. What one gains is a model that is considerably more tractable analytically. The detailed theoretical analyses of such models are of value both for what they suggest about the more realistic Heisenberg model, and also as a preliminary testing ground for various approximation techniques.

By far the most important simplification of the Heisenberg model is the Ising

---

22 When this analysis is generalized to more complex crystal structures (which is straightforward), the result (33.51) provides a way to distinguish simple ferromagnetic solids from simple ferrimagnetic ones; if a spontaneous magnetization (below T_c) is due to a positive exchange interaction (ferromagnetism), then the term in $1/T^2$ in the high-temperature susceptibility should be positive; if it is due to a negative (antiferromagnetic) coupling between unlike spins, then the term in $1/T^2$ in the high-temperature susceptibility should be negative.

23 This is the most important content of a phenomenological modification of Curie's law, known as the Curie-Weiss law. See the discussion of mean field theory below.

24 This is reviewed by M. E. Fisher, *Rep. Progr. Phys.* **30** (pt. II), 615 (1967).

25 The most important is the method of Padé approximants. This is reviewed by G. A. Baker, *Advances in Theoretical Physics I*, K. A. Brueckner, ed., Academic Press, New York, 1965.

model, in which the terms in $\mathbf{S}_+$ and $\mathbf{S}_-$ are simply dropped from the Heisenberg Hamiltonian (33.9), leaving

$$\mathcal{H}^{\text{Ising}} = -\frac{1}{2}\sum_{\mathbf{R},\mathbf{R}'} J(\mathbf{R}-\mathbf{R}')\mathbf{S}_z(\mathbf{R})\mathbf{S}_z(\mathbf{R}') - g\mu_B H \sum_{\mathbf{R}} \mathbf{S}_z(\mathbf{R}). \tag{33.52}$$

Since all the $\mathbf{S}_z(\mathbf{R})$ commute, $\mathcal{H}^{\text{Ising}}$ is explicitly diagonal in the representation in which each individual $\mathbf{S}_z(\mathbf{R})$ is diagonal; i.e., all the eigenfunctions and eigenvalues of the Hamiltonian are known. In spite of this, calculating the partition function is *still* a task of formidable difficulty. However, the high-temperature expansion is more easily evaluated and can be carried out to more terms than in the Heisenberg model, and the profound difficulties in the low-temperature expansion disappear (together, unfortunately, with the Bloch $T^{3/2}$ law).

Near the critical point, however, one can still do little more than extrapolate the high-temperature (and low-temperature) expansions, except in the case of the two-dimensional Ising model with only nearest-neighbor interactions.[26] In that one case, for several simple lattices (e.g., square, triangular, honeycomb) the exact free energy is known[27] in zero magnetic field, together with the spontaneous magnetization. It is sobering to note that the calculation of these results is among the most impressive *tours de force* achieved by theoretical physicists, in spite of the extensive oversimplifications that have had to be made to construct a model even this tractable.

According to the exact Onsager solution, the specific heat of the two-dimensional Ising model has a logarithmic singularity in zero magnetic field, as the critical temperature T_c is approached from either above or below. The spontaneous magnetization vanishes as $(T_c - T)^{1/8}$, and the susceptibility diverges as $(T - T_c)^{-7/4}$. Note that these exponents are quite different from the observed values described on page 699, except, perhaps, for the specific heat singularity (a very small power law divergence being hard to distinguish from a logarithmic singularity). This is a consequence of the two-dimensional structure of the model. Series expansions in three dimensions indicate singularities much closer to those observed.

Finally, we note another approach to the critical region arising from the hypothesis[28] that in the neighborhood of $T = T_c$ and $H = 0$ the magnetic equation of state should have the form

$$\frac{H}{|T_c - T|^{\beta+\gamma}} = f_{\pm}\left(\frac{M}{|T_c - T|^{\beta}}\right), \qquad T \gtrless T_c, \tag{33.53}$$

which is known as a scaling equation of state. Given such a form, one can deduce certain relations between the exponents describing the critical point singularities—for example (see Eqs. (33.1), (33.2), and (33.3)), $\alpha + 2\beta + \gamma = 2$—that can only be

[26] See, however, the remark on renormalization group methods in footnote 7. The model can also be analyzed completely in one dimension, but for any finite range of interaction there is no magnetic ordering at any temperature.

[27] The solution was found by L. Onsager, *Phys. Rev.* **65**, 117 (1944). The first published calculation of the spontaneous magnetization (Onsager reported the result but never published his calculation) is by C. N. Yang, *Phys. Rev.* **85**, 808 (1952). A relatively accessible version of the Onsager calculation of the free energy has been given by T. Schultz et al., *Rev. Mod. Phys.* **36**, 856 (1964).

[28] B. Widom, *J. Chem. Phys.* **43**, 3898 (1965); L. P. Kadanoff, *Physics* **2**, 263 (1966).

proved[29] as inequalities, but appear to be satisfied in real systems as strict equalities. The scaling concept has been applied to the static correlation function[30] and even to the time-dependent correlation function.[31] It has set the direction to many experiments on the critical point, which have, in turn, confirmed the original conjecture (see, for example, Figure 33.10). It is only with the recent theoretical work of K. G. Wilson, however, that a firm basis has been given to the scaling hypothesis.[32]

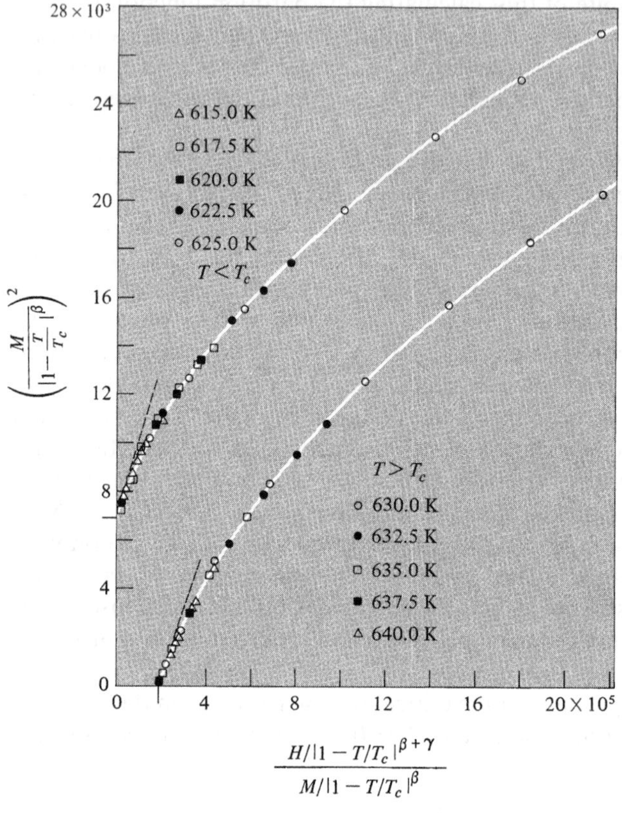

Figure 33.10
The magnetic equation of state of nickel near $T_c = 627.4$ K. If the scaling hypothesis is correct, there should be two temperature-independent exponents β and γ such that $H/|T - T_c|^{\beta+\gamma}$ depends on M and T only in the combination $M/|T - T_c|^{\beta}$. (The functional relations will not be the same, however, above and below T_c.) By plotting $[M/|1 - (T/T_c)|^{\beta}]^2$ vs. $[H/|1 - (T/T_c)|^{\beta+\gamma}]/[M/|1 - (T/T_c)|^{\beta}]$, one can demonstrate the extent to which the hypothesis is satisfied. For five different temperatures above T_c the points so plotted all lie on one universal curve; the same behavior is found for five different temperatures below T_c. The exponents used are $\beta = 0.378$ and $\gamma = 1.34$. (The scales are based on H in gauss and M in emu/gm.) (From J. S. Kouvel and J. B. Comly, *Phys. Rev. Lett.* **20**, 1237 (1968).)

---

[29] R. B. Griffiths (*J. Chem. Phys.* **43**, 1958 (1965)) gives a large number of thermodynamic inequalities that can be proved about singular quantities near the critical point.

[30] In its simplest form, scaling asserts (M. E. Fisher, *J. Math. Phys.* **5**, 944 (1964)) that the correlation function has the form

$$\Gamma(\mathbf{R}) = \frac{1}{R^p} f\left(\frac{R}{\xi}\right),$$

where $\xi(T)$, known as the correlation length, diverges at the critical temperature. It is clear, from the fact that the susceptibility diverges at the critical point, that the correlation function should acquire a very long spatial range at T_c (see Eqs. (33.35) and (33.38)). The scaling hypothesis makes the additional assumptions that the correlation function falls off as a simple power of R at T_c, and that it depends on temperature only through the variable $R/\xi(T)$.

[31] B. I. Halperin and P. C. Hohenberg, *Phys. Rev. Lett.* **19**, 700 (1967).

[32] See footnote 7, and also F. J. Wegner, *Phys. Rev.* **B5**, 4529 (1972).

MEAN FIELD THEORY

The earliest attempt at a quantitative analysis of the ferromagnetic transition was put forward by P. Weiss and is known as mean (or molecular) field theory.[33] Mean field theory gives a grossly inadequate picture of the critical region, fails to predict spin waves at low temperatures, and even at high temperatures reproduces without error only the leading correction to Curie's law. We nevertheless mention it here because (a) the theory has been so widely used and quoted that one must learn to recognize it and be aware of its inadequacies; (b) when one is confronted with a new situation (e.g., a particular complicated arrangement of spins on a crystal structure with several types of coupling), mean field theory probably offers the simplest rough way of sorting out the types of structures one might expect to arise; and (c) mean field theory is sometimes taken as a starting point for more sophisticated calculations.

Suppose that in the Heisenberg Hamiltonian (33.4) we focus our attention on a particular site $\mathbf{R}$ and isolate from $\mathcal{H}$ those terms containing $\mathbf{S}(\mathbf{R})$:

$$\Delta\mathcal{H} = -\mathbf{S}(\mathbf{R})\cdot\left(\sum_{\mathbf{R}\neq\mathbf{R}'} J(\mathbf{R}-\mathbf{R}')\mathbf{S}(\mathbf{R}') + g\mu_B\mathbf{H}\right). \tag{33.54}$$

This has the form of the energy of a spin in an effective external field:

$$\mathbf{H}_{\text{eff}} = \mathbf{H} + \frac{1}{g\mu_B}\sum_{\mathbf{R}'} J(\mathbf{R}-\mathbf{R}')\mathbf{S}(\mathbf{R}'), \tag{33.55}$$

but the "field" $\mathbf{H}_{\text{eff}}$ is an operator, depending in a complicated way on the detailed configuration of all the other spins at sites different from $\mathbf{R}$. The mean field approximation evades this complexity by replacing $\mathbf{H}_{\text{eff}}$ with its thermal equilibrium mean value. In the case of a ferromagnet[34] every spin has the same mean value, which can be represented in terms of the total magnetization density as

$$\langle\mathbf{S}(\mathbf{R})\rangle = \frac{V}{N}\frac{\mathbf{M}}{g\mu_B}. \tag{33.56}$$

If we replace each spin in Eq. (33.55) by its mean value (33.56) we arrive at the effective field,

$$\mathbf{H}_{\text{eff}} = \mathbf{H} + \lambda\mathbf{M}, \tag{33.57}$$

where

$$\lambda = \frac{V}{N}\frac{J_0}{(g\mu_B)^2}, \qquad J_0 = \sum_{\mathbf{R}} J(\mathbf{R}). \tag{33.58}$$

The mean field theory of a ferromagnet assumes that the only effect of interactions is to replace the field each spin feels by $\mathbf{H}_{\text{eff}}$. This is rarely justified in cases of practical

---

[33] The theory is readily generalized to describe all varieties of magnetic ordering, is closely analogous to the van der Waals theory of the liquid-vapor transition, and is a particular example of a very general theory of phase transitions due to Landau.

[34] For other cases see Problem 7. In general, one makes an initial *ansatz* for the equilibrium average of each $\mathbf{S}(\mathbf{R})$, uses that to construct the average mean field, and then requires (self-consistency) that the equilibrium average of each spin $\mathbf{S}(\mathbf{R})$, computed as if it were a free spin in the average mean field, agree with the initial *ansatz*.

interest, since it requires either that individual spin directions do not deviate drastically from their average values, or that the exchange interaction is of such long range that many spins contribute to (33.55), with individual spin fluctuations about the average canceling among themselves.

If we nevertheless make the mean field approximation, then the magnetization density is given by the solution to

$$M = M_0\left(\frac{H_{\text{eff}}}{T}\right), \tag{33.59}$$

where M_0 is the magnetization density in the field H at temperature T, calculated in the absence of magnetic interactions. We computed M_0 in Chapter 31, where we found (Eq. 31.44) that it depended on H and T only through their ratio, as Eq. (33.59) takes explicitly into account. If there is a spontaneous magnetization $M(T)$ at a temperature T, then it will be given by a nonzero solution to (33.59) when the applied field vanishes. Since $H_{\text{eff}} = \lambda M$ when $H = 0$, we must have

$$M(T) = M_0\left(\frac{\lambda M}{T}\right). \tag{33.60}$$

The possibility of solutions to Eq. (33.60) is most easily investigated graphically. If we write it as the pair of equations

$$M(T) = M_0(x),$$
$$M(T) = \frac{T}{\lambda}x, \tag{33.61}$$

then solutions will occur whenever the graph of $M_0(x)$ intersects the straight line $(T/\lambda)x$ (see Figure 33.11). This will happen at a nonzero value of x if and only if the slope of the straight line, T/λ, is less than the slope of $M_0(x)$ at the origin, $M_0'(0)$. The latter slope, however, can be expressed in terms of the zero-field susceptibility χ_0 calculated in the absence of interactions, for

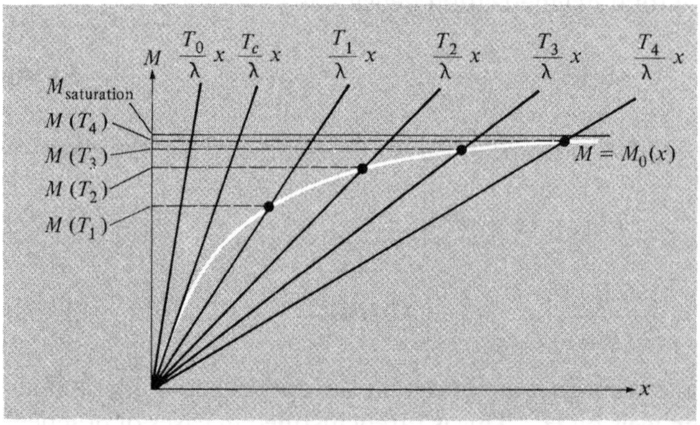

Figure 33.11
Graphical solution to the mean field equations (33.61). When T has a value exceeding T_c (for example, $T = T_0$) there is no solution except $M = 0$. When T is less than T_c (for example, $T = T_1, \ldots T_4$) there are solutions with nonzero M. The critical value of T, T_c, is determined by the geometrical condition that the slope of $M_0(x)$ at the origin be T_c/λ.

$$\chi_0 = \left(\frac{\partial M_0}{\partial H}\right)_{H=0} = \frac{M_0'(0)}{T}. \tag{33.62}$$

Comparing this with the explicit form (31.47) of Curie's law, we can read off the value of $M_0'(0)$, and conclude that the critical temperature T_c below which a nonzero spontaneous magnetization can be found is given by

$$T_c = \frac{N}{V}\frac{(g\mu_B)^2}{3k_B}S(S+1)\lambda = \frac{S(S+1)}{3k_B}J_0. \tag{33.63}$$

In Table 33.4 this prediction is compared with the exact critical temperatures for several two- and three-dimensional Ising models.[35] The actual critical temperatures are lower than the mean field prediction by as much as a factor of 2. However, agreement does improve with increasing lattice dimensionality and coordination number, as one might expect.

Table 33.4
RATIO OF THE EXACT CRITICAL TEMPERATURES TO THOSE PREDICTED BY MEAN FIELD THEORY (MFT) FOR SEVERAL NEAREST NEIGHBOR ISING MODELS[a]

LATTICE	DIMENSIONALITY	COORDINATION NUMBER	T_c/T_c^{mft}
Honeycomb	2	3	0.5062173
Square	2	4	0.5672963
Triangular	2	6	0.6068256
Diamond	3	4	0.67601
Simple cubic	3	6	0.75172
Body-centered cubic	3	8	0.79385
Face-centered cubic	3	12	0.8162

[a] The two-dimensional values of T_c are known in closed form; the values in three dimensions have been computed by extrapolation techniques to the accuracy quoted.
Source: M. E. Fisher, *Repts. Prog. Phys.* **30** (pt. II), 615 (1967).

Just below T_c, Eq. (33.60) gives a spontaneous magnetization that varies as $(T_c - T)^{1/2}$, regardless of the dimensionality of the lattice (see Problem 6). This is in striking contrast to the known results $M \sim (T_c - T)^\beta$, with $\beta = \frac{1}{8}$ for the two-dimensional Ising model, and $\beta \simeq \frac{1}{3}$ for most three-dimensional physical and model systems. Note, though, that the agreement with mean field theory again improves as the dimensionality increases.[36]

Near zero temperature mean field theory predicts that the spontaneous magnetization deviates from its saturation value by a term of order e^{-J_0S/k_BT} (Problem 9). This is in striking contrast to the $T^{3/2}$ dependence that a more accurate analysis of the isotropic[37] Heisenberg model predicts and experiments have confirmed.

[35] When applying (33.63) to the Ising model, $\frac{1}{3}S(S+1)$ must be replaced by the term from which it arose: the mean value of $\mathbf{S}_z^2$ for a randomly oriented spin.

[36] It is believed that in more than four dimensions the mean field critical indices are correct.

[37] The spontaneous magnetization of the anisotropic Heisenberg model does deviate only exponentially from saturation. However, J_0/k_BT is replaced by $\Delta J/k_BT$, where ΔJ is a measure of the anisotropy in the exchange coupling, which is very much smaller than J_0 when the anisotropy is weak. See Problem 5.

The susceptibility in the mean field approximation is given by differentiating (33.59):

$$\chi = \frac{\partial M}{\partial H} = \frac{\partial M_0}{\partial H_{\text{eff}}}\frac{\partial H_{\text{eff}}}{\partial H} = \chi_0(1 + \lambda\chi). \tag{33.64}$$

Hence

$$\chi = \frac{\chi_0}{1 - \lambda\chi_0}, \tag{33.65}$$

where χ_0 is evaluated in the field H_{eff}. Above T_c in the limit of zero applied field, H_{eff} vanishes and the susceptibility χ_0 assumes the Curie's law form (31.47). Equation (33.65) then gives a zero-field susceptibility

$$\chi = \frac{\chi_0}{1 - (T_c/T)}. \tag{33.66}$$

This result is identical in form to Curie's law for an ideal paramagnet (Eq. (31.47)), except that the T in the denominator has been replaced by $T - T_c$, a modification known as the Curie-Weiss law. The term "law" is an unfortunate one, since near T_c the measured and calculated susceptibilities of three-dimensional ferromagnets diverge as an inverse power of $T - T_c$ somewhere between $\frac{5}{4}$ and $\frac{4}{3}$, rather than as the simple pole that (33.66) predicts.[38] However, the dominant (order $1/T^2$) correction to the high-temperature Curie's law susceptibility given by (33.66) does agree with the exact result (33.50), and this is the only real content of the Curie-Weiss law: the high-temperature correction to the susceptibility of a ferromagnet makes it larger than the value predicted by Curie's law.[39] Corrections beyond the dominant one at high temperatures disagree with the prediction of (33.66), and therefore once one leaves the high-temperature regime the Curie-Weiss law is little more than a particularly simple and not very reliable way of extrapolating the high-temperature susceptibility series down to lower T.

CONSEQUENCES OF DIPOLAR INTERACTIONS IN FERROMAGNETS: DOMAINS

Although the critical temperature of iron is over 1000 K, a piece of iron picked from the shelf normally appears to be "unmagnetized." The same piece of iron, however, is attracted by magnetic fields far more strongly than a paramagnetic substance, and can be "magnetized" by stroking it with a "permanent magnet."

To explain these phenomena it is necessary to consider the hitherto neglected magnetic dipolar interactions between the spins. We stressed in Chapter 32 that this interaction is very weak, the dipolar coupling between nearest neighbors being typically a thousand times smaller than the exchange coupling. However, the exchange

---

[38] In the two-dimensional Ising model the susceptibility diverges as $(T - T_c)^{-7/4}$, an even wider departure from the Curie-Weiss prediction. Note once again, however, that the mean field prediction improves with increasing dimensionality.

[39] In antiferromagnets mean field theory leads above T_c to a susceptibility of the form (33.66), but with a pole at negative T (see Problem 7). Here, again, this result is not reliable except for its prediction about the sign of the high-temperature correction to Curie's law.

interaction is quite short-ranged (falling off exponentially with spin separation in a ferromagnetic insulator) whereas the dipolar interaction is not (falling off only as the inverse cube of the separation). As a result, the magnetic configuration of a macroscopic sample can be quite complex, for the dipolar energies become significant when enormous numbers of spins are involved, and can then considerably alter the spin configuration favored by the short-range exchange interactions.

In particular, a uniformly magnetized configuration such as we have used to characterize the ferromagnetic state is exceedingly uneconomical in dipolar energy. The dipolar energy can be substantially reduced (Figure 33.12) by dividing the specimen into uniformly magnetized *domains* of macroscopic size, whose magnetization vectors point in widely different directions. Such a subdivision is paid for in exchange energy, for the spins near the boundary of a domain will experience unfavorable exchange interactions with the nearby spins in the neighboring misaligned domain. Because, however, the exchange interaction is short-ranged, it is only the spins near the domain boundaries that will have their exchange energies raised. In contrast, the gain in magnetic dipolar energy is a bulk effect: because of the long range of the interaction, the dipolar energy of *every* spin drops when domains are formed. Therefore, provided that the domains are not too small, domain formation will be favored in spite of the vastly greater strength of the exchange interaction. Every spin can lower its (small) dipolar energy, but only a few (those near domain boundaries) have their (large) exchange energy raised.

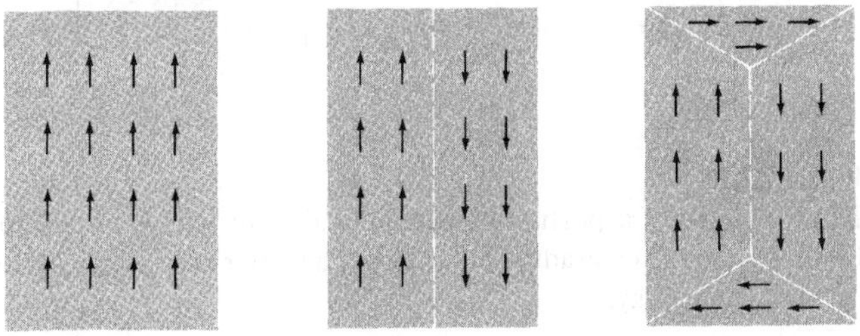

(a) (b) (c)

Figure 33.12
A ferromagnetically ordered solid can reduce its magnetic dipolar energy by breaking up into a complex structure of domains. Thus the single-domain structure (a) has a much higher dipolar energy than the structure (b) consisting of two domains. (To see this think of the two halves of (b) as being two bar magnets. To form the single domain (a), one of the magnets in (b) must be reversed, thereby changing a configuration in which opposite poles are near one another to one in which like poles are near one another.) The two-domain structure (b) can lower its dipolar energy still further by producing the additional domains shown in (c).

The ease with which a ferromagnet below T_c retains or loses (by breaking up into domains) its spontaneous magnetization, as well as the process by which application of a field forces the spontaneous magnetization to reappear, is intimately tied to the physics of how domains alter their size and orientation. The structure of the boundary between two domains (known as the domain wall, or Bloch wall) plays an important role in these processes. An abrupt boundary (Figure 33.13a) between two domains is unnecessarily costly in exchange energy. One can lower the surface energy of a domain wall by spreading out the reversal of spin direction over many spins.[40] For if the spin reversal is spread over n spins, then as one passes through the wall each spin will be seen to differ in orientation from its neighbor by an angle π/n (Figure 33.13b). In a crude classical picture the exchange energy of successive pairs will therefore be not the minimum value $-JS^2$, but rather $-JS^2 \cos(\pi/n) \approx -JS^2[1 - \frac{1}{2}(\pi/n)^2]$. Since it takes n steps to reverse the spin, the cost of achieving a 180° spin reversal down a line of n spins will be

$$\Delta E = n\left[-JS^2 \cos\left(\frac{\pi}{n}\right) - (-JS^2)\right] = \frac{\pi^2}{2n} JS^2, \tag{33.67}$$

which is lower than the cost of an abrupt (one-step) reversal by the factor $\pi^2/2n$.

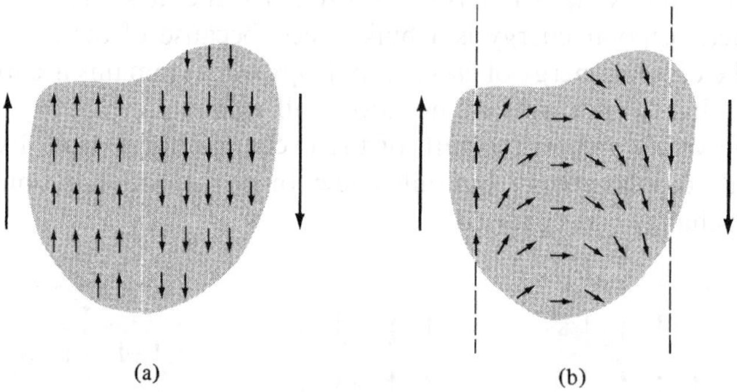

Figure 33.13
Detailed view of a portion of domain wall showing (a) an abrupt boundary and (b) a gradual boundary. The latter type is less costly in exchange energy.

If this were the only consideration, the domain wall would broaden out to a thickness limited only by dipolar interactions. However, the above analysis assumed that the exchange coupling between neighboring spins was perfectly isotropic, depending only on the angle between them. Although interactions in the Heisenberg Hamiltonian (33.4) have this isotropy, this is only because spin-orbit coupling was neglected in its derivation. In a real solid the spins will be coupled to the electronic charge density via spin-orbit coupling, and their energy will therefore depend to some extent on their absolute orientation with respect to the crystal axes, as well as on their relative orientation with respect to one another. Although this dependence of the spin energy on absolute orientation (known as *anisotropy energy*) may be quite weak, it will, on the average, contribute a fixed energy per spin to the energy of a line of deviant spins, and will therefore eventually outweigh the increasingly small reductions in exchange energy made possible by successive extensions of the domain wall

---

40 We consider the case where the wall is not so very thick that the dipolar energy of the boundary itself is consequential.

thickness. Thus in practice the thickness of a domain wall is determined by a balance between exchange and anisotropy energies.[41]

The "magnetization" of a piece of "unmagnetized" iron by application of a field (well below T_c) is a process in which domains are rearranged and reoriented. When a *weak* field is applied, domains oriented along the field can grow at the expense of adversely oriented domains by smooth motion of the domain walls (Figure 33.14).[42] The magnetization process in weak fields is reversible: As the aligning field returns to zero, the domains revert to their original forms (with zero bulk magnetization for the entire specimen). If, however, the aligning field is not weak, favorably aligned domains may also extend themselves by irreversible processes. For example, the

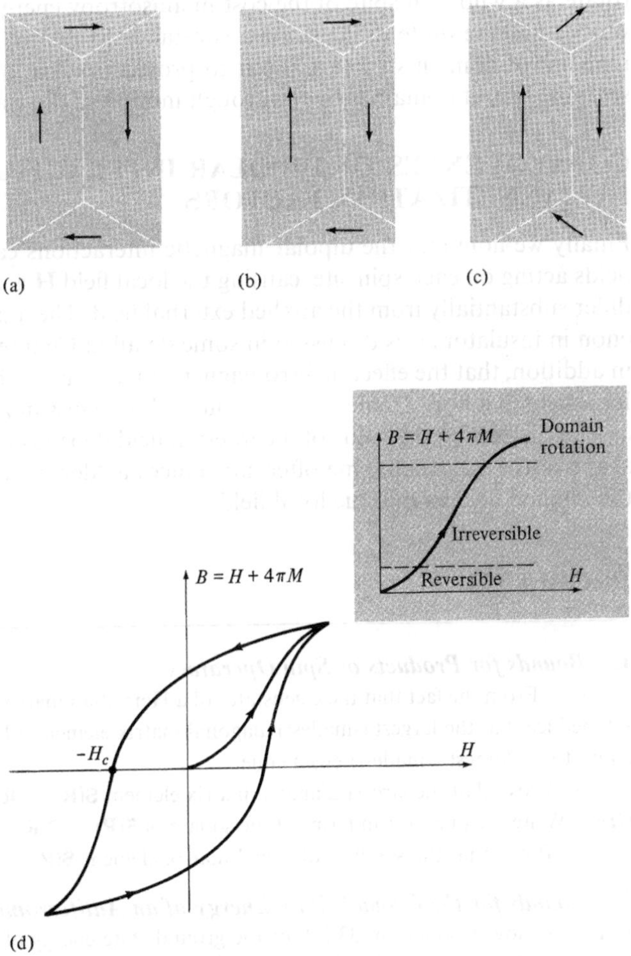

Figure 33.14
The magnetization process. (a) An unmagnetized specimen. (b) The specimen in a weak field that favors spin up. The domain of up spins has grown at the expense of the domain of down spins by motion of the domain wall to the right. In (c) the applied field is stronger, and domain rotation is starting to take place. The magnetization curve (conventionally plotted as $B = H + 4\pi M$ vs. H) from zero magnetization (configuration (a) in zero field) up to saturation is shown in the inset. If one subsequently reduces the field, the magnetization does not return to zero with the field, and a hysteresis curve (d) results. At the field $-H_c$, B vanishes. Sometimes this is taken as an alternative definition of the coercive force.

---

41 Anisotropy energy is also responsible for the observed phenomenon of "easy" and "hard" axes of magnetization.

42 The gradual reversal of spins in the wall is important for the smoothness of the wall's motion. To move an abrupt wall by a series of 180° flips of individual spins would require moving each spin through a large (exchange) energy barrier.

reversible low-field motion of domain walls may be hindered by crystalline imperfections through which the wall will pass only if the gain in external field energy is sufficiently large. When the aligning field is removed, these defects may prevent the domain walls from returning to their original unmagnetized configuration. It then becomes necessary to apply a rather strong field in the opposite direction to restore the unmagnetized configuration. This phenomenon is known as *hysteresis*, and the field necessary to restore zero magnetization (usually from saturation) is known as the *coercive force*. Evidently the value of the coercive force depends on the state of preparation of the specimen.

At very large fields it may become energetically favorable for entire domains to rotate as a whole, in spite of the cost in anisotropy energy. Once magnetized in this way, it may be quite difficult for a substance to reform into domains unless some remnant of domain structure is left to provide nucleation centers for the less catastrophic sort of domain growth through motion of the walls.

CONSEQUENCES OF DIPOLAR INTERACTIONS: DEMAGNETIZATION FACTORS

Finally we note that the dipolar magnetic interactions can result in strong internal fields acting on each spin site, causing the local field **H** a spin actually experiences to differ substantially from the applied external field. The analogous electrical phenomenon in insulators was discussed in some detail in Chapter 27. Here we merely note, in addition, that the effect in ferromagnetic materials can be quite large: The internal local field in a ferromagnet can be thousands of gauss in zero external field. As in the case of dielectrics, the value of the internal field depends in a complicated way on the shape of the specimen. One often introduces a "demagnetization factor" to convert the applied field to the true local field.

PROBLEMS

---

1. *Bounds for Products of Spin Operators*

(a) From the fact that the eigenstates of a Hermitian matrix form a complete orthonormal set, deduce that the largest (smallest) diagonal matrix element a Hermitian operator can have is equal to its largest (smallest) eigenvalue.

(b) Prove that the largest diagonal matrix element $\mathbf{S}(\mathbf{R}) \cdot \mathbf{S}(\mathbf{R}')$ can have when $\mathbf{R} \neq \mathbf{R}'$ is S^2. (*Hint:* Write the operator in terms of the square of $\mathbf{S}(\mathbf{R}) + \mathbf{S}(\mathbf{R}')$.)

(c) Prove that the smallest diagonal matrix element $\mathbf{S}(\mathbf{R}) \cdot \mathbf{S}(\mathbf{R}')$ can have is $-S(S+1)$.

2. *Bounds for the Ground-State Energy of an Antiferromagnet*

Derive the lower bound in (33.17) for the ground state energy of a Heisenberg antiferromagnet from one of the results in Problem 1. Derive the upper bound in (33.17) from a variational argument, using as a trial ground state the one described on page 704.

3. *Exact Ground-State Energy of a Simple "Antiferromagnet"*

Show that the ground-state energy of the four spin antiferromagnetic nearest-neighbor Heisenberg linear chain,

$$\mathcal{H} = J(\mathbf{S}_1 \cdot \mathbf{S}_2 + \mathbf{S}_2 \cdot \mathbf{S}_3 + \mathbf{S}_3 \cdot \mathbf{S}_4 + \mathbf{S}_4 \cdot \mathbf{S}_1), \tag{33.68}$$

is

$$E_0 = -4JS^2\left[1 + \frac{1}{2S}\right]. \tag{33.69}$$

(*Hint:* Write the Hamiltonian in the form

$$\mathcal{H} = \tfrac{1}{2}J[(\mathbf{S}_1 + \mathbf{S}_2 + \mathbf{S}_3 + \mathbf{S}_4)^2 - (\mathbf{S}_1 + \mathbf{S}_3)^2 - (\mathbf{S}_2 + \mathbf{S}_4)^2].) \tag{33.70}$$

4. *Properties of Spin Wave States*

(a) Confirm the normalization in Eqs. (33.19) and (33.20).

(b) Derive Eq. (33.27).

(c) Show that $\langle \mathbf{k}|\mathbf{S}_\perp(\mathbf{R})|\mathbf{k}\rangle = 0$, i.e., that the phase of the spin wave is unspecified in the state $|\mathbf{k}\rangle$.

5. *Anisotropic Heisenberg Model*

Consider the anisotropic Heisenberg spin Hamiltonian

$$\mathcal{H} = -\tfrac{1}{2}\sum_{\mathbf{RR}'}[J_z(\mathbf{R}-\mathbf{R}')S_z(\mathbf{R})S_z(\mathbf{R}') + J(\mathbf{R}-\mathbf{R}')\mathbf{S}_\perp(\mathbf{R})\cdot\mathbf{S}_\perp(\mathbf{R}')] \tag{33.71}$$

with $J_z(\mathbf{R}-\mathbf{R}') > J(\mathbf{R}-\mathbf{R}') > 0$.

(a) Show that the ground state (33.5) and one–spin–wave states (33.23) remain eigenstates of $\mathcal{H}$, but that the spin wave excitation energies are raised by

$$S\sum_{\mathbf{R}}[J_z(\mathbf{R}) - J(\mathbf{R})]. \tag{33.72}$$

(b) Show that the low-temperature spontaneous magnetization now deviates from saturation only exponentially in $-1/T$.

(c) Show that the argument on page 708, that there can be no spontaneous magnetization in two dimensions, no longer works.

6. *Mean Field Theory Near the Critical Point*

For small x, the Brillouin function $B_J(x)$ has the form $Ax - Bx^3$, where A and B are positive.

(a) Deduce that as T approaches T_c from below, the spontaneous magnetization of a ferromagnet vanishes as $(T_c - T)^{1/2}$ according to mean field theory.

(b) Deduce that at T_c, the magnetization density $M(H, T_c)$ vanishes as $H^{1/3}$ in mean field theory. (Observations and calculations indicate an exponent closer to 1/5 for three-dimensional systems. The exponent for the two-dimensional Ising model is 1/15.)

7. *Mean Field Theory of Ferrimagnetism and Antiferromagnetism*

Consider a magnetic structure made up of two types of spins that occupy two interpenetrating sublattices. Let spins within sublattice 1 be coupled by exchange constants J_1, within sublattice 2, by J_2, and between sublattices 1 and 2, by J_3.

(a) Generalize the mean field theory of a simple ferromagnet to this structure, showing that Eq. (33.59) for the spontaneous magnetization generalizes to two coupled equations for the two sublattice magnetizations of the form:

$$\begin{aligned} M_1 &= M_0[(H + \lambda_1 M_1 + \lambda_3 M_2)/T], \\ M_2 &= M_0[(H + \lambda_2 M_2 + \lambda_3 M_1)/T]. \end{aligned} \tag{33.73}$$

(b) Deduce from this that above T_c the zero-field susceptibility is the ratio of a polynomial linear in T to one quadratic in T.

(c) Verify that the susceptibility reduces back to the Curie-Weiss form when the ions in the two sublattices are identical and ferromagnetically coupled ($\lambda_1 = \lambda_2 > 0$, $\lambda_3 > 0$).

(d) Verify that when the ions in the two sublattices are identical ($\lambda_1 = \lambda^2 > 0$) and antiferromagnetically coupled ($\lambda_3 < 0$) with $|\lambda_3| > |\lambda_1|$, the temperature in the Curie-Weiss "law" becomes negative.

8. *High-Temperature Susceptibility of Ferrimagnets and Antiferromagnets*
Generalize the high-temperature susceptibility expansion to the case of the structure described in Problem 7, and compare the exact leading ($O(1/T^2)$) correction to Curie's law to the mean field result.

9. *Low-Temperature Spontaneous Magnetization in Mean Field Theory*
Show that when T is far below T_c, the mean field theory of a ferromagnet predicts a spontaneous magnetization that differs from its saturation value exponentially in $-1/T$.

34
Superconductivity

In Chapter 32 we found that the independent electron approximation cannot adequately describe most magnetically ordered solids. In many metals without any magnetic ordering a still more spectacular failure of the independent electron approximation sets in abruptly at very low temperatures, where another kind of electronically ordered state is established, known as the superconducting state. Superconductivity is not peculiar to a few metals. More than 20 metallic elements can become superconductors (Table 34.1). Even certain semiconductors can be made superconducting under suitable conditions,[1] and the list of alloys whose superconducting properties have been measured stretches into the thousands.[2]

Table 34.1
SUPERCONDUCTING ELEMENTS[a]

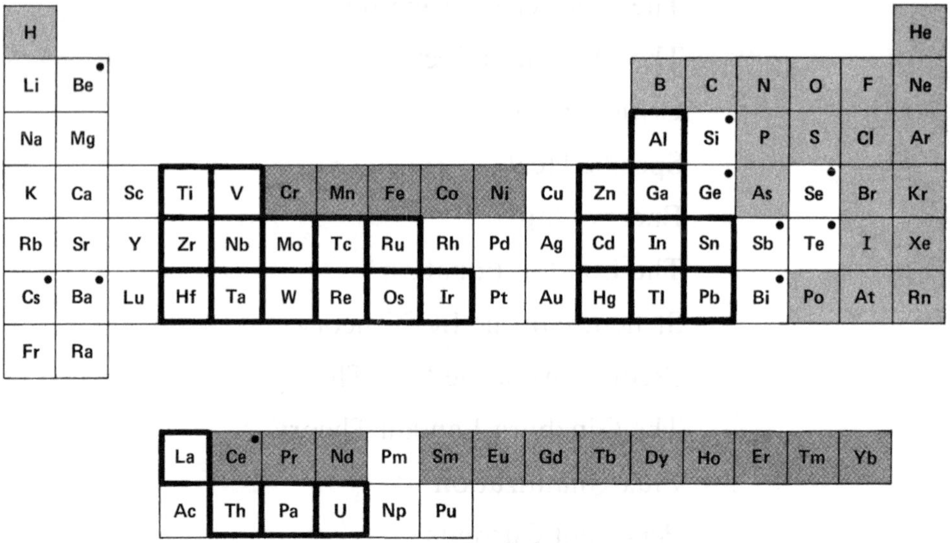

[a]Elements that are superconducting only under special conditions are indicated separately. Note the incompatibility of superconducting and magnetic order. After G. Gladstone, et al. Parks *op. cit*, note 6.

Legend:

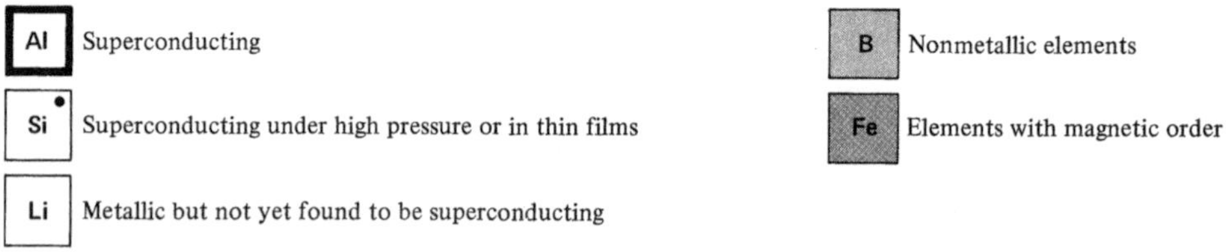

The characteristic properties of metals in the superconducting state appear highly anomalous when regarded from the point of view of the independent electron approximation. The most striking features of a superconductor are:

---

[1] Such as application of high pressure, or preparation of the specimen in very thin films. A striking example of the unexpected ways in which superconductivity can be enhanced is provided by bismuth: Amorphous bismuth is a superconductor at *higher* temperatures than crystalline bismuth, which makes no sense at all in the independent electron approximation.

[2] See B. W. Roberts, *Progr. Cryog.* **4**, 161 (1964).

1. A superconductor can behave as if it had no measurable DC electrical resistivity. Currents have been established in superconductors which, in the absence of any driving field, have nevertheless shown no discernible decay for as long as people have had the patience to watch.[3]
2. A superconductor can behave as a perfect diamagnet. A sample in thermal equilibrium in an applied magnetic field, provided the field is not too strong, carries electrical surface currents. These currents give rise to an additional magnetic field that precisely cancels the applied magnetic field in the interior of the superconductor.
3. A superconductor usually behaves as if there were a gap in energy of width 2Δ centered about the Fermi energy, in the set of allowed one-electron levels.[4] Thus an electron of energy $\mathcal{E}$ can be accommodated by (or extracted from) a superconductor[5] only if $\mathcal{E} - \mathcal{E}_F$ (or $\mathcal{E}_F - \mathcal{E}$) exceeds Δ. The energy gap Δ increases in size as the temperature drops, leveling off to a maximum value $\Delta(0)$ at very low temperatures.

The theory of superconductivity is quite extensive and highly specialized. Like the theories we have described elsewhere in this book, it is based on the nonrelativistic quantum mechanics of electrons and ions, but beyond that its similarity to the other models and theories we have examined diminishes rapidly. The microscopic theory of superconductivity cannot be described in the language of the independent electron approximation. Even comparatively elementary microscopic calculations for superconductors rely on formal techniques (field theoretic methods) which, while conceptually no more sophisticated than the ordinary methods of quantum mechanics, require considerable experience and practice before they can be used with confidence and understanding.

Consequently, to a greater degree than in other chapters we shall limit our survey of the theory of superconductivity to qualitative descriptions of some of the major concepts, together with statements of a few of the simpler predictions. The reader who wishes to acquire even an elementary working knowledge of the subject must consult one of the many available books.[6]

---

[3] The record appears to be $2\frac{1}{2}$ years; S. C. Collins, quoted in E. A. Lynton, *Superconductivity*, Wiley, New York, 1969.

[4] Under a variety of special conditions superconductivity can also occur without an energy gap. Gapless superconductivity can be produced, for example, by introducing a suitable concentration of magnetic impurities. A review is given by K. Maki in *Superconductivity*, R. D. Parks, ed., Dekker, New York, 1969. In the context of superconductivity, the term "energy gap" always refers to the quantity Δ.

[5] This is most directly observed in electron tunneling experiments, which are described below along with other manifestations of the energy gap.

[6] Two fundamental references on the phenomenological theory are F. London, *Superfluids*, vol. 1, Wiley, New York, 1954, and Dover, New York, 1954, and D. Shoenberg, *Superconductivity*, Cambridge, 1962. A very brief survey is given by E. A. Lynton, *Superconductivity*, Methuen, London, 1969. The microscopic theory is expounded in J. R. Schrieffer, *Superconductivity*, W. A. Benjamin, New York, 1964, and in the final chapter of A. A. Abrikosov, L. P. Gorkov, and I. E. Dzyaloshinski, *Methods of Quantum Field Theory in Statistical Physics*, Prentice-Hall, Englewood Cliffs, N.J., 1963. A detailed survey of the theoretical aspects of the subject has been given by G. Rickayzen, *Theory of Superconductivity*, Interscience, New York, 1965, and in somewhat less detail by P. de Gennes, *Superconductivity of Metals and Alloys*, W. A. Benjamin, Menlo Park, Calif., 1966. A survey of all aspects of the subject, theoretical and experimental, by many of the leading experts in the field is *Superconductivity*, R. D. Parks, ed., Dekker, New York, 1969.

This chapter is organized as follows:

1. A survey of the basic empirical facts about superconductivity.
2. A description of the phenomenological London equation and its relation to perfect diamagnetism.
3. A qualitative description of the microscopic theory of Bardeen, Cooper, and Schrieffer.
4. A summary of some of the fundamental equilibrium predictions of the microscopic theory and how they compare with experiment.
5. A qualitative discussion of the relationship between the microscopic theory, the concept of an "order parameter," and the transport properties of superconductors.
6. A description of the remarkable tunneling phenomena between superconductors predicted by B. D. Josephson.

CRITICAL TEMPERATURE

The transition to the superconducting state is a sharp one in bulk specimens. Above a critical temperature[7] T_c the properties of the metal are completely normal; below T_c superconducting properties are displayed, the most dramatic of which is the absence of any measurable DC electrical resistance. Measured critical temperatures range from a few millidegrees Kelvin[8] up to a little over 20 K. The corresponding thermal energy $k_B T_c$ varies from about 10^{-7} eV up to a few thousandths of an electron volt. This is quite minute compared with the energies we have become accustomed to regarding as significant in solids.[9] Transition temperatures of the superconducting elements are listed in Table 34.2.

PERSISTENT CURRENTS

Figure 34.1 displays the resistivity of a superconducting metal vs. temperature as the critical temperature T_c is crossed. Above T_c the resistivity has the form characteristic of a normal metal, $\rho(T) = \rho_0 + BT^5$, the constant term arising from impurity[10] and defect scattering, and the term in T^5 arising from phonon scattering. Below T_c these mechanisms lose the power to degrade an electric current and the resistivity drops abruptly to zero. Currents can flow in a superconductor with no discernible dissipation of energy.[11] There are, however, some limitations:

1. Superconductivity is destroyed by application of a sufficiently large magnetic field (see below).

---

[7] The critical temperature is that at which the transition occurs in the absence of an applied magnetic field. When a magnetic field is present (see below) the transition occurs at a lower temperature, and the nature of the transition changes from second order to first order; i.e., there is a latent heat in nonzero field.

[8] The lowest temperatures at which superconductivity has been sought, to date.

[9] Thus $\mathcal{E}_F \sim 10$ eV, $\hbar\omega_D \sim 0.1$ eV.

[10] We assume there are no magnetic impurities present; see page 687.

[11] When Ampère first proposed that magnetism could be understood in terms of electric currents flowing in individual molecules, it was objected that no currents were known to flow without dissipation. Ampère persisted in his view and was vindicated by the quantum theory, which permits stationary molecular states in which a net current flows (see Chapter 31). A solid in the superconducting state is behaving like one enormous molecule. The presence of an electric current without dissipation in a superconductor is a dramatic macroscopic manifestation of quantum mechanics.

Table 34.2
VALUES OF T_c AND H_c FOR THE SUPERCONDUCTING ELEMENTS[a]

ELEMENT		T_c (K)	H_c (GAUSS)[b]
Al		1.196	99
Cd		0.56	30
Ga		1.091	51
Hf		0.09	—
Hg	α (rhomb)	4.15	411
	β	3.95	339
In		3.40	293
Ir		0.14	19
La	α (hcp)	4.9	798
	β (fcc)	6.06	1096
Mo		0.92	98
Nb		9.26	1980
Os		0.655	65
Pa		1.4	—
Pb		7.19	803
Re		1.698	198
Ru		0.49	66
Sn		3.72	305
Ta		4.48	830
Tc		7.77	1410
Th		1.368	162
Ti		0.39	100
Tl		2.39	171
U	α	0.68	—
	γ	1.80	—
V		5.30	1020
W		0.012	1
Zn		0.875	53
Zr		0.65	47

[a] For type II superconductors, the zero-temperature critical field quoted is obtained from an equal-area construction: The low-field ($H < H_{c1}$) magnetization is extrapolated linearly to a field H_c chosen to give an enclosed area equal to the area under the actual magnetization curve.

[b] At $T = 0$ (K).

Sources: B. W. Roberts, *Progr. Cryog.* **4**, 161 (1964); G. Gladstone, M. A. Jensen, and J. R. Schrieffer, *Superconductivity*, R. D. Parks, ed., Dekker, New York, 1969; *Handbook of Chemistry and Physics*, 55th ed., Chemical Rubber Publishing Co., Cleveland, 1974–1975.

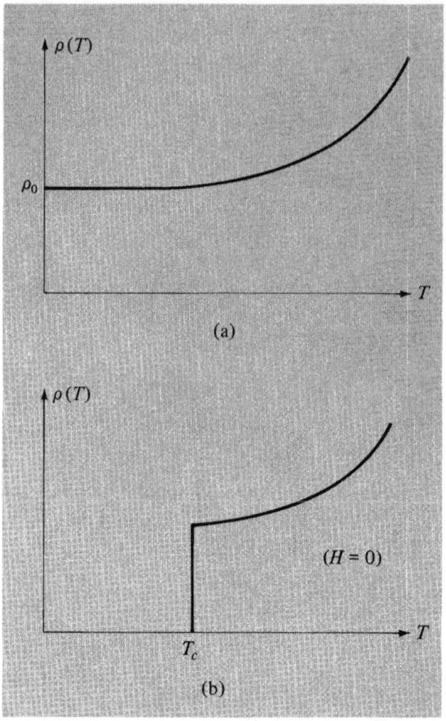

Figure 34.1
(a) Low-temperature resistivity of a normal metal ($\rho(T) = \rho_0 + BT^5$) containing nonmagnetic impurities (b) Low-temperature resistivity of a superconductor (in zero magnetic field) containing nonmagnetic impurities. At T_c, ρ drops abruptly to zero.

2. If the current exceeds a "critical current," the superconducting state will be destroyed (Silsbee effect). The size of the critical current (which can be as large as 100 amp in a 1-mm wire) depends on the nature and geometry of the specimen, and is related to whether the magnetic field produced by the current exceeds the critical field at the surface of the superconductor.[12]
3. A superconductor well below its transition temperature will also respond without dissipation to an AC electric field provided that the frequency is not too large. The change from dissipationless to normal response occurs at a frequency ω of order $\Delta/\hbar$, where Δ is the energy gap.

THERMOELECTRIC PROPERTIES

In the independent electron approximation good electrical conductors are also good conductors of heat, since the conduction electrons transport entropy as well as electric charge.[13] Superconductors, contrary to this, are poor thermal conductors (Figure 34.2).[14] They also exhibit no Peltier effect; i.e., an electric current at uniform temperature in a superconductor is not accompanied by a thermal current, as it would be in a normal metal. The absence of a Peltier effect indicates that those electrons that participate in the persistent current carry no entropy. The poor thermal

[12] See Problem 3.
[13] See page 253.
[14] This property is exploited to make thermal switches.

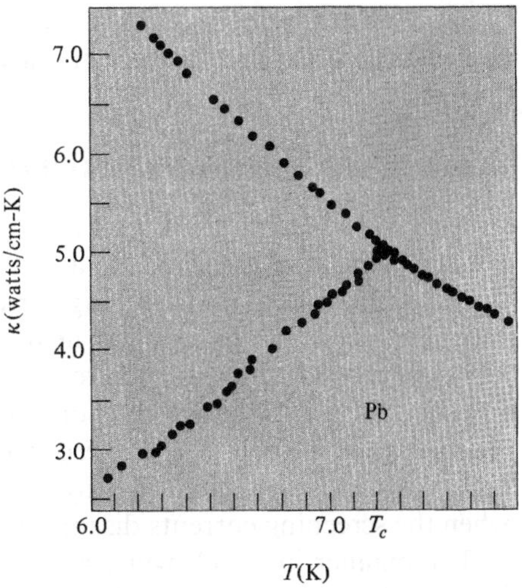

Figure 34.2
The thermal conductivity of lead. Below T_c the lower curve gives the thermal conductivity in the superconducting state, and the upper curve, in the normal state. The normal sample is produced below T_c by application of a magnetic field, which is assumed otherwise to have no appreciable affect on the thermal conductivity. (Reproduced by permission of the National Research Council of Canada from J. H. P. Watson and G. M. Graham, *Can. J. Phys.* **41**, 1738 (1963).)

conductivity indicates that even when a superconductor is not carrying an electric current, only a fraction of its conduction electrons are capable of transporting entropy.[15]

MAGNETIC PROPERTIES: PERFECT DIAMAGNETISM

A magnetic field (provided that it is not too strong) cannot penetrate into the interior of a superconductor. This is most dramatically illustrated by the Meissner-Ochsenfeld effect: If a normal metal in a magnetic field[16] is cooled below its superconducting transition temperature, the magnetic flux is abruptly expelled. Thus the transition, when it occurs in a magnetic field, is accompanied by the appearance of whatever surface currents are required to cancel the magnetic field in the interior of the specimen.

Note that this is not implied by perfect conductivity (i.e., $\sigma = \infty$) alone, even though perfect conductivity does imply a somewhat related property: If a perfect conductor, initially in zero magnetic field, is moved into a region of nonzero field (or if a field is turned on), then Faraday's law of induction gives rise to eddy currents that cancel the magnetic field in the interior. If, however, a magnetic field were established in a perfect conductor, its expulsion would be equally resisted. Eddy currents would be induced to maintain the field if the sample were moved into a field-free region (or if the applied field were turned off). Thus perfect conductivity implies a time-independent magnetic field in the interior, but is noncommittal as to the value that field must have. In a superconductor, the field is not only independent of time, but also zero.

---

[15] Presumably the efficacy of the phonons in conducting heat remains undiminished, but this is generally a less important contribution to the thermal conductivity than that of the conduction electrons.

[16] A normal metal is only weakly paramagnetic or diamagnetic (no magnetically ordered metals are superconductors) and an applied magnetic field can penetrate it.

We shall examine the relation between perfect conductivity and the Meissner effect somewhat more quantitatively in our discussion of the London equation below.

MAGNETIC PROPERTIES: THE CRITICAL FIELD

Consider a superconductor at a temperature T below its critical temperature T_c. As a magnetic field H is turned on, a certain amount of energy is expended to establish the magnetic field of the screening currents that cancels the field in the interior of the superconductor. If the applied field is large enough it will become energetically advantageous for the specimen to revert back to the normal state, allowing the field to penetrate. For although the normal state has a higher free energy than the superconducting state below T_c in zero field, at high enough fields this increase in free energy will be more than offset by the lowering of magnetic field energy that occurs when the screening currents disappear and the field is allowed to enter the specimen.

The manner in which penetration occurs with increasing field strength depends in general on the geometry of the specimen. However, for the simplest geometry—long, thin, cylindrically shaped samples with their axes parallel to the applied magnetic field—there are two clearly distinguishable kinds of behavior:

Type I Below a *critical field* $H_c(T)$ that increases as T falls below T_c, there is no penetration of flux; when the applied field exceeds $H_c(T)$ the entire specimen reverts to the normal state and the field penetrates perfectly.[17] The resulting phase diagram in the H-T plane is pictured in Figure 34.3.[18] One often describes this type of field penetration by plotting the macroscopic diamagnetic magnetization density M vs. the applied field H (Figure 34.4a).

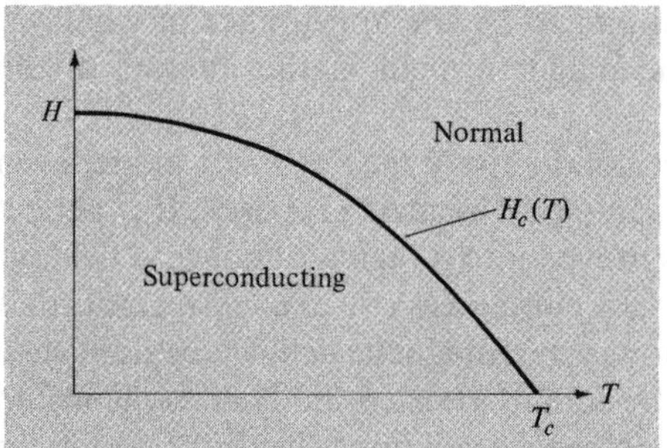

Figure 34.3
The phase boundary between the superconducting and normal states of a type I superconductor in the H-T plane. The boundary is given by the curve $H_c(T)$.

Type II Below a *lower critical field* $H_{c1}(T)$ there is no penetration of flux; when the applied field exceeds an *upper critical field* $H_{c2}(T) > H_{c1}(T)$, the entire specimen reverts to the normal state and the field penetrates perfectly. When the applied field strength is between $H_{c1}(T)$ and $H_{c2}(T)$, there is partial penetration of flux, and the

[17] Except for the small diamagnetic and paramagnetic effects characteristic of normal metals.
[18] Some quantitative thermodynamic consequences of this behavior are explored in Problem 1.

Figure 34.4
(a) Magnetization curve of a type I superconductor. Below H_c no field penetrates: $B = 0$ (or $M = -H/4\pi$). (See footnote 30 for the distinction between B and H in a superconductor.) (b) Magnetization curve of a type II superconductor. Below H_{c1} behavior is as in the type I case. Between H_{c1} and H_{c2}, M falls smoothly to zero, and B rises smoothly to H.

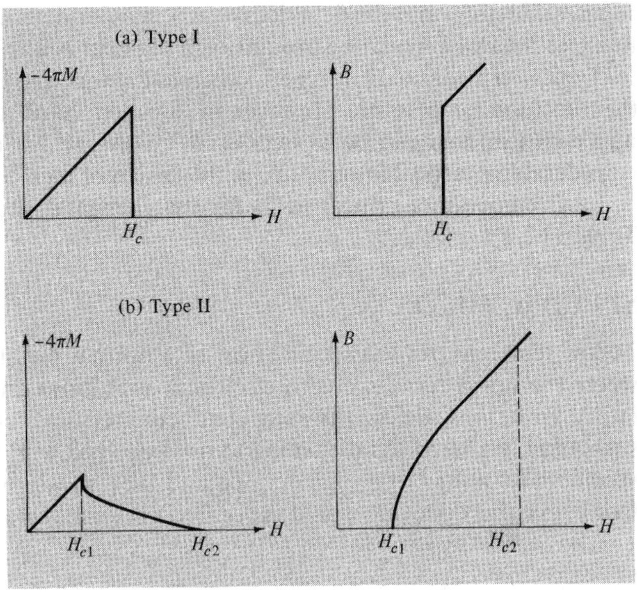

sample develops a rather complicated microscopic structure of both normal and superconducting regions, known as the *mixed state.*[19] The magnetization curve corresponding to type II behavior is shown in Figure 34.4b.

It was proposed by A. A. Abrikosov, and subsequently confirmed by experiment (Figure 34.5), that in the mixed state the field partially penetrates the sample in the form of thin filaments of flux. Within each filament the field is high, and the material is not superconducting. Outside of the core of the filaments, the material remains

Figure 34.5
Triangular array of vortex lines emerging through the surface of a $Pb_{.98}In_{.02}$ superconducting foil in a field of 80 gauss normal to the surface. (Courtesy of J. Silcox and G. Dolan.) The vortices are revealed by the coagulation of fine ferromagnetic particles. Neighboring vortices are about half a micron apart.

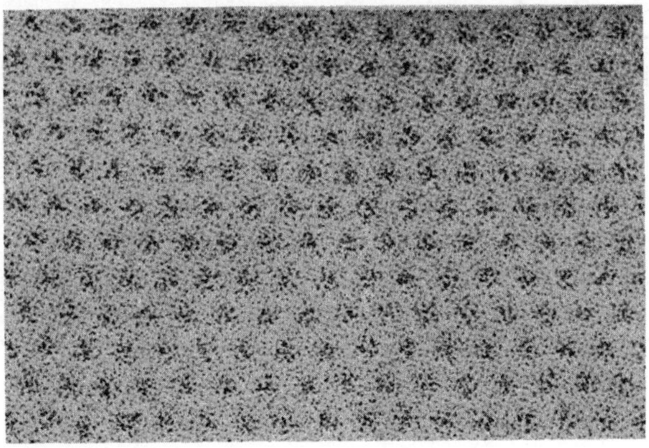

[19] Not to be confused with the *intermediate state*, a configuration a type I superconductor may assume when its shape is more complex than a cylinder parallel to the field, in which macroscopic superconducting and normal regions are interleaved in such a way as to lower the magnetic field energy by more than the cost in free energy of the normal regions.

superconducting, and the field decays in a manner determined by the London equation (see below). Circulating around each filament is a vortex of screening current.[20]

Typical critical fields in type I superconductors are about 10^2 gauss well below the transition temperature. However, in so-called "hard" type II superconductors the upper critical field can be as high as 10^5 gauss, which makes type II materials of considerable practical importance in the design of high-field magnets.

Low temperature critical fields for the elemental superconductors are given in Table 34.2.

SPECIFIC HEAT

At low temperatures the specific heat of a normal metal has the form $AT + BT^3$, where the linear term is due to electronic excitations and the cubic term is due to lattice vibrations. Below the superconducting critical temperature this behavior is substantially altered. As the temperature drops below T_c (in zero magnetic field) the specific heat jumps to a higher value and then slowly decreases, eventually falling well below the value one would expect for a normal metal (Figure 34.6). By applying

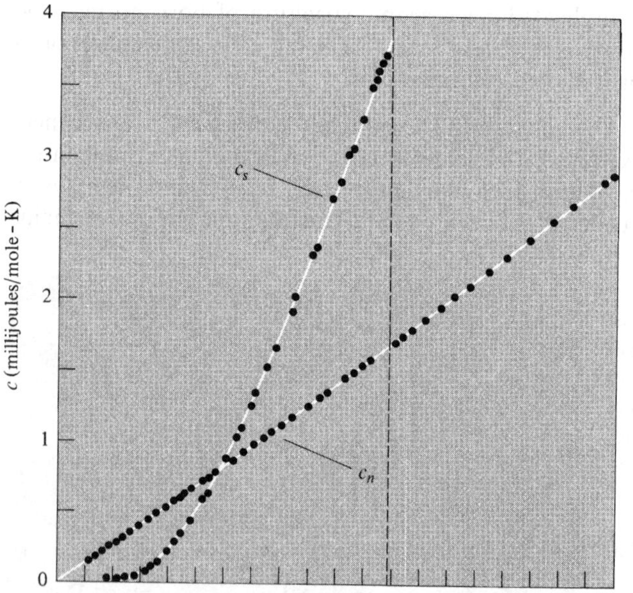

Figure 34.6
Low-temperature specific heat of normal and superconducting aluminum. The normal phase is produced below T_c by application of a weak (300-gauss) magnetic field, which destroys the superconducting ordering but has otherwise negligible effect on the specific heat. The Debye temperature is quite high in aluminum, so the specific heat is dominated by the electronic contribution throughout this temperature range (as can be seen from the fact that the normal-state curve is quite close to being linear). The discontinuity at T_c agrees well with the theoretical prediction (34.22) $[c_s - c_n]/c_n = 1.43$. Well below T_c, c_s drops far below c_n, suggesting the existence of an energy gap. (N. E. Phillips, *Phys. Rev.* **114**, 676 (1959).)

[20] The term "vortex" is often used to refer to the filaments themselves, as well as to the structure of the current in the vicinity of each filament. It can be shown that the magnetic flux enclosed by each vortex is just equal to the magnetic flux quantum, $hc/2e$ (see footnote 60).

a magnetic field to drive the metal into the normal state, one can compare the specific heats of the superconducting and normal states below the critical temperature.[21] Such an analysis reveals that in the superconducting state the linear electronic contribution to the specific heat is replaced by a term that vanishes much more rapidly at very low temperatures, having a dominant low-temperature behavior of the form $\exp(-\Delta/k_B T)$. This is the characteristic thermal behavior of a system whose excited levels are separated from the ground state by an energy 2Δ.[22] Both theory (see Eq. (34.19)) and experiment (see Table 34.3) indicate that the energy gap Δ is of order $k_B T_c$.

OTHER MANIFESTATIONS OF THE ENERGY GAP

Normal Tunneling

The conduction electrons in a superconductor and a normal metal can be brought into thermal equilibrium with one another by placing the metals into such close contact that they are separated only by a thin insulating layer,[23] which the electrons can cross by quantum-mechanical tunneling. In thermal equilibrium enough electrons have passed from one metal to the other to make the chemical potentials of electrons in both metals equal.[24] When both metals are normal, application of a potential difference then raises the chemical potential of one metal with respect to the other, and further electrons tunnel through the insulating layer. Such "tunneling currents" at normal metal junctions have been observed to obey Ohm's law. However, when one of the metals is a superconductor well below its critical temperature, then no current is observed to flow until the potential V reaches a threshold value, $eV = \Delta$ (see Figure 34.7). The size of Δ is in good agreement with the value inferred from low-temperature specific heat measurements, confirming the picture of a gap in the density of one-electron levels in the superconductor. As the temperature is raised toward T_c, the threshold voltage declines,[25] indicating that the energy gap itself is declining with increasing temperature.

Frequency Dependent Electromagnetic Behavior

The response of a metal to electromagnetic radiation (for example the transmission through thin films or the reflection from bulk samples) is determined by the frequency dependent conductivity. This in turn depends on the available mechanisms for energy absorption by the conduction electrons at the given frequency. Because the electronic excitation spectrum in the superconducting state is characterized by an energy gap Δ, one would expect the AC conductivity to differ substantially from its normal state form at frequencies small compared with $\Delta/\hbar$, and to be essentially the same in the superconducting and normal states at frequencies large compared with $\Delta/\hbar$. Except

---

[21] The normal specific heat is not appreciably affected by the presence of a magnetic field.

[22] See point 3, page 727.

[23] For example, the thin layer of oxide on the surfaces of the two specimens.

[24] See page 360.

[25] The threshold also becomes blurred, due to the presence of thermally excited electrons, which require less energy to tunnel.

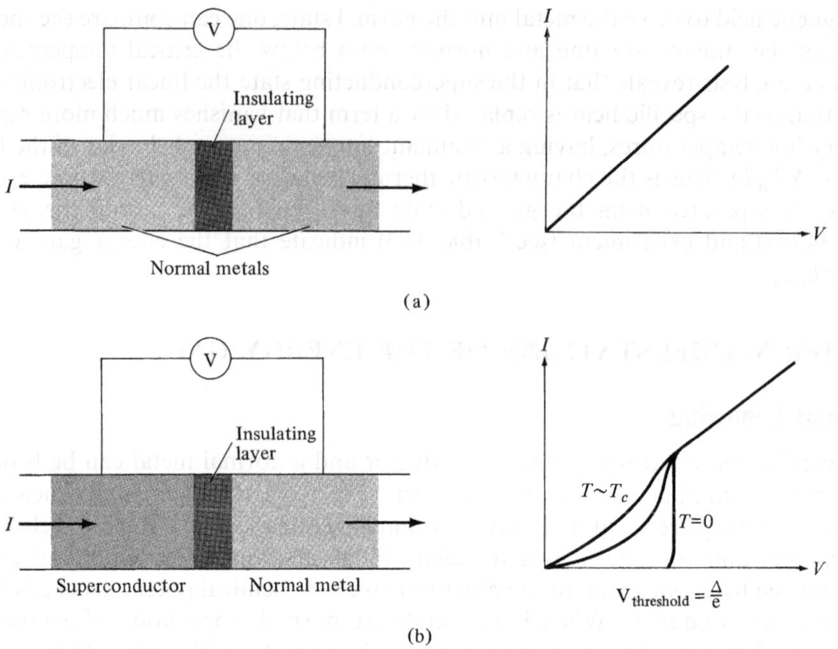

Figure 34.7
(a) Current-voltage relation for electron tunneling through a thin insulating barrier between two normal metals. For small currents and voltages the relation is linear. (b) Current-voltage relation for electron tunneling through a thin insulating barrier between a superconductor and a normal metal. The relation is strongly temperature-dependent. At $T = 0$ there is a sharp threshold, which is blurred at higher temperatures due to the thermal excitation of electrons across the energy gap within the superconductor.

quite near the critical temperature (see p. 744), $\Delta/\hbar$ is typically in the range between microwave and infrared frequencies. In the superconducting state an AC behavior is observed which is indistinguishable from that in the normal state at optical frequencies. Deviations from normal state behavior first appear in the infrared, and only at microwave frequencies does AC behavior fully displaying the lack of electronic absorption characteristic of an energy gap become completely developed.

Acoustic Attenuation

When a sound wave propagates through a metal the microscopic electric fields due to the displacement of the ions can impart energy to electrons near the Fermi level, thereby removing energy from the wave.[26] Well below T_c the rate of attenuation is markedly lower in a superconductor than a normal metal, as one would expect for sound waves, where $\hbar\omega < 2\Delta$.

---

[26] See pages 275–277.

THE LONDON EQUATION

F. London and H. London first examined in a quantitative way the fundamental fact that a metal in the superconducting state permits no magnetic field in its interior.[27] Their analysis starts with the two-fluid model of Gorter and Casimir.[28] The only crucial assumption of this model that we shall use is that in a superconductor at temperature $T < T_c$, only a fraction $n_s(T)/n$ of the total number of conduction electrons are capable of participating in a supercurrent. The quantity $n_s(T)$ is known as the density of superconducting electrons. It approaches the full electronic density n as T falls well below T_c, but it drops to zero as T rises to T_c. The remaining fraction of electrons are assumed to constitute a "normal fluid" of density $n - n_s$ that cannot carry an electric current without normal dissipation. The normal current and the supercurrent are assumed to flow in parallel; since the latter flows with no resistance whatever, it will carry the entire current induced by any small transitory electric field, and the normal electrons will remain quite inert. Normal electrons are therefore ignored in the discussion that follows.

Suppose that an electric field momentarily arises within a superconductor. The superconducting electrons will be freely accelerated without dissipation so that their mean velocity $\mathbf{v}_s$ will satisfy[29]

$$m \frac{d\mathbf{v}_s}{dt} = -e\mathbf{E}. \tag{34.1}$$

Since the current density carried by these electrons is $\mathbf{j} = -e\mathbf{v}_s n_s$, Eq. (34.1) can be written as

$$\frac{d}{dt}\mathbf{j} = \frac{n_s e^2}{m}\mathbf{E}. \tag{34.2}$$

Note that the Fourier transform of (34.2) gives the ordinary AC conductivity for an electron gas of density n_s in the Drude model, Eq. (1.29), when the relaxation time τ becomes infinitely large:

$$\mathbf{j}(\omega) = \sigma(\omega)\mathbf{E}(\omega),$$
$$\sigma(\omega) = i\frac{n_s e^2}{m\omega}. \tag{34.3}$$

Substituting (34.2) into Faraday's law of induction,

$$\nabla \times \mathbf{E} = -\frac{1}{c}\frac{\partial \mathbf{B}}{\partial t}, \tag{34.4}$$

[27] F. London and H. London, *Proc. Roy. Soc.* (London), **A149**, 71 (1935), *Physica* **2**, 341 (1935); F. London, *Superfluids* vol. 1, Wiley, New York, 1954, and Dover, New York, 1954.

[28] The two-fluid model is also used to describe superfluid helium-4, and is described in both of the volumes by F. London, *Superfluids*, vols. 1 and 2, *Ibid.*

[29] We ignore band structure effects throughout this chapter and describe the electrons with free electron dynamics.

gives the following relation between current density and magnetic field:

$$\frac{\partial}{\partial t}\left(\nabla \times \mathbf{j} + \frac{n_s e^2}{mc}\mathbf{B}\right) = 0. \tag{34.5}$$

This relation, together with the Maxwell equation[30]

$$\nabla \times \mathbf{B} = \frac{4\pi}{c}\mathbf{j}, \tag{34.6}$$

determines the magnetic fields and current densities that can exist within a perfect conductor.

Note in particular that any static field **B** determines a static current density **j** through Eq. (34.6). Since any time-independent **B** and **j** are trivially solutions to (34.5), the two equations are consistent with an arbitrary static magnetic field. This is incompatible with the observed behavior of superconductors, which permit *no* fields in their interior. F. London and H. London discovered that this characteristic behavior of superconductors could be obtained by restricting the full set of solutions of (34.5) to those that obey[31]

$$\nabla \times \mathbf{j} = -\frac{n_s e^2}{mc}\mathbf{B}, \tag{34.7}$$

which is known as the London equation. Equation (34.5), which characterizes any medium that conducts electricity without dissipation, requires that $\nabla \times \mathbf{j} + (n_s e^2/mc)\mathbf{B}$ be independent of time; the more restrictive London equation, which specifically characterizes superconductors and distinguishes them from mere "perfect conductors," requires in addition that the time-independent value be zero.

---

[30] We assume that the rate of time variation is so slow that the displacement current can be neglected. We also take the field in (34.6) to be **B** rather than **H**; this is because **j** represents the mean *microscopic* current flowing in the superconductor. The field **H** would appear only if we represented **j** by an effective magnetization density satisfying $\nabla \times \mathbf{M} = \mathbf{j}/c$, and defined **H** in the usual way as $\mathbf{H} = \mathbf{B} - 4\pi\mathbf{M}$. In that case Eq. (34.6) would be replaced by the equation $\nabla \times \mathbf{H} = 0$. Given the definitions of **H** and **M**, this would be a completely equivalent formulation.

[31] This is a local relation; i.e., the current at the point **r** is related to the field at the same point. A. B. Pippard pointed out that, more generally, the current at **r** should be determined by the field within a neighborhood of the point **r** according to a relation of the form

$$\nabla \times \mathbf{j}(\mathbf{r}) = -\int d\mathbf{r}' K(\mathbf{r} - \mathbf{r}')\mathbf{B}(\mathbf{r}'),$$

where the kernel $K(\mathbf{r})$ is appreciable only for r less than a length ξ_0. The distance ξ_0 is one of several fundamental lengths characterizing a superconductor, all of which, unfortunately, are indiscriminately referred to as "the coherence length." In pure materials well below the critical temperature all such coherence lengths are the same, but near T_c or in materials with short impurity mean free paths, the "coherence length" may vary from one context to another. We shall avoid this tangle of coherence lengths by restricting our comments on its significance to the low-temperature pure case, where all coherence lengths agree. It turns out that in such cases the criterion for whether a superconductor is type I or type II is that the coherence length be large (type I) or small (type II) compared with the London penetration depth Λ (Eq. (34.9)).

The reason for replacing (34.5) by the more restrictive London equation is that the latter leads directly to the Meissner effect.[32] Equations (34.6) and (34.7) imply that

$$\nabla^2 \mathbf{B} = \frac{4\pi n_s e^2}{mc^2} \mathbf{B},$$

$$\nabla^2 \mathbf{j} = \frac{4\pi n_s e^2}{mc^2} \mathbf{j}. \tag{34.8}$$

These equations, in turn, predict that currents and magnetic fields in superconductors can exist only within a layer of thickness Λ of the surface, where Λ, known as the London penetration depth, is given by[33]

$$\Lambda = \left(\frac{mc^2}{4\pi n_s e^2}\right)^{1/2} = 41.9 \left(\frac{r_s}{a_0}\right)^{3/2} \left(\frac{n}{n_s}\right)^{1/2} \text{Å}. \tag{34.9}$$

Thus the London equation implies the Meissner effect, along with a specific picture of the surface currents that screen out the applied field. These currents occur within a surface layer of thickness 10^2–10^3 Å (well below T_c—the thickness can be considerably greater near the critical temperature, where n_s approaches zero). Within this same surface layer the field drops continuously to zero. These predictions are confirmed by the fact that the field penetration is not complete in superconducting films as thin as or thinner than the penetration depth Λ.

MICROSCOPIC THEORY: QUALITATIVE FEATURES

The microscopic theory of superconductivity was put forth by Bardeen, Cooper, and Schrieffer in 1957.[34] In a broad survey such as this we cannot develop the formalism necessary for an adequate description of their theory, and can only describe in a qualitative way the underlying physical principles and the major theoretical predictions.

The theory of superconductivity requires, to begin with, a net *attractive* interaction between electrons in the neighborhood of the Fermi surface. Although the direct electrostatic interaction is repulsive, it is possible for the ionic motion to "overscreen" the Coulomb interaction, leading to a net attraction.[35] We described this possibility

---

[32] We shall see below that the London equation is also suggested by certain features of the microscopic electronic ordering.

[33] Consider, for example, the case of a semiinfinite superconductor occupying the half space $x > 0$. Then Eq. (34.8) implies that the physical solutions decay exponentially:

$$B(x) = B(0)e^{-x/\Lambda}.$$

Other geometries are examined in Problem 2.

[34] J. Bardeen, L. N. Cooper, and J. R. Schrieffer, *Phys. Rev.* **108**, 1175 (1957). The theory is generally referred to as the BCS theory.

[35] Direct evidence that the ionic motion plays a role in establishing superconductivity is provided by the *isotope effect*: The critical temperature of different isotopes of a given metallic element varies from one isotope to another, frequently (but not always) as the inverse square root of the ionic mass. The fact that there is any dependence on ionic mass demonstrates that the ions cannot play a merely static role in the transition, but must be dynamically involved.

in Chapter 26, where we found, in a simplified model, that allowing the ions to move in response to motions of the electrons led to a net interaction between electrons with wave vectors **k** and **k**′ of the form[36]

$$v^{\text{eff}}_{\mathbf{k},\mathbf{k}'}(\mathbf{k}, \mathbf{k}') = \frac{4\pi e^2}{q^2 + k_0{}^2} \cdot \frac{\omega^2}{\omega^2 - \omega_q{}^2}, \tag{34.10}$$

where $\hbar\omega$ is the difference in electronic energies, k_0 is the Thomas-Fermi wave vector (17.50), $\mathbf{q}$ is the difference in electron wave vectors, and $\omega_{\mathbf{q}}$ is the frequency of a phonon of wave vector $\mathbf{q}$.

Thus screening by the ionic motion can yield a net attractive interaction between electrons with energies sufficiently close together (roughly, separated by less than $\hbar\omega_D$, a measure of the typical phonon energy). This attraction[37] underlies the theory of superconductivity.

Given that electrons whose energies differ by $O(\hbar\omega_D)$ can experience a net attraction, the possibility arises that such electrons might form bound pairs.[38] This would appear to be doubtful, since in three dimensions two particles must interact with a certain minimum strength to form a bound state, a condition that the rather limited effective attraction would be unlikely to meet. However, Cooper[39] argued that this apparently implausible possibility was made quite likely by the influence of the remaining $N - 2$ electrons on the interacting pair, through the Pauli exclusion principle.

Cooper considered the problem of two electrons with an attractive interaction that would be far too weak to bind them if they were in isolation. He demonstrated, however, that in the presence of a Fermi sphere of additional electrons[40] the exclusion principle radically altered the two-electron problem so that a bound state existed no matter how weak the attraction. Aside from indicating that the net attraction need not have a minimum strength to bind a pair, Cooper's calculation also indicated how the superconducting transition temperature could be so low compared with all other characteristic temperatures of the solid. This followed from the form of his solution, which gave a binding energy that was very small compared with the potential energy of attraction when the attraction was weak.

Cooper's argument applies to a single pair of electrons in the presence of a normal Fermi distribution of additional electrons. The theory of Bardeen, Cooper, and

---

36 See pages 518–519. That such an attraction was possible and might be the source of superconductivity was first emphasized by H. Fröhlich.

37 Any other mechanism leading to a net attractive interaction between electrons near the Fermi surface would also lead to a superconducting state at low enough temperature. However, no cases of superconductivity due to other mechanisms have been convincingly established in metals.

38 More generally, one might inquire into the possibility of n electrons binding together, but the weak interaction and the Pauli exclusion principle make the case $n = 2$ the most promising.

39 L. N. Cooper, *Phys. Rev.* **104**, 1189 (1956).

40 The degenerate Fermi distribution of additional electrons was taken to play no role other than prohibiting the two electrons from occupying any levels with wave vectors less than k_F. Thus the Cooper calculation was basically a two-electron calculation except that analysis was restricted to states built out of one-electron levels from which all plane waves with wave vectors less than k_F had been excluded. See Problem 4.

Schrieffer took an essential further step, constructing a ground state in which *all* electrons form bound pairs. This is a considerable extension of the Cooper model, for each electron now plays two roles: It provides the necessary restriction on allowed wave vectors (via the exclusion principle) that makes possible the binding of other pairs in spite of the weakness of the attraction; at the same time, the electron itself is participating in one of the bound pairs.

The BCS approximation to the electronic ground state wave function can be described as follows: Group the N conduction electrons into $N/2$ pairs[41] and let each pair be described by a bound-state wave function $\phi(\mathbf{r}s, \mathbf{r}'s')$, where $\mathbf{r}$ is the electronic position and s is the spin quantum number. Then consider the N-electron wave function that is just the product of $N/2$ *identical* such two-electron wave functions:

$$\Psi(\mathbf{r}_1 s_1, \ldots, \mathbf{r}_N s_N) = \phi(\mathbf{r}_1 s_1, \mathbf{r}_2 s_2) \ldots \phi(\mathbf{r}_{N-1} s_{N-1}, \mathbf{r}_N s_N). \qquad \textbf{(34.11)}$$

This describes a state in which all electrons are bound, in pairs, into identical two-electron states. However, it lacks the symmetry required by the Pauli principle. To construct a state that changes sign whenever the space and spin coordinates of any two electrons are interchanged, we must antisymmetrize the state (34.11). This leads to the BCS ground state:[42]

$$\Psi_{\text{BCS}} = \mathcal{A}\Psi. \qquad \textbf{(34.12)}$$

It may seem surprising that the state (34.12) satisfies the Pauli principle even though all the pair wave functions ϕ appearing in it are identical. Indeed, if we had constructed a product state analogous to (34.11) out of N identical *one*-electron levels, subsequent antisymmetrization would cause it to vanish. The fundamental requirement of antisymmetry implies that no one-electron level can be doubly occupied when the states are antisymmetrized products of one-electron levels. However, the requirement of antisymmetry does not imply a corresponding restriction on the occupancy of two-electron levels in states that are antisymmetrized products of two-electron levels.[43]

It can be demonstrated that if the state (34.12) is taken as a trial state in a variational estimate of the ground-state energy, then the optimum choice of ϕ must lead to a lower energy than the best choice of Slater determinants (i.e., the best independent electron trial function) for any attractive interaction, no matter how weak.

In the BCS theory the pair wave functions ϕ are taken to be singlet states;[44] i.e.,

---

[41] The odd electron (if N is odd) is of no significance in the limit of a large system.

[42] The antisymmetrizer $\mathcal{A}$ simply adds to the function it acts upon each of the $N! - 1$ other functions obtained by all possible permutations of the arguments, weighted with $+1$ or -1 according to whether the permutation is constructed out of an even or odd number of pair interchanges.

[43] This is why it is possible for a pair of fermions to behave statistically like a boson. Indeed, if the binding energy of each pair were so strong that the size of the pair were small compared with the interparticle spacing r_s, then the ground state would consist of $N/2$ bosons, all condensed into the same two-electron level. As we shall see, however, the size of a Cooper pair is large compared with r_s, and it can be highly misleading to view the Cooper pairs as independent bosons.

[44] If the pair states were triplets (spin 1) this would imply characteristic magnetic properties that are not observed. Triplet pairing has, however, been observed in liquid helium-3, a degenerate Fermi liquid that bears many resemblances to the electron gas in metals. See, for example, *Nobel Symposium* 24, *Collective Properties of Physical Systems*, B. Lundqvist and S. Lundqvist, eds., Academic Press, New York, 1973, pages 84–120.

the two electrons in the pair have opposite spin and the orbital part of the wave function, $\phi(\mathbf{r}, \mathbf{r}')$ is symmetric. If the pair state is chosen to be translationally invariant (ignoring possible complications due to the periodic potential of the lattice) so that $\phi(\mathbf{r}, \mathbf{r}')$ has the form $\chi(\mathbf{r} - \mathbf{r}')$, then one can write:

$$\chi(\mathbf{r} - \mathbf{r}') = \frac{1}{V} \sum_{\mathbf{k}} \chi_{\mathbf{k}} e^{i\mathbf{k}\cdot\mathbf{r}} e^{-i\mathbf{k}\cdot\mathbf{r}'}. \tag{34.13}$$

Thus χ can be viewed as a superposition of products of one-electron levels in each term of which electrons with equal and opposite wave vectors are paired.[45]

One result of the variational calculation of Ψ_{BCS} is that the spatial range ξ_0 of the pair wave function[46] is very large compared with the spacing between electrons r_s. A crude estimate of ξ_0 can be constructed as follows: The pair wave function $\phi(\mathbf{r})$ is presumably a superposition of one-electron levels with energies within $O(\Delta)$ of $\mathcal{E}_F$, since outside of that energy range tunneling experiments indicate that the one-electron level density is little altered from the form it has in a normal metal. The spread in momenta of the one-electron levels making up the pair state is therefore fixed by the condition

$$\Delta = \delta\mathcal{E} = \delta\left(\frac{p^2}{2m}\right) = \left(\frac{p_F}{m}\right)\delta p \approx v_F\,\delta p. \tag{34.14}$$

The spatial range of $\phi(\mathbf{r})$ is thus of order

$$\xi_0 \sim \frac{\hbar}{\delta p} \sim \frac{\hbar v_F}{\Delta} \sim \frac{1}{k_F}\frac{\mathcal{E}_F}{\Delta}. \tag{34.15}$$

Since $\mathcal{E}_F$ is typically 10^3–10^4 times Δ, and k_F is of order 10^8 cm^{-1}, ξ_0 is typically 10^3 Å.

Thus within the region occupied by any given pair will be found the centers of many (millions, or more) pairs. This is a very crucial feature of the superconducting state: The pairs cannot be thought of as independent particles, but are spatially interlocked in a very intricate manner, which is essential to the stability of the state.

The above description summarizes the essential features of the electronic ground state in a superconductor. To describe the excited states, or the thermal or transport properties of a superconductor, one must resort to more sophisticated formalisms. We shall not go into these here, except to emphasize that the underlying physical picture remains that of a system of paired electrons. In nonequilibrium processes the pair state can be more complex. At nonzero temperatures a fraction of the pairs are thermally dissociated, and the density of superconducting electrons n_s is determined by the fraction that remain paired. Furthermore, because of the intricate self-consistent nature of the pairing, the thermal dissociation of some of the pairs at nonzero

---

[45] This aspect of the ground state is often emphasized, and the assertion is made that electrons with opposite spins and wave vectors are bound in pairs. This is no more (or less) accurate than the assertion that any translationally invariant bound state of two identical particles pairs them with equal and opposite momenta; i.e., the assertion correctly focuses attention on the fact that the total momentum of the bound pair is zero, but it misleadingly distracts attention from the fact that the state is a superposition of such pairs, and therefore localized in the relative position coordinate (unlike a single product of plane waves).

[46] In pure superconductors well below T_c this turns out to be the same as the coherence length, described in footnote 31. It is therefore denoted by the same symbol.

temperatures results in a temperature dependence in the characteristic properties (for example, the range of the pair function) of those pairs that remain bound. As T rises through T_c all pairs become dissociated, and the ground state reverts continuously back to the normal ground state of the independent electron approximation.

QUANTITATIVE PREDICTIONS OF THE ELEMENTARY MICROSCOPIC THEORY

In its simplest form the BCS theory makes two gross oversimplifications in the basic Hamiltonian that describes the conduction electrons:

1. The conduction electrons are treated in the free electron approximation; band structure effects are ignored.
2. The rather complicated net attractive interaction[47] (34.10) between electrons near the Fermi energy is further simplified to an effective interaction V. The matrix element of V between a two-electron state with electronic wave vectors $\mathbf{k}_1$ and $\mathbf{k}_2$, and another with wave vectors $\mathbf{k}_3$ and $\mathbf{k}_4$, is taken in a volume Ω to be

$$\begin{aligned} \langle \mathbf{k}_1\mathbf{k}_2|V|\mathbf{k}_3\mathbf{k}_4\rangle &= -V_0/\Omega, \text{ when } \mathbf{k}_1 + \mathbf{k}_2 = \mathbf{k}_3 + \mathbf{k}_4, \\ &\qquad |\mathcal{E}(\mathbf{k}_i) - \mathcal{E}_F| < \hbar\omega,\ i = 1, \ldots, 4, \\ &= 0, \text{ otherwise.} \end{aligned} \tag{34.16}$$

The restriction on wave vectors is required for any translationally invariant potential; the significant aspect of the interaction (34.16) is the attraction experienced whenever all four free electron energies are within an amount $\hbar\omega$ (usually taken to be of order $\hbar\omega_D$) of the Fermi energy.

Equation (34.16) is a gross oversimplification of the actual net interaction, and any results depending on its detailed features are to be viewed with suspicion. Fortunately, the theory predicts a number of relations from which the two phenomenological parameters V_0 and $\hbar\omega$ are absent. These relations are rather well obeyed by a large class of superconductors, with certain notable exceptions (such as lead and mercury). Even these exceptions, known as "strong coupling superconductors" have been convincingly brought into the more general framework of the BCS theory, provided that the simplifications inherent in the approximate interaction (34.16) are abandoned, along with certain other overly simple representations of the effects of phonons.[48]

From the model Hamiltonian (34.16), the BCS theory deduces the following major equilibrium predictions:

---

[47] One must not forget that even Eq. (34.10) is a comparatively crude representation of the detailed dynamic interaction induced among the electrons by the phonons. In so-called strong coupling superconductors (see below) even Eq. (34.10) is inadequate.

[48] In the theory of strong-coupling superconductors one treats the full electron-phonon system, without at the start trying to eliminate the phonons in favor of an effective interaction of the form (34.16), or even (34.10). As a result, the net interaction between electrons is more complicated and no longer instantaneous but retarded. Furthermore, the lifetimes due to electron-phonon scattering of the electronic levels within $\hbar\omega_D$ from the Fermi level may be so short that the picture of well-defined one-electron levels out of which pairs are formed also requires modification.

Critical Temperature

In zero magnetic field, superconducting ordering sets in at a critical temperature given by

$$k_B T_c = 1.13\hbar\omega e^{-1/N_0 V_0}, \tag{34.17}$$

where N_0 is the density of electronic levels for a single spin population in the normal metal[49] and ω and V_0 are the parameters of the model Hamiltonian (34.16). Because of the exponential dependence, the effective coupling V_0 cannot be determined precisely enough to permit very accurate computations of the critical temperature from (34.17). However, this same exponential dependence accounts for the very low critical temperatures (typically one to three orders of magnitude below the Debye temperature), for although $\hbar\omega$ is of order $k_B\Theta_D$, the strong dependence on $N_0 V_0$ can lead to the observed range of critical temperatures with $N_0 V_0$ in the range from 0.1 to 0.5, i.e., with $V_0 n$ in the range[50] $0.1\mathcal{E}_F$ to $0.5\mathcal{E}_F$. Note also that no matter how weak the coupling V_0, the theory predicts a transition, though the transition temperature (34.17) may be unobservably low.

Energy Gap

A formula similar to (34.17) is predicted for the zero-temperature energy gap:

$$\Delta(0) = 2\hbar\omega e^{-1/N_0 V_0}. \tag{34.18}$$

The ratio of (34.18) to (34.17) gives a fundamental formula independent of the phenomenological parameters:

$$\frac{\Delta(0)}{k_B T_c} = 1.76. \tag{34.19}$$

This result appears to hold for a large number of superconductors to within about 10 percent (Table 34.3). Those for which it fails (for example, lead and mercury, where the discrepancy is closer to 30 percent) tend systematically to deviate from other predictions of the simple theory as well, and can be brought closer into line with theoretical predictions by using the more elaborate analysis of the strong-coupling theory.

The elementary theory also predicts that near the critical temperature (in zero field) the energy gap vanishes according to the universal law[51]

$$\frac{\Delta(T)}{\Delta(0)} = 1.74\left(1 - \frac{T}{T_c}\right)^{1/2}, \qquad T \approx T_c. \tag{34.20}$$

---

[49] The quantity N_0 is simply $g(\mathcal{E}_F)/2$. This notation for the density of levels is widely used in the literature of superconductivity.

[50] The quantity N_0 is of order $n/\mathcal{E}_F$. See Eq. (2.65).

[51] Equation (34.20) is a characteristic result of mean field theory (cf. the prediction of mean field theory that the spontaneous magnetization vanishes as $(T_c - T)^{1/2}$, Chapter 33, Problem 6). Mean field theory is known to be wrong in ferromagnets sufficiently near the critical temperature. Presumably it fails sufficiently near T_c in a superconductor as well, but arguments have been advanced that the region inside of which mean field theory fails is exceedingly small (typically $(T_c - T)/T_c \approx 10^{-8}$). Superconductors provide a rare example of a phase transition that is well described by a mean field theory quite near the critical point.

Table 34.3
MEASURED VALUES[a] OF $2\Delta(0)/k_B T_c$

ELEMENT	$2\Delta(0)/k_B T_c$
Al	3.4
Cd	3.2
Hg (α)	4.6
In	3.6
Nb	3.8
Pb	4.3
Sn	3.5
Ta	3.6
Tl	3.6
V	3.4
Zn	3.2

[a] $\Delta(0)$ is taken from tunneling experiments. Note that the BCS value for this ratio is 3.53. Most of the values listed have an uncertainty of ± 0.1.

Source: R. Mersevey and B. B. Schwartz, *Superconductivity*, R. D. Parks, ed., Dekker, New York, 1969.

Critical Field

The elementary BCS prediction for $H_c(T)$ is often expressed in terms of the deviation from the empirical law:[52]

$$\frac{H_c(T)}{H_c(0)} \approx 1 - \left(\frac{T}{T_c}\right)^2. \tag{34.21}$$

The quantity $[H_c(T)/H_c(0)] - [1 - (T/T_c)^2]$ is shown for several superconductors in Figure 34.8, along with the BCS prediction. The departure is small in all cases, but note that the strong-coupling superconductors lead and mercury are more out of line than the others.

Specific Heat

At the critical temperature (in zero magnetic field) the elementary BCS theory predicts a discontinuity in the specific heat that can also be put in a form independent of the parameters in the model Hamiltonian (34.16):[53]

$$\left.\frac{c_s - c_n}{c_n}\right|_{T_c} = 1.43. \tag{34.22}$$

[52] The BCS prediction can again be cast in a parameter independent form. At low temperatures it is $H_c(T)/H_c(0) \approx 1 - 1.06(T/T_c)^2$, while near T_c it is $H_c(T)/H_c(0) = 1.74[1 - (T/T_c)]$.

[53] A specific heat discontinuity at T_c is also a characteristic mean field theory result. Presumably, *very* close to T_c the specific heat may diverge.

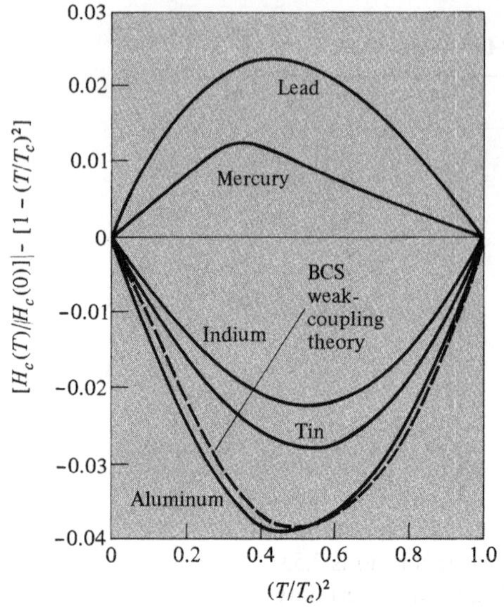

Figure 34.8
The deviation from the crude empirical relation $H_c(T)/H_c(0) \approx 1 - [T/T_c]^2$, as measured in several metals and as predicted by the simple BCS theory. Deviations from the simple BCS prediction are more pronounced in the "strong-coupling" superconductors, lead and mercury. (J. C. Swihart et al., *Phys. Rev. Lett.* **14**, 106 (1965).)

The agreement of this prediction with experiment is again good to about 10 percent except for the strong-coupling superconductors (Table 34.4).

The low-temperature electronic specific heat can also be cast in a parameter independent form,

$$\frac{c_s}{\gamma T_c} = 1.34 \left(\frac{\Delta(0)}{T}\right)^{3/2} e^{-\Delta(0)/T}, \tag{34.23}$$

where γ is the coefficient of the linear term in the specific heat of the metal in the normal state (Eq. (2.80)). Note the exponential drop, on a scale determined by the energy gap $\Delta(0)$.

MICROSCOPIC THEORY AND THE MEISSNER EFFECT

In the presence of a magnetic field, diamagnetic currents will flow in the equilibrium state of a metal, whether it is normal or superconducting, although the currents are vastly greater in a superconductor. In a free electron model the current will be determined to first order in the field by an equation of the form[54]

$$\nabla \times \mathbf{j}(\mathbf{r}) = -\int d\mathbf{r}'\, K(\mathbf{r} - \mathbf{r}')\mathbf{B}(\mathbf{r}'). \tag{34.24}$$

If it happens that the kernel $K(\mathbf{r})$ satisfies

$$\int d\mathbf{r}\, K(\mathbf{r}) = K_0 \neq 0, \tag{34.25}$$

then in the limit of magnetic fields that vary slowly over the range of $K(\mathbf{r})$, Eq. (34.24) reduces to

$$\nabla \times \mathbf{j}(\mathbf{r}) = -K_0\mathbf{B}(\mathbf{r}), \tag{34.26}$$

[54] This is the same kernel K mentioned in footnote 31.

Table 34.4
MEASURED VALUES OF THE RATIO[a]
$[(c_s - c_n)/c_n]_{T_c}$

ELEMENT	$\left[\dfrac{c_s - c_n}{c_n}\right]_{T_c}$
Al	1.4
Cd	1.4
Ga	1.4
Hg	2.4
In	1.7
La (HCP)	1.5
Nb	1.9
Pb	2.7
Sn	1.6
Ta	1.6
Tl	1.5
V	1.5
Zn	1.3

[a] The simple BCS prediction is $[(c_s - c_n)/c_n]_{T_c} = 1.43$.
Source: R. Mersevey and B. B. Schwartz, *Superconductivity*, R. D. Parks, ed., Dekker, New York, 1969.

which is nothing but the London equation (34.7), with n_s given by

$$n_s = \frac{mc}{e^2} K_0. \tag{34.27}$$

Since the London equation implies the Meissner effect, it follows that in normal metals the constant K_0 must vanish. To demonstrate that BCS theory implies the Meissner effect, one calculates the kernel $K(\mathbf{r})$ by perturbation theory in the applied field, and verifies explicitly that $K_0 \neq 0$.

The actual demonstration that $K_0 \neq 0$ is a fairly complex application of BCS theory. However, a more intuitive explanation for the London equation was offered at the time the equation was first put forth, by the Londons. This explanation can be made somewhat more compelling through a phenomenological theory of V. L. Ginzburg and L. D. Landau,[55] which, though proposed seven years prior to the BCS theory, can be quite naturally described in terms of some of the fundamental notions of the microscopic theory.

THE GINZBURG–LANDAU THEORY

Ginzburg and Landau asserted that the superconducting state could be characterized by a complex "order parameter" $\psi(\mathbf{r})$, which vanishes above T_c and whose

[55] V. L. Ginzburg and L. D. Landau, *Zh. Eksp. Teor. Fiz.* **20**, 1064 (1950).

magnitude measures the degree of superconducting order at position **r** below T_c.[56] From the perspective of the BCS theory the order parameter can be viewed as a one-particle wave function describing the position of the center of mass of a Cooper pair. Since all Cooper pairs are in the same two-electron state, a single function suffices. Because the order parameter does not refer to the relative coordinate of the two electrons in the pair, the description of a superconductor in terms of $\psi(\mathbf{r})$ is valid only for phenomena that vary slowly on the scale[57] of the dimensions of the pair.

In the ground state of the superconductor each pair is in a translationally invariant state that does not depend on the center of mass coordinate; i.e., the order parameter is a constant. The order parameter develops interesting structure when currents flow, or when an applied field appears. A fundamental assumption of the Ginzburg-Landau theory is that the current flowing in a superconductor characterized by order parameter $\psi(\mathbf{r})$ in the presence of a magnetic field given by the vector potential $\mathbf{A}(\mathbf{r})$ is given by the ordinary quantum-mechanical formula for the current due to a particle of charge $-2e$ and mass $2m$ (i.e., the Cooper pair itself) described by a wave function $\psi(\mathbf{r})$, namely

$$\mathbf{j} = -\frac{e}{2m}\left[\psi^*\left\{\left(\frac{\hbar}{i}\nabla + \frac{2e}{c}\mathbf{A}\right)\psi\right\} + \left\{\left(\frac{\hbar}{i}\nabla + \frac{2e}{c}\mathbf{A}\right)\psi\right\}^*\psi\right]. \tag{34.28}$$

The London equation (34.7) follows from (34.28), provided one also assumes that the significant spatial variation of the order parameter $\psi = |\psi|e^{i\phi}$ is through the phase ϕ, and not the magnitude $|\psi|$. Since the magnitude of the order parameter measures the degree of superconducting ordering, this assumption restricts consideration to disturbances in which the density of Cooper pairs is not appreciably altered from its uniform thermal equilibrium value. This should be the case in phenomena in which the pairs can flow, but not accumulate or be destroyed.[58]

Given this assumption, the relation (34.28) for the current simplifies to

$$\mathbf{j} = -\left[\frac{2e^2}{mc}\mathbf{A} + \frac{e\hbar}{m}\nabla\phi\right]|\psi|^2. \tag{34.29}$$

---

[56] It is sometimes helpful to bear in mind an analogy to a Heisenberg ferromagnet, where the order parameter may be viewed as the mean value of the local spin $\mathbf{s}(\mathbf{r})$. Above T_c, $\mathbf{s}(\mathbf{r})$ vanishes; below, it gives the local value of the spontaneous magnetization. In the ground state $\mathbf{s}(\mathbf{r})$ is independent of **r** (and, correspondingly, in a uniform superconductor that carries no current, $\psi(\mathbf{r})$ is constant). However, one can consider more complicated configurations of the ferromagnet, in which, for example, the magnetization is constrained by applied fields to point in different directions at two ends of a bar. A position-dependent $\mathbf{s}(\mathbf{r})$ is also useful in investigating aspects of domain structure. In a similar way, a position-dependent $\psi(\mathbf{r})$ is used to investigate current-carrying configurations of a superconductor.

[57] Well below T_c this is just the length ξ_0, described on page 742.

[58] More generally, when the degree of superconducting order does have significant spatial variation, one must use a second Ginzburg-Landau equation in conjunction with (34.28) to determine both ψ and the current. The second equation relates the spatial rate of change of the order parameter to the vector potential, and bears a (to some extent deceptive) resemblance to a one-particle Schrödinger equation. The use of the full set of Ginzburg-Landau equations is essential, for example, in the description of vortices in type II superconductors, for in the core of the vortex the magnitude of the order parameter drops rapidly to zero, yielding a region in which the magnetic flux is appreciable.

Since the curl of any gradient vanishes, and since $|\psi|^2$ is essentially constant, we immediately deduce the London equation (34.7) provided that we identify the superfluid density n_s with $2|\psi|^2$, which is reasonable in view of the interpretation of ψ as a wave function characterizing particles of charge $2e$.

FLUX QUANTIZATION

Equation (34.29) has an even more striking implication than the London equation. Consider a superconductor in the shape of a ring (Figure 34.9). If we integrate (34.29)

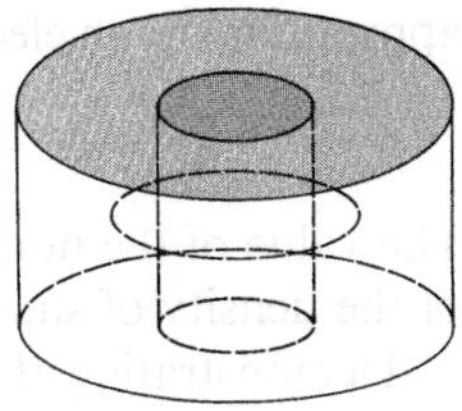

Figure 34.9
A ring of superconducting material, showing a path encircling the aperture and lying well within the interior of the superconductor.

over a path deep inside the superconducting material enclosing the hole in the ring, then, since appreciable currents can flow only near the surface of the superconductor, we find that

$$0 = \oint \mathbf{j} \cdot d\boldsymbol{\ell} = \oint \left(\frac{2e^2}{mc} \mathbf{A} + \frac{e\hbar}{m} \nabla \phi \right) \cdot d\boldsymbol{\ell}. \tag{34.30}$$

Stokes theorem gives

$$\int \mathbf{A} \cdot d\boldsymbol{\ell} = \int \nabla \times \mathbf{A} \cdot d\mathbf{S} = \int \mathbf{B} \cdot d\mathbf{S} = \Phi, \tag{34.31}$$

where Φ is the flux enclosed by the ring.[59] Furthermore, since the order parameter is single-valued, its phase must change by 2π times an integer n when the ring is encircled:

$$\oint \nabla \phi \cdot d\boldsymbol{\ell} = \Delta \phi = 2\pi n. \tag{34.32}$$

Combining these results, we conclude that the magnetic flux enclosed by the ring must be quantized:

$$|\Phi| = \frac{nhc}{2e} = n\Phi_0. \tag{34.33}$$

The quantity $\Phi_0 = hc/2e = 2.0679 \times 10^{-7}$ gauss-cm^2 is known as the fluxoid, or flux quantum. Flux quantization has been observed and is one of the most compelling pieces of evidence for the validity of the description of a superconductor by means of a complex order parameter.[60]

---

[59] Since the magnetic field cannot penetrate the superconducting material, the enclosed flux will not depend on the choice of path, as long as the path is deep inside the material.

[60] B. S. Deaver and W. M. Fairbank, *Phys. Rev. Lett.* **7**, 43 (1961); R. Doll and M. Näbauer, *Phys. Rev. Lett.* **7**, 51 (1961). The Ginzburg-Landau theory also predicts, and experiments have confirmed, that each vortex in a type II superconductor contains a single quantum of magnetic flux.

MICROSCOPIC THEORY AND PERSISTENT CURRENTS

The property for which superconductors are named is unfortunately one of the most difficult to extract from the microscopic theory. In a sense perfect conductivity is implied by the Meissner effect, for macroscopic currents must flow without dissipation in order to screen macroscopic magnetic fields in equilibrium. Indeed, the direct microscopic derivation of persistent currents is not dissimilar to that of the Meissner effect. One calculates to linear order the current induced by an electric field, and demonstrates that there is a piece in the AC conductivity that has the form (34.3) appropriate to an electron gas without dissipation. For this it suffices to prove that[61]

$$\lim_{\omega \to 0} \omega \operatorname{Im} \sigma(\omega) \neq 0. \tag{34.34}$$

The value of the nonzero constant determines, by comparison with (34.3), the value of the density of superconducting electrons n_s.

Demonstrating that (34.34) holds is more complicated than demonstrating the condition (34.25) for the Meissner effect, because it is essential to include the effects of scattering: If there were no scattering any metal would obey (34.34), but even in the absence of scattering a calculation of diamagnetism in a normal metal reveals no Meissner effect. However, the calculation has been done[62] and the value of n_s deduced from the low-frequency conductivity is found to agree with that deduced from the calculation of the Meissner effect. The calculation is unfortunately quite formal, and does not provide any intuitive explanation for the remarkable fact that none of the familiar scattering mechanisms are effective in degrading a current, once it has been established in a superconducting metal. Such an intuitive explanation is at least suggested by the following line of reasoning:[63]

Suppose that we use an electric field to establish a current in a metal, and then turn off the field and ask how the current can decay. In a normal metal the current can be degraded one electron at a time; i.e., scattering processes can reduce the total momentum of the electronic system by a series of collisions of individual electrons with impurities, phonons, imperfections, etc., each of which, on the average, drives the momentum distribution back to its equilibrium form, in which the total current vanishes. When a current is established in a superconductor, all the Cooper pairs move together: The single two-electron state describing each of the pairs is one with

---

[61] Equation (34.34) is not dissimilar in structure to the condition (34.25) for a Meissner effect. The integral over all space of the kernel K is equal to the $k = 0$ limit of its spatial Fourier transform. In both cases one must establish the failure of a certain electromagnetic response function to vanish, in an appropriate long-wavelength or low-frequency limit.

[62] See, for example, A. A. Abrikosov, L. P. Gorkov, and I. E. Dzyaloshinski, *Methods of Quantum Field Theory in Statistical Physics*, Prentice-Hall, Englewood Cliffs, N.J., 1963, pp. 334–341.

[63] A variety of "intuitive" arguments have been offered on this point, many of them quite spurious. There is, for example, an argument (based on an old argument of Landau's to explain superfluidity in ^{4}He) purporting to deduce persistent currents from the existence of a gap in the one-electron excitation spectrum. But this merely explains why currents cannot be degraded by one-electron excitations, leaving open the possibility of degrading the current pair by pair. The argument we indicate here can be found under many guises, associated with the notions of "rigidity of the wave function," "off-diagonal long-range order," or "long-range phase coherence."

a nonzero center of mass momentum.[64] One might expect that such a current could be degraded by single-pair collisions, analogous to the one-electron collisions in a normal metal, in which individual pairs have their center of mass momenta reduced back to zero by collisions. Such a suggestion, however, fails to take into account the delicate interdependence of the pairs.[65] Essential to the stability of the paired state is the fact that all the other pairs exist and are described by identical pair wave functions. Thus one cannot change the pair wave functions individually without destroying the paired state altogether, at an enormous cost in free energy.

The supercurrent-degrading transition that is least costly in free energy depends, in general, on the geometry of the specimen, but it usually requires destruction of the pairing in some macroscopic portion of the sample. Such processes are possible, but the cost in free energy will in general be so large that the lifetime of the supercurrent is infinite on any practical time scale.[66]

SUPERCURRENT TUNNELING; THE JOSEPHSON EFFECTS

We have described (page 735) the tunneling of single electrons from a superconducting metal through a thin insulating barrier into a normal metal, and indicated how measurements of the tunneling current give information about the density of one-electron levels in the superconductor. Measurements can also be made of the tunneling currents when both metals are superconductors, and the results are fitted well by assuming that both metals have one-electron-level densities of the form predicted by the BCS theory (Figure 34.10). In 1962 Josephson[67] predicted that in addition to this "normal tunneling" of single electrons there should be another component to the tunneling current carried by paired electrons: Provided that the barrier is not too thick, electron pairs can traverse the junction from one superconductor to the other without dissociating.

One immediate consequence of this observation is that a supercurrent of pairs should flow across the junction in the absence of any applied electric field (the DC Josephson effect). Because the two superconductors are only weakly coupled (i.e., because the paired electrons must cross a gap of nonsuperconducting material), the typical tunneling current across the junction will be far smaller than typical critical currents for single specimens.

---

[64] That a supercurrent is very well described by regarding all paired electrons as occupying a single quantum state is confirmed by the absence of thermoelectric effects in a superconductor (see page 730). If a supercurrent resembled the disorderly flow of electrons that constitutes a current in a normal metal, then there would be an accompanying thermal current (Peltier effect).

[65] Recall (page 742) that within the radius of a given pair will be found the centers of millions of other pairs.

[66] Thus, in principle, supercurrent-carrying states are only metastable. For suitable geometries (i.e., specimens that are very small in one spatial dimension or more) the fluctuations necessary to destroy a supercurrent need not be overwhelmingly improbable, and one can observe the decay of the "persistent current." A very appealing microscopic picture of such processes has been given by V. Ambegaokar and J. S. Langer, *Phys. Rev.* **164**, 498 (1967).

[67] B. D. Josephson, *Phys. Lett.* **1**, 251 (1962). See also the articles by Josephson and Mercereau in *Superconductivity*, R. D. Parks, ed., Dekker, New York, 1969.

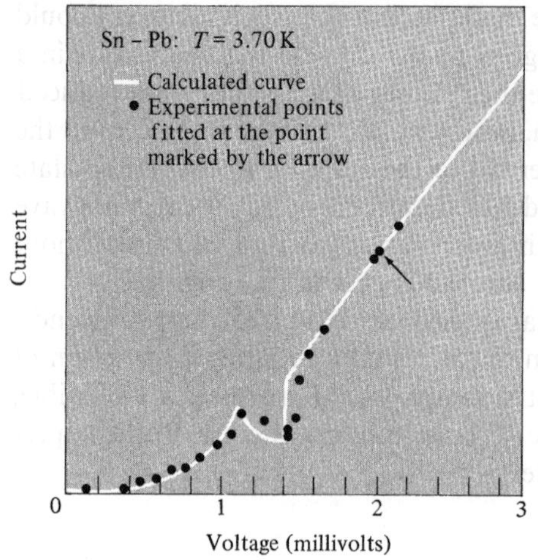

Figure 34.10
Normal tunneling current between two superconductors (tin and lead). The solid curve is the prediction of BCS theory. (S. Shapiro et al., *IBM J. Res. Develop.* **6**, 34 (1962).)

Josephson predicted a variety of further effects by assuming that the superconducting ordering on both sides of the junction could be described by a single order parameter $\psi(\mathbf{r})$. He showed that the tunneling current would be determined by the change in phase of the order parameter across the junction. Furthermore, by using gauge invariance to relate the phase of the order parameter to the value of an applied vector potential, he was able to show that the tunneling current would depend in a sensitive way upon any magnetic field present in the junction. Specifically, the tunneling current in the presence of a magnetic field should have the form

$$I = I_0 \frac{\sin \pi\Phi/\Phi_0}{\pi\Phi/\Phi_0}, \tag{34.35}$$

where Φ is the total magnetic flux in the junction, Φ_0 is the flux quantum $hc/2e$, and I_0 depends on the temperature and the structure of the junction, but not upon magnetic field. Such effects were subsequently observed (Figure 34.11), thereby providing impressive further confirmation of the fundamental validity of the order parameter description of the superconducting state, as well as vindicating Josephson's highly imaginative application of the theory.[68]

Similar considerations led Josephson to predict further that if a DC electric potential were applied across such a junction the induced supercurrent would be oscillatory (the AC Josephson effect) with angular frequency

$$\omega_J = \frac{2eV}{\hbar}. \tag{34.36}$$

This remarkable result—that a DC electric field should induce an alternating current—was not only observed, but has been made the basis for highly accurate techniques for measuring voltages as well as the precise value of the fundamental constant e/h.[69]

---

[68] The value of the flux quantum is exceedingly small, making the effect of practical importance as an highly sensitive way to measure magnetic field strengths.

[69] W. H. Parker et al., *Phys. Rev.* **177**, 639 (1969).

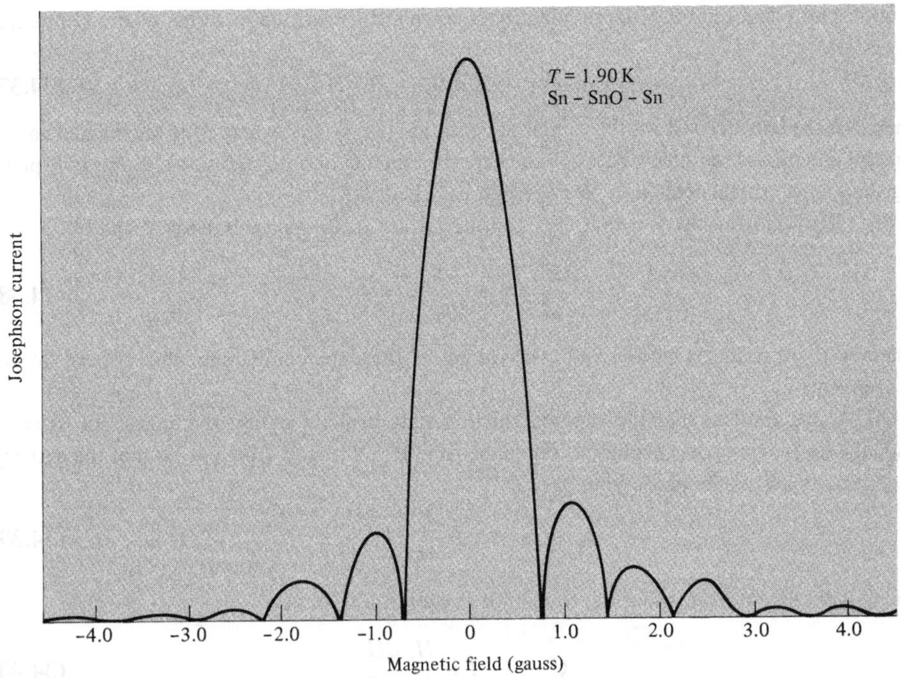

Figure 34.11
Josephson tunneling current as a function of magnetic field in an Sn-SnO-Sn junction. (R. C. Jaklevic, quoted in James E. Mercereau, *Superconductivity*, vol. 1, R. D. Parks, ed., Dekker, New York, 1969, p. 393.)

It is fitting that this volume should conclude with this sketchy and tantalizingly incomplete survey of superconductivity. The rich and highly original theories, both microscopic and phenomenological, that have successfully evolved over the past two decades to cope with superconducting phenomena are indicative of the fundamental health and future promise of the contemporary theory of solids. In spite of the novelty and, at times, the forbidding complexity of the concepts upon which the theory of superconductivity is based, one must not forget that it rests on a broad foundation extending over almost all the important areas of the theory of solids we have examined in earlier chapters. In no other subject are the two fundamental branches of solid state physics—the dynamics of electrons and the vibrations of the lattice of ions—so intimately fused, with such spectacular consequences.

PROBLEMS

1. *Thermodynamics of the Superconducting State*

The equilibrium state of a superconductor in a uniform magnetic field is determined by the temperature T and the magnitude of the field H. (Assume that the pressure P is fixed, and that the superconductor is a long cylinder parallel to the field so that demagnetization effects are unimportant.) The thermodynamic identity is conveniently written in terms of the Gibbs free energy G:

$$dG = -S\,dT - \mathfrak{M}\,dH \tag{34.37}$$

where S is the entropy and $\mathfrak{M}$, the total magnetization ($\mathfrak{M} = MV$, where M is the magnetization density). The phase boundary between the superconducting and normal states in the H-T plane is given by the critical field curve, $H_c(T)$ (Figure 34.3).

(a) Deduce, from the fact that G is continuous across the phase boundary, that

$$\frac{dH_c(T)}{dT} = \frac{S_n - S_s}{\mathfrak{M}_s - \mathfrak{M}_n}, \tag{34.38}$$

(where the subscripts s and n indicate values in the superconducting and normal phase, respectively).

(b) Using the fact that the superconducting state displays perfect diamagnetism ($B = 0$), while the normal state has negligible diamagnetism ($M \approx 0$), show from (34.38) that the entropy discontinuity across the phase boundary is

$$S_n - S_s = -\frac{V}{4\pi} H_c \frac{dH_c}{dT}, \tag{34.39}$$

and thus the latent heat, when the transition occurs in a field, is

$$Q = -TV \frac{H_c}{4\pi} \frac{dH_c}{dT}. \tag{34.40}$$

(c) Show that when the transition occurs at zero field (i.e., at the critical point) there is a specific heat discontinuity given by

$$(c_p)_n - (c_p)_s = -\frac{T}{4\pi}\left(\frac{dH_c}{dT}\right)^2. \tag{34.41}$$

2. *The London Equation for a Superconducting Slab*

Consider an infinite superconducting slab bounded by two parallel planes perpendicular to the y-axis at $y = \pm d$. Let a uniform magnetic field of strength H_0 be applied along the z-axis.

(a) Taking as a boundary condition that the parallel component of $\mathbf{B}$ be continuous at the surface, deduce from the London equation (34.7) and the Maxwell equation (34.6) that within the superconductor

$$\mathbf{B} = B(y)\hat{\mathbf{z}}, \quad B(y) = H_0 \frac{\cosh(y/\Lambda)}{\cosh(d/\Lambda)}. \tag{34.42}$$

(b) Show that the diamagnetic current density flowing in equilibrium is

$$\mathbf{j} = j(y)\hat{\mathbf{x}}, \quad j(y) = \frac{c}{4\pi\Lambda} H_0 \frac{\sinh(y/\Lambda)}{\cosh(d/\Lambda)}.$$

(c) The magnetization density at a point within the slab is $\mathbf{M(y)} = (\mathbf{B(y)} - \mathbf{H}_0)/4\pi$. Show that the average magnetization density (averaged over the thickness of the slab) is

$$\overline{M} = -\frac{H_0}{4\pi}\left(1 - \frac{\Lambda}{d}\tanh\frac{d}{\Lambda}\right), \tag{34.43}$$

and give the limiting form for the susceptibility when the slab is thick ($d \gg \Lambda$) and thin ($d \ll \Lambda$).

3. *Critical Current in a Cylindrical Wire*

A current of I amperes flows in a cylindrical superconducting wire of radius r cm. Show that when the field produced by the current immediately outside the wire is H_c (in gauss), then

$$I = 5rH_c. \tag{34.44}$$

4. *The Cooper Problem*

Consider a pair of electrons in a singlet state, described by the symmetric spatial wave function

$$\phi(\mathbf{r} - \mathbf{r}') = \int \frac{d\mathbf{k}}{(2\pi)^3} \chi(\mathbf{k}) e^{i\mathbf{k}\cdot(\mathbf{r}-\mathbf{r}')}. \tag{34.45}$$

In the momentum representation the Schrödinger equation has the form

$$\left(E - 2\frac{\hbar^2 k^2}{2m}\right)\chi(\mathbf{k}) = \int \frac{d\mathbf{k}'}{(2\pi)^3} V(\mathbf{k}, \mathbf{k}')\chi(\mathbf{k}'). \tag{34.46}$$

We assume that the two electrons interact in the presence of a degenerate free electron gas, whose existence is felt only via the exclusion principle: Electron levels with $k < k_F$ are forbidden to each of the two electrons, which gives the constraint:

$$\chi(\mathbf{k}) = 0, \quad k < k_F. \tag{34.47}$$

We take the interaction of the pair to have the simple attractive form (cf. Eq. (34.16)):

$$\begin{aligned} V(\mathbf{k}_1\, \mathbf{k}_2) &\equiv -V, \qquad \mathcal{E}_F \leqslant \frac{\hbar^2 k_i^2}{2m} \leqslant \mathcal{E}_F + \hbar\omega, \quad i = 1, 2; \\ &= 0, \qquad \text{otherwise,} \end{aligned} \tag{34.48}$$

and look for a bound-state solution to the Schrödinger equation (34.46) consistent with the constraint (34.47). Since we are considering only one-electron levels which in the absence of the attraction have energies in excess of $2\mathcal{E}_F$, a bound state will be one with energy E less than $2\mathcal{E}_F$, and the binding energy will be

$$\Delta = 2\mathcal{E}_F - E. \tag{34.49}$$

(a) Show that a bound state of energy E exists provided that

$$1 = V \int_{\mathcal{E}_F}^{\mathcal{E}_F + \hbar\omega} \frac{N(\mathcal{E})\, d\mathcal{E}}{2\mathcal{E} - E}, \tag{34.50}$$

where $N(\mathcal{E})$ is the density of one-electron levels of a given spin.

(b) Show that Eq. (34.50) has a solution with $E < 2\mathcal{E}_F$ for arbitrarily weak V, provided that $N(\mathcal{E}_F) \neq 0$. (Note the crucial role played by the exclusion principle: If the lower cutoff were not $\mathcal{E}_F$, but 0, then since $N(0) = 0$, there would *not* be a solution for arbitrarily weak coupling).

(c) Assuming that $N(\mathcal{E})$ differs negligibly from $N(\mathcal{E}_F)$ in the range $\mathcal{E}_F < \mathcal{E} < \mathcal{E}_F + \hbar\omega$, show that the binding energy is given by

$$\Delta = 2\hbar\omega \frac{e^{-2/N(\mathcal{E}_F)V}}{1 - e^{-2/N(\mathcal{E}_F)V}}, \tag{34.51}$$

or, in the weak-coupling limit:

$$\Delta = 2\hbar\omega e^{-2/N(\mathcal{E}_F)V}. \tag{34.52}$$

Appendix A

Summary of Important Numerical Relations in The Free Electron Theory of Metals

We collect here those results of free electron theory from Chapters 1 and 2 that are useful in making rough numerical estimates of metallic properties. We use the following values of the fundamental constants:[1]

Electronic charge:	e	$= 1.60219 \times 10^{-19}$ coulomb
		$= 4.80324 \times 10^{-10}$ esu
Speed of light:	c	$= 2.997925 \times 10^{10}$ cm/sec
Planck's constant:	h	$= 6.6262 \times 10^{-27}$ erg-sec
	$h/2\pi = \hbar$	$= 1.05459 \times 10^{-27}$ erg-sec
Electronic mass:	m	$= 9.1095 \times 10^{-28}$ gm
Boltzmann's constant:	k_B	$= 1.3807 \times 10^{-16}$ erg/K
		$= 0.8617 \times 10^{-4}$ eV/K
Bohr radius:	$\hbar^2/me^2 = a_0$	$= 0.529177$ Å
Rydberg:	$e^2/2a_0$	$= 13.6058$ eV
Electron volt:	1 eV	$= 1.60219 \times 10^{-12}$ erg
		$= 1.1604 \times 10^4$ K

IDEAL FERMI GAS:

$$k_F = [3.63 \text{ Å}^{-1}] \times [r_s/a_0]^{-1} \tag{2.23}$$
$$v_F = [4.20 \times 10^8 \text{ cm/sec}] \times [r_s/a_0]^{-1} \tag{2.24}$$
$$\mathcal{E}_F = [50.1 \text{ eV}] \times [r_s/a_0]^{-2} \tag{2.26}$$
$$T_F = [58.2 \times 10^4 \text{ K}] \times [r_s/a_0]^{-2} \tag{2.33}$$

The quantity r_s (Eq. (1.2)) is listed for selected metals in Table 1.1, and is numerically given by

$$\frac{r_s}{a_0} = 5.44[n_{22}]^{-1/3},$$

where the electronic density is $n = n_{22} \times 10^{22}/\text{cm}^3$.

RELAXATION TIME AND MEAN FREE PATH:

$$\tau = [2.2 \times 10^{-15} \text{ sec}] \times [(r_s/a_0)^3/\rho_\mu] \tag{1.8}$$
$$l = [92 \text{ Å}] \times [(r_s/a_0)^2/\rho_\mu] \tag{2.91}$$

Here ρ_μ is the resistivity in microhm centimeters, listed for selected metals in Table 1.2.

[1] B. N. Taylor, W. H. Parker, and D. N. Langenberg, *Rev. Mod. Phys.* **41**, 375 (1969). We quote the values to far greater precision than it is meaningful to use in free electron calculations. Other constants are given on the back inside cover.

CYCLOTRON FREQUENCY:

$$\begin{aligned} \nu_c = \omega_c/2\pi &= 2.80H \times 10^6 \text{ Hz} \\ \hbar\omega_c &= 1.16H \times 10^{-8} \text{ eV} \\ &= 1.34H \times 10^{-4} \text{ K} \end{aligned} \tag{1.22}$$

Here $\omega_c = eH/mc$ (Eq. (1.18)) and H (above) is the magnetic field in gauss.

PLASMA FREQUENCY:

$$\begin{aligned} \nu_p = \omega_p/2\pi &= [11.4 \times 10^{15} \text{ Hz}] \times (r_s/a_0)^{-3/2} \\ \hbar\omega_p &= [47.1 \text{ eV}] \times (r_s/a_0)^{-3/2} \end{aligned} \tag{1.40}$$

Here $\omega_p = [4\pi ne^2/m]^{1/2}$ (Eq. (1.38)).

Appendix B

The Chemical Potential

One believes[1] that in the limit of a large system the Helmholtz free energy per unit volume approaches a smooth function of density and temperature:

$$\lim_{\substack{N,V\to\infty \\ N/V\to n}} \frac{1}{V} F(N, V, T) = f(n, T), \tag{B.1}$$

or, to an excellent approximation for large N and V,

$$F(N, V, T) = Vf(n, T). \tag{B.2}$$

Since (Eq. (2.45)) the chemical potential is defined by

$$\mu = F(N + 1, V, T) - F(N, V, T), \tag{B.3}$$

we have, for a large system,

$$\begin{aligned}\mu = V\left[f\left(\frac{N+1}{V}, T\right) - f\left(\frac{N}{V}, T\right)\right] &= V\left[f\left(n + \frac{1}{V}, T\right) - f(n, T)\right], \\ &\underset{V\to\infty}{\to} \left(\frac{\partial f}{\partial n}\right)_T.\end{aligned} \tag{B.4}$$

The pressure is given by $P = -(\partial F/\partial V)_T$, which (B.2) and (B.4) reduce to $P = -f + \mu n$. Since $F = U - TS$, where U is the internal energy and S is the entropy, it follows that the chemical potential is just the Gibbs free energy per particle:

$$\mu = \frac{G}{N}, \quad G = U - TS + PV. \tag{B.5}$$

Since $T = (\partial u/\partial s)_n$ (where the energy density $u = U/V$ and entropy density $s = S/V$ are defined in the same way as the free energy density f was defined in (B.1)), it follows from (B.4) that μ can also be written in the forms

$$\mu = \left(\frac{\partial u}{\partial n}\right)_s, \tag{B.6}$$

or

$$\mu = -T\left(\frac{\partial s}{\partial n}\right)_u. \tag{B.7}$$

---

[1] In many cases it can be proved. See, for example, J. L. Lebowitz and E. H. Lieb, *Phys. Rev. Lett.* **22**, 631 (1969).

Appendix C

The Sommerfeld Expansion

The Sommerfeld expansion is applied to integrals of the form

$$\int_{-\infty}^{\infty} d\mathcal{E}\, H(\mathcal{E}) f(\mathcal{E}), \quad f(\mathcal{E}) = \frac{1}{e^{(\mathcal{E}-\mu)/k_B T} + 1}, \tag{C.1}$$

where $H(\mathcal{E})$ vanishes as $\mathcal{E} \to -\infty$ and diverges no more rapidly than some power of $\mathcal{E}$ as $\mathcal{E} \to +\infty$. If one defines

$$K(\mathcal{E}) = \int_{-\infty}^{\mathcal{E}} H(\mathcal{E}')\, d\mathcal{E}', \tag{C.2}$$

so that

$$H(\mathcal{E}) = \frac{dK(\mathcal{E})}{d\mathcal{E}}, \tag{C.3}$$

then one can integrate by parts[1] in (C.1) to get

$$\int_{-\infty}^{\infty} H(\mathcal{E}) f(\mathcal{E})\, d\mathcal{E} = \int_{-\infty}^{\infty} K(\mathcal{E}) \left(-\frac{\partial f}{\partial \mathcal{E}} \right) d\mathcal{E}. \tag{C.4}$$

Since f is indistinguishable from zero when $\mathcal{E}$ is more than a few $k_B T$ greater than μ, and indistinguishable from unity when $\mathcal{E}$ is more than a few $k_B T$ less than μ, its $\mathcal{E}$-derivative will be appreciable only within a few $k_B T$ of μ. Provided that H is nonsingular and not too rapidly varying in the neighborhood of $\mathcal{E} = \mu$, it is very reasonable to evaluate (C.4) by expanding $K(\mathcal{E})$ in a Taylor series about $\mathcal{E} = \mu$, with the expectation that only the first few terms will be of importance:

$$K(\mathcal{E}) = K(\mu) + \sum_{n=1}^{\infty} \left[\frac{(\mathcal{E} - \mu)^n}{n!} \right] \left[\frac{d^n K(\mathcal{E})}{d\mathcal{E}^n} \right]_{\mathcal{E}=\mu}. \tag{C.5}$$

When we substitute (C.5) in (C.4), the leading term gives just $K(\mu)$, since

$$\int_{-\infty}^{\infty} (-\partial f / \partial \mathcal{E})\, d\mathcal{E} = 1.$$

Furthermore, since $\partial f / \partial \mathcal{E}$ is an even function of $\mathcal{E} - \mu$, only terms with even n in (C.5) contribute to (C.4), and if we reexpress K in terms of the original function H through (C.2), we find that:

$$\int_{-\infty}^{\infty} d\mathcal{E}\, H(\mathcal{E}) f(\mathcal{E}) = \int_{-\infty}^{\mu} H(\mathcal{E})\, d\mathcal{E} + \sum_{n=1}^{\infty} \int_{-\infty}^{\infty} \frac{(\mathcal{E} - \mu)^{2n}}{(2n)!} \left(-\frac{\partial f}{\partial \mathcal{E}} \right) d\mathcal{E}\, \frac{d^{2n-1}}{d\mathcal{E}^{2n-1}} H(\mathcal{E}) \Big|_{\mathcal{E}=\mu}. \tag{C.6}$$

[1] The integrated term vanishes at ∞ because the Fermi function vanishes more rapidly than K diverges, and at $-\infty$ because the Fermi function approaches unity while K approaches zero.

Finally, making the substitution $(\mathcal{E} - \mu)/k_B T = x$, we find that

$$\int_{-\infty}^{\infty} H(\mathcal{E}) f(\mathcal{E})\, d\mathcal{E} = \int_{-\infty}^{\mu} H(\mathcal{E})\, d\mathcal{E} + \sum_{n=1}^{\infty} a_n (k_B T)^{2n} \frac{d^{2n-1}}{d\mathcal{E}^{2n-1}} H(\mathcal{E})\Big|_{\mathcal{E}=\mu}, \tag{C.7}$$

where the a_n are dimensionless numbers given by

$$a_n = \int_{-\infty}^{\infty} \frac{x^{2n}}{(2n)!}\left(-\frac{d}{dx}\frac{1}{e^x + 1}\right) dx. \tag{C.8}$$

By elementary manipulations one can show that

$$a_n = 2\left(1 - \frac{1}{2^{2n}} + \frac{1}{3^{2n}} - \frac{1}{4^{2n}} + \frac{1}{5^{2n}} - \cdots\right). \tag{C.9}$$

This is usually written in terms of the Riemann zeta function, $\zeta(n)$, as

$$a_n = \left(2 - \frac{1}{2^{2(n-1)}}\right)\zeta(2n), \tag{C.10}$$

where

$$\zeta(n) = 1 + \frac{1}{2^n} + \frac{1}{3^n} + \frac{1}{4^n} + \cdots. \tag{C.11}$$

For the first few n, $\zeta(2n)$ has the values[2]

$$\zeta(2n) = 2^{2n-1} \frac{\pi^{2n}}{(2n)!} B_n \tag{C.12}$$

where the B_n are known as Bernoulli numbers, and

$$B_1 = \frac{1}{6}, \quad B_2 = \frac{1}{30}, \quad B_3 = \frac{1}{42}, \quad B_4 = \frac{1}{30}, \quad B_5 = \frac{5}{66}. \tag{C.13}$$

In most practical calculations in metals physics, one rarely needs to know more than $\zeta(2) = \pi^2/6$, and never goes beyond $\zeta(4) = \pi^4/90$. Nevertheless, if one should wish to carry the Sommerfeld expansion (2.70) beyond $n = 5$ (and hence past the values of the B_n listed in (C.13)), by the time $2n$ is as large as 12 the a_n can be evaluated to five-place accuracy by retaining only the first two terms in the alternating series (C.9).

---

[2] See, for example, E. Jahnke and F. Emde, *Tables of Functions*, 4th ed., Dover, New York, 1945, p. 272.

Appendix D

Plane-Wave Expansions of Periodic Functions in More Than One Dimension

We start from the general observation that the plane waves $e^{i\mathbf{k}\cdot\mathbf{r}}$ form a complete set of functions in which any function (subject to suitable conditions of regularity) can be expanded.[1] If a function $f(\mathbf{r})$ has the periodicity of a Bravais lattice, i.e., if $f(\mathbf{r} + \mathbf{R}) = f(\mathbf{r})$ for all $\mathbf{r}$ and all $\mathbf{R}$ in the Bravais lattice, then only plane waves with the periodicity of the Bravais lattice can occur in the expansion. Since the set of wave vectors for plane waves with the periodicity of the lattice is just the reciprocal lattice, a function periodic in the direct lattice will have a plane wave expansion of the form

$$\boxed{f(\mathbf{r}) = \sum_{\mathbf{K}} f_{\mathbf{K}} e^{i\mathbf{K}\cdot\mathbf{r}},} \tag{D.1}$$

where the sum is over all reciprocal lattice vectors **K**.

The Fourier coefficients $f_{\mathbf{K}}$ are given by

$$\boxed{f_{\mathbf{K}} = \frac{1}{v}\int_C d\mathbf{r}\, e^{-i\mathbf{K}\cdot\mathbf{r}} f(\mathbf{r})} \tag{D.2}$$

where the integral is over any direct lattice primitive cell C, and v is the volume of the primitive cell.[2] This follows from multiplying (D.1) by $e^{-i\mathbf{K}\cdot\mathbf{r}}/v$ and integrating over the primitive cell C, provided that[3] we can show

$$\int_C d\mathbf{r}\, e^{i\mathbf{K}\cdot\mathbf{r}} = 0 \tag{D.3}$$

for any nonvanishing reciprocal lattice vector **K**.

To establish (D.3) we merely note that since $e^{i\mathbf{K}\cdot\mathbf{r}}$ has the periodicity of the lattice (**K** being a reciprocal lattice vector) its integral over a primitive cell is independent of the choice of cell.[2] In particular, it will not be changed if we translate the primitive cell C through a vector $\mathbf{d}$ (not necessarily a Bravais lattice vector). However, the integral over the translated cell C' can be written as the integral of $e^{i\mathbf{K}\cdot(\mathbf{r}+\mathbf{d})}$ over the

---

[1] No attempt at mathematical rigor is made here. The mathematical subtleties in three dimensions are no more difficult than they are in one, since the function can be considered one variable at a time. Our concern is more one of bookkeeping and notation: how most compactly to express the basic formulas of three-dimensional Fourier series, quite aside from how they can be rigorously derived.

[2] The choice of cell is immaterial, since the integrand is periodic. That the integral of a periodic function over a primitive cell does not depend on the choice of cells is most easily seen by recalling that any primitive cell can be cut up and reassembled into any other, by translating its pieces through Bravais lattice vectors. But translating a periodic function through a Bravais lattice vector leaves it unchanged.

[3] In our mathematically lax manner, we ignore possible difficulties in interchanging orders of integration and summation.

original cell. Thus

$$\int_C d\mathbf{r}\, e^{i\mathbf{K}\cdot(\mathbf{r}+\mathbf{d})} = \int_{C'} d\mathbf{r}\, e^{i\mathbf{K}\cdot\mathbf{r}} = \int_C d\mathbf{r}\, e^{i\mathbf{K}\cdot\mathbf{r}} \tag{D.4}$$

or

$$(e^{i\mathbf{K}\cdot\mathbf{d}} - 1)\int_C d\mathbf{r}\, e^{i\mathbf{K}\cdot\mathbf{r}} = 0. \tag{D.5}$$

Since $e^{i\mathbf{K}\cdot\mathbf{d}} - 1$ can vanish for arbitrary $\mathbf{d}$ only if $\mathbf{K}$ itself vanishes, this establishes (D.3) for nonzero $\mathbf{K}$.

These formulas have several kinds of application. They can be applied directly to functions with the real-space periodicity of a crystal Bravais lattice. They can also be applied to functions that are periodic in k-space with the periodicity of the reciprocal lattice. In that case, noting that the reciprocal of the reciprocal is the direct lattice and that the volume of a reciprocal lattice primitive cell is $(2\pi)^3/v$, we can transcribe (D.1) and (D.2) to read:

$$\boxed{\phi(\mathbf{k}) = \sum_{\mathbf{R}} e^{+i\mathbf{R}\cdot\mathbf{k}}\phi_{\mathbf{R}}} \tag{D.6}$$

for any $\phi(\mathbf{k})$ with the periodicity of the reciprocal lattice ($\phi(\mathbf{k} + \mathbf{K}) = \phi(\mathbf{k})$ for all $\mathbf{k}$ and all reciprocal lattice vectors $\mathbf{K}$), where the sum is over all direct lattice vectors $\mathbf{R}$ and

$$\boxed{\phi_{\mathbf{R}} = v\int \frac{d\mathbf{k}}{(2\pi)^3}\, e^{-i\mathbf{R}\cdot\mathbf{k}}\phi(\mathbf{k}),} \tag{D.7}$$

where v is the volume of the direct lattice primitive cell, and the integral is over any reciprocal lattice primitive cell (such as the first Brillouin zone).

Another important application is to functions in real space whose only periodicity is that imposed by the Born-von Karman boundary condition (Eq. (8.22)):

$$f(\mathbf{r} + N_i\mathbf{a}_i) = f(\mathbf{r}), \qquad i = 1, 2, 3. \tag{D.8}$$

Such functions are periodic in a very large (and unphysical) Bravais lattice generated by the three primitive vectors $N_i\mathbf{a}_i$, $i = 1, 2, 3$. The reciprocal to this lattice has primitive vectors $\mathbf{b}_i/N_i$, where the $\mathbf{b}_i$ are related to the $\mathbf{a}_i$ by Eq. (5.3). A vector of this reciprocal lattice has the form

$$\mathbf{k} = \sum_{i=1}^{3} \frac{m_i}{N_i}\mathbf{b}_i, \qquad m_i \text{ integral.} \tag{D.9}$$

Since the volume of the primitive cell associated with the Born-von Karman periodicity is the volume V of the entire crystal, (D.1) and (D.2) now become

$$\boxed{f(\mathbf{r}) = \sum_{\mathbf{k}} f_{\mathbf{k}} e^{i\mathbf{k}\cdot\mathbf{r}}} \tag{D.10}$$

for any f satisfying the Born-von Karman boundary condition (D.8), where the sum is over all **k** of the form (D.9) and

$$f_{\mathbf{k}} = \frac{1}{V} \int d\mathbf{r}\, e^{-i\mathbf{k}\cdot\mathbf{r}} f(\mathbf{r}), \tag{D.11}$$

where the integral is over the entire crystal. Note also the analogue of (D.3):

$$\int_V d\mathbf{r}\, e^{i\mathbf{k}\cdot\mathbf{r}} = 0 \tag{D.12}$$

for any **k** of the form (D.9) different from 0.

Appendix E

The Velocity and Effective Mass of Bloch Electrons

One may evaluate the derivatives $\partial \mathcal{E}_n/\partial k_i$ and $\partial^2 \mathcal{E}_n/\partial k_i \partial k_j$ by noting that they are the coefficients of the linear and quadratic terms in $\mathbf{q}$, in the expansion

$$\mathcal{E}_n(\mathbf{k} + \mathbf{q}) = \mathcal{E}_n(\mathbf{k}) + \sum_i \frac{\partial \mathcal{E}_n}{\partial k_i} q_i + \frac{1}{2} \sum_{ij} \frac{\partial^2 \mathcal{E}_n}{\partial k_i \, \partial k_j} q_i q_j + O(q^3). \tag{E.1}$$

Since, however, $\mathcal{E}_n(\mathbf{k} + \mathbf{q})$ is the eigenvalue of $H_{\mathbf{k}+\mathbf{q}}$ (Eq. (8.48)), we can calculate the required terms from the fact that

$$H_{\mathbf{k}+\mathbf{q}} = H_{\mathbf{k}} + \frac{\hbar^2}{m} \mathbf{q} \cdot \left(\frac{1}{i} \nabla + \mathbf{k} \right) + \frac{\hbar^2}{2m} q^2, \tag{E.2}$$

as an exercise in perturbation theory.

Perturbation theory asserts that if $H = H_0 + V$ and the normalized eigenvectors and eigenvalues of H_0 are

$$H_0 \psi_n = E_n^0 \psi_n, \tag{E.3}$$

then to second order in V, the corresponding eigenvalues of H are

$$E_n = E_n^0 + \int d\mathbf{r} \, \psi_n^* V \psi_n + \sum_{n' \neq n} \frac{|\int d\mathbf{r} \, \psi_n^* V \psi_{n'}|^2}{(E_n^0 - E_{n'}^0)} + \cdots. \tag{E.4}$$

To calculate to linear order in $\mathbf{q}$ we need only keep the term linear in $\mathbf{q}$ in (E.2) and insert it into the first-order term in (E.4). In this way, we find that

$$\sum_i \frac{\partial \mathcal{E}_n}{\partial k_i} q_i = \sum_i \int d\mathbf{r} \, u_{n\mathbf{k}}^* \frac{\hbar^2}{m} \left(\frac{1}{i} \nabla + \mathbf{k} \right)_i q_i u_{n\mathbf{k}}, \tag{E.5}$$

(where the integrations are either over a primitive cell or over the entire crystal, depending on whether the normalization integral $\int d\mathbf{r} \, |u_{n\mathbf{k}}|^2$ has been taken equal to unity over a primitive cell or over the entire crystal). Therefore

$$\frac{\partial \mathcal{E}_n}{\partial \mathbf{k}} = \frac{\hbar^2}{m} \int d\mathbf{r} \, u_{n\mathbf{k}}^* \left(\frac{1}{i} \nabla + \mathbf{k} \right) u_{n\mathbf{k}}. \tag{E.6}$$

If we express this in terms of the Bloch functions $\psi_{n\mathbf{k}}$ via (8.3), it can be written as

$$\boxed{\frac{\partial \mathcal{E}_n}{\partial \mathbf{k}} = \frac{\hbar^2}{m} \int d\mathbf{r} \, \psi_{n\mathbf{k}}^* \frac{1}{i} \nabla \psi_{n\mathbf{k}}.} \tag{E.7}$$

Since $(1/m)(\hbar/i)\nabla$ is the velocity operator,[1] this establishes that $(1/\hbar)(\partial \mathcal{E}_n(\mathbf{k})/\partial \mathbf{k})$ is the mean velocity of an electron in the Bloch level given by n, $\mathbf{k}$.

---

[1] The velocity operator is $\mathbf{v} = d\mathbf{r}/dt = (1/i\hbar)[\mathbf{r}, \mathbf{H}] = \mathbf{p}/m = \hbar\nabla/mi$.

To calculate $\partial^2 \mathcal{E}_n/\partial k_i\, \partial k_j$ we need $\mathcal{E}_n(\mathbf{k} + \mathbf{q})$ to second order in q. Equations (E.2) and (E.4) give[2]

$$\sum_{ij} \frac{1}{2} \frac{\partial^2 \mathcal{E}_n}{\partial k_i\, \partial k_j} q_i q_j = \frac{\hbar^2}{2m} q^2 + \sum_{n' \neq n} \frac{\left| \int d\mathbf{r}\, u_{n\mathbf{k}}^* \frac{\hbar^2}{m} \mathbf{q} \cdot \left(\frac{1}{i} \nabla + \mathbf{k} \right) u_{n'\mathbf{k}} \right|^2}{\mathcal{E}_{n\mathbf{k}} - \mathcal{E}_{n'\mathbf{k}}}. \quad \text{(E.8)}$$

Again using (8.3), we can express (E.8) in terms of Bloch functions:

$$\sum_{ij} \frac{1}{2} \frac{\partial^2 \mathcal{E}_n}{\partial k_i\, \partial k_j} q_i q_j = \frac{\hbar^2}{2m} q^2 + \sum_{n' \neq n} \frac{\left| \langle n\mathbf{k} | \frac{\hbar^2}{mi} \mathbf{q} \cdot \nabla | n'\mathbf{k} \rangle \right|^2}{\mathcal{E}_{n\mathbf{k}} - \mathcal{E}_{n'\mathbf{k}}}, \quad \text{(E.9)}$$

where we have used the notation:

$$\int d\mathbf{r}\, \psi_{n\mathbf{k}}^* X \psi_{n'\mathbf{k}} = \langle n\mathbf{k} | X | n'\mathbf{k} \rangle. \quad \text{(E.10)}$$

Consequently,

$$\frac{\partial^2 \mathcal{E}_n(\mathbf{k})}{\partial k_i\, \partial k_j} = \frac{\hbar^2}{m} \delta_{ij} + \left(\frac{\hbar^2}{m} \right)^2 \sum_{n' \neq n} \frac{\langle n\mathbf{k} | \frac{1}{i} \nabla_i | n'\mathbf{k} \rangle \langle n'\mathbf{k} | \frac{1}{i} \nabla_j | n\mathbf{k} \rangle + \langle n\mathbf{k} | \frac{1}{i} \nabla_j | n'\mathbf{k} \rangle \langle n'\mathbf{k} | \frac{1}{i} \nabla_i | n\mathbf{k} \rangle}{\mathcal{E}_n(\mathbf{k}) - \mathcal{E}_{n'}(\mathbf{k})}. \quad \text{(E.11)}$$

The quantity on the right side of (E.11) (times a factor $1/\hbar^2$) is the inverse "effective mass tensor" (page 228) and the formula (E.11) is often called "the effective mass theorem."

---

[2] The first term on the right of (E.8) comes from placing the second-order term in $H_{\mathbf{k}+\mathbf{q}}$(Eq. (E.2)) into the first-order term in the perturbation theory formula (Eq. (E.4)). The second term on the right comes from placing the first-order term in $H_{\mathbf{k}+\mathbf{q}}$into the second-order term in the perturbation theory formula.

Appendix F

Some Identities Related to Fourier Analysis of Periodic Systems

To derive Fourier inversion formulas it suffices to establish the identity

$$\boxed{\sum_{\mathbf{R}} e^{i\mathbf{k}\cdot\mathbf{R}} = N\delta_{\mathbf{k},\mathbf{0}},} \tag{F.1}$$

where $\mathbf{R}$ runs through the N sites of the Bravais lattice

$$\mathbf{R} = \sum_{i=1}^{3} n_i\mathbf{a}_i, \qquad 0 \leqslant n_i < N_i, \quad N_1N_2N_3 = N, \tag{F.2}$$

and $\mathbf{k}$ is any vector in the first Brillouin zone consistent with the Born-von Karman boundary condition appropriate to the N points specified by (F.2).

The identity is most simply proved by noting that because $\mathbf{k}$ is consistent with the Born-von Karman periodic boundary condition, the value of the sum in (F.1) is unchanged if every $\mathbf{R}$ is displaced by the same $\mathbf{R}_0$, where $\mathbf{R}_0$ is itself any vector of the form (F.2):

$$\sum_{\mathbf{R}} e^{i\mathbf{k}\cdot\mathbf{R}} = \sum_{\mathbf{R}} e^{i\mathbf{k}\cdot(\mathbf{R}+\mathbf{R}_0)} = e^{i\mathbf{k}\cdot\mathbf{R}_0}\sum_{\mathbf{R}} e^{i\mathbf{k}\cdot\mathbf{R}}. \tag{F.3}$$

Consequently the sum must vanish unless $e^{i\mathbf{k}\cdot\mathbf{R}_0} = 1$, for all $\mathbf{R}_0$ of the form (F.2), i.e., for all vectors $\mathbf{R}_0$ of the Bravais lattice. This is possible only if $\mathbf{k}$ is a reciprocal lattice vector. But $\mathbf{k} = \mathbf{0}$ is the only reciprocal lattice vector in the first Brillouin zone.[1] Hence the left side of (F.1) does indeed vanish if $\mathbf{k} \neq \mathbf{0}$, and is trivially equal to N, when $\mathbf{k} = \mathbf{0}$.

A closely related identity of similar importance is

$$\boxed{\sum_{\mathbf{k}} e^{i\mathbf{k}\cdot\mathbf{R}} = N\delta_{\mathbf{R},\mathbf{0}},} \tag{F.4}$$

where $\mathbf{R}$ is any vector of the form (F.2), and the sum on $\mathbf{k}$ runs through all sites in the first Brillouin zone consistent with the Born-von Karman boundary condition The sum in (F.4) is now unchanged if every $\mathbf{k}$ is translated by the same vector $\mathbf{k}_0$ lying in the first zone and consistent with the Born-von Karman boundary condition, for the primitive cell constructed by shifting the entire first zone by $\mathbf{k}_0$, can be re-assembled into the first zone by shifting appropriate pieces of it through reciprocal lattice vectors. Since no term of the form $e^{i\mathbf{k}\cdot\mathbf{R}}$ is changed when $\mathbf{k}$ is shifted by a

[1] If $\mathbf{k}$ is not restricted to the first zone, the sum in (F.1) will therefore vanish unless $\mathbf{k}$ is a reciprocal lattice vector $\mathbf{K}$, in which case it will be equal to N.

reciprocal lattice vector, the sum over the shifted zone is identical to the sum over the original zone. Thus:

$$\sum_{\mathbf{k}} e^{i\mathbf{k}\cdot\mathbf{R}} = \sum_{\mathbf{k}} e^{i(\mathbf{k}+\mathbf{k}_0)\cdot\mathbf{R}} = e^{i\mathbf{k}_0\cdot\mathbf{R}} \sum_{\mathbf{k}} e^{i\mathbf{k}\cdot\mathbf{R}}, \tag{F.5}$$

and therefore the sum on the left side of (F.4) must vanish unless $e^{i\mathbf{k}_0\cdot\mathbf{R}}$ is unity for all $\mathbf{k}_0$ consistent with the Born-von Karman boundary condition. The only $\mathbf{R}$ of the form (F.2) for which this is possible is $\mathbf{R} = \mathbf{0}$. And when $\mathbf{R} = \mathbf{0}$, the sum in (F.4) is trivially equal to N.

Appendix G

The Variational Principle for Schrödinger's Equation

We wish to show that the functional $E[\psi]$ (Eq. (11.17)) is made stationary over all differentiable functions ψ satisfying the Bloch condition with wave vector $\mathbf{k}$, by the $\psi_{\mathbf{k}}$ that satisfy the Schrödinger equation:

$$-\frac{\hbar^2}{2m}\nabla^2\psi_{\mathbf{k}} + U(\mathbf{r})\psi_{\mathbf{k}} = \mathcal{E}_{\mathbf{k}}\psi_{\mathbf{k}}. \tag{G.1}$$

By this we mean the following: Let ψ be close to one of the $\psi_{\mathbf{k}}$, so that

$$\psi = \psi_{\mathbf{k}} + \delta\psi, \tag{G.2}$$

where $\delta\psi$ is small. Let ψ satisfy the Bloch condition with wave vector $\mathbf{k}$, so that $\delta\psi$ does as well. Then

$$E[\psi] = E[\psi_{\mathbf{k}}] + O(\delta\psi)^2. \tag{G.3}$$

To keep notation simple in the proof it helps to define

$$F[\phi, \chi] = \int d\mathbf{r}\left(\frac{\hbar^2}{2m}\nabla\phi^* \cdot \nabla\chi + U(\mathbf{r})\phi^*\chi\right), \tag{G.4}$$

and to use the standard notation

$$(\phi, \chi) = \int d\mathbf{r}\, \phi^*\chi. \tag{G.5}$$

Note that $E[\psi]$ is then given by

$$E[\psi] = \frac{F[\psi, \psi]}{(\psi, \psi)}. \tag{G.6}$$

The variational principle follows directly from the fact that

$$\begin{aligned} F[\phi, \psi_{\mathbf{k}}] &= \mathcal{E}_{\mathbf{k}}(\phi, \psi_{\mathbf{k}}), \\ F[\psi_{\mathbf{k}}, \phi] &= \mathcal{E}_{\mathbf{k}}(\psi_{\mathbf{k}}, \phi), \end{aligned} \tag{G.7}$$

for arbitrary ϕ that satisfy the Bloch condition with wave vector $\mathbf{k}$. This is because the Bloch condition requires the integrands in (G.7) to have the periodicity of the lattice. Consequently, one can use the integration-by-parts formulas of Appendix I to transfer both gradients onto $\psi_{\mathbf{k}}$; (G.7) then follows at once from the fact that $\psi_{\mathbf{k}}$ satisfies the Schrödinger equation (G.1).

We can now write

$$\begin{aligned} F[\psi, \psi] &= F[\psi_{\mathbf{k}} + \delta\psi, \psi_{\mathbf{k}} + \delta\psi] \\ &= F[\psi_{\mathbf{k}}, \psi_{\mathbf{k}}] + F[\delta\psi, \psi_{\mathbf{k}}] + F[\psi_{\mathbf{k}}, \delta\psi] + O(\delta\psi)^2 \\ &= \mathcal{E}_{\mathbf{k}}\{(\psi_{\mathbf{k}}, \psi_{\mathbf{k}}) + (\psi_{\mathbf{k}}, \delta\psi) + (\delta\psi, \psi_{\mathbf{k}})\} + O(\delta\psi)^2. \end{aligned} \tag{G.8}$$

Furthermore,

$$(\psi, \psi) = (\psi_{\mathbf{k}}, \psi_{\mathbf{k}}) + (\psi_{\mathbf{k}}, \delta\psi) + (\delta\psi, \psi_{\mathbf{k}}) + O(\delta\psi)^2. \quad \textbf{(G.9)}$$

Dividing (G.8) by (G.9), we have

$$E[\psi] = \frac{F[\psi, \psi]}{(\psi, \psi)} = \mathcal{E}_{\mathbf{k}} + O(\delta\psi)^2, \quad \textbf{(G.10)}$$

which establishes the variational principle. (The fact that $E[\psi_{\mathbf{k}}] = \mathcal{E}_{\mathbf{k}}$ follows at once from (G.10) by setting $\delta\psi$ equal to zero.)

Note that nothing in the derivation requires ψ to have a continuous first derivative.

Appendix H

Hamiltonian Formulation of the Semiclassical Equations of Motion, and Liouville's Theorem

The semiclassical equations of motion (12.6a) and (12.6b) can be written in the canonical Hamiltonian form

$$\dot{\mathbf{r}} = \frac{\partial H}{\partial \mathbf{p}}, \quad \dot{\mathbf{p}} = -\frac{\partial H}{\partial \mathbf{r}}, \tag{H.1}$$

where the Hamiltonian for electrons in the nth band is

$$H(\mathbf{r}, \mathbf{p}) = \mathcal{E}_n\left(\frac{1}{\hbar}\left[\mathbf{p} + \frac{e}{c}\mathbf{A}(\mathbf{r}, t)\right]\right) - e\phi(\mathbf{r}, t), \tag{H.2}$$

the fields are given in terms of the vector and scalar potentials by

$$\mathbf{H} = \nabla \times \mathbf{A}, \quad \mathbf{E} = -\nabla\phi - \frac{1}{c}\frac{\partial \mathbf{A}}{\partial t}, \tag{H.3}$$

and the variable $\mathbf{k}$ appearing in (12.6a) and (12.6b) is defined to be

$$\hbar\mathbf{k} = \mathbf{p} + \frac{e}{c}\mathbf{A}(\mathbf{r}, t). \tag{H.4}$$

To verify that (12.6a) and (12.6b) follow from (H.1) through (H.4) is a somewhat complicated, but conceptually straightforward, exercise in differentiation (just as it is in the free electron case).

Note that the canonical crystal momentum (the variable that plays the role of the canonical momentum in the Hamiltonian formulation) is not $\hbar\mathbf{k}$, but (from Eq. (H.4)),

$$\mathbf{p} = \hbar\mathbf{k} - \frac{e}{c}\mathbf{A}(\mathbf{r}, t). \tag{H.5}$$

Because the semiclassical equations for each band have the canonical Hamiltonian form, Liouville's theorem[1] implies that regions of six-dimensional rp-space evolve in time in such a way as to preserve their volume. Because, however, $\mathbf{k}$ differs from $\mathbf{p}$ only by an additive vector that does not depend on $\mathbf{p}$, any region in rp-space has the same volume as the corresponding region in rk-space.[2] This establishes Liouville's theorem in the form used in Chapters 12 and 13.

---

[1] The proof depends only on the equations of motion having the form (H.1). See, for example, Keith R. Symon, *Mechanics*, 3rd ed., Addison-Wesley, Reading, Mass, 1971, p. 395.

[2] Formally, the Jacobian $\partial(\mathbf{r}, \mathbf{p})/\partial(\mathbf{r}, \mathbf{k})$ is unity.

Appendix I

Green's Theorem for Periodic Functions

If $u(\mathbf{r})$ and $v(\mathbf{r})$ both have the periodicity of a Bravais lattice,[1] then the following identities hold for integrals taken over a primitive cell C:

$$\int_C d\mathbf{r}\; u\, \nabla v = -\int_C d\mathbf{r}\; v\, \nabla u, \tag{I.1}$$

$$\int_C d\mathbf{r}\; u\, \nabla^2 v = \int_C d\mathbf{r}\; v\, \nabla^2 u. \tag{I.2}$$

These are proved as follows:

Let $f(\mathbf{r})$ be any function with the periodicity of the Bravais lattice. Since C is a primitive cell, the integral

$$I(\mathbf{r}') = \int_C d\mathbf{r}\; f(\mathbf{r} + \mathbf{r}') \tag{I.3}$$

is independent of $\mathbf{r}'$. Therefore, in particular,

$$\nabla' I(\mathbf{r}') = \int_C d\mathbf{r}\; \nabla' f(\mathbf{r} + \mathbf{r}') = \int_C d\mathbf{r}\; \nabla f(\mathbf{r} + \mathbf{r}') = 0, \tag{I.4}$$

$$\nabla'^2 I(\mathbf{r}') = \int_C d\mathbf{r}\; \nabla'^2 f(\mathbf{r} + \mathbf{r}') = \int_C d\mathbf{r}\; \nabla^2 f(\mathbf{r} + \mathbf{r}') = 0. \tag{I.5}$$

Evaluating these at $\mathbf{r}' = 0$ we find that any periodic f satisfies

$$\int_C d\mathbf{r}\; \nabla f(\mathbf{r}) = 0, \tag{I.6}$$

$$\int_C d\mathbf{r}\; \nabla^2 f(\mathbf{r}) = 0. \tag{I.7}$$

Equation (I.1) follows directly from (I.6) applied to the case $f = uv$. To derive (I.2) set $f = uv$ in (I.7) to find

$$\int_C d\mathbf{r}\; (\nabla^2 u)v + \int_C d\mathbf{r}\; u(\nabla^2 v) + 2\int_C d\mathbf{r}\; \nabla u \cdot \nabla v = 0. \tag{I.8}$$

We can apply (I.1) to the last term in (I.8), taking the two periodic functions in (I.1) to be v and the various components of the gradient of u. This gives

$$2\int_C d\mathbf{r}\; \nabla u \cdot \nabla v = -2\int_C d\mathbf{r}\; u \nabla^2 v, \tag{I.9}$$

which reduces (I.8) to (I.2).

---

[1] We use a real-space notation, although the theorem, of course, also holds for periodic functions in k-space.

Appendix J

Conditions for the Absence of Interband Transitions in Uniform Electric or Magnetic Fields

The theories of electric breakdown and magnetic breakthrough that underly the conditions (12.8) and (12.9) are quite intricate. In this appendix we present some very crude ways of understanding the conditions.

In the limit of vanishingly small periodic potential, electric breakdown occurs whenever the wave vector of an electron crosses a Bragg plane (page 219). When the periodic potential is weak, but not zero, we can ask why breakdown might still be expected to occur near a Bragg plane, and how strong the potential must be for this possibility to be excluded.

In a weak periodic potential, points near Bragg planes are characterized by $\mathcal{E}(\mathbf{k})$ having a large curvature (see, for example, Figure 9.3). As a result, near a Bragg plane a small spread in wave vector can cause a large spread in velocity, since

$$\Delta v(\mathbf{k}) = \frac{\partial v}{\partial \mathbf{k}} \cdot \Delta \mathbf{k} \approx \frac{1}{\hbar}\left(\frac{\partial^2 \mathcal{E}}{\partial k^2}\right) \Delta k. \tag{J.1}$$

For the semiclassical picture to remain valid, the uncertainty in velocity must remain small compared with a typical electronic velocity v_F. This sets an upper limit on Δk:

$$\Delta k \ll \frac{\hbar v_F}{\partial^2 \mathcal{E}/\partial k^2}. \tag{J.2}$$

Since the periodic potential is weak, we can estimate the maximum value of $\partial^2 \mathcal{E}/\partial k^2$ by differentiating the nearly free electron result (9.26) in a direction normal to the Bragg plane, and evaluating the result on the plane:

$$\frac{\partial^2 \mathcal{E}}{\partial k^2} \approx \frac{(\hbar^2 \mathbf{K}/m)^2}{|U_{\mathbf{K}}|}. \tag{J.3}$$

Since $\hbar K/m \approx v_F$, and $\mathcal{E}_{\text{gap}} \approx |U_{\mathbf{K}}|$ (see Eq. (9.27)), Eq. (J.2) becomes

$$\Delta k \ll \frac{\mathcal{E}_{\text{gap}}}{\hbar v_F}, \tag{J.4}$$

which sets a lower limit on the uncertainty in the electron's position,

$$\Delta x \sim \frac{1}{\Delta k} \gg \frac{\hbar v_F}{\mathcal{E}_{\text{gap}}}. \tag{J.5}$$

As a result of this uncertainty in position the potential energy of the electron in the applied field is uncertain by

$$e\,\Delta\phi = eE\,\Delta x \gg \frac{eE\hbar v_F}{\mathcal{E}_{\text{gap}}}. \tag{J.6}$$

If this uncertainty in potential energy becomes comparable to the band gap, an interband transition can occur without violating energy conservation. To prohibit this we must have

$$\mathcal{E}_{\text{gap}} \gg \frac{eE\hbar v_F}{\mathcal{E}_{\text{gap}}}. \tag{J.7}$$

Since $\hbar v_F/a \approx \mathcal{E}_F$, when a is the lattice constant, (J.7) can also be written in the form

$$\frac{\mathcal{E}_{\text{gap}}^2}{\mathcal{E}_F} \gg eEa. \tag{J.8}$$

A similarly rough argument gives the condition for magnetic breakthrough. Since energy cannot be gained from a magnetic field, breakthrough requires a lack of definition in an electron's wave vector comparable to the distance between points of equal energy on two different branches of the Fermi surface. The k-space distance to be bridged is of order $\mathcal{E}_{\text{gap}}/|\partial\mathcal{E}/\partial\mathbf{k}|$ (Fig. J1), which is of order $\mathcal{E}_{\text{gap}}/\hbar v_F$. A condition

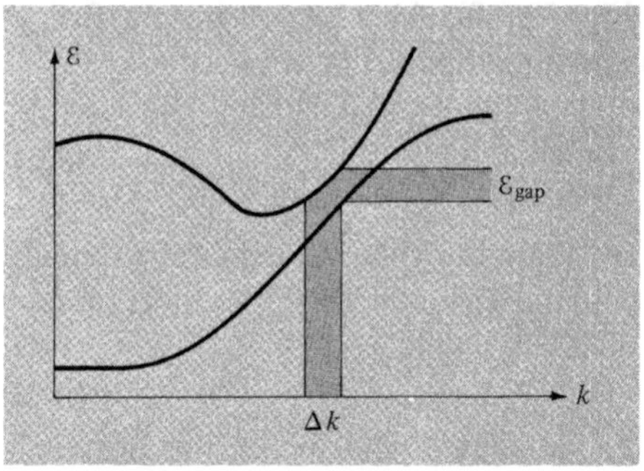

Figure J.1
When two bands come close together, the separation in k-space of points with the same energy is roughly $\Delta k = \mathcal{E}_{\text{gap}}/|\partial\mathcal{E}/\partial\mathbf{k}|$. Here $\mathcal{E}_{\text{gap}}$ is the minimum vertical separation between the two bands.

prohibiting breakthrough is therefore

$$\Delta k \ll \frac{\mathcal{E}_{\text{gap}}}{\hbar v_F}. \tag{J.9}$$

On page 230 we found that in a magnetic field the semiclassical orbit in real space is given by rotating the k-space orbit through 90° about the field direction and scaling it with the factor $\hbar c/eH$. Consequently the uncertainty relation

$$\Delta k_y\, \Delta y > 1 \tag{J.10}$$

implies a k-space uncertainty relation:

$$\Delta k_y\, \Delta k_x > \frac{eH}{\hbar c}. \tag{J.11}$$

Thus in a magnetic field an electron cannot be localized in k-space to better than a region of dimensions

$$\Delta k \approx \left(\frac{eH}{\hbar c}\right)^{1/2} \tag{J.12}$$

In conjunction with (J.9) this means that to avoid breakthrough it is necessary that

$$\left(\frac{eH}{\hbar c}\right)^{1/2} \ll \frac{\mathcal{E}_{\text{gap}}}{\hbar v_F}. \tag{J.13}$$

Taking $\mathcal{E}_F = \frac{1}{2}mv_F^2$, we can rewrite this condition as

$$\hbar\omega_c \ll \frac{\mathcal{E}_{\text{gap}}^2}{\mathcal{E}_F}. \tag{J.14}$$

Appendix K

Optical Properties of Solids

Consider a plane electromagnetic wave with angular frequency ω propagating along the z-axis through a medium with conductivity $\sigma(\omega)$ and dielectric constant[1] $\epsilon^0(\omega)$. (We ignore in this discussion the case of magnetic media, assuming that the magnetic permeability μ is unity; i.e., we take **B** equal to **H** in Maxwell's equations.) If we define $\mathbf{D}(z, \omega)$, $\mathbf{E}(z, \omega)$, and $\mathbf{j}(z, \omega)$ in the usual way:

$$\mathbf{j}(z, t) = \text{Re}\,[\mathbf{j}(z, \omega)e^{-i\omega t}], \quad \text{etc.,} \tag{K.1}$$

then the electric displacement and current density are related to the electric field by:

$$\mathbf{j}(z, \omega) = \sigma(\omega)\mathbf{E}(z, \omega); \quad \mathbf{D}(z, \omega) = \epsilon^0(\omega)\,\mathbf{E}(z, \omega). \tag{K.2}$$

ASSUMPTION OF LOCALITY

Equation (K.2) is a local relation; i.e., the current or displacement at a point is entirely determined by the value of the electric field at the same point. This assumption is valid (see pages 16–17) provided that the spatial variation of the fields is small compared with the electronic mean free path in the medium. Once the fields have been calculated assuming locality, the assumption is easily checked and found to be valid at optical frequencies.

ASSUMPTION OF ISOTROPY

For simplicity we assume the medium is sufficiently simple that $\sigma_{ij}(\omega) = \sigma(\omega)\delta_{ij}$, $\epsilon^0_{ij}(\omega) = \epsilon^0(\omega)\delta_{ij}$; i.e., **D** and **j** are parallel to **E**. This will be true, for example, in any crystal with cubic symmetry or in any polycrystalline specimen. To study birefringent crystals the assumption must be dropped.

CONVENTIONAL NATURE OF THE DISTINCTION BETWEEN $\epsilon^0(\omega)$ AND $\sigma(\omega)$

The dielectric constant and conductivity enter into a determination of the optical properties of a solid only in the combination:

$$\epsilon(\omega) = \epsilon^0(\omega) + \frac{4\pi i\sigma(\omega)}{\omega}. \tag{K.3}$$

As a result, one is free to redefine ϵ^0 by adding to it an arbitrary function of frequency, provided one makes a corresponding redefinition of σ so that the combination (K.3) is preserved:

$$\epsilon^0(\omega) \to \epsilon^0(\omega) + \delta\epsilon(\omega), \; \sigma(\omega) \to \sigma(\omega) - \frac{\omega}{4\pi i}\,\delta\epsilon(\omega). \tag{K.4}$$

---

[1] We use Gaussian units, in which the dielectric constant of empty space is unity. We use ϵ^0 for the dielectric constant of the medium.

This freedom of choice reflects a genuine ambiguity in the physical definitions of ϵ^0 and σ, which describe distinguishable physical processes only in the DC case, where σ describes the "free charges" (those that can move freely over arbitrary distances in response to the DC field) and ϵ^0 describes the "bound charges" (those that are bound to equilibrium positions and only stretched to new equilibrium positions by the DC field; see Figure K.1).

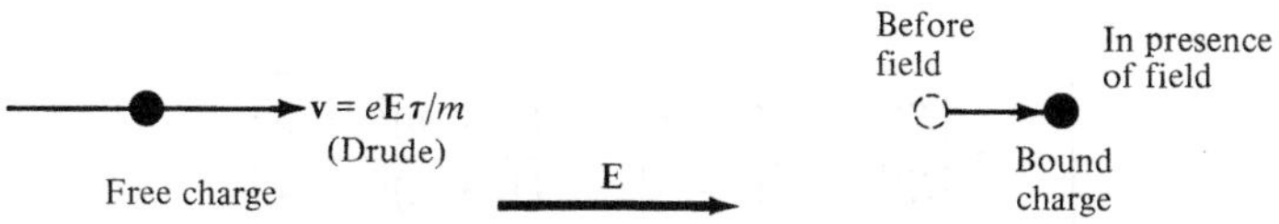

Figure K.1
The response of "free" and "bound" charges to a DC electric field. The free charge moves for as long as the field acts, but the bound charge is constrained by restoring forces, and can be displaced ("polarized") only to a new equilibrium position.

In the case of an AC field the distinction blurs. The free charges do not move arbitrarily far, but oscillate back and forth with the frequency of the field, whereas the bound charges no longer come to rest at new equilibrium positions, but also oscillate at the field frequency.

If the frequency of the field is sufficiently low ($\omega \ll 1/\tau$) the distinction can still be preserved, but on rather different grounds: The free charge velocities will respond in phase with the field (i.e., $\sigma(\omega)$ will be predominantly real) while the bound charge velocities will respond out of phase with the field (i.e., $\epsilon^0(\omega)$ will be predominantly real). At higher frequencies even this distinction disappears: The free charges can have a substantial out-of-phase response (indeed in metals, at optical frequencies, $\sigma(\omega)$ is predominantly imaginary) and the bound charges can have a considerable in-phase response (which may or may not be present at optical frequencies, depending on the material).

Thus at high frequencies the distinction between free and bound charges (and hence between $\sigma(\omega)$ and $\epsilon^0(\omega)$) is entirely a conventional one. In discussing metals, at least two different conventions are widely used:

1. σ is reserved for the response of electrons in partially filled bands (i.e., the conduction electrons) and ϵ^0 describes the response of the electrons in completely filled bands (e.g., the core electrons).[2]
2. The response of all electrons (core and conduction) is lumped into the single dielectric constant,

$$\epsilon(\omega) = \epsilon^0(\omega) + \frac{4\pi i\sigma(\omega)}{\omega}, \tag{K.5}$$

where $\epsilon(\omega)$ now includes the contributions of both partially filled and filled bands to the current. This convention enables the optical theories of metals to

[2] From the point of view of the semiclassical theory of Chapters 12 and 13, the filled bands are inert, and do not respond to the fields. However, corrections to the semiclassical theory do predict some core polarization.

use the same notation as is used for insulators, where, by universal convention, all charge is considered bound.

THE REFLECTIVITY

When a plane wave is normally incident from vacuum on a medium with dielectric constant ϵ (in the sense of Eq. (K.5)) the fraction r of power reflected (the reflectivity) is given by[3]

$$r = \frac{|E^r|^2}{|E^i|^2} = \left|\frac{1-\kappa}{1+\kappa}\right|^2 = \frac{(1-n)^2 + k^2}{(1+n)^2 + k^2} \tag{K.6}$$

where

$$\kappa = \sqrt{\epsilon}, \quad n = \text{Re}\,\kappa, \quad k = \text{Im}\,\kappa. \tag{K.7}$$

DETERMINATION OF $\epsilon(\omega)$ FROM THE MEASURED REFLECTIVITY

To determine n and k (and hence $\epsilon = (n + ik)^2$) from the reflectivity (K.6), more information is required. Two approaches are followed:

1. One can exploit the fact that n and k are related by the Kramers-Kronig relations:[4]

$$n(\omega) = 1 + P\int_{-\infty}^{\infty} \frac{d\omega'}{\pi}\frac{k(\omega')}{\omega' - \omega}, \quad k(\omega) = -P\int_{-\infty}^{\infty} \frac{d\omega'}{\pi}\frac{n(\omega') - 1}{\omega' - \omega} \tag{K.8}$$

   Either of these equations, plus a knowledge of $r(\omega)$ at *all* frequencies, permits one, in principle, to disentangle the separate values of $n(\omega)$ and $k(\omega)$. In practice the numerical analysis can be quite complicated, and the method has the disadvantage of requiring measurements to be made at enough frequencies to give reliable extrapolations to the entire frequency range, as required in (K.8).
2. One can use the generalization of (K.6) to nonnormal angles of incidence.[5] One then obtains a second expression for the reflectivity at angle of incidence θ, involving θ, $n(\omega)$, $k(\omega)$, and the polarization of the incident radiation. By comparing this with the measured reflectivity at angle θ, one obtains a second equation involving $n(\omega)$ and $k(\omega)$, and the two can then be extracted.

THE RELATION BETWEEN ϵ AND THE INTERBAND ABSORPTION IN A METAL

In a metal the core polarization generally leads to a real contribution to ϵ^0, and Eq. (K.5) gives, for the imaginary part,

$$\text{Im}\,\epsilon = \frac{4\pi}{\omega}\,\text{Re}\,\sigma. \tag{K.8}$$

---

[3] This result is proved in most texts on electromagnetism. See, for example, L. D. Landau and E. M. Lifshitz, *Electrodynamics of Continuous Media*, Addison-Wesley, Reading, Mass., 1960, p. 274. The quantities E^i and E^r are the amplitudes of the incident and reflected electric fields. The square root in (K.7) is the one yielding non-negative n.

[4] Landau and Lifshitz, *op. cit.*, p. 259. (The "P" indicates a principal value integral.)

[5] *Ibid.*, pp. 272–277.

In the semiclassical theory the AC conductivity (13.34) has a real part only by virtue of collisions. In the absence of collisions the semiclassical conductivity is purely imaginary and ϵ entirely real, signifying no dissipation of electromagnetic energy within the metal. If, however, we use the correction to the semiclassical result (13.37), then the conductivity has a real part even in the absence of collisions, and we find that[6]

$$\operatorname{Im} \epsilon = \frac{4\pi^2 e^2}{\omega} \int \frac{d\mathbf{k}}{(2\pi)^3} \sum_{nn'} D_{nn'}(\mathbf{k}) \delta\left(\frac{\mathcal{E}_n(\mathbf{k}) - \mathcal{E}_{n'}(\mathbf{k})}{\hbar} - \omega\right), \tag{K.9}$$

where

$$D_{nn'}(\mathbf{k}) = \frac{f(\mathcal{E}_{n'}(\mathbf{k})) - f(\mathcal{E}_n(\mathbf{k}))}{\mathcal{E}_n(\mathbf{k}) - \mathcal{E}_{n'}(k)} \frac{1}{3} \sum_i |\langle n\mathbf{k}|v_i|n'\mathbf{k}\rangle|^2. \tag{K.10}$$

The quantity $D_{nn'}(\mathbf{k})$ is nonnegative, and negligibly small (when $k_B T \ll \mathcal{E}_F$) unless one of the pair of levels $n\mathbf{k}$, $n'\mathbf{k}$ is occupied and the other is unoccupied. Therefore the additional contribution to the real (absorptive) part of the conductivity arises whenever $\hbar\omega$ is equal to the energy difference between two levels with the same $\mathbf{k}$, one occupied and the other empty. This is precisely the condition for interband absorption suggested by intuitive arguments about photons (see page 293).

---

[6] We specialize to the case of a metal with cubic symmetry. The frequency appearing in the denominator of (13.37) is to be evaluated in the limit of a vanishingly small positive imaginary part η, and we use the identity:

$$\lim_{\eta \to 0} \operatorname{Im} \frac{1}{x + i\eta} = -\pi\delta(x).$$

Appendix L

Quantum Theory of the Harmonic Crystal

We first summarize the quantum theory of a single (one-dimensional) harmonic oscillator with Hamiltonian

$$h = \frac{p^2}{2m} + \tfrac{1}{2}m\omega^2 q^2. \tag{L.1}$$

The structure of this Hamiltonian is simplified by defining the "lowering" operator

$$a = \sqrt{\frac{m\omega}{2\hbar}}\, q + i\sqrt{\frac{1}{2\hbar m\omega}}\, p, \tag{L.2}$$

and its adjoint, the "raising" operator

$$a^\dagger = \sqrt{\frac{m\omega}{2\hbar}}\, q - i\sqrt{\frac{1}{2\hbar m\omega}}\, p. \tag{L.3}$$

The canonical commutation relations $[q, p] = i\hbar$ imply that

$$[a, a^\dagger] = 1. \tag{L.4}$$

If the Hamiltonian is expressed in terms of a and $a^\dagger$ rather than q and p, it takes on the simple form:

$$h = \hbar\omega(a^\dagger a + \tfrac{1}{2}). \tag{L.5}$$

It is not difficult to show[1] that the commutation relations (L.4) imply that the eigenvalues of (L.5) are of the form $(n + \frac{1}{2})\hbar\omega$, where $n = 0, 1, 2, \ldots$. If the ground state of h is denoted by $|0\rangle$, then the nth excited state $|n\rangle$ is

$$|n\rangle = \frac{1}{\sqrt{n!}}(a^\dagger)^n|0\rangle, \tag{L.6}$$

and satisfies

$$a^\dagger a|n\rangle = n|n\rangle, \qquad h|n\rangle = (n + \tfrac{1}{2})\hbar\omega|n\rangle. \tag{L.7}$$

The matrix elements of a and $a^\dagger$ in this complete set of states are given by:

$$\begin{aligned} \langle n'|a|n\rangle &= 0, \qquad n' \neq n - 1, \\ \langle n-1|a|n\rangle &= \sqrt{n}, \\ \langle n'|a^\dagger|n\rangle &= \langle n|a|n'\rangle. \end{aligned} \tag{L.8}$$

All these results follow in a straightforward way from Eqs. (L.4) and (L.5).

The procedure in the case of a harmonic crystal is similar. The Hamiltonian is now given by Eq. (23.2).[2] Let $\omega_s(\mathbf{k})$ and $\boldsymbol{\epsilon}_s(\mathbf{k})$ be the frequency and polarization vector for

---

[1] See, for example, D. Park, *Introduction to the Quantum Theory*, McGraw-Hill, New York, 1964, p. 110.

[2] We summarize the derivation only for monatomic Bravais lattices, stating below how the conclusions are to be generalized when there is a polyatomic basis.

the classical normal mode with polarization s and wave vector $\mathbf{k}$, as described on page 439. In analogy to (L.2) we now define[3] the "phonon annihilation operator"

$$a_{\mathbf{k}s} = \frac{1}{\sqrt{N}} \sum_{\mathbf{R}} e^{-i\mathbf{k}\cdot\mathbf{R}} \boldsymbol{\epsilon}_s(\mathbf{k}) \cdot \left[\sqrt{\frac{M\omega_s(\mathbf{k})}{2\hbar}}\, \mathbf{u}(\mathbf{R}) + i \sqrt{\frac{1}{2\hbar M\omega_s(\mathbf{k})}}\, \mathbf{P}(\mathbf{R}) \right], \tag{L.9}$$

and its adjoint, the "phonon creation operator"

$$a^{\dagger}_{\mathbf{k}s} = \frac{1}{\sqrt{N}} \sum_{\mathbf{R}} e^{i\mathbf{k}\cdot\mathbf{R}} \boldsymbol{\epsilon}_s(\mathbf{k}) \cdot \left[\sqrt{\frac{M\omega_s(\mathbf{k})}{2\hbar}}\, \mathbf{u}(\mathbf{R}) - i \sqrt{\frac{1}{2\hbar M\omega_s(\mathbf{k})}}\, \mathbf{P}(\mathbf{R}) \right]. \tag{L.10}$$

The canonical commutation relations

$$\begin{aligned} [u_\mu(\mathbf{R}), P_\nu(\mathbf{R}')] &= i\hbar\, \delta_{\mu\nu}\, \delta_{\mathbf{R},\mathbf{R}'}, \\ [u_\mu(\mathbf{R}), u_\nu(\mathbf{R}')] &= [P_\mu(\mathbf{R}), P_\nu(\mathbf{R}')] = 0, \end{aligned} \tag{L.11}$$

the identity[4]

$$\sum_{\mathbf{R}} e^{i\mathbf{k}\cdot\mathbf{R}} = \left\{ \begin{array}{ll} 0, & \mathbf{k} \text{ not a reciprocal lattice vector,} \\ N, & \mathbf{k} \text{ a reciprocal lattice vector,} \end{array} \right\} \tag{L.12}$$

and the orthonormality of the polarization vectors (Eq. (22.61)) yield the commutation relations

$$\begin{aligned} [a_{\mathbf{k}s}, a^{\dagger}_{\mathbf{k}'s'}] &= \delta_{\mathbf{k}\mathbf{k}'}\, \delta_{ss'}, \\ [a_{\mathbf{k}s}, a_{\mathbf{k}'s'}] &= [a^{\dagger}_{\mathbf{k}s}, a^{\dagger}_{\mathbf{k}'s'}] = 0, \end{aligned} \tag{L.13}$$

which are analogous to (L.4).

One can invert (L.9) to express the original coordinates and momenta in terms of the $a_{\mathbf{k}s}$ and $a^{\dagger}_{\mathbf{k}s}$:

$$\begin{aligned} \mathbf{u}(\mathbf{R}) &= \frac{1}{\sqrt{N}} \sum_{\mathbf{k}s} \sqrt{\frac{\hbar}{2M\omega_s(\mathbf{k})}}\, (a_{\mathbf{k}s} + a^{\dagger}_{-\mathbf{k}s}) \boldsymbol{\epsilon}_s(\mathbf{k}) e^{i\mathbf{k}\cdot\mathbf{R}}, \\ \mathbf{P}(\mathbf{R}) &= \frac{-i}{\sqrt{N}} \sum_{\mathbf{k}s} \sqrt{\frac{\hbar M\omega_s(\mathbf{k})}{2}}\, (a_{\mathbf{k}s} - a^{\dagger}_{-\mathbf{k}s}) \boldsymbol{\epsilon}_s(\mathbf{k}) e^{i\mathbf{k}\cdot\mathbf{R}}. \end{aligned} \tag{L.14}$$

Equation (L.14) can be verified by direct substitution of (L.9) and (L.10), and by use of the "completeness relation" that holds for any complete set of real orthogonal vectors,

$$\sum_{s=1}^{3} [\boldsymbol{\epsilon}_s(\mathbf{k})]_\mu [\boldsymbol{\epsilon}_s(\mathbf{k})]_\nu = \delta_{\mu\nu}, \tag{L.15}$$

together with the identity[4,5]

$$\sum_{\mathbf{k}} e^{i\mathbf{k}\cdot\mathbf{R}} = 0, \qquad \mathbf{R} \neq 0. \tag{L.16}$$

---

[3] If $\omega_s(\mathbf{k}) = 0$, this definition fails. The problem occurs for only three of the normal modes out of N (the $\mathbf{k} = 0$ acoustic modes) and can usually be ignored. It is a reflection of the fact that the three degrees of freedom describing translations of the entire crystal as a whole cannot be described as oscillator degrees of freedom. Only in problems in which one wishes to consider translations of the crystal as a whole, or the total momentum of the crystal, does it become important to treat these degrees of freedom correctly as well. The problem is discussed further in Appendix M.

[4] See Appendix F.

[5] Use has also been made of the fact that in a monatomic Bravais lattice $\omega_s(\mathbf{k}) = \omega_s(-\mathbf{k})$, and $\boldsymbol{\epsilon}_s(\mathbf{k}) = \boldsymbol{\epsilon}_s(-\mathbf{k})$.

We can express the harmonic Hamiltonian in terms of the new oscillator variables by substituting (L.14) into (23.2). If the identity (L.16) and the orthonormality of the polarization vectors of a given **k**, are used, it can be shown that the kinetic energy is given by:

$$\frac{1}{2M}\sum_{\mathbf{R}} \mathbf{P}(\mathbf{R})^2 = \frac{1}{4}\sum_{\mathbf{k}s} \hbar\omega_s(\mathbf{k})(a_{\mathbf{k}s} - a^\dagger_{-\mathbf{k}s})(a^\dagger_{\mathbf{k}s} - a_{-\mathbf{k}s}). \tag{L.17}$$

The potential energy takes on a similar structure when one exploits the fact that the polarization vectors are eigenvectors of the dynamical matrix $\mathsf{D}(\mathbf{k})$ (see (22.57):

$$U = \frac{1}{4}\sum_{\mathbf{k}s} \hbar\omega_s(\mathbf{k})(a_{\mathbf{k}s} + a^\dagger_{-\mathbf{k}s})(a_{-\mathbf{k}s} + a^\dagger_{\mathbf{k}s}). \tag{L.18}$$

Adding these together, we find that

$$H = \frac{1}{2}\sum \hbar\omega_s(\mathbf{k})(a_{\mathbf{k}s}a^\dagger_{\mathbf{k}s} + a^\dagger_{\mathbf{k}s}a_{\mathbf{k}s}), \tag{L.19}$$

or, using the commutation relations (L.13),

$$H = \sum \hbar\omega_s(\mathbf{k})(a^\dagger_{\mathbf{k}s}a_{\mathbf{k}s} + \tfrac{1}{2}). \tag{L.20}$$

This is nothing more than the sum of $3N$ independent oscillator Hamiltonians, one for each wave vector and polarization. When a Hamiltonian separates into a sum of commuting sub-Hamiltonians, its eigenstates are simply all products of the eigenstates of each of the separate sub-Hamiltonians, and the corresponding eigenvalues are the sum of the individual eigenvalues of the sub-Hamiltonians. We can therefore specify an eigenstate of H by giving a set of $3N$ quantum numbers $n_{\mathbf{k}s}$, one for each of the $3N$ independent oscillator Hamiltonians $\hbar\omega_s(\mathbf{k})(a^\dagger_{\mathbf{k}s}a_{\mathbf{k}s} + \frac{1}{2})$. The energy of such a state is just

$$E = \sum (n_{\mathbf{k}s} + \tfrac{1}{2})\hbar\omega_s(\mathbf{k}). \tag{L.21}$$

In many applications (such as those in Chapter 23) it is necessary to know only the form (L.21) of the eigenvalues of H. However, in problems involving the interaction of lattice vibrations with external radiation or with each other (i.e., in problems where anharmonic terms are important) it is essential to use the relations (L.14), for it is the **u**'s and **P**'s in terms of which physical interactions are simply expressed, but the a's and $a^\dagger$'s that have simple matrix elements in the harmonic stationary states.

A similar procedure is used to transform the Hamiltonian for a lattice with a polyatomic basis. We quote here only the final result:

The definitions (L.9) and (L.10) (which now define the $a_{\mathbf{k}s}$ and $a^\dagger_{\mathbf{k}s}$ for $s = 1, \ldots 3p$, where p is the number of ions in the basis) remain valid provided one makes the replacements

$$\begin{aligned}
\mathbf{u}(\mathbf{R}) &\to \mathbf{u}^i(\mathbf{R}),\\
\mathbf{P}(\mathbf{R}) &\to \mathbf{P}^i(\mathbf{R}),\\
M &\to M^i,\\
\boldsymbol{\epsilon}_s(\mathbf{k}) &\to \sqrt{M^i}\boldsymbol{\epsilon}_s{}^i(\mathbf{k}) \text{ in (L.9)},\\
\boldsymbol{\epsilon}_s(\mathbf{k}) &\to \sqrt{M^i}\boldsymbol{\epsilon}_s{}^i(\mathbf{k})^* \text{ in (L.10)},
\end{aligned} \tag{L.22}$$

and sums over the index i (which specifies the type of basis ion). The $\boldsymbol{\epsilon}_s{}^i(\mathbf{k})$ are now the polarization vectors of the classical normal modes as defined in Eq. (22.67), obeying the orthonormality relation (22.68), the completeness relation

$$\sum_{s=1}^{3p} [\boldsymbol{\epsilon}_s{}^i(\mathbf{k})^*]_\mu \, [\boldsymbol{\epsilon}_s{}^j(\mathbf{k})]_\nu = \frac{1}{M_i} \delta_{ij} \, \delta_{\mu\nu}, \tag{L.23}$$

and the condition[6]

$$\boldsymbol{\epsilon}_s{}^i(-\mathbf{k}) = \boldsymbol{\epsilon}_s{}^i(\mathbf{k})^*. \tag{L.24}$$

The inversion formulas (L.14) remain valid provided the replacements[7] specified in (L.22) are made, and the commutation relations (L.13) and the form (L.20) of the harmonic Hamiltonian are unchanged.

---

6 Together with the condition $\omega_s(\mathbf{k}) = \omega_s(-\mathbf{k})$ this holds quite generally. However in the polyatomic case the polarization vectors are not, in general, real.

7 The first of the two prescriptions for the polarization vectors is to be used in generalizing both of equations (L.14).

Appendix M

Conservation of Crystal Momentum

Associated with every symmetry of a Hamiltonian is a conservation law. The Hamiltonian of a crystal possesses a symmetry, closely related to the translational symmetry of the Bravais lattice, that leads to a very general conservation law known as the conservation of *crystal momentum.* Consider the Hamiltonian:

$$H = \sum_{\mathbf{R}} \frac{\mathbf{P}(\mathbf{R})^2}{2M} + \frac{1}{2}\sum_{\mathbf{R},\mathbf{R}'} \phi[\mathbf{R} + \mathbf{u}(\mathbf{R}) - \mathbf{R}' - \mathbf{u}(\mathbf{R}')]$$
$$+ \sum_{i=1}^{n} \frac{p_i^2}{2m_i} + \frac{1}{2}\sum_{i \neq j} v_{ij}(\mathbf{r}_i - \mathbf{r}_j)$$
$$+ \sum_{\mathbf{R},i} w_i(\mathbf{r}_i - \mathbf{R} - \mathbf{u}(\mathbf{R})). \qquad \textbf{(M.1)}$$

The first two terms are the Hamiltonian of the ion cores. Note that we have not made the harmonic approximation,[1] but have represented the ion-ion interaction by a general sum of pair potentials.[2] The next two terms are the Hamiltonian of n additional particles, and the last term gives the interaction of these particles with the ions. To keep the discussion general, we do not specify the nature of the n particles, though the following possibilities are of interest:

1. If $n = 0$, we are discussing an isolated insulating crystal.
2. If $n = 1$, we can apply the discussion to the scattering of a single particle, say a neutron, by an insulating crystal.
3. If we wish to discuss an isolated metal, we can let the n particles be the conduction electrons ($n \approx 10^{23}$), in which case all the m_i would be the electron mass m and the $v_{ij}(\mathbf{r})$ would all be the same function of $\mathbf{r}$.
4. We can let the n particles be the conduction electrons and an incident external particle, if we wish to discuss neutron scattering by a metal.

Note that the Hamiltonian (M.1) is invariant under the uniform translation:

$$\mathbf{r}_i \to \mathbf{r}_i + \mathbf{r}_0, \quad i = 1, \ldots, n,$$
$$\mathbf{u}(\mathbf{R}) \to \mathbf{u}(\mathbf{R}) + \mathbf{r}_0, \quad \text{for all } \mathbf{R}. \qquad \textbf{(M.2)}$$

This familiar symmetry leads to conservation of the total *momentum* of the ions and particles, and is not the symmetry we are interested in. *Crystal momentum* conservation arises from the fact that when the translation vector $\mathbf{r}_0$ is a Bravais lattice vector $\mathbf{R}_0$, then it is possible to simulate the translation of the ions by a simple permutation of the

[1] We shall, however, be introducing operators defined in terms of the phonon operators a and $a^\dagger$ (Appendix L). These will not be well defined, if the system is so unlike a harmonic crystal with Bravais lattice $\{\mathbf{R}\}$ as to lie in a completely different Hilbert space. Thus although the procedure we shall follow makes formal sense for any system (e.g., a liquid, or a crystal with a Bravais lattice different from that specified by the $\{\mathbf{R}\}$), the conclusions can be meaningfully applied only to the case of a crystal with the Bravais lattice $\{\mathbf{R}\}$.

[2] Even the assumption of pair interactions is unnecessary. It is made only to give H a concrete form that is not too complex to obscure the argument.

ionic variables; i.e., the Hamiltonian (M.1) is also invariant under the transformation:

$$\begin{aligned} &\mathbf{r}_i \to \mathbf{r}_i + \mathbf{R}_0, \qquad i = 1, \ldots, n, \\ &\mathbf{u}(\mathbf{R}) \to \mathbf{u}(\mathbf{R} - \mathbf{R}_0), \quad P(\mathbf{R}) \to P(\mathbf{R} - \mathbf{R}_0), \quad \text{for all } \mathbf{R}, \end{aligned} \tag{M.3}$$

as can be verified explicitly by direct substitution into (M.1).

To emphasize the difference between symmetry (M.2) (with $\mathbf{r}_0 = \mathbf{R}_0$) and symmetry (M.3), consider the following two symmetry-breaking terms one might add to the Hamiltonian (M.1):

1. We might add a term:

$$\frac{1}{2} K \sum_{\mathbf{R}} [\mathbf{u}(\mathbf{R})]^2, \tag{M.4}$$

which ties each ion to its equilibrium site with a harmonic spring. This term destroys the translational symmetry (M.2) of the Hamiltonian, and momentum will not be conserved in its presence. However, the term (M.4) is invariant under the permutation symmetry (M.3), so its addition will not destroy the conservation of crystal momentum.

2. Suppose, on the other hand, we alter the Hamiltonian (M.1) so as to maintain the translational symmetry but destroy the permutation symmetry. We might, for example, give each ion a different mass, replacing the ionic kinetic-energy term by:

$$\sum_{\mathbf{R}} \frac{\mathbf{P}(\mathbf{R})^2}{2M(\mathbf{R})}. \tag{M.5}$$

The resulting Hamiltonian will continue to be invariant under the spatial translation (M.2) so total momentum will continue to be conserved. However, it will no longer be invariant under the permutation (M.3), and *crystal* momentum will not be conserved.

Of the two conservation laws, crystal momentum conservation is by far the more important. In practice, crystals are not free to recoil as a whole, and even if they were, the minute changes in the total momentum of the entire crystal produced by the scattering of a single neutron would be impossible to measure directly.

DERIVATION OF THE CONSERVATION LAW

To derive the conservation law implied by the symmetry (M.3) we must describe the quantum-mechanical operators that produce this transformation. That part of the transformation affecting the particle coordinates, $\mathbf{r}_i \to \mathbf{r}_i + \mathbf{R}_0$, is effected by the particle translational operator $T_{\mathbf{R}_0}$ (see Chapter 8). It is a fundamental result of quantum mechanics that this transformation can be written as a unitary transformation involving the total momentum operator for the particles:[3]

$$\begin{aligned} &\mathbf{r}_i \to \mathbf{r}_i + \mathbf{R}_0 = T_{\mathbf{R}_0} \mathbf{r}_i T_{\mathbf{R}_0}^{-1} = e^{(i/\hbar)\mathbf{P} \cdot \mathbf{R}_0} \mathbf{r}_i e^{-(i/\hbar)\mathbf{P} \cdot \mathbf{R}_0}, \\ &\mathbf{P} = \sum_{i=1}^{n} \mathbf{P}_i. \end{aligned} \tag{M.6}$$

[3] See, for example, K. Gottfried, *Quantum Mechanics*, vol. I, W. A. Benjamin, Menlo Park, Calif., 1966, p. 245.

In addition, we need the operator that produces the transformation (M.3) on the ionic variables. Crystal momentum conservation takes on a simple form primarily because this transformation has a structure very much like Eq. (M.6):

$$\begin{aligned} \mathbf{u}(\mathbf{R}) \to \mathbf{u}(\mathbf{R} - \mathbf{R}_0) &= \mathfrak{I}_{\mathbf{R}_0}\, \mathbf{u}(\mathbf{R})\, \mathfrak{I}_{\mathbf{R}_0}{}^{-1} = e^{i\mathcal{K}\cdot\mathbf{R}_0}\, \mathbf{u}(\mathbf{R})\, e^{-i\mathcal{K}\cdot\mathbf{R}_0}, \\ \mathbf{P}(\mathbf{R}) \to \mathbf{P}(\mathbf{R} - \mathbf{R}_0) &= \mathfrak{I}_{\mathbf{R}_0}\, \mathbf{P}(\mathbf{R})\, \mathfrak{I}_{\mathbf{R}_0}{}^{-1} = e^{i\mathcal{K}\cdot\mathbf{R}_0}\, \mathbf{P}(\mathbf{R})\, e^{-i\mathcal{K}\cdot\mathbf{R}_0}. \end{aligned} \tag{M.7}$$

The operator $\mathcal{K}$ is not related in any way to the total momentum operator $\mathbf{P} = \Sigma\mathbf{P}(\mathbf{R})$ for the ions, but is specified[4] by taking the eigenstates of $\mathcal{K}$ to be the eigenstates of the harmonic part of the ion-ion Hamiltonian, and its eigenvalue in a state with phonon occupation numbers $n_{\mathbf{k}s}$ to be given by

$$\mathcal{K}|\{n_{\mathbf{k}s}\}\rangle = \left(\sum_{\mathbf{k}s} \mathbf{k} n_{\mathbf{k}s}\right) |\{n_{\mathbf{k}s}\}\rangle. \tag{M.8}$$

To verify that the operator $\mathcal{K}$ defined by (M.8) does produce the transformation (M.7), we must use the representation (L.14) that gives $\mathbf{u}(\mathbf{R})$ and $\mathbf{P}(\mathbf{R})$ in terms of the oscillator raising and lowering operators for each normal mode:

$$\mathbf{u}(\mathbf{R}) = \frac{1}{\sqrt{N}} \sum_{\mathbf{k}s} \sqrt{\frac{\hbar}{2M\omega_s(\mathbf{k})}}\, (a_{\mathbf{k}s} + a^{\dagger}_{-\mathbf{k}s}) e^{i\mathbf{k}\cdot\mathbf{R}} \boldsymbol{\epsilon}_s(\mathbf{k}). \tag{M.9}$$

(We consider explicitly only the $\mathbf{u}(\mathbf{R})$. The argument for the $\mathbf{P}(\mathbf{R})$ is virtually the same.) Since the only operators in (M.9) are $a_{\mathbf{k}s}$ and $a^{\dagger}_{-\mathbf{k}s}$, we have

$$\begin{aligned} &e^{i\mathcal{K}\cdot\mathbf{R}_0}\, \mathbf{u}(\mathbf{R})\, e^{-i\mathcal{K}\cdot\mathbf{R}_0} \\ &= \frac{1}{\sqrt{N}} \sum_{\mathbf{k}s} \sqrt{\frac{\hbar}{2M\omega_s(\mathbf{k})}}\, (e^{i\mathcal{K}\cdot\mathbf{R}_0} a_{\mathbf{k}s} e^{-i\mathcal{K}\cdot\mathbf{R}_0} + e^{i\mathcal{K}\cdot\mathbf{R}_0} a^{\dagger}_{-\mathbf{k}s} e^{-i\mathcal{K}\cdot\mathbf{R}_0}) e^{i\mathbf{k}\cdot\mathbf{R}} \boldsymbol{\epsilon}_s(\mathbf{k}). \end{aligned} \tag{M.10}$$

We will therefore have established (M.7) when we prove that

$$\begin{aligned} e^{i\mathcal{K}\cdot\mathbf{R}_0}\, a_{\mathbf{k}s}\, e^{-i\mathcal{K}\cdot\mathbf{R}_0} &= e^{-i\mathbf{k}\cdot\mathbf{R}_0} a_{\mathbf{k}s}, \\ e^{i\mathcal{K}\cdot\mathbf{R}_0}\, a^{\dagger}_{-\mathbf{k}s}\, e^{-i\mathcal{K}\cdot\mathbf{R}_0} &= e^{-i\mathbf{k}\cdot\mathbf{R}_0} a^{\dagger}_{-\mathbf{k}s}, \end{aligned} \tag{M.11}$$

for substitution of (M.11) into (M.10) reduces it back to the form (M.9) with $\mathbf{R}$ replaced by $\mathbf{R} - \mathbf{R}_0$.

Both results in (M.11) will follow if we can establish the single identity:

$$e^{i\mathcal{K}\cdot\mathbf{R}_0}\, a^{\dagger}_{\mathbf{k}s}\, e^{-i\mathcal{K}\cdot\mathbf{R}_0} = e^{i\mathbf{k}\cdot\mathbf{R}_0} a^{\dagger}_{\mathbf{k}s}, \tag{M.12}$$

for the first of (M.11) is just the adjoint of (M.12) and the second follows from letting $\mathbf{k} \to -\mathbf{k}$. Equation (M.12) will be established if we can demonstrate that the operators on both sides of the equation have the same effect on a complete set of states, for they will then have the same effect on any linear combination of states from the complete set, and thus on any state whatsoever. We again choose the complete set to be the eigenstates of the harmonic Hamiltonian. The operator $a^{\dagger}_{\mathbf{k}s}$, being an oscillator-

---

[4] An operator is defined by giving a complete set of eigenstates and the corresponding eigenvalues, since any state can be written as a linear combination of eigenstates. Note that here the subtle assumption that the solid is actually crystalline with Bravais lattice $\{\mathbf{R}\}$ appears; if it were not, its states could not be written as linear combinations of eigenstates of a harmonic crystal with Bravais lattice $\{\mathbf{R}\}$.

raising operator for the **k**s normal mode, simply acts on a state with a particular set of phonon occupation numbers to produce (to within a normalization constant) a state in which the occupation number for the **k**s normal mode is increased by one, and all other occupation numbers remain the same. We may, to begin with, write

$$e^{i\mathcal{K}\cdot\mathbf{R}_0} a^\dagger_{\mathbf{k}s}\, e^{-i\mathcal{K}\cdot\mathbf{R}_0}|\{n_{\mathbf{k}s}\}\rangle = \exp\left(-i\sum_{\mathbf{k}'s}\mathbf{k}'n_{\mathbf{k}'s}\cdot\mathbf{R}_0\right) e^{i\mathcal{K}\cdot\mathbf{R}_0} a^\dagger_{\mathbf{k}s}|\{n_{\mathbf{k}s}\}\rangle, \tag{M.13}$$

which simply exploits (M.8). Since the state $a^\dagger_{\mathbf{k}s}|\{n_{\mathbf{k}s}\}\rangle$ differs from $|\{n_{\mathbf{k}s}\}\rangle$ only in that the occupation number of mode **k**s is increased by one, it will also be an eigenstate of $\varkappa$ with eigenvalue $\Sigma\mathbf{k}'n_{\mathbf{k}'s} + \mathbf{k}$. Therefore every term in the eigenvalue of $e^{i\mathbf{k}\cdot\mathbf{R}_0}$ cancels the corresponding term in the eigenvalue of $e^{-i\mathbf{k}\cdot\mathbf{R}_0}$ except for the single extra term $e^{i\mathbf{k}\cdot\mathbf{R}_0}$, and we have:

$$\exp\left(-i\sum_{\mathbf{k}'s}\mathbf{k}'n_{\mathbf{k}'s}\cdot\mathbf{R}_0\right) e^{i\mathcal{K}\cdot\mathbf{R}_0} a^\dagger_{\mathbf{k}s}|\{n_{\mathbf{k}s}\}\rangle = e^{i\mathbf{k}\cdot\mathbf{R}_0} a^\dagger_{\mathbf{k}s}|\{n_{\mathbf{k}s}\}\rangle. \tag{M.14}$$

Equations (M.14) and (M.13) establish (M.12), and therefore $\mathcal{K}$ gives the desired transformation. The operator $\hbar\mathcal{K}$ is called the crystal momentum operator.

APPLICATIONS

We illustrate the workings of crystal momentum conservation in several cases:

1. ***Isolated Insulator*** If only the ions are present, then the invariance (M.3) implies that the Hamiltonian commutes with the operator $e^{i\mathcal{K}\cdot\mathbf{R}_0}$, since

$$\begin{aligned} e^{i\mathcal{K}\cdot\mathbf{R}_0} H(\{\mathbf{u}(\mathbf{R}), \mathbf{P}(\mathbf{R})\}) e^{-i\mathcal{K}\cdot\mathbf{R}_0} &= H(\{\mathbf{u}(\mathbf{R}-\mathbf{R}_0), \mathbf{P}(\mathbf{R}-\mathbf{R}_0)\}) \equiv H, \\ \text{or}\quad e^{i\mathcal{K}\cdot\mathbf{R}_0} H &= H e^{i\mathcal{K}\cdot\mathbf{R}_0}. \end{aligned} \tag{M.15}$$

This means that the operator $\mathfrak{I}_{\mathbf{R}_0} = e^{i\mathcal{K}\cdot\mathbf{R}_0}$ is a constant of the motion; i.e., if the crystal is in an eigenstate of $\mathfrak{I}_{\mathbf{R}_0}$ at time $t = 0$, it will remain in an eigenstate at all subsequent times. In particular, suppose that the crystal is in an eigenstate of the harmonic Hamiltonian at $t = 0$ with phonon occupation numbers $n_{\mathbf{k}s}$. Because the full Hamiltonian is not harmonic this will not be a stationary state. However, crystal momentum conservation requires that it remain an eigenstate of $\mathfrak{I}_{\mathbf{R}_0}$ for all Bravais lattice vectors $\mathbf{R}_0$. This means that the state at future times can only be a linear combination of eigenstates of the harmonic Hamiltonian with phonon occupation numbers $n'_{\mathbf{k}s}$ leading to the same eigenvalue of $\mathfrak{I}_{\mathbf{R}_0}$ as the original state:

$$\exp\left(i\sum\mathbf{k}n'_{\mathbf{k}s}\cdot\mathbf{R}_0\right) = \exp\left(i\sum\mathbf{k}n_{\mathbf{k}s}\cdot\mathbf{R}_0\right). \tag{M.16}$$

Since this must hold for *arbitrary* Bravais lattice vectors $\mathbf{R}_0$ (the $\mathfrak{I}_{\mathbf{R}_0}$ commute with one another for different $\mathbf{R}_0$) we have

$$\exp\left\{i\left[\sum(\mathbf{k}n_{\mathbf{k}s} - \mathbf{k}n'_{\mathbf{k}s})\cdot\mathbf{R}_0\right]\right\} = 1 \qquad \text{for all } \mathbf{R}_0, \tag{M.17}$$

which requires that

$$\sum\mathbf{k}n_{\mathbf{k}s} = \sum\mathbf{k}n'_{\mathbf{k}s} + \text{reciprocal lattice vector}. \tag{M.18}$$

Thus *the total phonon wave vector in an anharmonic crystal is conserved to within an additive reciprocal lattice vector.*

2. ***Scattering of a Neutron by an Insulator*** Suppose that at the start of the experiment the crystal is in an eigenstate of the harmonic Hamiltonian with phonon occupation numbers $n_{\mathbf{k}s}$ and the neutron is in a state with *real* momentum $\mathbf{p}$, satisfying:

$$T_{\mathbf{R}_0}\psi(\mathbf{r}) = e^{(i/\hbar)\mathbf{p}\cdot\mathbf{R}_0}\psi(r) \qquad (\text{i.e., } \psi(\mathbf{r}) = e^{(i/\hbar)\mathbf{p}\cdot\mathbf{R}_0}), \tag{M.19}$$

where $T_{\mathbf{R}_0}$ is the neutron translation operator. The invariance of the total neutron-ion Hamiltonian under (M.3) implies that the product of the neutron translation and ion permutation operators commutes with H for any $\mathbf{R}_0$:

$$[T_{\mathbf{R}_0}\mathfrak{I}_{\mathbf{R}_0}, H] = 0. \tag{M.20}$$

In the initial state Φ we have:

$$T_{\mathbf{R}_0}\mathfrak{I}_{\mathbf{R}_0}\Phi = \exp\left[i(\mathbf{p}/\hbar + \Sigma\mathbf{k}n_{\mathbf{k}s})\cdot\mathbf{R}_0\right]\Phi, \tag{M.21}$$

and therefore subsequent states must continue to be eigenstates with the same eigenvalue. They can therefore be represented as linear combinations of states in which the neutron has momentum $\mathbf{p}'$ and the crystal has occupation numbers $n'_{\mathbf{k}s}$ with the restriction that

$$\mathbf{p}' + \hbar\sum\mathbf{k}n'_{\mathbf{k}s} = \mathbf{p} + \hbar\sum\mathbf{k}n_{\mathbf{k}s} + \hbar\times\text{reciprocal lattice vector}. \tag{M.22}$$

Thus *the change in the momentum of the neutron must be balanced by a change in the crystal momentum*[5] *of the phonons, to within an additive reciprocal lattice vector times* $\hbar$.

3. ***Isolated Metal*** If the particles are conduction electrons then we can consider at $t = 0$ a state in which the electrons are in a specified set of Bloch levels. Now each Bloch level (see Eq. (8.21)) is an eigenstate of the electron translation operator:

$$T_{\mathbf{R}_0}\psi_{n\mathbf{k}}(\mathbf{r}) = e^{i\mathbf{k}\cdot\mathbf{R}_0}\psi_{n\mathbf{k}}(\mathbf{r}). \tag{M.23}$$

If, in addition, the crystal is in an eigenstate of the harmonic Hamiltonian at $t = 0$, then the combined electron translation and ion permutation operator $T_{\mathbf{R}_0}\mathfrak{I}_{\mathbf{R}_0}$ will have the eigenvalue

$$\exp\left[i(\mathbf{k}_e + \Sigma\mathbf{k}n_{\mathbf{k}s})\cdot\mathbf{R}_0\right], \tag{M.24}$$

where $\mathbf{k}_e$ is the sum of the electronic wave vectors of all the occupied Bloch levels (i.e., $\hbar\mathbf{k}_e$ is the total electronic crystal momentum). Since this operator commutes with the electron-ion Hamiltonian, the metal must remain in an eigenstate at all subsequent times. Therefore *the change in the total electronic crystal momentum must be compensated for by a change in the total ionic crystal momentum to within an additive reciprocal lattice vector.*

4. ***Scattering of a Neutron by a Metal*** In the same way, we can deduce that *when neutrons are scattered by a metal, the change in the neutron momentum must be*

---

[5] The eigenvalue $\hbar\Sigma\mathbf{k}n_{\mathbf{k}s}$ of the crystal momentum operator $\hbar\mathfrak{K}$, is called the crystal momentum. (cf. page 472).

balanced by a change in the total crystal momentum of the electrons and ions, to within an additive reciprocal lattice vector times ħ. Neutrons, however, interact only weakly with electrons, and in practice it is only the lattice crystal momentum that changes. This case is therefore essentially the same as case 2. Note, however, that X rays do interact strongly with electrons, so crystal momentum can be lost to the electronic system in X-ray scattering.

Appendix N

Theory of the Scattering of Neutrons by a Crystal

Let a neutron with momentum **p** be scattered by a crystal and emerge with momentum **p**′. We assume that the only degrees of freedom of the crystal are those associated with ionic motion, that before the scattering the ions are in an eigenstate of the crystal Hamiltonian with energy E_i, and that after the scattering the ions are in an eigenstate of the crystal Hamiltonian with energy E_f. We describe the initial and final states and energies of the composite neutron-ion system as follows:

Before scattering:

$$\Psi_i = \psi_{\mathbf{p}}(\mathbf{r})\Phi_i, \qquad \psi_{\mathbf{p}} = \frac{1}{\sqrt{V}} e^{i\mathbf{p}\cdot\mathbf{r}/\hbar},$$

$$\mathcal{E}_i = E_i + p^2/2M_n;$$

After scattering:

$$\Psi_f = \psi_{\mathbf{p}'}(\mathbf{r})\Phi_f, \qquad \psi_{\mathbf{p}'} = \frac{1}{\sqrt{V}} e^{i\mathbf{p}'\cdot\mathbf{r}/\hbar},$$

$$\mathcal{E}_f = E_f + p'^2/2M_n. \tag{N.1}$$

It is convenient to define variables ω and **q** in terms of the neutron energy gain and momentum transfer:

$$\hbar\omega = \frac{p'^2}{2M_n} - \frac{p^2}{2M_n},$$

$$\hbar\mathbf{q} = \mathbf{p}' - \mathbf{p}. \tag{N.2}$$

We describe the neutron-ion interaction by

$$V(\mathbf{r}) = \sum_{\mathbf{R}} v(\mathbf{r} - \mathbf{r}(\mathbf{R})) = \frac{1}{V}\sum_{\mathbf{k},\mathbf{R}} v_{\mathbf{k}} e^{i\mathbf{k}\cdot[\mathbf{r}-\mathbf{r}(\mathbf{R})]}. \tag{N.3}$$

Because the range of v is of order 10^{-13} cm (a typical nuclear dimension), its Fourier components will vary on the scale of $k \approx 10^{13}\ \text{cm}^{-1}$, and therefore be essentially independent of k for wave vectors of order $10^8\ \text{cm}^{-1}$, the relevant range for experiments that measure phonon spectra. The constant v_0 is conventionally written in terms of a length a, known as the scattering length, defined so that the total cross section for scattering of a neutron by a single isolated ion is given in Born approximation by $4\pi a^2$.[1] Eq. (N.3) is thus written

$$V(\mathbf{r}) = \frac{2\pi\hbar^2 a}{M_n V}\sum_{\mathbf{k},\mathbf{R}} e^{i\mathbf{k}\cdot[\mathbf{r}-\mathbf{r}(\mathbf{R})]}. \tag{N.4}$$

[1] We assume that the nuclei have spin zero and are of a single isotope. In general, one must consider the possibility of a depending on the nuclear state. This leads to two types of terms in the cross section: a *coherent* term, which has the form of the cross section we derive below, but with a replaced by its mean value, and an additional piece, known as the *incoherent* term, which has no striking energy dependence and contributes, along with the multiphonon processes, to the diffuse background.

The probability per unit time for a neutron to scatter from $\mathbf{p}$ to $\mathbf{p}'$ by virtue of its interaction with the ions is almost always calculated with the "golden rule" of lowest-order time-dependent perturbation theory:[2]

$$
\begin{aligned}
P &= \sum_f \frac{2\pi}{\hbar}\,\delta(\mathcal{E}_i - \mathcal{E}_f)|(\Psi_i, V\Psi_f)|^2 \\
&= \sum_f \frac{2\pi}{\hbar}\,\delta(E_f - E_i + \hbar\omega)\left|\frac{1}{V}\int d\mathbf{r}\, e^{i\mathbf{q}\cdot\mathbf{r}}(\Phi_i, V(\mathbf{r})\Phi_f)\right|^2 \\
&= \frac{(2\pi\hbar)^3}{(M_n V)^2}\, a^2 \sum_f \delta(E_f - E_i + \hbar\omega)\left|\sum_{\mathbf{R}} (\Phi_i, e^{i\mathbf{q}\cdot\mathbf{r}(\mathbf{R})}\Phi_f)\right|^2 .
\end{aligned}
\tag{N.5}
$$

The transition rate P is related to the measured cross section, $d\sigma/d\Omega\, dE$, by the fact that the cross section, transition rate, and incident neutron flux, $j = (p/M_n)|\psi_{\mathbf{p}}|^2 = (1/V)(p/M_n)$ satisfy[3]

$$
\begin{aligned}
j\,\frac{d\sigma}{d\Omega\, dE}\, d\Omega\, dE &= \frac{p}{M_n V}\frac{d\sigma}{d\Omega\, dE}\, d\Omega\, dE = \frac{PV\, d\mathbf{p}'}{(2\pi\hbar)^3} \\
&= \frac{PVp'^2\, dp'\, d\Omega}{(2\pi\hbar)^3} = \frac{PVM_n p'\, dE\, d\Omega}{(2\pi\hbar)^3}.
\end{aligned}
\tag{N.6}
$$

For a given initial state i (and all final states f compatible with the energy-conserving δ-function), Eqs. (N.5) and (N.6) give

$$
\frac{d\sigma}{d\Omega\, dE} = \frac{p'}{p}\frac{Na^2}{\hbar} S_i(\mathbf{q}, \omega), \tag{N.7}
$$

where

$$
S_i(\mathbf{q}, \omega) = \frac{1}{N}\sum_f \delta\left(\frac{E_f - E_i}{\hbar} + \omega\right)\left|\sum_{\mathbf{R}} (\Phi_i, e^{i\mathbf{q}\cdot\mathbf{r}(\mathbf{R})}\Phi_f)\right|^2 . \tag{N.8}
$$

To evaluate S_i we use the representation

$$
\delta(\omega) = \int_{-\infty}^{\infty} \frac{dt}{2\pi}\, e^{i\omega t}, \tag{N.9}
$$

and note that any operator A obeys the relation $e^{i(E_f - E_i)t/\hbar}(\Phi_f, A\Phi_i) = (\Phi_f, A(t)\Phi_i)$, where $A(t) = e^{iHt/\hbar}Ae^{-iHt/\hbar}$. Furthermore for any pair of operators A and B,

$$
\sum_f (\Phi_i, A\Phi_f)(\Phi_f, B\Phi_i) = (\Phi_i, AB\Phi_i). \tag{N.10}
$$

Therefore

$$
S_i(\mathbf{q}, \omega) = \frac{1}{N}\int \frac{dt}{2\pi}\, e^{i\omega t} \sum_{\mathbf{RR}'} e^{-i\mathbf{q}\cdot(\mathbf{R}-\mathbf{R}')}(\Phi_i, \exp[i\mathbf{q}\cdot\mathbf{u}(\mathbf{R}')]\exp[-i\mathbf{q}\cdot\mathbf{u}(\mathbf{R}, t)]\Phi_i). \tag{N.11}
$$

---

[2] See, for example, D. Park *Introduction to the Quantum Theory* McGraw-Hill, New York, 1964, p. 244. The analysis of neutron scattering data relies heavily on this use of lowest-order perturbation theory—i.e., on the Born approximation. Higher order perturbation theory produces so-called *multiple scattering* corrections.

[3] We use the fact that a volume element $d\mathbf{p}'$ contains $V\, d\mathbf{p}'/(2\pi\hbar)^3$ neutron states of a given spin. (The argument is identical to that given in Chapter 2 for electrons.)

Generally the crystal will be in thermal equilibrium, and we must therefore average the cross section for the given i over a Maxwell-Boltzmann distribution of equilibrium states. This requires us to replace S_i by its thermal average

$$S(\mathbf{q}, \omega) = \frac{1}{N} \sum_{\mathbf{RR}'} e^{-i\mathbf{q}\cdot(\mathbf{R}-\mathbf{R}')} \int \frac{dt}{2\pi} e^{i\omega t} \langle \exp[i\mathbf{q}\cdot\mathbf{u}(\mathbf{R}')] \exp[-i\mathbf{q}\cdot\mathbf{u}(\mathbf{R}, t)]\rangle, \tag{N.12}$$

where

$$\langle A \rangle = \frac{\sum e^{-E_i/k_BT}(\Phi_i, A\Phi_i)}{\sum e^{-E_i/k_BT}}. \tag{N.13}$$

Finally,

$$\frac{d\sigma}{d\Omega\, dE} = \frac{p'}{p} \frac{Na^2}{\hbar} S(\mathbf{q}, \omega). \tag{N.14}$$

$S(\mathbf{q}, \omega)$ is known as the *dynamical structure factor* of the crystal, and is entirely determined by the crystal itself without reference to any properties of the neutrons.[4] Furthermore, our result (N.14) has not even exploited the harmonic approximation, and is therefore quite general, applying (with the appropriate changes in notation) even to the scattering of neutrons by liquids. To extract the peculiar characteristics of neutron scattering by a lattice of ions, we now make the harmonic approximation.

In a harmonic crystal the position of any ion at time t is a linear function of the positions and momenta of all the ions at time zero. It can be proved,[5] however, that if A and B are operators linear in the $\mathbf{u}(\mathbf{R})$ and $\mathbf{P}(\mathbf{R})$ of a *harmonic crystal*, then

$$\langle e^A e^B \rangle = e^{(1/2)\langle A^2 + 2AB + B^2 \rangle}. \tag{N.15}$$

This result is directly applicable to (N.12):

$$\langle \exp[i\mathbf{q}\cdot\mathbf{u}(\mathbf{R}')] \exp[-i\mathbf{q}\cdot\mathbf{u}(\mathbf{R}, t)]\rangle = \exp\left(-\tfrac{1}{2}\langle[\mathbf{q}\cdot\mathbf{u}(\mathbf{R}')]^2\rangle - \tfrac{1}{2}\langle[\mathbf{q}\cdot\mathbf{u}(\mathbf{R}, t)]^2\rangle + \langle[\mathbf{q}\cdot\mathbf{u}(\mathbf{R}')][\mathbf{q}\cdot\mathbf{u}(\mathbf{R}, t)]\rangle\right). \tag{N.16}$$

This can be further simplified from the observation that the operator products depend only on the relative positions and times

$$\begin{aligned} \langle[\mathbf{q}\cdot\mathbf{u}(\mathbf{R}')]^2\rangle &= \langle[\mathbf{q}\cdot\mathbf{u}(\mathbf{R}, t)]^2\rangle = \langle[\mathbf{q}\cdot\mathbf{u}(0)]^2\rangle \equiv 2W, \\ \langle[\mathbf{q}\cdot\mathbf{u}(\mathbf{R}')][\mathbf{q}\cdot\mathbf{u}(\mathbf{R}, t)]\rangle &= \langle[\mathbf{q}\cdot\mathbf{u}(0)][\mathbf{q}\cdot\mathbf{u}(\mathbf{R}-\mathbf{R}', t)]\rangle, \end{aligned} \tag{N.17}$$

and therefore:

$$S(\mathbf{q}, \omega) = e^{-2W} \int \frac{dt}{2\pi} e^{i\omega t} \sum_{\mathbf{R}} e^{-i\mathbf{q}\cdot\mathbf{R}} \exp\langle[\mathbf{q}\cdot\mathbf{u}(0)][\mathbf{q}\cdot\mathbf{u}(\mathbf{R}, t)]\rangle. \tag{N.18}$$

Equation (N.18) is an *exact* evaluation of $S(\mathbf{q}, \omega)$, Eq. (N.12), *provided that the crystal is harmonic.*

---

[4] It is simply the Fourier transform of the density autocorrelation function.

[5] N. D. Mermin, *J. Math. Phys.* 7, 1038 (1966) gives a particularly compact proof.

In Chapter 24 we classified neutron scatterings according to the number, m, of phonons emitted and/or absorbed by the neutron. If one expands the exponential occurring in the integrand of S,

$$\exp\langle[\mathbf{q}\cdot\mathbf{u}(0)][\mathbf{q}\cdot\mathbf{u}(\mathbf{R},t)]\rangle = \sum_{m=0}^{\infty}\frac{1}{m!}\Big(\langle[\mathbf{q}\cdot\mathbf{u}(0)][\mathbf{q}\cdot\mathbf{u}(\mathbf{R},t)]\rangle\Big)^m, \quad \text{(N.19)}$$

then it can be shown that the mth term in this expansion gives precisely the contribution of the m-phonon processes to the total cross section. We limit ourselves here to showing that the $m = 0$ and $m = 1$ terms give the structure we deduced on less precise grounds for the zero- and one-phonon processes of Chapter 24.

1. *Zero-Phonon Contribution* ($m = 0$) If the exponential on the extreme right of (N.18) is replaced by unity, then the sum over $\mathbf{R}$ can be evaluated with Eq. (L.12), the time integral reduces to a δ-function as in (N.9), and the no-phonon contribution to $S(\mathbf{q}, \omega)$ is just

$$\boxed{S_{(0)}(\mathbf{q},\omega) = e^{-2W}\,\delta(\omega)\,N\sum_{\mathbf{K}}\delta_{\mathbf{q},\mathbf{K}}.} \quad \text{(N.20)}$$

The δ-function requires the scattering to be elastic. Integrating over final energies, we find that:

$$\frac{d\sigma}{d\Omega} = \int dE\,\frac{d\sigma}{d\Omega\,dE} = e^{-2W}(Na)^2\sum_{\mathbf{K}}\delta_{\mathbf{q},\mathbf{K}}. \quad \text{(N.21)}$$

This is precisely what one expects for Bragg-reflected neutrons: The scattering is elastic and occurs only for momentum transfers equal to $\hbar$ times a reciprocal lattice vector. The fact that Bragg scattering is a coherent process is reflected in the cross section being proportional to N^2 times the cross section a^2 for a single scatterer, rather than merely to N times the single ion cross section. Thus the *amplitudes* combine additively (rather than the cross sections). The effect of the thermal vibrations of the ions about their equilibrium positions is entirely contained in the factor e^{-2W}, which is known as the Debye-Waller factor. Since the mean square ionic displacement from equilibrium $\langle[\mathbf{u}(0)]^2\rangle$ will increase with temperature, we find that the thermal vibrations of the ions diminish the intensity of the Bragg peaks but do *not* (as was feared in the early days of X-ray scattering) eliminate the peaks altogether.[6]

2. *One-Phonon Contribution* ($m = 1$) To evaluate the contribution to $d\sigma/d\Omega\,dE$ from the $m = 1$ term in (N.19) one requires the form of

$$\langle[\mathbf{q}\cdot\mathbf{u}(0)][\mathbf{q}\cdot\mathbf{u}(\mathbf{R},t)]\rangle, \quad \text{(N.22)}$$

This is readily evaluated from Eq. (L.14) and the fact that[7]

$$\begin{aligned}
a_{\mathbf{k}s}(t) &= a_{\mathbf{k}s}e^{-i\omega_s(\mathbf{k})t}, & a_{\mathbf{k}s}^{\dagger}(t) &= a_{\mathbf{k}s}^{\dagger}e^{i\omega_s(\mathbf{k})t},\\
\langle a_{\mathbf{k}'s'}^{\dagger}a_{\mathbf{k}s}\rangle &= n_s(\mathbf{k})\delta_{\mathbf{k}\mathbf{k}'}\delta_{ss'}, & \langle a_{\mathbf{k}s}^{\dagger}a_{\mathbf{k}'s'}^{\dagger}\rangle &= 0,\\
\langle a_{\mathbf{k}s}a_{\mathbf{k}'s'}^{\dagger}\rangle &= [1+n_s(\mathbf{k})]\,\delta_{\mathbf{k}\mathbf{k}'}\delta_{ss'}, & \langle a_{\mathbf{k}s}a_{\mathbf{k}'s'}\rangle &= 0.
\end{aligned} \quad \text{(N.23)}$$

---

[6] This is a mark of the long-range order that always persists in a true crystal.

[7] Here, as in (23.10), $n_s(\mathbf{q})$ is the Bose-Einstein occupation factor for phonons in mode s with wave vector $\mathbf{q}$ and energy $\hbar\omega_s(\mathbf{q})$.

We then find that

$$S_{(1)}(\mathbf{q}, \omega) = e^{-2W} \sum_s \frac{\hbar}{2M\omega_s(\mathbf{q})} [\mathbf{q} \cdot \boldsymbol{\epsilon}_s(\mathbf{q})]^2 \Big([1 + n_s(\mathbf{q})]\, \delta[\omega + \omega_s(\mathbf{q})] + n_s(\mathbf{q})\, \delta[\omega - \omega_s(\mathbf{q})] \Big). \quad \textbf{(N.24)}$$

Substituting this into (N.14) we find that the one-phonon cross section is:

$$\frac{d\sigma}{d\Omega\, dE} = Ne^{-2W} \frac{p'}{p} a^2 \sum_s \frac{1}{2M\omega_s(\mathbf{q})} [\mathbf{q} \cdot \boldsymbol{\epsilon}_s(\mathbf{q})]^2 \Big([1 + n_s(\mathbf{q})]\, \delta[\omega + \omega_s(\mathbf{q})] + n_s(\mathbf{q})\, \delta[\omega - \omega_s(\mathbf{q})] \Big). \quad \textbf{(N.25)}$$

Note that this does indeed vanish unless the one-phonon conservation laws (24.9) or (24.10) are satisfied; thus, as a function of energy, $d\sigma/d\Omega\, dE$ is a series of sharp delta-function peaks at the allowed final neutron energies.

This structure makes it possible to distinguish the one-phonon processes from all the remaining terms in the multiphonon expansion of S or the cross section, all of which can be shown to be smooth functions of the final neutron energy. Note that the intensity in the one-phonon peaks is also modulated by the same Debye-Waller factor that diminishes the intensity of the Bragg peaks. Note also the factor $[\mathbf{q} \cdot \boldsymbol{\epsilon}_s(\mathbf{q})]^2$, which enables one to extract information about the phonon polarization vectors. Finally, the thermal factors $n_s(\mathbf{q})$ and $1 + n_s(\mathbf{q})$ are for processes in which phonons are absorbed or emitted, respectively. These factors are typical for processes involving the creation and absorption of Bose-Einstein particles, and indicate (as is reasonable) that at very low temperatures processes in which phonons are emitted will be the dominant ones (when they are allowed by the conservation laws).

APPLICATION TO X-RAY SCATTERING

Aside from the factor $(p'/p)a^2$, peculiar to the dynamics of neutrons, the inelastic scattering cross section for X rays should have precisely the same form as (N.14). However, one can generally not resolve the small (compared with X-ray energies) energy losses or gains occurring in one-phonon processes, and must therefore, in effect, integrate the cross section over all final energies:

$$\frac{d\sigma}{d\Omega} \propto \int d\omega\, S(\mathbf{q}, \omega) \propto e^{-2W} \sum_{\mathbf{R}} e^{-i\mathbf{q} \cdot \mathbf{R}} \exp \langle [\mathbf{q} \cdot \mathbf{u}(0)][\mathbf{q} \cdot \mathbf{u}(\mathbf{R})] \rangle. \quad \textbf{(N.26)}$$

This result is simply related to our discussion of X-ray scattering in Chapter 6, in which we relied on the static lattice model. In that chapter we found that the scattering for a monatomic Bravais lattice was proportional to a factor:

$$\Big|\sum_{\mathbf{R}} e^{i\mathbf{q} \cdot \mathbf{R}}\Big|^2. \quad \textbf{(N.27)}$$

Equation (N.26) generalizes this result by allowing for the ions to be displaced from their equilibrium positions, $\mathbf{R} \to \mathbf{R} + \mathbf{u}(\mathbf{R})$, and taking a thermal equilibrium average over ionic configurations.

Making the multiphonon expansion in (N.26) will yield the frequency integrals of the individual terms in the multiphonon expansion we made in the neutron case. The no-phonon terms continue to give the Bragg peaks, diminished by the Debye-Waller factor – an aspect of the intensity of the Bragg peaks that was not taken into account in our discussion of Chapter 6. The one-phonon term yields a scattering cross section proportional to

$$\int d\omega\, S_{(1)}(\mathbf{q}, \omega) = e^{-2W} \sum_s \frac{\hbar}{2M\omega_s(\mathbf{q})} (\mathbf{q} \cdot \boldsymbol{\epsilon}_s(\mathbf{q}))^2 \coth \tfrac{1}{2}\beta\hbar\omega_s(\mathbf{q}), \qquad \textbf{(N.28)}$$

where $\mathbf{q}$ is the change in X-ray wave vector. Since the change in photon energy is minute, $\mathbf{q}$ is entirely determined by the incident X-ray energy and the direction of observation. The contribution to (N.28) of the terms arising from the various branches of the phonon spectrum may be disentangled, by doing the experiment at several values of $\mathbf{q}$ differing by reciprocal lattice vectors. However the major problem is distinguishing the contribution (N.28) of the one-phonon processes to the total scattering cross section, from the contribution of the multiphonon terms, since the characteristic structure of the one-phonon terms lies entirely in their singular energy dependence, which is lost once the integral over ω has been performed. In practice, one can do little better than to try to estimate the multiphonon contribution from the general result (N.26). Alternatively, one can work at temperatures so low, and momentum transfers q sufficiently small, as to make the expansion (N.19) rapidly convergent. If crystals were strictly classical this would always be possible, since the deviations from equilibrium of the ions would vanish as $T \to 0$. Unfortunately, however, the zero-point ionic vibrations are present even at $T = 0$, and there is therefore an intrinsic limit to the rate at which the multiphonon expansion can converge.

Appendix O

Anharmonic Terms and n-Phonon Processes

An anharmonic term of nth degree in the ionic displacements $\mathbf{u}$ can be written in terms of the normal mode lowering and raising operators a and $a^\dagger$, through Eq. (L.14). Such a term will consist of linear combinations of products containing m lowering operators, $a_{\mathbf{k}_1 s_1}, \ldots a_{\mathbf{k}_m s_m}$, and $n - m$ raising operators, $a^\dagger_{\mathbf{k}_{m+1} s_{m+1}}, \ldots a^\dagger_{\mathbf{k}_n s_n}$ $(0 \leqslant m \leqslant n)$. Each such product, when acting on a state characterized by a definite set of phonon occupation numbers, produces a state in which the $n_{\mathbf{k}s}$ are all unchanged except for being reduced by one when $\mathbf{k}s = \mathbf{k}_1 s_1, \ldots \mathbf{k}_m s_m$, and increased by one when $\mathbf{k}s = \mathbf{k}_{m+1} s_{m+1}, \ldots \mathbf{k}_n s_n$. Thus an anharmonic term of the nth degree has nonvanishing matrix elements only between states in which just n phonon occupation numbers differ.[1]

[1] The overwhelming majority of terms in the expansion of any anharmonic term will have all the $\mathbf{k}_i s_i$ different, unless the crystal is microscopically small. Terms in which the occupation number of a given normal mode changes by two or more make a negligible contribution in the limit of a large crystal.

Appendix P

Evaluation of the Landé g-Factor

We take the vector product of both sides of (31.34) with $\langle JLSJ_z'|\mathbf{J}|JLSJ_z\rangle$ and sum over J_z', for fixed J. Since the matrix elements of $\mathbf{J}$ vanish between states with differing values of J, we may as well sum over all states in the $(2L + 1)(2S + 1)$ dimensional manifold with given L and S. Having noted this, however, we can use the completeness relation

$$\sum_{JJ_z'} |JLSJ_z'\rangle\langle JLSJ_z'| = \mathbf{1}, \tag{P.1}$$

to replace the sums of products of matrix elements by matrix elements of the operator products:

$$\langle JLSJ_z|(\mathbf{L} + g_0\mathbf{S})\cdot\mathbf{J}|JLSJ_z\rangle = g(JLS)\langle JLSJ_z|\mathbf{J}^2|JLSJ_z\rangle. \tag{P.2}$$

Now we simply observe that the operator identities:

$$\begin{gathered}
\mathbf{S}^2 = (\mathbf{J} - \mathbf{L})^2 = \mathbf{J}^2 + \mathbf{L}^2 - 2\mathbf{L}\cdot\mathbf{J},\\
\mathbf{L}^2 = (\mathbf{J} - \mathbf{S})^2 = \mathbf{J}^2 + \mathbf{S}^2 - 2\mathbf{S}\cdot\mathbf{J},\\
\langle JLSJ_z|\begin{Bmatrix}\mathbf{J}^2\\ \mathbf{L}^2\\ \mathbf{S}^2\end{Bmatrix}|JLSJ_z\rangle = \begin{Bmatrix}J(J+1)\\ L(L+1)\\ S(S+1)\end{Bmatrix},
\end{gathered} \tag{P.3}$$

permit one to evaluate (P.2) as

$$\begin{aligned}
g(JLS)J(J + 1) &= \langle JLSJ_z|(\mathbf{L}\cdot\mathbf{J})|JLSJ_z\rangle + g_0\langle JLSJ_z|(\mathbf{S}\cdot\mathbf{J})|JLSJ_z\rangle\\
&= \frac{1}{2}[J(J + 1) + L(L + 1) - S(S + 1)]\\
&\quad + \frac{g_0}{2}[J(J + 1) + S(S + 1) - L(L + 1)],
\end{aligned} \tag{P.4}$$

which is equivalent to the result (31.37) quoted in the text.

Index